Machine Learning: A Probabilistic Perspective

Adaptive Computation and Machine Learning

Thomas Dietterich, Editor

Christopher Bishop, David Heckerman, Michael Jordan, and Michael Kearns, Associate Editors

Machine Learning
A Probabilistic Perspective

Kevin P. Murphy

The MIT Press
Cambridge, Massachusetts
London, England

For information about special quantity discounts, please email special_sales@mitpress.mit.edu

This book was set in the LaTeX programming language by the author. Printed and bound in China.

Library of Congress Cataloging-in-Publication Information

Murphy, Kevin P.
Machine learning : a probabilistic perspective / Kevin P. Murphy.
p. cm. — (Adaptive computation and machine learning series)
Includes bibliographical references and index.
ISBN 978-0-262-01802-9 (hardcover : alk. paper)
1. Machine learning. 2. Probabilities. I. Title.
Q325.5.M87 2012
006.3'1—dc23
2012004558

15 14 13 12 11

This book is dedicated to Alessandro, Michael and Stefano, and to the memory of Gerard Joseph Murphy.

Contents

Preface

Introduction

With the ever increasing amounts of data in electronic form, the need for automated methods for data analysis continues to grow. The goal of machine learning is to develop methods that can automatically detect patterns in data, and then to use the uncovered patterns to predict future data or other outcomes of interest. Machine learning is thus closely related to the fields of statistics and data mining, but differs slightly in terms of its emphasis and terminology. This book provides a detailed introduction to the field, and includes worked examples drawn from application domains such as molecular biology, text processing, computer vision, and robotics.

Target audience

This book is suitable for upper-level undergraduate students and beginning graduate students in computer science, statistics, electrical engineering, econometrics, or anyone else who has the appropriate mathematical background. Specifically, the reader is assumed to already be familiar with basic multivariate calculus, probability, linear algebra, and computer programming. Prior exposure to statistics is helpful but not necessary.

A probabilistic approach

This books adopts the view that the best way to make machines that can learn from data is to use the tools of probability theory, which has been the mainstay of statistics and engineering for centuries. Probability theory can be applied to any problem involving uncertainty. In machine learning, uncertainty comes in many forms: what is the best prediction (or decision) given some data? what is the best model given some data? what measurement should I perform next? etc.

The systematic application of probabilistic reasoning to all inferential problems, including inferring parameters of statistical models, is sometimes called a Bayesian approach. However, this term tends to elicit very strong reactions (either positive or negative, depending on who you ask), so we prefer the more neutral term "probabilistic approach". Besides, we will often use techniques such as maximum likelihood estimation, which are not Bayesian methods, but certainly fall within the probabilistic paradigm.

Rather than describing a cookbook of different heuristic methods, this book stresses a principled model-based approach to machine learning. For any given model, a variety of algorithms can often be applied. Conversely, any given algorithm can often be applied to a variety of models. This kind of modularity, where we distinguish model from algorithm, is good pedagogy and good engineering.

We will often use the language of graphical models to specify our models in a concise and intuitive way. In addition to aiding comprehension, the graph structure aids in developing efficient algorithms, as we will see. However, this book is not primarily about graphical models; it is about probabilistic modeling in general.

A practical approach

Nearly all of the methods described in this book have been implemented in a MATLAB software package called PMTK, which stands for probabilistic modeling toolkit. This is freely available from pmtk3.googlecode.com (the digit 3 refers to the third edition of the toolkit, which is the one used in this version of the book). There are also a variety of supporting files, written by other people, available at pmtksupport.googlecode.com. These will be downloaded automatically, if you follow the setup instructions described on the PMTK website.

MATLAB is a high-level, interactive scripting language ideally suited to numerical computation and data visualization, and can be purchased from www.mathworks.com. Some of the code requires the Statistics toolbox, which needs to be purchased separately. There is also a free version of Matlab called Octave, available at http://www.gnu.org/software/octave/, which supports most of the functionality of MATLAB. Some (but not all) of the code in this book also works in Octave. See the PMTK website for details.

PMTK was used to generate many of the figures in this book; the source code for these figures is included on the PMTK website, allowing the reader to easily see the effects of changing the data or algorithm or parameter settings. The book refers to files by name, e.g., naiveBayesFit. In order to find the corresponding file, you can use two methods: within Matlab you can type which naiveBayesFit and it will return the full path to the file; or, if you do not have Matlab but want to read the source code anyway, you can use your favorite search engine, which should return the corresponding file from the pmtk3.googlecode.com website.

Details on *how to use* PMTK can be found on its website. Details on the *underlying theory* behind these methods can be found in this book.

Acknowledgments

A book this large is obviously a team effort. I would especially like to thank the following people: my wife Margaret, for keeping the home fires burning as I toiled away in my office for the last six years; Matt Dunham, who created many of the figures in this book, and who wrote much of the code in PMTK; Baback Moghaddam (RIP), who gave extremely detailed feedback on every page of an earlier draft of the book; Chris Williams, who also gave very detailed feedback; Cody Severinski and Wei-Lwun Lu, who assisted with figures; generations of UBC students, who gave helpful comments on earlier drafts; Daphne Koller, Nir Friedman, and Chris Manning, for letting me use their latex style files; Stanford University, Google Research and Skyline College for hosting me during part of my sabbatical; and various Canadian funding agencies (NSERC, CRC and CIFAR) who have supported me financially over the years.

In addition, I would like to thank the following people for giving me helpful feedback on parts of the book, and/or for sharing figures, code, exercises or even (in some cases) text: David Blei, Sebastien Bratieres, Hannes Bretschneider, Greg Corrado, Jutta Degener, Arnaud Doucet, Mario Figueiredo, Nando de Freitas, Mark Girolami, Gabriel Goh, Tom Griffiths, Katherine Heller, Geoff Hinton, Aapo Hyvarinen, Tommi Jaakkola, Mike Jordan, Charles Kemp, Emtiyaz Khan, Bonnie Kirkpatrick, Daphne Koller, Zico Kolter, Honglak Lee, Julien Mairal, Andrew McPherson, Tom Minka, Ian Nabney, Robert Piche, Arthur Pope, Carl Rassmussen, Ryan Rifkin, Ruslan Salakhutdinov, Mark Schmidt, Daniel Selsam, David Sontag, Erik Sudderth, Josh Tenenbaum, Martin Wainwright, Yair Weiss, Kai Yu.

Kevin Patrick Murphy
Palo Alto, California
June 2012

First printing: August 2012
Second printing: November 2012 (same as first)
Third printing: February 2013 (fixed some typos)
Fourth printing: August 2013 (fixed many typos)

1 *Introduction*

1.1 Machine learning: what and why?

> We are drowning in information and starving for knowledge. — John Naisbitt.

We are entering the era of **big data**. For example, there are about 40 billion indexed web pages[1]; 100 hours of video are uploaded to YouTube every minute[2]; the genomes of 1000s of people, each of which has a length of 3.8×10^9 base pairs, have been sequenced by various labs; Walmart handles more than 1M transactions per hour and has databases containing more than 2.5 petabytes (2.5×10^{15}) of information (Cukier 2010); and so on.

This deluge of data calls for automated methods of data analysis, which is what **machine learning** provides. In particular, we define machine learning as a set of methods that can automatically detect patterns in data, and then use the uncovered patterns to predict future data, or to perform other kinds of decision making under uncertainty (such as planning how to collect more data!).

This books adopts the view that the best way to solve such problems is to use the tools of probability theory. Probability theory can be applied to any problem involving uncertainty. In machine learning, uncertainty comes in many forms: what is the best prediction about the future given some past data? what is the best model to explain some data? what measurement should I perform next? etc. The probabilistic approach to machine learning is closely related to the field of statistics, but differs slightly in terms of its emphasis and terminology[3].

We will describe a wide variety of probabilistic models, suitable for a wide variety of data and tasks. We will also describe a wide variety of algorithms for learning and using such models. The goal is not to develop a cook book of ad hoc techniques, but instead to present a unified view of the field through the lens of probabilistic modeling and inference. Although we will pay attention to computational efficiency, details on how to scale these methods to truly massive datasets are better described in other books, such as (Rajaraman and Ullman 2011; Bekkerman et al. 2011).

---

1. See http://www.worldwidewebsize.com/. There are over 1 trillion unique URLs (see http://googleblog.blogspot.com/2008/07/we-knew-web-was-big.html), but many of these point to the same page, or are not part of standard search engine indexes.
2. Source: http://www.youtube.com/t/press_statistics.
3. Rob Tibshirani, a statistician at Stanford university, has created an amusing comparison between machine learning and statistics, available at http://www-stat.stanford.edu/~tibs/stat315a/glossary.pdf. See also (van Iterson et al. 2012).

It should be noted, however, that even when one has an apparently massive data set, the effective number of data points for certain cases of interest might be quite small. In fact, data across a variety of domains exhibits a property known as the **long tail**, which means that a few things are very common, but most things are quite rare. For example, some words (such as "the" and "and") are very common, but most (such as "pareidolia") are very rare. Similarly, some movies and books are very popular, but most are not. See Section 2.4.7 for further details on the long tail. One consequence of the long tail is that understanding or predicting the behavior of most items requires learning from small amounts of data, even if the total amount of data is large (Jordan 2011). In this book, we will discuss techniques that can handle data sets of this kind.

1.1.1 Types of machine learning

Machine learning is usually divided into two main types. In the **predictive** or **supervised learning** approach, the goal is to learn a mapping from inputs $\mathbf{x}$ to outputs y, given a labeled set of input-output pairs $\mathcal{D} = \{(\mathbf{x}_i, y_i)\}_{i=1}^{N}$. Here $\mathcal{D}$ is called the **training set**, and N is the number of training examples.

In the simplest setting, each training input $\mathbf{x}_i$ is a D-dimensional vector of numbers, representing, say, the height and weight of a person. These are called **features**, **attributes** or **covariates**. They are often stored in an $N \times D$ **design matrix** $\mathbf{X}$, such as the one in Figure 1.1(b). In general, however, $\mathbf{x}_i$ could be a complex structured object, such as an image, a sentence, an email message, a time series, a molecular shape, a graph, etc.

Similarly the form of the output or **response variable** can in principle be anything, but most methods assume that y_i is a **categorical** or **nominal** variable from some finite set, $y_i \in \{1, \ldots, C\}$ (such as male or female), or that y_i is a real-valued scalar (such as income level). When y_i is categorical, the problem is known as **classification** or **pattern recognition**, and when y_i is real-valued, the problem is known as **regression**. Another variant, known as **ordinal regression**, occurs where label space $\mathcal{Y}$ has some natural ordering, such as grades A–F.

The second main type of machine learning is the **descriptive** or **unsupervised learning** approach. Here we are only given inputs, $\mathcal{D} = \{\mathbf{x}_i\}_{i=1}^{N}$, and the goal is to find "interesting patterns" in the data. This is sometimes called **knowledge discovery**. This is a much less well-defined problem, since we are not told what kinds of patterns to look for, and there is no obvious error metric to use (unlike supervised learning, where we can compare our prediction of y for a given $\mathbf{x}$ to the observed value).

There is a third type of machine learning, known as **reinforcement learning**, which is somewhat less commonly used. This is useful for learning how to act or behave when given occasional reward or punishment signals. (For example, consider how a baby learns to walk.) Unfortunately, RL is beyond the scope of this book, although we do discuss decision theory in Section 5.7, which is the basis of RL. See e.g., (Kaelbling et al. 1996; Sutton and Barto 1998; Russell and Norvig 2010; Szepesvari 2010; Wiering and van Otterlo 2012) for more information on RL.

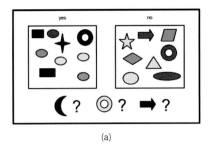
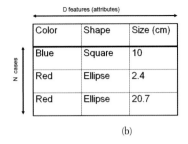

D features (attributes)				
Color	Shape	Size (cm)		Label
Blue	Square	10		1
Red	Ellipse	2.4		1
Red	Ellipse	20.7		0

(a)   (b)

Figure 1.1 Left: Some labeled training examples of colored shapes, along with 3 unlabeled test cases. Right: Representing the training data as an $N \times D$ design matrix. Row i represents the feature vector $\mathbf{x}_i$. The last column is the label, $y_i \in \{0, 1\}$. Based on a figure by Leslie Kaelbling.

1.2 Supervised learning

We begin our investigation of machine learning by discussing supervised learning, which is the form of ML most widely used in practice.

1.2.1 Classification

In this section, we discuss classification. Here the goal is to learn a mapping from inputs $\mathbf{x}$ to outputs y, where $y \in \{1, \ldots, C\}$, with C being the number of classes. If $C = 2$, this is called **binary classification** (in which case we often assume $y \in \{0, 1\}$); if $C > 2$, this is called **multiclass classification**. If the class labels are not mutually exclusive (e.g., somebody may be classified as tall and strong), we call it **multi-label classification**, but this is best viewed as predicting multiple related binary class labels (a so-called **multiple output model**). When we use the term "classification", we will mean multiclass classification with a single output, unless we state otherwise.

One way to formalize the problem is as **function approximation**. We assume $y = f(\mathbf{x})$ for some unknown function f, and the goal of learning is to estimate the function f given a labeled training set, and then to make predictions using $\hat{y} = \hat{f}(\mathbf{x})$. (We use the hat symbol to denote an estimate.) Our main goal is to make predictions on novel inputs, meaning ones that we have not seen before (this is called **generalization**), since predicting the response on the training set is easy (we can just look up the answer).

1.2.1.1 Example

As a simple toy example of classification, consider the problem illustrated in Figure 1.1(a). We have two classes of object which correspond to labels 0 and 1. The inputs are colored shapes. These have been described by a set of D features, which are stored in an $N \times D$ design matrix $\mathbf{X}$, shown in Figure 1.1(b). The input features $\mathbf{x}$ can be discrete, continuous or a combination of the two. In addition to the inputs, we have a vector of training labels $\mathbf{y}$.

In Figure 1.1, the test cases are a blue crescent, a yellow circle and a blue arrow. None of these have been seen before. Thus we are required to **generalize** beyond the training set. A reasonable guess is that blue crescent should be $y = 1$, since all blue shapes are labeled 1 in the training set. The yellow circle is harder to classify, since some yellow things are labeled $y = 1$ and some are labeled $y = 0$, and some circles are labeled $y = 1$ and some $y = 0$. Consequently it is not clear what the right label should be in the case of the yellow circle. Similarly, the correct label for the blue arrow is unclear, since although all the previously seen blue things have a positive label, all the previously seen arrows have a negative label.

1.2.1.2   The need for probabilistic predictions

To handle ambiguous cases, such as the yellow circle above, it is desirable to return a probability. The reader is assumed to already have some familiarity with basic concepts in probability. If not, please consult Chapter 2 for a refresher, if necessary.

We will denote the probability distribution over possible labels, given the input vector $\mathbf{x}$ and training set $\mathcal{D}$ by $p(y|\mathbf{x}, \mathcal{D})$. In general, this represents a vector of length C. (If there are just two classes, it is sufficient to return the single number $p(y = 1|\mathbf{x}, \mathcal{D})$, since $p(y = 1|\mathbf{x}, \mathcal{D}) + p(y = 0|\mathbf{x}, \mathcal{D}) = 1$.) In our notation, we make explicit that the probability is conditional on the test input $\mathbf{x}$, as well as the training set $\mathcal{D}$, by putting these terms on the right hand side of the conditioning bar $|$. We are also implicitly conditioning on the form of model that we use to make predictions. When choosing between different models, we will make this assumption explicit by writing $p(y|\mathbf{x}, \mathcal{D}, M)$, where M denotes the model. However, if the model is clear from context, we will drop M from our notation for brevity.

Given a probabilistic output, we can always compute our "best guess" as to the "true label" using

$$\hat{y} = \hat{f}(\mathbf{x}) = \operatorname*{argmax}_{c=1}^{C} p(y = c|\mathbf{x}, \mathcal{D}) \tag{1.1}$$

This corresponds to the most probable class label, and is called the **mode** of the distribution $p(y|\mathbf{x}, \mathcal{D})$; it is also known as a **MAP estimate** (MAP stands for **maximum a posteriori**). Using the most probable label makes intuitive sense, but we will give a more formal justification for this procedure in Section 5.7.

Now consider a case such as the yellow circle, where $p(\hat{y}|\mathbf{x}, \mathcal{D})$ is far from 1.0. If we are not very confident of our answer, it might be better to say "I don't know" instead of returning an answer that we don't really trust. This is particularly important in domains such as medicine and finance where we may be risk averse, as we explain in Section 5.7.

One interesting application of the "I don't know" option arises when playing the TV game show Jeopardy. In this game, contestants have to solve various word puzzles and answer a variety of trivia questions, but if they answer incorrectly, they lose money. In 2011, IBM unveiled a computer system called Watson which beat the top human Jeopardy champion. Watson uses a variety of interesting techniques (Ferrucci et al. 2010), but the most pertinent one for our present discussion is that it contains a module that estimates how confident it is of its answer. The system only chooses to "buzz in" its answer if sufficiently confident it is correct.

Another application where estimating uncertainty is important is in online advertising. Google has a system known as SmartASS (ad selection system) that predicts the probability you will click on an ad based on your search history and other user and ad-specific features (Metz 2010). This probability is known as the **click-through rate** or **CTR**, and can be used to maximize expected

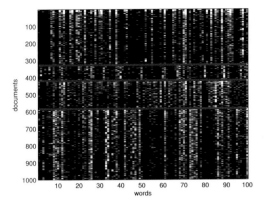

Figure 1.2 Subset of size 16242 x 100 of the 20-newsgroups data. We only show 1000 rows, for clarity. Each row is a document (represented as a bag-of-words bit vector), each column is a word. The red lines separate the 4 classes, which are (in descending order) comp, rec, sci, talk (these are the titles of USENET groups). We can see that there are subsets of words whose presence or absence is indicative of the class. The data is available from `http://cs.nyu.edu/~roweis/data.html`. Figure generated by `newsgroupsVisualize`.

profit. We will discuss some of the basic principles behind systems such as SmartASS later in this book.

The recent best selling book *The Signal and the Noise* (Silver 2012) gives many more examples of why it is important to use probabilistic methods when performing predictions, whether of political races or of the weather.

1.2.1.3 Real-world applications

Classification is probably the most widely used form of machine learning, and has been used to solve many interesting and often difficult real-world problems. We have already mentioned some important applications. We give a few more examples below.

Document classification and email spam filtering

In **document classification**, the goal is to classify a document, such as a web page or email message, into one of C classes, that is, to compute $p(y = c|\mathbf{x}, \mathcal{D})$, where $\mathbf{x}$ is some representation of the text. A special case of this is **email spam filtering**, where the classes are spam $y = 1$ or ham $y = 0$.

Most classifiers assume that the input vector $\mathbf{x}$ has a fixed size. A common way to represent variable-length documents in feature-vector format is to use a **bag of words** representation. This is explained in detail in Section 3.4.4.1, but the basic idea is to define $x_{ij} = 1$ iff word j occurs in document i. If we apply this transformation to every document in our data set, we get a binary document × word co-occurrence matrix: see Figure 1.2 for an example. Essentially the document classification problem has been reduced to one that looks for subtle changes in the pattern of bits. For example, we may notice that most spam messages have a high probability of

(a)   (b)   (c)

Figure 1.3   Three types of iris flowers: setosa, versicolor and virginica.   Source: `http://www.statlab.u` `ni-heidelberg.de/data/iris/` . Used with kind permission of Dennis Kramb and SIGNA.

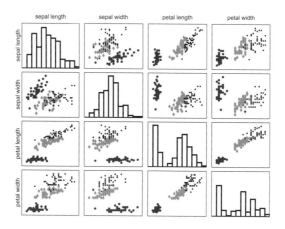

Figure 1.4   Visualization of the Iris data as a pairwise scatter plot. The diagonal plots the marginal histograms of the 4 features. The off diagonals contain scatterplots of all possible pairs of features. Red circle = setosa, green diamond = versicolor, blue star = virginica. Figure generated by `fisheririsDemo`.

containing the words "buy", "cheap", "viagra", etc. In Exercise 8.1 and Exercise 8.2, you will get hands-on experience applying various classification techniques to the spam filtering problem.

Classifying flowers

Figure 1.3 gives another example of classification, due to the statistician Ronald Fisher. The goal is to learn to distinguish three different kinds of iris flower, called setosa, versicolor and virginica. Fortunately, rather than working directly with images, a botanist has already extracted 4 useful features or characteristics: sepal length and width, and petal length and width. (Such **feature extraction** is an important, but difficult, task. Most machine learning methods use features

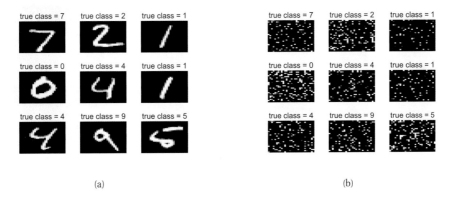

(a)   (b)

Figure 1.5 (a) First 9 test MNIST gray-scale images. (b) Same as (a), but with the features permuted randomly. Classification performance is identical on both versions of the data (assuming the training data is permuted in an identical way). Figure generated by `shuffledDigitsDemo`.

chosen by some human. Later we will discuss some methods that can learn good features from the data.) If we make a **scatter plot** of the iris data, as in Figure 1.4, we see that it is easy to distinguish setosas (red circles) from the other two classes by just checking if their petal length or width is below some threshold. However, distinguishing versicolor from virginica is slightly harder; any decision will need to be based on at least two features. (It is always a good idea to perform **exploratory data analysis**, such as plotting the data, before applying a machine learning method.)

Image classification and handwriting recognition

Now consider the harder problem of classifying images directly, where a human has not pre-processed the data. We might want to classify the image as a whole, e.g., is it an indoors or outdoors scene? is it a horizontal or vertical photo? does it contain a dog or not? This is called **image classification**.

In the special case that the images consist of isolated handwritten letters and digits, for example, in a postal or ZIP code on a letter, we can use classification to perform **handwriting recognition**. A standard dataset used in this area is known as **MNIST**, which stands for "Modified National Institute of Standards"[4]. (The term "modified" is used because the images have been preprocessed to ensure the digits are mostly in the center of the image.) This dataset contains 60,000 training images and 10,000 test images of the digits 0 to 9, as written by various people. The images are size 28×28 and have grayscale values in the range $0 : 255$. See Figure 1.5(a) for some example images.

Many generic classification methods ignore any structure in the input features, such as spatial layout. Consequently, they can also just as easily handle data that looks like Figure 1.5(b), which is the same data except we have randomly permuted the order of all the features. (You will verify this in Exercise 1.1.) This flexibility is both a blessing (since the methods are general

---

4. Available from `http://yann.lecun.com/exdb/mnist/`.

(a)   (b)

Figure 1.6 Example of face detection. (a) Input image (Murphy family, photo taken 5 August 2010). Used with kind permission of Bernard Diedrich of Sherwood Studios. (b) Output of classifier, which detected 5 faces at different poses. This was produced using the online demo at `http://demo.pittpatt.com/`. The classifier was trained on 1000s of manually labeled images of faces and non-faces, and then was applied to a dense set of overlapping patches in the test image. Only the patches whose probability of containing a face was sufficiently high were returned. Used with kind permission of Pittpatt.com

purpose) and a curse (since the methods ignore an obviously useful source of information). We will discuss methods for exploiting structure in the input features later in the book.

Face detection and recognition

A harder problem is to find objects within an image; this is called **object detection** or **object localization**. An important special case of this is **face detection**. One approach to this problem is to divide the image into many small overlapping patches at different locations, scales and orientations, and to classify each such patch based on whether it contains face-like texture or not. This is called a **sliding window detector**. The system then returns those locations where the probability of face is sufficiently high. See Figure 1.6 for an example. Such face detection systems are built-in to most modern digital cameras; the locations of the detected faces are used to determine the center of the auto-focus. Another application is automatically blurring out faces in Google's StreetView system.

Having found the faces, one can then proceed to perform **face recognition**, which means estimating the identity of the person (see Figure 1.10(a)). In this case, the number of class labels might be very large. Also, the features one should use are likely to be different than in the face detection problem: for recognition, subtle differences between faces such as hairstyle may be important for determining identity, but for detection, it is important to be **invariant** to such details, and to just focus on the differences between faces and non-faces. For more information about visual object detection, see e.g., (Szeliski 2010).

1.2.2 Regression

Regression is just like classification except the response variable is continuous. Figure 1.7 shows a simple example: we have a single real-valued input $x_i \in \mathbb{R}$, and a single real-valued response

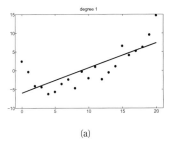
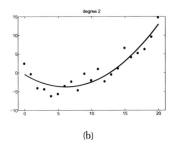

(a)   (b)

Figure 1.7   (a) Linear regression on some 1d data. (b) Same data with polynomial regression (degree 2). Figure generated by `linregPolyVsDegree`.

$y_i \in \mathbb{R}$. We consider fitting two models to the data: a straight line and a quadratic function. (We explain how to fit such models below.) Various extensions of this basic problem can arise, such as having high-dimensional inputs, outliers, non-smooth responses, etc. We will discuss ways to handle such problems later in the book.

Here are some examples of real-world regression problems.

- Predict tomorrow's stock market price given current market conditions and other possible side information.

- Predict the age of a viewer watching a given video on YouTube.

- Predict the location in 3d space of a robot arm end effector, given control signals (torques) sent to its various motors.

- Predict the amount of prostate specific antigen (PSA) in the body as a function of a number of different clinical measurements.

- Predict the temperature at any location inside a building using weather data, time, door sensors, etc.

1.3   Unsupervised learning

We now consider **unsupervised learning**, where we are just given output data, without any inputs. The goal is to discover "interesting structure" in the data; this is sometimes called **knowledge discovery**. Unlike supervised learning, we are not told what the desired output is for each input. Instead, we will formalize our task as one of **density estimation**, that is, we want to build models of the form $p(\mathbf{x}_i|\boldsymbol{\theta})$. There are two differences from the supervised case. First, we have written $p(\mathbf{x}_i|\boldsymbol{\theta})$ instead of $p(y_i|\mathbf{x}_i, \boldsymbol{\theta})$; that is, supervised learning is conditional density estimation, whereas unsupervised learning is unconditional density estimation. Second, $\mathbf{x}_i$ is a vector of features, so we need to create multivariate probability models. By contrast, in supervised learning, y_i is usually just a single variable that we are trying to predict. This means that for most supervised learning problems, we can use univariate probability models (with input-dependent parameters), which significantly simplifies the problem. (We will discuss

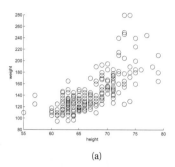
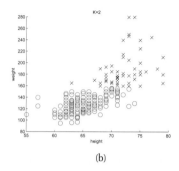

(a)   (b)

Figure 1.8   (a) The height and weight of some people. (b) A possible clustering using $K = 2$ clusters. Figure generated by `kmeansHeightWeight`.

multi-output classification in Chapter 19, where we will see that it also involves multivariate probability models.)

Unsupervised learning is arguably more typical of human and animal learning. It is also more widely applicable than supervised learning, since it does not require a human expert to manually label the data. Labeled data is not only expensive to acquire[5], but it also contains relatively little information, certainly not enough to reliably estimate the parameters of complex models. Geoff Hinton, who is a famous professor of ML at the University of Toronto, has said:

> When we're learning to see, nobody's telling us what the right answers are — we just look. Every so often, your mother says "that's a dog", but that's very little information. You'd be lucky if you got a few bits of information — even one bit per second — that way. The brain's visual system has 10^{14} neural connections. And you only live for 10^9 seconds. So it's no use learning one bit per second. You need more like 10^5 bits per second. And there's only one place you can get that much information: from the input itself. — Geoffrey Hinton, 1996 (quoted in (Gorder 2006)).

Below we describe some canonical examples of unsupervised learning.

1.3.1   Discovering clusters

As a canonical example of unsupervised learning, consider the problem of **clustering** data into groups. For example, Figure 1.8(a) plots some 2d data, representing the height and weight of a group of 210 people. It seems that there might be various clusters, or subgroups, although it is not clear how many. Let K denote the number of clusters. Our first goal is to estimate the distribution over the number of clusters, $p(K|\mathcal{D})$; this tells us if there are subpopulations within the data. For simplicity, we often approximate the distribution $p(K|\mathcal{D})$ by its mode, $K^* = \arg\max_K p(K|\mathcal{D})$. In the supervised case, we were told that there are two classes (male

---

5. The advent of **crowd sourcing** web sites such as Mechanical Turk, (`https://www.mturk.com/mturk/welcome`), which outsource data processing tasks to humans all over the world, has reduced the cost of labeling data. Nevertheless, the amount of unlabeled data is still orders of magnitude larger than the amount of labeled data.

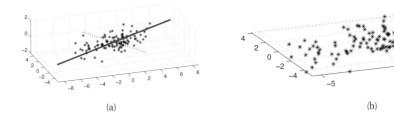

(a)   (b)

Figure 1.9   (a) A set of points that live on a 2d linear subspace embedded in 3d. The solid red line is the first principal component direction. The dotted black line is the second PC direction. (b) 2D representation of the data. Figure generated by `pcaDemo3d`.

and female), but in the unsupervised case, we are free to choose as many or few clusters as we like. Picking a model of the "right" complexity is called model selection, and will be discussed in detail below.

Our second goal is to estimate which cluster each point belongs to. Let $z_i \in \{1, \ldots, K\}$ represent the cluster to which data point i is assigned. (z_i is an example of a **hidden** or **latent** variable, since it is never observed in the training set.) We can infer which cluster each data point belongs to by computing $z_i^* = \mathrm{argmax}_k \, p(z_i = k | \mathbf{x}_i, \mathcal{D})$. This is illustrated in Figure 1.8(b), where we use different colors to indicate the assignments, assuming $K = 2$.

In this book, we focus on **model based clustering**, which means we fit a probabilistic model to the data, rather than running some ad hoc algorithm. The advantages of the model-based approach are that one can compare different kinds of models in an objective way (in terms of the likelihood they assign to the data), we can combine them together into larger systems, etc.

Here are some real world applications of clustering.

- In astronomy, the **autoclass** system (Cheeseman et al. 1988) discovered a new type of star, based on clustering astrophysical measurements.

- In e-commerce, it is common to cluster users into groups, based on their purchasing or web-surfing behavior, and then to send customized targeted advertising to each group (see e.g., (Berkhin 2006)).

- In biology, it is common to cluster flow-cytometry data into groups, to discover different sub-populations of cells (see e.g., (Lo et al. 2009)).

1.3.2   Discovering latent factors

When dealing with high dimensional data, it is often useful to reduce the dimensionality by projecting the data to a lower dimensional subspace which captures the "essence" of the data. This is called **dimensionality reduction**. A simple example is shown in Figure 1.9, where we project some 3d data down to a 2d plane. The 2d approximation is quite good, since most points lie close to this subspace. Reducing to 1d would involve projecting points onto the red line in Figure 1.9(a); this would be a rather poor approximation. (We will make this notion precise in Chapter 12.)

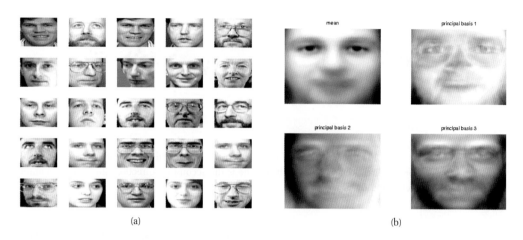

(a)   (b)

Figure 1.10   a) 25 randomly chosen 64×64 pixel images from the Olivetti face database. (b) The mean and the first three principal component basis vectors (eigenfaces). Figure generated by `pcaImageDemo`.

The motivation behind this technique is that although the data may appear high dimensional, there may only be a small number of degrees of variability, corresponding to **latent factors**. For example, when modeling the appearance of face images, there may only be a few underlying latent factors which describe most of the variability, such as lighting, pose, identity, etc, as illustrated in Figure 1.10.

When used as input to other statistical models, such low dimensional representations often result in better predictive accuracy, because they focus on the "essence" of the object, filtering out inessential features. Also, low dimensional representations are useful for enabling fast nearest neighbor searches and two dimensional projections are very useful for **visualizing** high dimensional data.

The most common approach to dimensionality reduction is called **principal components analysis** or **PCA**. This can be thought of as an unsupervised version of (multi-output) linear regression, where we observe the high-dimensional response $\mathbf{y}$, but not the low-dimensional "cause" $\mathbf{z}$. Thus the model has the form $\mathbf{z} \rightarrow \mathbf{y}$; we have to "invert the arrow", and infer the latent low-dimensional $\mathbf{z}$ from the observed high-dimensional $\mathbf{y}$. See Section 12.1 for details.

Dimensionality reduction, and PCA in particular, has been applied in many different areas. Some examples include the following:

- In biology, it is common to use PCA to interpret gene microarray data, to account for the fact that each measurement is usually the result of many genes which are correlated in their behavior by the fact that they belong to different biological pathways.

- In natural language processing, it is common to use a variant of PCA called latent semantic analysis for document retrieval (see Section 27.2.2).

- In signal processing (e.g., of acoustic or neural signals), it is common to use ICA (which is a variant of PCA) to separate signals into their different sources (see Section 12.6).

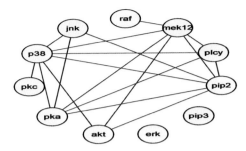

Figure 1.11 A sparse undirected Gaussian graphical model learned using graphical lasso (Section 26.7.2) applied to some flow cytometry data (from (Sachs et al. 2005)), which measures the phosphorylation status of 11 proteins. Figure generated by `ggmLassoDemo`.

- In computer graphics, it is common to project motion capture data to a low dimensional space, and use it to create animations. See Section 15.5 for one way to tackle such problems.

1.3.3 Discovering graph structure

Sometimes we measure a set of correlated variables, and we would like to discover which ones are most correlated with which others. This can be represented by a graph G, in which nodes represent variables, and edges represent direct dependence between variables (we will make this precise in Chapter 10, when we discuss graphical models). We can then learn this graph structure from data, i.e., we compute $\hat{G} = \operatorname{argmax} p(G|\mathcal{D})$.

As with unsupervised learning in general, there are two main applications for learning sparse graphs: to discover new knowledge, and to get better joint probability density estimators. We now give somes example of each.

- Much of the motivation for learning sparse graphical models comes from the systems biology community. For example, suppose we measure the phosphorylation status of some proteins in a cell (Sachs et al. 2005). Figure 1.11 gives an example of a graph structure that was learned from this data (using methods discussed in Section 26.7.2). As another example, Smith et al. (2006) showed that one can recover the neural "wiring diagram" of a certain kind of bird from time-series EEG data. The recovered structure closely matched the known functional connectivity of this part of the bird brain.

- In some cases, we are not interested in interpreting the graph structure, we just want to use it to model correlations and to make predictions. One example of this is in financial portfolio management, where accurate models of the covariance between large numbers of different stocks is important. Carvahlo and West (2007) show that by learning a sparse graph, and then using this as the basis of a trading strategy, it is possible to outperform (i.e., make more money than) methods that do not exploit sparse graphs. Another example is predicting traffic jams on the freeway. Horvitz et al. (2005) describe a deployed system called JamBayes for predicting traffic flow in the Seattle area; predictions are made using a graphical model whose structure was learned from data.

(a)   (b)

Figure 1.12   (a) A noisy image with an occluder. (b) An estimate of the underlying pixel intensities, based on a pairwise MRF model.   Source: Figure 8 of (Felzenszwalb and Huttenlocher 2006).   Used with kind permission of Pedro Felzenszwalb.

1.3.4   Matrix completion

Sometimes we have missing data, that is, variables whose values are unknown. For example, we might have conducted a survey, and some people might not have answered certain questions. Or we might have various sensors, some of which fail. The corresponding design matrix will then have "holes" in it; these missing entries are often represented by **NaN**, which stands for "not a number". The goal of **imputation** is to infer plausible values for the missing entries. This is sometimes called **matrix completion**. Below we give some example applications.

1.3.4.1   Image inpainting

An interesting example of an imputation-like task is known as **image inpainting**. The goal is to "fill in" holes (e.g., due to scratches or occlusions) in an image with realistic texture. This is illustrated in Figure 1.12, where we denoise the image, as well as impute the pixels hidden behind the occlusion. This can be tackled by building a joint probability model of the pixels, given a set of clean images, and then inferring the unknown variables (pixels) given the known variables (pixels). This is somewhat like market basket analysis, except the data is real-valued and spatially structured, so the kinds of probability models we use are quite different. See Sections 19.6.2.7 and 13.8.4 for some possible choices.

1.3.4.2   Collaborative filtering

Another interesting example of an imputation-like task is known as **collaborative filtering**. A common example of this concerns predicting which movies people will want to watch based on how they, and other people, have rated movies which they have already seen. The key idea is that the prediction is not based on features of the movie or user (although it could be), but merely on a ratings matrix. More precisely, we have a matrix $\mathbf{X}$ where $X(m, u)$ is the rating

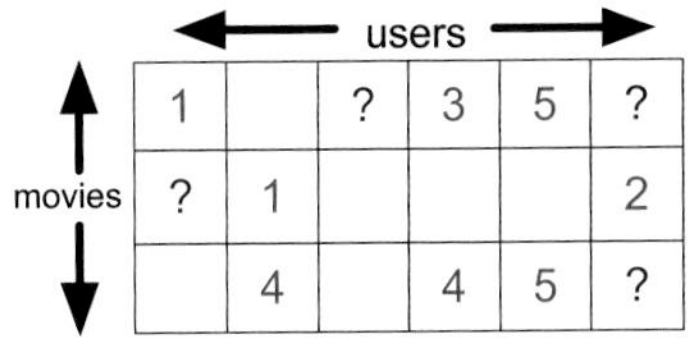

Figure 1.13 Example of movie-rating data. Training data is in red, test data is denoted by ?, empty cells are unknown.

(say an integer between 1 and 5, where 1 is dislike and 5 is like) by user u of movie m. Note that most of the entries in $\mathbf{X}$ will be missing or unknown, since most users will not have rated most movies. Hence we only observe a tiny subset of the $\mathbf{X}$ matrix, and we want to predict a different subset. In particular, for any given user u, we might want to predict which of the unrated movies he/she is most likely to want to watch.

In order to encourage research in this area, the DVD rental company Netflix created a competition, launched in 2006, with a \$1M USD prize (see `http://netflixprize.com/`). In particular, they provided a large matrix of ratings, on a scale of 1 to 5, for $\sim 18k$ movies created by $\sim 500k$ users. The full matrix would have $\sim 9 \times 10^9$ entries, but only about 1% of the entries are observed, so the matrix is extremely **sparse**. A subset of these are used for training, and the rest for testing, as shown in Figure 1.13. The goal of the competition was to predict more accurately than Netflix's existing system. On 21 September 2009, the prize was awarded to a team of researchers known as "BellKor's Pragmatic Chaos". Section 27.6.2 discusses some of their methodology. Further details on the teams and their methods can be found at `http://www.netflixprize.com/community/viewtopic.php?id=1537`.

1.3.4.3 Market basket analysis

In commercial data mining, there is much interest in a task called **market basket analysis**. The data consists of a (typically very large but sparse) binary matrix, where each column represents an item or product, and each row represents a transaction. We set $x_{ij} = 1$ if item j was purchased on the i'th transaction. Many items are purchased together (e.g., bread and butter), so there will be correlations amongst the bits. Given a new partially observed bit vector, representing a subset of items that the consumer has bought, the goal is to predict which other bits are likely to turn on, representing other items the consumer might be likely to buy. (Unlike collaborative filtering, we often assume there is no missing data in the training data, since we know the past shopping behavior of each customer.)

This task arises in other domains besides modeling purchasing patterns. For example, similar techniques can be used to model dependencies between files in complex software systems. In this case, the task is to predict, given a subset of files that have been changed, which other ones need to be updated to ensure consistency (see e.g., (Hu et al. 2010)).

It is common to solve such tasks using **frequent itemset mining**, which create association rules (see e.g., (Hastie et al. 2009, sec 14.2) for details). Alternatively, we can adopt a probabilistic approach, and fit a joint density model $p(x_1, \ldots, x_D)$ to the bit vectors, see e.g., (Hu et al.

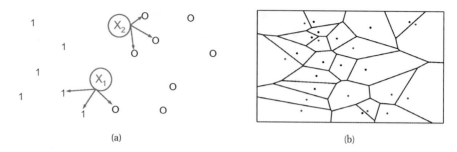

(a)   (b)

Figure 1.14   (a) Illustration of a K-nearest neighbors classifier in 2d for $K = 3$. The 3 nearest neighbors of test point $\mathbf{x}_1$ have labels 1, 1 and 0, so we predict $p(y = 1|\mathbf{x}_1, \mathcal{D}, K = 3) = 2/3$. The 3 nearest neighbors of test point x_2 have labels 0, 0, and 0, so we predict $p(y = 1|\mathbf{x}_2, \mathcal{D}, K = 3) = 0/3$. (b) Illustration of the Voronoi tesselation induced by 1-NN. Based on Figure 4.13 of (Duda et al. 2001). Figure generated by `knnVoronoi`.

2010). Such models often have better predictive accuracy than association rules, although they may be less interpretible. This is typical of the difference between data mining and machine learning: in data mining, there is more emphasis on interpretable models, whereas in machine learning, there is more emphasis on accurate models.

1.4   Some basic concepts in machine learning

In this Section, we provide an introduction to some key ideas in machine learning. We will expand on these concepts later in the book, but we introduce them briefly here, to give a flavor of things to come.

1.4.1   Parametric vs non-parametric models

In this book, we will be focussing on probabilistic models of the form $p(y|\mathbf{x})$ or $p(\mathbf{x})$, depending on whether we are interested in supervised or unsupervised learning respectively. There are many ways to define such models, but the most important distinction is this: does the model have a fixed number of parameters, or does the number of parameters grow with the amount of training data? The former is called a **parametric model**, and the latter is called a **non-parametric model**. Parametric models have the advantage of often being faster to use, but the disadvantage of making stronger assumptions about the nature of the data distributions. Non-parametric models are more flexible, but often computationally intractable for large datasets. We will give examples of both kinds of models in the sections below. We focus on supervised learning for simplicity, although much of our discussion also applies to unsupervised learning.

1.4.2   A simple non-parametric classifier: K-nearest neighbors

A simple example of a non-parametric classifier is the K **nearest neighbor** (**KNN**) classifier. This simply "looks at" the K points in the training set that are nearest to the test input $\mathbf{x}$,

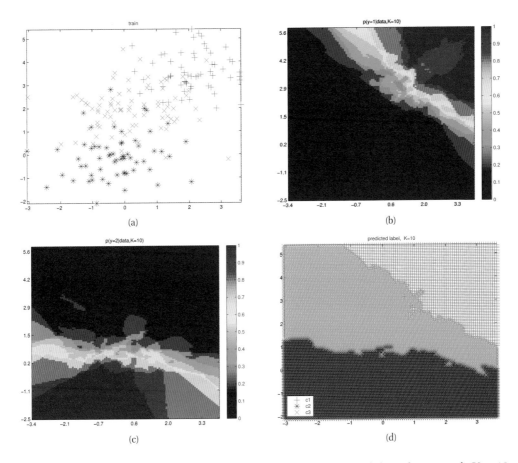

Figure 1.15 (a) Some synthetic 3-class training data in 2d. (b) Probability of class 1 for KNN with $K = 10$. (c) Probability of class 2. (d) MAP estimate of class label. Figure generated by `knnClassifyDemo`.

counts how many members of each class are in this set, and returns that empirical fraction as the estimate, as illustrated in Figure 1.14. More formally,

$$p(y = c | \mathbf{x}, \mathcal{D}, K) = \frac{1}{K} \sum_{i \in N_K(\mathbf{x}, \mathcal{D})} \mathbb{I}(y_i = c) \tag{1.2}$$

where $N_K(\mathbf{x}, \mathcal{D})$ are the (indices of the) K nearest points to $\mathbf{x}$ in $\mathcal{D}$ and $\mathbb{I}(e)$ is the **indicator function** defined as follows:

$$\mathbb{I}(e) = \begin{cases} 1 & \text{if } e \text{ is true} \\ 0 & \text{if } e \text{ is false} \end{cases} \tag{1.3}$$

This method is an example of **memory-based learning** or **instance-based learning**. It can be derived from a probabilistic framework as explained in Section 14.7.3. The most common

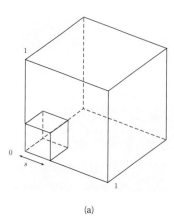
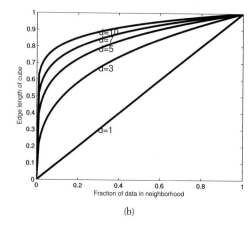

(a)   (b)

Figure 1.16 Illustration of the curse of dimensionality. (a) We embed a small cube of side s inside a larger unit cube. (b) We plot the edge length of a cube needed to cover a given volume of the unit cube as a function of the number of dimensions. Based on Figure 2.6 from (Hastie et al. 2009). Figure generated by `curseDimensionality`.

distance metric to use is Euclidean distance (which limits the applicability of the technique to data which is real-valued), although other metrics can be used.

Figure 1.15 gives an example of the method in action, where the input is two dimensional, we have three classes, and $K = 10$. (We discuss the effect of K below.) Panel (a) plots the training data. Panel (b) plots $p(y = 1|\mathbf{x}, \mathcal{D})$ where $\mathbf{x}$ is evaluated on a grid of points. Panel (c) plots $p(y = 2|\mathbf{x}, \mathcal{D})$. We do not need to plot $p(y = 3|\mathbf{x}, \mathcal{D})$, since probabilities sum to one. Panel (d) plots the MAP estimate $\hat{y}(\mathbf{x}) = \mathrm{argmax}_c(y = c|\mathbf{x}, \mathcal{D})$.

A KNN classifier with $K = 1$ induces a **Voronoi tessellation** of the points (see Figure 1.14(b)). This is a partition of space which associates a region $V(\mathbf{x}_i)$ with each point $\mathbf{x}_i$ in such a way that all points in $V(\mathbf{x}_i)$ are closer to $\mathbf{x}_i$ than to any other point. Within each cell, the predicted label is the label of the corresponding training point.

1.4.3 The curse of dimensionality

The KNN classifier is simple and can work quite well, provided it is given a good distance metric and has enough labeled training data. In fact, it can be shown that the KNN classifier can come within a factor of 2 of the best possible performance if $N \to \infty$ (Cover and Hart 1967).

However, the main problem with KNN classifiers is that they do not work well with high dimensional inputs. The poor performance in high dimensional settings is due to the **curse of dimensionality**.

To explain the curse, we give some examples from (Hastie et al. 2009, p22). Consider applying a KNN classifier to data where the inputs are uniformly distributed in the D-dimensional unit cube. Suppose we estimate the density of class labels around a test point $\mathbf{x}$ by "growing" a hyper-cube around $\mathbf{x}$ until it contains a desired fraction f of the data points. The expected edge length of this cube will be $e_D(f) = f^{1/D}$. If $D = 10$, and we want to base our estimate on 10%

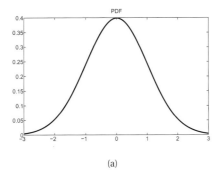
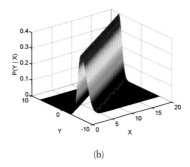

(a)   (b)

Figure 1.17 (a) A Gaussian pdf with mean 0 and variance 1. Figure generated by `gaussPlotDemo`. (b) Visualization of the conditional density model $p(y|x, \boldsymbol{\theta}) = \mathcal{N}(y|w_0 + w_1x, \sigma^2)$. The density falls off exponentially fast as we move away from the regression line. Figure generated by `linregWedgeDemo2`.

of the data, we have $e_{10}(0.1) = 0.8$, so we need to extend the cube 80% along each dimension around $\mathbf{x}$. Even if we only use 1% of the data, we find $e_{10}(0.01) = 0.63$: see Figure 1.16. Since the entire range of the data is only 1 along each dimension, we see that the method is no longer very local, despite the name "nearest neighbor". The trouble with looking at neighbors that are so far away is that they may not be good predictors about the behavior of the input-output function at a given point.

1.4.4 Parametric models for classification and regression

The main way to combat the curse of dimensionality is to make some assumptions about the nature of the data distribution (either $p(y|\mathbf{x})$ for a supervised problem or $p(\mathbf{x})$ for an unsupervised problem). These assumptions, known as **inductive bias**, are often embodied in the form of a **parametric model**, which is a statistical model with a fixed number of parameters. Below we briefly describe two widely used examples; we will revisit these and other models in much greater depth later in the book.

1.4.5 Linear regression

One of the most widely used models for regression is known as **linear regression**. This asserts that the response is a linear function of the inputs. This can be written as follows:

$$y(\mathbf{x}) = \mathbf{w}^T\mathbf{x} + \epsilon = \sum_{j=1}^{D} w_j x_j + \epsilon \tag{1.4}$$

where $\mathbf{w}^T\mathbf{x}$ represents the inner or **scalar product** between the input vector $\mathbf{x}$ and the model's **weight vector** $\mathbf{w}$[6], and ϵ is the **residual error** between our linear predictions and the true response.

---

6. In statistics, it is more common to denote the regression weights by $\boldsymbol{\beta}$.

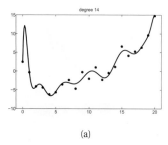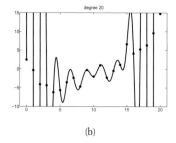

(a)   (b)

Figure 1.18 Polynomial of degrees 14 and 20 fit by least squares to 21 data points. Figure generated by `linregPolyVsDegree`.

We often assume that ϵ has a **Gaussian**[7] or **normal** distribution. We denote this by $\epsilon \sim \mathcal{N}(\mu, \sigma^2)$, where μ is the mean and σ^2 is the variance (see Chapter 2 for details). When we plot this distribution, we get the well-known **bell curve** shown in Figure 1.17(a).

To make the connection between linear regression and Gaussians more explicit, we can rewrite the model in the following form:

$$p(y|\mathbf{x}, \boldsymbol{\theta}) = \mathcal{N}(y|\mu(\mathbf{x}), \sigma^2(\mathbf{x})) \tag{1.5}$$

This makes it clear that the model is a conditional probability density. In the simplest case, we assume μ is a linear function of $\mathbf{x}$, so $\mu = \mathbf{w}^T\mathbf{x}$, and that the noise is fixed, $\sigma^2(x) = \sigma^2$. In this case, $\boldsymbol{\theta} = (\mathbf{w}, \sigma^2)$ are the parameters of the model.

For example, suppose the input is one-dimensional. We can represent the expected response as follows:

$$\mu(\mathbf{x}) = w_0 + w_1 x = \mathbf{w}^T\mathbf{x} \tag{1.6}$$

where w_0 is the intercept or **bias** term, w_1 is the slope, and where we have defined the vector $\mathbf{x} = (1, x)$. (Prepending a constant 1 term to an input vector is a common notational trick which allows us to combine the intercept term with the other terms in the model.) If w_1 is positive, it means we expect the output to increase as the input increases. This is illustrated in 1d in Figure 1.17(b); a more conventional plot, of the mean response vs x, is shown in Figure 1.7(a).

Linear regression can be made to model non-linear relationships by replacing $\mathbf{x}$ with some non-linear function of the inputs, $\phi(\mathbf{x})$. That is, we use

$$p(y|\mathbf{x}, \boldsymbol{\theta}) = \mathcal{N}(y|\mathbf{w}^T\phi(\mathbf{x}), \sigma^2) \tag{1.7}$$

This is known as **basis function expansion**. For example, Figure 1.18 illustrates the case where $\phi(\mathbf{x}) = [1, x, x^2, \ldots, x^d]$, for $d = 14$ and $d = 20$; this is known as **polynomial regression**. We will consider other kinds of basis functions later in the book. In fact, many popular machine learning methods — such as support vector machines, neural networks, classification and regression trees, etc. — can be seen as just different ways of estimating basis functions from data, as we discuss in Chapters 14 and 16.

---

7. Carl Friedrich Gauss (1777–1855) was a German mathematician and physicist.

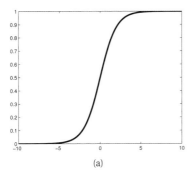
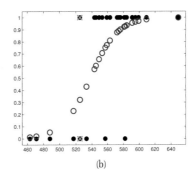

(a)   (b)

Figure 1.19 (a) The sigmoid or logistic function. We have $\text{sigm}(-\infty) = 0$, $\text{sigm}(0) = 0.5$, and $\text{sigm}(\infty) = 1$. Figure generated by `sigmoidPlot`. (b) Logistic regression for SAT scores. Solid black dots are the data. The open red circles are the predicted probabilities. The green crosses denote two students with the same SAT score of 525 (and hence same input representation x) but with different training labels (one student passed, $y = 1$, the other failed, $y = 0$). Hence this data is not perfectly separable using just the SAT feature. Figure generated by `logregSATdemo`.

1.4.6 Logistic regression

We can generalize linear regression to the (binary) classification setting by making two changes. First we replace the Gaussian distribution for y with a **Bernoulli** distribution[8],which is more appropriate for the case when the response is binary, $y \in \{0, 1\}$. That is, we use

$$p(y|\mathbf{x}, \mathbf{w}) = \text{Ber}(y|\mu(\mathbf{x})) \tag{1.8}$$

where $\mu(\mathbf{x}) = \mathbb{E}[y|\mathbf{x}] = p(y = 1|\mathbf{x})$. Second, we compute a linear combination of the inputs, as before, but then we pass this through a function that ensures $0 \le \mu(\mathbf{x}) \le 1$ by defining

$$\mu(\mathbf{x}) = \text{sigm}(\mathbf{w}^T \mathbf{x}) \tag{1.9}$$

where $\text{sigm}(\eta)$ refers to the **sigmoid** function, also known as the **logistic** or **logit** function. This is defined as

$$\text{sigm}(\eta) \triangleq \frac{1}{1 + \exp(-\eta)} = \frac{e^\eta}{e^\eta + 1} \tag{1.10}$$

The term "sigmoid" means S-shaped: see Figure 1.19(a) for a plot. It is also known as a **squashing function**, since it maps the whole real line to $[0, 1]$, which is necessary for the output to be interpreted as a probability.

Putting these two steps together we get

$$p(y|\mathbf{x}, \mathbf{w}) = \text{Ber}(y|\text{sigm}(\mathbf{w}^T \mathbf{x})) \tag{1.11}$$

This is called **logistic regression** due to its similarity to linear regression (although it is a form of classification, not regression!).

---

8. Jacob Bernoulli (1654–1705) was a Swiss mathematician.

A simple example of logistic regression is shown in Figure 1.19(b), where we plot

$$p(y_i = 1|x_i, \mathbf{w}) = \text{sigm}(w_0 + w_1 x_i) \tag{1.12}$$

where x_i is the SAT[9] score of student i and y_i is whether they passed or failed a class. The solid black dots show the training data, and the red circles plot $p(y = 1|\mathbf{x}_i, \hat{\mathbf{w}})$, where $\hat{\mathbf{w}}$ are the parameters estimated from the training data (we discuss how to compute these estimates in Section 8.3.4).

If we threshold the output probability at 0.5, we can induce a **decision rule** of the form

$$\hat{y}(x) = 1 \iff p(y = 1|\mathbf{x}) > 0.5 \tag{1.13}$$

By looking at Figure 1.19(b), we see that $\text{sigm}(w_0 + w_1 x) = 0.5$ for $x \approx 545 = x^*$. We can imagine drawing a vertical line at $x = x^*$; this is known as a decision boundary. Everything to the left of this line is classified as a 0, and everything to the right of the line is classified as a 1.

We notice that this decision rule has a non-zero error rate even on the training set. This is because the data is not **linearly separable**, i.e., there is no straight line we can draw to separate the 0s from the 1s. We can create models with non-linear decision boundaries using basis function expansion, just as we did with non-linear regression. We will see many examples of this later in the book.

1.4.7 Overfitting

When we fit highly flexible models, we need to be careful that we do not **overfit** the data, that is, we should avoid trying to model every minor variation in the input, since this is more likely to be noise than true signal. This is illustrated in Figure 1.18(b), where we see that using a high degree polynomial results in a curve that is very "wiggly". It is unlikely that the true function has such extreme oscillations. Thus using such a model might result in inaccurate predictions of future outputs.

As another example, consider the KNN classifier. The value of K can have a large effect on the behavior of this model. When $K = 1$, the method makes no errors on the training set (since we just return the labels of the original training points), but the resulting prediction surface is very "wiggly" (see Figure 1.20(a)). Therefore the method may not work well at predicting future data. In Figure 1.20(b), we see that using $K = 5$ results in a smoother prediction surface, because we are averaging over a larger neighborhood. As K increases, the predictions becomes smoother until, in the limit of $K = N$, we end up predicting the majority label of the whole data set. Below we discuss how to pick the "right" value of K.

1.4.8 Model selection

When we have a variety of models of different complexity (e.g., linear or logistic regression models with different degree polynomials, or KNN classifiers with different values of K), how should we pick the right one? A natural approach is to compute the **misclassification rate** on

---

9. SAT stands for "Scholastic Aptitude Test". This is a standardized test for college admissions used in the United States (the data in this example is from (Johnson and Albert 1999, p87)).

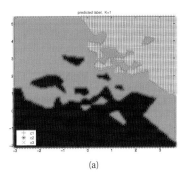
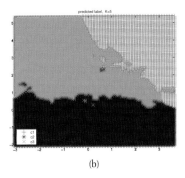

(a)   (b)

Figure 1.20 Prediction surface for KNN on the data in Figure 1.15(a). (a) K=1. (b) K=5. Figure generated by `knnClassifyDemo`.

the training set for each method. This is defined as follows:

$$\text{err}(f, \mathcal{D}) = \frac{1}{N} \sum_{i=1}^{N} \mathbb{I}(f(\mathbf{x}_i) \neq y_i) \tag{1.14}$$

where $f(\mathbf{x})$ is our classifier. In Figure 1.21(a), we plot this error rate vs K for a KNN classifier (dotted blue line). We see that increasing K *increases* our error rate on the training set, because we are over-smoothing. As we said above, we can get minimal error on the training set by using $K = 1$, since this model is just memorizing the data.

However, what we care about is **generalization error**, which is the expected value of the misclassification rate when averaged over future data (see Section 6.3 for details). This can be approximated by computing the misclassification rate on a large independent *test set*, not used during model training. We plot the test error vs K in Figure 1.21(a) in solid red (upper curve). Now we see a **U-shaped curve**: for complex models (small K), the method overfits, and for simple models (big K), the method **underfits**. Therefore, an obvious way to pick K is to pick the value with the minimum error on the test set (in this example, any value between 10 and 100 should be fine).

Unfortunately, when training the model, we don't have access to the test set (by assumption), so we cannot use the test set to pick the model of the right complexity.[10] However, we can create a test set by partitioning the training set into two: the part used for training the model, and a second part, called the **validation set**, used for selecting the model complexity. We then fit all the models on the training set, and evaluate their performance on the validation set, and pick the best. Once we have picked the best, we can refit it to all the available data. If we have a separate test set, we can evaluate performance on this, in order to estimate the accuracy of our method. (We discuss this in more detail in Section 6.5.3.)

Often we use about 80% of the data for the training set, and 20% for the validation set. But if the number of training cases is small, this technique runs into problems, because the model

---

10. In academic settings, we usually do have access to the test set, but we should not use it for model fitting or model selection, otherwise we will get an unrealistically optimistic estimate of performance of our method. This is one of the "golden rules" of machine learning research.

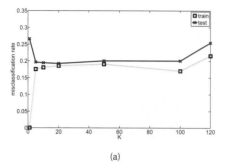
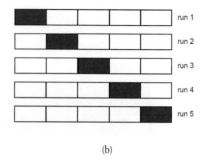

| (a) | (b) |

Figure 1.21   (a) Misclassification rate vs K in a K-nearest neighbor classifier. On the left, where K is small, the model is complex and hence we overfit. On the right, where K is large, the model is simple and we underfit. Dotted blue line: training set (size 200). Solid red line: test set (size 500). (b) Schematic of 5-fold cross validation. Figure generated by `knnClassifyDemo`.

won't have enough data to train on, and we won't have enough data to make a reliable estimate of the future performance.

A simple but popular solution to this is to use **cross validation** (**CV**). The idea is simple: we split the training data into K **folds**; then, for each fold $k \in \{1, \dots, K\}$, we train on all the folds but the k'th, and test on the k'th, in a round-robin fashion, as sketched in Figure 1.21(b). We then compute the error averaged over all the folds, and use this as a proxy for the test error. (Note that each point gets predicted only once, although it will be used for training $K - 1$ times.) It is common to use $K = 5$; this is called 5-fold CV. If we set $K = N$, then we get a method called **leave-one out cross validation**, or **LOOCV**, since in fold i, we train on all the data cases except for i, and then test on i. Exercise 1.3 asks you to compute the 5-fold CV estimate of the test error vs K, and to compare it to the empirical test error in Figure 1.21(a).

Choosing K for a KNN classifier is a special case of a more general problem known as **model selection**, where we have to choose between models with different degrees of flexibility. Cross-validation is widely used for solving such problems, although we will discuss other approaches later in the book.

1.4.9   No free lunch theorem

> All models are wrong, but some models are useful. — George Box (Box and Draper 1987, p424).[11]

Much of machine learning is concerned with devising different models, and different algorithms to fit them. We can use methods such as cross validation to empirically choose the best method for our particular problem. However, there is no single best model that works optimally for all kinds of problems — this is sometimes called the **no free lunch theorem** (Wolpert 1996). The reason for this is that a set of assumptions that works well in one domain may work poorly in another.

---

11. George Box is a retired statistics professor at the University of Wisconsin.

As a consequence of the no free lunch theorem, we need to develop many different types of models, to cover the wide variety of data that occurs in the real world. And for each model, there may be many different algorithms we can use to train the model, which make different speed-accuracy-complexity tradeoffs. It is this combination of data, models and algorithms that we will be studying in the subsequent chapters.

Exercises

Exercise 1.1 KNN classifier on shuffled MNIST data

Run `mnist1NNdemo` and verify that the misclassification rate (on the first 1000 test cases) of MNIST of a 1-NN classifier is 3.8%. (If you run it all on all 10,000 test cases, the error rate is 3.09%.) Modify the code so that you first randomly permute the features (columns of the training and test design matrices), as in `shuffledDigitsDemo`, and then apply the classifier. Verify that the error rate is not changed.

Exercise 1.2 Approximate KNN classifiers

Use the Matlab/C++ code at `http://people.cs.ubc.ca/~mariusm/index.php/FLANN/FLANN` to perform approximate nearest neighbor search, and combine it with `mnist1NNdemo` to classify the MNIST data set. How much speedup do you get, and what is the drop (if any) in accuracy?

Exercise 1.3 CV for KNN

Use `knnClassifyDemo` to plot the CV estimate of the misclassification rate on the test set. Compare this to Figure 1.21(a). Discuss the similarities and differences to the test error rate.

2 *Probability*

2.1 Introduction

> Probability theory is nothing but common sense reduced to calculation. — Pierre Laplace, 1812

In the previous chapter, we saw how probability can play a useful role in machine learning. In this chapter, we discuss probability theory in more detail. We do not have space to go into great detail — for that, you are better off consulting some of the excellent textbooks available on this topic, such as (Jaynes 2003; Bertsekas and Tsitsiklis 2008; Wasserman 2004). But we will briefly review many of the key ideas you will need in later chapters.

Before we start with the more technical material, let us pause and ask: what is probability? We are all familiar with phrases such as "the probability that a coin will land heads is 0.5". But what does this mean? There are actually at least two different interpretations of probability. One is called the **frequentist** interpretation. In this view, probabilities represent long run frequencies of events. For example, the above statement means that, if we flip the coin many times, we expect it to land heads about half the time.[1]

The other interpretation is called the **Bayesian** interpretation of probability. In this view, probability is used to quantify our **uncertainty** about something; hence it is fundamentally related to information rather than repeated trials (Jaynes 2003; Lindley 2006). In the Bayesian view, the above statement means we believe the coin is equally likely to land heads or tails on the next toss.

One big advantage of the Bayesian interpretation is that it can be used to model our uncertainty about events that do not have long term frequencies. For example, we might want to compute the probability that the polar ice cap will melt by 2020 CE. This event will happen zero or one times, but cannot happen repeatedly. Nevertheless, we ought to be able to quantify our uncertainty about this event; based on how probable we think this event is, we will (hopefully!) take appropriate actions (see Section 5.7 for a discussion of optimal decision making under uncertainty). To give some more machine learning oriented examples, we might have received a specific email message, and want to compute the probability it is spam. Or we might have observed a "blip" on our radar screen, and want to compute the probability distribution over the location of the corresponding target (be it a bird, plane, or missile). In all these cases, the

---

1. Actually, the Stanford statistician (and former professional magician) Persi Diaconis has shown that a coin is about 51% likely to land facing the same way up as it started, due to the physics of the problem (Diaconis et al. 2007).

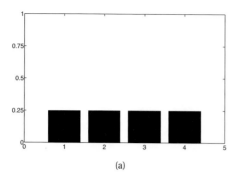 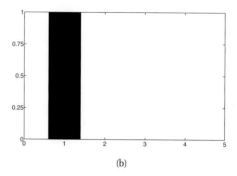

Figure 2.1   (A) a uniform distribution on $\{1, 2, 3, 4\}$, with $p(x = k) = 1/4$. (b) a degenerate distribution $p(x) = 1$ if $x = 1$ and $p(x) = 0$ if $x \in \{2, 3, 4\}$. Figure generated by `discreteProbDistFig`.

idea of repeated trials does not make sense, but the Bayesian interpretation is valid and indeed quite natural. We shall therefore adopt the Bayesian interpretation in this book. Fortunately, the basic rules of probability theory are the same, no matter which interpretation is adopted.

2.2   A brief review of probability theory

This section is a very brief review of the basics of probability theory, and is merely meant as a refresher for readers who may be "rusty". Please consult some of the excellent textbooks available on this topic, such as (Rice 1995; Jaynes 2003; Bertsekas and Tsitsiklis 2008; Wasserman 2004; Lindley 2006), for further details, if necessary. Readers who are already familiar with these basics may safely skip this section.

2.2.1   Discrete random variables

The expression $p(A)$ denotes the probability that the event A is true. For example, A might be the logical expression "it will rain tomorrow". We require that $0 \leq p(A) \leq 1$, where $p(A) = 0$ means the event definitely will not happen, and $p(A) = 1$ means the event definitely will happen. We write $p(\overline{A})$ to denote the probability of the event not A; this is defined to $p(\overline{A}) = 1 - p(A)$. We will often write $A = 1$ to mean the event A is true, and $A = 0$ to mean the event A is false.

We can extend the notion of binary events by defining a **discrete random variable** (**rv** for short) X, which can take on any value from a finite or countably infinite set $\mathcal{X}$. We denote the probability of the event that $X = x$ by $p(X = x)$, or just $p(x)$ for short. Here $p()$ is called a **probability mass function** or **pmf**. This satisfies the properties $0 \leq p(x) \leq 1$ and $\sum_{x \in \mathcal{X}} p(x) = 1$. Figure 2.1 shows two pmf's defined on the finite **state space** $\mathcal{X} = \{1, 2, 3, 4\}$. On the left we have a uniform distribution, $p(x) = 1/4$, and on the right, we have a degenerate distribution, $p(x) = \mathbb{I}(x = 1)$, where $\mathbb{I}()$ is the binary **indicator function**. This distribution represents the fact that X is always equal to the value 1, in other words, it is a constant.

2.2.2 Fundamental rules

In this section, we review the basic rules of probability.

2.2.2.1 Probability of a union of two events

Given two events, A and B, we define the probability of A or B as follows:

$$p(A \vee B) \quad = \quad p(A) + p(B) - p(A \wedge B) \tag{2.1}$$

$$= \quad p(A) + p(B) \text{ if } A \text{ and } B \text{ are mutually exclusive} \tag{2.2}$$

2.2.2.2 Joint probabilities

We define the probability of the joint event A and B as follows:

$$p(A, B) = p(A \wedge B) = p(A|B)p(B) \tag{2.3}$$

This is sometimes called the **product rule**. Given a **joint distribution** on two events $p(A, B)$, we define the **marginal distribution** as follows:

$$p(A) = \sum_b p(A, B) = \sum_b p(A|B = b)p(B = b) \tag{2.4}$$

where we are summing over all possible states of B. We can define $p(B)$ similarly. This is sometimes called the **sum rule** or the **rule of total probability**.

The product rule can be applied multiple times to yield the **chain rule** of probability:

$$p(X_{1:D}) = p(X_1)p(X_2|X_1)p(X_3|X_2, X_1)p(X_4|X_1, X_2, X_3) \ldots p(X_D|X_{1:D-1}) \tag{2.5}$$

where we introduce the Matlab-like notation $1 : D$ to denote the set $\{1, 2, \ldots, D\}$.

2.2.2.3 Conditional probability

We define the **conditional probability** of event A, given that event B is true, as follows:

$$p(A|B) = \frac{p(A, B)}{p(B)} \text{ if } p(B) > 0 \tag{2.6}$$

2.2.3 Bayes' rule

Combining the definition of conditional probability with the product and sum rules yields **Bayes' rule**, also called **Bayes' Theorem**[2]:

$$p(X = x|Y = y) = \frac{p(X = x, Y = y)}{p(Y = y)} = \frac{p(X = x)p(Y = y|X = x)}{\sum_{x'} p(X = x')p(Y = y|X = x')} \tag{2.7}$$

Sir Harold Jeffreys wrote that "Bayes's theorem is to the theory of probability what Pythagoras's theorem is to geometry" (Jeffreys 1973). We show two simple applications of the theorem below, but we will encounter many, many more throughout the book. (See also (McGrayne 2011) for an interesting historical account of its applications to many other real-world problems.)

---

2. Thomas Bayes (1702–1761) was an English mathematician and Presbyterian minister. Technically speaking, we should write "Bayes' rule", but often we will just write "Bayes rule" instead, dropping the apostrophe, since it is less fussy.

2.2.3.1   Example: medical diagnosis

As an example of how to use this rule, consider the following medical diagnosis problem. Suppose you are a woman in your 40s, and you decide to have a medical test for breast cancer called a **mammogram**. If the test is positive, what is the probability you have cancer? That obviously depends on how reliable the test is. Suppose you are told the test has a **sensitivity** of 80%, which means, if you have cancer, the test will be positive with probability 0.8. In other words,

$$p(x = 1|y = 1) = 0.8 \tag{2.8}$$

where $x = 1$ is the event the mammogram is positive, and $y = 1$ is the event you have breast cancer. Many people conclude they are therefore 80% likely to have cancer. But this is false! It ignores the prior probability of having breast cancer, which fortunately is quite low:

$$p(y = 1) = 0.004 \tag{2.9}$$

Ignoring this prior is called the **base rate fallacy**. We also need to take into account the fact that the test may be a **false positive** or **false alarm**. Unfortunately, such false positives are quite likely (with current screening technology):

$$p(x = 1|y = 0) = 0.1 \tag{2.10}$$

Combining these three terms using Bayes rule, we can compute the correct answer as follows:

$$p(y = 1|x = 1) = \frac{p(x = 1|y = 1)p(y = 1)}{p(x = 1|y = 1)p(y = 1) + p(x = 1|y = 0)p(y = 0)} \tag{2.11}$$

$$= \frac{0.8 \times 0.004}{0.8 \times 0.004 + 0.1 \times 0.996} = 0.031 \tag{2.12}$$

where $p(y = 0) = 1 - p(y = 1) = 0.996$. In other words, if you test positive, you only have about a 3% chance of actually having breast cancer![3]

2.2.3.2   Example: Generative classifiers

We can generalize the medical diagnosis example to classify feature vectors $\mathbf{x}$ of arbitrary type as follows:

$$p(y = c|\mathbf{x}) = \frac{p(y = c)p(\mathbf{x}|y = c)}{\sum_{c'} p(y = c'|\boldsymbol{\theta})p(\mathbf{x}|y = c')} \tag{2.13}$$

This is called a **generative classifier**, since it specifies how to generate the data using the **class-conditional density** $p(\mathbf{x}|y = c)$ and the class prior $p(y = c)$. We discuss such models in detail in Chapters 3 and 4. An alternative approach is to directly fit the class posterior, $p(y = c|\mathbf{x})$; this is known as a discriminative classifier. We discuss the pros and cons of the two approaches in Section 8.6.

---

3. These numbers are from (McGrayne 2011, p257). Based on this analysis, the US government decided not to recommend annual mammogram screening to women in their 40s: the number of false alarms would cause needless worry and stress amongst women, and result in unnecesssary, expensive, and potentially harmful followup tests. See Section 5.7 for the optimal way to trade off risk versus reward in the face of uncertainty.

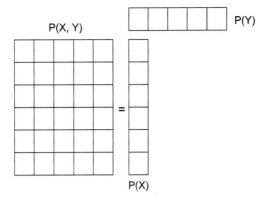

Figure 2.2   Computing $p(x, y) = p(x)p(y)$, where $X \perp Y$. Here X and Y are discrete random variables; X has 6 possible states (values) and Y has 5 possible states. A general joint distribution on two such variables would require $(6 \times 5) - 1 = 29$ parameters to define it (we subtract 1 because of the sum-to-one constraint). By assuming (unconditional) independence, we only need $(6 - 1) + (5 - 1) = 9$ parameters to define $p(x, y)$.

2.2.4   Independence and conditional independence

We say X and Y are **unconditionally independent** or **marginally independent**, denoted $X \perp Y$, if we can represent the joint as the product of the two marginals (see Figure 2.2), i.e.,

$$X \perp Y \iff p(X, Y) = p(X)p(Y) \tag{2.14}$$

In general, we say a set of variables is mutually independent if the joint can be written as a product of marginals.

Unfortunately, unconditional independence is rare, because most variables can influence most other variables. However, usually this influence is mediated via other variables rather than being direct. We therefore say X and Y are **conditionally independent** (CI) given Z iff the conditional joint can be written as a product of conditional marginals:

$$X \perp Y | Z \iff p(X, Y | Z) = p(X | Z)p(Y | Z) \tag{2.15}$$

When we discuss graphical models in Chapter 10, we will see that we can write this assumption as a graph $X - Z - Y$, which captures the intuition that all the dependencies between X and Y are mediated via Z. For example, the probability it will rain tomorrow (event X) is independent of whether the ground is wet today (event Y), given knowledge of whether it is raining today (event Z). Intuitively, this is because Z "causes" both X and Y, so if we know Z, we do not need to know about Y in order to predict X or vice versa. We shall expand on this concept in Chapter 10.

Another characterization of CI is this:

Theorem 2.2.1. $X \perp Y | Z$ *iff there exist functions g and h such that*

$$p(x, y | z) = g(x, z)h(y, z) \tag{2.16}$$

for all x, y, z such that $p(z) > 0$.

See Exercise 2.8 for the proof.

CI assumptions allow us to build large probabilistic models from small pieces. We will see many examples of this throughout the book. In particular, in Section 3.5, we discuss naive Bayes classifiers, in Section 17.2, we discuss Markov models, and in Chapter 10 we discuss graphical models; all of these models heavily exploit CI properties.

2.2.5   Continuous random variables

So far, we have only considered reasoning about uncertain discrete quantities. We will now show (following (Jaynes 2003, p107)) how to extend probability to reason about uncertain continuous quantities.

Suppose X is some uncertain continuous quantity. The probability that X lies in any interval $a \leq X \leq b$ can be computed as follows. Define the events $A = (X \leq a)$, $B = (X \leq b)$ and $W = (a < X \leq b)$. We have that $B = A \vee W$, and since A and W are mutually exclusive, the sum rules gives

$$p(B) = p(A) + p(W) \tag{2.17}$$

and hence

$$p(W) = p(B) - p(A) \tag{2.18}$$

Define the function $F(q) \triangleq p(X \leq q)$. This is called the **cumulative distribution function** or **cdf** of X. This is a monotonically non-decreasing function. See Figure 2.3(a) for an example. Using this notation we have

$$p(a < X \leq b) = F(b) - F(a) \tag{2.19}$$

Now define $f(x) = \frac{d}{dx}F(x)$ (we assume this derivative exists); this is called the **probability density function** or **pdf**. See Figure 2.3(b) for an example. Given a pdf, we can compute the probability of a continuous variable being in a finite interval as follows:

$$p(a < X \leq b) = \int_a^b f(x)dx \tag{2.20}$$

As the size of the interval gets smaller, we can write

$$p(x \leq X \leq x + dx) \approx p(x)dx \tag{2.21}$$

We require $p(x) \geq 0$, but it is possible for $p(x) > 1$ for any given x, so long as the density integrates to 1. As an example, consider the **uniform distribution** $\mathrm{Unif}(a, b)$:

$$\mathrm{Unif}(x|a, b) = \frac{1}{b - a} \mathbb{I}(a \leq x \leq b) \tag{2.22}$$

If we set $a = 0$ and $b = \frac{1}{2}$, we have $p(x) = 2$ for any $x \in [0, \frac{1}{2}]$.

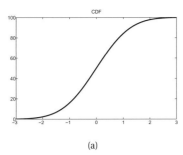
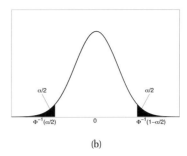

(a)   (b)

Figure 2.3 (a) Plot of the cdf for the standard normal, $\mathcal{N}(0,1)$. (b) Corresponding pdf. The shaded regions each contain $\alpha/2$ of the probability mass. Therefore the nonshaded region contains $1 - \alpha$ of the probability mass. If the distribution is Gaussian $\mathcal{N}(0,1)$, then the leftmost cutoff point is $\Phi^{-1}(\alpha/2)$, where Φ is the cdf of the Gaussian. By symmetry, the rightmost cutoff point is $\Phi^{-1}(1 - \alpha/2) = -\Phi^{-1}(\alpha/2)$. If $\alpha = 0.05$, the central interval is 95%, and the left cutoff is -1.96 and the right is 1.96. Figure generated by `quantileDemo`.

2.2.6 Quantiles

If the cdf F is a monotonically increasing function, it has an inverse; let us denote this by F^{-1}. If F is the cdf of X, then $F^{-1}(\alpha)$ is the value of x_α such that $p(X \leq x_\alpha) = \alpha$; this is called the α **quantile** of F. The value $F^{-1}(0.5)$ is the **median** of the distribution, with half of the probability mass on the left, and half on the right. The values $F^{-1}(0.25)$ and $F^{-1}(0.75)$ are the lower and upper **quartiles**.

We can also use the inverse cdf to compute **tail area probabilities**. For example, if Φ is the cdf of the Gaussian distribution $\mathcal{N}(0,1)$, then points to the left of $\Phi^{-1}(\alpha/2)$ contain $\alpha/2$ of the probability mass, as illustrated in Figure 2.3(b). By symmetry, points to the right of $\Phi^{-1}(1-\alpha/2)$ also contain $\alpha/2$ of the mass. Hence the central interval $(\Phi^{-1}(\alpha/2), \Phi^{-1}(1 - \alpha/2))$ contains $1 - \alpha$ of the mass. If we set $\alpha = 0.05$, the central 95% interval is covered by the range

$$(\Phi^{-1}(0.025), \Phi^{-1}(0.975)) = (-1.96, 1.96) \tag{2.23}$$

If the distribution is $\mathcal{N}(\mu, \sigma^2)$, then the 95% interval becomes $(\mu - 1.96\sigma, \mu + 1.96\sigma)$. This is sometimes approximated by writing $\mu \pm 2\sigma$.

2.2.7 Mean and variance

The most familiar property of a distribution is its **mean**, or **expected value**, denoted by μ. For discrete rv's, it is defined as $\mathbb{E}[X] \triangleq \sum_{x \in \mathcal{X}} x\, p(x)$, and for continuous rv's, it is defined as $\mathbb{E}[X] \triangleq \int_{\mathcal{X}} x\, p(x) dx$. (If $\mathbb{E}[X] \triangleq \int_{\mathcal{X}} |x|\, p(x) dx$ is not finite, the mean is not defined; we will see some examples of this later.)

The **variance** is a measure of the "spread" of a distribution, denoted by σ^2. This is defined as follows:

$$\text{var}[X] \triangleq \mathbb{E}\left[(X-\mu)^2\right] = \int (x-\mu)^2 p(x)dx \tag{2.24}$$

$$= \int x^2 p(x)dx + \mu^2 \int p(x)dx - 2\mu \int x p(x)dx = \mathbb{E}\left[X^2\right] - \mu^2 \tag{2.25}$$

from which we derive the useful result

$$\mathbb{E}\left[X^2\right] = \mu^2 + \sigma^2 \tag{2.26}$$

The **standard deviation** is defined as

$$\text{std}[X] \triangleq \sqrt{\text{var}[X]} \tag{2.27}$$

This is useful since it has the same units as X itself.

2.3   Some common discrete distributions

In this section, we review some commonly used parametric distributions defined on discrete state spaces, both finite and countably infinite.

2.3.1   The binomial and Bernoulli distributions

Suppose we toss a coin n times. Let $X \in \{0,\dots,n\}$ be the number of heads. If the probability of heads is θ, then we say X has a **binomial** distribution, written as $X \sim \text{Bin}(n,\theta)$. The pmf is given by

$$\text{Bin}(k|n,\theta) \triangleq \binom{n}{k}\theta^k (1-\theta)^{n-k} \tag{2.28}$$

where

$$\binom{n}{k} \triangleq \frac{n!}{(n-k)!k!} \tag{2.29}$$

is the number of ways to choose k items from n (this is known as the **binomial coefficient**, and is pronounced "n choose k"). See Figure 2.4 for some examples of the binomial distribution. This distribution has the following mean and variance:

$$\text{mean} = n\theta, \quad \text{var} = n\theta(1-\theta) \tag{2.30}$$

Now suppose we toss a coin only once. Let $X \in \{0,1\}$ be a binary random variable, with probability of "success" or "heads" of θ. We say that X has a **Bernoulli** distribution. This is written as $X \sim \text{Ber}(\theta)$, where the pmf is defined as

$$\text{Ber}(x|\theta) = \theta^{\mathbb{I}(x=1)}(1-\theta)^{\mathbb{I}(x=0)} \tag{2.31}$$

In other words,

$$\text{Ber}(x|\theta) = \begin{cases} \theta & \text{if } x=1 \\ 1-\theta & \text{if } x=0 \end{cases} \tag{2.32}$$

This is obviously just a special case of a Binomial distribution with $n=1$.

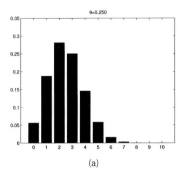
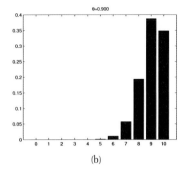

(a)   (b)

Figure 2.4 Illustration of the binomial distribution with $n = 10$ and $\theta \in \{0.25, 0.9\}$. Figure generated by `binomDistPlot`.

2.3.2 The multinomial and multinoulli distributions

The binomial distribution can be used to model the outcomes of coin tosses. To model the outcomes of tossing a K-sided die, we can use the **multinomial** distribution. This is defined as follows: let $\mathbf{x} = (x_1, \ldots, x_K)$ be a random vector, where x_j is the number of times side j of the die occurs. Then $\mathbf{x}$ has the following pmf:

$$\text{Mu}(\mathbf{x}|n, \boldsymbol{\theta}) \triangleq \binom{n}{x_1 \ldots x_K} \prod_{j=1}^{K} \theta_j^{x_j} \tag{2.33}$$

where θ_j is the probability that side j shows up, and

$$\binom{n}{x_1 \ldots x_K} \triangleq \frac{n!}{x_1! x_2! \cdots x_K!} \tag{2.34}$$

is the **multinomial coefficient** (the number of ways to divide a set of size $n = \sum_{k=1}^{K} x_k$ into subsets with sizes x_1 up to x_K).

Now suppose $n = 1$. This is like rolling a K-sided dice once, so $\mathbf{x}$ will be a vector of 0s and 1s (a bit vector), in which only one bit can be turned on. Specifically, if the dice shows up as face k, then the k'th bit will be on. In this case, we can think of x as being a scalar categorical random variable with K states (values), and $\mathbf{x}$ is its **dummy encoding**, that is, $\mathbf{x} = [\mathbb{I}(x = 1), \ldots, \mathbb{I}(x = K)]$. For example, if $K = 3$, we encode the states 1, 2 and 3 as $(1, 0, 0)$, $(0, 1, 0)$, and $(0, 0, 1)$. This is also called a **one-hot encoding**, since we imagine that only one of the K "wires" is "hot" or on. In this case, the pmf becomes

$$\text{Mu}(\mathbf{x}|1, \boldsymbol{\theta}) = \prod_{j=1}^{K} \theta_j^{\mathbb{I}(x_j=1)} \tag{2.35}$$

See Figure 2.1 for some examples. This very common special case is known as a **categorical** or **discrete** distribution. (Gustavo Lacerda suggested we call it the **multinoulli distribution**, by analogy with the Binomial/ Bernoulli distinction, a term which we shall adopt in this book.) We

Name	n	K	x
Multinomial	-	-	$\mathbf{x} \in \{0, 1, \ldots, n\}^K$, $\sum_{k=1}^{K} x_k = n$
Multinoulli	1	-	$\mathbf{x} \in \{0, 1\}^K$, $\sum_{k=1}^{K} x_k = 1$ (1-of-K encoding)
Binomial	-	1	$x \in \{0, 1, \ldots, n\}$
Bernoulli	1	1	$x \in \{0, 1\}$

Table 2.1 Summary of the multinomial and related distributions.

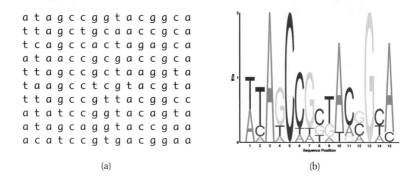

```
a t a g c c g g t a c g g c a
t t a g c t g c a a c c g c a
t c a g c c a c t a g a g c a
a t a a c c g c g a c c g c a
t t a g c c g c t a a g g t a
t a a g c c t c g t a c g t a
t t a g c c g t t a c g g c c
a t a t c c g g t a c a g t a
a t a g c a g g t a c c g a a
a c a t c c g t g a c g g a a
```

(a)   (b)

Figure 2.5 (a) Some aligned DNA sequences. (b) The corresponding sequence logo. The vertical axis represents $2 - H$, where H is the entropy of the distribution for that column (measured in bits). Thus deterministic distributions (with an entropy of 0) have height 2, and uniform distributions (with an entropy of 2) have height 0. Figure generated by `seqlogoDemo`.

will use the following notation for this case:

$$\mathrm{Cat}(x|\boldsymbol{\theta}) \triangleq \mathrm{Mu}(\mathbf{x}|1, \boldsymbol{\theta}) \tag{2.36}$$

In otherwords, if $x \sim \mathrm{Cat}(\boldsymbol{\theta})$, then $p(x = j|\boldsymbol{\theta}) = \theta_j$. See Table 2.1 for a summary.

2.3.2.1   Application: DNA sequence motifs

An interesting application of multinomial models arises in **biosequence analysis**. Suppose we have a set of (aligned) DNA sequences, such as in Figure 2.5(a), where there are 10 rows (sequences) and 15 columns (locations along the genome). We see that several locations are conserved by evolution (e.g., because they are part of a gene coding region), since the corresponding columns tend to be "pure". For example, column 13 is all G's.

One way to visually summarize the data is by using a **sequence logo**: see Figure 2.5(b). We plot the letters A, C, G and T, with the most probable letter on the top, and with a fontsize related to their empirical probability (for details of the vertical scaling, see Section 2.8.1). The empirical probability distribution at location t, $\hat{\boldsymbol{\theta}}_t$, is obtained by normalizing the vector of

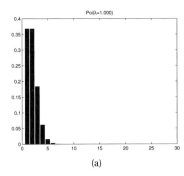
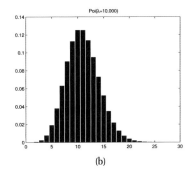

(a)   (b)

Figure 2.6   Illustration of some Poisson distributions for $\lambda \in \{1, 10\}$. We have truncated the x-axis to 30 for clarity, but the support of the distribution is over all the non-negative integers. Figure generated by `poissonPlotDemo`.

counts (see Equation 3.48):

$$\mathbf{N}_t = \left(\sum_{i=1}^{N} \mathbb{I}(X_{it} = 1), \sum_{i=1}^{N} \mathbb{I}(X_{it} = 2), \sum_{i=1}^{N} \mathbb{I}(X_{it} = 3), \sum_{i=1}^{N} \mathbb{I}(X_{it} = 4) \right) \tag{2.37}$$

$$\hat{\boldsymbol{\theta}}_t = \mathbf{N}_t / N \tag{2.38}$$

This distribution is known as a **motif**. We can also compute the most probable letter in each location; this is called the **consensus sequence**.

2.3.3   The Poisson distribution

We say that $X \in \{0, 1, 2, \ldots\}$ has a **Poisson** distribution with parameter $\lambda > 0$, written $X \sim \text{Poi}(\lambda)$, if its pmf is

$$\text{Poi}(x|\lambda) = e^{-\lambda} \frac{\lambda^x}{x!} \tag{2.39}$$

The first term is just the normalization constant, required to ensure the distribution sums to 1.

The Poisson distribution is often used as a model for counts of rare events like radioactive decay and traffic accidents. See Figure 2.6 for some plots.

2.3.4   The empirical distribution

Given a set of data, $\mathcal{D} = \{x_1, \ldots, x_N\}$, we define the **empirical distribution**, also called the **empirical measure**, as follows:

$$p_{\text{emp}}(A) \triangleq \frac{1}{N} \sum_{i=1}^{N} \delta_{x_i}(A) \tag{2.40}$$

where $\delta_x(A)$ is the **Dirac measure**, defined by

$$\delta_x(A) = \begin{cases} 0 & \text{if } x \notin A \\ 1 & \text{if } x \in A \end{cases} \tag{2.41}$$

In general, we can associate "weights" with each sample:

$$p(x) = \sum_{i=1}^{N} w_i \delta_{x_i}(x) \tag{2.42}$$

where we require $0 \le w_i \le 1$ and $\sum_{i=1}^{N} w_i = 1$. We can think of this as a histogram, with "spikes" at the data points x_i, where w_i determines the height of spike i. This distribution assigns 0 probability to any point not in the data set.

2.4   Some common continuous distributions

In this section we present some commonly used univariate (one-dimensional) continuous probability distributions.

2.4.1   Gaussian (normal) distribution

The most widely used distribution in statistics and machine learning is the Gaussian or normal distribution. Its pdf is given by

$$\mathcal{N}(x|\mu, \sigma^2) \triangleq \frac{1}{\sqrt{2\pi\sigma^2}} e^{-\frac{1}{2\sigma^2}(x-\mu)^2} \tag{2.43}$$

Here $\mu = \mathbb{E}[X]$ is the mean (and mode), and $\sigma^2 = \text{var}[X]$ is the variance. $\sqrt{2\pi\sigma^2}$ is the normalization constant needed to ensure the density integrates to 1 (see Exercise 2.11).

We write $X \sim \mathcal{N}(\mu, \sigma^2)$ to denote that $p(X = x) = \mathcal{N}(x|\mu, \sigma^2)$. If $X \sim \mathcal{N}(0, 1)$, we say X follows a **standard normal** distribution. See Figure 2.3(b) for a plot of this pdf; this is sometimes called the **bell curve**.

We will often talk about the **precision** of a Gaussian, by which we mean the inverse variance: $\lambda = 1/\sigma^2$. A high precision means a narrow distribution (low variance) centered on μ.[4]

Note that, since this is a pdf, we can have $p(x) > 1$. To see this, consider evaluating the density at its center, $x = \mu$. We have $\mathcal{N}(\mu|\mu, \sigma^2) = (\sigma\sqrt{2\pi})^{-1}e^0$, so if $\sigma < 1/\sqrt{2\pi}$, we have $p(x) > 1$.

The cumulative distribution function or cdf of the Gaussian is defined as

$$\Phi(x; \mu, \sigma^2) \triangleq \int_{-\infty}^{x} \mathcal{N}(z|\mu, \sigma^2)dz \tag{2.44}$$

See Figure 2.3(a) for a plot of this cdf when $\mu = 0$, $\sigma^2 = 1$. This integral has no closed form expression, but is built in to most software packages. In particular, we can compute it in terms of the **error function (erf)**:

$$\Phi(x; \mu, \sigma) = \frac{1}{2}[1 + \text{erf}(z/\sqrt{2})] \tag{2.45}$$

---

4. The symbol λ will have many different meanings in this book, in order to be consistent with the rest of the literature. The intended meaning should be clear from context.

where $z = (x - \mu)/\sigma$ and

$$\text{erf}(x) \triangleq \frac{2}{\sqrt{\pi}} \int_0^x e^{-t^2} dt \qquad (2.46)$$

The Gaussian distribution is the most widely used distribution in statistics. There are several reasons for this. First, it has two parameters which are easy to interpret, and which capture some of the most basic properties of a distribution, namely its mean and variance. Second, the central limit theorem (Section 2.6.3) tells us that sums of independent random variables have an approximately Gaussian distribution, making it a good choice for modeling residual errors or "noise". Third, the Gaussian distribution makes the least number of assumptions (has maximum entropy), subject to the constraint of having a specified mean and variance, as we show in Section 9.2.6; this makes it a good default choice in many cases. Finally, it has a simple mathematical form, which results in easy to implement, but often highly effective, methods, as we will see. See (Jaynes 2003, ch 7) for a more extensive discussion of why Gaussians are so widely used.

2.4.2 Degenerate pdf

In the limiting case where $\sigma^2 \to 0$, the Gaussian becomes an infinitely tall and infinitely thin "spike" centered at μ:

$$\lim_{\sigma^2 \to 0} \mathcal{N}(x|\mu, \sigma^2) = \delta(x - \mu) \qquad (2.47)$$

where δ is called a **Dirac delta function**, and is defined as

$$\delta(x) = \begin{cases} \infty & \text{if } x = 0 \\ 0 & \text{if } x \neq 0 \end{cases} \qquad (2.48)$$

such that

$$\int_{-\infty}^{\infty} \delta(x) dx = 1 \qquad (2.49)$$

A useful property of delta functions is the **sifting property**, which selects out a single term from a sum or integral:

$$\int_{-\infty}^{\infty} f(x)\delta(x - \mu) dx = f(\mu) \qquad (2.50)$$

since the integrand is only non-zero if $x - \mu = 0$.

2.4.3 The Student's t distribution

One problem with the Gaussian distribution is that it is sensitive to outliers, since the log-probability only decays quadratically with distance from the center. A more robust distribution is the **Student's t distribution**, which we shall call the Student distribution for short (although

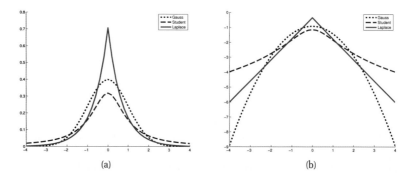

(a)   (b)

Figure 2.7   (a) The pdf's for a $\mathcal{N}(0,1)$, $\mathcal{T}(0,1,1)$ and $\text{Lap}(0,1/\sqrt{2})$. The mean is 0 and the variance is 1 for both the Gaussian and Laplace. The mean and variance of the Student is undefined when $\nu = 1$. (b) Log of these pdf's. Note that the Student distribution is not log-concave for any parameter value, unlike the Laplace distribution, which is always log-concave (and log-convex...) Nevertheless, both are unimodal. Figure generated by `studentLaplacePdfPlot`.

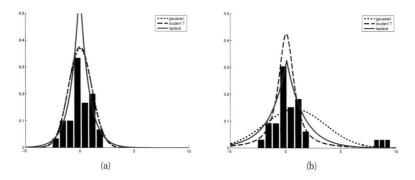

(a)   (b)

Figure 2.8   Illustration of the effect of outliers on fitting Gaussian, Student and Laplace distributions. (a) No outliers (the Gaussian and Student curves are on top of each other). (b) With outliers. We see that the Gaussian is more affected by outliers than the Student and Laplace distributions. Based on Figure 2.16 of (Bishop 2006). Figure generated by `robustDemo`.

it is more commonly called the t-distribution).[5] Its pdf is as follows:

$$\mathcal{T}(x|\mu,\sigma^2,\nu) \quad \propto \quad \left[1 + \frac{1}{\nu}\left(\frac{x-\mu}{\sigma}\right)^2\right]^{-\left(\frac{\nu+1}{2}\right)} \tag{2.51}$$

---

5. This distribution has a colourful etymology. It was first published in 1908 by William Sealy Gosset, who worked at the Guinness brewery in Dublin, Ireland. Since his employer would not allow him to use his own name, he called it the "Student" distribution. The origin of the term t seems to have arisen in the context of tables of the Student distribution, used by Fisher when developing the basis of classical statistical inference. See `http://jeff560.tripod.com/s.html` for more historical details.

where μ is the mean, $\sigma^2 > 0$ is the scale parameter, and $\nu > 0$ is called the **degrees of freedom**. See Figure 2.7 for some plots. For later reference, we note that the distribution has the following properties:

$$\text{mean} = \mu, \text{mode} = \mu, \ \text{var} = \frac{\nu\sigma^2}{(\nu - 2)} \tag{2.52}$$

The variance is only defined if $\nu > 2$. The mean is only defined if $\nu > 1$.

As an illustration of the robustness of the Student distribution, consider Figure 2.8. On the left, we show a Gaussian and a Student fit to some data with no outliers. On the right, we add some outliers. We see that the Gaussian is affected a lot, whereas the Student distribution hardly changes. This is because the Student has heavier tails, at least for small ν (see Figure 2.7).

If $\nu = 1$, this distribution is known as the **Cauchy** or **Lorentz** distribution. This is notable for having such heavy tails that the integral that defines the mean does not converge.

To ensure finite variance, we require $\nu > 2$. It is common to use $\nu = 4$, which gives good performance in a range of problems (Lange et al. 1989). For $\nu \gg 5$, the Student distribution rapidly approaches a Gaussian distribution and loses its robustness properties.

2.4.4 The Laplace distribution

Another distribution with heavy tails is the **Laplace distribution**[6], also known as the **double sided exponential** distribution. This has the following pdf:

$$\text{Lap}(x|\mu, b) \ \triangleq \ \frac{1}{2b} \exp\left(-\frac{|x - \mu|}{b}\right) \tag{2.53}$$

Here μ is a location parameter and $b > 0$ is a scale parameter. See Figure 2.7 for a plot. This distribution has the following properties:

$$\text{mean} = \mu, \ \text{mode} = \mu, \ \text{var} = 2b^2 \tag{2.54}$$

Its robustness to outliers is illustrated in Figure 2.8. It also puts more probability density at 0 than the Gaussian. This property is a useful way to encourage sparsity in a model, as we will see in Section 13.3.

2.4.5 The gamma distribution

The **gamma distribution** is a flexible distribution for positive real valued rv's, $x > 0$. It is defined in terms of two parameters, called the shape $a > 0$ and the rate $b > 0$:[7]

$$\text{Ga}(T|\text{shape} = a, \text{rate} = b) \ \triangleq \ \frac{b^a}{\Gamma(a)} T^{a-1} e^{-Tb} \tag{2.55}$$

---

6. Pierre-Simon Laplace (1749–1827) was a French mathematician, who played a key role in creating the field of Bayesian statistics.

7. There is an alternative parameterization, where we use the scale parameter instead of the rate: $\text{Ga}_s(T|a, b) \triangleq \text{Ga}(T|a, 1/b)$. This version is the one used by Matlab's `gampdf`, although in this book will use the rate parameterization unless otherwise specified.

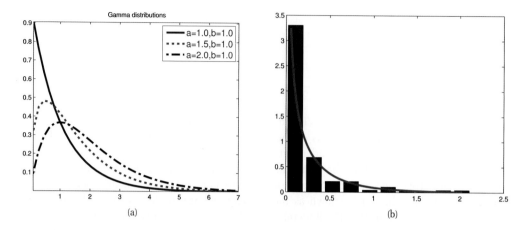

Figure 2.9 (a) Some $\text{Ga}(a, b = 1)$ distributions. If $a \leq 1$, the mode is at 0, otherwise it is > 0. As we increase the rate b, we reduce the horizontal scale, thus squeezing everything leftwards and upwards. Figure generated by `gammaPlotDemo`. (b) An empirical pdf of some rainfall data, with a fitted Gamma distribution superimposed. Figure generated by `gammaRainfallDemo`.

where $\Gamma(a)$ is the gamma function:

$$\Gamma(x) \triangleq \int_0^\infty u^{x-1} e^{-u} du \tag{2.56}$$

See Figure 2.9 for some plots. For later reference, we note that the distribution has the following properties:

$$\text{mean} = \frac{a}{b}, \;\; \text{mode} = \frac{a-1}{b}, \;\; \text{var} = \frac{a}{b^2} \tag{2.57}$$

There are several distributions which are just special cases of the Gamma, which we discuss below.

- **Exponential distribution** This is defined by $\text{Expon}(x|\lambda) \triangleq \text{Ga}(x|1, \lambda)$, where λ is the rate parameter. This distribution describes the times between events in a Poisson process, i.e. a process in which events occur continuously and independently at a constant average rate λ.
- **Erlang distribution** This is the same as the Gamma distribution where a is an integer. It is common to fix $a = 2$, yielding the one-parameter Erlang distribution, $\text{Erlang}(x|\lambda) = \text{Ga}(x|2, \lambda)$, where λ is the rate parameter.
- **Chi-squared distribution** This is defined by $\chi^2(x|\nu) \triangleq \text{Ga}(x|\frac{\nu}{2}, \frac{1}{2})$. This is the distribution of the sum of squared Gaussian random variables. More precisely, if $Z_i \sim \mathcal{N}(0, 1)$, and $S = \sum_{i=1}^{\nu} Z_i^2$, then $S \sim \chi_\nu^2$.

Another useful result is the following: If $X \sim \text{Ga}(a, b)$, then one can show (Exercise 2.10) that $\frac{1}{X} \sim \text{IG}(a, b)$, where IG is the **inverse gamma** distribution defined by

$$\text{IG}(x|\text{shape} = a, \text{scale} = b) \;\; \triangleq \;\; \frac{b^a}{\Gamma(a)} x^{-(a+1)} e^{-b/x} \tag{2.58}$$

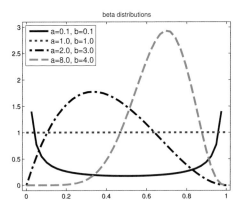

Figure 2.10 Some beta distributions. Figure generated by `betaPlotDemo`.

The distribution has these properties

$$\text{mean} = \frac{b}{a-1}, \quad \text{mode} = \frac{b}{a+1}, \text{var} = \frac{b^2}{(a-1)^2(a-2)}, \tag{2.59}$$

The mean only exists if $a > 1$. The variance only exists if $a > 2$.

We will see applications of these distributions later on.

2.4.6 The beta distribution

The **beta distribution** has support over the interval $[0, 1]$ and is defined as follows:

$$\text{Beta}(x|a,b) = \frac{1}{B(a,b)} x^{a-1}(1-x)^{b-1} \tag{2.60}$$

Here $B(p, q)$ is the beta function,

$$B(a,b) \triangleq \frac{\Gamma(a)\Gamma(b)}{\Gamma(a+b)} \tag{2.61}$$

See Figure 2.10 for plots of some beta distributions. We require $a, b > 0$ to ensure the distribution is integrable (i.e., to ensure $B(a,b)$ exists). If $a = b = 1$, we get the uniform distribution. If a and b are both less than 1, we get a bimodal distribution with "spikes" at 0 and 1; if a and b are both greater than 1, the distribution is unimodal. For later reference, we note that the distribution has the following properties (Exercise 2.16):

$$\text{mean} = \frac{a}{a+b}, \quad \text{mode} = \frac{a-1}{a+b-2}, \quad \text{var} = \frac{ab}{(a+b)^2(a+b+1)} \tag{2.62}$$

2.4.7 Pareto distribution

The **Pareto distribution** is defined as follows:

$$\text{Pareto}(x|k,m) = km^k x^{-(k+1)}\mathbb{I}(x \geq m) \tag{2.63}$$

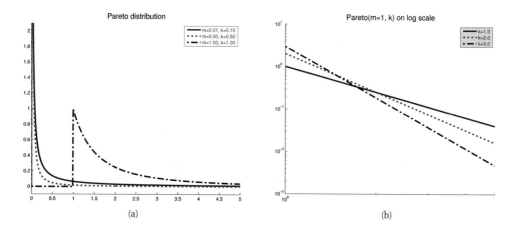

Figure 2.11 (a) The Pareto distribution Pareto$(x|m, k)$. (b) The pdf on a log-log scale. Figure generated by `paretoPlot`.

This density asserts that x must be greater than some constant m, but not too much greater, where k controls what is "too much". As $k \rightarrow \infty$, the distribution approaches $\delta(x - m)$. See Figure 2.11(a) for some plots. This distribution has the following properties

$$\text{mean} = \frac{km}{k - 1} \text{ if } k > 1 \ , \quad \text{mode} = m, \quad \text{var} = \frac{m^2 k}{(k - 1)^2 (k - 2)} \text{ if } k > 2 \qquad (2.64)$$

When $m = 0$, the distribution has the form $p(x) = kx^a$, where $a = -(k + 1)$. If we plot the distribution on a log-log scale, it forms a straight line, of the form $\log p(x) = a \log x + k$. See Figure 2.11(b) for an illustration; this is known as a **power law**. This is useful for modeling the distribution of quantities that exhibit **long tails**, also called **heavy tails**. For example, it has been observed that the most frequent word in English ("the") occurs approximately twice as often as the second most frequent word ("of"), which occurs twice as often as the fourth most frequent word, etc. If we plot the frequency of words vs their rank, we will get a power law; this is known as **Zipf's law**. Wealth has a similarly skewed distribution, especially in plutocracies such as the USA.[8]

2.5 Joint probability distributions

So far, we have been mostly focusing on modeling univariate probability distributions. In this section, we start our discussion of the more challenging problem of building joint probability distributions on multiple related random variables; this will be a central topic in this book.

A **joint probability distribution** has the form $p(x_1, \ldots, x_D)$ for a set of $D > 1$ variables, and models the (stochastic) relationships between the variables. If all the variables are discrete,

---

8. In the USA, 400 Americans have more wealth than half of all Americans combined. (Source: `http://www.politifact.com/wisconsin/statements/2011/mar/10/michael-moore/michael-moore-s ays-400-americans-have-more-wealth-.`) See (Hacker and Pierson 2010) for a political analysis of how such an extreme distribution of income has arisen in a democratic country.

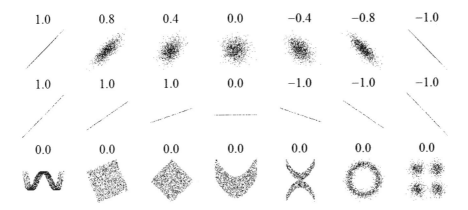

Figure 2.12 Several sets of (x, y) points, with the correlation coefficient of x and y for each set. Note that the correlation reflects the noisiness and direction of a linear relationship (top row), but not the slope of that relationship (middle), nor many aspects of nonlinear relationships (bottom). N.B.: the figure in the center has a slope of 0 but in that case the correlation coefficient is undefined because the variance of Y is zero. Source: http://en.wikipedia.org/wiki/File:Correlation_examples.png

we can represent the joint distribution as a big multi-dimensional array, with one variable per dimension. However, the number of parameters needed to define such a model is $O(K^D)$, where K is the number of states for each variable.

We can define high dimensional joint distributions using fewer parameters by making conditional independence assumptions, as we explain in Chapter 10. In the case of continuous distributions, an alternative approach is to restrict the form of the pdf to certain functional forms, some of which we will examine below.

2.5.1 Covariance and correlation

The **covariance** between two rv's X and Y measures the degree to which X and Y are (linearly) related. Covariance is defined as

$$\operatorname{cov}[X, Y] \triangleq \mathbb{E}\left[(X - \mathbb{E}[X])(Y - \mathbb{E}[Y])\right] = \mathbb{E}[XY] - \mathbb{E}[X]\mathbb{E}[Y] \tag{2.65}$$

If $\mathbf{x}$ is a d-dimensional random vector, its **covariance matrix** is defined to be the following symmetric, positive definite matrix:

$$\operatorname{cov}[\mathbf{x}] \triangleq \mathbb{E}\left[(\mathbf{x} - \mathbb{E}[\mathbf{x}])(\mathbf{x} - \mathbb{E}[\mathbf{x}])^T\right] \tag{2.66}$$

$$= \begin{pmatrix} \operatorname{var}[X_1] & \operatorname{cov}[X_1, X_2] & \cdots & \operatorname{cov}[X_1, X_d] \\ \operatorname{cov}[X_2, X_1] & \operatorname{var}[X_2] & \cdots & \operatorname{cov}[X_2, X_d] \\ \vdots & \vdots & \ddots & \vdots \\ \operatorname{cov}[X_d, X_1] & \operatorname{cov}[X_d, X_2] & \cdots & \operatorname{var}[X_d] \end{pmatrix} \tag{2.67}$$

Covariances can be between 0 and infinity. Sometimes it is more convenient to work with a normalized measure, with a finite upper bound. The (Pearson) **correlation coefficient** between

X and Y is defined as

$$\text{corr}\,[X, Y] \triangleq \frac{\text{cov}\,[X, Y]}{\sqrt{\text{var}\,[X]\,\text{var}\,[Y]}} \tag{2.68}$$

A **correlation matrix** has the form

$$\mathbf{R} = \begin{pmatrix} \text{corr}\,[X_1, X_1] & \text{corr}\,[X_1, X_2] & \cdots & \text{corr}\,[X_1, X_d] \\ \vdots & \vdots & \ddots & \vdots \\ \text{corr}\,[X_d, X_1] & \text{corr}\,[X_d, X_2] & \cdots & \text{corr}\,[X_d, X_d] \end{pmatrix} \tag{2.69}$$

One can show (Exercise 4.3) that $-1 \leq \text{corr}\,[X, Y] \leq 1$. Hence in a correlation matrix, each entry on the diagonal is 1, and the other entries are between -1 and 1.

One can also show that $\text{corr}\,[X, Y] = 1$ if and only if $Y = aX + b$ for some parameters a and b, i.e., if there is a *linear* relationship between X and Y (see Exercise 4.4). Intuitively one might expect the correlation coefficient to be related to the slope of the regression line, i.e., the coefficient a in the expression $Y = aX + b$. However, as we show in Equation 7.99 later, the regression coefficient is in fact given by $a = \text{cov}\,[X, Y]\,/\text{var}\,[X]$. A better way to think of the correlation coefficient is as a degree of linearity: see Figure 2.12.

If X and Y are independent, meaning $p(X, Y) = p(X)p(Y)$ (see Section 2.2.4), then $\text{cov}\,[X, Y] = 0$, and hence $\text{corr}\,[X, Y] = 0$ so they are uncorrelated. However, the converse is not true: *uncorrelated does not imply independent*. For example, let $X \sim U(-1, 1)$ and $Y = X^2$. Clearly Y is dependent on X (in fact, Y is uniquely determined by X), yet one can show (Exercise 4.1) that $\text{corr}\,[X, Y] = 0$. Some striking examples of this fact are shown in Figure 2.12. This shows several data sets where there is clear dependence between X and Y, and yet the correlation coefficient is 0. A more general measure of dependence between random variables is mutual information, discussed in Section 2.8.3. This is only zero if the variables truly are independent.

2.5.2 The multivariate Gaussian

The **multivariate Gaussian** or **multivariate normal** (**MVN**) is the most widely used joint probability density function for continuous variables. We discuss MVNs in detail in Chapter 4; here we just give some definitions and plots.

The pdf of the MVN in D dimensions is defined by the following:

$$\mathcal{N}(\mathbf{x}|\boldsymbol{\mu}, \boldsymbol{\Sigma}) \triangleq \frac{1}{(2\pi)^{D/2}|\boldsymbol{\Sigma}|^{1/2}} \exp\left[-\frac{1}{2}(\mathbf{x} - \boldsymbol{\mu})^T \boldsymbol{\Sigma}^{-1}(\mathbf{x} - \boldsymbol{\mu})\right] \tag{2.70}$$

where $\boldsymbol{\mu} = \mathbb{E}\,[\mathbf{x}] \in \mathbb{R}^D$ is the mean vector, and $\boldsymbol{\Sigma} = \text{cov}\,[\mathbf{x}]$ is the $D \times D$ covariance matrix. Sometimes we will work in terms of the **precision matrix** or **concentration matrix** instead. This is just the inverse covariance matrix, $\boldsymbol{\Lambda} = \boldsymbol{\Sigma}^{-1}$. The normalization constant $(2\pi)^{-D/2}|\boldsymbol{\Lambda}|^{1/2}$ just ensures that the pdf integrates to 1 (see Exercise 4.5).

Figure 2.13 plots some MVN densities in 2d for three different kinds of covariance matrices. A full covariance matrix has $D(D + 1)/2$ parameters (we divide by 2 since $\boldsymbol{\Sigma}$ is symmetric). A diagonal covariance matrix has D parameters, and has 0s in the off-diagonal terms. A **spherical** or **isotropic** covariance, $\boldsymbol{\Sigma} = \sigma^2 \mathbf{I}_D$, has one free parameter.

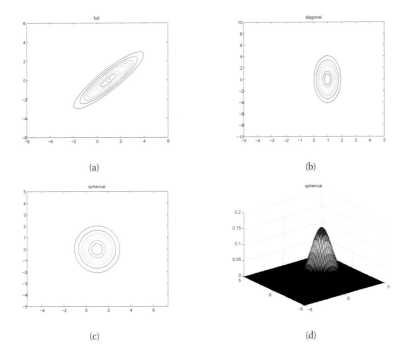

(a)

(b)

(c)

(d)

Figure 2.13 We show the level sets for 2d Gaussians. (a) A full covariance matrix has elliptical contours. (b) A diagonal covariance matrix is an **axis aligned** ellipse. (c) A spherical covariance matrix has a circular shape. (d) Surface plot for the spherical Gaussian in (c). Figure generated by `gaussPlot2Ddemo`.

2.5.3 Multivariate Student t distribution

A more robust alternative to the MVN is the **multivariate Student t** distribution, whose pdf is given by

$$\mathcal{T}(\mathbf{x}|\boldsymbol{\mu}, \boldsymbol{\Sigma}, \nu) = \frac{\Gamma(\nu/2 + D/2)}{\Gamma(\nu/2)} \frac{|\boldsymbol{\Sigma}|^{-1/2}}{\nu^{D/2}\pi^{D/2}} \times \left[1 + \frac{1}{\nu}(\mathbf{x} - \boldsymbol{\mu})^T\boldsymbol{\Sigma}^{-1}(\mathbf{x} - \boldsymbol{\mu})\right]^{-(\frac{\nu+D}{2})} \tag{2.71}$$

$$= \frac{\Gamma(\nu/2 + D/2)}{\Gamma(\nu/2)}|\pi\mathbf{V}|^{-1/2} \times \left[1 + (\mathbf{x} - \boldsymbol{\mu})^T\mathbf{V}^{-1}(\mathbf{x} - \boldsymbol{\mu})\right]^{-(\frac{\nu+D}{2})} \tag{2.72}$$

where $\boldsymbol{\Sigma}$ is called the scale matrix (since it is not exactly the covariance matrix) and $\mathbf{V} = \nu\boldsymbol{\Sigma}$. This has fatter tails than a Gaussian. The smaller ν is, the fatter the tails. As $\nu \to \infty$, the distribution tends towards a Gaussian. The distribution has the following properties

$$\text{mean} = \boldsymbol{\mu}, \ \text{mode} = \boldsymbol{\mu}, \ \text{Cov} = \frac{\nu}{\nu - 2}\boldsymbol{\Sigma} \tag{2.73}$$

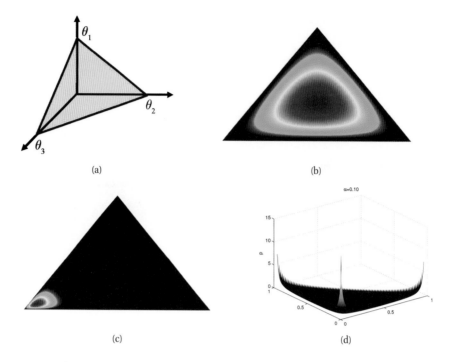

(a)   (b)

(c)   (d)

Figure 2.14 (a) The Dirichlet distribution when $K = 3$ defines a distribution over the simplex, which can be represented by the triangular surface. Points on this surface satisfy $0 \leq \theta_k \leq 1$ and $\sum_{k=1}^{3} \theta_k = 1$. (b) Plot of the Dirichlet density when $\boldsymbol{\alpha} = (2, 2, 2)$. (c) $\boldsymbol{\alpha} = (20, 2, 2)$. Figure generated by `visDirichletGui`, by Jonathan Huang. (d) $\boldsymbol{\alpha} = (0.1, 0.1, 0.1)$. (The comb-like structure on the edges is a plotting artifact.) Figure generated by `dirichlet3dPlot`.

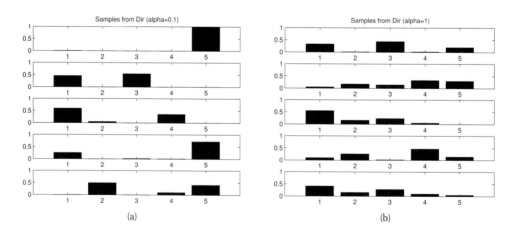

(a)   (b)

Figure 2.15 Samples from a 5-dimensional symmetric Dirichlet distribution for different parameter values. (a) $\boldsymbol{\alpha} = (0.1, \ldots, 0.1)$. This results in very sparse distributions, with many 0s. (b) $\boldsymbol{\alpha} = (1, \ldots, 1)$. This results in more uniform (and dense) distributions. Figure generated by `dirichletHistogramDemo`.

2.5.4 Dirichlet distribution

A multivariate generalization of the beta distribution is the **Dirichlet**[9] distribution, which has support over the **probability simplex**, defined by

$$S_K = \{\mathbf{x} : 0 \le x_k \le 1, \sum_{k=1}^K x_k = 1\} \tag{2.74}$$

The pdf is defined as follows:

$$\text{Dir}(\mathbf{x}|\boldsymbol{\alpha}) \triangleq \frac{1}{B(\boldsymbol{\alpha})} \prod_{k=1}^K x_k^{\alpha_k - 1} \mathbb{I}(\mathbf{x} \in S_K) \tag{2.75}$$

where $B(\alpha_1, \ldots, \alpha_K)$ is the natural generalization of the beta function to K variables:

$$B(\boldsymbol{\alpha}) \triangleq \frac{\prod_{k=1}^K \Gamma(\alpha_k)}{\Gamma(\alpha_0)} \tag{2.76}$$

where $\alpha_0 \triangleq \sum_{k=1}^K \alpha_k$.

Figure 2.14 shows some plots of the Dirichlet when $K = 3$, and Figure 2.15 for some sampled probability vectors. We see that $\alpha_0 = \sum_{k=1}^K \alpha_k$ controls the strength of the distribution (how peaked it is), and the α_k control where the peak occurs. For example, $\text{Dir}(1, 1, 1)$ is a uniform distribution, $\text{Dir}(2, 2, 2)$ is a broad distribution centered at $(1/3, 1/3, 1/3)$, and $\text{Dir}(20, 20, 20)$ is a narrow distribution centered at $(1/3, 1/3, 1/3)$. If $\alpha_k < 1$ for all k, we get "spikes" at the corners of the simplex.

For future reference, the distribution has these properties

$$\mathbb{E}[x_k] = \frac{\alpha_k}{\alpha_0}, \quad \text{mode}[x_k] = \frac{\alpha_k - 1}{\alpha_0 - K}, \quad \text{var}[x_k] = \frac{\alpha_k(\alpha_0 - \alpha_k)}{\alpha_0^2(\alpha_0 + 1)} \tag{2.77}$$

where $\alpha_0 = \sum_k \alpha_k$. Often we use a symmetric Dirichlet prior of the form $\alpha_k = \alpha/K$. In this case, the mean becomes $1/K$, and the variance becomes $\text{var}[x_k] = \frac{K-1}{K^2(\alpha+1)}$. So increasing α increases the precision (decreases the variance) of the distribution.

2.6 Transformations of random variables

If $\mathbf{x} \sim p()$ is some random variable, and $\mathbf{y} = f(\mathbf{x})$, what is the distribution of $\mathbf{y}$? This is the question we address in this section.

2.6.1 Linear transformations

Suppose $f()$ is a linear function:

$$\mathbf{y} = f(\mathbf{x}) = \mathbf{A}\mathbf{x} + \mathbf{b} \tag{2.78}$$

9. Johann Dirichlet was a German mathematician, 1805–1859.

In this case, we can easily derive the mean and covariance of $\mathbf{y}$ as follows. First, for the mean, we have

$$\mathbb{E}[\mathbf{y}] = \mathbb{E}[\mathbf{Ax} + \mathbf{b}] = \mathbf{A}\boldsymbol{\mu} + \mathbf{b} \tag{2.79}$$

where $\boldsymbol{\mu} = \mathbb{E}[\mathbf{x}]$. This is called the **linearity of expectation**. If $f()$ is a scalar-valued function, $f(\mathbf{x}) = \mathbf{a}^T\mathbf{x} + b$, the corresponding result is

$$\mathbb{E}[\mathbf{a}^T\mathbf{x} + b] = \mathbf{a}^T\boldsymbol{\mu} + b \tag{2.80}$$

For the covariance, we have

$$\text{cov}[\mathbf{y}] = \text{cov}[\mathbf{Ax} + \mathbf{b}] = \mathbf{A}\boldsymbol{\Sigma}\mathbf{A}^T \tag{2.81}$$

where $\boldsymbol{\Sigma} = \text{cov}[\mathbf{x}]$. We leave the proof of this as an exercise. If $f()$ is scalar valued, the result becomes

$$\text{var}[y] = \text{var}[\mathbf{a}^T\mathbf{x} + b] = \mathbf{a}^T\boldsymbol{\Sigma}\mathbf{a} \tag{2.82}$$

We will use both of these results extensively in later chapters. Note, however, that the mean and covariance only completely define the distribution of $\mathbf{y}$ if $\mathbf{x}$ is Gaussian. In general we must use the techniques described below to derive the full distribution of $\mathbf{y}$, as opposed to just its first two moments.

2.6.2 General transformations

If X is a discrete rv, we can derive the pmf for y by simply summing up the probability mass for all the x's such that $f(x) = y$:

$$p_y(y) = \sum_{x:f(x)=y} p_x(x) \tag{2.83}$$

For example, if $f(X) = 1$ if X is even and $f(X) = 0$ otherwise, and $p_x(X)$ is uniform on the set $\{1, \ldots, 10\}$, then $p_y(1) = \sum_{x \in \{2,4,6,8,10\}} p_x(x) = 0.5$, and $p_y(0) = 0.5$ similarly. Note that in this example, f is a many-to-one function.

If X is continuous, we cannot use Equation 2.83 since $p_x(x)$ is a density, not a pmf, and we cannot sum up densities. Instead, we work with cdf's, and write

$$P_y(y) \triangleq P(Y \leq y) = P(f(X) \leq y) = P(X \in \{x | f(x) \leq y\}) \tag{2.84}$$

We can derive the pdf of y by differentiating the cdf.

In the case of monotonic and hence invertible functions, we can write

$$P_y(y) = P(f(X) \leq y) = P(X \leq f^{-1}(y)) = P_x(f^{-1}(y)) \tag{2.85}$$

Taking derivatives we get

$$p_y(y) \triangleq \frac{d}{dy}P_y(y) = \frac{d}{dy}P_x(f^{-1}(y)) = \frac{dx}{dy}\frac{d}{dx}P_x(x) = \frac{dx}{dy}p_x(x) \tag{2.86}$$

where $x = f^{-1}(y)$. We can think of dx as a measure of volume in the x-space; similarly dy measures volume in y space. Thus $\frac{dx}{dy}$ measures the change in volume. Since the sign of this change is not important, we take the absolute value to get the general expression:

$$p_y(y) = p_x(x)\left|\frac{dx}{dy}\right| \tag{2.87}$$

This is called **change of variables** formula. We can understand this result more intuitively as follows. Observations falling in the range $(x, x+\delta x)$ will get transformed into $(y, y+\delta y)$, where $p_x(x)\delta x \approx p_y(y)\delta y$. Hence $p_y(y) \approx p_x(x)|\frac{\delta x}{\delta y}|$. For example, suppose $X \sim U(-1, 1)$, and $Y = X^2$. Then $p_y(y) = \frac{1}{2}y^{-\frac{1}{2}}$. See also Exercise 2.10.

2.6.2.1 Multivariate change of variables *

We can extend the previous results to multivariate distributions as follows. Let f be a function that maps $\mathbb{R}^n$ to $\mathbb{R}^n$, and let $\mathbf{y} = f(\mathbf{x})$. Then its **Jacobian matrix J** is given by

$$\mathbf{J}_{\mathbf{x}\to\mathbf{y}} \triangleq \frac{\partial(y_1, \ldots, y_n)}{\partial(x_1, \ldots, x_n)} \triangleq \begin{pmatrix} \frac{\partial y_1}{\partial x_1} & \cdots & \frac{\partial y_1}{\partial x_n} \\ \vdots & \ddots & \vdots \\ \frac{\partial y_n}{\partial x_1} & \cdots & \frac{\partial y_n}{\partial x_n} \end{pmatrix} \tag{2.88}$$

$|\det \mathbf{J}|$ measures how much a unit cube changes in volume when we apply f.

If f is an invertible mapping, we can define the pdf of the transformed variables using the Jacobian of the inverse mapping $\mathbf{y} \to \mathbf{x}$:

$$p_y(\mathbf{y}) = p_x(\mathbf{x})\left|\det\left(\frac{\partial\mathbf{x}}{\partial\mathbf{y}}\right)\right| = p_x(\mathbf{x})|\det \mathbf{J}_{\mathbf{y}\to\mathbf{x}}| \tag{2.89}$$

In Exercise 4.5 you will use this formula to derive the normalization constant for a multivariate Gaussian.

As a simple example, consider transforming a density from Cartesian coordinates $\mathbf{x} = (x_1, x_2)$ to polar coordinates $\mathbf{y} = (r, \theta)$, where $x_1 = r\cos\theta$ and $x_2 = r\sin\theta$. Then

$$\mathbf{J}_{\mathbf{y}\to\mathbf{x}} = \begin{pmatrix} \frac{\partial x_1}{\partial r} & \frac{\partial x_1}{\partial \theta} \\ \frac{\partial x_2}{\partial r} & \frac{\partial x_2}{\partial \theta} \end{pmatrix} = \begin{pmatrix} \cos\theta & -r\sin\theta \\ \sin\theta & r\cos\theta \end{pmatrix} \tag{2.90}$$

and

$$|\det \mathbf{J}| = |r\cos^2\theta + r\sin^2\theta| = |r| \tag{2.91}$$

Hence

$$\begin{align} p_\mathbf{y}(\mathbf{y}) &= p_\mathbf{x}(\mathbf{x})|\det \mathbf{J}| \tag{2.92} \\ p_{r,\theta}(r,\theta) &= p_{x_1,x_2}(x_1, x_2)r = p_{x_1,x_2}(r\cos\theta, r\sin\theta)r \tag{2.93} \end{align}$$

To see this geometrically, notice that the area of the shaded patch in Figure 2.16 is given by

$$P(r \leq R \leq r+dr, \theta \leq \Theta \leq \theta+d\theta) = p_{r,\theta}(r,\theta)drd\theta \tag{2.94}$$

In the limit, this is equal to the density at the center of the patch, $p(r,\theta)$, times the size of the patch, $r\,dr\,d\theta$. Hence

$$p_{r,\theta}(r,\theta)drd\theta = p_{x_1,x_2}(r\cos\theta, r\sin\theta)r\,dr\,d\theta \tag{2.95}$$

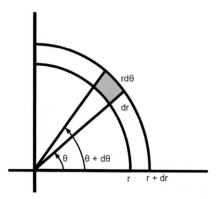

Figure 2.16 Change of variables from polar to Cartesian. The area of the shaded patch is $r \, dr \, d\theta$. Based on (Rice 1995) Figure 3.16.

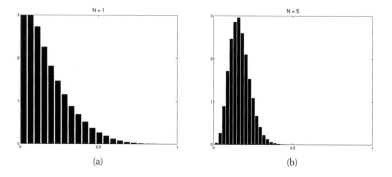

Figure 2.17 The central limit theorem in pictures. We plot a histogram of $\frac{1}{N} \sum_{i=1}^{N} x_{ij}$, where $x_{ij} \sim$ Beta$(1, 5)$, for $j = 1 : 10000$. As $N \to \infty$, the distribution tends towards a Gaussian. (a) $N = 1$. (b) $N = 5$. Based on Figure 2.6 of (Bishop 2006). Figure generated by `centralLimitDemo`.

2.6.3 Central limit theorem

Now consider N random variables with pdf's (not necessarily Gaussian) $p(x_i)$, each with mean μ and variance σ^2. We assume each variable is **independent and identically distributed** or **iid** for short. Let $S_N = \sum_{i=1}^{N} X_i$ be the sum of the rv's. This is a simple but widely used transformation of rv's. One can show that, as N increases, the distribution of this sum approaches

$$p(S_N = s) = \frac{1}{\sqrt{2\pi N \sigma^2}} \exp\left(-\frac{(s - N\mu)^2}{2N\sigma^2}\right) \tag{2.96}$$

Hence the distribution of the quantity

$$Z_N \triangleq \frac{S_N - N\mu}{\sigma\sqrt{N}} = \frac{\overline{X} - \mu}{\sigma/\sqrt{N}} \tag{2.97}$$

converges to the standard normal, where $\overline{X} = \frac{1}{N} \sum_{i=1}^{N} x_i$ is the sample mean. This is called the **central limit theorem**. See e.g., (Jaynes 2003, p222) or (Rice 1995, p169) for a proof.

In Figure 2.17 we give an example in which we compute the mean of rv's drawn from a beta distribution. We see that the sampling distribution of the mean value rapidly converges to a Gaussian distribution.

2.7 Monte Carlo approximation

In general, computing the distribution of a function of an rv using the change of variables formula can be difficult. One simple but powerful alternative is as follows. First we generate S samples from the distribution, call them $x_1, \ldots, x_S$. (There are many ways to generate such samples; one popular method, for high dimensional distributions, is called Markov chain Monte Carlo or MCMC; this will be explained in Chapter 24.) Given the samples, we can approximate the distribution of $f(X)$ by using the empirical distribution of $\{f(x_s)\}_{s=1}^{S}$. This is called a **Monte Carlo** approximation, named after a city in Europe known for its plush gambling casinos. Monte Carlo techniques were first developed in the area of statistical physics — in particular, during development of the atomic bomb — but are now widely used in statistics and machine learning as well.

We can use Monte Carlo to approximate the expected value of any function of a random variable. We simply draw samples, and then compute the arithmetic mean of the function applied to the samples. This can be written as follows:

$$\mathbb{E}\left[f(X)\right] = \int f(x)p(x)dx \approx \frac{1}{S} \sum_{s=1}^{S} f(x_s) \tag{2.98}$$

where $x_s \sim p(X)$. This is called Monte Carlo integration, and has the advantage over numerical integration (which is based on evaluating the function at a fixed grid of points) that the function is only evaluated in places where there is non-negligible probability.

By varying the function $f()$, we can approximate many quantities of interest, such as

- $\overline{x} = \frac{1}{S} \sum_{s=1}^{S} x_s \to \mathbb{E}\left[X\right]$
- $\frac{1}{S} \sum_{s=1}^{S} (x_s - \overline{x})^2 \to \mathrm{var}\left[X\right]$
- $\frac{1}{S}|\{x_s \le c\}| \to P(X \le c)$
- $\mathrm{median}\{x_1, \ldots, x_S\} \to \mathrm{median}(X)$

We give some examples below, and will see many more in later chapters.

2.7.1 Example: change of variables, the MC way

In Section 2.6.2, we discussed how to analytically compute the distribution of a function of a random variable, $y = f(x)$. A much simpler approach is to use a Monte Carlo approximation. For example, suppose $x \sim \mathrm{Unif}(-1, 1)$ and $y = x^2$. We can approximate $p(y)$ by drawing many samples from $p(x)$, squaring them, and computing the resulting empirical distribution. See Figure 2.18 for an illustration. We will use this technique extensively in later chapters. See also Figure 5.2.

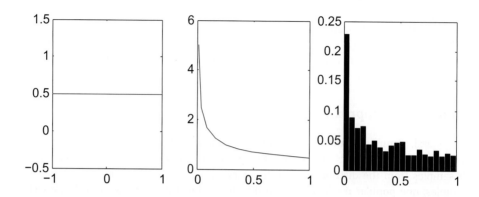

Figure 2.18 Computing the distribution of $y = x^2$, where $p(x)$ is uniform (left). The analytic result is shown in the middle, and the Monte Carlo approximation is shown on the right. Figure generated by `changeOfVarsDemo1d`.

2.7.2 Example: estimating π by Monte Carlo integration

MC approximation can be used for many applications, not just statistical ones. Suppose we want to estimate π. We know that the area of a circle with radius r is πr^2, but it is also equal to the following definite integral:

$$I = \int_{-r}^{r} \int_{-r}^{r} \mathbb{I}(x^2 + y^2 \leq r^2) dx dy \tag{2.99}$$

Hence $\pi = I/(r^2)$. Let us approximate this by Monte Carlo integration. Let $f(x, y) = \mathbb{I}(x^2 + y^2 \leq r^2)$ be an indicator function that is 1 for points inside the circle, and 0 outside, and let $p(x)$ and $p(y)$ be uniform distributions on $[-r, r]$, so $p(x) = p(y) = 1/(2r)$. Then

$$I = (2r)(2r) \int \int f(x, y) p(x) p(y) dx dy \tag{2.100}$$

$$= 4r^2 \int \int f(x, y) p(x) p(y) dx dy \tag{2.101}$$

$$\approx 4r^2 \frac{1}{S} \sum_{s=1}^{S} f(x_s, y_s) \tag{2.102}$$

We find $\hat{\pi} = 3.1416$ with standard error 0.09 (see Section 2.7.3 for a discussion of standard errors). We can plot the points that are accepted or rejected as in Figure 2.19.

2.7.3 Accuracy of Monte Carlo approximation

The accuracy of an MC approximation increases with sample size. This is illustrated in Figure 2.20, On the top line, we plot a histogram of samples from a Gaussian distribution. On the bottom line, we plot a smoothed version of these samples, created using a kernel density

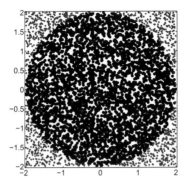

Figure 2.19 Estimating π by Monte Carlo integration. Blue points are inside the circle, red crosses are outside. Figure generated by `mcEstimatePi`.

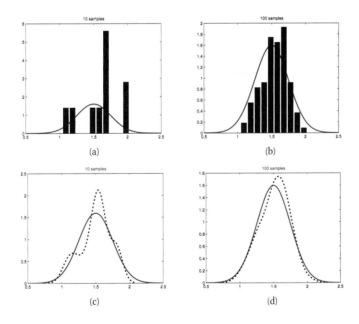

Figure 2.20 10 and 100 samples from a Gaussian distribution, $\mathcal{N}(\mu = 1.5, \sigma^2 = 0.25)$. Solid red line is true pdf. Top line: histogram of samples. Bottom line: kernel density estimate derived from samples in dotted blue, solid red line is true pdf. Based on Figure 4.1 of (Hoff 2009). Figure generated by `mcAccuracyDemo`.

estimate (Section 14.7.2). This smoothed distribution is then evaluated on a dense grid of points and plotted. Note that this smoothing is just for the purposes of plotting, it is not used for the Monte Carlo estimate itself.

If we denote the exact mean by $\mu = \mathbb{E}\left[f(X)\right]$, and the MC approximation by $\hat{\mu}$, one can show that, with independent samples,

$$(\hat{\mu} - \mu) \rightarrow \mathcal{N}(0, \frac{\sigma^2}{S}) \tag{2.103}$$

where

$$\sigma^2 = \text{var}\left[f(X)\right] = \mathbb{E}\left[f(X)^2\right] - \mathbb{E}\left[f(X)\right]^2 \tag{2.104}$$

This is a consequence of the central-limit theorem. Of course, σ^2 is unknown in the above expression, but it can also be estimated by MC:

$$\hat{\sigma}^2 \quad = \quad \frac{1}{S} \sum_{s=1}^{S} (f(x_s) - \hat{\mu})^2 \tag{2.105}$$

Then we have

$$P\left\{ \mu - 1.96 \frac{\hat{\sigma}}{\sqrt{S}} \leq \hat{\mu} \leq \mu + 1.96 \frac{\hat{\sigma}}{\sqrt{S}} \right\} \approx 0.95 \tag{2.106}$$

The term $\sqrt{\frac{\hat{\sigma}^2}{S}}$ is called the (numerical or empirical) **standard error**, and is an estimate of our uncertainty about our estimate of μ. (See Section 6.2 for more discussion on standard errors.)

If we want to report an answer which is accurate to within $\pm\epsilon$ with probability at least 95%, we need to use a number of samples S which satisfies $1.96\sqrt{\hat{\sigma}^2/S} \leq \epsilon$. We can approximate the 1.96 factor by 2, yielding $S \geq \frac{4\hat{\sigma}^2}{\epsilon^2}$.

2.8   Information theory

Information theory is concerned with representing data in a compact fashion (a task known as **data compression** or **source coding**), as well as with transmitting and storing it in a way that is robust to errors (a task known as **error correction** or **channel coding**). At first, this seems far removed from the concerns of probability theory and machine learning, but in fact there is an intimate connection. To see this, note that compactly representing data requires allocating short codewords to highly probable bit strings, and reserving longer codewords to less probable bit strings. This is similar to the situation in natural language, where common words (such as "a", "the", "and") are generally much shorter than rare words. Also, decoding messages sent over noisy channels requires having a good probability model of the kinds of messages that people tend to send. In both cases, we need a model that can predict which kinds of data are likely and which unlikely, which is also a central problem in machine learning (see (MacKay 2003) for more details on the connection between information theory and machine learning).

Obviously we cannot go into the details of information theory here (see e.g., (Cover and Thomas 2006) if you are interested to learn more). However, we will introduce a few basic concepts that we will need later in the book.

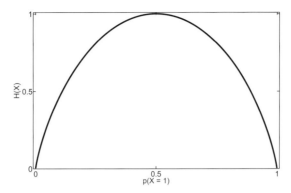

Figure 2.21 Entropy of a Bernoulli random variable as a function of θ. The maximum entropy is $\log_2 2 = 1$. Figure generated by `bernoulliEntropyFig`.

2.8.1 Entropy

The **entropy** of a random variable X with distribution p, denoted by $\mathbb{H}(X)$ or sometimes $\mathbb{H}(p)$, is a measure of its uncertainty. In particular, for a discrete variable with K states, it is defined by

$$\mathbb{H}(X) \triangleq -\sum_{k=1}^{K} p(X = k) \log_2 p(X = k) \tag{2.107}$$

Usually we use log base 2, in which case the units are called **bits** (short for binary digits). If we use log base e, the units are called **nats**. For example, if $X \in \{1, \ldots, 5\}$ with histogram distribution $p = [0.25, 0.25, 0.2, 0.15, 0.15]$, we find $H = 2.2855$. The discrete distribution with maximum entropy is the uniform distribution (see Section 9.2.6 for a proof). Hence for a K-ary random variable, the entropy is maximized if $p(x = k) = 1/K$; in this case, $\mathbb{H}(X) = \log_2 K$. Conversely, the distribution with minimum entropy (which is zero) is any delta-function that puts all its mass on one state. Such a distribution has no uncertainty. In Figure 2.5(b), where we plotted a DNA sequence logo, the height of each bar is defined to be $2 - H$, where H is the entropy of that distribution, and 2 is the maximum possible entropy. Thus a bar of height 0 corresponds to a uniform distribution, whereas a bar of height 2 corresponds to a deterministic distribution.

For the special case of binary random variables, $X \in \{0, 1\}$, we can write $p(X = 1) = \theta$ and $p(X = 0) = 1 - \theta$. Hence the entropy becomes

$$\begin{aligned} \mathbb{H}(X) &= -[p(X = 1) \log_2 p(X = 1) + p(X = 0) \log_2 p(X = 0)] & \text{(2.108)} \\ &= -[\theta \log_2 \theta + (1 - \theta) \log_2 (1 - \theta)] & \text{(2.109)} \end{aligned}$$

This is called the **binary entropy function**, and is also written $\mathbb{H}(\theta)$. We plot this in Figure 2.21. We see that the maximum value of 1 occurs when the distribution is uniform, $\theta = 0.5$.

2.8.2   KL divergence

One way to measure the dissimilarity of two probability distributions, p and q, is known as the **Kullback-Leibler divergence** (**KL divergence**) or **relative entropy**. This is defined as follows:

$$\mathbb{KL}\left(p||q\right) \triangleq \sum_{k=1}^{K} p_k \log \frac{p_k}{q_k} \tag{2.110}$$

where the sum gets replaced by an integral for pdfs.[10] We can rewrite this as

$$\mathbb{KL}\left(p||q\right) = \sum_{k} p_k \log p_k - \sum_{k} p_k \log q_k = -\mathbb{H}\left(p\right) + \mathbb{H}\left(p, q\right) \tag{2.111}$$

where $\mathbb{H}\left(p, q\right)$ is called the **cross entropy**,

$$\mathbb{H}\left(p, q\right) \triangleq -\sum_{k} p_k \log q_k \tag{2.112}$$

One can show (Cover and Thomas 2006) that the cross entropy is the average number of bits needed to encode data coming from a source with distribution p when we use model q to define our codebook. Hence the "regular" entropy $\mathbb{H}\left(p\right) = \mathbb{H}\left(p, p\right)$, defined in Section 2.8.1, is the expected number of bits if we use the true model, so the KL divergence is the difference between these. In other words, the KL divergence is the average number of *extra* bits needed to encode the data, due to the fact that we used distribution q to encode the data instead of the true distribution p.

The "extra number of bits" interpretation should make it clear that $\mathbb{KL}\left(p||q\right) \geq 0$, and that the KL is only equal to zero iff $q = p$. We now give a proof of this important result.

Theorem 2.8.1. (**Information inequality**) $\mathbb{KL}\left(p||q\right) \geq 0$ *with equality iff $p = q$.*

Proof. To prove the theorem, we need to use **Jensen's inequality**. This states that, for any convex function f, we have that

$$f\left(\sum_{i=1}^{n} \lambda_i \mathbf{x}_i\right) \leq \sum_{i=1}^{n} \lambda_i f(\mathbf{x}_i) \tag{2.113}$$

where $\lambda_i \geq 0$ and $\sum_{i=1}^{n} \lambda_i = 1$. This is clearly true for $n = 2$ (by definition of convexity), and can be proved by induction for $n > 2$.

Let us now prove the main theorem, following (Cover and Thomas 2006, p28). Let $A = \{x : p(x) > 0\}$ be the support of $p(x)$. Then

$$-\mathbb{KL}\left(p||q\right) \quad = \quad -\sum_{x \in A} p(x) \log \frac{p(x)}{q(x)} = \sum_{x \in A} p(x) \log \frac{q(x)}{p(x)} \tag{2.114}$$

$$\leq \quad \log \sum_{x \in A} p(x) \frac{q(x)}{p(x)} = \log \sum_{x \in A} q(x) \tag{2.115}$$

$$\leq \quad \log \sum_{x \in \mathcal{X}} q(x) = \log 1 = 0 \tag{2.116}$$

---

10. The KL divergence is not a distance, since it is asymmetric. One symmetric version of the KL divergence is the **Jensen-Shannon divergence**, defined as $JS(p_1, p_2) = 0.5\mathbb{KL}\left(p_1||q\right) + 0.5\mathbb{KL}\left(p_2||q\right)$, where $q = 0.5p_1 + 0.5p_2$.

where the first inequality follows from Jensen's. Since $\log(x)$ is a strictly concave function, we have equality in Equation 2.115 iff $p(x) = cq(x)$ for some c. We have equality in Equation 2.116 iff $\sum_{x \in A} q(x) = \sum_{x \in \mathcal{X}} q(x) = 1$, which implies $c = 1$. Hence $\mathbb{KL}(p||q) = 0$ iff $p(x) = q(x)$ for all x. $\qquad\square$

One important consequence of this result is that the *discrete distribution with the maximum entropy is the uniform distribution*. More precisely, $\mathbb{H}(X) \leq \log|\mathcal{X}|$, where $|\mathcal{X}|$ is the number of states for X, with equality iff $p(x)$ is uniform. To see this, let $u(x) = 1/|\mathcal{X}|$. Then

$$0 \quad \leq \quad \mathbb{KL}(p||u) = \sum_x p(x) \log \frac{p(x)}{u(x)} \tag{2.117}$$

$$= \quad \sum_x p(x) \log p(x) - \sum_x p(x) \log u(x) = -\mathbb{H}(X) + \log|\mathcal{X}| \tag{2.118}$$

This is a formulation of Laplace's **principle of insufficient reason**, which argues in favor of using uniform distributions when there are no other reasons to favor one distribution over another. See Section 9.2.6 for a discussion of how to create distributions that satisfy certain constraints, but otherwise are as least-commital as possible. (For example, the Gaussian satisfies first and second moment constraints, but otherwise has maximum entropy.)

2.8.3 Mutual information

Consider two random variables, X and Y. Suppose we want to know how much knowing one variable tells us about the other. We could compute the correlation coefficient, but this is only defined for real-valued random variables, and furthermore, this is a very limited measure of dependence, as we saw in Figure 2.12. A more general approach is to determine how similar the joint distribution $p(X, Y)$ is to the factored distribution $p(X)p(Y)$. This is called the **mutual information** or **MI**, and is defined as follows:

$$\mathbb{I}(X; Y) \triangleq \mathbb{KL}(p(X, Y)||p(X)p(Y)) = \sum_x \sum_y p(x, y) \log \frac{p(x, y)}{p(x)p(y)} \tag{2.119}$$

We have $\mathbb{I}(X; Y) \geq 0$ with equality iff $p(X, Y) = p(X)p(Y)$. That is, the MI is zero iff the variables are independent.

To gain insight into the meaning of MI, it helps to re-express it in terms of joint and conditional entropies. One can show (Exercise 2.12) that the above expression is equivalent to the following:

$$\mathbb{I}(X; Y) = \mathbb{H}(X) - \mathbb{H}(X|Y) = \mathbb{H}(Y) - \mathbb{H}(Y|X) \tag{2.120}$$

where $\mathbb{H}(Y|X)$ is the **conditional entropy**, defined as $\mathbb{H}(Y|X) = \sum_x p(x)\mathbb{H}(Y|X = x)$. Thus we can interpret the MI between X and Y as the reduction in uncertainty about X after observing Y, or, by symmetry, the reduction in uncertainty about Y after observing X. We will encounter several applications of MI later in the book. See also Exercises 2.13 and 2.14 for the connection between MI and correlation coefficients.

A quantity which is closely related to MI is the **pointwise mutual information** or **PMI**. For two events (not random variables) x and y, this is defined as

$$\mathrm{PMI}(x, y) \triangleq \log \frac{p(x, y)}{p(x)p(y)} = \log \frac{p(x|y)}{p(x)} = \log \frac{p(y|x)}{p(y)} \tag{2.121}$$

This measures the discrepancy between these events occuring together compared to what would be expected by chance. Clearly the MI of X and Y is just the expected value of the PMI. Interestingly, we can rewrite the PMI as follows:

$$\text{PMI}(x, y) = \log \frac{p(x|y)}{p(x)} = \log \frac{p(y|x)}{p(y)} \tag{2.122}$$

This is the amount we learn from updating the prior $p(x)$ into the posterior $p(x|y)$, or equivalently, updating the prior $p(y)$ into the posterior $p(y|x)$.

2.8.3.1   Mutual information for continuous random variables *

The above formula for MI is defined for discrete random variables. For continuous random variables, it is common to first **discretize** or **quantize** them, by dividing the ranges of each variable into bins, and computing how many values fall in each histogram bin (Scott 1979). We can then easily compute the MI using the formula above (see `mutualInfoAllPairsMixed` for some code, and `miMixedDemo` for a demo).

Unfortunately, the number of bins used, and the location of the bin boundaries, can have a significant effect on the results. One way around this is to try to estimate the MI directly, without first performing density estimation (Learned-Miller 2004). Another approach is to try many different bin sizes and locations, and to compute the maximum MI achieved. This statistic, appropriately normalized, is known as the **maximal information coefficient** (MIC) (Reshef et al. 2011). More precisely, define

$$m(x, y) = \frac{\max_{G \in \mathcal{G}(x,y)} \mathbb{I}\left(X(G); Y(G)\right)}{\log \min(x, y)} \tag{2.123}$$

where $\mathcal{G}(x, y)$ is the set of 2d grids of size $x \times y$, and $X(G), Y(G)$ represents a discretization of the variables onto this grid. (The maximization over bin locations can be performed efficiently using dynamic programming (Reshef et al. 2011).) Now define the MIC as

$$\text{MIC} \triangleq \max_{x,y:xy<B} m(x, y) \tag{2.124}$$

where B is some sample-size dependent bound on the number of bins we can use and still reliably estimate the distribution ((Reshef et al. 2011) suggest $B = N^{0.6}$). It can be shown that the MIC lies in the range $[0, 1]$, where 0 represents no relationship between the variables, and 1 represents a noise-free relationship of any form, not just linear.

Figure 2.22 gives an example of this statistic in action. The data consists of 357 variables measuring a variety of social, economic, health and political indicators, collected by the World Health Organization (WHO). On the left of the figure, we see the correlation coefficient (CC) plotted against the MIC for all 63,566 variable pairs. On the right of the figure, we see scatter plots for particular pairs of variables, which we now discuss:

- The point marked C has a low CC and a low MIC. The corresponding scatter plot makes it clear that there is no relationship between these two variables (percentage of lives lost to injury and density of dentists in the population).

- The points marked D and H have high CC (in absolute value) and high MIC, because they represent nearly linear relationships.

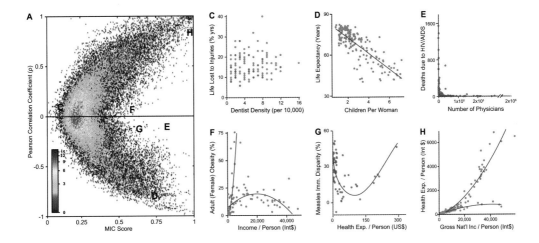

Figure 2.22 Left: Correlation coefficient vs maximal information criterion (MIC) for all pairwise relationships in the WHO data. Right: scatter plots of certain pairs of variables. The red lines are non-parametric smoothing regressions (Section 15.4.6) fit separately to each trend. Source: Figure 4 of (Reshef et al. 2011). Used with kind permission of David Reshef and the American Association for the Advancement of Science.

- The points marked E, F, and G have low CC but high MIC. This is because they correspond to non-linear (and sometimes, as in the case of E and F, non-functional, i.e., one-to-many) relationships between the variables.

In summary, we see that statistics (such as MIC) based on mutual information can be used to discover interesting relationships between variables in a way that simpler measures, such as correlation coefficients, cannot. For this reason, the MIC has been called "a correlation for the 21st century" (Speed 2011).

Exercises

Exercise 2.1 Probabilities are sensitive to the form of the question that was used to generate the answer

(Source: Minka.) My neighbor has two children. Assuming that the gender of a child is like a coin flip, it is most likely, a priori, that my neighbor has one boy and one girl, with probability 1/2. The other possibilities—two boys or two girls—have probabilities 1/4 and 1/4.

a. Suppose I ask him whether he has any boys, and he says yes. What is the probability that one child is a girl?
b. Suppose instead that I happen to see one of his children run by, and it is a boy. What is the probability that the other child is a girl?

Exercise 2.2 Legal reasoning

(Source: Peter Lee.) Suppose a crime has been committed. Blood is found at the scene for which there is no innocent explanation. It is of a type which is present in 1% of the population.

a. The prosecutor claims: "There is a 1% chance that the defendant would have the crime blood type if he were innocent. Thus there is a 99% chance that he is guilty". This is known as the **prosecutor's fallacy**. What is wrong with this argument?

b. The defender claims: "The crime occurred in a city of 800,000 people. The blood type would be found in approximately 8000 people. The evidence has provided a probability of just 1 in 8000 that the defendant is guilty, and thus has no relevance." This is known as the **defender's fallacy**. What is wrong with this argument?

Exercise 2.3 Variance of a sum

Show that the variance of a sum is $\text{var}[X+Y] = \text{var}[X] + \text{var}[Y] + 2\text{cov}[X,Y]$, where $\text{cov}[X,Y]$ is the covariance between X and Y.

Exercise 2.4 Bayes rule for medical diagnosis

(Source: Koller.) After your yearly checkup, the doctor has bad news and good news. The bad news is that you tested positive for a serious disease, and that the test is 99% accurate (i.e., the probability of testing positive given that you have the disease is 0.99, as is the probability of testing negative given that you don't have the disease). The good news is that this is a rare disease, striking only one in 10,000 people. What are the chances that you actually have the disease? (Show your calculations as well as giving the final result.)

Exercise 2.5 The Monty Hall problem

(Source: Mackay.) On a game show, a contestant is told the rules as follows:

> There are three doors, labelled 1, 2, 3. A single prize has been hidden behind one of them. You get to select one door. Initially your chosen door will *not* be opened. Instead, the gameshow host will open one of the other two doors, and *he will do so in such a way as not to reveal the prize.* For example, if you first choose door 1, he will then open one of doors 2 and 3, and it is guaranteed that he will choose which one to open so that the prize will not be revealed.
>
> At this point, you will be given a fresh choice of door: you can either stick with your first choice, or you can switch to the other closed door. All the doors will then be opened and you will receive whatever is behind your final choice of door.

Imagine that the contestant chooses door 1 first; then the gameshow host opens door 3, revealing nothing behind the door, as promised. Should the contestant (a) stick with door 1, or (b) switch to door 2, or (c) does it make no difference? You may assume that initially, the prize is equally likely to be behind any of the 3 doors. Hint: use Bayes rule.

Exercise 2.6 Conditional independence

(Source: Koller.)

a. Let $H \in \{1, \ldots, K\}$ be a discrete random variable, and let e_1 and e_2 be the observed values of two other random variables E_1 and E_2. Suppose we wish to calculate the vector

$$\vec{P}(H|e_1, e_2) = (P(H = 1|e_1, e_2), \ldots, P(H = K|e_1, e_2))$$

Which of the following sets of numbers are sufficient for the calculation?

   i. $P(e_1, e_2), P(H), P(e_1|H), P(e_2|H)$
   ii. $P(e_1, e_2), P(H), P(e_1, e_2|H)$
   iii. $P(e_1|H), P(e_2|H), P(H)$

b. Now suppose we now assume $E_1 \perp E_2|H$ (i.e., E_1 and E_2 are conditionally independent given H). Which of the above 3 sets are sufficient now?

Show your calculations as well as giving the final result. Hint: use Bayes rule.

Exercise 2.7 Pairwise independence does not imply mutual independence

We say that two random variables are pairwise independent if

$$p(X_2|X_1) = p(X_2) \tag{2.125}$$

and hence

$$p(X_2, X_1) = p(X_1)p(X_2|X_1) = p(X_1)p(X_2) \tag{2.126}$$

We say that n random variables are mutually independent if

$$p(X_i|X_S) = p(X_i) \quad \forall S \subseteq \{1, \ldots, n\} \setminus \{i\} \tag{2.127}$$

and hence

$$p(X_{1:n}) = \prod_{i=1}^{n} p(X_i) \tag{2.128}$$

Show that pairwise independence between all pairs of variables does not necessarily imply mutual independence. It suffices to give a counter example.

Exercise 2.8 Conditional independence iff joint factorizes

In the text we said $X \perp Y|Z$ iff

$$p(x, y|z) = p(x|z)p(y|z) \tag{2.129}$$

for all x, y, z such that $p(z) > 0$. Now prove the following alternative definition: $X \perp Y|Z$ iff there exist functions g and h such that

$$p(x, y|z) = g(x, z)h(y, z) \tag{2.130}$$

for all x, y, z such that $p(z) > 0$.

Exercise 2.9 Conditional independence

(Source: Koller.) Are the following properties true? Prove or disprove. Note that we are not restricting attention to distributions that can be represented by a graphical model.

a. True or false? $(X \perp W|Z, Y) \wedge (X \perp Y|Z) \Rightarrow (X \perp Y, W|Z)$

b. True or false? $(X \perp Y|Z) \wedge (X \perp Y|W) \Rightarrow (X \perp Y|Z, W)$

Exercise 2.10 Deriving the inverse gamma density

Let $X \sim \text{Ga}(a, b)$, i.e.

$$\text{Ga}(x|a, b) = \frac{b^a}{\Gamma(a)} x^{a-1} e^{-xb} \tag{2.131}$$

Let $Y = 1/X$. Show that $Y \sim \text{IG}(a, b)$, i.e.,

$$\text{IG}(x|\text{shape} = a, \text{scale} = b) = \frac{b^a}{\Gamma(a)} x^{-(a+1)} e^{-b/x} \tag{2.132}$$

Hint: use the change of variables formula.

Exercise 2.11 Normalization constant for a 1D Gaussian

The normalization constant for a zero-mean Gaussian is given by

$$Z = \int_a^b \exp\left(-\frac{x^2}{2\sigma^2}\right) dx \tag{2.133}$$

where $a = -\infty$ and $b = \infty$. To compute this, consider its square

$$Z^2 = \int_a^b \int_a^b \exp\left(-\frac{x^2 + y^2}{2\sigma^2}\right) dxdy \tag{2.134}$$

Let us change variables from cartesian (x, y) to polar (r, θ) using $x = r\cos\theta$ and $y = r\sin\theta$. Since $dxdy = rdrd\theta$, and $cos^2\theta + \sin^2\theta = 1$, we have

$$Z^2 = \int_0^{2\pi} \int_0^\infty r\exp\left(-\frac{r^2}{2\sigma^2}\right) drd\theta \tag{2.135}$$

Evaluate this integral and hence show $Z = \sqrt{\sigma^2 2\pi}$. Hint 1: separate the integral into a product of two terms, the first of which (involving $d\theta$) is constant, so is easy. Hint 2: if $u = e^{-r^2/2\sigma^2}$ then $du/dr = -\frac{1}{\sigma^2}re^{-r^2/2\sigma^2}$, so the second integral is also easy (since $\int u'(r)dr = u(r)$).

Exercise 2.12 Expressing mutual information in terms of entropies

Show that

$$I(X, Y) = H(X) - H(X|Y) = H(Y) - H(Y|X) \tag{2.136}$$

Exercise 2.13 Mutual information for correlated normals

(Source: (Cover and Thomas 1991, Q9.3).) Find the mutual information $I(X_1, X_2)$ where $\mathbf{X}$ has a bivariate normal distribution:

$$\begin{pmatrix} X_1 \\ X_2 \end{pmatrix} \sim \mathcal{N}\left(\mathbf{0}, \begin{pmatrix} \sigma^2 & \rho\sigma^2 \\ \rho\sigma^2 & \sigma^2 \end{pmatrix}\right) \tag{2.137}$$

Evaluate $I(X_1, X_2)$ for $\rho = 1$, $\rho = 0$ and $\rho = -1$ and comment. Hint: The (differential) entropy of a d-dimensional Gaussian is

$$h(\mathbf{X}) = \frac{1}{2}\log_2\left[(2\pi e)^d \det \Sigma\right] \tag{2.138}$$

In the 1d case, this becomes

$$h(X) = \frac{1}{2}\log_2\left[2\pi e\sigma^2\right] \tag{2.139}$$

Hint: $\log(0) = \infty$.

Exercise 2.14 A measure of correlation (normalized mutual information)

(Source: (Cover and Thomas 1991, Q2.20).) Let X and Y be discrete random variables which are identically distributed (so $H(X) = H(Y)$) but not necessarily independent. Define

$$r = 1 - \frac{H(Y|X)}{H(X)} \tag{2.140}$$

a. Show $r = \frac{I(X,Y)}{H(X)}$

b. Show $0 \leq r \leq 1$

c. When is $r = 0$?

d. When is $r = 1$?

Exercise 2.15 MLE minimizes KL divergence to the empirical distribution

Let $p_{\text{emp}}(x)$ be the empirical distribution, and let $q(x|\boldsymbol{\theta})$ be some model. Show that $\text{argmin}_q \, \mathbb{KL}\left(p_{\text{emp}}||q\right)$ is obtained by $q(x) = q(x; \hat{\boldsymbol{\theta}})$, where $\hat{\boldsymbol{\theta}}$ is the MLE. Hint: use non-negativity of the KL divergence.

Exercise 2.16 Mean, mode, variance for the beta distribution

Suppose $\theta \sim \text{Beta}(a, b)$. Derive the mean, mode and variance.

Exercise 2.17 Expected value of the minimum

Suppose X, Y are two points sampled independently and uniformly at random from the interval $[0, 1]$. What is the expected location of the leftmost point?

3 *Generative models for discrete data*

3.1 Introduction

In Section 2.2.3.2, we discussed how to classify a feature vector $\mathbf{x}$ by applying Bayes rule to a generative classifier of the form

$$p(y = c|\mathbf{x}, \boldsymbol{\theta}) \propto p(\mathbf{x}|y = c, \boldsymbol{\theta})p(y = c|\boldsymbol{\theta}) \tag{3.1}$$

The key to using such models is specifying a suitable form for the class-conditional density $p(\mathbf{x}|y = c, \boldsymbol{\theta})$, which defines what kind of data we expect to see in each class. In this chapter, we focus on the case where the observed data are discrete symbols. We also discuss how to infer the unknown parameters $\boldsymbol{\theta}$ of such models.

3.2 Bayesian concept learning

Consider how a child learns to understand the meaning of a word, such as "dog". Presumably the child's parents point out positive examples of this concept, saying such things as, "look at the cute dog!", or "mind the doggy", etc. However, it is very unlikely that they provide negative examples, by saying "look at that non-dog". Certainly, negative examples may be obtained during an active learning process — the child says "look at the dog" and the parent says "that's a cat, dear, not a dog" — but psychological research has shown that people can learn concepts from positive examples alone (Xu and Tenenbaum 2007).

We can think of learning the meaning of a word as equivalent to **concept learning**, which in turn is equivalent to binary classification. To see this, define $f(x) = 1$ if x is an example of the concept C, and $f(x) = 0$ otherwise. Then the goal is to learn the indicator function f, which just defines which elements are in the set C. By allowing for uncertainty about the definition of f, or equivalently the elements of C, we can emulate **fuzzy set theory**, but using standard probability calculus. Note that standard binary classification techniques require positive and negative examples. By contrast, we will devise a way to learn from positive examples alone.

For pedagogical purposes, we will consider a very simple example of concept learning called the **number game**, based on part of Josh Tenenbaum's PhD thesis (Tenenbaum 1999). The game proceeds as follows. I choose some simple arithmetical concept C, such as "prime number" or "a number between 1 and 10". I then give you a series of randomly chosen positive examples $\mathcal{D} = \{x_1, \ldots, x_N\}$ drawn from C, and ask you whether some new test case $\tilde{x}$ belongs to C, i.e., I ask you to classify $\tilde{x}$.

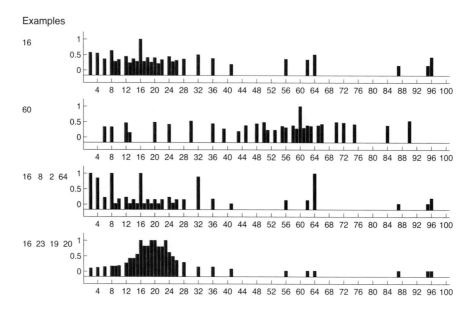

Figure 3.1 Empirical predictive distribution averaged over 8 humans in the number game. First two rows: after seeing $\mathcal{D} = \{16\}$ and $\mathcal{D} = \{60\}$. This illustrates diffuse similarity. Third row: after seeing $\mathcal{D} = \{16, 8, 2, 64\}$. This illustrates rule-like behavior (powers of 2). Bottom row: after seeing $\mathcal{D} = \{16, 23, 19, 20\}$. This illustrates focussed similarity (numbers near 20). Source: Figure 5.5 of (Tenenbaum 1999). Used with kind permission of Josh Tenenbaum.

Suppose, for simplicity, that all numbers are integers between 1 and 100. Now suppose I tell you "16" is a positive example of the concept. What other numbers do you think are positive? 17? 6? 32? 99? It's hard to tell with only one example, so your predictions will be quite vague. Presumably numbers that are similar in some sense to 16 are more likely. But similar in what way? 17 is similar, because it is "close by", 6 is similar because it has a digit in common, 32 is similar because it is also even and a power of 2, but 99 does not seem similar. Thus some numbers are more likely than others. We can represent this as a probability distribution, $p(\tilde{x}|\mathcal{D})$, which is the probability that $\tilde{x} \in C$ given the data $\mathcal{D}$ for any $\tilde{x} \in \{1, \ldots, 100\}$. This is called the **posterior predictive distribution**. Figure 3.1(top) shows the predictive distribution of people derived from a lab experiment. We see that people predict numbers that are similar to 16, under a variety of kinds of similarity.

Now suppose I tell you that 8, 2 and 64 are *also* positive examples. Now you may guess that the hidden concept is "powers of two". This is an example of **induction**. Given this hypothesis, the predictive distribution is quite specific, and puts most of its mass on powers of 2, as shown in Figure 3.1(third row). If instead I tell you the data is $\mathcal{D} = \{16, 23, 19, 20\}$, you will get a different kind of **generalization gradient**, as shown in Figure 3.1(bottom).

How can we explain this behavior and emulate it in a machine? The classic approach to induction is to suppose we have a **hypothesis space** of concepts, $\mathcal{H}$, such as: odd numbers, even numbers, all numbers between 1 and 100, powers of two, all numbers ending in j (for

$0 \leq j \leq 9$), etc. The subset of $\mathcal{H}$ that is consistent with the data D is called the **version space**. As we see more examples, the version space shrinks and we become increasingly certain about the concept (Mitchell 1997).

However, the version space is not the whole story. After seeing $\mathcal{D} = \{16\}$, there are many consistent rules; how do you combine them to predict if $\tilde{x} \in C$? Also, after seeing $\mathcal{D} = \{16, 8, 2, 64\}$, why did you choose the rule "powers of two" and not, say, "all even numbers", or "powers of two except for 32", both of which are equally consistent with the evidence? We will now provide a Bayesian explanation for this.

3.2.1 Likelihood

We must explain why we chose $h_{\text{two}} \triangleq$"powers of two", and not, say, $h_{\text{even}} \triangleq$ "even numbers" after seeing $\mathcal{D} = \{16, 8, 2, 64\}$, given that both hypotheses are consistent with the evidence. The key intuition is that we want to avoid **suspicious coincidences**. If the true concept was even numbers, how come we only saw numbers that happened to be powers of two?

To formalize this, let us assume that examples are sampled uniformly at random from the **extension** of a concept. (The extension of a concept is just the set of numbers that belong to it, e.g., the extension of h_{even} is $\{2, 4, 6, \ldots, 98, 100\}$; the extension of "numbers ending in 9" is $\{9, 19, \ldots, 99\}$.) Tenenbaum calls this the **strong sampling assumption**. Given this assumption, the probability of independently sampling N items (with replacement) from h is given by

$$p(\mathcal{D}|h) = \left[\frac{1}{\text{size}(h)}\right]^N = \left[\frac{1}{|h|}\right]^N \tag{3.2}$$

This crucial equation embodies what Tenenbaum calls the **size principle**, which means the model favors the simplest (smallest) hypothesis consistent with the data. This is more commonly known as **Occam's razor**.[1]

To see how it works, let $\mathcal{D} = \{16\}$. Then $p(\mathcal{D}|h_{\text{two}}) = 1/6$, since there are only 6 powers of two less than 100, but $p(\mathcal{D}|h_{\text{even}}) = 1/50$, since there are 50 even numbers. So the likelihood that $h = h_{\text{two}}$ is higher than if $h = h_{\text{even}}$. After 4 examples, the likelihood of h_{two} is $(1/6)^4 = 7.7 \times 10^{-4}$, whereas the likelihood of h_{even} is $(1/50)^4 = 1.6 \times 10^{-7}$. This is a **likelihood ratio** of almost 5000:1 in favor of h_{two}. This quantifies our earlier intuition that $D = \{16, 8, 2, 64\}$ would be a very suspicious coincidence if generated by h_{even}.

3.2.2 Prior

Suppose $D = \{16, 8, 2, 64\}$. Given this data, the concept $h' =$"powers of two except 32" is more likely than $h =$"powers of two", since h' does not need to explain the coincidence that 32 is missing from the set of examples.

However, the hypothesis $h' =$"powers of two except 32" seems "conceptually unnatural". We can capture such intuition by assigning low prior probability to unnatural concepts. Of course, your prior might be different than mine. This **subjective** aspect of Bayesian reasoning is a source of much controversy, since it means, for example, that a child and a math professor

---

1. William of Occam (also spelt Ockham) was an English monk and philosopher, 1288–1348.

will reach different answers. In fact, they presumably not only have different priors, but also different hypothesis spaces. However, we can finesse that by defining the hypothesis space of the child and the math professor to be the same, and then setting the child's prior weight to be zero on certain "advanced" concepts. Thus there is no sharp distinction between the prior and the hypothesis space.

Although the subjectivity of the prior is controversial, it is actually quite useful. If you are told the numbers are from some arithmetic rule, then given 1200, 1500, 900 and 1400, you may think 400 is likely but 1183 is unlikely. But if you are told that the numbers are examples of healthy cholesterol levels, you would probably think 400 is unlikely and 1183 is likely. Thus we see that the prior is the mechanism by which background knowledge can be brought to bear on a problem. Without this, rapid learning (i.e., from small samples sizes) is impossible.

So, what prior should we use? For illustration purposes, let us use a simple prior which puts uniform probability on 30 simple arithmetical concepts, such as "even numbers", "odd numbers", "prime numbers", "numbers ending in 9", etc. To make things more interesting, we make the concepts even and odd more likely apriori. We also include two "unnatural" concepts, namely "powers of 2, plus 37" and "powers of 2, except 32", but give them low prior weight. See Figure 3.2(a) for a plot of this prior. We will consider a slightly more sophisticated prior later on.

3.2.3 Posterior

The posterior is simply the likelihood times the prior, normalized. In this context we have

$$p(h|\mathcal{D}) \;=\; \frac{p(\mathcal{D}|h)p(h)}{\sum_{h' \in \mathcal{H}} p(\mathcal{D}, h')} = \frac{p(h)\mathbb{I}(\mathcal{D} \in h)/|h|^N}{\sum_{h' \in \mathcal{H}} p(h')\mathbb{I}(\mathcal{D} \in h')/|h'|^N} \tag{3.3}$$

where $\mathbb{I}(\mathcal{D} \in h)$ is 1 **iff** (if and only if) all the data are in the extension of the hypothesis h. Figure 3.2 plots the prior, likelihood and posterior after seeing $\mathcal{D} = \{16\}$. We see that the posterior is a combination of prior and likelihood. In the case of most of the concepts, the prior is uniform, so the posterior is proportional to the likelihood. However, the "unnatural" concepts of "powers of 2, plus 37" and "powers of 2, except 32" have low posterior support, despite having high likelihood, due to the low prior. Conversely, the concept of odd numbers has low posterior support, despite having a high prior, due to the low likelihood.

Figure 3.3 plots the prior, likelihood and posterior after seeing $\mathcal{D} = \{16, 8, 2, 64\}$. Now the likelihood is much more peaked on the powers of two concept, so this dominates the posterior. Essentially the learner has an **aha** moment, and figures out the true concept. (Here we see the need for the low prior on the unnatural concepts, otherwise we would have overfit the data and picked "powers of 2, except for 32".)

In general, when we have enough data, the posterior $p(h|\mathcal{D})$ becomes peaked on a single concept, namely the MAP estimate, i.e.,

$$p(h|\mathcal{D}) \rightarrow \delta_{\hat{h}^{MAP}}(h) \tag{3.4}$$

where $\hat{h}^{MAP} = \operatorname{argmax}_h p(h|\mathcal{D})$ is the posterior mode, and where δ is the **Dirac measure** defined by

$$\delta_x(A) = \begin{cases} 1 & \text{if } x \in A \\ 0 & \text{if } x \notin A \end{cases} \tag{3.5}$$

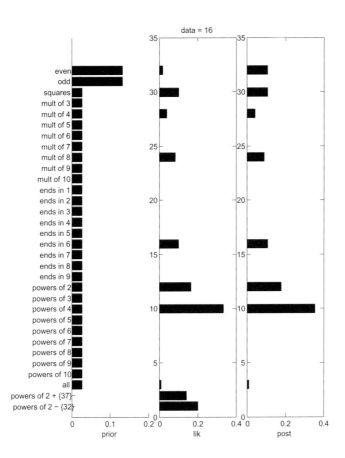

Figure 3.2   Prior, likelihood and posterior for $\mathcal{D} = \{16\}$. Based on (Tenenbaum 1999). Figure generated by `numbersGame`.

Note that the MAP estimate can be written as

$$\hat{h}^{MAP} = \underset{h}{\operatorname{argmax}}\, p(\mathcal{D}|h)p(h) = \underset{h}{\operatorname{argmax}}\, [\log p(\mathcal{D}|h) + \log p(h)] \tag{3.6}$$

Since the likelihood term depends exponentially on N, and the prior stays constant, as we get more and more data, the MAP estimate converges towards the **maximum likelihood estimate** or **MLE**:

$$\hat{h}^{mle} \triangleq \underset{h}{\operatorname{argmax}}\, p(\mathcal{D}|h) = \underset{h}{\operatorname{argmax}}\, \log p(\mathcal{D}|h) \tag{3.7}$$

In other words, if we have enough data, we see that the **data overwhelms the prior**. In this

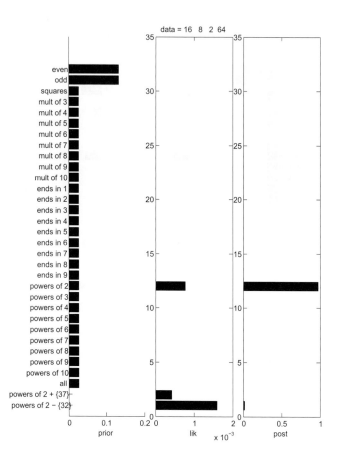

Figure 3.3   Prior, likelihood and posterior for $\mathcal{D} = \{16, 8, 2, 64\}$. Based on (Tenenbaum 1999). Figure generated by `numbersGame`.

case, the MAP estimate converges towards the MLE.

If the true hypothesis is in the hypothesis space, then the MAP/ ML estimate will converge upon this hypothesis. Thus we say that Bayesian inference (and ML estimation) are consistent estimators (see Section 6.4.1 for details). We also say that the hypothesis space is **identifiable in the limit**, meaning we can recover the truth in the limit of infinite data. If our hypothesis class is not rich enough to represent the "truth" (which will usually be the case), we will converge on the hypothesis that is as close as possible to the truth. However, formalizing this notion of "closeness" is beyond the scope of this chapter.

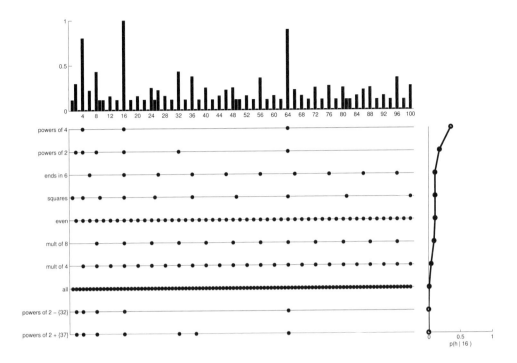

Figure 3.4 Posterior over hypotheses and the corresponding predictive distribution after seeing one example, $\mathcal{D} = \{16\}$. A dot means this number is consistent with this hypothesis. The graph $p(h|\mathcal{D})$ on the right is the weight given to hypothesis h. By taking a weighed sum of dots, we get $p(\tilde{x} \in C|\mathcal{D})$ (top). Based on Figure 2.9 of (Tenenbaum 1999). Figure generated by `numbersGame`.

3.2.4 Posterior predictive distribution

The posterior is our internal **belief state** about the world. The way to test if our beliefs are justified is to use them to predict objectively observable quantities (this is the basis of the scientific method). Specifically, the posterior predictive distribution in this context is given by

$$p(\tilde{x} \in C|\mathcal{D}) = \sum_h p(y = 1|\tilde{x}, h)p(h|\mathcal{D}) \tag{3.8}$$

This is just a weighted average of the predictions of each individual hypothesis and is called **Bayes model averaging** (Hoeting et al. 1999). This is illustrated in Figure 3.4. The dots at the bottom show the predictions from each hypothesis; the vertical curve on the right shows the weight associated with each hypothesis. If we multiply each row by its weight and add up, we get the distribution at the top.

When we have a small and/or ambiguous dataset, the posterior $p(h|\mathcal{D})$ is vague, which induces a broad predictive distribution. However, once we have "figured things out", the posterior becomes a delta function centered at the MAP estimate. In this case, the predictive distribution becomes

$$p(\tilde{x} \in C | \mathcal{D}) = \sum_h p(\tilde{x}|h)\delta_{\hat{h}}(h) = p(\tilde{x}|\hat{h}) \tag{3.9}$$

This is called a **plug-in approximation** to the predictive density and is very widely used, due to its simplicity. However, in general, this under-represents our uncertainty, and our predictions will not be as "smooth" as when using BMA. We will see more examples of this later in the book.

Although MAP learning is simple, it cannot explain the gradual shift from similarity-based reasoning (with uncertain posteriors) to rule-based reasoning (with certain posteriors). For example, suppose we observe $\mathcal{D} = \{16\}$. If we use the simple prior above, the minimal consistent hypothesis is "all powers of 4", so only 4 and 16 get a non-zero probability of being predicted. This is of course an example of overfitting. Given $\mathcal{D} = \{16, 8, 2, 64\}$, the MAP hypothesis is "all powers of two". Thus the plug-in predictive distribution gets broader (or stays the same) as we see more data: it starts narrow, but is forced to broaden as it sees more data. In contrast, in the Bayesian approach, we start broad and then narrow down as we learn more, which makes more intuitive sense. In particular, given $\mathcal{D} = \{16\}$, there are many hypotheses with non-negligible posterior support, so the predictive distribution is broad. However, when we see $\mathcal{D} = \{16, 8, 2, 64\}$, the posterior concentrates its mass on one hypothesis, so the predictive distribution becomes narrower. So the predictions made by a plug-in approach and a Bayesian approach are quite different in the small sample regime, although they converge to the same answer as we see more data.

3.2.5 A more complex prior

To model human behavior, Tenenbaum used a slightly more sophisticated prior which was derived by analysing some experimental data of how people measure similarity between numbers; see (Tenenbaum 1999, p208) for details. The result is a set of arithmetical concepts similar to those mentioned above, plus all intervals between n and m for $1 \leq n, m \leq 100$. (Note that these hypotheses are not mutually exclusive.) Thus the prior is a **mixture** of two priors, one over arithmetical rules, and one over intervals:

$$p(h) = \pi_0 p_{\text{rules}}(h) + (1 - \pi_0)p_{\text{interval}}(h) \tag{3.10}$$

The only free parameter in the model is the relative weight, π_0, given to these two parts of the prior. The results are not very sensitive to this value, so long as $\pi_0 > 0.5$, reflecting the fact that people are more likely to think of concepts defined by rules. The predictive distribution of the model, using this larger hypothesis space, is shown in Figure 3.5. It is strikingly similar to the human predictive distribution, shown in Figure 3.1, even though it was not fit to human data (modulo the choice of hypothesis space).

3.3 The beta-binomial model

The number game involved inferring a distribution over a discrete variable drawn from a finite hypothesis space, $h \in \mathcal{H}$, given a series of discrete observations. This made the computations particularly simple: we just needed to sum, multiply and divide. However, in many applications, the unknown parameters are continuous, so the hypothesis space is (some subset) of $\mathbb{R}^K$, where

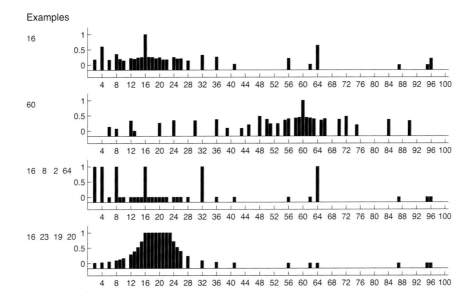

Examples

Figure 3.5 Predictive distributions for the model using the full hypothesis space. Compare to Figure 3.1. The predictions of the Bayesian model are only plotted for those values of $\tilde{x}$ for which human data is available; this is why the top line looks sparser than Figure 3.4. Source: Figure 5.6 of (Tenenbaum 1999). Used with kind permission of Josh Tenenbaum.

K is the number of parameters. This complicates the mathematics, since we have to replace sums with integrals. However, the basic ideas are the same.

We will illustrate this by considering the problem of inferring the probability that a coin shows up heads, given a series of observed coin tosses. Although this might seem trivial, it turns out that this model forms the basis of many of the methods we will consider later in this book, including naive Bayes classifiers, Markov models, etc. It is historically important, since it was the example which was analyzed in Bayes' original paper of 1763. (Bayes' analysis was subsequently generalized by Pierre-Simon Laplace, creating what we now call "Bayes rule" — see (Stigler 1986) for further historical details.)

We will follow our now-familiar recipe of specifying the likelihood and prior, and deriving the posterior and posterior predictive.

3.3.1 Likelihood

Suppose $X_i \sim \text{Ber}(\theta)$, where $X_i = 1$ represents "heads", $X_i = 0$ represents "tails", and $\theta \in [0, 1]$ is the rate parameter (probability of heads). If the data are iid, the likelihood has the form

$$p(\mathcal{D}|\theta) = \theta^{N_1}(1 - \theta)^{N_0} \tag{3.11}$$

where we have $N_1 = \sum_{i=1}^{N} \mathbb{I}(x_i = 1)$ heads and $N_0 = \sum_{i=1}^{N} \mathbb{I}(x_i = 0)$ tails. These two counts are called the **sufficient statistics** of the data, since this is all we need to know about $\mathcal{D}$ to infer θ. (An alternative set of sufficient statistics are N_1 and $N = N_0 + N_1$.)

More formally, we say $\mathbf{s}(\mathcal{D})$ is a sufficient statistic for data $\mathcal{D}$ if $p(\boldsymbol{\theta}|\mathcal{D}) = p(\boldsymbol{\theta}|\mathbf{s}(\mathcal{D}))$. If we use a uniform prior, this is equivalent to saying $p(\mathcal{D}|\boldsymbol{\theta}) \propto p(\mathbf{s}(\mathcal{D})|\boldsymbol{\theta})$. Consequently, if we have two datasets with the same sufficient statistics, we will infer the same value for $\boldsymbol{\theta}$.

Now suppose the data consists of the count of the number of heads N_1 observed in a fixed number $N = N_1 + N_0$ of trials. In this case, we have $N_1 \sim \text{Bin}(N, \theta)$, where Bin represents the binomial distribution, which has the following pmf:

$$\text{Bin}(k|n, \theta) \triangleq \binom{n}{k} \theta^k (1 - \theta)^{n-k} \tag{3.12}$$

Since $\binom{n}{k}$ is a constant independent of θ, the likelihood for the binomial sampling model is the same as the likelihood for the Bernoulli model. So any inferences we make about θ will be the same whether we observe the counts, $\mathcal{D} = (N_1, N)$, or a sequence of trials, $\mathcal{D} = \{x_1, \ldots, x_N\}$.

3.3.2 Prior

We need a prior which has support over the interval $[0, 1]$. To make the math easier, it would be convenient if the prior had the same form as the likelihood, i.e., if the prior looked like

$$p(\theta) \propto \theta^{\gamma_1} (1 - \theta)^{\gamma_2} \tag{3.13}$$

for some prior parameters γ_1 and γ_2. If this were the case, then we could easily evaluate the posterior by simply adding up the exponents:

$$p(\theta|\mathcal{D}) \propto p(\mathcal{D}|\theta)p(\theta) = \theta^{N_1}(1 - \theta)^{N_0} \theta^{\gamma_1}(1 - \theta)^{\gamma_2} = \theta^{N_1 + \gamma_1}(1 - \theta)^{N_0 + \gamma_2} \tag{3.14}$$

When the prior and the posterior have the same form, we say that the prior is a **conjugate prior** for the corresponding likelihood. Conjugate priors are widely used because they simplify computation, and are easy to interpret, as we see below.

In the case of the Bernoulli, the conjugate prior is the beta distribution, which we encountered in Section 2.4.6:

$$\text{Beta}(\theta|a, b) \propto \theta^{a-1}(1 - \theta)^{b-1} \tag{3.15}$$

The parameters of the prior are called **hyper-parameters**. We can set them in order to encode our prior beliefs. For example, to encode our beliefs that θ has mean 0.7 and standard deviation 0.2, we set $a = 2.975$ and $b = 1.275$ (Exercise 3.15). Or to encode our beliefs that θ has mean 0.15 and that we think it lives in the interval $(0.05, 0.30)$ with probability 0.95, then we find $a = 4.5$ and $b = 25.5$ (Exercise 3.16).

If we know "nothing" about θ, except that it lies in the interval $[0, 1]$, we can use a uniform prior, which is a kind of uninformative prior (see Section 5.4.2 for details). The uniform distribution can be represented by a beta distribution with $a = b = 1$.

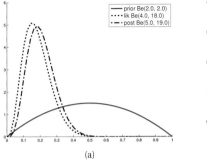 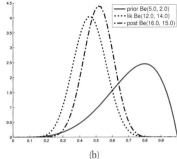

(a) (b)

Figure 3.6 (a) Updating a Beta(2, 2) prior with a Binomial likelihood with sufficient statistics $N_1 = 3, N_0 = 17$ to yield a Beta(5,19) posterior. (b) Updating a Beta(5, 2) prior with a Binomial likelihood with sufficient statistics $N_1 = 11, N_0 = 13$ to yield a Beta(16, 15) posterior. Figure generated by `binomialBetaPosteriorDemo`.

3.3.3 Posterior

If we multiply the likelihood by the beta prior we get the following posterior (following Equation 3.14):

$$p(\theta|\mathcal{D}) \quad \propto \quad \text{Bin}(N_1|\theta, N_0 + N_1)\text{Beta}(\theta|a, b) \propto \text{Beta}(\theta|N_1 + a, N_0 + b) \tag{3.16}$$

In particular, the posterior is obtained by adding the prior hyper-parameters to the empirical counts. For this reason, the hyper-parameters are known as **pseudo counts**. The strength of the prior, also known as the **equivalent sample size** of the prior, is the sum of the pseudo counts, $\alpha_0 = a + b$; this plays a role analogous to the data set size, $N_1 + N_0 = N$.

Figure 3.6(a) gives an example where we update a weak Beta(2,2) prior with a peaked likelihood function, corresponding to a large sample size; we see that the posterior is essentially identical to the likelihood, since the data has overwhelmed the prior. Figure 3.6(b) gives an example where we update a strong Beta(5,2) prior with a peaked likelihood function; now we see that the posterior is a "compromise" between the prior and likelihood.

Note that updating the posterior sequentially is equivalent to updating in a single batch. To see this, suppose we have two data sets $\mathcal{D}'$ and $\mathcal{D}''$ with sufficient statistics N_1', N_0' and N_1'', N_0''. Let $N_1 = N_1' + N_1''$ and $N_0 = N_0' + N_0''$ be the sufficient statistics of the combined datasets. In batch mode we have

$$p(\theta|\mathcal{D}', \mathcal{D}'') \quad \propto \quad \text{Bin}(N_1|\theta, N_1 + N_0)\text{Beta}(\theta|a, b) \propto \text{Beta}(\theta|N_1 + a, N_0 + b) \tag{3.17}$$

In sequential mode, we get the same final result:

$$p(\theta|\mathcal{D}', \mathcal{D}'') \quad \propto \quad p(\mathcal{D}''|\theta)p(\theta|\mathcal{D}') \tag{3.18}$$

$$\propto \quad \text{Bin}(N_1''|\theta, N_1'' + N_0'')\text{Beta}(\theta|N_1' + a, N_0' + b) \tag{3.19}$$

$$\propto \quad \text{Beta}(\theta|\ N_1' + N_1'' + a, N_0' + N_0'' + b) = \text{Beta}(\theta|N_1 + a, N_0 + b) \tag{3.20}$$

This makes Bayesian inference particularly well-suited to **online learning**, as we will see in Section 8.5.5.

3.3.3.1   Posterior mean and mode

From Equation 2.62, the MAP estimate is given by

$$\hat{\theta}_{MAP} \quad = \quad \frac{a + N_1 - 1}{a + b + N - 2} \tag{3.21}$$

If we use a uniform prior, then the MAP estimate reduces to the MLE, which is just the empirical fraction of heads:

$$\hat{\theta}_{MLE} = \frac{N_1}{N} \tag{3.22}$$

This makes intuitive sense, but it can also be derived by applying elementary calculus to maximize the likelihood function in Equation 3.11. (Exercise 3.1).

By contrast, the posterior mean is given by,

$$\bar{\theta} \quad = \quad \frac{a + N_1}{a + b + N} \tag{3.23}$$

This difference between the mode and the mean will prove important later.

We will now show that the posterior mean is a convex combination of the prior mean and the MLE, which captures the notion that the posterior is a compromise between what we previously believed and what the data is telling us.

Let $\alpha_0 = a + b$ be the equivalent sample size of the prior, which controls its strength, and let the prior mean be $m_1 = a/\alpha_0$. Then the posterior mean is given by

$$\mathbb{E}\left[\theta | \mathcal{D}\right] \quad = \quad \frac{\alpha_0 m_1 + N_1}{N + \alpha_0} = \frac{\alpha_0}{N + \alpha_0} m_1 + \frac{N}{N + \alpha_0} \frac{N_1}{N} = \lambda m_1 + (1 - \lambda)\hat{\theta}_{MLE} \tag{3.24}$$

where $\lambda = \frac{\alpha_0}{N + \alpha_0}$ is the ratio of the prior to posterior equivalent sample size. So the weaker the prior, the smaller is λ, and hence the closer the posterior mean is to the MLE as $N \to \infty$. One can show similarly that the posterior mode is a convex combination of the prior mode and the MLE, and that it too converges to the MLE.

3.3.3.2   Posterior variance

The mean and mode are point estimates, but it is useful to know how much we can trust them. The variance of the posterior is one way to measure this. The variance of the Beta posterior is given by

$$\text{var}\left[\theta | \mathcal{D}\right] = \frac{(a + N_1)(b + N_0)}{(a + N_1 + b + N_0)^2 (a + N_1 + b + N_0 + 1)} \tag{3.25}$$

We can simplify this formidable expression in the case that $N \gg a, b$, to get

$$\text{var}\left[\theta | \mathcal{D}\right] \approx \frac{N_1 N_0}{N N N} = \frac{\hat{\theta}(1 - \hat{\theta})}{N} \tag{3.26}$$

where $\hat{\theta}$ is the MLE. Hence the "**error bar**" in our estimate (i.e., the posterior standard deviation), is given by

$$\sigma = \sqrt{\text{var}\left[\theta | \mathcal{D}\right]} \approx \sqrt{\frac{\hat{\theta}(1 - \hat{\theta})}{N}} \tag{3.27}$$

We see that the uncertainty goes down at a rate of $1/\sqrt{N}$. Note, however, that the uncertainty (variance) is maximized when $\hat{\theta} = 0.5$, and is minimized when $\hat{\theta}$ is close to 0 or 1. This means it is easier to be sure that a coin is biased than to be sure that it is fair.

3.3.4 Posterior predictive distribution

So far, we have been focusing on inference of the unknown parameter(s). Let us now turn our attention to prediction of future observable data.

Consider predicting the probability of heads in a single future trial under a $\mathrm{Beta}(a, b)$ posterior. We have

$$p(\tilde{x} = 1|\mathcal{D}) = \int_0^1 p(x = 1|\theta)p(\theta|\mathcal{D})d\theta \qquad (3.28)$$

$$= \int_0^1 \theta\, \mathrm{Beta}(\theta|a, b)d\theta = \mathbb{E}\left[\theta|\mathcal{D}\right] = \frac{a}{a + b} \qquad (3.29)$$

Thus we see that the mean of the posterior predictive distribution is equivalent (in this case) to plugging in the posterior mean parameters: $p(\tilde{x}|\mathcal{D}) = \mathrm{Ber}(\tilde{x}|\mathbb{E}\left[\theta|\mathcal{D}\right])$.

3.3.4.1 Overfitting and the black swan paradox

Suppose instead that we plug-in the MLE, i.e., we use $p(\tilde{x}|\mathcal{D}) \approx \mathrm{Ber}(\tilde{x}|\hat{\theta}_{MLE})$. Unfortunately, this approximation can perform quite poorly when the sample size is small. For example, suppose we have seen $N = 3$ tails in a row. The MLE is $\hat{\theta} = 0/3 = 0$, since this makes the observed data as probable as possible. However, using this estimate, we predict that heads are impossible. This is called the **zero count problem** or the **sparse data problem**, and frequently occurs when estimating counts from small amounts of data. One might think that in the era of "big data", such concerns are irrelevant, but note that once we partition the data based on certain criteria — such as the number of times a *specific person* has engaged in a *specific activity* — the sample sizes can become much smaller. This problem arises, for example, when trying to perform personalized recommendation of web pages. Thus Bayesian methods are still useful, even in the big data regime (Jordan 2011).

The zero-count problem is analogous to a problem in philosophy called the **black swan paradox**. This is based on the ancient Western conception that all swans were white. In that context, a black swan was a metaphor for something that could not exist. (Black swans were discovered in Australia by European explorers in the 17th Century.) The term "black swan paradox" was first coined by the famous philosopher of science Karl Popper; the term has also been used as the title of a recent popular book (Taleb 2007). This paradox was used to illustrate the problem of **induction**, which is the problem of how to draw general conclusions about the future from specific observations from the past.

Let us now derive a simple Bayesian solution to the problem. We will use a uniform prior, so $a = b = 1$. In this case, plugging in the posterior mean gives **Laplace's rule of succession**

$$p(\tilde{x} = 1|\mathcal{D}) = \frac{N_1 + 1}{N_1 + N_0 + 2} \qquad (3.30)$$

This justifies the common practice of adding 1 to the empirical counts, normalizing and then plugging them in, a technique known as **add-one smoothing**. (Note that plugging in the MAP parameters would not have this smoothing effect, since the mode has the form $\hat{\theta} = \frac{N_1+a-1}{N+a+b-2}$, which becomes the MLE if $a = b = 1$.)

3.3.4.2 Predicting the outcome of multiple future trials

Suppose now we were interested in predicting the number of heads, x, in M future trials. This is given by

$$p(x|\mathcal{D}, M) \quad = \quad \int_0^1 \mathrm{Bin}(x|\theta, M)\mathrm{Beta}(\theta|a, b)d\theta \tag{3.31}$$

$$= \quad \binom{M}{x} \frac{1}{B(a, b)} \int_0^1 \theta^x (1 - \theta)^{M-x}\theta^{a-1}(1 - \theta)^{b-1}d\theta \tag{3.32}$$

We recognize the integral as the normalization constant for a $\mathrm{Beta}(a+x, M-x+b)$ distribution. Hence

$$\int_0^1 \theta^x (1 - \theta)^{M-x}\theta^{a-1}(1 - \theta)^{b-1}d\theta = B(x + a, M - x + b) \tag{3.33}$$

Thus we find that the posterior predictive is given by the following, known as the (compound) **beta-binomial** distribution:

$$Bb(x|a, b, M) \quad \triangleq \quad \binom{M}{x} \frac{B(x + a, M - x + b)}{B(a, b)} \tag{3.34}$$

This distribution has the following mean and variance

$$\mathbb{E}[x] = M\frac{a}{a + b}, \quad \mathrm{var}[x] = \frac{Mab}{(a + b)^2}\frac{(a + b + M)}{a + b + 1} \tag{3.35}$$

If $M = 1$, and hence $x \in \{0, 1\}$, we see that the mean becomes $\mathbb{E}[x|\mathcal{D}] = p(x = 1|\mathcal{D}) = \frac{a}{a+b}$, which is consistent with Equation 3.29.

This process is illustrated in Figure 3.7(a). We start with a Beta(2,2) prior, and plot the posterior predictive density after seeing $N_1 = 3$ heads and $N_0 = 17$ tails. Figure 3.7(b) plots a plug-in approximation using a MAP estimate. We see that the Bayesian prediction has longer tails, spreading its probablity mass more widely, and is therefore less prone to overfitting and black-swan type paradoxes.

3.4 The Dirichlet-multinomial model

In the previous section, we discussed how to infer the probability that a coin comes up heads. In this section, we generalize these results to infer the probability that a die with K sides comes up as face k. This might seem like another toy exercise, but the methods we will study are widely used to analyse text data, biosequence data, etc., as we will see later.

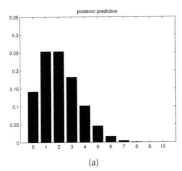
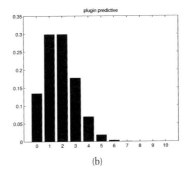

(a)   (b)

Figure 3.7   (a) Posterior predictive distributions after seeing $N_1 = 3, N_0 = 17$. (b) Plugin approximation. Figure generated by `betaBinomPostPredDemo`.

3.4.1 Likelihood

Suppose we observe N dice rolls, $\mathcal{D} = \{x_1, \ldots, x_N\}$, where $x_i \in \{1, \ldots, K\}$. If we assume the data is iid, the likelihood has the form

$$p(\mathcal{D}|\boldsymbol{\theta}) \quad = \quad \prod_{k=1}^{K} \theta_k^{N_k} \tag{3.36}$$

where $N_k = \sum_{i=1}^{N} \mathbb{I}(y_i = k)$ is the number of times event k occured (these are the sufficient statistics for this model). The likelihood for the multinomial model has the same form, up to an irrelevant constant factor.

3.4.2 Prior

Since the parameter vector lives in the K-dimensional probability simplex, we need a prior that has support over this simplex. Ideally it would also be conjugate. Fortunately, the Dirichlet distribution (Section 2.5.4) satisfies both criteria. So we will use the following prior:

$$\text{Dir}(\boldsymbol{\theta}|\boldsymbol{\alpha}) \quad = \quad \frac{1}{B(\boldsymbol{\alpha})} \prod_{k=1}^{K} \theta_k^{\alpha_k - 1} \mathbb{I}(\mathbf{x} \in S_K) \tag{3.37}$$

3.4.3 Posterior

Multiplying the likelihood by the prior, we find that the posterior is also Dirichlet:

$$p(\boldsymbol{\theta}|\mathcal{D}) \quad \propto \quad p(\mathcal{D}|\boldsymbol{\theta})p(\boldsymbol{\theta}) \tag{3.38}$$

$$\propto \quad \prod_{k=1}^{K} \theta_k^{N_k} \theta_k^{\alpha_k - 1} = \prod_{k=1}^{K} \theta_k^{\alpha_k + N_k - 1} \tag{3.39}$$

$$= \quad \text{Dir}(\boldsymbol{\theta}|\alpha_1 + N_1, \ldots, \alpha_K + N_K) \tag{3.40}$$

We see that the posterior is obtained by adding the prior hyper-parameters (pseudo-counts) α_k to the empirical counts N_k.

We can derive the mode of this posterior (i.e., the MAP estimate) by using calculus. However, we must enforce the constraint that $\sum_k \theta_k = 1$.[2]. We can do this by using a **Lagrange multiplier**. The constrained objective function, or **Lagrangian**, is given by the log likelihood plus log prior plus the constraint:

$$\ell(\boldsymbol{\theta}, \lambda) = \sum_k N_k \log \theta_k + \sum_k (\alpha_k - 1) \log \theta_k + \lambda \left(1 - \sum_k \theta_k \right) \tag{3.41}$$

To simplify notation, we define $N_k' \triangleq N_k + \alpha_k - 1$. Taking derivatives with respect to λ yields the original constraint:

$$\frac{\partial \ell}{\partial \lambda} = \left(1 - \sum_k \theta_k \right) = 0 \tag{3.42}$$

Taking derivatives with respect to θ_k yields

$$\frac{\partial \ell}{\partial \theta_k} = \frac{N_k'}{\theta_k} - \lambda = 0 \tag{3.43}$$

$$N_k' = \lambda \theta_k \tag{3.44}$$

We can solve for λ using the sum-to-one constraint:

$$\sum_k N_k' = \lambda \sum_k \theta_k \tag{3.45}$$

$$N + \alpha_0 - K = \lambda \tag{3.46}$$

where $\alpha_0 \triangleq \sum_{k=1}^K \alpha_k$ is the equivalent sample size of the prior. Thus the MAP estimate is given by

$$\hat{\theta}_k = \frac{N_k + \alpha_k - 1}{N + \alpha_0 - K} \tag{3.47}$$

which is consistent with Equation 2.77. If we use a uniform prior, $\alpha_k = 1$, we recover the MLE:

$$\hat{\theta}_k = N_k / N \tag{3.48}$$

This is just the empirical fraction of times face k shows up.

---

2. We do not need to explicitly enforce the constraint that $\theta_k \geq 0$ since the gradient of the objective has the form $N_k / \theta_k - \lambda$; so negative values would reduce the objective, rather than maximize it. (Of course, this does not preclude setting $\theta_k = 0$, and indeed this is the optimal solution if $N_k = 0$ and $\alpha_k = 1$.)

3.4.4 Posterior predictive

The posterior predictive distribution for a single multinoulli trial is given by the following expression:

$$p(X = j|\mathcal{D}) = \int p(X = j|\boldsymbol{\theta})p(\boldsymbol{\theta}|\mathcal{D})d\boldsymbol{\theta} \tag{3.49}$$

$$= \int p(X = j|\theta_j) \left[\int p(\boldsymbol{\theta}_{-j}, \theta_j|\mathcal{D})d\boldsymbol{\theta}_{-j}\right] d\theta_j \tag{3.50}$$

$$= \int \theta_j p(\theta_j|\mathcal{D})d\theta_j = \mathbb{E}\left[\theta_j|\mathcal{D}\right] = \frac{\alpha_j + N_j}{\sum_k (\alpha_k + N_k)} = \frac{\alpha_j + N_j}{\alpha_0 + N} \tag{3.51}$$

where $\boldsymbol{\theta}_{-j}$ are all the components of $\boldsymbol{\theta}$ except θ_j. See also Exercise 3.13.

The above expression avoids the zero-count problem, just as we saw in Section 3.3.4.1. In fact, this form of Bayesian smoothing is even more important in the multinomial case than the binary case, since the likelihood of data sparsity increases once we start partitioning the data into many categories.

3.4.4.1 Worked example: language models using bag of words

One application of Bayesian smoothing using the Dirichlet-multinomial model is to **language modeling**, which means predicting which words might occur next in a sequence. Here we will take a very simple-minded approach, and assume that the i'th word, $X_i \in \{1, \ldots, K\}$, is sampled independently from all the other words using a $\text{Cat}(\boldsymbol{\theta})$ distribution. This is called the **bag of words** model. Given a past sequence of words, how can we predict which one is likely to come next?

For example, suppose we observe the following sequence (part of a children's nursery rhyme):

Mary had a little lamb, little lamb, little lamb,
Mary had a little lamb, its fleece as white as snow

Furthermore, suppose our vocabulary consists of the following words:

mary lamb little big fleece white black snow rain unk
1   2   3   4   5   6   7   8   9   10

Here **unk** stands for unknown, and represents all other words that do not appear elsewhere on the list. To encode each line of the nursery rhyme, we first strip off punctuation, and remove any **stop words** such as "a", "as", "the", etc. We can also perform **stemming**, which means reducing words to their base form, such as stripping off the final s in plural words, or the *ing* from verbs (e.g., *raining* becomes *rain*). In this example, no words need stemming. Finally, we replace each word by its index into the vocabulary to get:

1 10 3 2 3 2 3 2
1 10 3 2 10 5 6 8

We now ignore the word order, and count how often each word occurred, resulting in a histogram of word counts:

Token	1	2	3	4	5	6	7	8	9	10
Word	mary	lamb	little	big	fleece	white	black	snow	rain	unk
Count	2	4	4	0	1	1	0	1	0	3

Denote the above counts by N_j. If we use a $\text{Dir}(\boldsymbol{\alpha})$ prior for $\boldsymbol{\theta}$, the posterior predictive is just

$$p(\tilde{X} = j|D) = E[\theta_j|D] = \frac{\alpha_j + N_j}{\sum_{j'} \alpha_{j'} + N_{j'}} = \frac{1 + N_j}{10 + 17} \tag{3.52}$$

If we set $\alpha_j = 1$, we get

$$p(\tilde{X} = j|D) = (3/27, 5/27, 5/27, 1/27, 2/27, 2/27, 1/27, 2/27, 1/27, 5/27) \tag{3.53}$$

The modes of the predictive distribution are $X = 2$ ("lamb") and $X = 10$ ("unk"). Note that the words "big", "black" and "rain" are predicted to occur with non-zero probability in the future, even though they have never been seen before. Later on we will see more sophisticated language models.

3.5 Naive Bayes classifiers

In this section, we discuss how to classify vectors of discrete-valued features, $\mathbf{x} \in \{1, \dots, K\}^D$, where K is the number of values for each feature, and D is the number of features. We will use a generative approach. This requires us to specify the class-conditional distribution, $p(\mathbf{x}|y = c)$. The simplest approach is to assume the features are **conditionally independent** given the class label. This allows us to write the class-conditional density as a product of one-dimensional densities:

$$p(\mathbf{x}|y = c, \boldsymbol{\theta}) = \prod_{j=1}^{D} p(x_j|y = c, \boldsymbol{\theta}_{jc}) \tag{3.54}$$

The resulting model is called a **naive Bayes classifier** (NBC).

The model is called "naive" since we do not expect the features to be independent, even conditional on the class label. However, even if the naive Bayes assumption is not true, it often results in classifiers that work well (Domingos and Pazzani 1997). One reason for this is that the model is quite simple (it only has $O(CD)$ parameters, for C classes and D features), and hence it is relatively immune to overfitting.

The form of the class-conditional density depends on the type of each feature. We give some possibilities below:

- In the case of real-valued features, we can use the Gaussian distribution: $p(\mathbf{x}|y = c, \boldsymbol{\theta}) = \prod_{j=1}^{D} \mathcal{N}(x_j|\mu_{jc}, \sigma_{jc}^2)$, where μ_{jc} is the mean of feature j in objects of class c, and σ_{jc}^2 is its variance.

- In the case of binary features, $x_j \in \{0, 1\}$, we can use the Bernoulli distribution: $p(\mathbf{x}|y = c, \boldsymbol{\theta}) = \prod_{j=1}^{D} \text{Ber}(x_j|\mu_{jc})$, where μ_{jc} is the probability that feature j occurs in class c. This is sometimes called the **multivariate Bernoulli naive Bayes** model. We will see an application of this below.

- In the case of categorical features, $x_j \in \{1, \ldots, K\}$, we can use the multinoulli distribution: $p(\mathbf{x}|y = c, \boldsymbol{\theta}) = \prod_{j=1}^{D} \mathrm{Cat}(x_j|\boldsymbol{\mu}_{jc})$, where $\boldsymbol{\mu}_{jc}$ is a histogram over the K possible values for x_j in class c.

Obviously we can handle other kinds of features, or use different distributional assumptions. Also, it is easy to mix and match features of different types.

3.5.1 Model fitting

We now discuss how to "train" a naive Bayes classifier. This usually means computing the MLE or the MAP estimate for the parameters. However, we will also discuss how to compute the full posterior, $p(\boldsymbol{\theta}|\mathcal{D})$.

3.5.1.1 MLE for NBC

The probability for a single data case is given by

$$p(\mathbf{x}_i, y_i|\boldsymbol{\theta}) = p(y_i|\boldsymbol{\pi}) \prod_j p(x_{ij}|\boldsymbol{\theta}_j) = \prod_c \pi_c^{\mathbb{I}(y_i=c)} \prod_j \prod_c p(x_{ij}|\boldsymbol{\theta}_{jc})^{\mathbb{I}(y_i=c)} \tag{3.55}$$

Hence the log-likelihood is given by

$$\log p(\mathcal{D}|\boldsymbol{\theta}) = \sum_{c=1}^{C} N_c \log \pi_c + \sum_{j=1}^{D} \sum_{c=1}^{C} \sum_{i:y_i=c} \log p(x_{ij}|\boldsymbol{\theta}_{jc}) \tag{3.56}$$

We see that this expression decomposes into a series of terms, one concerning $\boldsymbol{\pi}$, and DC terms containing the $\boldsymbol{\theta}_{jc}$'s. Hence we can optimize all these parameters separately.

From Equation 3.48, the MLE for the class prior is given by

$$\hat{\pi}_c = \frac{N_c}{N} \tag{3.57}$$

where $N_c \triangleq \sum_i \mathbb{I}(y_i = c)$ is the number of examples in class c.

The MLE for the likelihood depends on the type of distribution we choose to use for each feature. For simplicity, let us suppose all features are binary, so $x_j|y = c \sim \mathrm{Ber}(\theta_{jc})$. In this case, the MLE becomes

$$\hat{\theta}_{jc} = \hat{P}(x_j = 1|y = c) = \frac{N\hat{P}(x_j = 1, y = c)}{N\hat{P}(y = c)} = \frac{N_{jc}}{N_c} \tag{3.58}$$

where $N_{jc} \triangleq \sum_i \mathbb{I}(x_{ij} = 1, y_i = c)$ is the number of examples in class c where feature j turns on.

It is extremely simple to implement this model fitting procedure: See Algorithm 3.1 for some pseudo-code (and `naiveBayesFit` for some Matlab code). This algorithm obviously takes $O(ND)$ time. The method is easily generalized to handle features of mixed type. This simplicity is one reason the method is so widely used.

Figure 3.8 gives an example where we have 2 classes and 600 binary features, representing the presence or absence of words in a bag-of-words model. The plot visualizes the $\boldsymbol{\theta}_c$ vectors for the

---

Algorithm 3.1: Fitting a naive Bayes classifier to binary features

1   $N_c = 0$, $N_{jc} = 0$;
2   **for** $i = 1 : N$ **do**
3       $c = y_i$ // Class label of i'th example;
4       $N_c := N_c + 1$;
5   **for** $j = 1 : D$ **do**
6   **if** $x_{ij} = 1$ **then**
7               $N_{jc} := N_{jc} + 1$
8   $\hat{\pi}_c = \frac{N_c}{N}$, $\hat{\theta}_{jc} = \frac{N_{jc}}{N_c}$

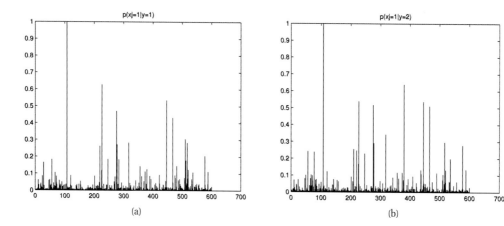

Figure 3.8   Class conditional densities $p(x_j = 1 | y = c)$ for two document classes, corresponding to "X windows" and "MS windows". Figure generated by `naiveBayesBowDemo`.

two classes. The big spike at index 107 corresponds to the word "subject", which occurs in both classes with probability 1. (In Section 3.5.4, we discuss how to "filter out" such uninformative features.)

3.5.1.2   Bayesian naive Bayes

The trouble with maximum likelihood is that it can overfit. For example, consider the example in Figure 3.8: the feature corresponding to the word "subject" (call it feature j) always occurs in both classes, so we estimate $\hat{\theta}_{jc} = 1$. What will happen if we encounter a new email which does not have this word in it? Our algorithm will crash and burn, since we will find that $p(y = c|\mathbf{x}, \hat{\boldsymbol{\theta}}) = 0$ for both classes! This is another manifestation of the black swan paradox discussed in Section 3.3.4.1.

A simple solution to overfitting is to be Bayesian. For simplicity, we will use a factored prior:

$$p(\boldsymbol{\theta}) = p(\boldsymbol{\pi}) \prod_{j=1}^{D} \prod_{c=1}^{C} p(\theta_{jc}) \tag{3.59}$$

We will use a $\text{Dir}(\boldsymbol{\alpha})$ prior for $\boldsymbol{\pi}$ and a $\text{Beta}(\beta_0, \beta_1)$ prior for each θ_{jc}. Often we just take $\boldsymbol{\alpha} = 1$ and $\beta_0 = \beta_1 = 1$, corresponding to add-one or Laplace smoothing.

Combining the factored likelihood in Equation 3.56 with the factored prior above gives the following factored posterior:

$$p(\boldsymbol{\theta}|\mathcal{D}) = p(\boldsymbol{\pi}|\mathcal{D}) \prod_{j=1}^{D} \prod_{c=}^{C} p(\theta_{jc}|\mathcal{D}) \tag{3.60}$$

$$p(\boldsymbol{\pi}|\mathcal{D}) = \text{Dir}(N_1 + \alpha_1 \dots, N_C + \alpha_C) \tag{3.61}$$

$$p(\theta_{jc}|\mathcal{D}) = \text{Beta}((N_c - N_{jc}) + \beta_0, N_{jc} + \beta_1) \tag{3.62}$$

In other words, to compute the posterior, we just update the prior counts with the empirical counts from the likelihood. It is straightforward to modify algorithm 3.1 to handle this version of model "fitting".

3.5.2 Using the model for prediction

At test time, the goal is to compute

$$p(y = c|\mathbf{x}, \mathcal{D}) \propto p(y = c|\mathcal{D}) \prod_{j=1}^{D} p(x_j|y = c, \mathcal{D}) \tag{3.63}$$

The correct Bayesian procedure is to integrate out the unknown parameters:

$$p(y = c|\mathbf{x}, \mathcal{D}) \propto \left[\int \text{Cat}(y = c|\boldsymbol{\pi}) p(\boldsymbol{\pi}|\mathcal{D}) d\boldsymbol{\pi} \right] \tag{3.64}$$

$$\prod_{j=1}^{D} \left[\int \text{Ber}(x_j|y = c, \theta_{jc}) p(\boldsymbol{\theta}_{jc}|\mathcal{D}) \right] \tag{3.65}$$

Fortunately, this is easy to do, at least if the posterior is Dirichlet. In particular, from Equation 3.51, we know the posterior predictive density can be obtained by simply plugging in the posterior mean parameters $\overline{\boldsymbol{\theta}}$. Hence

$$p(y = c|\mathbf{x}, \mathcal{D}) \propto \overline{\pi}_c \prod_{j=1}^{D} (\overline{\theta}_{jc})^{\mathbb{I}(x_j=1)} (1 - \overline{\theta}_{jc})^{\mathbb{I}(x_j=0)} \tag{3.66}$$

$$\overline{\theta}_{jc} = \frac{N_{jc} + \beta_1}{N_c + \beta_0 + \beta_1} \tag{3.67}$$

$$\overline{\pi}_c = \frac{N_c + \alpha_c}{N + \alpha_0} \tag{3.68}$$

where $\alpha_0 = \sum_c \alpha_c$.

If we have approximated the posterior by a single point, $p(\boldsymbol{\theta}|\mathcal{D}) \approx \delta_{\hat{\boldsymbol{\theta}}}(\boldsymbol{\theta})$, where $\hat{\boldsymbol{\theta}}$ may be the ML or MAP estimate, then the posterior predictive density is obtained by simply plugging in the parameters, to yield a virtually identical rule:

$$p(y = c|\mathbf{x}, \mathcal{D}) \quad \propto \quad \hat{\pi}_c \prod_{j=1}^{D} (\hat{\theta}_{jc})^{\mathbb{I}(x_j=1)} (1 - \hat{\theta}_{jc})^{\mathbb{I}(x_j=0)} \tag{3.69}$$

The only difference is we replaced the posterior mean $\overline{\boldsymbol{\theta}}$ with the posterior mode or MLE $\hat{\boldsymbol{\theta}}$. However, this small difference can be important in practice, since the posterior mean will result in less overfitting (see Section 3.4.4.1).

3.5.3   The log-sum-exp trick

We now discuss one important practical detail that arises when using generative classifiers of any kind. We can compute the posterior over class labels using Equation 2.13, using the appropriate class-conditional density (and a plug-in approximation). Unfortunately a naive implementation of Equation 2.13 can fail due to **numerical underflow**. The problem is that $p(\mathbf{x}|y = c)$ is often a very small number, especially if $\mathbf{x}$ is a high-dimensional vector. This is because we require that $\sum_{\mathbf{x}} p(\mathbf{x}|y) = 1$, so the probability of observing any particular high-dimensional vector is small. The obvious solution is to take logs when applying Bayes rule, as follows:

$$\log p(y = c|\mathbf{x}) = b_c - \log \left[\sum_{c'=1}^{C} e^{b_{c'}} \right] \tag{3.70}$$

$$b_c \triangleq \log p(\mathbf{x}|y = c) + \log p(y = c) \tag{3.71}$$

However, this requires evaluating the following expression

$$\log[\sum_{c'} e^{b_{c'}}] = \log \sum_{c'} p(y = c', \mathbf{x}) = \log p(\mathbf{x}) \tag{3.72}$$

and we can't add up in the log domain. Fortunately, we can factor out the largest term, and just represent the remaining numbers relative to that. For example,

$$\log(e^{-120} + e^{-121}) = \log \left(e^{-120}(e^0 + e^{-1}) \right) = \log(e^0 + e^{-1}) - 120 \tag{3.73}$$

In general, we have

$$\log \sum_c e^{b_c} = \log \left[(\sum_c e^{b_c - B}) e^B \right] = \left[\log(\sum_c e^{b_c - B}) \right] + B \tag{3.74}$$

where $B = \max_c b_c$. This is called the **log-sum-exp** trick, and is widely used. (See the function `logsumexp` for an implementation.)

This trick is used in Algorithm 3.2 which gives pseudo-code for using an NBC to compute $p(y_i|\mathbf{x}_i, \hat{\boldsymbol{\theta}})$. See `naiveBayesPredict` for the Matlab code. Note that we do not need the log-sum-exp trick if we only want to compute $\hat{y}_i$, since we can just maximize the unnormalized quantity $\log p(y_i = c) + \log p(\mathbf{x}_i|y = c)$.

Algorithm 3.2: Predicting with a naive bayes classifier for binary features

1 **for** $i = 1 : N$ **do**
2   **for** $c = 1 : C$ **do**
3          $L_{ic} = \log \hat{\pi}_c$;
4   **for** $j = 1 : D$ **do**
5   **if** $x_{ij} = 1$ **then** $L_{ic} := L_{ic} + \log \hat{\theta}_{jc}$;
6   **else** $L_{ic} := L_{ic} + \log(1 - \hat{\theta}_{jc})$;
7      $p_{ic} = \exp(L_{ic} - \text{logsumexp}(L_{i,:}))$;
8      $\hat{y}_i = \text{argmax}_c \, p_{ic}$;

3.5.4 Feature selection using mutual information

Since an NBC is fitting a joint distribution over potentially many features, it can suffer from overfitting. In addition, the run-time cost is $O(D)$, which may be too high for some applications.

One common approach to tackling both of these problems is to perform **feature selection**, to remove "irrelevant" features that do not help much with the classification problem. The simplest approach to feature selection is to evaluate the relevance of each feature separately, and then take the top K, where K is chosen based on some tradeoff between accuracy and complexity. This approach is known as variable **ranking**, **filtering**, or **screening**.

One way to measure relevance is to use mutual information (Section 2.8.3) between feature X_j and the class label Y:

$$I(X, Y) = \sum_{x_j} \sum_y p(x_j, y) \log \frac{p(x_j, y)}{p(x_j)p(y)} \tag{3.75}$$

The mutual information can be thought of as the reduction in entropy on the label distribution once we observe the value of feature j. If the features are binary, it is easy to show (Exercise 3.21) that the MI can be computed as follows

$$I_j = \sum_c \left[\theta_{jc}\pi_c \log \frac{\theta_{jc}}{\theta_j} + (1 - \theta_{jc})\pi_c \log \frac{1 - \theta_{jc}}{1 - \theta_j} \right] \tag{3.76}$$

where $\pi_c = p(y = c)$, $\theta_{jc} = p(x_j = 1|y = c)$, and $\theta_j = p(x_j = 1) = \sum_c \pi_c \theta_{jc}$. (All of these quantities can be computed as a by-product of fitting a naive Bayes classifier.)

Table 3.1 illustrates what happens if we apply this to the binary bag of words dataset used in Figure 3.8. We see that the words with highest mutual information are much more discriminative than the words which are most probable. For example, the most probable word in both classes is "subject", which always occurs because this is newsgroup data, which always has a subject line. But obviously this is not very discriminative. The words with highest MI with the class label are (in decreasing order) "windows", "microsoft", "DOS" and "motif", which makes sense, since the classes correspond to Microsoft Windows and X Windows.

class 1	prob	class 2	prob	highest MI	MI
subject	0.998	subject	0.998	windows	0.215
this	0.628	windows	0.639	microsoft	0.095
with	0.535	this	0.540	dos	0.092
but	0.471	with	0.538	motif	0.078
you	0.431	but	0.518	window	0.067

Table 3.1 We list the 5 most likely words for class 1 (X windows) and class 2 (MS windows). We also show the 5 words with highest mutual information with class label. Produced by `naiveBayesBowDemo`

3.5.5 Classifying documents using bag of words

Document classification is the problem of classifying text documents into different categories. One simple approach is to represent each document as a binary vector, which records whether each word is present or not, so $x_{ij} = 1$ iff word j occurs in document i, otherwise $x_{ij} = 0$. We can then use the following class conditional density:

$$p(\mathbf{x}_i|y_i = c, \boldsymbol{\theta}) = \prod_{j=1}^{D} \text{Ber}(x_{ij}|\theta_{jc}) = \prod_{j=1}^{D} \theta_{jc}^{\mathbb{I}(x_{ij})}(1 - \theta_{jc})^{\mathbb{I}(1-x_{ij})} \tag{3.77}$$

This is called the **Bernoulli product model**, or the **binary independence model**.

However, ignoring the number of times each word occurs in a document loses some information (McCallum and Nigam 1998). A more accurate representation counts the number of occurrences of each word. Specifically, let $\mathbf{x}_i$ be a vector of counts for document i, so $x_{ij} \in \{0, 1, \ldots, N_i\}$, where N_i is the number of terms in document i (so $\sum_{j=1}^{D} x_{ij} = N_i$). For the class conditional densities, we can use a multinomial distribution:

$$p(\mathbf{x}_i|y_i = c, \boldsymbol{\theta}) = \text{Mu}(\mathbf{x}_i|N_i, \boldsymbol{\theta}_c) = \frac{N_i!}{\prod_{j=1}^{D} x_{ij}!} \prod_{j=1}^{D} \theta_{jc}^{x_{ij}} \tag{3.78}$$

where we have implicitly assumed that the document length N_i is independent of the class. Here θ_{jc} is the probability of generating word j in documents of class c; these parameters satisfy the constraint that $\sum_{j=1}^{D} \theta_{jc} = 1$ for each class c.[3]

Although the multinomial classifier is easy to train and easy to use at test time, it does not work particularly well for document classification. One reason for this is that it does not take into account the **burstiness** of word usage. This refers to the phenomenon that most words never appear in any given document, but if they do appear once, they are likely to appear more than once, i.e., words occur in bursts.

The multinomial model cannot capture the burstiness phenomenon. To see why, note that Equation 3.78 has the form $\theta_{jc}^{N_{ij}}$, and since $\theta_{jc} \ll 1$ for rare words, it becomes increasingly unlikely to generate many of them. For more frequent words, the decay rate is not as fast. To see why intuitively, note that the most frequent words are function words which are not specific

---

3. Since Equation 3.78 models each word independently, this model is often called a naive Bayes classifier, although technically the features x_{ij} are not independent, because of the constraint $\sum_j x_{ij} = N_i$.

to the class, such as "and", "the", and "but"; the chance of the word "and" occuring is pretty much the same no matter how many time it has previously occurred (modulo document length), so the independence assumption is more reasonable for common words. However, since rare words are the ones that matter most for classification purposes, these are the ones we want to model the most carefully.

Various ad hoc heuristics have been proposed to improve the performance of the multinomial document classifier (Rennie et al. 2003). We now present an alternative class conditional density that performs as well as these ad hoc methods, yet is probabilistically sound (Madsen et al. 2005).

Suppose we simply replace the multinomial class conditional density with the **Dirichlet Compound Multinomial** or **DCM** density, defined as follows:

$$p(\mathbf{x}_i|y_i = c, \boldsymbol{\alpha}) = \int \mathrm{Mu}(\mathbf{x}_i|N_i, \boldsymbol{\theta}_c)\mathrm{Dir}(\boldsymbol{\theta}_c|\boldsymbol{\alpha}_c)d\boldsymbol{\theta}_c = \frac{N_i!}{\prod_{j=1}^{D} x_{ij}!} \frac{B(\mathbf{x}_i + \boldsymbol{\alpha}_c)}{B(\boldsymbol{\alpha}_c)} \quad (3.79)$$

(This equation is derived in Equation 5.24.) Surprisingly this simple change is all that is needed to capture the burstiness phenomenon. The intuitive reason for this is as follows: After seeing one occurence of a word, say word j, the posterior counts on θ_j gets updated, making another occurence of word j more likely. By contrast, if θ_j is fixed, then the occurences of each word are independent. The multinomial model corresponds to drawing a ball from an urn with K colors of ball, recording its color, and then replacing it. By contrast, the DCM model corresponds to drawing a ball, recording its color, and then replacing it with one additional copy; this is called the **Polya urn**.

Using the DCM as the class conditional density gives much better results than using the multinomial, and has performance comparable to state of the art methods, as described in (Madsen et al. 2005). The only disadvantage is that fitting the DCM model is more complex; see (Minka 2000e; Elkan 2006) for the details.

Exercises

Exercise 3.1 MLE for the Bernoulli/ binomial model

Derive Equation 3.22 by optimizing the log of the likelihood in Equation 3.11.

Exercise 3.2 Marginal likelihood for the Beta-Bernoulli model

In Equation 5.23, we showed that the marginal likelihood is the ratio of the normalizing constants:

$$p(D) = \frac{Z(\alpha_1 + N_1, \alpha_0 + N_0)}{Z(\alpha_1, \alpha_0)} = \frac{\Gamma(\alpha_1 + N_1)\Gamma(\alpha_0 + N_0)}{\Gamma(\alpha_1 + \alpha_0 + N)} \frac{\Gamma(\alpha_1 + \alpha_0)}{\Gamma(\alpha_1)\Gamma(\alpha_0)} \quad (3.80)$$

We will now derive an alternative derivation of this fact. By the chain rule of probability,

$$p(x_{1:N}) = p(x_1)p(x_2|x_1)p(x_3|x_{1:2})\ldots \quad (3.81)$$

In Section 3.3.4, we showed that the posterior predictive distribution is

$$p(X = k|D_{1:N}) = \frac{N_k + \alpha_k}{\sum_i N_i + \alpha_i} \triangleq \frac{N_k + \alpha_k}{N + \alpha} \quad (3.82)$$

where $k \in \{0,1\}$ and $D_{1:N}$ is the data seen so far. Now suppose $D = H, T, T, H, H$ or $D = 1, 0, 0, 1, 1$. Then

$$p(D) \quad = \quad \frac{\alpha_1}{\alpha} \cdot \frac{\alpha_0}{\alpha+1} \cdot \frac{\alpha_0+1}{\alpha+2} \cdot \frac{\alpha_1+1}{\alpha+3} \cdot \frac{\alpha_1+2}{\alpha+4} \tag{3.83}$$

$$= \quad \frac{[\alpha_1(\alpha_1+1)(\alpha_1+2)]\,[\alpha_0(\alpha_0+1)]}{\alpha(\alpha+1)\cdots(\alpha+4)} \tag{3.84}$$

$$= \quad \frac{[(\alpha_1)\cdots(\alpha_1+N_1-1)]\,[(\alpha_0)\cdots(\alpha_0+N_0-1)]}{(\alpha)\cdots(\alpha+N-1)} \tag{3.85}$$

Show how this reduces to Equation 3.80 by using the fact that, for integers, $(\alpha-1)! = \Gamma(\alpha)$.

Exercise 3.3 Posterior predictive for Beta-Binomial model

Recall from Equation 3.32 that the posterior predictive for the Beta-Binomial is given by

$$p(x|n, D) \quad = \quad Bb(x|\alpha_0', \alpha_1', n) \tag{3.86}$$

$$= \quad \frac{B(x+\alpha_1', n-x+\alpha_0')}{B(\alpha_1', \alpha_0')} \binom{n}{x} \tag{3.87}$$

Prove that this reduces to

$$p(\tilde{x} = 1|D) = \frac{\alpha_1'}{\alpha_0' + \alpha_1'} \tag{3.88}$$

when $n = 1$ (and hence $x \in \{0, 1\}$). i.e., show that

$$Bb(1|\alpha_1', \alpha_0', 1) \quad = \quad \frac{\alpha_1'}{\alpha_1' + \alpha_0'} \tag{3.89}$$

Hint: use the fact that

$$\Gamma(\alpha_0 + \alpha_1 + 1) = (\alpha_0 + \alpha_1)\Gamma(\alpha_0 + \alpha_1) \tag{3.90}$$

Exercise 3.4 Beta updating from censored likelihood

(Source: Gelman.) Suppose we toss a coin $n = 5$ times. Let X be the number of heads. We observe that there are fewer than 3 heads, but we don't know exactly how many. Let the prior probability of heads be $p(\theta) = \text{Beta}(\theta|1, 1)$. Compute the posterior $p(\theta|X < 3)$ up to normalization constants, i.e., derive an expression proportional to $p(\theta, X < 3)$. Hint: the answer is a mixture distribution.

Exercise 3.5 Uninformative prior for log-odds ratio

Let

$$\phi = \text{logit}(\theta) = \log \frac{\theta}{1-\theta} \tag{3.91}$$

Show that if $p(\phi) \propto 1$, then $p(\theta) \propto \text{Beta}(\theta|0, 0)$. Hint: use the change of variables formula.

Exercise 3.6 MLE for the Poisson distribution

The Poisson pmf is defined as $\text{Poi}(x|\lambda) = e^{-\lambda}\frac{\lambda^x}{x!}$, for $x \in \{0, 1, 2, \ldots\}$ where $\lambda > 0$ is the rate parameter. Derive the MLE.

Exercise 3.7 Bayesian analysis of the Poisson distribution

In Exercise 3.6, we defined the Poisson distribution with rate λ and derived its MLE. Here we perform a conjugate Bayesian analysis.

a. Derive the posterior $p(\lambda|D)$ assuming a conjugate prior $p(\lambda) = \text{Ga}(\lambda|a, b) \propto \lambda^{a-1}e^{-\lambda b}$. Hint: the posterior is also a Gamma distribution.

b. What does the posterior mean tend to as $a \to 0$ and $b \to 0$? (Recall that the mean of a $\text{Ga}(a, b)$ distribution is a/b.)

Exercise 3.8 MLE for the uniform distribution

(Source: Kaelbling.) Consider a uniform distribution centered on 0 with width $2a$. The density function is given by

$$p(x) = \frac{1}{2a} I(x \in [-a, a]) \tag{3.92}$$

a. Given a data set $x_1, \ldots, x_n$, what is the maximum likelihood estimate of a (call it $\hat{a}$)?

b. What probability would the model assign to a new data point x_{n+1} using $\hat{a}$?

c. Do you see any problem with the above approach? Briefly suggest (in words) a better approach.

Exercise 3.9 Bayesian analysis of the uniform distribution

Consider the uniform distribution $\text{Unif}(0, \theta)$. The maximum likelihood estimate is $\hat{\theta} = \max(D)$, as we saw in Exercise 3.8, but this is unsuitable for predicting future data since it puts zero probability mass outside the training data. In this exercise, we will perform a Bayesian analysis of the uniform distribution (following (Minka 2001a)). The conjugate prior is the Pareto distribution, $p(\theta) = \text{Pareto}(\theta|K, b)$, defined in Section 2.4.7. Given a Pareto prior, the joint distribution of θ and $D = (x_1, \ldots, x_N)$ is

$$p(D, \theta) = \frac{Kb^K}{\theta^{N+K+1}} \mathbb{I}(\theta \geq \max(D, b)) \tag{3.93}$$

Let $m = \max(D)$. The evidence (the probability that all N samples came from the same uniform distribution) is

$$
\begin{aligned}
p(D) &= \int_m^\infty \frac{Kb^K}{\theta^{N+K+1}} d\theta & \text{(3.94)} \\
&= \begin{cases} \frac{K}{(N+K)b^N} & \text{if } m \leq b \\ \frac{Kb^K}{(N+K)m^{N+K}} & \text{if } m > b \end{cases} & \text{(3.95)}
\end{aligned}
$$

Derive the posterior $p(\theta|D)$, and show that it can be expressed as a Pareto distribution.

Exercise 3.10 Taxicab (tramcar) problem

Suppose you arrive in a new city and see a taxi numbered 100. How many taxis are there in this city? Let us assume taxis are numbered sequentially as integers starting from 0 up to some unknown upper bound θ. (We number taxis from 0 for simplicity; we can also count from 1 without changing the analysis.) Hence the likelihood function is $p(x) = U(0, \theta)$, the uniform distribution. The goal is to estimate θ. We will use the Bayesian analysis from Exercise 3.9.

a. Suppose we see one taxi numbered 100, so $D = \{100\}$, $m = 100$, $N = 1$. Using an (improper) non-informative prior on θ of the form $p(\theta) = \text{Pareto}(\theta|0, 0) \propto 1/\theta$, what is the posterior $p(\theta|D)$? (For simplicity, ignore the difference between real-values and integers.)

b. Compute the posterior mean, mode and median number of taxis in the city, if such quantities exist.

c. Rather than trying to compute a point estimate of the number of taxis, we can compute the predictive density over the next taxicab number using

$$p(D'|D, \alpha) = \int p(D'|\theta)p(\theta|D, \alpha)d\theta = p(D'|\beta) \qquad (3.96)$$

where $\alpha = (b, K)$ are the hyper-parameters, $\beta = (c, N + K)$ are the updated hyper-parameters. Now consider the case $D = \{m\}$, and $D' = \{x\}$. Using Equation 3.95, write down an expression for

$$p(x|D, \alpha) \qquad (3.97)$$

As above, use a non-informative prior $b = K = 0$.

d. Use the predictive density formula to compute the probability that the next taxi you will see (say, the next day) has number 100, 50 or 150, i.e., compute $p(x = 100|D, \alpha)$, $p(x = 50|D, \alpha)$, $p(x = 150|D, \alpha)$.

e. Briefly describe (1-2 sentences) some ways we might make the model more accurate at prediction.

Exercise 3.11 Bayesian analysis of the exponential distribution

A lifetime X of a machine is modeled by an exponential distribution with unknown parameter θ. The likelihood is $p(x|\theta) = \theta e^{-\theta x}$ for $x \geq 0$, $\theta > 0$.

a. Show that the MLE is $\hat{\theta} = 1/\bar{x}$, where $\bar{x} = \frac{1}{N} \sum_{i=1}^{N} x_i$.

b. Suppose we observe $X_1 = 5, X_2 = 6, X_3 = 4$ (the lifetimes (in years) of 3 different iid machines). What is the MLE given this data?

c. Assume that an expert believes θ should have a prior distribution that is also exponential

$$p(\theta) = \text{Expon}(\theta|\lambda) \qquad (3.98)$$

Choose the prior parameter, call it $\hat{\lambda}$, such that $\mathbb{E}[\theta] = 1/3$. Hint: recall that the Gamma distribution has the form

$$\text{Ga}(\theta|a, b) \quad \propto \quad \theta^{a-1}e^{-\theta b} \qquad (3.99)$$

and its mean is a/b.

d. What is the posterior, $p(\theta|\mathcal{D}, \hat{\lambda})$?

e. Is the exponential prior conjugate to the exponential likelihood?

f. What is the posterior mean, $\mathbb{E}\left[\theta|\mathcal{D}, \hat{\lambda}\right]$?

g. Explain why the MLE and posterior mean differ. Which is more reasonable in this example?

Exercise 3.12 MAP estimation for the Bernoulli with non-conjugate priors

(Source: Jaakkola.) In the book, we discussed Bayesian inference of a Bernoulli rate parameter with the prior $p(\theta) = \text{Beta}(\theta|\alpha, \beta)$. We know that, with this prior, the MAP estimate is given by

$$\hat{\theta} = \frac{N_1 + \alpha - 1}{N + \alpha + \beta - 2} \qquad (3.100)$$

where N_1 is the number of heads, N_0 is the number of tails, and $N = N_0 + N_1$ is the total number of trials.

a. Now consider the following prior, that believes the coin is fair, or is slightly biased towards tails:

$$p(\theta) \quad = \quad \begin{cases} 0.5 & \text{if } \theta = 0.5 \\ 0.5 & \text{if } \theta = 0.4 \\ 0 & \text{otherwise} \end{cases} \qquad (3.101)$$

Derive the MAP estimate under this prior as a function of N_1 and N.

b. Suppose the true parameter is $\theta = 0.41$. Which prior leads to a better estimate when N is small? Which prior leads to a better estimate when N is large?

Exercise 3.13 Posterior predictive distribution for a batch of data with the Dirichlet-multinomial model

In Equation 3.51, we gave the the posterior predictive distribution for a single multinomial trial using a Dirichlet prior. Now consider predicting a *batch* of new data, $\tilde{\mathcal{D}} = (X_1, \ldots, X_m)$, consisting of m single multinomial trials (think of predicting the next m words in a sentence, assuming they are drawn iid). Derive an expression for

$$p(\tilde{\mathcal{D}}|\mathcal{D}, \alpha) \tag{3.102}$$

Your answer should be a function of $\boldsymbol{\alpha}$, and the old and new counts (sufficient statistics), defined as

$$N_k^{old} = \sum_{i \in \mathcal{D}} I(x_i = k) \tag{3.103}$$

$$N_k^{new} = \sum_{i \in \tilde{\mathcal{D}}} I(x_i = k) \tag{3.104}$$

Hint: recall that, for a vector of counts $N_{1:K}$, the marginal likelihood (evidence) is given by

$$p(\mathcal{D}|\boldsymbol{\alpha}) = \frac{\Gamma(\alpha)}{\Gamma(N + \alpha)} \prod_k \frac{\Gamma(N_k + \alpha_k)}{\Gamma(\alpha_k)} \tag{3.105}$$

where $\alpha = \sum_k \alpha_k$ and $N = \sum_k N_k$.

Exercise 3.14 Posterior predictive for Dirichlet-multinomial

(Source: Koller.)

a. Suppose we compute the empirical distribution over letters of the Roman alphabet plus the space character (a distribution over 27 values) from 2000 samples. Suppose we see the letter "e" 260 times. What is $p(x_{2001} = e|\mathcal{D})$, if we assume $\boldsymbol{\theta} \sim \text{Dir}(\alpha_1, \ldots, \alpha_{27})$, where $\alpha_k = 10$ for all k?

b. Suppose, in the 2000 samples, we saw "e" 260 times, "a" 100 times, and "p" 87 times. What is $p(x_{2001} = p, x_{2002} = a|\mathcal{D})$, if we assume $\boldsymbol{\theta} \sim \text{Dir}(\alpha_1, \ldots, \alpha_{27})$, where $\alpha_k = 10$ for all k? Show your work.

Exercise 3.15 Setting the beta hyper-parameters

Suppose $\theta \sim \beta(\alpha_1, \alpha_2)$ and we believe that $\mathbb{E}[\theta] = m$ and $\text{var}[\theta] = v$. Using Equation 2.62, solve for α_1 and α_2 in terms of m and v. What values do you get if $m = 0.7$ and $v = 0.2^2$?

Exercise 3.16 Setting the beta hyper-parameters II

(Source: Draper.) Suppose $\theta \sim \beta(\alpha_1, \alpha_2)$ and we believe that $\mathbb{E}[\theta] = m$ and $p(\ell < \theta < u) = 0.95$. Write a program that can solve for α_1 and α_2 in terms of m, ℓ and u. Hint: write α_2 as a function of α_1 and m, so the pdf only has one unknown; then write down the probability mass contained in the interval as an integral, and minimize its squared discrepancy from 0.95. What values do you get if $m = 0.15$, $\ell = 0.05$ and $u = 0.3$? What is the equivalent sample size of this prior?

Exercise 3.17 Marginal likelihood for beta-binomial under uniform prior

Suppose we toss a coin N times and observe N_1 heads. Let $N_1 \sim \text{Bin}(N, \theta)$ and $\theta \sim \text{Beta}(1, 1)$. Show that the marginal likelihood is $p(N_1|N) = 1/(N + 1)$. Hint: $\Gamma(x + 1) = x!$ if x is an integer.

Exercise 3.18 Bayes factor for coin tossing

Suppose we toss a coin $N = 10$ times and observe $N_1 = 9$ heads. Let the null hypothesis be that the coin is fair, and the alternative be that the coin can have any bias, so $p(\theta) = \text{Unif}(0, 1)$. Derive the Bayes factor $BF_{1,0}$ in favor of the biased coin hypothesis. What if $N = 100$ and $N_1 = 90$? Hint: see Exercise 3.17.

Exercise 3.19 Irrelevant features with naive Bayes

(Source: Jaakkola.) Let $x_{iw} = 1$ if word w occurs in document i and $x_{iw} = 0$ otherwise. Let θ_{cw} be the estimated probability that word w occurs in documents of class c. Then the log-likelihood that document $\mathbf{x}$ belongs to class c is

$$\log p(\mathbf{x}_i | c, \theta) = \log \prod_{w=1}^{W} \theta_{cw}^{x_{iw}} (1 - \theta_{cw})^{1-x_{iw}} \tag{3.106}$$

$$= \sum_{w=1}^{W} x_{iw} \log \theta_{cw} + (1 - x_{iw}) \log(1 - \theta_{cw}) \tag{3.107}$$

$$= \sum_{w=1}^{W} x_{iw} \log \frac{\theta_{cw}}{1 - \theta_{cw}} + \sum_{w} \log(1 - \theta_{cw}) \tag{3.108}$$

where W is the number of words in the vocabulary. We can write this more succinctly as

$$\log p(\mathbf{x}_i | c, \theta) = \phi(\mathbf{x}_i)^T \boldsymbol{\beta}_c \tag{3.109}$$

where $\mathbf{x}_i = (x_{i1}, \dots, x_{iW})$ is a bit vector, $\phi(\mathbf{x}_i) = (\mathbf{x}_i, 1)$, and

$$\boldsymbol{\beta}_c = (\log \frac{\theta_{c1}}{1 - \theta_{c1}}, \dots, \log \frac{\theta_{cW}}{1 - \theta_{cW}}, \sum_{w} \log(1 - \theta_{cw}))^T \tag{3.110}$$

We see that this is a linear classifier, since the class-conditional density is a linear function (an inner product) of the parameters $\boldsymbol{\beta}_c$.

a. Assuming $p(C = 1) = p(C = 2) = 0.5$, write down an expression for the log posterior odds ratio, $\log_2 \frac{p(c=1|\mathbf{x}_i)}{p(c=2|\mathbf{x}_i)}$, in terms of the features $\phi(\mathbf{x}_i)$ and the parameters $\boldsymbol{\beta}_1$ and $\boldsymbol{\beta}_2$.

b. Intuitively, words that occur in both classes are not very "discriminative", and therefore should not affect our beliefs about the class label. Consider a particular word w. State the conditions on $\theta_{1,w}$ and $\theta_{2,w}$ (or equivalently the conditions on $\beta_{1,w}, \beta_{2,w}$) under which the presence or absence of w in a test document will have no effect on the class posterior (such a word will be ignored by the classifier). Hint: using your previous result, figure out when the posterior odds ratio is 0.5/0.5.

c. The posterior mean estimate of θ, using a Beta(1,1) prior, is given by

$$\hat{\theta}_{cw} = \frac{1 + \sum_{i \in c} x_{iw}}{2 + n_c} \tag{3.111}$$

where the sum is over the n_c documents in class c. Consider a particular word w, and suppose it always occurs in every document (regardless of class). Let there be n_1 documents of class 1 and n_2 be the number of documents in class 2, where $n_1 \neq n_2$ (since e.g., we get much more non-spam than spam; this is an example of class imbalance). If we use the above estimate for θ_{cw}, will word w be ignored by our classifier? Explain why or why not.

d. What other ways can you think of which encourage "irrelevant" words to be ignored?

Exercise 3.20 Class conditional densities for binary data

Consider a generative classifier for C classes with class conditional density $p(\mathbf{x}|y)$ and uniform class prior $p(y)$. Suppose all the D features are binary, $x_j \in \{0, 1\}$. If we assume all the features are conditionally independent (the naive Bayes assumption), we can write

$$p(\mathbf{x}|y = c) = \prod_{j=1}^{D} \mathrm{Ber}(x_j|\theta_{jc}) \tag{3.112}$$

This requires DC parameters.

a. Now consider a different model, which we will call the "full" model, in which all the features are fully dependent (i.e., we make no factorization assumptions). How might we represent $p(\mathbf{x}|y = c)$ in this case? How many parameters are needed to represent $p(\mathbf{x}|y = c)$?

b. Assume the number of features D is fixed. Let there be N training cases. If the sample size N is very small, which model (naive Bayes or full) is likely to give lower test set error, and why?

c. If the sample size N is very large, which model (naive Bayes or full) is likely to give lower test set error, and why?

d. What is the computational complexity of fitting the full and naive Bayes models as a function of N and D? Use big-Oh notation. (Fitting the model here means computing the MLE or MAP parameter estimates. You may assume you can convert a D-bit vector to an array index in $O(D)$ time.)

e. What is the computational complexity of applying the full and naive Bayes models at test time to a single test case?

f. Suppose the test case has missing data. Let $\mathbf{x}_v$ be the visible features of size v, and $\mathbf{x}_h$ be the hidden (missing) features of size h, where $v + h = D$. What is the computational complexity of computing $p(y|\mathbf{x}_v, \boldsymbol{\theta})$ for the full and naive Bayes models, as a function of v and h?

Exercise 3.21 Mutual information for naive Bayes classifiers with binary features

Derive Equation 3.76.

Exercise 3.22 Fitting a naive Bayes spam filter by hand

(Source: Daphne Koller.). Consider a Naive Bayes model (multivariate Bernoulli version) for spam classification with the vocabulary V="secret", "offer", "low", "price", "valued", "customer", "today", "dollar", "million", "sports", "is", "for", "play", "healthy", "pizza". We have the following example spam messages "million dollar offer", "secret offer today", "secret is secret" and normal messages, "low price for valued customer", "play secret sports today", "sports is healthy", "low price pizza". Give the MLEs for the following parameters: θ_{spam}, $\theta_{\mathrm{secret}|\mathrm{spam}}$, $\theta_{\mathrm{secret}|\mathrm{non-spam}}$, $\theta_{\mathrm{sports}|\mathrm{non-spam}}$, $\theta_{\mathrm{dollar}|\mathrm{spam}}$.

4 Gaussian models

4.1 Introduction

In this chapter, we discuss the **multivariate Gaussian** or **multivariate normal** (**MVN**), which is the most widely used joint probability density function for continuous variables. It will form the basis for many of the models we will encounter in later chapters.

Unfortunately, the level of mathematics in this chapter is higher than in many other chapters. In particular, we rely heavily on linear algebra and matrix calculus. This is the price one must pay in order to deal with high-dimensional data. Beginners may choose to skip sections marked with a *. In addition, since there are so many equations in this chapter, we have put a box around those that are particularly important.

4.1.1 Notation

Let us briefly say a few words about notation. We denote vectors by boldface lower case letters, such as $\mathbf{x}$. We denote matrices by boldface upper case letters, such as $\mathbf{X}$. We denote entries in a matrix by non-bold upper case letters, such as X_{ij}.

All vectors are assumed to be column vectors unless noted otherwise. We use $[x_1, \ldots, x_D]$ to denote a column vector created by stacking D scalars. Similarly, if we write $\mathbf{x} = [\mathbf{x}_1, \ldots, \mathbf{x}_D]$, where the left hand side is a tall column vector, we mean to stack the $\mathbf{x}_i$ along the rows; this is usually written as $\mathbf{x} = (\mathbf{x}_1^T, \ldots, \mathbf{x}_D^T)^T$, but that is rather ugly. If we write $\mathbf{X} = [\mathbf{x}_1, \ldots, \mathbf{x}_D]$, where the left hand side is a matrix, we mean to stack the $\mathbf{x}_i$ along the columns, creating a matrix.

4.1.2 Basics

Recall from Section 2.5.2 that the pdf for an MVN in D dimensions is defined by the following:

$$\mathcal{N}(\mathbf{x}|\boldsymbol{\mu}, \boldsymbol{\Sigma}) \triangleq \frac{1}{(2\pi)^{D/2}|\boldsymbol{\Sigma}|^{1/2}} \exp\left[-\frac{1}{2}(\mathbf{x} - \boldsymbol{\mu})^T \boldsymbol{\Sigma}^{-1}(\mathbf{x} - \boldsymbol{\mu})\right] \tag{4.1}$$

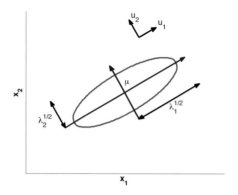

Figure 4.1 Visualization of a 2 dimensional Gaussian density. The major and minor axes of the ellipse are defined by the first two eigenvectors of the covariance matrix, namely $\mathbf{u}_1$ and $\mathbf{u}_2$. Based on Figure 2.7 of (Bishop 2006).

The expression inside the exponent is the Mahalanobis distance between a data vector $\mathbf{x}$ and the mean vector $\boldsymbol{\mu}$, We can gain a better understanding of this quantity by performing an **eigendecomposition** of $\boldsymbol{\Sigma}$. That is, we write $\boldsymbol{\Sigma} = \mathbf{U}\boldsymbol{\Lambda}\mathbf{U}^T$, where $\mathbf{U}$ is an orthonormal matrix of eigenvectors satsifying $\mathbf{U}^T\mathbf{U} = \mathbf{I}$, and $\boldsymbol{\Lambda}$ is a diagonal matrix of eigenvalues.

Using the eigendecomposition, we have that

$$\boldsymbol{\Sigma}^{-1} = \mathbf{U}^{-T}\boldsymbol{\Lambda}^{-1}\mathbf{U}^{-1} = \mathbf{U}\boldsymbol{\Lambda}^{-1}\mathbf{U}^T = \sum_{i=1}^{D} \frac{1}{\lambda_i}\mathbf{u}_i\mathbf{u}_i^T \tag{4.2}$$

where $\mathbf{u}_i$ is the i'th column of $\mathbf{U}$, containing the i'th eigenvector. Hence we can rewrite the Mahalanobis distance as follows:

$$(\mathbf{x} - \boldsymbol{\mu})^T\boldsymbol{\Sigma}^{-1}(\mathbf{x} - \boldsymbol{\mu}) \quad = \quad (\mathbf{x} - \boldsymbol{\mu})^T\left(\sum_{i=1}^{D} \frac{1}{\lambda_i}\mathbf{u}_i\mathbf{u}_i^T\right)(\mathbf{x} - \boldsymbol{\mu}) \tag{4.3}$$

$$= \quad \sum_{i=1}^{D} \frac{1}{\lambda_i}(\mathbf{x} - \boldsymbol{\mu})^T\mathbf{u}_i\mathbf{u}_i^T(\mathbf{x} - \boldsymbol{\mu}) = \sum_{i=1}^{D} \frac{y_i^2}{\lambda_i} \tag{4.4}$$

where $y_i \triangleq \mathbf{u}_i^T(\mathbf{x} - \boldsymbol{\mu})$. Recall that the equation for an ellipse in 2d is

$$\frac{y_1^2}{\lambda_1} + \frac{y_2^2}{\lambda_2} = 1 \tag{4.5}$$

Hence we see that the contours of equal probability density of a Gaussian lie along ellipses. This is illustrated in Figure 4.1. The eigenvectors determine the orientation of the ellipse, and the eigenvalues determine how elongated it is.

In general, we see that the Mahalanobis distance corresponds to Euclidean distance in a transformed coordinate system, where we shift by $\boldsymbol{\mu}$ and rotate by $\mathbf{U}$.

4.1.3 MLE for an MVN

We now describe one way to estimate the parameters of an MVN, using MLE. In later sections, we will discuss Bayesian inference for the parameters, which can mitigate overfitting, and can provide a measure of confidence in our estimates.

Theorem 4.1.1 (MLE for a Gaussian). *If we have N iid samples $\mathbf{x}_i \sim \mathcal{N}(\boldsymbol{\mu}, \boldsymbol{\Sigma})$, then the MLE for the parameters is given by*

$$\hat{\boldsymbol{\mu}}_{mle} = \frac{1}{N}\sum_{i=1}^{N} \mathbf{x}_i \triangleq \overline{\mathbf{x}} \tag{4.6}$$

$$\hat{\boldsymbol{\Sigma}}_{mle} = \frac{1}{N}\sum_{i=1}^{N}(\mathbf{x}_i - \overline{\mathbf{x}})(\mathbf{x}_i - \overline{\mathbf{x}})^T = \frac{1}{N}\left(\sum_{i=1}^{N} \mathbf{x}_i\mathbf{x}_i^T\right) - \overline{\mathbf{x}}\,\overline{\mathbf{x}}^T \tag{4.7}$$

That is, the MLE is just the empirical mean and empirical covariance. In the univariate case, we get the following familiar results:

$$\hat{\mu} = \frac{1}{N}\sum_i x_i = \overline{x} \tag{4.8}$$

$$\hat{\sigma}^2 = \frac{1}{N}\sum_i (x_i - \overline{x})^2 = \left(\frac{1}{N}\sum_i x_i^2\right) - (\overline{x})^2 \tag{4.9}$$

4.1.3.1 Proof *

To prove this result, we will need several results from matrix algebra, which we summarize below. In the equations, $\mathbf{a}$ and $\mathbf{b}$ are vectors, and $\mathbf{A}$ and $\mathbf{B}$ are matrices. Also, the notation $\text{tr}(\mathbf{A})$ refers to the **trace** of a matrix, which is the sum of its diagonals: $\text{tr}(\mathbf{A}) = \sum_i A_{ii}$.

$$\begin{aligned}
\frac{\partial(\mathbf{b}^T\mathbf{a})}{\partial\mathbf{a}} &= \mathbf{b} \\
\frac{\partial(\mathbf{a}^T\mathbf{A}\mathbf{a})}{\partial\mathbf{a}} &= (\mathbf{A} + \mathbf{A}^T)\mathbf{a} \\
\frac{\partial}{\partial\mathbf{A}}\text{tr}(\mathbf{B}\mathbf{A}) &= \mathbf{B}^T \\
\frac{\partial}{\partial\mathbf{A}}\log|\mathbf{A}| &= \mathbf{A}^{-T} \triangleq (\mathbf{A}^{-1})^T \\
\text{tr}(\mathbf{A}\mathbf{B}\mathbf{C}) = \text{tr}&(\mathbf{C}\mathbf{A}\mathbf{B}) = \text{tr}(\mathbf{B}\mathbf{C}\mathbf{A})
\end{aligned} \tag{4.10}$$

The last equation is called the **cyclic permutation property** of the trace operator. Using this, we can derive the widely used **trace trick**, which reorders the scalar inner product $\mathbf{x}^T\mathbf{A}\mathbf{x}$ as follows

$$\mathbf{x}^T\mathbf{A}\mathbf{x} = \text{tr}(\mathbf{x}^T\mathbf{A}\mathbf{x}) = \text{tr}(\mathbf{x}\mathbf{x}^T\mathbf{A}) = \text{tr}(\mathbf{A}\mathbf{x}\mathbf{x}^T) \tag{4.11}$$

Proof. We can now begin with the proof. The log-likelihood (dropping additive constants) is given by

$$\ell(\boldsymbol{\mu}, \boldsymbol{\Sigma}) = \log p(\mathcal{D}|\boldsymbol{\mu}, \boldsymbol{\Sigma}) \quad = \quad \frac{N}{2} \log |\boldsymbol{\Lambda}| - \frac{1}{2} \sum_{i=1}^{N} (\mathbf{x}_i - \boldsymbol{\mu})^T \boldsymbol{\Lambda} (\mathbf{x}_i - \boldsymbol{\mu}) \tag{4.12}$$

where $\boldsymbol{\Lambda} = \boldsymbol{\Sigma}^{-1}$ is the precision matrix.

Using the substitution $\mathbf{y}_i = \mathbf{x}_i - \boldsymbol{\mu}$ and the chain rule of calculus, we have

$$\frac{\partial}{\partial \boldsymbol{\mu}} (\mathbf{x}_i - \boldsymbol{\mu})^T \boldsymbol{\Sigma}^{-1} (\mathbf{x}_i - \boldsymbol{\mu}) \quad = \quad \frac{\partial}{\partial \mathbf{y}_i} \mathbf{y}_i^T \boldsymbol{\Sigma}^{-1} \mathbf{y}_i \frac{\partial \mathbf{y}_i}{\partial \boldsymbol{\mu}} \tag{4.13}$$

$$= \quad -1(\boldsymbol{\Sigma}^{-1} + \boldsymbol{\Sigma}^{-T}) \mathbf{y}_i \tag{4.14}$$

Hence

$$\frac{\partial}{\partial \boldsymbol{\mu}} \ell(\boldsymbol{\mu}, \boldsymbol{\Sigma}) \quad = \quad -\frac{1}{2} \sum_{i=1}^{N} -2\boldsymbol{\Sigma}^{-1} (\mathbf{x}_i - \boldsymbol{\mu}) = \boldsymbol{\Sigma}^{-1} \sum_{i=1}^{N} (\mathbf{x}_i - \boldsymbol{\mu}) = 0 \tag{4.15}$$

$$\hat{\boldsymbol{\mu}} \quad = \quad \frac{1}{N} \sum_{i=1}^{N} \mathbf{x}_i = \overline{\mathbf{x}} \tag{4.16}$$

So the MLE of $\boldsymbol{\mu}$ is just the empirical mean.

Now we can use the trace-trick to rewrite the log-likelihood for $\boldsymbol{\Lambda}$ as follows:

$$\ell(\boldsymbol{\Lambda}) \quad = \quad \frac{N}{2} \log |\boldsymbol{\Lambda}| - \frac{1}{2} \sum_{i} \text{tr}[(\mathbf{x}_i - \boldsymbol{\mu})(\mathbf{x}_i - \boldsymbol{\mu})^T \boldsymbol{\Lambda}] \tag{4.17}$$

$$= \quad \frac{N}{2} \log |\boldsymbol{\Lambda}| - \frac{1}{2} \text{tr}[\mathbf{S}_{\mu} \boldsymbol{\Lambda}] \tag{4.18}$$

$$\tag{4.19}$$

where

$$\mathbf{S}_{\mu} \triangleq \sum_{i=1}^{N} (\mathbf{x}_i - \boldsymbol{\mu})(\mathbf{x}_i - \boldsymbol{\mu})^T \tag{4.20}$$

is the scatter matrix centered on $\boldsymbol{\mu}$. Taking derivatives of this expression with respect to $\boldsymbol{\Lambda}$ yields

$$\frac{\partial \ell(\boldsymbol{\Lambda})}{\partial \boldsymbol{\Lambda}} \quad = \quad \frac{N}{2} \boldsymbol{\Lambda}^{-T} - \frac{1}{2} \mathbf{S}_{\mu}^T = 0 \tag{4.21}$$

$$\boldsymbol{\Lambda}^{-T} \quad = \quad \boldsymbol{\Lambda}^{-1} = \boldsymbol{\Sigma} = \frac{1}{N} \mathbf{S}_{\mu} \tag{4.22}$$

so

$$\hat{\boldsymbol{\Sigma}} = \frac{1}{N} \sum_{i=1}^{N} (\mathbf{x}_i - \boldsymbol{\mu})(\mathbf{x}_i - \boldsymbol{\mu})^T \tag{4.23}$$

which is just the empirical covariance matrix centered on $\boldsymbol{\mu}$. If we plug-in the MLE $\boldsymbol{\mu} = \overline{\mathbf{x}}$ (since both parameters must be simultaneously optimized), we get the standard equation for the MLE of a covariance matrix. $\square$

4.1.4 Maximum entropy derivation of the Gaussian *

In this section, we show that the multivariate Gaussian is the distribution with maximum entropy subject to having a specified mean and covariance (see also Section 9.2.6). This is one reason the Gaussian is so widely used: the first two moments are usually all that we can reliably estimate from data, so we want a distribution that captures these properties, but otherwise makes as few additional assumptions as possible.

To simplify notation, we will assume the mean is zero. The pdf has the form

$$p(\mathbf{x}) = \frac{1}{Z} \exp(-\frac{1}{2}\mathbf{x}^T\boldsymbol{\Sigma}^{-1}\mathbf{x}) \tag{4.24}$$

If we define $f_{ij}(\mathbf{x}) = x_i x_j$ and $\lambda_{ij} = \frac{1}{2}(\boldsymbol{\Sigma}^{-1})_{ij}$, for $i, j \in \{1, \dots, D\}$, we see that this is in the same form as Equation 9.74. The (differential) entropy of this distribution (using log base e) is given by

$$h(\mathcal{N}(\boldsymbol{\mu}, \boldsymbol{\Sigma})) = \frac{1}{2} \ln \left[(2\pi e)^D |\boldsymbol{\Sigma}| \right] \tag{4.25}$$

We now show the MVN has maximum entropy amongst all distributions with a specified covariance $\boldsymbol{\Sigma}$.

Theorem 4.1.2. *Let $q(\mathbf{x})$ be any density satisfying $\int q(\mathbf{x})x_i x_j d\mathbf{x} = \Sigma_{ij}$. Let $p = \mathcal{N}(\mathbf{0}, \boldsymbol{\Sigma})$. Then $h(q) \le h(p)$.*

Proof. (From (Cover and Thomas 1991, p234).) We have

$$0 \quad \le \quad \mathbb{KL}\left(q||p\right) = \int q(\mathbf{x}) \log \frac{q(\mathbf{x})}{p(\mathbf{x})} d\mathbf{x} \tag{4.26}$$

$$= \quad -h(q) - \int q(\mathbf{x}) \log p(\mathbf{x}) d\mathbf{x} \tag{4.27}$$

$$=^* \quad -h(q) - \int p(\mathbf{x}) \log p(\mathbf{x}) d\mathbf{x} \tag{4.28}$$

$$= \quad -h(q) + h(p) \tag{4.29}$$

where the key step in Equation 4.28 (marked with a *) follows since q and p yield the same moments for the quadratic form encoded by $\log p(\mathbf{x})$. $\square$

4.2 Gaussian discriminant analysis

One important application of MVNs is to define the class conditional densities in a generative classifier, i.e.,

$$p(\mathbf{x}|y = c, \boldsymbol{\theta}) = \mathcal{N}(\mathbf{x}|\boldsymbol{\mu}_c, \boldsymbol{\Sigma}_c) \tag{4.30}$$

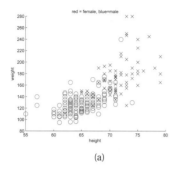
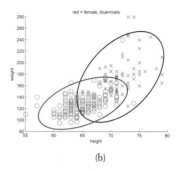

(a)   (b)

Figure 4.2   (a) Height/weight data. (b) Visualization of 2d Gaussians fit to each class. 95% of the probability mass is inside the ellipse. Figure generated by `gaussHeightWeight`.

The resulting technique is called (Gaussian) **discriminant analysis** or GDA (even though it is a generative, not discriminative, classifier — see Section 8.6 for more on this distinction). If $\boldsymbol{\Sigma}_c$ is diagonal, this is equivalent to naive Bayes.

We can classify a feature vector using the following decision rule, derived from Equation 2.13:

$$\hat{y}(\mathbf{x}) = \operatorname*{argmax}_{c} \left[\log p(y = c | \boldsymbol{\pi}) + \log p(\mathbf{x} | \boldsymbol{\theta}_c) \right] \tag{4.31}$$

When we compute the probability of $\mathbf{x}$ under each class conditional density, we are measuring the distance from $\mathbf{x}$ to the center of each class, $\boldsymbol{\mu}_c$, using Mahalanobis distance. This can be thought of as a **nearest centroids classifier**.

As an example, Figure 4.2 shows two Gaussian class-conditional densities in 2d, representing the height and weight of men and women. We can see that the features are correlated, as is to be expected (tall people tend to weigh more). The ellipses for each class contain 95% of the probability mass. If we have a uniform prior over classes, we can classify a new test vector as follows:

$$\hat{y}(\mathbf{x}) = \operatorname*{argmin}_{c} (\mathbf{x} - \boldsymbol{\mu}_c)^T \boldsymbol{\Sigma}_c^{-1} (\mathbf{x} - \boldsymbol{\mu}_c) \tag{4.32}$$

4.2.1   Quadratic discriminant analysis (QDA)

The posterior over class labels is given by Equation 2.13. We can gain further insight into this model by plugging in the definition of the Gaussian density, as follows:

$$p(y = c | \mathbf{x}, \boldsymbol{\theta}) = \frac{\pi_c |2\pi \boldsymbol{\Sigma}_c|^{-\frac{1}{2}} \exp\left[-\frac{1}{2} (\mathbf{x} - \boldsymbol{\mu}_c)^T \boldsymbol{\Sigma}_c^{-1} (\mathbf{x} - \boldsymbol{\mu}_c) \right]}{\sum_{c'} \pi_{c'} |2\pi \boldsymbol{\Sigma}_{c'}|^{-\frac{1}{2}} \exp\left[-\frac{1}{2} (\mathbf{x} - \boldsymbol{\mu}_{c'})^T \boldsymbol{\Sigma}_{c'}^{-1} (\mathbf{x} - \boldsymbol{\mu}_{c'}) \right]} \tag{4.33}$$

Thresholding this results in a quadratic function of $\mathbf{x}$. The result is known as **quadratic discriminant analysis** (QDA). Figure 4.3 gives some examples of what the decision boundaries look like in 2D.[1]

---

1. See `http://home.comcast.net/~tom.fawcett/public_html/ML-gallery/pages/` for an interesting visualization of decision boundaires in 2d for a variety of classificatino methods.

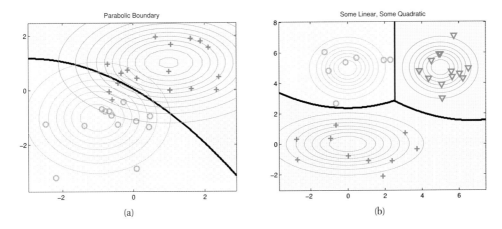

Figure 4.3 Quadratic decision boundaries in 2D for the 2 and 3 class case. Figure generated by discrimAnalysisDboundariesDemo.

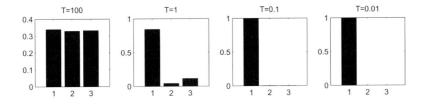

Figure 4.4 Softmax distribution $\mathcal{S}(\boldsymbol{\eta}/T)$, where $\boldsymbol{\eta} = (3, 0, 1)$, at different temperatures T. When the temperature is high (left), the distribution is uniform, whereas when the temperature is low (right), the distribution is "spiky", with all its mass on the largest element. Figure generated by softmaxDemo2.

4.2.2 Linear discriminant analysis (LDA)

We now consider a special case in which the covariance matrices are **tied** or **shared** across classes, $\boldsymbol{\Sigma}_c = \boldsymbol{\Sigma}$. In this case, we can simplify Equation 4.33 as follows:

$$p(y = c | \mathbf{x}, \boldsymbol{\theta}) \quad \propto \quad \pi_c \exp\left[\boldsymbol{\mu}_c^T \boldsymbol{\Sigma}^{-1} \mathbf{x} - \frac{1}{2}\mathbf{x}^T \boldsymbol{\Sigma}^{-1} \mathbf{x} - \frac{1}{2}\boldsymbol{\mu}_c^T \boldsymbol{\Sigma}^{-1} \boldsymbol{\mu}_c\right] \tag{4.34}$$

$$= \quad \exp\left[\boldsymbol{\mu}_c^T \boldsymbol{\Sigma}^{-1} \mathbf{x} - \frac{1}{2}\boldsymbol{\mu}_c^T \boldsymbol{\Sigma}^{-1} \boldsymbol{\mu}_c + \log \pi_c\right] \exp[-\frac{1}{2}\mathbf{x}^T \boldsymbol{\Sigma}^{-1} \mathbf{x}] \tag{4.35}$$

Since the quadratic term $\mathbf{x}^T \boldsymbol{\Sigma}^{-1} \mathbf{x}$ is independent of c, it will cancel out in the numerator and denominator. If we define

$$\gamma_c \quad = \quad -\frac{1}{2}\boldsymbol{\mu}_c^T \boldsymbol{\Sigma}^{-1} \boldsymbol{\mu}_c + \log \pi_c \tag{4.36}$$

$$\boldsymbol{\beta}_c \quad = \quad \boldsymbol{\Sigma}^{-1} \boldsymbol{\mu}_c \tag{4.37}$$

then we can write

$$p(y = c|\mathbf{x}, \boldsymbol{\theta}) = \frac{e^{\boldsymbol{\beta}_c^T \mathbf{x} + \gamma_c}}{\sum_{c'} e^{\boldsymbol{\beta}_{c'}^T \mathbf{x} + \gamma_{c'}}} = \mathcal{S}(\boldsymbol{\eta})_c \tag{4.38}$$

where $\boldsymbol{\eta} = [\boldsymbol{\beta}_1^T \mathbf{x} + \gamma_1, \dots, \boldsymbol{\beta}_C^T \mathbf{x} + \gamma_C]$, and $\mathcal{S}$ is the **softmax** function, defined as follows:

$$\mathcal{S}(\boldsymbol{\eta})_c = \frac{e^{\eta_c}}{\sum_{c'=1}^{C} e^{\eta_{c'}}} \tag{4.39}$$

The softmax function is so-called since it acts a bit like the max function. To see this, let us divide each η_c by a constant T called the **temperature**. Then as $T \to 0$, we find

$$\mathcal{S}(\boldsymbol{\eta}/T)_c = \begin{cases} 1.0 & \text{if } c = \operatorname{argmax}_{c'} \eta_{c'} \\ 0.0 & \text{otherwise} \end{cases} \tag{4.40}$$

In other words, at low temperatures, the distribution spends essentially all of its time in the most probable state, whereas at high temperatures, it visits all states uniformly. See Figure 4.4 for an illustration. Note that this terminology comes from the area of statistical physics, where it is common to use the **Boltzmann distribution**, which has the same form as the softmax function.

An interesting property of Equation 4.38 is that, if we take logs, we end up with a linear function of $\mathbf{x}$. (The reason it is linear is because the $\mathbf{x}^T \boldsymbol{\Sigma}^{-1} \mathbf{x}$ cancels from the numerator and denominator.) Thus the decision boundary between any two classes, say c and c', will be a straight line. Hence this technique is called **linear discriminant analysis** or **LDA**. [2] We can derive the form of this line as follows:

$$p(y = c|\mathbf{x}, \boldsymbol{\theta}) = p(y = c'|\mathbf{x}, \boldsymbol{\theta}) \tag{4.41}$$

$$\boldsymbol{\beta}_c^T \mathbf{x} + \gamma_c = \boldsymbol{\beta}_{c'}^T \mathbf{x} + \gamma_{c'} \tag{4.42}$$

$$\mathbf{x}^T (\boldsymbol{\beta}_{c'} - \boldsymbol{\beta}) = \gamma_{c'} - \gamma_c \tag{4.43}$$

See Figure 4.5 for some examples.

An alternative to fitting an LDA model and then deriving the class posterior is to directly fit $p(y|\mathbf{x}, \mathbf{W}) = \text{Cat}(y|\mathbf{W}\mathbf{x})$ for some $C \times D$ weight matrix $\mathbf{W}$. This is called **multi-class logistic regression**, or **multinomial logistic regression**.[3] We will discuss this model in detail in Section 8.2. The difference between the two approaches is explained in Section 8.6.

4.2.3 Two-class LDA

To gain further insight into the meaning of these equations, let us consider the binary case. In this case, the posterior is given by

$$p(y = 1|\mathbf{x}, \boldsymbol{\theta}) = \frac{e^{\boldsymbol{\beta}_1^T \mathbf{x} + \gamma_1}}{e^{\boldsymbol{\beta}_1^T \mathbf{x} + \gamma_1} + e^{\boldsymbol{\beta}_0^T \mathbf{x} + \gamma_0}} \tag{4.44}$$

$$= \frac{1}{1 + e^{(\boldsymbol{\beta}_0 - \boldsymbol{\beta}_1)^T \mathbf{x} + (\gamma_0 - \gamma_1)}} = \text{sigm}\left((\boldsymbol{\beta}_1 - \boldsymbol{\beta}_0)^T \mathbf{x} + (\gamma_1 - \gamma_0)\right) \tag{4.45}$$

---

2. The abbreviation "LDA", could either stand for "linear discriminant analysis" or "latent Dirichlet allocation" (Section 27.3). We hope the meaning is clear from text.
3. In the language modeling community, this model is called a **maximum entropy** model, for reasons explained in Section 9.2.6.

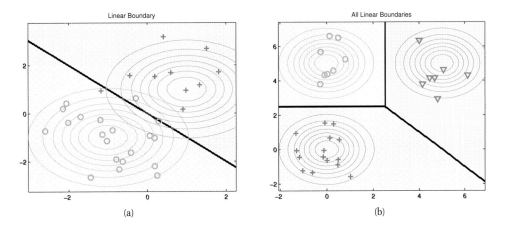

Figure 4.5 Linear decision boundaries in 2D for the 2 and 3 class case. Figure generated by discrimAnalysisDboundariesDemo.

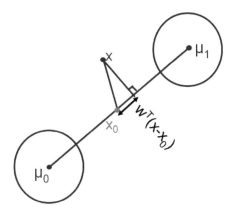

Figure 4.6 Geometry of LDA in the 2 class case where $\boldsymbol{\Sigma}_1 = \boldsymbol{\Sigma}_2 = \mathbf{I}$.

where $\mathrm{sigm}(\eta)$ refers to the sigmoid function (Equation 1.10).

Now

$$\gamma_1 - \gamma_0 = -\frac{1}{2}\boldsymbol{\mu}_1^T \boldsymbol{\Sigma}^{-1} \boldsymbol{\mu}_1 + \frac{1}{2}\boldsymbol{\mu}_0^T \boldsymbol{\Sigma}^{-1} \boldsymbol{\mu}_0 + \log(\pi_1/\pi_0) \tag{4.46}$$

$$= -\frac{1}{2}(\boldsymbol{\mu}_1 - \boldsymbol{\mu}_0)^T \boldsymbol{\Sigma}^{-1}(\boldsymbol{\mu}_1 + \boldsymbol{\mu}_0) + \log(\pi_1/\pi_0) \tag{4.47}$$

So if we define

$$\mathbf{w} = \boldsymbol{\beta}_1 - \boldsymbol{\beta}_0 = \boldsymbol{\Sigma}^{-1}(\boldsymbol{\mu}_1 - \boldsymbol{\mu}_0) \tag{4.48}$$

$$\mathbf{x}_0 = \frac{1}{2}(\boldsymbol{\mu}_1 + \boldsymbol{\mu}_0) - (\boldsymbol{\mu}_1 - \boldsymbol{\mu}_0)\frac{\log(\pi_1/\pi_0)}{(\boldsymbol{\mu}_1 - \boldsymbol{\mu}_0)^T \boldsymbol{\Sigma}^{-1}(\boldsymbol{\mu}_1 - \boldsymbol{\mu}_0)} \tag{4.49}$$

then we have $\mathbf{w}^T\mathbf{x}_0 = -(\gamma_1 - \gamma_0)$, and hence

$$p(y = 1|\mathbf{x}, \boldsymbol{\theta}) = \text{sigm}(\mathbf{w}^T(\mathbf{x} - \mathbf{x}_0)) \tag{4.50}$$

(This is closely related to logistic regression, which we will discuss in Section 8.2.) So the final decision rule is as follows: shift $\mathbf{x}$ by $\mathbf{x}_0$, project onto the line $\mathbf{w}$, and see if the result is positive or negative.

If $\boldsymbol{\Sigma} = \sigma^2\mathbf{I}$, then $\mathbf{w}$ is in the direction of $\boldsymbol{\mu}_1 - \boldsymbol{\mu}_0$. So we classify the point based on whether its projection is closer to $\boldsymbol{\mu}_0$ or $\boldsymbol{\mu}_1$. This is illustrated in Figure 4.6. Furthermore, if $\pi_1 = \pi_0$, then $\mathbf{x}_0 = \frac{1}{2}(\boldsymbol{\mu}_1 + \boldsymbol{\mu}_0)$, which is half way between the means. If we make $\pi_1 > \pi_0$, then $\mathbf{x}_0$ gets closer to $\boldsymbol{\mu}_0$, so more of the line belongs to class 1 *a priori*. Conversely if $\pi_1 < \pi_0$, the boundary shifts right. Thus we see that the class prior, π_c, just changes the decision threshold, and not the overall geometry, as we claimed above. (A similar argument applies in the multi-class case.)

The magnitude of $\mathbf{w}$ determines the steepness of the logistic function, and depends on how well-separated the means are, relative to the variance. In psychology and signal detection theory, it is common to define the **discriminability** of a signal from the background noise using a quantity called **d-prime**:

$$d' \triangleq \frac{\mu_1 - \mu_0}{\sigma} \tag{4.51}$$

where μ_1 is the mean of the signal and μ_0 is the mean of the noise, and σ is the standard deviation of the noise. If d' is large, the signal will be easier to discriminate from the noise.

4.2.4 MLE for discriminant analysis

We now discuss how to fit a discriminant analysis model. The simplest way is to use maximum likelihood. The log-likelihood function is as follows:

$$\log p(\mathcal{D}|\boldsymbol{\theta}) = \left[\sum_{i=1}^{N}\sum_{c=1}^{C}\mathbb{I}(y_i = c)\log\pi_c\right] + \sum_{c=1}^{C}\left[\sum_{i:y_i=c}\log\mathcal{N}(\mathbf{x}|\boldsymbol{\mu}_c, \boldsymbol{\Sigma}_c)\right] \tag{4.52}$$

We see that this factorizes into a term for $\boldsymbol{\pi}$, and C terms for each $\boldsymbol{\mu}_c$ and $\boldsymbol{\Sigma}_c$. Hence we can estimate these parameters separately. For the class prior, we have $\hat{\pi}_c = \frac{N_c}{N}$, as with naive Bayes. For the class-conditional densities, we just partition the data based on its class label, and compute the MLE for each Gaussian:

$$\hat{\boldsymbol{\mu}}_c = \frac{1}{N_c}\sum_{i:y_i=c}\mathbf{x}_i, \quad \hat{\boldsymbol{\Sigma}}_c = \frac{1}{N_c}\sum_{i:y_i=c}(\mathbf{x}_i - \hat{\boldsymbol{\mu}}_c)(\mathbf{x}_i - \hat{\boldsymbol{\mu}}_c)^T \tag{4.53}$$

See `discrimAnalysisFit` for a Matlab implementation. Once the model has been fit, you can make predictions using `discrimAnalysisPredict`, which uses a plug-in approximation.

4.2.5 Strategies for preventing overfitting

The speed and simplicity of the MLE method is one of its greatest appeals. However, the MLE can badly overfit in high dimensions. In particular, the MLE for a full covariance matrix is singular if $N_c < D$. And even when $N_c > D$, the MLE can be ill-conditioned, meaning it is close to singular. There are several possible solutions to this problem:

- Use a diagonal covariance matrix for each class, which assumes the features are conditionally independent; this is equivalent to using a naive Bayes classifier (Section 3.5).

- Use a full covariance matrix, but force it to be the same for all classes, $\boldsymbol{\Sigma}_c = \boldsymbol{\Sigma}$. This is an example of **parameter tying** or **parameter sharing**, and is equivalent to LDA (Section 4.2.2).

- Use a diagonal covariance matrix *and* force it to be shared. This is called diagonal covariance LDA, and is discussed in Section 4.2.7.

- Use a full covariance matrix, but impose a prior and then integrate it out. If we use a conjugate prior, this can be done in closed form, using the results from Section 4.6.3; this is analogous to the "Bayesian naive Bayes" method in Section 3.5.1.2. See (Minka 2000f) for details.

- Fit a full or diagonal covariance matrix by MAP estimation. We discuss two different kinds of prior below.

- Project the data into a low-dimensional subspace and fit the Gaussians there. See Section 8.6.3.3 for a way to find the best (most discriminative) linear projection.

We discuss some of these options below.

4.2.6 Regularized LDA *

Suppose we tie the covariance matrices, so $\boldsymbol{\Sigma}_c = \boldsymbol{\Sigma}$, as in LDA, and furthermore we perform MAP estimation of $\boldsymbol{\Sigma}$ using an inverse Wishart prior of the form $\mathrm{IW}(\mathrm{diag}(\hat{\boldsymbol{\Sigma}}_{mle}), \nu_0)$ (see Section 4.5.1). Then we have

$$\hat{\boldsymbol{\Sigma}} = \lambda \mathrm{diag}(\hat{\boldsymbol{\Sigma}}_{mle}) + (1 - \lambda)\hat{\boldsymbol{\Sigma}}_{mle} \tag{4.54}$$

where λ controls the amount of regularization, which is related to the strength of the prior, ν_0 (see Section 4.6.2.1 for details). This technique is known as **regularized discriminant analysis** or RDA (Hastie et al. 2009, p656).

When we evaluate the class conditional densities, we need to compute $\hat{\boldsymbol{\Sigma}}^{-1}$, and hence $\hat{\boldsymbol{\Sigma}}_{mle}^{-1}$, which is impossible to compute if $D > N$. However, we can use the SVD of $\mathbf{X}$ (Section 12.2.3) to get around this, as we show below. (Note that this trick cannot be applied to QDA, which is a nonlinear function of $\mathbf{x}$.)

Let $\mathbf{X} = \mathbf{U}\mathbf{D}\mathbf{V}^T$ be the SVD of the design matrix, where $\mathbf{V}$ is $D \times N$, $\mathbf{U}$ is an $N \times N$ orthogonal matrix, and $\mathbf{D}$ is a diagonal matrix of size N. Furthermore, define the $N \times N$ matrix $\mathbf{Z} = \mathbf{U}\mathbf{D}$; this is like a design matrix in a lower dimensional space (since we assume $N < D$). Also, define $\boldsymbol{\mu}_z = \mathbf{V}^T\boldsymbol{\mu}$ as the mean of the data in this reduced space; we can recover the original mean using $\boldsymbol{\mu} = \mathbf{V}\boldsymbol{\mu}_z$, since $\mathbf{V}^T\mathbf{V} = \mathbf{V}\mathbf{V}^T = \mathbf{I}$. With these definitions, we can rewrite the MLE as follows:

$$\hat{\boldsymbol{\Sigma}}_{mle} = \frac{1}{N}\mathbf{X}^T\mathbf{X} - \boldsymbol{\mu}\boldsymbol{\mu}^T \tag{4.55}$$

$$= \frac{1}{N}(\mathbf{Z}\mathbf{V}^T)^T(\mathbf{Z}\mathbf{V}^T) - (\mathbf{V}\boldsymbol{\mu}_z)(\mathbf{V}\boldsymbol{\mu}_z)^T \tag{4.56}$$

$$= \frac{1}{N}\mathbf{V}\mathbf{Z}^T\mathbf{Z}\mathbf{V}^T - \mathbf{V}\boldsymbol{\mu}_z\boldsymbol{\mu}_z^T\mathbf{V}^T \tag{4.57}$$

$$= \mathbf{V}(\frac{1}{N}\mathbf{Z}^T\mathbf{Z} - \boldsymbol{\mu}_z\boldsymbol{\mu}_z^T)\mathbf{V}^T \tag{4.58}$$

$$= \mathbf{V}\hat{\boldsymbol{\Sigma}}_z\mathbf{V}^T \tag{4.59}$$

where $\hat{\boldsymbol{\Sigma}}_z$ is the empirical covariance of $\mathbf{Z}$. Hence we can rewrite the MAP estimate as

$$\hat{\boldsymbol{\Sigma}}_{map} = \mathbf{V}\tilde{\boldsymbol{\Sigma}}_z\mathbf{V}^T \tag{4.60}$$

$$\tilde{\boldsymbol{\Sigma}}_z = \lambda\text{diag}(\hat{\boldsymbol{\Sigma}}_z) + (1-\lambda)\hat{\boldsymbol{\Sigma}}_z \tag{4.61}$$

Note, however, that we never need to actually compute the $D \times D$ matrix $\hat{\boldsymbol{\Sigma}}_{map}$. This is because Equation 4.38 tells us that to classify using LDA, all we need to compute is $p(y = c|\mathbf{x}, \boldsymbol{\theta}) \propto \exp(\delta_c)$, where

$$\delta_c = \mathbf{x}^T\boldsymbol{\beta}_c + \gamma_c, \quad \boldsymbol{\beta}_c = \hat{\boldsymbol{\Sigma}}^{-1}\boldsymbol{\mu}_c, \quad \gamma_c - \frac{1}{2}\boldsymbol{\mu}_c^T\boldsymbol{\beta}_c + \log\pi_c \tag{4.62}$$

We can compute the crucial $\boldsymbol{\beta}_c$ term for RDA without inverting the $D \times D$ matrix as follows:

$$\boldsymbol{\beta}_c = \hat{\boldsymbol{\Sigma}}_{map}^{-1}\boldsymbol{\mu}_c = (\mathbf{V}\tilde{\boldsymbol{\Sigma}}_z\mathbf{V}^T)^{-1}\boldsymbol{\mu}_c = \mathbf{V}\tilde{\boldsymbol{\Sigma}}_z^{-1}\mathbf{V}^T\boldsymbol{\mu}_c = \mathbf{V}\tilde{\boldsymbol{\Sigma}}_z^{-1}\boldsymbol{\mu}_{z,c} \tag{4.63}$$

where $\boldsymbol{\mu}_{z,c} = \mathbf{V}^T\boldsymbol{\mu}_c$ is the mean of the $\mathbf{Z}$ matrix for data belonging to class c. See `rdaFit` for the code.

4.2.7   Diagonal LDA

A simple alternative to RDA is to tie the covariance matrices, so $\boldsymbol{\Sigma}_c = \boldsymbol{\Sigma}$ as in LDA, and then to use a diagonal covariance matrix for each class. This is called the **diagonal LDA** model, and is equivalent to RDA with $\lambda = 1$. The corresponding discriminant function is as follows (compare to Equation 4.33):

$$\delta_c(\mathbf{x}) = \log p(\mathbf{x}, y = c|\boldsymbol{\theta}) = -\sum_{j=1}^{D}\frac{(x_j - \mu_{cj})^2}{2\sigma_j^2} + \log\pi_c \tag{4.64}$$

Typically we set $\hat{\mu}_{cj} = \overline{x}_{cj}$ and $\hat{\sigma}_j^2 = s_j^2$, which is the **pooled empirical variance** of feature j (pooled across classes) defined by

$$s_j^2 = \frac{\sum_{c=1}^{C}\sum_{i:y_i=c}(x_{ij} - \overline{x}_{cj})^2}{N - C} \tag{4.65}$$

In high-dimensional settings, this model can work much better than LDA and RDA (Bickel and Levina 2004).

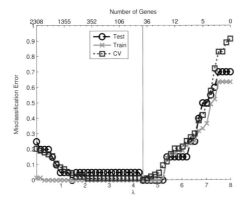

Figure 4.7 Error versus amount of shrinkage for nearest shrunken centroid classifier applied to the SRBCT gene expression data. Based on Figure 18.4 of (Hastie et al. 2009). Figure generated by `shrunkenCentroidsSRBCTdemo`.

4.2.8 Nearest shrunken centroids classifier *

One drawback of diagonal LDA is that it depends on all of the features. In high-dimensional problems, we might prefer a method that only depends on a subset of the features, for reasons of accuracy and interpretability. One approach is to use a screening method, perhaps based on mutual information, as in Section 3.5.4. We now discuss another approach to this problem known as the **nearest shrunken centroids** classifier (Hastie et al. 2009, p652).

The basic idea is to perform MAP estimation for diagonal LDA with a sparsity-promoting (Laplace) prior (see Section 13.3). More precisely, define the class-specific feature mean, μ_{cj}, in terms of the class-independent feature mean, m_j, and a class-specific offset, Δ_{cj}. Thus we have

$$\mu_{cj} = m_j + \Delta_{cj} \tag{4.66}$$

We will then put a prior on the Δ_{cj} terms to encourage them to be strictly zero and compute a MAP estimate. If, for feature j, we find that $\Delta_{cj} = 0$ for all c, then feature j will play no role in the classification decision (since μ_{cj} will be independent of c). Thus features that are not discriminative are automatically ignored. The details can be found in (Hastie et al. 2009, p652) and (Greenshtein and Park 2009). See `shrunkenCentroidsFit` for some code.

Let us give an example of the method in action, based on (Hastie et al. 2009, p652). Consider the problem of classifying a gene expression dataset, which 2308 genes, 4 classes, 63 training samples and 20 test samples. Using a diagonal LDA classifier produces 5 errors on the test set. Using the nearest shrunken centroids classifier produced 0 errors on the test set, for a range of λ values: see Figure 4.7. More importantly, the model is sparse and hence more interpretable: Figure 4.8 plots an unpenalized estimate of the difference, d_{cj}, in gray, as well as the shrunken estimates Δ_{cj} in blue. (These estimates are computed using the value of λ estimated by CV.) We see that only 39 genes are used, out of the original 2308.

Now consider an even harder problem, with 16,603 genes, a training set of 144 patients, a test set of 54 patients, and 14 different types of cancer (Ramaswamy et al. 2001). Hastie et al. (Hastie et al. 2009, p656) report that nearest shrunken centroids produced 17 errors on the test

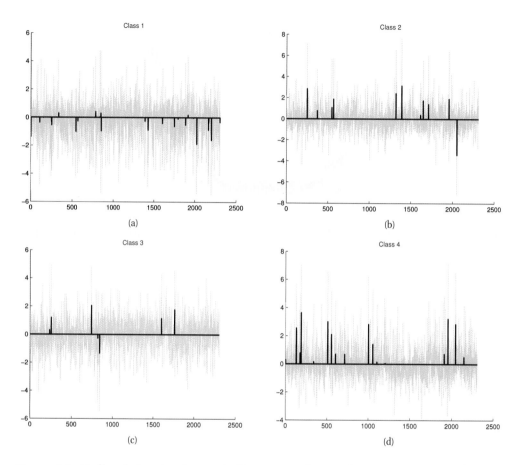

Figure 4.8 Profile of the shrunken centroids corresponding to $\lambda = 4.4$ (CV optimal in Figure 4.7). This selects 39 genes. Based on Figure 18.4 of (Hastie et al. 2009). Figure generated by `shrunkenCentroidsSRBCTdemo`.

set, using 6,520 genes, and that RDA (Section 4.2.6) produced 12 errors on the test set, using all 16,603 genes. The PMTK function `cancerHighDimClassifDemo` can be used to reproduce these numbers.

4.3 Inference in jointly Gaussian distributions

Given a joint distribution, $p(\mathbf{x}_1, \mathbf{x}_2)$, it is useful to be able to compute marginals $p(\mathbf{x}_1)$ and conditionals $p(\mathbf{x}_1|\mathbf{x}_2)$. We discuss how to do this below, and then give some applications. These operations take $O(D^3)$ time in the worst case. See Section 20.4.3 for faster methods.

4.3.1 Statement of the result

Theorem 4.3.1 (Marginals and conditionals of an MVN). *Suppose* $\mathbf{x} = (\mathbf{x}_1, \mathbf{x}_2)$ *is jointly Gaussian with parameters*

$$\boldsymbol{\mu} = \begin{pmatrix} \boldsymbol{\mu}_1 \\ \boldsymbol{\mu}_2 \end{pmatrix}, \quad \boldsymbol{\Sigma} = \begin{pmatrix} \boldsymbol{\Sigma}_{11} & \boldsymbol{\Sigma}_{12} \\ \boldsymbol{\Sigma}_{21} & \boldsymbol{\Sigma}_{22} \end{pmatrix}, \quad \boldsymbol{\Lambda} = \boldsymbol{\Sigma}^{-1} = \begin{pmatrix} \boldsymbol{\Lambda}_{11} & \boldsymbol{\Lambda}_{12} \\ \boldsymbol{\Lambda}_{21} & \boldsymbol{\Lambda}_{22} \end{pmatrix} \tag{4.67}$$

Then the marginals are given by

$$\begin{aligned} p(\mathbf{x}_1) &= \mathcal{N}(\mathbf{x}_1 | \boldsymbol{\mu}_1, \boldsymbol{\Sigma}_{11}) \\ p(\mathbf{x}_2) &= \mathcal{N}(\mathbf{x}_2 | \boldsymbol{\mu}_2, \boldsymbol{\Sigma}_{22}) \end{aligned} \tag{4.68}$$

and the posterior conditional is given by

$$\begin{aligned} p(\mathbf{x}_1 | \mathbf{x}_2) &= \mathcal{N}(\mathbf{x}_1 | \boldsymbol{\mu}_{1|2}, \boldsymbol{\Sigma}_{1|2}) \\ \boldsymbol{\mu}_{1|2} &= \boldsymbol{\mu}_1 + \boldsymbol{\Sigma}_{12} \boldsymbol{\Sigma}_{22}^{-1} (\mathbf{x}_2 - \boldsymbol{\mu}_2) \\ &= \boldsymbol{\mu}_1 - \boldsymbol{\Lambda}_{11}^{-1} \boldsymbol{\Lambda}_{12} (\mathbf{x}_2 - \boldsymbol{\mu}_2) \\ &= \boldsymbol{\Sigma}_{1|2} \left(\boldsymbol{\Lambda}_{11} \boldsymbol{\mu}_1 - \boldsymbol{\Lambda}_{12} (\mathbf{x}_2 - \boldsymbol{\mu}_2) \right) \\ \boldsymbol{\Sigma}_{1|2} &= \boldsymbol{\Sigma}_{11} - \boldsymbol{\Sigma}_{12} \boldsymbol{\Sigma}_{22}^{-1} \boldsymbol{\Sigma}_{21} = \boldsymbol{\Lambda}_{11}^{-1} \end{aligned} \tag{4.69}$$

Equation 4.69 is of such crucial importance in this book that we have put a box around it, so you can easily find it. For the proof, see Section 4.3.4.

We see that both the marginal and conditional distributions are themselves Gaussian. For the marginals, we just extract the rows and columns corresponding to $\mathbf{x}_1$ or $\mathbf{x}_2$. For the conditional, we have to do a bit more work. However, it is not that complicated: the conditional mean is just a linear function of $\mathbf{x}_2$, and the conditional covariance is just a constant matrix that is independent of $\mathbf{x}_2$. We give three different (but equivalent) expressions for the posterior mean, and two different (but equivalent) expressions for the posterior covariance; each one is useful in different circumstances.

4.3.2 Examples

Below we give some examples of these equations in action, which will make them seem more intuitive.

4.3.2.1 Marginals and conditionals of a 2d Gaussian

Let us consider a 2d example. The covariance matrix is

$$\boldsymbol{\Sigma} = \begin{pmatrix} \sigma_1^2 & \rho \sigma_1 \sigma_2 \\ \rho \sigma_1 \sigma_2 & \sigma_2^2 \end{pmatrix} \tag{4.70}$$

The marginal $p(x_1)$ is a 1D Gaussian, obtained by projecting the joint distribution onto the x_1 line:

$$p(x_1) = \mathcal{N}(x_1 | \mu_1, \sigma_1^2) \tag{4.71}$$

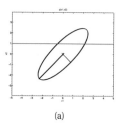 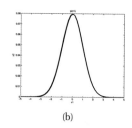 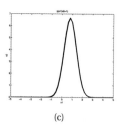

(a)   (b)   (c)

Figure 4.9 (a) A joint Gaussian distribution $p(x_1, x_2)$ with a correlation coefficient of 0.8. We plot the 95% contour and the principal axes. (b) The unconditional marginal $p(x_1)$. (c) The conditional $p(x_1|x_2) = \mathcal{N}(x_1|0.8, 0.36)$, obtained by slicing (a) at height $x_2 = 1$. Figure generated by `gaussCondition2Ddemo2`.

Suppose we observe $X_2 = x_2$; the conditional $p(x_1|x_2)$ is obtained by "slicing" the joint distribution through the $X_2 = x_2$ line (see Figure 4.9):

$$p(x_1|x_2) \;=\; \mathcal{N}\left(x_1 \Big| \mu_1 + \frac{\rho\sigma_1\sigma_2}{\sigma_2^2}(x_2 - \mu_2),\ \sigma_1^2 - \frac{(\rho\sigma_1\sigma_2)^2}{\sigma_2^2}\right) \tag{4.72}$$

If $\sigma_1 = \sigma_2 = \sigma$, we get

$$p(x_1|x_2) \;=\; \mathcal{N}\left(x_1 | \mu_1 + \rho(x_2 - \mu_2),\ \sigma^2(1 - \rho^2)\right) \tag{4.73}$$

In Figure 4.9 we show an example where $\rho = 0.8$, $\sigma_1 = \sigma_2 = 1$, $\mu_1 = \mu_2 = 0$, and $x_2 = 1$. We see that $\mathbb{E}[x_1|x_2 = 1] = 0.8$, which makes sense, since $\rho = 0.8$ means that we believe that if x_2 increases by 1 (beyond its mean), then x_1 increases by 0.8. We also see $\mathrm{var}[x_1|x_2 = 1] = 1 - 0.8^2 = 0.36$. This also makes sense: our uncertainty about x_1 has gone down, since we have learned something about x_1 (indirectly) by observing x_2. If $\rho = 0$, we get $p(x_1|x_2) = \mathcal{N}(x_1|\mu_1,\ \sigma_1^2)$, since x_2 conveys no information about x_1 if they are uncorrelated (and hence independent).

4.3.2.2   Interpolating noise-free data

Suppose we want to estimate a 1d function, defined on the interval $[0, T]$, such that $y_i = f(t_i)$ for N observed points t_i. We assume for now that the data is noise-free, so we want to **interpolate** it, that is, fit a function that goes exactly through the data. (See Section 4.4.2.3 for the noisy data case.) The question is: how does the function behave in between the observed data points? It is often reasonable to assume that the unknown function is smooth. In Chapter 15, we shall see how to encode *priors over functions*, and how to update such a prior with observed values to get a posterior over functions. But in this section, we take a simpler approach, which is adequate for MAP estimation of functions defined on 1d inputs. We follow the presentation of (Calvetti and Somersalo 2007, p135).

We start by discretizing the problem. First we divide the support of the function into D equal subintervals. We then define

$$x_j = f(s_j),\quad s_j = jh,\quad h = \frac{T}{D},\ 1 \le j \le D \tag{4.74}$$

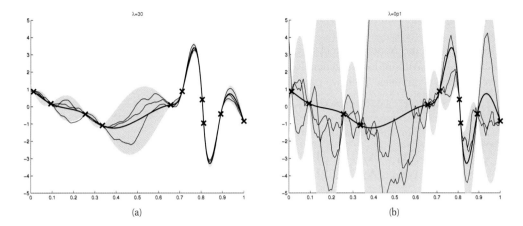

Figure 4.10 Interpolating noise-free data using a Gaussian with prior precision λ. (a) $\lambda = 30$. (b) $\lambda = 0.01$. See also Figure 4.15. Based on Figure 7.1 of (Calvetti and Somersalo 2007). Figure generated by `gaussInterpDemo`.

We can encode our smoothness prior by assuming that x_j is an average of its neighbors, x_{j-1} and x_{j+1}, plus some Gaussian noise:

$$x_j = \frac{1}{2}(x_{j-1} + x_{j+1}) + \epsilon_j, \quad 2 \le j \le D-1 \tag{4.75}$$

where $\epsilon \sim \mathcal{N}(\mathbf{0}, (1/\lambda)\mathbf{I})$. The precision term λ controls how much we think the function will vary: a large λ corresponds to a belief that the function is very smooth, a small λ corresponds to a belief that the function is quite "wiggly". In vector form, the above equation can be written as follows:

$$\mathbf{L}\mathbf{x} = \epsilon \tag{4.76}$$

where $\mathbf{L}$ is the $(D-2) \times D$ second order **finite difference matrix**

$$\mathbf{L} = \frac{1}{2} \begin{pmatrix} -1 & 2 & -1 & & \\ & -1 & 2 & -1 & \\ & & \ddots & & \\ & & -1 & 2 & -1 \end{pmatrix} \tag{4.77}$$

The corresponding prior has the form

$$p(\mathbf{x}) = \mathcal{N}(\mathbf{x}|\mathbf{0}, (\lambda \mathbf{L}^T \mathbf{L})^{-1}) \propto \exp\left(-\frac{\lambda}{2}||\mathbf{L}\mathbf{x}||_2^2\right) \tag{4.78}$$

We will henceforth assume we have scaled $\mathbf{L}$ by λ so we can ignore the λ term, and just write $\mathbf{\Lambda} = \mathbf{L}^T \mathbf{L}$ for the precision matrix.

Note that although $\mathbf{x}$ is D-dimensional, the precision matrix $\mathbf{\Lambda}$ only has rank $D-2$. Thus this is an improper prior, known as an intrinsic Gaussian random field (see Section 19.4.4 for more information). However, providing we observe $N \geq 2$ data points, the posterior will be proper.

Now let $\mathbf{x}_2$ be the N noise-free observations of the function, and $\mathbf{x}_1$ be the $D - N$ unknown function values. Without loss of generality, assume that the unknown variables are ordered first, then the known variables. Then we can partition the $\mathbf{L}$ matrix as follows:

$$\mathbf{L} = [\mathbf{L}_1, \ \mathbf{L}_2], \ \mathbf{L}_1 \in \mathbb{R}^{(D-2) \times (D-N)}, \ \mathbf{L}_2 \in \mathbb{R}^{(D-2) \times (N)} \tag{4.79}$$

We can also partition the precision matrix of the joint distribution:

$$\mathbf{\Lambda} = \mathbf{L}^T \mathbf{L} = \begin{pmatrix} \mathbf{\Lambda}_{11} & \mathbf{\Lambda}_{12} \\ \mathbf{\Lambda}_{21} & \mathbf{\Lambda}_{22} \end{pmatrix} = \begin{pmatrix} \mathbf{L}_1^T \mathbf{L}_1 & \mathbf{L}_1^T \mathbf{L}_2 \\ \mathbf{L}_2^T \mathbf{L}_1 & \mathbf{L}_2^T \mathbf{L}_2 \end{pmatrix} \tag{4.80}$$

Using Equation 4.69, we can write the conditional distribution as follows:

$$p(\mathbf{x}_1 | \mathbf{x}_2) \quad = \quad \mathcal{N}(\boldsymbol{\mu}_{1|2}, \mathbf{\Sigma}_{1|2}) \tag{4.81}$$

$$\boldsymbol{\mu}_{1|2} \quad = \quad -\mathbf{\Lambda}_{11}^{-1} \mathbf{\Lambda}_{12} \mathbf{x}_2 = -\mathbf{L}_1^{-1} \mathbf{L}_2 \mathbf{x}_2 \tag{4.82}$$

$$\mathbf{\Sigma}_{1|2} \quad = \quad \mathbf{\Lambda}_{11}^{-1} \tag{4.83}$$

Note that we can compute the mean by solving the following system of linear equations:

$$\mathbf{L}_1 \boldsymbol{\mu}_{1|2} = -\mathbf{L}_2 \mathbf{x}_2 \tag{4.84}$$

This is efficient since $\mathbf{L}_1$ is tridiagonal. Figure 4.10 gives an illustration of these equations. We see that the posterior mean $\boldsymbol{\mu}_{1|2}$ equals the observed data at the specified points, and smoothly interpolates in between, as desired.

It is also interesting to plot the 95% **pointwise marginal credibility intervals**, $\mu_j \pm 2\sqrt{\Sigma_{1|2,jj}}$, shown in grey. We see that the variance goes up as we move away from the data. We also see that the variance goes up as we decrease the precision of the prior, λ. Interestingly, λ has no effect on the posterior mean, since it cancels out when multiplying $\mathbf{\Lambda}_{11}$ and $\mathbf{\Lambda}_{12}$. By contrast, when we consider noisy data in Section 4.4.2.3, we will see that the prior precision affects the smoothness of posterior mean estimate.

The marginal credibility intervals do not capture the fact that neighboring locations are correlated. We can represent that by drawing complete functions (i.e., vectors $\mathbf{x}$) from the posterior, and plotting them. These are shown by the thin lines in Figure 4.10. These are not quite as smooth as the posterior mean itself. This is because the prior only penalizes first-order differences. See Section 4.4.2.3 for further discussion of this point.

4.3.2.3   Data imputation

Suppose we are missing some entries in a design matrix. If the columns are correlated, we can use the observed entries to predict the missing entries. Figure 4.11 shows a simple example. We sampled some data from a 20-dimensional Gaussian, and then deliberately "hid" 50% of the data in each row. We then inferred the missing entries given the observed entries, using the true (generating) model. More precisely, for each row i, we compute $p(\mathbf{x}_{\mathbf{h}_i} | \mathbf{x}_{\mathbf{v}_i}, \boldsymbol{\theta})$, where $\mathbf{h}_i$ and $\mathbf{v}_i$ are the indices of the hidden and visible entries in case i. From this, we compute the marginal distribution of each missing variable, $p(x_{h_{ij}} | \mathbf{x}_{\mathbf{v}_i}, \boldsymbol{\theta})$. We then plot the mean of this distribution, $\hat{x}_{ij} = \mathbb{E}[x_j | \mathbf{x}_{\mathbf{v}_i}, \boldsymbol{\theta}]$; this represents our "best guess" about the true value of that entry, in the

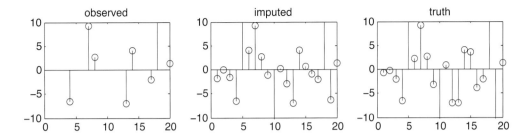

Figure 4.11 Illustration of data imputation. Left column: visualization of three rows of the data matrix with missing entries. Middle column: mean of the posterior predictive, based on partially observed data in that row, but the true model parameters. Right column: true values. Figure generated by gaussImputationDemo.

sense that it minimizes our expected squared error (see Section 5.7 for details). Figure 4.11 shows that the estimates are quite close to the truth. (Of course, if $j \in \mathbf{v}_i$, the expected value is equal to the observed value, $\hat{x}_{ij} = x_{ij}$.)

We can use $\text{var}\left[x_{h_{ij}} | \mathbf{x}_{\mathbf{v}_i}, \boldsymbol{\theta} \right]$ as a measure of confidence in this guess, although this is not shown. Alternatively, we could draw multiple samples from $p(\mathbf{x}_{\mathbf{h}_i} | \mathbf{x}_{\mathbf{v}_i}, \boldsymbol{\theta})$; this is called **multiple imputation**.

In addition to imputing the missing entries, we may be interested in computing the likelihood of each partially observed row in the table, $p(\mathbf{x}_{\mathbf{v}_i} | \boldsymbol{\theta})$, which can be computed using Equation 4.68. This is useful for detecting outliers (atypical observations).

4.3.3 Information form

Suppose $\mathbf{x} \sim \mathcal{N}(\boldsymbol{\mu}, \boldsymbol{\Sigma})$. One can show that $\mathbb{E}[\mathbf{x}] = \boldsymbol{\mu}$ is the mean vector, and $\text{cov}[\mathbf{x}] = \boldsymbol{\Sigma}$ is the covariance matrix. These are called the **moment parameters** of the distribution. However, it is sometimes useful to use the **canonical parameters** or **natural parameters**, defined as

$$\boldsymbol{\Lambda} \triangleq \boldsymbol{\Sigma}^{-1}, \ \ \boldsymbol{\xi} \triangleq \boldsymbol{\Sigma}^{-1} \boldsymbol{\mu} \tag{4.85}$$

We can convert back to the moment parameters using

$$\boldsymbol{\mu} = \boldsymbol{\Lambda}^{-1} \boldsymbol{\xi}, \ \ \boldsymbol{\Sigma} = \boldsymbol{\Lambda}^{-1} \tag{4.86}$$

(These two parameterizations are explained in more detail in Section 9.2, when we discuss the exponential family.)

Using the canonical parameters, we can write the MVN in **information form** (also called **canonical form**) as follows:

$$\mathcal{N}_c(\mathbf{x} | \boldsymbol{\xi}, \boldsymbol{\Lambda}) = (2\pi)^{-D/2} |\boldsymbol{\Lambda}|^{\frac{1}{2}} \exp\left[-\frac{1}{2}(\mathbf{x}^T \boldsymbol{\Lambda} \mathbf{x} + \boldsymbol{\xi}^T \boldsymbol{\Lambda}^{-1} \boldsymbol{\xi} - 2\mathbf{x}^T \boldsymbol{\xi}) \right] \tag{4.87}$$

where we use the notation $\mathcal{N}_c()$ to distinguish it from the moment parameterization $\mathcal{N}()$.

It is also possible to derive the marginalization and conditioning formulas in information form. We find

$$p(\mathbf{x}_2) \quad = \quad \mathcal{N}_c(\mathbf{x}_2|\boldsymbol{\xi}_2 - \boldsymbol{\Lambda}_{21}\boldsymbol{\Lambda}_{11}^{-1}\boldsymbol{\xi}_1, \boldsymbol{\Lambda}_{22} - \boldsymbol{\Lambda}_{21}\boldsymbol{\Lambda}_{11}^{-1}\boldsymbol{\Lambda}_{12}) \tag{4.88}$$

$$p(\mathbf{x}_1|\mathbf{x}_2) \quad = \quad \mathcal{N}_c(\mathbf{x}_1|\boldsymbol{\xi}_1 - \boldsymbol{\Lambda}_{12}\mathbf{x}_2, \boldsymbol{\Lambda}_{11}) \tag{4.89}$$

Thus we see that marginalization is easier in moment form, and conditioning is easier in information form.

Another operation that is significantly easier in information form is multiplying two Gaussians. One can show that

$$\mathcal{N}_c(\xi_f, \lambda_f)\mathcal{N}_c(\xi_g, \lambda_g) \quad \propto \quad \mathcal{N}_c(\xi_f + \xi_g, \lambda_f + \lambda_g) \tag{4.90}$$

However, in moment form, things are much messier:

$$\mathcal{N}(\mu_f, \sigma_f^2)\mathcal{N}(\mu_g, \sigma_g^2) \propto \mathcal{N}\left(\frac{\mu_f\sigma_g^2 + \mu_g\sigma_f^2}{\sigma_g^2 + \sigma_g^2}, \frac{\sigma_f^2\sigma_g^2}{\sigma_g^2 + \sigma_g^2}\right) \tag{4.91}$$

4.3.4   Proof of the result *

We now prove Theorem 4.3.1. Readers who are intimidated by heavy matrix algebra can safely skip this section. We first derive some results that we will need here and elsewhere in the book. We will return to the proof at the end.

4.3.4.1   Inverse of a partitioned matrix using Schur complements

The key tool we need is a way to invert a partitioned matrix. This can be done using the following result.

Theorem 4.3.2 (Inverse of a partitioned matrix). *Consider a general partitioned matrix*

$$\mathbf{M} = \begin{pmatrix} \mathbf{E} & \mathbf{F} \\ \mathbf{G} & \mathbf{H} \end{pmatrix} \tag{4.92}$$

where we assume $\mathbf{E}$ *and* $\mathbf{H}$ *are invertible. We have*

$$\mathbf{M}^{-1} \quad = \quad \begin{pmatrix} (\mathbf{M}/\mathbf{H})^{-1} & -(\mathbf{M}/\mathbf{H})^{-1}\mathbf{F}\mathbf{H}^{-1} \\ -\mathbf{H}^{-1}\mathbf{G}(\mathbf{M}/\mathbf{H})^{-1} & \mathbf{H}^{-1} + \mathbf{H}^{-1}\mathbf{G}(\mathbf{M}/\mathbf{H})^{-1}\mathbf{F}\mathbf{H}^{-1} \end{pmatrix} \tag{4.93}$$

$$= \quad \begin{pmatrix} \mathbf{E}^{-1} + \mathbf{E}^{-1}\mathbf{F}(\mathbf{M}/\mathbf{E})^{-1}\mathbf{G}\mathbf{E}^{-1} & -\mathbf{E}^{-1}\mathbf{F}(\mathbf{M}/\mathbf{E})^{-1} \\ -(\mathbf{M}/\mathbf{E})^{-1}\mathbf{G}\mathbf{E}^{-1} & (\mathbf{M}/\mathbf{E})^{-1} \end{pmatrix} \tag{4.94}$$

where

$$\mathbf{M}/\mathbf{H} \quad \triangleq \quad \mathbf{E} - \mathbf{F}\mathbf{H}^{-1}\mathbf{G} \tag{4.95}$$

$$\mathbf{M}/\mathbf{E} \quad \triangleq \quad \mathbf{H} - \mathbf{G}\mathbf{E}^{-1}\mathbf{F} \tag{4.96}$$

We say that $\mathbf{M}/\mathbf{H}$ *is the* **Schur complement** *of* $\mathbf{M}$ *wrt* $\mathbf{H}$. *Equation 4.93 is called the* **partitioned inverse formula**.

Proof. If we could block diagonalize $\mathbf{M}$, it would be easier to invert. To zero out the top right block of $\mathbf{M}$ we can pre-multiply as follows

$$\begin{pmatrix} \mathbf{I} & -\mathbf{F}\mathbf{H}^{-1} \\ \mathbf{0} & \mathbf{I} \end{pmatrix} \begin{pmatrix} \mathbf{E} & \mathbf{F} \\ \mathbf{G} & \mathbf{H} \end{pmatrix} = \begin{pmatrix} \mathbf{E} - \mathbf{F}\mathbf{H}^{-1}\mathbf{G} & \mathbf{0} \\ \mathbf{G} & \mathbf{H} \end{pmatrix} \tag{4.97}$$

Similarly, to zero out the bottom left we can post-multiply as follows

$$\begin{pmatrix} \mathbf{E} - \mathbf{F}\mathbf{H}^{-1}\mathbf{G} & \mathbf{0} \\ \mathbf{G} & \mathbf{H} \end{pmatrix} \begin{pmatrix} \mathbf{I} & \mathbf{0} \\ -\mathbf{H}^{-1}\mathbf{G} & \mathbf{I} \end{pmatrix} = \begin{pmatrix} \mathbf{E} - \mathbf{F}\mathbf{H}^{-1}\mathbf{G} & \mathbf{0} \\ \mathbf{0} & \mathbf{H} \end{pmatrix} \tag{4.98}$$

Putting it all together we get

$$\underbrace{\begin{pmatrix} \mathbf{I} & -\mathbf{F}\mathbf{H}^{-1} \\ \mathbf{0} & \mathbf{I} \end{pmatrix}}_{\mathbf{X}} \underbrace{\begin{pmatrix} \mathbf{E} & \mathbf{F} \\ \mathbf{G} & \mathbf{H} \end{pmatrix}}_{\mathbf{M}} \underbrace{\begin{pmatrix} \mathbf{I} & \mathbf{0} \\ -\mathbf{H}^{-1}\mathbf{G} & \mathbf{I} \end{pmatrix}}_{\mathbf{Z}} = \underbrace{\begin{pmatrix} \mathbf{E} - \mathbf{F}\mathbf{H}^{-1}\mathbf{G} & \mathbf{0} \\ \mathbf{0} & \mathbf{H} \end{pmatrix}}_{\mathbf{W}} \tag{4.99}$$

Taking the inverse of both sides yields

$$\mathbf{Z}^{-1}\mathbf{M}^{-1}\mathbf{X}^{-1} = \mathbf{W}^{-1} \tag{4.100}$$

and hence

$$\mathbf{M}^{-1} = \mathbf{Z}\mathbf{W}^{-1}\mathbf{X} \tag{4.101}$$

Substituting in the definitions we get

$$\begin{pmatrix} \mathbf{E} & \mathbf{F} \\ \mathbf{G} & \mathbf{H} \end{pmatrix}^{-1} = \begin{pmatrix} \mathbf{I} & \mathbf{0} \\ -\mathbf{H}^{-1}\mathbf{G} & \mathbf{I} \end{pmatrix} \begin{pmatrix} (\mathbf{M}/\mathbf{H})^{-1} & \mathbf{0} \\ \mathbf{0} & \mathbf{H}^{-1} \end{pmatrix} \begin{pmatrix} \mathbf{I} & -\mathbf{F}\mathbf{H}^{-1} \\ \mathbf{0} & \mathbf{I} \end{pmatrix} \tag{4.102}$$

$$= \begin{pmatrix} (\mathbf{M}/\mathbf{H})^{-1} & \mathbf{0} \\ -\mathbf{H}^{-1}\mathbf{G}(\mathbf{M}/\mathbf{H})^{-1} & \mathbf{H}^{-1} \end{pmatrix} \begin{pmatrix} \mathbf{I} & -\mathbf{F}\mathbf{H}^{-1} \\ \mathbf{0} & \mathbf{I} \end{pmatrix} \tag{4.103}$$

$$= \begin{pmatrix} (\mathbf{M}/\mathbf{H})^{-1} & -(\mathbf{M}/\mathbf{H})^{-1}\mathbf{F}\mathbf{H}^{-1} \\ -\mathbf{H}^{-1}\mathbf{G}(\mathbf{M}/\mathbf{H})^{-1} & \mathbf{H}^{-1} + \mathbf{H}^{-1}\mathbf{G}(\mathbf{M}/\mathbf{H})^{-1}\mathbf{F}\mathbf{H}^{-1} \end{pmatrix} \tag{4.104}$$

Alternatively, we could have decomposed the matrix $\mathbf{M}$ in terms of $\mathbf{E}$ and $\mathbf{M}/\mathbf{E} = (\mathbf{H} - \mathbf{G}\mathbf{E}^{-1}\mathbf{F})$, yielding

$$\begin{pmatrix} \mathbf{E} & \mathbf{F} \\ \mathbf{G} & \mathbf{H} \end{pmatrix}^{-1} = \begin{pmatrix} \mathbf{E}^{-1} + \mathbf{E}^{-1}\mathbf{F}(\mathbf{M}/\mathbf{E})^{-1}\mathbf{G}\mathbf{E}^{-1} & -\mathbf{E}^{-1}\mathbf{F}(\mathbf{M}/\mathbf{E})^{-1} \\ -(\mathbf{M}/\mathbf{E})^{-1}\mathbf{G}\mathbf{E}^{-1} & (\mathbf{M}/\mathbf{E})^{-1} \end{pmatrix} \tag{4.105}$$

$\square$

4.3.4.2   The matrix inversion lemma

We now derive some useful corollaries of the above result.

Corollary 4.3.1 (Matrix inversion lemma). *Consider a general partitioned matrix* $\mathbf{M} = \begin{pmatrix} \mathbf{E} & \mathbf{F} \\ \mathbf{G} & \mathbf{H} \end{pmatrix}$, *where we assume* $\mathbf{E}$ *and* $\mathbf{H}$ *are invertible. We have*

$$
\begin{aligned}
(\mathbf{E} - \mathbf{F}\mathbf{H}^{-1}\mathbf{G})^{-1} &= \mathbf{E}^{-1} + \mathbf{E}^{-1}\mathbf{F}(\mathbf{H} - \mathbf{G}\mathbf{E}^{-1}\mathbf{F})^{-1}\mathbf{G}\mathbf{E}^{-1} & (4.106) \\
(\mathbf{E} - \mathbf{F}\mathbf{H}^{-1}\mathbf{G})^{-1}\mathbf{F}\mathbf{H}^{-1} &= \mathbf{E}^{-1}\mathbf{F}(\mathbf{H} - \mathbf{G}\mathbf{E}^{-1}\mathbf{F})^{-1} & (4.107) \\
|\mathbf{E} - \mathbf{F}\mathbf{H}^{-1}\mathbf{G}| &= |\mathbf{H} - \mathbf{G}\mathbf{E}^{-1}\mathbf{F}||\mathbf{H}^{-1}||\mathbf{E}| & (4.108)
\end{aligned}
$$

The first two equations are known as the **matrix inversion lemma** or the **Sherman-Morrison-Woodbury formula**. The third equation is known as the **matrix determinant lemma**. A typical application in machine learning / statistics is the following. Let $\mathbf{E} = \boldsymbol{\Sigma}$ be a $N \times N$ diagonal matrix, let $\mathbf{F} = \mathbf{G}^T = \mathbf{X}$ of size $N \times D$, where $N \gg D$, and let $\mathbf{H}^{-1} = -\mathbf{I}$. Then we have

$$
(\boldsymbol{\Sigma} + \mathbf{X}\mathbf{X}^T)^{-1} = \boldsymbol{\Sigma}^{-1} - \boldsymbol{\Sigma}^{-1}\mathbf{X}(\mathbf{I} + \mathbf{X}^T\boldsymbol{\Sigma}^{-1}\mathbf{X})^{-1}\mathbf{X}^T\boldsymbol{\Sigma}^{-1} \tag{4.109}
$$

The LHS takes $O(N^3)$ time to compute, the RHS takes time $O(D^3)$ to compute.

Another application concerns computing a **rank one update** of an inverse matrix. Let $H = -1$ (a scalar), $\mathbf{F} = \mathbf{u}$ (a column vector), and $\mathbf{G} = \mathbf{v}^T$ (a row vector). Then we have

$$
\begin{aligned}
(\mathbf{E} + \mathbf{u}\mathbf{v}^T)^{-1} &= \mathbf{E}^{-1} + \mathbf{E}^{-1}\mathbf{u}(-1 - \mathbf{v}^T\mathbf{E}^{-1}\mathbf{u})^{-1}\mathbf{v}^T\mathbf{E}^{-1} & (4.110) \\
&= \mathbf{E}^{-1} - \frac{\mathbf{E}^{-1}\mathbf{u}\mathbf{v}^T\mathbf{E}^{-1}}{1 + \mathbf{v}^T\mathbf{E}^{-1}\mathbf{u}} & (4.111)
\end{aligned}
$$

This is useful when we incrementally add a data vector to a design matrix, and want to update our sufficient statistics. (One can derive an analogous formula for removing a data vector.)

Proof. To prove Equation 4.106, we simply equate the top left block of Equation 4.93 and Equation 4.94. To prove Equation 4.107, we simple equate the top right blocks of Equations 4.93 and 4.94. The proof of Equation 4.108 is left as an exercise. $\qquad\square$

4.3.4.3   Proof of Gaussian conditioning formulas

We can now return to our original goal, which is to derive Equation 4.69. Let us factor the joint $p(\mathbf{x}_1, \mathbf{x}_2)$ as $p(\mathbf{x}_2)p(\mathbf{x}_1|\mathbf{x}_2)$ as follows:

$$
E = \exp\left\{ -\frac{1}{2} \begin{pmatrix} \mathbf{x}_1 - \boldsymbol{\mu}_1 \\ \mathbf{x}_2 - \boldsymbol{\mu}_2 \end{pmatrix}^T \begin{pmatrix} \boldsymbol{\Sigma}_{11} & \boldsymbol{\Sigma}_{12} \\ \boldsymbol{\Sigma}_{21} & \boldsymbol{\Sigma}_{22} \end{pmatrix}^{-1} \begin{pmatrix} \mathbf{x}_1 - \boldsymbol{\mu}_1 \\ \mathbf{x}_2 - \boldsymbol{\mu}_2 \end{pmatrix} \right\} \tag{4.112}
$$

Using Equation 4.102 the above exponent becomes

$$
E = \exp\left\{-\frac{1}{2}\begin{pmatrix}\mathbf{x}_1-\boldsymbol{\mu}_1\\\mathbf{x}_2-\boldsymbol{\mu}_2\end{pmatrix}^T\begin{pmatrix}\mathbf{I} & \mathbf{0}\\-\boldsymbol{\Sigma}_{22}^{-1}\boldsymbol{\Sigma}_{21} & \mathbf{I}\end{pmatrix}\begin{pmatrix}(\boldsymbol{\Sigma}/\boldsymbol{\Sigma}_{22})^{-1} & \mathbf{0}\\\mathbf{0} & \boldsymbol{\Sigma}_{22}^{-1}\end{pmatrix}\right. \tag{4.113}
$$

$$
\left.\times\begin{pmatrix}\mathbf{I} & -\boldsymbol{\Sigma}_{12}\boldsymbol{\Sigma}_{22}^{-1}\\\mathbf{0} & \mathbf{I}\end{pmatrix}\begin{pmatrix}\mathbf{x}_1-\boldsymbol{\mu}_1\\\mathbf{x}_2-\boldsymbol{\mu}_2\end{pmatrix}\right\} \tag{4.114}
$$

$$
= \exp\left\{-\frac{1}{2}(\mathbf{x}_1-\boldsymbol{\mu}_1-\boldsymbol{\Sigma}_{12}\boldsymbol{\Sigma}_{22}^{-1}(\mathbf{x}_2-\boldsymbol{\mu}_2))^T(\boldsymbol{\Sigma}/\boldsymbol{\Sigma}_{22})^{-1}\right. \tag{4.115}
$$

$$
\left.(\mathbf{x}_1-\boldsymbol{\mu}_1-\boldsymbol{\Sigma}_{12}\boldsymbol{\Sigma}_{22}^{-1}(\mathbf{x}_2-\boldsymbol{\mu}_2))\right\}\times\exp\left\{-\frac{1}{2}(\mathbf{x}_2-\boldsymbol{\mu}_2)^T\boldsymbol{\Sigma}_{22}^{-1}(\mathbf{x}_2-\boldsymbol{\mu}_2)\right\} \tag{4.116}
$$

This is of the form

$$
\exp(\text{quadratic form in }\mathbf{x}_1,\mathbf{x}_2)\times\exp(\text{quadratic form in }\mathbf{x}_2) \tag{4.117}
$$

Hence we have successfully factorized the joint as

$$
\begin{aligned}
p(\mathbf{x}_1,\mathbf{x}_2) &= p(\mathbf{x}_1|\mathbf{x}_2)p(\mathbf{x}_2) &\tag{4.118}\\
&= \mathcal{N}(\mathbf{x}_1|\boldsymbol{\mu}_{1|2},\boldsymbol{\Sigma}_{1|2})\mathcal{N}(\mathbf{x}_2|\boldsymbol{\mu}_2,\boldsymbol{\Sigma}_{22}) &\tag{4.119}
\end{aligned}
$$

where the parameters of the conditional distribution can be read off from the above equations using

$$
\begin{aligned}
\boldsymbol{\mu}_{1|2} &= \boldsymbol{\mu}_1+\boldsymbol{\Sigma}_{12}\boldsymbol{\Sigma}_{22}^{-1}(\mathbf{x}_2-\boldsymbol{\mu}_2) &\tag{4.120}\\
\boldsymbol{\Sigma}_{1|2} &= \boldsymbol{\Sigma}/\boldsymbol{\Sigma}_{22}=\boldsymbol{\Sigma}_{11}-\boldsymbol{\Sigma}_{12}\boldsymbol{\Sigma}_{22}^{-1}\boldsymbol{\Sigma}_{21} &\tag{4.121}
\end{aligned}
$$

We can also use the fact that $|\mathbf{M}|=|\mathbf{M}/\mathbf{H}||\mathbf{H}|$ to check the normalization constants are correct:

$$
\begin{aligned}
(2\pi)^{(d_1+d_2)/2}|\boldsymbol{\Sigma}|^{\frac{1}{2}} &= (2\pi)^{(d_1+d_2)/2}(|\boldsymbol{\Sigma}/\boldsymbol{\Sigma}_{22}|\,|\boldsymbol{\Sigma}_{22}|)^{\frac{1}{2}} &\tag{4.122}\\
&= (2\pi)^{d_1/2}|\boldsymbol{\Sigma}/\boldsymbol{\Sigma}_{22}|^{\frac{1}{2}}\,(2\pi)^{d_2/2}|\boldsymbol{\Sigma}_{22}|^{\frac{1}{2}} &\tag{4.123}
\end{aligned}
$$

where $d_1=\dim(\mathbf{x}_1)$ and $d_2=\dim(\mathbf{x}_2)$.

We leave the proof of the other forms of the result in Equation 4.69 as an exercise.

4.4 Linear Gaussian systems

Suppose we have two variables, $\mathbf{x}$ and $\mathbf{y}$. Let $\mathbf{x}\in\mathbb{R}^{D_x}$ be a hidden variable, and $\mathbf{y}\in\mathbb{R}^{D_y}$ be a noisy observation of $\mathbf{x}$. Let us assume we have the following prior and likelihood:

$$
\boxed{\begin{aligned}
p(\mathbf{x}) &= \mathcal{N}(\mathbf{x}|\boldsymbol{\mu}_x,\boldsymbol{\Sigma}_x)\\
p(\mathbf{y}|\mathbf{x}) &= \mathcal{N}(\mathbf{y}|\mathbf{A}\mathbf{x}+\mathbf{b},\boldsymbol{\Sigma}_y)
\end{aligned}} \tag{4.124}
$$

where $\mathbf{A}$ is a matrix of size $D_y \times D_x$. This is an example of a **linear Gaussian system**. We can represent this schematically as $\mathbf{x} \to \mathbf{y}$, meaning $\mathbf{x}$ generates $\mathbf{y}$. In this section, we show how to "invert the arrow", that is, how to infer $\mathbf{x}$ from $\mathbf{y}$. We state the result below, then give several examples, and finally we derive the result. We will see many more applications of these results in later chapters.

4.4.1   Statement of the result

Theorem 4.4.1 (Bayes rule for linear Gaussian systems). *Given a linear Gaussian system, as in Equation 4.124, the posterior $p(\mathbf{x}|\mathbf{y})$ is given by the following:*

$$
\begin{aligned}
p(\mathbf{x}|\mathbf{y}) &= \mathcal{N}(\mathbf{x}|\boldsymbol{\mu}_{x|y}, \boldsymbol{\Sigma}_{x|y}) \\
\boldsymbol{\Sigma}_{x|y}^{-1} &= \boldsymbol{\Sigma}_x^{-1} + \mathbf{A}^T \boldsymbol{\Sigma}_y^{-1} \mathbf{A} \\
\boldsymbol{\mu}_{x|y} &= \boldsymbol{\Sigma}_{x|y}[\mathbf{A}^T \boldsymbol{\Sigma}_y^{-1} (\mathbf{y} - \mathbf{b}) + \boldsymbol{\Sigma}_x^{-1} \boldsymbol{\mu}_x]
\end{aligned}
\tag{4.125}
$$

In addition, the normalization constant $p(\mathbf{y})$ is given by

$$
p(\mathbf{y}) = \mathcal{N}(\mathbf{y}|\mathbf{A}\boldsymbol{\mu}_x + \mathbf{b}, \boldsymbol{\Sigma}_y + \mathbf{A}\boldsymbol{\Sigma}_x\mathbf{A}^T)
\tag{4.126}
$$

For the proof, see Section 4.4.3.

4.4.2   Examples

In this section, we give some example applications of the above result.

4.4.2.1   Inferring an unknown scalar from noisy measurements

Suppose we make N noisy measurements y_i of some underlying quantity x; let us assume the measurement noise has fixed precision $\lambda_y = 1/\sigma^2$, so the likelihood is

$$
p(y_i|x) \quad = \quad \mathcal{N}(y_i|x, \lambda_y^{-1})
\tag{4.127}
$$

Now let us use a Gaussian prior for the value of the unknown source:

$$
p(x) \quad = \quad \mathcal{N}(x|\mu_0, \lambda_0^{-1})
\tag{4.128}
$$

We want to compute $p(x|y_1, \ldots, y_N, \sigma^2)$. We can convert this to a form that lets us apply Bayes rule for Gaussians by defining $\mathbf{y} = (y_1, \ldots, y_N)$, $\mathbf{A} = \mathbf{1}_N^T$ (an $1 \times N$ row vector of 1's), and $\boldsymbol{\Sigma}_y^{-1} = \mathrm{diag}(\lambda_y \mathbf{I})$. Then we get

$$
p(x|\mathbf{y}) \quad = \quad \mathcal{N}(x|\mu_N, \lambda_N^{-1})
\tag{4.129}
$$

$$
\lambda_N \quad = \quad \lambda_0 + N\lambda_y
\tag{4.130}
$$

$$
\mu_N \quad = \quad \frac{N\lambda_y \overline{y} + \lambda_0 \mu_0}{\lambda_N} = \frac{N\lambda_y}{N\lambda_y + \lambda_0}\overline{y} + \frac{\lambda_0}{N\lambda_y + \lambda_0}\mu_0
\tag{4.131}
$$

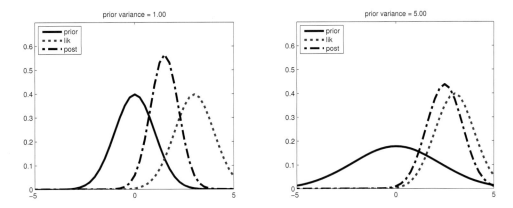

Figure 4.12 Inference about x given a noisy observation $y = 3$. (a) Strong prior $\mathcal{N}(0, 1)$. The posterior mean is "shrunk" towards the prior mean, which is 0. (a) Weak prior $\mathcal{N}(0, 5)$. The posterior mean is similar to the MLE. Figure generated by `gaussInferParamsMean1d`.

These equations are quite intuitive: the posterior precision λ_N is the prior precision λ_0 plus N units of measurement precision λ_y. Also, the posterior mean μ_N is a convex combination of the MLE $\overline{y}$ and the prior mean μ_0. This makes it clear that the posterior mean is a compromise between the MLE and the prior. If the prior is weak relative to the signal strength (λ_0 is small relative to λ_y), we put more weight on the MLE. If the prior is strong relative to the signal strength (λ_0 is large relative to λ_y), we put more weight on the prior. This is illustrated in Figure 4.12, which is very similar to the analogous results for the beta-binomial model in Figure 3.6.

Note that the posterior mean is written in terms of $N\lambda_y\overline{y}$, so having N measurements each of precision λ_y is like having one measurement with value $\overline{y}$ and precision $N\lambda_y$.

We can rewrite the results in terms of the posterior variance, rather than posterior precision, as follows:

$$p(x|\mathcal{D}, \sigma^2) = \mathcal{N}(x|\mu_N, \tau_N^2) \tag{4.132}$$

$$\tau_N^2 = \frac{1}{\frac{N}{\sigma^2} + \frac{1}{\tau_0^2}} = \frac{\sigma^2 \tau_0^2}{N\tau_0^2 + \sigma^2} \tag{4.133}$$

$$\mu_N = \tau_N^2 \left(\frac{\mu_0}{\tau_0^2} + \frac{N\overline{y}}{\sigma^2} \right) = \frac{\sigma^2}{N\tau_0^2 + \sigma^2} \mu_0 + \frac{N\tau_0^2}{N\tau_0^2 + \sigma^2} \overline{y} \tag{4.134}$$

where $\tau_0^2 = 1/\lambda_0$ is the prior variance and $\tau_N^2 = 1/\lambda_N$ is the posterior variance.

We can also compute the posterior sequentially, by updating after each observation. If $N = 1$, we can rewrite the posterior after seeing a single observation as follows (where we define $\Sigma_y = \sigma^2$, $\Sigma_0 = \tau_0^2$ and $\Sigma_1 = \tau_1^2$ to be the variances of the likelihood, prior and posterior):

$$p(x|y) \quad = \quad \mathcal{N}(x|\mu_1, \Sigma_1) \tag{4.135}$$

$$\Sigma_1 \quad = \quad \left(\frac{1}{\Sigma_0} + \frac{1}{\Sigma_y} \right)^{-1} = \frac{\Sigma_y \Sigma_0}{\Sigma_0 + \Sigma_y} \tag{4.136}$$

$$\mu_1 \quad = \quad \Sigma_1 \left(\frac{\mu_0}{\Sigma_0} + \frac{y}{\Sigma_y} \right) \tag{4.137}$$

We can rewrite the posterior mean in 3 different ways:

$$\mu_1 \quad = \quad \frac{\Sigma_y}{\Sigma_y + \Sigma_0} \mu_0 + \frac{\Sigma_0}{\Sigma_y + \Sigma_0} y \tag{4.138}$$

$$= \quad \mu_0 + (y - \mu_0) \frac{\Sigma_0}{\Sigma_y + \Sigma_0} \tag{4.139}$$

$$= \quad y - (y - \mu_0) \frac{\Sigma_y}{\Sigma_y + \Sigma_0} \tag{4.140}$$

The first equation is a convex combination of the prior and the data. The second equation is the prior mean adjusted towards the data. The third equation is the data adjusted towards the prior mean; this is called **shrinkage**. These are all equivalent ways of expressing the tradeoff between likelihood and prior. If Σ_0 is small relative to Σ_Y, corresponding to a strong prior, the amount of shrinkage is large (see Figure 4.12(a)), whereas if Σ_0 is large relative to Σ_y, corresponding to a weak prior, the amount of shrinkage is small (see Figure 4.12(b)).

Another way to quantify the amount of shrinkage is in terms of the **signal-to-noise ratio**, which is defined as follows:

$$\text{SNR} \triangleq \frac{\mathbb{E}\left[X^2\right]}{\mathbb{E}\left[\epsilon^2\right]} = \frac{\Sigma_0 + \mu_0^2}{\Sigma_y} \tag{4.141}$$

where $x \sim \mathcal{N}(\mu_0, \Sigma_0)$ is the true signal, $y = x + \epsilon$ is the observed signal, and $\epsilon \sim \mathcal{N}(0, \Sigma_y)$ is the noise term.

4.4.2.2   Inferring an unknown vector from noisy measurements

Now consider N vector-valued observations, $\mathbf{y}_i \sim \mathcal{N}(\mathbf{x}, \boldsymbol{\Sigma}_y)$, and a Gaussian prior, $\mathbf{x} \sim \mathcal{N}(\boldsymbol{\mu}_0, \boldsymbol{\Sigma}_0)$. Setting $\mathbf{A} = \mathbf{I}$, $\mathbf{b} = \mathbf{0}$, and using $\overline{\mathbf{y}}$ for the effective observation with precision $N\boldsymbol{\Sigma}_y^{-1}$, we have

$$p(\mathbf{x}|\mathbf{y}_1, \ldots, \mathbf{y}_N) \quad = \quad \mathcal{N}(\mathbf{x}|\boldsymbol{\mu}_N, \boldsymbol{\Sigma}_N) \tag{4.142}$$

$$\boldsymbol{\Sigma}_N^{-1} \quad = \quad \boldsymbol{\Sigma}_0^{-1} + N\boldsymbol{\Sigma}_y^{-1} \tag{4.143}$$

$$\boldsymbol{\mu}_N \quad = \quad \boldsymbol{\Sigma}_N(\boldsymbol{\Sigma}_y^{-1}(N\overline{\mathbf{y}}) + \boldsymbol{\Sigma}_0^{-1}\boldsymbol{\mu}_0) \tag{4.144}$$

See Figure 4.13 for a 2d example. We can think of $\mathbf{x}$ as representing the true, but unknown, location of an object in 2d space, such as a missile or airplane, and the $\mathbf{y}_i$ as being noisy observations, such as radar "blips". As we receive more blips, we are better able to localize the source. In Section 18.3.1, we will see how to extend this example to track moving objects using the famous Kalman filter algorithm.

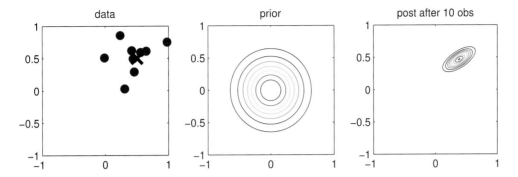

Figure 4.13 Illustration of Bayesian inference for the mean of a 2d Gaussian. (a) The data is generated from $\mathbf{y}_i \sim \mathcal{N}(\mathbf{x}, \boldsymbol{\Sigma}_y)$, where $\mathbf{x} = [0.5, 0.5]^T$ and $\boldsymbol{\Sigma}_y = 0.1[2, 1; 1, 1]$). We assume the sensor noise covariance $\boldsymbol{\Sigma}_y$ is known but $\mathbf{x}$ is unknown. The black cross represents $\mathbf{x}$. (b) The prior is $p(\mathbf{x}) = \mathcal{N}(\mathbf{x}|\mathbf{0}, 0.1\mathbf{I}_2)$. (c) We show the posterior after 10 data points have been observed. Figure generated by `gaussInferParamsMean2d`.

Now suppose we have multiple measuring devices, and we want to combine them together; this is known as **sensor fusion**. If we have multiple observations with different covariances (corresponding to sensors with different reliabilities), the posterior will be an appropriate weighted average of the data. Consider the example in Figure 4.14. We use an uninformative prior on $\mathbf{x}$, namely $p(\mathbf{x}) = \mathcal{N}(\boldsymbol{\mu}_0, \boldsymbol{\Sigma}_0) = \mathcal{N}(\mathbf{0}, 10^{10}\mathbf{I}_2)$. We get 2 noisy observations, $\mathbf{y}_1 \sim \mathcal{N}(\mathbf{x}, \boldsymbol{\Sigma}_{y,1})$ and $\mathbf{y}_2 \sim \mathcal{N}(\mathbf{x}, \boldsymbol{\Sigma}_{y,2})$. We then compute $p(\mathbf{x}|\mathbf{y}_1, \mathbf{y}_2)$.

In Figure 4.14(a), we set $\boldsymbol{\Sigma}_{y,1} = \boldsymbol{\Sigma}_{y,2} = 0.01\mathbf{I}_2$, so both sensors are equally reliable. In this case, the posterior mean is halfway between the two observations, $\mathbf{y}_1$ and $\mathbf{y}_2$. In Figure 4.14(b), we set $\boldsymbol{\Sigma}_{y,1} = 0.05\mathbf{I}_2$ and $\boldsymbol{\Sigma}_{y,2} = 0.01\mathbf{I}_2$, so sensor 2 is more reliable than sensor 1. In this case, the posterior mean is closer to $\mathbf{y}_2$. In Figure 4.14(c), we set

$$\boldsymbol{\Sigma}_{y,1} = 0.01 \begin{pmatrix} 10 & 1 \\ 1 & 1 \end{pmatrix}, \quad \boldsymbol{\Sigma}_{y,2} = 0.01 \begin{pmatrix} 1 & 1 \\ 1 & 10 \end{pmatrix} \tag{4.145}$$

so sensor 1 is more reliable in the y_2 component (vertical direction), and sensor 2 is more reliable in the y_1 component (horizontal direction). In this case, the posterior mean uses $\mathbf{y}_1$'s vertical component and $\mathbf{y}_2$'s horizontal component.

Note that this technique crucially relies on modeling our uncertainty of each sensor; computing an unweighted average would give the wrong result. However, we have assumed the sensor precisions are known. When they are not, we should model out uncertainty about $\boldsymbol{\Sigma}_1$ and $\boldsymbol{\Sigma}_2$ as well. See Section 4.6.4 for details.

4.4.2.3 Interpolating noisy data

We now revisit the example of Section 4.3.2.2. This time we no longer assume noise-free observations. Instead, let us assume that we obtain N noisy observations y_i; without loss of generality, assume these correspond to $x_1, \ldots, x_N$. We can model this setup as a linear

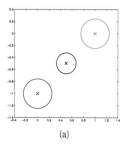 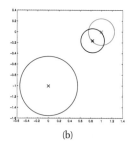 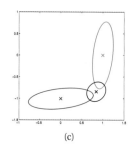

(a)   (b)   (c)

Figure 4.14   We observe $\mathbf{y}_1 = (0, -1)$ (red cross) and $\mathbf{y}_2 = (1, 0)$ (green cross) and infer $E(\boldsymbol{\mu}|\mathbf{y}_1, \mathbf{y}_2, \boldsymbol{\theta})$ (black cross). (a) Equally reliable sensors, so the posterior mean estimate is in between the two circles. (b) Sensor 2 is more reliable, so the estimate shifts more towards the green circle. (c) Sensor 1 is more reliable in the vertical direction, Sensor 2 is more reliable in the horizontal direction. The estimate is an appropriate combination of the two measurements. Figure generated by `sensorFusion2d`.

Gaussian system:

$$\mathbf{y} = \mathbf{A}\mathbf{x} + \boldsymbol{\epsilon} \tag{4.146}$$

where $\boldsymbol{\epsilon} \sim \mathcal{N}(\mathbf{0}, \boldsymbol{\Sigma}_y)$, $\boldsymbol{\Sigma}_y = \sigma^2 \mathbf{I}$, σ^2 is the observation noise, and $\mathbf{A}$ is a $N \times D$ projection matrix that selects out the observed elements. For example, if $N = 2$ and $D = 4$ we have

$$\mathbf{A} = \begin{pmatrix} 1 & 0 & 0 & 0 \\ 0 & 1 & 0 & 0 \end{pmatrix} \tag{4.147}$$

Using the same improper prior as before, $\boldsymbol{\Sigma}_x = (\mathbf{L}^T\mathbf{L})^{-1}$, we can easily compute the posterior mean and variance. In Figure 4.15, we plot the posterior mean, posterior variance, and some posterior samples. Now we see that the prior precision λ effects the posterior mean as well as the posterior variance. In particular, for a strong prior (large λ), the estimate is very smooth, and the uncertainty is low. but for a weak prior (small λ), the estimate is wiggly, and the uncertainty (away from the data) is high.

The posterior mean can also be computed by solving the following optimization problem:

$$\min_{\mathbf{x}} \frac{1}{2\sigma^2} \sum_{i=1}^{N} (x_i - y_i)^2 + \frac{\lambda}{2} \sum_{j=1}^{D} \left[(x_j - x_{j-1})^2 + (x_j - x_{j+1})^2 \right] \tag{4.148}$$

where we have defined $x_0 = x_1$ and $x_{D+1} = x_D$ for notational simplicity. We recognize this as a discrete approximation to the following problem:

$$\min_{f} \frac{1}{2\sigma^2} \int (f(t) - y(t))^2 dt + \frac{\lambda}{2} \int [f'(t)]^2 dt \tag{4.149}$$

where $f'(t)$ is the first derivative of f. The first term measures fit to the data, and the second term penalizes functions that are "too wiggly". This is an example of **Tikhonov regularization**, which is a popular approach to **functional data analysis**. See Chapter 15 for more sophisticated approaches, which enforce higher order smoothness (so the resulting samples look less "jagged").

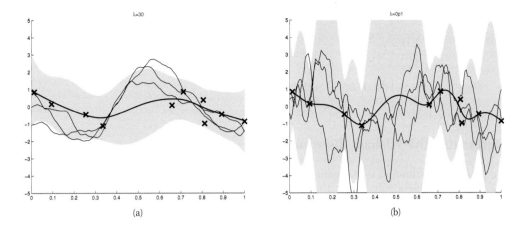

Figure 4.15 Interpolating noisy data (noise variance $\sigma^2 = 1$) using a Gaussian with prior precision λ. (a) $\lambda = 30$. (b) $\lambda = 0.01$. See also Figure 4.10. Based on Figure 7.1 of (Calvetti and Somersalo 2007). Figure generated by gaussInterpNoisyDemo. See also splineBasisDemo.

4.4.3 Proof of the result *

We now derive Equation 4.125. The basic idea is to derive the joint distribution, $p(\mathbf{x}, \mathbf{y}) = p(\mathbf{x})p(\mathbf{y}|\mathbf{x})$, and then to use the results from Section 4.3.1 for computing $p(\mathbf{x}|\mathbf{y})$.

In more detail, we proceed as follows. The log of the joint distribution is as follows (dropping irrelevant constants):

$$\log p(\mathbf{x}, \mathbf{y}) = -\frac{1}{2}(\mathbf{x} - \boldsymbol{\mu}_x)^T \boldsymbol{\Sigma}_x^{-1}(\mathbf{x} - \boldsymbol{\mu}_x) - \frac{1}{2}(\mathbf{y} - \mathbf{A}\mathbf{x} - \mathbf{b})^T \boldsymbol{\Sigma}_y^{-1}(\mathbf{y} - \mathbf{A}\mathbf{x} - \mathbf{b}) \quad (4.150)$$

This is clearly a joint Gaussian distribution, since it is the exponential of a quadratic form.

Expanding out the quadratic terms involving $\mathbf{x}$ and $\mathbf{y}$, and ignoring linear and constant terms, we have

$$
\begin{align}
Q &= -\frac{1}{2}\mathbf{x}^T \boldsymbol{\Sigma}_x^{-1}\mathbf{x} - \frac{1}{2}\mathbf{y}^T \boldsymbol{\Sigma}_y^{-1}\mathbf{y} - \frac{1}{2}(\mathbf{A}\mathbf{x})^T \boldsymbol{\Sigma}_y^{-1}(\mathbf{A}\mathbf{x}) + \mathbf{y}^T \boldsymbol{\Sigma}_y^{-1}\mathbf{A}\mathbf{x} \quad &(4.151)\\
&= -\frac{1}{2}\begin{pmatrix}\mathbf{x}\\\mathbf{y}\end{pmatrix}^T \begin{pmatrix}\boldsymbol{\Sigma}_x^{-1} + \mathbf{A}^T \boldsymbol{\Sigma}_y^{-1}\mathbf{A} & -\mathbf{A}^T \boldsymbol{\Sigma}_y^{-1}\\ -\boldsymbol{\Sigma}_y^{-1}\mathbf{A} & \boldsymbol{\Sigma}_y^{-1}\end{pmatrix}\begin{pmatrix}\mathbf{x}\\\mathbf{y}\end{pmatrix} \quad &(4.152)\\
&= -\frac{1}{2}\begin{pmatrix}\mathbf{x}\\\mathbf{y}\end{pmatrix}^T \boldsymbol{\Sigma}^{-1}\begin{pmatrix}\mathbf{x}\\\mathbf{y}\end{pmatrix} \quad &(4.153)
\end{align}
$$

where the precision matrix of the joint is defined as

$$\boldsymbol{\Sigma}^{-1} = \begin{pmatrix}\boldsymbol{\Sigma}_x^{-1} + \mathbf{A}^T \boldsymbol{\Sigma}_y^{-1}\mathbf{A} & -\mathbf{A}^T \boldsymbol{\Sigma}_y^{-1}\\ -\boldsymbol{\Sigma}_y^{-1}\mathbf{A} & \boldsymbol{\Sigma}_y^{-1}\end{pmatrix} \triangleq \boldsymbol{\Lambda} = \begin{pmatrix}\boldsymbol{\Lambda}_{xx} & \boldsymbol{\Lambda}_{xy}\\ \boldsymbol{\Lambda}_{yx} & \boldsymbol{\Lambda}_{yy}\end{pmatrix} \quad (4.154)$$

From Equation 4.69, and using the fact that $\boldsymbol{\mu}_y = \mathbf{A}\boldsymbol{\mu}_x + \mathbf{b}$, we have

$$p(\mathbf{x}|\mathbf{y}) = \mathcal{N}(\boldsymbol{\mu}_{x|y}, \boldsymbol{\Sigma}_{x|y}) \tag{4.155}$$

$$\boldsymbol{\Sigma}_{x|y} = \boldsymbol{\Lambda}_{xx}^{-1} = (\boldsymbol{\Sigma}_x^{-1} + \mathbf{A}^T \boldsymbol{\Sigma}_y^{-1} \mathbf{A})^{-1} \tag{4.156}$$

$$\boldsymbol{\mu}_{x|y} = \boldsymbol{\Sigma}_{x|y} \left(\boldsymbol{\Lambda}_{xx}\boldsymbol{\mu}_x - \boldsymbol{\Lambda}_{xy}(\mathbf{y} - \boldsymbol{\mu}_y) \right) \tag{4.157}$$

$$= \boldsymbol{\Sigma}_{x|y} \left(\boldsymbol{\Sigma}_x^{-1}\boldsymbol{\mu}_x + \mathbf{A}^T \boldsymbol{\Sigma}_y^{-1}(\mathbf{y} - \mathbf{b}) \right) \tag{4.158}$$

4.5   Digression: The Wishart distribution *

The **Wishart** distribution is the generalization of the Gamma distribution to positive definite matrices. Press (Press 2005, p107) has said "The Wishart distribution ranks next to the (multivariate) normal distribution in order of importance and usefulness in multivariate statistics". We will mostly use it to model our uncertainty in covariance matrices, $\boldsymbol{\Sigma}$, or their inverses, $\boldsymbol{\Lambda} = \boldsymbol{\Sigma}^{-1}$.

The pdf of the Wishart is defined as follows:

$$\text{Wi}(\boldsymbol{\Lambda}|\mathbf{S}, \nu) = \frac{1}{Z_{\text{Wi}}} |\boldsymbol{\Lambda}|^{(\nu-D-1)/2} \exp\left(-\frac{1}{2}\text{tr}(\boldsymbol{\Lambda}\mathbf{S}^{-1}) \right) \tag{4.159}$$

Here ν is called the "degrees of freedom" and $\mathbf{S}$ is the "scale matrix". (We shall get more intuition for these parameters shortly.) The normalization constant for this distribution (which requires integrating over all symmetric pd matrices) is the following formidable expression

$$Z_{\text{Wi}} = 2^{\nu D/2} \Gamma_D(\nu/2) |\mathbf{S}|^{\nu/2} \tag{4.160}$$

where $\Gamma_D(a)$ is the **multivariate gamma function**:

$$\Gamma_D(x) = \pi^{D(D-1)/4} \prod_{i=1}^{D} \Gamma\left(x + (1-i)/2 \right) \tag{4.161}$$

Hence $\Gamma_1(a) = \Gamma(a)$ and

$$\Gamma_D(\nu_0/2) = \pi^{D(D-1)/4} \prod_{i=1}^{D} \Gamma(\frac{\nu_0 + 1 - i}{2}) \tag{4.162}$$

The normalization constant only exists (and hence the pdf is only well defined) if $\nu > D - 1$.

There is a connection between the Wishart distribution and the Gaussian. In particular, let $\mathbf{x}_i \sim \mathcal{N}(0, \boldsymbol{\Sigma})$. Then the scatter matrix $\mathbf{S} = \sum_{i=1}^{N} \mathbf{x}_i \mathbf{x}_i^T$ has a Wishart distribution: $\mathbf{S} \sim \text{Wi}(\boldsymbol{\Sigma}, N)$. Hence $\mathbb{E}[\mathbf{S}] = N\boldsymbol{\Sigma}$. More generally, one can show that the mean and mode of $\text{Wi}(\mathbf{S}, \nu)$ are given by

$$\text{mean} = \nu\mathbf{S}, \ \text{mode} = (\nu - D - 1)\mathbf{S} \tag{4.163}$$

where the mode only exists if $\nu > D + 1$.

If $D = 1$, the Wishart reduces to the Gamma distribution:

$$\text{Wi}(\lambda|s^{-1}, \nu) = \text{Ga}(\lambda|\frac{\nu}{2}, \frac{s}{2}) \tag{4.164}$$

4.5.1 Inverse Wishart distribution

Recall that we showed (Exercise 2.10) that if $\lambda \sim \text{Ga}(a, b)$, then that $\frac{1}{\lambda} \sim \text{IG}(a, b)$. Similarly, if $\mathbf{\Sigma}^{-1} \sim \text{Wi}(\mathbf{S}, \nu)$ then $\mathbf{\Sigma} \sim \text{IW}(\mathbf{S}^{-1}, \nu + D + 1)$, where IW is the **inverse Wishart**, the multidimensional generalization of the inverse Gamma. It is defined as follows, for $\nu > D - 1$ and $\mathbf{S} \succ 0$:

$$\text{IW}(\mathbf{\Sigma}|\mathbf{S}, \nu) = \frac{1}{Z_{IW}} |\mathbf{\Sigma}|^{-(\nu+D+1)/2} \exp\left(-\frac{1}{2}\text{tr}(\mathbf{S}^{-1}\mathbf{\Sigma}^{-1})\right) \tag{4.165}$$

$$Z_{IW} = |\mathbf{S}|^{-\nu/2} 2^{\nu D/2} \Gamma_D(\nu/2) \tag{4.166}$$

One can show that the distribution has these properties

$$\text{mean} = \frac{\mathbf{S}^{-1}}{\nu - D - 1}, \quad \text{mode} = \frac{\mathbf{S}^{-1}}{\nu + D + 1} \tag{4.167}$$

If $D = 1$, this reduces to the inverse Gamma:

$$\text{IW}(\sigma^2|S^{-1}, \nu) = \text{IG}(\sigma^2|\nu/2, S/2) \tag{4.168}$$

4.5.2 Visualizing the Wishart distribution *

Since the Wishart is a distribution over matrices, it is hard to plot as a density function. However, we can easily sample from it, and in the 2d case, we can use the eigenvectors of the resulting matrix to define an ellipse, as explained in Section 4.1.2. See Figure 4.16 for some examples.

For higher dimensional matrices, we can plot marginals of the distribution. The diagonals of a Wishart distributed matrix have Gamma distributions, so are easy to plot. It is hard in general to work out the distribution of the off-diagonal elements, but we can sample matrices from the distribution, and then compute the distribution empirically. In particular, we can convert each sampled matrix to a correlation matrix, and thus compute a Monte Carlo approximation (Section 2.7) to the expected correlation coefficients:

$$\mathbb{E}[R_{ij}] \approx \frac{1}{S} \sum_{s=1}^{S} \mathbf{R}(\mathbf{\Sigma}^{(s)})_{ij} \tag{4.169}$$

where $\mathbf{\Sigma}^{(s)} \sim \text{Wi}(\mathbf{\Sigma}, \nu)$ and $\mathbf{R}(\mathbf{\Sigma})$ converts matrix $\mathbf{\Sigma}$ into a correlation matrix:

$$R_{ij} = \frac{\mathbf{\Sigma}_{ij}}{\sqrt{\mathbf{\Sigma}_{ii}\mathbf{\Sigma}_{jj}}} \tag{4.170}$$

We can then use kernel density estimation (Section 14.7.2) to produce a smooth approximation to the univariate density $\mathbb{E}[R_{ij}]$ for plotting purposes. See Figure 4.16 for some examples.

4.6 Inferring the parameters of an MVN

So far, we have discussed inference in a Gaussian assuming the parameters $\boldsymbol{\theta} = (\boldsymbol{\mu}, \mathbf{\Sigma})$ are known. We now discuss how to infer the parameters themselves. We will assume the data has the form $\mathbf{x}_i \sim \mathcal{N}(\boldsymbol{\mu}, \mathbf{\Sigma})$ for $i = 1 : N$ and is fully observed, so we have no missing data (see

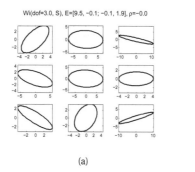
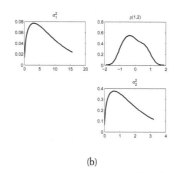

(a)   (b)

Figure 4.16   Visualization of the Wishart distribution. Left: Some samples from the Wishart distribution, $\boldsymbol{\Sigma} \sim \text{Wi}(\mathbf{S}, \nu)$, where $\mathbf{S} = [3.1653, -0.0262; -0.0262, 0.6477]$ and $\nu = 3$. Right: Plots of the marginals (which are Gamma), and the approximate (sample-based) marginal on the correlation coefficient. If $\nu = 3$ there is a lot of uncertainty about the value of the correlation coefficient ρ (see the almost uniform distribution on $[-1, 1]$). The sampled matrices are highly variable, and some are nearly singular. As ν increases, the sampled matrices are more concentrated on the prior $\mathbf{S}$. Figure generated by `wiPlotDemo`.

Section 11.6.1 for how to estimate parameters of an MVN in the presence of missing values). To simplify the presentation, we derive the posterior in three parts: first we compute $p(\boldsymbol{\mu}|\mathcal{D}, \boldsymbol{\Sigma})$; then we compute $p(\boldsymbol{\Sigma}|\mathcal{D}, \boldsymbol{\mu})$; finally we compute the joint $p(\boldsymbol{\mu}, \boldsymbol{\Sigma}|\mathcal{D})$.

4.6.1   Posterior distribution of μ

We have discussed how to compute the MLE for $\boldsymbol{\mu}$; we now discuss how to compute its posterior, which is useful for modeling our uncertainty about its value.

The likelihood has the form

$$p(\mathcal{D}|\boldsymbol{\mu}) = \mathcal{N}(\overline{\mathbf{x}}|\boldsymbol{\mu}, \frac{1}{N}\boldsymbol{\Sigma}) \tag{4.171}$$

For simplicity, we will use a conjugate prior, which in this case is a Gaussian. In particular, if $p(\boldsymbol{\mu}) = \mathcal{N}(\boldsymbol{\mu}|\mathbf{m}_0, \mathbf{V}_0)$ then we can derive a Gaussian posterior for $\boldsymbol{\mu}$ based on the results in Section 4.4.2.2. We get

$$
\begin{align}
p(\boldsymbol{\mu}|\mathcal{D}, \boldsymbol{\Sigma}) &= \mathcal{N}(\boldsymbol{\mu}|\mathbf{m}_N, \mathbf{V}_N) \tag{4.172}\\
\mathbf{V}_N^{-1} &= \mathbf{V}_0^{-1} + N\boldsymbol{\Sigma}^{-1} \tag{4.173}\\
\mathbf{m}_N &= \mathbf{V}_N(\boldsymbol{\Sigma}^{-1}(N\overline{\mathbf{x}}) + \mathbf{V}_0^{-1}\mathbf{m}_0) \tag{4.174}
\end{align}
$$

This is exactly the same process as inferring the location of an object based on noisy radar "blips", except now we are inferring the mean of a distribution based on noisy samples. (To a Bayesian, there is no difference between uncertainty about parameters and uncertainty about anything else.)

We can model an uninformative prior by setting $\mathbf{V}_0 = \infty \mathbf{I}$. In this case we have $p(\boldsymbol{\mu}|\mathcal{D}, \boldsymbol{\Sigma}) = \mathcal{N}(\overline{\mathbf{x}}, \frac{1}{N}\boldsymbol{\Sigma})$, so the posterior mean is equal to the MLE. We also see that the posterior variance goes down as $1/N$, which is a standard result from frequentist statistics.

4.6.2 Posterior distribution of Σ *

We now discuss how to compute $p(\Sigma|\mathcal{D}, \boldsymbol{\mu})$. The likelihood has the form

$$p(\mathcal{D}|\boldsymbol{\mu}, \Sigma) \quad \propto \quad |\Sigma|^{-\frac{N}{2}} \exp\left(-\frac{1}{2}\text{tr}(\mathbf{S}_{\mu}\Sigma^{-1})\right) \tag{4.175}$$

The corresponding conjugate prior is known as the inverse Wishart distribution (Section 4.5.1). Recall that this has the following pdf:

$$\text{IW}(\Sigma|\mathbf{S}_0^{-1}, \nu_0) \quad \propto \quad |\Sigma|^{-(\nu_0+D+1)/2} \exp\left(-\frac{1}{2}\text{tr}(\mathbf{S}_0\Sigma^{-1})\right) \tag{4.176}$$

Here $\nu_0 > D - 1$ is the degrees of freedom (dof), and $\mathbf{S}_0$ is a symmetric pd matrix. We see that $\mathbf{S}_0^{-1}$ plays the role of the prior scatter matrix, and $N_0 \triangleq \nu_0 + D + 1$ controls the strength of the prior, and hence plays a role analogous to the sample size N.

Multiplying the likelihood and prior we find that the posterior is also inverse Wishart:

$$p(\Sigma|\mathcal{D}, \boldsymbol{\mu}) \quad \propto \quad |\Sigma|^{-\frac{N}{2}} \exp\left(-\frac{1}{2}\text{tr}(\Sigma^{-1}\mathbf{S}_{\mu})\right) |\Sigma|^{-(\nu_0+D+1)/2}$$

$$\exp\left(-\frac{1}{2}\text{tr}(\Sigma^{-1}\mathbf{S}_0)\right) \tag{4.177}$$

$$= \quad |\Sigma|^{-\frac{N+(\nu_0+D+1)}{2}} \exp\left(-\frac{1}{2}\text{tr}\left[\Sigma^{-1}(\mathbf{S}_{\mu} + \mathbf{S}_0)\right]\right) \tag{4.178}$$

$$= \quad \text{IW}(\Sigma|\mathbf{S}_N, \nu_N) \tag{4.179}$$

$$\nu_N \quad = \quad \nu_0 + N \tag{4.180}$$

$$\mathbf{S}_N \quad = \quad \mathbf{S}_0 + \mathbf{S}_{\mu} \tag{4.181}$$

In words, this says that the posterior strength ν_N is the prior strength ν_0 plus the number of observations N, and the posterior scatter matrix $\mathbf{S}_N$ is the prior scatter matrix $\mathbf{S}_0$ plus the data scatter matrix $\mathbf{S}_{\mu}$.

4.6.2.1 MAP estimation

We see from Equation 4.7 that $\hat{\Sigma}_{mle}$ is a rank $\min(N, D)$ matrix. If $N < D$, this is not full rank, and hence will be uninvertible. And even if $N > D$, it may be the case that $\hat{\Sigma}$ is ill-conditioned (meaning it is nearly singular).

To solve these problems, we can use the posterior mode (or mean). One can show (using techniques analogous to the derivation of the MLE) that the MAP estimate is given by

$$\hat{\Sigma}_{map} = \frac{\mathbf{S}_N}{\nu_N + D + 1} = \frac{\mathbf{S}_0 + \mathbf{S}_{\mu}}{N_0 + N} \tag{4.182}$$

If we use an improper uniform prior, corresponding to $N_0 = 0$ and $\mathbf{S}_0 = \mathbf{0}$, we recover the MLE.

Let us now consider the use of a proper informative prior, which is necessary whenever D/N is large (say bigger than 0.1). Let $\boldsymbol{\mu} = \overline{\mathbf{x}}$, so $\mathbf{S}_{\mu} = \mathbf{S}_{\overline{x}}$. Then we can rewrite the MAP estimate

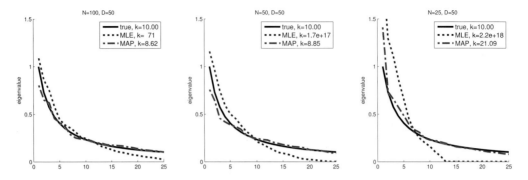

Figure 4.17   Estimating a covariance matrix in $D = 50$ dimensions using $N \in \{100, 50, 25\}$ samples. We plot the eigenvalues in descending order for the true covariance matrix (solid black), the MLE (dotted blue) and the MAP estimate (dashed red), using Equation 4.184 with $\lambda = 0.9$. We also list the condition number of each matrix in the legend. Based on Figure 1 of (Schaefer and Strimmer 2005). Figure generated by `shrinkcovDemo`.

as a convex combination of the prior mode and the MLE. To see this, let $\boldsymbol{\Sigma}_0 \triangleq \frac{\mathbf{S}_0}{N_0}$ be the prior mode. Then the posterior mode can be rewritten as

$$\hat{\boldsymbol{\Sigma}}_{map} = \frac{\mathbf{S}_0 + \mathbf{S}_{\overline{x}}}{N_0 + N} = \frac{N_0}{N_0 + N} \frac{\mathbf{S}_0}{N_0} + \frac{N}{N_0 + N} \frac{\mathbf{S}}{N} = \lambda \boldsymbol{\Sigma}_0 + (1 - \lambda)\hat{\boldsymbol{\Sigma}}_{mle} \tag{4.183}$$

where $\lambda = \frac{N_0}{N_0+N}$, controls the amount of shrinkage towards the prior.

This invites the question: where do the parameters of the prior come from? It is common to set λ by cross validation. Alternatively, we can use the closed-form formula provided in (Ledoit and Wolf 2004b,a; Schaefer and Strimmer 2005), which is the optimal frequentist estimate if we use squared loss. This is arguably not the most natural loss function for covariance matrices (because it ignores the postive definite constraint), but it results in a simple estimator, which is implemented in the PMTK function `shrinkcov`. We discuss Bayesian ways of estimating λ later.

As for the prior covariance matrix, $\mathbf{S}_0$, it is common to use the following (data dependent) prior: $\mathbf{S}_0 = \mathrm{diag}(\hat{\boldsymbol{\Sigma}}_{mle})$. In this case, the MAP estimate is given by

$$\hat{\boldsymbol{\Sigma}}_{map}(i, j) = \begin{cases} \hat{\boldsymbol{\Sigma}}_{mle}(i, j) & \text{if } i = j \\ (1 - \lambda)\hat{\boldsymbol{\Sigma}}_{mle}(i, j) & \text{otherwise} \end{cases} \tag{4.184}$$

Thus we see that the diagonal entries are equal to their ML estimates, and the off-diagonal elements are "shrunk" somewhat towards 0. This technique is therefore called **shrinkage estimation**, or **regularized estimation**.

The benefits of MAP estimation are illustrated in Figure 4.17. We consider fitting a 50-dimensional Gaussian to $N = 100$, $N = 50$ and $N = 25$ data points. We see that the MAP estimate is always well-conditioned, unlike the MLE. In particular, we see that the **eigenvalue spectrum** of the MAP estimate is much closer to that of the true matrix than the MLE's. The eigenvectors, however, are unaffected.

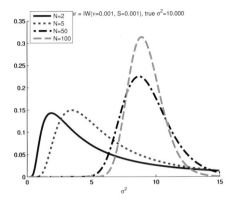

Figure 4.18 Sequential updating of the posterior for σ^2 starting from an uninformative prior. The data was generated from a Gaussian with known mean $\mu = 5$ and unknown variance $\sigma^2 = 10$. Figure generated by `gaussSeqUpdateSigma1D`.

The importance of regularizing the estimate of Σ will become apparent in later chapters, when we consider fitting covariance matrices to high-dimensional data.

4.6.2.2 Univariate posterior

In the 1d case, the likelihood has the form

$$p(\mathcal{D}|\sigma^2) \quad \propto \quad (\sigma^2)^{-N/2} \exp\left(-\frac{1}{2\sigma^2}\sum_{i=1}^{N}(x_i - \mu)^2\right) \tag{4.185}$$

The standard conjugate prior is the inverse Gamma distribution, which is just the scalar version of the inverse Wishart:

$$\text{IG}(\sigma^2|a_0, b_0) \propto (\sigma^2)^{-(a_0+1)} \exp(-\frac{b_0}{\sigma^2}) \tag{4.186}$$

Multiplying the likelihood and the prior, we see that the posterior is also IG:

$$p(\sigma^2|\mathcal{D}) \quad = \quad \text{IG}(\sigma^2|a_N, b_N) \tag{4.187}$$

$$a_N \quad = \quad a_0 + N/2 \tag{4.188}$$

$$b_N \quad = \quad b_0 + \frac{1}{2}\sum_{i=1}^{N}(x_i - \mu)^2 \tag{4.189}$$

See Figure 4.18 for an illustration.

The form of the posterior is not quite as pretty as the multivariate case, because of the factors of $\frac{1}{2}$. This arises because $\text{IW}(\sigma^2|s_0, \nu_0) = \text{IG}(\sigma^2|\frac{s_0}{2}, \frac{\nu_0}{2})$. Another problem with using the $\text{IG}(a_0, b_0)$ distribution is that the strength of the prior is encoded in both a_0 and b_0. To avoid both of these problems, it is common (in the statistics literature) to use an alternative parameterization of the IG distribution, known as the (scaled) **inverse chi-squared distribution**. This is defined as follows:

$$\chi^{-2}(\sigma^2|\nu_0, \sigma_0^2) = \text{IG}(\sigma^2|\frac{\nu_0}{2}, \frac{\nu_0\sigma_0^2}{2}) \propto (\sigma^2)^{-\nu_0/2-1} \exp(-\frac{\nu_0\sigma_0^2}{2\sigma^2}) \tag{4.190}$$

Here ν_0 controls the strength of the prior, and σ_0^2 encodes the value of the prior. With this prior, the posterior becomes

$$p(\sigma^2|\mathcal{D}, \mu) = \chi^{-2}(\sigma^2|\nu_N, \sigma_N^2) \tag{4.191}$$

$$\nu_N = \nu_0 + N \tag{4.192}$$

$$\sigma_N^2 = \frac{\nu_0\sigma_0^2 + \sum_{i=1}^{N}(x_i - \mu)^2}{\nu_N} \tag{4.193}$$

We see that the posterior dof ν_N is the prior dof ν_0 plus N, and the posterior sum of squares $\nu_N\sigma_N^2$ is the prior sum of squares $\nu_0\sigma_0^2$ plus the data sum of squares.

We can emulate an uninformative prior, $p(\sigma^2) \propto \sigma^{-2}$, by setting $\nu_0 = 0$, which makes intuitive sense (since it corresponds to a zero virtual sample size).

4.6.3 Posterior distribution of μ and Σ *

We now discuss how to compute $p(\mu, \Sigma|\mathcal{D})$. These results are a bit complex, but will prove useful later on in this book. Feel free to skip this section on a first reading.

4.6.3.1 Likelihood

The likelihood is given by

$$p(\mathcal{D}|\mu, \Sigma) = (2\pi)^{-ND/2}|\Sigma|^{-\frac{N}{2}} \exp\left(-\frac{1}{2}\sum_{i=1}^{N}(\mathbf{x}_i - \mu)^T\Sigma^{-1}(\mathbf{x}_i - \mu)\right) \tag{4.194}$$

Now one can show that

$$\sum_{i=1}^{N}(\mathbf{x}_i - \mu)^T\Sigma^{-1}(\mathbf{x}_i - \mu) = \text{tr}(\Sigma^{-1}\mathbf{S}_{\overline{x}}) + N(\overline{\mathbf{x}} - \mu)^T\Sigma^{-1}(\overline{\mathbf{x}} - \mu) \tag{4.195}$$

Hence we can rewrite the likelihood as follows:

$$p(\mathcal{D}|\mu, \Sigma) = (2\pi)^{-ND/2}|\Sigma|^{-\frac{N}{2}} \exp\left(-\frac{N}{2}(\mu - \overline{\mathbf{x}})^T\Sigma^{-1}(\mu - \overline{\mathbf{x}})\right) \tag{4.196}$$

$$\exp\left(-\frac{1}{2}\text{tr}(\Sigma^{-1}\mathbf{S}_{\overline{x}})\right) \tag{4.197}$$

We will use this form below.

4.6.3.2 Prior

The obvious prior to use is the following

$$p(\mu, \Sigma) = \mathcal{N}(\mu|\mathbf{m}_0, \mathbf{V}_0)\text{IW}(\Sigma|\mathbf{S}_0, \nu_0) \tag{4.198}$$

Unfortunately, this is not conjugate to the likelihood. To see why, note that $\boldsymbol{\mu}$ and $\boldsymbol{\Sigma}$ appear together in a non-factorized way in the likelihood; hence they will also be coupled together in the posterior.

The above prior is sometimes called **semi-conjugate** or **conditionally conjugate**, since both conditionals, $p(\boldsymbol{\mu}|\boldsymbol{\Sigma})$ and $p(\boldsymbol{\Sigma}|\boldsymbol{\mu})$, are individually conjugate. To create a full conjugate prior, we need to use a prior where $\boldsymbol{\mu}$ and $\boldsymbol{\Sigma}$ are dependent on each other. We will use a joint distribution of the form

$$p(\boldsymbol{\mu}, \boldsymbol{\Sigma}) = p(\boldsymbol{\Sigma})p(\boldsymbol{\mu}|\boldsymbol{\Sigma}) \tag{4.199}$$

Looking at the form of the likelihood equation, Equation 4.197, we see that a natural conjugate prior has the form of a **Normal-inverse-Wishart** or **NIW** distribution, defined as follows:

$$\text{NIW}(\boldsymbol{\mu}, \boldsymbol{\Sigma}|\mathbf{m}_0, \kappa_0, \nu_0, \mathbf{S}_0) \triangleq \tag{4.200}$$

$$\mathcal{N}(\boldsymbol{\mu}|\mathbf{m}_0, \frac{1}{\kappa_0}\boldsymbol{\Sigma}) \times \text{IW}(\boldsymbol{\Sigma}|\mathbf{S}_0, \nu_0) \tag{4.201}$$

$$= \frac{1}{Z_{NIW}}|\boldsymbol{\Sigma}|^{-\frac{1}{2}}\exp\left(-\frac{\kappa_0}{2}(\boldsymbol{\mu}-\mathbf{m}_0)^T\boldsymbol{\Sigma}^{-1}(\boldsymbol{\mu}-\mathbf{m}_0)\right) \tag{4.202}$$

$$\times|\boldsymbol{\Sigma}|^{-\frac{\nu_0+D+1}{2}}\exp\left(-\frac{1}{2}\text{tr}(\boldsymbol{\Sigma}^{-1}\mathbf{S}_0)\right) \tag{4.203}$$

$$= \frac{1}{Z_{NIW}}|\boldsymbol{\Sigma}|^{-\frac{\nu_0+D+2}{2}} \tag{4.204}$$

$$\times\exp\left(-\frac{\kappa_0}{2}(\boldsymbol{\mu}-\mathbf{m}_0)^T\boldsymbol{\Sigma}^{-1}(\boldsymbol{\mu}-\mathbf{m}_0)-\frac{1}{2}\text{tr}(\boldsymbol{\Sigma}^{-1}\mathbf{S}_0)\right) \tag{4.205}$$

$$Z_{NIW} = 2^{\nu_0 D/2}\Gamma_D(\nu_0/2)(2\pi/\kappa_0)^{D/2}|\mathbf{S}_0|^{-\nu_0/2} \tag{4.206}$$

where $\Gamma_D(a)$ is the multivariate Gamma function.

The parameters of the NIW can be interpreted as follows: $\mathbf{m}_0$ is our prior mean for $\boldsymbol{\mu}$, and κ_0 is how strongly we believe this prior; and $\mathbf{S}_0$ is (proportional to) our prior mean for $\boldsymbol{\Sigma}$, and ν_0 is how strongly we believe this prior.[4]

One can show (Minka 2000f) that the (improper) uninformative prior has the form

$$\lim_{k\to 0}\mathcal{N}(\mu|\mathbf{m}_0, \boldsymbol{\Sigma}/k)\text{IW}(\boldsymbol{\Sigma}|\mathbf{S}_0, k) \quad \propto \quad |2\pi\boldsymbol{\Sigma}|^{-\frac{1}{2}}|\boldsymbol{\Sigma}|^{-(D+1)/2} \tag{4.207}$$

$$\propto \quad |\boldsymbol{\Sigma}|^{-(\frac{D}{2}+1)} \propto \text{NIW}(\boldsymbol{\mu}, \boldsymbol{\Sigma}|\mathbf{0}, 0, 0, 0\mathbf{I}) \tag{4.208}$$

In practice, it is often better to use a weakly informative data-dependent prior. A common choice (see e.g., (Chipman et al. 2001, p81), (Fraley and Raftery 2007, p6)) is to use $\mathbf{S}_0 = \text{diag}(\mathbf{S}_{\overline{x}})/N$, and $\nu_0 = D + 2$, to ensure $\mathbb{E}[\boldsymbol{\Sigma}] = \mathbf{S}_0$, and to set $\boldsymbol{\mu}_0 = \overline{\mathbf{x}}$ and κ_0 to some small number, such as 0.01.

---

4. Although this prior has four parameters, there are really only three free parameters, since our uncertainty in the mean is proportional to the variance. In particular, if we believe that the variance is large, then our uncertainty in μ must be large too. This makes sense intuitively, since if the data has large spread, it may be hard to pin down its mean. See also Exercise 9.1, where we will see the three free parameters more explicitly. If we want separate "control" over our confidence in $\boldsymbol{\mu}$ and $\boldsymbol{\Sigma}$, we must use a semi-conjugate prior.

4.6.3.3   Posterior

The posterior can be shown (Exercise 4.11) to be NIW with updated parameters:

$$p(\boldsymbol{\mu}, \boldsymbol{\Sigma}|\mathcal{D}) \quad = \quad \text{NIW}(\boldsymbol{\mu}, \boldsymbol{\Sigma}|\mathbf{m}_N, \kappa_N, \nu_N, \mathbf{S}_N) \tag{4.209}$$

$$\mathbf{m}_N \quad = \quad \frac{\kappa_0 \mathbf{m}_0 + N\overline{\mathbf{x}}}{\kappa_N} = \frac{\kappa_0}{\kappa_0 + N}\mathbf{m}_0 + \frac{N}{\kappa_0 + N}\overline{\mathbf{x}} \tag{4.210}$$

$$\kappa_N \quad = \quad \kappa_0 + N \tag{4.211}$$

$$\nu_N \quad = \quad \nu_0 + N \tag{4.212}$$

$$\mathbf{S}_N \quad = \quad \mathbf{S}_0 + \mathbf{S}_{\overline{x}} + \frac{\kappa_0 N}{\kappa_0 + N}(\overline{\mathbf{x}} - \mathbf{m}_0)(\overline{\mathbf{x}} - \mathbf{m}_0)^T \tag{4.213}$$

$$= \quad \mathbf{S}_0 + \mathbf{S} + \kappa_0 \mathbf{m}_0 \mathbf{m}_0^T - \kappa_N \mathbf{m}_N \mathbf{m}_N^T \tag{4.214}$$

where we have defined $\mathbf{S} \triangleq \sum_{i=1}^{N} \mathbf{x}_i \mathbf{x}_i^T$ as the uncentered sum-of-squares matrix (this is easier to update incrementally than the centered version).

This result is actually quite intuitive: the posterior mean is a convex combination of the prior mean and the MLE, with "strength" $\kappa_0 + N$; and the posterior scatter matrix $\mathbf{S}_N$ is the prior scatter matrix $\mathbf{S}_0$ plus the empirical scatter matrix $\mathbf{S}_{\overline{x}}$ plus an extra term due to the uncertainty in the mean (which creates its own virtual scatter matrix).

4.6.3.4   Posterior mode

The mode of the joint distribution has the following form:

$$\operatorname{argmax} p(\boldsymbol{\mu}, \boldsymbol{\Sigma}|\mathcal{D}) \quad = \quad (\mathbf{m}_N, \frac{\mathbf{S}_N}{\nu_N + D + 2}) \tag{4.215}$$

If we set $\kappa_0 = 0$, this reduces to

$$\operatorname{argmax} p(\boldsymbol{\mu}, \boldsymbol{\Sigma}|\mathcal{D}) \quad = \quad (\overline{\mathbf{x}}, \frac{\mathbf{S}_0 + \mathbf{S}_{\overline{x}}}{\nu_0 + N + D + 2}) \tag{4.216}$$

The corresponding estimate $\hat{\boldsymbol{\Sigma}}$ is almost the same as Equation 4.183, but differs by 1 in the denominator, because this is the mode of the joint, not the mode of the marginal.

4.6.3.5   Posterior marginals

The posterior marginal for $\boldsymbol{\Sigma}$ is simply

$$p(\boldsymbol{\Sigma}|\mathcal{D}) \quad = \quad \int p(\boldsymbol{\mu}, \boldsymbol{\Sigma}|\mathcal{D})d\boldsymbol{\mu} = \text{IW}(\boldsymbol{\Sigma}|\mathbf{S}_N, \nu_N) \tag{4.217}$$

The mode and mean of this marginal are given by

$$\hat{\boldsymbol{\Sigma}}_{map} = \frac{\mathbf{S}_N}{\nu_N + D + 1}, \quad \mathbb{E}\left[\boldsymbol{\Sigma}\right] = \frac{\mathbf{S}_N}{\nu_N - D - 1} \tag{4.218}$$

One can show that the posterior marginal for $\boldsymbol{\mu}$ has a multivariate Student's t distribution:

$$p(\boldsymbol{\mu}|\mathcal{D}) \quad = \quad \int p(\boldsymbol{\mu}, \boldsymbol{\Sigma}|\mathcal{D})d\boldsymbol{\Sigma} = \mathcal{T}(\boldsymbol{\mu}|\mathbf{m}_N, \frac{1}{\kappa_N(\nu_N - D + 1)}\mathbf{S}_N, \nu_N - D + 1) \tag{4.219}$$

This follows from the fact that the Student distribution can be represented as a scaled mixture of Gaussians (see Equation 11.61).

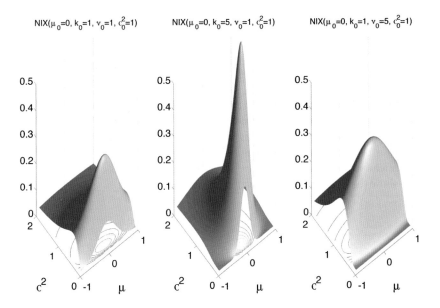

Figure 4.19 The $NI\chi^2(m_0, \kappa_0, \nu_0, \sigma_0^2)$ distribution. m_0 is the prior mean and κ_0 is how strongly we believe this; σ_0^2 is the prior variance and ν_0 is how strongly we believe this. (a) $m_0 = 0, \kappa_0 = 1, \nu_0 = 1, \sigma_0^2 = 1$. Notice that the contour plot (underneath the surface) is shaped like a "squashed egg". (b) We increase the strength of our belief in the mean, so it gets narrower: $m_0 = 0, \kappa_0 = 5, \nu_0 = 1, \sigma_0^2 = 1$. (c) We increase the strength of our belief in the variance, so it gets narrower: $m_0 = 0, \kappa_0 = 1, \nu_0 = 5, \sigma_0^2 = 1$. Figure generated by NIXdemo2.

4.6.3.6 Posterior predictive

The posterior predictive is given by

$$p(\mathbf{x}|\mathcal{D}) = \frac{p(\mathbf{x}, \mathcal{D})}{p(\mathcal{D})} \tag{4.220}$$

so it can be easily evaluated in terms of a ratio of marginal likelihoods.

It turns out that this ratio has the form of a multivariate Student-T distribution:

$$p(\mathbf{x}|\mathcal{D}) = \int\int \mathcal{N}(\mathbf{x}|\boldsymbol{\mu}, \boldsymbol{\Sigma})\text{NIW}(\boldsymbol{\mu}, \boldsymbol{\Sigma}|\mathbf{m}_N, \kappa_N, \nu_N, \mathbf{S}_N)d\boldsymbol{\mu}d\boldsymbol{\Sigma} \tag{4.221}$$

$$= \mathcal{T}(\mathbf{x}|\mathbf{m}_N, \frac{\kappa_N + 1}{\kappa_N(\nu_N - D + 1)}\mathbf{S}_N, \nu_N - D + 1) \tag{4.222}$$

The Student-T has wider tails than a Gaussian, which takes into account the fact that $\boldsymbol{\Sigma}$ is unknown. However, this rapidly becomes Gaussian-like.

4.6.3.7 Posterior for scalar data

We now specialise the above results to the case where x_i is 1D. These results are widely used in the statistics literature. As in Section 4.6.2.2, it is conventional not to use the normal inverse

Wishart, but to use the **normal inverse chi-squared** or **NIX** distribution, defined by

$$NI\chi^2(\mu,\sigma^2|m_0,\kappa_0,\nu_0,\sigma_0^2) \quad \triangleq \quad \mathcal{N}(\mu|m_0,\sigma^2/\kappa_0) \; \chi^{-2}(\sigma^2|\nu_0,\sigma_0^2) \tag{4.223}$$

$$\propto \quad (\frac{1}{\sigma^2})^{(\nu_0+3)/2} \exp\left(-\frac{\nu_0\sigma_0^2 + \kappa_0(\mu-m_0)^2}{2\sigma^2}\right) \tag{4.224}$$

See Figure 4.19 for some plots. Along the μ axis, the distribution is shaped like a Gaussian, and along the σ^2 axis, the distribution is shaped like a χ^{-2}; the contours of the joint density have a "squashed egg" appearance. Interestingly, we see that the contours for μ are more peaked for small values of σ^2, which makes sense, since if the data is low variance, we will be able to estimate its mean more reliably.

One can show that the posterior is given by

$$p(\mu,\sigma^2|\mathcal{D}) \quad = \quad NI\chi^2(\mu,\sigma^2|m_N,\kappa_N,\nu_N,\sigma_N^2) \tag{4.225}$$

$$m_N \quad = \quad \frac{\kappa_0 m_0 + N\bar{x}}{\kappa_N} \tag{4.226}$$

$$\kappa_N \quad = \quad \kappa_0 + N \tag{4.227}$$

$$\nu_N \quad = \quad \nu_0 + N \tag{4.228}$$

$$\nu_N\sigma_N^2 \quad = \quad \nu_0\sigma_0^2 + \sum_{i=1}^{N}(x_i - \bar{x})^2 + \frac{N\kappa_0}{\kappa_0 + N}(m_0 - \bar{x})^2 \tag{4.229}$$

The posterior marginal for σ^2 is just

$$p(\sigma^2|\mathcal{D}) \quad = \quad \int p(\mu,\sigma^2|\mathcal{D})d\mu = \chi^{-2}(\sigma^2|\nu_N,\sigma_N^2) \tag{4.230}$$

with the posterior mean given by $\mathbb{E}\left[\sigma^2|\mathcal{D}\right] = \frac{\nu_N}{\nu_N-2}\sigma_N^2$.

The posterior marginal for μ has a Student distribution, which follows from the fact that the Student distribution is a (scaled) mixture of Gaussians:

$$p(\mu|\mathcal{D}) \quad = \quad \int p(\mu,\sigma^2|D)d\sigma^2 = \mathcal{T}(\mu|m_N,\sigma_N^2/\kappa_N,\nu_N) \tag{4.231}$$

with the posterior mean given by $\mathbb{E}\left[\mu|\mathcal{D}\right] = m_N$.

Let us see how these results look if we use the following uninformative prior:

$$p(\mu,\sigma^2) \propto p(\mu)p(\sigma^2) \propto \sigma^{-2} \propto NI\chi^2(\mu,\sigma^2|\mu_0=0,\kappa_0=0,\nu_0=-1,\sigma_0^2=0) \tag{4.232}$$

With this prior, the posterior has the form

$$p(\mu,\sigma^2|\mathcal{D}) \quad = \quad NI\chi^2(\mu,\sigma^2|m_N=\bar{x},\kappa_N=N,\nu_N=N-1,\sigma_N^2=s^2) \tag{4.233}$$

where

$$s^2 \quad \triangleq \quad \frac{1}{N-1}\sum_{i=1}^{N}(x_i-\bar{x})^2 = \frac{N}{N-1}\hat{\sigma}_{mle}^2 \tag{4.234}$$

is the the **sample standard deviation**. (In Section 6.4.2, we show that this is an unbiased estimate of the variance.) Hence the marginal posterior for the mean is given by

$$p(\mu|\mathcal{D}) = \mathcal{T}(\mu|\bar{x}, \frac{s^2}{N}, N-1) \tag{4.235}$$

and the posterior variance of μ is

$$\text{var}\left[\mu|\mathcal{D}\right] = \frac{\nu_N}{\nu_N - 2}\sigma_N^2 = \frac{N-1}{N-3}\frac{s^2}{N} \rightarrow \frac{s^2}{N} \tag{4.236}$$

The square root of this is called the **standard error of the mean**:

$$\sqrt{\text{var}\left[\mu|\mathcal{D}\right]} \approx \frac{s}{\sqrt{N}} \tag{4.237}$$

Thus an approximate 95% posterior **credible interval** for the mean is

$$I_{.95}(\mu|\mathcal{D}) = \overline{x} \pm 2\frac{s}{\sqrt{N}} \tag{4.238}$$

(Bayesian credible intervals are discussed in more detail in Section 5.2.2; they are contrasted with frequentist confidence intervals in Section 6.6.1.)

4.6.3.8 Bayesian t-test

Suppose we want to test the hypothesis that $\mu \neq \mu_0$ for some known value μ_0 (often 0), given values $x_i \sim \mathcal{N}(\mu, \sigma^2)$. This is called a two-sided, one-sample **t-test**. A simple way to perform such a test is just to check if $\mu_0 \in I_{0.95}(\mu|\mathcal{D})$. If it is not, then we can be 95% sure that $\mu \neq \mu_0$.[5] A more common scenario is when we want to test if two paired samples have the same mean. More precisely, suppose $y_i \sim \mathcal{N}(\mu_1, \sigma^2)$ and $z_i \sim \mathcal{N}(\mu_2, \sigma^2)$. We want to determine if $\mu = \mu_1 - \mu_2 > 0$, using $x_i = y_i - z_i$ as our data. We can evaluate this quantity as follows:

$$p(\mu > \mu_0|\mathcal{D}) = \int_{\mu_0}^{\infty} p(\mu|\mathcal{D})d\mu \tag{4.239}$$

This is called a one-sided, **paired t-test**. (For a similar approach to unpaired tests, comparing the difference in binomial proportions, see Section 5.2.3.)

To calculate the posterior, we must specify a prior. Suppose we use an uninformative prior. As we showed above, we find that the posterior marginal on μ has the form

$$p(\mu|\mathcal{D}) = \mathcal{T}(\mu|\overline{x}, \frac{s^2}{N}, N-1) \tag{4.240}$$

Now let us define the following **t statistic**:

$$t \triangleq \frac{\overline{x} - \mu_0}{s/\sqrt{N}} \tag{4.241}$$

where the denominator is the standard error of the mean. We see that

$$p(\mu|\mathcal{D}) = 1 - F_{N-1}(t) \tag{4.242}$$

where $F_{\nu}(t)$ is the cdf of the standard Student t distribution $\mathcal{T}(0, 1, \nu)$.

---

5. A more complex approach is to perform Bayesian model comparison. That is, we compute the Bayes factor (described in Section 5.3.3) $p(\mathcal{D}|H_0)/p(\mathcal{D}|H_1)$, where H_0 is the point null hypothesis that $\mu = \mu_0$, and H_1 is the alternative hypothesis that $\mu \neq \mu_0$. See (Gonen et al. 2005; Rouder et al. 2009) for details.

4.6.3.9   Connection with frequentist statistics *

If we use an uninformative prior, it turns out that the above Bayesian analysis gives the same result as derived using frequentist methods. (We discuss frequentist statistics in Chapter 6.) Specifically, from the above results, we see that

$$\frac{\mu - \overline{x}}{\sqrt{s/N}} | \mathcal{D} \sim t_{N-1} \tag{4.243}$$

This has the same form as the sampling distribution of the MLE:

$$\frac{\mu - \overline{X}}{\sqrt{s/N}} | \mu \sim t_{N-1} \tag{4.244}$$

The reason is that the Student distribution is symmetric in its first two arguments, so $\mathcal{T}(\overline{x}|\mu, \sigma^2, \nu) = \mathcal{T}(\mu|\overline{x}, \sigma^2, \nu)$; hence statements about the posterior for μ have the same form as statements about the sampling distribution of $\overline{x}$. Consequently, the (one-sided) p-value (defined in Section 6.6.2) returned by a frequentist test is the same as $p(\mu > \mu_0|\mathcal{D})$ returned by the Bayesian method. See bayesTtestDemo for an example.

Despite the superficial similarity, these two results have a different interpretation: in the Bayesian approach, μ is unknown and $\overline{x}$ is fixed, whereas in the frequentist approach, $\overline{X}$ is unknown and μ is fixed. More equivalences between frequentist and Bayesian inference in simple models using uninformative priors can be found in (Box and Tiao 1973). See also Section 7.6.3.3.

4.6.4   Sensor fusion with unknown precisions *

In this section, we apply the results in Section 4.6.3 to the problem of sensor fusion in the case where the precision of each measurement device is unknown. This generalizes the results of Section 4.4.2.2, where the measurement model was assumed to be Gaussian with known precision. The unknown precision case turns out to give qualitatively different results, yielding a potentially multi-modal posterior as we will see. Our presentation is based on (Minka 2001e).

Suppose we want to pool data from multiple sources to estimate some quantity $\mu \in \mathbb{R}$, but the reliability of the sources is unknown. Specifically, suppose we have two different measurement devices, x and y, with different precisions: $x_i|\mu \sim \mathcal{N}(\mu, \lambda_x^{-1})$ and $y_i|\mu \sim \mathcal{N}(\mu, \lambda_y^{-1})$. We make two independent measurements with each device, which turn out to be

$$x_1 = 1.1, x_2 = 1.9, y_1 = 2.9, y_2 = 4.1 \tag{4.245}$$

We will use a non-informative prior for μ, $p(\mu) \propto 1$, which we can emulate using an infinitely broad Gaussian, $p(\mu) = \mathcal{N}(\mu|m_0 = 0, \lambda_0^{-1} = \infty)$. If the λ_x and λ_y terms were known, then the posterior would be Gaussian:

$$p(\mu|\mathcal{D}, \lambda_x, \lambda_y) = \mathcal{N}(\mu|m_N, \lambda_N^{-1}) \tag{4.246}$$

$$\lambda_N = \lambda_0 + N_x \lambda_x + N_y \lambda_y \tag{4.247}$$

$$m_N = \frac{\lambda_x N_x \overline{x} + \lambda_y N_y \overline{y}}{N_x \lambda_x + N_y \lambda_y} \tag{4.248}$$

where $N_x = 2$ is the number of x measurements, $N_y = 2$ is the number of y measurements, $\overline{x} = \frac{1}{N_x} \sum_{i=1}^{N_x} x_i = 1.5$ and $\overline{y} = \frac{1}{N_y} \sum_{i=1}^{N_y} y_i = 3.5$. This result follows because the posterior precision is the sum of the measurement precisions, and the posterior mean is a weighted sum of the prior mean (which is 0) and the data means.

However, the measurement precisions are not known. Initially we will estimate them by maximum likelihood. The log-likelihood is given by

$$\ell(\mu, \lambda_x, \lambda_y) \quad = \quad \log \lambda_x - \frac{\lambda_x}{2} \sum_i (x_i - \mu)^2 + \log \lambda_y - \frac{\lambda_y}{2} \sum_i (y_i - \mu)^2 \tag{4.249}$$

The MLE is obtained by solving the following simultaneous equations:

$$\frac{\partial \ell}{\partial \mu} \quad = \quad \lambda_x N_x (\overline{x} - \mu) + \lambda_y N_y (\overline{y} - \mu) = 0 \tag{4.250}$$

$$\frac{\partial \ell}{\partial \lambda_x} \quad = \quad \frac{1}{\lambda_x} - \frac{1}{N_x} \sum_{i=1}^{N_x} (x_i - \mu)^2 = 0 \tag{4.251}$$

$$\frac{\partial \ell}{\partial \lambda_y} \quad = \quad \frac{1}{\lambda_y} - \frac{1}{N_y} \sum_{i=1}^{N_y} (y_i - \mu)^2 = 0 \tag{4.252}$$

This gives

$$\hat{\mu} \quad = \quad \frac{N_x \hat{\lambda}_x \overline{x} + N_y \hat{\lambda}_y \overline{y}}{N_x \hat{\lambda}_x + N_y \hat{\lambda}_y} \tag{4.253}$$

$$1/\hat{\lambda}_x \quad = \quad \frac{1}{N_x} \sum_i (x_i - \hat{\mu})^2 \tag{4.254}$$

$$1/\hat{\lambda}_y \quad = \quad \frac{1}{N_y} \sum_i (y_i - \hat{\mu})^2 \tag{4.255}$$

We notice that the MLE for μ has the same form as the posterior mean, m_N.

We can solve these equations by fixed point iteration. Let us initialize by estimating $\lambda_x = 1/s_x^2$ and $\lambda_y = 1/s_y^2$, where $s_x^2 = \frac{1}{N_x} \sum_{i=1}^{N_x} (x_i - \overline{x})^2 = 0.16$ and $s_y^2 = \frac{1}{N_y} \sum_{i=1}^{N_y} (y_i - \overline{y})^2 = 0.36$. Using this, we get $\hat{\mu} = 2.1154$, so $p(\mu | \mathcal{D}, \hat{\lambda}_x, \hat{\lambda}_y) = \mathcal{N}(\mu | 2.1154, 0.0554)$. If we now iterate, we converge to $\hat{\lambda}_x = 1/0.1662$, $\hat{\lambda}_y = 1/4.0509$, $p(\mu | \mathcal{D}, \hat{\lambda}_x, \hat{\lambda}_y) = \mathcal{N}(\mu | 1.5788, 0.0798)$.

The plug-in approximation to the posterior is plotted in Figure 4.20(a). This weights each sensor according to its estimated precision. Since sensor y was estimated to be much less reliable than sensor x, we have $\mathbb{E}\left[\mu | \mathcal{D}, \hat{\lambda}_x, \hat{\lambda}_y\right] \approx \overline{x}$, so we effectively ignore the y sensor.

Now we will adopt a Bayesian approach and integrate out the unknown precisions, rather than trying to estimate them. That is, we compute

$$p(\mu | \mathcal{D}) \propto p(\mu) \left[\int p(\mathcal{D}_x | \mu, \lambda_x) p(\lambda_x | \mu) d\lambda_x \right] \left[\int p(\mathcal{D}_y | \mu, \lambda_y) p(\lambda_y | \mu) d\lambda_y \right] \tag{4.256}$$

We will use uninformative Jeffrey's priors, $p(\mu) \propto 1$, $p(\lambda_x | \mu) \propto 1/\lambda_x$ and $p(\lambda_y | \mu) \propto 1/\lambda_y$.

Since the x and y terms are symmetric, we will just focus on one of them. The key integral is

$$I = \int p(\mathcal{D}_x | \mu, \lambda_x) p(\lambda_x | \mu) d\lambda_x \quad \propto \quad \int \lambda_x^{-1} (N_x \lambda_x)^{N_x/2} \tag{4.257}$$

$$\exp\left(-\frac{N_x}{2} \lambda_x (\overline{x} - \mu)^2 - \frac{N_x}{2} s_x^2 \lambda_x \right) d\lambda_x \tag{4.258}$$

Exploiting the fact that $N_x = 2$ this simplifies to

$$I = \int \lambda_x^{-1} \lambda_x^1 \exp(-\lambda_x [(\overline{x} - \mu)^2 + s_x^2]) d\lambda_x \tag{4.259}$$

We recognize this as proportional to the integral of an unnormalized Gamma density

$$\text{Ga}(\lambda | a, b) \propto \lambda^{a-1} e^{-\lambda b} \tag{4.260}$$

where $a = 1$ and $b = (\overline{x} - \mu)^2 + s_x^2$. Hence the integral is proportional to the normalizing constant of the Gamma distribution, $\Gamma(a) b^{-a}$, so we get

$$I \propto \int p(\mathcal{D}_x | \mu, \lambda_x) p(\lambda_x | \mu) d\lambda_x \quad \propto \quad \left((\overline{x} - \mu)^2 + s_x^2 \right)^{-1} \tag{4.261}$$

and the posterior becomes

$$p(\mu | \mathcal{D}) \propto \frac{1}{(\overline{x} - \mu)^2 + s_x^2} \; \frac{1}{(\overline{y} - \mu)^2 + s_y^2} \tag{4.262}$$

The exact posterior is plotted in Figure 4.20(b). We see that it has two modes, one near $\overline{x} = 1.5$ and one near $\overline{y} = 3.5$. These correspond to the beliefs that the x sensor is more reliable than the y one, and vice versa. The weight of the first mode is larger, since the data from the x sensor agree more with each other, so it seems slightly more likely that the x sensor is the reliable one. (They obviously cannot both be reliable, since they disagree on the values that they are reporting.) However, the Bayesian solution keeps open the possibility that the y sensor is the more reliable one; from two measurements, we cannot tell, and choosing just the x sensor, as the plug-in approximation does, results in overconfidence (a posterior that is too narrow).

Exercises

Exercise 4.1 Uncorrelated does not imply independent

Let $X \sim U(-1, 1)$ and $Y = X^2$. Clearly Y is dependent on X (in fact, Y is uniquely determined by X). However, show that $\rho(X, Y) = 0$. Hint: if $X \sim U(a, b)$ then $E[X] = (a + b)/2$ and $\text{var}[X] = (b - a)^2/12$.

Exercise 4.2 Uncorrelated and Gaussian does not imply independent unless *jointly* Gaussian

Let $X \sim \mathcal{N}(0, 1)$ and $Y = WX$, where $p(W = -1) = p(W = 1) = 0.5$. It is clear that X and Y are not independent, since Y is a function of X.

a. Show $Y \sim \mathcal{N}(0, 1)$.

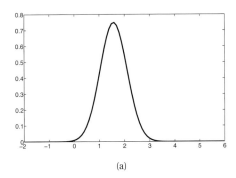 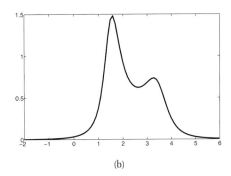

(a) (b)

Figure 4.20 Posterior for μ. (a) Plug-in approximation. (b) Exact posterior. Figure generated by `sensorFusionUnknownPrec`.

b. Show $\operatorname{cov}[X, Y] = 0$. Thus X and Y are uncorrelated but dependent, even though they are Gaussian. Hint: use the definition of covariance

$$\operatorname{cov}[X, Y] = \mathbb{E}[XY] - \mathbb{E}[X]\mathbb{E}[Y] \tag{4.263}$$

and the **rule of iterated expectation**

$$\mathbb{E}[XY] = \mathbb{E}[\mathbb{E}[XY|W]] \tag{4.264}$$

Exercise 4.3 Correlation coefficient is between -1 and +1

Prove that $-1 \leq \rho(X, Y) \leq 1$

Exercise 4.4 Correlation coefficient for linearly related variables is ± 1

Show that, if $Y = aX + b$ for some parameters $a > 0$ and b, then $\rho(X, Y) = 1$. Similarly show that if $a < 0$, then $\rho(X, Y) = -1$.

Exercise 4.5 Normalization constant for a multidimensional Gaussian

Prove that the normalization constant for a d-dimensional Gaussian is given by

$$(2\pi)^{d/2}|\boldsymbol{\Sigma}|^{\frac{1}{2}} = \int \exp(-\frac{1}{2}(\mathbf{x} - \mu)^T\boldsymbol{\Sigma}^{-1}(\mathbf{x} - \boldsymbol{\mu}))d\mathbf{x} \tag{4.265}$$

Hint: diagonalize $\boldsymbol{\Sigma}$ and use the fact that $|\boldsymbol{\Sigma}| = \prod_i \lambda_i$ to write the joint pdf as a product of d one-dimensional Gaussians in a transformed coordinate system. (You will need the change of variables formula.) Finally, use the normalization constant for univariate Gaussians.

Exercise 4.6 Bivariate Gaussian

Let $\mathbf{x} \sim \mathcal{N}(\boldsymbol{\mu}, \boldsymbol{\Sigma})$ where $\mathbf{x} \in \mathbb{R}^2$ and

$$\boldsymbol{\Sigma} = \begin{pmatrix} \sigma_1^2 & \rho\sigma_1\sigma_2 \\ \rho\sigma_1\sigma_2 & \sigma_2^2 \end{pmatrix} \tag{4.266}$$

where ρ is the correlation coefficient. Show that the pdf is given by

$$p(x_1, x_2) = \frac{1}{2\pi\sigma_1\sigma_2\sqrt{1 - \rho^2}} \exp\left(-\frac{1}{2(1 - \rho^2)}\right. \tag{4.267}$$

$$\left.\left[\frac{(x_1 - \mu_1)^2}{\sigma_1^2} + \frac{(x_2 - \mu_2)^2}{\sigma_2^2} - 2\rho\frac{(x_1 - \mu_1)}{\sigma_1}\frac{(x_2 - \mu_2)}{\sigma_2}\right]\right) \tag{4.268}$$

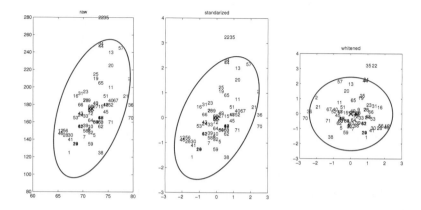

Figure 4.21 (a) Height/weight data for the men. (b) Standardized. (c) Whitened.

Exercise 4.7 Conditioning a bivariate Gaussian

Consider a bivariate Gaussian distribution $p(x_1, x_2) = \mathcal{N}(x|\mu, \Sigma)$ where

$$\Sigma = \begin{pmatrix} \sigma_1^2 & \sigma_{12} \\ \sigma_{21} & \sigma_2^2 \end{pmatrix} = \sigma_1 \sigma_2 \begin{pmatrix} \frac{\sigma_1}{\sigma_2} & \rho \\ \rho & \frac{\sigma_2}{\sigma_1} \end{pmatrix} \tag{4.269}$$

where the correlation coefficient is given by

$$\rho \triangleq \frac{\sigma_{12}}{\sigma_1 \sigma_2} \tag{4.270}$$

a. What is $P(X_2|x_1)$? Simplify your answer by expressing it in terms of ρ, σ_2, σ_1, μ_1, μ_2 and x_1.

b. Assume $\sigma_1 = \sigma_2 = 1$. What is $P(X_2|x_1)$ now?

Exercise 4.8 Whitening vs standardizing

a. Load the height/weight data using `rawdata = dlmread('heightWeightData.txt')`. The first column is the class label (1=male, 2=female), the second column is height, the third weight. Extract the height/weight data corresponding to the males. Fit a 2d Gaussian to the male data, using the empirical mean and covariance. Plot your Gaussian as an ellipse (use `gaussPlot2d`), superimposing on your scatter plot. It should look like Figure 4.21(a), where we have labeled each datapoint by its index. Turn in your figure and code.

b. **Standardizing** the data means ensuring the empirical variance along each dimension is 1. This can be done by computing $\frac{x_{ij} - \overline{x}_j}{\sigma_j}$, where σ_j is the empirical std of dimension j. Standardize the data and replot. It should look like Figure 4.21(b). (Use `axis('equal')`.) Turn in your figure and code.

c. **Whitening** or **sphereing** the data means ensuring its empirical covariance matrix is proportional to **I**, so the data is uncorrelated and of equal variance along each dimension. This can be done by computing $\Lambda^{-\frac{1}{2}} U^T x$ for each data vector **x**, where **U** are the eigenvectors and Λ the eigenvalues of the covariance matrix $X^T X$. Whiten the data and replot. It should look like Figure 4.21(c). Note that whitening rotates the data, so people move to counter-intuitive locations in the new coordinate system (see e.g., person 2, who moves from the right hand side to the left).

Exercise 4.9 Sensor fusion with known variances in 1d

Suppose we have two sensors with known (and different) variances v_1 and v_2, but unknown (and the same) mean μ. Suppose we observe n_1 observations $y_i^{(1)} \sim \mathcal{N}(\mu, v_1)$ from the first sensor and n_2 observations

$y_i^{(2)} \sim \mathcal{N}(\mu, v_2)$ from the second sensor. (For example, suppose μ is the true temperature outside, and sensor 1 is a precise (low variance) digital thermosensing device, and sensor 2 is an imprecise (high variance) mercury thermometer.) Let $\mathcal{D}$ represent all the data from both sensors. What is the posterior $p(\mu|\mathcal{D})$, assuming a non-informative prior for μ (which we can simulate using a Gaussian with a precision of 0)? Give an explicit expression for the posterior mean and variance.

Exercise 4.10 Derivation of information form formulae for marginalizing and conditioning

Derive the information form results of Section 4.3.1.

Exercise 4.11 Derivation of the NIW posterior

Derive Equation 4.209. Hint: one can show that

$$N(\overline{\mathbf{x}} - \boldsymbol{\mu})(\overline{\mathbf{x}} - \boldsymbol{\mu})^T + \kappa_0(\boldsymbol{\mu} - \mathbf{m}_0)(\boldsymbol{\mu} - \mathbf{m}_0)^T \tag{4.271}$$

$$= \kappa_N(\boldsymbol{\mu} - \mathbf{m}_N)(\boldsymbol{\mu} - \mathbf{m}_N)^T + \frac{\kappa_0 N}{\kappa_N}(\overline{\mathbf{x}} - \mathbf{m}_0)(\overline{\mathbf{x}} - \mathbf{m}_0)^T \tag{4.272}$$

This is a matrix generalization of an operation called **completing the square**.[6]

Derive the corresponding result for the normal-Wishart model.

Exercise 4.12 BIC for Gaussians

(Source: Jaakkola.)

The Bayesian information criterion (BIC) is a penalized log-likelihood function that can be used for model selection (see Section 5.3.2.4). It is defined as

$$BIC = \log p(\mathcal{D}|\hat{\boldsymbol{\theta}}_{ML}) - \frac{d}{2}\log(N) \tag{4.273}$$

where d is the number of free parameters in the model and N is the number of samples. In this question, we will see how to use this to choose between a full covariance Gaussian and a Gaussian with a diagonal covariance. Obviously a full covariance Gaussian has higher likelihood, but it may not be "worth" the extra parameters if the improvement over a diagonal covariance matrix is too small. So we use the BIC score to choose the model.

Following Section 4.1.3, we can write

$$\log p(\mathcal{D}|\hat{\boldsymbol{\Sigma}}, \hat{\boldsymbol{\mu}}) = -\frac{N}{2}\mathrm{tr}\left(\hat{\boldsymbol{\Sigma}}^{-1}\hat{\mathbf{S}}\right) - \frac{N}{2}\log(|\hat{\boldsymbol{\Sigma}}|) \tag{4.274}$$

$$\hat{\mathbf{S}} = \frac{1}{N}\sum_{i=1}^{N}(\mathbf{x}_i - \overline{\mathbf{x}})(\mathbf{x}_i - \overline{\mathbf{x}})^T \tag{4.275}$$

where $\hat{\mathbf{S}}$ is the scatter matrix (empirical covariance), the trace of a matrix is the sum of its diagonals, and we have used the trace trick.

a. Derive the BIC score for a Gaussian in D dimensions with full covariance matrix. Simplify your answer as much as possible, exploiting the form of the MLE. Be sure to specify the number of free parameters d.

b. Derive the BIC score for a Gaussian in D dimensions with a *diagonal* covariance matrix. Be sure to specify the number of free parameters d. Hint: for the digaonal case, the ML estimate of $\boldsymbol{\Sigma}$ is the same as $\hat{\boldsymbol{\Sigma}}_{ML}$ except the off-diagonal terms are zero:

$$\hat{\boldsymbol{\Sigma}}_{diag} = \mathrm{diag}(\hat{\boldsymbol{\Sigma}}_{ML}(1,1), \ldots, \hat{\boldsymbol{\Sigma}}_{ML}(D,D)) \tag{4.276}$$

---

6. In the scalar case, completing the square means rewriting $c_2 x^2 + c_1 x + c_0$ as $-a(x-b)^2 + w$ where $a = -c_2$, $b = \frac{c_1}{2c_2}$ and $w = \frac{c_1^2}{4c_2} + c_0$.

Exercise 4.13 Gaussian posterior credible interval

(Source: DeGroot.)

Let $X \sim \mathcal{N}(\mu, \sigma^2 = 4)$ where μ is unknown but has prior $\mu \sim \mathcal{N}(\mu_0, \sigma_0^2 = 9)$. The posterior after seeing n samples is $\mu \sim \mathcal{N}(\mu_n, \sigma_n^2)$. (This is called a credible interval, and is the Bayesian analog of a confidence interval.) How big does n have to be to ensure

$$p(\ell \leq \mu_n \leq u|D) \geq 0.95 \tag{4.277}$$

where (ℓ, u) is an interval (centered on μ_n) of width 1 and D is the data? Hint: recall that 95% of the probability mass of a Gaussian is within $\pm 1.96\sigma$ of the mean.

Exercise 4.14 MAP estimation for 1D Gaussians

(Source: Jaakkola.)

Consider samples $x_1, \ldots, x_n$ from a Gaussian random variable with known variance σ^2 and unknown mean μ. We further assume a prior distribution (also Gaussian) over the mean, $\mu \sim \mathcal{N}(m, s^2)$, with fixed mean m and fixed variance s^2. Thus the only unknown is μ.

a. Calculate the MAP estimate $\hat{\mu}_{MAP}$. You can state the result without proof. Alternatively, with a lot more work, you can compute derivatives of the log posterior, set to zero and solve.

b. Show that as the number of samples n increase, the MAP estimate converges to the maximum likelihood estimate.

c. Suppose n is small and fixed. What does the MAP estimator converge to if we increase the prior variance s^2?

d. Suppose n is small and fixed. What does the MAP estimator converge to if we decrease the prior variance s^2?

Exercise 4.15 Sequential (recursive) updating of $\hat{\Sigma}$

(Source: (Duda et al. 2001, Q3.35,3.36).)

The unbiased estimate for the covariance of a d-dimensional Gaussian based on n samples is given by

$$\hat{\Sigma} = \mathbf{C}_n = \frac{1}{n-1} \sum_{i=1}^{n} (\mathbf{x}_i - \mathbf{m}_n)(\mathbf{x}_i - \mathbf{m}_n)^T \tag{4.278}$$

It is clear that it takes $O(nd^2)$ time to compute $\mathbf{C}_n$. If the data points arrive one at a time, it is more efficient to incrementally update these estimates than to recompute from scratch.

a. Show that the covariance can be sequentially updated as follows

$$\mathbf{C}_{n+1} = \frac{n-1}{n}\mathbf{C}_n + \frac{1}{n+1}(\mathbf{x}_{n+1} - \mathbf{m}_n)(\mathbf{x}_{n+1} - \mathbf{m}_n)^T \tag{4.279}$$

b. How much time does it take per sequential update? (Use big-O notation.)

c. Show that we can sequentially update the precision matrix using

$$\mathbf{C}_{n+1}^{-1} = \frac{n}{n-1}\left[\mathbf{C}_n^{-1} - \frac{\mathbf{C}_n^{-1}(\mathbf{x}_{n+1} - \mathbf{m}_n)(\mathbf{x}_{n+1} - \mathbf{m}_n)^T\mathbf{C}_n^{-1}}{\frac{n^2-1}{n} + (\mathbf{x}_{n+1} - \mathbf{m}_n)^T\mathbf{C}_n^{-1}(\mathbf{x}_{n+1} - \mathbf{m}_n)} \right] \tag{4.280}$$

Hint: notice that the update to $\mathbf{C}_{n+1}$ consists of adding a rank-one matrix, namely $\mathbf{u}\mathbf{u}^T$, where $\mathbf{u} = \mathbf{x}_{n+1} - \mathbf{m}_n$. Use the matrix inversion lemma for rank-one updates (Equation 4.111), which we repeat here for convenience:

$$(\mathbf{E} + \mathbf{u}\mathbf{v}^T)^{-1} = \mathbf{E}^{-1} - \frac{\mathbf{E}^{-1}\mathbf{u}\mathbf{v}^T\mathbf{E}^{-1}}{1 + \mathbf{v}^T\mathbf{E}^{-1}\mathbf{u}} \tag{4.281}$$

d. What is the time complexity per update?

Exercise 4.16 Likelihood ratio for Gaussians

(Source: Alpaydin p103 ex 4.). Consider a binary classifier where the K class-conditional densities are MVN $p(x|y = j) = \mathcal{N}(x|\mu_j, \Sigma_j)$. By Bayes rule, we have

$$\log \frac{p(y = 1|x)}{p(y = 0|x)} = \log \frac{p(x|y = 1)}{p(x|y = 0)} + \log \frac{p(y = 1)}{p(y = 0)} \tag{4.282}$$

In other words, the log posterior ratio is the log likelihood ratio plus the log prior ratio. For each of the 4 cases in the table below, derive an expression for the log likelihood ratio $\log \frac{p(x|y=1)}{p(x|y=0)}$, simplifying as much as possible.

Form of Σ_j	Cov	Num parameters
Arbitrary	Σ_j	$Kd(d+1)/2$
Shared	$\Sigma_j = \Sigma$	$d(d+1)/2$
Shared, axis-aligned	$\Sigma_j = \Sigma$ with $\Sigma_{ij} = 0$ for $i \neq j$	d
Shared, spherical	$\Sigma_j = \sigma^2 I$	1

Exercise 4.17 LDA/QDA on height/weight data

The function `discrimAnalysisHeightWeightDemo` fits an LDA and QDA model to the height/weight data. Compute the misclassification rate of both of these models on the training set. Turn in your numbers and code.

Exercise 4.18 Naive Bayes with mixed features

Consider a 3-class naive Bayes classifier with one binary feature and one Gaussian feature:

$$y \sim \text{Mu}(y|\boldsymbol{\pi}, 1), \quad x_1|y = c \sim \text{Ber}(x_1|\theta_c), \quad x_2|y = c \sim \mathcal{N}(x_2|\mu_c, \sigma_c^2) \tag{4.283}$$

Let the parameter vectors be as follows:

$$\boldsymbol{\pi} = (0.5, 0.25, 0.25), \quad \boldsymbol{\theta} = (0.5, 0.5, 0.5), \quad \boldsymbol{\mu} = (-1, 0, 1), \quad \boldsymbol{\sigma}^2 = (1, 1, 1) \tag{4.284}$$

a. Compute $p(y|x_1 = 0, x_2 = 0)$ (the result should be a vector of 3 numbers that sums to 1).
b. Compute $p(y|x_1 = 0)$.
c. Compute $p(y|x_2 = 0)$.
d. Explain any interesting patterns you see in your results. Hint: look at the parameter vector θ.

Exercise 4.19 Decision boundary for LDA with semi-tied covariances

Consider a generative classifier with class-conditional densities of the form $\mathcal{N}(\mathbf{x}|\boldsymbol{\mu}_c, \boldsymbol{\Sigma}_c)$. In LDA, we assume $\boldsymbol{\Sigma}_c = \boldsymbol{\Sigma}$, and in QDA, each $\boldsymbol{\Sigma}_c$ is arbitrary. Here we consider the 2-class case in which $\boldsymbol{\Sigma}_1 = k\boldsymbol{\Sigma}_0$, for $k > 1$. That is, the Gaussian ellipsoids have the same "shape", but the one for class 1 is "wider". Derive an expression for $p(y = 1|\mathbf{x}, \boldsymbol{\theta})$, simplifying as much as possible. Give a geometric interpretation of your result, if possible.

Exercise 4.20 Logistic regression vs LDA/QDA

(Source: Jaakkola.) Suppose we train the following binary classifiers via maximum likelihood.

a. GaussI: A generative classifier, where the class-conditional densities are Gaussian, with both covariance matrices set to $\mathbf{I}$ (identity matrix), i.e., $p(\mathbf{x}|y = c) = \mathcal{N}(\mathbf{x}|\boldsymbol{\mu}_c, \mathbf{I})$. We assume $p(y)$ is uniform.
b. GaussX: as for GaussI, but the covariance matrices are unconstrained, i.e., $p(\mathbf{x}|y = c) = \mathcal{N}(\mathbf{x}|\boldsymbol{\mu}_c, \boldsymbol{\Sigma}_c)$.

c. LinLog: A logistic regression model with linear features.

d. QuadLog: A logistic regression model, using linear and quadratic features (i.e., polynomial basis function expansion of degree 2).

After training we compute the performance of each model M on the training set as follows:

$$L(M) = \frac{1}{n} \sum_{i=1}^{n} \log p(y_i | \mathbf{x}_i, \hat{\boldsymbol{\theta}}, M) \tag{4.285}$$

(Note that this is the *conditional* log-likelihood $p(y|\mathbf{x}, \hat{\boldsymbol{\theta}})$ and not the joint log-likelihood $p(y, \mathbf{x}|\hat{\boldsymbol{\theta}})$.) We now want to compare the performance of each model. We will write $L(M) \leq L(M')$ if model M *must* have lower (or equal) log likelihood (on the training set) than M', for any training set (in other words, M is worse than M', at least as far as training set logprob is concerned). For each of the following model pairs, state whether $L(M) \leq L(M')$, $L(M) \geq L(M')$, or whether no such statement can be made (i.e., M might sometimes be better than M' and sometimes worse); also, for each question, briefly (1-2 sentences) explain why.

a. GaussI, LinLog.

b. GaussX, QuadLog.

c. LinLog, QuadLog.

d. GaussI, QuadLog.

e. Now suppose we measure performance in terms of the average misclassification rate on the training set:

$$R(M) = \frac{1}{n} \sum_{i=1}^{n} I(y_i \neq \hat{y}(\mathbf{x}_i)) \tag{4.286}$$

Is it true in general that $L(M) > L(M')$ implies that $R(M) < R(M')$? Explain why or why not.

Exercise 4.21 Gaussian decision boundaries

(Source: (Duda et al. 2001, Q3.7).) Let $p(x|y = j) = \mathcal{N}(x|\mu_j, \sigma_j)$ where $j = 1, 2$ and $\mu_1 = 0, \sigma_1^2 = 1, \mu_2 = 1, \sigma_2^2 = 10^6$. Let the class priors be equal, $p(y = 1) = p(y = 2) = 0.5$.

a. Find the decision region

$$R_1 = \{x : p(x|\mu_1, \sigma_1) \geq p(x|\mu_2, \sigma_2)\} \tag{4.287}$$

Sketch the result. Hint: draw the curves and find where they intersect. Find *both* solutions of the equation

$$p(x|\mu_1, \sigma_1) = p(x|\mu_2, \sigma_2) \tag{4.288}$$

Hint: recall that to solve a quadratic equation $ax^2 + bx + c = 0$, we use

$$x = \frac{-b \pm \sqrt{b^2 - 4ac}}{2a} \tag{4.289}$$

b. Now suppose $\sigma_2 = 1$ (and all other parameters remain the same). What is R_1 in this case?

Exercise 4.22 QDA with 3 classes

Consider a three category classification problem. Let the prior probabilites:

$$P(Y = 1) = P(Y = 2) = P(Y = 3) = 1/3 \qquad (4.290)$$

The class-conditional densities are multivariate normal densities with parameters:

$$\mu_1 = [0, 0]^T, \mu_2 = [1, 1]^T, \mu_3 = [-1, 1]^T \qquad (4.291)$$

$$\Sigma_1 = \begin{bmatrix} 0.7 & 0 \\ 0 & 0.7 \end{bmatrix}, \Sigma_2 = \begin{bmatrix} 0.8 & 0.2 \\ 0.2 & 0.8 \end{bmatrix}, \Sigma_3 = \begin{bmatrix} 0.8 & 0.2 \\ 0.2 & 0.8 \end{bmatrix} \qquad (4.292)$$

Classify the following points:

a. $\mathbf{x} = [-0.5, 0.5]$

b. $\mathbf{x} = [0.5, 0.5]$

Exercise 4.23 Scalar QDA

[Note: you can solve this exercise by hand or using a computer (matlab, R, whatever). In either case, show your work.] Consider the following training set of heights x (in inches) and gender y (male/female) of some US college students: $\mathbf{x} = (67, 79, 71, 68, 67, 60)$, $\mathbf{y} = (m, m, m, f, f, f)$.

a. Fit a Bayes classifier to this data, using maximum likelihood estimation, i.e., estimate the parameters of the class-conditional likelihoods

$$p(x|y = c) = \mathcal{N}(x; \mu_c, \sigma_c) \qquad (4.293)$$

and the class prior

$$p(y = c) = \pi_c \qquad (4.294)$$

What are your values of μ_c, σ_c, π_c for $c = m, f$? Show your work (so you can get partial credit if you make an arithmetic error).

b. Compute $p(y = m|x, \hat{\theta})$, where $x = 72$, and $\hat{\theta}$ are the MLE parameters. (This is called a plug-in prediction.)

c. What would be a simple way to extend this technique if you had multiple attributes per person, such as height and weight? Write down your proposed model as an equation.

5 *Bayesian statistics*

5.1 Introduction

We have now seen a variety of different probability models, and we have discussed how to fit them to data, i.e., we have discussed how to compute MAP parameter estimates $\hat{\boldsymbol{\theta}} = \operatorname{argmax} p(\boldsymbol{\theta}|\mathcal{D})$, using a variety of different priors. We have also discussed how to compute the full posterior $p(\boldsymbol{\theta}|\mathcal{D})$, as well as the posterior predictive density, $p(\mathbf{x}|\mathcal{D})$, for certain special cases (and in later chapters, we will discuss algorithms for the general case).

Using the posterior distribution to summarize everything we know about a set of unknown variables is at the core of **Bayesian statistics**. In this chapter, we discuss this approach to statistics in more detail. In Chapter 6, we discuss an alternative approach to statistics known as frequentist or classical statistics.

5.2 Summarizing posterior distributions

The posterior $p(\boldsymbol{\theta}|\mathcal{D})$ summarizes everything we know about the unknown quantities $\boldsymbol{\theta}$. In this section, we discuss some simple quantities that can be derived from a probability distribution, such as a posterior. These summary statistics are often easier to understand and visualize than the full joint.

5.2.1 MAP estimation

We can easily compute a **point estimate** of an unknown quantity by computing the posterior mean, median or mode. In Section 5.7, we discuss how to use decision theory to choose between these methods. Typically the posterior mean or median is the most appropriate choice for a real-valued quantity, and the vector of posterior marginals is the best choice for a discrete quantity. However, the posterior mode, aka the MAP estimate, is the most popular choice because it reduces to an optimization problem, for which efficient algorithms often exist. Futhermore, MAP estimation can be interpreted in non-Bayesian terms, by thinking of the log prior as a regularizer (see Section 6.5 for more details).

Although this approach is computationally appealing, it is important to point out that there are various drawbacks to MAP estimation, which we briefly discuss below. This will provide motivation for the more thoroughly Bayesian approach which we will study later in this chapter (and elsewhere in this book).

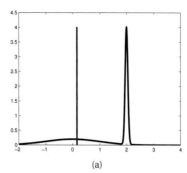
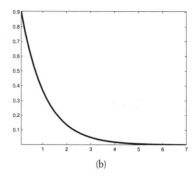

(a)   (b)

Figure 5.1   (a) A bimodal distribution in which the mode is very untypical of the distribution. The thin blue vertical line is the mean, which is arguably a better summary of the distribution, since it is near the majority of the probability mass. Figure generated by `bimodalDemo`. (b) A skewed distribution in which the mode is quite different from the mean. Figure generated by `gammaPlotDemo`.

5.2.1.1   No measure of uncertainty

The most obvious drawback of MAP estimation, and indeed of any other **point estimate** such as the posterior mean or median, is that it does not provide any measure of uncertainty. In many applications, it is important to know how much one can trust a given estimate. We can derive such confidence measures from the posterior, as we discuss in Section 5.2.2.

5.2.1.2   Plugging in the MAP estimate can result in overfitting

In machine learning, we often care more about predictive accuracy than in interpreting the parameters of our models. However, if we don't model the uncertainty in our parameters, then our predictive distribution will be overconfident. We saw several examples of this in Chapter 3, and we will see more examples later. Overconfidence in predictions is particularly problematic in situations where we may be risk averse; see Section 5.7 for details.

5.2.1.3   The mode is an untypical point

The mode of a posterior distribution is often a very poor choice as a summary, since the mode is usually quite untypical of the distribution, unlike the mean or median. This is illustrated in Figure 5.1(a) for a 1D continuous space. The basic problem is that the mode is a single point, the mean and median take the volume of the space into account. Another example is shown in Figure 5.1(b): here the mode is 0, but the mean is non-zero. Such skewed distributions often arise when inferring variance parameters, especially in hierarchical models. In such cases the MAP estimate (and hence the MLE) is obviously a very bad estimate.

How should we summarize a posterior if the mode is not a good choice? The answer is to use decision theory, which we discuss in Section 5.7. The basic idea is to specify a loss function, where $L(\theta, \hat{\theta})$ is the loss you incur if the truth is θ and your estimate is $\hat{\theta}$. If we use 0-1 loss, $L(\theta, \hat{\theta}) = \mathbb{I}(\theta \neq \hat{\theta})$, then the optimal estimate is the posterior mode. 0-1 loss means you only get "points" if you make no errors, otherwise you get nothing: there is no "partial credit" under

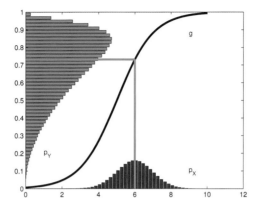

Figure 5.2 Example of the transformation of a density under a nonlinear transform. Note how the mode of the transformed distribution is not the transform of the original mode. Based on Exercise 1.4 of (Bishop 2006). Figure generated by `bayesChangeOfVar`.

this loss function! For continuous-valued quantities, we often prefer to use squared error loss, $L(\theta, \hat{\theta}) = (\theta - \hat{\theta})^2$; the corresponding optimal estimator is then the posterior mean, as we show in Section 5.7. Or we can use a more robust loss function, $L(\theta, \hat{\theta}) = |\theta - \hat{\theta}|$, which gives rise to the posterior median.

5.2.1.4 MAP estimation is not invariant to reparameterization *

A more subtle problem with MAP estimation is that the result we get depends on how we parameterize the probability distribution. Changing from one representation to another equivalent representation changes the result, which is not very desirable, since the units of measurement are arbitrary (e.g., when measuring distance, we can use centimetres or inches).

To understand the problem, suppose we compute the posterior for x. If we define $y = f(x)$, the distribution for y is given by Equation 2.87, which we repeat here for convenience:

$$p_y(y) = p_x(x)\left|\frac{dx}{dy}\right| \tag{5.1}$$

The $\left|\frac{dx}{dy}\right|$ term is called the Jacobian, and it measures the change in size of a unit volume passed through f. Let $\hat{x} = \operatorname{argmax}_x p_x(x)$ be the MAP estimate for x. In general it is not the case that $\hat{y} = \operatorname{argmax}_y p_y(y)$ is given by $f(\hat{x})$. For example, let $x \sim \mathcal{N}(6, 1)$ and $y = f(x)$, where

$$f(x) = \frac{1}{1 + \exp(-x + 5)} \tag{5.2}$$

We can derive the distribution of y using Monte Carlo simulation (see Section 2.7.1). The result is shown in Figure 5.2. We see that the original Gaussian has become "squashed" by the sigmoid nonlinearity. In particular, we see that the mode of the transformed distribution is not equal to the transform of the original mode.

To see how this problem arises in the context of MAP estimation, consider the following example, due to Michael Jordan. The Bernoulli distribution is typically parameterized by its mean μ, so $p(y = 1|\mu) = \mu$, where $y \in \{0, 1\}$. Suppose we have a uniform prior on the unit interval: $p_\mu(\mu) = 1 \; \mathbb{I}(0 \leq \mu \leq 1)$. If there is no data, the MAP estimate is just the mode of the prior, which can be anywhere between 0 and 1. We will now show that different parameterizations can pick different points in this interval arbitrarily.

First let $\theta = \sqrt{\mu}$ so $\mu = \theta^2$. The new prior is

$$p_\theta(\theta) = p_\mu(\mu)\Big|\frac{d\mu}{d\theta}\Big| = 2\theta \tag{5.3}$$

for $\theta \in [0, 1]$ so the new mode is

$$\hat{\theta}_{MAP} = \arg\max_{\theta \in [0,1]} 2\theta = 1 \tag{5.4}$$

Now let $\phi = 1 - \sqrt{1 - \mu}$. The new prior is

$$p_\phi(\phi) = p_\mu(\mu)\Big|\frac{d\mu}{d\phi}\Big| = 2(1 - \phi) \tag{5.5}$$

for $\phi \in [0, 1]$, so the new mode is

$$\hat{\phi}_{MAP} = \arg\max_{\phi \in [0,1]} 2 - 2\phi = 0 \tag{5.6}$$

Thus the MAP estimate depends on the parameterization. The MLE does not suffer from this since the likelihood is a function, not a probability density. Bayesian inference does not suffer from this problem either, since the change of measure is taken into account when integrating over the parameter space.

One solution to the problem is to optimize the following objective function:

$$\hat{\boldsymbol{\theta}} = \underset{\boldsymbol{\theta}}{\mathrm{argmax}}\, p(\mathcal{D}|\boldsymbol{\theta})p(\boldsymbol{\theta})|\mathbf{I}(\boldsymbol{\theta})|^{-\frac{1}{2}} \tag{5.7}$$

Here $\mathbf{I}(\boldsymbol{\theta})$ is the Fisher information matrix associated with $p(\mathbf{x}|\boldsymbol{\theta})$ (see Section 6.2.2). This estimate is parameterization independent, for reasons explained in (Jermyn 2005; Druilhet and Marin 2007). Unfortunately, optimizing Equation 5.7 is often difficult, which reduces the appeal of the whole approach.

5.2.2 Credible intervals

In addition to point estimates, we often want a measure of confidence. A standard measure of confidence in some (scalar) quantity θ is the "width" of its posterior distribution. This can be measured using a $100(1 - \alpha)\%$ **credible interval**, which is a (contiguous) region $C = (\ell, u)$ (standing for lower and upper) which contains $1 - \alpha$ of the posterior probability mass, i.e.,

$$C_\alpha(\mathcal{D}) = (\ell, u) : P(\ell \leq \theta \leq u|\mathcal{D}) = 1 - \alpha \tag{5.8}$$

There may be many such intervals, so we choose one such that there is $(1 - \alpha)/2$ mass in each tail; this is called a **central interval**.

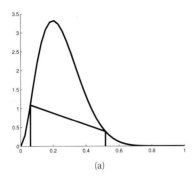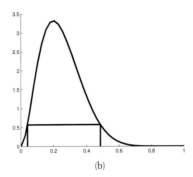

(a)   (b)

Figure 5.3   (a) Central interval and (b) HPD region for a Beta(3,9) posterior. The CI is (0.06, 0.52) and the HPD is (0.04, 0.48). Based on Figure 3.6 of (Hoff 2009). Figure generated by betaHPD.

If the posterior has a known functional form, we can compute the posterior central interval using $\ell = F^{-1}(\alpha/2)$ and $u = F^{-1}(1-\alpha/2)$, where F is the cdf of the posterior. For example, if the posterior is Gaussian, $p(\theta|\mathcal{D}) = \mathcal{N}(0, 1)$, and $\alpha = 0.05$, then we have $\ell = \Phi(\alpha/2) = -1.96$, and $u = \Phi(1 - \alpha/2) = 1.96$, where Φ denotes the cdf of the Gaussian. This is illustrated in Figure 2.3(c). This justifies the common practice of quoting a credible interval in the form of $\mu \pm 2\sigma$, where μ represents the posterior mean, σ represents the posterior standard deviation, and 2 is a good approximation to 1.96.

Of course, the posterior is not always Gaussian. For example, in our coin example, if we use a uniform prior and we observe $N_1 = 47$ heads out of $N = 100$ trials, then the posterior is a beta distribution, $p(\theta|\mathcal{D}) = \text{Beta}(48, 54)$. We find the 95% posterior credible interval is $(0.3749, 0.5673)$ (see betaCredibleInt for the one line of Matlab code we used to compute this).

If we don't know the functional form, but we can draw samples from the posterior, then we can use a Monte Carlo approximation to the posterior quantiles: we simply sort the S samples, and find the one that occurs at location α/S along the sorted list. As $S \to \infty$, this converges to the true quantile. See mcQuantileDemo for a demo.

People often confuse Bayesian credible intervals with frequentist confidence intervals. However, they are not the same thing, as we discuss in Section 6.6.1. In general, credible intervals are usually what people want to compute, but confidence intervals are usually what they actually compute, because most people are taught frequentist statistics but not Bayesian statistics. Fortunately, the mechanics of computing a credible interval are just as easy as computing a confidence interval (see e.g., betaCredibleInt for how to do it in Matlab).

5.2.2.1   Highest posterior density regions *

A problem with central intervals is that there might be points outside the CI which have higher probability density. This is illustrated in Figure 5.3(a), where we see that points outside the left-most CI boundary have higher density than those just inside the right-most CI boundary.

This motivates an alternative quantity known as the **highest posterior density** or **HPD** region. This is defined as the (set of) most probable points that in total constitute $100(1 - \alpha)\%$ of the

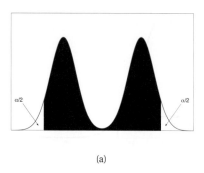
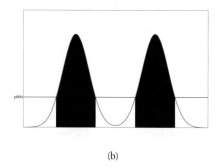

(a)   (b)

Figure 5.4 (a) Central interval and (b) HPD region for a hypothetical multimodal posterior. Based on Figure 2.2 of (Gelman et al. 2004). Figure generated by `postDensityIntervals`.

probability mass. More formally, we find the threshold p^* on the pdf such that

$$1 - \alpha = \int_{\theta:p(\theta|\mathcal{D})>p^*} p(\theta|\mathcal{D})d\theta \tag{5.9}$$

and then define the HPD as

$$C_\alpha(\mathcal{D}) = \{\theta : p(\theta|\mathcal{D}) \geq p^*\} \tag{5.10}$$

In 1d, the HPD region is sometimes called a **highest density interval** or **HDI**. For example, Figure 5.3(b) shows the 95% HDI of a $\mathrm{Beta}(3,9)$ distribution, which is $(0.04, 0.48)$. We see that this is narrower than the CI, even though it still contains 95% of the mass; furthermore, every point inside of it has higher density than every point outside of it.

For a unimodal distribution, the HDI will be the narrowest interval around the mode containing 95% of the mass. To see this, imagine "water filling" in reverse, where we lower the level until 95% of the mass is revealed, and only 5% is submerged. This gives a simple algorithm for computing HDIs in the 1d case: simply search over points such that the interval contains 95% of the mass and has minimal width. This can be done by 1d numerical optimization if we know the inverse CDF of the distribution, or by search over the sorted data points if we have a bag of samples (see `betaHPD` for a demo).

If the posterior is multimodal, the HDI may not even be a connected region: see Figure 5.4(b) for an example. However, summarizing multimodal posteriors is always difficult.

5.2.3   Inference for a difference in proportions

Sometimes we have multiple parameters, and we are interested in computing the posterior distribution of some function of these parameters. For example, suppose you are about to buy something from Amazon.com, and there are two sellers offering it for the same price. Seller 1 has 90 positive reviews and 10 negative reviews. Seller 2 has 2 positive reviews and 0 negative reviews. Who should you buy from?[1]

---

1. This example is from `www.johndcook.com/blog/2011/09/27/bayesian-amazon`. See also `lingpipe-blog.com/2009/10/13/bayesian-counterpart-to-fisher-exact-test-on-contingency-tables`.

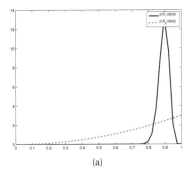
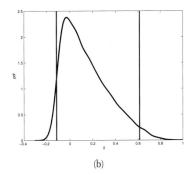

(a)   (b)

Figure 5.5   (a) Exact posteriors $p(\theta_i|\mathcal{D}_i)$. (b) Monte Carlo approximation to $p(\delta|\mathcal{D})$. We use kernel density estimation to get a smooth plot. The vertical lines enclose the 95% central interval. Figure generated by `amazonSellerDemo`.

On the face of it, you should pick seller 2, but we cannot be very confident that seller 2 is better since it has had so few reviews. In this section, we sketch a Bayesian analysis of this problem. Similar methodology can be used to compare rates or proportions across groups for a variety of other settings.

Let θ_1 and θ_2 be the unknown reliabilities of the two sellers. Since we don't know much about them, we'll endow them both with uniform priors, $\theta_i \sim \text{Beta}(1,1)$. The posteriors are $p(\theta_1|\mathcal{D}_1) = \text{Beta}(91, 11)$ and $p(\theta_2|\mathcal{D}_2) = \text{Beta}(3, 1)$.

We want to compute $p(\theta_1 > \theta_2|\mathcal{D})$. For convenience, let us define $\delta = \theta_1 - \theta_2$ as the difference in the rates. (Alternatively we might want to work in terms of the log-odds ratio.) We can compute the desired quantity using numerical integration:

$$
\begin{aligned}
p(\delta > 0|\mathcal{D}) &= \int_0^1 \int_0^1 \mathbb{I}(\theta_1 > \theta_2)\text{Beta}(\theta_1|y_1 + 1, N_1 - y_1 + 1) \\
&\quad \text{Beta}(\theta_2|y_2 + 1, N_2 - y_2 + 1)d\theta_1 d\theta_2
\end{aligned}
\tag{5.11}
$$

We find $p(\delta > 0|\mathcal{D}) = 0.710$, which means you are better off buying from seller 1! See `amazonSellerDemo` for the code. (It is also possible to solve the integral analytically (Cook 2005).)

A simpler way to solve the problem is to approximate the posterior $p(\delta|\mathcal{D})$ by Monte Carlo sampling. This is easy, since θ_1 and θ_2 are independent in the posterior, and both have beta distributions, which can be sampled from using standard methods. The distributions $p(\theta_i|\mathcal{D}_i)$ are shown in Figure 5.5(a), and a MC approximation to $p(\delta|\mathcal{D})$, together with a 95% HPD, is shown Figure 5.5(b). An MC approximation to $p(\delta > 0|\mathcal{D})$ is obtained by counting the fraction of samples where $\theta_1 > \theta_2$; this turns out to be 0.718, which is very close to the exact value. (See `amazonSellerDemo` for the code.)

5.3   Bayesian model selection

In Figure 1.18, we saw that using too high a degree polynomial results in overfitting, and using too low a degree results in underfitting. Similarly, in Figure 7.8(a), we will see that using too small a regularization parameter results in overfitting, and too large a value results in underfitting. In general, when faced with a set of models (i.e., families of parametric distributions) of different complexity, how should we choose the best one? This is called the **model selection** problem.

One approach is to use cross-validation to estimate the generalization error of all the candidate models, and then to pick the model that seems the best. However, this requires fitting each model K times, where K is the number of CV folds. A more efficient approach is to compute the posterior over models,

$$p(m|\mathcal{D}) = \frac{p(\mathcal{D}|m)p(m)}{\sum_{m\in\mathcal{M}} p(m,\mathcal{D})} \tag{5.12}$$

From this, we can easily compute the MAP model, $\hat{m} = \operatorname{argmax} p(m|\mathcal{D})$. This is called **Bayesian model selection**.

If we use a uniform prior over models, $p(m) \propto 1$, this amounts to picking the model which maximizes

$$p(\mathcal{D}|m) = \int p(\mathcal{D}|\boldsymbol{\theta})p(\boldsymbol{\theta}|m)d\boldsymbol{\theta} \tag{5.13}$$

This quantity is called the **marginal likelihood**, the **integrated likelihood**, or the **evidence** for model m. The details on how to perform this integral will be discussed in Section 5.3.2. But first we give an intuitive interpretation of what this quantity means.

5.3.1   Bayesian Occam's razor

One might think that using $p(\mathcal{D}|m)$ to select models would always favor the model with the most parameters. This is true if we use $p(\mathcal{D}|\hat{\boldsymbol{\theta}}_m)$ to select models, where $\hat{\boldsymbol{\theta}}_m$ is the MLE or MAP estimate of the parameters for model m, because models with more parameters will fit the data better, and hence achieve higher likelihood. However, if we integrate out the parameters, rather than maximizing them, we are automatically protected from overfitting: models with more parameters do not necessarily have higher *marginal* likelihood. This is called the **Bayesian Occam's razor** effect (MacKay 1995b; Murray and Ghahramani 2005), named after the principle known as **Occam's razor**, which says one should pick the simplest model that adequately explains the data.

One way to understand the Bayesian Occam's razor is to notice that the marginal likelihood can be rewritten as follows, based on the chain rule of probability (Equation 2.5):

$$p(\mathcal{D}) = p(y_1)p(y_2|y_1)p(y_3|y_{1:2})\dots p(y_N|y_{1:N-1}) \tag{5.14}$$

where we have dropped the conditioning on $\mathbf{x}$ for brevity. This is similar to a leave-one-out cross-validation estimate (Section 1.4.8) of the likelihood, since we predict each future point given all the previous ones. (Of course, the order of the data does not matter in the above expression.) If a model is too complex, it will overfit the "early" examples and will then predict the remaining ones poorly.

Another way to understand the Bayesian Occam's razor effect is to note that probabilities must sum to one. Hence $\sum_{\mathcal{D}'} p(\mathcal{D}'|m) = 1$, where the sum is over all possible data sets. Complex models, which can predict many things, must spread their probability mass thinly, and hence will not obtain as large a probability for any given data set as simpler models. This is sometimes

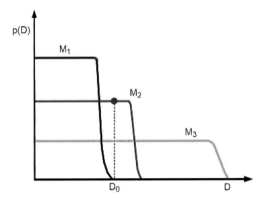

Figure 5.6 A schematic illustration of the Bayesian Occam's razor. The broad (green) curve corresponds to a complex model, the narrow (blue) curve to a simple model, and the middle (red) curve is just right. Based on Figure 3.13 of (Bishop 2006). See also (Murray and Ghahramani 2005, Figure 2) for a similar plot produced on real data.

called the **conservation of probability mass** principle, and is illustrated in Figure 5.6. On the horizontal axis we plot all possible data sets in order of increasing complexity (measured in some abstract sense). On the vertical axis we plot the predictions of 3 possible models: a simple one, M_1; a medium one, M_2; and a complex one, M_3. We also indicate the actually observed data $\mathcal{D}_0$ by a vertical line. Model 1 is too simple and assigns low probability to $\mathcal{D}_0$. Model 3 also assigns $\mathcal{D}_0$ relatively low probability, because it can predict many data sets, and hence it spreads its probability quite widely and thinly. Model 2 is "just right": it predicts the observed data with a reasonable degree of confidence, but does not predict too many other things. Hence model 2 is the most probable model.

As a concrete example of the Bayesian Occam's razor, consider the data in Figure 5.7. We plot polynomials of degrees 1, 2 and 3 fit to $N = 5$ data points. It also shows the posterior over models, where we use a Gaussian prior (see Section 7.6 for details). There is not enough data to justify a complex model, so the MAP model is $d = 1$. Figure 5.8 shows what happens when $N = 30$. Now it is clear that $d = 2$ is the right model (the data was in fact generated from a quadratic).

As another example, Figure 7.8(c) plots $\log p(\mathcal{D}|\lambda)$ vs $\log(\lambda)$, for the polynomial ridge regression model, where λ ranges over the same set of values used in the CV experiment. We see that the maximum evidence occurs at roughly the same point as the minimum of the test MSE, which also corresponds to the point chosen by CV.

When using the Bayesian approach, we are not restricted to evaluating the evidence at a finite grid of values. Instead, we can use numerical optimization to find $\lambda^* = \mathrm{argmax}_\lambda\, p(\mathcal{D}|\lambda)$. This technique is called **empirical Bayes** or **type II maximum likelihood** (see Section 5.6 for details). An example is shown in Figure 7.8(b): we see that the curve has a similar shape to the CV estimate, but it can be computed more efficiently.

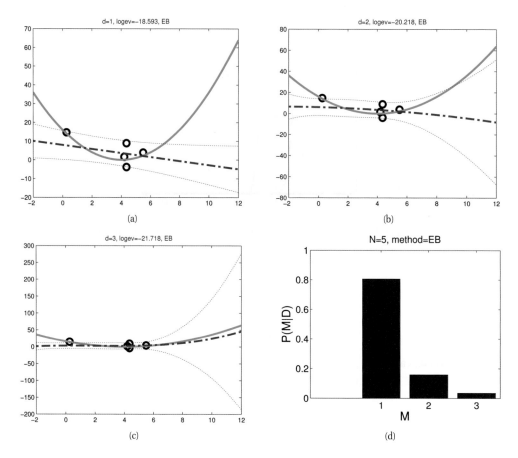

Figure 5.7 (a-c) We plot polynomials of degrees 1, 2 and 3 fit to $N = 5$ data points using empirical Bayes. The solid green curve is the true function, the dashed red curve is the prediction (dotted blue lines represent $\pm\sigma$ around the mean). (d) We plot the posterior over models, $p(d|\mathcal{D})$, assuming a uniform prior $p(d) \propto 1$. Based on a figure by Zoubin Ghahramani. Figure generated by `linregEbModelSelVsN`.

5.3.2   Computing the marginal likelihood (evidence)

When discussing parameter inference for a fixed model, we often wrote

$$p(\boldsymbol{\theta}|\mathcal{D}, m) \propto p(\boldsymbol{\theta}|m)p(\mathcal{D}|\boldsymbol{\theta}, m) \tag{5.15}$$

thus ignoring the normalization constant $p(\mathcal{D}|m)$. This is valid since $p(\mathcal{D}|m)$ is constant wrt $\boldsymbol{\theta}$. However, when comparing models, we need to know how to compute the marginal likelihood, $p(\mathcal{D}|m)$. In general, this can be quite hard, since we have to integrate over all possible parameter values, but when we have a conjugate prior, it is easy to compute, as we now show.

Let $p(\boldsymbol{\theta}) = q(\boldsymbol{\theta})/Z_0$ be our prior, where $q(\boldsymbol{\theta})$ is an unnormalized distribution, and Z_0 is the normalization constant of the prior. Let $p(\mathcal{D}|\boldsymbol{\theta}) = q(\mathcal{D}|\boldsymbol{\theta})/Z_\ell$ be the likelihood, where Z_ℓ contains any constant factors in the likelihood. Finally let $p(\boldsymbol{\theta}|\mathcal{D}) = q(\boldsymbol{\theta}|\mathcal{D})/Z_N$ be our poste-

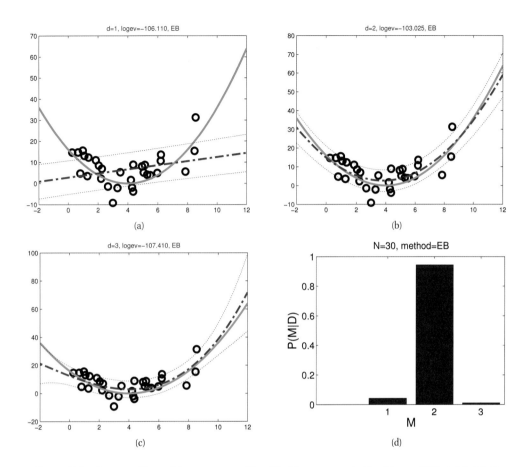

Figure 5.8 Same as Figure 5.7 except now $N = 30$. Figure generated by `linregEbModelSelVsN`.

rior, where $q(\boldsymbol{\theta}|\mathcal{D}) = q(\mathcal{D}|\boldsymbol{\theta})q(\boldsymbol{\theta})$ is the unnormalized posterior, and Z_N is the normalization constant of the posterior. We have

$$p(\boldsymbol{\theta}|\mathcal{D}) = \frac{p(\mathcal{D}|\boldsymbol{\theta})p(\boldsymbol{\theta})}{p(\mathcal{D})} \tag{5.16}$$

$$\frac{q(\boldsymbol{\theta}|\mathcal{D})}{Z_N} = \frac{q(\mathcal{D}|\boldsymbol{\theta})q(\boldsymbol{\theta})}{Z_\ell Z_0 p(\mathcal{D})} \tag{5.17}$$

$$p(\mathcal{D}) = \frac{Z_N}{Z_0 Z_\ell} \tag{5.18}$$

So assuming the relevant normalization constants are tractable, we have an easy way to compute the marginal likelihood. We give some examples below.

5.3.2.1   Beta-binomial model

Let us apply the above result to the Beta-binomial model. Since we know $p(\theta|\mathcal{D}) = \text{Beta}(\theta|a', b')$, where $a' = a + N_1$ and $b' = b + N_0$, we know the normalization constant of the posterior is $B(a', b')$. Hence

$$p(\theta|\mathcal{D}) = \frac{p(\mathcal{D}|\theta)p(\theta)}{p(\mathcal{D})} \tag{5.19}$$

$$= \frac{1}{p(\mathcal{D})} \left[\frac{1}{B(a,b)} \theta^{a-1}(1-\theta)^{b-1} \right] \left[\binom{N}{N_1} \theta^{N_1}(1-\theta)^{N_0} \right] \tag{5.20}$$

$$= \binom{N}{N_1} \frac{1}{p(\mathcal{D})} \frac{1}{B(a,b)} \left[\theta^{a+N_1-1}(1-\theta)^{b+N_0-1} \right] \tag{5.21}$$

So

$$\frac{1}{B(a+N_1, b+N_0)} = \binom{N}{N_1} \frac{1}{p(\mathcal{D})} \frac{1}{B(a,b)} \tag{5.22}$$

$$p(\mathcal{D}) = \binom{N}{N_1} \frac{B(a+N_1, b+N_0)}{B(a,b)} \tag{5.23}$$

The marginal likelihood for the Beta-Bernoulli model is the same as above, except it is missing the $\binom{N}{N_1}$ term.

5.3.2.2   Dirichlet-multinoulli model

By the same reasoning as the Beta-Bernoulli case, one can show that the marginal likelihood for the Dirichlet-multinoulli model is given by

$$p(\mathcal{D}) = \frac{B(\mathbf{N} + \boldsymbol{\alpha})}{B(\boldsymbol{\alpha})} \tag{5.24}$$

where

$$B(\boldsymbol{\alpha}) = \frac{\prod_{k=1}^{K} \Gamma(\alpha_k)}{\Gamma(\sum_k \alpha_k)} \tag{5.25}$$

Hence we can rewrite the above result in the following form, which is what is usually presented in the literature:

$$p(\mathcal{D}) = \frac{\Gamma(\sum_k \alpha_k)}{\Gamma(N + \sum_k \alpha_k)} \prod_k \frac{\Gamma(N_k + \alpha_k)}{\Gamma(\alpha_k)} \tag{5.26}$$

We will see many applications of this equation later.

5.3.2.3   Gaussian-Wishart-Gaussian model

Consider the case of an MVN likelihood with a conjugate NIW prior (this is known as the Gaussian-Wishart-Gaussian model, because we combine a Gaussian-Wishart prior with a Gaussian likelihood). Let Z_0 be the normalizer for the prior, Z_N be normalizer for the posterior, and let $Z_l = (2\pi)^{ND/2}$ be the normalizer for the likelihood. Then it is easy to see that

$$p(\mathcal{D}) = \frac{Z_N}{Z_0 Z_l} \tag{5.27}$$

$$= \frac{1}{\pi^{ND/2}} \frac{1}{2^{ND/2}} \frac{\left(\frac{2\pi}{\kappa_N}\right)^{D/2} |\mathbf{S}_N|^{-\nu_N/2} 2^{(\nu_0+N)D/2} \Gamma_D(\nu_N/2)}{\left(\frac{2\pi}{\kappa_0}\right)^{D/2} |\mathbf{S}_0|^{-\nu_0/2} 2^{\nu_0 D/2} \Gamma_D(\nu_0/2)} \tag{5.28}$$

$$= \frac{1}{\pi^{ND/2}} \left(\frac{\kappa_0}{\kappa_N}\right)^{D/2} \frac{|\mathbf{S}_0|^{\nu_0/2}}{|\mathbf{S}_N|^{\nu_N/2}} \frac{\Gamma_D(\nu_N/2)}{\Gamma_D(\nu_0/2)} \tag{5.29}$$

This equation will prove useful later, such as when we consider mixtures of Gaussian models.[2]

5.3.2.4 BIC approximation to log marginal likelihood

In general, computing the integral in Equation 5.13 can be quite difficult. One simple but popular approximation is known as the **Bayesian information criterion** or **BIC**, which has the following form (Schwarz 1978):

$$\text{BIC} \triangleq \log p(\mathcal{D}|\hat{\boldsymbol{\theta}}) - \frac{\text{dof}(\hat{\boldsymbol{\theta}})}{2} \log N \approx \log p(\mathcal{D}) \tag{5.30}$$

where $\text{dof}(\hat{\boldsymbol{\theta}})$ is the number of **degrees of freedom** in the model, and $\hat{\boldsymbol{\theta}}$ is the MLE for the model.[3] We see that this has the form of a **penalized log likelihood**, where the penalty term depends on the model's complexity. See Section 8.4.2 for the derivation of the BIC score.

As an example, consider linear regression. As we show in Section 7.3, the MLE is given by $\hat{\mathbf{w}} = (\mathbf{X}^T\mathbf{X})^{-1}\mathbf{X}^T\mathbf{y}$ and $\hat{\sigma}^2 = \text{RSS}/N$, where $\text{RSS} = \sum_{i=1}^N (y_i - \hat{\mathbf{w}}_{mle}^T \mathbf{x}_i)^2$. The corresponding log likelihood is given by

$$\log p(\mathcal{D}|\hat{\boldsymbol{\theta}}) = -\frac{N}{2} \log(2\pi\hat{\sigma}^2) - \frac{N}{2} \tag{5.31}$$

Hence the BIC score is as follows (dropping constant terms)

$$\text{BIC} = -\frac{N}{2} \log(\hat{\sigma}^2) - \frac{D}{2} \log(N) \tag{5.32}$$

where D is the number of variables in the model. In the statistics literature, it is common to use an alternative definition of BIC, which we call the BIC *cost* (since we want to minimize it):

$$\text{BIC-cost} \triangleq -2\log p(\mathcal{D}|\hat{\boldsymbol{\theta}}) + \text{dof}(\hat{\boldsymbol{\theta}}) \log N \approx -2\log p(\mathcal{D}) \tag{5.33}$$

In the context of linear regression, this becomes

$$\text{BIC-cost} = N \log(\hat{\sigma}^2) + D \log(N) \tag{5.34}$$

---

2. It is often useful to consider ratios of marginal likelihoods, known as Bayes factors. In this case, a lot of the computational effort of scoring two models can be shared. For details, see http://www.stats.ox.ac.uk/~teh/res earch/notes/GaussianInverseWishart.pdf

3. Traditionally the BIC score is defined using the ML estimate $\hat{\boldsymbol{\theta}}$, so it is independent of the prior. However, for models such as mixtures of Gaussians, the ML estimate can be poorly behaved, so it is better to evaluate the BIC score using the MAP estimate, as in (Fraley and Raftery 2007).

The BIC method is very closely related to the **minimum description length** or **MDL** principle, which characterizes the score for a model in terms of how well it fits the data, minus how complex the model is to define. See (Hansen and Yu 2001) for details.

There is a very similar expression to BIC / MDL called the **Akaike information criterion** or **AIC**, defined as

$$\text{AIC}(m, \mathcal{D}) \triangleq \log p(\mathcal{D}|\hat{\boldsymbol{\theta}}_{MLE}) - \text{dof}(m) \tag{5.35}$$

This is derived from a frequentist framework, and cannot be interpreted as an approximation to the marginal likelihood. Nevertheless, the form of this expression is very similar to BIC. We see that the penalty for AIC is less than for BIC. This causes AIC to pick more complex models. However, this can result in better predictive accuracy. See e.g., (Clarke et al. 2009, sec 10.2) for further discussion on such information criteria.

5.3.2.5   Effect of the prior

Sometimes it is not clear how to set the prior. When we are performing posterior inference, the details of the prior may not matter too much, since the likelihood often overwhelms the prior anyway. But when computing the marginal likelihood, the prior plays a much more important role, since we are averaging the likelihood over all possible parameter settings, as weighted by the prior.

In Figures 5.7 and 5.8, where we demonstrated model selection for linear regression, we used a prior of the form $p(\mathbf{w}) = \mathcal{N}(\mathbf{0}, \alpha^{-1}\mathbf{I})$. Here α is a tuning parameter that controls how strong the prior is. This parameter can have a large effect, as we discuss in Section 7.5. Intuitively, if α is large, the weights are "forced" to be small, so we need to use a complex model with many small parameters (e.g., a high degree polynomial) to fit the data. Conversely, if α is small, we will favor simpler models, since each parameter is "allowed" to vary in magnitude by a lot.

If the prior is unknown, the correct Bayesian procedure is to put a prior on the prior. That is, we should put a prior on the hyper-parameter α as well as the parameters $\mathbf{w}$. To compute the marginal likelihood, we should integrate out all unknowns, i.e., we should compute

$$p(\mathcal{D}|m) = \int \int p(\mathcal{D}|\mathbf{w})p(\mathbf{w}|\alpha, m)p(\alpha|m)d\mathbf{w}d\alpha \tag{5.36}$$

Of course, this requires specifying the hyper-prior. Fortunately, the higher up we go in the Bayesian hierarchy, the less sensitive are the results to the prior settings. So we can usually make the hyper-prior uninformative.

A computational shortcut is to optimize α rather than integrating it out. That is, we use

$$p(\mathcal{D}|m) \approx \int p(\mathcal{D}|\mathbf{w})p(\mathbf{w}|\hat{\alpha}, m)d\mathbf{w} \tag{5.37}$$

where

$$\hat{\alpha} = \underset{\alpha}{\text{argmax}}\, p(\mathcal{D}|\alpha, m) = \underset{\alpha}{\text{argmax}} \int p(\mathcal{D}|\mathbf{w})p(\mathbf{w}|\alpha, m)d\mathbf{w} \tag{5.38}$$

This approach is called empirical Bayes (EB), and is discussed in more detail in Section 5.6. This is the method used in Figures 5.7 and 5.8.

Bayes factor $BF(1,0)$	Interpretation
$BF < \frac{1}{100}$	Decisive evidence for M_0
$BF < \frac{1}{10}$	Strong evidence for M_0
$\frac{1}{10} < BF < \frac{1}{3}$	Moderate evidence for M_0
$\frac{1}{3} < BF < 1$	Weak evidence for M_0
$1 < BF < 3$	Weak evidence for M_1
$3 < BF < 10$	Moderate evidence for M_1
$BF > 10$	Strong evidence for M_1
$BF > 100$	Decisive evidence for M_1

Table 5.1 Jeffreys scale of evidence for interpreting Bayes factors.

5.3.3 Bayes factors

Suppose our prior on models is uniform, $p(m) \propto 1$. Then model selection is equivalent to picking the model with the highest marginal likelihood. Now suppose we just have two models we are considering, call them the **null hypothesis**, M_0, and the **alternative hypothesis**, M_1. Define the **Bayes factor** as the ratio of marginal likelihoods:

$$BF_{1,0} \triangleq \frac{p(\mathcal{D}|M_1)}{p(\mathcal{D}|M_0)} = \frac{p(M_1|\mathcal{D})}{p(M_0|\mathcal{D})} / \frac{p(M_1)}{p(M_0)} \tag{5.39}$$

(This is like a **likelihood ratio**, except we integrate out the parameters, which allows us to compare models of different complexity.) If $BF_{1,0} > 1$ then we prefer model 1, otherwise we prefer model 0.

Of course, it might be that $BF_{1,0}$ is only slightly greater than 1. In that case, we are not very confident that model 1 is better. Jeffreys (1961) proposed a scale of evidence for interpreting the magnitude of a Bayes factor, which is shown in Table 5.1. This is a Bayesian alternative to the frequentist concept of a p-value (see Section 6.6.2). Alternatively, we can just convert the Bayes factor to a posterior over models. If $p(M_1) = p(M_0) = 0.5$, we have

$$p(M_0|\mathcal{D}) = \frac{BF_{0,1}}{1 + BF_{0,1}} = \frac{1}{BF_{1,0} + 1} \tag{5.40}$$

5.3.3.1 Example: Testing if a coin is fair

Suppose we observe some coin tosses, and want to decide if the data was generated by a fair coin, $\theta = 0.5$, or a potentially biased coin, where θ could be any value in $[0, 1]$. Let us denote the first model by M_0 and the second model by M_1. The marginal likelihood under M_0 is simply

$$p(\mathcal{D}|M_0) = \left(\frac{1}{2}\right)^N \tag{5.41}$$

where N is the number of coin tosses. The marginal likelihood under M_1, using a Beta prior, is

$$p(\mathcal{D}|M_1) \quad = \quad \int p(\mathcal{D}|\theta)p(\theta)d\theta = \frac{B(\alpha_1 + N_1, \alpha_0 + N_0)}{B(\alpha_1, \alpha_0)} \tag{5.42}$$

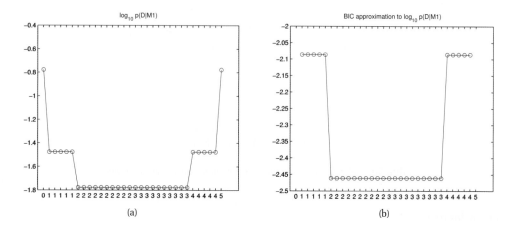

Figure 5.9   (a) Log marginal likelihood for the coins example. (b) BIC approximation. Figure generated by `coinsModelSelDemo`.

We plot $\log p(\mathcal{D}|M_1)$ vs the number of heads N_1 in Figure 5.9(a), assuming $N = 5$ and $\alpha_1 = \alpha_0 = 1$. (The shape of the curve is not very sensitive to α_1 and α_0, as long as $\alpha_0 = \alpha_1$.) If we observe 2 or 3 heads, the unbiased coin hypothesis M_0 is more likely than M_1, since M_0 is a simpler model (it has no free parameters) — it would be a suspicious coincidence if the coin were biased but happened to produce almost exactly 50/50 heads/tails. However, as the counts become more extreme, we favor the biased coin hypothesis. Note that, if we plot the log Bayes factor, $\log BF_{1,0}$, it will have exactly the same shape, since $\log p(\mathcal{D}|M_0)$ is a constant. See also Exercise 3.18.

Figure 5.9(b) shows the BIC approximation to $\log p(\mathcal{D}|M_1)$ for our biased coin example from Section 5.3.3.1. We see that the curve has approximately the same shape as the exact log marginal likelihood, which is all that matters for model selection purposes, since the absolute scale is irrelevant. In particular, it favors the simpler model unless the data is overwhelmingly in support of the more complex model.

5.3.4   Jeffreys-Lindley paradox *

Problems can arise when we use improper priors (i.e., priors that do not integrate to 1) for model selection / hypothesis testing, even though such priors may be acceptable for other purposes. For example, consider testing the hypotheses $M_0 : \theta \in \Theta_0$ vs $M_1 : \theta \in \Theta_1$. To define the marginal density on θ, we use the following mixture model

$$p(\theta) \quad = \quad p(\theta|M_0)p(M_0) + p(\theta|M_1)p(M_1) \tag{5.43}$$

This is only meaningful if $p(\theta|M_0)$ and $p(\theta|M_1)$ are proper (normalized) density functions. In this case, the posterior is given by

$$p(M_0|\mathcal{D}) = \frac{p(M_0)p(\mathcal{D}|M_0)}{p(M_0)p(\mathcal{D}|M_0) + p(M_1)p(\mathcal{D}|M_1)} \quad (5.44)$$

$$= \frac{p(M_0)\int_{\Theta_0} p(\mathcal{D}|\theta)p(\theta|M_0)d\theta}{p(M_0)\int_{\Theta_0} p(\mathcal{D}|\theta)p(\theta|M_0)d\theta + p(M_1)\int_{\Theta_1} p(\mathcal{D}|\theta)p(\theta|M_1)d\theta} \quad (5.45)$$

Now suppose we use improper priors, $p(\theta|M_0) \propto c_0$ and $p(\theta|M_1) \propto c_1$. Then

$$p(M_0|\mathcal{D}) = \frac{p(M_0)c_0\int_{\Theta_0} p(\mathcal{D}|\theta)d\theta}{p(M_0)c_0\int_{\Theta_0} p(\mathcal{D}|\theta)d\theta + p(M_1)c_1\int_{\Theta_1} p(\mathcal{D}|\theta)d\theta} \quad (5.46)$$

$$= \frac{p(M_0)c_0\ell_0}{p(M_0)c_0\ell_0 + p(M_1)c_1\ell_1} \quad (5.47)$$

where $\ell_i = \int_{\Theta_i} p(\mathcal{D}|\theta)d\theta$ is the integrated or marginal likelihood for model i. Now let $p(M_0) = p(M_1) = \frac{1}{2}$. Hence

$$p(M_0|\mathcal{D}) = \frac{c_0\ell_0}{c_0\ell_0 + c_1\ell_1} = \frac{\ell_0}{\ell_0 + (c_1/c_0)\ell_1} \quad (5.48)$$

Thus we can change the posterior arbitrarily by choosing c_1 and c_0 as we please. Note that using proper, but very vague, priors can cause similar problems. In particular, the Bayes factor will always favor the simpler model, since the probability of the observed data under a complex model with a very diffuse prior will be very small. This is called the **Jeffreys-Lindley paradox**.

Thus it is important to use proper priors when performing model selection. Note, however, that, if M_0 and M_1 share the same prior over a subset of the parameters, this part of the prior can be improper, since the corresponding normalization constant will cancel out.

5.4 Priors

The most controversial aspect of Bayesian statistics is its reliance on priors. Bayesians argue this is unavoidable, since nobody is a **tabula rasa** or **blank slate**: all inference must be done conditional on certain assumptions about the world. Nevertheless, one might be interested in minimizing the impact of one's prior assumptions. We briefly discuss some ways to do this below.

5.4.1 Uninformative priors

If we don't have strong beliefs about what θ should be, it is common to use an **uninformative** or **non-informative** prior, and to "let the data speak for itself".

The issue of designing uninformative priors is actually somewhat tricky. As an example of the difficulty, consider a Bernoulli parameter, $\theta \in [0,1]$. One might think that the most uninformative prior would be the uniform distribution, $\text{Beta}(1,1)$. But the posterior mean in this case is $\mathbb{E}[\theta|\mathcal{D}] = \frac{N_1+1}{N_1+N_0+2}$, whereas the MLE is $\frac{N_1}{N_1+N_0}$. Hence one could argue that the prior wasn't completely uninformative after all.

Clearly by decreasing the magnitude of the pseudo counts, we can lessen the impact of the prior. By the above argument, the most non-informative prior is

$$\lim_{c \to 0} \text{Beta}(c, c) = \text{Beta}(0, 0) \tag{5.49}$$

which is a mixture of two equal point masses at 0 and 1 (see (Zhu and Lu 2004)). This is also called the **Haldane prior**. Note that the Haldane prior is an improper prior, meaning it does not integrate to 1. However, as long as we see at least one head and at least one tail, the posterior will be proper.

In Section 5.4.2.1 we will argue that the "right" uninformative prior is in fact $\text{Beta}(\frac{1}{2}, \frac{1}{2})$. Clearly the difference in practice between these three priors is very likely negligible. In general, it is advisable to perform some kind of **sensitivity analysis**, in which one checks how much one's conclusions or predictions change in response to change in the modeling assumptions, which includes the choice of prior, but also the choice of likelihood and any kind of data pre-processing. If the conclusions are relatively insensitive to the modeling assumptions, one can have more confidence in the results.

5.4.2   Jeffreys priors *

Recall the example of the uniform prior for the probability of heads. This means the probability of a heads could be anywhere between 0 and 1. But what about the probability of two heads in a row, which is given by $\phi = \theta^2$? That has prior distribution $p(\phi) = 1/(2\sqrt{\phi}) = \text{Beta}(0.5, 1)$, which is far from uniform.

Harold Jeffreys[4] designed a general purpose technique for creating non-informative priors. The result is known as the **Jeffreys prior**. The key observation is that if $p(\phi)$ is non-informative, then any re-parameterization of the prior, such as $\theta = h(\phi)$ for some function h, should also be non-informative. Now, by the change of variables formula,

$$p_\theta(\theta) = p_\phi(\phi) \left| \frac{d\phi}{d\theta} \right| \tag{5.50}$$

so the prior will in general change. However, let us pick

$$p_\phi(\phi) \propto (I(\phi))^{\frac{1}{2}} \tag{5.51}$$

where $I(\phi)$ is the **Fisher information**:

$$I(\phi) \triangleq -\mathbb{E}\left[\left(\frac{d \log p(X|\phi)}{d\phi} \right) 2 \right] \tag{5.52}$$

This is a measure of curvature of the expected negative log likelihood and hence a measure of stability of the MLE (see Section 6.2.2). Now

$$\frac{d \log p(x|\theta)}{d\theta} = \frac{d \log p(x|\phi)}{d\phi} \frac{d\phi}{d\theta} \tag{5.53}$$

---

4. Harold Jeffreys, 1891 – 1989, was an English mathematician, statistician, geophysicist, and astronomer.

Squaring and taking expectations over x, we have

$$I(\theta) = -\mathbb{E}\left[\left(\frac{d\log p(X|\theta)}{d\theta}\right)^2\right] = I(\phi)\left(\frac{d\phi}{d\theta}\right)^2 \tag{5.54}$$

$$I(\theta)^{\frac{1}{2}} = I(\phi)^{\frac{1}{2}}\left|\frac{d\phi}{d\theta}\right| \tag{5.55}$$

so we find the transformed prior is

$$p_\theta(\theta) = p_\phi(\phi)\left|\frac{d\phi}{d\theta}\right| \propto (I(\phi))^{\frac{1}{2}}\left|\frac{d\phi}{d\theta}\right| = I(\theta)^{\frac{1}{2}} \tag{5.56}$$

So $p_\theta(\theta)$ and $p_\phi(\phi)$ are the same.

Intuitively, we can understand the Jeffreys prior as follows: It puts less prior weight on parameter values where the likelihood function is "flat" (since the Fisher information measures curvature of the log likelihood), thus preventing the prior from having undue influence.

Below we give some examples of Jeffreys priors.

5.4.2.1   Example: Jeffreys prior for the Bernoulli and multinoulli

Suppose $X \sim \text{Ber}(\theta)$. The log likelihood for a single sample is

$$\log p(X|\theta) = X\log\theta + (1-X)\log(1-\theta) \tag{5.57}$$

The **score function** is just the gradient of the log-likelihood:

$$s(\theta) \triangleq \frac{d}{d\theta}\log p(X|\theta) = \frac{X}{\theta} - \frac{1-X}{1-\theta} \tag{5.58}$$

The **observed information** is the second derivative of the log-likelihood:

$$J(\theta) = -\frac{d^2}{d\theta^2}\log p(X|\theta) = -s'(\theta|X) = \frac{X}{\theta^2} + \frac{1-X}{(1-\theta)^2} \tag{5.59}$$

The Fisher information is the expected information:

$$I(\theta) = E[J(\theta|X)|X \sim \theta] = \frac{\theta}{\theta^2} + \frac{1-\theta}{(1-\theta)^2} = \frac{1}{\theta(1-\theta)} \tag{5.60}$$

Hence Jeffreys prior is

$$p(\theta) \propto \theta^{-\frac{1}{2}}(1-\theta)^{-\frac{1}{2}} = \frac{1}{\sqrt{\theta(1-\theta)}} \propto \text{Beta}(\frac{1}{2},\frac{1}{2}) \tag{5.61}$$

Now consider a multinoulli random variable with K states. One can show that the Jeffreys prior is given by

$$p(\boldsymbol{\theta}) \propto \text{Dir}(\frac{1}{2},\ldots,\frac{1}{2}) \tag{5.62}$$

Note that this is different from the more obvious choices of $\text{Dir}(\frac{1}{K},\ldots,\frac{1}{K})$ or $\text{Dir}(1,\ldots,1)$.

5.4.2.2   Example: Jeffreys prior for location and scale parameters

One can show that the Jeffreys prior for a location parameter, such as the Gaussian mean, is $p(\mu) \propto 1$. Thus is an example of a **translation invariant prior**, which satisfies the property that the probability mass assigned to any interval, $[A, B]$ is the same as that assigned to any other shifted interval of the same width, such as $[A - c, B - c]$. That is,

$$\int_{A-c}^{B-c} p(\mu)d\mu = (A - c) - (B - c) = (A - B) = \int_A^B p(\mu)d\mu \tag{5.63}$$

This can be achieved using $p(\mu) \propto 1$, which we can approximate by using a Gaussian with infinite variance, $p(\mu) = \mathcal{N}(\mu|0, \infty)$. Note that this is an **improper prior**, since it does not integrate to 1. Using improper priors is fine as long as the posterior is proper, which will be the case provided we have seen $N \geq 1$ data points, since we can "nail down" the location as soon as we have seen a single data point.

Similarly, one can show that the Jeffreys prior for a scale parameter, such as the Gaussian variance, is $p(\sigma^2) \propto 1/\sigma^2$. This is an example of a **scale invariant prior**, which satisfies the property that the probability mass assigned to any interval $[A, B]$ is the same as that assigned to any other interval $[A/c, B/c]$ which is scaled in size by some constant factor $c > 0$. (For example, if we change units from meters to feet we do not want that to affect our inferences.) This can be achieved by using

$$p(s) \propto 1/s \tag{5.64}$$

To see this, note that

$$\int_{A/c}^{B/c} p(s)ds \quad = \quad [\log s]_{A/c}^{B/c} = \log(B/c) - \log(A/c) \tag{5.65}$$

$$= \quad \log(B) - \log(A) = \int_A^B p(s)ds \tag{5.66}$$

We can approximate this using a degenerate Gamma distribution (Section 2.4.5), $p(s) = \text{Ga}(s|0, 0)$. The prior $p(s) \propto 1/s$ is also improper, but the posterior is proper as soon as we have seen $N \geq 2$ data points (since we need at least two data points to estimate a variance).

5.4.3   Robust priors

In many cases, we are not very confident in our prior, so we want to make sure it does not have an undue influence on the result. This can be done by using **robust priors** (Insua and Ruggeri 2000), which typically have heavy tails, which avoids forcing things to be too close to the prior mean.

Let us consider an example from (Berger 1985, p7). Suppose $x \sim \mathcal{N}(\theta, 1)$. We observe that $x = 5$ and we want to estimate θ. The MLE is of course $\hat{\theta} = 5$, which seems reasonable. The posterior mean under a uniform prior is also $\bar{\theta} = 5$. But now suppose we know that the prior median is 0, and the prior quantiles are at -1 and 1, so $p(\theta \leq -1) = p(-1 < \theta \leq 0) = p(0 < \theta \leq 1) = p(1 < \theta) = 0.25$. Let us also assume the prior is smooth and unimodal.

It is easy to show that a Gaussian prior of the form $\mathcal{N}(\theta|0, 2.19^2)$ satisfies these prior constraints. But in this case the posterior mean is given by 3.43, which doesn't seem very satisfactory.

Now suppose we use as a Cauchy prior $\mathcal{T}(\theta|0, 1, 1)$. This also satisfies the prior constraints of our example. But this time we find (using numerical method integration: see `robustPriorDemo` for the code) that the posterior mean is about 4.6, which seems much more reasonable.

5.4.4 Mixtures of conjugate priors

Robust priors are useful, but can be computationally expensive to use. Conjugate priors simplify the computation, but are often not robust, and not flexible enough to encode our prior knowledge. However, it turns out that a **mixture of conjugate priors** is also conjugate (Exercise 5.1), and can approximate any kind of prior (Dallal and Hall 1983; Diaconis and Ylvisaker 1985). Thus such priors provide a good compromise between computational convenience and flexibility.

For example, suppose we are modeling coin tosses, and we think the coin is either fair, or is biased towards heads. This cannot be represented by a beta distribution. However, we can model it using a mixture of two beta distributions. For example, we might use

$$p(\theta) = 0.5\ \mathrm{Beta}(\theta|20, 20) + 0.5\ \mathrm{Beta}(\theta|30, 10) \tag{5.67}$$

If θ comes from the first distribution, the coin is fair, but if it comes from the second, it is biased towards heads.

We can represent a mixture by introducing a latent indicator variable z, where $z = k$ means that θ comes from mixture component k. The prior has the form

$$p(\theta) = \sum_k p(z = k)p(\theta|z = k) \tag{5.68}$$

where each $p(\theta|z = k)$ is conjugate, and $p(z = k)$ are called the (prior) mixing weights. One can show (Exercise 5.1) that the posterior can also be written as a mixture of conjugate distributions as follows:

$$p(\theta|\mathcal{D}) = \sum_k p(z = k|\mathcal{D})p(\theta|\mathcal{D}, z = k) \tag{5.69}$$

where $p(Z = k|\mathcal{D})$ are the posterior mixing weights given by

$$p(Z = k|\mathcal{D}) = \frac{p(Z = k)p(\mathcal{D}|Z = k)}{\sum_{k'} p(Z = k')p(\mathcal{D}|Z = k')} \tag{5.70}$$

Here the quantity $p(\mathcal{D}|Z = k)$ is the marginal likelihood for mixture component k (see Section 5.3.2.1).

5.4.4.1 Example

Suppose we use the mixture prior

$$p(\theta) = 0.5\mathrm{Beta}(\theta|a_1, b_1) + 0.5\mathrm{Beta}(\theta|a_2, b_2) \tag{5.71}$$

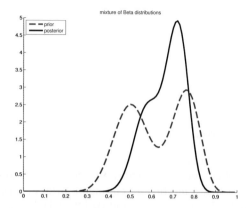

Figure 5.10   A mixture of two Beta distributions. Figure generated by `mixBetaDemo`.

where $a_1 = b_1 = 20$ and $a_2 = b_2 = 10$. and we observe N_1 heads and N_0 tails. The posterior becomes

$$p(\theta|\mathcal{D}) = p(Z = 1|\mathcal{D})\text{Beta}(\theta|a_1 + N_1, b_1 + N_0) + p(Z = 2|\mathcal{D})\text{Beta}(\theta|a_2 + N_1, b_2 + N_0) \quad (5.72)$$

If $N_1 = 20$ heads and $N_0 = 10$ tails, then, using Equation 5.23, the posterior becomes

$$p(\theta|\mathcal{D}) = 0.346\,\text{Beta}(\theta|40, 30) + 0.654\,\text{Beta}(\theta|50, 20) \quad (5.73)$$

See Figure 5.10 for an illustration.

5.4.4.2   Application: Finding conserved regions in DNA and protein sequences

We mentioned that Dirichlet-multinomial models are widely used in biosequence analysis. Let us give a simple example to illustrate some of the machinery that has developed. Specifically, consider the sequence logo discussed in Section 2.3.2.1. Now suppose we want to find locations which represent coding regions of the genome. Such locations often have the same letter across all sequences, because of evolutionary pressure. So we need to find columns which are "pure", or nearly so, in the sense that they are mostly all As, mostly all Ts, mostly all Cs, or mostly all Gs. One approach is to look for low-entropy columns; these will be ones whose distribution is nearly deterministic (pure).

But suppose we want to associate a confidence measure with our estimates of purity. This can be useful if we believe adjacent locations are conserved together. In this case, we can let $Z_1 = 1$ if location t is conserved, and let $Z_t = 0$ otherwise. We can then add a dependence between adjacent Z_t variables using a Markov chain; see Chapter 17 for details.

In any case, we need to define a likelihood model, $p(\mathbf{N}_t|Z_t)$, where $\mathbf{N}_t$ is the vector of (A,C,G,T) counts for column t. It is natural to make this be a multinomial distribution with parameter $\boldsymbol{\theta}_t$. Since each column has a different distribution, we will want to integrate out $\boldsymbol{\theta}_t$ and thus compute the marginal likelihood

$$p(\mathbf{N}_t|Z_t) = \int p(\mathbf{N}_t|\boldsymbol{\theta}_t)p(\boldsymbol{\theta}_t|Z_t)d\boldsymbol{\theta}_t \quad (5.74)$$

But what prior should we use for $\boldsymbol{\theta}_t$? When $Z_t = 0$ we can use a uniform prior, $p(\boldsymbol{\theta}|Z_t = 0) = \text{Dir}(1,1,1,1)$, but what should we use if $Z_t = 1$? After all, if the column is conserved, it could be a (nearly) pure column of As, Cs, Gs, or Ts. A natural approach is to use a mixture of Dirichlet priors, each one of which is "tilted" towards the appropriate corner of the 4-dimensional simplex, e.g.,

$$p(\boldsymbol{\theta}|Z_t = 1) = \frac{1}{4}\text{Dir}(\boldsymbol{\theta}|(10,1,1,1)) + \cdots + \frac{1}{4}\text{Dir}(\boldsymbol{\theta}|(1,1,1,10)) \qquad (5.75)$$

Since this is conjugate, we can easily compute $p(\mathbf{N}_t|Z_t)$. See (Brown et al. 1993) for an application of these ideas to a real bio-sequence problem.

5.5 Hierarchical Bayes

A key requirement for computing the posterior $p(\boldsymbol{\theta}|\mathcal{D})$ is the specification of a prior $p(\boldsymbol{\theta}|\boldsymbol{\eta})$, where $\boldsymbol{\eta}$ are the hyper-parameters. What if we don't know how to set $\boldsymbol{\eta}$? In some cases, we can use uninformative priors, we we discussed above. A more Bayesian approach is to put a prior on our priors! In terms of graphical models (Chapter 10), we can represent the situation as follows:

$$\boldsymbol{\eta} \rightarrow \boldsymbol{\theta} \rightarrow \mathcal{D} \qquad (5.76)$$

This is an example of a **hierarchical Bayesian model**, also called a **multi-level model**, since there are multiple levels of unknown quantities. We give a simple example below, and we will see many others later in the book.

5.5.1 Example: modeling related cancer rates

Consider the problem of predicting cancer rates in various cities (this example is from (Johnson and Albert 1999, p24)). In particular, suppose we measure the number of people in various cities, N_i, and the number of people who died of cancer in these cities, x_i. We assume $x_i \sim \text{Bin}(N_i, \theta_i)$, and we want to estimate the cancer rates θ_i. One approach is to estimate them all separately, but this will suffer from the sparse data problem (underestimation of the rate of cancer due to small N_i). Another approach is to assume all the θ_i are the same; this is called **parameter tying**. The resulting pooled MLE is just $\hat{\theta} = \frac{\sum_i x_i}{\sum_i N_i}$. But the assumption that all the cities have the same rate is a rather strong one. A compromise approach is to assume that the θ_i are similar, but that there may be city-specific variations. This can be modeled by assuming the θ_i are drawn from some common distribution, say $\theta_i \sim \text{Beta}(a, b)$. The full joint distribution can be written as

$$p(\mathcal{D}, \boldsymbol{\theta}, \boldsymbol{\eta}|\mathbf{N}) = p(\boldsymbol{\eta}) \prod_{i=1}^{N} \text{Bin}(x_i|N_i, \theta_i)\text{Beta}(\theta_i|\boldsymbol{\eta}) \qquad (5.77)$$

where $\boldsymbol{\eta} = (a, b)$.

Note that it is crucial that we infer $\boldsymbol{\eta} = (a, b)$ from the data; if we just clamp it to a constant, the θ_i will be conditionally independent, and there will be no information flow between them. By contrast, by treating $\boldsymbol{\eta}$ as an unknown (hidden variable), we allow the data-poor cities to **borrow statistical strength** from data-rich ones.

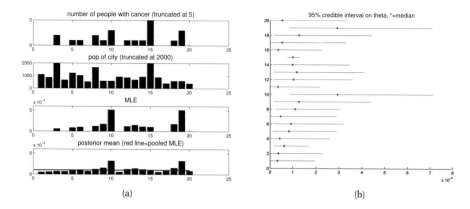

Figure 5.11 (a) Results of fitting the model using the data from (Johnson and Albert 1999, p24). First row: Number of cancer incidents x_i in 20 cities in Missouri. Second row: population size N_i. The largest city (number 15) has a population of $N_{15} = 53637$ and $x_{15} = 54$ incidents, but we truncate the vertical axes of the first two rows so that the differences between the other cities are visible. Third row: MLE $\hat{\theta}_i$. The red line is the pooled MLE. Fourth row: posterior mean $\mathbb{E}[\theta_i|\mathcal{D}]$. The red line is $\mathbb{E}[a/(a+b)|\mathcal{D}]$, the population-level mean. (b) Posterior 95% credible intervals on the cancer rates. Figure generated by `cancerRatesEb`

Suppose we compute the joint posterior $p(\boldsymbol{\eta}, \boldsymbol{\theta}|\mathcal{D})$. From this we can get the posterior marginals $p(\theta_i|\mathcal{D})$. In Figure 5.11(a), we plot the posterior means, $\mathbb{E}[\theta_i|\mathcal{D}]$, as blue bars, as well as the population level mean, $\mathbb{E}[a/(a+b)|\mathcal{D}]$, shown as a red line (this represents the average of the θ_i's). We see that the posterior mean is shrunk towards the pooled estimate more strongly for cities with small sample sizes N_i. For example, city 1 and city 20 both have a 0 observed cancer incidence rate, but city 20 has a smaller population, so its rate is shrunk more towards the population-level estimate (i.e., it is closer to the horizontal red line) than city 1.

Figure 5.11(b) shows the 95% posterior credible intervals for θ_i. We see that city 15, which has a very large population (53,637 people), has small posterior uncertainty. Consequently this city has the largest impact on the posterior estimate of $\boldsymbol{\eta}$, which in turn will impact the estimate of the cancer rates for other cities. Cities 10 and 19, which have the highest MLE, also have the highest posterior uncertainty, reflecting the fact that such a high estimate is in conflict with the prior (which is estimated from all the other cities).

In the above example, we have one parameter per city, modeling the probability the response is on. By making the Bernoulli rate parameter be a function of covariates, $\theta_i = \text{sigm}(\mathbf{w}_i^T\mathbf{x})$, we can model multiple correlated logistic regression tasks. This is called **multi-task learning**, and will be discussed in more detail in Section 9.5.

5.6 Empirical Bayes

In hierarchical Bayesian models, we need to compute the posterior on multiple levels of latent variables. For example, in a two-level model, we need to compute

$$p(\boldsymbol{\eta}, \boldsymbol{\theta}|\mathcal{D}) \propto p(\mathcal{D}|\boldsymbol{\theta})p(\boldsymbol{\theta}|\boldsymbol{\eta})p(\boldsymbol{\eta}) \tag{5.78}$$

In some cases, we can analytically marginalize out $\boldsymbol{\theta}$; this leaves is with the simpler problem of just computing $p(\boldsymbol{\eta}|\mathcal{D})$.

As a computational shortcut, we can approximate the posterior on the hyper-parameters with a point-estimate, $p(\boldsymbol{\eta}|\mathcal{D}) \approx \delta_{\hat{\boldsymbol{\eta}}}(\boldsymbol{\eta})$, where $\hat{\boldsymbol{\eta}} = \operatorname{argmax} p(\boldsymbol{\eta}|\mathcal{D})$. Since $\boldsymbol{\eta}$ is typically much smaller than $\boldsymbol{\theta}$ in dimensionality, it is less prone to overfitting, so we can safely use a uniform prior on $\boldsymbol{\eta}$. Then the estimate becomes

$$\hat{\boldsymbol{\eta}} = \operatorname{argmax} p(\mathcal{D}|\boldsymbol{\eta}) = \operatorname{argmax} \left[\int p(\mathcal{D}|\boldsymbol{\theta})p(\boldsymbol{\theta}|\boldsymbol{\eta})d\boldsymbol{\theta} \right] \tag{5.79}$$

where the quantity inside the brackets is the marginal or integrated likelihood, sometimes called the evidence. This overall approach is called **empirical Bayes** (**EB**) or **type-II maximum likelihood**. In machine learning, it is sometimes called the **evidence procedure**.

Empirical Bayes violates the principle that the prior should be chosen independently of the data. However, we can just view it as a computationally cheap approximation to inference in a hierarchical Bayesian model, just as we viewed MAP estimation as an approximation to inference in the one level model $\boldsymbol{\theta} \to \mathcal{D}$. In fact, we can construct a hierarchy in which the more integrals one performs, the "more Bayesian" one becomes:

Method	Definition
Maximum likelihood	$\hat{\boldsymbol{\theta}} = \operatorname{argmax}_{\boldsymbol{\theta}} p(\mathcal{D}|\boldsymbol{\theta})$
MAP estimation	$\hat{\boldsymbol{\theta}} = \operatorname{argmax}_{\boldsymbol{\theta}} p(\mathcal{D}|\boldsymbol{\theta})p(\boldsymbol{\theta}|\boldsymbol{\eta})$
ML-II (Empirical Bayes)	$\hat{\boldsymbol{\eta}} = \operatorname{argmax}_{\boldsymbol{\eta}} \int p(\mathcal{D}|\boldsymbol{\theta})p(\boldsymbol{\theta}|\boldsymbol{\eta})d\boldsymbol{\theta} = \operatorname{argmax}_{\boldsymbol{\eta}} p(\mathcal{D}|\boldsymbol{\eta})$
MAP-II	$\hat{\boldsymbol{\eta}} = \operatorname{argmax}_{\boldsymbol{\eta}} \int p(\mathcal{D}|\boldsymbol{\theta})p(\boldsymbol{\theta}|\boldsymbol{\eta})p(\boldsymbol{\eta})d\boldsymbol{\theta} = \operatorname{argmax}_{\boldsymbol{\eta}} p(\mathcal{D}|\boldsymbol{\eta})p(\boldsymbol{\eta})$
Full Bayes	$p(\boldsymbol{\theta}, \boldsymbol{\eta}|\mathcal{D}) \propto p(\mathcal{D}|\boldsymbol{\theta})p(\boldsymbol{\theta}|\boldsymbol{\eta})p(\boldsymbol{\eta})$

Note that EB can be shown to have good frequentist properties (see e.g., (Carlin and Louis 1996; Efron 2010)), so it is widely used by non-Bayesians. For example, the popular James-Stein estimator, discussed in Section 6.3.3.2, can be derived using EB.

5.6.1 Example: beta-binomial model

Let us return to the cancer rates model. We can analytically integrate out the θ_i's, and write down the marginal likelihood directly, as follows:

$$p(\mathcal{D}|a,b) = \prod_i \int \operatorname{Bin}(x_i|N_i, \theta_i)\operatorname{Beta}(\theta_i|a,b)d\theta_i \tag{5.80}$$

$$= \prod_i \frac{B(a + x_i, b + N_i - x_i)}{B(a,b)} \tag{5.81}$$

Various ways of maximizing this wrt a and b are discussed in (Minka 2000e).

Having estimated a and b, we can plug in the hyper-parameters to compute the posterior $p(\theta_i|\hat{a}, \hat{b}, \mathcal{D})$ in the usual way, using conjugate analysis. The net result is that the posterior mean of each θ_i is a weighted average of its local MLE and the prior means, which depends on $\boldsymbol{\eta} = (a, b)$; but since $\boldsymbol{\eta}$ is estimated based on all the data, each θ_i is influenced by all the data.

5.6.2 Example: Gaussian-Gaussian model

We now study another example that is analogous to the cancer rates example, except the data is real-valued. We will use a Gaussian likelihood and a Gaussian prior. This will allow us to write down the solution analytically.

In particular, suppose we have data from multiple related groups. For example, x_{ij} could be the test score for student i in school j, for $j = 1 : D$ and $i = 1 : N_j$. We want to estimate the mean score for each school, θ_j. However, since the sample size, N_j, may be small for some schools, we can regularize the problem by using a hierarchical Bayesian model, where we assume θ_j come from a common prior, $\mathcal{N}(\mu, \tau^2)$.

The joint distribution has the following form:

$$p(\boldsymbol{\theta}, \mathcal{D}|\boldsymbol{\eta}, \sigma^2) = \prod_{j=1}^{D} \mathcal{N}(\theta_j|\mu, \tau^2) \prod_{i=1}^{N_j} \mathcal{N}(x_{ij}|\theta_j, \sigma^2) \tag{5.82}$$

where we assume σ^2 is known for simplicity. (We relax this assumption in Exercise 24.4.) We explain how to estimate $\boldsymbol{\eta}$ below. Once we have estimated $\boldsymbol{\eta} = (\mu, \tau)$, we can compute the posteriors over the θ_j's. To do that, it simplifies matters to rewrite the joint distribution in the following form, exploiting the fact that N_j Gaussian measurements with values x_{ij} and variance σ^2 are equivalent to one measurement of value $\overline{x}_j \triangleq \frac{1}{N_j} \sum_{i=1}^{N_j} x_{ij}$ with variance $\sigma_j^2 \triangleq \sigma^2/N_j$. This yields

$$p(\boldsymbol{\theta}, \mathcal{D}|\hat{\boldsymbol{\eta}}, \sigma^2) = \prod_{j=1}^{D} \mathcal{N}(\theta_j|\hat{\mu}, \hat{\tau}^2) \mathcal{N}(\overline{x}_j|\theta_j, \sigma_j^2) \tag{5.83}$$

From this, it follows from the results of Section 4.4.1 that the posteriors are given by

$$p(\theta_j|\mathcal{D}, \hat{\mu}, \hat{\tau}^2) = \mathcal{N}(\theta_j|\hat{B}_j\hat{\mu} + (1 - \hat{B}_j)\overline{x}_j, (1 - \hat{B}_j)\sigma_j^2) \tag{5.84}$$

$$\hat{B}_j \triangleq \frac{\sigma_j^2}{\sigma_j^2 + \hat{\tau}^2} \tag{5.85}$$

where $\hat{\mu} = \overline{x}$ and $\hat{\tau}^2$ will be defined below.

The quantity $0 \leq \hat{B}_j \leq 1$ controls the degree of **shrinkage** towards the overall mean, μ. If the data is reliable for group j (e.g., because the sample size N_j is large), then σ_j^2 will be small relative to τ^2; hence $\hat{B}_j$ will be small, and we will put more weight on $\overline{x}_j$ when we estimate θ_j. However, groups with small sample sizes will get regularized (shrunk towards the overall mean μ) more heavily. We will see an example of this below.

If $\sigma_j = \sigma$ for all groups j, the posterior mean becomes

$$\hat{\theta}_j = \hat{B}\overline{x} + (1 - \hat{B})\overline{x}_j = \overline{x} + (1 - \hat{B})(\overline{x}_j - \overline{x}) \tag{5.86}$$

This has exactly the same form as the James Stein estimator discussed in Section 6.3.3.2.

5.6.2.1 Example: predicting baseball scores

We now give an example of shrinkage applied to baseball batting averages, from (Efron and Morris 1975). We observe the number of hits for $D = 18$ players during the first $T = 45$ games.

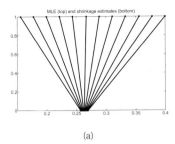
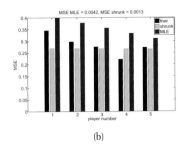

(a)   (b)

Figure 5.12 (a) MLE parameters (top) and corresponding shrunken estimates (bottom). (b) We plot the true parameters (blue), the posterior mean estimate (green), and the MLEs (red) for 5 of the players. Figure generated by `shrinkageDemoBaseball`.

Call the number of hits b_i. We assume $b_j \sim \text{Bin}(T, \theta_j)$, where θ_j is the "true" batting average for player j. The goal is to estimate the θ_j. The MLE is of course $\hat{\theta}_j = x_j$, where $x_j = b_j/T$ is the empirical batting average. However, we can use an EB approach to do better.

To apply the Gaussian shrinkage approach described above, we require that the likelihood be Gaussian, $x_j \sim \mathcal{N}(\theta_j, \sigma^2)$ for known σ^2. (We drop the i subscript since we assume $N_j = 1$, since x_j already represents the average for player j.) However, in this example we have a binomial likelihood. While this has the right mean, $\mathbb{E}[x_j] = \theta_j$, the variance is not constant:

$$\text{var}[x_j] = \frac{1}{T^2} \text{var}[b_j] = \frac{T\theta_j(1 - \theta_j)}{T^2} \tag{5.87}$$

So let us apply a **variance stabilizing transform**[5] to x_j to better match the Gaussian assumption:

$$y_j = f(y_j) = \sqrt{T} \arcsin(2y_j - 1) \tag{5.88}$$

Now we have approximately $y_j \sim \mathcal{N}(f(\theta_j), 1) = \mathcal{N}(\mu_j, 1)$. We use Gaussian shrinkage to estimate the μ_j using Equation 5.86 with $\sigma^2 = 1$, and we then transform back to get

$$\hat{\theta}_j = 0.5(\sin(\hat{\mu}_j/\sqrt{T}) + 1) \tag{5.89}$$

The results are shown in Figure 5.12(a-b). In (a), we plot the MLE $\hat{\theta}_j$ and the posterior mean $\overline{\theta}_j$. We see that all the estimates have shrunk towards the global mean, 0.265. In (b), we plot the true value θ_j, the MLE $\hat{\theta}_j$ and the posterior mean $\overline{\theta}_j$. (The "true" values of θ_j are estimated from a large number of independent games.) We see that, on average, the shrunken estimate is much closer to the true parameters than the MLE is. Specifically, the mean squared error, defined by MSE $= \frac{1}{N} \sum_{j=1}^{D} (\theta_j - \overline{\theta}_j)^2$, is over three times smaller using the shrinkage estimates $\overline{\theta}_j$ than using the MLEs $\hat{\theta}_j$.

---

5. Suppose $\mathbb{E}[X] = \mu$ and $\text{var}[X] = \sigma^2(\mu)$. Let $Y = f(X)$. Then a Taylor series expansions gives $Y \approx f(\mu) + (X - \mu)f'(\mu)$. Hence $\text{var}[Y] \approx f'(\mu)^2\text{var}[X - \mu] = f'(\mu)^2\sigma^2(\mu)$. A variance stabilizing transformation is a function f such that $f'(\mu)^2\sigma^2(\mu)$ is independent of μ.

5.6.2.2   Estimating the hyper-parameters

In this section, we give an algorithm for estimating η. Suppose initially that $\sigma_j^2 = \sigma^2$ is the same for all groups. In this case, we can derive the EB estimate in closed form, as we now show. From Equation 4.126, we have

$$p(\overline{x}_j|\mu,\tau^2,\sigma^2) \;=\; \int \mathcal{N}(\overline{x}_j|\theta_j,\sigma^2)\mathcal{N}(\theta_j|\mu,\tau^2)d\theta_j = \mathcal{N}(\overline{x}_j|\mu,\tau^2+\sigma^2) \tag{5.90}$$

Hence the marginal likelihood is

$$p(\mathcal{D}|\mu,\tau^2,\sigma^2) = \prod_{j=1}^{D} \mathcal{N}(\overline{x}_j|\mu,\tau^2+\sigma^2) \tag{5.91}$$

Thus we can estimate the hyper-parameters using the usual MLEs for a Gaussian. For μ, we have

$$\hat{\mu} = \frac{1}{D}\sum_{j=1}^{D} \overline{x}_j = \overline{x} \tag{5.92}$$

which is the overall mean.

For the variance, we can use moment matching (which is equivalent to the MLE for a Gaussian): we simply equate the model variance to the empirical variance:

$$\hat{\tau}^2 + \sigma^2 = \frac{1}{D}\sum_{j=1}^{D}(\overline{x}_j - \overline{x})^2 \triangleq s^2 \tag{5.93}$$

so $\hat{\tau}^2 = s^2 - \sigma^2$. Since we know τ^2 must be positive, it is common to use the following revised estimate:

$$\hat{\tau}^2 = \max\{0, s^2 - \sigma^2\} = (s^2 - \sigma^2)_+ \tag{5.94}$$

Hence the shrinkage factor is

$$\hat{B} = \frac{\sigma^2}{\sigma^2 + \hat{\tau}^2} = \frac{\sigma^2}{\sigma^2 + (s^2 - \sigma^2)_+} \tag{5.95}$$

In the case where the σ_j^2's are different, we can no longer derive a solution in closed form. Exercise 11.13 discusses how to use the EM algorithm to derive an EB estimate, and Exercise 24.4 discusses how to perform full Bayesian inference in this hierarchical model.

5.7   Bayesian decision theory

We have seen how probability theory can be used to represent and update our beliefs about the state of the world. However, ultimately our goal is to convert our beliefs into actions. In this section, we discuss the optimal way to do this.

We can formalize any given statistical decision problem as a game against nature (as opposed to a game against other strategic players, which is the topic of game theory, see e.g., (Shoham and Leyton-Brown 2009) for details). In this game, nature picks a state or parameter or label, $y \in \mathcal{Y}$, unknown to us, and then generates an observation, $\mathbf{x} \in \mathcal{X}$, which we get to see. We then have to make a decision, that is, we have to choose an action a from some **action space** $\mathcal{A}$. Finally we incur some **loss**, $L(y, a)$, which measures how compatible our action a is with nature's hidden state y. For example, we might use misclassification loss, $L(y, a) = \mathbb{I}(y \neq a)$, or squared loss, $L(y, a) = (y - a)^2$. We will see some other examples below.

Our goal is to devise a **decision procedure** or **policy**, $\delta : \mathcal{X} \rightarrow \mathcal{A}$, which specifies the optimal action for each possible input. By optimal, we mean the action that minimizes the expected loss:

$$\delta(\mathbf{x}) = \underset{a \in \mathcal{A}}{\operatorname{argmin}} \, \mathbb{E}\left[L(y, a)\right] \tag{5.96}$$

In economics, it is more common to talk of a **utility function**; this is just negative loss, $U(y, a) = -L(y, a)$. Thus the above rule becomes

$$\delta(\mathbf{x}) = \underset{a \in \mathcal{A}}{\operatorname{argmax}} \, \mathbb{E}\left[U(y, a)\right] \tag{5.97}$$

This is called the **maximum expected utility principle**, and is the essence of what we mean by **rational behavior**.

Note that there are two different interpretations of what we mean by "expected". In the Bayesian version, which we discuss below, we mean the expected value of y given the data we have seen so far. In the frequentist version, which we discuss in Section 6.3, we mean the expected value of y and $\mathbf{x}$ that we expect to see in the future.

In the Bayesian approach to decision theory, the optimal action, having observed $\mathbf{x}$, is defined as the action a that minimizes the **posterior expected loss**:

$$\rho(a|\mathbf{x}) \triangleq \mathbb{E}_{p(y|\mathbf{x})}\left[L(y, a)\right] = \sum_y L(y, a) p(y|\mathbf{x}) \tag{5.98}$$

(If y is continuous (e.g., when we want to estimate a parameter vector), we should replace the sum with an integral.) Hence the **Bayes estimator**, also called the **Bayes decision rule**, is given by

$$\delta(\mathbf{x}) = \underset{a \in \mathcal{A}}{\operatorname{argmin}} \, \rho(\mathbf{a}|\mathbf{x}) \tag{5.99}$$

5.7.1 Bayes estimators for common loss functions

In this section we show how to construct Bayes estimators for the loss functions most commonly arising in machine learning.

5.7.1.1 MAP estimate minimizes 0-1 loss

The **0-1 loss** is defined by

$$L(y, a) = \mathbb{I}(y \neq a) = \begin{cases} 0 & \text{if } a = y \\ 1 & \text{if } a \neq y \end{cases} \tag{5.100}$$

This is commonly used in classification problems where y is the true class label and $a = \hat{y}$ is the estimate.

For example, in the two class case, we can write the loss matrix as follows:

	$\hat{y} = 1$	$\hat{y} = 0$
$y = 1$	0	1
$y = 0$	1	0

(In Section 5.7.2, we generalize this loss function so it penalizes the two kinds of errors on the off-diagonal differently.)

The posterior expected loss is

$$\rho(a|\mathbf{x}) = p(a \neq y|\mathbf{x}) = 1 - p(y|\mathbf{x}) \tag{5.101}$$

Hence the action that minimizes the expected loss is the posterior mode or MAP estimate

$$y^*(\mathbf{x}) = \underset{y \in \mathcal{Y}}{\operatorname{argmax}} \, p(y|\mathbf{x}) \tag{5.102}$$

5.7.1.2 Reject option

In classification problems where $p(y|\mathbf{x})$ is very uncertain, we may prefer to choose a **reject action**, in which we refuse to classify the example as any of the specified classes, and instead say "don't know". Such ambiguous cases can be handled by e.g., a human expert. See Figure 5.13 for an illustration. This is useful in **risk averse** domains such as medicine and finance.

We can formalize the reject option as follows. Let choosing $a = C + 1$ correspond to picking the reject action, and choosing $a \in \{1, \ldots, C\}$ correspond to picking one of the classes. Suppose we define the loss function as

$$L(y = j, a = i) = \begin{cases} 0 & \text{if } i = j \text{ and } i, j \in \{1, \ldots, C\} \\ \lambda_r & \text{if } i = C + 1 \\ \lambda_s & \text{otherwise} \end{cases} \tag{5.103}$$

where λ_r is the cost of the reject action, and λ_s is the cost of a substitution error. In Exercise 5.3, you will show that the optimal action is to pick the reject action if the most probable class has a probability below $1 - \frac{\lambda_r}{\lambda_s}$; otherwise you should just pick the most probable class.

5.7.1.3 Posterior mean minimizes ℓ_2 (quadratic) loss

For continuous parameters, a more appropriate loss function is **squared error**, ℓ_2 **loss**, or **quadratic loss**, defined as

$$L(y, a) = (y - a)^2 \tag{5.104}$$

The posterior expected loss is given by

$$\rho(a|\mathbf{x}) = \mathbb{E}\left[(y - a)^2|\mathbf{x}\right] = \mathbb{E}\left[y^2|\mathbf{x}\right] - 2a\mathbb{E}\left[y|\mathbf{x}\right] + a^2 \tag{5.105}$$

Hence the optimal estimate is the posterior mean:

$$\frac{\partial}{\partial a}\rho(a|\mathbf{x}) = -2\mathbb{E}\left[y|\mathbf{x}\right] + 2a = 0 \implies \hat{y} = \mathbb{E}\left[y|\mathbf{x}\right] = \int y p(y|\mathbf{x}) dy \tag{5.106}$$

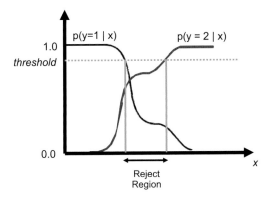

Figure 5.13 For some regions of input space, where the class posteriors are uncertain, we may prefer not to choose class 1 or 2; instead we may prefer the reject option. Based on Figure 1.26 of (Bishop 2006).

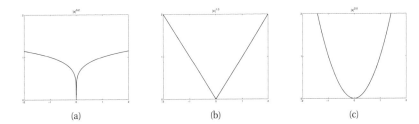

Figure 5.14 (a-c). Plots of the $L(y, a) = |y - a|^q$ vs $|y - a|$ for $q = 0.2$, $q = 1$ and $q = 2$. Figure generated by `lossFunctionFig`.

This is often called the **minimum mean squared error** estimate or **MMSE** estimate.

In a linear regression problem, we have

$$p(y|\mathbf{x}, \boldsymbol{\theta}) = \mathcal{N}(y|\mathbf{x}^T \mathbf{w}, \sigma^2) \tag{5.107}$$

In this case, the optimal estimate given some training data $\mathcal{D}$ is given by

$$\mathbb{E}[y|\mathbf{x}, \mathcal{D}] = \mathbf{x}^T \mathbb{E}[\mathbf{w}|\mathcal{D}] \tag{5.108}$$

That is, we just plug-in the posterior mean parameter estimate. Note that this is the optimal thing to do no matter what prior we use for $\mathbf{w}$.

5.7.1.4 Posterior median minimizes ℓ_1 (absolute) loss

The ℓ_2 loss penalizes deviations from the truth quadratically, and thus is sensitive to outliers. A more robust alternative is the absolute or ℓ_1 **loss**, $L(y, a) = |y - a|$ (see Figure 5.14). The optimal estimate is the posterior median, i.e., a value a such that $P(y < a|\mathbf{x}) = P(y \geq a|\mathbf{x}) = 0.5$. See Exercise 5.9 for a proof.

5.7.1.5 Supervised learning

Consider a prediction function $\delta : \mathcal{X} \to \mathcal{Y}$, and suppose we have some cost function $\ell(y, y')$ which gives the cost of predicting y' when the truth is y. We can define the loss incurred by taking action δ (i.e., using this predictor) when the unknown state of nature is $\boldsymbol{\theta}$ (the parameters of the data generating mechanism) as follows:

$$L(\boldsymbol{\theta}, \delta) \triangleq \mathbb{E}_{(\mathbf{x},y) \sim p(\mathbf{x},y|\boldsymbol{\theta})}\left[\ell(y, \delta(\mathbf{x})) = \sum_{\mathbf{x}}\sum_{y} L(y, \delta(\mathbf{x}))p(\mathbf{x}, y|\boldsymbol{\theta})\right] \tag{5.109}$$

This is known as the **generalization error**. Our goal is to minimize the posterior expected loss, given by

$$\rho(\delta|\mathcal{D}) = \int p(\boldsymbol{\theta}|\mathcal{D})L(\boldsymbol{\theta}, \delta)d\boldsymbol{\theta} \tag{5.110}$$

This should be contrasted with the frequentist risk which is defined in Equation 6.47.

5.7.2 The false positive vs false negative tradeoff

In this section, we focus on binary decision problems, such as hypothesis testing, two-class classification, object / event detection, etc. There are two types of error we can make: a **false positive** (aka **false alarm**), which arises when we estimate $\hat{y} = 1$ but the truth is $y = 0$; or a **false negative** (aka **missed detection**), which arises when we estimate $\hat{y} = 0$ but the truth is $y = 1$. The 0-1 loss treats these two kinds of errors equivalently. However, we can consider the following more general loss matrix:

	$\hat{y} = 1$	$\hat{y} = 0$
$y = 1$	0	L_{FN}
$y = 0$	L_{FP}	0

where L_{FN} is the cost of a false negative, and L_{FP} is the cost of a false positive. The posterior expected loss for the two possible actions is given by

$$\rho(\hat{y} = 0|\mathbf{x}) = L_{FN}\, p(y = 1|\mathbf{x}) \tag{5.111}$$
$$\rho(\hat{y} = 1|\mathbf{x}) = L_{FP}\, p(y = 0|\mathbf{x}) \tag{5.112}$$

Hence we should pick class $\hat{y} = 1$ iff

$$\rho(\hat{y} = 0|\mathbf{x}) > \rho(\hat{y} = 1|\mathbf{x}) \tag{5.113}$$
$$\frac{p(y = 1|\mathbf{x})}{p(y = 0|\mathbf{x})} > \frac{L_{FP}}{L_{FN}} \tag{5.114}$$

If $L_{FN} = cL_{FP}$, it is easy to show (Exercise 5.10) that we should pick $\hat{y} = 1$ iff $p(y = 1|\mathbf{x})/p(y = 0|\mathbf{x}) > \tau$, where $\tau = c/(1 + c)$ (see also (Muller et al. 2004)). For example, if a false negative costs twice as much as false positive, so $c = 2$, then we use a decision threshold of $2/3$ before declaring a positive.

Below we discuss ROC curves, which provide a way to study the FP-FN tradeoff without having to choose a specific threshold.

		Truth		
		1	0	Σ
Estimate	1	TP	FP	$\hat{N}_+ = TP + FP$
	0	FN	TN	$\hat{N}_- = FN + TN$
	Σ	$N_+ = TP + FN$	$N_- = FP + TN$	$N = TP + FP + FN + TN$

Table 5.2   Quantities derivable from a confusion matrix. N_+ is the true number of positives, $\hat{N}_+$ is the "called" number of positives, N_- is the true number of negatives, $\hat{N}_-$ is the "called" number of negatives.

5.7.2.1   ROC curves and all that

Suppose we are solving a binary decision problem, such as classification, hypothesis testing, object detection, etc. Also, assume we have a labeled data set, $\mathcal{D} = \{(\mathbf{x}_i, y_i)\}$. Let $\delta(\mathbf{x}) = \mathbb{I}(f(\mathbf{x}) > \tau)$ be our decision rule, where $f(\mathbf{x})$ is a measure of confidence that $y = 1$ (this should be monotonically related to $p(y = 1|\mathbf{x})$, but does not need to be a probability), and τ is some threshold parameter. For each given value of τ, we can apply our decision rule and count the number of true positives, false positives, true negatives, and false negatives that occur, as shown in Table 5.2. This table of errors is called a **confusion matrix**.

From this table, we can compute the **true positive rate** (TPR), also known as the **sensitivity**, **recall** or **hit rate**, by using $TPR = TP/N_+ \approx p(\hat{y} = 1|y = 1)$. We can also compute the **false positive rate** (FPR), also called the **false alarm rate**, or the **type I error rate**, by using $FPR = FP/N_- \approx p(\hat{y} = 1|y = 0)$. These and other definitions are summarized in Tables 5.3 and 5.4. We can combine these errors in any way we choose to compute a loss function.

However, rather than than computing the TPR and FPR for a fixed threshold τ, we can run our detector for a set of thresholds, and then plot the TPR vs FPR as an implicit function of τ. This is called a **receiver operating characteristic** or **ROC** curve. See Figure 5.15(a) for an example. Any system can achieve the point on the bottom left, $(FPR = 0, TPR = 0)$, by setting $\tau = 1$ and thus classifying everything as negative; similarly any system can achieve the point on the top right, $(FPR = 1, TPR = 1)$, by setting $\tau = 0$ and thus classifying everything as positive. If a system is performing at chance level, then we can achieve any point on the diagonal line $TPR = FPR$ by choosing an appropriate threshold. A system that perfectly separates the positives from negatives has a threshold that can achieve the top left corner, $(FPR = 0, TPR = 1)$; by varying the threshold such a system will "hug" the left axis and then the top axis, as shown in Figure 5.15(a).

The quality of a ROC curve is often summarized as a single number using the **area under the curve** or **AUC**. Higher AUC scores are better; the maximum is obviously 1.[6] Another summary statistic that is used is the **equal error rate** or **EER**, also called the **cross over rate**, defined as the value which satisfies $FPR = FNR$. Since $FNR = 1 - TPR$, we can compute the EER by drawing a line from the top left to the bottom right and seeing where it intersects the ROC curve (see points A and B in Figure 5.15(a)). Lower EER scores are better; the minimum is obviously 0.

---

6. One can show (Fawcett 2003) that the AUC is equal to the probabillity that a randomly drawn positive example has a higher score than a randomly drawn negative example, $p(s(\mathbf{x}_1) > s(\mathbf{x}_2)|y_1 = 1, y_2 = 0)$. Thus the AUC measures the ability to rank items correctly.

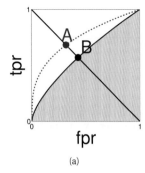
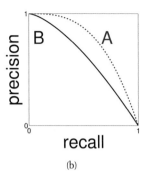

(a)   (b)

Figure 5.15 (a) ROC curves for two hypothetical classification systems. A is better than B. We plot the true positive rate (TPR) vs the false positive rate (FPR) as we vary the threshold τ. We also indicate the equal error rate (EER) with the red and blue dots, and the area under the curve (AUC) for classifier B by the shaded blue area. Figure generated by ROChand. (b) A precision-recall curve for two hypothetical classification systems. A is better than B. Figure generated by PRhand.

	$y = 1$	$y = 0$
$\hat{y} = 1$	TP/N_+=TPR=sensitivity=recall	FP/N_-=FPR=type I
$\hat{y} = 0$	FN/N_+=FNR=miss rate=type II	TN/N_-=TNR=specifity

Table 5.3 Estimating $p(\hat{y}|y)$ from a confusion matrix. Abbreviations: FNR = false negative rate, FPR = false positive rate, TNR = true negative rate, TPR = true positive rate.

	$y = 1$	$y = 0$
$\hat{y} = 1$	$TP/\hat{N}_+$=precision=PPV	$FP/\hat{N}_+$=FDP
$\hat{y} = 0$	$FN/\hat{N}_-$	$TN/\hat{N}_-$=NPV

Table 5.4 Estimating $p(y|\hat{y})$ from a confusion matrix. Abbreviations: FDP = false discovery probability, NPV = negative predictive value, PPV = positive predictive value,

5.7.2.2   Precision recall curves

When trying to detect a rare event (such as retrieving a relevant document or finding a face in an image), the number of negatives is very large. Hence comparing $TPR = TP/N_+$ to $FPR = FP/N_-$ is not very informative, since the FPR will be very small. Hence all the "action" in the ROC curve will occur on the extreme left. In such cases, it is common to plot the TPR versus the number of false positives, rather than vs the false positive rate.

However, in some cases, the very notion of "negative" is not well-defined. For example, when detecting objects in images (see Section 1.2.1.3), if the detector works by classifying patches, then the number of patches examined — and hence the number of true negatives — is a parameter of the algorithm, not part of the problem definition. So we would like to use a measure that only talks about positives.

The **precision** is defined as $TP/\hat{N}_+ = p(y = 1|\hat{y} = 1)$ and the **recall** is defined as $TP/N_+ = p(\hat{y} = 1|y = 1)$. Precision measures what fraction of our detections are actually positive, and recall measures what fraction of the positives we actually detected. If $\hat{y}_i \in \{0, 1\}$ is the predicted label, and $y_i \in \{0, 1\}$ is the true label, we can estimate precision and recall using

$$P = \frac{\sum_i y_i \hat{y}_i}{\sum_i \hat{y}_i}, \quad R = \frac{\sum_i y_i \hat{y}_i}{\sum_i y_i} \tag{5.115}$$

A **precision recall curve** is a plot of precision vs recall as we vary the threshold τ. See Figure 5.15(b). Hugging the top right is the best one can do.

This curve can be summarized as a single number using the mean precision (averaging over recall values), which approximates the area under the curve. Alternatively, one can quote the precision for a fixed recall level, such as the precision of the first $K = 10$ entities recalled. This is called the **average precision at K** score. This measure is widely used when evaluating information retrieval systems.

5.7.2.3 F-scores *

For a fixed threshold, one can compute a single precision and recall value. These are often combined into a single statistic called the **F score**, or **F1 score**, which is the harmonic mean of precision and recall:

$$F_1 \triangleq \frac{2}{1/P + 1/R} = \frac{2PR}{R + P} \tag{5.116}$$

Using Equation 5.115, we can write this as

$$F_1 = \frac{2 \sum_{i=1}^{N} y_i \hat{y}_i}{\sum_{i=1}^{N} y_i + \sum_{i=1}^{N} \hat{y}_i} \tag{5.117}$$

This is a widely used measure in information retrieval systems.

To understand why we use the harmonic mean instead of the arithmetic mean, $(P + R)/2$, consider the following scenario. Suppose we recall all entries, so $R = 1$. The precision will be given by the **prevalence**, $p(y = 1)$. Suppose the prevalence is low, say $p(y = 1) = 10^{-4}$. The arithmetic mean of P and R is given by $(P + R)/2 = (10^{-4} + 1)/2 \approx 50\%$. By contrast, the harmonic mean of this strategy is only $\frac{2 \times 10^{-4} \times 1}{1 + 10^{-4}} \approx 0.2\%$.

In the multi-class case (e.g., for document classification problems), there are two ways to generalize F_1 scores. The first is called **macro-averaged F1**, and is defined as $\sum_{c=1}^{C} F_1(c)/C$, where $F_1(c)$ is the F_1 score obtained on the task of distinguishing class c from all the others. The other is called **micro-averaged F1**, and is defined as the F_1 score where we pool all the counts from each class's contingency table.

Table 5.5 gives an example that illustrates the difference. We see that the precision of class 1 is 0.5, and of class 2 is 0.9. The macro-averaged precision is therefore 0.7, whereas the micro-averaged precision is 0.83. The latter is much closer to the precision of class 2 than to the precision of class 1, since class 2 is five times larger than class 1. To give equal weight to each class, use macro-averaging.

	Class 1			Class 2			Pooled	
	$y = 1$	$y = 0$		$y = 1$	$y = 0$		$y = 1$	$y = 0$
$\hat{y} = 1$	10	10	$\hat{y} = 1$	90	10	$\hat{y} = 1$	100	20
$\hat{y} = 0$	10	970	$\hat{y} = 0$	10	890	$\hat{y} = 0$	20	1860

Table 5.5 Illustration of the difference between macro- and micro-averaging. y is the true label, and $\hat{y}$ is the called label. In this example, the macro-averaged precision is $[10/(10 + 10) + 90/(10 + 90)]/2 = (0.5 + 0.9)/2 = 0.7$. The micro-averaged precision is $100/(100 + 20) \approx 0.83$. Based on Table 13.7 of (Manning et al. 2008).

5.7.2.4   False discovery rates *

Suppose we are trying to discover a rare phenomenon using some kind of high throughput measurement device, such as a gene expression micro array, or a radio telescope. We will need to make many binary decisions of the form $p(y_i = 1|\mathcal{D}) > \tau$, where $\mathcal{D} = \{\mathbf{x}_i\}_{i=1}^{N}$ and N may be large. This is called **multiple hypothesis testing**. Note that the difference from standard binary classification is that we are classifying y_i based on all the data, not just based on $\mathbf{x}_i$. So this is a simultaneous classification problem, where we might hope to do better than a series of individual classification problems.

How should we set the threshold τ? A natural approach is to try to minimize the expected number of false positives. In the Bayesian approach, this can be computed as follows:

$$FD(\tau, \mathcal{D}) \triangleq \sum_i \underbrace{(1 - p_i)}_{\text{pr. error}} \underbrace{\mathbb{I}(p_i > \tau)}_{\text{discovery}} \tag{5.118}$$

where $p_i \triangleq p(y_i = 1|\mathcal{D})$ is your belief that this object exhibits the phenomenon in question. We then define the posterior expected **false discovery rate** as follows:

$$FDR(\tau, \mathcal{D}) \triangleq FD(\tau, \mathcal{D})/N(\tau, \mathcal{D}) \tag{5.119}$$

where $N(\tau, \mathcal{D}) = \sum_i \mathbb{I}(p_i > \tau)$ is the number of discovered items. Given a desired FDR tolerance, say $\alpha = 0.05$, one can then adapt τ to achieve this; this is called the **direct posterior probability approach** to controlling the FDR (Newton et al. 2004; Muller et al. 2004).

In order to control the FDR it is very helpful to estimate the p_i's jointly (e.g., using a hierarchical Bayesian model, as in Section 5.5), rather than independently. This allows the pooling of statistical strength, and thus lower FDR. See e.g., (Berry and Hochberg 1999) for more information.

5.7.3   Other topics *

In this section, we briefly mention a few other topics related to Bayesian decision theory. We do not have space to go into detail, but we include pointers to the relevant literature.

5.7.3.1   Contextual bandits

A **one-armed bandit** is a colloquial term for a slot machine, found in casinos around the world. The game is this: you insert some money, pull an arm, and wait for the machine to stop; if you're lucky, you win some money. Now imagine there is a bank of K such machines to choose from. Which one should you use? This is called a **multi-armed bandit**, and can be modeled using Bayesian decision theory: there are K possible actions, and each action has an unknown reward (payoff function) r_k. By maintaining a belief state, $p(r_{1:K}|\mathcal{D}) = \prod_k p(r_k|\mathcal{D})$, one can devise an optimal policy; this can be compiled into a series of **Gittins Indices** (Gittins 1989). This optimally solves the **exploration-exploitation** tradeoff, which specifies how many times one should try each action before deciding to go with the winner.

Now consider an extension where each arm, and the player, has an associated feature vector; call all these features $\mathbf{x}$. This is called a **contextual bandit** (see e.g., (Sarkar 1991; Scott 2010; Li et al. 2011)). For example, the "arms" could represent ads or news articles which we want to show to the user, and the features could represent properties of these ads or articles, such as a bag of words, as well as properties of the user, such as demographics. If we assume a linear model for reward, $r_k = \boldsymbol{\theta}_k^T \mathbf{x}$, we can maintain a distribution over the parameters of each arm, $p(\boldsymbol{\theta}_k|\mathcal{D})$, where $\mathcal{D}$ is a series of tuples of the form $(a, \mathbf{x}, r)$, which specifies which arm was pulled, what its features were, and what the resulting outcome was (e.g., $r = 1$ if the user clicked on the ad, and $r = 0$ otherwise). We discuss ways to compute $p(\boldsymbol{\theta}_k|\mathcal{D})$ from linear and logistic regression models in later chapters.

Given the posterior, we must decide what action to take. One common heuristic, known as **UCB** (which stands for "upper confidence bound") is to take the action which maximizes

$$k^* = \underset{k=1}{\overset{K}{\operatorname{argmax}}} \mu_k + \lambda \sigma_k \tag{5.120}$$

where $\mu_k = \mathbb{E}[r_k|\mathcal{D}]$, $\sigma_k^2 = \operatorname{var}[r_k|\mathcal{D}]$ and λ is a tuning parameter that trades off exploration and exploitation. The intuition is that we should pick actions which we believe are good (μ_k is large), and/ or actions about which we are uncertain (σ_k is large).

An even simpler method, known as **Thompson sampling**, is as follows. At each step, we pick action k with a probability that is equal to its probability of being the optimal action:

$$p_k = \int \mathbb{I}(\mathbb{E}[r|a, \mathbf{x}, \boldsymbol{\theta}] = \max_{a'} \mathbb{E}[r|a', \mathbf{x}, \boldsymbol{\theta}]) p(\boldsymbol{\theta}|\mathcal{D}) d\boldsymbol{\theta} \tag{5.121}$$

We can approximate this by drawing a single sample from the posterior, $\boldsymbol{\theta}^t \sim p(\boldsymbol{\theta}|\mathcal{D})$, and then choosing $k^* = \operatorname{argmax}_k \mathbb{E}[r|\mathbf{x}, k, \boldsymbol{\theta}^t]$. Despite its simplicity, this has been shown to work quite well (Chapelle and Li 2011).

5.7.3.2 Utility theory

Suppose we are a doctor trying to decide whether to operate on a patient or not. We imagine there are 3 states of nature: the patient has no cancer, the patient has lung cancer, or the patient has breast cancer. Since the action and state space is discrete, we can represent the loss function $L(\theta, a)$ as a **loss matrix**, such as the following:

	Surgery	No surgery
No cancer	20	0
Lung cancer	10	50
Breast cancer	10	60

These numbers reflects the fact that not performing surgery when the patient has cancer is very bad (loss of 50 or 60, depending on the type of cancer), since the patient might die; not performing surgery when the patient does not have cancer incurs no loss (0); performing surgery when the patient does not have cancer is wasteful (loss of 20); and performing surgery when the patient does have cancer is painful but necessary (10).

It is natural to ask where these numbers come from. Ultimately they represent the personal **preferences** or values of a fictitious doctor, and are somewhat arbitrary: just as some people prefer chocolate ice cream and others prefer vanilla, there is no such thing as the "right" loss/ utility function. However, it can be shown (see e.g., (DeGroot 1970)) that any set of consistent preferences can be converted to a scalar loss/ utility function. Note that utility can be measured on an arbitrary scale, such as dollars, since it is only relative values that matter.[7]

5.7.3.3 Sequential decision theory

So far, we have concentrated on **one-shot decision problem**s, where we only have to make one decision and then the game ends. In Section 10.6, we will generalize this to multi-stage or sequential decision problems. Such problems frequently arise in many business and engineering settings. This is closely related to the problem of reinforcement learning. However, further discussion of this point is beyond the scope of this book.

Exercises

Exercise 5.1 Proof that a mixture of conjugate priors is indeed conjugate

Derive Equation 5.69.

Exercise 5.2 Optimal threshold on classification probability

Consider a case where we have learned a conditional probability distribution $P(y|\mathbf{x})$. Suppose there are only two classes, and let $p_0 = P(Y = 0|\mathbf{x})$ and $p_1 = P(Y = 1|\mathbf{x})$. Consider the loss matrix below:

predicted label $\hat{y}$	true label y	
	0	1
0	0	λ_{01}
1	λ_{10}	0

a. Show that the decision $\hat{y}$ that minimizes the expected loss is equivalent to setting a probability threshold θ and predicting $\hat{y} = 0$ if $p_1 < \theta$ and $\hat{y} = 1$ if $p_1 \geq \theta$. What is θ as a function of λ_{01} and λ_{10}? (Show your work.)

b. Show a loss matrix where the threshold is 0.1. (Show your work.)

---

7. People are often squeamish about talking about human lives in monetary terms, but all decision making requires tradeoffs, and one needs to use some kind of "currency" to compare different courses of action. Insurance companies do this all the time. Ross Schachter, a decision theorist at Stanford University, likes to tell a story of a school board who rejected a study on asbestos removal from schools because it performed a **cost-benefit analysis**, which was considered "inhumane" because they put a dollar value on children's health; the result of rejecting the report was that the asbestos was not removed, which is surely more "inhumane". In medical domains, one often measures utility in terms of **QALY**, or quality-adjusted life-years, instead of dollars, but it's the same idea. Of course, even if you do not explicitly specify how much you value different people's lives, your *behavior* will reveal your implicit values / preferences, and these preferences can then be converted to a real-valued scale, such as dollars or QALY. Inferring a utility function from behavior is called **inverse reinforcement learning**.

Exercise 5.3 Reject option in classifiers

(Source: (Duda et al. 2001, Q2.13).)

In many classification problems one has the option either of assigning $\mathbf{x}$ to class j or, if you are too uncertain, of choosing the **reject option**. If the cost for rejects is less than the cost of falsely classifying the object, it may be the optimal action. Let α_i mean you choose action i, for $i = 1 : C + 1$, where C is the number of classes and $C + 1$ is the reject action. Let $Y = j$ be the true (but unknown) **state of nature**. Define the loss function as follows

$$\lambda(\alpha_i | Y = j) = \begin{cases} 0 & \text{if } i = j \text{ and } i, j \in \{1, \dots, C\} \\ \lambda_r & \text{if } i = C + 1 \\ \lambda_s & \text{otherwise} \end{cases} \tag{5.122}$$

In otherwords, you incur 0 loss if you correctly classify, you incur λ_r loss (cost) if you choose the reject option, and you incur λ_s loss (cost) if you make a substitution error (misclassification).

a. Show that the minimum risk is obtained if we decide $Y = j$ if $p(Y = j|\mathbf{x}) \geq p(Y = k|\mathbf{x})$ for all k (i.e., j is the most probable class) *and* if $p(Y = j|\mathbf{x}) \geq 1 - \frac{\lambda_r}{\lambda_s}$; otherwise we decide to reject.

b. Describe qualitatively what happens as λ_r/λ_s is increased from 0 to 1 (i.e., the relative cost of rejection increases).

Exercise 5.4 More reject options

In many applications, the classifier is allowed to "reject" a test example rather than classifying it into one of the classes. Consider, for example, a case in which the cost of a misclassification is \$10 but the cost of having a human manually make the decison is only \$3. We can formulate this as the following loss matrix:

Decision	true label y	
$\hat{y}$	0	1
predict 0	0	10
predict 1	10	0
reject	3	3

a. Suppose $P(y = 1|\mathbf{x})$ is predicted to be 0.2. Which decision minimizes the expected loss?

b. Now suppose $P(y = 1|\mathbf{x}) = 0.4$. Now which decision minimizes the expected loss?

c. Show that in general, for this loss matrix, but for any posterior distribution, there will be two thresholds θ_0 and θ_1 such that the optimal decision is to predict 0 if $p_1 < \theta_0$, reject if $\theta_0 \leq p_1 \leq \theta_1$, and predict 1 if $p_1 > \theta_1$ (where $p_1 = p(y = 1|\mathbf{x})$). What are these thresholds?

Exercise 5.5 Newsvendor problem

Consider the following classic problem in decision theory / economics. Suppose you are trying to decide how much quantity Q of some product (e.g., newspapers) to buy to maximize your profits. The optimal amount will depend on how much demand D you think there is for your product, as well as its cost to you C and its selling price P. Suppose D is unknown but has pdf $f(D)$ and cdf $F(D)$. We can evaluate the expected profit by considering two cases: if $D > Q$, then we sell all Q items, and make profit $\pi = (P - C)Q$; but if $D < Q$, we only sell D items, at profit $(P - C)D$, but have wasted $C(Q - D)$ on the unsold items. So the expected profit if we buy quantity Q is

$$E\pi(Q) = \int_Q^\infty (P - C)Qf(D)dD + \int_0^Q (P - C)Df(D) - \int_0^Q C(Q - D)f(D)dD \tag{5.123}$$

Simplify this expression, and then take derivatives wrt Q to show that the optimal quantity Q^* (which maximizes the expected profit) satisfies

$$F(Q^*) = \frac{P - C}{P} \tag{5.124}$$

Exercise 5.6 Bayes factors and ROC curves

Let $B = p(D|H_1)/p(D|H_0)$ be the bayes factor in favor of model 1. Suppose we plot two ROC curves, one computed by thresholding B, and the other computed by thresholding $p(H_1|D)$. Will they be the same or different? Explain why.

Exercise 5.7 Bayes model averaging helps predictive accuracy

Let Δ be a quantity that we want to predict, let $\mathcal{D}$ be the observed data and $\mathcal{M}$ be a finite set of models. Suppose our action is to provide a probabilistic prediction $p()$, and the loss function is $L(\Delta, p()) = -\log p(\Delta)$. We can either perform Bayes model averaging and predict using

$$p^{BMA}(\Delta) = \sum_{m \in \mathcal{M}} p(\Delta|m, \mathcal{D})p(m|\mathcal{D}) \tag{5.125}$$

or we could predict using any single model (a plugin approximation)

$$p^m(\Delta) = p(\Delta|m, \mathcal{D}) \tag{5.126}$$

Show that, for all models $m \in \mathcal{M}$, the posterior expected loss using BMA is lower, i.e.,

$$\mathbb{E}\left[L(\Delta, p^{BMA})\right] \le \mathbb{E}\left[L(\Delta, p^m)\right] \tag{5.127}$$

where the expectation over Δ is with respect to

$$p(\Delta|\mathcal{D}) = \sum_{m \in \mathcal{M}} p(\Delta|m, \mathcal{D})p(m|\mathcal{D}) \tag{5.128}$$

Hint: use the non-negativity of the KL divergence.

Exercise 5.8 MLE and model selection for a 2d discrete distribution

(Source: Jaakkola.)

Let $x \in \{0, 1\}$ denote the result of a coin toss ($x = 0$ for tails, $x = 1$ for heads). The coin is potentially biased, so that heads occurs with probability θ_1. Suppose that someone else observes the coin flip and reports to you the outcome, y. But this person is unreliable and only reports the result correctly with probability θ_2; i.e., $p(y|x, \theta_2)$ is given by

	$y = 0$	$y = 1$
$x = 0$	θ_2	$1 - \theta_2$
$x = 1$	$1 - \theta_2$	θ_2

Assume that θ_2 is independent of x and θ_1.

a. Write down the joint probability distribution $p(x, y|\boldsymbol{\theta})$ as a 2×2 table, in terms of $\boldsymbol{\theta} = (\theta_1, \theta_2)$.

b. Suppose have the following dataset: $\mathbf{x} = (1, 1, 0, 1, 1, 0, 0)$, $\mathbf{y} = (1, 0, 0, 0, 1, 0, 1)$. What are the MLEs for θ_1 and θ_2? Justify your answer. Hint: note that the likelihood function factorizes,

$$p(x, y|\boldsymbol{\theta}) = p(y|x, \theta_2)p(x|\theta_1) \tag{5.129}$$

What is $p(\mathcal{D}|\hat{\boldsymbol{\theta}}, M_2)$ where M_2 denotes this 2-parameter model? (You may leave your answer in fractional form if you wish.)

c. Now consider a model with 4 parameters, $\boldsymbol{\theta} = (\theta_{0,0}, \theta_{0,1}, \theta_{1,0}, \theta_{1,1})$, representing $p(x, y|\boldsymbol{\theta}) = \theta_{x,y}$. (Only 3 of these parameters are free to vary, since they must sum to one.) What is the MLE of $\boldsymbol{\theta}$? What is $p(\mathcal{D}|\hat{\boldsymbol{\theta}}, M_4)$ where M_4 denotes this 4-parameter model?

d. Suppose we are not sure which model is correct. We compute the leave-one-out cross validated log likelihood of the 2-parameter model and the 4-parameter model as follows:

$$L(m) = \sum_{i=1}^{n} \log p(x_i, y_i | m, \hat{\theta}(\mathcal{D}_{-i})) \tag{5.130}$$

and $\hat{\theta}(\mathcal{D}_{-i}))$ denotes the MLE computed on $\mathcal{D}$ excluding row i. Which model will CV pick and why? Hint: notice how the table of counts changes when you omit each training case one at a time.

e. Recall that an alternative to CV is to use the BIC score, defined as

$$\mathrm{BIC}(M, \mathcal{D}) \triangleq \log p(\mathcal{D}|\hat{\boldsymbol{\theta}}_{MLE}) - \frac{\mathrm{dof}(M)}{2} \log N \tag{5.131}$$

where $\mathrm{dof}(M)$ is the number of free parameters in the model, Compute the BIC scores for both models (use log base e). Which model does BIC prefer?

Exercise 5.9 Posterior median is optimal estimate under L1 loss

Prove that the posterior median is optimal estimate under L1 loss.

Exercise 5.10 Decision rule for trading off FPs and FNs

If $L_{FN} = cL_{FP}$, show that we should pick $\hat{y} = 1$ iff $p(y = 1|\mathbf{x})/p(y = 0|\mathbf{x}) > \tau$, where $\tau = c/(1 + c)$

6 *Frequentist statistics*

6.1 Introduction

The approach to statistical inference that we described in Chapter 5 is known as Bayesian statistics. Perhaps surprisingly, this is considered controversial by some people, whereas the application of Bayes rule to non-statistical problems — such as medical diagnosis (Section 2.2.3.1), spam filtering (Section 3.4.4.1), or airplane tracking (Section 18.2.1) — is not controversial. The reason for the objection has to do with a misguided distinction between parameters of a statistical model and other kinds of unknown quantities.[1]

Attempts have been made to devise approaches to statistical inference that avoid treating parameters like random variables, and which thus avoid the use of priors and Bayes rule. Such approaches are known as **frequentist statistics**, **classical statistics** or **orthodox statistics**. Instead of being based on the posterior distribution, they are based on the concept of a sampling distribution. This is the distribution that an estimator has when applied to multiple data sets sampled from the true but unknown distribution; see Section 6.2 for details. It is this notion of variation across repeated trials that forms the basis for modeling uncertainty used by the frequentist approach.

By contrast, in the Bayesian approach, we only ever condition on the actually observed data; there is no notion of repeated trials. This allows the Bayesian to compute the probability of one-off events, as we discussed in Section 2.1. Perhaps more importantly, the Bayesian approach avoids certain paradoxes that plague the frequentist approach (see Section 6.6). Nevertheless, it is important to be familiar with frequentist statistics (especially Section 6.5), since it is widely used in machine learning.

6.2 Sampling distribution of an estimator

In frequentist statistics, a parameter estimate $\hat{\theta}$ is computed by applying an **estimator** δ to some data $\mathcal{D}$, so $\hat{\theta} = \delta(\mathcal{D})$. The parameter is viewed as fixed and the data as random, which is the exact opposite of the Bayesian approach. The uncertainty in the parameter estimate can be measured by computing the **sampling distribution** of the estimator. To understand this

---

1. Parameters are sometimes considered to represent true (but unknown) physical quantities, which are therefore not random. However, we have seen that it is perfectly reasonable to use a probability distribution to represent one's uncertainty about an unknown constant.

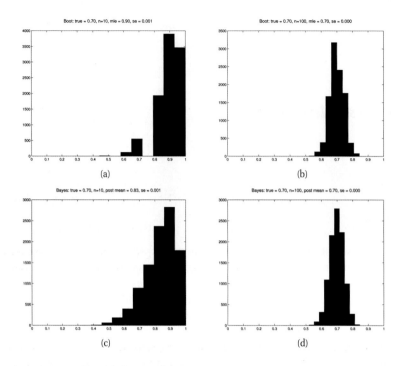

Figure 6.1 Bootstrap (top row) vs Bayes (bottom row). The N data cases were generated from $\text{Ber}(\theta = 0.7)$. Left column: $N = 10$. Right column: $N = 100$. (a-b) A bootstrap approximation to the sampling distribution of the MLE for a Bernoulli distribution. We show the histogram derived from $B = 10,000$ bootstrap samples. (c-d) Histogram of 10,000 samples from the posterior distirbution using a uniform prior. Figure generated by `bootstrapDemoBer`.

concept, imagine sampling many different data sets $\mathcal{D}^{(s)}$ from some true model, $p(\cdot|\theta^*)$, i.e., let $\mathcal{D}^{(s)} = \{x_i^{(s)}\}_{i=1}^N$, where $x_i^s \sim p(\cdot|\theta^*)$, and θ^* is the true parameter. Here $s = 1 : S$ indexes the sampled data set, and N is the size of each such dataset. Now apply the estimator $\hat{\theta}(\cdot)$ to each $\mathcal{D}^{(s)}$ to get a set of estimates, $\{\hat{\theta}(\mathcal{D}^{(s)})\}$. As we let $S \to \infty$, the distribution induced on $\hat{\theta}(\cdot)$ is the sampling distribution of the estimator. We will discuss various ways to use the sampling distribution in later sections. But first we sketch two approaches for computing the sampling distribution itself.

6.2.1 Bootstrap

The **bootstrap** is a simple Monte Carlo technique to approximate the sampling distribution. This is particularly useful in cases where the estimator is a complex function of the true parameters.

The idea is simple. If we knew the true parameters θ^*, we could generate many (say S) fake datasets, each of size N, from the true distribution, $x_i^s \sim p(\cdot|\theta^*)$, for $s = 1 : S$, $i = 1 : N$. We could then compute our estimator from each sample, $\hat{\theta}^s = f(x_{1:N}^s)$ and use the empirical distribution of the resulting samples as our estimate of the sampling distribution. Since θ is unknown, the idea of the **parametric bootstrap** is to generate the samples using $\hat{\theta}(\mathcal{D})$ instead. An alternative, called the **non-parametric bootstrap**, is to sample the x_i^s (with replacement) from the original data $\mathcal{D}$, and then compute the induced distribution as before. Some methods for speeding up the bootstrap when applied to massive data sets are discussed in (Kleiner et al. 2011).

Figure 6.1(a-b) shows an example where we compute the sampling distribution of the MLE for a Bernoulli using the parametric bootstrap. (Results using the non-parametric bootstrap are essentially the same.) We see that the sampling distribution is asymmetric, and therefore quite far from Gaussian, when $N = 10$; when $N = 100$, the distribution looks more Gaussian, as theory suggests (see below).

A natural question is: what is the connection between the parameter estimates $\hat{\theta}^s = \hat{\theta}(x_{1:N}^s)$ computed by the bootstrap and parameter values sampled from the posterior, $\theta^s \sim p(\cdot|\mathcal{D})$? Conceptually they are quite different. But in the common case that that the prior is not very strong, they can be quite similar. For example, Figure 6.1(c-d) shows an example where we compute the posterior using a uniform Beta(1,1) prior, and then sample from it. We see that the posterior and the sampling distribution are quite similar. So one can think of the bootstrap distribution as a "poor man's" posterior; see (Hastie et al. 2001, p235) for details.

However, perhaps surprisingly, bootstrap can be slower than posterior sampling. The reason is that the bootstrap has to fit the model S times, whereas in posterior sampling, we usually only fit the model once (to find a local mode), and then perform local exploration around the mode. Such local exploration is usually much faster than fitting a model from scratch.

6.2.2 Large sample theory for the MLE *

In some cases, the sampling distribution for some estimators can be computed analytically. In particular, it can be shown that, under certain conditions, as the sample size tends to infinity, the sampling distribution of the MLE becomes Gaussian. Informally, the requirement for this result to hold is that each parameter in the model gets to "see" an infinite amount of data, and that the model be identifiable. Unfortunately this excludes many of the models of interest to machine learning. Nevertheless, let us assume we are in a simple setting where the theorem holds.

The center of the Gaussian will be the MLE $\hat{\boldsymbol{\theta}}$. But what about the variance of this Gaussian? Intuitively the variance of the estimator will be (inversely) related to the amount of curvature of the likelihood surface at its peak. If the curvature is large, the peak will be "sharp", and the variance low; in this case, the estimate is "well determined". By contrast, if the curvature is small, the peak will be nearly "flat", so the variance is high.

Let us now formalize this intuition. Define the **score function** as the gradient of the log likelihood evaluated at some point $\hat{\boldsymbol{\theta}}$:

$$\mathbf{s}(\hat{\boldsymbol{\theta}}) \triangleq \nabla \log p(\mathcal{D}|\boldsymbol{\theta})|_{\hat{\boldsymbol{\theta}}} \tag{6.1}$$

Define the **observed information matrix** as the gradient of the negative score function, or equivalently, the Hessian of the NLL:

$$\mathbf{J}(\hat{\boldsymbol{\theta}}(\mathcal{D})) \triangleq -\nabla \mathbf{s}(\hat{\boldsymbol{\theta}}) = -\nabla_{\boldsymbol{\theta}}^2 \log p(\mathcal{D}|\boldsymbol{\theta})|_{\hat{\boldsymbol{\theta}}} \tag{6.2}$$

In 1D, this becomes

$$J(\hat{\theta}(\mathcal{D})) = -\frac{d}{d\theta^2} \log p(\mathcal{D}|\theta)|_{\hat{\theta}} \tag{6.3}$$

This is just a measure of curvature of the negative log-likelihood function at $\hat{\theta}$.

Since we are studying the sampling distribution, $\mathcal{D} = (\mathbf{x}_1, \ldots, \mathbf{x}_N)$ is a set of random variables. The **Fisher information matrix** is defined to be the expected value of the observed information matrix:[2]

$$\mathbf{I}_N(\hat{\boldsymbol{\theta}}|\boldsymbol{\theta}^*) \triangleq \mathbb{E}_{\boldsymbol{\theta}^*}\left[\mathbf{J}(\hat{\boldsymbol{\theta}}(\mathcal{D}))\right] \tag{6.4}$$

where $\mathbb{E}_{\boldsymbol{\theta}^*}[\mathbf{f}(\mathcal{D})] \triangleq \frac{1}{N}\sum_{i=1}^N \mathbf{f}(\mathbf{x}_i)p(\mathbf{x}_i|\boldsymbol{\theta}^*)$ is the expected value of the function $\mathbf{f}$ when applied to data sampled from $\boldsymbol{\theta}^*$. Often $\boldsymbol{\theta}^*$, representing the "true parameter" that generated the data, is assumed known, so we just write $\mathbf{I}_N(\hat{\boldsymbol{\theta}}) \triangleq \mathbf{I}_N(\hat{\boldsymbol{\theta}}|\boldsymbol{\theta}^*)$ for short. Furthermore, it is easy to see that $\mathbf{I}_N(\hat{\boldsymbol{\theta}}) = N\mathbf{I}_1(\hat{\boldsymbol{\theta}})$, because the log-likelihood for a sample of size N is just N times "steeper" than the log-likelihood for a sample of size 1. So we can just write $\mathbf{I}(\hat{\boldsymbol{\theta}}) \triangleq \mathbf{I}_1(\hat{\boldsymbol{\theta}})$. This is the notation that is usually used.

Now let $\hat{\boldsymbol{\theta}} \triangleq \hat{\boldsymbol{\theta}}_{mle}(\mathcal{D})$ be the MLE, where $\mathcal{D} \sim \boldsymbol{\theta}^*$. It can be shown that

$$\hat{\boldsymbol{\theta}} \to \mathcal{N}(\boldsymbol{\theta}^*, \mathbf{I}_N(\boldsymbol{\theta}^*)^{-1}) \tag{6.5}$$

as $N \to \infty$ (see e.g., (Rice 1995, p265) for a proof). We say that the sampling distribution of the MLE is **asymptotically normal**.

What about the variance of the MLE, which can be used as some measure of confidence in the MLE? Unfortunately, $\boldsymbol{\theta}^*$ is unknown, so we can't evaluate the variance of the sampling distribution. However, we can approximate the sampling distribution by replacing $\boldsymbol{\theta}^*$ with $\hat{\boldsymbol{\theta}}$. Consequently, the approximate **standard errors** of $\hat{\theta}_k$ are given by

$$\hat{se}_k \triangleq \mathbf{I}_N(\hat{\boldsymbol{\theta}})_{k,k}^{-\frac{1}{2}} \tag{6.6}$$

For example, from Equation 5.60 we know that the Fisher information for a binomial sampling model is

$$I(\theta) = \frac{1}{\theta(1-\theta)} \tag{6.7}$$

So the approximate standard error of the MLE is

$$\hat{se} = \frac{1}{\sqrt{I_N(\hat{\theta})}} = \frac{1}{\sqrt{NI(\hat{\theta})}} = \left(\frac{\hat{\theta}(1-\hat{\theta})}{N}\right)^{\frac{1}{2}} \tag{6.8}$$

---

2. This is not the usual definition, but is equivalent to it under standard assumptions. More precisely, the standard definition is as follows (we just give the scalar case to simplify notation): $I(\hat{\theta}|\theta^*) \triangleq \text{var}_{\theta^*}\left[\frac{d}{d\theta}\log p(X|\theta)|_{\hat{\theta}}\right]$, that is, the variance of the score function. If $\hat{\theta}$ is the MLE, it is easy to see that $\mathbb{E}_{\theta^*}\left[\frac{d}{d\theta}\log p(X|\theta)|_{\hat{\theta}}\right] = 0$ (since the gradient must be zero at a maximum), so the variance reduces to the expected square of the score function: $I(\hat{\theta}|\theta^*) = \mathbb{E}_{\theta^*}\left[(\frac{d}{d\theta}\log p(X|\theta))^2\right]$. It can be shown (e.g., (Rice 1995, p263)) that $\mathbb{E}_{\theta^*}\left[(\frac{d}{d\theta}\log p(X|\theta))^2\right] = -\mathbb{E}_{\theta^*}\left[\frac{d^2}{d\theta^2}\log p(X|\theta)\right]$, so now the Fisher information reduces to the expected second derivative of the NLL, which is a much more intuitive quantity than the variance of the score.

where $\hat{\theta} = \frac{1}{N} \sum_i X_i$. Compare this to Equation 3.27, which is the posterior standard deviation under a uniform prior.

6.3 Frequentist decision theory

In frequentist or classical decision theory, there is a loss function and a likelihood, but there is no prior and hence no posterior or posterior expected loss. Thus there is no automatic way of deriving an optimal estimator, unlike the Bayesian case. Instead, in the frequentist approach, we are free to choose any estimator or decision procedure $\delta : \mathcal{X} \to \mathcal{A}$ we want.[3]

Having chosen an estimator, we define its expected loss or **risk** as follows:

$$R(\theta^*, \delta) \triangleq \mathbb{E}_{p(\tilde{\mathcal{D}}|\theta^*)}\left[L(\theta^*, \delta(\tilde{\mathcal{D}}))\right] = \int L(\theta^*, \delta(\tilde{\mathcal{D}}))p(\tilde{\mathcal{D}}|\theta^*)d\tilde{\mathcal{D}} \tag{6.9}$$

where $\tilde{\mathcal{D}}$ is data sampled from "nature's distribution", which is represented by parameter θ^*. In other words, the expectation is wrt the sampling distribution of the estimator. Compare this to the Bayesian posterior expected loss:

$$\rho(a|\mathcal{D}, \pi) \triangleq \mathbb{E}_{p(\theta|\mathcal{D}, \pi)}\left[L(\theta, a)\right] = \int_\Theta L(\theta, \mathbf{a})p(\theta|\mathcal{D}, \pi)d\theta \tag{6.10}$$

We see that the Bayesian approach averages over θ (which is unknown) and conditions on $\mathcal{D}$ (which is known), whereas the frequentist approach averages over $\tilde{\mathcal{D}}$ (thus ignoring the observed data), and conditions on θ^* (which is unknown).

Not only is the frequentist definition unnatural, it cannot even be computed, because θ^* is unknown. Consequently, we cannot compare different estimators in terms of their frequentist risk. We discuss various solutions to this below.

6.3.1 Bayes risk

How do we choose amongst estimators? We need some way to convert $R(\boldsymbol{\theta}^*, \delta)$ into a single measure of quality, $R(\delta)$, which does not depend on knowing $\boldsymbol{\theta}^*$. One approach is to put a prior on θ^*, and then to define **Bayes risk** or **integrated risk** of an estimator as follows:

$$R_B(\delta) \triangleq \mathbb{E}_{p(\boldsymbol{\theta}^*)}\left[R(\boldsymbol{\theta}^*, \delta)\right] = \int R(\boldsymbol{\theta}^*, \delta)p(\boldsymbol{\theta}^*)d\boldsymbol{\theta}^* \tag{6.11}$$

A **Bayes estimator** or **Bayes decision rule** is one which minimizes the expected risk:

$$\delta_B \triangleq \underset{\delta}{\operatorname{argmin}}\, R_B(\delta) \tag{6.12}$$

Note that the integrated risk is also called the **preposterior risk**, since it is before we have seen the data. Minimizing this can be useful for experiment design.

We will now prove a very important theorem that connects the Bayesian and frequentist approaches to decision theory.

---

3. In practice, the frequentist approach is usually only applied to one-shot statistical decision problems — such as classification, regression and parameter estimation — since its non-constructive nature makes it difficult to apply to sequential decision problems, which adapt to data online.

Theorem 6.3.1. *A Bayes estimator can be obtained by minimizing the posterior expected loss for each* $\mathbf{x}$.

Proof. By switching the order of integration, we have

$$R_B(\delta) = \int \left[\sum_{\mathbf{x}} \sum_{y} L(y, \delta(\mathbf{x})) p(\mathbf{x}, y | \boldsymbol{\theta}^*) \right] p(\boldsymbol{\theta}^*) d\boldsymbol{\theta}^* \tag{6.13}$$

$$= \sum_{\mathbf{x}} \sum_{y} \int_{\Theta} L(y, \delta(\mathbf{x})) p(\mathbf{x}, y, \boldsymbol{\theta}^*) d\boldsymbol{\theta}^* \tag{6.14}$$

$$= \sum_{\mathbf{x}} \left[\sum_{y} L(y, \delta(\mathbf{x})) p(y | \mathbf{x}) dy \right] p(\mathbf{x}) \tag{6.15}$$

$$= \sum_{\mathbf{x}} \rho(\delta(\mathbf{x}) | \mathbf{x}) \; p(\mathbf{x}) \tag{6.16}$$

To minimize the overall expectation, we just minimize the term inside for each $\mathbf{x}$, so our decision rule is to pick

$$\delta_B(\mathbf{x}) = \operatorname*{argmin}_{a \in \mathcal{A}} \rho(a | \mathbf{x}) \tag{6.17}$$

$\square$

Hence we see that the picking the optimal action on a case-by-case basis (as in the Bayesian approach) is optimal on average (as in the frequentist approach). In other words, the Bayesian approach provides a good way of achieving frequentist goals. In fact, one can go further and prove the following.

Theorem 6.3.2 (Wald, 1950). *Every admissable decision rule is a Bayes decision rule with respect to some, possibly improper, prior distribution.*

This theorem shows that *the best way to minimize frequentist risk is to be Bayesian!* See (Bernardo and Smith 1994, p448) for further discussion of this point.

6.3.2   Minimax risk

Obviously some frequentists dislike using Bayes risk since it requires the choice of a prior (although this is only in the evaluation of the estimator, not necessarily as part of its construction). An alternative approach is as follows. Define the **maximum risk** of an estimator as

$$R_{max}(\delta) \triangleq \max_{\boldsymbol{\theta}^*} R(\boldsymbol{\theta}^*, \delta) \tag{6.18}$$

A **minimax rule** is one which minimizes the maximum risk:

$$\delta_{MM} \triangleq \operatorname*{argmin}_{\delta} R_{max}(\delta) \tag{6.19}$$

For example, in Figure 6.2, we see that δ_1 has lower worst-case risk than δ_2, ranging over all possible values of θ^*, so it is the minimax estimator (see Section 6.3.3.1 for an explanation of how to compute a risk function for an actual model).

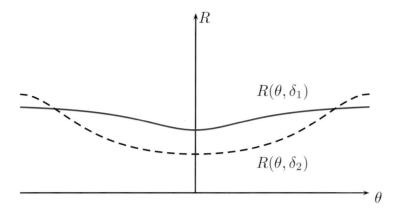

Figure 6.2 Risk functions for two decision procedures, δ_1 and δ_2. Since δ_1 has lower worst case risk, it is the minimax estimator, even though δ_2 has lower risk for most values of θ. Thus minimax estimators are overly conservative.

Minimax estimators have a certain appeal. However, computing them can be hard. And furthermore, they are very pessimistic. In fact, one can show that all minimax estimators are equivalent to Bayes estimators under a **least favorable prior**. In most statistical situations (excluding game theoretic ones), assuming nature is an adversary is not a reasonable assumption.

6.3.3 Admissible estimators

The basic problem with frequentist decision theory is that it relies on knowing the true distribution $p(\cdot|\theta^*)$ in order to evaluate the risk. However, it might be the case that some estimators are worse than others regardless of the value of θ^*. In particular, if $R(\theta, \delta_1) \leq R(\theta, \delta_2)$ for all $\theta \in \Theta$, then we say that δ_1 **dominates** δ_2. The domination is said to be strict if the inequality is strict for some θ. An estimator is said to be **admissible** if it is not strictly dominated by any other estimator.

6.3.3.1 Example

Let us give an example, based on (Bernardo and Smith 1994). Consider the problem of estimating the mean of a Gaussian. We assume the data is sampled from $x_i \sim \mathcal{N}(\theta^*, \sigma^2 = 1)$ and use quadratic loss, $L(\theta, \hat{\theta}) = (\theta - \hat{\theta})^2$. The corresponding risk function is the MSE. Some possible decision rules or estimators $\hat{\theta}(\mathbf{x}) = \delta(\mathbf{x})$ are as follows:

- $\delta_1(\mathbf{x}) = \overline{x}$, the sample mean

- $\delta_2(\mathbf{x}) = \tilde{\mathbf{x}}$, the sample median

- $\delta_3(\mathbf{x}) = \theta_0$, a fixed value

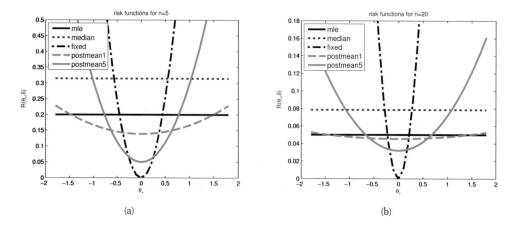

Figure 6.3 Risk functions for estimating the mean of a Gaussian using data sampled $\mathcal{N}(\theta^*, \sigma^2 = 1)$. The solid dark blue horizontal line is the MLE, the solid light blue curved line is the posterior mean when $\kappa = 5$. Left: $N = 5$ samples. Right: $N = 20$ samples. Based on Figure B.1 of (Bernardo and Smith 1994). Figure generated by `riskFnGauss`.

- $\delta_\kappa(\mathbf{x})$, the posterior mean under a $\mathcal{N}(\theta|\theta_0, \sigma^2/\kappa)$ prior:

$$\delta_\kappa(\mathbf{x}) = \frac{N}{N+\kappa}\overline{x} + \frac{\kappa}{N+\kappa}\theta_0 = w\overline{x} + (1-w)\theta_0 \tag{6.20}$$

For δ_κ, we consider a weak prior, $\kappa = 1$, and a stronger prior, $\kappa = 5$. The prior mean is θ_0, some fixed value. We assume σ^2 is known. (Thus $\delta_3(\mathbf{x})$ is the same as $\delta_\kappa(\mathbf{x})$ with an infinitely strong prior, $\kappa = \infty$.)

Let us now derive the risk functions analytically. (We can do this since in this toy example, we know the true parameter θ^*.) In Section 6.4.4, we show that the MSE can be decomposed into squared bias plus variance:

$$MSE(\hat{\theta}(\cdot)|\theta^*) = \text{var}\left[\hat{\theta}\right] + \text{bias}^2(\hat{\theta}) \tag{6.21}$$

The sample mean is unbiased, so its risk is

$$MSE(\delta_1|\theta^*) = \text{var}\left[\overline{x}\right] = \frac{\sigma^2}{N} \tag{6.22}$$

The sample median is also unbiased. One can show that the variance is approximately $\pi/(2N)$, so

$$MSE(\delta_2|\theta^*) = \frac{\pi}{2N} \tag{6.23}$$

For $\delta_3(\mathbf{x}) = \theta_0$, the variance is zero, so

$$MSE(\delta_3|\theta^*) = (\theta^* - \theta_0)^2 \tag{6.24}$$

Finally, for the posterior mean, we have

$$MSE(\delta_\kappa|\theta^*) \quad = \quad \mathbb{E}\left[(w\overline{x} + (1-w)\theta_0 - \theta^*)^2\right] \tag{6.25}$$

$$= \quad \mathbb{E}\left[(w(\overline{x} - \theta^*) + (1-w)(\theta_0 - \theta^*))^2\right] \tag{6.26}$$

$$= \quad w^2\frac{\sigma^2}{N} + (1-w)^2(\theta_0 - \theta^*)^2 \tag{6.27}$$

$$= \quad \frac{1}{(N+\kappa)^2}\left(N\sigma^2 + \kappa^2(\theta_0 - \theta^*)^2\right) \tag{6.28}$$

These functions are plotted in Figure 6.3 for $N \in \{5, 20\}$. We see that in general, the best estimator depends on the value of θ^*, which is unknown. If θ^* is very close to θ_0, then δ_3 (which just predicts θ_0) is best. If θ^* is within some reasonable range around θ_0, then the posterior mean, which combines the prior guess of θ_0 with the actual data, is best. If θ^* is far from θ_0, the MLE is best. None of this should be suprising: a small amount of shrinkage (using the posterior mean with a weak prior) is usually desirable, assuming our prior mean is sensible.

What is more surprising is that the risk of decision rule δ_2 (sample median) is always higher than that of δ_1 (sample mean) for every value of θ^*. Consequently the sample median is an inadmissible estimator for this particular problem (where the data is assumed to come from a Gaussian).

In practice, the sample median is often better than the sample mean, because it is more robust to outliers. One can show (Minka 2000d) that the median is the Bayes estimator (under squared loss) if we assume the data comes from a Laplace distribution, which has heavier tails than a Gaussian. More generally, we can construct robust estimators by using flexible models of our data, such as mixture models or non-parametric density estimators (Section 14.7.2), and then computing the posterior mean or median.

6.3.3.2 Stein's paradox *

Suppose we have N iid random variables $X_i \sim \mathcal{N}(\theta_i, 1)$, and we want to estimate the θ_i. The obvious estimator is the MLE, which in this case sets $\hat{\theta}_i = x_i$. It turns out that this is an inadmissible estimator under quadratic loss, when $N \geq 4$.

To show this, it suffices to construct an estimator that is better. The James-Stein estimator is one such estimator, and is defined as follows:

$$\hat{\theta}_i \quad = \quad \hat{B}\overline{x} + (1 - \hat{B})x_i = \overline{x} + (1 - \hat{B})(x_i - \overline{x}) \tag{6.29}$$

where $\overline{x} = \frac{1}{N}\sum_{i=1}^{N} x_i$ and $0 < B < 1$ is some tuning constant. This estimate "shrinks" the θ_i towards the overall mean. (We derive this estimator using an empirical Bayes approach in Section 5.6.2.)

It can be shown that this shrinkage estimator has lower frequentist risk (MSE) than the MLE (sample mean) for $N \geq 4$. This is known as **Stein's paradox**. The reason it is called a paradox is illustrated by the following example. Suppose θ_i is the "true" IQ of student i and X_i is his test score. Why should my estimate of θ_i depend on the global mean $\overline{x}$, and hence on some other student's scores? One can create even more paradoxical examples by making the different dimensions be qualitatively different, e.g., θ_1 is my IQ, θ_2 is the average rainfall in Vancouver, etc.

The solution to the paradox is the following. If your goal is to estimate just θ_i, you cannot do better than using x_i, but if the goal is to estimate the whole vector $\boldsymbol{\theta}$, and you use squared error as your loss function, then shrinkage helps. To see this, suppose we want to estimate $||\boldsymbol{\theta}||_2^2$ from a single sample $\mathbf{x} \sim \mathcal{N}(\boldsymbol{\theta}, \mathbf{I})$. A simple estimate is $||\mathbf{x}||_2^2$, but this will overestimate the result, since

$$\mathbb{E}\left[||\mathbf{x}||_2^2\right] = \mathbb{E}\left[\sum_i x_i^2\right] = \sum_{i=1}^{N}\left(1 + \theta_i^2\right) = N + ||\boldsymbol{\theta}||_2^2 \tag{6.30}$$

Consequently we can reduce our risk by pooling information, even from unrelated sources, and shrinking towards the overall mean. In Section 5.6.2, we give a Bayesian explanation for this. See also (Efron and Morris 1975).

6.3.3.3 Admissibility is not enough

It seems clear that we can restrict our search for good estimators to the class of admissible estimators. But in fact it is easy to construct admissible estimators, as we show in the following example.

Theorem 6.3.3. *Let $X \sim \mathcal{N}(\theta, 1)$, and consider estimating θ under squared loss. Let $\delta_1(x) = \theta_0$, a constant independent of the data. This is an admissible estimator.*

Proof. Suppose not. Then there is some other estimator δ_2 with smaller risk, so $R(\theta^*, \delta_2) \leq R(\theta^*, \delta_1)$, where the inequality must be strict for some θ^*. Suppose the true parameter is $\theta^* = \theta_0$. Then $R(\theta^*, \delta_1) = 0$, and

$$R(\theta^*, \delta_2) = \int (\delta_2(x) - \theta_0)^2 p(x|\theta_0) dx \tag{6.31}$$

Since $0 \leq R(\theta^*, \delta_2) \leq R(\theta^*, \delta_1)$ for all θ^*, and $R(\theta_0, \delta_1) = 0$, we have $R(\theta_0, \delta_2) = 0$ and hence $\delta_2(x) = \theta_0 = \delta_1(x)$. Thus the only way δ_2 can avoid having higher risk than δ_1 at some specific point θ_0 is by being equal to δ_1. Hence there is no other estimator δ_2 with strictly lower risk, so δ_2 is admissible. $\qquad\square$

6.4 Desirable properties of estimators

Since frequentist decision theory does not provide an automatic way to choose the best estimator, we need to come up with other heuristic selection principles. In this section, we discuss some properties we would like estimators to have. Unfortunately, we will see that we cannot achieve all of these properties at the same time.

6.4.1 Consistent estimators

An estimator is said to be **consistent** if it eventually recovers the true parameters that generated the data as the sample size goes to infinity, i.e., $\hat{\theta}(\mathcal{D}) \rightarrow \theta^*$ as $|\mathcal{D}| \rightarrow \infty$ (where the arrow denotes convergence in probability). Of course, this concept only makes sense if the data actually comes from the specified model with parameters θ^*, which is not usually the case with real data. Nevertheless, it can be a useful theoretical property.

It can be shown that the MLE is a consistent estimator. The intuitive reason is that maximizing likelihood is equivalent to minimizing $\mathbb{KL}\left(p(\cdot|\boldsymbol{\theta}^*)||p(\cdot|\hat{\boldsymbol{\theta}})\right)$, where $p(\cdot|\boldsymbol{\theta}^*)$ is the true distribution and $p(\cdot|\hat{\boldsymbol{\theta}})$ is our estimate. We can achieve 0 KL divergence iff $\hat{\boldsymbol{\theta}} = \boldsymbol{\theta}^*$.[4]

6.4.2 Unbiased estimators

The **bias** of an estimator is defined as

$$\text{bias}(\hat{\theta}(\cdot)) = \mathbb{E}_{p(\mathcal{D}|\theta_*)}\left[\hat{\theta}(\mathcal{D}) - \theta_*\right] \tag{6.32}$$

where θ_* is the true parameter value. If the bias is zero, the estimator is called **unbiased**. This means the sampling distribution is centered on the true parameter. For example, the MLE for a Gaussian mean is unbiased:

$$\text{bias}(\hat{\mu}) = \mathbb{E}\left[\overline{x}\right] - \mu = \mathbb{E}\left[\frac{1}{N}\sum_{i=1}^{N} x_i\right] - \mu = \frac{N\mu}{N} - \mu = 0 \tag{6.33}$$

However, the MLE for a Gaussian variance, $\hat{\sigma}^2$, is not an unbiased estimator of σ^2. In fact, one can show (Exercise 6.3) that

$$\mathbb{E}\left[\hat{\sigma}^2\right] = \frac{N-1}{N}\sigma^2 \tag{6.34}$$

However, the following estimator

$$\hat{\sigma}_{N-1}^2 = \frac{N}{N-1}\hat{\sigma}^2 = \frac{1}{N-1}\sum_{i=1}^{N}(x_i - \overline{x})^2 \tag{6.35}$$

is an unbiased estimator, which we can easily prove as follows:

$$\mathbb{E}\left[\hat{\sigma}_{N-1}^2\right] = \mathbb{E}\left[\frac{N}{N-1}\hat{\sigma}^2\right] = \frac{N}{N-1}\frac{N-1}{N}\sigma^2 = \sigma^2 \tag{6.36}$$

In Matlab, `var(X)` returns $\hat{\sigma}_{N-1}^2$, whereas `var(X,1)` returns $\hat{\sigma}^2$ (the MLE). For large enough N, the difference will be negligible.

Although the MLE may sometimes be a biased estimator, one can show that asymptotically, it is always unbiased. (This is necessary for the MLE to be a consistent estimator.)

Although being unbiased sounds like a desirable property, this is not always true. See Section 6.4.4 and (Lindley 1972) for discussion of this point.

6.4.3 Minimum variance estimators

It seems intuitively reasonable that we want our estimator to be unbiased (although we shall give some arguments against this claim below). However, being unbiased is not enough. For

---

4. If the model is unidentifiable, the MLE may select a set of parameters that is different from the true parameters but for which the induced distribution, $p(\cdot|\hat{\boldsymbol{\theta}})$, is the same as the exact distribution. Such parameters are said to be likelihood equivalent.

example, suppose we want to estimate the mean of a Gaussian from $\mathcal{D} = \{x_1, \ldots, x_N\}$. The estimator that just looks at the first data point, $\hat{\theta}(\mathcal{D}) = x_1$, is an unbiased estimator, but will generally be further from θ_* than the empirical mean $\bar{x}$ (which is also unbiased). So the variance of an estimator is also important.

A natural question is: how low can the variance go? A famous result, called the **Cramer-Rao lower bound**, provides a lower bound on the variance of any unbiased estimator. More precisely,

Theorem 6.4.1 (Cramer-Rao inequality). *Let $X_1, \ldots, X_n \sim p(X|\theta_0)$ and $\hat{\theta} = \hat{\theta}(x_1, \ldots, x_n)$ be an unbiased estimator of θ_0. Then, under various smoothness assumptions on $p(X|\theta_0)$, we have*

$$\text{var}\left[\hat{\theta}\right] \geq \frac{1}{nI(\theta_0)} \tag{6.37}$$

where $I(\theta_0)$ is the Fisher information matrix (see Section 6.2.2).

A proof can be found e.g., in (Rice 1995, p275).

It can be shown that the MLE achieves the Cramer Rao lower bound, and hence has the smallest asymptotic variance of any unbiased estimator. Thus MLE is said to be **asymptotically optimal**.

6.4.4 The bias-variance tradeoff

Although using an unbiased estimator seems like a good idea, this is not always the case. To see why, suppose we use quadratic loss. As we showed above, the corresponding risk is the MSE. We now derive a very useful decomposition of the MSE. (All expectations and variances are wrt the true distribution $p(\mathcal{D}|\theta^*)$, but we drop the explicit conditioning for notational brevity.) Let $\hat{\theta} = \hat{\theta}(\mathcal{D})$ denote the estimate, and $\bar{\theta} = \mathbb{E}\left[\hat{\theta}\right]$ denote the expected value of the estimate (as we vary $\mathcal{D}$). Then we have

$$\mathbb{E}\left[(\hat{\theta} - \theta^*)^2\right] = \mathbb{E}\left[\left[(\hat{\theta} - \bar{\theta}) + (\bar{\theta} - \theta^*)\right]^2\right] \tag{6.38}$$

$$= \mathbb{E}\left[\left(\hat{\theta} - \bar{\theta}\right)^2\right] + 2(\bar{\theta} - \theta^*)\mathbb{E}\left[\hat{\theta} - \bar{\theta}\right] + (\bar{\theta} - \theta^*)^2 \tag{6.39}$$

$$= \mathbb{E}\left[\left(\hat{\theta} - \bar{\theta}\right)^2\right] + (\bar{\theta} - \theta^*)^2 \tag{6.40}$$

$$= \text{var}\left[\hat{\theta}\right] + \text{bias}^2(\hat{\theta}) \tag{6.41}$$

In words,

$$\boxed{\text{MSE} = \text{variance} + \text{bias}^2} \tag{6.42}$$

This is called the **bias-variance tradeoff** (see e.g., (Geman et al. 1992)). What it means is that it might be wise to use a biased estimator, so long as it reduces our variance, assuming our goal is to minimize squared error.

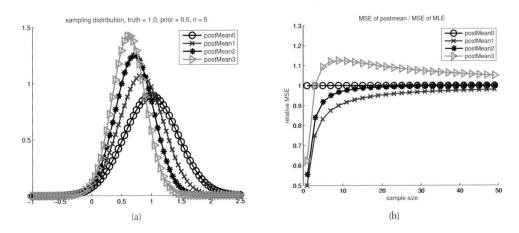

Figure 6.4 Left: Sampling distribution of the MAP estimate with different prior strengths κ_0. (The MLE corresponds to $\kappa_0 = 0$.) Right: MSE relative to that of the MLE versus sample size. Based on Figure 5.6 of (Hoff 2009). Figure generated by `samplingDistGaussShrinkage`.

6.4.4.1 Example: estimating a Gaussian mean

Let us give an example, based on (Hoff 2009, p79). Suppose we want to estimate the mean of a Gaussian from $\mathbf{x} = (x_1, \ldots, x_N)$. We assume the data is sampled from $x_i \sim \mathcal{N}(\theta^* = 1, \sigma^2)$. An obvious estimate is the MLE. This has a bias of 0 and a variance of

$$\text{var}\left[\overline{x}|\theta^*\right] = \frac{\sigma^2}{N} \tag{6.43}$$

But we could also use a MAP estimate. In Section 4.6.1, we show that the MAP estimate under a Gaussian prior of the form $\mathcal{N}(\theta_0, \sigma^2/\kappa_0)$ is given by

$$\tilde{x} \triangleq \frac{N}{N + \kappa_0}\overline{x} + \frac{\kappa_0}{N + \kappa_0}\theta_0 = w\overline{x} + (1 - w)\theta_0 \tag{6.44}$$

where $0 \leq w \leq 1$ controls how much we trust the MLE compared to our prior. (This is also the posterior mean, since the mean and mode of a Gaussian are the same.) The bias and variance are given by

$$\mathbb{E}\left[\tilde{x}\right] - \theta^* = w\theta_0 + (1 - w)\theta_0 - \theta^* = (1 - w)(\theta_0 - \theta^*) \tag{6.45}$$

$$\text{var}\left[\tilde{x}\right] = w^2\frac{\sigma^2}{N} \tag{6.46}$$

So although the MAP estimate is biased (assuming $w < 1$), it has lower variance.

Let us assume that our prior is slightly misspecified, so we use $\theta_0 = 0$, whereas the truth is $\theta^* = 1$. In Figure 6.4(a), we see that the sampling distribution of the MAP estimate for $\kappa_0 > 0$ is biased away from the truth, but has lower variance (is narrower) than that of the MLE.

In Figure 6.4(b), we plot $\text{mse}(\tilde{x})/\text{mse}(\overline{x})$ vs N. We see that the MAP estimate has lower MSE than the MLE, especially for small sample size, for $\kappa_0 \in \{1, 2\}$. The case $\kappa_0 = 0$ corresponds to

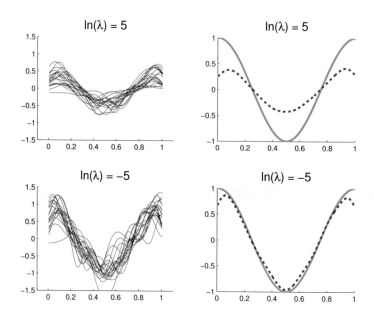

Figure 6.5 Illustration of bias-variance tradeoff for ridge regression. We generate 100 data sets from the true function, shown in solid green. Left: we plot the regularized fit for 20 different data sets. We use linear regression with a Gaussian RBF expansion, with 25 centers evenly spread over the $[0, 1]$ interval. Right: we plot the average of the fits, averaged over all 100 datasets. Top row: strongly regularized: we see that the individual fits are similar to each other (low variance), but the average is far from the truth (high bias). Bottom row: lightly regularized: we see that the individual fits are quite different from each other (high variance), but the average is close to the truth (low bias). Based on (Bishop 2006) Figure 3.5. Figure generated by `biasVarModelComplexity3`.

the MLE, and the case $\kappa_0 = 3$ corresponds to a strong prior, which hurts performance because the prior mean is wrong. It is clearly important to "tune" the strength of the prior, a topic we discuss later.

6.4.4.2   Example: ridge regression

Another important example of the bias variance tradeoff arises in ridge regression, which we discuss in Section 7.5. In brief, this corresponds to MAP estimation for linear regression under a Gaussian prior, $p(\mathbf{w}) = \mathcal{N}(\mathbf{w}|\mathbf{0}, \lambda^{-1}\mathbf{I})$ The zero-mean prior encourages the weights to be small, which reduces overfitting; the precision term, λ, controls the strength of this prior. Setting $\lambda = 0$ results in the MLE; using $\lambda > 0$ results in a biased estimate. To illustrate the effect on the variance, consider a simple example. Figure 6.5 on the left plots each individual fitted curve, and on the right plots the average fitted curve. We see that as we increase the strength of the regularizer, the variance decreases, but the bias increases.

6.4.4.3 Bias-variance tradeoff for classification

If we use 0-1 loss instead of squared error, the above analysis breaks down, since the frequentist risk is no longer expressible as squared bias plus variance. In fact, one can show (Exercise 7.2 of (Hastie et al. 2009)) that the bias and variance combine multiplicatively. If the estimate is on the correct side of the decision boundary, then the bias is negative, and decreasing the variance will decrease the misclassification rate. But if the estimate is on the wrong side of the decision boundary, then the bias is positive, so it pays to *increase* the variance (Friedman 1997a). This little known fact illustrates that the bias-variance tradeoff is not very useful for classification. It is better to focus on expected loss (see below), not directly on bias and variance. We can approximate the expected loss using cross validation, as we discuss in Section 6.5.3.

6.5 Empirical risk minimization

Frequentist decision theory suffers from the fundamental problem that one cannot actually compute the risk function, since it relies on knowing the true data distribution. (By contrast, the Bayesian posterior expected loss can always be computed, since it conditions on the the data rather than conditioning on θ^*.) However, there is one setting which avoids this problem, and that is where the task is to predict observable quantities, as opposed to estimating hidden variables or parameters. That is, instead of looking at loss functions of the form $L(\theta, \delta(\mathcal{D}))$, where θ is the true but unknown parameter, and $\delta(\mathcal{D})$ is our estimator, let us look at loss functions of the form $L(y, \delta(\mathbf{x}))$, where y is the true but unknown response, and $\delta(\mathbf{x})$ is our prediction given the input $\mathbf{x}$. In this case, the frequentist risk becomes

$$R(p_*, \delta) \triangleq \mathbb{E}_{(\mathbf{x},y)\sim p_*} [L(y, \delta(\mathbf{x}))] = \sum_{\mathbf{x}} \sum_{y} L(y, \delta(\mathbf{x}))p_*(\mathbf{x}, y) \tag{6.47}$$

where p_* represents "nature's distribution". Of course, this distribution is unknown, but a simple approach is to use the empirical distribution, derived from some training data $\mathcal{D}$, to approximate p_*, i.e.,

$$p_*(\mathbf{x}, y) \approx p_{\text{emp}}(\mathbf{x}, y|\mathcal{D}) \triangleq \frac{1}{N} \sum_{i=1}^{N} \delta_{\mathbf{x}_i}(\mathbf{x})\delta_{y_i}(y) \tag{6.48}$$

We then define the **empirical risk** as follows:

$$R_{emp}(\mathcal{D}, \delta) \triangleq R(p_{\text{emp}}(\cdot|\mathcal{D}), \delta) = \frac{1}{N} \sum_{i=1}^{N} L(y_i, \delta(\mathbf{x}_i)) \tag{6.49}$$

In the case of 0-1 loss, $L(y, \delta(\mathbf{x})) = \mathbb{I}(y \neq \delta(\mathbf{x}))$, this becomes the **misclassification rate**. In the case of squared error loss, $L(y, \delta(\mathbf{x})) = (y - \delta(\mathbf{x}))^2$, this becomes the **mean squared error**. We define the task of **empirical risk minimization** or **ERM** as finding a decision procedure (typically a classification rule) to minimize the empirical risk:

$$\delta_{ERM}(\mathcal{D}) = \underset{\delta}{\operatorname{argmin}} \, R_{emp}(\mathcal{D}, \delta) \tag{6.50}$$

In the unsupervised case, we eliminate all references to y, and replace $L(y, \delta(\mathbf{x}))$ with $L(\mathbf{x}, \delta(\mathbf{x}))$, where, for example, $L(\mathbf{x}, \delta(\mathbf{x})) = ||\mathbf{x} - \delta(\mathbf{x})||_2^2$, which measures the reconstruction error. We can define the decision rule using $\delta(\mathbf{x}) = \text{decode}(\text{encode}(\mathbf{x}))$, as in vector quantization (Section 11.4.2.6) or PCA (section 12.2). Finally, we define the empirical risk as

$$R_{emp}(\mathcal{D}, \delta) = \frac{1}{N} \sum_{i=1}^{N} L(\mathbf{x}_i, \delta(\mathbf{x}_i)) \tag{6.51}$$

Of course, we can always trivially minimize this risk by setting $\delta(\mathbf{x}) = \mathbf{x}$, so it is critical that the encoder-decoder go via some kind of bottleneck.

6.5.1   Regularized risk minimization

Note that the empirical risk is equal to the Bayes risk if our prior about "nature's distribution" is that it is exactly equal to the empirical distribution (Minka 2001b):

$$\mathbb{E}\left[R(p_*, \delta)|p_* = p_{\text{emp}}\right] = R_{emp}(\mathcal{D}, \delta) \tag{6.52}$$

Therefore minimizing the empirical risk will typically result in overfitting. It is therefore often necessary to add a complexity penalty to the objective function:

$$R'(\mathcal{D}, \delta) = R_{emp}(\mathcal{D}, \delta) + \lambda C(\delta) \tag{6.53}$$

where $C(\delta)$ measures the complexity of the prediction function $\delta(\mathbf{x})$ and λ controls the strength of the complexity penalty. This approach is known as **regularized risk minimization** (RRM). Note that if the loss function is negative log likelihood, and the regularizer is a negative log prior, this is equivalent to MAP estimation.

The two key issues in RRM are: how do we measure complexity, and how do we pick λ. For a linear model, we can define the complexity of in terms of its degrees of freedom, discussed in Section 7.5.3. For more general models, we can use the VC dimension, discussed in Section 6.5.4. To pick λ, we can use the methods discussed in Section 6.5.2.

6.5.2   Structural risk minimization

The regularized risk minimization principle says that we should fit the model, for a given complexity penalty, by using

$$\hat{\delta}_\lambda = \underset{\delta}{\operatorname{argmin}} \left[R_{emp}(\mathcal{D}, \delta) + \lambda C(\delta)\right] \tag{6.54}$$

But how should we pick λ? We cannot using the training set, since this will underestimate the true risk, a problem known as **optimism of the training error**. As an alternative, we can use the following rule, known as the **structural risk minimization** principle: (Vapnik 1998):

$$\hat{\lambda} = \underset{\lambda}{\operatorname{argmin}} \hat{R}(\hat{\delta}_\lambda) \tag{6.55}$$

where $\hat{R}(\delta)$ is an estimate of the risk. There are two widely used estimates: cross validation and theoretical upper bounds on the risk. We discuss both of these below.

6.5.3 Estimating the risk using cross validation

We can estimate the risk of some estimator using a validation set. If we don't have a separate validation set, we can use **cross validation** (CV), as we briefly discussed in Section 1.4.8. More precisely, CV is defined as follows. Let there be $N = |\mathcal{D}|$ data cases in the training set. Denote the data in the k'th test fold by $\mathcal{D}_k$ and all the other data by $\mathcal{D}_{-k}$. (In **stratified CV**, these folds are chosen so the class proportions (if discrete labels are present) are roughly equal in each fold.) Let $\mathcal{F}$ be a learning algorithm or fitting function that takes a dataset and a model index m (this could a discrete index, such as the degree of a polynomial, or a continuous index, such as the strength of a regularizer) and returns a parameter vector:

$$\hat{\boldsymbol{\theta}}_m = \mathcal{F}(\mathcal{D}, m) \tag{6.56}$$

Finally, let $\mathcal{P}$ be a prediction function that takes an input and a parameter vector and returns a prediction:

$$\hat{y} = \mathcal{P}(\mathbf{x}, \hat{\boldsymbol{\theta}}) = f(\mathbf{x}, \hat{\boldsymbol{\theta}}) \tag{6.57}$$

Thus the combined **fit-predict cycle** is denoted as

$$f_m(\mathbf{x}, \mathcal{D}) = \mathcal{P}(\mathbf{x}, \mathcal{F}(\mathcal{D}, m)) \tag{6.58}$$

The K-fold CV estimate of the risk of f_m is defined by

$$R(m, \mathcal{D}, K) \triangleq \frac{1}{N} \sum_{k=1}^{K} \sum_{i \in \mathcal{D}_k} L\left(y_i, \mathcal{P}(\mathbf{x}_i, \mathcal{F}(\mathcal{D}_{-k}, m))\right) \tag{6.59}$$

Note that we can call the fitting algorithm once per fold. Let $f_m^k(\mathbf{x}) = \mathcal{P}(\mathbf{x}, \mathcal{F}(\mathcal{D}_{-k}, m))$ be the function that was trained on all the data except for the test data in fold k. Then we can rewrite the CV estimate as

$$R(m, \mathcal{D}, K) = \frac{1}{N} \sum_{k=1}^{K} \sum_{i \in \mathcal{D}_k} L\left(y_i, f_m^k(\mathbf{x}_i)\right) = \frac{1}{N} \sum_{i=1}^{N} L\left(y_i, f_m^{k(i)}(\mathbf{x}_i)\right) \tag{6.60}$$

where $k(i)$ is the fold in which i is used as test data. In other words, we predict y_i using a model that was trained on data that does not contain $\mathbf{x}_i$.

If $K = N$, the method is known as **leave one out cross validation** or LOOCV. In this case, n the estimated risk becomes

$$R(m, \mathcal{D}, N) = \frac{1}{N} \sum_{i=1}^{N} L\left(y_i, f_m^{-i}(\mathbf{x}_i)\right) \tag{6.61}$$

where $f_m^i(\mathbf{x}) = \mathcal{P}(\mathbf{x}, \mathcal{F}(\mathcal{D}_{-i}, m))$. This requires fitting the model N times, where for f_m^{-i} we omit the i'th training case. Fortunately, for some model classes and loss functions (namely linear models and quadratic loss), we can fit the model once, and analytically "remove" the effect of the i'th training case. This is known as **generalized cross validation** or GCV.

6.5.3.1 Example: using CV to pick λ for ridge regression

As a concrete example, consider picking the strength of the ℓ_2 regularizer in penalized linear regression. We use the following rule:

$$\hat{\lambda} = \arg \min_{\lambda \in [\lambda_{min}, \lambda_{max}]} R(\lambda, \mathcal{D}_{\text{train}}, K) \tag{6.62}$$

where $[\lambda_{min}, \lambda_{max}]$ is a finite range of λ values that we search over, and $R(\lambda, \mathcal{D}_{\text{train}}, K)$ is the K-fold CV estimate of the risk of using λ, given by

$$R(\lambda, \mathcal{D}_{\text{train}}, K) = \frac{1}{|\mathcal{D}_{\text{train}}|} \sum_{k=1}^{K} \sum_{i \in \mathcal{D}_k} L(y_i, f_\lambda^k(\mathbf{x}_i)) \tag{6.63}$$

where $f_\lambda^k(\mathbf{x}) = \mathbf{x}^T \hat{\mathbf{w}}_\lambda(\mathcal{D}_{-k})$ is the prediction function trained on data excluding fold k, and $\hat{\mathbf{w}}_\lambda(\mathcal{D}) = \arg \min_{\mathbf{w}} NLL(\mathbf{w}, \mathcal{D}) + \lambda ||\mathbf{w}||_2^2$ is the MAP estimate. Figure 6.6(b) gives an example of a CV estimate of the risk vs $\log(\lambda)$, where the loss function is squared error.

When performing classification, we usually use 0-1 loss. In this case, we optimize a convex upper bound on the empirical risk to estimate $\mathbf{w}_\lambda$, but we optimize (the CV estimate of) the risk itself to estimate λ. We can handle the non-smooth 0-1 loss function when estimating λ because we are using brute-force search over the entire (one-dimensional) space.

When we have more than one or two tuning parameters, this approach becomes infeasible. In such cases, one can use empirical Bayes, which allows one to optimize large numbers of hyper-parameters using gradient-based optimizers instead of brute-force search. See Section 5.6 for details.

6.5.3.2 The one standard error rule

The above procedure estimates the risk, but does not give any measure of uncertainty. A standard frequentist measure of uncertainty of an estimate is the standard error of the mean, defined by

$$se = \frac{\hat{\sigma}}{\sqrt{N}} = \sqrt{\frac{\hat{\sigma}^2}{N}} \tag{6.64}$$

where $\hat{\sigma}^2$ is an estimate of the variance of the loss:

$$\hat{\sigma}^2 = \frac{1}{N} \sum_{i=1}^{N} (L_i - \overline{L})^2, \quad L_i = L(y_i, f_m^{k(i)}(\mathbf{x}_i)) \quad \overline{L} = \frac{1}{N} \sum_{i=1}^{N} L_i \tag{6.65}$$

Note that σ measures the intrinsic variability of L_i across samples, whereas se measures our uncertainty about the mean $\overline{L}$.

Suppose we apply CV to a set of models and compute the mean and se of their estimated risks. A common heuristic for picking a model from these noisy estimates is to pick the value which corresponds to the simplest model whose risk is no more than one standard error above the risk of the best model; this is called the **one-standard error rule** (Hastie et al. 2001, p216). For example, in Figure 6.6, we see that this heuristic does not choose the lowest point on the curve, but one that is slightly to its right, since that corresponds to a more heavily regularized model with essentially the same empirical performance.

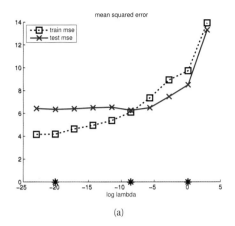
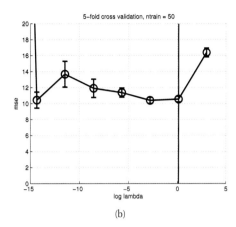

      (a)   (b)

Figure 6.6 (a) Mean squared error for ℓ_2 penalized degree 14 polynomial regression vs log regularizer. Same as in Figures 7.8, except now we have $N = 50$ training points instead of 21. The stars correspond to the values used to plot the functions in Figure 7.7. (b) CV estimate. The vertical scale is truncated for clarity. The blue line corresponds to the value chosen by the one standard error rule. Figure generated by `linregPolyVsRegDemo`.

6.5.3.3 CV for model selection in non-probabilistic unsupervised learning

If we are performing unsupervised learning, we must use a loss function such as $L(\mathbf{x}, \delta(\mathbf{x})) = ||\mathbf{x} - \delta(\mathbf{x})||_2$, which measures reconstruction error. Here $\delta(\mathbf{x})$ is some encode-decode scheme. However, as we discussed in Section 11.5.2, we cannot use CV to determine the complexity of δ, since we will always get lower loss with a more complex model, even if evaluated on the test set. This is because more complex models will compress the data less, and induce less distortion. Consequently, we must either use probabilistic models, or invent other heuristics.

6.5.4 Upper bounding the risk using statistical learning theory *

The principle problem with cross validation is that it is slow, since we have to fit the model multiple times. A faster approach is to compute analytic approximations or bounds to the generalization error. This is the studied in the field of **statistical learning theory** (SLT). More precisely, SLT tries to bound the risk $R(p_*, h)$ for any data distribution p_* and hypothesis $h \in \mathcal{H}$ in terms of the empirical risk $R_{emp}(\mathcal{D}, h)$, the sample size $N = |\mathcal{D}|$, and the size of the hypothesis space $\mathcal{H}$.

Let us initially consider the case where the hypothesis space is finite, with size $\dim(\mathcal{H}) = |\mathcal{H}|$. In other words, we are selecting a model/ hypothesis from a finite list, rather than optimizing real-valued parameters, Then we can prove the following.

Theorem 6.5.1. *For any data distribution p_*, and any dataset $\mathcal{D}$ of size N drawn from p_*, the probability that our estimate of the error rate will be more than ϵ wrong, in the worst case, is upper* *bounded as follows:*

$$P\left(\max_{h \in \mathcal{H}} |R_{emp}(\mathcal{D}, h) - R(p_*, h)| > \epsilon\right) \leq 2 \dim(\mathcal{H}) e^{-2N\epsilon^2} \tag{6.66}$$

Proof. To prove this, we need two useful results. First, **Hoeffding's inequality**, which states that if $X_1, \ldots, X_N \sim \text{Ber}(\theta)$, then, for any $\epsilon > 0$,

$$P(|\overline{x} - \theta| > \epsilon) \leq 2e^{-2N\epsilon^2} \tag{6.67}$$

where $\overline{x} = \frac{1}{N} \sum_{i=1}^{N} x_i$. Second, the **union bound**, which says that if $A_1, \ldots, A_d$ are a set of events, then $P(\cup_{i=1}^{d} A_i) \leq \sum_{i=1}^{d} P(A_i)$.

Finally, for notational brevity, let $R(h) = R(h, p_*)$ be the true risk, and $\hat{R}_N(h) = R_{emp}(\mathcal{D}, h)$ be the empirical risk.

Using these results we have

$$P\left(\max_{h \in \mathcal{H}} |\hat{R}_N(h) - R(h)| > \epsilon\right) = P\left(\bigcup_{h \in \mathcal{H}} |\hat{R}_N(h) - R(h)| > \epsilon\right) \tag{6.68}$$

$$\leq \sum_{h \in \mathcal{H}} P\left(|\hat{R}_N(h) - R(h)| > \epsilon\right) \tag{6.69}$$

$$\leq \sum_{h \in \mathcal{H}} 2e^{-2N\epsilon^2} = 2\dim(\mathcal{H})e^{-2N\epsilon^2} \tag{6.70}$$

$\square$

Ths bound tells us that the optimism of the training error increases with $\dim(\mathcal{H})$[5], but decreases with $N = |\mathcal{D}|$, as is to be expected.

If the hypothesis space $\mathcal{H}$ is infinite (e.g., we have real-valued parameters), we cannot use $\dim(\mathcal{H}) = |\mathcal{H}|$. Instead, we can use a quantity called the **Vapnik-Chervonenkis** or **VC** dimension of the hypothesis class. See (Vapnik 1998) for details.

Stepping back from all the theory, the key intuition behind statistical learning theory is quite simple. Suppose we find a model with low empirical risk. If the hypothesis space $\mathcal{H}$ is very big, relative to the data size, then it is quite likely that we just got "lucky" and were given a data set that is well-modeled by our chosen function by chance. However, this does not mean that such a function will have low generalization error. But if the hypothesis class is sufficiently constrained in size, and/or the training set is sufficiently large, then we are unlikely to get lucky in this way, so a low empirical risk is evidence of a low true risk.

The advantage of statistical learning theory compared to CV is that the bounds on the risk are quicker to compute than using CV. The disadvantage is that it is hard to compute the VC dimension for many interesting models, and the upper bounds are usually very loose (although see (Kaariainen and Langford 2005)).

One can extend statistical learning theory by taking computational complexity of the learner into account. This field is called **computational learning theory** or **COLT**. Most of this work

---

5. Note that optimism of the training error does not necessarily increase with model complexity, but it does increase with the number of different models that are being searched over.

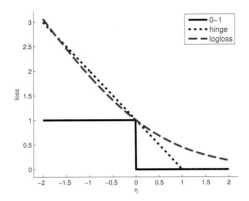

Figure 6.7 Illustration of various loss functions for binary classification. The horizontal axis is the margin $y\eta$, the vertical axis is the loss. The log loss uses log base 2. Figure generated by `hingeLossPlot`.

focuses on the case where h is a binary classifier, and the loss function is 0-1 loss. If we observe a low empirical risk, and the hypothesis space is suitably "small", then we can say that our estimated function is **probably approximately correct** or **PAC**. A hypothesis space is said to be **efficiently PAC-learnable** if there is a polynomial time algorithm that can identify a function that is PAC. See (Kearns and Vazirani 1994) for details.

6.5.5 Surrogate loss functions

Minimizing the loss in the ERM/ RRM framework is not always easy. For example, we might want to optimize the AUC or F1 scores. Or more simply, we might just want to minimize the 0-1 loss, as is common in classification. Unfortunately, the 0-1 risk is a very non-smooth objective and hence is hard to optimize. One alternative is to use maximum likelihood estimation instead, since log-likelihood is a smooth convex upper bound on the 0-1 risk, as we show below.

To see this, consider binary logistic regression, and let $y_i \in \{-1, +1\}$. Suppose our decision function computes the log-odds ratio,

$$f(\mathbf{x}_i) = \log \frac{p(y = 1|\mathbf{x}_i, \mathbf{w})}{p(y = -1|\mathbf{x}_i, \mathbf{w})} = \mathbf{w}^T \mathbf{x}_i = \eta_i \tag{6.71}$$

Then the corresponding probability distribution on the output label is

$$p(y_i|\mathbf{x}_i, \mathbf{w}) = \text{sigm}(y_i \eta_i) \tag{6.72}$$

Let us define the **log-loss** as as

$$L_{\text{nll}}(y, \eta) = -\log p(y|\mathbf{x}, \mathbf{w}) = \log(1 + e^{-y\eta}) \tag{6.73}$$

It is clear that minimizing the average log-loss is equivalent to maximizing the likelihood.

Now consider computing the most probable label, which is equivalent to using $\hat{y} = -1$ if $\eta_i < 0$ and $\hat{y} = +1$ if $\eta_i \geq 0$. The 0-1 loss of our function becomes

$$L_{01}(y, \eta) = \mathbb{I}(y \neq \hat{y}) = \mathbb{I}(y\eta < 0) \tag{6.74}$$

Figure 6.7 plots these two loss functions. We see that the NLL is indeed an upper bound on the 0-1 loss.

Log-loss is an example of a **surrogate loss function**. Another example is the **hinge loss**:

$$L_{\text{hinge}}(y, \eta) = \max(0, 1 - y\eta) \tag{6.75}$$

See Figure 6.7 for a plot. We see that the function looks like a door hinge, hence its name. This loss function forms the basis of a popular classification method known as support vector machines (SVM), which we will discuss in Section 14.5.

The surrogate is usually chosen to be a convex upper bound, since convex functions are easy to minimize. See e.g., (Bartlett et al. 2006) for more information.

6.6   Pathologies of frequentist statistics *

> I believe that it would be very difficult to persuade an intelligent person that current [frequentist] statistical practice was sensible, but that there would be much less difficulty with an approach via likelihood and Bayes' theorem. — George Box, 1962.

Frequentist statistics exhibits various forms of weird and undesirable behaviors, known as **pathologies**. We give a few examples below, in order to caution the reader; these and other examples are explained in more detail in (Lindley 1972; Lindley and Phillips 1976; Lindley 1982; Berger 1985; Jaynes 2003; Minka 1999).

6.6.1   Counter-intuitive behavior of confidence intervals

A **confidence interval** is an interval derived from the sampling distribution of an estimator (whereas a Bayesian credible interval is derived from the posterior of a parameter, as we discussed in Section 5.2.2). More precisely, a frequentist confidence interval for some parameter θ is defined by the following (rather un-natural) expression:

$$C'_\alpha(\theta) = (\ell, u) : P(\ell(\tilde{\mathcal{D}}) \le \theta \le u(\tilde{\mathcal{D}}) | \tilde{\mathcal{D}} \sim \theta) = 1 - \alpha \tag{6.76}$$

That is, if we sample hypothetical future data $\tilde{\mathcal{D}}$ from θ, then $(\ell(\tilde{\mathcal{D}}), u(\tilde{\mathcal{D}}))$, is a confidence interval if the parameter θ lies inside this interval $1 - \alpha$ percent of the time.

Let us step back for a moment and think about what is going on. In Bayesian statistics, we condition on what is known — namely the observed data, $\mathcal{D}$ — and average over what is not known, namely the parameter θ. In frequentist statistics, we do exactly the opposite: we condition on what is unknown — namely the true parameter value θ — and average over hypothetical future data sets $\tilde{\mathcal{D}}$.

This counter-intuitive definition of confidence intervals can lead to bizarre results. Consider the following example from (Berger 1985, p11). Suppose we draw two integers $\mathcal{D} = (x_1, x_2)$ from

$$p(x|\theta) = \begin{cases} 0.5 & \text{if } x = \theta \\ 0.5 & \text{if } x = \theta + 1 \\ 0 & \text{otherwise} \end{cases} \tag{6.77}$$

If $\theta = 39$, we would expect the following outcomes each with probability 0.25:

$$(39, 39), (39, 40), (40, 39), (40, 40) \tag{6.78}$$

Let $m = \min(x_1, x_2)$ and define the following confidence interval:

$$[\ell(\mathcal{D}), u(\mathcal{D})] = [m, m] \tag{6.79}$$

For the above samples this yields

$$[39, 39], \quad [39, 39], \quad [39, 39], \quad [40, 40] \tag{6.80}$$

Hence Equation 6.79 is clearly a 75% CI, since 39 is contained in 3/4 of these intervals. However, if $\mathcal{D} = (39, 40)$ then $p(\theta = 39|\mathcal{D}) = 1.0$, so we know that θ must be 39, yet we only have 75% "confidence" in this fact.

Another, less contrived example, is as follows. Suppose we want to estimate the parameter θ of a Bernoulli distribution. Let $\overline{x} = \frac{1}{N} \sum_{i=1}^{N} x_i$ be the sample mean. The MLE is $\hat{\theta} = \overline{x}$. An approximate 95% confidence interval for a Bernoulli parameter is $\overline{x} \pm 1.96\sqrt{\overline{x}(1 - \overline{x})/N}$ (this is called a **Wald interval** and is based on a Gaussian approximation to the Binomial distribution; compare to Equation 3.27). Now consider a single trial, where $N = 1$ and $x_1 = 0$. The MLE is 0, which overfits, as we saw in Section 3.3.4.1. But our 95% confidence interval is also $(0, 0)$, which seems even worse. It can be argued that the above flaw is because we approximated the true sampling distribution with a Gaussian, or because the sample size was to small, or the parameter "too extreme". However, the Wald interval can behave badly even for large N, and non-extreme parameters (Brown et al. 2001).

6.6.2 p-values considered harmful

Suppose we want to decide whether to accept or reject some baseline model, which we will call the **null hypothesis**. We need to define some decision rule. In frequentist statistics, it is standard to first compute a quantity called the **p-value**, which is defined as the probability (under the null) of observing some **test statistic** $f(\mathcal{D})$ (such as the chi-squared statistic) that is as large *or larger* than that actually observed:[6]

$$\text{pvalue}(\mathcal{D}) \triangleq P(f(\tilde{\mathcal{D}}) \geq f(\mathcal{D})|\tilde{\mathcal{D}} \sim H_0) \tag{6.81}$$

This quantity relies on computing a **tail area probability** of the sampling distribution; we give an example of how to do this below.

Given the p-value, we define our decision rule as follows: we reject the null hypothesis iff the p-value is less than some threshold, such as $\alpha = 0.05$. If we do reject it, we say the difference between the observed test statistic and the expected test statistic is **statistically significant** at level α. This approach is known as **null hypothesis significance testing**, or **NHST**.

This procedure guarantees that our expected type I (false positive) error rate is at most α. This is sometimes interpreted as saying that frequentist hypothesis testing is very conservative, since it is unlikely to accidently reject the null hypothesis. But in fact the opposite is the case: because this method only worries about trying to reject the null, it can never gather evidence in favor of the null, no matter how large the sample size. Because of this, p-values tend to overstate the evidence against the null, and are thus very "trigger happy".

---

6. The reason we cannot just compute the probability of the observed value of the test statistic is that this will have probability zero under a pdf. The p-value is defined in terms of the cdf, so is always a number between 0 and 1.

In general there can be huge differences between p-values and the quantity that we really care about, which is the posterior probability of the null hypothesis given the data, $p(H_0|\mathcal{D})$. In particular, Sellke et al. (2001) show that even if the p-value is as low as 0.05, the posterior probability of H_0 is at least 30%, and often much higher. So frequentists often claim to have "significant" evidence of an effect that cannot be explained by the null hypothesis, whereas Bayesians are usually more conservative in their claims. For example, p-values have been used to "prove" that ESP (extra-sensory perception) is real (Wagenmakers et al. 2011), even though ESP is clearly very improbable. For this reason, p-values have been banned from certain medical journals (Matthews 1998).

Another problem with p-values is that their computation depends on decisions you make about when to stop collecting data, even if these decisions don't change the data you actually observed. For example, suppose I toss a coin $n = 12$ times and observe $s = 9$ successes (heads) and $f = 3$ failures (tails), so $n = s + f$. In this case, n is fixed and s (and hence f) is random. The relevant sampling model is the binomial

$$\text{Bin}(s|n, \theta) = \binom{n}{s} \theta^s (1 - \theta)^{n-s} \tag{6.82}$$

Let the null hypothesis be that the coin is fair, $\theta = 0.5$, where θ is the probability of success (heads). The one-sided p-value, using test statistic $t(s) = s$, is

$$p_1 = P(S \geq 9|H_0) = \sum_{s=9}^{12} \text{Bin}(s|12, 0.5) = \sum_{s=9}^{12} \binom{12}{s} 0.5^{12} = 0.073 \tag{6.83}$$

The two-sided p-value is

$$p_2 = \sum_{s=9}^{12} \text{Bin}(s|12, 0.5) + \sum_{s=0}^{3} \text{Bin}(s|12, 0.5) = 0.073 + 0.073 = 0.146 \tag{6.84}$$

In either case, the p-value is larger than the magical 5% threshold, so a frequentist would not reject the null hypothesis.

Now suppose I told you that I actually kept tossing the coin until I observed $f = 3$ tails. In this case, f is fixed and n (and hence $s = n - f$) is random. The probability model becomes the **negative binomial distribution**, given by

$$\text{NegBinom}(s|f, \theta) = \binom{s + f - 1}{f - 1} \theta^s (1 - \theta)^f \tag{6.85}$$

where $f = n - s$.

Note that the term which depends on θ is the same in Equations 6.82 and 6.85, so the posterior over θ would be the same in both cases. However, these two interpretations of the same data give different p-values. In particular, under the negative binomial model we get

$$p_3 = P(S \geq 9|H_0) = \sum_{s=9}^{\infty} \binom{3 + s - 1}{2} (1/2)^s (1/2)^3 = 0.0327 \tag{6.86}$$

So the p-value is 3%, and suddenly there seems to be significant evidence of bias in the coin! Obviously this is ridiculous: the data is the same, so our inferences about the coin should be the same. After all, I could have chosen the experimental protocol at random. It is the outcome of the experiment that matters, not the details of how I decided which one to run.

Although this might seem like just a mathematical curiosity, this also has significant practical implications. In particular, the fact that the **stopping rule** affects the computation of the p-value means that frequentists often do not terminate experiments early, even when it is obvious what the conclusions are, lest it adversely affect their statistical analysis. If the experiments are costly or harmful to people, this is obviously a bad idea. Perhaps it is not surprising, then, that the US Food and Drug Administration (FDA), which regulates clinical trials of new drugs, has recently become supportive of Bayesian methods[7], since Bayesian methods are not affected by the stopping rule.

6.6.3 The likelihood principle

The fundamental reason for many of these pathologies is that frequentist inference violates the **likelihood principle**, which says that inference should be based on the likelihood of the observed data, not based on hypothetical future data that you have not observed. Bayes obviously satisfies the likelihood principle, and consequently does not suffer from these pathologies.

A compelling argument in favor of the likelihood principle was presented in (Birnbaum 1962), who showed that it followed automatically from two simpler principles. The first of these is the **sufficiency principle**, which says that a sufficient statistic contains all the relevant information about an unknown parameter (arguably this is true by definition). The second principle is known as **weak conditionality**, which says that inferences should be based on the events that happened, not which might have happened. To motivate this, consider an example from (Berger 1985). Suppose we need to analyse a substance, and can send it either to a laboratory in New York or in California. The two labs seem equally good, so a fair coin is used to decide between them. The coin comes up heads, so the California lab is chosen. When the results come back, should it be taken into account that the coin could have come up tails and thus the New York lab could have been used? Most people would argue that the New York lab is irrelevant, since the tails event didn't happen. This is an example of weak conditionality. Given this principle, one can show that all inferences should only be based on what was observed, which is in contrast to standard frequentist procedures. See (Berger and Wolpert 1988) for further details on the likelihood principle.

6.6.4 Why isn't everyone a Bayesian?

Given these fundamental flaws of frequentist statistics, and the fact that Bayesian methods do not have such flaws, an obvious question to ask is: "Why isn't everyone a Bayesian?" The (frequentist) statistician Bradley Efron wrote a paper with exactly this title (Efron 1986). His short paper is well worth reading for anyone interested in this topic. Below we quote his opening section:

> The title is a reasonable question to ask on at least two counts. First of all, everyone used to be a Bayesian. Laplace wholeheartedly endorsed Bayes's formulation of the inference

---

7. See http://yamlb.wordpress.com/2006/06/19/the-us-fda-is-becoming-progressively-more-bayes ian/.

problem, and most 19th-century scientists followed suit. This included Gauss, whose statistical work is usually presented in frequentist terms.

A second and more important point is the cogency of the Bayesian argument. Modern statisticians, following the lead of Savage and de Finetti, have advanced powerful theoretical arguments for preferring Bayesian inference. A byproduct of this work is a disturbing catalogue of inconsistencies in the frequentist point of view.

Nevertheless, everyone is not a Bayesian. The current era (1986) is the first century in which statistics has been widely used for scientific reporting, and in fact, 20th-century statistics is mainly non-Bayesian. However, Lindley (1975) predicts a change for the 21st century.

Time will tell whether Lindley was right....

Exercises

Exercise 6.1 Pessimism of LOOCV

(Source: (Witten and Frank 2005), p152.). Suppose we have a completely random labeled dataset (i.e., the features $\mathbf{x}$ tell us nothing about the class labels y) with N_1 examples of class 1, and N_2 examples of class 2, where $N_1 = N_2$. What is the best misclassification rate any method can achieve? What is the estimated misclassification rate of the same method using LOOCV?

Exercise 6.2 James Stein estimator for Gaussian means

Consider the 2 stage model $Y_i|\theta_i \sim \mathcal{N}(\theta_i, \sigma^2)$ and $\theta_i|\mu \sim \mathcal{N}(m_0, \tau_0^2)$. Suppose $\sigma^2 = 500$ is known and we observe the following 6 data points, $i = 1 : 6$:

```
1505, 1528, 1564, 1498, 1600, 1470
```

a. Find the ML-II estimates of m_0 and τ_0^2.

b. Find the posterior estimates $\mathbb{E}[\theta_i|y_i, m_0, \tau_0]$ and $\text{var}[\theta_i|y_i, m_0, \tau_0]$ for $i = 1$. (The other terms, $i = 2 : 6$, are computed similarly.)

c. Give a 95% credible interval for $p(\theta_i|y_i, m_0, \tau_0)$ for $i = 1$. Do you trust this interval (assuming the Gaussian assumption is reasonable)? i.e. is it likely to be too large or too small, or just right?

d. What do you expect would happen to your estimates if σ^2 were much smaller (say $\sigma^2 = 1$)? You do not need to compute the numerical answer; just briefly explain what would happen qualitatively, and why.

Exercise 6.3 $\hat{\sigma}_{MLE}^2$ is biased

Show that $\hat{\sigma}_{MLE}^2 = \frac{1}{N}\sum_{n=1}^N (x_n - \hat{\mu})^2$ is a biased estimator of σ^2, i.e., show

$$\mathbb{E}_{X_1,\dots,X_n \sim \mathcal{N}(\mu,\sigma)}[\hat{\sigma}^2(X_1,\dots,X_n)] \neq \sigma^2$$

Hint: note that $X_1,\dots,X_N$ are independent, and use the fact that the expectation of a product of independent random variables is the product of the expectations.

Exercise 6.4 Estimation of σ^2 when μ is known

Suppose we sample $x_1,\dots,x_N \sim \mathcal{N}(\mu,\sigma^2)$ where μ is a *known* constant. Derive an expression for the MLE for σ^2 in this case. Is it unbiased?

7 *Linear regression*

7.1 Introduction

Linear regression is the "work horse" of statistics and (supervised) machine learning. When augmented with kernels or other forms of basis function expansion, it can model also non-linear relationships, as we will see.

7.2 Model specification

As we discussed in Section 1.4.5, linear regression is a model of the form

$$p(y|\mathbf{x}, \boldsymbol{\theta}) = \mathcal{N}(y|\mathbf{w}^T\mathbf{x}, \sigma^2) \tag{7.1}$$

Linear regression can be made to model non-linear relationships by replacing $\mathbf{x}$ with some non-linear function of the inputs, $\phi(\mathbf{x})$. That is, we use

$$p(y|\mathbf{x}, \boldsymbol{\theta}) = \mathcal{N}(y|\mathbf{w}^T\phi(\mathbf{x}), \sigma^2) \tag{7.2}$$

This is known as **basis function expansion**. (Note that the model is still linear in the parameters $\mathbf{w}$, so it is still called linear regression; the importance of this will become clear below.) A simple example are polynomial basis functions, where the model has the form

$$\phi(x) = [1, x, x^2, \dots, x^d] \tag{7.3}$$

Figure 1.18 illustrates the effect of changing d: increasing the degree d allows us to create increasingly complex functions. We can also apply linear regression to more than 1 input. For example, consider modeling temperature as a function of location. Figure 7.1(a) plots $\mathbb{E}[y|\mathbf{x}] = w_0 + w_1 x_1 + w_2 x_2$, and Figure 7.1(b) plots $\mathbb{E}[y|\mathbf{x}] = w_0 + w_1 x_1 + w_2 x_2 + w_3 x_1^2 + w_4 x_2^2$.

If we have a categorical input variable with K possible values, we need to use a dummy encoding for it (see Section 2.3.2). If all of the inputs are categorical, the model is known as **anova**, which stands for **analysis of variance**.

7.3 Maximum likelihood estimation (least squares)

The most common way to estimate the parameters of a statistical model is to compute the maximum likelihood estimate or MLE (see Section 3.2.3), which is defined by

$$\hat{\boldsymbol{\theta}} \triangleq \arg\max_{\boldsymbol{\theta}} \log p(\mathcal{D}|\boldsymbol{\theta}) \tag{7.4}$$

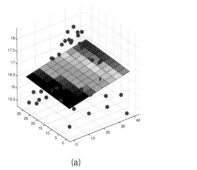 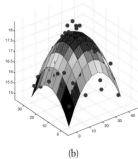

(a)   (b)

Figure 7.1 Linear regression applied to 2d data. Vertical axis is temperature, horizontal axes are location within a room. Data was collected by some remote sensing motes at Intel's lab in Berkeley, CA (data courtesy of Romain Thibaux). (a) The fitted plane has the form $\hat{f}(\mathbf{x}) = w_0 + w_1 x_1 + w_2 x_2$. (b) Temperature data is fitted with a quadratic of the form $\hat{f}(\mathbf{x}) = w_0 + w_1 x_1 + w_2 x_2 + w_3 x_1^2 + w_4 x_2^2$. Produced by `surfaceFitDemo`.

It is common to assume the training examples are independent and identically distributed, commonly abbreviated to **iid**. This means we can write the log-likelihood as follows:

$$\ell(\boldsymbol{\theta}) \triangleq \log p(\mathcal{D}|\boldsymbol{\theta}) = \sum_{i=1}^{N} \log p(y_i|\mathbf{x}_i, \boldsymbol{\theta}) \tag{7.5}$$

Instead of maximizing the log-likelihood, we can equivalently minimize the **negative log likelihood** or **NLL**:

$$\text{NLL}(\boldsymbol{\theta}) \triangleq -\sum_{i=1}^{N} \log p(y_i|\mathbf{x}_i, \boldsymbol{\theta}) \tag{7.6}$$

The NLL formulation is sometimes more convenient, since many optimization software packages are designed to find the minima of functions, rather than maxima.

Now let us apply the method of MLE to the linear regression setting. Inserting the definition of the Gaussian into the above, we find that the log likelihood is given by

$$\ell(\boldsymbol{\theta}) = \sum_{i=1}^{N} \log \left[\left(\frac{1}{2\pi\sigma^2} \right)^{\frac{1}{2}} \exp \left(-\frac{1}{2\sigma^2} (y_i - \mathbf{w}^T \mathbf{x}_i)^2 \right) \right] \tag{7.7}$$

$$= \frac{-1}{2\sigma^2} RSS(\mathbf{w}) - \frac{N}{2} \log(2\pi\sigma^2) \tag{7.8}$$

RSS stands for **residual sum of squares** and is defined by

$$\text{RSS}(\mathbf{w}) \triangleq \sum_{i=1}^{N} (y_i - \mathbf{w}^T \mathbf{x}_i)^2 \tag{7.9}$$

The RSS is also called the **sum of squared errors**, or SSE, and SSE/N is called the **mean squared error** or **MSE**. It can also be written as the square of the ℓ_2 **norm** of the vector of

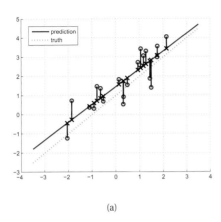
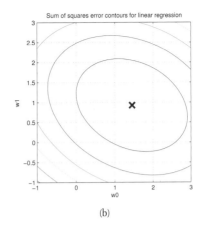

(a)   (b)

Figure 7.2 (a) In linear least squares, we try to minimize the sum of squared distances from each training point (denoted by a red circle) to its approximation (denoted by a blue cross), that is, we minimize the sum of the lengths of the little vertical blue lines. The red diagonal line represents $\hat{y}(x) = w_0 + w_1 x$, which is the least squares regression line. Note that these residual lines are not perpendicular to the least squares line, in contrast to Figure 12.5. Figure generated by `residualsDemo`. (b) Contours of the RSS error surface for the same example. The red cross represents the MLE, $\mathbf{w} = (1.45, 0.93)$. Figure generated by `contoursSSEdemo`.

residual errors:

$$\text{RSS}(\mathbf{w}) = ||\boldsymbol{\epsilon}||_2^2 = \sum_{i=1}^{N} \epsilon_i^2 \tag{7.10}$$

where $\epsilon_i = (y_i - \mathbf{w}^T \mathbf{x}_i)$.

We see that the MLE for $\mathbf{w}$ is the one that minimizes the RSS, so this method is known as **least squares**. This method is illustrated in Figure 7.2(a). The training data (x_i, y_i) are shown as red circles, the estimated values $(x_i, \hat{y}_i)$ are shown as blue crosses, and the residuals $\epsilon_i = y_i - \hat{y}_i$ are shown as vertical blue lines. The goal is to find the setting of the parameters (the slope w_1 and intercept w_0) such that the resulting red line minimizes the sum of squared residuals (the lengths of the vertical blue lines).

In Figure 7.2(b), we plot the NLL surface for our linear regression example. We see that it is a quadratic "bowl" with a unique minimum, which we now derive. (Importantly, this is true even if we use basis function expansion, such as polynomials, because the NLL is still *linear in the parameters* $\mathbf{w}$, even if it is not linear in the inputs $\mathbf{x}$.)

7.3.1 Derivation of the MLE

First, we rewrite the objective in a form that is more amenable to differentiation:

$$\text{NLL}(\mathbf{w}) = \frac{1}{2}(\mathbf{y} - \mathbf{X}\mathbf{w})^T(\mathbf{y} - \mathbf{X}\mathbf{w}) = \frac{1}{2}\mathbf{w}^T(\mathbf{X}^T\mathbf{X})\mathbf{w} - \mathbf{w}^T(\mathbf{X}^T\mathbf{y}) \tag{7.11}$$

where

$$\mathbf{X}^T\mathbf{X} = \sum_{i=1}^{N} \mathbf{x}_i \mathbf{x}_i^T = \sum_{i=1}^{N} \begin{pmatrix} x_{i,1}^2 & \cdots & x_{i,1}x_{i,D} \\ & \ddots & \\ x_{i,D}x_{i,1} & \cdots & x_{i,D}^2 \end{pmatrix} \tag{7.12}$$

is the **sum of squares** matrix and

$$\mathbf{X}^T\mathbf{y} = \sum_{i=1}^{N} \mathbf{x}_i y_i. \tag{7.13}$$

Using results from Equation 4.10, we see that the gradient of this is given by

$$\mathbf{g}(\mathbf{w}) = [\mathbf{X}^T\mathbf{X}\mathbf{w} - \mathbf{X}^T\mathbf{y}] = \sum_{i=1}^{N} \mathbf{x}_i(\mathbf{w}^T\mathbf{x}_i - y_i) \tag{7.14}$$

Equating to zero we get

$$\mathbf{X}^T\mathbf{X}\mathbf{w} = \mathbf{X}^T\mathbf{y} \tag{7.15}$$

This is known as the **normal equation**. The corresponding solution $\hat{\mathbf{w}}$ to this linear system of equations is called the **ordinary least squares** or **OLS** solution, which is given by

$$\boxed{\hat{\mathbf{w}}_{OLS} = (\mathbf{X}^T\mathbf{X})^{-1}\mathbf{X}^T\mathbf{y}} \tag{7.16}$$

7.3.2 Geometric interpretation

This equation has an elegant geometrical intrepretation, as we now explain. We assume $N > D$, so we have more examples than features. The columns of $\mathbf{X}$ define a linear subspace of dimensionality D which is embedded in N dimensions. Let the j'th column be $\tilde{\mathbf{x}}_j$, which is a vector in $\mathbb{R}^N$. (This should not be confused with $\mathbf{x}_i \in \mathbb{R}^D$, which represents the i'th data case.) Similarly, $\mathbf{y}$ is a vector in $\mathbb{R}^N$. For example, suppose we have $N = 3$ examples in $D = 2$ dimensions:

$$\mathbf{X} = \begin{pmatrix} 1 & 2 \\ 1 & -2 \\ 1 & 2 \end{pmatrix}, \quad \mathbf{y} = \begin{pmatrix} 8.8957 \\ 0.6130 \\ 1.7761 \end{pmatrix} \tag{7.17}$$

These vectors are illustrated in Figure 7.3.

We seek a vector $\hat{\mathbf{y}} \in \mathbb{R}^N$ that lies in this linear subspace and is as close as possible to $\mathbf{y}$, i.e., we want to find

$$\underset{\hat{\mathbf{y}} \in \text{span}(\{\tilde{\mathbf{x}}_1, \ldots, \tilde{\mathbf{x}}_D\})}{\operatorname{argmin}} \|\mathbf{y} - \hat{\mathbf{y}}\|_2. \tag{7.18}$$

Since $\hat{\mathbf{y}} \in \text{span}(\mathbf{X})$, there exists some weight vector $\mathbf{w}$ such that

$$\hat{\mathbf{y}} = w_1\tilde{\mathbf{x}}_1 + \cdots + w_D\tilde{\mathbf{x}}_D = \mathbf{X}\mathbf{w} \tag{7.19}$$

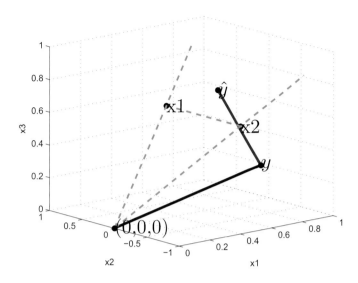

Figure 7.3 Graphical interpretation of least squares for $N = 3$ examples and $D = 2$ features. $\tilde{\mathbf{x}}_1$ and $\tilde{\mathbf{x}}_2$ are vectors in $\mathbb{R}^3$; together they define a 2D plane. $\mathbf{y}$ is also a vector in $\mathbb{R}^3$ but does not lie on this 2D plane. The orthogonal projection of $\mathbf{y}$ onto this plane is denoted $\hat{\mathbf{y}}$. The red line from $\mathbf{y}$ to $\hat{\mathbf{y}}$ is the residual, whose norm we want to minimize. For visual clarity, all vectors have been converted to unit norm. Figure generated by `leastSquaresProjection`.

To minimize the norm of the residual, $\mathbf{y} - \hat{\mathbf{y}}$, we want the residual vector to be orthogonal to every column of $\mathbf{X}$, so $\tilde{\mathbf{x}}_j^T(\mathbf{y} - \hat{\mathbf{y}}) = 0$ for $j = 1 : D$. Hence

$$\tilde{\mathbf{x}}_j^T(\mathbf{y} - \hat{\mathbf{y}}) = 0 \Rightarrow \mathbf{X}^T(\mathbf{y} - \mathbf{Xw}) = \mathbf{0} \Rightarrow \mathbf{w} = (\mathbf{X}^T\mathbf{X})^{-1}\mathbf{X}^T\mathbf{y} \tag{7.20}$$

Hence our projected value of $\mathbf{y}$ is given by

$$\hat{\mathbf{y}} = \mathbf{X}\hat{\mathbf{w}} = \mathbf{X}(\mathbf{X}^T\mathbf{X})^{-1}\mathbf{X}^T\mathbf{y} \tag{7.21}$$

This corresponds to an **orthogonal projection** of $\mathbf{y}$ onto the column space of $\mathbf{X}$. The projection matrix $\mathbf{P} \triangleq \mathbf{X}(\mathbf{X}^T\mathbf{X})^{-1}\mathbf{X}^T$ is called the **hat matrix**, since it "puts the hat on $\mathbf{y}$".

7.3.3 Convexity

When discussing least squares, we noted that the NLL had a bowl shape with a unique minimum. The technical term for functions like this is **convex**. Convex functions play a very important role in machine learning.

Let us define this concept more precisely. We say a *set* $\mathcal{S}$ is **convex** if for any $\boldsymbol{\theta}, \boldsymbol{\theta}' \in \mathcal{S}$, we have

$$\lambda\boldsymbol{\theta} + (1 - \lambda)\boldsymbol{\theta}' \in \mathcal{S}, \quad \forall \lambda \in [0, 1] \tag{7.22}$$

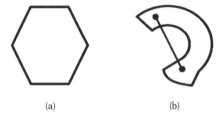

(a)   (b)

Figure 7.4   (a) Illustration of a convex set. (b) Illustration of a nonconvex set.

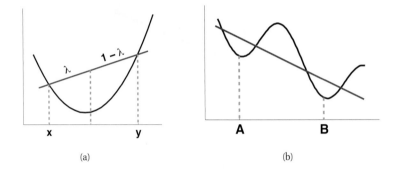

(a)   (b)

Figure 7.5   (a) Illustration of a convex function. We see that the chord joining $(x, f(x))$ to $(y, f(y))$ lies above the function. (b) A function that is neither convex nor concave. **A** is a local minimum, **B** is a global minimum. Figure generated by `convexFnHand`.

That is, if we draw a line from $\boldsymbol{\theta}$ to $\boldsymbol{\theta}'$, all points on the line lie inside the set. See Figure 7.4(a) for an illustration of a convex set, and Figure 7.4(b) for an illustration of a non-convex set.

A *function* $f(\boldsymbol{\theta})$ is called convex if its **epigraph** (the set of points above the function) defines a convex set. Equivalently, a function $f(\boldsymbol{\theta})$ is called convex if it is defined on a convex set and if, for any $\boldsymbol{\theta}, \boldsymbol{\theta}' \in \mathcal{S}$, and for any $0 \le \lambda \le 1$, we have

$$f(\lambda\boldsymbol{\theta} + (1 - \lambda)\boldsymbol{\theta}') \le \lambda f(\boldsymbol{\theta}) + (1 - \lambda)f(\boldsymbol{\theta}') \tag{7.23}$$

See Figure 7.5 for a 1d example. A function is called **strictly convex** if the inequality is strict. A function $f(\boldsymbol{\theta})$ is **concave** if $-f(\boldsymbol{\theta})$ is convex. Examples of scalar convex functions include θ^2, e^θ, and $\theta \log \theta$ (for $\theta > 0$). Examples of scalar concave functions include $\log(\theta)$ and $\sqrt{\theta}$.

Intuitively, a (strictly) convex function has a "bowl shape", and hence has a unique global minimum θ^* corresponding to the bottom of the bowl. Hence its second derivative must be positive everywhere, $\frac{d^2}{d\theta^2} f(\theta) > 0$. A twice-continuously differentiable, multivariate function f is convex iff its Hessian is positive definite for all $\boldsymbol{\theta}$.[1] In the machine learning context, the function f often corresponds to the NLL.

---

1. Recall that the Hessian is the matrix of second partial derivatives, defined by $H_{jk} = \frac{\partial f^2(\theta)}{\partial\theta_j \partial\theta_k}$. Also, recall that a matrix $\mathbf{H}$ is **positive definite** iff $\mathbf{v}^T \mathbf{H} \mathbf{v} > 0$ for any non-zero vector $\mathbf{v}$.

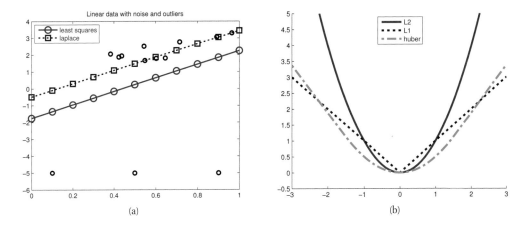

Figure 7.6 (a) Illustration of robust linear regression. Figure generated by `linregRobustDemoCombined`. (b) Illustration of ℓ_2, ℓ_1, and Huber loss functions. Figure generated by `huberLossDemo`.

Models where the NLL is convex are desirable, since this means we can always find the globally optimal MLE. We will see many examples of this later in the book. However, many models of interest will not have concave likelihoods. In such cases, we will discuss ways to derive locally optimal parameter estimates.

7.4 Robust linear regression *

It is very common to model the noise in regression models using a Gaussian distribution with zero mean and constant variance, $\epsilon_i \sim \mathcal{N}(0, \sigma^2)$, where $\epsilon_i = y_i - \mathbf{w}^T \mathbf{x}_i$. In this case, maximizing likelihood is equivalent to minimizing the sum of squared residuals, as we have seen. However, if we have **outliers** in our data, this can result in a poor fit, as illustrated in Figure 7.6(a). (The outliers are the points on the bottom of the figure.) This is because squared error penalizes deviations quadratically, so points far from the line have more affect on the fit than points near to the line.

One way to achieve **robustness** to outliers is to replace the Gaussian distribution for the response variable with a distribution that has **heavy tails**. Such a distribution will assign higher likelihood to outliers, without having to perturb the straight line to "explain" them.

One possibility is to use the Laplace distribution, introduced in Section 2.4.4. If we use this as our observation model for regression, we get the following likelihood:

$$p(y|\mathbf{x}, \mathbf{w}, b) = \text{Lap}(y|\mathbf{w}^T\mathbf{x}, b) \propto \exp(-\frac{1}{b}|y - \mathbf{w}^T\mathbf{x}|) \tag{7.24}$$

The robustness arises from the use of $|y - \mathbf{w}^T\mathbf{x}|$ instead of $(y - \mathbf{w}^T\mathbf{x})^2$. For simplicity, we will assume b is fixed. Let $r_i \triangleq y_i - \mathbf{w}^T\mathbf{x}_i$ be the i'th residual. The NLL has the form

$$\ell(\mathbf{w}) = \sum_i |r_i(\mathbf{w})| \tag{7.25}$$

Likelihood	Prior	Name	Section
Gaussian	Uniform	Least squares	7.3
Gaussian	Gaussian	Ridge	7.5
Gaussian	Laplace	Lasso	13.3
Laplace	Uniform	Robust regression	7.4
Student	Uniform	Robust regression	Exercise 11.12

Table 7.1 Summary of various likelihoods and priors used for linear regression. The likelihood refers to the distributional form of $p(y|\mathbf{x}, \mathbf{w}, \sigma^2)$, and the prior refers to the distributional form of $p(\mathbf{w})$. MAP estimation with a uniform distribution corresponds to MLE.

Unfortunately, this is a non-linear objective function, which is hard to optimize. Fortunately, we can convert the NLL to a linear objective, subject to linear constraints, using the following **split variable** trick. First we define

$$r_i \triangleq r_i^+ - r_i^- \tag{7.26}$$

and then we impose the linear inequality constraints that $r_i^+ \geq 0$ and $r_i^- \geq 0$. Now the constrained objective becomes

$$\min_{\mathbf{w}, \mathbf{r}^+, \mathbf{r}^-} \sum_i (r_i^+ + r_i^-) \qquad \text{s.t.} \quad r_i^+ \geq 0, r_i^- \geq 0, \mathbf{w}^T \mathbf{x}_i + r_i^+ - r_i^- = y_i \tag{7.27}$$

This is an example of a **linear program** with $D + 2N$ unknowns and $3N$ constraints.

Since this is a convex optimization problem, it has a unique solution. To solve an LP, we must first write it in standard form, which as follows:

$$\min_{\boldsymbol{\theta}} \mathbf{f}^T \boldsymbol{\theta} \quad \text{s.t.} \quad \mathbf{A}\boldsymbol{\theta} \leq \mathbf{b}, \ \mathbf{A}_{eq}\boldsymbol{\theta} = \mathbf{b}_{eq}, \ \mathbf{l} \leq \boldsymbol{\theta} \leq \mathbf{u} \tag{7.28}$$

In our current example, $\boldsymbol{\theta} = (\mathbf{w}, \mathbf{r}^+, \mathbf{r}^-)$, $\mathbf{f} = [\mathbf{0}, \mathbf{1}, \mathbf{1}]$, $\mathbf{A} = []$, $\mathbf{b} = []$, $\mathbf{A}_{eq} = [\mathbf{X}, \mathbf{I}, -\mathbf{I}]$, $\mathbf{b}_{eq} = \mathbf{y}$, $\mathbf{l} = [-\infty \mathbf{1}, \mathbf{0}, \mathbf{0}]$, $\mathbf{u} = []$. This can be solved by any LP solver (see e.g., (Boyd and Vandenberghe 2004)). See Figure 7.6(a) for an example of the method in action.

An alternative to using NLL under a Laplace likelihood is to minimize the **Huber loss** function (Huber 1964), defined as follows:

$$L_H(r, \delta) = \begin{cases} r^2/2 & \text{if } |r| \leq \delta \\ \delta|r| - \delta^2/2 & \text{if } |r| > \delta \end{cases} \tag{7.29}$$

This is equivalent to ℓ_2 for errors that are smaller than δ, and is equivalent to ℓ_1 for larger errors. See Figure 7.6(b). The advantage of this loss function is that it is everywhere differentiable, using the fact that $\frac{d}{dr}|r| = \text{sign}(r)$ if $r \neq 0$. We can also check that the function has a continuous first derivative, since the gradients of the two parts of the function match at $r = \pm\delta$, namely $\frac{d}{dr}L_H(r, \delta)|_{r=\delta} = \delta$. Consequently optimizing the Huber loss is much faster than using the Laplace likelihood, since we can use standard smooth optimization methods (such as quasi-Newton) instead of linear programming.

Figure 7.6(a) gives an illustration of the Huber loss function. The results are qualitatively similiar to the probabilistic methods. (In fact, it turns out that the Huber method also has a probabilistic interpretation, although it is rather unnatural (Pontil et al. 1998).)

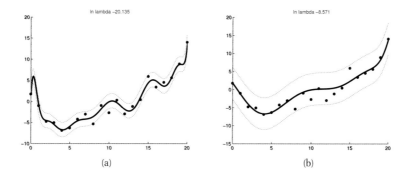

Figure 7.7 Degree 14 Polynomial fit to $N = 21$ data points with increasing amounts of ℓ_2 regularization. Data was generated from noise with variance $\sigma^2 = 4$. The error bars, representing the noise variance σ^2, get wider as the fit gets smoother, since we are ascribing more of the data variation to the noise. Figure generated by `linregPolyVsRegDemo`.

7.5   Ridge regression

One problem with ML estimation is that it can result in overfitting. In this section, we discuss a way to ameliorate this problem by using MAP estimation with a Gaussian prior. For simplicity, we assume a Gaussian likelihood, rather than a robust likelihood.

7.5.1   Basic idea

The reason that the MLE can overfit is that it is picking the parameter values that are the best for modeling the training data; but if the data is noisy, such parameters often result in complex functions. As a simple example, suppose we fit a degree 14 polynomial to $N = 21$ data points using least squares. The resulting curve is very "wiggly", as shown in Figure 7.7(a). The corresponding least squares coefficients (excluding w_0) are as follows:

```
6.560, -36.934, -109.255, 543.452, 1022.561, -3046.224, -3768.013,
8524.540, 6607.897, -12640.058, -5530.188, 9479.730, 1774.639, -2821.526
```

We see that there are many large positive and negative numbers. These balance out exactly to make the curve "wiggle" in just the right way so that it almost perfectly interpolates the data. But this situation is unstable: if we changed the data a little, the coefficients would change a lot.

We can encourage the parameters to be small, thus resulting in a smoother curve, by using a zero-mean Gaussian prior:

$$p(\mathbf{w}) = \prod_j \mathcal{N}(w_j | 0, \tau^2) \tag{7.30}$$

where $1/\tau^2$ controls the strength of the prior. The corresponding MAP estimation problem becomes

$$\underset{\mathbf{w}}{\operatorname{argmax}} \sum_{i=1}^N \log \mathcal{N}(y_i | w_0 + \mathbf{w}^T \mathbf{x}_i, \sigma^2) + \sum_{j=1}^D \log \mathcal{N}(w_j | 0, \tau^2) \tag{7.31}$$

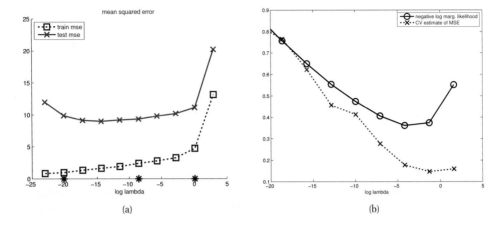

Figure 7.8 (a) Training error (dotted blue) and test error (solid red) for a degree 14 polynomial fit by ridge regression, plotted vs $\log(\lambda)$. Data was generated from noise with variance $\sigma^2 = 4$ (training set has size $N = 21$). Note: Models are ordered from complex (small regularizer) on the left to simple (large regularizer) on the right. The stars correspond to the values used to plot the functions in Figure 7.7. (b) Estimate of performance using training set. Dotted blue: 5-fold cross-validation estimate of future MSE. Solid black: negative log marginal likelihood, $-\log p(\mathcal{D}|\lambda)$. Both curves have been vertically rescaled to [0,1] to make them comparable. Figure generated by `linregPolyVsRegDemo`.

It is a simple exercise to show that this is equivalent to minimizing the following:

$$J(\mathbf{w}) = \frac{1}{N}\sum_{i=1}^{N}(y_i - (w_0 + \mathbf{w}^T\mathbf{x}_i))^2 + \lambda||\mathbf{w}||_2^2 \qquad (7.32)$$

where $\lambda \triangleq \sigma^2/\tau^2$ and $||\mathbf{w}||_2^2 = \sum_j w_j^2 = \mathbf{w}^T\mathbf{w}$ is the squared two-norm. Here the first term is the MSE/ NLL as usual, and the second term, $\lambda \geq 0$, is a complexity penalty. The corresponding solution is given by

$$\hat{\mathbf{w}}_{ridge} = (\lambda\mathbf{I}_D + \mathbf{X}^T\mathbf{X})^{-1}\mathbf{X}^T\mathbf{y} \qquad (7.33)$$

This technique is known as **ridge regression**, or **penalized least squares**. In general, adding a Gaussian prior to the parameters of a model to encourage them to be small is called ℓ_2 **regularization** or **weight decay**. Note that the offset term w_0 is not regularized, since this just affects the height of the function, not its complexity. By penalizing the sum of the magnitudes of the weights, we ensure the function is simple (since $\mathbf{w} = \mathbf{0}$ corresponds to a straight line, which is the simplest possible function, corresponding to a constant.)

We illustrate this idea in Figure 7.7, where we see that increasing λ results in smoother functions. The resulting coefficients also become smaller. For example, using $\lambda = 10^{-3}$, we have

2.128, 0.807, 16.457, 3.704, -24.948, -10.472, -2.625, 4.360, 13.711,
10.063, 8.716, 3.966, -9.349, -9.232

In Figure 7.8(a), we plot the MSE on the training and test sets vs $\log(\lambda)$. We see that, as we increase λ (so the model becomes more constrained), the error on the training set increases. For the test set, we see the characteristic U-shaped curve, where the model overfits and then underfits. It is common to use cross validation to pick λ, as shown in Figure 7.8(b). In Section 7.6.4 we will discuss a more probabilistic approach.

We will consider a variety of different priors in this book. Each of these corresponds to a different form of **regularization**. This technique is very widely used to prevent overfitting.

7.5.2 Numerically stable computation *

Interestingly, ridge regression, which works better statistically, is also easier to fit numerically, since $(\lambda \mathbf{I}_D + \mathbf{X}^T \mathbf{X})$ is much better conditioned (and hence more likely to be invertible) than $\mathbf{X}^T \mathbf{X}$, at least for suitable large λ.

Nevertheless, inverting matrices is still best avoided, for reasons of numerical stability. (Indeed, if you write w=inv(X' * X)*X'*y in Matlab, it will give you a warning.) We now describe a useful trick for fitting ridge regression models (and hence by extension, computing vanilla OLS estimates) that is more numerically robust. We assume the prior has the form $p(\mathbf{w}) = \mathcal{N}(\mathbf{0}, \mathbf{\Lambda}^{-1})$, where $\mathbf{\Lambda}$ is the precision matrix. In the case of ridge regression, $\mathbf{\Lambda} = (1/\tau^2)\mathbf{I}$. To avoid penalizing the w_0 term, we should center the data first, as explained in Exercise 7.5.

First let us augment the original data with some "virtual data" coming from the prior:

$$\tilde{\mathbf{X}} = \begin{pmatrix} \mathbf{X}/\sigma \\ \sqrt{\mathbf{\Lambda}} \end{pmatrix}, \quad \tilde{\mathbf{y}} = \begin{pmatrix} \mathbf{y}/\sigma \\ \mathbf{0}_{D \times 1} \end{pmatrix} \tag{7.34}$$

where $\mathbf{\Lambda} = \sqrt{\mathbf{\Lambda}}\sqrt{\mathbf{\Lambda}}^T$ is a **Cholesky decomposition** of $\mathbf{\Lambda}$. We see that $\tilde{\mathbf{X}}$ is $(N + D) \times D$, where the extra rows represent pseudo-data from the prior.

We now show that the NLL on this expanded data is equivalent to *penalized* NLL on the original data:

$$f(\mathbf{w}) = (\tilde{\mathbf{y}} - \tilde{\mathbf{X}}\mathbf{w})^T(\tilde{\mathbf{y}} - \tilde{\mathbf{X}}\mathbf{w}) \tag{7.35}$$

$$= \left(\begin{pmatrix} \mathbf{y}/\sigma \\ \mathbf{0} \end{pmatrix} - \begin{pmatrix} \mathbf{X}/\sigma \\ \sqrt{\mathbf{\Lambda}} \end{pmatrix} \mathbf{w} \right)^T \left(\begin{pmatrix} \mathbf{y}/\sigma \\ \mathbf{0} \end{pmatrix} - \begin{pmatrix} \mathbf{X}/\sigma \\ \sqrt{\mathbf{\Lambda}} \end{pmatrix} \mathbf{w} \right) \tag{7.36}$$

$$= \begin{pmatrix} \frac{1}{\sigma}(\mathbf{y} - \mathbf{X}\mathbf{w}) \\ -\sqrt{\mathbf{\Lambda}}\mathbf{w} \end{pmatrix}^T \begin{pmatrix} \frac{1}{\sigma}(\mathbf{y} - \mathbf{X}\mathbf{w}) \\ -\sqrt{\mathbf{\Lambda}}\mathbf{w} \end{pmatrix} \tag{7.37}$$

$$= \frac{1}{\sigma^2}(\mathbf{y} - \mathbf{X}\mathbf{w})^T(\mathbf{y} - \mathbf{X}\mathbf{w}) + (\sqrt{\mathbf{\Lambda}}\mathbf{w})^T(\sqrt{\mathbf{\Lambda}}\mathbf{w}) \tag{7.38}$$

$$= \frac{1}{\sigma^2}(\mathbf{y} - \mathbf{X}\mathbf{w})^T(\mathbf{y} - \mathbf{X}\mathbf{w}) + \mathbf{w}^T \mathbf{\Lambda} \mathbf{w} \tag{7.39}$$

Hence the MAP estimate is given by

$$\hat{\mathbf{w}}_{ridge} = (\tilde{\mathbf{X}}^T \tilde{\mathbf{X}})^{-1} \tilde{\mathbf{X}}^T \tilde{\mathbf{y}} \tag{7.40}$$

as we claimed.

Now let

$$\tilde{\mathbf{X}} = \mathbf{QR} \tag{7.41}$$

be the **QR decomposition** of $\mathbf{X}$, where $\mathbf{Q}$ is orthonormal (meaning $\mathbf{Q}^T\mathbf{Q} = \mathbf{QQ}^T = \mathbf{I}$), and $\mathbf{R}$ is upper triangular. Then

$$(\tilde{\mathbf{X}}^T\tilde{\mathbf{X}})^{-1} = (\mathbf{R}^T\mathbf{Q}^T\mathbf{QR})^{-1} = (\mathbf{R}^T\mathbf{R})^{-1} = \mathbf{R}^{-1}\mathbf{R}^{-T} \tag{7.42}$$

Hence

$$\hat{\mathbf{w}}_{ridge} = \mathbf{R}^{-1}\mathbf{R}^{-T}\mathbf{R}^T\mathbf{Q}^T\tilde{\mathbf{y}} = \mathbf{R}^{-1}\mathbf{Q}\tilde{\mathbf{y}} \tag{7.43}$$

Note that $\mathbf{R}$ is easy to invert since it is upper triangular. This gives us a way to compute the ridge estimate while avoiding having to invert $(\mathbf{\Lambda} + \mathbf{X}^T\mathbf{X})$.

We can use this technique to find the MLE, by simply computing the QR decomposition of the unaugmented matrix $\mathbf{X}$, and using the original $\mathbf{y}$. This is the method of choice for solving least squares problems. (In fact, it is so common that it can be implemented in one line of Matlab, using the **backslash operator**: w=X\y.) Note that computing the QR decomposition of an $N \times D$ matrix takes $O(ND^2)$ time, and is numerically very stable.

If $D \gg N$, we should first perform an SVD decomposition. In particular, let $\mathbf{X} = \mathbf{USV}^T$ be the SVD of $\mathbf{X}$, where $\mathbf{V}^T\mathbf{V} = \mathbf{I}_N$, $\mathbf{UU}^T = \mathbf{U}^T\mathbf{U} = \mathbf{I}_N$, and $\mathbf{S}$ is a diagonal $N \times N$ matrix. Now let $\mathbf{Z} = \mathbf{US}$ be an $N \times N$ matrix. Then we can rewrite the ridge estimate thus:

$$\hat{\mathbf{w}}_{ridge} = \mathbf{V}(\mathbf{Z}^T\mathbf{Z} + \lambda\mathbf{I}_N)^{-1}\mathbf{Z}^T\mathbf{y} \tag{7.44}$$

In other words, we can replace the D-dimensional vectors $\mathbf{x}_i$ with the N-dimensional vectors $\mathbf{z}_i$ and perform our penalized fit as before. We then transform the N-dimensional solution to the D-dimensional solution by multiplying by $\mathbf{V}$. Geometrically, we are rotating to a new coordinate system in which all but the first N coordinates are zero. This does not affect the solution since the spherical Gaussian prior is rotationally invariant. The overall time is now $O(DN^2)$ operations.

7.5.3 Connection with PCA *

In this section, we discuss an interesting connection between ridge regression and PCA (Section 12.2), which gives further insight into why ridge regression works well. Our discussion is based on (Hastie et al. 2009, p66).

Let $\mathbf{X} = \mathbf{USV}^T$ be the SVD of $\mathbf{X}$. From Equation 7.44, we have

$$\hat{\mathbf{w}}_{ridge} = \mathbf{V}(\mathbf{S}^2 + \lambda\mathbf{I})^{-1}\mathbf{SU}^T\mathbf{y} \tag{7.45}$$

Hence the ridge predictions on the training set are given by

$$\hat{\mathbf{y}} = \mathbf{X}\hat{\mathbf{w}}_{ridge} = \mathbf{USV}^T\mathbf{V}(\mathbf{S}^2 + \lambda\mathbf{I})^{-1}\mathbf{SU}^T\mathbf{y} \tag{7.46}$$

$$= \mathbf{U}\tilde{\mathbf{S}}\mathbf{U}^T\mathbf{y} = \sum_{j=1}^{D}\mathbf{u}_j\tilde{S}_{jj}\mathbf{u}_j^T\mathbf{y} \tag{7.47}$$

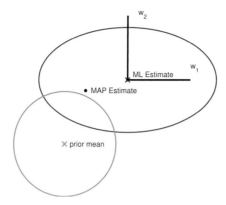

Figure 7.9 Geometry of ridge regression. The likelihood is shown as an ellipse, and the prior is shown as a circle centered on the origin. Based on Figure 3.15 of (Bishop 2006). Figure generated by `geomRidge`

where

$$\tilde{S}_{jj} \triangleq [\mathbf{S}(\mathbf{S}^2 + \lambda I)^{-1}\mathbf{S}]_{jj} = \frac{\sigma_j^2}{\sigma_j^2 + \lambda} \tag{7.48}$$

and σ_j are the singular values of $\mathbf{X}$. Hence

$$\hat{\mathbf{y}} = \mathbf{X}\hat{\mathbf{w}}_{ridge} = \sum_{j=1}^{D} \mathbf{u}_j \frac{\sigma_j^2}{\sigma_j^2 + \lambda} \mathbf{u}_j^T \mathbf{y} \tag{7.49}$$

In contrast, the least squares prediction is

$$\hat{\mathbf{y}} = \mathbf{X}\hat{\mathbf{w}}_{ls} = (\mathbf{U}\mathbf{S}\mathbf{V}^T)(\mathbf{V}\mathbf{S}^{-1}\mathbf{U}^T\mathbf{y}) = \mathbf{U}\mathbf{U}^T\mathbf{y} = \sum_{j=1}^{D} \mathbf{u}_j\mathbf{u}_j^T \mathbf{y} \tag{7.50}$$

If σ_j^2 is small compared to λ, then direction $\mathbf{u}_j$ will not have much effect on the prediction. In view of this, we *define* the effective number of **degrees of freedom** of the model as follows:

$$\text{dof}(\lambda) = \sum_{j=1}^{D} \frac{\sigma_j^2}{\sigma_j^2 + \lambda} \tag{7.51}$$

When $\lambda = 0$, $\text{dof}(\lambda) = D$, and as $\lambda \to \infty$, $\text{dof}(\lambda) \to 0$.

Let us try to understand why this behavior is desirable. In Section 7.6, we show that $\text{cov}\,[\mathbf{w}|\mathcal{D}] = \sigma^2(\mathbf{X}^T\mathbf{X})^{-1}$, if we use a uniform prior for $\mathbf{w}$. Thus the directions in which we are most uncertain about $\mathbf{w}$ are determined by the eigenvectors of this matrix with the smallest eigenvalues, as shown in Figure 4.1. Furthermore, in Section 12.2.3, we show that the squared singular values σ_j^2 are equal to the eigenvalues of $\mathbf{X}^T\mathbf{X}$. Hence small singular values σ_j correspond to directions with high posterior variance. It is these directions which ridge shrinks the most.

This process is illustrated in Figure 7.9. The horizontal w_1 parameter is not-well determined by the data (has high posterior variance), but the vertical w_2 parameter is well-determined. Hence w_2^{map} is close to w_2^{mle}, but w_1^{map} is shifted strongly towards the prior mean, which is 0. (Compare to Figure 4.14(c), which illustrated sensor fusion with sensors of different reliabilities.) In this way, ill-determined parameters are reduced in size towards 0. This is called **shrinkage**.

There is a related, but different, technique called **principal components regression**. The idea is this: first use PCA to reduce the dimensionality to K dimensions, and then use these low dimensional features as input to regression. However, this technique does not work as well as ridge in terms of predictive accuracy (Hastie et al. 2001, p70). The reason is that in PC regression, only the first K (derived) dimensions are retained, and the remaining $D - K$ dimensions are entirely ignored. By contrast, ridge regression uses a "soft" weighting of all the dimensions.

7.5.4   Regularization effects of big data

Regularization is the most common way to avoid overfitting. However, another effective approach — which is not always available — is to use lots of data. It should be intuitively obvious that the more training data we have, the better we will be able to learn.[2] So we expect the test set error to decrease to some plateau as N increases.

This is illustrated in Figure 7.10, where we plot the mean squared error incurred on the test set achieved by polynomial regression models of different degrees vs N (a plot of error vs training set size is known as a **learning curve**). The level of the plateau for the test error consists of two terms: an irreducible component that all models incur, due to the intrinsic variability of the generating process (this is called the **noise floor**); and a component that depends on the discrepancy between the generating process (the "truth") and the model: this is called **structural error**.

In Figure 7.10, the truth is a degree 2 polynomial, and we try fitting polynomials of degrees 1, 2 and 25 to this data. Call the 3 models $\mathcal{M}_1$, $\mathcal{M}_2$ and $\mathcal{M}_{25}$. We see that the structural error for models $\mathcal{M}_2$ and $\mathcal{M}_{25}$ is zero, since both are able to capture the true generating process. However, the structural error for $\mathcal{M}_1$ is substantial, which is evident from the fact that the plateau occurs high above the noise floor.

For any model that is expressive enough to capture the truth (i.e., one with small structural error), the test error will go to the noise floor as $N \rightarrow \infty$. However, it will typically go to zero faster for simpler models, since there are fewer parameters to estimate. In particular, for finite training sets, there will be some discrepancy between the parameters that we estimate and the best parameters that we could estimate given the particular model class. This is called **approximation error**, and goes to zero as $N \rightarrow \infty$, but it goes to zero faster for simpler models. This is illustrated in Figure 7.10. See also Exercise 7.1.

In domains with lots of data, simple methods can work surprisingly well (Halevy et al. 2009). However, there are still reasons to study more sophisticated learning methods, because there will always be problems for which we have little data. For example, even in such a data-rich domain as web search, as soon as we want to start personalizing the results, the amount of data available for any given user starts to look small again (relative to the complexity of the problem).

---

2. This assumes the training data is randomly sampled, and we don't just get repetitions of the same examples. Having informatively sampled data can help even more; this is the motivation for an approach known as active learning, where you get to choose your training data.

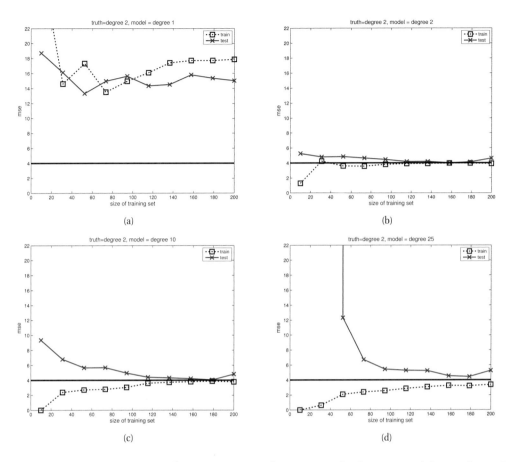

Figure 7.10 MSE on training and test sets vs size of training set, for data generated from a degree 2 polynomial with Gaussian noise of variance $\sigma^2 = 4$. We fit polynomial models of varying degree to this data. (a) Degree 1. (b) Degree 2. (c) Degree 10. (d) Degree 25. Note that for small training set sizes, the test error of the degree 25 polynomial is higher than that of the degree 2 polynomial, due to overfitting, but this difference vanishes once we have enough data. Note also that the degree 1 polynomial is too simple and has high test error even given large amounts of training data. Figure generated by `linregPolyVsN`.

In such cases, we may want to learn multiple related models at the same time, which is known as multi-task learning. This will allow us to "borrow statistical strength" from tasks with lots of data and to share it with tasks with little data. We will discuss ways to do later in the book.

7.6 Bayesian linear regression

Although ridge regression is a useful way to compute a point estimate, sometimes we want to compute the full posterior over $\mathbf{w}$ and σ^2. For simplicity, we will initially assume the noise variance σ^2 is known, so we focus on computing $p(\mathbf{w}|\mathcal{D}, \sigma^2)$. Then in Section 7.6.3 we consider the general case, where we compute $p(\mathbf{w}, \sigma^2 | \mathcal{D})$. We assume throughout a Gaussian likelihood model. Performing Bayesian inference with a robust likelihood is also possible, but requires more advanced techniques (see Exercise 24.5).

7.6.1 Computing the posterior

In linear regression, the likelihood is given by

$$p(\mathbf{y} | \mathbf{X}, \mathbf{w}, \mu, \sigma^2) \quad = \quad \mathcal{N}(\mathbf{y} | \mu + \mathbf{X}\mathbf{w}, \sigma^2 \mathbf{I}_N) \tag{7.52}$$

$$\propto \quad \exp\left(-\frac{1}{2\sigma^2}(\mathbf{y} - \mu\mathbf{1}_N - \mathbf{X}\mathbf{w})^T(\mathbf{y} - \mu\mathbf{1}_N - \mathbf{X}\mathbf{w})\right) \tag{7.53}$$

where μ is an offset term. If the inputs are centered, so $\sum_i x_{ij} = 0$ for each j, the mean of the output is equally likely to be positive or negative. So let us put an improper prior on μ of the form $p(\mu) \propto 1$, and then integrate it out to get

$$p(\mathbf{y} | \mathbf{X}, \mathbf{w}, \sigma^2) \propto \exp\left(-\frac{1}{2\sigma^2} ||\mathbf{y} - \overline{y}\mathbf{1}_N - \mathbf{X}\mathbf{w}||_2^2\right) \tag{7.54}$$

where $\overline{y} = \frac{1}{N}\sum_{i=1}^N y_i$ is the empirical mean of the output. For notational simplicity, we shall assume the output has been centered, and write $\mathbf{y}$ for $\mathbf{y} - \overline{y}\mathbf{1}_N$.

The conjugate prior to the above Gaussian likelihood is also a Gaussian, which we will denote by $p(\mathbf{w}) = \mathcal{N}(\mathbf{w} | \mathbf{w}_0, \mathbf{V}_0)$. Using Bayes rule for Gaussians, Equation 4.125, the posterior is given by

$$p(\mathbf{w} | \mathbf{X}, \mathbf{y}, \sigma^2) \quad \propto \quad \mathcal{N}(\mathbf{w} | \mathbf{w}_0, \mathbf{V}_0)\mathcal{N}(\mathbf{y} | \mathbf{X}\mathbf{w}, \sigma^2 \mathbf{I}_N) = \mathcal{N}(\mathbf{w} | \mathbf{w}_N, \mathbf{V}_N) \tag{7.55}$$

$$\mathbf{w}_N \quad = \quad \mathbf{V}_N \mathbf{V}_0^{-1} \mathbf{w}_0 + \frac{1}{\sigma^2} \mathbf{V}_N \mathbf{X}^T \mathbf{y} \tag{7.56}$$

$$\mathbf{V}_N^{-1} \quad = \quad \mathbf{V}_0^{-1} + \frac{1}{\sigma^2} \mathbf{X}^T \mathbf{X} \tag{7.57}$$

$$\mathbf{V}_N \quad = \quad \sigma^2 (\sigma^2 \mathbf{V}_0^{-1} + \mathbf{X}^T \mathbf{X})^{-1} \tag{7.58}$$

If $\mathbf{w}_0 = \mathbf{0}$ and $\mathbf{V}_0 = \tau^2 \mathbf{I}$, then the posterior mean reduces to the ridge estimate, if we define $\lambda = \frac{\sigma^2}{\tau^2}$. This is because the mean and mode of a Gaussian are the same.

To gain insight into the posterior distribution (and not just its mode), let us consider a 1D example:

$$y(x, \mathbf{w}) = w_0 + w_1 x + \epsilon \tag{7.59}$$

where the "true" parameters are $w_0 = -0.3$ and $w_1 = 0.5$. In Figure 7.11 we plot the prior, the likelihood, the posterior, and some samples from the posterior predictive. In particular, the right hand column plots the function $y(x, \mathbf{w}^{(s)})$ where x ranges over $[-1, 1]$, and $\mathbf{w}^{(s)} \sim \mathcal{N}(\mathbf{w} | \mathbf{w}_N, \mathbf{V}_N)$ is a sample from the parameter posterior. Initially, when we sample from the prior (first row), our predictions are "all over the place", since our prior is uniform. After we see one data point (second row), our posterior becomes constrained by the corresponding likelihood, and our predictions pass close to the observed data. However, we see that the posterior has a ridge-like shape, reflecting the fact that there are many possible solutions, with different

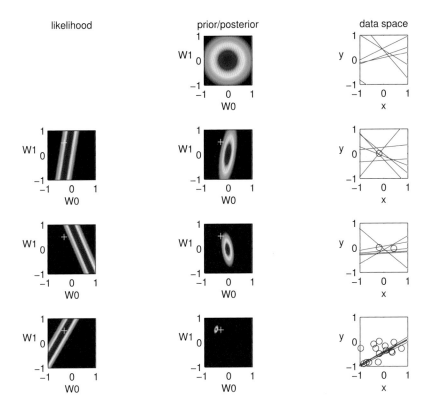

Figure 7.11 Sequential Bayesian inference of the parameters of a linear regression model $p(y|\mathbf{x}) = \mathcal{N}(y|w_0x_0 + w_1x_1, \sigma^2)$. Left column: likelihood function for current data point. Middle column: posterior given first n data points, $p(w_0, w_1|\mathbf{x}_{1:n}, y_{1:n}, \sigma^2)$. Right column: samples from the current posterior predictive distribution. Row 1: prior distribution ($n = 0$). Row 2: after 1 data point. Row 3: after 2 data points. Row 4: after 20 data points. The white cross in columns 1 and 2 represents the true parameter value; we see that the mode of the posterior rapidly converges to this point. The blue circles in column 3 are the observed data points. Based on Figure 3.7 of (Bishop 2006). Figure generated by `bayesLinRegDemo2d`.

slopes/intercepts. This makes sense since we cannot uniquely infer two parameters from one observation. After we see two data points (third row), the posterior becomes much narrower, and our predictions all have similar slopes and intercepts. After we observe 20 data points (last row), the posterior is essentially a delta function centered on the true value, indicated by a white cross. (The estimate converges to the truth since the data was generated from this model, and because Bayes is a consistent estimator; see Section 6.4.1 for discussion of this point.)

7.6.2 Computing the posterior predictive

It's tough to make predictions, especially about the future. — Yogi Berra

In machine learning, we often care more about predictions than about interpreting the parameters. Using Equation 4.126, we can easily show that the posterior predictive distribution at a test point $\mathbf{x}$ is also Gaussian:

$$p(y|\mathbf{x}, \mathcal{D}, \sigma^2) = \int \mathcal{N}(y|\mathbf{x}^T\mathbf{w}, \sigma^2)\mathcal{N}(\mathbf{w}|\mathbf{w}_N, \mathbf{V}_N)d\mathbf{w} \tag{7.60}$$

$$= \mathcal{N}(y|\mathbf{w}_N^T\mathbf{x}, \sigma_N^2(\mathbf{x})) \tag{7.61}$$

$$\sigma_N^2(\mathbf{x}) = \sigma^2 + \mathbf{x}^T\mathbf{V}_N\mathbf{x} \tag{7.62}$$

The variance in this prediction, $\sigma_N^2(\mathbf{x})$, depends on two terms: the variance of the observation noise, σ^2, and the variance in the parameters, $\mathbf{V}_N$. The latter translates into variance about observations in a way which depends on how close $\mathbf{x}$ is to the training data $\mathcal{D}$. This is illustrated in Figure 7.12(b), where we see that the error bars get larger as we move away from the training points, representing increased uncertainty. This is important for applications such as active learning, where we want to model what we don't know as well as what we do.[3] By contrast, the plugin approximation has constant sized error bars (as illustrated in Figure 7.12(a)), since

$$p(y|\mathbf{x}, \mathcal{D}, \sigma^2) \approx \int \mathcal{N}(y|\mathbf{x}^T\mathbf{w}, \sigma^2)\delta_{\hat{\mathbf{w}}}(\mathbf{w})d\mathbf{w} = p(y|\mathbf{x}, \hat{\mathbf{w}}, \sigma^2). \tag{7.63}$$

7.6.3 Bayesian inference when σ^2 is unknown *

In this section, we apply the results in Section 4.6.3 to the problem of computing $p(\mathbf{w}, \sigma^2|\mathcal{D})$ for a linear regression model. This generalizes the results from Section 7.6.1 where we assumed σ^2 was known. In the case where we use an uninformative prior, we will see some interesting connections to frequentist statistics.

7.6.3.1 Conjugate prior

As usual, the likelihood has the form

$$p(\mathbf{y}|\mathbf{X}, \mathbf{w}, \sigma^2) = \mathcal{N}(\mathbf{y}|\mathbf{X}\mathbf{w}, \sigma^2\mathbf{I}_N) \tag{7.64}$$

By analogy to Section 4.6.3, one can show that the natural conjugate prior has the following form:

$$p(\mathbf{w}, \sigma^2) = \text{NIG}(\mathbf{w}, \sigma^2|\mathbf{w}_0, \mathbf{V}_0, a_0, b_0) \tag{7.65}$$

$$\triangleq \mathcal{N}(\mathbf{w}|\mathbf{w}_0, \sigma^2\mathbf{V}_0)\text{IG}(\sigma^2|a_0, b_0) \tag{7.66}$$

$$= \frac{b_0^{a_0}}{(2\pi)^{D/2}|\mathbf{V}_0|^{\frac{1}{2}}\Gamma(a_0)} (\sigma^2)^{-(a_0+(D/2)+1)} \tag{7.67}$$

$$\times \exp\left[-\frac{(\mathbf{w} - \mathbf{w}_0)^T\mathbf{V}_0^{-1}(\mathbf{w} - \mathbf{w}_0) + 2b_0}{2\sigma^2}\right] \tag{7.68}$$

---

3. Nate Silver (Silver 2012, Ch. 6) gives an interesting example of the importance of reporting confidence intervals together with point predictions. In 1997, the National Weather Service forecast that the Red River in North Dakota would reach flood levels of 49 feet, plus or minus 9 feet. However, to "avoid confusing the public", they only reported the point estimate of 49 feet, and omitted their error bars. When the officials of the town of Grand Forks, ND heard this forecast, they were relieved, since their flood walls were 51 feet tall. They thought they were safe from flooding, since the walls were slightly taller than the predicted river levels. In reality, the river crested to 54 feet, which was well within the margin of error forecast by the NWS. Unfortunately this meant that the town was flooded, and nearly all of the town's 50,000 residents had to be evacuated.

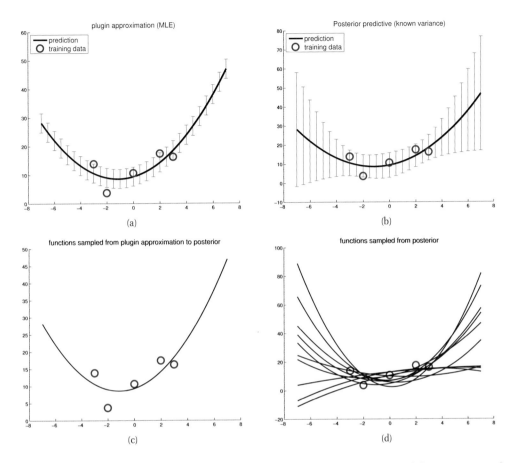

Figure 7.12 (a) Plug-in approximation to predictive density (we plug in the MLE of the parameters). (b) Posterior predictive density, obtained by integrating out the parameters. Black curve is posterior mean, error bars are 2 standard deviations of the posterior predictive density. (c) 10 samples from the plugin approximation to posterior predictive. (d) 10 samples from the posterior predictive. Figure generated by `linregPostPredDemo`.

With this prior and likelihood, one can show that the posterior has the following form:

$$p(\mathbf{w}, \sigma^2 | \mathcal{D}) = \text{NIG}(\mathbf{w}, \sigma^2 | \mathbf{w}_N, \mathbf{V}_N, a_N, b_N) \tag{7.69}$$

$$\mathbf{w}_N = \mathbf{V}_N(\mathbf{V}_0^{-1}\mathbf{w}_0 + \mathbf{X}^T\mathbf{y}) \tag{7.70}$$

$$\mathbf{V}_N = (\mathbf{V}_0^{-1} + \mathbf{X}^T\mathbf{X})^{-1} \tag{7.71}$$

$$a_N = a_0 + n/2 \tag{7.72}$$

$$b_N = b_0 + \frac{1}{2}\left(\mathbf{w}_0^T\mathbf{V}_0^{-1}\mathbf{w}_0 + \mathbf{y}^T\mathbf{y} - \mathbf{w}_N^T\mathbf{V}_N^{-1}\mathbf{w}_N\right) \tag{7.73}$$

The expressions for $\mathbf{w}_N$ and $\mathbf{V}_N$ are similar to the case where σ^2 is known. The expression for a_N is also intuitive, since it just updates the counts. The expression for b_N can be interpreted as follows: it is the prior sum of squares, b_0, plus the empirical sum of squares, $\mathbf{y}^T\mathbf{y}$, plus a term due to the error in the prior on $\mathbf{w}$.

The posterior marginals are as follows:

$$p(\sigma^2|\mathcal{D}) \;=\; \mathrm{IG}(a_N, b_N) \tag{7.74}$$

$$p(\mathbf{w}|\mathcal{D}) \;=\; \mathcal{T}(\mathbf{w}_N, \frac{b_N}{a_N}\mathbf{V}_N, 2a_N) \tag{7.75}$$

We give a worked example of using these equations in Section 7.6.3.3.

By analogy to Section 4.6.3.6, the posterior predictive distribution is a Student t distribution. In particular, given m new test inputs $\tilde{\mathbf{X}}$, we have

$$p(\tilde{\mathbf{y}}|\tilde{\mathbf{X}}, \mathcal{D}) \;=\; \mathcal{T}(\tilde{\mathbf{y}}|\tilde{\mathbf{X}}\mathbf{w}_N, \frac{b_N}{a_N}(\mathbf{I}_m + \tilde{\mathbf{X}}\mathbf{V}_N\tilde{\mathbf{X}}^T), 2a_N) \tag{7.76}$$

The predictive variance has two components: $(b_N/a_N)\mathbf{I}_m$ due to the measurement noise, and $(b_N/a_N)\tilde{\mathbf{X}}\mathbf{V}_N\tilde{\mathbf{X}}^T$ due to the uncertainty in $\mathbf{w}$. This latter term varies depending on how close the test inputs are to the training data.

It is common to set $a_0 = b_0 = 0$, corresponding to an uninformative prior for σ^2, and to set $\mathbf{w}_0 = \mathbf{0}$ and $\mathbf{V}_0 = g(\mathbf{X}^T\mathbf{X})^{-1}$ for any positive value g. This is called Zellner's **g-prior** (Zellner 1986). Here g plays a role analogous to $1/\lambda$ in ridge regression. However, the prior covariance is proportional to $(\mathbf{X}^T\mathbf{X})^{-1}$ rather than $\mathbf{I}$. This ensures that the posterior is invariant to scaling of the inputs (Minka 2000b). See also Exercise 7.10.

We will see below that if we use an uninformative prior, the posterior precision given N measurements is $\mathbf{V}_N^{-1} = \mathbf{X}^T\mathbf{X}$. The **unit information prior** is defined to contain as much information as one sample (Kass and Wasserman 1995). To create a unit information prior for linear regression, we need to use $\mathbf{V}_0^{-1} = \frac{1}{N}\mathbf{X}^T\mathbf{X}$, which is equivalent to the g-prior with $g = N$.

7.6.3.2 Uninformative prior

An uninformative prior can be obtained by considering the uninformative limit of the conjugate g-prior, which corresponds to setting $g = \infty$. This is equivalent to an improper NIG prior with $\mathbf{w}_0 = 0$, $\mathbf{V}_0 = \infty\mathbf{I}$, $a_0 = 0$ and $b_0 = 0$, which gives $p(\mathbf{w}, \sigma^2) \propto \sigma^{-(D+2)}$.

Alternatively, we can start with the semi-conjugate prior $p(\mathbf{w}, \sigma^2) = p(\mathbf{w})p(\sigma^2)$, and take each term to its uninformative limit individually, which gives $p(\mathbf{w}, \sigma^2) \propto \sigma^{-2}$. This is equivalent to an improper NIG prior with $\mathbf{w}_0 = \mathbf{0}, \mathbf{V} = \infty\mathbf{I}$, $a_0 = -D/2$ and $b_0 = 0$. The corresponding posterior is given by

$$p(\mathbf{w}, \sigma^2|\mathcal{D}) \;=\; \mathrm{NIG}(\mathbf{w}, \sigma^2|\mathbf{w}_N, \mathbf{V}_N, a_N, b_N) \tag{7.77}$$

$$\mathbf{w}_N \;=\; \hat{\mathbf{w}}_{mle} = (\mathbf{X}^T\mathbf{X})^{-1}\mathbf{X}^T\mathbf{y} \tag{7.78}$$

$$\mathbf{V}_N \;=\; (\mathbf{X}^T\mathbf{X})^{-1} \tag{7.79}$$

$$a_N \;=\; \frac{N-D}{2} \tag{7.80}$$

$$b_N \;=\; \frac{s^2}{2} \tag{7.81}$$

$$s^2 \;\triangleq\; (\mathbf{y} - \mathbf{X}\hat{\mathbf{w}}_{mle})^T(\mathbf{y} - \mathbf{X}\hat{\mathbf{w}}_{mle}) \tag{7.82}$$

w_j	$\mathbb{E}\left[w_j \mid \mathcal{D}\right]$	$\sqrt{\mathrm{var}\left[w_j \mid \mathcal{D}\right]}$	95% CI	sig
w0	10.998	3.06027	[4.652, 17.345]	*
w1	-0.004	0.00156	[-0.008, -0.001]	*
w2	-0.054	0.02190	[-0.099, -0.008]	*
w3	0.068	0.09947	[-0.138, 0.274]	
w4	-1.294	0.56381	[-2.463, -0.124]	*
w5	0.232	0.10438	[0.015, 0.448]	*
w6	-0.357	1.56646	[-3.605, 2.892]	
w7	-0.237	1.00601	[-2.324, 1.849]	
w8	0.181	0.23672	[-0.310, 0.672]	
w9	-1.285	0.86485	[-3.079, 0.508]	
w10	-0.433	0.73487	[-1.957, 1.091]	

Table 7.2 Posterior mean, standard deviation and credible intervals for a linear regression model with an uninformative prior fit to the caterpillar data. Produced by `linregBayesCaterpillar`.

The marginal distribution of the weights is given by

$$p(\mathbf{w}|\mathcal{D}) = \mathcal{T}(\mathbf{w}|\hat{\mathbf{w}}, \frac{s^2}{N-D}\mathbf{C}, N-D) \tag{7.83}$$

where $\mathbf{C} = (\mathbf{X}^T\mathbf{X})^{-1}$ and $\hat{\mathbf{w}}$ is the MLE. We discuss the implications of these equations below.

7.6.3.3 An example where Bayesian and frequentist inference coincide *

The use of a (semi-conjugate) uninformative prior is interesting because the resulting posterior turns out to be equivalent to the results from frequentist statistics (see also Section 4.6.3.9). In particular, from Equation 7.83 we have

$$p(w_j|\mathcal{D}) = T(w_j|\hat{w}_j, \frac{C_{jj}s^2}{N-D}, N-D) \tag{7.84}$$

This is equivalent to the sampling distribution of the MLE which is given by the following (see e.g., (Rice 1995, p542), (Casella and Berger 2002, p554)):

$$\frac{w_j - \hat{w}_j}{s_j} \sim t_{N-D} \tag{7.85}$$

where

$$s_j = \sqrt{\frac{s^2 C_{jj}}{N-D}} \tag{7.86}$$

is the standard error of the estimated parameter. (See Section 6.2 for a discussion of sampling distributions.) Consequently, the frequentist confidence interval and the Bayesian marginal credible interval for the parameters are the same in this case.

As a worked example of this, consider the caterpillar dataset from (Marin and Robert 2007). (The details of what the data mean don't matter for our present purposes.) We can compute the posterior mean and standard deviation, and the 95% credible intervals (CI) for the regression coefficients using Equation 7.84. The results are shown in Table 7.2. It is easy to check that these 95% credible intervals are identical to the 95% confidence intervals computed using standard frequentist methods (see `linregBayesCaterpillar` for the code).

We can also use these marginal posteriors to compute if the coefficients are "significantly" different from 0. An informal way to do this (without using decision theory) is to check if its 95% CI excludes 0. From Table 7.2, we see that the CIs for coefficients 0, 1, 2, 4, 5 are all significant by this measure, so we put a little star by them. It is easy to check that these results are the same as those produced by standard frequentist software packages which compute p-values at the 5% level.

Although the correspondence between the Bayesian and frequentist results might seem appealing to some readers, recall from Section 6.6 that frequentist inference is riddled with pathologies. Also, note that the MLE does not even exist when $N < D$, so standard frequentist inference theory breaks down in this setting. Bayesian inference theory still works, although it requires the use of proper priors. (See (Maruyama and George 2008) for one extension of the g-prior to the case where $D > N$.)

7.6.4 EB for linear regression (evidence procedure)

So far, we have assumed the prior is known. In this section, we describe an empirical Bayes procedure for picking the hyper-parameters. More precisely, we choose $\boldsymbol{\eta} = (\alpha, \lambda)$ to maximize the marginal likelihood, where $\lambda = 1/\sigma^2$ is the precision of the observation noise and $\boldsymbol{\alpha}$ is the precision of the prior, $p(\mathbf{w}) = \mathcal{N}(\mathbf{w}|\mathbf{0}, \alpha^{-1}\mathbf{I})$. This is known as the **evidence procedure** (MacKay 1995b).[4] See Section 13.7.4 for the algorithmic details.

The evidence procedure provides an alternative to using cross validation. For example, in Figure 7.13(b), we plot the log marginal likelihood for different values of α, as well as the maximum value found by the optimizer. We see that, in this example, we get the same result as 5-CV, shown in Figure 7.13(a). (We kept $\lambda = 1/\sigma^2$ fixed in both methods, to make them comparable.)

The principal practical advantage of the evidence procedure over CV will become apparent in Section 13.7, where we generalize the prior by allowing a different α_j for every feature. This can be used to perform feature selection, using a technique known as automatic relevancy determination or ARD. By contrast, it would not be possible to use CV to tune D different hyper-parameters.

The evidence procedure is also useful when comparing different kinds of models, since it provides a good approximation to the evidence:

$$p(\mathcal{D}|m) = \int\int p(\mathcal{D}|\mathbf{w}, m)p(\mathbf{w}|m, \boldsymbol{\eta})p(\boldsymbol{\eta}|m)d\mathbf{w}d\boldsymbol{\eta} \tag{7.87}$$

$$\approx \max_{\boldsymbol{\eta}} \int p(\mathcal{D}|\mathbf{w}, m)p(\mathbf{w}|m, \boldsymbol{\eta})p(\boldsymbol{\eta}|m)d\mathbf{w} \tag{7.88}$$

It is important to (at least approximately) integrate over $\boldsymbol{\eta}$ rather than setting it arbitrarily, for reasons discussed in Section 5.3.2.5. Indeed, this is the method we used to evaluate the marginal

---

4. Alternatively, we could integrate out λ analytically, as shown in Section 7.6.3, and just optimize α (Buntine and Weigend 1991). However, it turns out that this is less accurate than optimizing both α and λ (MacKay 1999).

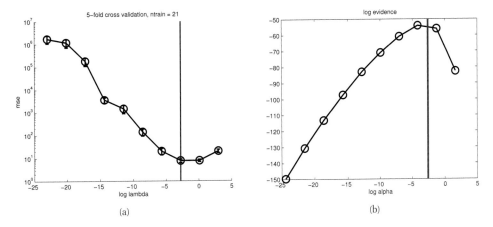

Figure 7.13   (a) Estimate of test MSE produced by 5-fold cross-validation vs $\log(\lambda)$. The smallest value is indicated by the vertical line. Note the vertical scale is in log units. (c) Log marginal likelihood vs $\log(\alpha)$. The largest value is indicated by the vertical line. Figure generated by linregPolyVsRegDemo.

likelihood for the polynomial regression models in Figures 5.7 and 5.8. For a "more Bayesian" approach, in which we model our uncertainty about $\boldsymbol{\eta}$ rather than computing point estimates, see Section 21.5.2.

Exercises

Exercise 7.1 Behavior of training set error with increasing sample size

The error on the test will always decrease as we get more training data, since the model will be better estimated. However, as shown in Figure 7.10, for sufficiently complex models, the error on the training set can increase we we get more training data, until we reach some plateau. Explain why.

Exercise 7.2 Multi-output linear regression

(Source: Jaakkola.)

When we have multiple independent outputs in linear regression, the model becomes

$$p(\mathbf{y}|\mathbf{x}, \mathbf{W}) = \prod_{j=1}^{M} \mathcal{N}(y_j | \mathbf{w}_j^T \mathbf{x}_i, \sigma_j^2) \tag{7.89}$$

Since the likelihood factorizes across dimensions, so does the MLE. Thus

$$\hat{\mathbf{W}} = [\hat{\mathbf{w}}_1, \dots, \hat{\mathbf{w}}_M] \tag{7.90}$$

where $\hat{\mathbf{w}}_j = (\mathbf{X}^T \mathbf{X})^{-1} \mathbf{Y}_{:,j}$.

In this exercise we apply this result to a model with 2 dimensional response vector $\mathbf{y}_i \in \mathbb{R}^2$. Suppose we have some binary input data, $x_i \in \{0, 1\}$. The training data is as follows:

x	y
0	$(-1, -1)^T$
0	$(-1, -2)^T$
0	$(-2, -1)^T$
1	$(1, 1)^T$
1	$(1, 2)^T$
1	$(2, 1)^T$

Let us embed each x_i into 2d using the following basis function:

$$\phi(0) = (1, 0)^T, \quad \phi(1) = (0, 1)^T \tag{7.91}$$

The model becomes

$$\hat{\mathbf{y}} = \mathbf{W}^T \phi(x) \tag{7.92}$$

where $\mathbf{W}$ is a 2×2 matrix. Compute the MLE for $\mathbf{W}$ from the above data.

Exercise 7.3 Centering and ridge regression

Assume that $\bar{\mathbf{x}} = 0$, so the input data has been centered. Show that the optimizer of

$$J(\mathbf{w}, w_0) \quad = \quad (\mathbf{y} - \mathbf{X}\mathbf{w} - w_0 \mathbf{1})^T (\mathbf{y} - \mathbf{X}\mathbf{w} - w_0 \mathbf{1}) + \lambda \mathbf{w}^T \mathbf{w} \tag{7.93}$$

is

$$\hat{w}_0 \quad = \quad \bar{y} \tag{7.94}$$

$$\mathbf{w} \quad = \quad (\mathbf{X}^T \mathbf{X} + \lambda \mathbf{I})^{-1} \mathbf{X}^T \mathbf{y} \tag{7.95}$$

Exercise 7.4 MLE for σ^2 for linear regression

Show that the MLE for the error variance in linear regression is given by

$$\hat{\sigma}^2 = \frac{1}{N} \sum_{i=1}^{N} (y_i - \mathbf{x}_i^T \hat{\mathbf{w}})^2 \tag{7.96}$$

This is just the empirical variance of the residual errors when we plug in our estimate of $\hat{\mathbf{w}}$.

Exercise 7.5 MLE for the offset term in linear regression

Linear regression has the form $\mathbb{E}[y|\mathbf{x}] = w_0 + \mathbf{w}^T \mathbf{x}$. It is common to include a column of 1's in the design matrix, so we can solve for the offset term w_0 term and the other parameters $\mathbf{w}$ at the same time using the normal equations. However, it is also possible to solve for $\mathbf{w}$ and w_0 separately. Show that

$$\hat{w}_0 \quad = \quad \frac{1}{N} \sum_i y_i - \frac{1}{N} \sum_i \mathbf{x}_i^T \mathbf{w} = \bar{y} - \bar{\mathbf{x}}^T \mathbf{w} \tag{7.97}$$

So $\hat{w}_0$ models the difference in the average output from the average predicted output. Also, show that

$$\hat{\mathbf{w}} = (\mathbf{X}_c^T \mathbf{X}_c)^{-1} \mathbf{X}_c^T \mathbf{y}_c = \left[\sum_{i=1}^{N} (\mathbf{x}_i - \bar{\mathbf{x}})(\mathbf{x}_i - \bar{\mathbf{x}})^T \right]^{-1} \left[\sum_{i=1}^{N} (y_i - \bar{y})(\mathbf{x}_i - \bar{\mathbf{x}}) \right] \tag{7.98}$$

where $\mathbf{X}_c$ is the centered input matrix containing $\mathbf{x}_i^c = \mathbf{x}_i - \bar{\mathbf{x}}$ along its rows, and $\mathbf{y}_c = \mathbf{y} - \bar{\mathbf{y}}$ is the centered output vector. Thus we can first compute $\hat{\mathbf{w}}$ on centered data, and then estimate w_0 using $\bar{y} - \bar{\mathbf{x}}^T \hat{\mathbf{w}}$.

Exercise 7.6 MLE for simple linear regression

Simple linear regression refers to the case where the input is scalar, so $D = 1$. Show that the MLE in this case is given by the following equations, which may be familiar from basic statistics classes:

$$w_1 = \frac{\sum_i (x_i - \bar{x})(y_i - \bar{y})}{\sum_i (x_i - \bar{x})^2} = \frac{\sum_i x_i y_i - N \bar{x} \, \bar{y}}{\sum_i x_i^2 - N \bar{x}^2} \approx \frac{\text{cov}\,[X, Y]}{\text{var}\,[X]} \tag{7.99}$$

$$w_0 = \bar{y} - w_1 \bar{x} \approx \mathbb{E}\,[Y] - w_1 \mathbb{E}\,[X] \tag{7.100}$$

See `linregDemo1` for a demo.

Exercise 7.7 Sufficient statistics for online linear regression

(Source: Jaakkola.) Consider fitting the model $\hat{y} = w_0 + w_1 x$ using least squares. Unfortunately we did not keep the original data, x_i, y_i, but we do have the following functions (statistics) of the data:

$$\bar{x}^{(n)} = \frac{1}{n} \sum_{i=1}^{n} x_i, \quad \bar{y}^{(n)} = \frac{1}{n} \sum_{i=1}^{n} y_i \tag{7.101}$$

$$C_{xx}^{(n)} = \frac{1}{n} \sum_{i=1}^{n} (x_i - \bar{x})^2, \quad C_{xy}^{(n)} = \frac{1}{n} \sum_{i=1}^{n} (x_i - \bar{x})(y_i - \bar{y}), \quad C_{yy}^{(n)} = \frac{1}{n} \sum_{i=1}^{n} (y_i - \bar{y})^2 \tag{7.102}$$

a. What are the minimal set of statistics that we need to estimate w_1? (Hint: see Equation 7.99.)

b. What are the minimal set of statistics that we need to estimate w_0? (Hint: see Equation 7.97.)

c. Suppose a new data point, x_{n+1}, y_{n+1} arrives, and we want to update our sufficient statistics without looking at the old data, which we have not stored. (This is useful for online learning.) Show that we can this for $\bar{x}$ as follows.

$$\bar{x}^{(n+1)} \triangleq \frac{1}{n+1} \sum_{i=1}^{n+1} x_i = \frac{1}{n+1} \left(n \bar{x}^{(n)} + x_{n+1} \right) \tag{7.103}$$

$$= \bar{x}^{(n)} + \frac{1}{n+1} (x_{n+1} - \bar{x}^{(n)}) \tag{7.104}$$

This has the form: new estimate is old estimate plus correction. We see that the size of the correction diminishes over time (i.e., as we get more samples). Derive a similar expression to update $\bar{y}$

d. Show that one can update $C_{xy}^{(n+1)}$ recursively using

$$C_{xy}^{(n+1)} = \frac{1}{n+1} \left[x_{n+1} y_{n+1} + n C_{xy}^{(n)} + n \bar{x}^{(n)} \bar{y}^{(n)} - (n+1) \bar{x}^{(n+1)} \bar{y}^{(n+1)} \right] \tag{7.105}$$

Derive a similar expression to update C_{xx}.

e. Implement the online learning algorithm, i.e., write a function of the form `[w,ss] = linregUpdateSS(ss, x, y)`, where x and y are scalars and ss is a structure containing the sufficient statistics.

f. Plot the coefficients over "time", using the dataset in `linregDemo1`. (Specifically, use `[x,y] = polyDataMake('sampling','thibaux')`.) Check that they converge to the solution given by the batch (offline) learner (i.e, ordinary least squares). Your result should look like Figure 7.14.

Turn in your derivation, code and plot.

Exercise 7.8 Bayesian linear regression in 1d with known σ^2

(Source: Bolstad.) Consider fitting a model of the form

$$p(y|x, \boldsymbol{\theta}) = \mathcal{N}(y|w_0 + w_1 x, \sigma^2) \tag{7.106}$$

to the data shown below:

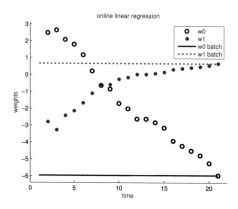

Figure 7.14 Regression coefficients over time. Produced by Exercise 7.7.

```
x = [94,96,94,95,104,106,108,113,115,121,131];
y = [0.47, 0.75, 0.83, 0.98, 1.18, 1.29, 1.40, 1.60, 1.75, 1.90, 2.23];
```

a. Compute an unbiased estimate of σ^2 using

$$\hat{\sigma}^2 = \frac{1}{N-2} \sum_{i=1}^{N} (y_i - \hat{y}_i)^2 \tag{7.107}$$

(The denominator is $N-2$ since we have 2 inputs, namely the offset term and x.) Here $\hat{y}_i = \hat{w}_0 + \hat{w}_1 x_i$, and $\hat{\mathbf{w}} = (\hat{w}_0, \hat{w}_1)$ is the MLE.

b. Now assume the following prior on $\mathbf{w}$:

$$p(\mathbf{w}) = p(w_0)p(w_1) \tag{7.108}$$

Use an (improper) uniform prior on w_0 and a $\mathcal{N}(0,1)$ prior on w_1. Show that this can be written as a Gaussian prior of the form $p(\mathbf{w}) = \mathcal{N}(\mathbf{w}|\mathbf{w}_0, \mathbf{V}_0)$. What are $\mathbf{w}_0$ and $\mathbf{V}_0$?

c. Compute the marginal posterior of the slope, $p(w_1|\mathcal{D}, \sigma^2)$, where $\mathcal{D}$ is the data above, and σ^2 is the unbiased estimate computed above. What is $\mathbb{E}\left[w_1|\mathcal{D}, \sigma^2\right]$ and $\text{var}\left[w_1|\mathcal{D}, \sigma^2\right]$ Show your work. (You can use Matlab if you like.) Hint: the posterior variance is a very small number!

d. What is a 95% credible interval for w_1?

Exercise 7.9 Generative model for linear regression

Linear regression is the problem of estimating $E[Y|\mathbf{x}]$ using a linear function of the form $w_0 + \mathbf{w}^T\mathbf{x}$. Typically we assume that the conditional distribution of Y given $\mathbf{X}$ is Gaussian. We can either estimate this conditional Gaussian directly (a discriminative approach), or we can fit a Gaussian to the joint distribution of $\mathbf{X}, Y$ and then derive $E[Y|\mathbf{X} = \mathbf{x}]$.

In Exercise 7.5 we showed that the discriminative approach leads to these equations

$$E[Y|\mathbf{x}] = w_0 + \mathbf{w}^T\mathbf{x} \tag{7.109}$$

$$w_0 = \bar{y} - \bar{\mathbf{x}}^T\mathbf{w} \tag{7.110}$$

$$\mathbf{w} = (\mathbf{X}_c^T\mathbf{X}_c)^{-1}\mathbf{X}_c^T\mathbf{y}_c \tag{7.111}$$

where $\mathbf{X}_c = \mathbf{X} - \bar{\mathbf{X}}$ is the centered input matrix, and $\bar{\mathbf{X}} = \mathbf{1}_n \bar{\mathbf{x}}^T$ replicates $\bar{\mathbf{x}}$ across the rows. Similarly, $\mathbf{y}_c = \mathbf{y} - \bar{\mathbf{y}}$ is the centered output vector, and $\bar{\mathbf{y}} = \mathbf{1}_n \bar{y}$ replicates $\bar{y}$ across the rows.

a. By finding the maximum likelihood estimates of $\boldsymbol{\Sigma}_{XX}$, $\boldsymbol{\Sigma}_{XY}$, $\boldsymbol{\mu}_X$ and $\boldsymbol{\mu}_Y$, derive the above equations by fitting a joint Gaussian to $\mathbf{X}, Y$ and using the formula for conditioning a Gaussian (see Section 4.3.1). Show your work.

b. What are the advantages and disadvantages of this approach compared to the standard discriminative approach?

Exercise 7.10 Bayesian linear regression using the g-prior

Show that when we use the g-prior, $p(\mathbf{w}, \sigma^2) = \text{NIG}(\mathbf{w}, \sigma^2 | \mathbf{0}, g(\mathbf{X}^T \mathbf{X})^{-1}, 0, 0)$, the posterior has the following form:

$$
\begin{align}
p(\mathbf{w}, \sigma^2 | \mathcal{D}) &= \text{NIG}(\mathbf{w}, \sigma^2 | \mathbf{w}_N, \mathbf{V}_N, a_N, b_N) && (7.112) \\
\mathbf{V}_N &= \frac{g}{g+1} (\mathbf{X}^T \mathbf{X})^{-1} && (7.113) \\
\mathbf{w}_N &= \frac{g}{g+1} \hat{\mathbf{w}}_{mle} && (7.114) \\
a_N &= N/2 && (7.115) \\
b_N &= \frac{s^2}{2} + \frac{1}{2(g+1)} \hat{\mathbf{w}}_{mle}^T \mathbf{X}^T \mathbf{X} \hat{\mathbf{w}}_{mle} && (7.116) \\
&&& (7.117)
\end{align}
$$

8 *Logistic regression*

8.1 Introduction

One way to build a probabilistic classifier is to create a joint model of the form $p(y, \mathbf{x}) = p(y)p(\mathbf{x}|y)$, and then to condition on the observed features $\mathbf{x}$, thereby deriving the class posterior, $p(y|\mathbf{x})$. This is called a **generative classifier**, since it specifies how to generate the observed features $\mathbf{x}$ for each class y. See Sections 3.5 and 4.2 for details. A more popular alternative, which we discuss in this chapter, is to fit a model of the form $p(y|\mathbf{x})$, which directly models the mapping from inputs $\mathbf{x}$ to output y. This is called a **discriminative classifier**, since it discriminates between the class labels, but cannot generate examples of each class. We compare the generative and discriminative approaches, in Section 8.6.

In this chapter, we focus on discriminative classifiers which are linear in the parameters, extending the machinery of Chapter 7 to the classification setting. This will turn out to significantly simplify model fitting, as we will see. In later chapters, we will consider non-linear and non-parametric discriminative models.

8.2 Model specification

As we discussed in Section 1.4.6, logistic regression corresponds to the following binary classification model:

$$p(y|\mathbf{x}, \mathbf{w}) = \text{Ber}(y|\text{sigm}(\mathbf{w}^T \mathbf{x})) \tag{8.1}$$

A 1d example is shown in Figure 1.19(b). Logistic regression can easily be extended to higher-dimensional inputs. For example, Figure 8.1 shows plots of $p(y = 1|\mathbf{x}, \mathbf{w}) = \text{sigm}(\mathbf{w}^T \mathbf{x})$ for 2d input and different weight vectors $\mathbf{w}$. If we threshold these probabilities at 0.5, we induce a linear decision boundary, whose normal (perpendicular) is given by $\mathbf{w}$.

Logistic regression is one of the most popular approaches to classification, for several reasons, including the following:

- LR models are easy to fit to data, as we discuss in Section 8.3. By "easy" we mean that the algorithms are simple to implement, and are very fast. In particular, there are methods that take time linear in the number of non-zeros in the dataset, which is the minimal possible time, since we need to "touch" each data point at least once. By contrast, more complex classifiers (such as SVMs, neural nets, decision trees, etc.) are often much slower to train.

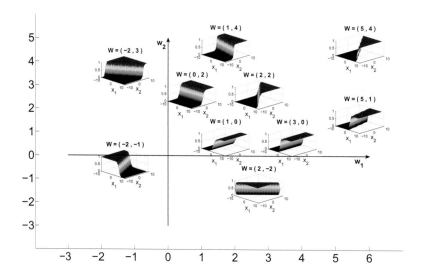

Figure 8.1 Plots of $\text{sigm}(w_1 x_1 + w_2 x_2)$. Here $\mathbf{w} = (w_1, w_2)$ defines the normal to the decision boundary. Points to the right of this have $\text{sigm}(\mathbf{w}^T \mathbf{x}) > 0.5$, and points to the left have $\text{sigm}(\mathbf{w}^T \mathbf{x}) < 0.5$. Based on Figure 39.3 of (MacKay 2003). Figure generated by `sigmoidplot2D`.

- LR models are easy to interpret. In particular, let us define the **log odds** as follows: $LO \triangleq \log \frac{p(y=1|\mathbf{x})}{p(y=0|\mathbf{x})}$. It can be shown that this is a linear function of $\mathbf{w}$, i.e., $LO = \mathbf{w}^T \mathbf{x}$. To see this significance of this, let us consider a simple example. Suppose we have a feature vector where the first component is the number of cigarettes you smoke per day, and the second is the number of minutes you exercise. The goal is to predict the probability you will get lung cancer. If we estimate the parameters to be $\hat{\mathbf{w}} = (1.3, -1.1)$, this means that for every extra cigarette you smoke, you increase your risk of getting cancer by a factor of $e^{1.3}$.

- LR models are easy to extend to multi-class classification, as we discuss in Section 8.3.7, as well as to other kinds of output data, such as counts (see Section 9.3).

- LR models can easily be extended to handle non-linear decision boundaries by using kernels (see Section 14.3), or by learning the features from data (see Chapter 16).

8.3   Model fitting

In this section, we discuss algorithms for estimating the parameters of a logistic regression model.

8.3.1 MLE

The negative log-likelihood for logistic regression is given by

$$\text{NLL}(\mathbf{w}) \quad = \quad -\sum_{i=1}^{N} \log[\mu_i^{\mathbb{I}(y_i=1)} \times (1-\mu_i)^{\mathbb{I}(y_i=0)}] \tag{8.2}$$

$$= \quad -\sum_{i=1}^{N} [y_i \log \mu_i + (1-y_i) \log(1-\mu_i)] \tag{8.3}$$

where $\mu_i = \text{sigm}(\mathbf{w}^T \mathbf{x}_i)$. This is also called the **cross-entropy** error function (see Section 2.8.2).

Another way of writing this is as follows. Suppose $\tilde{y}_i \in \{-1, +1\}$ instead of $y_i \in \{0, 1\}$. We have $p(\tilde{y} = 1) = \frac{1}{1 + \exp(-\mathbf{w}^T \mathbf{x})}$ and $p(\tilde{y} = -1) = \frac{1}{1 + \exp(+\mathbf{w}^T \mathbf{x})}$. Hence

$$NLL(\mathbf{w}) = \sum_{i=1}^{N} \log(1 + \exp(-\tilde{y}_i \mathbf{w}^T \mathbf{x}_i)) \tag{8.4}$$

Unlike linear regression, we can no longer write down the MLE in closed form. Instead, we need to use an optimization algorithm to compute it. For this, we need to derive the gradient and Hessian.

In the case of logistic regression, one can show (Exercise 8.3) that the gradient and Hessian of this are given by the following

$$\mathbf{g} \quad = \quad \frac{d}{d\mathbf{w}} NLL(\mathbf{w}) = \sum_{i} (\mu_i - y_i)\mathbf{x}_i = \mathbf{X}^T(\boldsymbol{\mu} - \mathbf{y}) \tag{8.5}$$

$$\mathbf{H} \quad = \quad \frac{d}{d\mathbf{w}} \mathbf{g}(\mathbf{w})^T = \sum_{i} (\nabla_{\mathbf{w}} \mu_i)\mathbf{x}_i^T = \sum_{i} \mu_i(1-\mu_i)\mathbf{x}_i \mathbf{x}_i^T \tag{8.6}$$

$$= \quad \mathbf{X}^T \mathbf{S} \mathbf{X} \tag{8.7}$$

where $\mathbf{S} \triangleq \text{diag}(\mu_i(1-\mu_i))$. One can also show (Exercise 8.3) that $\mathbf{H}$ is positive definite. Hence the NLL is convex and has a unique global minimum. Below we discuss some methods for finding this minimum.

8.3.2 Steepest descent

Perhaps the simplest algorithm for unconstrained optimization is **gradient descent**, also known as **steepest descent**. This can be written as follows:

$$\boldsymbol{\theta}_{k+1} = \boldsymbol{\theta}_k - \eta_k \mathbf{g}_k \tag{8.8}$$

where η_k is the **step size** or **learning rate**. The main issue in gradient descent is: how should we set the step size? This turns out to be quite tricky. If we use a constant learning rate, but make it too small, convergence will be very slow, but if we make it too large, the method can fail to converge at all. This is illustrated in Figure 8.2, where we plot the following function[1]

$$f(\boldsymbol{\theta}) = 0.5(\theta_1^2 - \theta_2)^2 + 0.5(\theta_1 - 1)^2 \tag{8.9}$$

---

1. This function, from (Aoki 1971, 106), is not convex, as is easily verified by computing the Hessian and evaluating its determinant; if $2(\theta_1^2 - \theta_2) + 1 < 0$, the determinant is negative. Nevertheless, the function has a positive definite Hessian in the neighborhood of its global minimum at (1,1).

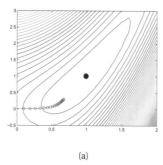
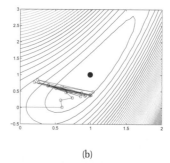

(a)   (b)

Figure 8.2 Gradient descent on a simple function, starting from $(0,0)$, for 20 steps, using a fixed learning rate (step size) η. The global minimum is at $(1,1)$. (a) $\eta = 0.1$. (b) $\eta = 0.6$. Figure generated by `steepestDescentDemo`.

By inspection, the global minimum is at $(1, 1)$. We arbitrarily decide to start from $(0, 0)$. In Figure 8.2(a), we use a fixed step size of $\eta = 0.1$; we see that it moves slowly along the valley. In Figure 8.2(b), we use a fixed step size of $\eta = 0.6$; we see that the algorithm starts oscillating up and down the sides of the valley and never converges to the optimum.

Let us develop a more stable method for picking the step size, so that the method is guaranteed to converge to a local optimum no matter where we start. (This property is called **global convergence**, which should not be confused with convergence to the global optimum!) By Taylor's theorem, we have

$$f(\boldsymbol{\theta} + \eta \mathbf{d}) \approx f(\boldsymbol{\theta}) + \eta \mathbf{g}^T \mathbf{d} \tag{8.10}$$

where $\mathbf{d}$ is our descent direction. So if η is chosen small enough, then $f(\boldsymbol{\theta} + \eta \mathbf{d}) < f(\boldsymbol{\theta})$, since the gradient will be negative. But we don't want to choose the step size η too small, or we will move very slowly and may not reach the minimum. So let us pick η to minimize

$$\phi(\eta) = f(\boldsymbol{\theta}_k + \eta \mathbf{d}_k) \tag{8.11}$$

This is called **line minimization** or **line search**. There are various methods for solving this 1d optimization problem; see (Nocedal and Wright 2006) for details.

Figure 8.3(a) demonstrates that line search does indeed work for our simple problem. However, we see that the steepest descent path with exact line searches exhibits a characteristic **zig-zag** behavior. To see why, note that an exact line search satisfies $\eta_k = \arg\min_{\eta>0} \phi(\eta)$. A necessary condition for the optimum is $\phi'(\eta) = 0$. By the chain rule, $\phi'(\eta) = \mathbf{d}^T \mathbf{g}$, where $\mathbf{g} = f'(\boldsymbol{\theta} + \eta \mathbf{d})$ is the gradient at the end of the step. So we either have $\mathbf{g} = \mathbf{0}$, which means we have found a stationary point, or $\mathbf{g} \perp \mathbf{d}$, which means that exact search stops at a point where the local gradient is perpendicular to the search direction. Hence consecutive directions will be orthogonal (see Figure 8.3(b)). This explains the zig-zag behavior.

One simple heuristic to reduce the effect of zig-zagging is to add a **momentum** term, $(\boldsymbol{\theta}_k - \boldsymbol{\theta}_{k-1})$, as follows:

$$\boldsymbol{\theta}_{k+1} = \boldsymbol{\theta}_k - \eta_k \mathbf{g}_k + \mu_k(\boldsymbol{\theta}_k - \boldsymbol{\theta}_{k-1}) \tag{8.12}$$

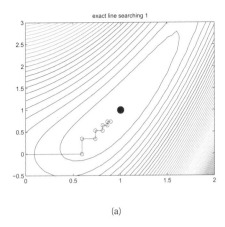

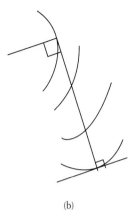

(a) (b)

Figure 8.3 (a) Steepest descent on the same function as Figure 8.2, starting from $(0, 0)$, using line search. Figure generated by `steepestDescentDemo`. (b) Illustration of the fact that at the end of a line search (top of picture), the local gradient of the function will be perpendicular to the search direction. Based on Figure 10.6.1 of (Press et al. 1988).

where $0 \leq \mu_k \leq 1$ controls the importance of the momentum term. In the optimization community, this is known as the **heavy ball method** (see e.g., (Bertsekas 1999)).

An alternative way to minimize "zig-zagging" is to use the method of **conjugate gradients** (see e.g., (Nocedal and Wright 2006, ch 5) or (Golub and van Loan 1996, Sec 10.2)). This is the method of choice for quadratic objectives of the form $f(\boldsymbol{\theta}) = \boldsymbol{\theta}^T \mathbf{A} \boldsymbol{\theta}$, which arise when solving linear systems. However, non-linear CG is less popular.

8.3.3 Newton's method

Algorithm 8.1: Newton's method for minimizing a strictly convex function

---
1 Initialize $\boldsymbol{\theta}_0$;
2 **for** $k = 1, 2, \ldots$ *until convergence* **do**
3   Evaluate $\mathbf{g}_k = \nabla f(\boldsymbol{\theta}_k)$;
4   Evaluate $\mathbf{H}_k = \nabla^2 f(\boldsymbol{\theta}_k)$;
5   Solve $\mathbf{H}_k \mathbf{d}_k = -\mathbf{g}_k$ for $\mathbf{d}_k$;
6   Use line search to find stepsize η_k along $\mathbf{d}_k$;
7     $\boldsymbol{\theta}_{k+1} = \boldsymbol{\theta}_k + \eta_k \mathbf{d}_k$;

One can derive faster optimization methods by taking the curvature of the space (i.e., the Hessian) into account. These are called **second order** optimization metods. The primary example is **Newton's algorithm**. This is an iterative algorithm which consists of updates of the

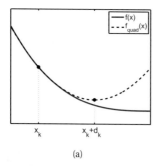
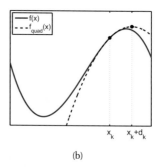

(a)   (b)

Figure 8.4 Illustration of Newton's method for minimizing a 1d function. (a) The solid curve is the function $f(x)$. The dotted line $f_{quad}(x)$ is its second order approximation at x_k. The Newton step d_k is what must be added to x_k to get to the minimum of $f_{quad}(x)$. Based on Figure 13.4 of (Vandenberghe 2006). Figure generated by `newtonsMethodMinQuad`. (b) Illustration of Newton's method applied to a nonconvex function. We fit a quadratic around the current point x_k and move to its stationary point, $x_{k+1} = x_k + d_k$. Unfortunately, this is a local maximum, not minimum. This means we need to be careful about the extent of our quadratic approximation. Based on Figure 13.11 of (Vandenberghe 2006). Figure generated by `newtonsMethodNonConvex`.

form

$$\boldsymbol{\theta}_{k+1} = \boldsymbol{\theta}_k - \eta_k \mathbf{H}_k^{-1} \mathbf{g}_k \tag{8.13}$$

The full pseudo-code is given in Algorithm 8.1.

This algorithm can be derived as follows. Consider making a second-order Taylor series approximation of $f(\boldsymbol{\theta})$ around $\boldsymbol{\theta}_k$:

$$f_{quad}(\boldsymbol{\theta}) = f_k + \mathbf{g}_k^T(\boldsymbol{\theta} - \boldsymbol{\theta}_k) + \frac{1}{2}(\boldsymbol{\theta} - \boldsymbol{\theta}_k)^T \mathbf{H}_k(\boldsymbol{\theta} - \boldsymbol{\theta}_k) \tag{8.14}$$

Let us rewrite this as

$$f_{quad}(\boldsymbol{\theta}) = \boldsymbol{\theta}^T \mathbf{A} \boldsymbol{\theta} + \mathbf{b}^T \boldsymbol{\theta} + c \tag{8.15}$$

where

$$\mathbf{A} = \frac{1}{2}\mathbf{H}_k, \quad \mathbf{b} = \mathbf{g}_k - \mathbf{H}_k \boldsymbol{\theta}_k, \quad c = f_k - \mathbf{g}_k^T \boldsymbol{\theta}_k + \frac{1}{2}\boldsymbol{\theta}_k^T \mathbf{H}_k \boldsymbol{\theta}_k \tag{8.16}$$

The minimum of f_{quad} is at

$$\boldsymbol{\theta} = -\frac{1}{2}\mathbf{A}^{-1}\mathbf{b} = \boldsymbol{\theta}_k - \mathbf{H}_k^{-1}\mathbf{g}_k \tag{8.17}$$

Thus the Newton step $\mathbf{d}_k = -\mathbf{H}_k^{-1}\mathbf{g}_k$ is what should be added to $\boldsymbol{\theta}_k$ to minimize the second order approximation of f around $\boldsymbol{\theta}_k$. See Figure 8.4(a) for an illustration.

In its simplest form (as listed), Newton's method requires that $\mathbf{H}_k$ be positive definite, which will hold if the function is strictly convex. If the objective function is not convex, then $\mathbf{H}_k$ may not be positive definite, so $\mathbf{d}_k = -\mathbf{H}_k^{-1}\mathbf{g}_k$ may not be a descent direction (see Figure 8.4(b) for an example). In this case, one simple strategy is to revert to steepest descent, $\mathbf{d}_k = -\mathbf{g}_k$. The **Levenberg Marquardt** algorithm is an adaptive way to blend between Newton steps and steepest descent steps. This method is widely used when solving nonlinear least squares problems. An alternative approach is this: Rather than computing $\mathbf{d}_k = -\mathbf{H}_k^{-1}\mathbf{g}_k$ directly, we can solve the linear system of equations $\mathbf{H}_k \mathbf{d}_k = -\mathbf{g}_k$ for $\mathbf{d}_k$ using conjugate gradient (CG). If $\mathbf{H}_k$ is not positive definite, we can simply truncate the CG iterations as soon as negative curvature is detected; this is called **truncated Newton**.

8.3.4 Iteratively reweighted least squares (IRLS)

Let us now apply Newton's algorithm to find the MLE for binary logistic regression. The Newton update at iteration $k + 1$ for this model is as follows (using $\eta_k = 1$, since the Hessian is exact):

$$
\begin{align}
\mathbf{w}_{k+1} &= \mathbf{w}_k - \mathbf{H}^{-1}\mathbf{g}_k \tag{8.18} \\
&= \mathbf{w}_k + (\mathbf{X}^T\mathbf{S}_k\mathbf{X})^{-1}\mathbf{X}^T(\mathbf{y} - \boldsymbol{\mu}_k) \tag{8.19} \\
&= (\mathbf{X}^T\mathbf{S}_k\mathbf{X})^{-1}\left[(\mathbf{X}^T\mathbf{S}_k\mathbf{X})\mathbf{w}_k + \mathbf{X}^T(\mathbf{y} - \boldsymbol{\mu}_k)\right] \tag{8.20} \\
&= (\mathbf{X}^T\mathbf{S}_k\mathbf{X})^{-1}\mathbf{X}^T\left[\mathbf{S}_k\mathbf{X}\mathbf{w}_k + \mathbf{y} - \boldsymbol{\mu}_k\right] \tag{8.21} \\
&= (\mathbf{X}^T\mathbf{S}_k\mathbf{X})^{-1}\mathbf{X}^T\mathbf{S}_k\mathbf{z}_k \tag{8.22}
\end{align}
$$

where we have defined the **working response** as

$$
\mathbf{z}_k \triangleq \mathbf{X}\mathbf{w}_k + \mathbf{S}_k^{-1}(\mathbf{y} - \boldsymbol{\mu}_k) \tag{8.23}
$$

Equation 8.22 is an example of a **weighted least squares problem**, which is a minimizer of

$$
\sum_{i=1}^{N} S_{ki}(z_{ki} - \mathbf{w}^T\mathbf{x}_i)^2 \tag{8.24}
$$

Since $\mathbf{S}_k$ is a diagonal matrix, we can rewrite the targets in component form (for each case $i = 1 : N$) as

$$
z_{ki} = \mathbf{w}_k^T\mathbf{x}_i + \frac{y_i - \mu_{ki}}{\mu_{ki}(1 - \mu_{ki})} \tag{8.25}
$$

This algorithm is known as **iteratively reweighted least squares** or **IRLS** for short, since at each iteration, we solve a weighted least squares problem, where the weight matrix $\mathbf{S}_k$ changes at each iteration. See Algorithm 8.2 for some pseudocode.

8.3.5 Quasi-Newton (variable metric) methods

The mother of all second-order optimization algorithm is Newton's algorithm, which we discussed in Section 8.3.3. Unfortunately, it may be too expensive to compute $\mathbf{H}$ explicitly. **Quasi-Newton** methods iteratively build up an approximation to the Hessian using information gleaned from the gradient vector at each step. The most common method is called **BFGS** (named after its inventors, Broyden, Fletcher, Goldfarb and Shanno), which updates the approximation to the

---

Algorithm 8.2: Iteratively reweighted least squares (IRLS)

1 $\mathbf{w} = \mathbf{0}_D$;
2 $w_0 = \log(\overline{y}/(1 - \overline{y}))$;
3 **repeat**
4 $\quad \eta_i = w_0 + \mathbf{w}^T \mathbf{x}_i$;
5 $\quad \mu_i = \text{sigm}(\eta_i)$;
6 $\quad s_i = \mu_i(1 - \mu_i)$;
7 $\quad z_i = \eta_i + \frac{y_i - \mu_i}{s_i}$;
8 $\quad \mathbf{S} = \text{diag}(s_{1:N})$;
9 $\quad \mathbf{w} = (\mathbf{X}^T \mathbf{S} \mathbf{X})^{-1} \mathbf{X}^T \mathbf{S} \mathbf{z}$;
10 **until** *converged*;

Hessian $\mathbf{B}_k \approx \mathbf{H}_k$ as follows:

$$\mathbf{B}_{k+1} = \mathbf{B}_k + \frac{\mathbf{y}_k \mathbf{y}_k^T}{\mathbf{y}_k^T \mathbf{s}_k} - \frac{(\mathbf{B}_k \mathbf{s}_k)(\mathbf{B}_k \mathbf{s}_k)^T}{\mathbf{s}_k^T \mathbf{B}_k \mathbf{s}_k} \tag{8.26}$$

$$\mathbf{s}_k = \boldsymbol{\theta}_k - \boldsymbol{\theta}_{k-1} \tag{8.27}$$

$$\mathbf{y}_k = \mathbf{g}_k - \mathbf{g}_{k-1} \tag{8.28}$$

This is a rank-two update to the matrix, and ensures that the matrix remains positive definite (under certain restrictions on the step size). We typically start with a diagonal approximation, $\mathbf{B}_0 = \mathbf{I}$. Thus BFGS can be thought of as a "diagonal plus low-rank" approximation to the Hessian.

Alternatively, BFGS can iteratively update an approximation to the inverse Hessian, $\mathbf{C}_k \approx \mathbf{H}_k^{-1}$, as follows:

$$\mathbf{C}_{k+1} = \left(\mathbf{I} - \frac{\mathbf{s}_k \mathbf{y}_k^T}{\mathbf{y}_k^T \mathbf{s}_k}\right) \mathbf{C}_k \left(\mathbf{I} - \frac{\mathbf{y}_k \mathbf{s}_k^T}{\mathbf{y}_k^T \mathbf{s}_k}\right) + \frac{\mathbf{s}_k \mathbf{s}_k^T}{\mathbf{y}_k^T \mathbf{s}_k} \tag{8.29}$$

Since storing the Hessian takes $O(D^2)$ space, for very large problems, one can use **limited memory BFGS**, or **L-BFGS**, where $\mathbf{H}_k$ or $\mathbf{H}_k^{-1}$ is approximated by a diagonal plus low rank matrix. In particular, the product $\mathbf{H}_k^{-1} \mathbf{g}_k$ can be obtained by performing a sequence of inner products with $\mathbf{s}_k$ and $\mathbf{y}_k$, using only the m most recent $(\mathbf{s}_k, \mathbf{y}_k)$ pairs, and ignoring older information. The storage requirements are therefore $O(mD)$. Typically $m \sim 20$ suffices for good performance. See (Nocedal and Wright 2006, p177) for more information. L-BFGS is often the method of choice for most unconstrained smooth optimization problems that arise in machine learning (although see Section 8.5).

8.3.6 ℓ_2 regularization

Just as we prefer ridge regression to linear regression, so we should prefer MAP estimation for logistic regression to computing the MLE. In fact, regularization is important in the classification setting even if we have lots of data. To see why, suppose the data is linearly separable. In this case, the MLE is obtained when $||\mathbf{w}|| \to \infty$, corresponding to an infinitely steep sigmoid function, $\mathbb{I}(\mathbf{w}^T\mathbf{x} > w_0)$, also known as a **linear threshold unit**. This assigns the maximal amount of probability mass to the training data. However, such a solution is very brittle and will not generalize well.

To prevent this, we can use ℓ_2 regularization, just as we did with ridge regression. We note that the new objective, gradient and Hessian have the following forms:

$$f'(\mathbf{w}) = \text{NLL}(\mathbf{w}) + \lambda\mathbf{w}^T\mathbf{w} \tag{8.30}$$

$$\mathbf{g}'(\mathbf{w}) = \mathbf{g}(\mathbf{w}) + 2\lambda\mathbf{w} \tag{8.31}$$

$$\mathbf{H}'(\mathbf{w}) = \mathbf{H}(\mathbf{w}) + 2\lambda\mathbf{I} \tag{8.32}$$

It is a simple matter to pass these modified equations into any gradient-based optimizer.

8.3.7 Multi-class logistic regression

Now we consider **multinomial logistic regression**, sometimes called a **maximum entropy classifier**. This is a model of the form

$$p(y = c|\mathbf{x}, \mathbf{W}) = \frac{\exp(\mathbf{w}_c^T\mathbf{x})}{\sum_{c'=1}^C \exp(\mathbf{w}_{c'}^T\mathbf{x})} \tag{8.33}$$

where $\mathbf{w}_c$ is the c'th column of $\mathbf{W}$. A slight variant, known as a **conditional logit model**, normalizes over a different set of classes for each data case; this can be useful for modeling choices that users make between different sets of items that are offered to them.

Let us now introduce some notation. Let $\mu_{ic} = p(y_i = c|\mathbf{x}_i, \mathbf{W}) = \mathcal{S}(\boldsymbol{\eta}_i)_c$, where $\boldsymbol{\eta}_i = \mathbf{W}^T\mathbf{x}_i$ is a $C \times 1$ vector. Also, let $y_{ic} = \mathbb{I}(y_i = c)$ be the one-of-C encoding of y_i; thus $\mathbf{y}_i$ is a bit vector, in which the c'th bit turns on iff $y_i = c$. Following (Krishnapuram et al. 2005), let us set $\mathbf{w}_C = \mathbf{0}$, to ensure identifiability.

With this, the log-likelihood can be written as

$$\ell(\mathbf{W}) = \log \prod_{i=1}^N \prod_{c=1}^C \mu_{ic}^{y_{ic}} = \sum_{i=1}^N \sum_{c=1}^C y_{ic} \log \mu_{ic} \tag{8.34}$$

$$= \sum_{i=1}^N \left[\left(\sum_{c=1}^C y_{ic}\mathbf{w}_c^T\mathbf{x}_i \right) - \log\left(\sum_{c'=1}^C \exp(\mathbf{w}_{c'}^T\mathbf{x}_i) \right) \right] \tag{8.35}$$

The NLL is the cross entropy loss function:

$$f(\mathbf{W}) = -\ell(\mathbf{W}) \tag{8.36}$$

We now proceed to compute the gradient and Hessian of this expression. Since $\mathbf{W}$ is a matrix, the notation gets a bit heavy, but the ideas are simple. It helps to define $\mathbf{A} \otimes \mathbf{B}$ be the **Kronecker product** of matrices $\mathbf{A}$ and $\mathbf{B}$. If $\mathbf{A}$ is an $m \times n$ matrix and $\mathbf{B}$ is a $p \times q$ matrix, then $\mathbf{A} \times \mathbf{B}$ is the $mp \times nq$ block matrix

$$\mathbf{A} \otimes \mathbf{B} = \begin{bmatrix} a_{11}\mathbf{B} & \cdots & a_{1n}\mathbf{B} \\ \vdots & \ddots & \vdots \\ a_{m1}\mathbf{B} & \cdots & a_{mn}\mathbf{B} \end{bmatrix} \tag{8.37}$$

Returning to the task at hand, one can show (Exercise 8.4) that the gradient is given by

$$\mathbf{g}(\mathbf{W}) = \nabla f(\mathbf{W}) = \sum_{i=1}^{N} (\boldsymbol{\mu}_i - \mathbf{y}_i) \otimes \mathbf{x}_i \tag{8.38}$$

where $\mathbf{y}_i = (\mathbb{I}(y_i = 1), \ldots, \mathbb{I}(y_i = C - 1))$ and $\boldsymbol{\mu}_i = [p(y_i = 1|\mathbf{x}_i, \mathbf{w}), \ldots, p(y_i = C - 1|\mathbf{x}_i, \mathbf{W})]$ are column vectors of length $C-1$, For example, if we have $D = 3$ feature dimensions and $C = 3$ classes, this becomes

$$\mathbf{g}(\mathbf{W}) = \sum_i \begin{pmatrix} (\mu_{i1} - y_{i1})x_{i1} \\ (\mu_{i1} - y_{i1})x_{i2} \\ (\mu_{i1} - y_{i1})x_{i3} \\ (\mu_{i2} - y_{i2})x_{i1} \\ (\mu_{i2} - y_{i2})x_{i2} \\ (\mu_{i2} - y_{i2})x_{i3} \end{pmatrix} \tag{8.39}$$

In other words, for each class c, the derivative for the weights in the c'th column is

$$\nabla_{\mathbf{w}_c} f(\mathbf{W}) = \sum_i (\mu_{ic} - y_{ic})\mathbf{x}_i \tag{8.40}$$

This has the same form as in the binary logistic regression case, namely an error term times $\mathbf{x}_i$. (This turns out to be a general property of distributions in the exponential family, as we will see in Section 9.3.2.)

One can also show (Exercise 8.4) that the Hessian is the following block structured $D(C - 1) \times D(C - 1)$ matrix:

$$\mathbf{H}(\mathbf{W}) = \nabla^2 f(\mathbf{w}) = \sum_{i=1}^{N} (\text{diag}(\boldsymbol{\mu}_i) - \boldsymbol{\mu}_i \boldsymbol{\mu}_i^T) \otimes (\mathbf{x}_i \mathbf{x}_i^T) \tag{8.41}$$

For example, if we have 3 features and 3 classes, this becomes

$$\mathbf{H}(\mathbf{W}) = \sum_i \begin{pmatrix} \mu_{i1} - \mu_{i1}^2 & -\mu_{i1}\mu_{i2} \\ -\mu_{i1}\mu_{i2} & \mu_{i2} - \mu_{i2}^2 \end{pmatrix} \otimes \begin{pmatrix} x_{i1}x_{i1} & x_{i1}x_{i2} & x_{i1}x_{i3} \\ x_{i2}x_{i1} & x_{i2}x_{i2} & x_{i2}x_{i3} \\ x_{i3}x_{i1} & x_{i3}x_{i2} & x_{i3}x_{i3} \end{pmatrix} \tag{8.42}$$

$$= \sum_i \begin{pmatrix} (\mu_{i1} - \mu_{i1}^2)\mathbf{X}_i & -\mu_{i1}\mu_{i2}\mathbf{X}_i \\ -\mu_{i1}\mu_{i2}\mathbf{X}_i & (\mu_{i2} - \mu_{i2}^2)\mathbf{X}_i \end{pmatrix} \tag{8.43}$$

where $\mathbf{X}_i = \mathbf{x}_i \mathbf{x}_i^T$. In other words, the block c, c' submatrix is given by

$$\mathbf{H}_{c,c'}(\mathbf{W}) = \sum_i \mu_{ic}(\delta_{c,c'} - \mu_{i,c'})\mathbf{x}_i \mathbf{x}_i^T \tag{8.44}$$

This is also a positive definite matrix, so there is a unique MLE.

Now consider minimizing

$$f'(\mathbf{W}) \triangleq -\log p(\mathcal{D}|\mathbf{w}) - \log p(\mathbf{W}) \tag{8.45}$$

where $p(\mathbf{W}) = \prod_c \mathcal{N}(\mathbf{w}_c | \mathbf{0}, \mathbf{V}_0)$. The new objective, its gradient and Hessian are given by

$$f'(\mathbf{W}) = f(\mathbf{W}) + \frac{1}{2}\sum_c \mathbf{w}_c \mathbf{V}_0^{-1}\mathbf{w}_c \tag{8.46}$$

$$\mathbf{g}'(\mathbf{W}) = \mathbf{g}(\mathbf{W}) + \mathbf{V}_0^{-1}(\sum_c \mathbf{w}_c) \tag{8.47}$$

$$\mathbf{H}'(\mathbf{W}) = \mathbf{H}(\mathbf{W}) + \mathbf{I}_C \otimes \mathbf{V}_0^{-1} \tag{8.48}$$

This can be passed to any gradient-based optimizer to find the MAP estimate. Note, however, that the Hessian has size $O((CD) \times (CD))$, which is C times more row and columns than in the binary case, so limited memory BFGS is more appropriate than Newton's method. See `logregFit` for some Matlab code.

8.4 Bayesian logistic regression

It is natural to want to compute the full posterior over the parameters, $p(\mathbf{w}|\mathcal{D})$, for logistic regression models. This can be useful for any situation where we want to associate confidence intervals with our predictions (e.g., this is necessary when solving contextual bandit problems, discussed in Section 5.7.3.1).

Unfortunately, unlike the linear regression case, this cannot be done exactly, since there is no convenient conjugate prior for logistic regression. We discuss one simple approximation below; some other approaches include MCMC (Section 24.3.3.1), variational inference (Section 21.8.1.1), expectation propagation (Kuss and Rasmussen 2005), etc. For notational simplicity, we stick to binary logistic regression.

8.4.1 Laplace approximation

In this section, we discuss how to make a Gaussian approximation to a posterior distribution. (In physics, there is an analogous technique known as a **saddle point approximation**.) The approximation works as follows. Suppose $\boldsymbol{\theta} \in \mathbb{R}^D$. Let

$$p(\boldsymbol{\theta}|\mathcal{D}) = \frac{1}{Z}e^{-E(\boldsymbol{\theta})} \tag{8.49}$$

where $E(\boldsymbol{\theta})$ is called an **energy function**, and is equal to the negative log of the unnormalized log posterior, $E(\boldsymbol{\theta}) = -\log p(\boldsymbol{\theta}, \mathcal{D})$, with $Z = p(\mathcal{D})$ being the normalization constant. Performing a Taylor series expansion around the mode $\boldsymbol{\theta}^*$ (i.e., the lowest energy state) we get

$$E(\boldsymbol{\theta}) \approx E(\boldsymbol{\theta}^*) + (\boldsymbol{\theta} - \boldsymbol{\theta}^*)^T\mathbf{g} + \frac{1}{2}(\boldsymbol{\theta} - \boldsymbol{\theta}^*)^T\mathbf{H}(\boldsymbol{\theta} - \boldsymbol{\theta}^*) \tag{8.50}$$

where $\mathbf{g}$ is the gradient and $\mathbf{H}$ is the Hessian of the energy function evaluated at the mode:

$$\mathbf{g} \triangleq \nabla E(\boldsymbol{\theta})|_{\boldsymbol{\theta}^*}, \quad \mathbf{H} \triangleq \frac{\partial^2 E(\boldsymbol{\theta})}{\partial \boldsymbol{\theta} \partial \boldsymbol{\theta}^T}|_{\boldsymbol{\theta}^*} \tag{8.51}$$

Since $\boldsymbol{\theta}^*$ is the mode, the gradient term is zero. Hence

$$\hat{p}(\boldsymbol{\theta}, \mathcal{D}) = e^{-E(\boldsymbol{\theta}^*)} \exp\left[-\frac{1}{2}(\boldsymbol{\theta} - \boldsymbol{\theta}^*)^T \mathbf{H}(\boldsymbol{\theta} - \boldsymbol{\theta}^*)\right] \tag{8.52}$$

$$\hat{p}(\boldsymbol{\theta}|\mathcal{D}) = \frac{1}{Z}\hat{p}(\boldsymbol{\theta}, \mathcal{D}) \propto \mathcal{N}(\boldsymbol{\theta}|\boldsymbol{\theta}^*, \mathbf{H}^{-1}) \tag{8.53}$$

$$Z = e^{-E(\boldsymbol{\theta}^*)}(2\pi)^{D/2}|\mathbf{H}|^{-\frac{1}{2}} \tag{8.54}$$

The last line follows from normalization constant of the multivariate Gaussian.

Equation 8.54 is known as the **Laplace approximation** to the marginal likelihood. Therefore Equation 8.53 is sometimes called a Laplace approximation to the posterior. However, in the statistics community, the term "Laplace approximation" refers to a more sophisticated method (see e.g. (Rue et al. 2009) for details). It may therefore be better to use the term "Gaussian approximation" to refer to Equation 8.53. In any case, a Gaussian approximation is often a reasonable approximation, since posteriors often become more "Gaussian-like" as the sample size increases, for reasons analogous to the central limit theorem.

8.4.2 Derivation of the Bayesian information criterion (BIC)

We can rewrite the Laplace approximation to the log marginal likelihood as follows, dropping irrelevant constants:

$$\log p(\mathcal{D}) \approx \log p(\mathcal{D}|\boldsymbol{\theta}^*) + \log p(\boldsymbol{\theta}^*) - \frac{1}{2}\log|\mathbf{H}| \tag{8.55}$$

We see that this is the log likelihood plus some penalty terms. If we have a uniform prior, $p(\boldsymbol{\theta}) \propto 1$, we can drop the second term, and replace $\boldsymbol{\theta}^*$ with the MLE, $\hat{\boldsymbol{\theta}}$, yielding $\log p(\mathcal{D}) \approx \log p(\mathcal{D}|\hat{\boldsymbol{\theta}}) - \frac{1}{2}\log|\mathbf{H}|$.

We now focus on approximating the $\log|\mathbf{H}|$ term, which is sometimes called the **Occam factor**, since it is a measure of model complexity (volume of the posterior distribution). We have $\mathbf{H} = \sum_{i=1}^{N}\mathbf{H}_i$, where $\mathbf{H}_i = \nabla\nabla \log p(\mathcal{D}_i|\boldsymbol{\theta})$. Let us approximate each $\mathbf{H}_i$ by a fixed matrix $\hat{\mathbf{H}}$. Then we have

$$\log|\mathbf{H}| = \log|N\hat{\mathbf{H}}| = \log(N^d|\hat{\mathbf{H}}|) = D\log N + \log|\hat{\mathbf{H}}| \tag{8.56}$$

where $D = \dim(\boldsymbol{\theta})$ and we have assumed $\mathbf{H}$ is full rank. We can drop the $\log|\hat{\mathbf{H}}|$ term, since it is independent of N, and thus will get overwhelmed by the likelihood. Putting all the pieces together, we recover the BIC score (Section 5.3.2.4):

$$\log p(\mathcal{D}) \approx \log p(\mathcal{D}|\hat{\boldsymbol{\theta}}) - \frac{D}{2}\log N \tag{8.57}$$

8.4.3 Gaussian approximation for logistic regression

Now let us apply the Gaussian approximation to logistic regression. We will use a Gaussian prior of the form $p(\mathbf{w}) = \mathcal{N}(\mathbf{w}|\mathbf{0}, \mathbf{V}_0)$, just as we did in MAP estimation. The approximate posterior is given by

$$p(\mathbf{w}|\mathcal{D}) \approx \mathcal{N}(\mathbf{w}|\hat{\mathbf{w}}, \mathbf{H}^{-1}) \tag{8.58}$$

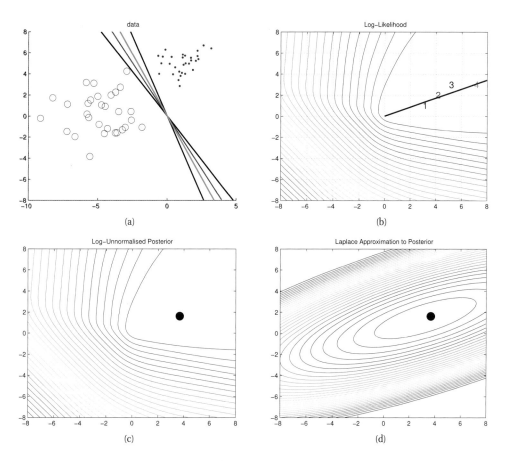

Figure 8.5 (a) Two-class data in 2d. (b) Log-likelihood for a logistic regression model. The line is drawn from the origin in the direction of the MLE (which is at infinity). The numbers correspond to 4 points in parameter space, corresponding to the lines in (a). (c) Unnormalized log posterior (assuming vague spherical prior). (d) Laplace approximation to posterior. Based on a figure by Mark Girolami. Figure generated by `logregLaplaceGirolamiDemo`.

where $\hat{\mathbf{w}} = \arg\min_{\mathbf{w}} E(\mathbf{w})$, $E(\mathbf{w}) = -(\log p(\mathcal{D}|\mathbf{w}) + \log p(\mathbf{w}))$, and $\mathbf{H} = \nabla^2 E(\mathbf{w})|_{\hat{\mathbf{w}}}$.

As an example, consider the linearly separable 2D data in Figure 8.5(a). There are many parameter settings that correspond to lines that perfectly separate the training data; we show 4 examples. The likelihood surface is shown in Figure 8.5(b), where we see that the likelihood is unbounded as we move up and to the right in parameter space, along a ridge where $w_2/w_1 = 2.35$ (this is indicated by the diagonal line). The reasons for this is that we can maximize the likelihood by driving $||\mathbf{w}||$ to infinity (subject to being on this line), since large regression weights make the sigmoid function very steep, turning it into a step function. Consequently the MLE is not well defined when the data is linearly separable.

To regularize the problem, let us use a vague spherical prior centered at the origin, $\mathcal{N}(\mathbf{w}|\mathbf{0}, 100\mathbf{I})$.

Multiplying this spherical prior by the likelihood surface results in a highly skewed posterior, shown in Figure 8.5(c). The posterior is skewed because the likelihood function "chops off" regions of parameter space (in a "soft" fashion) which disagree with the data. The MAP estimate is shown by the blue dot. Unlike the MLE, this is not at infinity.

The Gaussian approximation to this posterior is shown in Figure 8.5(d). We see that this is a symmetric distribution, and therefore not a great approximation. Of course, it gets the mode correct (by construction), and it at least represents the fact that there is more uncertainty along the southwest-northeast direction (which corresponds to uncertainty about the orientation of separating lines) than perpendicular to this. Although a crude approximation, this is surely better than approximating the posterior by a delta function, which is what MAP estimation does.

8.4.4   Approximating the posterior predictive

Given the posterior, we can compute credible intervals, perform hypothesis tests, etc., just as we did in Section 7.6.3.3 in the case of linear regression. But in machine learning, interest usually focusses on prediction. The posterior predictive distribution has the form

$$p(y|\mathbf{x}, \mathcal{D}) \quad = \quad \int p(y|\mathbf{x}, \mathbf{w})p(\mathbf{w}|\mathcal{D})d\mathbf{w} \tag{8.59}$$

Unfortunately this integral is intractable.

The simplest approximation is the plug-in approximation, which, in the binary case, takes the form

$$p(y = 1|\mathbf{x}, \mathcal{D}) \quad \approx \quad p(y = 1|\mathbf{x}, \mathbb{E}[\mathbf{w}]) \tag{8.60}$$

where $\mathbb{E}[\mathbf{w}]$ is the posterior mean. In this context, $\mathbb{E}[\mathbf{w}]$ is called the **Bayes point**. Of course, such a plug-in estimate underestimates the uncertainty. We discuss some better approximations below.

8.4.4.1   Monte Carlo approximation

A better approach is to use a Monte Carlo approximation, as follows:

$$p(y = 1|\mathbf{x}, \mathcal{D}) \quad \approx \quad \frac{1}{S} \sum_{s=1}^{S} \text{sigm}((\mathbf{w}^s)^T \mathbf{x}) \tag{8.61}$$

where $\mathbf{w}^s \sim p(\mathbf{w}|\mathcal{D})$ are samples from the posterior. (This technique can be trivially extended to the multi-class case.) If we have approximated the posterior using Monte Carlo, we can reuse these samples for prediction. If we made a Gaussian approximation to the posterior, we can draw *independent* samples from the Gaussian using standard methods.

Figure 8.6(b) shows samples from the posterior predictive for our 2d example. Figure 8.6(c) shows the average of these samples. By averaging over multiple predictions, we see that the uncertainty in the decision boundary "splays out" as we move further from the training data. So although the decision boundary is linear, the posterior predictive density is not linear. Note also that the posterior mean decision boundary is roughly equally far from both classes; this is the Bayesian analog of the large margin principle discussed in Section 14.5.2.2.

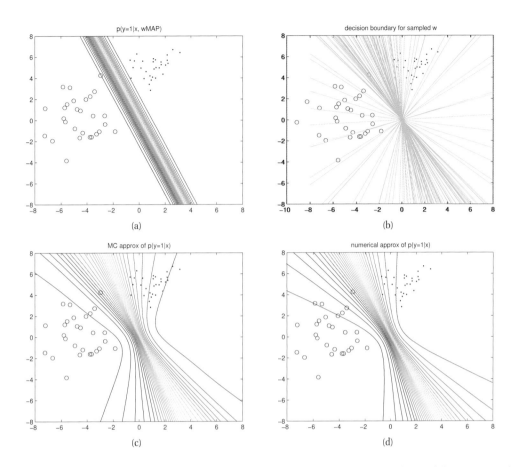

Figure 8.6 Posterior predictive distribution for a logistic regression model in 2d. Top left: contours of $p(y = 1|\mathbf{x}, \hat{\mathbf{w}}_{map})$. Top right: samples from the posterior predictive distribution. Bottom left: Averaging over these samples. Bottom right: moderated output (probit approximation). Based on a figure by Mark Girolami. Figure generated by `logregLaplaceGirolamiDemo`.

Figure 8.7(a) shows an example in 1d. The red dots denote the mean of the posterior predictive evaluated at the training data. The vertical blue lines denote 95% credible intervals for the posterior predictive; the small blue star is the median. We see that, with the Bayesian approach, we are able to model our uncertainty about the probability a student will pass the exam based on his SAT score, rather than just getting a point estimate.

8.4.4.2 Probit approximation (moderated output) *

If we have a Gaussian approximation to the posterior $p(\mathbf{w}|\mathcal{D}) \approx \mathcal{N}(\mathbf{w}|\mathbf{m}_N, \mathbf{V}_N)$, we can also compute a deterministic approximation to the posterior predictive distribution, at least in the

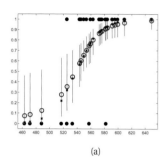

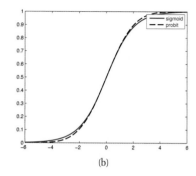

(a)   (b)

Figure 8.7 (a) Posterior predictive density for SAT data. The red circle denotes the posterior mean, the blue cross the posterior median, and the blue lines denote the 5th and 95th percentiles of the predictive distribution. Figure generated by `logregSATdemoBayes`. (b) The logistic (sigmoid) function $\text{sigm}(x)$ in solid red, with the rescaled probit function $\Phi(\lambda x)$ in dotted blue superimposed. Here $\lambda = \sqrt{\pi/8}$, which was chosen so that the derivatives of the two curves match at $x = 0$. Based on Figure 4.9 of (Bishop 2006). Figure generated by `probitPlot`. Figure generated by `probitRegDemo`.

binary case. We proceed as follows:

$$p(y = 1|\mathbf{x}, \mathcal{D}) \approx \int \text{sigm}(\mathbf{w}^T\mathbf{x})p(\mathbf{w}|\mathcal{D})d\mathbf{w} = \int \text{sigm}(a)\mathcal{N}(a|\mu, \sigma^2)da \tag{8.62}$$

$$a \triangleq \mathbf{w}^T\mathbf{x} \tag{8.63}$$

$$\mu \triangleq \mathbb{E}[a] = \mathbf{m}_N^T\mathbf{x} \tag{8.64}$$

$$\sigma^2 \triangleq \text{var}[a] = \int p(a|\mathcal{D})[a^2 - \mathbb{E}[a]^2]da \tag{8.65}$$

$$= \int p(\mathbf{w}|\mathcal{D})[(\mathbf{w}^T\mathbf{x})^2 - (\mathbf{m}_N^T\mathbf{x})^2]d\mathbf{w} = \mathbf{x}^T\mathbf{V}_N\mathbf{x} \tag{8.66}$$

Thus we see that we need to evaluate the expectation of a sigmoid with respect to a Gaussian. This can be approximated by exploiting the fact that the sigmoid function is similar to the **probit** function, which is given by the cdf of the standard normal:

$$\Phi(a) \triangleq \int_{-\infty}^{a} \mathcal{N}(x|0, 1)dx \tag{8.67}$$

Figure 8.7(b) plots the sigmoid and probit functions. We have rescaled the axes so that $\text{sigm}(a)$ has the same slope as $\Phi(\lambda a)$ at the origin, where $\lambda^2 = \pi/8$.

The advantage of using the probit is that one can convolve it with a Gaussian analytically:

$$\int \Phi(\lambda a)\mathcal{N}(a|\mu, \sigma^2)da = \Phi\left(\frac{\mu}{(\lambda^{-2} + \sigma^2)^{\frac{1}{2}}}\right) \tag{8.68}$$

We now plug in the approximation $\text{sigm}(a) \approx \Phi(\lambda a)$ to both sides of this equation to get

$$\int \text{sigm}(a)\mathcal{N}(a|\mu,\sigma^2)da \quad \approx \quad \text{sigm}(\kappa(\sigma^2)\mu) \tag{8.69}$$

$$\kappa(\sigma^2) \quad \triangleq \quad (1 + \pi\sigma^2/8)^{-\frac{1}{2}} \tag{8.70}$$

Applying this to the logistic regression model we get the following expression (first suggested in (Spiegelhalter and Lauritzen 1990)):

$$p(y = 1|\mathbf{x}, \mathcal{D}) \quad \approx \quad \text{sigm}(\kappa(\sigma^2)\mu) \tag{8.71}$$

Figure 8.6(d) indicates that this gives very similar results to the Monte Carlo approximation.

Using Equation 8.71 is sometimes called a **moderated output**, since it is less extreme than the plug-in estimate. To see this, note that $0 \leq \kappa(\sigma^2) \leq 1$ and hence

$$\text{sigm}(\kappa(\sigma^2)\mu) \leq \text{sigm}(\mu) = p(y = 1|\mathbf{x}, \hat{\mathbf{w}}) \tag{8.72}$$

where the inequality is strict if $\mu \neq 0$. If $\mu > 0$, we have $p(y = 1|\mathbf{x}, \hat{\mathbf{w}}) > 0.5$, but the moderated prediction is always closer to 0.5, so it is less confident. However, the decision boundary occurs whenever $p(y = 1|\mathbf{x}, \mathcal{D}) = \text{sigm}(\kappa(\sigma^2)\mu) = 0.5$, which implies $\mu = \hat{\mathbf{w}}^T\mathbf{x} = 0$. Hence the decision boundary for the moderated approximation is the same as for the plug-in approximation. So the number of misclassifications will be the same for the two methods, but the log-likelihood will not. (Note that in the multiclass case, taking into account posterior covariance gives different answers than the plug-in approach: see Exercise 3.10.3 of (Rasmussen and Williams 2006).)

8.4.5 Residual analysis (outlier detection) *

It is sometimes useful to detect data cases which are "outliers". This is called **residual analysis** or **case analysis**. In a regression setting, this can be performed by computing $r_i = y_i - \hat{y}_i$, where $\hat{y}_i = \hat{\mathbf{w}}^T\mathbf{x}_i$. These values should follow a $\mathcal{N}(0, \sigma^2)$ distribution, if the modelling assumptions are correct. This can be assessed by creating a **qq-plot**, where we plot the N theoretical quantiles of a Gaussian distribution against the N empirical quantiles of the r_i. Points that deviate from the straightline are potential outliers.

Classical methods, based on residuals, do not work well for binary data, because they rely on asymptotic normality of the test statistics. However, adopting a Bayesian approach, we can just define outliers to be points which which $p(y_i|\hat{y}_i)$ is small, where we typically use $\hat{y}_i = \text{sigm}(\hat{\mathbf{w}}^T\mathbf{x}_i)$. Note that $\hat{\mathbf{w}}$ was estimated from all the data. A better method is to exclude $(\mathbf{x}_i, y_i)$ from the estimate of $\mathbf{w}$ when predicting y_i. That is, we define outliers to be points which have low probability under the cross-validated posterior predictive distribution, defined by

$$p(y_i|\mathbf{x}_i, \mathbf{x}_{-i}, \mathbf{y}_{-i}) = \int p(y_i|\mathbf{x}_i, \mathbf{w}) \prod_{i' \neq i} p(y_{i'}|\mathbf{x}_{i'}, \mathbf{w})p(\mathbf{w})d\mathbf{w} \tag{8.73}$$

This can be efficiently approximated by sampling methods (Gelfand 1996). For further discussion of residual analysis in logistic regression models, see e.g.,(Johnson and Albert 1999, Sec 3.4).

8.5   Online learning and stochastic optimization

Traditionally machine learning is performed **offline**, which means we have a **batch** of data, and we optimize an equation of the following form

$$f(\boldsymbol{\theta}) = \frac{1}{N} \sum_{i=1}^{N} f(\boldsymbol{\theta}, \mathbf{z}_i) \tag{8.74}$$

where $\mathbf{z}_i = (\mathbf{x}_i, y_i)$ in the supervised case, or just $\mathbf{x}_i$ in the unsupervised case, and $f(\boldsymbol{\theta}, \mathbf{z}_i)$ is some kind of loss function. For example, we might use

$$f(\boldsymbol{\theta}, \mathbf{z}_i) = -\log p(y_i|\mathbf{x}_i, \boldsymbol{\theta}) \tag{8.75}$$

in which case we are trying to maximize the likelihood. Alternatively, we might use

$$f(\boldsymbol{\theta}, \mathbf{z}_i) = L(y_i, h(\mathbf{x}_i, \boldsymbol{\theta})) \tag{8.76}$$

where $h(\mathbf{x}_i, \boldsymbol{\theta})$ is a prediction function, and $L(y, \hat{y})$ is some other loss function such as squared error or the Huber loss. In frequentist decision theory, the average loss is called the risk (see Section 6.3), so this overall approach is called empirical risk minimization or ERM (see Section 6.5 for details).

However, if we have **streaming data**, we need to perform **online learning**, so we can update our estimates as each new data point arrives rather than waiting until "the end" (which may never occur). And even if we have a batch of data, we might want to treat it like a stream if it is too large to hold in main memory. Below we discuss learning methods for this kind of scenario.[2]

8.5.1   Online learning and regret minimization

Suppose that at each step, "nature" presents a sample $\mathbf{z}_k$ and the "learner" must respond with a parameter estimate $\boldsymbol{\theta}_k$. In the theoretical machine learning community, the objective used in online learning is the **regret**, which is the averaged loss incurred relative to the best we could have gotten in hindsight using a single fixed parameter value:

$$\text{regret}_k \triangleq \frac{1}{k} \sum_{t=1}^{k} f(\boldsymbol{\theta}_t, \mathbf{z}_t) - \min_{\boldsymbol{\theta}_* \in \Theta} \frac{1}{k} \sum_{t=1}^{k} f(\boldsymbol{\theta}_*, \mathbf{z}_t) \tag{8.77}$$

For example, imagine we are investing in the stock-market. Let θ_j be the amount we invest in stock j, and let z_j be the return on this stock. Our loss function is $f(\boldsymbol{\theta}, \mathbf{z}) = -\boldsymbol{\theta}^T \mathbf{z}$. The regret is how much better (or worse) we did by trading at each step, rather than adopting a "buy and hold" strategy using an oracle to choose which stocks to buy.

---

2. A simple implementation trick can be used to speed up batch learning algorithms when applied to data sets that are too large to hold in memory. First note that the naive implementation makes a pass over the data file, from the beginning to end, accumulating the sufficient statistics and gradients as it goes; then an update is performed and the process repeats. Unfortunately, at the end of each pass, the data from the beginning of the file will have been evicted from the cache (since are are assuming it cannot all fit into memory). Rather than going back to the beginning of the file and reloading it, we can simply work backwards from the end of the file, which is already in memory. We then repeat this forwards-backwards pattern over the data. This simple trick is known as **rocking**.

One simple algorithm for online learning is **online gradient descent** (Zinkevich 2003), which is as follows: at each step k, update the parameters using

$$\boldsymbol{\theta}_{k+1} = \text{proj}_{\Theta}(\boldsymbol{\theta}_k - \eta_k \mathbf{g}_k) \tag{8.78}$$

where $\text{proj}_{\mathcal{V}}(\mathbf{v}) = \text{argmin}_{\mathbf{w} \in \mathbf{V}} ||\mathbf{w} - \mathbf{v}||_2$ is the projection of vector $\mathbf{v}$ onto space $\mathcal{V}$, $\mathbf{g}_k = \nabla f(\boldsymbol{\theta}_k, \mathbf{z}_k)$ is the gradient, and η_k is the step size. (The projection step is only needed if the parameter must be constrained to live in a certain subset of $\mathbb{R}^D$. See Section 13.4.3 for details.) Below we will see how this approach to regret minimization relates to more traditional objectives, such as MLE.

There are a variety of other approaches to regret minimization which are beyond the scope of this book (see e.g., Cesa-Bianchi and Lugosi (2006) for details).

8.5.2 Stochastic optimization and risk minimization

Now suppose that instead of minimizing regret with respect to the past, we want to minimize expected loss in the future, as is more common in (frequentist) statistical learning theory. That is, we want to minimize

$$f(\boldsymbol{\theta}) = \mathbb{E}\left[f(\boldsymbol{\theta}, \mathbf{z})\right] \tag{8.79}$$

where the expectation is taken over future data. Optimizing functions where some of the variables in the objective are random is called **stochastic optimization**.[3]

Suppose we receive an infinite stream of samples from the distribution. One way to optimize stochastic objectives such as Equation 8.79 is to perform the update in Equation 8.78 at each step. This is called **stochastic gradient descent** or **SGD** (Nemirovski and Yudin 1978). Since we typically want a single parameter estimate, we can use a running average:

$$\overline{\boldsymbol{\theta}}_k = \frac{1}{k} \sum_{t=1}^{k} \boldsymbol{\theta}_t \tag{8.80}$$

This is called **Polyak-Ruppert averaging**, and can be implemented recursively as follows:

$$\overline{\boldsymbol{\theta}}_k = \overline{\boldsymbol{\theta}}_{k-1} - \frac{1}{k}(\overline{\boldsymbol{\theta}}_{k-1} - \boldsymbol{\theta}_k) \tag{8.81}$$

See e.g., (Spall 2003; Kushner and Yin 2003) for details.

8.5.2.1 Setting the step size

We now discuss some sufficient conditions on the learning rate to guarantee convergence of SGD. These are known as the **Robbins-Monro** conditions:

$$\sum_{k=1}^{\infty} \eta_k = \infty, \ \sum_{k=1}^{\infty} \eta_k^2 < \infty. \tag{8.82}$$

---

3. Note that in stochastic optimization, the objective is stochastic, and therefore the algorithms will be, too. However, it is also possible to apply stochastic optimization algorithms to deterministic objectives. Examples include simulated annealing (Section 24.6.1) and stochastic gradient descent applied to the empirical risk minimization problem. There are some interesting theoretical connections between online learning and stochastic optimization (Cesa-Bianchi and Lugosi 2006), but this is beyond the scope of this book.

The set of values of η_k over time is called the learning rate **schedule**. Various formulas are used, such as $\eta_k = 1/k$, or the following (Bottou 1998; Bach and Moulines 2011):

$$\eta_k = (\tau_0 + k)^{-\kappa} \tag{8.83}$$

where $\tau_0 \geq 0$ slows down early iterations of the algorithm, and $\kappa \in (0.5, 1]$ controls the rate at which old values are forgotten.

The need to adjust these tuning parameters is one of the main drawbacks of stochastic optimization. One simple heuristic (Bottou 2007) is as follows: store an initial subset of the data, and try a range of η values on this subset; then choose the one that results in the fastest decrease in the objective and apply it to all the rest of the data. Note that this may not result in convergence, but the algorithm can be terminated when the performance improvement on a hold-out set plateaus (this is called **early stopping**).

8.5.2.2 Per-parameter step sizes

One drawback of SGD is that it uses the same step size for all parameters. We now briefly present a method known as **adagrad** (short for adaptive gradient) (Duchi et al. 2010), which is similar in spirit to a diagonal Hessian approximation. (See also (Schaul et al. 2012) for a similar approach.) In particular, if $\theta_i(k)$ is parameter i at time k, and $g_i(k)$ is its gradient, then we make an update as follows:

$$\theta_i(k+1) = \theta_i(k) - \eta \frac{g_i(k)}{\tau_0 + \sqrt{s_i(k)}} \tag{8.84}$$

where $s_i(k) = \sum_{t=1}^{k} g_i(t)^2$. This can be recursively updated as follows:

$$s_i(k) = s_i(k-1) + g_i(k)^2 \tag{8.85}$$

The result is a per-parameter step size that adapts to the curvature of the loss function. This method was original derived for the regret minimization case, but it can be applied more generally.

8.5.2.3 SGD compared to batch learning

If we don't have an infinite data stream, we can "simulate" one by sampling data points at random from our training set. Essentially we are optimizing Equation 8.74 by treating it as an expectation with respect to the empirical distribution.

In theory, we should sample with replacement, although in practice it is usually better to randomly permute the data and sample without replacement, and then to repeat. A single such pass over the entire data set is called an **epoch**. See Algorithm 8.3 for some pseudocode.

In this offline case, it is often better to compute the gradient of a **mini-batch** of B data cases. If $B = 1$, this is standard SGD, and if $B = N$, this is standard **steepest descent**. Typically $B \sim 100$ is used.

Although a simple first-order method, SGD performs surprisingly well on some problems, especially ones with large data sets (Bottou 2007). The intuitive reason for this is that one can get a fairly good estimate of the gradient by looking at just a few examples. Carefully evaluating

---

Algorithm 8.3: Stochastic gradient descent

---

1 Initialize $\boldsymbol{\theta}$, η;
2 **repeat**
3   Randomly permute data;
4   **for** $i = 1 : N$ **do**
5         $\mathbf{g} = \nabla f(\boldsymbol{\theta}, \mathbf{z}_i)$;
6         $\boldsymbol{\theta} \leftarrow \text{proj}_{\Theta}(\boldsymbol{\theta} - \eta\mathbf{g})$;
7   Update η;
8 **until** *converged*;

---

precise gradients using large datasets is often a waste of time, since the algorithm will have to recompute the gradient again anyway at the next step. It is often a better use of computer time to have a noisy estimate and to move rapidly through parameter space. As an extreme example, suppose we double the training set by duplicating every example. Batch methods will take twice as long, but online methods will be unaffected, since the direction of the gradient has not changed (doubling the size of the data changes the magnitude of the gradient, but that is irrelevant, since the gradient is being scaled by the step size anyway).

In addition to enhanced speed, SGD is often less prone to getting stuck in shallow local minima, because it adds a certain amount of "noise". Consequently it is quite popular in the machine learning community for fitting models with non-convex objectives, such as neural networks (Section 16.5) and deep belief networks (Section 28.1).

Recently, it has become popular to initialize with SGD, and then switch to batch optimization methods, such as L-BFGS (see e.g., (Agarwal et al. 2011)). The intuition is that SGD rapidly finds a good basin of attraction, and then L-BFGS, which forms a quadratic approximation to the basin, can quickly find the (possibly local) minimum.

8.5.3 The LMS algorithm

As an example of SGD, let us consider how to compute the MLE for linear regression in an online fashion. We derived the batch gradient in Equation 7.14. The online gradient at iteration k is given by

$$\mathbf{g}_k = \mathbf{x}_i(\boldsymbol{\theta}_k^T \mathbf{x}_i - y_i) \tag{8.86}$$

where $i = i(k)$ is the training example to use at iteration k. If the data set is streaming, we use $i(k) = k$; we shall assume this from now on, for notational simplicity. Equation 8.86 is easy to interpret: it is the feature vector $\mathbf{x}_k$ weighted by the difference between what we predicted, $\hat{y}_k = \boldsymbol{\theta}_k^T \mathbf{x}_k$, and the true response, y_k; hence the gradient acts like an error signal.

After computing the gradient, we take a step along it as follows:

$$\boldsymbol{\theta}_{k+1} = \boldsymbol{\theta}_k - \eta_k(\hat{y}_k - y_k)\mathbf{x}_k \tag{8.87}$$

(There is no need for a projection step, since this is an unconstrained optimization problem.) This algorithm is called the **least mean squares** or **LMS** algorithm, and is also known as the **delta rule**, or the **Widrow-Hoff rule**.

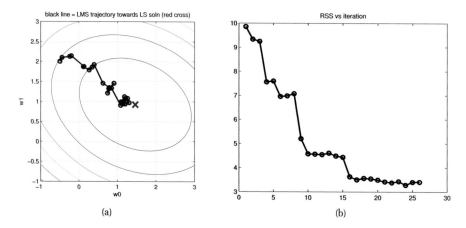

Figure 8.8 Illustration of the LMS algorithm. Left: we start from $\boldsymbol{\theta} = (-0.5, 2)$ and slowly converging to the least squares solution of $\hat{\boldsymbol{\theta}} = (1.45, 0.92)$ (red cross). Right: plot of objective function over time. Note that it does not decrease monotonically. Figure generated by LMSdemo.

Figure 8.8 shows the results of applying this algorithm to the data shown in Figure 7.2. We start at $\boldsymbol{\theta} = (-0.5, 2)$ and converge (in the sense that $||\boldsymbol{\theta}_k - \boldsymbol{\theta}_{k-1}||_2^2$ drops below a threshold of 10^{-2}) in about 26 iterations.

Note that LMS may require multiple passes through the data to find the optimum. By contrast, the recursive least squares algorithm, which is based on the Kalman filter and which uses second-order information, finds the optimum in a single pass (see Section 18.2.3). See also Exercise 7.7.

8.5.4 The perceptron algorithm

Now let us consider how to fit a binary logistic regression model in an online manner. The batch gradient was given in Equation 8.5. In the online case, the weight update has the simple form

$$\boldsymbol{\theta}_k = \boldsymbol{\theta}_{k-1} - \eta_k \mathbf{g}_i = \boldsymbol{\theta}_{k-1} - \eta_k (\mu_i - y_i) \mathbf{x}_i \tag{8.88}$$

where $\mu_i = p(y_i = 1 | \mathbf{x}_i, \boldsymbol{\theta}_k) = \mathbb{E}[y_i | \mathbf{x}_i, \boldsymbol{\theta}_k]$. We see that this has exactly the same form as the LMS algorithm. Indeed, this property holds for all generalized linear models (Section 9.3).

We now consider an approximation to this algorithm. Specifically, let

$$\hat{y}_i = \arg \max_{y \in \{0,1\}} p(y | \mathbf{x}_i, \boldsymbol{\theta}) \tag{8.89}$$

represent the most probable class label. We replace $\mu_i = p(y = 1 | \mathbf{x}_i, \boldsymbol{\theta}) = \text{sigm}(\boldsymbol{\theta}^T \mathbf{x}_i)$ in the gradient expression with $\hat{y}_i$. Thus the approximate gradient becomes

$$\mathbf{g}_i \approx (\hat{y}_i - y_i) \mathbf{x}_i \tag{8.90}$$

It will make the algebra simpler if we assume $y \in \{-1, +1\}$ rather than $y \in \{0, 1\}$. In this case, our prediction becomes

$$\hat{y}_i = \text{sign}(\boldsymbol{\theta}^T \mathbf{x}_i) \tag{8.91}$$

Then if $\hat{y}_i y_i = -1$, we have made an error, but if $\hat{y}_i y_i = +1$, we guessed the right label.

At each step, we update the weight vector by adding on the gradient. The key observation is that, if we predicted correctly, then $\hat{y}_i = y_i$, so the (approximate) gradient is zero and we do not change the weight vector. But if $\mathbf{x}_i$ is misclassified, we update the weights as follows: If $\hat{y}_i = 1$ but $y_i = -1$, then the negative gradient is $-(\hat{y}_i - y_i)\mathbf{x}_i = -2\mathbf{x}_i$; and if $\hat{y}_i = -1$ but $y_i = 1$, then the negative gradient is $-(\hat{y}_i - y_i)\mathbf{x}_i = 2\mathbf{x}_i$. We can absorb the factor of 2 into the learning rate η and just write the update, in the case of a misclassification, as

$$\boldsymbol{\theta}_k = \boldsymbol{\theta}_{k-1} + \eta_k y_i \mathbf{x}_i \tag{8.92}$$

Since it is only the sign of the weights that matter, not the magnitude, we will set $\eta_k = 1$. See Algorithm 8.4 for the pseudocode.

One can show that this method, known as the **perceptron algorithm** (Rosenblatt 1958), will converge, provided the data is linearly separable, i.e., that there exist parameters $\boldsymbol{\theta}$ such that predicting with $\text{sign}(\boldsymbol{\theta}^T \mathbf{x})$ achieves 0 error on the training set. However, if the data is not linearly separable, the algorithm will not converge, and even if it does converge, it may take a long time. There are much better ways to train logistic regression models (such as using proper SGD, without the gradient approximation, or IRLS, discussed in Section 8.3.4). However, the perceptron algorithm is historically important: it was one of the first machine learning algorithms ever derived (by Frank Rosenblatt in 1957), and was even implemented in analog hardware. In addition, the algorithm can be used to fit models where computing marginals $p(y_i|\mathbf{x}, \boldsymbol{\theta})$ is more expensive than computing the MAP output, $\arg\max_{\mathbf{y}} p(\mathbf{y}|\mathbf{x}, \boldsymbol{\theta})$; this arises in some structured-output classification problems. See Section 19.7 for details.

Algorithm 8.4: Perceptron algorithm

1 Input: linearly separable data set $\mathbf{x}_i \in \mathbb{R}^D$, $y_i \in \{-1, +1\}$ for $i = 1 : N$;
2 Initialize $\boldsymbol{\theta}_0$;
3 $k \leftarrow 0$;
4 **repeat**
5     $k \leftarrow k + 1$;
6     $i \leftarrow k \bmod N$;
7   **if** $\hat{y}_i \neq y_i$ **then**
8         $\boldsymbol{\theta}_{k+1} \leftarrow \boldsymbol{\theta}_k + y_i \mathbf{x}_i$
9   **else**
10   no-op
11 **until** *converged*;

8.5.5   A Bayesian view

Another approach to online learning is to adopt a Bayesian view. This is conceptually quite simple: we just apply Bayes rule recursively:

$$p(\boldsymbol{\theta}|\mathcal{D}_{1:k}) \propto p(\mathcal{D}_k|\boldsymbol{\theta})p(\boldsymbol{\theta}|\mathcal{D}_{1:k-1}) \tag{8.93}$$

This has the obvious advantage of returning a posterior instead of just a point estimate. It also allows for the online adaptation of hyper-parameters, which is important since cross-validation cannot be used in an online setting. Finally, it has the (less obvious) advantage that it can be quicker than SGD. To see why, note that by modeling the posterior variance of each parameter in addition to its mean, we effectively associate a different learning rate for each parameter (de Freitas et al. 2000), which is a simple way to model the curvature of the space. These variances can then be adapted using the usual rules of probability theory. By contrast, getting second-order optimization methods to work online is more tricky (see e.g., (Schraudolph et al. 2007; Sunehag et al. 2009; Bordes et al. 2009, 2010)).

As a simple example, in Section 18.2.3 we show how to use the Kalman filter to fit a linear regression model online. Unlike the LMS algorithm, this converges to the optimal (offline) answer in a single pass over the data. An extension which can learn a robust non-linear regression model in an online fashion is described in (Ting et al. 2010). For the GLM case, we can use an assumed density filter (Section 18.5.3), where we approximate the posterior by a Gaussian with a diagonal covariance; the variance terms serve as a per-parameter step-size. See Section 18.5.3.2 for details. Another approach is to use particle filtering (Section 23.5); this was used in (Andrieu et al. 2000) for sequentially learning a kernelized linear/logistic regression model.

8.6   Generative vs discriminative classifiers

In Section 4.2.2, we showed that the posterior over class labels induced by Gaussian discriminant analysis (GDA) has exactly the same form as logistic regression, namely $p(y = 1|\mathbf{x}) = \text{sigm}(\mathbf{w}^T\mathbf{x})$. The decision boundary is therefore a linear function of $\mathbf{x}$ in both cases. Note, however, that many generative models can give rise to a logistic regression posterior, e.g., if each class-conditional density is Poisson, $p(x|y = c) = \text{Poi}(x|\lambda_c)$. So the assumptions made by GDA are much stronger than the assumptions made by logistic regression.

A further difference between these models is the way they are trained. When fitting a discriminative model, we usually maximize the conditional log likelihood $\sum_{i=1}^{N} \log p(y_i|\mathbf{x}_i, \boldsymbol{\theta})$, whereas when fitting a generative model, we usually maximize the joint log likelihood, $\sum_{i=1}^{N} \log p(y_i, \mathbf{x}_i|\boldsymbol{\theta})$. It is clear that these can, in general, give different results (see Exercise 4.20).

When the Gaussian assumptions made by GDA are correct, the model will need less training data than logistic regression to achieve a certain level of performance, but if the Gaussian assumptions are incorrect, logistic regression will do better (Ng and Jordan 2002). This is because discriminative models do not need to model the distribution of the features. This is illustrated in Figure 8.10. We see that the class conditional densities are rather complex; in particular, $p(x|y = 1)$ is a multimodal distribution, which might be hard to estimate. However, the class posterior, $p(y = c|x)$, is a simple sigmoidal function, centered on the threshold value of 0.55. This suggests that, in general, discriminative methods will be more accurate, since their "job" is in some sense easier. However, accuracy is not the only important factor when choosing a method. Below we discuss some other advantages and disadvantages of each approach.

8.6.1 Pros and cons of each approach

- **Easy to fit?** As we have seen, it is usually very easy to fit generative classifiers. For example, in Sections 3.5.1.1 and 4.2.4, we show that we can fit a naive Bayes model and an LDA model by simple counting and averaging. By contrast, logistic regression requires solving a convex optimization problem (see Section 8.3.4 for the details), which is much slower.

- **Fit classes separately?** In a generative classifier, we estimate the parameters of each class conditional density independently, so we do not have to retrain the model when we add more classes. In contrast, in discriminative models, all the parameters interact, so the whole model must be retrained if we add a new class. (This is also the case if we train a generative model to maximize a discriminative objective (Salojarvi et al. 2005).)

- **Handle missing features easily?** Sometimes some of the inputs (components of $\mathbf{x}$) are not observed. In a generative classifier, there is a simple method for dealing with this, as we discuss in Section 8.6.2. However, in a discriminative classifier, there is no principled solution to this problem, since the model assumes that $\mathbf{x}$ is always available to be conditioned on (although see (Marlin 2008) for some heuristic approaches).

- **Can handle unlabeled training data?** There is much interest in **semi-supervised learning**, which uses unlabeled data to help solve a supervised task. This is fairly easy to do using generative models (see e.g., (Lasserre et al. 2006; Liang et al. 2007)), but is much harder to do with discriminative models.

- **Symmetric in inputs and outputs?** We can run a generative model "backwards", and infer probable inputs given the output by computing $p(\mathbf{x}|y)$. This is not possible with a discriminative model. The reason is that a generative model defines a joint distribution on $\mathbf{x}$ and y, and hence treats both inputs and outputs symmetrically.

- **Can handle feature preprocessing?** A big advantage of discriminative methods is that they allow us to preprocess the input in arbitrary ways, e.g., we can replace $\mathbf{x}$ with $\phi(\mathbf{x})$, which could be some basis function expansion, as illustrated in Figure 8.9. It is often hard to define a generative model on such pre-processed data, since the new features are correlated in complex ways.

- **Well-calibrated probabilities?** Some generative models, such as naive Bayes, make strong independence assumptions which are often not valid. This can result in very extreme posterior class probabilities (very near 0 or 1). Discriminative models, such as logistic regression, are usually better calibrated in terms of their probability estimates.

We see that there are arguments for and against both kinds of models. It is therefore useful to have both kinds in your "toolbox". See Table 8.1 for a summary of the classification and regression techniques we cover in this book.

8.6.2 Dealing with missing data

Sometimes some of the inputs (components of $\mathbf{x}$) are not observed; this could be due to a sensor failure, or a failure to complete an entry in a survey, etc. This is called the **missing data**

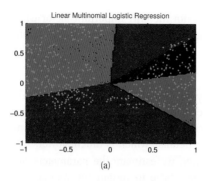
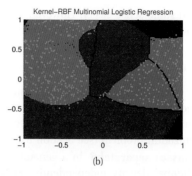

Figure 8.9　(a) Multinomial logistic regression for 5 classes in the original feature space. (b) After basis function expansion, using RBF kernels with a bandwidth of 1, and using all the data points as centers. Figure generated by `logregMultinomKernelDemo`.

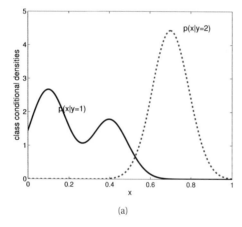
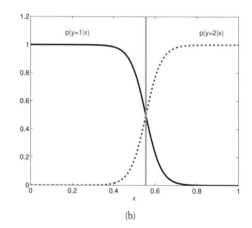

Figure 8.10　The class-conditional densities $p(x|y = c)$ (left) may be more complex than the class posteriors $p(y = c|x)$ (right). Based on Figure 1.27 of (Bishop 2006). Figure generated by `generativeVsDiscrim`.

problem (Little and Rubin 1987). The ability to handle missing data in a principled way is one of the biggest advantages of generative models.

To formalize our assumptions, we can associate a binary response variable $r_i \in \{0, 1\}$, that specifies whether each value $\mathbf{x}_i$ is observed or not. The joint model has the form $p(\mathbf{x}_i, r_i|\boldsymbol{\theta}, \boldsymbol{\phi}) = p(r_i|\mathbf{x}_i, \boldsymbol{\phi})p(\mathbf{x}_i|\boldsymbol{\theta})$, where $\boldsymbol{\phi}$ are the parameters controlling whether the item is observed or not. If we assume $p(r_i|\mathbf{x}_i, \boldsymbol{\phi}) = p(r_i|\boldsymbol{\phi})$, we say the data is **missing completely at random** or **MCAR**. If we assume $p(r_i|\mathbf{x}_i, \boldsymbol{\phi}) = p(r_i|\mathbf{x}_i^o, \boldsymbol{\phi})$, where $\mathbf{x}_i^o$ is the observed part of $\mathbf{x}_i$, we say the data is **missing at random** or **MAR**. If neither of these assumptions hold, we say the data is **not missing at random** or **NMAR**. In this case, we have to model the missing data mechanism, since the pattern of missingness is informative about the values of the missing data and the corresponding parameters. This is the case in most collaborative filtering problems, for

Model	Classif/regr	Gen/Discr	Param/Non	Section
Discriminant analysis	Classif	Gen	Param	Sec. 4.2.2, 4.2.4
Naive Bayes classifier	Classif	Gen	Param	Sec. 3.5, 3.5.1.2
Tree-augmented Naive Bayes classifier	Classif	Gen	Param	Sec. 10.2.1
Linear regression	Regr	Discrim	Param	Sec. 1.4.5, 7.3, 7.6,
Logistic regression	Classif	Discrim	Param	Sec. 1.4.6, 8.3.4, 8.4.3, 21.8.1.1
Sparse linear/ logistic regression	Both	Discrim	Param	Ch. 13
Mixture of experts	Both	Discrim	Param	Sec. 11.2.4
Multilayer perceptron (MLP)/ Neural network	Both	Discrim	Param	Ch. 16
Conditional random field (CRF)	Classif	Discrim	Param	Sec. 19.6
K nearest neighbor classifier	Classif	Gen	Non	Sec. 1.4.2, 14.7.3
(Infinite) Mixture Discriminant analysis	Classif	Gen	Non	Sec. 14.7.3
Classification and regression trees (CART)	Both	Discrim	Non	Sec. 16.2
Boosted model	Both	Discrim	Non	Sec. 16.4
Sparse kernelized lin/logreg (SKLR)	Both	Discrim	Non	Sec. 14.3.2
Relevance vector machine (RVM)	Both	Discrim	Non	Sec. 14.3.2
Support vector machine (SVM)	Both	Discrim	Non	Sec. 14.5
Gaussian processes (GP)	Both	Discrim	Non	Ch. 15
Smoothing splines	Regr	Discrim	Non	Section 15.4.6

Table 8.1 List of various models for classification and regression which we discuss in this book. Columns are as follows: Model name; is the model suitable for classification, regression, or both; is the model generative or discriminative; is the model parametric or non-parametric; list of sections in book which discuss the model. See also `http://pmtk3.googlecode.com/svn/trunk/docs/tutorial/html/tutSupervised.html` for the PMTK equivalents of these models. Any generative probabilistic model (e.g., HMMs, Boltzmann machines, Bayesian networks, etc.) can be turned into a classifier by using it as a class conditional density.

example. See e.g., (Marlin 2008) for further discussion. We will henceforth assume the data is MAR.

When dealing with missing data, it is helpful to distinguish the cases when there is missingness only at test time (so the training data is **complete data**), from the harder case when there is missingness also at training time. We will discuss these two cases below. Note that the class label is always missing at test time, by definition; if the class label is also sometimes missing at training time, the problem is called semi-supervised learning.

8.6.2.1 Missing data at test time

In a generative classifier, we can handle features that are MAR by marginalizing them out. For example, if we are missing the value of x_1, we can compute

$$p(y = c|\mathbf{x}_{2:D}, \boldsymbol{\theta}) \quad \propto \quad p(y = c|\boldsymbol{\theta})p(\mathbf{x}_{2:D}|y = c, \boldsymbol{\theta}) \tag{8.94}$$

$$= \quad p(y = c|\boldsymbol{\theta}) \sum_{x_1} p(x_1, \mathbf{x}_{2:D}|y = c, \boldsymbol{\theta}) \tag{8.95}$$

If we make the naive Bayes assumption, the marginalization can be performed as follows:

$$\sum_{x_1} p(x_1, x_{2:D}|y = c, \boldsymbol{\theta}) = \left[\sum_{x_1} p(x_1|\boldsymbol{\theta}_{1c})\right] \prod_{j=2}^{D} p(x_j|\boldsymbol{\theta}_{jc}) = \prod_{j=2}^{D} p(x_j|\boldsymbol{\theta}_{jc}) \tag{8.96}$$

where we exploited the fact that $\sum_{x_1} p(x_1|y = c, \boldsymbol{\theta}) = 1$. Hence in a naive Bayes classifier, we can simply ignore missing features at test time. Similarly, in discriminant analysis, no matter what regularization method was used to estimate the parameters, we can always analytically marginalize out the missing variables (see Section 4.3):

$$p(\mathbf{x}_{2:D}|y = c, \boldsymbol{\theta}) = \mathcal{N}(\mathbf{x}_{2:D}|\boldsymbol{\mu}_{c,2:D}, \boldsymbol{\Sigma}_{c,2:D,2:D}) \tag{8.97}$$

8.6.2.2 Missing data at training time

Missing data at training time is harder to deal with. In particular, computing the MLE or MAP estimate is no longer a simple optimization problem, for reasons discussed in Section 11.3.2. However, soon we will study a variety of more sophisticated algorithms (such as EM algorithm, in Section 11.4) for finding approximate ML or MAP estimates in such cases.

8.6.3 Fisher's linear discriminant analysis (FLDA) *

Discriminant analysis is a generative approach to classification, which requires fitting an MVN to the features. As we have discussed, this can be problematic in high dimensions. An alternative approach is to reduce the dimensionality of the features $\mathbf{x} \in \mathbb{R}^D$ and then fit an MVN to the resulting low-dimensional features $\mathbf{z} \in \mathbb{R}^L$. The simplest approach is to use a linear projection matrix, $\mathbf{z} = \mathbf{Wx}$, where $\mathbf{W}$ is a $L \times D$ matrix. One approach to finding $\mathbf{W}$ would be to use PCA (Section 12.2); the result would be very similar to RDA (Section 4.2.6), since SVD and PCA are essentially equivalent. However, PCA is an unsupervised technique that does not take class labels into account. Thus the resulting low dimensional features are not necessarily optimal for classification, as illustrated in Figure 8.11. An alternative approach is to find the matrix $\mathbf{W}$ such that the low-dimensional data can be classified as well as possible using a Gaussian class-conditional density model. The assumption of Gaussianity is reasonable since we are computing linear combinations of (potentially non-Gaussian) features. This approach is called **Fisher's linear discriminant analysis**, or **FLDA**.

FLDA is an interesting hybrid of discriminative and generative techniques. The drawback of this technique is that it is restricted to using $L \leq C - 1$ dimensions, regardless of D, for reasons that we will explain below. In the two-class case, this means we are seeking a single vector $\mathbf{w}$ onto which we can project the data. Below we derive the optimal $\mathbf{w}$ in the two-class case. We then generalize to the multi-class case, and finally we give a probabilistic interpretation of this technique.

8.6.3.1 Derivation of the optimal 1d projection

We now derive this optimal direction $\mathbf{w}$, for the two-class case, following the presentation of (Bishop 2006, Sec 4.1.4). Define the class-conditional means as

$$\boldsymbol{\mu}_1 = \frac{1}{N_1} \sum_{i:y_i=1} \mathbf{x}_i, \ \ \boldsymbol{\mu}_2 = \frac{1}{N_2} \sum_{i:y_i=2} \mathbf{x}_i \tag{8.98}$$

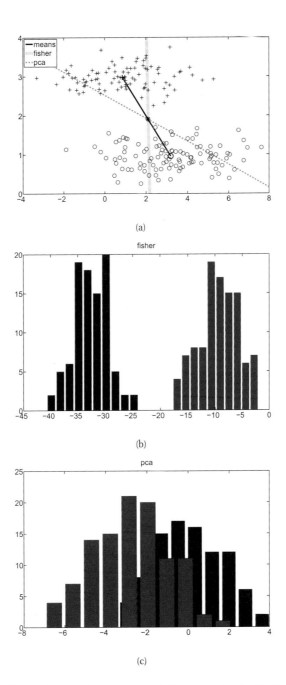

(a)

(b)

(c)

Figure 8.11 Example of Fisher's linear discriminant. (a) Two class data in 2D. Dashed green line = first principal basis vector. Dotted red line = Fisher's linear discriminant vector. Solid black line joins the class-conditional means. (b) Projection of points onto Fisher's vector shows good class separation. (c) Projection of points onto PCA vector shows poor class separation. Figure generated by `fisherLDAdemo`.

Let $m_k = \mathbf{w}^T \boldsymbol{\mu}_k$ be the projection of each mean onto the line $\mathbf{w}$. Also, let $z_i = \mathbf{w}^T \mathbf{x}_i$ be the projection of the data onto the line. The variance of the projected points is proportional to

$$s_k^2 = \sum_{i:y_i=k} (z_i - m_k)^2 \tag{8.99}$$

The goal is to find $\mathbf{w}$ such that we maximize the distance between the means, $m_2 - m_1$, while also ensuring the projected clusters are "tight":

$$J(\mathbf{w}) = \frac{(m_2 - m_1)^2}{s_1^2 + s_2^2} \tag{8.100}$$

We can rewrite the right hand side of the above in terms of $\mathbf{w}$ as follows

$$J(\mathbf{w}) = \frac{\mathbf{w}^T \mathbf{S}_B \mathbf{w}}{\mathbf{w}^T \mathbf{S}_W \mathbf{w}} \tag{8.101}$$

where $\mathbf{S}_B$ is the between-class scatter matrix given by

$$\mathbf{S}_B = (\boldsymbol{\mu}_2 - \boldsymbol{\mu}_1)(\boldsymbol{\mu}_2 - \boldsymbol{\mu}_1)^T \tag{8.102}$$

and $\mathbf{S}_W$ is the within-class scatter matrix, given by

$$\mathbf{S}_W = \sum_{i:y_i=1} (\mathbf{x}_i - \boldsymbol{\mu}_1)(\mathbf{x}_i - \boldsymbol{\mu}_1)^T + \sum_{i:y_i=2} (\mathbf{x}_i - \boldsymbol{\mu}_2)(\mathbf{x}_i - \boldsymbol{\mu}_2)^T \tag{8.103}$$

To see this, note that

$$\mathbf{w}^T \mathbf{S}_B \mathbf{w} = \mathbf{w}^T (\boldsymbol{\mu}_2 - \boldsymbol{\mu}_1)(\boldsymbol{\mu}_2 - \boldsymbol{\mu}_1)^T \mathbf{w} = (m_2 - m_1)(m_2 - m_1) \tag{8.104}$$

and

$$\begin{aligned}
\mathbf{w}^T \mathbf{S}_W \mathbf{w} &= \sum_{i:y_i=1} \mathbf{w}^T (\mathbf{x}_i - \boldsymbol{\mu}_1)(\mathbf{x}_i - \boldsymbol{\mu}_1)^T \mathbf{w} + \\
&\quad \sum_{i:y_i=2} \mathbf{w}^T (\mathbf{x}_i - \boldsymbol{\mu}_2)(\mathbf{x}_i - \boldsymbol{\mu}_2)^T \mathbf{w} \tag{8.105} \\
&= \sum_{i:y_i=1} (z_i - m_1)^2 + \sum_{i:y_i=2} (z_i - m_2)^2 \tag{8.106}
\end{aligned}$$

Equation 8.101 is a ratio of two scalars; we can take its derivative with respect to $\mathbf{w}$ and equate to zero. One can show (Exercise 12.6) that $J(\mathbf{w})$ is maximized when

$$\mathbf{S}_B \mathbf{w} = \lambda \mathbf{S}_W \mathbf{w} \tag{8.107}$$

where

$$\lambda = \frac{\mathbf{w}^T \mathbf{S}_B \mathbf{w}}{\mathbf{w}^T \mathbf{S}_W \mathbf{w}} \tag{8.108}$$

Equation 8.107 is called a **generalized eigenvalue** problem. If $\mathbf{S}_W$ is invertible, we can convert it to a regular eigenvalue problem:

$$\mathbf{S}_W^{-1} \mathbf{S}_B \mathbf{w} = \lambda \mathbf{w} \tag{8.109}$$

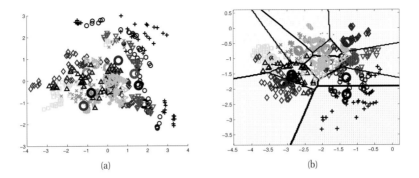

(a)   (b)

Figure 8.12 (a) PCA projection of vowel data to 2d. (b) FLDA projection of vowel data to 2d. We see there is better class separation in the FLDA case. Based on Figure 4.11 of (Hastie et al. 2009). Figure generated by `fisherDiscrimVowelDemo`, by Hannes Bretschneider.

However, in the two class case, there is a simpler solution. In particular, since

$$\mathbf{S}_B \mathbf{w} = (\boldsymbol{\mu}_2 - \boldsymbol{\mu}_1)(\boldsymbol{\mu}_2 - \boldsymbol{\mu}_1)^T \mathbf{w} = (\boldsymbol{\mu}_2 - \boldsymbol{\mu}_1)(m_2 - m_1) \tag{8.110}$$

then, from Equation 8.109 we have

$$\lambda \, \mathbf{w} = \mathbf{S}_W^{-1}(\boldsymbol{\mu}_2 - \boldsymbol{\mu}_1)(m_2 - m_1) \tag{8.111}$$
$$\mathbf{w} \propto \mathbf{S}_W^{-1}(\boldsymbol{\mu}_2 - \boldsymbol{\mu}_1) \tag{8.112}$$

Since we only care about the directionality, and not the scale factor, we can just set

$$\mathbf{w} = \mathbf{S}_W^{-1}(\boldsymbol{\mu}_2 - \boldsymbol{\mu}_1) \tag{8.113}$$

This is the optimal solution in the two-class case. If $\mathbf{S}_W \propto \mathbf{I}$, meaning the pooled covariance matrix is isotropic, then $\mathbf{w}$ is proportional to the vector that joins the class means. This is an intuitively reasonable direction to project onto, as shown in Figure 8.11.

8.6.3.2   Extension to higher dimensions and multiple classes

We can extend the above idea to multiple classes, and to higher dimensional subspaces, by finding a projection *matrix* $\mathbf{W}$ which maps from D to L. Let $\mathbf{z}_i = \mathbf{W}\mathbf{x}_i$ be the low dimensional projection of the i'th data point. Let $\mathbf{m}_c = \frac{1}{N_c} \sum_{i:y_i=c} \mathbf{z}_i$ be the corresponding mean for the c'th class and $\mathbf{m} = \frac{1}{N} \sum_{c=1}^{C} \mathbf{m}_c$ be the overall mean, both in the low dimensional space. We define the following scatter matrices:

$$\tilde{\mathbf{S}}_W = \sum_{c=1}^{C} \sum_{i:y_i=c} (\mathbf{z}_i - \mathbf{m}_c)(\mathbf{z}_i - \mathbf{m}_c)^T \tag{8.114}$$

$$\tilde{\mathbf{S}}_B = \sum_{c=1}^{C} N_c (\mathbf{z}_i - \mathbf{m})(\mathbf{z}_i - \mathbf{m})^T \tag{8.115}$$

Finally, we define the objective function as maximizing the following:[4]

$$J(\mathbf{W}) = \frac{|\tilde{\mathbf{S}}_B|}{|\tilde{\mathbf{S}}_W|} = \frac{|\mathbf{W}^T \mathbf{S}_B \mathbf{W}|}{\mathbf{W}^T \mathbf{S}_W \mathbf{W}|} \tag{8.116}$$

where $\mathbf{S}_W$ and $\mathbf{S}_B$ are defined in the original high dimensional space in the obvious way (namely using $\mathbf{x}_i$ instead of $\mathbf{z}_i$, $\boldsymbol{\mu}_c$ instead of $\mathbf{m}_c$, and $\boldsymbol{\mu}$ instead of $\mathbf{m}$). The solution can be shown (Duda et al. 2001) to be

$$\mathbf{W} = \mathbf{S}_W^{-\frac{1}{2}} \mathbf{U} \tag{8.117}$$

where $\mathbf{U}$ are the L leading eigenvectors of $\mathbf{S}_W^{-\frac{1}{2}} \mathbf{S}_B \mathbf{S}_W^{-\frac{1}{2}}$, assuming $\mathbf{S}_W$ is non-singular. (If it is singular, we can first perform PCA on all the data.)

Figure 8.12 gives an example of this method applied to some $D = 10$ dimensional speech data, representing $C = 11$ different vowel sounds. We see that FLDA gives better class separation than PCA.

Note that FLDA is restricted to finding at most a $L \leq C - 1$ dimensional linear subspace, no matter how large D, because the rank of the between class scatter matrix $\mathbf{S}_B$ is $C - 1$. (The -1 term arises because of the $\boldsymbol{\mu}$ term, which is a linear function of the $\boldsymbol{\mu}_c$.) This is a rather severe restriction which limits the usefulness of FLDA.

As an aside, we mention a method for feature selection based on the FLDA method. We can define the quality of a given subset of features, $\mathbf{Z}$, using $J(\mathbf{Z}) = \text{tr}\left\{\tilde{\mathbf{S}}_B \tilde{\mathbf{S}}_T^{-1}\right\}$, where $\tilde{\mathbf{S}}_T = \tilde{\mathbf{S}}_W + \tilde{\mathbf{S}}_B = \sum_{i=1}^N (\mathbf{z}_i - \mathbf{m})(\mathbf{z}_i - \mathbf{m})^T$ is the total scatter matrix in the reduced space. This is called the **Fisher score** (Duda et al. 2001). Picking the optimal subset takes exponential time, so a standard heuristic is to pick features one at a time. In the 1d case, for dimension j, the Fisher score becomes $J(\mathbf{z}^j) = \frac{\sum_{c=1}^C N_c (\mu_c^j - \mu^j)^2}{(s^j)^2}$, where $(s^j)^2 = \sum_{c=1}^C N_c (s_c^j)^2$ is the overall variance of feature j. This can be extended to select subsets of features as described in (Gu et al. 2011).

8.6.3.3 Probabilistic interpretation of FLDA *

To find a valid probabilistic interpretation of FLDA, we follow the approach of (Kumar and Andreo 1998; Zhou et al. 2009). They proposed a model known as **heteroscedastic LDA** (HLDA), which works as follows. Let $\mathbf{W}$ be a $D \times D$ invertible matrix, and let $\mathbf{z}_i = \mathbf{W}\mathbf{x}_i$ be a transformed version of the data. We now fit full covariance Gaussians to the transformed data, one per class, but with the constraint that only the first L components will be class-specific; the remaining $H = D - L$ components will be shared across classes, and will thus not be discriminative. That is, we use

$$p(\mathbf{z}_i | \boldsymbol{\theta}, y_i = c) = \mathcal{N}(\mathbf{z}_i | \boldsymbol{\mu}_c, \boldsymbol{\Sigma}_c) \tag{8.118}$$

$$\boldsymbol{\mu}_c \triangleq (\mathbf{m}_c; \mathbf{m}_0) \tag{8.119}$$

$$\boldsymbol{\Sigma}_c \triangleq \begin{pmatrix} \mathbf{S}_c & \mathbf{0} \\ \mathbf{0} & \mathbf{S}_0 \end{pmatrix} \tag{8.120}$$

---

4. An alternative criterion that is sometimes used (Fukunaga 1990) is $J(\mathbf{W}) = \text{tr}\left\{\tilde{\mathbf{S}}_W^{-1} \tilde{\mathbf{S}}_B\right\} = \text{tr}\left\{(\mathbf{W}\mathbf{S}_W\mathbf{W}^T)^{-1}(\mathbf{W}\mathbf{S}_B\mathbf{W}^T)\right\}$.

where $\mathbf{m}_0$ is the shared H dimensional mean and $\mathbf{S}_0$ is the shared $H \times H$ covariance. The pdf of the original (untransformed) data is given by

$$
\begin{aligned}
p(\mathbf{x}_i|y_i = c, \mathbf{W}, \boldsymbol{\theta}) &= |\mathbf{W}| \, \mathcal{N}(\mathbf{W}\mathbf{x}_i|\boldsymbol{\mu}_c, \boldsymbol{\Sigma}_c) && (8.121) \\
&= |\mathbf{W}| \, \mathcal{N}(\mathbf{W}_L\mathbf{x}_i|\mathbf{m}_c, \mathbf{S}_c) \, \mathcal{N}(\mathbf{W}_H\mathbf{x}_i|\mathbf{m}_0, \mathbf{S}_0) && (8.122)
\end{aligned}
$$

where $\mathbf{W} = \begin{pmatrix} \mathbf{W}_L \\ \mathbf{W}_H \end{pmatrix}$. For fixed $\mathbf{W}$, it is easy to derive the MLE for $\boldsymbol{\theta}$. One can then optimize $\mathbf{W}$ using gradient methods.

In the special case that the $\boldsymbol{\Sigma}_c$ are diagonal, there is a closed-form solution for $\mathbf{W}$ (Gales 1999). And in the special case the $\boldsymbol{\Sigma}_c$ are all equal, we recover classical LDA (Zhou et al. 2009).

In view of this result, it should be clear that HLDA will outperform LDA if the class covariances are not equal within the discriminative subspace (i.e., if the assumption that $\boldsymbol{\Sigma}_c$ is independent of c is a poor assumption). This is easy to demonstrate on synthetic data, and is also the case on more challenging tasks such as speech recognition (Kumar and Andreo 1998). Furthermore, we can extend the model by allowing each class to use its own projection matrix; this is known as **multiple LDA** (Gales 2002).

Exercises

Exercise 8.1 Spam classification using logistic regression

Consider the email spam data set discussed on p300 of (Hastie et al. 2009). This consists of 4601 email messages, from which 57 features have been extracted. These are as follows:

- 48 features, in $[0, 100]$, giving the percentage of words in a given message which match a given word on the list. The list contains words such as "business", "free", "george", etc. (The data was collected by George Forman, so his name occurs quite a lot.)

- 6 features, in $[0, 100]$, giving the percentage of characters in the email that match a given character on the list. The characters are   ;   ([!   \$ #

- Feature 55: The average length of an uninterrupted sequence of capital letters (max is 40.3, mean is 4.9)

- Feature 56: The length of the longest uninterrupted sequence of capital letters (max is 45.0, mean is 52.6)

- Feature 57: The sum of the lengts of uninterrupted sequence of capital letters (max is 25.6, mean is 282.2)

Load the data from `spamData.mat`, which contains a training set (of size 3065) and a test set (of size 1536).

One can imagine performing several kinds of preprocessing to this data. Try each of the following separately:

a. Standardize the columns so they all have mean 0 and unit variance.

b. Transform the features using $\log(x_{ij} + 0.1)$.

c. Binarize the features using $\mathbb{I}(x_{ij} > 0)$.

For each version of the data, fit a logistic regression model. Use cross validation to choose the strength of the ℓ_2 regularizer. Report the mean error rate on the training and test sets. You should get numbers similar to this:

method	train	test
stnd	0.082	0.079
log	0.052	0.059
binary	0.065	0.072

(The precise values will depend on what regularization value you choose.) Turn in your code and numerical results.

(See also Exercise 8.2.

Exercise 8.2 Spam classification using naive Bayes

We will re-examine the dataset from Exercise 8.1.

a. Use `naiveBayesFit` and `naiveBayesPredict` on the binarized spam data. What is the training and test error? (You can try different settings of the pseudocount α if you like (this corresponds to the Beta(α, α) prior each θ_{jc}), although the default of $\alpha = 1$ is probably fine.) Turn in your error rates.

b. Modify the code so it can handle real-valued features. Use a Gaussian density for each feature; fit it with maximum likelihood. What are the training and test error rates on the standardized data and the log transformed data? Turn in your 4 error rates and code.

Exercise 8.3 Gradient and Hessian of log-likelihood for logistic regression

a. Let $\sigma(a) = \frac{1}{1+e^{-a}}$ be the sigmoid function. Show that

$$\frac{d\sigma(a)}{da} = \sigma(a)(1 - \sigma(a)) \tag{8.123}$$

b. Using the previous result and the chain rule of calculus, derive an expression for the gradient of the log likelihood (Equation 8.5).

c. The Hessian can be written as $\mathbf{H} = \mathbf{X}^T \mathbf{S} \mathbf{X}$, where $\mathbf{S} \triangleq \mathrm{diag}(\mu_1(1 - \mu_1), \ldots, \mu_n(1 - \mu_n))$. Show that $\mathbf{H}$ is positive definite. (You may assume that $0 < \mu_i < 1$, so the elements of $\mathbf{S}$ will be strictly positive, and that $\mathbf{X}$ is full rank.)

Exercise 8.4 Gradient and Hessian of log-likelihood for multinomial logistic regression

a. Let $\mu_{ik} = \mathcal{S}(\boldsymbol{\eta}_i)_k$. Prove that the Jacobian of the softmax is

$$\frac{\partial \mu_{ik}}{\partial \eta_{ij}} = \mu_{ik}(\delta_{kj} - \mu_{ij}) \tag{8.124}$$

where $\delta_{kj} = I(k = j)$.

b. Hence show that

$$\nabla_{\mathbf{w}_c} \ell \;\;=\;\; \sum_i (y_{ic} - \mu_{ic})\mathbf{x}_i \tag{8.125}$$

Hint: use the chain rule and the fact that $\sum_c y_{ic} = 1$.

c. Show that the block submatrix of the Hessian for classes c and c' is given by

$$\mathbf{H}_{c,c'} \;\;=\;\; -\sum_i \mu_{ic}(\delta_{c,c'} - \mu_{i,c'})\mathbf{x}_i \mathbf{x}_i^T \tag{8.126}$$

Exercise 8.5 Symmetric version of ℓ_2 regularized multinomial logistic regression

(Source: Ex 18.3 of (Hastie et al. 2009).)

Multiclass logistic regression has the form

$$p(y = c | \mathbf{x}, \mathbf{W}) = \frac{\exp(w_{c0} + \mathbf{w}_c^T \mathbf{x})}{\sum_{k=1}^{C} \exp(w_{k0} + \mathbf{w}_k^T \mathbf{x})} \tag{8.127}$$

where $\mathbf{W}$ is a $(D+1) \times C$ weight matrix. We can arbitrarily define $\mathbf{w}_c = \mathbf{0}$ for one of the classes, say $c = C$, since $p(y = C | \mathbf{x}, \mathbf{W}) = 1 - \sum_{c=1}^{C-1} p(y = c | \mathbf{x}, \mathbf{w})$. In this case, the model has the form

$$p(y = c | \mathbf{x}, \mathbf{W}) = \frac{\exp(w_{c0} + \mathbf{w}_c^T \mathbf{x})}{1 + \sum_{k=1}^{C-1} \exp(w_{k0} + \mathbf{w}_k^T \mathbf{x})} \tag{8.128}$$

If we don't "clamp" one of the vectors to some constant value, the parameters will be unidentifiable. However, suppose we don't clamp $\mathbf{w}_c = \mathbf{0}$, so we are using Equation 8.127, but we add ℓ_2 regularization by optimizing

$$\sum_{i=1}^{N} \log p(y_i | \mathbf{x}_i, \mathbf{W}) - \lambda \sum_{c=1}^{C} ||\mathbf{w}_c||_2^2 \tag{8.129}$$

Show that at the optimum we have $\sum_{c=1}^{C} \hat{w}_{cj} = 0$ for $j = 1 : D$. (For the unregularized $\hat{w}_{c0}$ terms, we still need to enforce that $w_{0C} = 0$ to ensure identifiability of the offset.)

Exercise 8.6 Elementary properties of ℓ_2 regularized logistic regression

(Source: Jaakkola.). Consider minimizing

$$J(\mathbf{w}) = -\ell(\mathbf{w}, \mathcal{D}_{\text{train}}) + \lambda ||\mathbf{w}||_2^2 \tag{8.130}$$

where

$$\ell(\mathbf{w}, \mathcal{D}) = \frac{1}{|\mathcal{D}|} \sum_{i \in \mathcal{D}} \log \sigma(y_i \mathbf{x}_i^T \mathbf{w}) \tag{8.131}$$

is the average log-likelihood on data set $\mathcal{D}$, for $y_i \in \{-1, +1\}$. Answer the following true/ false questions.

a. $J(\mathbf{w})$ has multiple locally optimal solutions: T/F?

b. Let $\hat{\mathbf{w}} = \arg\min_{\mathbf{w}} J(\mathbf{w})$ be a global optimum. $\hat{\mathbf{w}}$ is sparse (has many zero entries): T/F?

c. If the training data is linearly separable, then some weights w_j might become infinite if $\lambda = 0$: T/F?

d. $\ell(\hat{\mathbf{w}}, \mathcal{D}_{\text{train}})$ always increases as we increase λ: T/F?

e. $\ell(\hat{\mathbf{w}}, \mathcal{D}_{\text{test}})$ always increases as we increase λ: T/F?

Exercise 8.7 Regularizing separate terms in 2d logistic regression

(Source: Jaakkola.)

a. Consider the data in Figure 8.13, where we fit the model $p(y = 1 | \mathbf{x}, \mathbf{w}) = \sigma(w_0 + w_1 x_1 + w_2 x_2)$. Suppose we fit the model by maximum likelihood, i.e., we minimize

$$J(\mathbf{w}) = -\ell(\mathbf{w}, \mathcal{D}_{\text{train}}) \tag{8.132}$$

where $\ell(\mathbf{w}, \mathcal{D}_{\text{train}})$ is the log likelihood on the training set. Sketch a possible decision boundary corresponding to $\hat{\mathbf{w}}$. (Copy the figure first (a rough sketch is enough), and then superimpose your answer on your copy, since you will need multiple versions of this figure). Is your answer (decision boundary) unique? How many classification errors does your method make on the training set?

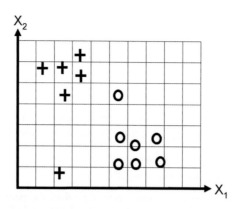

Figure 8.13 Data for logistic regression question.

b. Now suppose we regularize only the w_0 parameter, i.e., we minimize

$$J_0(\mathbf{w}) = -\ell(\mathbf{w}, \mathcal{D}_{\text{train}}) + \lambda w_0^2 \tag{8.133}$$

Suppose λ is a very large number, so we regularize w_0 all the way to 0, but all other parameters are unregularized. Sketch a possible decision boundary. How many classification errors does your method make on the training set? Hint: consider the behavior of simple linear regression, $w_0 + w_1 x_1 + w_2 x_2$ when $x_1 = x_2 = 0$.

c. Now suppose we heavily regularize only the w_1 parameter, i.e., we minimize

$$J_1(\mathbf{w}) = -\ell(\mathbf{w}, \mathcal{D}_{\text{train}}) + \lambda w_1^2 \tag{8.134}$$

Sketch a possible decision boundary. How many classification errors does your method make on the training set?

d. Now suppose we heavily regularize only the w_2 parameter. Sketch a possible decision boundary. How many classification errors does your method make on the training set?

9 Generalized linear models and the exponential family

9.1 Introduction

We have now encountered a wide variety of probability distributions: the Gaussian, the Bernoulli, the Student t, the uniform, the gamma, etc. It turns out that most of these are members of a broader class of distributions known as the **exponential family**.[1] In this chapter, we discuss various properties of this family. This allows us to derive theorems and algorithms with very broad applicability.

We will see how we can easily use any member of the exponential family as a class-conditional density in order to make a generative classifier. In addition, we will discuss how to build discriminative models, where the response variable has an exponential family distribution, whose mean is a linear function of the inputs; this is known as a generalized linear model, and generalizes the idea of logistic regression to other kinds of response variables.

9.2 The exponential family

Before defining the exponential family, we mention several reasons why it is important:

- It can be shown that, under certain regularity conditions, the exponential family is the only family of distributions with finite-sized sufficient statistics, meaning that we can compress the data into a fixed-sized summary without loss of information. This is particularly useful for online learning, as we will see later.
- The exponential family is the only family of distributions for which conjugate priors exist, which simplifies the computation of the posterior (see Section 9.2.5).
- The exponential family can be shown to be the family of distributions that makes the least set of assumptions subject to some user-chosen constraints (see Section 9.2.6).
- The exponential family is at the core of generalized linear models, as discussed in Section 9.3.
- The exponential family is at the core of variational inference, as discussed in Section 21.2.

---

1. The exceptions are the Student t, which does not have the right form, and the uniform distribution, which does not have fixed support independent of the parameter values.

9.2.1   Definition

A pdf or pmf $p(\mathbf{x}|\boldsymbol{\theta})$, for $\mathbf{x} = (x_1, \ldots, x_m) \in \mathcal{X}^m$ and $\boldsymbol{\theta} \in \Theta \subseteq \mathbb{R}^d$, is said to be in the **exponential family** if it is of the form

$$p(\mathbf{x}|\boldsymbol{\theta}) = \frac{1}{Z(\boldsymbol{\theta})} h(\mathbf{x}) \exp[\boldsymbol{\theta}^T \phi(\mathbf{x})] \tag{9.1}$$

$$= h(\mathbf{x}) \exp[\boldsymbol{\theta}^T \phi(\mathbf{x}) - A(\boldsymbol{\theta})] \tag{9.2}$$

where

$$Z(\boldsymbol{\theta}) = \int_{\mathcal{X}^m} h(\mathbf{x}) \exp[\boldsymbol{\theta}^T \phi(\mathbf{x})] d\mathbf{x} \tag{9.3}$$

$$A(\boldsymbol{\theta}) = \log Z(\boldsymbol{\theta}) \tag{9.4}$$

Here $\boldsymbol{\theta}$ are called the **natural parameters** or **canonical parameters**, $\phi(\mathbf{x}) \in \mathbb{R}^d$ is called a vector of **sufficient statistics**, $Z(\boldsymbol{\theta})$ is called the **partition function**, $A(\boldsymbol{\theta})$ is called the **log partition function** or **cumulant function**, and $h(\mathbf{x})$ is a scaling constant, often 1. If $\phi(\mathbf{x}) = \mathbf{x}$, we say it is a **natural exponential family**.

Equation 9.2 can be generalized by writing

$$p(\mathbf{x}|\boldsymbol{\theta}) = h(\mathbf{x}) \exp[\eta(\boldsymbol{\theta})^T \phi(\mathbf{x}) - A(\eta(\boldsymbol{\theta}))] \tag{9.5}$$

where η is a function that maps the parameters $\boldsymbol{\theta}$ to the canonical parameters $\boldsymbol{\eta} = \eta(\boldsymbol{\theta})$. If $\dim(\boldsymbol{\theta}) < \dim(\eta(\boldsymbol{\theta}))$, it is called a **curved exponential family**, which means we have more sufficient statistics than parameters. If $\eta(\boldsymbol{\theta}) = \boldsymbol{\theta}$, the model is said to be in **canonical form**. We will assume models are in canonical form unless we state otherwise.

9.2.2   Examples

Let us consider some examples to make things clearer.

9.2.2.1   Bernoulli

The Bernoulli for $x \in \{0, 1\}$ can be written in exponential family form as follows:

$$\text{Ber}(x|\mu) = \mu^x (1-\mu)^{1-x} = \exp[x \log(\mu) + (1-x) \log(1-\mu)] = \exp[\phi(x)^T \boldsymbol{\theta}] \tag{9.6}$$

where $\phi(x) = [\mathbb{I}(x = 0), \mathbb{I}(x = 1)]$ and $\boldsymbol{\theta} = [\log(\mu), \log(1-\mu)]$. However, this representation is **over-complete** since there is a linear dependence between the features:

$$\mathbf{1}^T \phi(x) = \mathbb{I}(x = 0) + \mathbb{I}(x = 1) = 1 \tag{9.7}$$

Consequently $\boldsymbol{\theta}$ is not uniquely identifiable. It is common to require that the representation be **minimal**, which means there is a unique $\boldsymbol{\theta}$ associated with the distribution. In this case, we can just define

$$\text{Ber}(x|\mu) = (1-\mu) \exp\left[x \log\left(\frac{\mu}{1-\mu}\right)\right] \tag{9.8}$$

Now we have $\phi(x) = x$, $\theta = \log\left(\frac{\mu}{1-\mu}\right)$, which is the log-odds ratio, and $Z = 1/(1 - \mu)$. We can recover the mean parameter μ from the canonical parameter using

$$\mu = \text{sigm}(\theta) = \frac{1}{1 + e^{-\theta}} \tag{9.9}$$

9.2.2.2 Multinoulli

We can represent the multinoulli as a minimal exponential family as follows (where $x_k = \mathbb{I}(x = k)$):

$$\text{Cat}(x|\boldsymbol{\mu}) = \prod_{k=1}^{K} \mu_k^{x_k} = \exp\left[\sum_{k=1}^{K} x_k \log \mu_k\right] \tag{9.10}$$

$$= \exp\left[\sum_{k=1}^{K-1} x_k \log \mu_k + \left(1 - \sum_{k=1}^{K-1} x_k\right)\log\left(1 - \sum_{k=1}^{K-1} \mu_k\right)\right] \tag{9.11}$$

$$= \exp\left[\sum_{k=1}^{K-1} x_k \log\left(\frac{\mu_k}{1 - \sum_{j=1}^{K-1} \mu_j}\right) + \log\left(1 - \sum_{k=1}^{K-1} \mu_k\right)\right] \tag{9.12}$$

$$= \exp\left[\sum_{k=1}^{K-1} x_k \log\left(\frac{\mu_k}{\mu_K}\right) + \log \mu_K\right] \tag{9.13}$$

where $\mu_K = 1 - \sum_{k=1}^{K-1} \mu_k$. We can write this in exponential family form as follows:

$$\text{Cat}(x|\boldsymbol{\theta}) = \exp(\boldsymbol{\theta}^T \phi(\mathbf{x}) - A(\boldsymbol{\theta})) \tag{9.14}$$

$$\boldsymbol{\theta} = [\log \frac{\mu_1}{\mu_K}, \dots, \log \frac{\mu_{K-1}}{\mu_K}] \tag{9.15}$$

$$\phi(x) = [\mathbb{I}(x = 1), \dots, \mathbb{I}(x = K - 1)] \tag{9.16}$$

We can recover the mean parameters from the canonical parameters using

$$\mu_k = \frac{e^{\theta_k}}{1 + \sum_{j=1}^{K-1} e^{\theta_j}} \tag{9.17}$$

From this, we find

$$\mu_K = 1 - \frac{\sum_{j=1}^{K-1} e^{\theta_j}}{1 + \sum_{j=1}^{K-1} e^{\theta_j}} = \frac{1}{1 + \sum_{j=1}^{K-1} e^{\theta_j}} \tag{9.18}$$

and hence

$$A(\boldsymbol{\theta}) = \log\left(1 + \sum_{k=1}^{K-1} e^{\theta_k}\right) \tag{9.19}$$

If we define $\theta_K = 0$, we can write $\boldsymbol{\mu} = \mathcal{S}(\boldsymbol{\theta})$ and $A(\boldsymbol{\theta}) = \log \sum_{k=1}^{K} e^{\theta_k}$, where $\mathcal{S}$ is the softmax function in Equation 4.39.

9.2.2.3   Univariate Gaussian

The univariate Gaussian can be written in exponential family form as follows:

$$\mathcal{N}(x|\mu, \sigma^2) \;=\; \frac{1}{(2\pi\sigma^2)^{\frac{1}{2}}} \exp[-\frac{1}{2\sigma^2}(x-\mu)^2] \tag{9.20}$$

$$=\; \frac{1}{(2\pi\sigma^2)^{\frac{1}{2}}} \exp[-\frac{1}{2\sigma^2}x^2 + \frac{\mu}{\sigma^2}x - \frac{1}{2\sigma^2}\mu^2] \tag{9.21}$$

$$=\; \frac{1}{Z(\boldsymbol{\theta})} \exp(\boldsymbol{\theta}^T \boldsymbol{\phi}(x)) \tag{9.22}$$

where

$$\boldsymbol{\theta} \;=\; \begin{pmatrix} \mu/\sigma^2 \\ \frac{-1}{2\sigma^2} \end{pmatrix} \tag{9.23}$$

$$\boldsymbol{\phi}(x) \;=\; \begin{pmatrix} x \\ x^2 \end{pmatrix} \tag{9.24}$$

$$Z(\mu, \sigma^2) \;=\; \sqrt{2\pi}\sigma \exp[\frac{\mu^2}{2\sigma^2}] \tag{9.25}$$

$$A(\boldsymbol{\theta}) \;=\; \frac{-\theta_1^2}{4\theta_2} - \frac{1}{2}\log(-2\theta_2) - \frac{1}{2}\log(2\pi) \tag{9.26}$$

9.2.2.4   Non-examples

Not all distributions of interest belong to the exponential family. For example, the uniform distribution, $X \sim \text{Unif}(a, b)$, does not, since the support of the distribution depends on the parameters. Also, the Student T distribution (Section 11.4.5) does not belong, since it does not have the required form.

9.2.3   Log partition function

An important property of the exponential family is that derivatives of the log partition function can be used to generate **cumulants** of the sufficient statistics.[2] For this reason, $A(\boldsymbol{\theta})$ is sometimes called a **cumulant function**. We will prove this for a 1-parameter distribution; this can be generalized to a K-parameter distribution in a straightforward way. For the first

---

2. The first and second cumulants of a distribution are its mean $\mathbb{E}[X]$ and variance $\text{var}[X]$, whereas the first and second moments are its mean $\mathbb{E}[X]$ and $\mathbb{E}[X^2]$.

derivative we have

$$\frac{dA}{d\theta} = \frac{d}{d\theta} \left(\log \int \exp(\theta\phi(x))h(x)dx \right) \tag{9.27}$$

$$= \frac{\frac{d}{d\theta} \int \exp(\theta\phi(x))h(x)dx}{\int \exp(\theta\phi(x))h(x)dx} \tag{9.28}$$

$$= \frac{\int \phi(x)\exp(\theta\phi(x))h(x)dx}{\exp(A(\theta))} \tag{9.29}$$

$$= \int \phi(x)\exp(\theta\phi(x) - A(\theta))h(x)dx \tag{9.30}$$

$$= \int \phi(x)p(x)dx = \mathbb{E}\left[\phi(x)\right] \tag{9.31}$$

For the second derivative we have

$$\frac{d^2 A}{d\theta^2} = \int \phi(x)\exp\left(\theta\phi(x) - A(\theta)\right)h(x)(\phi(x) - A'(\theta))dx \tag{9.32}$$

$$= \int \phi(x)p(x)(\phi(x) - A'(\theta))dx \tag{9.33}$$

$$= \int \phi^2(x)p(x)dx - A'(\theta)\int \phi(x)p(x)dx \tag{9.34}$$

$$= \mathbb{E}\left[\phi^2(X)\right] - \mathbb{E}\left[\phi(x)\right]^2 = \text{var}\left[\phi(x)\right] \tag{9.35}$$

where we used the fact that $A'(\theta) = \frac{dA}{d\theta} = \mathbb{E}\left[\phi(x)\right]$.

In the multivariate case, we have that

$$\frac{\partial^2 A}{\partial\theta_i\partial\theta_j} = \mathbb{E}\left[\phi_i(x)\phi_j(x)\right] - \mathbb{E}\left[\phi_i(x)\right]\mathbb{E}\left[\phi_j(x)\right] \tag{9.36}$$

and hence

$$\nabla^2 A(\boldsymbol{\theta}) = \text{cov}\left[\boldsymbol{\phi}(\mathbf{x})\right] \tag{9.37}$$

Since the covariance is positive definite, we see that $A(\boldsymbol{\theta})$ is a convex function (see Section 7.3.3).

9.2.3.1 Example: the Bernoulli distribution

For example, consider the Bernoulli distribution. We have $A(\theta) = \log(1 + e^\theta)$, so the mean is given by

$$\frac{dA}{d\theta} = \frac{e^\theta}{1 + e^\theta} = \frac{1}{1 + e^{-\theta}} = \text{sigm}(\theta) = \mu \tag{9.38}$$

The variance is given by

$$\frac{d^2 A}{d\theta^2} = \frac{d}{d\theta}(1 + e^{-\theta})^{-1} = (1 + e^{-\theta})^{-2}.e^{-\theta} \tag{9.39}$$

$$= \frac{e^{-\theta}}{1 + e^{-\theta}}\frac{1}{1 + e^{-\theta}} = \frac{1}{e^\theta + 1}\frac{1}{1 + e^{-\theta}} = (1 - \mu)\mu \tag{9.40}$$

9.2.4   MLE for the exponential family

The likelihood of an exponential family model has the form

$$p(\mathcal{D}|\boldsymbol{\theta}) = \left[\prod_{i=1}^{N} h(\mathbf{x}_i)\right] g(\boldsymbol{\theta})^N \exp\left(\boldsymbol{\eta}(\boldsymbol{\theta})^T [\sum_{i=1}^{N} \boldsymbol{\phi}(\mathbf{x}_i)]\right) \tag{9.41}$$

We see that the sufficient statistics are N and

$$\boldsymbol{\phi}(\mathcal{D}) = [\sum_{i=1}^{N} \phi_1(\mathbf{x}_i), \ldots, \sum_{i=1}^{N} \phi_K(\mathbf{x}_i)] \tag{9.42}$$

For example, for the Bernoulli model we have $\phi = [\sum_i \mathbb{I}(x_i = 1)]$, and for the univariate Gaussian, we have $\phi = [\sum_i x_i, \sum_i x_i^2]$. (We also need to know the sample size, N.)

The **Pitman-Koopman-Darmois theorem** states that, under certain regularity conditions, the exponential family is the only family of distributions with finite sufficient statistics. (Here, finite means of a size independent of the size of the data set.)

One of the conditions required in this theorem is that the support of the distribution not be dependent on the parameter. For a simple example of such a distribution, consider the uniform distribution

$$p(x|\theta) = U(x|\theta) = \frac{1}{\theta}\mathbb{I}(0 \le x \le \theta) \tag{9.43}$$

The likelihood is given by

$$p(\mathcal{D}|\boldsymbol{\theta}) = \theta^{-N}\mathbb{I}(0 \le \max\{x_i\} \le \theta) \tag{9.44}$$

So the sufficient statistics are N and $s(\mathcal{D}) = \max_i x_i$. This is finite in size, but the uniform distribution is not in the exponential family because its support set, $\mathcal{X}$, depends on the parameters.

We now descibe how to compute the MLE for a canonical exponential family model. Given N iid data points $\mathcal{D} = (x_1, \ldots, x_N)$, the log-likelihood is

$$\log p(\mathcal{D}|\boldsymbol{\theta}) = \boldsymbol{\theta}^T \boldsymbol{\phi}(\mathcal{D}) - NA(\boldsymbol{\theta}) \tag{9.45}$$

Since $-A(\boldsymbol{\theta})$ is concave in $\boldsymbol{\theta}$, and $\boldsymbol{\theta}^T\boldsymbol{\phi}(\mathcal{D})$ is linear in $\boldsymbol{\theta}$, we see that the log likelihood is concave, and hence has a unique global maximum. To derive this maximum, we use the fact that the derivative of the log partition function yields the expected value of the sufficient statistic vector (Section 9.2.3):

$$\nabla_{\boldsymbol{\theta}} \log p(\mathcal{D}|\boldsymbol{\theta}) = \boldsymbol{\phi}(\mathcal{D}) - N\mathbb{E}[\boldsymbol{\phi}(\mathbf{X})] \tag{9.46}$$

Setting this gradient to zero, we see that at the MLE, the empirical average of the sufficient statistics must equal the model's theoretical expected sufficient statistics, i.e., $\hat{\boldsymbol{\theta}}$ must satisfy

$$\mathbb{E}[\boldsymbol{\phi}(\mathbf{X})] = \frac{1}{N}\sum_{i=1}^{N} \boldsymbol{\phi}(\mathbf{x}_i) \tag{9.47}$$

This is called **moment matching**. For example, in the Bernoulli distribution, we have $\phi(X) = \mathbb{I}(X = 1)$, so the MLE satisfies

$$\mathbb{E}\left[\phi(X)\right] = p(X = 1) = \hat{\mu} = \frac{1}{N}\sum_{i=1}^{N}\mathbb{I}(x_i = 1) \tag{9.48}$$

9.2.5 Bayes for the exponential family *

We have seen that exact Bayesian analysis is considerably simplified if the prior is conjugate to the likelihood. Informally this means that the prior $p(\boldsymbol{\theta}|\boldsymbol{\tau})$ has the same form as the likelihood $p(\mathcal{D}|\boldsymbol{\theta})$. For this to make sense, we require that the likelihood have finite sufficient statistics, so that we can write $p(\mathcal{D}|\boldsymbol{\theta}) = p(\mathbf{s}(\mathcal{D})|\boldsymbol{\theta})$. This suggests that the only family of distributions for which conjugate priors exist is the exponential family. We will derive the form of the prior and posterior below.

9.2.5.1 Likelihood

The likelihood of the exponential family is given by

$$p(\mathcal{D}|\boldsymbol{\theta}) \propto g(\boldsymbol{\theta})^N \exp\left(\boldsymbol{\eta}(\boldsymbol{\theta})^T \mathbf{s}_N\right) \tag{9.49}$$

where $\mathbf{s}_N = \sum_{i=1}^{N}\mathbf{s}(\mathbf{x}_i)$. In terms of the canonical parameters this becomes

$$p(\mathcal{D}|\boldsymbol{\eta}) \propto \exp(N\boldsymbol{\eta}^T\bar{\mathbf{s}} - NA(\boldsymbol{\eta})) \tag{9.50}$$

where $\bar{\mathbf{s}} = \frac{1}{N}\mathbf{s}_N$.

9.2.5.2 Prior

The natural conjugate prior has the form

$$p(\boldsymbol{\theta}|\nu_0, \boldsymbol{\tau}_0) \propto g(\boldsymbol{\theta})^{\nu_0} \exp\left(\boldsymbol{\eta}(\boldsymbol{\theta})^T \boldsymbol{\tau}_0\right) \tag{9.51}$$

Let us write $\boldsymbol{\tau}_0 = \nu_0\bar{\boldsymbol{\tau}}_0$, to separate out the size of the prior pseudo-data, ν_0, from the mean of the sufficient statistics on this pseudo-data, $\bar{\boldsymbol{\tau}}_0$. In canonical form, the prior becomes

$$p(\boldsymbol{\eta}|\nu_0, \bar{\boldsymbol{\tau}}_0) \propto \exp(\nu_0\boldsymbol{\eta}^T\bar{\boldsymbol{\tau}}_0 - \nu_0 A(\boldsymbol{\eta})) \tag{9.52}$$

9.2.5.3 Posterior

The posterior is given by

$$p(\boldsymbol{\theta}|\mathcal{D}) = p(\boldsymbol{\theta}|\nu_N, \boldsymbol{\tau}_N) = p(\boldsymbol{\theta}|\nu_0 + N, \boldsymbol{\tau}_0 + \mathbf{s}_N) \tag{9.53}$$

So we see that we just update the hyper-parameters by adding. In canonical form, this becomes

$$p(\boldsymbol{\eta}|\mathcal{D}) \propto \exp\left(\boldsymbol{\eta}^T(\nu_0\bar{\boldsymbol{\tau}}_0 + N\bar{\mathbf{s}}) - (\nu_0 + N)A(\boldsymbol{\eta})\right) \tag{9.54}$$

$$= p(\boldsymbol{\eta}|\nu_0 + N, \frac{\nu_0\bar{\boldsymbol{\tau}}_0 + N\bar{\mathbf{s}}}{\nu_0 + N}) \tag{9.55}$$

So we see that the posterior hyper-parameters are a convex combination of the prior mean hyper-parameters and the average of the sufficient statistics.

9.2.5.4   Posterior predictive density

Let us derive a generic expression for the predictive density for future observables $\mathcal{D}' = (\tilde{\mathbf{x}}_1, \ldots, \tilde{\mathbf{x}}_{N'})$ given past data $\mathcal{D} = (\mathbf{x}_1, \ldots, \mathbf{x}_N)$ as follows. For notational brevity, we will combine the sufficient statistics with the size of the data, as follows: $\tilde{\boldsymbol{\tau}}_0 = (\nu_0, \boldsymbol{\tau}_0)$, $\tilde{\mathbf{s}}(\mathcal{D}) = (N, \mathbf{s}(\mathcal{D}))$, and $\tilde{\mathbf{s}}(\mathcal{D}') = (N', \mathbf{s}(\mathcal{D}'))$. So the prior becomes

$$p(\boldsymbol{\theta}|\tilde{\boldsymbol{\tau}}_0) = \frac{1}{Z(\tilde{\boldsymbol{\tau}}_0)} g(\boldsymbol{\theta})^{\nu_0} \exp(\boldsymbol{\eta}(\boldsymbol{\theta})^T \boldsymbol{\tau}_0) \tag{9.56}$$

The likelihood and posterior have a similar form. Hence

$$
\begin{aligned}
p(\mathcal{D}'|\mathcal{D}) &= \int p(\mathcal{D}'|\boldsymbol{\theta}) p(\boldsymbol{\theta}|\mathcal{D}) d\boldsymbol{\theta} && (9.57) \\[2mm]
&= \left[\prod_{i=1}^{N'} h(\tilde{\mathbf{x}}_i) \right] Z(\tilde{\boldsymbol{\tau}}_0 + \tilde{\mathbf{s}}(\mathcal{D}))^{-1} \int g(\boldsymbol{\theta})^{\nu_0 + N + N'} d\boldsymbol{\theta} && (9.58) \\[2mm]
&\quad \times \exp\left(\sum_k \eta_k(\boldsymbol{\theta})(\tau_k + \sum_{i=1}^{N} s_k(\mathbf{x}_i) + \sum_{i=1}^{N'} s_k(\tilde{\mathbf{x}}_i) \right) d\boldsymbol{\theta} && (9.59) \\[2mm]
&= \left[\prod_{i=1}^{N'} h(\tilde{\mathbf{x}}_i) \right] \frac{Z(\tilde{\boldsymbol{\tau}}_0 + \tilde{\mathbf{s}}(\mathcal{D}) + \tilde{\mathbf{s}}(\mathcal{D}'))}{Z(\tilde{\boldsymbol{\tau}}_0 + \tilde{\mathbf{s}}(\mathcal{D}))} && (9.60)
\end{aligned}
$$

If $N = 0$, this becomes the marginal likelihood of $\mathcal{D}'$, which reduces to the familiar form of normalizer of the posterior divided by the normalizer of the prior, multiplied by a constant.

9.2.5.5   Example: Bernoulli distribution

As a simple example, let us revisit the Beta-Bernoulli model in our new notation.

The likelihood is given by

$$p(\mathcal{D}|\theta) = (1-\theta)^N \exp\left(\log(\frac{\theta}{1-\theta}) \sum_i x_i \right) \tag{9.61}$$

Hence the conjugate prior is given by

$$
\begin{aligned}
p(\theta|\nu_0, \tau_0) &\propto (1-\theta)^{\nu_0} \exp\left(\log(\frac{\theta}{1-\theta}) \tau_0 \right) && (9.62) \\[2mm]
&= \theta^{\tau_0} (1-\theta)^{\nu_0 - \tau_0} && (9.63)
\end{aligned}
$$

If we define $\alpha = \tau_0 + 1$ and $\beta = \nu_0 - \tau_0 + 1$, we see that this is a beta distribution.

We can derive the posterior as follows, where $s = \sum_i \mathbb{I}(x_i = 1)$ is the sufficient statistic:

$$
\begin{aligned}
p(\theta|\mathcal{D}) &\propto \theta^{\tau_0 + s} (1-\theta)^{\nu_0 - \tau_0 + n - s} && (9.64) \\[2mm]
&= \theta^{\tau_n} (1-\theta)^{\nu_n - \tau_n} && (9.65)
\end{aligned}
$$

We can derive the posterior predictive distribution as follows. Assume $p(\theta) = \text{Beta}(\theta|\alpha, \beta)$, and let $s = s(\mathcal{D})$ be the number of heads in the past data. We can predict the probability of a given sequence of future heads, $\mathcal{D}' = (\tilde{x}_1, \ldots, \tilde{x}_m)$, with sufficient statistic $s' = \sum_{i=1}^{m} \mathbb{I}(\tilde{x}_i = 1)$, as follows:

$$p(\mathcal{D}'|\mathcal{D}) = \int_0^1 p(\mathcal{D}'|\theta) \text{Beta}(\theta|\alpha_n, \beta_n) d\theta \tag{9.66}$$

$$= \frac{\Gamma(\alpha_n + \beta_n)}{\Gamma(\alpha_n)\Gamma(\beta_n)} \int_0^1 \theta^{\alpha_n + t' - 1}(1 - \theta)^{\beta_n + m - t' - 1} d\theta \tag{9.67}$$

$$= \frac{\Gamma(\alpha_n + \beta_n)}{\Gamma(\alpha_n)\Gamma(\beta_n)} \frac{\Gamma(\alpha_{n+m})\Gamma(\beta_{n+m})}{\Gamma(\alpha_{n+m} + \beta_{n+m})} \tag{9.68}$$

where

$$\alpha_{n+m} = \alpha_n + s' = \alpha + s + s' \tag{9.69}$$

$$\beta_{n+m} = \beta_n + (m - s') = \beta + (n - s) + (m - s') \tag{9.70}$$

9.2.6 Maximum entropy derivation of the exponential family *

Although the exponential family is convenient, is there any deeper justification for its use? It turns out that there is: it is the distribution that makes the least number of assumptions about the data, subject to a specific set of user-specified constraints, as we explain below. In particular, suppose all we know is the expected values of certain features or functions:

$$\sum_{\mathbf{x}} f_k(\mathbf{x}) p(\mathbf{x}) = F_k \tag{9.71}$$

where F_k are known constants, and $f_k(\mathbf{x})$ is an arbitrary function. The principle of **maximum entropy** or **maxent** says we should pick the distribution with maximum entropy (closest to uniform), subject to the constraints that the moments of the distribution match the empirical moments of the specified functions.

To maximize entropy subject to the constraints in Equation 9.71, and the constraints that $p(\mathbf{x}) \geq 0$ and $\sum_{\mathbf{x}} p(\mathbf{x}) = 1$, we need to use Lagrange multipliers. The Lagrangian is given by

$$J(p, \boldsymbol{\lambda}) = -\sum_{\mathbf{x}} p(\mathbf{x}) \log p(\mathbf{x}) + \lambda_0 (1 - \sum_{\mathbf{x}} p(\mathbf{x})) + \sum_k \lambda_k (F_k - \sum_{\mathbf{x}} p(\mathbf{x}) f_k(\mathbf{x})) \tag{9.72}$$

We can use the calculus of variations to take derivatives wrt the function p, but we will adopt a simpler approach and treat p as a fixed length vector (since we are assuming $\mathbf{x}$ is discrete). Then we have

$$\frac{\partial J}{\partial p(\mathbf{x})} = -1 - \log p(\mathbf{x}) - \lambda_0 - \sum_k \lambda_k f_k(\mathbf{x}) \tag{9.73}$$

Setting $\frac{\partial J}{\partial p(\mathbf{x})} = 0$ yields

$$p(\mathbf{x}) = \frac{1}{Z} \exp(-\sum_k \lambda_k f_k(\mathbf{x})) \tag{9.74}$$

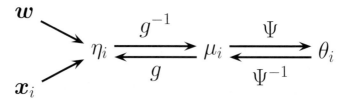

Figure 9.1   A visualization of the various features of a GLM. Based on Figure 8.3 of (Jordan 2007).

where $Z = e^{1+\lambda_0}$. Using the sum to one constraint, we have

$$1 \quad = \quad \sum_{\mathbf{x}} p(\mathbf{x}) = \frac{1}{Z} \sum_{\mathbf{x}} \exp(-\sum_k \lambda_k f_k(\mathbf{x})) \tag{9.75}$$

Hence the normalization constant is given by

$$Z \quad = \quad \sum_{\mathbf{x}} \exp(-\sum_k \lambda_k f_k(\mathbf{x})) \tag{9.76}$$

Thus the maxent distribution $p(\mathbf{x})$ has the form of the exponential family (Section 9.2), also known as the **Gibbs distribution**.

9.3   Generalized linear models (GLMs)

Linear and logistic regression are examples of **generalized linear model**s, or **GLM**s (McCullagh and Nelder 1989). These are models in which the output density is in the exponential family (Section 9.2), and in which the mean parameters are a linear combination of the inputs, passed through a possibly nonlinear function, such as the logistic function. We describe GLMs in more detail below. We focus on scalar outputs for notational simplicity. (This excludes multinomial logistic regression, but this is just to simplify the presentation.)

9.3.1   Basics

To understand GLMs, let us first consider the case of an unconditional distribution for a scalar response variable:

$$p(y_i|\theta, \sigma^2) \quad = \quad \exp\left[\frac{y_i\theta - A(\theta)}{\sigma^2} + c(y_i, \sigma^2) \right] \tag{9.77}$$

where σ^2 is the **dispersion parameter** (often set to 1), θ is the natural parameter, A is the partition function, and c is a normalization constant. For example, in the case of logistic regression, θ is the log-odds ratio, $\theta = \log(\frac{\mu}{1-\mu})$, where $\mu = \mathbb{E}[y] = p(y = 1)$ is the mean parameter (see Section 9.2.2.1). To convert from the mean parameter to the natural parameter

Distrib.	Link $g(\mu)$	$\theta = \psi(\mu)$	$\mu = \psi^{-1}(\theta) = \mathbb{E}[y]$
$\mathcal{N}(\mu, \sigma^2)$	identity	$\theta = \mu$	$\mu = \theta$
$\mathrm{Bin}(N, \mu)$	logit	$\theta = \log(\frac{\mu}{1-\mu})$	$\mu = \mathrm{sigm}(\theta)$
$\mathrm{Poi}(\mu)$	log	$\theta = \log(\mu)$	$\mu = e^\theta$

Table 9.1 Canonical link functions ψ and their inverses for some common GLMs.

we can use a function ψ, so $\theta = \Psi(\mu)$. This function is uniquely determined by the form of the exponential family distribution. In fact, this is an invertible mapping, so we have $\mu = \Psi^{-1}(\theta)$. Furthermore, we know from Section 9.2.3 that the mean is given by the derivative of the partition function, so we have $\mu = \Psi^{-1}(\theta) = A'(\theta)$.

Now let us add inputs/ covariates. We first define a linear function of the inputs:

$$\eta_i = \mathbf{w}^T \mathbf{x}_i \tag{9.78}$$

We now make the mean of the distribution be some invertible monotonic function of this linear combination. By convention, this function, known as the **mean function**, is denoted by g^{-1}, so

$$\mu_i = g^{-1}(\eta_i) = g^{-1}(\mathbf{w}^T \mathbf{x}_i) \tag{9.79}$$

See Figure 9.1 for a summary of the basic model.

The inverse of the mean function, namely $g()$, is called the **link function**. We are free to choose almost any function we like for g, so long as it is invertible, and so long as g^{-1} has the appropriate range. For example, in logistic regression, we set $\mu_i = g^{-1}(\eta_i) = \mathrm{sigm}(\eta_i)$.

One particularly simple form of link function is to use $g = \psi$; this is called the **canonical link function**. In this case, $\theta_i = \eta_i = \mathbf{w}^T \mathbf{x}_i$, so the model becomes

$$p(y_i | \mathbf{x}_i, \mathbf{w}, \sigma^2) = \exp\left[\frac{y_i \mathbf{w}^T \mathbf{x}_i - A(\mathbf{w}^T \mathbf{x}_i)}{\sigma^2} + c(y_i, \sigma^2)\right] \tag{9.80}$$

In Table 9.1, we list some distributions and their canonical link functions. We see that for the Bernoulli/ binomial distribution, the canonical link is the logit function, $g(\mu) = \log(\eta/(1-\eta))$, whose inverse is the logistic function, $\mu = \mathrm{sigm}(\eta)$.

Based on the results in Section 9.2.3, we can show that the mean and variance of the response variable are as follows:

$$\mathbb{E}[y | \mathbf{x}_i, \mathbf{w}, \sigma^2] = \mu_i = A'(\theta_i) \tag{9.81}$$
$$\mathrm{var}[y | \mathbf{x}_i, \mathbf{w}, \sigma^2] = \sigma_i^2 = A''(\theta_i)\sigma^2 \tag{9.82}$$

To make the notation clearer, let us consider some simple examples.

- For linear regression, we have

$$\log p(y_i | \mathbf{x}_i, \mathbf{w}, \sigma^2) = \frac{y_i \mu_i - \frac{\mu_i^2}{2}}{\sigma^2} - \frac{1}{2}\left(\frac{y_i^2}{\sigma^2} + \log(2\pi\sigma^2)\right) \tag{9.83}$$

where $y_i \in \mathbb{R}$, and $\theta_i = \mu_i = \mathbf{w}^T \mathbf{x}_i$ Here $A(\theta) = \theta^2/2$, so $\mathbb{E}[y_i] = \mu_i$ and $\mathrm{var}[y_i] = \sigma^2$.

- For binomial regression, we have

$$
\log p(y_i|\mathbf{x}_i, \mathbf{w}) \quad = \quad y_i \log(\frac{\pi_i}{1 - \pi_i}) + N_i \log(1 - \pi_i) + \log \binom{N_i}{y_i} \tag{9.84}
$$

where $y_i \in \{0, 1, \ldots, N_i\}$, $\pi_i = \mathrm{sigm}(\mathbf{w}^T\mathbf{x}_i)$, $\theta_i = \log(\pi_i/(1 - \pi_i)) = \mathbf{w}^T\mathbf{x}_i$, and $\sigma^2 = 1$. Here $A(\theta) = N_i \log(1 + e^\theta)$, so $\mathbb{E}[y_i] = N_i\pi_i = \mu_i$, $\mathrm{var}[y_i] = N_i\pi_i(1 - \pi_i)$.

- For **poisson regression**, we have

$$
\log p(y_i|\mathbf{x}_i, \mathbf{w}) \quad = \quad y_i \log \mu_i - \mu_i - \log(y_i!) \tag{9.85}
$$

where $y_i \in \{0, 1, 2, \ldots\}$, $\mu_i = \exp(\mathbf{w}^T\mathbf{x}_i)$, $\theta_i = \log(\mu_i) = \mathbf{w}^T\mathbf{x}_i$, and $\sigma^2 = 1$. Here $A(\theta) = e^\theta$, so $\mathbb{E}[y_i] = \mathrm{var}[y_i] = \mu_i$. Poisson regression is widely used in bio-statistical applications, where y_i might represent the number of diseases of a given person or place, or the number of reads at a genomic location in a high-throughput sequencing context (see e.g., (Kuan et al. 2009)).

9.3.2 ML and MAP estimation

One of the appealing properties of GLMs is that they can be fit using exactly the same methods that we used to fit logistic regression. In particular, the log-likelihood has the following form:

$$
\ell(\mathbf{w}) = \log p(\mathcal{D}|\mathbf{w}) \quad = \quad \frac{1}{\sigma^2} \sum_{i=1}^{N} \ell_i \tag{9.86}
$$

$$
\ell_i \quad \triangleq \quad \theta_i y_i - A(\theta_i) \tag{9.87}
$$

We can compute the gradient vector using the chain rule as follows:

$$
\frac{d\ell_i}{dw_j} \quad = \quad \frac{d\ell_i}{d\theta_i} \frac{d\theta_i}{d\mu_i} \frac{d\mu_i}{d\eta_i} \frac{d\eta_i}{dw_j} \tag{9.88}
$$

$$
= \quad (y_i - A'(\theta_i)) \frac{d\theta_i}{d\mu_i} \frac{d\mu_i}{d\eta_i} x_{ij} \tag{9.89}
$$

$$
= \quad (y_i - \mu_i) \frac{d\theta_i}{d\mu_i} \frac{d\mu_i}{d\eta_i} x_{ij} \tag{9.90}
$$

If we use a canonical link, $\theta_i = \eta_i$, this simplifies to

$$
\nabla_{\mathbf{w}}\ell(\mathbf{w}) = \frac{1}{\sigma^2} \left[\sum_{i=1}^{N} (y_i - \mu_i)\mathbf{x}_i \right] \tag{9.91}
$$

which is a sum of the input vectors, weighted by the errors. This can be used inside a (stochastic) gradient descent procedure, discussed in Section 8.5.2. However, for improved efficiency, we should use a second-order method. If we use a canonical link, the Hessian is given by

$$
\mathbf{H} = -\frac{1}{\sigma^2} \sum_{i=1}^{N} \frac{d\mu_i}{d\theta_i} \mathbf{x}_i \mathbf{x}_i^T = -\frac{1}{\sigma^2} \mathbf{X}^T \mathbf{S} \mathbf{X} \tag{9.92}
$$

Name	Formula
Logistic	$g^{-1}(\eta) = \text{sigm}(\eta) = \frac{e^\eta}{1+e^\eta}$
Probit	$g^{-1}(\eta) = \Phi(\eta)$
Log-log	$g^{-1}(\eta) = \exp(-\exp(-\eta))$
Complementary log-log	$g^{-1}(\eta) = 1 - \exp(-\exp(\eta))$

Table 9.2 Summary of some possible mean functions for binary regression.

where $\mathbf{S} = \text{diag}(\frac{d\mu_1}{d\theta_1}, \ldots, \frac{d\mu_N}{d\theta_N})$ is a diagonal weighting matrix. This can be used inside the IRLS algorithm (Section 8.3.4). Specifically, we have the following Newton update:

$$\mathbf{w}_{t+1} = (\mathbf{X}^T \mathbf{S}_t \mathbf{X})^{-1} \mathbf{X}^T \mathbf{S}_t \mathbf{z}_t \tag{9.93}$$

$$\mathbf{z}_t = \boldsymbol{\theta}_t + \mathbf{S}_t^{-1}(\mathbf{y} - \boldsymbol{\mu}_t) \tag{9.94}$$

where $\boldsymbol{\theta}_t = \mathbf{X}\mathbf{w}_t$ and $\boldsymbol{\mu}_t = g^{-1}(\boldsymbol{\eta}_t)$.

If we extend the derivation to handle non-canonical links, we find that the Hessian has another term. However, it turns out that the expected Hessian is the same as in Equation 9.92; using the expected Hessian (known as the Fisher information matrix) instead of the actual Hessian is known as the **Fisher scoring method**.

It is straightforward to modify the above procedure to perform MAP estimation with a Gaussian prior: we just modify the objective, gradient and Hessian, just as we added ℓ_2 regularization to logistic regression in Section 8.3.6.

9.3.3 Bayesian inference

Bayesian inference for GLMs is usually conducted using MCMC (Chapter 24). Possible methods include Metropolis Hastings with an IRLS-based proposal (Gamerman 1997), Gibbs sampling using adaptive rejection sampling (ARS) for each full-conditional (Dellaportas and Smith 1993), etc. See e.g., (Dey et al. 2000) for futher information. It is also possible to use the Gaussian approximation (Section 8.4.1) or variational inference (Section 21.8.1.1).

9.4 Probit regression

In (binary) logistic regression, we use a model of the form $p(y = 1|\mathbf{x}_i, \mathbf{w}) = \text{sigm}(\mathbf{w}^T \mathbf{x}_i)$. In general, we can write $p(y = 1|\mathbf{x}_i, \mathbf{w}) = g^{-1}(\mathbf{w}^T \mathbf{x}_i)$, for any function g^{-1} that maps $[-\infty, \infty]$ to $[0, 1]$. Several possible mean functions are listed in Table 9.2.

In this section, we focus on the case where $g^{-1}(\eta) = \Phi(\eta)$, where $\Phi(\eta)$ is the cdf of the standard normal. This is known as **probit regression**. The probit function is very similar to the logistic function, as shown in Figure 8.7(b). However, this model has some advantages over logistic regression, as we will see.

9.4.1 ML/MAP estimation using gradient-based optimization

We can find the MLE for probit regression using standard gradient methods. Let $\mu_i = \mathbf{w}^T \mathbf{x}_i$, and let $\tilde{y}_i \in \{-1, +1\}$. Then the gradient of the log-likelihod for a specific case is given by

$$\mathbf{g}_i \triangleq \frac{d}{d\mathbf{w}} \log p(\tilde{y}_i | \mathbf{w}^T \mathbf{x}_i) = \frac{d\mu_i}{d\mathbf{w}} \frac{d}{d\mu_i} \log p(\tilde{y}_i | \mathbf{w}^T \mathbf{x}_i) = \mathbf{x}_i \frac{\tilde{y}_i \phi(\mu_i)}{\Phi(\tilde{y}_i \mu_i)} \tag{9.95}$$

where ϕ is the standard normal pdf, and Φ is its cdf. Similarly, the Hessian for a single case is given by

$$\mathbf{H}_i = \frac{d}{d\mathbf{w}^2} \log p(\tilde{y}_i | \mathbf{w}^T \mathbf{x}_i) = -\mathbf{x}_i \left(\frac{\phi(\mu_i)^2}{\Phi(\tilde{y}_i \mu_i)^2} + \frac{\tilde{y}_i \mu_i \phi(\mu_i)}{\Phi(\tilde{y}_i \mu_i)} \right) \mathbf{x}_i^T \tag{9.96}$$

We can modify these expressions to compute the MAP estimate in a straightforward manner. In particular, if we use the prior $p(\mathbf{w}) = \mathcal{N}(\mathbf{0}, \mathbf{V}_0)$, the gradient and Hessian of the penalized log likelihood have the form $\sum_i \mathbf{g}_i + 2\mathbf{V}_0^{-1}\mathbf{w}$ and $\sum_i \mathbf{H}_i + 2\mathbf{V}_0^{-1}$. These expressions can be passed to any gradient-based optimizer. See `probitRegDemo` for a demo.

9.4.2 Latent variable interpretation

We can interpret the probit (and logistic) model as follows. First, let us associate each item $\mathbf{x}_i$ with two latent utilities, u_{0i} and u_{1i}, corresponding to the possible choices of $y_i = 0$ and $y_i = 1$. We then assume that the observed choice is whichever action has larger utility. More precisely, the model is as follows:

$$u_{0i} \triangleq \mathbf{w}_0^T \mathbf{x}_i + \delta_{0i} \tag{9.97}$$
$$u_{1i} \triangleq \mathbf{w}_1^T \mathbf{x}_i + \delta_{1i} \tag{9.98}$$
$$y_i = \mathbb{I}(u_{1i} > u_{10}) \tag{9.99}$$

where $\mathbf{w} = \mathbf{w}_1 - \mathbf{w}_0$, and δ's are error terms, representing all the other factors that might be relevant in decision making that we have chosen not to (or are unable to) model. This is called a **random utility model** or **RUM** (McFadden 1974; Train 2009).

Since it is only the difference in utilities that matters, let us define $z_i = u_{1i} - u_{0i} = \mathbf{w}^T \mathbf{x}_i + \epsilon_i$, where $\epsilon_i = \delta_{1i} - \delta_{0i}$. If the δ's have a Gaussian distribution, then so does ϵ_i. Thus we can write

$$z_i \triangleq \mathbf{w}^T \mathbf{x}_i + \epsilon_i \tag{9.100}$$
$$\epsilon_i \sim \mathcal{N}(0, 1) \tag{9.101}$$
$$y_i = 1 = \mathbb{I}(z_i \geq 0) \tag{9.102}$$

Following (Fruhwirth-Schnatter and Fruhwirth 2010), we call this the difference RUM or **dRUM** model.

When we marginalize out z_i, we recover the probit model:

$$p(y_i = 1 | \mathbf{x}_i, \mathbf{w}) = \int \mathbb{I}(z_i \geq 0) \mathcal{N}(z_i | \mathbf{w}^T \mathbf{x}_i, 1) dz_i \tag{9.103}$$
$$= p(\mathbf{w}^T \mathbf{x}_i + \epsilon \geq 0) = p(\epsilon \geq -\mathbf{w}^T \mathbf{x}_i) \tag{9.104}$$
$$= 1 - \Phi(-\mathbf{w}^T \mathbf{x}_i) = \Phi(\mathbf{w}^T \mathbf{x}_i) \tag{9.105}$$

where we used the symmetry of the Gaussian.[3] This latent variable interpretation provides an alternative way to fit the model, as discussed in Section 11.4.6.

Interestingly, if we use a Gumbel distribution for the δ's, we induce a logistic distribution for ϵ_i, and the model reduces to logistic regression. See Section 24.5.1 for further details.

9.4.3 Ordinal probit regression *

One advantage of the latent variable interpretation of probit regression is that it is easy to extend to the case where the response variable is ordinal, that is, it can take on C discrete values which can be ordered in some way, such as low, medium and high. This is called **ordinal regression**. The basic idea is as follows. We introduce $C + 1$ thresholds γ_j and set

$$y_i = j \quad \text{if} \quad \gamma_{j-1} < z_i \leq \gamma_j \tag{9.106}$$

where $\gamma_0 \leq \cdots \leq \gamma_C$. For identifiability reasons, we set $\gamma_0 = -\infty$, $\gamma_1 = 0$ and $\gamma_C = \infty$. For example, if $C = 2$, this reduces to the standard binary probit model, whereby $z_i < 0$ produces $y_i = 0$ and $z_i \geq 0$ produces $y_i = 1$. If $C = 3$, we partition the real line into 3 intervals: $(-\infty, 0]$, $(0, \gamma_2]$, (γ_2, ∞). We can vary the parameter γ_2 to ensure the right relative amount of probability mass falls in each interval, so as to match the empirical frequencies of each class label.

Finding the MLEs for this model is a bit trickier than for binary probit regression, since we need to optimize for $\mathbf{w}$ and $\boldsymbol{\gamma}$, and the latter must obey an ordering constraint. See e.g., (Kawakatsu and Largey 2009) for an approach based on EM. It is also possible to derive a simple Gibbs sampling algorithm for this model (see e.g., (Hoff 2009, p216)).

9.4.4 Multinomial probit models *

Now consider the case where the response variable can take on C unordered categorical values, $y_i \in \{1, \ldots, C\}$. The **multinomial probit** model is defined as follows:

$$z_{ic} = \mathbf{w}^T \mathbf{x}_{ic} + \epsilon_{ic} \tag{9.107}$$

$$\boldsymbol{\epsilon} \sim \mathcal{N}(\mathbf{0}, \mathbf{R}) \tag{9.108}$$

$$y_i = \arg\max_c z_{ic} \tag{9.109}$$

See e.g., (Dow and Endersby 2004; Scott 2009; Fruhwirth-Schnatter and Fruhwirth 2010) for more details on the model and its connection to multinomial logistic regression. (By defining $\mathbf{w} = [\mathbf{w}_1, \ldots, \mathbf{w}_C]$, and $\mathbf{x}_{ic} = [\mathbf{0}, \ldots, \mathbf{0}, \mathbf{x}_i, \mathbf{0}, \ldots, \mathbf{0}]$, we can recover the more familiar formulation $z_{ic} = \mathbf{x}_i^T \mathbf{w}_c$.) Since only relative utilities matter, we constrain $\mathbf{R}$ to be a correlation matrix. If instead of setting $y_i = \arg\max_c z_{ic}$ we use $y_{ic} = \mathbb{I}(z_{ic} > 0)$, we get a model known as **multivariate probit**, which is one way to model C correlated binary outcomes (see e.g., (Talhouk et al. 2012)).

---

3. Note that the assumption that the Gaussian noise term is zero mean and unit variance is made without loss of generality. To see why, suppose we used some other mean μ and variance σ^2. Then we could easily rescale $\mathbf{w}$ and add an offset term without changing the likelihood. since $P(\mathcal{N}(0, 1) \geq -\mathbf{w}^T \mathbf{x}) = P(\mathcal{N}(\mu, \sigma^2) \geq -(\mathbf{w}^T \mathbf{x} + \mu)/\sigma)$.

9.5   Multi-task learning

Sometimes we want to fit many related classification or regression models. It is often reasonable to assume the input-output mapping is similar across these different models, so we can get better performance by fitting all the parameters at the same time. In machine learning, this setup is often called **multi-task learning** (Caruana 1998), **transfer learning** (e.g., (Raina et al. 2005)), or **learning to learn** (Thrun and Pratt 1997). In statistics, this is usually tackled using hierarchical Bayesian models (Bakker and Heskes 2003), as we discuss below, although there are other possible methods, such as using Gaussian processes (see e.g., (Chai 2010)).

9.5.1   Hierarchical Bayes for multi-task learning

Let y_{ij} be the response of the i'th item in group j, for $i = 1 : N_j$ and $j = 1 : J$. For example, j might index schools, i might index students within a school, and y_{ij} might be the test score, as in Section 5.6.2. Or j might index people, and i might index purchases, and y_{ij} might be the identity of the item that was purchased (this is known as **discrete choice modeling** (Train 2009)). Let $\mathbf{x}_{ij}$ be a feature vector associated with y_{ij}. The goal is to fit the models $p(y_j|\mathbf{x}_j)$ for all j.

Although some groups may have lots of data, there is often a long tail, where the majority of groups have little data. Thus we can't reliably fit each model separately, but we don't want to use the same model for all groups. As a compromise, we can fit a separate model for each group, but encourage the model parameters to be similar across groups. More precisely, suppose $\mathbb{E}[y_{ij}|\mathbf{x}_{ij}] = g(\mathbf{x}_{ij}^T \boldsymbol{\beta}_j)$, where g is the link function for the GLM. Furthermore, suppose $\boldsymbol{\beta}_j \sim \mathcal{N}(\boldsymbol{\beta}_*, \sigma_j^2 \mathbf{I})$, and that $\boldsymbol{\beta}_* \sim \mathcal{N}(\boldsymbol{\mu}, \sigma_*^2 \mathbf{I})$. In this model, groups with small sample size borrow statistical strength from the groups with larger sample size, because the $\boldsymbol{\beta}_j$'s are correlated via the latent common parents $\boldsymbol{\beta}_*$ (see Section 5.5 for further discussion of this point). The term σ_j^2 controls how much group j depends on the common parents and the σ_*^2 term controls the strength of the overall prior.

Suppose, for simplicity, that $\boldsymbol{\mu} = \mathbf{0}$, and that σ_j^2 and σ_*^2 are all known (e.g., they could be set by cross validation). The overall log probability has the form

$$\log p(\mathcal{D}|\boldsymbol{\beta}) + \log p(\boldsymbol{\beta}) = \sum_j \left[\log p(\mathcal{D}_j|\boldsymbol{\beta}_j) - \frac{||\boldsymbol{\beta}_j - \boldsymbol{\beta}_*||^2}{2\sigma_j^2} \right] - \frac{||\boldsymbol{\beta}_*||^2}{2\sigma_*^2} \tag{9.110}$$

We can perform MAP estimation of $\boldsymbol{\beta} = (\boldsymbol{\beta}_{1:J}, \boldsymbol{\beta}_*)$ using standard gradient methods. Alternatively, we can perform an iterative optimization scheme, alternating between optimizing the $\boldsymbol{\beta}_j$ and the $\boldsymbol{\beta}_*$; since the likelihood and prior are convex, this is guaranteed to converge to the global optimum. Note that once the models are trained, we can discard $\boldsymbol{\beta}_*$, and use each model separately.

9.5.2   Application to personalized email spam filtering

An interesting application of multi-task learning is **personalized spam filtering**. Suppose we want to fit one classifier per user, $\boldsymbol{\beta}_j$. Since most users do not label their email as spam or not, it will be hard to estimate these models independently. So we will let the $\boldsymbol{\beta}_j$ have a common prior $\boldsymbol{\beta}_*$, representing the parameters of a generic user.

In this case, we can emulate the behavior of the above model with a simple trick (Daume 2007b; Attenberg et al. 2009; Weinberger et al. 2009): we make two copies of each feature $\mathbf{x}_i$, one concatenated with the user id, and one not. The effect will be to learn a predictor of the form

$$\mathbb{E}\left[y_i|\mathbf{x}_i, u\right] = (\boldsymbol{\beta}_*, \mathbf{w}_1, \cdots, \mathbf{w}_J)^T[\mathbf{x}_i, \mathbb{I}(u=1)\mathbf{x}_i, \cdots, \mathbb{I}(u=J)\mathbf{x}_i] \tag{9.111}$$

where u is the user id and $\mathbf{w}_j = \boldsymbol{\beta}_j - \boldsymbol{\beta}_*$. In other words,

$$\mathbb{E}\left[y_i|\mathbf{x}_i, u=j\right] = (\boldsymbol{\beta}_*^T + \mathbf{w}_j)^T\mathbf{x}_i \tag{9.112}$$

Thus $\boldsymbol{\beta}_*$ will be estimated from everyone's email, whereas $\mathbf{w}_j$ will just be estimated from user j's email.

To see the correspondence with the above hierarchical Bayesian model, note that the log probability of the original model can be rewritten as

$$\sum_j \left[\log p(\mathcal{D}_j|\boldsymbol{\beta}_* + \mathbf{w}_j) - \frac{||\mathbf{w}_j||^2}{2\sigma_j^2}\right] - \frac{||\boldsymbol{\beta}_*||^2}{2\sigma_*^2} \tag{9.113}$$

If we assume $\sigma_j^2 = \sigma_*^2$, the effect is the same as using the augmented feature trick, with the same regularizer strength for both $\mathbf{w}_j$ and $\boldsymbol{\beta}_*$. However, one typically gets better performance by not requiring that σ_j^2 be equal to σ_*^2 (Finkel and Manning 2009).

9.5.3   Application to domain adaptation

Domain adaptation is the problem of training a set of classifiers on data drawn from different distributions, such as email and newswire text. This problem is obviously a special case of multi-task learning, where the tasks are the same.

(Finkel and Manning 2009) used the above hierarchical Bayesian model to perform domain adaptation for two NLP tasks, namely named entity recognition and parsing. They report reasonably large improvements over fitting separate models to each dataset, and small improvements over the approach of pooling all the data and fitting a single model.

9.5.4   Other kinds of prior

In multi-task learning, it is common to assume that the prior is Gaussian. However, sometimes other priors are more suitable. For example, consider the task of **conjoint analysis**, which requires figuring out which features of a product customers like best. This can be modelled using the same hierarchical Bayesian setup as above, but where we use a sparsity-promoting prior on $\boldsymbol{\beta}_j$, rather than a Gaussian prior. This is called **multi-task feature selection**. See e.g., (Lenk et al. 1996; Argyriou et al. 2008) for some possible approaches.

It is not always reasonable to assume that all tasks are all equally similar. If we pool the parameters across tasks that are qualitatively different, the performance will be worse than not using pooling, because the inductive bias of our prior is wrong. Indeed, it has been found experimentally that sometimes multi-task learning does worse than solving each task separately (this is called **negative transfer**).

One way around this problem is to use a more flexible prior, such as a mixture of Gaussians. Such flexible priors can provide robustness against prior mis-specification. See e.g., (Xue et al. 2007; Jacob et al. 2008) for details. One can of course combine mixtures with sparsity-promoting priors (Ji et al. 2009). Many other variants are possible.

9.6   Generalized linear mixed models *

Suppose we generalize the multi-task learning scenario to allow the response to include information at the group level, $\mathbf{x}_j$, as well as at the item level, $\mathbf{x}_{ij}$. Similarly, we can allow the parameters to vary across groups, $\boldsymbol{\beta}_j$, or to be tied across groups, $\boldsymbol{\alpha}$. This gives rise to the following model:

$$\mathbb{E}\left[y_{ij}|\mathbf{x}_{ij}, \mathbf{x}_j\right] = g\left(\phi_1(\mathbf{x}_{ij})^T\boldsymbol{\beta}_j + \phi_2(\mathbf{x}_j)^T\boldsymbol{\beta}'_j + \phi_3(\mathbf{x}_{ij})^T\boldsymbol{\alpha} + \phi_4(\mathbf{x}_j)^T\boldsymbol{\alpha}'\right) \quad (9.114)$$

where the ϕ_k are basis functions. This model can be represented pictorially as shown in Figure 9.2(a). (Such figures will be explained in Chapter 10.) Note that the number of $\boldsymbol{\beta}_j$ parameters grows with the number of groups, whereas the size of $\boldsymbol{\alpha}$ is fixed.

Frequentists call the terms $\boldsymbol{\beta}_j$ **random effects**, since they vary randomly across groups, but they call $\boldsymbol{\alpha}$ a **fixed effect**, since it is viewed as a fixed but unknown constant. A model with both fixed and random effects is called a **mixed model**. If $p(y|\mathbf{x})$ is a GLM, the overall model is called a **generalized linear mixed effects model** or **GLMM**. Such models are widely used in statistics.

9.6.1   Example: semi-parametric GLMMs for medical data

Consider the following example from (Wand 2009). Suppose y_{ij} is the amount of spinal bone mineral density (SBMD) for person j at measurement i. Let x_{ij} be the age of person, and let x_j be their ethnicity, which can be one of: White, Asian, Black, or Hispanic. The primary goal is to determine if there are significant differences in the mean SBMD among the four ethnic groups, after accounting for age. The data is shown in the light gray lines in Figure 9.2(b). We see that there is a nonlinear effect of SBMD vs age, so we will use a **semi-parametric model** which combines linear regression with non-parametric regression (Ruppert et al. 2003). We also see that there is variation across individuals within each group, so we will use a mixed effects model. Specifically, we will use $\phi_1(\mathbf{x}_{ij}) = 1$ to account for the random effect of each person; $\phi_2(\mathbf{x}_{ij}) = 0$ since no other coefficients are person-specific; $\phi_3(x_{ij}) = [b_k(x_{ij})]$, where b_k is the k'th spline basis functions (see Section 15.4.6.2), to account for the nonlinear effect of age; and $\phi_4(x_j) = [\mathbb{I}(x_j = w), \mathbb{I}(x_j = a), \mathbb{I}(x_j = b), \mathbb{I}(x_j = h)]$ to account for the effect of the different ethnicities. Furthermore, we use a linear link function. The overall model is therefore

$$\mathbb{E}\left[y_{ij}|x_{ij}, x_j\right] \quad = \quad \beta_j + \boldsymbol{\alpha}^T\mathbf{b}(x_{ij}) + \epsilon_{ij} \quad (9.115)$$
$$+\alpha'_w\mathbb{I}(x_j = w) + \alpha'_a\mathbb{I}(x_j = a) + \alpha'_b\mathbb{I}(x_j = b) + \alpha'_h\mathbb{I}(x_j = h) \quad (9.116)$$

where $\epsilon_{ij} \sim \mathcal{N}(0, \sigma_y^2)$. $\boldsymbol{\alpha}$ contains the non-parametric part of the model related to age, $\boldsymbol{\alpha}'$ contains the parametric part of the model related to ethnicity, and β_j is a random offset for person j. We endow all of these regression coefficients with separate Gaussian priors. We can then perform posterior inference to compute $p(\boldsymbol{\alpha}, \boldsymbol{\alpha}', \boldsymbol{\beta}, \sigma^2|\mathcal{D})$ (see Section 9.6.2 for

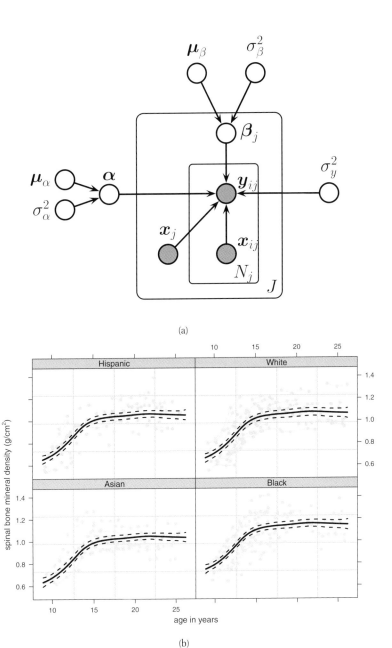

(a)

(b)

Figure 9.2 (a) Directed graphical model for generalized linear mixed effects model with J groups. (b) Spinal bone mineral density vs age for four different ethnic groups. Raw data is shown in the light gray lines. Fitted model shown in black (solid is the posterior predicted mean, dotted is the posterior predictive variance). From Figure 9 of (Wand 2009). Used with kind permission of Matt Wand

computational details). After fitting the model, we can compute the prediction for each group. See Figure 9.2(b) for the results. We can also perform significance testing, by computing $p(\alpha_g - \alpha_w | \mathcal{D})$ for each ethnic group g relative to some baseline (say, White), as we did in Section 5.2.3.

9.6.2 Computational issues

The principle problem with GLMMs is that they can be difficult to fit, for two reasons. First, $p(y_{ij} | \boldsymbol{\theta})$ may not be conjugate to the prior $p(\boldsymbol{\theta})$ where $\boldsymbol{\theta} = (\boldsymbol{\alpha}, \boldsymbol{\beta})$. Second, there are two levels of unknowns in the model, namely the regression coefficients $\boldsymbol{\theta}$ and the means and variances of the priors $\boldsymbol{\eta} = (\boldsymbol{\mu}, \boldsymbol{\sigma})$.

One approach is to adopt fully Bayesian inference methods, such as variational Bayes (Hall et al. 2011) or MCMC (Gelman and Hill 2007). We discuss VB in Section 21.5, and MCMC in Section 24.1.

An alternative approach is to use empirical Bayes, which we discuss in general terms in Section 5.6. In the context of a GLMM, we can use the EM algorithm (Section 11.4), where in the E step we compute $p(\boldsymbol{\theta} | \boldsymbol{\eta}, \mathcal{D})$, and in the M step we optimize $\boldsymbol{\eta}$. In the linear regression setting, the E step can be performed exactly, but in general we need to use approximations. Traditional methods use numerical quadrature or Monte Carlo (see e.g., (Breslow and Clayton 1993)). A faster approach is to use variational EM; see (Braun and McAuliffe 2010) for an application of variational EM to a multi-level discrete choice modeling problem.

In frequentist statistics, there is a popular method for fitting GLMMs called **generalized estimating equations** or **GEE** (Hardin and Hilbe 2003). However, we do not recommend this approach, since it is not as statistically efficient as likelihood-based methods (see Section 6.4.3). In addition, it can only provide estimates of the population parameters $\boldsymbol{\alpha}$, but not the random effects $\boldsymbol{\beta}_j$, which are sometimes of interest in themselves.

9.7 Learning to rank *

In this section, we discuss the **learning to rank** problem. That is, we want to learn a function that can rank order a set of items (we will be more precise below). The most common application is to information retrieval. Specifically, suppose we have a query q and a set of documents $d_1, \ldots, d_m$ that might be relevant to q (e.g., all documents that contain the string q). We would like to sort these documents in decreasing order of relevance and show the top k to the user. Similar problems arise in other areas, such as collaborative filtering. (Ranking players in a game or tournament setting is a slightly different kind of problem; see Section 22.5.5.)

Below we summarize some methods for solving this problem, following the presentation of (Liu 2009). This material is not based on GLMs, but we include it in this chapter anyway for lack of a better place.

A standard way to measure the relevance of a document d to a query q is to use a probabilistic language model based on a bag of words model. That is, we define $\text{rel}(d, q) \triangleq p(q|d) = \prod_{i=1}^{n} p(q_i|d)$, where q_i is the i'th word or term, and $p(q_i|d)$ is a multinoulli distribution estimated from document d. In practice, we need to smooth the estimated distribution, for example by using a Dirichlet prior, representing the overall frequency of each word. This can be estimated from all documents in the system. More precisely, we can use

$$p(t|d) = (1 - \lambda)\frac{\text{TF}(t, d)}{\text{LEN}(d)} + \lambda p(t|\text{background}) \tag{9.117}$$

where $\text{TF}(t, d)$ is the frequency of term t in document d, $\text{LEN}(d)$ is the number of words in d, and $0 < \lambda < 1$ is a smoothing parameter (see e.g., Zhai and Lafferty (2004) for details).

However, there might be many other signals that we can use to measure relevance. For example, the PageRank of a web document is a measure of its authoritativeness, derived from the web's link structure (see Section 17.2.4 for details). We can also compute how often and where the query occurs in the document. Below we discuss how to learn how to combine all these signals to come up with a good ranking[4]

9.7.1 The pointwise approach

Generalizing the above approach, suppose that for each query document pair, we define a feature vector, $\mathbf{x}(q, d)$. For example, this might contain the query-document similarity score and the page rank score of the document. Furthermore, suppose we have a set of labels y_j representing the degree of relevance of document d_j to query q. Such labels might be binary (e.g., relevant or irrelevant), or they may represent a degree of relevance (e.g., very relevant, somewhat relevant, irrelevant). Such labels can be obtained from query logs, by thresholding the number of times a document was clicked on for a given query.

If we have binary relevance labels, we can solve the problem using a standard binary classification scheme to estimate, $p(y = 1|\mathbf{x}(q, d))$. If we have ordered relevancy labels, we can use ordinal regression to predict the rating, $p(y = r|\mathbf{x}(q, d))$. In either case, we can then sort the documents by this scoring metric. This is called the **pointwise approach**, and is widely used because of its simplicity. However, this method does not take into account the location of each document in the list. Thus it penalizes errors at the end of the list just as much as errors at the beginning, which is often not the desired behavior. In addition, each decision about relevance is made very myopically.

9.7.2 The pairwise approach

There is evidence (e.g., (Carterette et al. 2008)) that people are better at judging the relative relevance of two items rather than absolute relevance. Consequently, the data might tell us that d_j is more relevant than d_k for a given query, or vice versa. We can model this kind of data using a binary classifier of the form $p(y_{jk}|\mathbf{x}(q, d_j), \mathbf{x}(q, d_k))$, where we set $y_{jk} = 1$ if $\text{rel}(d_j, q) > \text{rel}(d_k, q)$ and $y_{jk} = 0$ otherwise.

One way to model such a function is as follows:

$$p(y_{jk} = 1|\mathbf{x}_j, \mathbf{x}_k) = \text{sigm}(f(\mathbf{x}_j) - f(\mathbf{x}_k)) \tag{9.118}$$

where $f(\mathbf{x})$ is a scoring function, often taken to be linear, $f(\mathbf{x}) = \mathbf{w}^T\mathbf{x}$. This is a special kind of neural network known as **RankNet** (Burges et al. 2005) (see Section 16.5 for a general

---

4. Rather surprisingly, Google does not (or at least, did not as of 2008) using such learning methods in its search engine. Source: Peter Norvig, quoted in `http://anand.typepad.com/datawocky/2008/05/are-human-experts-less-p rone-to-catastrophic-errors-than-machine-learned-models.html`.

discussion of neural networks). We can find the MLE of $\mathbf{w}$ by maximizing the log likelihood, or equivalently, by minimizing the cross entropy loss, given by

$$
L \;=\; \sum_{i=1}^{N}\sum_{j=1}^{m_i}\sum_{k=j+1}^{m_i} L_{ijk} \tag{9.119}
$$

$$
-L_{ijk} \;=\; \mathbb{I}(y_{ijk}=1)\log p(y_{ijk}=1|\mathbf{x}_{ij},\mathbf{x}_{ik},\mathbf{w})
$$
$$
+\mathbb{I}(y_{ijk}=0)\log p(y_{ijk}=0|\mathbf{x}_{ij},\mathbf{x}_{ik},\mathbf{w}) \tag{9.120}
$$

This can be optimized using gradient descent. A variant of RankNet is used by Microsoft's Bing search engine.[5] In addition, this method has been extended to personalized ranking in (Rendle et al. 2009); in this setting, we learn a different ranking function for each user, and the features are latent variables derived from the user-item matrix, as in collaborative filtering.

9.7.3   The listwise approach

The pairwise approach suffers from the problem that decisions about relevance are made just based on a pair of items, rather than considering the full context. We now consider methods that look at the entire list of items at the same time.

We can define a total order on a list by specifying a permutation of its indices, $\boldsymbol{\pi}$. To model our uncertainty about $\boldsymbol{\pi}$, we can use the **Plackett-Luce** distribution, which derives its name from independent work by (Plackett 1975) and (Luce 1959). This has the following form:

$$
p(\boldsymbol{\pi}|\mathbf{s}) = \prod_{j=1}^{m} \frac{s_j}{\sum_{u=j}^{m} s_u} \tag{9.121}
$$

where $s_j = s(\pi^{-1}(j))$ is the score of the document ranked at the j'th position.

To understand Equation 9.121, let us consider a simple example. Suppose $\boldsymbol{\pi} = (A, B, C)$. Then we have that $p(\boldsymbol{\pi})$ is the probability of A being ranked first, times the probability of B being ranked second given that A is ranked first, times the probabilty of C being ranked third given that A and B are ranked first and second. In other words,

$$
p(\boldsymbol{\pi}|\mathbf{s}) = \frac{s_A}{s_A + s_B + s_C} \times \frac{s_B}{s_B + s_C} \times \frac{s_C}{s_C} \tag{9.122}
$$

To incorporate features, we can define $s(d) = f(\mathbf{x}(q,d))$, where we often take f to be a linear function, $f(\mathbf{x}) = \mathbf{w}^T\mathbf{x}$. This is known as the **ListNet** model (Cao et al. 2007). To train this model, let $\mathbf{y}_i$ be the relevance scores of the documents for query i. We then minimize the cross entropy term

$$
-\sum_i \sum_{\boldsymbol{\pi}} p(\boldsymbol{\pi}|\mathbf{y}_i)\log p(\boldsymbol{\pi}|\mathbf{s}_i) \tag{9.123}
$$

Of course, as stated, this is intractable, since the i'th term needs to sum over $m_i!$ permutations. To make this tractable, we can consider permutations over the top k positions only:

$$
p(\boldsymbol{\pi}_{1:k}|\mathbf{s}_{1:m}) = \prod_{j=1}^{k} \frac{s_j}{\sum_{u=1}^{m} s_u} \tag{9.124}
$$

---

5. Source: http://www.bing.com/community/site_blogs/b/search/archive/2009/06/01/user-needs-features-and-the-science-behind-bing.aspx.

There are only $m!/(m-k)!$ such permutations. If we set $k = 1$, we can evaluate each cross entropy term (and its derivative) in $O(m)$ time.

In the special case where only one document from the presented list is deemed relevant, say $y_i = c$, we can instead use multinomial logistic regression:

$$p(y_i = c|\mathbf{x}) = \frac{\exp(s_c)}{\sum_{c'=1}^{m} \exp(s_{c'})} \tag{9.125}$$

This often performs at least as well as ranking methods, at least in the context of collaborative filtering (Yang et al. 2011).

9.7.4 Loss functions for ranking

There are a variety of ways to measure the performance of a ranking system, which we summarize below.

- **Mean average precision (MAP)**. In the case of binary relevance labels, we can define the **precision at k** of some ordering as follows:

$$\text{P@k}(\boldsymbol{\pi}) \triangleq \frac{\text{num. relevant documents in the top } k \text{ positions of } \boldsymbol{\pi}}{k} \tag{9.126}$$

We then define the average precision as follows:

$$\text{AP}(\boldsymbol{\pi}) \triangleq \frac{\sum_k \text{P@k}(\boldsymbol{\pi}) \cdot I_k}{\text{num. relevant documents}} \tag{9.127}$$

where I_k is 1 iff document k is relevant. For example, if we have the relevancy labels $\mathbf{y} = (1, 0, 1, 0, 1)$, then the AP is $\frac{1}{3}(\frac{1}{1} + \frac{2}{3} + \frac{3}{5}) \approx 0.76$. Finally, we define the **mean average precision** as the AP averaged over all queries.
- **Mean reciprocal rank (MRR)**. For a query q, let the rank position of its first relevant document be denoted by $r(q)$. Then we define the **mean reciprocal rank** to be $1/r(q)$ averaged over all queries.[6]
- **Normalized discounted cumulative gain (NDCG)**. Suppose the relevance labels have multiple levels. We can define the **discounted cumulative gain** of the first k items in an ordering as follows:

$$\text{DCG@k}(\mathbf{r}) = r_1 + \sum_{i=2}^{k} \frac{r_i}{\log_2 i} \tag{9.128}$$

where r_i is the relevance of item i and the $\log_2$ term is used to discount items later in the list. Table 9.3 gives a simple numerical example. An alternative definition, that places stronger emphasis on retrieving relevant documents, uses

$$\text{DCG@k}(\mathbf{r}) = \sum_{i=1}^{k} \frac{2^{r_i} - 1}{\log_2(1 + i)} \tag{9.129}$$

---

6. To see the advantage of averaging $1/r(q)$ instead of $r(q)$, consider the effect of poorly rated items. They will have large rank, which will have a large impact on the mean of $r(q)$, but they will have small reciprocal rank, and thus will have minimal impact on the mean of $1/r(q)$.

i	1	2	3	4	5	6
r_i	3	2	3	0	1	2
$\log_2 i$	0	1	1.59	2.0	2.32	2.59
$\frac{r_i}{\log_2 i}$	N/A	2	1.887	0	0.431	0.772

Table 9.3 Illustration of how to compute NDCG, from `http://en.wikipedia.org/wiki/Discounted`
`_cumulative_gain`. The value r_i is the relevance score of the item in position i. From this, we see
that DCG@6 = $3 + (2 + 1.887 + 0 + 0.431 + 0.772) = 8.09$. The maximum DCG is obtained using the
ordering with scores 3, 3, 2, 2, 1, 0. Hence the ideal DCG is 8.693, and so the normalized DCG is 8.09 /
8.693 = 0.9306.

The trouble with DCG is that it varies in magnitude just because the length of a returned
list may vary. It is therefore common to normalize this measure by the ideal DCG, which is
the DCG obtained by using the optimal ordering: IDCG@k($\mathbf{r}$) = $\mathrm{argmax}_{\boldsymbol{\pi}}$ DCG@k($\mathbf{r}$). This
can be easily computed by sorting $r_{1:m}$ and then computing DCG@k. Finally, we define
the **normalized discounted cumulative gain** or **NDCG** as DCG/IDCG. Table 9.3 gives a
simple numerical example. The NDCG can be averaged over queries to give a measure of
performance.

- **Rank correlation**. We can measure the correlation between the ranked list, $\boldsymbol{\pi}$, and the
  relevance judgment, $\boldsymbol{\pi}^*$, using a variety of methods. One approach, known as the (weighted)
  Kendall's τ statistics, is defined in terms of the weighted pairwise inconsistency between the
  two lists:

$$\tau(\boldsymbol{\pi}, \boldsymbol{\pi}^*) = \frac{\sum_{u<v} w_{uv}\left[1 + \mathrm{sgn}(\pi_u - \pi_v)\mathrm{sgn}(\pi_u^* - \pi_v^*)\right]}{2\sum_{u<v} w_{uv}} \tag{9.130}$$

A variety of other measures are commonly used.

These loss functions can be used in different ways. In the Bayesian approach, we first fit the
model using posterior inference; this depends on the likelihood and prior, but not the loss. We
then choose our actions at test time to minimize the expected future loss. One way to do this is
to sample parameters from the posterior, $\boldsymbol{\theta}^s \sim p(\boldsymbol{\theta}|\mathcal{D})$, and then evaluate, say, the precision@k
for different thresholds, averaging over $\boldsymbol{\theta}^s$. See (Zhang et al. 2010) for an example of such an
approach.

In the frequentist approach, we try to minimize the empirical loss on the training set. The
problem is that these loss functions are not differentiable functions of the model parameters.
We can either use gradient-free optimization methods, or we can minimize a surrogate loss
function instead. Cross entropy loss (i.e., negative log likelihood) is an example of a widely used
surrogate loss function.

Another loss, known as **weighted approximate-rank pairwise** or **WARP** loss, proposed in
(Usunier et al. 2009) and extended in (Weston et al. 2010), provides a better approximation to

the precision@k loss. WARP is defined as follows:

$$\text{WARP}(\mathbf{f}(\mathbf{x},:), y) \triangleq L(\text{rank}(\mathbf{f}(\mathbf{x},:), y)) \tag{9.131}$$

$$\text{rank}(\mathbf{f}(\mathbf{x},:), y) = \sum_{y' \neq y} \mathbb{I}(f(\mathbf{x}, y') \geq f(\mathbf{x}, y)) \tag{9.132}$$

$$L(k) \triangleq \sum_{j=1}^{k} \alpha_j, \quad \text{with } \alpha_1 \geq \alpha_2 \geq \cdots \geq 0 \tag{9.133}$$

Here $\mathbf{f}(x,:) = [f(\mathbf{x}, 1), \ldots, f(\mathbf{x}, |y|)]$ is the vector of scores for each possible output label, or, in IR terms, for each possible document corresponding to input query $\mathbf{x}$. The expression $\text{rank}(\mathbf{f}(\mathbf{x},:), y)$ measures the rank of the true label y assigned by this scoring function. Finally, L transforms the integer rank into a real-valued penalty. Using $\alpha_1 = 1$ and $\alpha_{j>1} = 0$ would optimize the proportion of top-ranked correct labels. Setting $\alpha_{1:k}$ to be non-zero values would optimize the top k in the ranked list, which will induce good performance as measured by MAP or precision@k. As it stands, WARP loss is still hard to optimize, but it can be further approximated by Monte Carlo sampling, and then optimized by gradient descent, as described in (Weston et al. 2010).

Exercises

Exercise 9.1 Conjugate prior for univariate Gaussian in exponential family form

Derive the conjugate prior for μ and $\lambda = 1/\sigma^2$ for a univariate Gaussian using the exponential family, by analogy to Section 9.2.5.5. By suitable reparameterization, show that the prior has the form $p(\mu, \lambda) = \mathcal{N}(\mu|\gamma, \lambda(2\alpha - 1))\text{Ga}(\lambda|\alpha, \beta)$, and thus only has 3 free parameters.

Exercise 9.2 The MVN is in the exponential family

Show that we can write the MVN in exponential family form. Hint: use the information form defined in Section 4.3.3.

10 Directed graphical models (Bayes nets)

10.1 Introduction

> I basically know of two principles for treating complicated systems in simple ways: the first is the principle of modularity and the second is the principle of abstraction. I am an apologist for computational probability in machine learning because I believe that probability theory implements these two principles in deep and intriguing ways — namely through factorization and through averaging. Exploiting these two mechanisms as fully as possible seems to me to be the way forward in machine learning. — Michael Jordan, 1997 (quoted in (Frey 1998)).

Suppose we observe multiple correlated variables, such as words in a document, pixels in an image, or genes in a microarray. How can we compactly *represent* the **joint distribution** $p(\mathbf{x}|\boldsymbol{\theta})$? How can we use this distribution to *infer* one set of variables given another in a reasonable amount of computation time? And how can we *learn* the parameters of this distribution with a reasonable amount of data? These questions are at the core of probabilistic modeling, inference and learning, and form the topic of this chapter.

10.1.1 Chain rule

By the **chain rule** of probability, we can always represent a joint distribution as follows, using any ordering of the variables:

$$p(x_{1:V}) = p(x_1)p(x_2|x_1)p(x_3|x_2, x_1)p(x_4|x_1, x_2, x_3) \ldots p(x_V|x_{1:V-1}) \tag{10.1}$$

where V is the number of variables, the Matlab-like notation $1 : V$ denotes the set $\{1, 2, \ldots, V\}$, and where we have dropped the conditioning on the fixed parameters $\boldsymbol{\theta}$ for brevity. The problem with this expression is that it becomes more and more complicated to represent the conditional distributions $p(x_t|\mathbf{x}_{1:t-1})$ as t gets large.

For example, suppose all the variables have K states. We can represent $p(x_1)$ as a table of $O(K)$ numbers, representing a discrete distribution (there are actually only $K - 1$ free parameters, due to the sum-to-one constraint, but we write $O(K)$ for simplicity). Similarly, we can represent $p(x_2|x_1)$ as a table of $O(K^2)$ numbers by writing $p(x_2 = j|x_1 = i) = T_{ij}$; we say that $\mathbf{T}$ is a **stochastic matrix**, since it satisfies the constraint $\sum_j T_{ij} = 1$ for all rows i, and $0 \leq T_{ij} \leq 1$ for all entries. Similarly, we can represent $p(x_3|x_1, x_2)$ as a 3d table with

$O(K^3)$ numbers. These are called **conditional probability tables** or **CPTs**. We see that there are $O(K^V)$ parameters in the model. We would need an awful lot of data to learn so many parameters.

One solution is to replace each CPT with a more parsimonius **conditional probability distribution** or **CPD**, such as multinomial logistic regression, i.e., $p(x_t = k|\mathbf{x}_{1:t-1}) = \mathcal{S}(\mathbf{W}_t\mathbf{x}_{1:t-1})_k$. The total number of parameters is now only $O(K^2V^2)$, making this a compact density model (Neal 1992; Frey 1998). This is adequate if all we want to do is evaluate the probability of a fully observed vector $\mathbf{x}_{1:T}$. For example, we can use this model to define a class-conditional density, $p(\mathbf{x}|y = c)$, thus making a generative classifier (Bengio and Bengio 2000). However, this model is not useful for other kinds of prediction tasks, since each variable depends on all the previous variables. So we need another approach.

10.1.2 Conditional independence

The key to efficiently representing large joint distributions is to make some assumptions about conditional independence (**CI**). Recall from Section 2.2.4 that X and Y are conditionally independent given Z, denoted $X \perp Y|Z$, if and only if (iff) the conditional joint can be written as a product of conditional marginals, i.e.,

$$X \perp Y|Z \iff p(X, Y|Z) = p(X|Z)p(Y|Z) \tag{10.2}$$

Let us see why this might help. Suppose we assume that $x_{t+1} \perp \mathbf{x}_{1:t-1}|x_t$, or in words, "the future is independent of the past given the present". This is called the (first order) **Markov assumption**. Using this assumption, plus the chain rule, we can write the joint distribution as follows:

$$p(\mathbf{x}_{1:V}) = p(x_1) \prod_{t=1}^{V} p(x_t|x_{t-1}) \tag{10.3}$$

This is called a (first-order) **Markov chain**. They can be characterized by an initial distribution over states, $p(x_1 = i)$, plus a **state transition matrix** $p(x_t = j|x_{t-1} = i)$. See Section 17.2 for more information.

10.1.3 Graphical models

Although the first-order Markov assumption is useful for defining distributions on 1d sequences, how can we define distributions on 2d images, or 3d videos, or, in general, arbitrary collections of variables (such as genes belonging to some biological pathway)? This is where graphical models come in.

A **graphical model** (**GM**) is a way to represent a joint distribution by making CI assumptions. In particular, the nodes in the graph represent random variables, and the (lack of) edges represent CI assumptions. (A better name for these models would in fact be "independence diagrams", but the term "graphical models" is now entrenched.) There are several kinds of graphical model, depending on whether the graph is directed, undirected, or some combination of directed and undirected. In this chapter, we just study directed graphs. We consider undirected graphs in Chapter 19.

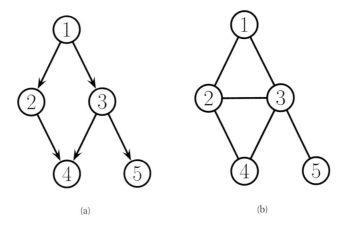

(a)   (b)

Figure 10.1   (a) A simple DAG on 5 nodes, numbered in topological order. Node 1 is the root, nodes 4 and 5 are the leaves. (b) A simple undirected graph, with the following maximal cliques: $\{1, 2, 3\}$, $\{2, 3, 4\}$, $\{3, 5\}$.

10.1.4   Graph terminology

Before we continue, we must define a few basic terms, most of which are very intuitive.

A **graph** $G = (\mathcal{V}, \mathcal{E})$ consists of a set of **nodes** or **vertices**, $\mathcal{V} = \{1, \ldots, V\}$, and a set of **edges**, $\mathcal{E} = \{(s, t) : s, t \in \mathcal{V}\}$. We can represent the graph by its **adjacency matrix**, in which we write $G(s, t) = 1$ to denote $(s, t) \in \mathcal{E}$, that is, if $s \rightarrow t$ is an edge in the graph. If $G(s, t) = 1$ iff $G(t, s) = 1$, we say the graph is **undirected**, otherwise it is **directed**. We usually assume $G(s, s) = 0$, which means there are no **self loops**.

Here are some other terms we will commonly use:

- **Parent** For a directed graph, the **parents** of a node is the set of all nodes that feed into it: $\mathrm{pa}(s) \triangleq \{t : G(t, s) = 1\}$.
- **Child** For a directed graph, the **children** of a node is the set of all nodes that feed out of it: $\mathrm{ch}(s) \triangleq \{t : G(s, t) = 1\}$.
- **Family** For a directed graph, the **family** of a node is the node and its parents, $\mathrm{fam}(s) = \{s\} \cup \mathrm{pa}(s)$.
- **Root** For a directed graph, a **root** is a node with no parents.
- **Leaf** For a directed graph, a **leaf** is a node with no children.
- **Ancestors** For a directed graph, the **ancestors** are the parents, grand-parents, etc of a node. That is, the ancestors of t is the set of nodes that connect to t via a trail: $\mathrm{anc}(t) \triangleq \{s : s \rightsquigarrow t\}$.
- **Descendants** For a directed graph, the **descendants** are the children, grand-children, etc of a node. That is, the descendants of s is the set of nodes that can be reached via trails from s: $\mathrm{desc}(s) \triangleq \{t : s \rightsquigarrow t\}$.
- **Neighbors** For any graph, we define the **neighbors** of a node as the set of all immediately connected nodes, $\mathrm{nbr}(s) \triangleq \{t : G(s, t) = 1 \vee G(t, s) = 1\}$. For an undirected graph, we

write $s \sim t$ to indicate that s and t are neighbors (so $(s, t) \in \mathcal{E}$ is an edge in the graph).

- **Degree** The **degree** of a node is the number of neighbors. For directed graphs, we speak of the **in-degree** and **out-degree**, which count the number of parents and children.

- **Cycle or loop** For any graph, we define a **cycle** or **loop** to be a series of nodes such that we can get back to where we started by following edges, $s_1 - s_2 \cdots - s_n - s_1$, $n \geq 2$. If the graph is directed, we may speak of a directed cycle. For example, in Figure 10.1(a), there are no directed cycles, but $1 \to 2 \to 4 \to 3 \to 1$ is an undirected cycle.

- **DAG** A **directed acyclic graph** or **DAG** is a directed graph with no directed cycles. See Figure 10.1(a) for an example.

- **Topological ordering** For a DAG, a **topological ordering** or **total ordering** is a numbering of the nodes such that parents have lower numbers than their children. For example, in Figure 10.1(a), we can use $(1, 2, 3, 4, 5)$, or $(1, 3, 2, 5, 4)$, etc.

- **Path or trail** A **path** or **trail** $s \rightsquigarrow t$ is a series of directed edges leading from s to t.

- **Tree** An undirected **tree** is an undirectecd graph with no cycles. A directed tree is a DAG in which there are no directed cycles. If we allow a node to have multiple parents, we call it a **polytree**, otherwise we call it a moral directed tree.

- **Forest** A **forest** is a set of trees.

- **Subgraph** A (node-induced) **subgraph** G_A is the graph created by using the nodes in A and their corresponding edges, $G_A = (\mathcal{V}_A, \mathcal{E}_A)$.

- **Clique** For an undirected graph, a **clique** is a set of nodes that are all neighbors of each other. A **maximal clique** is a clique which cannot be made any larger without losing the clique property. For example, in Figure 10.1(b), $\{1, 2\}$ is a clique but it is not maximal, since we can add 3 and still maintain the clique property. In fact, the maximal cliques are as follows: $\{1, 2, 3\}$, $\{2, 3, 4\}$, $\{3, 5\}$.

10.1.5 Directed graphical models

A **directed graphical model** or **DGM** is a GM whose graph is a DAG. These are more commonly known as **Bayesian networks**. However, there is nothing inherently "Bayesian" about Bayesian networks: they are just a way of defining probability distributions. These models are also called **belief networks**. The term "belief" here refers to subjective probability. Once again, there is nothing inherently subjective about the kinds of probability distributions represented by DGMs. Finally, these models are sometimes called **causal networks**, because the directed arrows are sometimes interpreted as representing causal relations. However, there is nothing inherently causal about DGMs. (See Section 26.6.1 for a discussion of causal DGMs.) For these reasons, we use the more neutral (but less glamorous) term DGM.

The key property of DAGs is that the nodes can be ordered such that parents come before children. This is called a topological ordering, and it can be constructed from any DAG. Given such an order, we define the **ordered Markov property** to be the assumption that a node only depends on its immediate parents, not on all predecessors in the ordering, i.e.,

$$x_s \perp \mathbf{x}_{\text{pred}(s) \backslash \text{pa}(s)} | \mathbf{x}_{\text{pa}(s)} \tag{10.4}$$

where $\text{pa}(s)$ are the parents of node s, and $\text{pred}(s)$ are the predecessors of node s in the ordering. This is a natural generalization of the first-order Markov property to from chains to general DAGs.

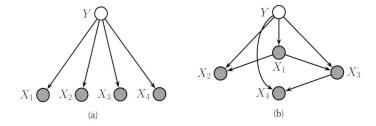

Figure 10.2   (a) A naive Bayes classifier represented as a DGM. We assume there are $D = 4$ features, for simplicity. Shaded nodes are observed, unshaded nodes are hidden. (b) Tree-augmented naive Bayes classifier for $D = 4$ features. In general, the tree topology can change depending on the value of y.

For example, the DAG in Figure 10.1(a) encodes the following joint distribution:

$$p(\mathbf{x}_{1:5}) \quad = \quad p(x_1)p(x_2|x_1)p(x_3|x_1,\cancel{x_2})p(x_4|\cancel{x_1},x_2,x_3)p(x_5|\cancel{x_1},\cancel{x_2},x_3,\cancel{x_4}) \qquad (10.5)$$

$$= \quad p(x_1)p(x_2|x_1)p(x_3|x_1)p(x_4|x_2,x_3)p(x_5|x_3) \qquad (10.6)$$

In general, we have

$$p(\mathbf{x}_{1:V}|G) = \prod_{t=1}^{V} p(x_t|\mathbf{x}_{\mathrm{pa}(t)}) \qquad (10.7)$$

where each term $p(x_t|\mathbf{x}_{\mathrm{pa}(t)})$ is a CPD. We have written the distribution as $p(\mathbf{x}|G)$ to emphasize that this equation only holds if the CI assumptions encoded in DAG G are correct. However, we will usual drop this explicit conditioning for brevity. If each node has $O(F)$ parents and K states, the number of parameters in the model is $O(VK^F)$, which is much less than the $O(K^V)$ needed by a model which makes no CI assumptions.

10.2   Examples

In this section, we show a wide variety of commonly used probabilistic models can be conveniently represented as DGMs.

10.2.1   Naive Bayes classifiers

In Section 3.5, we introduced the naive Bayes classifier. This assumes the features are conditionally independent given the class label. This assumption is illustrated in Figure 10.2(a). This allows us to write the joint distirbution as follows:

$$p(y, \mathbf{x}) = p(y) \prod_{j=1}^{D} p(x_j|y) \qquad (10.8)$$

The naive Bayes assumption is rather naive, since it assumes the features are conditionally independent. One way to capture correlation between the features is to use a graphical model. In particular, if the model is a tree, the method is known as a **tree-augmented naive Bayes**

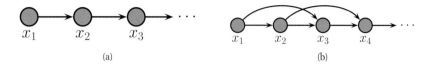

Figure 10.3   A first and second order Markov chain.

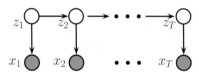

Figure 10.4   A first-order HMM.

classifier or **TAN** model (Friedman et al. 1997). This is illustrated in Figure 10.2(b). The reason to use a tree, as opposed to a generic graph, is two-fold. First, it is easy to find the optimal tree structure using the Chow-Liu algorithm, as explained in Section 26.3. Second, it is easy to handle missing features in a tree-structured model, as we explain in Section 20.2.

10.2.2   Markov and hidden Markov models

Figure 10.3(a) illustrates a first-order Markov chain as a DAG. Of course, the assumption that the immediate past, x_{t-1}, captures everything we need to know about the entire history, $\mathbf{x}_{1:t-2}$, is a bit strong. We can relax it a little by adding a dependence from x_{t-2} to x_t as well; this is called a **second order Markov chain**, and is illustrated in Figure 10.3(b). The corresponding joint has the following form:

$$p(\mathbf{x}_{1:T}) = p(x_1, x_2)p(x_3|x_1, x_2)p(x_4|x_2, x_3)\ldots = p(x_1, x_2)\prod_{t=3}^{T} p(x_t|x_{t-1}, x_{t-2}) \qquad (10.9)$$

We can create higher-order Markov models in a similar way. See Section 17.2 for a more detailed discussion of Markov models.

Unfortunately, even the second-order Markov assumption may be inadequate if there are long-range correlations amongst the observations. We can't keep building ever higher order models, since the number of parameters will blow up. An alternative approach is to assume that there is an underlying hidden process, that can be modeled by a first-order Markov chain, but that the data is a noisy observation of this process. The result is known as a **hidden Markov model** or **HMM**, and is illustrated in Figure 10.4. Here z_t is known as a **hidden variable** at "time" t, and x_t is the observed variable. (We put "time" in quotation marks, since these models can be applied to any kind of sequence data, such as genomics or language, where t represents location rather than time.) The CPD $p(z_t|z_{t-1})$ is the **transition model**, and the CPD $p(\mathbf{x}_t|z_t)$ is the **observation model**.

h_0	h_1	h_2	$P(v=0|h_1,h_2)$	$P(v=1|h_1,h_2)$
1	0	0	θ_0	$1-\theta_0$
1	1	0	$\theta_0\theta_1$	$1-\theta_0\theta_1$
1	0	1	$\theta_0\theta_2$	$1-\theta_0\theta_2$
1	1	1	$\theta_0\theta_1\theta_2$	$1-\theta_0\theta_1\theta_2$

Table 10.1   Noisy-OR CPD for 2 parents augmented with leak node. We have omitted the t subscript for brevity.

The hidden variables often represent quantities of interest, such as the identity of the word that someone is currently speaking. The observed variables are what we measure, such as the acoustic waveform. What we would like to do is estimate the hidden state given the data, i.e., to compute $p(z_t|\mathbf{x}_{1:t}, \boldsymbol{\theta})$. This is called **state estimation**, and is just another form of probabilistic inference. See Chapter 17 for further details on HMMs.

10.2.3   Medical diagnosis

Consider modeling the relationship between various variables that are measured in an intensive care unit (ICU), such as the breathing rate of a patient, their blood pressure, etc. The **alarm network** in Figure 10.5(a) is one way to represent these (in)dependencies (Beinlich et al. 1989). This model has 37 variables and 504 parameters.

Since this model was created by hand, by a process called **knowledge engineering**, it is known as a **probabilistic expert system**. In Section 10.4, we discuss how to learn the parameters of DGMs from data, assuming the graph structure is known, and in Chapter 26, we discuss how to learn the graph structure itself.

A different kind of medical diagnosis network, known as the **quick medical reference** or **QMR** network (Shwe et al. 1991), is shown in Figure 10.5(b). This was designed to model infectious diseases. The QMR model is a **bipartite graph** structure, with diseases (causes) at the top and symptoms or findings at the bottom. All nodes are binary. We can write the distribution as follows:

$$p(\mathbf{v}, \mathbf{h}) = \prod_s p(h_s) \prod_t p(v_t|\mathbf{h}_{\mathrm{pa}(t)}) \tag{10.10}$$

where h_s represent the **hidden nodes** (diseases), and v_t represent the **visible nodes** (symptoms). The CPD for the root nodes are just Bernoulli distributions, representing the prior probability of that disease. Representing the CPDs for the leaves (symptoms) using CPTs would require too many parameters, because the **fan-in** (number of parents) of many leaf nodes is very high. A natural alternative is to use logistic regression to model the CPD, $p(v_t = 1|\mathbf{h}_{\mathrm{pa}(t)}) = \mathrm{sigm}(\mathbf{w}_t^T\mathbf{h}_{\mathrm{pa}(t)})$. (A DGM in which the CPDs are logistic regression distributions is known as a **sigmoid belief net** (Neal 1992).) However, since the parameters of this model were created by hand, an alternative CPD, known as the **noisy-OR** model, was used.

The noisy-OR model assumes that if a parent is on, then the child will usually also be on (since it is an or-gate), but occasionally the "links" from parents to child may fail, independently at random. In this case, even if the parent is on, the child may be off. To model this more precisely, let $\theta_{st} = 1 - q_{st}$ be the probability that the $s \rightarrow t$ link fails, so $q_{st} = 1 - \theta_{st} = p(v_t =$

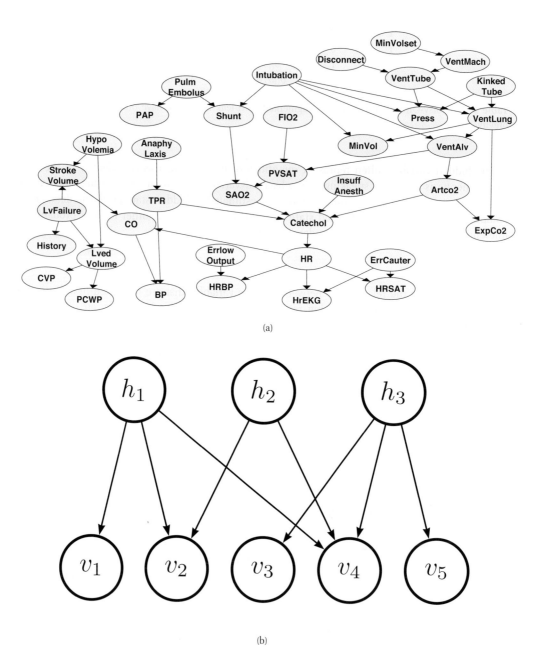

(a)

(b)

Figure 10.5 (a) The alarm network. Figure generated by `visualizeAlarmNetwork`. (b) The QMR network. All nodes are binary. The hidden nodes h_s represent diseases, and the visible nodes v_t represent symptoms. In the full network, there are 570 hidden (disease) nodes and 4075 visible (symptom) nodes.

G^p	G^m	$p(X = a)$	$p(X = b)$	$p(X = o)$	$p(X = ab)$
a	a	1	0	0	0
a	b	0	0	0	1
a	o	1	0	0	0
b	a	0	0	0	1
b	b	0	1	0	0
b	o	0	1	0	0
o	a	1	0	0	0
o	b	0	1	0	0
o	o	0	0	1	0

Table 10.2 CPT which encodes a mapping from genotype to phenotype (bloodtype). This is a deterministic, but many-to-one, mapping.

$1|h_s = 1, \mathbf{h}_{-s} = 0)$ is the probability that s can activate t on its own (its "causal power"). The only way for the child to be off is if all the links from all parents that are on fail independently at random. Thus

$$p(v_t = 0|\mathbf{h}) = \prod_{s \in pa(t)} \theta_{st}^{\mathbb{I}(h_s=1)} \qquad (10.11)$$

Obviously, $p(v_t = 1|\mathbf{h}) = 1 - p(v_t = 0|\mathbf{h})$.

If we observe that $v_t = 1$ but all its parents are off, then this contradicts the model. Such a data case would get probability zero under the model, which is problematic, because it is possible that someone exhibits a symptom but does not have any of the specified diseases. To handle this, we add a dummy **leak node** h_0, which is always on; this represents "all other causes". The parameter q_{0t} represents the probability that the background leak can cause the effect on its own. The modified CPD becomes $p(v_t = 0|\mathbf{h}) = \theta_{0t} \prod_{s \in pa(t)} \theta_{st}^{h_s}$. See Table 10.1 for a numerical example.

If we define $w_{st} \triangleq \log(\theta_{st})$, we can rewrite the CPD as

$$p(v_t = 1|\mathbf{h}) = 1 - \exp\left(w_{0t} + \sum_s h_s w_{st}\right) \qquad (10.12)$$

We see that this is similar to a logistic regression model.

Bipartite models with noisy-OR CPDs are called **BN2O** models. It is relatively easy to set the θ_{st} parameters by hand, based on domain expertise. However, it is also possible to learn them from data (see e.g, (Neal 1992; Meek and Heckerman 1997)). Noisy-OR CPDs have also proved useful in modeling human causal learning (Griffiths and Tenenbaum 2005), as well as general binary classification settings (Yuille and Zheng 2009).

10.2.4 Genetic linkage analysis *

Another important (and historically very early) application of DGMs is to the problem of **genetic linkage analysis**. We start with a **pedigree graph**, which is a DAG that representing the relationship between parents and children, as shown in Figure 10.6(a). We then convert this to a DGM, as we explain below. Finally we perform probabilistic inference in the resulting model.

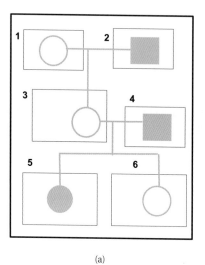

(a)

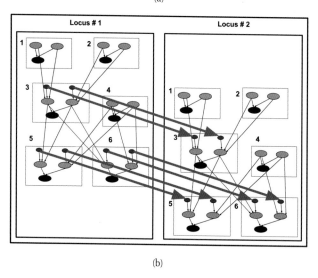

(b)

Figure 10.6 Left: family tree, circles are females, squares are males. Individuals with the disease of interest are highlighted. Right: DGM for two loci. Blue nodes X_{ij} is the observed phenotype for individual i at locus j. All other nodes are hidden. Orange nodes $G_{ij}^{p/m}$ is the paternal/ maternal allele. Small red nodes $z_{ijl}^{p/m}$ are the paternal/ maternal selection switching variables. These are linked across loci, $z_{ij}^m \rightarrow z_{i,j+1}^m$ and $z_{ij}^p \rightarrow z_{i,j+1}^p$. The founder (root) nodes do not have any parents, and hence do no need switching variables. Based on Figure 3 from (Friedman et al. 2000).

In more detail, for each person (or animal) i and location or locus j along the genome, we create three nodes: the observed **marker** X_{ij} (which can be a property such as blood type, or just a fragment of DNA that can be measured), and two hidden **alleles**, G_{ij}^m and G_{ij}^p, one inherited from i's mother (maternal allele) and the other from i's father (paternal allele). Together, the ordered pair $\mathbf{G}_{ij} = (G_{ij}^m, G_{ij}^p)$ constitutes i's hidden **genotype** at locus j.

Obviously we must add $G_{ij}^m \to X_{ij}$ and $G_{ij}^p \to X_{ij}$ arcs representing the fact that genotypes cause phenotypes (observed manifestations of genotypes). The CPD $p(X_{ij}|G_{ij}^m, G_{ij}^p)$ is called the **penetrance model**. As a very simple example, suppose $X_{ij} \in \{A, B, O, AB\}$ represents person i's observed bloodtype, and $G_{ij}^m, G_{ij}^p \in \{A, B, O\}$ is their genotype. We can represent the penetrance model using the deterministic CPD shown in Table 10.2. For example, A dominates O, so if a person has genotype AO or OA, their phenotype will be A.

In addition, we add arcs from i's mother and father into $\mathbf{G}_{ij}$, reflecting the **Mendelian inheritance** of genetic material from one's parents. More precisely, let $m_i = k$ be i's mother. Then G_{ij}^m could either be equal to G_{kj}^m or G_{kj}^p, that is, i's maternal allele is a copy of one of its mother's two alleles. Let Z_{ij}^m be a hidden variable than specifies the choice. We can model this using the following CPD, known as the **inheritance model**:

$$p(G_{ij}^m|G_{kj}^m, G_{kj}^p, Z_{ij}^m) = \begin{cases} \mathbb{I}(G_{ij}^m = G_{kj}^m) & \text{if } Z_{ij}^m = m \\ \mathbb{I}(G_{ij}^m = G_{kj}^p) & \text{if } Z_{ij}^m = p \end{cases} \tag{10.13}$$

We can define $p(G_{ij}^p|G_{kj}^m, G_{kj}^p, Z_{ij}^p)$ similarly, where $k = p_i$ is i's father. The values of the Z_{ij} are said to specify the **phase** of the genotype. The values of $G_{i,j}^p$, $G_{i,j}^m$, $Z_{i,j}^p$ and $Z_{i,j}^m$ constitute the **haplotype** of person i at locus j.[1]

Next, we need to specify the prior for the root nodes, $p(G_{ij}^m)$ and $p(G_{ij}^p)$. This is called the **founder model**, and represents the overall prevalence of difference kinds of alleles in the population. We usually assume independence between the loci for these founder alleles.

Finally, we need to specify priors for the switch variables that control the inheritance process. These variables are spatially correlated, since adjacent sites on the genome are typically inherited together (recombination events are rare). We can model this by imposing a two-state Markov chain on the Z's, where the probability of switching state at locus j is given by $\theta_j = \frac{1}{2}(1 - e^{-2d_j})$, where d_j is the distance between loci j and $j + 1$. This is called the **recombination model**.

The resulting DGM is shown in Figure 10.6(b): it is a series of replicated pedigree DAGs, augmented with switching Z variables, which are linked using Markov chains. (There is a related model known as **phylogenetic HMM** (Siepel and Haussler 2003), which is used to model evolution amongst phylogenies.)

As a simplified example of how this model can be used, suppose we only have one locus, corresponding to blood type. For brevity, we will drop the j index. Suppose we observe $x_i = A$. Then there are 3 possible genotypes: $\mathbf{G}_i$ is (A, A), (A, O) or (O, A). There is ambiguity because the genotype to phenotype mapping is many-to-one. We want to reverse this mapping. This is known as an **inverse problem**. Fortunately, we can use the blood types of relatives to help disambiguate the evidence. Information will "flow" from the other $x_{i'}$'s up to their $\mathbf{G}_{i'}$'s, then across to i's $\mathbf{G}_i$ via the pedigree DAG. Thus we can combine our **local evidence** $p(x_i|\mathbf{G}_i)$

---

1. Sometimes the observed marker is equal to the unphased genotype, which is the unordered set $\{G_{ij}^p, G_{ij}^m\}$; however, the phased or hidden genotype is not directly measurable.

with an informative prior, $p(\mathbf{G}_i|\mathbf{x}_{-i})$, conditioned on the other data, to get a less entropic local posterior, $p(G_i|\mathbf{x}) \propto p(x_i|G_i)p(G_i|\mathbf{x}_{-i})$.

In practice, the model is used to try to determine where along the genome a given disease-causing gene is assumed to lie — this is the genetic linkage analysis task. The method works as follows. First, suppose all the parameters of the model, including the distance between all the marker loci, are known. The only unknown is the location of the disease-causing gene. If there are L marker loci, we construct $L + 1$ models: in model ℓ, we postulate that the disease gene comes after marker ℓ, for $0 < \ell < L + 1$. We can estimate the Markov switching parameter $\hat{\theta}_\ell$, and hence the distance d_ℓ between the disease gene and its nearest known locus. We measure the quality of that model using its likelihood, $p(\mathcal{D}|\hat{\theta}_\ell)$. We then can then pick the model with highest likelihood (which is equivalent to the MAP model under a uniform prior).

Note, however, that computing the likelihood requires marginalizing out all the hidden Z and G variables. See (Fishelson and Geiger 2002) and the references therein for some exact methods for this task; these are based on the variable elimination algorithm, which we discuss in Section 20.3. Unfortunately, for reasons we explain in Section 20.5, exact methods can be computationally intractable if the number of individuals and/or loci is large. See (Albers et al. 2006) for an approximate method for computing the likelihood; this is based on a form of variational inference, which we will discuss in Section 22.4.1.

10.2.5 Directed Gaussian graphical models *

Consider a DGM where all the variables are real-valued, and all the CPDs have the following form:

$$p(x_t|\mathbf{x}_{\mathrm{pa}(t)}) = \mathcal{N}(x_t|\mu_t + \mathbf{w}_t^T \mathbf{x}_{\mathrm{pa}(t)}, \sigma_t^2) \tag{10.14}$$

This is called a **linear Gaussian** CPD. As we show below, multiplying all these CPDs together results in a large joint Gaussian distribution of the form $p(\mathbf{x}) = \mathcal{N}(\mathbf{x}|\boldsymbol{\mu}, \boldsymbol{\Sigma})$. This is called a directed GGM, or a **Gaussian Bayes net**.

We now explain how to derive $\boldsymbol{\mu}$ and $\boldsymbol{\Sigma}$ from the CPD parameters, following (Shachter and Kenley 1989, App. B). For convenience, we will rewrite the CPDs in the following form:

$$x_t = \mu_t + \sum_{s \in \mathrm{pa}(t)} w_{ts}(x_s - \mu_s) + \sigma_t z_t \tag{10.15}$$

where $z_t \sim \mathcal{N}(0, 1)$, σ_t is the conditional standard deviation of x_t given its parents, w_{ts} is the strength of the $s \to t$ edge, and μ_t is the local mean.[2]

It is easy to see that the global mean is just the concatenation of the local means, $\boldsymbol{\mu} = (\mu_1, \ldots, \mu_D)$. We now derive the global covariance, $\boldsymbol{\Sigma}$. Let $\mathbf{S} \triangleq \mathrm{diag}(\boldsymbol{\sigma})$ be a diagonal matrix containing the standard deviations. We can rewrite Equation 10.15 in matrix-vector form as follows:

$$(\mathbf{x} - \boldsymbol{\mu}) = \mathbf{W}(\mathbf{x} - \boldsymbol{\mu}) + \mathbf{S}\mathbf{z} \tag{10.16}$$

---

2. If we do not subtract off the parent's mean (i.e., if we use $x_t = \mu_t + \sum_{s \in \mathrm{pa}(t)} w_{ts}x_s + \sigma_t z_t$), the derivation of $\boldsymbol{\Sigma}$ is much messier, as can be seen by looking at (Bishop 2006, p370).

where $\mathbf{W}$ is the matrix of regression weights. Now let $\mathbf{e}$ be a vector of noise terms:

$$\mathbf{e} \triangleq \mathbf{S}\mathbf{z} \tag{10.17}$$

We can rearrange this to get

$$\mathbf{e} = (\mathbf{I} - \mathbf{W})(\mathbf{x} - \boldsymbol{\mu}) \tag{10.18}$$

Since $\mathbf{W}$ is lower triangular (because $w_{ts} = 0$ if $t > s$ in the topological ordering), we have that $\mathbf{I} - \mathbf{W}$ is lower triangular with 1s on the diagonal. Hence

$$\begin{pmatrix} e_1 \\ e_2 \\ \vdots \\ e_d \end{pmatrix} = \begin{pmatrix} 1 & & & & \\ -w_{21} & 1 & & & \\ -w_{32} & -w_{31} & 1 & & \\ \vdots & & & \ddots & \\ -w_{d1} & -w_{d2} & \dots & -w_{d,d-1} & 1 \end{pmatrix} \begin{pmatrix} x_1 - \mu_1 \\ x_2 - \mu_2 \\ \vdots \\ x_d - \mu_d \end{pmatrix} \tag{10.19}$$

Since $\mathbf{I} - \mathbf{W}$ is always invertible, we can write

$$\mathbf{x} - \boldsymbol{\mu} = (\mathbf{I} - \mathbf{W})^{-1}\mathbf{e} \triangleq \mathbf{U}\mathbf{e} = \mathbf{U}\mathbf{S}\mathbf{z} \tag{10.20}$$

where we defined $\mathbf{U} = (\mathbf{I} - \mathbf{W})^{-1}$. This matrix is part of the Cholesky decomposition of $\boldsymbol{\Sigma}$, as we now show:

$$\begin{aligned} \boldsymbol{\Sigma} &= \operatorname{cov}[\mathbf{x}] = \operatorname{cov}[\mathbf{x} - \boldsymbol{\mu}] \tag{10.21} \\ &= \operatorname{cov}[\mathbf{U}\mathbf{S}\mathbf{z}] = \mathbf{U}\mathbf{S}\operatorname{cov}[\mathbf{z}]\,\mathbf{S}\mathbf{U}^T = \mathbf{U}\mathbf{S}^2\mathbf{U}^T \tag{10.22} \end{aligned}$$

10.3 Inference

We have seen that graphical models provide a compact way to define joint probability distributions. Given such a joint distribution, what can we do with it? The main use for such a joint distribution is to perform **probabilistic inference**. This refers to the task of estimating unknown quantities from known quantities. For example, in Section 10.2.2, we introduced HMMs, and said that one of the goals is to estimate the hidden states (e.g., words) from the observations (e.g., speech signal). And in Section 10.2.4, we discussed genetic linkage analysis, and said that one of the goals is to estimate the likelihood of the data under various DAGs, corresponding to different hypotheses about the location of the disease-causing gene.

In general, we can pose the inference problem as follows. Suppose we have a set of correlated random variables with joint distribution $p(\mathbf{x}_{1:V}|\boldsymbol{\theta})$. (In this section, we are assuming the parameters $\boldsymbol{\theta}$ of the model are known. We discuss how to learn the parameters in Section 10.4.) Let us partition this vector into the **visible variables** $\mathbf{x}_v$, which are observed, and the **hidden variables**, $\mathbf{x}_h$, which are unobserved. Inference refers to computing the posterior distribution of the unknowns given the knowns:

$$p(\mathbf{x}_h|\mathbf{x}_v, \boldsymbol{\theta}) = \frac{p(\mathbf{x}_h, \mathbf{x}_v|\boldsymbol{\theta})}{p(\mathbf{x}_v|\boldsymbol{\theta})} = \frac{p(\mathbf{x}_h, \mathbf{x}_v|\boldsymbol{\theta})}{\sum_{\mathbf{x}_h'} p(\mathbf{x}_h', \mathbf{x}_v|\boldsymbol{\theta})} \tag{10.23}$$

Essentially we are **conditioning** on the data by **clamping** the visible variables to their observed values, $\mathbf{x}_v$, and then normalizing, to go from $p(\mathbf{x}_h, \mathbf{x}_v)$ to $p(\mathbf{x}_h|\mathbf{x}_v)$. The normalization constant $p(\mathbf{x}_v|\boldsymbol{\theta})$ is the likelihood of the data, also called the **probability of the evidence**.

Sometimes only some of the hidden variables are of interest to us. So let us partition the hidden variables into **query variables**, $\mathbf{x}_q$, whose value we wish to know, and the remaining **nuisance variables**, $\mathbf{x}_u$, which are "uninteresting". We can compute the desired marginal on the query nodes by **marginalizing out** the nuisance variables:

$$p(\mathbf{x}_q|\mathbf{x}_v, \boldsymbol{\theta}) = \sum_{\mathbf{x}_u} p(\mathbf{x}_q, \mathbf{x}_u|\mathbf{x}_v, \boldsymbol{\theta}) \tag{10.24}$$

In Section 4.3.1, we saw how to perform all these operations for a multivariate Gaussian in $O(V^3)$ time, where V is the number of variables. What if we have discrete random variables, with say K states each? If the joint distribution is represented as a multi-dimensional table, we can always perform these operations exactly, but this will take $O(K^V)$ time. In Chapter 20, we explain how to exploit the factorization encoded by the GM to perform these operations in $O(VK^{w+1})$ time, where w is a quantity known as the treewidth of the graph. This measures how "tree-like" the graph is. If the graph is a tree (or a chain), we have $w = 1$, so for these models, inference takes time linear in the number of nodes. Unfortunately, for more general graphs, exact inference can take time exponential in the number of nodes, as we explain in Section 20.5. We will therefore examine various approximate inference schemes later in the book.

10.4   Learning

In the graphical models literature, it is common to distinguish between inference and learning. Inference means computing (functions of) $p(\mathbf{x}_h|\mathbf{x}_v, \boldsymbol{\theta})$, where v are the visible nodes, h are the hidden nodes, and $\boldsymbol{\theta}$ are the parameters of the model, assumed to be known. Learning usually means computing a MAP estimate of the parameters given data:

$$\hat{\boldsymbol{\theta}} = \underset{\boldsymbol{\theta}}{\operatorname{argmax}} \sum_{i=1}^{N} \log p(\mathbf{x}_{i,v}|\boldsymbol{\theta}) + \log p(\boldsymbol{\theta}) \tag{10.25}$$

where $\mathbf{x}_{i,v}$ are the visible variables in case i. If we have a uniform prior, $p(\boldsymbol{\theta}) \propto 1$, this reduces to the MLE, as usual.

If we adopt a Bayesian view, the parameters are unknown variables and should also be inferred. Thus to a Bayesian, there is no distinction between inference and learning. In fact, we can just add the parameters as nodes to the graph, condition on $\mathcal{D}$, and then infer the values of all the nodes. (We discuss this in more detail below.)

In this view, the main difference between hidden variables and parameters is that the number of hidden variables grows with the amount of training data (since there is usually a set of hidden variables for each observed data case), whereas the number of parameters in usually fixed (at least in a parametric model). This means that we must integrate out the hidden variables to avoid overfitting, but we may be able to get away with point estimation techniques for parameters, which are fewer in number.

10.4.1   Plate notation

When inferring parameters from data, we often assume the data is iid. We can represent this assumption explicitly using a graphical model, as shown in Figure 10.7(a). This illustrates the

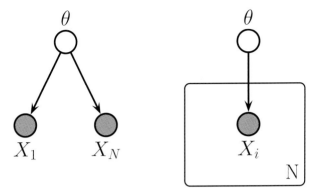

Figure 10.7 Left: data points x_i are conditionally independent given θ. Right: Plate notation. This represents the same model as the one on the left, except the repeated x_i nodes are inside a box, known as a plate; the number in the lower right hand corner, N, specifies the number of repetitions of the X_i node.

assumption that each data case was generated independently but from the same distribution. Notice that the data cases are only independent conditional on the parameters $\boldsymbol{\theta}$; marginally, the data cases are dependent. Nevertheless, we can see that, in this example, the order in which the data cases arrive makes no difference to our beliefs about $\boldsymbol{\theta}$, since all orderings will have the same sufficient statistics. Hence we say the data is **exchangeable**.

To avoid visual clutter, it is common to use a form of **syntactic sugar** called **plates**: we simply draw a little box around the repeated variables, with the convention that nodes within the box will get repeated when the model is **unrolled**. We often write the number of copies or repetitions in the bottom right corner of the box. See Figure 10.7(b) for a simple example. The corresponding joint distribution has the form

$$p(\boldsymbol{\theta}, \mathcal{D}) = p(\boldsymbol{\theta}) \left[\prod_{i=1}^{N} p(\mathbf{x}_i|\boldsymbol{\theta}) \right] \tag{10.26}$$

This DGM represents the CI assumptions behind the models we considered in Chapter 5.

A slightly more complex example is shown in Figure 10.8. On the left we show a naive Bayes classifier that has been "unrolled" for D features, but uses a plate to represent repetition over cases $i = 1 : N$. The version on the right shows the same model using **nested plate** notation. When a variable is inside two plates, it will have two sub-indices. For example, we write θ_{jc} to represent the parameter for feature j in class-conditional density c. Note that plates can be nested or crossing. Notational devices for modeling more complex parameter tying patterns can be devised (e.g., (Heckerman et al. 2004)), but these are not widely used. What is not clear from the figure is that θ_{jc} is used to generate x_{ij} iff $y_i = c$, otherwise it is ignored. This is an example of **context specific independence**, since the CI relationship $x_{ij} \perp \theta_{jc}$ only holds if $y_i \neq c$.

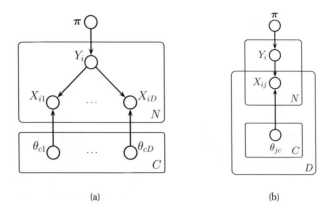

(a)   (b)

Figure 10.8   Naive Bayes classifier as a DGM. (a) With single plates. (b) WIth nested plates.

10.4.2   Learning from complete data

If all the variables are fully observed in each case, so there is no missing data and there are no
hidden variables, we say the data is **complete**. For a DGM with complete data, the likelihood is
given by

$$p(\mathcal{D}|\boldsymbol{\theta}) \quad = \quad \prod_{i=1}^{N} p(\mathbf{x}_i|\boldsymbol{\theta}) = \prod_{i=1}^{N}\prod_{t=1}^{V} p(x_{it}|\mathbf{x}_{i,\mathrm{pa}(t)}, \boldsymbol{\theta}_t) = \prod_{t=1}^{V} p(\mathcal{D}_t|\boldsymbol{\theta}_t) \tag{10.27}$$

where $\mathcal{D}_t$ is the data associated with node t and its parents, i.e., the t'th family. This is a
product of terms, one per CPD. We say that the likelihood **decomposes** according to the graph
structure.

Now suppose that the prior factorizes as well:

$$p(\boldsymbol{\theta}) = \prod_{t=1}^{V} p(\boldsymbol{\theta}_t) \tag{10.28}$$

Then clearly the posterior also factorizes:

$$p(\boldsymbol{\theta}|\mathcal{D}) \propto p(\mathcal{D}|\boldsymbol{\theta})p(\boldsymbol{\theta}) = \prod_{t=1}^{V} p(\mathcal{D}_t|\boldsymbol{\theta}_t)p(\boldsymbol{\theta}_t) \tag{10.29}$$

This means we can compute the posterior of each CPD independently. In other words,

factored prior plus factored likelihood implies factored posterior   (10.30)

Let us consider an example, where all CPDs are tabular, thus extending the earlier results of
Secion 3.5.1.2, where discussed Bayesian naive Bayes. We have a separate row (i.e., a separate
multinoulli distribution) for each **conditioning case**, i.e., for each combination of parent values,
as in Table 10.2. Formally, we can write the t'th CPT as $x_t|\mathbf{x}_{\mathrm{pa}(t)} = c \sim \mathrm{Cat}(\boldsymbol{\theta}_{tc})$, where
$\theta_{tck} \triangleq p(x_t = k|\mathbf{x}_{\mathrm{pa}(t)} = c)$, for $k = 1 : K_t$, $c = 1 : C_t$ and $t = 1 : T$. Here K_t is the number

of states for node t, $C_t \triangleq \prod_{s \in \mathrm{pa}(t)} K_s$ is the number of parent combinations, and T is the number of nodes. Obviously $\sum_k \theta_{tck} = 1$ for each row of each CPT.

Let us put a separate Dirichlet prior on each row of each CPT, i.e., $\boldsymbol{\theta}_{tc} \sim \mathrm{Dir}(\boldsymbol{\alpha}_{tc})$. Then we can compute the posterior by simply adding the pseudo counts to the empirical counts to get $\boldsymbol{\theta}_{tc}|\mathcal{D} \sim \mathrm{Dir}(\mathbf{N}_{tc} + \boldsymbol{\alpha}_{tc})$, where N_{tck} is the number of times that node t is in state k while its parents are in state c:

$$N_{tck} \triangleq \sum_{i=1}^{N} \mathbb{I}(x_{i,t} = k, x_{i,\mathrm{pa}(t)} = c) \tag{10.31}$$

From Equation 2.77, the mean of this distribution is given by the following:

$$\overline{\theta}_{tck} = \frac{N_{tck} + \alpha_{tck}}{\sum_{k'}(N_{tck'} + \alpha_{tck'})} \tag{10.32}$$

For example, consider the DGM in Figure 10.1(a). Suppose the training data consists of the following 5 cases:

x_1	x_2	x_3	x_4	x_5
0	0	1	0	0
0	1	1	1	1
1	1	0	1	0
0	1	1	0	0
0	1	1	1	0

Below we list all the sufficient statistics N_{tck}, and the posterior mean parameters $\overline{\theta}_{ick}$ under a Dirichlet prior with $\alpha_{ick} = 1$ (corresponding to add-one smoothing) for the $t = 4$ node:

x_2	x_3	$N_{tck=1}$	$N_{tck=0}$	$\overline{\theta}_{tck=1}$	$\overline{\theta}_{tck=0}$
0	0	0	0	1/2	1/2
1	0	1	0	2/3	1/3
0	1	0	1	1/3	2/3
1	1	2	1	3/5	2/5

It is easy to show that the MLE has the same form as Equation 10.32, except without the α_{tck} terms, i.e.,

$$\hat{\theta}_{tck} = \frac{N_{tck}}{\sum_{k'} N_{tck'}} \tag{10.33}$$

Of course, the MLE suffers from the zero-count problem discussed in Section 3.3.4.1, so it is important to use a prior to regularize the estimation problem.

10.4.3 Learning with missing and/or latent variables

If we have missing data and/or hidden variables, the likelihood no longer factorizes, and indeed it is no longer convex, as we explain in detail in Section 11.3. This means we will usually can only compute a locally optimal ML or MAP estimate. Bayesian inference of the parameters is even harder. We discuss suitable approximate inference techniques in later chapters.

10.5   Conditional independence properties of DGMs

At the heart of any graphical model is a set of conditional indepence (CI) assumptions. We write $\mathbf{x}_A \perp_G \mathbf{x}_B | \mathbf{x}_C$ if A is independent of B given C in the graph G, using the semantics to be defined below. Let $I(G)$ be the set of all such CI statements encoded by the graph.

We say that G is an **I-map** (independence map) for p, or that p is **Markov** wrt G, iff $I(G) \subseteq I(p)$, where $I(p)$ is the set of all CI statements that hold for distribution p. In other words, the graph is an I-map if it does not make any assertions of CI that are not true of the distribution. This allows us to use the graph as a safe proxy for p when reasoning about p's CI properties. This is helpful for designing algorithms that work for large classes of distributions, regardless of their specific numerical parameters $\boldsymbol{\theta}$.

Note that the fully connected graph is an I-map of all distributions, since it makes no CI assertions at all (since it is not missing any edges). We therefore say G is a **minimal I-map** of p if G is an I-map of p, and if there is no $G' \subseteq G$ which is an I-map of p.

It remains to specify how to determine if $\mathbf{x}_A \perp_G \mathbf{x}_B | \mathbf{x}_C$. Deriving these independencies for undirected graphs is easy (see Section 19.2), but the DAG situation is somewhat complicated, because of the need to respect the orientation of the directed edges. We give the details below.

10.5.1   d-separation and the Bayes Ball algorithm (global Markov properties)

First, we introduce some definitions. We say an *undirected path* P is **d-separated** by a set of nodes E (containing the evidence) iff at least one of the following conditions hold:

1. P contains a chain, $s \to m \to t$ or $s \leftarrow m \leftarrow t$, where $m \in E$

2. P contains a tent or fork, $s \swarrow^{m} \searrow t$, where $m \in E$

3. P contains a **collider** or **v-structure**, $s \searrow_{m} \swarrow t$, where m is not in E and nor is any descendant of m.

Next, we say that a *set of nodes* A is d-separated from a different set of nodes B given a third observed set E iff each undirected path from every node $a \in A$ to every node $b \in B$ is d-separated by E. Finally, we define the CI properties of a DAG as follows:

$$\mathbf{x}_A \perp_G \mathbf{x}_B | \mathbf{x}_E \iff \text{A is d-separated from B given E} \tag{10.34}$$

The **Bayes ball algorithm** (Shachter 1998) is a simple way to see if A is d-separated from B given E, based on the above definition. The idea is this. We "shade" all nodes in E, indicating that they are observed. We then place "balls" at each node in A, let them "bounce around" according to some rules, and then ask if any of the balls reach any of the nodes in B. The three main rules are shown in Figure 10.9. Notice that balls can travel opposite to edge directions. We see that a ball can pass through a chain, but not if it is shaded in the middle. Similarly, a ball can pass through a fork, but not if it is shaded in the middle. However, a ball cannot pass through a v-structure, unless it is shaded in the middle.

We can justify the 3 rules of Bayes ball as follows. First consider a chain structure $X \to Y \to Z$, which encodes

$$p(x, y, z) = p(x)p(y|x)p(z|y) \tag{10.35}$$

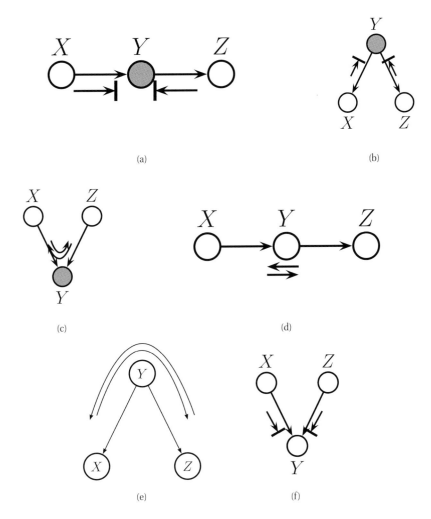

Figure 10.9 Bayes ball rules. A shaded node is one we condition on. If there is an arrow hitting a bar, it means the ball cannot pass through; otherwise the ball can pass through. Based on (Jordan 2007).

When we condition on y, are x and z independent? We have

$$p(x, z|y) \;=\; \frac{p(x)p(y|x)p(z|y)}{p(y)} = \frac{p(x, y)p(z|y)}{p(y)} = p(x|y)p(z|y) \qquad (10.36)$$

and therefore $x \perp z|y$. So observing the middle node of chain breaks it in two (as in a Markov chain).

Now consider the tent structure $X \leftarrow Y \rightarrow Z$. The joint is

$$p(x, y, z) = p(y)p(x|y)p(z|y) \qquad (10.37)$$

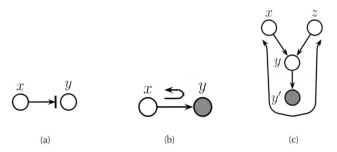

(a)   (b)   (c)

Figure 10.10   (a-b) Bayes ball boundary conditions. (c) Example of why we need boundary conditions. y' is an observed child of y, rendering y "effectively observed", so the ball bounces back up on its way from x to z.

When we condition on y, are x and z independent? We have

$$p(x, z|y) \quad = \quad \frac{p(x, y, z)}{p(y)} = \frac{p(y)p(x|y)p(z|y)}{p(y)} = p(x|y)p(z|y) \tag{10.38}$$

and therefore $x \perp z|y$. So observing a root node separates its children (as in a naive Bayes classifier: see Section 3.5).

Finally consider a v-structure $X \to Y \leftarrow Z$. The joint is

$$p(x, y, z) = p(x)p(z)p(y|x, z) \tag{10.39}$$

When we condition on y, are x and z independent? We have

$$p(x, z|y) = \frac{p(x)p(z)p(y|x, z)}{p(y)} \tag{10.40}$$

so $x \not\perp z|y$. However, in the unconditional distribution, we have

$$p(x, z) = p(x)p(z) \tag{10.41}$$

so we see that x and z are marginally independent. So we see that conditioning on a common child at the bottom of a v-structure makes its parents become dependent. This important effect is called **explaining away**, **inter-causal reasoning**, or **Berkson's paradox**. As an example of explaining away, suppose we toss two coins, representing the binary numbers 0 and 1, and we observe the "sum" of their values. A priori, the coins are independent, but once we observe their sum, they become coupled (e.g., if the sum is 1, and the first coin is 0, then we know the second coin is 1).

Finally, Bayes Ball also needs the "boundary conditions" shown in Figure 10.10(a-b). To understand where these rules come from, consider Figure 10.10(c). Suppose Y' is a noise-free copy of Y. Then if we observe Y', we effectively observe Y as well, so the parents X and Z have to compete to explain this. So if we send a ball down $X \to Y \to Y'$, it should "bounce back" up along $Y' \to Y \to Z$. However, if Y and all its children are hidden, the ball does not bounce back.

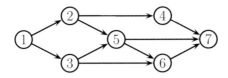

Figure 10.11 A DGM.

For example, in Figure 10.11, we see that $x_2 \perp x_6 | x_5, x_1$, since the $2 \to 5 \to 6$ path is blocked by x_5 (which is observed), the $2 \to 4 \to 7 \to 6$ path is blocked by x_7 (which is hidden), and the $2 \to 1 \to 3 \to 6$ path is blocked by x_1 (which is observed). However, we also see that $x_2 \not\perp x_6 | x_5, x_1, x_7$, since now the $2 \to 4 \to 7 \to 6$ path is no longer blocked by x_7 (which is now observed). Exercise 10.2 gives you some more practice in determining CI relationships for DGMs.

10.5.2 Other Markov properties of DGMs

From the d-separation criterion, one can conclude that

$$t \perp \mathrm{nd}(t) \setminus \mathrm{pa}(t) | \mathrm{pa}(t) \tag{10.42}$$

where the **non-descendants** of a node $\mathrm{nd}(t)$ are all the nodes except for its descendants, $\mathrm{nd}(t) = \mathcal{V} \setminus \{t \cup \mathrm{desc}(t)\}$. Equation 10.42 is called the **directed local Markov property**. For example, in Figure 10.11, we have $\mathrm{nd}(3) = \{2, 4\}$, and $\mathrm{pa}(3) = 1$, so $3 \perp 2, 4 | 1$.

A special case of this property is when we only look at predecessors of a node according to some topological ordering. We have

$$t \perp \mathrm{pred}(t) \setminus \mathrm{pa}(t) | \mathrm{pa}(t) \tag{10.43}$$

which follows since $\mathrm{pred}(t) \subseteq \mathrm{nd}(t)$. This is called the **ordered Markov property**, which justifies Equation 10.7. For example, in Figure 10.11, if we use the ordering $1, 2, \ldots, 7$. we find $\mathrm{pred}(3) = \{1, 2\}$ and $\mathrm{pa}(3) = 1$, so $3 \perp 2 | 1$.

We have now described three Markov properties for DAGs: the directed global Markov property G in Equation 10.34, the ordered Markov property O in Equation 10.43, and the directed local Markov property L in Equation 10.42. It is obvious that $G \implies L \implies O$. What is less obvious, but nevertheless true, is that $O \implies L \implies G$ (see e.g., (Koller and Friedman 2009) for the proof). Hence all these properties are equivalent.

Furthermore, any distribution p that is Markov wrt G can be factorized as in Equation 10.7; this is called the factorization property F. It is obvious that $O \implies F$, but one can show that the converse also holds (see e.g., (Koller and Friedman 2009) for the proof).

10.5.3 Markov blanket and full conditionals

The smallest set of nodes that renders a node t conditionally independent of all the other nodes in the graph is called t's **Markov blanket**; we will denote this by $\mathrm{mb}(t)$. One can show that the

Markov blanket of a node in a DGM is equal to the parents, the children, and the **co-parents**, i.e., other nodes who are also parents of its children:

$$mb(t) \triangleq ch(t) \cup pa(t) \cup copa(t) \tag{10.44}$$

For example, in Figure 10.11, we have

$$mb(5) = \{6, 7\} \cup \{2, 3\} \cup \{4\} = \{2, 3, 4, 6, 7\} \tag{10.45}$$

where 4 is a co-parent of 5 because they share a common child, namely 7.

To see why the co-parents are in the Markov blanket, note that when we derive $p(x_t|\mathbf{x}_{-t}) = p(x_t, \mathbf{x}_{-t})/p(\mathbf{x}_{-t})$, all the terms that do not involve x_t will cancel out between numerator and denominator, so we are left with a product of CPDs which contain x_t in their **scope**. Hence

$$p(x_t|\mathbf{x}_{-t}) \quad \propto \quad p(x_t|\mathbf{x}_{pa(t)}) \prod_{s \in ch(t)} p(x_s|\mathbf{x}_{pa(s)}) \tag{10.46}$$

For example, in Figure 10.11 we have

$$p(x_5|\mathbf{x}_{-5}) \propto p(x_5|x_2, x_3)p(x_6|x_3, x_5)p(x_7|x_4, x_5, x_6) \tag{10.47}$$

The resulting expression is called t's **full conditional**, and will prove to be important when we study Gibbs sampling (Section 24.2).

10.6 Influence (decision) diagrams *

We can represent multi-stage (Bayesian) decision problems by using a graphical notation known as a **decision diagram** or an **influence diagram** (Howard and Matheson 1981; Kjaerulff and Madsen 2008). This extends directed graphical models by adding **decision nodes** (also called **action nodes**), represented by rectangles, and **utility nodes** (also called **value nodes**), represented by diamonds. The original random variables are called **chance nodes**, and are represented by ovals, as usual.

Figure 10.12(a) gives a simple example, illustrating the famous **oil wild-catter** problem.[3] In this problem, you have to decide whether to drill an oil well or not. You have two possible actions: $d = 1$ means drill, $d = 0$ means don't drill. You assume there are 3 states of nature: $o = 0$ means the well is dry, $o = 1$ means it is wet (has some oil), and $o = 2$ means it is soaking (has a lot of oil). Suppose your prior beliefs are $p(o) = [0.5, 0.3, 0.2]$. Finally, you must specify the utility function $U(d, o)$. Since the states and actions are discrete, we can represent it as a table (analogous to a CPT in a DGM). Suppose we use the following numbers, in dollars:

	$o = 0$	$o = 1$	$o = 2$
$d = 0$	0	0	0
$d = 1$	-70	50	200

We see that if you don't drill, you incur no costs, but also make no money. If you drill a dry well, you lose $70; if you drill a wet well, you gain $50; and if you drill a soaking well, you gain $200. Your prior expected utility if you drill is given by

$$EU(d = 1) = \sum_{o=0}^{2} p(o)U(d, o) = 0.5 \cdot (-70) + 0.3 \cdot 50 + 0.2 \cdot 200 = 20 \tag{10.48}$$

---

3. This example is originally from (Raiffa 1968). Our presentation is based on some notes by Daphne Koller.

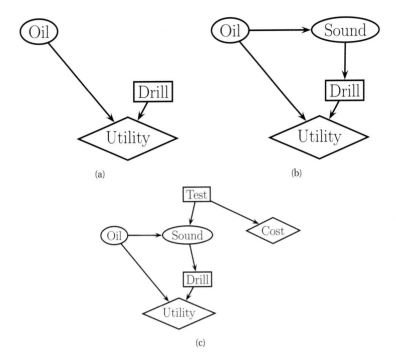

Figure 10.12 (a) Influence diagram for basic oil wild catter problem. (b) An extension in which we have an information arc from the Sound chance node to the Drill decision node. (c) An extension in which we get to decide whether to perform the test or not.

Your expected utility if you don't drill is 0. So your maximum expected utility is

$$MEU = \max\{EU(d = 0), EU(d = 1)\} = \max\{0, 20\} = 20 \tag{10.49}$$

and therefore the optimal action is to drill:

$$d^* = \arg\max\{EU(d = 0), EU(d = 1)\} = 1 \tag{10.50}$$

Now let us consider a slight extension to the model. Suppose you perform a sounding to estimate the state of the well. The sounding observation can be in one of 3 states: $s = 0$ is a diffuse reflection pattern, suggesting no oil; $s = 1$ is an open reflection pattern, suggesting some oil; and $s = 2$ is a closed reflection pattern, indicating lots of oil. Since S is caused by O, we add an $O \to S$ arc to our model. In addition, we assume that the outcome of the sounding test will be available before we decide whether to drill or not; hence we add an **information arc** from S to D. This is illustrated in Figure 10.12(b).

Let us model the reliability of our sensor using the following conditional distribution for $p(s|o)$:

	$s = 0$	$s = 1$	$s = 2$
$o = 0$	0.6	0.3	0.1
$o = 1$	0.3	0.4	0.3
$o = 2$	0.1	0.4	0.5

Suppose we do the sounding test and we observe $s = 0$. The posterior over the oil state is

$$p(o|s = 0) = [0.732, 0.219, 0.049] \tag{10.51}$$

Now your posterior expected utility of performing action d is

$$EU(d|s = 0) = \sum_{o=0}^{2} p(o|s = 0)U(o, d) \tag{10.52}$$

If $d = 1$, this gives

$$EU(d = 1|s = 0) = 0.732 \times (-70) + 0.219 \times 50 + 0.049 \times 200 = -30.5 \tag{10.53}$$

However, if $d = 0$, then $EU(d = 0|s = 0) = 0$, since not drilling incurs no cost. So if we observe $s = 0$, we are better off not drilling, which makes sense.

Now suppose we do the sounding test and we observe $s = 1$. By similar reasoning, one can show that $EU(d = 1|s = 1) = 32.9$, which is higher than $EU(d = 0|s = 1) = 0$. Similarly, if we observe $s = 2$, we have $EU(d = 1|s = 2) = 87.5$ which is much higher than $EU(d = 0|s = 2) = 0$. Hence the optimal policy $d^*(s)$ is as follows: if $s = 0$, choose $d^*(0) = 0$ and get \$0; if $s = 1$, choose $d^*(1) = 1$ and get \$32.9; and if $s = 2$, choose $d^*(2) = 1$ and get \$87.5.

You can compute your **expected profit** or maximum expected utility as follows:

$$MEU = \sum_{s} p(s)EU(d^*(s)|s) \tag{10.54}$$

This is the expected utility given possible outcomes of the sounding test, assuming you act optimally given the outcome. The prior marginal on the outcome of the test is

$$p(s) = \sum_{o} p(o)p(s|o) = [0.41, 0.35, 0.24] \tag{10.55}$$

Hence your maximum expected utility is

$$MEU = 0.41 \times 0 + 0.35 \times 32.9 + 0.24 \times 87.5 = 32.2 \tag{10.56}$$

Now suppose you can choose whether to do the test or not. This can be modelled as shown in Figure 10.12(c), where we add a new test node T. If $T = 1$, we do the test, and S can enter 1 of 3 states, determined by O, exactly as above. If $T = 0$, we don't do the test, and S enters a special unknown state. There is also some cost associated with performing the test.

Is it worth doing the test? This depends on how much our MEU changes if we know the outcome of the test (namely the state of S). If you don't do the test, we have $MEU = 20$ from Equation 10.49. If you do the test, you have $MEU = 32.2$ from Equation 10.56. So the improvement in utility if you do the test (and act optimally on its outcome) is \$12.2. This is

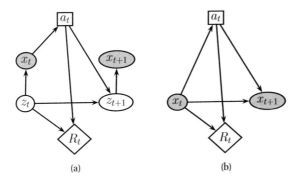

Figure 10.13 (a) A POMDP, shown as an influence diagram. z_t are hidden world states. We implicitly make the **no forgetting** assumption, which effectively means that a_t has arrows coming into it from all previous observations, $x_{1:t}$. (b) An MDP, shown as an influence diagram.

called the **value of perfect information** (VPI). So we should do the test as long as it costs less than $12.2.

In terms of graphical models, the VPI of a variable T can be determined by computing the MEU for the base influence diagram, I, and then computing the MEU for the same influence diagram where we add information arcs from T to the action nodes, and then computing the difference. In other words,

$$\text{VPI} = \text{MEU}(I + T \rightarrow D) - \text{MEU}(I) \tag{10.57}$$

where D is the decision node and T is the variable we are measuring.

It is possible to modify the variable elimination algorithm (Section 20.3) so that it computes the optimal policy given an influence diagram. These methods essentially work backwards from the final time-step, computing the optimal decision at each step assuming all following actions are chosen optimally. See e.g., (Lauritzen and Nilsson 2001; Kjaerulff and Madsen 2008) for details.

We could continue to extend the model in various ways. For example, we could imagine a dynamical system in which we test, observe outcomes, perform actions, move on to the next oil well, and continue drilling (and polluting) in this way. In fact, many problems in robotics, business, medicine, public policy, etc. can be usefully formulated as influence diagrams unrolled over time (Raiffa 1968; Lauritzen and Nilsson 2001; Kjaerulff and Madsen 2008).

A generic model of this form is shown in Figure 10.13(a). This is known as a **partially observed Markov decision process** or **POMDP** (pronounced "pom-d-p"). This is basically a hidden Markov model (Section 17.3) augmented with action and reward nodes. This can be used to model the **perception-action** cycle that all intelligent agents use (see e.g., (Kaelbling et al. 1998) for details).

A special case of a POMDP, in which the states are fully observed, is called a **Markov decision process** or **MDP**, shown in Figure 10.13(b). This is much easier to solve, since we only have to compute a mapping from observed states to actions. This can be solved using dynamic programming (see e.g., (Sutton and Barto 1998) for details).

In the POMDP case, the information arc from x_t to a_t is not sufficient to uniquely determine

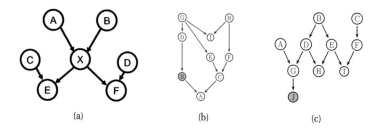

Figure 10.14   Some DGMs.

the best action, since the state is not fully observed. Instead, we need to choose actions based on our **belief state**, $p(z_t|\mathbf{x}_{1:t}, a_{1:t})$. Since the belief updating process is deterministic (see Section 17.4.2), we can compute a **belief state MDP**. For details on how to compute the policies for such models, see e.g., (Kaelbling et al. 1998; Spaan and Vlassis 2005).

Exercises

Exercise 10.1 Marginalizing a node in a DGM

(Source: Koller.)

Consider the DAG G in Figure 10.14(a). Assume it is a minimal I-map for $p(A, B, C, D, E, F, X)$. Now consider marginalizing out X. Construct a new DAG G' which is a minimal I-map for $p(A, B, C, D, E, F)$. Specify (and justify) which extra edges need to be added.

Exercise 10.2 Bayes Ball

(Source: Jordan.)

Here we compute some global independence statements from some directed graphical models. You can use the "Bayes ball" algorithm, the d-separation criterion, or the method of converting to an undirected graph (all should give the same results).

a. Consider the DAG in Figure 10.14(b). List all variables that are independent of A given evidence on B.

b. Consider the DAG in Figure 10.14(c). List all variables that are independent of A given evidence on J.

Exercise 10.3 Markov blanket for a DGM

Prove that the full conditional for node i in a DGM is given by

$$p(X_i|X_{-i}) \quad \propto \quad p(X_i|Pa(X_i)) \prod_{Y_j \in ch(X_i)} p(Y_j|Pa(Y_j)) \tag{10.58}$$

where $ch(X_i)$ are the children of X_i and $Pa(Y_j)$ are the parents of Y_j.

Exercise 10.4 Hidden variables in DGMs

Consider the DGMs in Figure 11.1 which both define $p(X_{1:6})$, where we number empty nodes left to right, top to bottom. The graph on the left defines the joint as

$$p(X_{1:6}) = \sum_h p(X_1)p(X_2)p(X_3)p(H = h|X_{1:3})p(X_4|H = h)p(X_5|H = h)p(X_6|H = h) \tag{10.59}$$

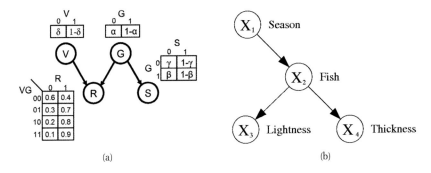

Figure 10.15   (a) Weather BN. (b) Fishing BN.

where we have marginalized over the hidden variable H. The graph on the right defines the joint as

$$p(X_{1:6}) = p(X_1)p(X_2)p(X_3)p(X_4|X_{1:3})p(X_5|X_{1:4})p(X_6|X_{1:5}) \tag{10.60}$$

a. (5 points) Assuming all nodes (including H) are binary and all CPDs are tabular, prove that the model on the left has 17 free parameters.

b. (5 points) Assuming all nodes are binary and all CPDs are tabular, prove that the model on the right has 59 free parameters.

c. (5 points) Suppose we have a data set $\mathcal{D} = X_{1:6}^n$ for $n = 1 : N$, where we observe the Xs but not H, and we want to estimate the parameters of the CPDs using maximum likelihood. For which model is this easier? Explain your answer.

Exercise 10.5 Bayes nets for a rainy day

(Source: Nando de Freitas.). In this question you must model a problem with 4 binary variables: G ="gray", V ="Vancouver", R ="rain" and S ="sad". Consider the directed graphical model describing the relationship between these variables shown in Figure 10.15(a).

a. Write down an expression for $P(S = 1|V = 1)$ in terms of $\alpha, \beta, \gamma, \delta$.

b. Write down an expression for $P(S = 1|V = 0)$. Is this the same or different to $P(S = 1|V = 1)$? Explain why.

c. Find maximum likelihood estimates of α, β, γ using the following data set, where each row is a training case. (You may state your answers without proof.)

V	G	R	S
1	1	1	1
1	1	0	1
1	0	0	0

$$\tag{10.61}$$

Exercise 10.6 Fishing nets

(Source: (Duda et al. 2001).) Consider the Bayes net shown in Figure 10.15(b). Here, the nodes represent the following variables

$$X_1 \in \{\text{winter, spring, summer, autumn}\}, \ X_2 \in \{\text{salmon, sea bass}\} \tag{10.62}$$
$$X_3 \in \{\text{light, medium, dark}\}, \ X_4 \in \{\text{wide, thin}\} \tag{10.63}$$

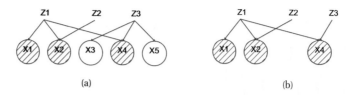

Figure 10.16   (a) A QMR-style network with some hidden leaves. (b) Removing the barren nodes.

The corresponding conditional probability tables are

$$p(x_1) = (\begin{array}{cccc} .25 & .25 & .25 & .25 \end{array}), \quad p(x_2|x_1) = \left(\begin{array}{cc} .9 & .1 \\ .3 & .7 \\ .4 & .6 \\ .8 & .2 \end{array} \right) \tag{10.64}$$

$$p(x_3|x_2) = \left(\begin{array}{ccc} .33 & .33 & .34 \\ .8 & .1 & .1 \end{array} \right), \quad p(x_4|x_2) = \left(\begin{array}{cc} .4 & .6 \\ .95 & .05 \end{array} \right) \tag{10.65}$$

Note that in $p(x_4|x_2)$, the rows represent x_2 and the columns x_4 (so each row sums to one and represents the child of the CPD). Thus $p(x_4 = \text{thin}|x_2 = \text{sea bass}) = 0.05$, $p(x_4 = \text{thin}|x_2 = \text{salmon}) = 0.6$, etc.

Answer the following queries. You may use matlab or do it by hand. In either case, show your work.

a. Suppose the fish was caught on December 20 — the end of autumn and the beginning of winter — and thus let $p(x_1) = (.5, 0, 0, .5)$ instead of the above prior. (This is called **soft evidence**, since we do not know the exact value of X_1, but we have a distribution over it.) Suppose the lightness has not been measured but it is known that the fish is thin. Classify the fish as salmon or sea bass.

b. Suppose all we know is that the fish is thin and medium lightness. What season is it now, most likely? Use $p(x_1) = (\begin{array}{cccc} .25 & .25 & .25 & .25 \end{array})$

Exercise 10.7 *Removing leaves in BN20 networks*

a. Consider the QMR network, where only some of the symtpoms are observed. For example, in Figure 10.16(a), X_4 and X_5 are hidden. Show that we can safely remove all the hidden leaf nodes without affecting the posterior over the disease nodes, i.e., prove that we can compute $p(\mathbf{z}_{1:3}|x_1, x_2, x_4)$ using the network in Figure 10.16(b). This is called barren node removal, and can be applied to any DGM.

b. Now suppose we partition the leaves into three groups: on, off and unknown. Clearly we can remove the unknown leaves, since they are hidden and do not affect their parents. Show that we can analytically remove the leaves that are in the "off state", by absorbing their effect into the prior of the parents. (This trick only works for noisy-OR CPDs.)

Exercise 10.8 *Handling negative findings in the QMR network*

Consider the QMR network. Let $\mathbf{d}$ be the hidden diseases, $\mathbf{f}^-$ be the negative findings (leaf nodes that are off), and $\mathbf{f}^-$ be the positive findings (leaf nodes that are on). We can compute the posterior $p(\mathbf{d}|\mathbf{f}^-\mathbf{f}^+)$ in two steps: first absorb the negative findings, $p(\mathbf{d}|\mathbf{f}^-) \propto p(\mathbf{d})p(\mathbf{f}^-|\mathbf{d})$, then absorb the positive findings, $p(\mathbf{d}|\mathbf{f}^-, \mathbf{f}^+) \propto p(\mathbf{d}|\mathbf{f}^-)p(\mathbf{f}^+|\mathbf{d})$. Show that the first step can be done in $O(|\mathbf{d}||\mathbf{f}^-|)$ time, where $|\mathbf{d}|$ is the number of dieases and $|\mathbf{f}^-|$ is the number of negative findings. For simplicity, you can ignore leak nodes. (Intuitively, the reason for this is that there is no correlation induced amongst the parents when the finding is off, since there is no explaining away.)

Exercise 10.9 Moralization does not introduce new independence statements

Recall that the process of moralizing a DAG means connecting together all "unmarried" parents that share a common child, and then dropping all the arrows. Let M be the moralization of DAG G. Show that $CI(M) \subseteq CI(G)$, where CI are the set of conditional independence statements implied by the model.

11 *Mixture models and the EM algorithm*

11.1 Latent variable models

In Chapter 10 we showed how graphical models can be used to define high-dimensional joint probability distributions. The basic idea is to model dependence between two variables by adding an edge between them in the graph. (Technically the graph represents conditional independence, but you get the point.)

An alternative approach is to assume that the observed variables are correlated because they arise from a hidden common "cause". Model with hidden variables are also known as **latent variable models** or **LVM**s. As we will see in this chapter, such models are harder to fit than models with no latent variables. However, they can have significant advantages, for two main reasons. First, LVMs often have fewer parameters than models that directly represent correlation in the visible space. This is illustrated in Figure 11.1. If all nodes (including H) are binary and all CPDs are tabular, the model on the left has 17 free parameters, whereas the model on the right has 59 free parameters.

Second, the hidden variables in an LVM can serve as a **bottleneck**, which computes a compressed representation of the data. This forms the basis of unsupervised learning, as we will see. Figure 11.2 illustrates some generic LVM structures that can be used for this purpose. In general there are L latent variables, $z_{i1}, \ldots, z_{IL}$, and D visible variables, $x_{i1}, \ldots, x_{iD}$, where usually $D \gg L$. If we have $L > 1$, there are many latent factors contributing to each observation, so we have a many-to-many mapping. If $L = 1$, we we only have a single latent variable; in this case, z_i is usually discrete, and we have a one-to-many mapping. We can also have a many-to-one mapping, representing different competing factors or causes for each observed variable; such models form the basis of probabilistic matrix factorization, discussed in Section 27.6.2. Finally, we can have a one-to-one mapping, which can be represented as $\mathbf{z}_i \to \mathbf{x}_i$. By allowing $\mathbf{z}_i$ and/or $\mathbf{x}_i$ to be vector-valued, this representation can subsume all the others. Depending on the form of the likelihood $p(\mathbf{x}_i|\mathbf{z}_i)$ and the prior $p(\mathbf{z}_i)$, we can generate a variety of different models, as summarized in Table 11.1.

11.2 Mixture models

The simplest form of LVM is when $z_i \in \{1, \ldots, K\}$, representing a discrete latent state. We will use a discrete prior for this, $p(z_i) = \text{Cat}(\boldsymbol{\pi})$. For the likelihood, we use $p(\mathbf{x}_i|z_i = k) = p_k(\mathbf{x}_i)$,

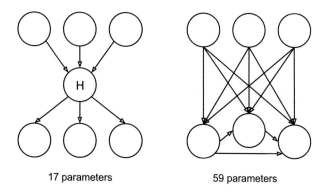

Figure 11.1 A DGM with and without hidden variables. The leaves represent medical symptoms. The roots represent primary causes, such as smoking, diet and exercise. The hidden variable can represent mediating factors, such as heart disease, which might not be directly visible.

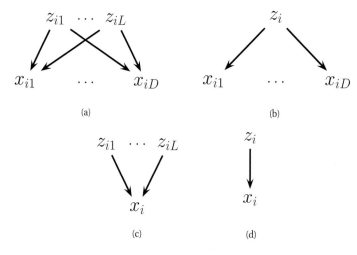

Figure 11.2 A latent variable model represented as a DGM. (a) Many-to-many. (b) One-to-many. (c) Many-to-one. (d) One-to-one.

where p_k is the k'th **base distribution** for the observations; this can be of any type. The overall model is known as a **mixture model**, since we are mixing together the K base distributions as follows:

$$p(\mathbf{x}_i|\boldsymbol{\theta}) = \sum_{k=1}^{K} \pi_k p_k(\mathbf{x}_i|\boldsymbol{\theta}) \tag{11.1}$$

This is a **convex combination** of the p_k's, since we are taking a weighted sum, where the **mixing weights** π_k satisfy $0 \leq \pi_k \leq 1$ and $\sum_{k=1}^{K} \pi_k = 1$. We give some examples below.

$p(\mathbf{x}_i|\mathbf{z}_i)$	$p(\mathbf{z}_i)$	Name	Section
MVN	Discrete	Mixture of Gaussians	11.2.1
Prod. Discrete	Discrete	Mixture of multinomials	11.2.2
Prod. Gaussian	Prod. Gaussian	Factor analysis/ probabilistic PCA	12.1.5
Prod. Gaussian	Prod. Laplace	Probabilistic ICA/ sparse coding	12.6
Prod. Discrete	Prod. Gaussian	Multinomial PCA	27.2.3
Prod. Discrete	Dirichlet	Latent Dirichlet allocation	27.3
Prod. Noisy-OR	Prod. Bernoulli	BN20/ QMR	10.2.3
Prod. Bernoulli	Prod. Bernoulli	Sigmoid belief net	27.7

Table 11.1 Summary of some popular directed latent variable models. Here "Prod" means product, so "Prod. Discrete" in the likelihood means a factored distribution of the form $\prod_j \mathrm{Cat}(x_{ij}|\mathbf{z}_i)$, and "Prod. Gaussian" means a factored distribution of the form $\prod_j \mathcal{N}(x_{ij}|\mathbf{z}_i)$. "PCA" stands for "principal components analysis". "ICA" stands for "indepedendent components analysis".

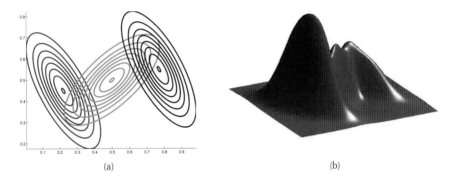

(a)   (b)

Figure 11.3 A mixture of 3 Gaussians in 2d. (a) We show the contours of constant probability for each component in the mixture. (b) A surface plot of the overall density. Based on Figure 2.23 of (Bishop 2006). Figure generated by `mixGaussPlotDemo`.

11.2.1 Mixtures of Gaussians

The most widely used mixture model is the **mixture of Gaussians** (MOG), also called a **Gaussian mixture model** or **GMM**. In this model, each base distribution in the mixture is a multivariate Gaussian with mean $\boldsymbol{\mu}_k$ and covariance matrix $\boldsymbol{\Sigma}_k$. Thus the model has the form

$$p(\mathbf{x}_i|\boldsymbol{\theta}) = \sum_{k=1}^{K} \pi_k \mathcal{N}(\mathbf{x}_i|\boldsymbol{\mu}_k, \boldsymbol{\Sigma}_k) \tag{11.2}$$

Figure 11.3 shows a mixture of 3 Gaussians in 2D. Each mixture component is represented by a different set of eliptical contours. Given a sufficiently large number of mixture components, a GMM can be used to approximate any density defined on $\mathbb{R}^D$.

11.2.2   Mixture of multinoullis

We can use mixture models to define density models on many kinds of data. For example, suppose our data consist of D-dimensional bit vectors. In this case, an appropriate class-conditional density is a product of Bernoullis:

$$p(\mathbf{x}_i|z_i = k, \boldsymbol{\theta}) = \prod_{j=1}^{D} \text{Ber}(x_{ij}|\mu_{jk}) = \prod_{j=1}^{D} \mu_{jk}^{x_{ij}}(1 - \mu_{jk})^{1-x_{ij}} \tag{11.3}$$

where μ_{jk} is the probability that bit j turns on in cluster k.

The latent variables do not have to any meaning, we might simply introduce latent variables in order to make the model more powerful. For example, one can show (Exercise 11.8) that the mean and covariance of the mixture distribution are given by

$$\mathbb{E}[\mathbf{x}] = \sum_k \pi_k \boldsymbol{\mu}_k \tag{11.4}$$

$$\text{cov}[\mathbf{x}] = \sum_k \pi_k [\boldsymbol{\Sigma}_k + \boldsymbol{\mu}_k \boldsymbol{\mu}_k^T] - \mathbb{E}[\mathbf{x}]\mathbb{E}[\mathbf{x}]^T \tag{11.5}$$

where $\boldsymbol{\Sigma}_k = \text{diag}(\mu_{jk}(1 - \mu_{jk}))$. So although the component distributions are factorized, the joint distribution is not. Thus the mixture distribution can capture correlations between variables, unlike a single product-of-Bernoullis model.

11.2.3   Using mixture models for clustering

There are two main applications of mixture models. The first is to use them as a **black-box** density model, $p(\mathbf{x}_i)$. This can be useful for a variety of tasks, such as data compression, outlier detection, and creating generative classifiers, where we model each class-conditional density $p(\mathbf{x}|y = c)$ by a mixture distribution (see Section 14.7.3).

The second, and more common, application of mixture models is to use them for clustering. We discuss this topic in detail in Chapter 25, but the basic idea is simple. We first fit the mixture model, and then compute $p(z_i = k|\mathbf{x}_i, \boldsymbol{\theta})$, which represents the posterior probability that point i belongs to cluster k. This is known as the **responsibility** of cluster k for point i, and can be computed using Bayes rule as follows:

$$r_{ik} \triangleq p(z_i = k|\mathbf{x}_i, \boldsymbol{\theta}) = \frac{p(z_i = k|\boldsymbol{\theta})p(\mathbf{x}_i|z_i = k, \boldsymbol{\theta})}{\sum_{k'=1}^{K} p(z_i = k'|\boldsymbol{\theta})p(\mathbf{x}_i|z_i = k', \boldsymbol{\theta})} \tag{11.6}$$

This procedure is called **soft clustering**, and is identical to the computations performed when using a generative classifier. The difference between the two models only arises at training time: in the mixture case, we never observe z_i, whereas with a generative classifier, we do observe y_i (which plays the role of z_i).

We can represent the amount of uncertainty in the cluster assignment by using $1 - \max_k r_{ik}$. Assuming this is small, it may be reasonable to compute a **hard clustering** using the MAP estimate, given by

$$z_i^* = \arg\max_k r_{ik} = \arg\max_k \log p(\mathbf{x}_i|z_i = k, \boldsymbol{\theta}) + \log p(\mathbf{z}_i = k|\boldsymbol{\theta}) \tag{11.7}$$

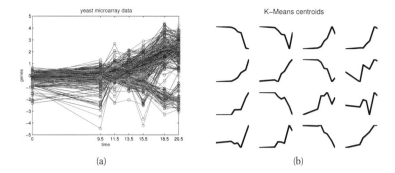

Figure 11.4 (a) Some yeast gene expression data plotted as a time series. (b) Visualizing the 16 cluster centers produced by K-means. Figure generated by `kmeansYeastDemo`.

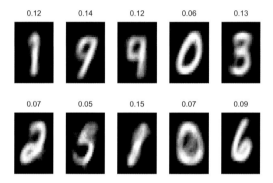

Figure 11.5 We fit a mixture of 10 Bernoullis to the binarized MNIST digit data. We show the MLE for the corresponding cluster means, $\boldsymbol{\mu}_k$. The numbers on top of each image represent the mixing weights $\hat{\pi}_k$. No labels were used when training the model. Figure generated by `mixBerMnistEM`.

Hard clustering using a GMM is illustrated in Figure 1.8, where we cluster some data representing the height and weight of people. The colors represent the hard assignments. Note that the identity of the labels (colors) used is immaterial; we are free to rename all the clusters, without affecting the partitioning of the data; this is called **label switching**.

Another example is shown in Figure 11.4. Here the data vectors $\mathbf{x}_i \in \mathbb{R}^7$ represent the expression levels of different genes at 7 different time points. We clustered them using a GMM. We see that there are several kinds of genes, such as those whose expression level goes up monotonically over time (in response to a given stimulus), those whose expression level goes down monotonically, and those with more complex response patterns. We have clustered the series into $K = 16$ groups. (See Section 11.5 for details on how to choose K.) For example, we can represent each cluster by a **prototype** or **centroid**. This is shown in Figure 11.4(b).

As an example of clustering binary data, consider a binarized version of the MNIST handwritten digit dataset (see Figure 1.5(a)), where we ignore the class labels. We can fit a mixture of

Bernoullis to this, using $K = 10$, and then visualize the resulting centroids, $\hat{\boldsymbol{\mu}}_k$, as shown in Figure 11.5. We see that the method correctly discovered some of the digit classes, but overall the results aren't great: it has created multiple clusters for some digits, and no clusters for others. There are several possible reasons for these "errors":

- The model is very simple and does not capture the relevant visual characteristics of a digit. For example, each pixel is treated independently, and there is no notion of shape or a stroke.

- Although we think there should be 10 clusters, some of the digits actually exhibit a fair degree of visual variety. For example, there are two ways of writing 7's (with and without the cross bar). Figure 1.5(a) illustrates some of the range in writing styles. Thus we need $K \gg 10$ clusters to adequately model this data. However, if we set K to be large, there is nothing in the model or algorithm preventing the extra clusters from being used to create multiple versions of the same digit, and indeed this is what happens. We can use model selection to prevent too many clusters from being chosen but what looks visually appealing and what makes a good density estimator may be quite different.

- The likelihood function is not convex, so we may be stuck in a local optimum, as we explain in Section 11.3.2.

This example is typical of mixture modeling, and goes to show one must be very cautious trying to "interpret" any clusters that are discovered by the method. (Adding a little bit of supervision, or using informative priors, can help a lot.)

11.2.4   Mixtures of experts

Section 14.7.3 described how to use mixture models in the context of generative classifiers. We can also use them to create discriminative models for classification and regression. For example, consider the data in Figure 11.6(a). It seems like a good model would be three different linear regression functions, each applying to a different part of the input space. We can model this by allowing the mixing weights and the mixture densities to be input-dependent:

$$p(y_i|\mathbf{x}_i, z_i = k, \boldsymbol{\theta}) = \mathcal{N}(y_i|\mathbf{w}_k^T\mathbf{x}_i, \sigma_k^2) \tag{11.8}$$

$$p(z_i|\mathbf{x}_i, \boldsymbol{\theta}) = \text{Cat}(z_i|\mathcal{S}(\mathbf{V}^T\mathbf{x}_i)) \tag{11.9}$$

See Figure 11.7(a) for the DGM.

This model is called a **mixture of experts** or MoE (Jordan and Jacobs 1994). The idea is that each submodel is considered to be an "expert" in a certain region of input space. The function $p(z_i = k|\mathbf{x}_i, \boldsymbol{\theta})$ is called a **gating function**, and decides which expert to use, depending on the input values. For example, Figure 11.6(b) shows how the three experts have "carved up" the 1d input space, Figure 11.6(a) shows the predictions of each expert individually (in this case, the experts are just linear regression models), and Figure 11.6(c) shows the overall prediction of the model, obtained using

$$p(y_i|\mathbf{x}_i, \boldsymbol{\theta}) = \sum_k p(z_i = k|\mathbf{x}_i, \boldsymbol{\theta})p(y_i|\mathbf{x}_i, z_i = k, \boldsymbol{\theta}) \tag{11.10}$$

We discuss how to fit this model in Section 11.4.3.

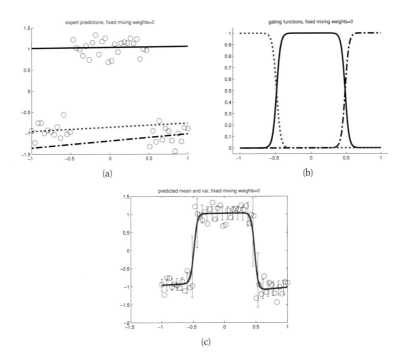

Figure 11.6 (a) Some data fit with three separate regression lines. (b) Gating functions for three different "experts". (c) The conditionally weighted average of the three expert predictions. Figure generated by `mixexpDemo`.

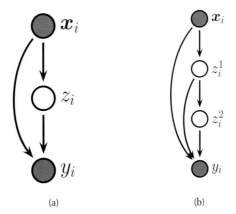

Figure 11.7 (a) A mixture of experts. (b) A hierarchical mixture of experts.

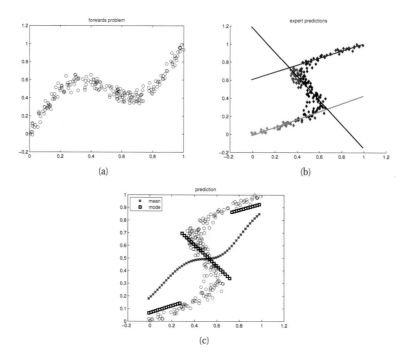

Figure 11.8 (a) Some data from a simple forwards model. (b) Some data from the inverse model, fit with a mixture of 3 linear regressions. Training points are color coded by their responsibilities. (c) The predictive mean (red cross) and mode (black square). Based on Figures 5.20 and 5.21 of (Bishop 2006). Figure generated by `mixexpDemoOneToMany`.

It should be clear that we can "plug in" any model for the expert. For example, we can use neural networks (Chapter 16) to represent both the gating functions and the experts. The result is known as a **mixture density network**. Such models are slower to train, but can be more flexible than mixtures of experts. See (Bishop 1994) for details.

It is also possible to make each expert be itself a mixture of experts. This gives rise to a model known as the **hierarchical mixture of experts**. See Figure 11.7(b) for the DGM, and Section 16.2.6 for further details.

11.2.4.1 Application to inverse problems

Mixtures of experts are useful in solving **inverse problems**. These are problems where we have to invert a many-to-one mapping. A typical example is in robotics, where the location of the end effector (hand) $\mathbf{y}$ is uniquely determined by the joint angles of the motors, $\mathbf{x}$. However, for any given location $\mathbf{y}$, there are many settings of the joints $\mathbf{x}$ that can produce it. Thus the inverse mapping $\mathbf{x} = f^{-1}(\mathbf{y})$ is not unique. Another example is **kinematic tracking** of people from video (Bo et al. 2008), where the mapping from image appearance to pose is not unique, due to self occlusion, etc.

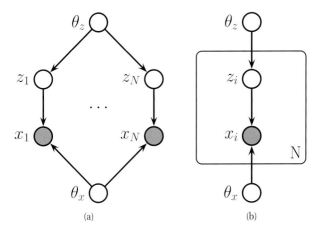

Figure 11.9 A LVM represented as a DGM. Left: Model is unrolled for N examples. Right: same model using plate notation.

A simpler example, for illustration purposes, is shown in Figure 11.8(a). We see that this defines a function, $y = f(x)$, since for every value x along the horizontal axis, there is a unique response y. This is sometimes called the **forwards model**. Now consider the problem of computing $x = f^{-1}(y)$. The corresponding inverse model is shown in Figure 11.8(b); this is obtained by simply interchanging the x and y axes. Now we see that for some values along the horizontal axis, there are multiple possible outputs, so the inverse is not uniquely defined. For example, if $y = 0.8$, then x could be 0.2 or 0.8. Consequently, the predictive distribution, $p(x|y, \boldsymbol{\theta})$ is multimodal.

We can fit a mixture of linear experts to this data. Figure 11.8(b) shows the prediction of each expert, and Figure 11.8(c) shows (a plugin approximation to) the posterior predictive mode and mean. Note that the posterior mean does not yield good predictions. In fact, any model which is trained to minimize mean squared error — even if the model is a flexible nonlinear model, such as neural network — will work poorly on inverse problems such as this. However, the posterior mode, where the mode is input dependent, provides a reasonable approximation.

11.3 Parameter estimation for mixture models

We have seen how to compute the posterior over the hidden variables given the observed variables, assuming the parameters are known. In this section, we discuss how to learn the parameters.

In Section 10.4.2, we showed that when we have complete data and a factored prior, the posterior over the parameters also factorizes, making computation very simple. Unfortunately this is no longer true if we have hidden variables and/or missing data. The reason is apparent from looking at Figure 11.9. If the z_i were observed, then by d-separation, we see that $\boldsymbol{\theta}_z \perp \boldsymbol{\theta}_x | \mathcal{D}$, and hence the posterior will factorize. But since, in an LVM, the z_i are hidden, the parameters are no longer independent, and the posterior does not factorize, making it much harder to

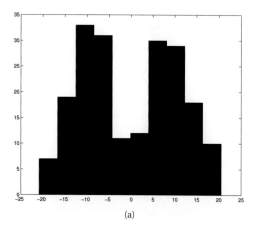
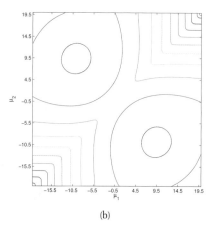

(a)   (b)

Figure 11.10   Left: $N = 200$ data points sampled from a mixture of 2 Gaussians in 1d, with $\pi_k = 0.5$, $\sigma_k = 5$, $\mu_1 = -10$ and $\mu_2 = 10$. Right: Likelihood surface $p(\mathcal{D}|\mu_1, \mu_2)$, with all other parameters set to their true values. We see the two symmetric modes, reflecting the unidentifiability of the parameters. Figure generated by `mixGaussLikSurfaceDemo`.

compute. This also complicates the computation of MAP and ML estimates, as we discus below.

11.3.1   Unidentifiability

The main problem with computing $p(\boldsymbol{\theta}|\mathcal{D})$ for an LVM is that the posterior may have multiple modes. To see why, consider a GMM. If the z_i were all observed, we would have a unimodal posterior for the parameters:

$$p(\boldsymbol{\theta}|\mathcal{D}) = \text{Dir}(\boldsymbol{\pi}|\mathcal{D}) \prod_{k=1}^{K} \text{NIW}(\boldsymbol{\mu}_k, \boldsymbol{\Sigma}_k|\mathcal{D}) \tag{11.11}$$

Consequently we can easily find the globally optimal MAP estimate (and hence globally optimal MLE).

But now suppose the z_i's are hidden. In this case, for each of the possible ways of "filling in" the z_i's, we get a different unimodal likelihood. Thus when we marginalize out over the z_i's, we get a multi-modal posterior for $p(\boldsymbol{\theta}|\mathcal{D})$.[1] These modes correspond to different labelings of the clusters. This is illustrated in Figure 11.10(b), where we plot the likelihood function, $p(\mathcal{D}|\mu_1, \mu_2)$, for a 2D GMM with $K = 2$ for the data is shown in Figure 11.10(a). We see two peaks, one corresponding to the case where $\mu_1 = -10$, $\mu_2 = 10$, and the other to the case where $\mu_1 = 10$, $\mu_2 = -10$. We say the parameters are not **identifiable**, since there is not a unique MLE. Therefore there cannot be a unique MAP estimate (assuming the prior does not rule out certain labelings), and hence the posterior must be multimodal. The question of how many modes there

---

1. Do not confuse multimodality of the parameter posterior, $p(\boldsymbol{\theta}|\mathcal{D})$, with the multimodality defined by the model, $p(\mathbf{x}|\boldsymbol{\theta})$. In the latter case, if we have K clusters, we would expect to only get K peaks, although it is theoretically possible to get more than K, at least if $D > 1$ (Carreira-Perpinan and Williams 2003).

are in the parameter posterior is hard to answer. There are $K!$ possible labelings, but some of the peaks might get merged. Nevertheless, there can be an exponential number, since finding the optimal MLE for a GMM is NP-hard (Aloise et al. 2009; Drineas et al. 2004).

Unidentifiability can cause a problem for Bayesian inference. For example, suppose we draw some samples from the posterior, $\boldsymbol{\theta}^{(s)} \sim p(\boldsymbol{\theta}|\mathcal{D})$, and then average them, to try to approximate the posterior mean, $\overline{\boldsymbol{\theta}} = \frac{1}{S}\sum_{s=1}^{S} \boldsymbol{\theta}^{(s)}$. (This kind of Monte Carlo approach is explained in more detail in Chapter 24.) If the samples come from different modes, the average will be meaningless. Note, however, that it is reasonable to average the posterior predictive distributions, $p(\mathbf{x}) \approx \frac{1}{S}\sum_{s=1}^{S} p(\mathbf{x}|\boldsymbol{\theta}^{(s)})$, since the likelihood function is invariant to which mode the parameters came from.

A variety of solutions have been proposed to the unidentifiability problem. These solutions depend on the details of the model and the inference algorithm that is used. For example, see (Stephens 2000) for an approach to handling unidentifiability in mixture models using MCMC.

The approach we will adopt in this chapter is much simpler: we just compute a single local mode, i.e., we perform approximate MAP estimation. (We say "approximate" since finding the globally optimal MLE, and hence MAP estimate, is NP-hard, at least for mixture models (Aloise et al. 2009).) This is by far the most common approach, because of its simplicity. It is also a reasonable approximation, at least if the sample size is sufficiently large. To see why, consider Figure 11.9(a). We see that there are N latent variables, each of which gets to "see" one data point each. However, there are only two latent parameters, each of which gets to see N data points. So the posterior uncertainty about the parameters is typically much less than the posterior uncertainty about the latent variables. This justifies the common strategy of computing $p(z_i|\mathbf{x}_i, \hat{\boldsymbol{\theta}})$, but not bothering to compute $p(\boldsymbol{\theta}|\mathcal{D})$. In Section 5.6, we will study hierarchical Bayesian models, which essentially put structure on top of the parameters. In such models, it is important to model $p(\boldsymbol{\theta}|\mathcal{D})$, so that the parameters can send information between themselves. If we used a point estimate, this would not be possible.

11.3.2 Computing a MAP estimate is non-convex

In the previous sections, we have argued, rather heuristically, that the likelihood function has multiple modes, and hence that finding an MAP or ML estimate will be hard. In this section, we show this result by more algebraic means, which sheds some additional insight into the problem. Our presentation is based in part on (Rennie 2004).

Consider the log-likelihood for an LVM:

$$\log p(\mathcal{D}|\boldsymbol{\theta}) = \sum_i \log \left[\sum_{\mathbf{z}_i} p(\mathbf{x}_i, \mathbf{z}_i|\boldsymbol{\theta}) \right] \tag{11.12}$$

Unfortunately, this objective is hard to maximize. since we cannot push the log inside the sum. This precludes certain algebraic simplifications, but does not prove the problem is hard.

Now suppose the joint probability distribution $p(\mathbf{z}_i, \mathbf{x}_i|\boldsymbol{\theta})$ is in the exponential family, which means it can be written as follows:

$$p(\mathbf{x}, \mathbf{z}|\boldsymbol{\theta}) = \frac{1}{Z(\boldsymbol{\theta})} \exp[\boldsymbol{\theta}^T \phi(\mathbf{x}, \mathbf{z})] \tag{11.13}$$

where $\phi(\mathbf{x}, \mathbf{z})$ are the sufficient statistics, and $Z(\boldsymbol{\theta})$ is the normalization constant (see Section 9.2 for more details). It can be shown (Exercise 9.2) that the MVN is in the exponential family, as are nearly all of the distributions we have encountered so far, including Dirichlet, multinomial, Gamma, Wishart, etc. (The Student distribution is a notable exception.) Furthermore, mixtures of exponential families are also in the exponential family, providing the mixing indicator variables are observed (Exercise 11.11).

With this assumption, the **complete data log likelihood** can be written as follows:

$$\ell_c(\boldsymbol{\theta}) = \sum_i \log p(\mathbf{x}_i, \mathbf{z}_i | \boldsymbol{\theta}) = \boldsymbol{\theta}^T \left(\sum_i \phi(\mathbf{x}_i, \mathbf{z}_i) \right) - N Z(\boldsymbol{\theta}) \tag{11.14}$$

The first term is clearly linear in $\boldsymbol{\theta}$. One can show that $Z(\boldsymbol{\theta})$ is a convex function (Boyd and Vandenberghe 2004), so the overall objective is concave (due to the minus sign), and hence has a unique maximum.

Now consider what happens when we have missing data. The **observed data log likelihood** is given by

$$\ell(\boldsymbol{\theta}) = \sum_i \log \sum_{\mathbf{z}_i} p(\mathbf{x}_i, \mathbf{z}_i | \boldsymbol{\theta}) = \sum_i \log \left[\sum_{\mathbf{z}_i} e^{\boldsymbol{\theta}^T \phi(\mathbf{z}_i, \mathbf{x}_i)} \right] - N \log Z(\boldsymbol{\theta}) \tag{11.15}$$

One can show that the log-sum-exp function is convex (Boyd and Vandenberghe 2004), and we know that $Z(\boldsymbol{\theta})$ is convex. However, the difference of two convex functions is not, in general, convex. So the objective is neither convex nor concave, and has local optima.

The disadvantage of non-convex functions is that it is usually hard to find their global optimum. Most optimization algorithms will only find a local optimum; which one they find depends on where they start. There are some algorithms, such as simulated annealing (Section 24.6.1) or genetic algorithms, that claim to always find the global optimum, but this is only under unrealistic assumptions (e.g., if they are allowed to be cooled "infinitely slowly", or allowed to run "infinitely long"). In practice, we will run a local optimizer, perhaps using **multiple random restarts** to increase out chance of finding a "good" local optimum. Of course, careful initialization can help a lot, too. We give examples of how to do this on a case-by-case basis.

Note that a convex method for fitting mixtures of Gaussians has been proposed. The idea is to assign one cluster per data point, and select from amongst them, using a convex ℓ_1-type penalty, rather than trying to optimize the locations of the cluster centers. See (Lashkari and Golland 2007) for details. This is essentially an unsupervised version of the approach used in sparse kernel logistic regression, which we will discuss in Section 14.3.2. Note, however, that the ℓ_1 penalty, although convex, is not necessarily a good way to promote sparsity, as discussed in Chapter 13. In fact, as we will see in that Chapter, some of the best sparsity-promoting methods use non-convex penalties, and use EM to optimie them! The moral of the story is: do not be afraid of non-convexity.

11.4 The EM algorithm

For many models in machine learning and statistics, computing the ML or MAP parameter estimate is easy provided we observe all the values of all the relevant random variables, i.e., if

Model	Section
Mix. Gaussians	11.4.2
Mix. experts	11.4.3
Factor analysis	12.1.5
Student T	11.4.5
Probit regression	11.4.6
DGM with hidden variables	11.4.4
MVN with missing data	11.6.1
HMMs	17.5.2
Shrinkage estimates of Gaussian means	Exercise 11.13

Table 11.2 Some models discussed in this book for which EM can be easily applied to find the ML/ MAP parameter estimate.

we have complete data. However, if we have missing data and/or latent variables, then computing the ML/MAP estimate becomes hard.

One approach is to use a generic gradient-based optimizer to find a local minimum of the **negative log likelihood** or **NLL**, given by

$$\text{NLL}(\boldsymbol{\theta}) = - \triangleq \frac{1}{N} \log p(\mathcal{D}|\boldsymbol{\theta}) \tag{11.16}$$

However, we often have to enforce constraints, such as the fact that covariance matrices must be positive definite, mixing weights must sum to one, etc., which can be tricky (see Exercise 11.5). In such cases, it is often much simpler (but not always faster) to use an algorithm called **expectation maximization**, or **EM** for short (Dempster et al. 1977; Meng and van Dyk 1997; McLachlan and Krishnan 1997). This is a simple iterative algorithm, often with closed-form updates at each step. Furthermore, the algorithm automatically enforce the required constraints.

EM exploits the fact that if the data were fully observed, then the ML/ MAP estimate would be easy to compute. In particular, EM is an iterative algorithm which alternates between inferring the missing values given the parameters (E step), and then optimizing the parameters given the "filled in" data (M step). We give the details below, followed by several examples. We end with a more theoretical discussion, where we put the algorithm in a larger context. See Table 11.2 for a summary of the applications of EM in this book.

11.4.1 Basic idea

Let $\mathbf{x}_i$ be the visible or observed variables in case i, and let $\mathbf{z}_i$ be the hidden or missing variables. The goal is to maximize the log likelihood of the observed data:

$$\ell(\boldsymbol{\theta}) = \sum_{i=1}^{N} \log p(\mathbf{x}_i|\boldsymbol{\theta}) = \sum_{i=1}^{N} \log \left[\sum_{\mathbf{z}_i} p(\mathbf{x}_i, \mathbf{z}_i|\boldsymbol{\theta}) \right] \tag{11.17}$$

Unfortunately this is hard to optimize, since the log cannot be pushed inside the sum.

EM gets around this problem as follows. Define the **complete data log likelihood** to be

$$\ell_c(\boldsymbol{\theta}) \triangleq \sum_{i=1}^{N} \log p(\mathbf{x}_i, \mathbf{z}_i | \boldsymbol{\theta}) \tag{11.18}$$

This cannot be computed, since $\mathbf{z}_i$ is unknown. So let us define the **expected complete data log likelihood** as follows:

$$Q(\boldsymbol{\theta}, \boldsymbol{\theta}^{t-1}) = \mathbb{E}\left[\ell_c(\boldsymbol{\theta}) | \mathcal{D}, \boldsymbol{\theta}^{t-1}\right] \tag{11.19}$$

where t is the current iteration number. Q is called the **auxiliary function**. The expectation is taken wrt the old parameters, $\boldsymbol{\theta}^{t-1}$, and the observed data $\mathcal{D}$. The goal of the **E step** is to compute $Q(\boldsymbol{\theta}, \boldsymbol{\theta}^{t-1})$, or rather, the terms inside of it which the MLE depends on; these are known as the **expected sufficient statistics** or ESS. In the **M step**, we optimize the Q function wrt $\boldsymbol{\theta}$:

$$\boldsymbol{\theta}^t = \arg\max_{\boldsymbol{\theta}} Q(\boldsymbol{\theta}, \boldsymbol{\theta}^{t-1}) \tag{11.20}$$

To perform MAP estimation, we modify the M step as follows:

$$\boldsymbol{\theta}^t = \operatorname*{argmax}_{\boldsymbol{\theta}} Q(\boldsymbol{\theta}, \boldsymbol{\theta}^{t-1}) + \log p(\boldsymbol{\theta}) \tag{11.21}$$

The E step remains unchanged.

In Section 11.4.7 we show that the EM algorithm monotonically increases the log likelihood of the observed data (plus the log prior, if doing MAP estimation), or it stays the same. So if the objective ever goes down, there must be a bug in our math or our code. (This is a surprisingly useful debugging tool!)

Below we explain how to perform the E and M steps for several simple models, that should make things clearer.

11.4.2   EM for GMMs

In this section, we discuss how to fit a mixture of Gaussians using EM. Fitting other kinds of mixture models requires a straightforward modification — see Exercise 11.3. We assume the number of mixture components, K, is known (see Section 11.5 for discussion of this point).

11.4.2.1 Auxiliary function

The expected complete data log likelihood is given by

$$Q(\boldsymbol{\theta}, \boldsymbol{\theta}^{(t-1)}) \triangleq \mathbb{E}\left[\sum_i \log p(\mathbf{x}_i, z_i | \boldsymbol{\theta})\right] \tag{11.22}$$

$$= \sum_i \mathbb{E}\left[\log\left[\prod_{k=1}^K (\pi_k p(\mathbf{x}_i | \boldsymbol{\theta}_k))^{\mathbb{I}(z_i = k)}\right]\right] \tag{11.23}$$

$$= \sum_i \sum_k \mathbb{E}\left[\mathbb{I}(z_i = k)\right] \log[\pi_k p(\mathbf{x}_i | \boldsymbol{\theta}_k)] \tag{11.24}$$

$$= \sum_i \sum_k p(z_i = k | \mathbf{x}_i, \boldsymbol{\theta}^{t-1}) \log[\pi_k p(\mathbf{x}_i | \boldsymbol{\theta}_k)] \tag{11.25}$$

$$= \sum_i \sum_k r_{ik} \log \pi_k + \sum_i \sum_k r_{ik} \log p(\mathbf{x}_i | \boldsymbol{\theta}_k) \tag{11.26}$$

where $r_{ik} \triangleq p(z_i = k | \mathbf{x}_i, \boldsymbol{\theta}^{(t-1)})$ is the **responsibility** that cluster k takes for data point i. This is computed in the E step, described below.

11.4.2.2 E step

The E step has the following simple form, which is the same for any mixture model:

$$r_{ik} = \frac{\pi_k p(\mathbf{x}_i | \boldsymbol{\theta}_k^{(t-1)})}{\sum_{k'} \pi_{k'} p(\mathbf{x}_i | \boldsymbol{\theta}_{k'}^{(t-1)})} \tag{11.27}$$

11.4.2.3 M step

In the M step, we optimize Q wrt $\boldsymbol{\pi}$ and the $\boldsymbol{\theta}_k$. For $\boldsymbol{\pi}$, we obviously have

$$\pi_k = \frac{1}{N} \sum_i r_{ik} = \frac{r_k}{N} \tag{11.28}$$

where $r_k \triangleq \sum_i r_{ik}$ is the weighted number of points assigned to cluster k.

To derive the M step for the $\boldsymbol{\mu}_k$ and $\boldsymbol{\Sigma}_k$ terms, we look at the parts of Q that depend on $\boldsymbol{\mu}_k$ and $\boldsymbol{\Sigma}_k$. We see that the result is

$$\ell(\boldsymbol{\mu}_k, \boldsymbol{\Sigma}_k) = \sum_k \sum_i r_{ik} \log p(\mathbf{x}_i | \boldsymbol{\theta}_k) \tag{11.29}$$

$$= -\frac{1}{2} \sum_i r_{ik} \left[\log |\boldsymbol{\Sigma}_k| + (\mathbf{x}_i - \boldsymbol{\mu}_k)^T \boldsymbol{\Sigma}_k^{-1} (\mathbf{x}_i - \boldsymbol{\mu}_k)\right] \tag{11.30}$$

This is just a weighted version of the standard problem of computing the MLEs of an MVN (see Section 4.1.3). One can show (Exercise 11.2) that the new parameter estimates are given by

$$\boldsymbol{\mu}_k = \frac{\sum_i r_{ik} \mathbf{x}_i}{r_k} \tag{11.31}$$

$$\boldsymbol{\Sigma}_k = \frac{\sum_i r_{ik} (\mathbf{x}_i - \boldsymbol{\mu}_k)(\mathbf{x}_i - \boldsymbol{\mu}_k)^T}{r_k} = \frac{\sum_i r_{ik} \mathbf{x}_i \mathbf{x}_i^T}{r_k} - \boldsymbol{\mu}_k \boldsymbol{\mu}_k^T \tag{11.32}$$

These equations make intuitive sense: the mean of cluster k is just the weighted average of all points assigned to cluster k, and the covariance is proportional to the weighted empirical scatter matrix.

After computing the new estimates, we set $\boldsymbol{\theta}^t = (\pi_k, \mu_k, \boldsymbol{\Sigma}_k)$ for $k = 1 : K$, and go to the next E step.

11.4.2.4   Example

An example of the algorithm in action is shown in Figure 11.11. We start with $\boldsymbol{\mu}_1 = (-1, 1)$, $\boldsymbol{\Sigma}_1 = \mathbf{I}$, $\boldsymbol{\mu}_2 = (1, -1)$, $\boldsymbol{\Sigma}_2 = \mathbf{I}$. We color code points such that blue points come from cluster 1 and red points from cluster 2. More precisely, we set the color to

$$\text{color}(i) = r_{i1}\text{blue} + r_{i2}\text{red} \tag{11.33}$$

so ambiguous points appear purple. After 20 iterations, the algorithm has converged on a good clustering. (The data was standardized, by removing the mean and dividing by the standard deviation, before processing. This often helps convergence.)

11.4.2.5   K-means algorithm

There is a popular variant of the EM algorithm for GMMs known as the **K-means algorithm**, which we now discuss. Consider a GMM in which we make the following assumptions: $\boldsymbol{\Sigma}_k = \sigma^2 \mathbf{I}_D$ is fixed, and $\pi_k = 1/K$ is fixed, so only the cluster centers, $\boldsymbol{\mu}_k \in \mathbb{R}^D$, have to be estimated. Now consider the following delta-function approximation to the posterior computed during the E step:

$$p(z_i = k|\mathbf{x}_i, \boldsymbol{\theta}) \approx \mathbb{I}(k = z_i^*) \tag{11.34}$$

where $z_i^* = \operatorname{argmax}_k p(z_i = k|\mathbf{x}_i, \boldsymbol{\theta})$. This is sometimes called **hard EM**, since we are making a hard assignment of points to clusters. Since we assumed an equal spherical covariance matrix for each cluster, the most probable cluster for $\mathbf{x}_i$ can be computed by finding the nearest prototype:

$$z_i^* = \arg\min_k ||\mathbf{x}_i - \boldsymbol{\mu}_k||_2^2 \tag{11.35}$$

Hence in each E step, we must find the Euclidean distance between N data points and K cluster centers, which takes $O(NKD)$ time. However, this can be sped up using various techniques, such as applying the triangle inequality to avoid some redundant computations (Elkan 2003). Given the hard cluster assignments, the M step updates each cluster center by computing the mean of all points assigned to it:

$$\boldsymbol{\mu}_k = \frac{1}{N_k} \sum_{i:z_i=k} \mathbf{x}_i \tag{11.36}$$

See Algorithm 11.1 for the pseudo-code.

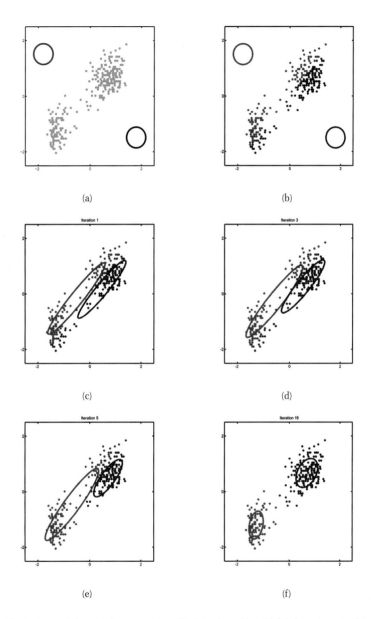

(a)   (b)

(c)   (d)

(e)   (f)

Figure 11.11   Illustration of the EM for a GMM applied to the Old Faithful data. (a) Initial (random) values of the parameters. (b) Posterior responsibility of each point computed in the first E step. The degree of redness indicates the degree to which the point belongs to the red cluster, and similarly for blue; this purple points have a roughly uniform posterior over clusters. (c) We show the updated parameters after the first M step. (d) After 3 iterations. (e) After 5 iterations. (f) After 16 iterations. Based on (Bishop 2006) Figure 9.8. Figure generated by `mixGaussDemoFaithful`.

Algorithm 11.1: K-means algorithm

1 *initialize* $\boldsymbol{\mu}_k$;
2 **repeat**
3   Assign each data point to its closest cluster center: $z_i = \arg\min_k ||\mathbf{x}_i - \boldsymbol{\mu}_k||_2^2$;
4   Update each cluster center by computing the mean of all points assigned to it:
    $\boldsymbol{\mu}_k = \frac{1}{N_k} \sum_{i:z_i=k} \mathbf{x}_i$;
5 **until** *converged*;

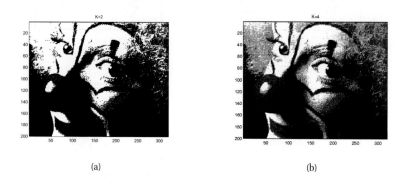

(a)   (b)

Figure 11.12 An image compressed using vector quantization with a codebook of size K. (a) $K = 2$. (b) $K = 4$. Figure generated by vqDemo.

11.4.2.6   Vector quantization

Since K-means is not a proper EM algorithm, it is not maximizing likelihood. Instead, it can be interpreted as a greedy algorithm for approximately minimizing a loss function related to data compression, as we now explain.

Suppose we want to perform lossy compression of some real-valued vectors, $\mathbf{x}_i \in \mathbb{R}^D$. A very simple approach to this is to use **vector quantization** or **VQ**. The basic idea is to replace each real-valued vector $\mathbf{x}_i \in \mathbb{R}^D$ with a discrete symbol $z_i \in \{1, \ldots, K\}$, which is an index into a **codebook** of K prototypes, $\boldsymbol{\mu}_k \in \mathbb{R}^D$. Each data vector is encoded by using the index of the most similar prototype, where similarity is measured in terms of Euclidean distance:

$$\text{encode}(\mathbf{x}_i) = \arg\min_k ||\mathbf{x}_i - \boldsymbol{\mu}_k||^2 \tag{11.37}$$

We can define a cost function that measures the quality of a codebook by computing the **reconstruction error** or **distortion** it induces:

$$J(\boldsymbol{\mu}, \mathbf{z} | K, \mathbf{X}) \triangleq \frac{1}{N} \sum_{i=1}^{N} ||\mathbf{x}_i - \text{decode}(\text{encode}(\mathbf{x}_i))||^2 = \frac{1}{N} \sum_{i=1}^{N} ||\mathbf{x}_i - \boldsymbol{\mu}_{z_i}||^2 \tag{11.38}$$

where $\text{decode}(k) = \boldsymbol{\mu}_k$. The K-means algorithm can be thought of as a simple iterative scheme for minimizing this objective.

Of course, we can achieve zero distortion if we assign one prototype to every data vector, but that takes $O(NDC)$ space, where N is the number of real-valued data vectors, each of length D, and C is the number of bits needed to represent a real-valued scalar (the quantization accuracy). However, in many data sets, we see similar vectors repeatedly, so rather than storing them many times, we can store them once and then create pointers to them. Hence we can reduce the space requirement to $O(N \log_2 K + KDC)$: the $O(N \log_2 K)$ term arises because each of the N data vectors needs to specify which of the K codewords it is using (the pointers); and the $O(KDC)$ term arises because we have to store each codebook entry, each of which is a D-dimensional vector. Typically the first term dominates the second, so we can approximate the **rate** of the encoding scheme (number of bits needed per object) as $O(\log_2 K)$, which is typically much less than $O(DC)$.

One application of VQ is to image compression. Consider the $N = 200 \times 320 = 64,000$ pixel image in Figure 11.12; this is gray-scale, so $D = 1$. If we use one byte to represent each pixel (a gray-scale intensity of 0 to 255), then $C = 8$, so we need $NC = 512,000$ bits to represent the image. For the compressed image, we need $N \log_2 K + KC$ bits. For $K = 4$, this is about 128kb, a factor of 4 compression. For $K = 8$, this is about 192kb, a factor of 2.6 compression, at negligible perceptual loss (see Figure 11.12(b)). Greater compression could be achieved if we modelled spatial correlation between the pixels, e.g., if we encoded 5x5 blocks (as used by JPEG). This is because the residual errors (differences from the model's predictions) would be smaller, and would take fewer bits to encode.

11.4.2.7 Initialization and avoiding local minima

Both K-means and EM need to be initialized. It is common to pick K data points at random, and to make these be the initial cluster centers. Or we can pick the centers sequentially so as to try to "cover" the data. That is, we pick the initial point uniformly at random. Then each subsequent point is picked from the remaining points with probability proportional to its squared distance to the points's closest cluster center. This is known as **farthest point clustering** (Gonzales 1985), or **k-means++** (Arthur and Vassilvitskii 2007; Bahmani et al. 2012). Surprisingly, this simple trick can be shown to guarantee that the distortion is never more than $O(\log K)$ worse than optimal (Arthur and Vassilvitskii 2007).

An heuristic that is commonly used in the speech recognition community is to incrementally "grow" GMMs: we initially give each cluster a score based on its mixture weight; after each round of training, we consider splitting the cluster with the highest score into two, with the new centroids being random perturbations of the original centroid, and the new scores being half of the old scores. If a new cluster has too small a score, or too narrow a variance, it is removed. We continue in this way until the desired number of clusters is reached. See (Figueiredo and Jain 2002) for a similar incremental approach.

11.4.2.8 MAP estimation

As usual, the MLE may overfit. The overfitting problem is particularly severe in the case of GMMs. To understand the problem, suppose for simplicity that $\Sigma_k = \sigma_k^2 I$, and that $K = 2$. It is possible to get an infinite likelihood by assigning one of the centers, say μ_2, to a single data point, say $\mathbf{x}_1$, since then the 1st term makes the following contribution to the likelihood:

$$\mathcal{N}(\mathbf{x}_1 | \mu_2, \sigma_2^2 I) = \frac{1}{\sqrt{2\pi\sigma_2^2}} e^0 \tag{11.39}$$

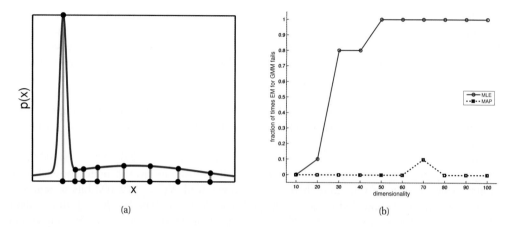

(a)   (b)

Figure 11.13 (a) Illustration of how singularities can arise in the likelihood function of GMMs. Based on (Bishop 2006) Figure 9.7. Figure generated by `mixGaussSingularity`. (b) Illustration of the benefit of MAP estimation vs ML estimation when fitting a Gaussian mixture model. We plot the fraction of times (out of 5 random trials) each method encounters numerical problems vs the dimensionality of the problem, for $N = 100$ samples. Solid red (upper curve): MLE. Dotted black (lower curve): MAP. Figure generated by `mixGaussMLvsMAP`.

Hence we can drive this term to infinity by letting $\sigma_2 \to 0$, as shown in Figure 11.13(a). We will call this the "collapsing variance problem".

An easy solution to this is to perform MAP estimation. The new auxiliary function is the expected complete data log-likelihood plus the log prior:

$$Q'(\boldsymbol{\theta}, \boldsymbol{\theta}^{old}) = \left[\sum_i \sum_k r_{ik} \log \pi_{ik} + \sum_i \sum_k r_{ik} \log p(\mathbf{x}_i | \boldsymbol{\theta}_k) \right] + \log p(\boldsymbol{\pi}) + \sum_k \log p(\boldsymbol{\theta}_k) \tag{11.40}$$

Note that the E step remains unchanged, but the M step needs to be modified, as we now explain.

For the prior on the mixture weights, it is natural to use a Dirichlet prior, $\boldsymbol{\pi} \sim \text{Dir}(\boldsymbol{\alpha})$, since this is conjugate to the categorical distribution. The MAP estimate is given by

$$\pi_k = \frac{r_k + \alpha_k - 1}{N + \sum_k \alpha_k - K} \tag{11.41}$$

If we use a uniform prior, $\alpha_k = 1$, this reduces to Equation 11.28.

The prior on the parameters of the class conditional densities, $p(\boldsymbol{\theta}_k)$, depends on the form of the class conditional densities. We discuss the case of GMMs below, and leave MAP estimation for mixtures of Bernoullis to Exercise 11.3.

For simplicity, let us consider a conjugate prior of the form

$$p(\boldsymbol{\mu}_k, \boldsymbol{\Sigma}_k) = \text{NIW}(\boldsymbol{\mu}_k, \boldsymbol{\Sigma}_k | \mathbf{m}_0, \kappa_0, \nu_0, \mathbf{S}_0) \tag{11.42}$$

From Section 4.6.3, the MAP estimate is given by

$$\hat{\boldsymbol{\mu}}_k = \frac{r_k \overline{\mathbf{x}}_k + \kappa_0 \mathbf{m}_0}{r_k + \kappa_0} \tag{11.43}$$

$$\tag{11.44}$$

$$\overline{\mathbf{x}}_k \triangleq \frac{\sum_i r_{ik} \mathbf{x}_i}{r_k} \tag{11.45}$$

$$\hat{\boldsymbol{\Sigma}}_k = \frac{\mathbf{S}_0 + \mathbf{S}_k + \frac{\kappa_0 r_k}{\kappa_0 + r_k}(\overline{\mathbf{x}}_k - \mathbf{m}_0)(\overline{\mathbf{x}}_k - \mathbf{m}_0)^T}{\nu_0 + r_k + D + 2} \tag{11.46}$$

$$\mathbf{S}_k \triangleq \sum_i r_{ik}(\mathbf{x}_i - \overline{\mathbf{x}}_k)(\mathbf{x}_i - \overline{\mathbf{x}}_k)^T \tag{11.47}$$

We now illustrate the benefits of using MAP estimation instead of ML estimation in the context of GMMs. We apply EM to some synthetic data in D dimensions, using either ML or MAP estimation. We count the trial as a "failure" if there are numerical issues involving singular matrices. For each dimensionality, we conduct 5 random trials. The results are illustrated in Figure 11.13(b) using $N = 100$. We see that as soon as D becomes even moderately large, ML estimation crashes and burns, whereas MAP estimation never encounters numerical problems.

When using MAP estimation, we need to specify the hyper-parameters. Here we mention some simple heuristics for setting them (Fraley and Raftery 2007, p163). We can set $\kappa_0 = 0$, so that the $\boldsymbol{\mu}_k$ are unregularized, since the numerical problems only arise from $\boldsymbol{\Sigma}_k$. In this case, the MAP estimates simplify to $\hat{\boldsymbol{\mu}}_k = \overline{\mathbf{x}}_k$ and $\hat{\boldsymbol{\Sigma}}_k = \frac{\mathbf{S}_0 + \mathbf{S}_k}{\nu_0 + r_k + D + 2}$, which is not quite so scary-looking.

Now we discuss how to set $\mathbf{S}_0$. One possibility is to use

$$\mathbf{S}_0 = \frac{1}{K^{1/D}} \text{diag}(s_1^2, \ldots, s_D^2) \tag{11.48}$$

where $s_j = (1/N) \sum_{i=1}^{N}(x_{ij} - \overline{x}_j)^2$ is the pooled variance for dimension j. (The reason for the $\frac{1}{K^{1/D}}$ term is that the resulting volume of each ellipsoid is then given by $|\mathbf{S}_0| = \frac{1}{K}|\text{diag}(s_1^2, \ldots, s_D^2)|$.) The parameter ν_0 controls how strongly we believe this prior. The weakest prior we can use, while still being proper, is to set $\nu_0 = D + 2$, so this is a common choice.

11.4.3   EM for mixture of experts

We can fit a mixture of experts model using EM in a straightforward manner. The expected complete data log likelihood is given by

$$Q(\boldsymbol{\theta}, \boldsymbol{\theta}^{old}) = \sum_{i=1}^{N} \sum_{k=1}^{K} r_{ik} \log[\pi_{ik} \mathcal{N}(y_i | \mathbf{w}_k^T \mathbf{x}_i, \sigma_k^2)] \tag{11.49}$$

$$\pi_{i,k} \triangleq \mathcal{S}(\mathbf{V}^T \mathbf{x}_i)_k \tag{11.50}$$

$$r_{ik} \propto \pi_{ik}^{old} \mathcal{N}(y_i | \mathbf{x}_i^T \mathbf{w}_k^{old}, (\sigma_k^{old})^2) \tag{11.51}$$

So the E step is the same as in a standard mixture model, except we have to replace π_k with $\pi_{i,k}$ when computing r_{ik}.

In the M step, we need to maximize $Q(\boldsymbol{\theta}, \boldsymbol{\theta}^{old})$ wrt $\mathbf{w}_k$, σ_k^2 and $\mathbf{V}$. For the regression parameters for model k, the objective has the form

$$Q(\boldsymbol{\theta}_k, \boldsymbol{\theta}^{old}) = \sum_{i=1}^{N} r_{ik} \left\{ -\frac{1}{\sigma_k^2} (y_i - \mathbf{w}_k^T \mathbf{x}_i) \right\} \tag{11.52}$$

We recognize this as a weighted least squares problem, which makes intuitive sense: if r_{ik} is small, then data point i will be downweighted when estimating model k's parameters. From Section 8.3.4 we can immediately write down the MLE as

$$\mathbf{w}_k = (\mathbf{X}^T \mathbf{R}_k \mathbf{X})^{-1} \mathbf{X}^T \mathbf{R}_k \mathbf{y} \tag{11.53}$$

where $\mathbf{R}_k = \text{diag}(r_{:,k})$. The MLE for the variance is given by

$$\sigma_k^2 = \frac{\sum_{i=1}^{N} r_{ik} (y_i - \mathbf{w}_k^T \mathbf{x}_i)^2}{\sum_{i=1}^{N} r_{ik}} \tag{11.54}$$

We replace the estimate of the unconditional mixing weights $\boldsymbol{\pi}$ with the estimate of the gating parameters, $\mathbf{V}$. The objective has the form

$$\ell(\mathbf{V}) = \sum_i \sum_k r_{ik} \log \pi_{i,k} \tag{11.55}$$

We recognize this as equivalent to the log-likelihood for multinomial logistic regression in Equation 8.34, except we replace the "hard" 1-of-C encoding $\mathbf{y}_i$ with the "soft" 1-of-K encoding $\mathbf{r}_i$. Thus we can estimate $\mathbf{V}$ by fitting a logistic regression model to soft target labels.

11.4.4   EM for DGMs with hidden variables

We can generalize the ideas behind EM for mixtures of experts to compute the MLE or MAP estimate for an arbitrary DGM. We could use gradient-based methods (Binder et al. 1997), but it is much simpler to use EM (Lauritzen 1995): in the E step, we just estimate the hidden variables, and in the M step, we will compute the MLE using these filled-in values. We give the details below.

For simplicity of presentation, we will assume all CPDs are tabular. Based on Section 10.4.2, let us write each CPT as follows:

$$p(x_{it} | \mathbf{x}_{i,\text{pa}(t)}, \boldsymbol{\theta}_t) = \prod_{c=1}^{K_{\text{pa}(t)}} \prod_{k=1}^{K_t} \theta_{tck}^{\mathbb{I}(x_{it}=i, \mathbf{x}_{i,\text{pa}(t)}=c)} \tag{11.56}$$

The log-likelihood of the complete data is given by

$$\log p(\mathcal{D} | \boldsymbol{\theta}) = \sum_{t=1}^{V} \sum_{c=1}^{K_{\text{pa}(t)}} \sum_{k=1}^{K_t} N_{tck} \log \theta_{tck} \tag{11.57}$$

where $N_{tck} = \sum_{i=1}^{N} \mathbb{I}(x_{it} = i, \mathbf{x}_{i,\text{pa}(t)} = c)$ are the empirical counts. Hence the expected complete data log-likelihood has the form

$$\mathbb{E}\left[\log p(\mathcal{D} | \boldsymbol{\theta}) \right] = \sum_t \sum_c \sum_k \overline{N}_{tck} \log \theta_{tck} \tag{11.58}$$

where

$$\overline{N}_{tck} = \sum_{i=1}^{N} \mathbb{E}\left[\mathbb{I}(x_{it} = i, \mathbf{x}_{i,\text{pa}(t)} = c)\right] = \sum_{i} p(x_{it} = k, \mathbf{x}_{i,\text{pa}(t)} = c|\mathcal{D}_i) \tag{11.59}$$

where $\mathcal{D}_i$ are all the visible variables in case i.

The quantity $p(x_{it}, \mathbf{x}_{i,\text{pa}(t)}|\mathcal{D}_i, \boldsymbol{\theta})$ is known as a **family marginal**, and can be computed using any GM inference algorithm. The $\overline{N}_{tjk}$ are the expected sufficient statistics, and constitute the output of the E step.

Given these ESS, the M step has the simple form

$$\hat{\theta}_{tck} = \frac{\overline{N}_{tck}}{\sum_{k'} \overline{N}_{tjk'}} \tag{11.60}$$

This can be proved by adding Lagrange multipliers (to enforce the constraint $\sum_k \theta_{tjk} = 1$) to the expected complete data log likelihood, and then optimizing each parameter vector $\boldsymbol{\theta}_{tc}$ separately. We can modify this to perform MAP estimation with a Dirichlet prior by simply adding pseudo counts to the expected counts.

11.4.5   EM for the Student distribution *

One problem with the Gaussian distribution is that it is sensitive to outliers, since the log-probability only decays quadratically with distance from the center. A more robust alternative is the Student t distribution, as discussed in Section 2.4.3.

Unlike the case of a Gaussian, there is no closed form formula for the MLE of a Student, even if we have no missing data, so we must resort to iterative optimization methods. The easiest one to use is EM, since it automatically enforces the constraints that ν is positive and that $\boldsymbol{\Sigma}$ is symmetric positive definite. In addition, the resulting algorithm turns out to have a simple intuitive form, as we see below.

At first blush, it might not be apparent why EM can be used, since there is no missing data. The key idea is to introduce an "artificial" hidden or auxiliary variable in order to simplify the algorithm. In particular, we will exploit the fact that a Student distribution can be written as a **Gaussian scale mixture**:

$$\mathcal{T}(\mathbf{x}_i|\boldsymbol{\mu}, \boldsymbol{\Sigma}, \nu) = \int \mathcal{N}(\mathbf{x}_i|\boldsymbol{\mu}, \boldsymbol{\Sigma}/z_i)\text{Ga}(z_i|\frac{\nu}{2}, \frac{\nu}{2})dz_i \tag{11.61}$$

(See Exercise 11.1 for a proof of this in the 1d case.) This can be thought of as an "infinite" mixture of Gaussians, each one with a slightly different covariance matrix.

Treating the z_i as missing data, we can write the complete data log likelihood as

$$\ell_c(\boldsymbol{\theta}) = \sum_{i=1}^{N} [\log \mathcal{N}(\mathbf{x}_i|\boldsymbol{\mu}, \boldsymbol{\Sigma}/z_i) + \log \text{Ga}(z_i|\nu/2, \nu/2)] \tag{11.62}$$

$$= \sum_{i=1}^{N} \left[-\frac{D}{2}\log(2\pi) - \frac{1}{2}\log|\boldsymbol{\Sigma}| - \frac{z_i}{2}\delta_i + \frac{\nu}{2}\log\frac{\nu}{2} - \log\Gamma(\frac{\nu}{2}) \right. \tag{11.63}$$

$$\left. +\frac{\nu}{2}(\log z_i - z_i) + (\frac{D}{2} - 1)\log z_i \right] \tag{11.64}$$

where we have defined the Mahalanobis distance to be

$$\delta_i = (\mathbf{x}_i - \boldsymbol{\mu})^T \boldsymbol{\Sigma}^{-1} (\mathbf{x}_i - \boldsymbol{\mu}) \tag{11.65}$$

We can partition this into two terms, one involving $\boldsymbol{\mu}$ and $\boldsymbol{\Sigma}$, and the other involving ν. We have, dropping irrelevant constants,

$$\ell_c(\boldsymbol{\theta}) = L_N(\boldsymbol{\mu}, \boldsymbol{\Sigma}) + L_G(\nu) \tag{11.66}$$

$$L_N(\boldsymbol{\mu}, \boldsymbol{\Sigma}) \triangleq -\frac{1}{2} N \log |\boldsymbol{\Sigma}| - \frac{1}{2} \sum_{i=1}^{N} z_i \delta_i \tag{11.67}$$

$$L_G(\nu) \triangleq -N \log \Gamma(\nu/2) + \frac{1}{2} N \nu \log(\nu/2) + \frac{1}{2} \nu \sum_{i=1}^{N} (\log z_i - z_i) \tag{11.68}$$

11.4.5.1   EM with ν known

Let us first derive the algorithm with ν assumed known, for simplicity. In this case, we can ignore the L_G term, so we only need to figure out how to compute $\mathbb{E}[z_i]$ wrt the old parameters.

From Section 4.6.2.2 we have

$$p(z_i | \mathbf{x}_i, \boldsymbol{\theta}) = \text{Ga}(z_i | \frac{\nu + D}{2}, \frac{\nu + \delta_i}{2}) \tag{11.69}$$

Now if $z_i \sim \text{Ga}(a, b)$, then $\mathbb{E}[z_i] = a/b$. Hence the E step at iteration t is

$$\overline{z}_i^{(t)} \triangleq \mathbb{E}\left[z_i | \mathbf{x}_i, \boldsymbol{\theta}^{(t)}\right] = \frac{\nu^{(t)} + D}{\nu^{(t)} + \delta_i^{(t)}} \tag{11.70}$$

The M step is obtained by maximizing $\mathbb{E}[L_N(\boldsymbol{\mu}, \boldsymbol{\Sigma})]$ to yield

$$\hat{\boldsymbol{\mu}}^{(t+1)} = \frac{\sum_i \overline{z}_i^{(t)} \mathbf{x}_i}{\sum_i \overline{z}_i^{(t)}} \tag{11.71}$$

$$\hat{\boldsymbol{\Sigma}}^{(t+1)} = \frac{1}{N} \sum_i \overline{z}_i^{(t)} (\mathbf{x}_i - \hat{\boldsymbol{\mu}}^{(t+1)})(\mathbf{x}_i - \hat{\boldsymbol{\mu}}^{(t+1)})^T \tag{11.72}$$

$$= \frac{1}{N} \left[\sum_i \overline{z}_i^{(t)} \mathbf{x}_i \mathbf{x}_i^T - \left(\sum_{i=1}^{N} \overline{z}_i^{(t)} \right) \hat{\boldsymbol{\mu}}^{(t+1)} (\hat{\boldsymbol{\mu}}^{(t+1)})^T \right] \tag{11.73}$$

These results are quite intuitive: the quantity $\overline{z}_i$ is the precision of measurement i, so if it is small, the corresponding data point is down-weighted when estimating the mean and covariance. This is how the Student achieves robustness to outliers.

11.4.5.2   EM with ν unknown

To compute the MLE for the degrees of freedom, we first need to compute the expectation of $L_G(\nu)$, which involves z_i and $\log z_i$. Now if $z_i \sim \text{Ga}(a, b)$, then one can show that

$$\overline{\ell}_i^{(t)} \triangleq \mathbb{E}\left[\log z_i | \boldsymbol{\theta}^{(t)}\right] = \Psi(a) - \log b \tag{11.74}$$

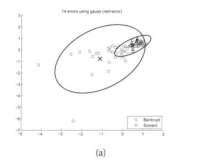
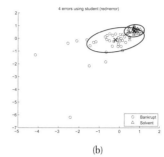

(a)   (b)

Figure 11.14 Mixture modeling on the bankruptcy data set. Left: Gaussian class conditional densities. Right: Student class conditional densities. Points that belong to class 1 are shown as triangles, points that belong to class 2 are shown as circles The estimated labels, based on the posterior probability of belonging to each mixture component, are computed. If these are incorrect, the point is colored red, otherwise it is colored blue. (Training data is in black.) Figure generated by `mixStudentBankruptcyDemo`.

where $\Psi(x) = \frac{d}{dx} \log \Gamma(x)$ is the digamma function. Hence, from Equation 11.69, we have

$$\bar{\ell}_i^{(t)} = \Psi(\frac{\nu^{(t)} + D}{2}) - \log(\frac{\nu^{(t)} + \delta_i^{(t)}}{2}) \tag{11.75}$$

$$= \log(\bar{z}_i^{(t)}) + \Psi(\frac{\nu^{(t)} + D}{2}) - \log(\frac{\nu^{(t)} + D}{2}) \tag{11.76}$$

Substituting into Equation 11.68, we have

$$\mathbb{E}\left[L_G(\nu)\right] = -N \log \Gamma(\nu/2) + \frac{N\nu}{2} \log(\nu/2) + \frac{\nu}{2} \sum_i (\bar{\ell}_i^{(t)} - \bar{z}_i^{(t)}) \tag{11.77}$$

The gradient of this expression is equal to

$$\frac{d}{d\nu} \mathbb{E}\left[L_G(\nu)\right] = -\frac{N}{2} \Psi(\nu/2) + \frac{N}{2} \log(\nu/2) + \frac{N}{2} + \frac{1}{2} \sum_i (\bar{\ell}_i^{(t)} - \bar{z}_i^{(t)}) \tag{11.78}$$

This has a unique solution in the interval $(0, +\infty]$ which can be found using a 1d constrained optimizer.

Performing a gradient-based optimization in the M step, rather than a closed-form update, is an example of what is known as the **generalized EM** algorithm. One can show that EM will still converge to a local optimum even if we only perform a "partial" improvement to the parameters in the M step.

11.4.5.3 Mixtures of Student distributions

It is easy to extend the above methods to fit a mixture of Student distributions. See Exercise 11.4 for the details.

Let us consider a small example from (Lo 2009, ch3). We have a $N = 66$, $D = 2$ data set regarding the bankrupty patterns of certain companies. The first feature specifies the ratio of retained earnings (RE) to total assets, and the second feature specifies the ratio of earnings before interests and taxes (EBIT) to total assets. We fit two models to this data, ignoring the class labels: a mixture of 2 Gaussians, and a mixture of 2 Students. We then use each fitted model to classify the data. We compute the most probable cluster membership and treat this as $\hat{y}_i$. We then compare $\hat{y}_i$ to the true labels y_i and compute an error rate. If this is more than 50%, we permute the latent labels (i.e., we consider cluster 1 to represent class 2 and vice versa), and then recompute the error rate. Points which are misclassified are then shown in red. The result is shown in Figure 11.14. We see that the Student model made 4 errors, the Gaussian model made 21. This is because the class-conditional densities contain some extreme values, causing the Gaussian to be a poor choice.

11.4.6   EM for probit regression *

In Section 9.4.2, we described the latent variable interpretation of probit regression. Recall that this has the form $p(y_i = 1|z_i) = \mathbb{I}(z_i > 0)$, where $z_i \sim \mathcal{N}(\mathbf{w}^T\mathbf{x}_i, 1)$ is latent. We now show how to fit this model using EM. (Although it is possible to fit probit regression models using gradient based methods, as shown in Section 9.4.1, this EM-based approach has the advantage that it generalized to many other kinds of models, as we will see later on.)

The complete data log likelihood has the following form, assuming a $\mathcal{N}(\mathbf{0}, \mathbf{V}_0)$ prior on $\mathbf{w}$:

$$\ell(\mathbf{z}, \mathbf{w}|\mathbf{V}_0) = \log p(\mathbf{y}|\mathbf{z}) + \log \mathcal{N}(\mathbf{z}|\mathbf{X}\mathbf{w}, \mathbf{I}) + \log \mathcal{N}(\mathbf{w}|\mathbf{0}, \mathbf{V}_0) \tag{11.79}$$

$$= \sum_i \log p(y_i|z_i) - \frac{1}{2}(\mathbf{z} - \mathbf{X}\mathbf{w})^T(\mathbf{z} - \mathbf{X}\mathbf{w}) - \frac{1}{2}\mathbf{w}^T\mathbf{V}_0^{-1}\mathbf{w} + \text{const} \tag{11.80}$$

The posterior in the E step is a **truncated Gaussian**:

$$p(z_i|y_i, \mathbf{x}_i, \mathbf{w}) = \begin{cases} \mathcal{N}(z_i|\mathbf{w}^T\mathbf{x}_i, 1)\mathbb{I}(z_i > 0) & \text{if } y_i = 1 \\ \mathcal{N}(z_i|\mathbf{w}^T\mathbf{x}_i, 1)\mathbb{I}(z_i < 0) & \text{if } y_i = 0 \end{cases} \tag{11.81}$$

In Equation 11.80, we see that $\mathbf{w}$ only depends linearly on $\mathbf{z}$, so we just need to compute $\mathbb{E}[z_i|y_i, \mathbf{x}_i, \mathbf{w}]$. Exercise 11.15 asks you to show that the posterior mean is given by

$$\mathbb{E}[z_i|\mathbf{w}, \mathbf{x}_i] = \begin{cases} \mu_i + \frac{\phi(\mu_i)}{1 - \Phi(-\mu_i)} = \mu_i + \frac{\phi(\mu_i)}{\Phi(\mu_i)} & \text{if } y_i = 1 \\ \mu_i - \frac{\phi(\mu_i)}{\Phi(-\mu_i)} = \mu_i - \frac{\phi(\mu_i)}{1 - \Phi(\mu_i)} & \text{if } y_i = 0 \end{cases} \tag{11.82}$$

where $\mu_i = \mathbf{w}^T\mathbf{x}_i$.

In the M step, we estimate $\mathbf{w}$ using ridge regression, where $\boldsymbol{\mu} = \mathbb{E}[\mathbf{z}]$ is the output we are trying to predict. Specifically, we have

$$\hat{\mathbf{w}} = (\mathbf{V}_0^{-1} + \mathbf{X}^T\mathbf{X})^{-1}\mathbf{X}^T\boldsymbol{\mu} \tag{11.83}$$

The EM algorithm is simple, but can be much slower than direct gradient methods, as illustrated in Figure 11.15. This is because the posterior entropy in the E step is quite high, since we only observe that z is positive or negative, but are given no information from the likelihood about its magnitude. Using a stronger regularizer can help speed convergence, because it constrains the range of plausible z values. In addition, one can use various speedup tricks, such as data augmentation (van Dyk and Meng 2001), but we do not discuss that here.

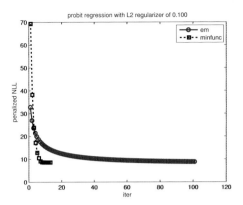

Figure 11.15 Fitting a probit regression model in 2d using a quasi-Newton method or EM. Figure generated by `probitRegDemo`.

11.4.7 Theoretical basis for EM *

In this section, we show that EM monotonically increases the observed data log likelihood until it reaches a local maximum (or saddle point, although such points are usually unstable). Our derivation will also serve as the basis for various generalizations of EM that we will discuss later.

11.4.7.1 Expected complete data log likelihood is a lower bound

Consider an arbitrary distribution $q(\mathbf{z}_i)$ over the hidden variables. The observed data log likelihood can be written as follows:

$$\ell(\boldsymbol{\theta}) \triangleq \sum_{i=1}^{N} \log \left[\sum_{\mathbf{z}_i} p(\mathbf{x}_i, \mathbf{z}_i | \boldsymbol{\theta}) \right] = \sum_{i=1}^{N} \log \left[\sum_{\mathbf{z}_i} q(\mathbf{z}_i) \frac{p(\mathbf{x}_i, \mathbf{z}_i | \boldsymbol{\theta})}{q(\mathbf{z}_i)} \right] \tag{11.84}$$

Now $\log(u)$ is a *concave* function, so from Jensen's inequality (Equation 2.113) we have the following *lower bound*:

$$\ell(\boldsymbol{\theta}) \geq \sum_{i} \sum_{\mathbf{z}_i} q_i(\mathbf{z}_i) \log \frac{p(\mathbf{x}_i, \mathbf{z}_i | \boldsymbol{\theta})}{q_i(\mathbf{z}_i)} \tag{11.85}$$

Let us denote this lower bound as follows:

$$Q(\boldsymbol{\theta}, q) \triangleq \sum_{i} \mathbb{E}_{q_i} \left[\log p(\mathbf{x}_i, \mathbf{z}_i | \boldsymbol{\theta}) \right] + \mathbb{H}\left(q_i \right) \tag{11.86}$$

where $\mathbb{H}\left(q_i \right)$ is the entropy of q_i.

The above argument holds for any positive distribution q. Which one should we choose? Intuitively we should pick the q that yields the tightest lower bound. The lower bound is a sum over i of terms of the following form:

$$L(\boldsymbol{\theta}, q_i) = \sum_{\mathbf{z}_i} q_i(\mathbf{z}_i) \log \frac{p(\mathbf{x}_i, \mathbf{z}_i | \boldsymbol{\theta})}{q_i(\mathbf{z}_i)} \tag{11.87}$$

$$= \sum_{\mathbf{z}_i} q_i(\mathbf{z}_i) \log \frac{p(\mathbf{z}_i | \mathbf{x}_i, \boldsymbol{\theta}) p(\mathbf{x}_i | \boldsymbol{\theta})}{q_i(\mathbf{z}_i)} \tag{11.88}$$

$$= \sum_{\mathbf{z}_i} q_i(\mathbf{z}_i) \log \frac{p(\mathbf{z}_i | \mathbf{x}_i, \boldsymbol{\theta})}{q_i(\mathbf{z}_i)} + \sum_{\mathbf{z}_i} q_i(\mathbf{z}_i) \log p(\mathbf{x}_i | \boldsymbol{\theta}) \tag{11.89}$$

$$= -\mathbb{KL}\left(q_i(\mathbf{z}_i) || p(\mathbf{z}_i | \mathbf{x}_i, \boldsymbol{\theta})\right) + \log p(\mathbf{x}_i | \boldsymbol{\theta}) \tag{11.90}$$

The $p(\mathbf{x}_i | \boldsymbol{\theta})$ term is independent of q_i, so we can maximize the lower bound by setting $q_i(\mathbf{z}_i) = p(\mathbf{z}_i | \mathbf{x}_i, \boldsymbol{\theta})$. Of course, $\boldsymbol{\theta}$ is unknown, so instead we use $q_i^t(\mathbf{z}_i) = p(\mathbf{z}_i | \mathbf{x}_i, \boldsymbol{\theta}^t)$, where $\boldsymbol{\theta}^t$ is our estimate of the parameters at iteration t. This is the output of the E step.

Plugging this in to the lower bound we get

$$Q(\boldsymbol{\theta}, q^t) = \sum_i \mathbb{E}_{q_i^t}\left[\log p(\mathbf{x}_i, \mathbf{z}_i | \boldsymbol{\theta})\right] + \mathbb{H}\left(q_i^t\right) \tag{11.91}$$

We recognize the first term as the expected complete data log likelihood. The second term is a constant wrt $\boldsymbol{\theta}$. So the M step becomes

$$\boldsymbol{\theta}^{t+1} = \arg\max_{\boldsymbol{\theta}} Q(\boldsymbol{\theta}, \boldsymbol{\theta}^t) = \arg\max_{\boldsymbol{\theta}} \sum_i \mathbb{E}_{q_i^t}\left[\log p(\mathbf{x}_i, \mathbf{z}_i | \boldsymbol{\theta})\right] \tag{11.92}$$

as usual.

Now comes the punchline. Since we used $q_i^t(\mathbf{z}_i) = p(\mathbf{z}_i | \mathbf{x}_i, \boldsymbol{\theta}^t)$, the KL divergence becomes zero, so $L(\boldsymbol{\theta}^t, q_i) = \log p(\mathbf{x}_i | \boldsymbol{\theta}^t)$, and hence

$$Q(\boldsymbol{\theta}^t, \boldsymbol{\theta}^t) = \sum_i \log p(\mathbf{x}_i | \boldsymbol{\theta}^t) = \ell(\boldsymbol{\theta}^t) \tag{11.93}$$

We see that the lower bound is tight after the E step. Since the lower bound "touches" the function, maximizing the lower bound will also "push up" on the function itself. That is, the M step is guaranteed to modify the parameters so as to increase the likelihood of the observed data (unless it is already at a local maximum).

This process is sketched in Figure 11.16. The dashed red curve is the original function (the observed data log-likelihood). The solid blue curve is the lower bound, evaluated at $\boldsymbol{\theta}^t$; this touches the objective function at $\boldsymbol{\theta}^t$. We then set $\boldsymbol{\theta}^{t+1}$ to the maximum of the lower bound (blue curve), and fit a new bound at that point (dotted green curve). The maximum of this new bound becomes $\boldsymbol{\theta}^{t+2}$, etc. (Compare this to Newton's method in Figure 8.4(a), which repeatedly fits and then optimizes a quadratic approximation.)

11.4.7.2   EM monotonically increases the observed data log likelihood

We now prove that EM monotonically increases the observed data log likelihood until it reaches a local optimum. We have

$$\ell(\boldsymbol{\theta}^{t+1}) \geq Q(\boldsymbol{\theta}^{t+1}, \boldsymbol{\theta}^t) \geq Q(\boldsymbol{\theta}^t, \boldsymbol{\theta}^t) = \ell(\boldsymbol{\theta}^t) \tag{11.94}$$

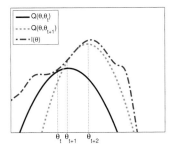

Figure 11.16   Illustration of EM as a bound optimization algorithm. Based on Figure 9.14 of (Bishop 2006). Figure generated by `emLogLikelihoodMax`.

where the first inequality follows since $Q(\boldsymbol{\theta}, \cdot)$ is a lower bound on $\ell(\boldsymbol{\theta})$; the second inequality follows since, by definition, $Q(\boldsymbol{\theta}^{t+1}, \boldsymbol{\theta}^t) = \max_{\boldsymbol{\theta}} Q(\boldsymbol{\theta}, \boldsymbol{\theta}^t) \geq Q(\boldsymbol{\theta}^t, \boldsymbol{\theta}^t)$; and the final equality follows Equation 11.93.

As a consequence of this result, if you do not observe monotonic increase of the observed data log likelihood, you must have an error in your math and/or code. (If you are performing MAP estimation, you must add on the log prior term to the objective.) This is a surprisingly powerful debugging tool.

11.4.8   Online EM

When dealing with large or streaming datasets, it is important to be able to learn online, as we discussed in Section 8.5. There are two main approaches to **online EM** in the literature. The first approach, known as **incremental EM** (Neal and Hinton 1998), optimizes the lower bound $Q(\boldsymbol{\theta}, q_1, \ldots, q_N)$ one q_i at a time; however, this requires storing the expected sufficient statistics for each data case. The second approach, known as **stepwise EM** (Sato and Ishii 2000; Cappe and Mouline 2009; Cappe 2010), is based on stochastic approximation theory, and only requires constant memory use. We explain both approaches in more detail below, following the presentation of (Liang and Klein 2009).

11.4.8.1   Batch EM review

Before explaining online EM, we review batch EM in a more abstract setting. Let $\phi(\mathbf{x}, \mathbf{z})$ be a vector of sufficient statistics for a single data case. (For example, for a mixture of multinoullis, this would be the count vector $a(j)$, which is the number of cluster j was used in $\mathbf{z}$, plus the matrix $B(j, v)$, which is of the number of times the hidden state was j and the observed letter was v.) Let $\mathbf{s}_i = \sum_{\mathbf{z}} p(\mathbf{z}|\mathbf{x}_i, \boldsymbol{\theta})\phi(\mathbf{x}_i, \mathbf{z})$ be the expected sufficient statistics for case i, and $\boldsymbol{\mu} = \sum_{i=1}^N \mathbf{s}_i$ be the sum of the ESS. Given $\boldsymbol{\mu}$, we can derive an ML or MAP estimate of the parameters in the M step; we will denote this operation by $\boldsymbol{\theta}(\boldsymbol{\mu})$. (For example, in the case of mixtures of multinoullis, we just need to normalize $\mathbf{a}$ and each row of $\mathbf{B}$.) With this notation under our belt, the pseudo code for batch EM is as shown in Algorithm 11.2.

Algorithm 11.2: Batch EM algorithm

---

1 *initialize* $\boldsymbol{\mu}$;
2 **repeat**
3      $\boldsymbol{\mu}^{new} = \mathbf{0}$;
4   **for** *each example* $i = 1 : N$ **do**
5          $\mathbf{s}_i := \sum_{\mathbf{z}} p(\mathbf{z}|\mathbf{x}_i, \boldsymbol{\theta}(\boldsymbol{\mu}))\phi(\mathbf{x}_i, \mathbf{z})$;
6          $\boldsymbol{\mu}^{new} := \boldsymbol{\mu}^{new} + \mathbf{s}_i$; ;
7      $\boldsymbol{\mu} := \boldsymbol{\mu}^{new}$;
8 **until** *converged*;

11.4.8.2 Incremental EM

In incremental EM (Neal and Hinton 1998), we keep track of $\boldsymbol{\mu}$ as well as the $\mathbf{s}_i$. When we come to a data case, we swap out the old $\mathbf{s}_i$ and replace it with the new $\mathbf{s}_i^{new}$, as shown in the code in Algorithm 11.3. Note that we can exploit the sparsity of $\mathbf{s}_i^{new}$ to speedup the computation of $\boldsymbol{\theta}$, since most components of $\boldsymbol{\mu}$ wil not have changed.

Algorithm 11.3: Incremental EM algorithm

---

1 *initialize* $\mathbf{s}_i$ for $i = 1 : N$;
2 $\boldsymbol{\mu} = \sum_i \mathbf{s}_i$;
3 **repeat**
4   **for** *each example* $i = 1 : N$ *in a random order* **do**
5          $\mathbf{s}_i^{new} := \sum_{\mathbf{z}} p(\mathbf{z}|\mathbf{x}_i, \boldsymbol{\theta}(\boldsymbol{\mu}))\phi(\mathbf{x}_i, \mathbf{z})$;
6          $\boldsymbol{\mu} := \boldsymbol{\mu} + \mathbf{s}_i^{new} - \mathbf{s}_i$;
7          $\mathbf{s}_i := \mathbf{s}_i^{new}$;
8 **until** *converged*;

This can be viewed as maximizing the lower bound $Q(\boldsymbol{\theta}, q_1, \ldots, q_N)$ by optimizing q_1, then $\boldsymbol{\theta}$, then q_2, then $\boldsymbol{\theta}$, etc. As such, this method is guaranteed to monotonically converge to a local maximum of the lower bound and to the log likelihood itself.

11.4.8.3 Stepwise EM

In stepwise EM, whenever we compute a new $\mathbf{s}_i$, we move $\boldsymbol{\mu}$ towards it, as shown in Algorithm 11.4.[2] At iteration k, the stepsize has value η_k, which must satisfy the Robbins-Monro conditions in Equation 8.82. For example, (Liang and Klein 2009) use $\eta_k = (2 + k)^{-\kappa}$ for $0.5 < \kappa \leq 1$. We can get somewhat better behavior by using a minibatch of size m before each update. It is possible to optimize m and κ to maximize the training set likelihood, by

---

2. A detail: As written the update for $\boldsymbol{\mu}$ does not exploit the sparsity of $\mathbf{s}_i$. We can fix this by storing $\mathbf{m} = \frac{\boldsymbol{\mu}}{\prod_{j < k}(1 - \eta_j)}$ instead of $\boldsymbol{\mu}$, and then using the sparse update $\mathbf{m} := \mathbf{m} + \frac{\eta_k}{\prod_{j < k}(1 - \eta_j)}\mathbf{s}_i$. This will not affect the results (i.e., $\boldsymbol{\theta}(\boldsymbol{\mu}) = \boldsymbol{\theta}(\mathbf{m})$), since scaling the counts by a global constant has no effect.

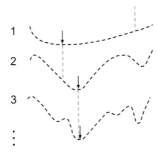

Figure 11.17 Illustration of deterministic annealing. Based on `http://en.wikipedia.org/wiki/Grad uated_optimization`.

trying different values in parallel for an initial trial period; this can significantly speed up the algorithm.

Algorithm 11.4: Stepwise EM algorithm

1 *initialize* $\boldsymbol{\mu}$; $k = 0$;
2 **repeat**
3   **for** *each example* $i = 1 : N$ *in a random order* **do**
4       $\mathbf{s}_i := \sum_{\mathbf{z}} p(\mathbf{z}|\mathbf{x}_i, \boldsymbol{\theta}(\boldsymbol{\mu}))\phi(\mathbf{x}_i, \mathbf{z})$;
5       $\boldsymbol{\mu} := (1 - \eta_k)\boldsymbol{\mu} + \eta_k \mathbf{s}_i$;
6       $k := k + 1$
7 **until** *converged*;

(Liang and Klein 2009) compare batch EM, incremental EM, and stepwise EM on four different unsupervised language modeling tasks. They found that stepwise EM (using $\kappa \approx 0.7$ and $m \approx 1000$) was faster than incremental EM, and both were much faster than batch EM. In terms of accuracy, stepwise EM was usually as good or sometimes even better than batch EM; incremental EM was often worse than either of the other methods.

11.4.9 Other EM variants *

EM is one of the most widely used algorithms in statistics and machine learning. Not surprisingly, many variations have been proposed. We briefly mention a few below, some of which we will use in later chapters. See (McLachlan and Krishnan 1997) for more information.

- **Annealed EM** In general, EM will only converge to a local maximum. To increase the chance of finding the global maximum, we can use a variety of methods. One approach is to use a method known as **deterministic annealing** (Rose 1998). The basic idea is to "smooth" the posterior "landscape" by raising it to a temperature, and then gradually cooling it, all the while slowly tracking the global maximum. See Figure 11.17. for a sketch. (A stochastic version

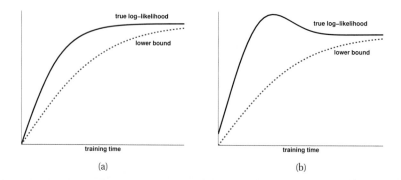

Figure 11.18   Illustration of possible behaviors of variational EM. (a) The lower bound increases at each iteration, and so does the likelihood. (b) The lower bound increases but the likelihood decreases. In this case, the algorithm is closing the gap between the approximate and true posterior. This can have a regularizing effect. Based on Figure 6 of (Saul et al. 1996). Figure generated by `varEMbound`.

of this algorithm is described in Section 24.6.1.) An annealed version of EM is described in (Ueda and Nakano 1998).

- **Variational EM** In Section 11.4.7, we showed that the optimal thing to do in the E step is to make q_i be the exact posterior over the latent variables, $q_i^t(\mathbf{z}_i) = p(\mathbf{z}_i|\mathbf{x}_i, \boldsymbol{\theta}^t)$. In this case, the lower bound on the log likelihood will be tight, so the M step will "push up" on the log-likelihood itself. However, sometimes it is computationally intractable to perform exact inference in the E step, but we may be able to perform approximate inference. If we can ensure that the E step is performing inference based on a a lower bound to the likelihood, then the M step can be seen as monotonically increasing this lower bound (see Figure 11.18). This is called **variational EM** (Neal and Hinton 1998). See Chapter 21 for some variational inference methods that can be used in the E step.

- **Monte Carlo EM** Another approach to handling an intractable E step is to use a Monte Carlo approximation to the expected sufficient statistics. That is, we draw samples from the posterior, $\mathbf{z}_i^s \sim p(\mathbf{z}_i|\mathbf{x}_i, \boldsymbol{\theta}^t)$, and then compute the sufficient statistics for each completed vector, $(\mathbf{x}_i, \mathbf{z}_i^s)$, and then average the results. This is called **Monte Carlo EM** or **MCEM** (Wei and Tanner 1990). (If we only draw a single sample, it is called **stochastic EM** (Celeux and Diebolt 1985).) One way to draw samples is to use MCMC (see Chapter 24). However, if we have to wait for MCMC to converge inside each E step, the method becomes very slow. An alternative is to use stochastic approximation, and only perform "brief" sampling in the E step, followed by a partial parameter update. This is called **stochastic approximation EM** (Delyon et al. 1999) and tends to work better than MCEM. Another alternative is to apply MCMC to infer the parameters as well as the latent variables (a fully Bayesian approach), thus eliminating the distinction between E and M steps. See Chapter 24 for details.

- **Generalized EM** Sometimes we can perform the E step exactly, but we cannot perform the M step exactly. However, we can still monotonically increase the log likelihood by performing a "partial" M step, in which we merely increase the expected complete data log likelihood, rather than maximizing it. For example, we might follow a few gradient steps. This is called

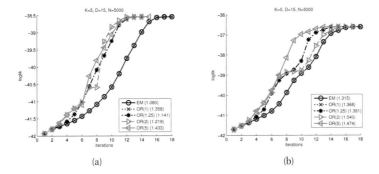

(a) (b)

Figure 11.19 Illustration of adaptive over-relaxed EM applied to a mixture of 5 Gaussians in 15 dimensions. We show the algorithm applied to two different datasets, randomly sampled from a mixture of 10 Gaussians. We plot the convergence for different update rates η. Using $\eta = 1$ gives the same results as regular EM. The actual running time is printed in the legend. Figure generated by `mixGaussOverRelaxedEmDemo`.

the **generalized EM** or **GEM** algorithm. (This is an unfortunate term, since there are many ways to generalize EM....)

- **ECM(E) algorithm** The **ECM** algorithm stands for "expectation conditional maximization", and refers to optimizing the parameters in the M step sequentially, if they turn out to be dependent. The **ECME** algorithm, which stands for "ECM either" (Liu and Rubin 1995), is a variant of ECM in which we maximize the expected complete data log likelihood (the Q function) as usual, or the observed data log likelihood, during one or more of the conditional maximization steps. The latter can be much faster, since it ignores the results of the E step, and directly optimizes the objective of interest. A standard example of this is when fitting the Student T distribution. For fixed ν, we can update Σ as usual, but then to update ν, we replace the standard update of the form $\nu^{t+1} = \arg\max_\nu Q((\boldsymbol{\mu}^{t+1}, \boldsymbol{\Sigma}^{t+1}, \nu), \boldsymbol{\theta}^t)$ with $\nu^{t+1} = \arg\max_\nu \log p(\mathcal{D}|\boldsymbol{\mu}^{t+1}, \boldsymbol{\Sigma}^{t+1}, \nu)$. See (McLachlan and Krishnan 1997) for more information.

- **Over-relaxed EM** Vanilla EM can be quite slow, especially if there is lots of missing data. The adaptive **overrelaxed EM algorithm** (Salakhutdinov and Roweis 2003) performs an update of the form $\boldsymbol{\theta}^{t+1} = \boldsymbol{\theta}^t + \eta(M(\boldsymbol{\theta}^t) - \boldsymbol{\theta}^t)$, where η is a step-size parameter, and $M(\boldsymbol{\theta}^t)$ is the usual update computed during the M step. Obviously this reduces to standard EM if $\eta = 1$, but using larger values of η can result in faster convergence. See Figure 11.19 for an illustration. Unfortunately, using too large a value of η can cause the algorithm to fail to converge.

Finally, note that EM is in fact just a special case of a larger class of algorithms known as **bound optimization** or **MM** algorithms (MM stands for **minorize-maximize**). See (Hunter and Lange 2004) for further discussion.

11.5   Model selection for latent variable models

When using LVMs, we must specify the number of latent variables, which controls the model complexity. In particuarl, in the case of mixture models, we must specify K, the number of clusters. Choosing these parameters is an example of model selection. We discuss some approaches below.

11.5.1   Model selection for probabilistic models

The optimal Bayesian approach, discussed in Section 5.3, is to pick the model with the largest marginal likelihood, $K^* = \text{argmax}_k\, p(\mathcal{D}|K)$.

There are two problems with this. First, evaluating the marginal likelihood for LVMs is quite difficult. In practice, simple approximations, such as BIC, can be used (see e.g., (Fraley and Raftery 2002)). Alternatively, we can use the cross-validated likelihood as a performance measure, although this can be slow, since it requires fitting each model F times, where F is the number of CV folds.

The second issue is the need to search over a potentially large number of models. The usual approach is to perform exhaustive search over all candidate values of K. However, sometimes we can set the model to its maximal size, and then rely on the power of the Bayesian Occam's razor to "kill off" unwanted components. An example of this will be shown in Section 21.6.1.6, when we discuss variational Bayes.

An alternative approach is to perform stochastic sampling in the space of models. Traditional approaches, such as (Green 1998, 2003; Lunn et al. 2009), are based on reversible jump MCMC, and use birth moves to propose new centers, and death moves to kill off old centers. However, this can be slow and difficult to implement. A simpler approach is to use a Dirichlet process mixture model, which can be fit using Gibbs sampling, but still allows for an unbounded number of mixture components; see Section 25.2 for details.

Perhaps surprisingly, these sampling-based methods can be faster than the simple approach of evaluating the quality of each K separately. The reason is that fitting the model for each K is often slow. By contrast, the sampling methods can often quickly determine that a certain value of K is poor, and thus they need not waste time in that part of the posterior.

11.5.2   Model selection for non-probabilistic methods

What if we are not using a probabilistic model? For example, how do we choose K for the K-means algorithm? Since this does not correspond to a probability model, there is no likelihood, so none of the methods described above can be used.

An obvious proxy for the likelihood is the reconstruction error. Define the squared reconstruction error of a data set $\mathcal{D}$, using model complexity K, as follows:

$$E(\mathcal{D}, K) = \frac{1}{|\mathcal{D}|} \sum_{i \in \mathcal{D}} ||\mathbf{x}_i - \hat{\mathbf{x}}_i||^2 \tag{11.95}$$

In the case of K-means, the reconstruction is given by $\hat{\mathbf{x}}_i = \boldsymbol{\mu}_{z_i}$, where $z_i = \text{argmin}_k ||\mathbf{x}_i - \boldsymbol{\mu}_k||_2^2$, as explained in Section 11.4.2.6.

Figure 11.20(a) plots the reconstruction error on the *test set* for K-means. We notice that the error decreases with increasing model complexity! The reason for this behavior is as follows:

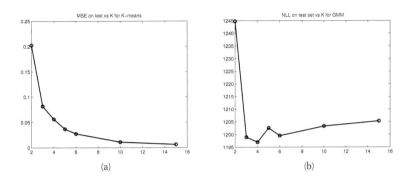

Figure 11.20   Test set performance vs K for data generated from a mixture of 3 Gaussians in 1d (data is shown in Figure 11.21(a)). (a) MSE on test set for K-means. (b) Negative log likelihood on test set for GMM. Figure generated by `kmeansModelSel1d`.

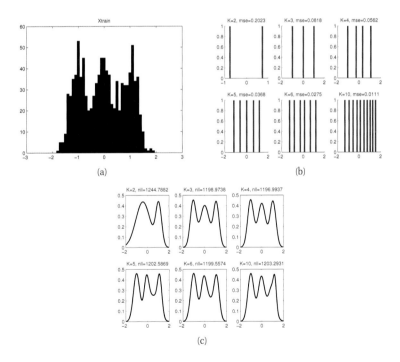

Figure 11.21   Synthetic data generated from a mixture of 3 Gaussians in 1d. (a) Histogram of training data. (Test data looks essentially the same.) (b) Centroids estimated by K-means for $K \in \{2, 3, 4, 5, 6, 10\}$. (c) GMM density model estimated by EM for for the same values of K. Figure generated by `kmeansModelSel1d`.

when we add more and more centroids to K-means, we can "tile" the space more densely, as shown in Figure 11.21(b). Hence any given test vector is more likely to find a close prototype to accurately represent it as K increases, thus decreasing reconstruction error. However, if we use a probabilistic model, such as the GMM, and plot the negative log-likelihood, we get the usual U-shaped curve on the test set, as shown in Figure 11.20(b).

In supervised learning, we can always use cross validation to select between non-probabilistic models of different complexity, but this is not the case with unsupervised learning. Although this is not a novel observation (e.g., it is mentioned in passing in (Hastie et al. 2009, p519), one of the standard references in this field), it is perhaps not as widely appreciated as it should be. In fact, it is one of the more compelling arguments in favor of probabilistic models.

Given that cross validation doesn't work, and supposing one is unwilling to use probabilistic models (for some bizarre reason...), how can one choose K? The most common approach is to plot the reconstruction error on the training set versus K, and to try to identify a **knee** or **kink** in the curve. The idea is that for $K < K^*$, where K^* is the "true" number of clusters, the rate of decrease in the error function will be high, since we are splitting apart things that should not be grouped together. However, for $K > K^*$, we are splitting apart "natural" clusters, which does not reduce the error by as much.

This kink-finding process can be automated by use of the **gap statistic** (Tibshirani et al. 2001). Nevertheless, identifying such kinks can be hard, as shown in Figure 11.20(a), since the loss function usually drops off gradually. A different approach to "kink finding" is described in Section 12.3.2.1.

11.6   Fitting models with missing data

Suppose we want to fit a joint density model by maximum likelihood, but we have "holes" in our data matrix, due to missing data (usually represented by NaNs). More formally, let $O_{ij} = 1$ if component j of data case i is observed, and let $O_{ij} = 0$ otherwise. Let $\mathbf{X}_v = \{x_{ij} : O_{ij} = 1\}$ be the visible data, and $\mathbf{X}_h = \{x_{ij} : O_{ij} = 0\}$ be the missing or hidden data. Our goal is to compute

$$\hat{\boldsymbol{\theta}} = \operatorname*{argmax}_{\boldsymbol{\theta}} p(\mathbf{X}_v | \boldsymbol{\theta}, \mathbf{O}) \tag{11.96}$$

Under the missing at random assumption (see Section 8.6.2), we have

$$p(\mathbf{X}_v | \boldsymbol{\theta}, \mathbf{O}) = \prod_{i=1}^{N} p(\mathbf{x}_{iv} | \boldsymbol{\theta}) \tag{11.97}$$

where $\mathbf{x}_{iv}$ is a vector created from row i and the columns indexed by the set $\{j : O_{ij} = 1\}$. Hence the log-likelihood has the form

$$\log p(\mathbf{X}_v | \boldsymbol{\theta}) = \sum_i \log p(\mathbf{x}_{iv} | \boldsymbol{\theta}) \tag{11.98}$$

where

$$p(\mathbf{x}_{iv} | \boldsymbol{\theta}) = \sum_{\mathbf{x}_{ih}} p(\mathbf{x}_{iv}, \mathbf{x}_{ih} | \boldsymbol{\theta}) \tag{11.99}$$

and $\mathbf{x}_{ih}$ is the vector of hidden variables for case i (assumed discrete for notational simplicity). Substituting in, we get

$$\log p(\mathbf{X}_v|\boldsymbol{\theta}) = \sum_i \log \left[\sum_{\mathbf{x}_{ih}} p(\mathbf{x}_{iv}, \mathbf{x}_{ih}|\boldsymbol{\theta}) \right] \tag{11.100}$$

Unfortunately, this objective is hard to maximize. since we cannot push the log inside the sum. However, we can use the EM algorithm to compute a local optimum. We give an example of this below.

11.6.1 EM for the MLE of an MVN with missing data

Suppose we want to fit an MVN by maximum likelihood, but we have missing data. We can use EM to find a local maximum of the objective, as we explain below.

11.6.1.1 Getting started

To get the algorithm started, we can compute the MLE based on those rows of the data matrix that are fully observed. If there are no such rows, we can use some ad-hoc imputation procedures, and then compute an initial MLE.

11.6.1.2 E step

Once we have $\boldsymbol{\theta}^{t-1}$, we can compute the expected complete data log likelihood at iteration t as follows:

$$
\begin{aligned}
Q(\boldsymbol{\theta}, \boldsymbol{\theta}^{t-1}) &= \mathbb{E}\left[\sum_{i=1}^N \log \mathcal{N}(\mathbf{x}_i|\boldsymbol{\mu}, \boldsymbol{\Sigma}) | \mathcal{D}, \boldsymbol{\theta}^{t-1} \right] \tag{11.101} \\
&= -\frac{N}{2}\log|2\pi\boldsymbol{\Sigma}| - \frac{1}{2}\sum_i \mathbb{E}\left[(\mathbf{x}_i - \boldsymbol{\mu})^T \boldsymbol{\Sigma}^{-1}(\mathbf{x}_i - \boldsymbol{\mu}) \right] \tag{11.102} \\
&= -\frac{N}{2}\log|2\pi\boldsymbol{\Sigma}| - \frac{1}{2}\text{tr}(\boldsymbol{\Sigma}^{-1}\sum_i \mathbb{E}\left[(\mathbf{x}_i - \boldsymbol{\mu})(\mathbf{x}_i - \boldsymbol{\mu})^T \right] \tag{11.103} \\
&= -\frac{N}{2}\log|\boldsymbol{\Sigma}| - \frac{ND}{2}\log(2\pi) - \frac{1}{2}\text{tr}(\boldsymbol{\Sigma}^{-1}\mathbb{E}\left[\mathbf{S}(\boldsymbol{\mu}) \right]) \tag{11.104}
\end{aligned}
$$

where

$$\mathbb{E}\left[\mathbf{S}(\boldsymbol{\mu}) \right] \triangleq \sum_i \left(\mathbb{E}\left[\mathbf{x}_i \mathbf{x}_i^T \right] + \boldsymbol{\mu}\boldsymbol{\mu}^T - 2\boldsymbol{\mu}\mathbb{E}\left[\mathbf{x}_i \right]^T \right) \tag{11.105}$$

(We drop the conditioning of the expectation on $\mathcal{D}$ and $\boldsymbol{\theta}^{t-1}$ for brevity.) We see that we need to compute $\sum_i \mathbb{E}\left[\mathbf{x}_i \right]$ and $\sum_i \mathbb{E}\left[\mathbf{x}_i \mathbf{x}_i^T \right]$; these are the expected sufficient statistics.

To compute these quantities, we use the results from Section 4.3.1. Specifically, consider case i, where components v are observed and components h are unobserved. We have

$$
\begin{aligned}
\mathbf{x}_{ih}|\mathbf{x}_{iv}, \boldsymbol{\theta} &\sim \mathcal{N}(\mathbf{m}_i, \mathbf{V}_i) \tag{11.106} \\
\mathbf{m}_i &\triangleq \boldsymbol{\mu}_h + \boldsymbol{\Sigma}_{hv}\boldsymbol{\Sigma}_{vv}^{-1}(\mathbf{x}_{iv} - \boldsymbol{\mu}_v) \tag{11.107} \\
\mathbf{V}_i &\triangleq \boldsymbol{\Sigma}_{hh} - \boldsymbol{\Sigma}_{hv}\boldsymbol{\Sigma}_{vv}^{-1}\boldsymbol{\Sigma}_{vh} \tag{11.108}
\end{aligned}
$$

Hence the expected sufficient statistics are

$$\mathbb{E}\left[\mathbf{x}_i\right] = \left(\mathbb{E}\left[\mathbf{x}_{ih}\right]; \mathbf{x}_{iv}\right) = \left(\mathbf{m}_i; \mathbf{x}_{iv}\right) \tag{11.109}$$

where we have assumed (without loss of generality) that the unobserved variables come before the observed variables in the node ordering.

To compute $\mathbb{E}\left[\mathbf{x}_i \mathbf{x}_i^T\right]$, we use the result that $\text{cov}\left[\mathbf{x}\right] = \mathbb{E}\left[\mathbf{x}\mathbf{x}^T\right] - \mathbb{E}\left[\mathbf{x}\right]\mathbb{E}\left[\mathbf{x}^T\right]$. Hence

$$\mathbb{E}\left[\mathbf{x}_i \mathbf{x}_i^T\right] = \mathbb{E}\left[\begin{pmatrix}\mathbf{x}_{ih}\\\mathbf{x}_{iv}\end{pmatrix}\begin{pmatrix}\mathbf{x}_{ih}^T & \mathbf{x}_{iv}^T\end{pmatrix}\right] = \begin{pmatrix}\mathbb{E}\left[\mathbf{x}_{ih}\mathbf{x}_{ih}^T\right] & \mathbb{E}\left[\mathbf{x}_{ih}\right]\mathbf{x}_{iv}^T\\\mathbf{x}_{iv}\mathbb{E}\left[\mathbf{x}_{ih}\right]^T & \mathbf{x}_{iv}\mathbf{x}_{iv}^T\end{pmatrix} \tag{11.110}$$

$$\mathbb{E}\left[\mathbf{x}_{ih}\mathbf{x}_{ih}^T\right] = \mathbb{E}\left[\mathbf{x}_{ih}\right]\mathbb{E}\left[\mathbf{x}_{ih}\right]^T + \mathbf{V}_i \tag{11.111}$$

11.6.1.3   M step

By solving $\nabla Q(\boldsymbol{\theta}, \boldsymbol{\theta}^{(t-1)}) = \mathbf{0}$, we can show that the M step is equivalent to plugging these ESS into the usual MLE equations to get

$$\boldsymbol{\mu}^t = \frac{1}{N}\sum_i \mathbb{E}\left[\mathbf{x}_i\right] \tag{11.112}$$

$$\boldsymbol{\Sigma}^t = \frac{1}{N}\sum_i \mathbb{E}\left[\mathbf{x}_i\mathbf{x}_i^T\right] - \boldsymbol{\mu}^t(\boldsymbol{\mu}^t)^T \tag{11.113}$$

Thus we see that EM is *not* equivalent to simply replacing variables by their expectations and applying the standard MLE formula; that would ignore the posterior variance and would result in an incorrect estimate. Instead we must compute the expectation of the sufficient statistics, and plug that into the usual equation for the MLE. We can easily modify the algorithm to perform MAP estimation, by plugging in the ESS into the equation for the MAP estimate. For an implementation, see `gaussMissingFitEm`.

11.6.1.4   Example

As an example of this procedure in action, let us reconsider the imputation problem from Section 4.3.2.3, which had $N = 100$ 10-dimensional data cases, with 50% missing data. Let us fit the parameters using EM. Call the resulting parameters $\hat{\boldsymbol{\theta}}$. We can use our model for predictions by computing $\mathbb{E}\left[\mathbf{x}_{ih}|\mathbf{x}_{iv}, \hat{\boldsymbol{\theta}}\right]$. Figure 11.22(a-b) indicates that the results obtained using the learned parameters are almost as good as with the true parameters. Not surprisingly, performance improves with more data, or as the fraction of missing data is reduced.

11.6.1.5   Extension to the GMM case

It is straightforward to fit a mixture of Gaussians in the presence of partially observed data vectors $\mathbf{x}_i$. We leave the details as an exercise.

Exercises

Exercise 11.1 Student T as infinite mixture of Gaussians

Derive Equation 11.61. For simplicity, assume a one-dimensional distribution.

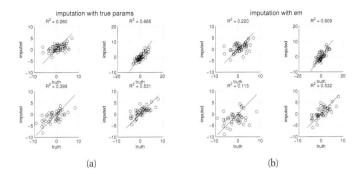

Figure 11.22 Illustration of data imputation. (a) Scatter plot of true values vs imputed values using true parameters. (b) Same as (a), but using parameters estimated with EM. Figure generated by gaussImputationDemo.

Exercise 11.2 EM for mixtures of Gaussians

Show that the M step for ML estimation of a mixture of Gaussians is given by

$$\boldsymbol{\mu}_k = \frac{\sum_i r_{ik}\mathbf{x}_i}{r_k} \tag{11.114}$$

$$\boldsymbol{\Sigma}_k = \frac{\sum_i r_{ik}(\mathbf{x}_i - \boldsymbol{\mu}_k)(\mathbf{x}_i - \boldsymbol{\mu}_k)^T}{r_k} = \frac{\sum_i r_{ik}\mathbf{x}_i\mathbf{x}_i^T - r_k\boldsymbol{\mu}_k\boldsymbol{\mu}_k^T}{r_k} \tag{11.115}$$

Exercise 11.3 EM for mixtures of Bernoullis

- Show that the M step for ML estimation of a mixture of Bernoullis is given by

$$\mu_{kj} = \frac{\sum_i r_{ik}x_{ij}}{\sum_i r_{ik}} \tag{11.116}$$

- Show that the M step for MAP estimation of a mixture of Bernoullis with a $\beta(\alpha, \beta)$ prior is given by

$$\mu_{kj} = \frac{\left(\sum_i r_{ik}x_{ij}\right) + \alpha - 1}{\left(\sum_i r_{ik}\right) + \alpha + \beta - 2} \tag{11.117}$$

Exercise 11.4 EM for mixture of Student distributions

Derive the EM algorithm for ML estimation of a mixture of multivariate Student T distributions.

Exercise 11.5 Gradient descent for fitting GMM

Consider the Gaussian mixture model

$$p(\mathbf{x}|\boldsymbol{\theta}) = \sum_k \pi_k \mathcal{N}(\mathbf{x}|\boldsymbol{\mu}_k, \boldsymbol{\Sigma}_k) \tag{11.118}$$

Define the log likelihood as

$$\ell(\boldsymbol{\theta}) = \sum_{n=1}^N \log p(\mathbf{x}_n|\boldsymbol{\theta}) \tag{11.119}$$

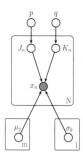

Figure 11.23   A mixture of Gaussians with two discrete latent indicators. J_n specifies which mean to use, and K_n specifies which variance to use.

Define the posterior responsibility that cluster k has for datapoint n as follows:

$$r_{nk} \triangleq p(z_n = k | \mathbf{x}_n, \boldsymbol{\theta}) = \frac{\pi_k \mathcal{N}(\mathbf{x}_n | \boldsymbol{\mu}_k, \boldsymbol{\Sigma}_k)}{\sum_{k'=1}^{K} \pi_{k'} \mathcal{N}(\mathbf{x}_n | \boldsymbol{\mu}_{k'}, \boldsymbol{\Sigma}_{k'})} \tag{11.120}$$

a. Show that the gradient of the log-likelihood wrt $\boldsymbol{\mu}_k$ is

$$\frac{d}{d\boldsymbol{\mu}_k} \ell(\boldsymbol{\theta}) = \sum_n r_{nk} \boldsymbol{\Sigma}_k^{-1} (\mathbf{x}_n - \boldsymbol{\mu}_k) \tag{11.121}$$

b. Derive the gradient of the log-likelihood wrt π_k. (For now, ignore any constraints on π_k.)

c. One way to handle the constraint that $\sum_{k=1}^{K} \pi_k = 1$ is to reparameterize using the softmax function:

$$\pi_k \triangleq \frac{e^{w_k}}{\sum_{k'=1}^{K} e^{w_{k'}}} \tag{11.122}$$

Here $w_k \in \mathbb{R}$ are unconstrained parameters. Show that

$$\frac{d}{dw_k} \ell(\boldsymbol{\theta}) = \sum_n r_{nk} - \pi_k \tag{11.123}$$

(There may be a constant factor missing in the above expression...) Hint: use the chain rule and the fact that

$$\frac{d\pi_j}{dw_k} = \begin{cases} \pi_j(1 - \pi_j) & \text{if } j = k \\ -\pi_j \pi_k & \text{if } j \neq k \end{cases} \tag{11.124}$$

which follows from Exercise 8.4(1).

d. Derive the gradient of the log-likelihood wrt $\boldsymbol{\Sigma}_k$. (For now, ignore any constraints on $\boldsymbol{\Sigma}_k$.)

e. One way to handle the constraint that $\boldsymbol{\Sigma}_k$ be a symmetric positive definite matrix is to reparameterize using a Cholesky decomposition, $\boldsymbol{\Sigma}_k = \mathbf{R}_k^T \mathbf{R}$, where $\mathbf{R}$ is an upper-triangular, but otherwise unconstrained matrix. Derive the gradient of the log-likelihood wrt $\mathbf{R}_k$.

Exercise 11.6 EM for a finite scale mixture of Gaussians

(Source: Jaakkola..) Consider the graphical model in Figure 11.23 which defines the following:

$$p(x_n | \theta) = \sum_{j=1}^{m} p_j \left[\sum_{k=1}^{l} q_k N(x_n | \mu_j, \sigma_k^2) \right] \tag{11.125}$$

where $\theta = \{p_1, \ldots, p_m, \mu_1, \ldots, \mu_m, q_1, \ldots, q_l, \sigma_1^2, \ldots, \sigma_l^2\}$ are all the parameters. Here $p_j \triangleq P(J_n = j)$ and $q_k \triangleq P(K_n = k)$ are the equivalent of mixture weights. We can think of this as a mixture of m non-Gaussian components, where each component distribution is a scale mixture, $p(x|j;\theta) = \sum_{k=1}^{l} q_k N(x; \mu_j, \sigma_k^2)$, combining Gaussians with different variances (scales).

We will now derive a generalized EM algorithm for this model. (Recall that in generalized EM, we do a partial update in the M step, rather than finding the exact maximum.)

a. Derive an expression for the responsibilities, $P(J_n = j, K_n = k | x_n, \theta)$, needed for the E step.

b. Write out a full expression for the expected complete log-likelihood

$$Q(\theta^{new}, \theta^{old}) = E_{\theta^{old}} \sum_{n=1}^{N} \log P(J_n, K_n, x_n | \theta^{new}) \tag{11.126}$$

c. Solving the M-step would require us to jointly optimize the means $\mu_1, \ldots, \mu_m$ and the variances $\sigma_1^2, \ldots, \sigma_l^2$. It will turn out to be simpler to first solve for the μ_j's given fixed σ_j^2's, and subsequently solve for σ_j^2's given the new values of μ_j's. For brevity, we will just do the first part. Derive an expression for the maximizing μ_j's given fixed $\sigma_{1:l}^2$, i.e., solve $\frac{\partial Q}{\partial \mu^{new}} = 0$.

Exercise 11.7 Manual calculation of the M step for a GMM

(Source: de Freitas.) In this question we consider clustering 1D data with a mixture of 2 Gaussians using the EM algorithm. You are given the 1-D data points $x = \begin{bmatrix} 1 & 10 & 20 \end{bmatrix}$. Suppose the output of the E step is the following matrix:

$$\mathbf{R} = \begin{bmatrix} 1 & 0 \\ 0.4 & 0.6 \\ 0 & 1 \end{bmatrix} \tag{11.127}$$

where entry $r_{i,c}$ is the probability of obervation x_i belonging to cluster c (the responsibility of cluster c for data point i). You just have to compute the M step. You may state the equations for maximum likelihood estimates of these quantities (which you should know) without proof; you just have to apply the equations to this data set. You may leave your answer in fractional form. Show your work.

a. Write down the likelihood function you are trying to optimize.

b. After performing the M step for the mixing weights π_1, π_2, what are the new values?

c. After performing the M step for the means μ_1 and μ_2, what are the new values?

Exercise 11.8 Moments of a mixture of Gaussians

Consider a mixture of K Gaussians

$$p(\mathbf{x}) = \sum_{k=1}^{K} \pi_k \mathcal{N}(\mathbf{x}|\boldsymbol{\mu}_k, \boldsymbol{\Sigma}_k) \tag{11.128}$$

a. Show that

$$\mathbb{E}[\mathbf{x}] = \sum_k \pi_k \boldsymbol{\mu}_k \tag{11.129}$$

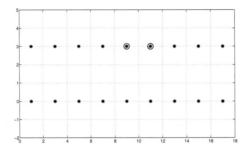

Figure 11.24   Some data points in 2d. Circles represent the initial guesses for $\mathbf{m}_1$ and $\mathbf{m}_2$.

b. Show that

$$\mathrm{cov}\left[\mathbf{x}\right] \quad = \quad \sum_k \pi_k [\mathbf{\Sigma}_k + \boldsymbol{\mu}_k \boldsymbol{\mu}_k^T] - \mathbb{E}\left[\mathbf{x}\right] \mathbb{E}\left[\mathbf{x}\right]^T \tag{11.130}$$

Hint: use the fact that $\mathrm{cov}\left[\mathbf{x}\right] = \mathbb{E}\left[\mathbf{x}\mathbf{x}^T\right] - \mathbb{E}\left[\mathbf{x}\right]\mathbb{E}\left[\mathbf{x}\right]^T$.

Exercise 11.9 K-means clustering by hand

(Source: Jaakkola.)

In Figure 11.24, we show some data points which lie on the integer grid. (Note that the x-axis has been compressed; distances should be measured using the actual grid coordinates.) Suppose we apply the K-means algorithm to this data, using $K = 2$ and with the centers initialized at the two circled data points. Draw the final clusters obtained after K-means converges (show the approximate location of the new centers and group together all the points assigned to each center). Hint: think about shortest Euclidean distance.

Exercise 11.10 Deriving the K-means cost function

Show that

$$J_W(\mathbf{z}) = \frac{1}{2} \sum_{k=1}^{K} \sum_{i:z_i=k} \sum_{i':z_{i'}=k} (x_i - x_{i'})^2 = \sum_{k=1}^{K} n_k \sum_{i:z_i=k} (x_i - \bar{x}_k)^2 \tag{11.131}$$

Hint: note that, for any μ,

$$\sum_i (x_i - \mu)^2 \quad = \quad \sum_i [(x_i - \bar{x}) - (\mu - \bar{x})]^2 \tag{11.132}$$

$$= \quad \sum_i (x_i - \bar{x})^2 + \sum_i (\bar{x} - \mu)^2 - 2 \sum_i (x_i - \bar{x})(\mu - \bar{x}) \tag{11.133}$$

$$= \quad ns^2 + n(\bar{x} - \mu)^2 \tag{11.134}$$

where $s^2 = \frac{1}{n} \sum_{i=1}^{n} (x_i - \bar{x})^2$, since

$$\sum_i (x_i - \bar{x})(\mu - \bar{x}) \quad = \quad (\mu - \bar{x})\left(\left(\sum_i x_i\right) - n\bar{x}\right) = (\mu - \bar{x})(n\bar{x} - n\bar{x}) = 0 \tag{11.135}$$

Exercise 11.11 Visible mixtures of Gaussians are in the exponential family

Show that the joint distribution $p(x, z|\boldsymbol{\theta})$ for a 1d GMM can be represented in exponential family form.

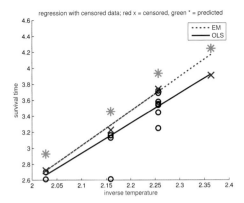

Figure 11.25 Example of censored linear regression. Black circles are observed training points, red crosses are observed but censored training points. Green stars are predicted values of the censored training points. We also show the lines fit by least squares (ignoring censoring) and by EM. Based on Figure 5.6 of (Tanner 1996). Figure generated by `linregCensoredSchmeeHahnDemo`, written by Hannes Bretschneider.

Exercise 11.12 *EM for robust linear regression with a Student t likelihood*

Consider a model of the form

$$p(y_i|\mathbf{x}_i, \mathbf{w}, \sigma^2, \nu) = \mathcal{T}(y_i|\mathbf{w}^T\mathbf{x}_i, \sigma^2, \nu) \tag{11.136}$$

Derive an EM algorithm to compute the MLE for $\mathbf{w}$. You may assume ν and σ^2 are fixed, for simplicity. Hint: see Section 11.4.5.

Exercise 11.13 *EM for EB estimation of Gaussian shrinkage model*

Extend the results of Section 5.6.2.2 to the case where the σ_j^2 are not equal (but are known). Hint: treat the θ_j as hidden variables, and then to integrate them out in the E step, and maximize $\boldsymbol{\eta} = (\mu, \tau^2)$ in the M step.

Exercise 11.14 *EM for censored linear regression*

Censored regression refers to the case where one knows the outcome is at least (or at most) a certain value, but the precise value is unknown. This arises in many different settings. For example, suppose one is trying to learn a model that can predict how long a program will take to run, for different settings of its parameters. One may abort certain runs if they seem to be taking too long; the resulting run times are said to be **right censored**. For such runs, all we know is that $y_i \geq c_i$, where c_i is the censoring time, that is, $y_i = \min(z_i, c_i)$, where z_i is the true running time and y_i is the observed running time. We can also define **left censored** and **interval censored** models.[3] Derive an EM algorithm for fitting a linear regression model to right-censored data. Hint: use the results from Exercise 11.15. See Figure 11.25 for an example, based on the data from (Schmee and Hahn 1979). We notice that the EM line is tilted upwards more, since the model takes into account the fact that the truncated values are actually higher than the observed values.

---

3. There is a closely related model in econometrics called the **Tobit model**, in which $y_i = \max(z_i, 0)$, so we only get to observe positive outcomes. An example of this is when z_i represents "desired investment", and y_i is actual investment. Probit regression (Section 9.4) is another example.

Exercise 11.15 Posterior mean and variance of a truncated Gaussian

Let $z_i = \mu_i + \sigma\epsilon_i$, where $\epsilon_i \sim \mathcal{N}(0,1)$. Sometimes, such as in probit regression or censored regression, we do not observe z_i, but we observe the fact that it is above some threshold, namely we observe the event $E = \mathbb{I}(z_i \geq c_i) = \mathbb{I}(\epsilon_i \geq \frac{c_i - \mu_i}{\sigma})$. (See Exercise 11.14 for details on censored regression, and Section 11.4.6 for probit regression.) Show that

$$\mathbb{E}\left[z_i | z_i \geq c_i\right] \quad = \quad \mu_i + \sigma H\left(\frac{c_i - \mu_i}{\sigma}\right) \tag{11.137}$$

and

$$\mathbb{E}\left[z_i^2 | z_i \geq c_i\right] \quad = \quad \mu_i^2 + \sigma^2 + \sigma(c_i + \mu_i) H\left(\frac{c_i - \mu_i}{\sigma}\right) \tag{11.138}$$

where we have defined

$$H(u) \triangleq \frac{\phi(u)}{1 - \Phi(u)} \tag{11.139}$$

and where $\phi(u)$ is the pdf of a standard Gaussian, and $\Phi(u)$ is its cdf.

Hint 1: we have $p(\epsilon_i | E) = \frac{p(\epsilon_i, E)}{p(E)}$, where E is some event of interest.

Hint 2: It can be shown that

$$\frac{d}{dw}\mathcal{N}(w|0,1) = -w\mathcal{N}(w|0,1) \tag{11.140}$$

and hence

$$\int_b^c w\mathcal{N}(w|0,1) = \mathcal{N}(b|0,1) - \mathcal{N}(c|0,1) \tag{11.141}$$

12 *Latent linear models*

12.1 Factor analysis

One problem with mixture models is that they only use a single latent variable to generate the observations. In particular, each observation can only come from one of K prototypes. One can think of a mixture model as using K hidden binary variables, representing a one-hot encoding of the cluster identity. But because these variables are mutually exclusive, the model is still limited in its representational power.

An alternative is to use a vector of real-valued latent variables, $\mathbf{z}_i \in \mathbb{R}^L$. The simplest prior to use is a Gaussian (we will consider other choices later):

$$p(\mathbf{z}_i) = \mathcal{N}(\mathbf{z}_i|\boldsymbol{\mu}_0, \boldsymbol{\Sigma}_0) \tag{12.1}$$

If the observations are also continuous, so $\mathbf{x}_i \in \mathbb{R}^D$, we may use a Gaussian for the likelihood. Just as in linear regression, we will assume the mean is a linear function of the (hidden) inputs, thus yielding

$$p(\mathbf{x}_i|\mathbf{z}_i, \boldsymbol{\theta}) = \mathcal{N}(\mathbf{W}\mathbf{z}_i + \boldsymbol{\mu}, \boldsymbol{\Psi}) \tag{12.2}$$

where $\mathbf{W}$ is a $D \times L$ matrix, known as the **factor loading matrix**, and $\boldsymbol{\Psi}$ is a $D \times D$ covariance matrix. We take $\boldsymbol{\Psi}$ to be diagonal, since the whole point of the model is to "force" $\mathbf{z}_i$ to explain the correlation, rather than "baking it in" to the observation's covariance. This overall model is called **factor analysis** or **FA**. The special case in which $\boldsymbol{\Psi} = \sigma^2 \mathbf{I}$ is called **probabilistic principal components analysis** or **PPCA**. The reason for this name will become apparent later.

The generative process, where $L = 1$, $D = 2$, $\boldsymbol{\Sigma}_0$ is spherical and $\boldsymbol{\Psi}$ is diagonal, is illustrated in Figure 12.1. We take an isotropic Gaussian "spray can", representing the prior on $\mathbf{z}_i$, and slide it along the 1d line defined by $\mathbf{w}z_i + \boldsymbol{\mu}$. This induces an elongated (and hence correlated) Gaussian in 2d, to which we add diagonal noise.

12.1.1 FA is a low rank parameterization of an MVN

FA can be thought of as a way of specifying a joint density model on $\mathbf{x}$ using a small number of parameters. To see this, note that from Equation 4.126, the induced marginal distribution

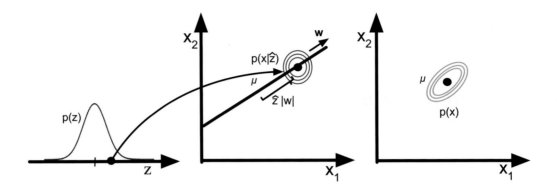

Figure 12.1 Illustration of the PPCA generative process, where we have $L = 1$ latent dimension generating $D = 2$ observed dimensions. Based on Figure 12.9 of (Bishop 2006).

$p(\mathbf{x}_i|\boldsymbol{\theta})$ is a Gaussian:

$$p(\mathbf{x}_i|\boldsymbol{\theta}) = \int \mathcal{N}(\mathbf{x}_i|\mathbf{W}\mathbf{z}_i + \boldsymbol{\mu}, \boldsymbol{\Psi})\mathcal{N}(\mathbf{z}_i|\boldsymbol{\mu}_0, \boldsymbol{\Sigma}_0)d\mathbf{z}_i \qquad (12.3)$$

$$= \mathcal{N}(\mathbf{x}_i|\mathbf{W}\boldsymbol{\mu}_0 + \boldsymbol{\mu}, \boldsymbol{\Psi} + \mathbf{W}\boldsymbol{\Sigma}_0\mathbf{W}^T) \qquad (12.4)$$

Hence $\mathbb{E}[\mathbf{x}] = \mathbf{W}\boldsymbol{\mu}_0 + \boldsymbol{\mu}$ and $\text{cov}[\mathbf{x}] = \mathbf{W}\mathbb{E}[\mathbf{z}\mathbf{z}^T]\mathbf{W}^T + \boldsymbol{\Psi} = \mathbf{W}\boldsymbol{\Sigma}_0\mathbf{W}^T + \boldsymbol{\Psi}$. From this, we see that we can set $\boldsymbol{\mu}_0 = \mathbf{0}$ without loss of generality, since we can always absorb $\mathbf{W}\boldsymbol{\mu}_0$ into $\boldsymbol{\mu}$. Similarly, we can set $\boldsymbol{\Sigma}_0 = \mathbf{I}$ without loss of generality, because we can always "emulate" a correlated prior by using defining a new weight matrix, $\tilde{\mathbf{W}} = \mathbf{W}\boldsymbol{\Sigma}_0^{-\frac{1}{2}}$ since

$$\text{cov}[\mathbf{x}] = \mathbf{W}\boldsymbol{\Sigma}_0\mathbf{W}^T + \boldsymbol{\Psi} = \tilde{\mathbf{W}}\tilde{\mathbf{W}}^T + \boldsymbol{\Psi} \qquad (12.5)$$

We thus see that FA approximates the covariance matrix of the visible vector using a low-rank decomposition:

$$\mathbf{C} \triangleq \text{cov}[\mathbf{x}] = \mathbf{W}\mathbf{W}^T + \boldsymbol{\Psi} \qquad (12.6)$$

This only uses $O(LD)$ parameters, which allows a flexible compromise between a full covariance Gaussian, with $O(D^2)$ parameters, and a diagonal covariance, with $O(D)$ parameters. Note that if we did not restrict $\boldsymbol{\Psi}$ to be diagonal, we could trivially set $\boldsymbol{\Psi}$ to a full covariance matrix; then we could set $\mathbf{W} = \mathbf{0}$, in which case the latent factors would not be required.

The marginal variance of each visible variable is given by $\text{var}[x_j] = \sum_{k=1}^{L} w_{jk}^2 + \psi_j$, where the first term is the variance due to the common factors, and the second ψ_j term is called the **uniqueness**, and is the variance term that is specific to that dimension.

12.1.2 Inference of the latent factors

Although FA can be thought of as just a way to define a density on $\mathbf{x}$, it is often used because we hope that the latent factors $\mathbf{z}$ will reveal something interesting about the data. To do this,

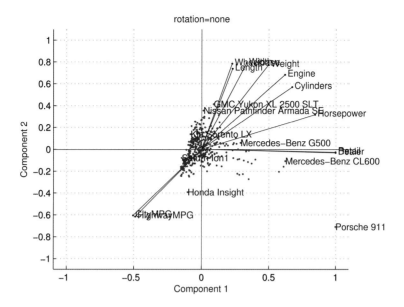

Figure 12.2 2D projection of 2004 cars data based on factor analysis. The blue text are the names of cars corresponding to certain chosen points. Figure generated by `faBiplotDemo`.

we need to compute the posterior over the latent factors. We can use Bayes rule for Gaussians to give

$$p(\mathbf{z}_i|\mathbf{x}_i, \boldsymbol{\theta}) = \mathcal{N}(\mathbf{z}_i|\mathbf{m}_i, \boldsymbol{\Sigma}_i) \tag{12.7}$$

$$\boldsymbol{\Sigma}_i \triangleq (\boldsymbol{\Sigma}_0^{-1} + \mathbf{W}^T \boldsymbol{\Psi}^{-1} \mathbf{W})^{-1} \tag{12.8}$$

$$\mathbf{m}_i \triangleq \boldsymbol{\Sigma}_i (\mathbf{W}^T \boldsymbol{\Psi}^{-1}(\mathbf{x}_i - \boldsymbol{\mu}) + \boldsymbol{\Sigma}_0^{-1} \boldsymbol{\mu}_0) \tag{12.9}$$

Note that in the FA model, $\boldsymbol{\Sigma}_i$ is actually independent of i, so we can denote it by $\boldsymbol{\Sigma}$. Computing this matrix takes $O(L^3 + L^2 D)$ time, and computing each $\mathbf{m}_i = \mathbb{E}[\mathbf{z}_i|\mathbf{x}_i, \boldsymbol{\theta}]$ takes $O(L^2 + LD)$ time. The $\mathbf{m}_i$ are sometimes called the latent **scores**, or latent **factors**.

Let us give a simple example, based (Shalizi 2009). We consider a dataset of $D = 11$ variables and $N = 387$ cases describing various aspects of cars, such as the engine size, the number of cylinders, the miles per gallon (MPG), the price, etc. We first fit a $L = 2$ dimensional model. We can plot the $\mathbf{m}_i$ scores as points in $\mathbb{R}^2$, to visualize the data, as shown in Figure 12.2.

To get a better understanding of the "meaning" of the latent factors, we can project unit vectors corresponding to each of the feature dimensions, $\mathbf{e}_1 = (1, 0, \dots, 0)$, $\mathbf{e}_2 = (0, 1, 0, \dots, 0)$, etc. into the low dimensional space. These are shown as blue lines in Figure 12.2; this is known as a **biplot**. We see that the horizontal axis represents price, corresponding to the features labeled "dealer" and "retail", with expensive cars on the right. The vertical axis represents fuel efficiency (measured in terms of MPG) versus size: heavy vehicles are less efficient and are higher up, whereas light vehicles are more efficient and are lower down. We can "verify" this interpretation by clicking on some points, and finding the closest exemplars in the training set, and printing their names, as in Figure 12.2. However, in general, interpreting latent variable models is fraught with difficulties, as we discuss in Section 12.1.3.

12.1.3   Unidentifiability

The parameters of an FA model are unidentifiable. To see this, suppose $\mathbf{R}$ is an arbitrary orthogonal rotation matrix, satisfying $\mathbf{R}\mathbf{R}^T = \mathbf{I}$. Let us define $\tilde{\mathbf{W}} = \mathbf{W}\mathbf{R}$; then the likelihood function of this modified matrix is the same as for the unmodified matrix, since

$$\text{cov}\,[\mathbf{x}] \;=\; \tilde{\mathbf{W}}\mathbb{E}\left[\mathbf{z}\mathbf{z}^T\right]\tilde{\mathbf{W}}^T + \mathbb{E}\left[\boldsymbol{\epsilon}\boldsymbol{\epsilon}^T\right] \tag{12.10}$$

$$\;=\; \mathbf{W}\mathbf{R}\mathbf{R}^T\mathbf{W}^T + \boldsymbol{\Psi} = \mathbf{W}\mathbf{W}^T + \boldsymbol{\Psi} \tag{12.11}$$

Geometrically, multiplying $\mathbf{W}$ by an orthogonal matrix is like rotating $\mathbf{z}$ before generating $\mathbf{x}$; but since $\mathbf{z}$ is drawn from an isotropic Gaussian, this makes no difference to the likelihood. Consequently, we cannot unique identify $\mathbf{W}$, and therefore cannot uniquely identify the latent factors, either.

To ensure a unique solution, we need to remove $L(L-1)/2$ degrees of freedom, since that is the number of orthonormal matrices of size $L \times L$.[1] In total, the FA model has $D + LD - L(L-1)/2$ free parameters (excluding the mean), where the first term arises from $\boldsymbol{\Psi}$. Obviously we require this to be less than or equal to $D(D+1)/2$, which is the number of parameters in an unconstrained (but symmetric) covariance matrix. This gives us an upper bound on L, as follows:

$$L_{max} = \lfloor D + 0.5(1 - \sqrt{1 + 8D}) \rfloor \tag{12.12}$$

For example, $D = 6$ implies $L \le 3$. But we usually never choose this upper bound, since it would result in overfitting (see discussion in Section 12.3 on how to choose L).

Unfortunately, even if we set $L < L_{max}$, we still cannot uniquely identify the parameters, since the rotational ambiguity still exists. Non-identifiability does not affect the predictive performance of the model. However, it does affect the loading matrix, and hence the interpretation of the latent factors. Since factor analysis is often used to uncover structure in the data, this problem needs to be addressed. Here are some commonly used solutions:

- **Forcing $\mathbf{W}$ to be orthonormal** Perhaps the cleanest solution to the identifiability problem is to force $\mathbf{W}$ to be orthonormal, and to order the columns by decreasing variance of the corresponding latent factors. This is the approach adopted by PCA, which we will discuss in Section 12.2. The result is not necessarily more interpretable, but at least it is unique.

- **Forcing $\mathbf{W}$ to be lower triangular** One way to achieve identifiability, which is popular in the Bayesian community (e.g., (Lopes and West 2004)), is to ensure that the first visible feature is only generated by the first latent factor, the second visible feature is only generated by the first two latent factors, and so on. For example, if $L = 3$ and $D = 4$, the correspond

---

1. To see this, note that there are $L - 1$ free parameters in $\mathbf{R}$ in the first column (since the column vector must be normalized to unit length), there are $L - 2$ free parameters in the second column (which must be orthogonal to the first), and so on.

factor loading matrix is given by

$$\mathbf{W} = \begin{pmatrix} w_{11} & 0 & 0 \\ w_{21} & w_{22} & 0 \\ w_{31} & w_{32} & w_{33} \\ w_{41} & w_{42} & w_{43} \end{pmatrix} \tag{12.13}$$

We also require that $w_{jj} > 0$ for $j = 1 : L$. The total number of parameters in this constrained matrix is $D + DL - L(L-1)/2$, which is equal to the number of uniquely identifiable parameters. The disadvantage of this method is that the first L visible variables, known as the **founder variables**, affect the interpretation of the latent factors, and so must be chosen carefully.[2]

- **Sparsity promoting priors on the weights** Instead of pre-specifying which entries in $\mathbf{W}$ are zero, we can encourage the entries to be zero, using ℓ_1 regularization (Zou et al. 2006), ARD (Bishop 1999; Archambeau and Bach 2008), or spike-and-slab priors (Rattray et al. 2009). This is called sparse factor analysis. This does not necessarily ensure a unique MAP estimate, but it does encourage interpretable solutions. See Section 13.8 on sparse coding for details.

- **Choosing an informative rotation matrix** There are a variety of heuristic methods that try to find rotation matrices $\mathbf{R}$ which can be used to modify $\mathbf{W}$ (and hence the latent factors) so as to try to increase the interpretability, typically by encouraging them to be (approximately) sparse. One popular method is known as **varimax** (Kaiser 1958).

- **Use of non-Gaussian priors for the latent factors** If we replace the prior on the latent variables, $p(\mathbf{z}_i)$, with a non-Gaussian distribution, we can sometimes uniquely identify $\mathbf{W}$, as well as the latent factors. This technique is known as independent components analysis or ICA, and is explained in Section 12.6.

12.1.4 Mixtures of factor analysers

The FA model assumes that the data lives on a low dimensional linear manifold. In reality, most data is better modeled by some form of low dimensional *curved* manifold. We can approximate a curved manifold by a piecewise linear manifold. This suggests the following model: let the k'th linear subspace of dimensionality L_k be represented by $\mathbf{W}_k$, for $k = 1 : K$. Suppose we have a latent indicator $q_i \in \{1, \ldots, K\}$ specifying which subspace we should use to generate the data. We then sample $\mathbf{z}_i$ from a Gaussian prior and pass it through the $\mathbf{W}_k$ matrix (where $k = q_i$), and add noise. More precisely, the model is as follows:

$$p(\mathbf{x}_i | \mathbf{z}_i, q_i = k, \boldsymbol{\theta}) = \mathcal{N}(\mathbf{x}_i | \boldsymbol{\mu}_k + \mathbf{W}_k \mathbf{z}_i, \boldsymbol{\Psi}) \tag{12.14}$$

$$p(\mathbf{z}_i | \boldsymbol{\theta}) = \mathcal{N}(\mathbf{z}_i | \mathbf{0}, \mathbf{I}) \tag{12.15}$$

$$p(q_i | \boldsymbol{\theta}) = \text{Cat}(q_i | \boldsymbol{\pi}) \tag{12.16}$$

This is called a **mixture of factor analysers** (MFA) (Ghahramani and Hinton 1996a). The CI assumptions are represented in Figure 12.3.

---

2. An additional complication arises if some prior theory specifies a known structural mapping from the latents to the observed variables (a situation known as **confirmatory factor analysis**, as opposed to the more common **exploratory factor analysis**, where nothing is known about $\mathbf{W}$). In this case, the sparsity pattern of $\mathbf{W}$ cannot be set as in Equation 12.13, but must conform to the structural prior. To ensure identifiability even in this case, one can post-process the samples from an MCMC run. For details, see (Erosheva and Curtis 2011).

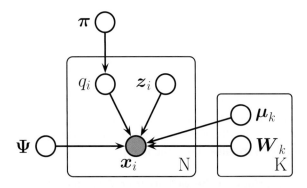

Figure 12.3   Mixture of factor analysers as a DGM.

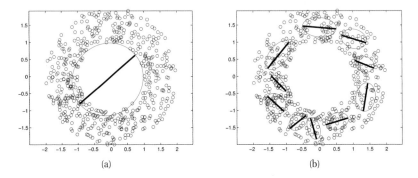

Figure 12.4   Mixture of 1d PPCAs fit to a dataset, for $K = 1, 10$.   Figure generated by `mixPpcaDemoNetlab`.

Another way to think about this model is as a low-rank version of a mixture of Gaussians. In particular, this model needs $O(KLD)$ parameters instead of the $O(KD^2)$ parameters needed for a mixture of full covariance Gaussians. This can reduce overfitting. In fact, MFA is a good generic density model for high-dimensional real-valued data.

12.1.5   EM for factor analysis models

Using the results from Chapter 4, it is straightforward to derive an EM algorithm to fit an FA model. With just a little more work, we can fit a mixture of FAs. Below we state the results without proof. The derivation can be found in (Ghahramani and Hinton 1996a); however, deriving these equations yourself is a useful exercise if you want to become proficient at the math.

To obtain the results for a single factor analyser, just set $r_{ic} = 1$ and $c = 1$ in the equations below. In Section 12.2.5 we will see a further simplification of these equations that arises when fitting a PPCA model, where the results will turn out to have a particularly simple and elegant intepretation.

Returning to the general case: in the E step, we compute the posterior responsibility of cluster c for data point i using

$$r_{ic} \triangleq p(q_i = c | \mathbf{x}_i, \boldsymbol{\theta}) \propto \pi_c \mathcal{N}(\mathbf{x}_i | \boldsymbol{\mu}_c, \mathbf{W}_c \mathbf{W}_c^T + \boldsymbol{\Psi}) \tag{12.17}$$

The conditional posterior for $\mathbf{z}_i$ is given by

$$p(\mathbf{z}_i | \mathbf{x}_i, q_i = c, \boldsymbol{\theta}) = \mathcal{N}(\mathbf{z}_i | \mathbf{m}_{ic}, \boldsymbol{\Sigma}_{ic}) \tag{12.18}$$
$$\boldsymbol{\Sigma}_{ic} \triangleq (\mathbf{I}_L + \mathbf{W}_c^T \boldsymbol{\Psi}_c^{-1} \mathbf{W}_c)^{-1} \tag{12.19}$$
$$\mathbf{m}_{ic} \triangleq \boldsymbol{\Sigma}_{ic}(\mathbf{W}_c^T \boldsymbol{\Psi}_c^{-1}(\mathbf{x}_i - \boldsymbol{\mu}_c)) \tag{12.20}$$

In the M step, it is easiest to estimate $\boldsymbol{\mu}_c$ and $\mathbf{W}_c$ at the same time, by defining $\tilde{\mathbf{W}}_c = (\mathbf{W}_c, \boldsymbol{\mu}_c)$, $\tilde{\mathbf{z}} = (\mathbf{z}, 1)$, Also, define

$$\mathbf{b}_{ic} \triangleq \mathbb{E}[\tilde{\mathbf{z}} | \mathbf{x}_i, q_i = c] = [\mathbf{m}_{ic}; 1] \tag{12.21}$$
$$\mathbf{C}_{ic} \triangleq \mathbb{E}[\tilde{\mathbf{z}}\tilde{\mathbf{z}}^T | \mathbf{x}_i, q_i = c] = \begin{pmatrix} \mathbb{E}[\mathbf{z}\mathbf{z}^T | \mathbf{x}_i, q_i = c] & \mathbb{E}[\mathbf{z} | \mathbf{x}_i, q_i = c] \\ \mathbb{E}[\mathbf{z} | \mathbf{x}_i, q_i = c]^T & 1 \end{pmatrix} \tag{12.22}$$

Then the M step is as follows:

$$\hat{\tilde{\mathbf{W}}}_c = \left[\sum_i r_{ic} \mathbf{x}_i \mathbf{b}_{ic}^T \right] \left[\sum_i r_{ic} \mathbf{C}_{ic} \right]^{-1} \tag{12.23}$$

$$\hat{\boldsymbol{\Psi}} = \frac{1}{N} \text{diag} \left\{ \sum_{ic} r_{ic} \left(\mathbf{x}_i - \hat{\tilde{\mathbf{W}}}_c \mathbf{b}_{ic} \right) \mathbf{x}_i^T \right\} \tag{12.24}$$

$$\hat{\pi}_c = \frac{1}{N} \sum_{i=1}^N r_{ic} \tag{12.25}$$

Note that these updates are for "vanilla" EM. A much faster version of this algorithm, based on ECM, is described in (Zhao and Yu 2008).

12.1.6 Fitting FA models with missing data

In many applications, such as collaborative filtering, we have missing data. One virtue of the EM approach to fitting an FA/PPCA model is that it is easy to extend to this case. However, overfitting can be a problem if there is a lot of missing data. Consequently it is important to perform MAP estimation or to use Bayesian inference. See e.g., (Ilin and Raiko 2010) for details.

12.2 Principal components analysis (PCA)

Consider the FA model where we constrain $\boldsymbol{\Psi} = \sigma^2 \mathbf{I}$, and $\mathbf{W}$ to be orthonormal. It can be shown (Tipping and Bishop 1999) that, as $\sigma^2 \to 0$, this model reduces to classical (non-probabilistic) **principal components analysis (PCA)**, also known as the **Karhunen Loeve** transform. The version where $\sigma^2 > 0$ is known as **probabilistic PCA (PPCA)** (Tipping and Bishop

1999), or **sensible PCA** (Roweis 1997). (An equivalent result was derived independently, from a different perspective, in (Moghaddam and Pentland 1995).)[3]

To make sense of this result, we first have to learn about classical PCA. We then connect PCA to the SVD. And finally we return to discuss PPCA.

12.2.1 Classical PCA: statement of the theorem

The **synthesis view** of classical PCA is summarized in the following theorem.

Theorem 12.2.1. *Suppose we want to find an orthogonal set of L linear basis vectors $\mathbf{w}_j \in \mathbb{R}^D$, and the corresponding scores $\mathbf{z}_i \in \mathbb{R}^L$, such that we minimize the average* **reconstruction error**

$$J(\mathbf{W}, \mathbf{Z}) = \frac{1}{N} \sum_{i=1}^{N} ||\mathbf{x}_i - \hat{\mathbf{x}}_i||^2 \tag{12.26}$$

where $\hat{\mathbf{x}}_i = \mathbf{W}\mathbf{z}_i$, subject to the constraint that $\mathbf{W}$ is orthonormal. Equivalently, we can write this objective as follows:

$$J(\mathbf{W}, \mathbf{Z}) = ||\mathbf{X} - \mathbf{W}\mathbf{Z}^T||_F^2 \tag{12.27}$$

where $\mathbf{Z}$ is an $N \times L$ matrix with the $\mathbf{z}_i$ in its rows, and $||\mathbf{A}||_F$ is the **Frobenius norm** *of matrix $\mathbf{A}$, defined by*

$$||\mathbf{A}||_F = \sqrt{\sum_{i=1}^{m} \sum_{j=1}^{n} a_{ij}^2} = \sqrt{\mathrm{tr}(\mathbf{A}^T \mathbf{A})} = ||\mathbf{A}(:)||_2 \tag{12.28}$$

The optimal solution is obtained by setting $\hat{\mathbf{W}} = \mathbf{V}_L$, where $\mathbf{V}_L$ contains the L eigenvectors with largest eigenvalues of the empirical covariance matrix, $\hat{\boldsymbol{\Sigma}} = \frac{1}{N} \sum_{i=1}^{N} \mathbf{x}_i \mathbf{x}_i^T$. (We assume the $\mathbf{x}_i$ have zero mean, for notational simplicity.) Furthermore, the optimal low-dimensional encoding of the data is given by $\hat{\mathbf{z}}_i = \mathbf{W}^T \mathbf{x}_i$, which is an orthogonal projection of the data onto the column space spanned by the eigenvectors.

An example of this is shown in Figure 12.5(a) for $D = 2$ and $L = 1$. The diagonal line is the vector $\mathbf{w}_1$; this is called the first principal component or principal direction. The data points $\mathbf{x}_i \in \mathbb{R}^2$ are orthogonally projected onto this line to get $z_i \in \mathbb{R}$. This is the best 1-dimensional approximation to the data. (We will discuss Figure 12.5(b) later.)

In general, it is hard to visualize higher dimensional data, but if the data happens to be a set of images, it is easy to do so. Figure 12.6 shows the first three principal vectors, reshaped as images, as well as the reconstruction of a specific image using a varying number of basis vectors. (We discuss how to choose L in Section 11.5.)

Below we will show that the principal directions are the ones along which the data shows maximal variance. This means that PCA can be "misled" by directions in which the variance is high merely because of the measurement scale. Figure 12.7(a) shows an example, where the vertical axis (weight) uses a larger range than the horizontal axis (height), resulting in a line that looks somewhat "unnatural". It is therefore standard practice to standardize the data first, or

---

3. In PPCA, we do not require $\mathbf{W}$ to be orthogonal.

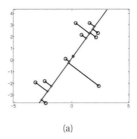
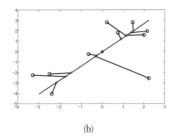

(a)   (b)

Figure 12.5 An illustration of PCA and PPCA where $D = 2$ and $L = 1$. Circles are the original data points, crosses are the reconstructions. The red star is the data mean. (a) PCA. The points are orthogonally projected onto the line. Figure generated by pcaDemo2d. (b) PPCA. The projection is no longer orthogonal: the reconstructions are shrunk towards the data mean (red star). Based on Figure 7.6 of (Nabney 2001). Figure generated by ppcaDemo2d.

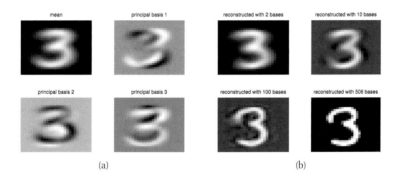

(a)   (b)

Figure 12.6 (a) The mean and the first three PC basis vectors (eigendigits) based on 25 images of the digit 3 (from the MNIST dataset). (b) Reconstruction of an image based on 2, 10, 100 and all the basis vectors. Figure generated by pcaImageDemo.

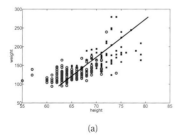
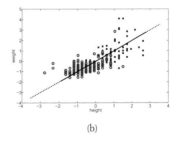

(a)   (b)

Figure 12.7 Effect of standardization on PCA applied to the height/ weight dataset. Left: PCA of raw data. Right: PCA of standardized data. Figure generated by pcaDemoHeightWeight.

equivalently, to work with correlation matrices instead of covariance matrices. The benefits of this are apparent from Figure 12.7(b).[4]

12.2.2   Proof *

Proof. We use $\mathbf{w}_j \in \mathbb{R}^D$ to denote the j'th principal direction, $\mathbf{x}_i \in \mathbb{R}^D$ to denote the i'th high-dimensional observation, $\mathbf{z}_i \in \mathbb{R}^L$ to denote the i'th low-dimensional representation, and $\tilde{\mathbf{z}}_j \in \mathbb{R}^N$ to denote the $[z_{1j}, \ldots, z_{Nj}]$, which is the j'th component of all the low-dimensional vectors.

Let us start by estimating the best 1d solution, $\mathbf{w}_1 \in \mathbb{R}^D$, and the corresponding projected points $\tilde{\mathbf{z}}_1 \in \mathbb{R}^N$. We will find the remaining bases $\mathbf{w}_2$, $\mathbf{w}_3$, etc. later. The reconstruction error is given by

$$J(\mathbf{w}_1, \mathbf{z}_1) \quad = \quad \frac{1}{N} \sum_{i=1}^{N} ||\mathbf{x}_i - z_{i1}\mathbf{w}_1||^2 = \frac{1}{N} \sum_{i=1}^{N} (\mathbf{x}_i - z_{i1}\mathbf{w}_1)^T (\mathbf{x}_i - z_{i1}\mathbf{w}_1) \tag{12.29}$$

$$= \quad \frac{1}{N} \sum_{i=1}^{N} [\mathbf{x}_i^T \mathbf{x}_i - 2z_{i1}\mathbf{w}_1^T \mathbf{x}_i + z_{i1}^2 \mathbf{w}_1^T \mathbf{w}_1] \tag{12.30}$$

$$= \quad \frac{1}{N} \sum_{i=1}^{N} [\mathbf{x}_i^T \mathbf{x}_i - 2z_{i1}\mathbf{w}_1^T \mathbf{x}_i + z_{i1}^2] \tag{12.31}$$

since $\mathbf{w}_1^T \mathbf{w}_1 = 1$ (by the orthonormality assumption). Taking derivatives wrt z_{i1} and equating to zero gives

$$\frac{\partial}{\partial z_{i1}} J(\mathbf{w}_1, \mathbf{z}_1) = \frac{1}{N} [-2\mathbf{w}_1^T \mathbf{x}_i + 2z_{i1}] = 0 \Rightarrow z_{i1} = \mathbf{w}_1^T \mathbf{x}_i \tag{12.32}$$

So the optimal reconstruction weights are obtained by orthogonally projecting the data onto the first principal direction, $\mathbf{w}_1$ (see Figure 12.5(a)). Plugging back in gives

$$J(\mathbf{w}_1) \quad = \quad \frac{1}{N} \sum_{i=1}^{N} [\mathbf{x}_i^T \mathbf{x}_i - z_{i1}^2] = \text{const} - \frac{1}{N} \sum_{i=1}^{N} z_{i1}^2 \tag{12.33}$$

Now the variance of the projected coordinates is given by

$$\text{var}[\tilde{\mathbf{z}}_1] = \mathbb{E}\left[\tilde{\mathbf{z}}_1^2\right] - (\mathbb{E}[\tilde{\mathbf{z}}_1])^2 = \frac{1}{N} \sum_{i=1}^{N} z_{i1}^2 - 0 \tag{12.34}$$

since

$$\mathbb{E}[z_{i1}] = \mathbb{E}\left[\mathbf{x}_i^T \mathbf{w}_1\right] = \mathbb{E}[\mathbf{x}_i]^T \mathbf{w}_1 = 0 \tag{12.35}$$

---

4. Note that this rescaling is not necessary for factor analysis, which allows each dimension to have its own variance, ψ_j. More formally, if $\mathbf{y}_i = \mathbf{C}\mathbf{x}_i$, and $\text{cov}[\mathbf{x}] = \mathbf{W}\mathbf{W}^T + \mathbf{\Psi}$, then $\text{cov}[\mathbf{y}] = \tilde{\mathbf{W}}\tilde{\mathbf{W}}^T + \tilde{\mathbf{\Psi}}$, where $\tilde{\mathbf{W}} = \mathbf{C}\mathbf{W}$ and $\tilde{\mathbf{\Psi}} = \mathbf{C}\mathbf{\Psi}\mathbf{C}^T$. So we see that the parameters learned for FA from the rescaled data are the same, up to a linear transform.

because the data has been centered. From this, we see that *minimizing* the reconstruction error is equivalent to *maximizing* the variance of the projected data, i.e.,

$$\arg\min_{\mathbf{w}_1} J(\mathbf{w}_1) = \arg\max_{\mathbf{w}_1} \text{var}\,[\tilde{\mathbf{z}}_1] \tag{12.36}$$

This is why it is often said that PCA finds the directions of maximal variance. This is called the **analysis view** of PCA.

The variance of the projected data can be written as

$$\frac{1}{N}\sum_{i=1}^{N} z_{i1}^2 = \frac{1}{N}\sum_{i=1}^{N} \mathbf{w}_1^T \mathbf{x}_i \mathbf{x}_i^T \mathbf{w}_1 = \mathbf{w}_1^T \hat{\boldsymbol{\Sigma}} \mathbf{w}_1 \tag{12.37}$$

where $\hat{\boldsymbol{\Sigma}} = \frac{1}{N}\sum_{i=1}^{N}\sum_i \mathbf{x}_i \mathbf{x}_i^T$ is the empirical covariance matrix (or correlation matrix if the data is standardized).

We can trivially maximize the variance of the projection (and hence minimize the reconstruction error) by letting $||\mathbf{w}_1|| \to \infty$, so we impose the constraint $||\mathbf{w}_1|| = 1$ and instead maximize

$$\tilde{J}(\mathbf{w}_1) = \mathbf{w}_1^T \hat{\boldsymbol{\Sigma}} \mathbf{w}_1 + \lambda_1(\mathbf{w}_1^T \mathbf{w}_1 - 1) \tag{12.38}$$

where λ_1 is the Lagrange multiplier. Taking derivatives and equating to zero we have

$$\frac{\partial}{\partial \mathbf{w}_1}\tilde{J}(\mathbf{w}_1) = 2\hat{\boldsymbol{\Sigma}}\mathbf{w}_1 - 2\lambda_1 \mathbf{w}_1 = 0 \tag{12.39}$$

$$\hat{\boldsymbol{\Sigma}}\mathbf{w}_1 = \lambda_1 \mathbf{w}_1 \tag{12.40}$$

Hence the direction that maximizes the variance is an eigenvector of the covariance matrix. Left multiplying by $\mathbf{w}_1$ (and using $\mathbf{w}_1^T \mathbf{w}_1 = 1$) we find that the variance of the projected data is

$$\mathbf{w}_1^T \hat{\boldsymbol{\Sigma}} \mathbf{w}_1 = \lambda_1 \tag{12.41}$$

Since we want to maximize the variance, we pick the eigenvector which corresponds to the largest eigenvalue.

Now let us find another direction $\mathbf{w}_2$ to further minimize the reconstruction error, subject to $\mathbf{w}_1^T \mathbf{w}_2 = 0$ and $\mathbf{w}_2^T \mathbf{w}_2 = 1$. The error is

$$J(\mathbf{w}_1, \mathbf{z}_1, \mathbf{w}_2, \mathbf{z}_2) = \frac{1}{N}\sum_{i=1}^{N} ||\mathbf{x}_i - z_{i1}\mathbf{w}_1 - z_{i2}\mathbf{w}_2||^2 \tag{12.42}$$

Optimizing wrt $\mathbf{w}_1$ and $\mathbf{z}_1$ gives the same solution as before. Exercise 12.4 asks you to show that $\frac{\partial J}{\partial \mathbf{z}_2} = 0$ yields $z_{i2} = \mathbf{w}_2^T \mathbf{x}_i$. In other words, the second principal encoding is gotten by projecting onto the second principal direction. Substituting in yields

$$J(\mathbf{w}_2) = \frac{1}{n}\sum_{i=1}^{N}[\mathbf{x}_i^T \mathbf{x}_i - \mathbf{w}_1^T \mathbf{x}_i \mathbf{x}_i^T \mathbf{w}_1 - \mathbf{w}_2^T \mathbf{x}_i \mathbf{x}_i^T \mathbf{w}_2] = \text{const} - \mathbf{w}_2^T \hat{\boldsymbol{\Sigma}} \mathbf{w}_2 \tag{12.43}$$

Dropping the constant term and adding the constraints yields

$$\tilde{J}(\mathbf{w}_2) = -\mathbf{w}_2^T \hat{\boldsymbol{\Sigma}} \mathbf{w}_2 + \lambda_2(\mathbf{w}_2^T \mathbf{w}_2 - 1) + \lambda_{12}(\mathbf{w}_2^T \mathbf{w}_1 - 0) \tag{12.44}$$

Exercise 12.4 asks you to show that the solution is given by the eigenvector with the second largest eigenvalue:

$$\hat{\mathbf{\Sigma}}\mathbf{w}_2 = \lambda_2\mathbf{w}_2 \tag{12.45}$$

The proof continues in this way. (Formally one can use induction.) $\qquad\square$

12.2.3 Singular value decomposition (SVD)

We have defined the solution to PCA in terms of eigenvectors of the covariance matrix. However, there is another way to obtain the solution, based on the **singular value decomposition**, or **SVD**. This basically generalizes the notion of eigenvectors from square matrices to any kind of matrix.

In particular, any (real) $N \times D$ matrix $\mathbf{X}$ can be decomposed as follows

$$\underbrace{\mathbf{X}}_{N \times D} = \underbrace{\mathbf{U}}_{N \times N}\underbrace{\mathbf{S}}_{N \times D}\underbrace{\mathbf{V}^T}_{D \times D} \tag{12.46}$$

where $\mathbf{U}$ is an $N \times N$ matrix whose columns are orthornormal (so $\mathbf{U}^T\mathbf{U} = \mathbf{I}_N$), $\mathbf{V}$ is $D \times D$ matrix whose rows and columns are orthonormal (so $\mathbf{V}^T\mathbf{V} = \mathbf{V}\mathbf{V}^T = \mathbf{I}_D$), and $\mathbf{S}$ is a $N \times D$ matrix containing the $r = \min(N, D)$ **singular values** $\sigma_i \geq 0$ on the main diagonal, with 0s filling the rest of the matrix. The columns of $\mathbf{U}$ are the left singular vectors, and the columns of $\mathbf{V}$ are the right singular vectors. See Figure 12.8(a) for an example.

Since there are at most D singular values (assuming $N > D$), the last $N - D$ columns of $\mathbf{U}$ are irrelevant, since they will be multiplied by 0. The **economy sized SVD**, or **thin SVD**, avoids computing these unnecessary elements. Let us denote this decomposition by $\hat{\mathbf{U}}\hat{\mathbf{S}}\hat{\mathbf{V}}$. If $N > D$, we have

$$\underbrace{\mathbf{X}}_{N \times D} = \underbrace{\hat{\mathbf{U}}}_{N \times D}\underbrace{\hat{\mathbf{S}}}_{D \times D}\underbrace{\hat{\mathbf{V}}^T}_{D \times D} \tag{12.47}$$

as in Figure 12.8(a). If $N < D$, we have

$$\underbrace{\mathbf{X}}_{N \times D} = \underbrace{\hat{\mathbf{U}}}_{N \times N}\underbrace{\hat{\mathbf{S}}}_{N \times N}\underbrace{\hat{\mathbf{V}}^T}_{N \times D} \tag{12.48}$$

Computing the economy-sized SVD takes $O(ND\min(N, D))$ time (Golub and van Loan 1996, p254).

The connection between eigenvectors and singular vectors is the following. For an arbitrary real matrix $\mathbf{X}$, if $\mathbf{X} = \mathbf{U}\mathbf{S}\mathbf{V}^T$, we have

$$\mathbf{X}^T\mathbf{X} = \mathbf{V}\mathbf{S}^T\mathbf{U}^T\,\mathbf{U}\mathbf{S}\mathbf{V}^T = \mathbf{V}(\mathbf{S}^T\mathbf{S})\mathbf{V}^T = \mathbf{V}\mathbf{D}\mathbf{V}^T \tag{12.49}$$

where $\mathbf{D} = \mathbf{S}^2$ is a diagonal matrix containing the squares singular values. Hence

$$(\mathbf{X}^T\mathbf{X})\mathbf{V} = \mathbf{V}\mathbf{D} \tag{12.50}$$

so the eigenvectors of $\mathbf{X}^T\mathbf{X}$ are equal to $\mathbf{V}$, the right singular vectors of $\mathbf{X}$, and the eigenvalues of $\mathbf{X}^T\mathbf{X}$ are equal to $\mathbf{D}$, the squared singular values. Similarly

$$\mathbf{X}\mathbf{X}^T = \mathbf{U}\mathbf{S}\mathbf{V}^T\,\mathbf{V}\mathbf{S}^T\mathbf{U}^T = \mathbf{U}(\mathbf{S}\mathbf{S}^T)\mathbf{U}^T \tag{12.51}$$

$$(\mathbf{X}\mathbf{X}^T)\mathbf{U} = \mathbf{U}(\mathbf{S}\mathbf{S}^T) = \mathbf{U}\mathbf{D} \tag{12.52}$$

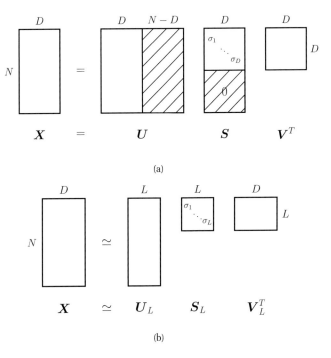

(a)

(b)

Figure 12.8 (a) SVD decomposition of non-square matrices $\mathbf{X} = \mathbf{USV}^T$. The shaded parts of $\mathbf{S}$, and all the off-diagonal terms, are zero. The shaded entries in $\mathbf{U}$ and $\mathbf{S}$ are not computed in the economy-sized version, since they are not needed. (b) Truncated SVD approximation of rank L.

so the eigenvectors of $\mathbf{XX}^T$ are equal to $\mathbf{U}$, the left singular vectors of $\mathbf{X}$. Also, the eigenvalues of $\mathbf{XX}^T$ are equal to the squared singular values. We can summarize all this as follows:

$$\mathbf{U} = \text{evec}(\mathbf{XX}^T), \quad \mathbf{V} = \text{evec}(\mathbf{X}^T\mathbf{X}), \quad \mathbf{S}^2 = \text{eval}(\mathbf{XX}^T) = \text{eval}(\mathbf{X}^T\mathbf{X}) \tag{12.53}$$

Since the eigenvectors are unaffected by linear scaling of a matrix, we see that the right singular vectors of $\mathbf{X}$ are equal to the eigenvectors of the empirical covariance $\hat{\mathbf{\Sigma}}$. Furthermore, the eigenvalues of $\hat{\mathbf{\Sigma}}$ are a scaled version of the squared singular values. This means we can perform PCA using just a few lines of code (see pcaPmtk).

However, the connection between PCA and SVD goes deeper. From Equation 12.46, we can represent a rank r matrix as follows:

$$\mathbf{X} = \sigma_1 \begin{pmatrix} | \\ \mathbf{u}_1 \\ | \end{pmatrix} \begin{pmatrix} - & \mathbf{v}_1^T & - \end{pmatrix} + \cdots + \sigma_r \begin{pmatrix} | \\ \mathbf{u}_r \\ | \end{pmatrix} \begin{pmatrix} - & \mathbf{v}_r^T & - \end{pmatrix} \tag{12.54}$$

If the singular values die off quickly as in Figure 12.10, we can produce a rank L approximation to the matrix as follows:

$$\mathbf{X} \approx \mathbf{U}_{:,1:L} \ \mathbf{S}_{1:L,1:L} \ \mathbf{V}_{:,1:L}^T \tag{12.55}$$

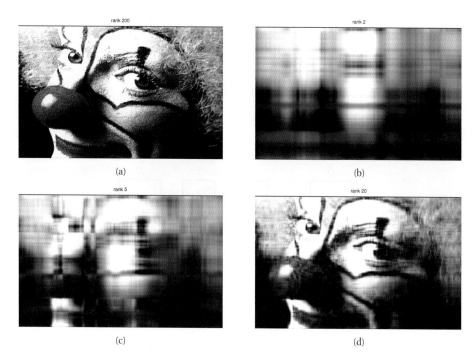

Figure 12.9 Low rank approximations to an image. Top left: The original image is of size 200×320, so has rank 200. Subsequent images have ranks 2, 5, and 20. Figure generated by `svdImageDemo`.

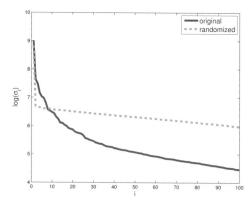

Figure 12.10 First 50 log singular values for the clown image (solid red line), and for a data matrix obtained by randomly shuffling the pixels (dotted green line). Figure generated by `svdImageDemo`.

This is called a **truncated SVD** (see Figure 12.8(b)). The total number of parameters needed to represent an $N \times D$ matrix using a rank L approximation is

$$NL + LD + L = L(N + D + 1) \tag{12.56}$$

As an example, consider the 200×320 pixel image in Figure 12.9(top left). This has 64,000 numbers in it. We see that a rank 20 approximation, with only $(200 + 320 + 1) \times 20 = 10,420$ numbers is a very good approximation.

One can show that the error in this approximation is given by

$$||\mathbf{X} - \mathbf{X}_L||_F \approx \sigma_{L+1} \tag{12.57}$$

Furthermore, one can show that the SVD offers the best rank L approximation to a matrix (best in the sense of minimizing the above Frobenius norm).

Let us connect this back to PCA. Let $\mathbf{X} = \mathbf{U}\mathbf{S}\mathbf{V}^T$ be a truncated SVD of $\mathbf{X}$. We know that $\hat{\mathbf{W}} = \mathbf{V}$, and that $\hat{\mathbf{Z}} = \mathbf{X}\hat{\mathbf{W}}$, so

$$\hat{\mathbf{Z}} = \mathbf{U}\mathbf{S}\mathbf{V}^T\mathbf{V} = \mathbf{U}\mathbf{S} \tag{12.58}$$

Furthermore, the optimal reconstruction is given by $\hat{\mathbf{X}} = \mathbf{Z}\hat{\mathbf{W}}^T$, so we find

$$\hat{\mathbf{X}} = \mathbf{U}\mathbf{S}\mathbf{V}^T \tag{12.59}$$

This is precisely the same as a truncated SVD approximation! This is another illustration of the fact that PCA is the best low rank approximation to the data.

12.2.4 Probabilistic PCA

We are now ready to revisit PPCA. One can show the following remarkable result.

Theorem 12.2.2 ((Tipping and Bishop 1999)). *Consider a factor analysis model in which* $\mathbf{\Psi} = \sigma^2\mathbf{I}$. *The observed data log likelihood is given by*

$$\log p(\mathbf{X}|\mathbf{W}, \sigma^2) = -\frac{N}{2}\ln|\mathbf{C}| - \frac{1}{2}\sum_{i=1}^{N}\mathbf{x}_i^T\mathbf{C}^{-1}\mathbf{x}_i = -\frac{N}{2}\ln|\mathbf{C}| + \text{tr}(\mathbf{C}^{-1}\hat{\mathbf{\Sigma}}) \tag{12.60}$$

where $\mathbf{C} = \mathbf{W}\mathbf{W}^T + \sigma^2\mathbf{I}$ *and* $\mathbf{S} = \frac{1}{N}\sum_{i=1}^{N}\mathbf{x}_i\mathbf{x}_i^T = (1/N)\mathbf{X}^T\mathbf{X}$. *(We are assuming centered data, for notational simplicity.) The maxima of the log-likelihood are given by*

$$\hat{\mathbf{W}} = \mathbf{V}(\mathbf{\Lambda} - \sigma^2\mathbf{I})^{\frac{1}{2}}\mathbf{R} \tag{12.61}$$

where $\mathbf{R}$ *is an arbitrary* $L \times L$ *orthogonal matrix,* $\mathbf{V}$ *is the* $D \times L$ *matrix whose columns are the first* L *eigenvectors of* $\mathbf{S}$, *and* $\mathbf{\Lambda}$ *is the corresponding diagonal matrix of eigenvalues. Without loss of generality, we can set* $\mathbf{R} = \mathbf{I}$. *Furthermore, the MLE of the noise variance is given by*

$$\hat{\sigma}^2 = \frac{1}{D - L}\sum_{j=L+1}^{D}\lambda_j \tag{12.62}$$

which is the average variance associated with the discarded dimensions.

Thus, as $\sigma^2 \to 0$, we have $\hat{\mathbf{W}} \to \mathbf{V}$, as in classical PCA. What about $\hat{\mathbf{Z}}$? It is easy to see that the posterior over the latent factors is given by

$$p(\mathbf{z}_i|\mathbf{x}_i, \hat{\boldsymbol{\theta}}) = \mathcal{N}(\mathbf{z}_i|\hat{\mathbf{F}}^{-1}\hat{\mathbf{W}}^T\mathbf{x}_i, \sigma^2\hat{\mathbf{F}}^{-1}) \tag{12.63}$$

$$\hat{\mathbf{F}} \triangleq \hat{\mathbf{W}}^T\hat{\mathbf{W}} + \hat{\sigma}^2\mathbf{I} \tag{12.64}$$

(Do not confuse $\mathbf{F} = \mathbf{W}^T\mathbf{W} + \sigma^2\mathbf{I}$ with $\mathbf{C} = \mathbf{W}\mathbf{W}^T + \sigma^2\mathbf{I}$.) Hence, as $\sigma^2 \to 0$, we find $\hat{\mathbf{W}} \to \mathbf{V}$, $\hat{\mathbf{F}} \to \mathbf{I}$ and $\hat{\mathbf{z}}_i \to \mathbf{V}^T\mathbf{x}_i$. Thus the posterior mean is obtained by an orthogonal projection of the data onto the column space of $\mathbf{V}$, as in classical PCA.

Note, however, that if $\sigma^2 > 0$, the posterior mean is not an orthogonal projection, since it is shrunk somewhat towards the prior mean, as illustrated in Figure 12.5(b). This sounds like an undesirable property, but it means that the reconstructions will be closer to the overall data mean, $\hat{\boldsymbol{\mu}} = \bar{\mathbf{x}}$.

12.2.5   EM algorithm for PCA

Although the usual way to fit a PCA model uses eigenvector methods, or the SVD, we can also use EM, which will turn out to have some advantages that we discuss below. EM for PCA relies on the probabilistic formulation of PCA. However the algorithm continues to work in the zero noise limit, $\sigma^2 = 0$, as shown by (Roweis 1997).

Let $\tilde{\mathbf{Z}}$ be a $L \times N$ matrix storing the posterior means (low-dimensional representations) along its columns. Similarly, let $\tilde{\mathbf{X}} = \mathbf{X}^T$ store the original data along its columns. From Equation 12.63, when $\sigma^2 = 0$, we have

$$\tilde{\mathbf{Z}} = (\mathbf{W}^T\mathbf{W})^{-1}\mathbf{W}^T\tilde{\mathbf{X}} \tag{12.65}$$

This constitutes the E step. Notice that this is just an orthogonal projection of the data.

From Equation 12.23, the M step is given by

$$\hat{\mathbf{W}} = \left[\sum_i \mathbf{x}_i \mathbb{E}\left[\mathbf{z}_i\right]^T\right]\left[\sum_i \mathbb{E}\left[\mathbf{z}_i\right]\mathbb{E}\left[\mathbf{z}_i\right]^T\right]^{-1} \tag{12.66}$$

where we exploited the fact that $\boldsymbol{\Sigma} = \text{cov}\left[\mathbf{z}_i|\mathbf{x}_i, \boldsymbol{\theta}\right] = 0\mathbf{I}$ when $\sigma^2 = 0$. It is worth comparing this expression to the MLE for multi-output linear regression (Equation 7.89), which has the form $\mathbf{W} = (\sum_i \mathbf{y}_i\mathbf{x}_i^T)(\sum_i \mathbf{x}_i\mathbf{x}_i^T)^{-1}$. Thus we see that the M step is like linear regression where we replace the observed inputs by the expected values of the latent variables.

In summary, here is the entire algorithm:

- **E step** $\tilde{\mathbf{Z}} = (\mathbf{W}^T\mathbf{W})^{-1}\mathbf{W}^T\tilde{\mathbf{X}}$
- **M step** $\mathbf{W} = \tilde{\mathbf{X}}\tilde{\mathbf{Z}}^T(\tilde{\mathbf{Z}}\tilde{\mathbf{Z}}^T)^{-1}$

(Tipping and Bishop 1999) showed that the only stable fixed point of the EM algorithm is the globally optimal solution. That is, the EM algorithm converges to a solution where $\mathbf{W}$ spans the same linear subspace as that defined by the first L eigenvectors. However, if we want $\mathbf{W}$ to be orthogonal, and to contain the eigenvectors in descending order of eigenvalue, we have to orthogonalize the resulting matrix (which can be done quite cheaply). Alternatively, we can modify EM to give the principal basis directly (Ahn and Oh 2003).

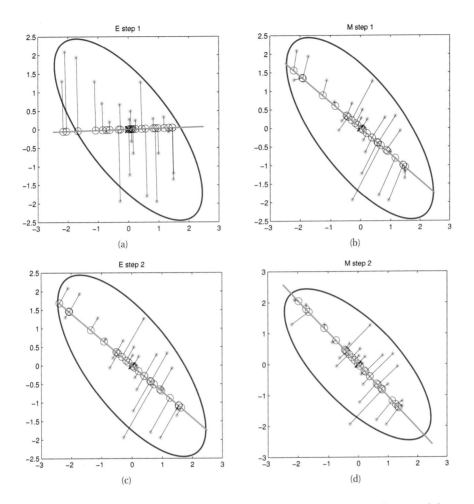

Figure 12.11 Illustration of EM for PCA when $D = 2$ and $L = 1$. Green stars are the original data points, black circles are their reconstructions. The weight vector $\mathbf{w}$ is represented by blue line. (a) We start with a random initial guess of $\mathbf{w}$. The E step is represented by the orthogonal projections. (b) We update the rod $\mathbf{w}$ in the M step, keeping the projections onto the rod (black circles) fixed. (c) Another E step. The black circles can 'slide' along the rod, but the rod stays fixed. (d) Another M step. Based on Figure 12.12 of (Bishop 2006). Figure generated by pcaEmStepByStep.

This algorithm has a simple physical analogy in the case $D = 2$ and $L = 1$ (Roweis 1997). Consider some points in $\mathbb{R}^2$ attached by springs to a rigid rod, whose orientation is defined by a vector $\mathbf{w}$. Let z_i be the location where the i'th spring attaches to the rod. In the E step, we hold the rod fixed, and let the attachment points slide around so as to minimize the spring energy (which is proportional to the sum of squared residuals). In the M step, we hold the attachment points fixed and let the rod rotate so as to minimize the spring energy. See Figure 12.11 for an illustration.

Apart from this pleasing intuitive interpretation, EM for PCA has the following advantages over eigenvector methods:

- EM can be faster. In particular, assuming $N, D \gg L$, the dominant cost of EM is the projection operation in the E step, so the overall time is $O(TLND)$, where T is the number of iterations. (Roweis 1997) showed experimentally that the number of iterations is usually very small (the mean was 3.6), regardless of N or D. (This results depends on the ratio of eigenvalues of the empirical covariance matrix.) This is much faster than the $O(\min(ND^2, DN^2))$ time required by straightforward eigenvector methods, although more sophisticated eigenvector methods, such as the Lanczos algorithm, have running times comparable to EM.

- EM can be implemented in an online fashion, i.e., we can update our estimate of $\mathbf{W}$ as the data streams in (see Section 11.4.8).

- EM can handle missing data in a simple way (see Section 12.1.6).

- EM can be extended to handle mixtures of PPCA/ FA models (see (Ghahramani and Hinton 1996a)).

- EM can be modified to variational EM or to variational Bayes EM to fit more complex models (see (Ghahramani and Beal 1999)).

12.3 Choosing the number of latent dimensions

In Section 11.5, we discussed how to choose the number of components K in a mixture model. In this section, we discuss how to choose the number of latent dimensions L in a FA/PCA model.

12.3.1 Model selection for FA/PPCA

If we use a probabilistic model, we can in principle compute $L^* = \operatorname{argmax}_L p(L|\mathcal{D})$. However, there are two problems with this. First, evaluating the marginal likelihood for LVMs is quite difficult. In practice, simple approximations, such as BIC or variational lower bounds (see Section 21.5), can be used (see also (Minka 2000a)). Alternatively, we can use the cross-validated likelihood as a performance measure, although this can be slow, since it requires fitting each model F times, where F is the number of CV folds.

The second issue is the need to search over a potentially large number of models. The usual approach is to perform exhaustive search over all candidate values of L. However, sometimes we can set the model to its maximal size, and then use a technique called automatic relevancy determination (Section 13.7), combined with EM, to automatically prune out irrelevant weights. This technique will be described in a supervised context in Chapter 13, but can be adapted to the (M)FA context as shown in (Bishop 1999; Ghahramani and Beal 2000).

Figure 12.12 illustrates this approach applied to a mixture of FAs fit to a small synthetic dataset. The figures visualize the weight matrices for each cluster, using **Hinton diagrams**, where where the size of the square is proportional to the value of the entry in the matrix.[5] We see that many of them are sparse. Figure 12.13 shows that the degree of sparsity depends on the amount

---

5. Geoff Hinton is an English professor of computer science at the University of Toronto.

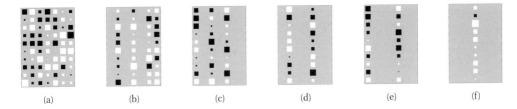

   (a)   (b)   (c)   (d)   (e)   (f)

Figure 12.12 Illustration of estimating the effective dimensionalities in a mixture of factor analysers using VBEM. The blank columns have been forced to 0 via the ARD mechanism. The data was generated from 6 clusters with intrinsic dimensionalities of $7, 4, 3, 2, 2, 1$, which the method has successfully estimated. Source: Figure 4.4 of (Beal 2003). Used with kind permission of Matt Beal.

number of points per cluster	intrinsic dimensionalities					
	1	7	4	3	2	2
8		2			1	
8	1		2			
16	1		4			2
32	1	6	3	3	2	2
64	1	7	4	3	2	2
128	1	7	4	3	2	2

Figure 12.13 We show the estimated number of clusters, and their estimated dimensionalities, as a function of sample size. The VBEM algorithm found two different solutions when $N = 8$. Note that more clusters, with larger effective dimensionalities, are discovered as the sample sizes increases. Source: Table 4.1 of (Beal 2003). Used with kind permission of Matt Beal.

of training data, in accord with the Bayesian Occam's razor. In particular, when the sample size is small, the method automatically prefers simpler models, but as the sample size gets sufficiently large, the method converges on the "correct" solution, which is one with 6 subspaces of dimensionality 1, 2, 2, 3, 4 and 7.

Although the ARD/ EM method is elegant, it still needs to perform search over K. This is done using "birth" and "death" moves (Ghahramani and Beal 2000). An alternative approach is to perform stochastic sampling in the space of models. Traditional approaches, such as (Lopes and West 2004), are based on reversible jump MCMC, and also use birth and death moves. However, this can be slow and difficult to implement. More recent approaches use non-parametric priors, combined with Gibbs sampling, see e.g., (Paisley and Carin 2009).

12.3.2 Model selection for PCA

Since PCA is not a probabilistic model, we cannot use any of the methods described above. An obvious proxy for the likelihood is the reconstruction error:

$$E(\mathcal{D}, L) = \frac{1}{|\mathcal{D}|} \sum_{i \in \mathcal{D}} ||\mathbf{x}_i - \hat{\mathbf{x}}_i||^2 \tag{12.67}$$

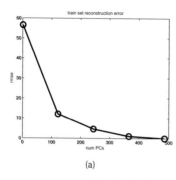
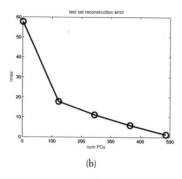

(a)   (b)

Figure 12.14   Reconstruction error on MNIST vs number of latent dimensions used by PCA. (a) Training set. (b) Test set. Figure generated by `pcaOverfitDemo`.

In the case of PCA, the reconstruction is given by by $\hat{\mathbf{x}}_i = \mathbf{W}\mathbf{z}_i + \boldsymbol{\mu}$, where $\mathbf{z}_i = \mathbf{W}^T(\mathbf{x}_i - \boldsymbol{\mu})$ and $\mathbf{W}$ and $\boldsymbol{\mu}$ are estimated from $\mathcal{D}_{\text{train}}$.

Figure 12.14(a) plots $E(\mathcal{D}_{\text{train}}, L)$ vs L on the MNIST training data in Figure 12.6. We see that it drops off quite quickly, indicating that we can capture most of the empirical correlation of the pixels with a small number of factors, as illustrated qualitatively in Figure 12.6.

Exercise 12.5 asks you to prove that the residual error from only using L terms is given by the sum of the discarded eigenvalues:

$$E(\mathcal{D}_{\text{train}}, L) \;=\; \sum_{j=L+1}^{D} \lambda_j \tag{12.68}$$

Therefore an alternative to plotting the error is to plot the retained eigenvalues, in decreasing order. This is called a **scree plot**, because "the plot looks like the side of a mountain, and 'scree' refers to the debris fallen from a mountain and lying at its base".[6] This will have the same shape as the residual error plot.

A related quantity is the **fraction of variance explained**, defined as

$$F(\mathcal{D}_{\text{train}}, L) = \frac{\sum_{j=1}^{L} \lambda_j}{\sum_{j'=1}^{L_{max}} \lambda_{j'}} \tag{12.69}$$

This captures the same information as the scree plot.

Of course, if we use $L = \text{rank}(\mathbf{X})$, we get zero reconstruction error on the training set. To avoid overfitting, it is natural to plot reconstruction error on the test set. This is shown in Figure 12.14(b). Here we see that the error continues to go down even as the model becomes more complex! Thus we do not get the usual U-shaped curve that we typically expect to see.

What is going on? The problem is that PCA is not a proper generative model of the data. It is merely a compression technique. If you give it more latent dimensions, it will be able to approximate the test data more accurately. By contrast, a probabilistic model enjoys a Bayesian

---

6. Quotation from `http://janda.org/workshop/factoranalysis/SPSSrun/SPSS08.htm`.

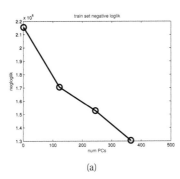
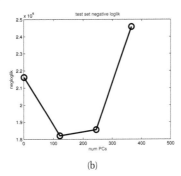

(a)   (b)

Figure 12.15 Negative log likelihood on MNIST vs number of latent dimensions used by PPCA. (a) Training set. (b) Test set. Figure generated by `pcaOverfitDemo`.

Occam's razor effect (Section 5.3.1), in that it gets "punished" if it wastes probability mass on parts of the space where there is little data. This is illustrated in Figure 12.15, which plots the negative log likelihood, computed using PPCA, vs L. Here, on the test set, we see the usual U-shaped curve.

These results are analogous to those in Section 11.5.2, where we discussed the issue of choosing K in the K-means algorithm vs using a GMM.

12.3.2.1 Profile likelihood

Although there is no U-shape, there is sometimes a "regime change" in the plots, from relatively large errors to relatively small. One way to automate the detection of this is described in (Zhu and Ghodsi 2006). The idea is this. Let λ_k be some measure of the error incurred by a model of size k, such that $\lambda_1 \geq \lambda_2 \geq \cdots \geq \lambda_{L_{max}}$. In PCA, these are the eigenvalues, but the method can also be applied to K-means. Now consider partitioning these values into two groups, depending on whether $k < L$ or $k > L$, where L is some threshold which we will determine. To measure the quality of L, we will use a simple change-point model, where $\lambda_k \sim \mathcal{N}(\mu_1, \sigma^2)$ if $k \leq L$, and $\lambda_k \sim \mathcal{N}(\mu_2, \sigma^2)$ if $k > L$. (It is important that σ^2 be the same in both models, to prevent overfitting in the case where one regime has less data than the other.) Within each of the two regimes, we assume the λ_k are iid, which is obviously incorrect, but is adequate for our present purposes. We can fit this model for each $L = 1 : L_{max}$ by partitioning the data and computing the MLEs, using a pooled estimate of the variance:

$$\mu_1(L) = \frac{\sum_{k \leq L} \lambda_k}{L}, \ \mu_2(L) = \frac{\sum_{k > L} \lambda_k}{N - L} \tag{12.70}$$

$$\sigma^2(L) = \frac{\sum_{k \leq L} (\lambda_k - \mu_1(L))^2 + \sum_{k > L} (\lambda_k - \mu_2(L))^2}{N} \tag{12.71}$$

We can then evaluate the **profile log likelihood**

$$\ell(L) = \sum_{k=1}^{L} \log \mathcal{N}(\lambda_k | \mu_1(L), \sigma^2(L)) + \sum_{k=L+1}^{K} \log \mathcal{N}(\lambda_k | \mu_2(L), \sigma^2(L)) \tag{12.72}$$

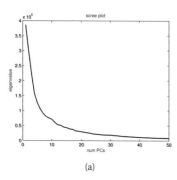
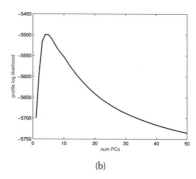

(a)   (b)

Figure 12.16   (a) Scree plot for training set, corresponding to Figure 12.14(a). (b) Profile likelihood. Figure generated by pcaOverfitDemo.

Finally, we choose $L^* = \arg\max \ell(L)$. This is illustrated in Figure 12.16. On the left, we plot the scree plot, which has the same shape as in Figure 12.14(a). On the right, we plot the profile likelihood. Rather miraculously, we see a fairly well-determined peak.

12.4   PCA for categorical data

In this section, we consider extending the factor analysis model to the case where the observed data is categorical rather than real-valued. That is, the data has the form $y_{ij} \in \{1, \ldots, C\}$, where $j = 1 : R$ is the number of observed response variables. We assume each y_{ij} is generated from a latent variable $\mathbf{z}_i \in \mathbb{R}^L$, with a Gaussian prior, which is passed through the softmax function as follows:

$$p(\mathbf{z}_i) = \mathcal{N}(\mathbf{0}, \mathbf{I}) \tag{12.73}$$

$$p(\mathbf{y}_i | \mathbf{z}_i, \boldsymbol{\theta}) = \prod_{r=1}^{R} \text{Cat}(y_{ir} | \mathcal{S}(\mathbf{W}_r^T \mathbf{z}_i + \mathbf{w}_{0r})) \tag{12.74}$$

where $\mathbf{W}_r \in \mathbb{R}^{L \times M}$ is the factor loading matrix for response j, and $\mathbf{w}_{0r} \in \mathbb{R}^M$ is the offset term for response r, and $\boldsymbol{\theta} = (\mathbf{W}_r, \mathbf{w}_{0r})_{r=1}^R$. (We need an explicit offset term, since clamping one element of $\mathbf{z}_i$ to 1 can cause problems when computing the posterior covariance.) As in factor analysis, we have defined the prior mean to be $\mathbf{m}_0 = \mathbf{0}$ and the prior covariance $\mathbf{V}_0 = \mathbf{I}$, since we can capture non-zero mean by changing $\mathbf{w}_{0j}$ and non-identity covariance by changing $\mathbf{W}_r$. We will call this **categorical PCA**. Versions of this model are used in psychometrics, where they are called **item response theory** models. See Chapter 27 for a discussion of related models.

It is interesting to study what kinds of distributions we can induce on the observed variables by varying the parameters. For simplicity, we assume there is a single ternary response variable, so y_i lives in the 3d probability simplex. Figure 12.17 shows what happens when we vary the parameters of the prior, $\mathbf{m}_0$ and $\mathbf{V}_0$, which is equivalent to varying the parameters of the likelihood, $\mathbf{W}_1$ and $\mathbf{w}_{01}$. We see that this can define fairly complex distributions over the simplex. This induced distribution is known as the **logistic normal** distribution (Aitchison 1982).

Figure 12.17 Some examples of the logistic normal distribution defined on the 3d simplex. (a) Diagonal covariance and non-zero mean. (b) Negative correlation between states 1 and 2. (c) Positive correlation between states 1 and 2. Source: Figure 1 of (Blei and Lafferty 2007). Used with kind permission of David Blei.

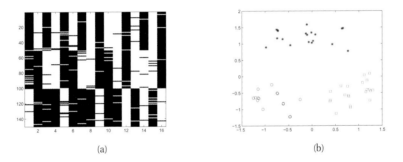

(a) (b)

Figure 12.18 Left: 150 synthetic 16 dimensional bit vectors. Right: the 2d embedding learned by binary PCA, using variational EM. We have color coded points by the identity of the true "prototype" that generated them. Figure generated by `binaryFaDemoTipping`.

We can fit this model to data using a modified version of EM. The basic idea is to infer a Gaussian approximation to the posterior $p(\mathbf{z}_i|\mathbf{y}_i, \boldsymbol{\theta})$ in the E step, and then to maximize $\boldsymbol{\theta}$ in the M step. The details for the multiclass case, can be found in (Khan et al. 2010) (see also Section 21.8.1.1). The details for the binary case for the the sigmoid link can be found in Exercise 21.9, and for the probit link in Exercise 21.10.

One application of such a model is to visualize high dimensional categorical data. Figure 12.18(a) shows a simple example where we have 150 6-dimensional bit vectors. It is clear that each sample is just a noisy copy of one of three binary prototypes. We fit a 2d catFA to this model, yielding approximate MLEs $\hat{\boldsymbol{\theta}}$. In Figure 12.18(b), we plot $\mathbb{E}\left[\mathbf{z}_i|\mathbf{x}_i, \hat{\boldsymbol{\theta}}\right]$. We see that there are three distinct clusters, as is to be expected.

In (Khan et al. 2010), we show that this model outperforms finite mixture models on the task of imputing missing entries in design matrices consisting of real and categorical data. This is useful for analysing social science survey data, which often has missing data and variables of mixed type.

12.5   PCA for paired and multi-view data

It is common to have a pair of related datasets, e.g., gene expression and gene copy number, or movie ratings by users and movie reviews. It is natural to want to combine these together into a low-dimensional embedding. This is an example of **data fusion**. A common task is to predict one element of the pair, say $\mathbf{y}_i$, from the other one, $\mathbf{x}_i$, via the low-dimensional "bottleneck". Below we discuss various latent Gaussian models for solving this task, following the presentation of (Virtanen 2010). (Most of the models easily generalize from modeling pairs to modeling sets of data, $\mathbf{x}_{im}$, for $m = 1 : M$, both we omit this case to simplify the notation.)

12.5.1   Supervised PCA (latent factor regression)

Consider the following model, illustrated in Figure 12.19(a):

$$
\begin{aligned}
p(\mathbf{z}_i) &= \mathcal{N}(\mathbf{0}, \mathbf{I}_L) & (12.75)\\
p(y_i|\mathbf{z}_i) &= \mathcal{N}(\mathbf{w}_y^T \mathbf{z}_i + \mu_y, \sigma_y^2) & (12.76)\\
p(\mathbf{x}_i|\mathbf{z}_i) &= \mathcal{N}(\mathbf{W}_x \mathbf{z}_i + \boldsymbol{\mu}_x, \sigma_x^2 \mathbf{I}_D) & (12.77)
\end{aligned}
$$

In (Yu et al. 2006), this is called **supervised PCA**. In (West 2003), this is called **Bayesian factor regression**. This model is like PCA, except that the target variable y_i is taken into account when learning the low dimensional embedding. Since the model is jointly Gaussian, we have

$$
y_i|\mathbf{x}_i \sim \mathcal{N}(\mathbf{x}_i^T \mathbf{w}, \mathbf{w}_y^T \mathbf{C} \mathbf{w}_y + \sigma_y^2) \tag{12.78}
$$

where $\mathbf{w} = \boldsymbol{\Psi}^{-1} \mathbf{W}_x \mathbf{C} \mathbf{w}_y$, $\boldsymbol{\Psi} = \sigma_x^2 \mathbf{I}_D$, and $\mathbf{C}^{-1} = \mathbf{I} + \mathbf{W}_x^T \boldsymbol{\Psi}^{-1} \mathbf{W}_x$. So although this is a joint density model of $(\mathbf{x}_i, y_i)$, we can compute the implied conditional distribution.[7]

The above discussion focussed on regression. (Guo 2009) generalizes the model to the exponential family, which is more appropriate if $\mathbf{x}_i$ and/or y_i are discrete. For example, if y_i is binary, we can write

$$
p(y_i|\mathbf{z}_i) = \text{Ber}(\text{sigm}(\mathbf{w}_y^T \mathbf{z}_i)) \tag{12.79}
$$

We can also easily generalize to the case where y_i is a vector of responses to be predicted as in multi-label classification. For example, (Ma et al. 2008; Williamson and Ghahramani 2008) used this model to perform collaborative filtering, where the goal is to predict $y_{ij} \in \{1, \ldots, 5\}$, the rating person i gives to movie j, where the "side information" $\mathbf{x}_i$ takes the form of a list of i's friends. Another extentions is to the semi-supervised case, where we do not observe $\mathbf{y}_i$ for all i (Yu et al. 2006).

In the classification case, we can no longer compute the conditional $p(\mathbf{y}_i|\mathbf{x}_i, \boldsymbol{\theta})$ in closed form, the model has a similar interpretation to the regression case, namely that we are predicting the response via a latent "bottleneck".

---

7. There is an interesting connection to Zellner's g-prior. Suppose $p(\mathbf{w}_y) = \mathcal{N}(\mathbf{0}, \frac{1}{g}\boldsymbol{\Sigma}^2)$, and let $\mathbf{X} = \mathbf{R}\mathbf{V}^T$ be the SVD of $\mathbf{X}$, where $\mathbf{V}^T \mathbf{V} = \mathbf{I}$ and $\mathbf{R}^T \mathbf{R} = \boldsymbol{\Sigma}^2 = \text{diag}(\sigma_j^2)$ contains the squared singular values. Then one can show (West 2003) that $p(\mathbf{w}) = \mathcal{N}(\mathbf{0}, g\mathbf{V}^{-T}\boldsymbol{\Sigma}^{-2}\mathbf{V}^{-1}) = \mathcal{N}(\mathbf{0}, g(\mathbf{X}^T\mathbf{X})^{-1})$. So the dependence of the prior for $\mathbf{w}$ on $\mathbf{X}$ arises from the fact that $\mathbf{w}$ is derived indirectly by a joint model of $\mathbf{X}$ and $\mathbf{y}$.

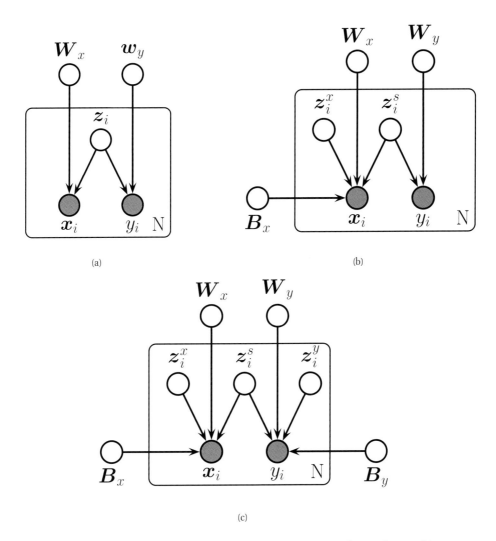

Figure 12.19 Gaussian latent factor models for paired data. (a) Supervised PCA. (b) Partial least squares. (c) Canonical correlation analysis.

The basic idea of compressing $\mathbf{x}_i$ to predict $\mathbf{y}_i$ can be formulated using information theory. In particular, we might want to find an encoding distribution $p(\mathbf{z}|\mathbf{x})$ such that we minimize

$$\mathbb{I}(X;Z) - \beta \mathbb{I}(X;Y) \tag{12.80}$$

where $\beta \geq 0$ is some parameter controlling the tradeoff between compression and predictive accuracy. This is known as the **information bottleneck** (Tishby et al. 1999). Often Z is taken to be discrete, as in clustering. However, in the Gaussian case, IB is closely related to the supervised PCA model (Chechik et al. 2005).

12.5.1.1 Discriminative supervised PCA

One problem with supervised PCA is that it puts as much weight on predicting the inputs $\mathbf{x}_i$ as the outputs $\mathbf{y}_i$. This can be partially alleviated by using a weighted objective of the following form (Rish et al. 2008):

$$\ell(\boldsymbol{\theta}) = \prod_i p(\mathbf{y}_i | \boldsymbol{\eta}_{iy})^{\alpha_y} p(\mathbf{x}_i | \boldsymbol{\eta}_{ix})^{\alpha_x} \tag{12.81}$$

where the α_m control the relative importance of the data sources, $\boldsymbol{\eta}_{ix} = \mathbf{W}_x \mathbf{z}_i$ and $\boldsymbol{\eta}_{iy} = \mathbf{W}_y \mathbf{z}_i$. For a Gaussian observation model, we can see that α_m just controls the noise variance:

$$\ell(\boldsymbol{\theta}) \propto \prod_i \exp(-\frac{1}{2}\alpha_x \|\mathbf{x}_i^T - \boldsymbol{\eta}_{ix}\|^2) \exp(-\frac{1}{2}\alpha_y \|\mathbf{y}_i^T - \boldsymbol{\eta}_{iy}\|^2) \tag{12.82}$$

This interpretation holds more generally for the exponential family. Note, however, that it is hard to estimate the α_m parameters, because changing them changes the normalization constant of the likelihood. We give an alternative approach to weighting $\mathbf{y}$ more heavily in Section 12.5.2.

12.5.2 Partial least squares

The technique of **partial least squares** (**PLS**) (Gustafsson 2001; Sun et al. 2009) is an asymmetric or more "discriminative" form of supervised PCA. The key idea is to allow some of the (co)variance in the input features to be explained by its own subspace, $\mathbf{z}_i^x$, and to let the rest of the subspace, $\mathbf{z}_i^s$, be shared between input and output. The model has the form

$$p(\mathbf{z}_i) \;=\; \mathcal{N}(\mathbf{z}_i^s | \mathbf{0}, \mathbf{I}_{L_s}) \mathcal{N}(\mathbf{z}_i^x | \mathbf{0}, \mathbf{I}_{L_x}) \tag{12.83}$$

$$p(\mathbf{y}_i | \mathbf{z}_i) \;=\; \mathcal{N}(\mathbf{W}_y \mathbf{z}_i^s + \boldsymbol{\mu}_y, \sigma^2 \mathbf{I}_{D_y}) \tag{12.84}$$

$$p(\mathbf{x}_i | \mathbf{z}_i) \;=\; \mathcal{N}(\mathbf{W}_x \mathbf{z}_i^s + \mathbf{B}_x \mathbf{z}_i^x + \boldsymbol{\mu}_x, \sigma^2 \mathbf{I}_{D_x}) \tag{12.85}$$

See Figure 12.19(b). The corresponding induced distribution on the visible variables has the form

$$p(\mathbf{v}_i | \boldsymbol{\theta}) = \int \mathcal{N}(\mathbf{v}_i | \mathbf{W}\mathbf{z}_i + \boldsymbol{\mu}, \sigma^2 \mathbf{I}) \mathcal{N}(\mathbf{z}_i | \mathbf{0}, \mathbf{I}) d\mathbf{z}_i = \mathcal{N}(\mathbf{v}_i | \boldsymbol{\mu}, \mathbf{W}\mathbf{W}^T + \sigma^2 \mathbf{I}) \tag{12.86}$$

where $\mathbf{v}_i = (\mathbf{x}_i; \mathbf{y}_i)$, $\boldsymbol{\mu} = (\boldsymbol{\mu}_y; \boldsymbol{\mu}_x)$ and

$$\mathbf{W} \;=\; \begin{pmatrix} \mathbf{W}_y & \mathbf{0} \\ \mathbf{W}_x & \mathbf{B}_x \end{pmatrix} \tag{12.87}$$

$$\mathbf{W}\mathbf{W}^T \;=\; \begin{pmatrix} \mathbf{W}_y\mathbf{W}_y^T & \mathbf{W}_x\mathbf{W}_x^T \\ \mathbf{W}_x\mathbf{W}_x^T & \mathbf{W}_x\mathbf{W}_x^T + \mathbf{B}_x\mathbf{B}_x^T \end{pmatrix} \tag{12.88}$$

We should choose L large enough so that the shared subspace does not capture covariate-specific variation.

This model can be easily generalized to discrete data using the exponential family (Virtanen 2010).

12.5.3 Canonical correlation analysis

Canonical correlation analysis or **CCA** is like a symmetric unsupervised version of PLS: it allows each view to have its own "private" subspace, but there is also a shared subspace. If we have two observed variables, $\mathbf{x}_i$ and $\mathbf{y}_i$, then we have three latent variables, $\mathbf{z}_i^s \in \mathbb{R}^{L_s}$ which is shared, $\mathbf{z}_i^x \in \mathbb{R}^{L_x}$ and $\mathbf{z}_i^y \in \mathbb{R}^{L_y}$ which are private. We can write the model as follows (Bach and Jordan 2005):

$$p(\mathbf{z}_i) = \mathcal{N}(\mathbf{z}_i^s|\mathbf{0}, \mathbf{I}_{L_s})\mathcal{N}(\mathbf{z}_i^x|\mathbf{0}, \mathbf{I}_{L_x})\mathcal{N}(\mathbf{z}_i^y|\mathbf{0}, \mathbf{I}_{L_y}) \tag{12.89}$$

$$p(\mathbf{x}_i|\mathbf{z}_i) = \mathcal{N}(\mathbf{x}_i|\mathbf{B}_x\mathbf{z}_i^x + \mathbf{W}_x\mathbf{z}_i^s + \boldsymbol{\mu}_x, \sigma^2\mathbf{I}_{D_x}) \tag{12.90}$$

$$p(\mathbf{y}_i|\mathbf{z}_i) = \mathcal{N}(\mathbf{y}_i|\mathbf{B}_y\mathbf{z}_i^y + \mathbf{W}_y\mathbf{z}_i^s + \boldsymbol{\mu}_y, \sigma^2\mathbf{I}_{D_y}) \tag{12.91}$$

See Figure 12.19(c). The corresponding observed joint distribution has the form

$$p(\mathbf{v}_i|\boldsymbol{\theta}) = \int \mathcal{N}(\mathbf{v}_i|\mathbf{W}\mathbf{z}_i + \boldsymbol{\mu}, \sigma^2\mathbf{I})\mathcal{N}(\mathbf{z}_i|\mathbf{0}, \mathbf{I})d\mathbf{z}_i = \mathcal{N}(\mathbf{v}_i|\boldsymbol{\mu}, \mathbf{W}\mathbf{W}^T + \sigma^2\mathbf{I}_D) \tag{12.92}$$

where

$$\mathbf{W} = \begin{pmatrix} \mathbf{W}_x & \mathbf{B}_x & \mathbf{0} \\ \mathbf{W}_y & \mathbf{0} & \mathbf{B}_y \end{pmatrix} \tag{12.93}$$

$$\mathbf{W}\mathbf{W}^T = \begin{pmatrix} \mathbf{W}_x\mathbf{W}_x^T + \mathbf{B}_x\mathbf{B}_x^T & \mathbf{W}_x\mathbf{W}_y^T \\ \mathbf{W}_y\mathbf{W}_y^T & \mathbf{W}_y\mathbf{W}_y^T + \mathbf{B}_y\mathbf{B}_y^T \end{pmatrix} \tag{12.94}$$

One can compute the MLE for this model using EM. (Bach and Jordan 2005) show that the resulting MLE is equivalent (up to rotation and scaling) to the classical, non-probabilistic view. However, the advantages of the probabilistic view are many: we can trivially generalize to more than two observed variables; we can create mixtures of CCA (Viinikanoja et al. 2010); we can create sparse versions of CCA using ARD (Archambeau and Bach 2008); we can generalize to the exponential family (Klami et al. 2010); we can perform Bayesian inference of the parameters (Wang 2007; Klami and Kaski 2008); we can handle non-parametric sparsity-promoting priors for $\mathbf{W}$ and $\mathbf{B}$ (Rai and Daume 2009); and so on.

12.6 Independent Component Analysis (ICA)

Consider the following situation. You are in a crowded room and many people are speaking. Your ears essentially act as two microphones, which are listening to a linear combination of the different speech signals in the room. Your goal is to deconvolve the mixed signals into their constituent parts. This is known as the **cocktail party problem**, and is an example of **blind signal separation** (BSS), or **blind source separation**, where "blind" means we know "nothing" about the source of the signals. Besides the obvious applications to acoustic signal processing, this problem also arises when analysing EEG and MEG signals, financial data, and any other dataset (not necessarily temporal) where latent sources or factors get mixed together in a linear way.

We can formalize the problem as follows. Let $\mathbf{x}_t \in \mathbb{R}^D$ be the observed signal at the sensors at "time" t, and $\mathbf{z}_t \in \mathbb{R}^L$ be the vector of source signals. We assume that

$$\mathbf{x}_t = \mathbf{W}\mathbf{z}_t + \boldsymbol{\epsilon}_t \tag{12.95}$$

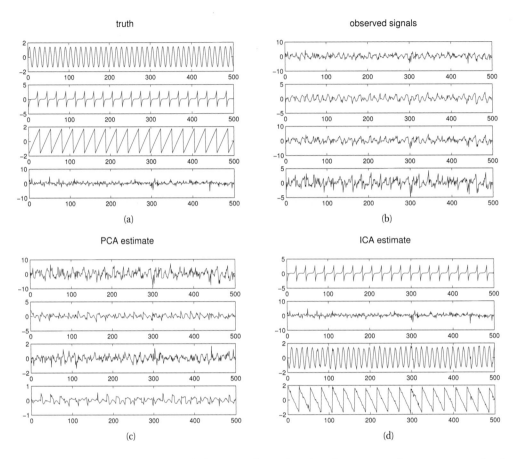

Figure 12.20   Illustration of ICA applied to 500 iid samples of a 4d source signal. (a) Latent signals. (b) Observations. (c) PCA estimate. (d) ICA estimate. Figure generated by icaDemo, written by Aapo Hyvarinen.

where $\mathbf{W}$ is an $D \times L$ matrix, and $\boldsymbol{\epsilon}_t \sim \mathcal{N}(\mathbf{0}, \boldsymbol{\Psi})$. In this section, we treat each time point as an independent observation, i.e., we do not model temporal correlation (so we could replace the t index with i, but we stick with t to be consistent with much of the ICA literature). The goal is to infer the source signals, $p(\mathbf{z}_t | \mathbf{x}_t, \boldsymbol{\theta})$, as illustrated in Figure 12.20. In this context, $\mathbf{W}$ is called the **mixing matrix**. If $L = D$ (number of sources = number of sensors), it will be a square matrix. Often we will assume the noise level, $|\boldsymbol{\Psi}|$, is zero, for simplicity.

So far, the model is identical to factor analysis (or PCA if there is no noise, except we don't in general require orthogonality of $\mathbf{W}$). However, we will use a different prior for $p(\mathbf{z}_t)$. In PCA, we assume each source is independent, and has a Gaussian distribution

$$p(\mathbf{z}_t) = \prod_{j=1}^{L} \mathcal{N}(z_{tj} | 0, 1) \tag{12.96}$$

We will now relax this Gaussian assumption and let the source distributions be any *non-Gaussian*

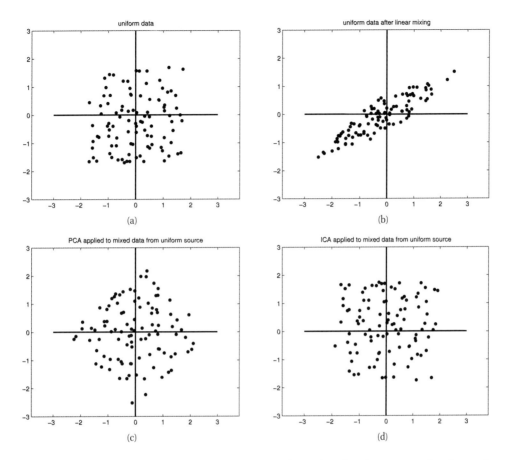

Figure 12.21 Illustration of ICA and PCA applied to 100 iid samples of a 2d source signal with a uniform distribution. (a) Latent signals. (b) Observations. (c) PCA estimate. (d) ICA estimate. Figure generated by `icaDemoUniform`, written by Aapo Hyvarinen.

distribution

$$p(\mathbf{z}_t) = \prod_{j=1}^{L} p_j(z_{tj}) \tag{12.97}$$

Without loss of generality, we can constrain the variance of the source distributions to be 1, because any other variance can be modelled by scaling the rows of $\mathbf{W}$ appropriately. The resulting model is known as **independent component analysis** or **ICA**.

The reason the Gaussian distribution is disallowed as a source prior in ICA is that it does not permit unique recovery of the sources, as illustrated in Figure 12.20(c). This is because the PCA likelihood is invariant to any orthogonal transformation of the sources $\mathbf{z}_t$ and mixing matrix $\mathbf{W}$. PCA can recover the best linear subspace in which the signals lie, but cannot uniquely recover the signals themselves.

To illustrate this, suppose we have two independent sources with uniform distributions, as shown in Figure 12.21(a). Now suppose we have the following mixing matrix

$$\mathbf{W} = \begin{pmatrix} 2 & 3 \\ 2 & 1 \end{pmatrix} \qquad (12.98)$$

Then we observe the data shown in Figure 12.21(b) (assuming no noise). If we apply PCA followed by scaling to this, we get the result in Figure 12.21(c). This corresponds to a whitening of the data (see Exercise 4.8). To uniquely recover the sources, we need to perform an additional rotation. The trouble is, there is no information in the symmetric Gaussian posterior to tell us which angle to rotate by. In a sense, PCA solves "half" of the problem, since it identifies the linear subspace; all that ICA has to do is then to identify the appropriate rotation. (Hence we see that ICA is not that different from methods such as varimax, which seek good rotations of the latent factors to enhance interpretability.)

Figure 12.21(d) shows that ICA can recover the source, up to a permutation of the indices and possible sign change. ICA requires that $\mathbf{W}$ is square and hence invertible. In the non-square case (e.g., where we have more sources than sensors), we cannot uniquely recover the true signal, but we can compute the posterior $p(\mathbf{z}_t | \mathbf{x}_t, \hat{\mathbf{W}})$, which represents our beliefs about the source. In both cases, we need to estimate $\mathbf{W}$ as well as the source distributions p_j. We discuss how to do this below.

12.6.1 Maximum likelihood estimation

In this section, we discuss ways to estimate square mixing matrices $\mathbf{W}$ for the noise-free ICA model. As usual, we will assume that the observations have been centered; hence we can also assume $\mathbf{z}$ is zero-mean. In addition, we assume the observations have been whitened, which can be done with PCA.

If the data is centered and whitened, we have $\mathbb{E}\left[\mathbf{x}\mathbf{x}^T\right] = \mathbf{I}$. But in the noise free case, we also have

$$\text{cov}\left[\mathbf{x}\right] = \mathbb{E}\left[\mathbf{x}\mathbf{x}^T\right] = \mathbf{W}\mathbb{E}\left[\mathbf{z}\mathbf{z}^T\right]\mathbf{W}^T = \mathbf{W}\mathbf{W}^T \qquad (12.99)$$

Hence we see that $\mathbf{W}$ must be orthogonal. This reduces the number of parameters we have to estimate from D^2 to $D(D-1)/2$. It will also simplify the math and the algorithms.

Let $\mathbf{V} = \mathbf{W}^{-1}$; these are often called the **recognition weights**, as opposed to $\mathbf{W}$, which are the **generative weights**.[8]

Since $\mathbf{x} = \mathbf{W}\mathbf{z}$, we have, from Equation 2.89,

$$p_x(\mathbf{W}\mathbf{z}_t) = p_z(\mathbf{z}_t)|\det(\mathbf{W}^{-1})| = p_z(\mathbf{V}\mathbf{x}_t)|\det(\mathbf{V})| \qquad (12.100)$$

Hence we can write the log-likelihood, assuming T iid samples, as follows:

$$\frac{1}{T}\log p(\mathcal{D}|\mathbf{V}) = \log|\det(\mathbf{V})| + \frac{1}{T}\sum_{j=1}^{L}\sum_{t=1}^{T}\log p_j(\mathbf{v}_j^T\mathbf{x}_t) \qquad (12.101)$$

---

8. In the literature, it is common to denote the generative weights by $\mathbf{A}$ and the recognition weights by $\mathbf{W}$, but we are trying to be consistent with the notation used earlier in this chapter.

where $\mathbf{v}_j$ is the j'th row of $\mathbf{V}$. Since we are constraining $\mathbf{V}$ to be orthogonal, the first term is a constant, so we can drop it. We can also replace the average over the data with an expectation operator to get the following objective

$$\text{NLL}(\mathbf{V}) = \sum_{j=1}^{L} \mathbb{E}\left[G_j(z_j)\right] \tag{12.102}$$

where $z_j = \mathbf{v}_j^T \mathbf{x}$ and $G_j(z) \triangleq -\log p_j(z)$. We want to minimize this subject to the constraint that the rows of $\mathbf{V}$ are orthogonal. We also want them to be unit norm, since this ensures that the variance of the factors is unity (since, with whitened data, $\mathbb{E}\left[\mathbf{v}_j^T \mathbf{x}\right] = ||\mathbf{v}_j||^2$), which is necessary to fix the scale of the weights. In otherwords, $\mathbf{V}$ should be an orthonormal matrix.

It is straightforward to derive a gradient descent algorithm to fit this model; however, it is rather slow. One can also derive a faster algorithm that follows the natural gradient; see e.g., (MacKay 2003, ch 34) for details. A popular alternative is to use an approximate Newton method, which we discuss in Section 12.6.2. Another approach is to use EM, which we discuss in Section 12.6.3.

12.6.2 The FastICA algorithm

We now describe the **fast ICA** algorithm, based on (Hyvarinen and Oja 2000), which we will show is an approximate Newton method for fitting ICA models.

For simplicity of presentation, we initially assume there is only one latent factor. In addition, we initially assume all source distributions are known and are the same, so we can just write $G(z) = -\log p(z)$. Let $g(z) = \frac{d}{dz}G(z)$. The constrained objective, and its gradient and Hessian, are given by

$$f(\mathbf{v}) = \mathbb{E}\left[G(\mathbf{v}^T \mathbf{x})\right] + \lambda(1 - \mathbf{v}^T \mathbf{v}) \tag{12.103}$$

$$\nabla f(\mathbf{v}) = \mathbb{E}\left[\mathbf{x}g(\mathbf{v}^T \mathbf{x})\right] - \beta \mathbf{v} \tag{12.104}$$

$$\mathbf{H}(\mathbf{v}) = \mathbb{E}\left[\mathbf{x}\mathbf{x}^T g'(\mathbf{v}^T \mathbf{x})\right] - \beta \mathbf{I} \tag{12.105}$$

where $\beta = 2\lambda$ is a Lagrange multiplier. Let us make the approximation

$$\mathbb{E}\left[\mathbf{x}\mathbf{x}^T g'(\mathbf{v}^T \mathbf{x})\right] \approx \mathbb{E}\left[\mathbf{x}\mathbf{x}^T\right] \mathbb{E}\left[g'(\mathbf{v}^T \mathbf{x})\right] = \mathbb{E}\left[g'(\mathbf{v}^T \mathbf{x})\right] \tag{12.106}$$

This makes the Hessian very easy to invert, giving rise to the following Newton update:

$$\mathbf{v}^* \triangleq \mathbf{v} - \frac{\mathbb{E}\left[\mathbf{x}g(\mathbf{v}^T \mathbf{x})\right] - \beta \mathbf{v}}{\mathbb{E}\left[g'(\mathbf{v}^T \mathbf{x})\right] - \beta} \tag{12.107}$$

One can rewrite this in the following way

$$\mathbf{v}^* \triangleq \mathbb{E}\left[\mathbf{x}g(\mathbf{v}^T \mathbf{x})\right] - \mathbb{E}\left[g'(\mathbf{v}^T \mathbf{x})\right] \mathbf{v} \tag{12.108}$$

(In practice, the expectations can be replaced by Monte Carlo estimates from the training set, which gives an efficient online learning algorithm.) After performing this update, one should project back onto the constraint surface using

$$\mathbf{v}^{new} \triangleq \frac{\mathbf{v}^*}{||\mathbf{v}^*||} \tag{12.109}$$

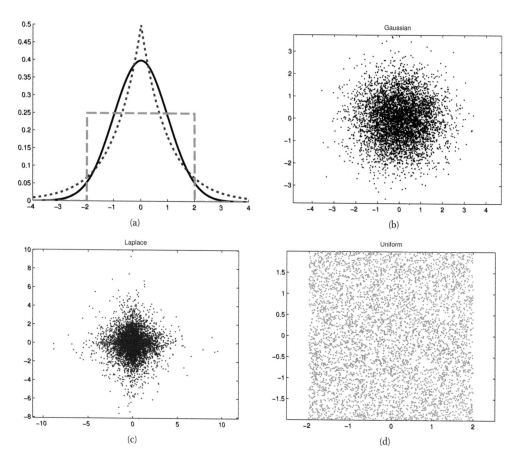

Figure 12.22 Illustration of Gaussian, sub-Gaussian (uniform) and super-Gaussian (Laplace) distributions in 1d and 2d. Figure generated by `subSuperGaussPlot`, written by Kevin Swersky.

One iterates this algorithm until convergence. (Due to the sign ambiguity of $\mathbf{v}$, the values of $\mathbf{v}$ may not converge, but the direction defined by this vector should converge, so one can assess convergence by monitoring $|\mathbf{v}^T\mathbf{v}^{new}|$, which should approach 1.)

Since the objective is not convex, there are multiple local optima. We can use this fact to learn multiple different weight vectors or latent features. We can either learn the features sequentially and then project out the part of $\mathbf{v}_j$ that lies in the subspace defined by earlier features, or we can learn them in parallel, and orthogonalize $\mathbf{V}$ in parallel. This latter approach is usually preferred, since, unlike PCA, the features are not ordered in any way. So the first feature is not "more important" than the second, and hence it is better to treat them symmetrically.

12.6.2.1 Modeling the source densities

So far, we have assumed that $G(z) = -\log p(z)$ is known. What kinds of models might be reasonable as signal priors? We know that using Gaussians (which correspond to quadratic functions for G) won't work. So we want some kind of non-Gaussian distribution. In general, there are several kinds of non-Gaussian distributions, such as the following:

- **Super-Gaussian distributions** These are distributions which have a big spike at the mean, and hence (in order to ensure unit variance) have heavy tails. The Laplace distribution is a classic example. See Figure 12.22. Formally, we say a distribution is **super-Gaussian** or **leptokurtic** ("lepto" coming from the Greek for "thin") if $\text{kurt}(z) > 0$, where $\text{kurt}(z)$ is the **kurtosis** of the distribution, defined by

$$\text{kurt}(z) \triangleq \frac{\mu_4}{\sigma^4} - 3 \tag{12.110}$$

where σ is the standard deviation, and μ_k is the k'th **central moment**, or moment about the mean:

$$\mu_k \triangleq \mathbb{E}\left[(X - \mathbb{E}[X])^k\right] \tag{12.111}$$

(So $\mu_1 = \mu$ is the mean, and $\mu_2 = \sigma^2$ is the variance.) It is conventional to subtract 3 in the definition of kurtosis to make the kurtosis of a Gaussian variable equal to zero.
- **Sub-Gaussian distributions** A **sub-Gaussian** or **platykurtic** ("platy" coming from the Greek for "broad") distribution has negative kurtosis. These are distributions which are much flatter than a Gaussian. The uniform distribution is a classic example. See Figure 12.22.
- **Skewed distributions** Another way to "be non-Gaussian" is to be asymmetric. One measure of this is **skewness**, defined by

$$\text{skew}(z) \triangleq \frac{\mu_3}{\sigma^3} \tag{12.112}$$

An example of a (right) skewed distribution is the gamma distribution (see Figure 2.9).

When one looks at the empirical distribution of many natural signals, such as images and speech, when passed through certain linear filters, they tend to be very super-Gaussian. This result holds both for the kind of linear filters found in certain parts of the brain, such as the simple cells found in the primary visual cortex, as well as for the kinds of linear filters used in signal processing, such as wavelet transforms. One obvious choice for modeling natural signals with ICA is therefore the Laplace distribution. For mean zero and variance 1, this has a log pdf given by

$$\log p(z) = -\sqrt{2}|z| - \log(\sqrt{2}) \tag{12.113}$$

Since the Laplace prior is not differentiable at the origin, it is more common to use other, smoother super-Gaussian distributions. One example is the logistic distribution. The corresponding log pdf, for the case where the mean is zero and the variance is 1 (so $\mu = 0$ and $s = \frac{\sqrt{3}}{\pi}$), is given by the following:

$$\log p(z) = -2\log\cosh(\frac{\pi}{2\sqrt{3}}z) - \log\frac{4\sqrt{3}}{\pi} \tag{12.114}$$

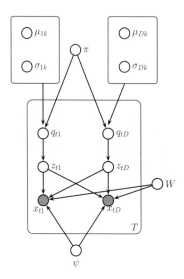

Figure 12.23 Modeling the source distributions using a mixture of univariate Gaussians (the independent factor analysis model of (Moulines et al. 1997; Attias 1999)).

Various ways of estimating $G(Z) = -\log p(z)$ are discussed in the seminal paper (Pham and Garrat 1997). However, when fitting ICA by maximum likelihood, it is not critical that the exact shape of the source distribution be known (although it is important to know whether it is sub or super Gaussian). Consequently, it is common to just use $G(z) = \sqrt{z}$ or $G(z) = \log \cosh(z)$ instead of the more complex expressions above.

12.6.3 Using EM

An alternative to assuming a particular form for $G(z)$, or equivalently for $p(z)$, is to use a flexible non-parametric density estimator, such as a mixture of (uni-variate) Gaussians:

$$p(q_j = k) = \pi_k \tag{12.115}$$
$$p(z_j | q_j = k) = \mathcal{N}(\mu_{j,k}, \sigma_{j,k}^2) \tag{12.116}$$
$$p(\mathbf{x}|\mathbf{z}) = \mathcal{N}(\mathbf{W}\mathbf{z}, \mathbf{\Psi}) \tag{12.117}$$

This approach was proposed in (Moulines et al. 1997; Attias 1999), and the corresponding graphical model is shown in Figure 12.23.

It is possible to derive an exact EM algorithm for this model. The key observation is that it is possible to compute $\mathbb{E}[\mathbf{z}_t | \mathbf{x}_t, \boldsymbol{\theta}]$ exactly by summing over all K^L combinations of the $\mathbf{q}_t$ variables, where K is the number of mixture components per source. (If this is too expensive, one can use a variational mean field approximation (Attias 1999).) We can then estimate all the source distributions in parallel by fitting a standard GMM to $\mathbb{E}[\mathbf{z}_t]$. When the source GMMs are known, we can compute the marginals $p_j(z_j)$ very easily, using

$$p_j(z_j) = \sum_{k=1}^{K} \pi_{j,k} \mathcal{N}(z_j | \mu_{j,k}, \sigma_{j,k}^2) \tag{12.118}$$

Given the p_j's, we can then use an ICA algorithm to estimate $\mathbf{W}$. Of course, these steps should be interleaved. The details can be found in (Attias 1999).

12.6.4 Other estimation principles *

It is quite common to estimate the parameters of ICA models using methods that seem different to maximum likelihood. We will review some of these methods below, because they give additional insight into ICA. However, we will also see that these methods in fact are equivalent to maximum likelihood after all. Our presentation is based on (Hyvarinen and Oja 2000).

12.6.4.1 Maximizing non-Gaussianity

An early approach to ICA was to find a matrix $\mathbf{V}$ such that the distribution $\mathbf{z} = \mathbf{V}\mathbf{x}$ is as far from Gaussian as possible. (There is a related approach in statistics called **projection pursuit**.) One measure of non-Gaussianity is kurtosis, but this can be sensitive to outliers. Another measure is the **negentropy**, defined as

$$\text{negentropy}(z) \triangleq \mathbb{H}\left(\mathcal{N}(\mu, \sigma^2)\right) - \mathbb{H}(z) \tag{12.119}$$

where $\mu = \mathbb{E}[z]$ and $\sigma^2 = \text{var}[z]$. Since the Gaussian is the maximum entropy distribution, this measure is always non-negative and becomes large for distributions that are highly non-Gaussian.

We can define our objective as maximizing

$$J(\mathbf{V}) = \sum_j \text{negentropy}(z_j) = \sum_j \mathbb{H}\left(\mathcal{N}(\mu_j, \sigma_j^2)\right) - \mathbb{H}(z_j) \tag{12.120}$$

where $\mathbf{z} = \mathbf{V}\mathbf{x}$. If we fix $\mathbf{V}$ to be orthogonal, and if we whiten the data, the covariance of $\mathbf{z}$ will be $\mathbf{I}$ independently of $\mathbf{V}$, so the first term is a constant. Hence

$$J(\mathbf{V}) = \sum_j -\mathbb{H}(z_j) + \text{const} = \sum_j \mathbb{E}[\log p(z_j)] + \text{const} \tag{12.121}$$

which we see is equal (up to a sign change, and irrelevant constants) to the log-likelihood in Equation 12.102.

12.6.4.2 Minimizing mutual information

One measure of dependence of a set of random variables is the **multi-information**, which is an extension of mutual information defined as follows:

$$I(\mathbf{z}) \triangleq \mathbb{KL}\left(p(\mathbf{z}) || \prod_j p(z_j)\right) = \sum_j \mathbb{H}(z_j) - \mathbb{H}(\mathbf{z}) \tag{12.122}$$

We would like to minimize this, since we are trying to find independent components. Put another way, we want the best possible factored approximation to the joint distribution.

Now since $\mathbf{z} = \mathbf{V}\mathbf{x}$, we have

$$I(\mathbf{z}) = \sum_j \mathbb{H}(z_j) - \mathbb{H}(\mathbf{V}\mathbf{x}) \tag{12.123}$$

If we constrain $\mathbf{V}$ to be orthogonal, we can drop the last term, since then $\mathbb{H}(\mathbf{V}\mathbf{x}) = \mathbb{H}(\mathbf{x})$ (since multiplying by $\mathbf{V}$ does not change the shape of the distribution), and $\mathbb{H}(\mathbf{x})$ is a constant which is is solely determined by the empirical distribution. Hence we have $I(\mathbf{z}) = \sum_j \mathbb{H}(z_j)$. Minimizing this is equivalent to maximizing the negentropy, which is equivalent to maximum likelihood.

12.6.4.3   Maximizing mutual information (infomax)

Instead of trying to minimize the mutual information between the components of $\mathbf{z}$, let us imagine a neural network where $\mathbf{x}$ is the input and $y_j = \phi(\mathbf{v}_j^T\mathbf{x}) + \epsilon$ is the noisy output, where ϕ is some nonlinear scalar function, and $\epsilon \sim \mathcal{N}(0,1)$. It seems reasonable to try to maximize the information flow through this system, a principle known as **infomax**. (Bell and Sejnowski 1995). That is, we want to maximize the mutual information between $\mathbf{y}$ (the internal neural representation) and $\mathbf{x}$ (the observed input signal). We have $\mathbb{I}(\mathbf{x}; \mathbf{y}) = \mathbb{H}(\mathbf{y}) - \mathbb{H}(\mathbf{y}|\mathbf{x})$, where the latter term is constant if we assume the noise has constant variance. One can show that we can approximate the former term as follows

$$\mathbb{H}(\mathbf{y}) = \sum_{j=1}^{L} \mathbb{E}\left[\log \phi'(\mathbf{v}_j^T\mathbf{x})\right] + \log|\det(\mathbf{V})| \tag{12.124}$$

where, as usual, we can drop the last term if $\mathbf{V}$ is orthogonal. If we define $\phi(z)$ to be a cdf, then $\phi'(z)$ is its pdf, and the above expression is equivalent to the log likelihood. In particular, if we use a logistic nonlinearity, $\phi(z) = \text{sigm}(z)$, then the corresponding pdf is the logistic distribution, and $\log \phi'(z) = \log \cosh(z)$ (ignoring irrelevant constants). Thus we see that infomax is equivalent to maximum likelihood.

Exercises

Exercise 12.1 M step for FA

For the FA model, show that the MLE in the M step for $\mathbf{W}$ is given by Equation 12.23.

Exercise 12.2 MAP estimation for the FA model

Derive the M step for the FA model using conjugate priors for the parameters.

Exercise 12.3 Heuristic for assessing applicability of PCA

(Source: (Press 2005, Q9.8).). Let the empirical covariance matrix Σ have eigenvalues $\lambda_1 \geq \lambda_2 \geq \cdots \geq \lambda_d > 0$. Explain why the variance of the evalues, $\sigma^2 = \frac{1}{d}\sum_{i=1}^{d}(\lambda_i - \bar{\lambda})^2$ is a good measure of whether or not PCA would be useful for analysing the data (the higher the value of σ^2 the more useful PCA).

Exercise 12.4 Deriving the second principal component

a. Let

$$J(\mathbf{v}_2, \mathbf{z}_2) = \frac{1}{n} \sum_{i=1}^{n} (\mathbf{x}_i - z_{i1}\mathbf{v}_1 - z_{i2}\mathbf{v}_2)^T (\mathbf{x}_i - z_{i1}\mathbf{v}_1 - z_{i2}\mathbf{v}_2) \tag{12.125}$$

Show that $\frac{\partial J}{\partial \mathbf{z}_2} = 0$ yields $z_{i2} = \mathbf{v}_2^T \mathbf{x}_i$.

b. Show that the value of $\mathbf{v}_2$ that minimizes

$$\tilde{J}(\mathbf{v}_2) = -\mathbf{v}_2^T \mathbf{C} \mathbf{v}_2 + \lambda_2 (\mathbf{v}_2^T \mathbf{v}_2 - 1) + \lambda_{12}(\mathbf{v}_2^T \mathbf{v}_1 - 0) \tag{12.126}$$

is given by the eigenvector of $\mathbf{C}$ with the second largest eigenvalue. Hint: recall that $\mathbf{C}\mathbf{v}_1 = \lambda_1 \mathbf{v}_1$ and $\frac{\partial \mathbf{x}^T \mathbf{A}\mathbf{x}}{\partial \mathbf{x}} = (\mathbf{A} + \mathbf{A}^T)\mathbf{x}$.

Exercise 12.5 Deriving the residual error for PCA

a. Prove that

$$||\mathbf{x}_i - \sum_{j=1}^{K} z_{ij}\mathbf{v}_j||^2 = \mathbf{x}_i^T \mathbf{x}_i - \sum_{j=1}^{K} \mathbf{v}_j^T \mathbf{x}_i \mathbf{x}_i^T \mathbf{v}_j \tag{12.127}$$

Hint: first consider the case $K = 2$. Use the fact that $\mathbf{v}_j^T \mathbf{v}_j = 1$ and $\mathbf{v}_j^T \mathbf{v}_k = 0$ for $k \neq j$. Also, recall $z_{ij} = \mathbf{x}_i^T \mathbf{v}_j$.

b. Now show that

$$J_K \triangleq \frac{1}{n} \sum_{i=1}^{n} \left(\mathbf{x}_i^T \mathbf{x}_i - \sum_{j=1}^{K} \mathbf{v}_j^T \mathbf{x}_i \mathbf{x}_i^T \mathbf{v}_j \right) = \frac{1}{n} \sum_{i=1}^{n} \mathbf{x}_i^T \mathbf{x}_i - \sum_{j=1}^{K} \lambda_j \tag{12.128}$$

Hint: recall $\mathbf{v}_j^T \mathbf{C} \mathbf{v}_j = \lambda_j \mathbf{v}_j^T \mathbf{v}_j = \lambda_j$.

c. If $K = d$ there is no truncation, so $J_d = 0$. Use this to show that the error from only using $K < d$ terms is given by

$$J_K = \sum_{j=K+1}^{d} \lambda_j \tag{12.129}$$

Hint: partition the sum $\sum_{j=1}^{d} \lambda_j$ into $\sum_{j=1}^{K} \lambda_j$ and $\sum_{j=K+1}^{d} \lambda_j$.

Exercise 12.6 Derivation of Fisher's linear discriminant

Show that the maximum of $J(\mathbf{w}) = \frac{\mathbf{w}^T \mathbf{S}_B \mathbf{w}}{\mathbf{w}^T \mathbf{S}_W \mathbf{w}}$ is given by $\mathbf{S}_B \mathbf{w} = \lambda \mathbf{S}_W \mathbf{w}$

where $\lambda = \frac{\mathbf{w}^T \mathbf{S}_B \mathbf{w}}{\mathbf{w}^T \mathbf{S}_W \mathbf{w}}$. Hint: recall that the derivative of a ratio of two scalars is given by $\frac{d}{dx} \frac{f(x)}{g(x)} = \frac{f'g - fg'}{g^2}$, where $f' = \frac{d}{dx} f(x)$ and $g' = \frac{d}{dx} g(x)$. Also, recall that $\frac{d}{d\mathbf{x}} \mathbf{x}^T \mathbf{A}\mathbf{x} = (\mathbf{A} + \mathbf{A}^T)\mathbf{x}$.

Exercise 12.7 PCA via successive deflation

Let $\mathbf{v}_1, \mathbf{v}_2, \ldots, \mathbf{v}_k$ be the first k eigenvectors with largest eigenvalues of $\mathbf{C} = \frac{1}{n}\mathbf{X}^T\mathbf{X}$, i.e., the principal basis vectors. These satisfy

$$\mathbf{v}_j^T \mathbf{v}_k = \begin{cases} 0 & \text{if } j \neq k \\ 1 & \text{if } j = k \end{cases} \tag{12.130}$$

We will construct a method for finding the $\mathbf{v}_j$ sequentially.

As we showed in class, $\mathbf{v}_1$ is the first principal eigenvector of $\mathbf{C}$, and satisfies $\mathbf{C}\mathbf{v}_1 = \lambda_1\mathbf{v}_1$. Now define $\tilde{\mathbf{x}}_i$ as the orthogonal projection of $\mathbf{x}_i$ onto the space orthogonal to $\mathbf{v}_1$:

$$\tilde{\mathbf{x}}_i = \mathbf{P}_{\perp\mathbf{v}_1}\,\mathbf{x}_i = (\mathbf{I} - \mathbf{v}_1\mathbf{v}_1^T)\mathbf{x}_i \tag{12.131}$$

Define $\tilde{\mathbf{X}} = [\tilde{\mathbf{x}}_1; ...; \tilde{\mathbf{x}}_n]$ as the **deflated matrix** of rank $d - 1$, which is obtained by removing from the d dimensional data the component that lies in the direction of the first principal direction:

$$\tilde{\mathbf{X}} = (\mathbf{I} - \mathbf{v}_1\mathbf{v}_1^T)^T\mathbf{X} = (\mathbf{I} - \mathbf{v}_1\mathbf{v}_1^T)\mathbf{X} \tag{12.132}$$

a. Using the facts that $\mathbf{X}^T\mathbf{X}\mathbf{v}_1 = n\lambda_1\mathbf{v}_1$ (and hence $\mathbf{v}_1^T\mathbf{X}^T\mathbf{X} = n\lambda_1\mathbf{v}_1^T$) and $\mathbf{v}_1^T\mathbf{v}_1 = 1$, show that the covariance of the deflated matrix is given by

$$\tilde{\mathbf{C}} \triangleq \frac{1}{n}\tilde{\mathbf{X}}^T\tilde{\mathbf{X}} = \frac{1}{n}\mathbf{X}^T\mathbf{X} - \lambda_1\mathbf{v}_1\mathbf{v}_1^T \tag{12.133}$$

b. Let $\mathbf{u}$ be the principal eigenvector of $\tilde{\mathbf{C}}$. Explain why $\mathbf{u} = \mathbf{v}_2$. (You may assume $\mathbf{u}$ is unit norm.)

c. Suppose we have a simple method for finding the leading eigenvector and eigenvalue of a pd matrix, denoted by $[\lambda, \mathbf{u}] = f(\mathbf{C})$. Write some pseudo code for finding the first K principal basis vectors of $\mathbf{X}$ that only uses the special f function and simple vector arithmetic, i.e., your code should not use SVD or the `eig` function. Hint: this should be a simple iterative routine that takes 2–3 lines to write. The input is $\mathbf{C}$, K and the function f, the output should be $\mathbf{v}_j$ and λ_j for $j = 1 : K$. Do not worry about being syntactically correct.

Exercise 12.8 Latent semantic indexing

(Source: de Freitas.). In this exercise, we study a technique called **latent semantic indexing**, which applies SVD to a document by term matrix, to create a low-dimensional embedding of the data that is designed to capture semantic similarity of words.

The file `lsiDocuments.pdf` contains 9 documents on various topics. A list of all the 460 unique words/terms that occur in these documents is in `lsiWords.txt`. A document by term matrix is in `lsiMatrix.txt`.

a. Let X be the transpose of `lsiMatrix`, so each column represents a document. Compute the SVD of X and make an approximation to it $\hat{X}$ using the first 2 singular values/ vectors. Plot the low dimensional representation of the 9 documents in 2D. You should get something like Figure 12.24.

b. Consider finding documents that are about alien abductions. If If you look at `lsiWords.txt`, there are 3 versions of this word, term 23 ("abducted"), term 24 ("abduction") and term 25 ("abductions"). Suppose we want to find documents containing the word "abducted". Documents 2 and 3 contain it, but document 1 does not. However, document 1 is clearly related to this topic. Thus LSI should also find document 1. Create a test document q containing the one word "abducted", and project it into the 2D subspace to make $\hat{q}$. Now compute the cosine similarity between $\hat{q}$ and the low dimensional representation of all the documents. What are the top 3 closest matches?

Exercise 12.9 Imputation in a FA model

Derive an expression for $p(\mathbf{x}_h|\mathbf{x}_v, \boldsymbol{\theta})$ for a FA model.

Exercise 12.10 Efficiently evaluating the PPCA density

Derive an expression for $p(\mathbf{x}|\hat{\mathbf{W}}, \hat{\sigma}^2)$ for the PPCA model based on plugging in the MLEs and using the matrix inversion lemma.

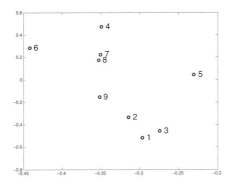

Figure 12.24 Projection of 9 documents into 2 dimensions. Figure generated by lsiCode.

Exercise 12.11 PPCA vs FA

(Source: Exercise 14.15 of (Hastie et al. 2009), due to Hinton.). Generate 200 observations from the following model, where $\mathbf{z}_i \sim \mathcal{N}(\mathbf{0}, \mathbf{I})$: $x_{i1} = z_{i1}$, $x_{i2} = z_{i1} + 0.001z_{i2}$, $x_{i3} = 10z_{i3}$. Fit a FA and PCA model with 1 latent factor. Hence show that the corresponding weight vector $\mathbf{w}$ aligns with the maximal variance direction (dimension 3) in the PCA case, but with the maximal correlation direction (dimensions 1+2) in the case of FA.

13 *Sparse linear models*

13.1 Introduction

We introduced the topic of feature selection in Section 3.5.4, where we discussed methods for finding input variables which had high mutual information with the output. The trouble with this approach is that it is based on a myopic strategy that only looks at one variable at a time. This can fail if there are interaction effects. For example, if $y = \text{xor}(x_1, x_2)$, then neither x_1 nor x_2 on its own can predict the response, but together they perfectly predict the response. For a real-world example of this, consider genetic association studies: sometimes two genes on their own may be harmless, but when present together they cause a recessive disease (Balding 2006).

In this chapter, we focus on selecting sets of variables at a time using a model-based approach. If the model is a generalized linear model, of the form $p(y|\mathbf{x}) = p(y|f(\mathbf{w}^T\mathbf{x}))$ for some link function f, then we can perform feature selection by encouraging the weight vector $\mathbf{w}$ to be **sparse**, i.e., to have lots of zeros. This approach turns out to offer significant computational advantages, as we will see below.

Here are some applications where feature selection/ sparsity is useful:

- In many problems, we have many more dimensions D than training cases N. The corresponding design matrix is short and fat, rather than tall and skinny. This is called the **small N, large D** problem. This is becoming increasingly prevalent as we develop more high throughput measurement devices, For example, with gene microarrays, it is common to measure the expression levels of $D \sim 10,000$ genes, but to only get $N \sim 100$ such examples. (It is perhaps a sign of the times that even our data seems to be getting fatter...) We may want to find the smallest set of features that can accurately predict the response (e.g., growth rate of the cell) in order to prevent overfitting, to reduce the cost of building a diagnostic device, or to help with scientific insight into the problem.

- In Chapter 14, we will use basis functions centered on the training examples, so $\phi(\mathbf{x}) = [\kappa(\mathbf{x}, \mathbf{x}_1), \ldots, \kappa(\mathbf{x}, \mathbf{x}_N)]$, where κ is a kernel function. The resulting design matrix has size $N \times N$. Feature selection in this context is equivalent to selecting a subset of the training examples, which can help reduce overfitting and computational cost. This is known as a sparse kernel machine.

- In signal processing, it is common to represent signals (images, speech, etc.) in terms of wavelet basis functions. To save time and space, it is useful to find a sparse representation of the signals, in terms of a small number of such basis functions. This allows us to estimate signals from a small number of measurements, as well as to compress the signal. See Section 13.8.3 for more information.

Note that the topic of feature selection and sparsity is currently one of the most active areas of machine learning/ statistics. In this chapter, we only have space to give an overview of the main results.

13.2   Bayesian variable selection

A natural way to pose the variable selection problem is as follows. Let $\gamma_j = 1$ if feature j is "relevant", and let $\gamma_j = 0$ otherwise. Our goal is to compute the posterior over models

$$p(\gamma|\mathcal{D}) = \frac{e^{-f(\gamma)}}{\sum_{\gamma'} e^{-f(\gamma')}} \tag{13.1}$$

where $f(\gamma)$ is the cost function:

$$f(\gamma) \triangleq -[\log p(\mathcal{D}|\gamma) + \log p(\gamma)] \tag{13.2}$$

For example, suppose we generate $N = 20$ samples from a $D = 10$ dimensional linear regression model, $y_i \sim \mathcal{N}(\mathbf{w}^T\mathbf{x}_i, \sigma^2)$, in which $K = 5$ elements of $\mathbf{w}$ are non-zero. In particular, we use $\mathbf{w} = (0.00, -1.67, 0.13, 0.00, 0.00, 1.19, 0.00, -0.04, 0.33, 0.00)$ and $\sigma^2 = 1$. We enumerate all $2^{10} = 1024$ models and compute $p(\gamma|\mathcal{D})$ for each one (we give the equations for this below). We order the models in **Gray code** order, which ensures consecutive vectors differ by exactly 1 bit (the reasons for this are computational, and are discussed in Section 13.2.3).

The resulting set of bit patterns is shown in Figure 13.1(a). The cost of each model, $f(\gamma)$, is shown in Figure 13.1(b). We see that this objective function is extremely "bumpy". The results are easier to interpret if we compute the posterior distribution over models, $p(\gamma|\mathcal{D})$. This is shown in Figure 13.1(c). The top 8 models are listed below:

model	prob	members
4	0.447	2,
61	0.241	2, 6,
452	0.103	2, 6, 9,
60	0.091	2, 3, 6,
29	0.041	2, 5,
68	0.021	2, 6, 7,
36	0.015	2, 5, 6,
5	0.010	2, 3,

The "true" model is $\{2, 3, 6, 8, 9\}$. However, the coefficients associated with features 3 and 8 are very small (relative to σ^2). so these variables are harder to detect. Given enough data, the method will converge on the true model (assuming the data is generated from a linear model), but for finite data sets, there will usually be considerable posterior uncertainty.

Interpreting the posterior over a large number of models is quite difficult, so we will seek various summary statistics. A natural one is the posterior mode, or MAP estimate

$$\hat{\gamma} = \operatorname{argmax} p(\gamma|\mathcal{D}) = \operatorname{argmin} f(\gamma) \tag{13.3}$$

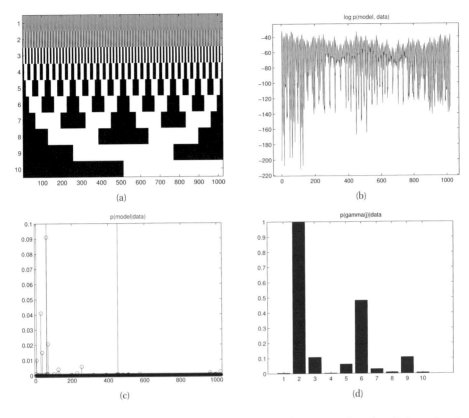

Figure 13.1 (a) All possible bit vectors of length 10 enumerated in Gray code order. (b) Score function for all possible models. (c) Posterior over all 1024 models. Vertical scale has been truncated at 0.1 for clarity. (d) Marginal inclusion probabilities. Figure generated by `linregAllsubsetsGraycodeDemo`.

However, the mode is often not representative of the full posterior mass (see Section 5.2.1.3). A better summary is the **median model** (Barbieri and Berger 2004; Carvahlo and Lawrence 2007), computed using

$$\hat{\gamma} = \{j : p(\gamma_j = 1|\mathcal{D}) > 0.5\} \tag{13.4}$$

This requires computing the posterior marginal **inclusion probabilities**, $p(\gamma_j = 1|\mathcal{D})$. These are shown in Figure 13.1(d). We see that the model is confident that variables 2 and 6 are included; if we lower the decision threshold to 0.1, we would add 3 and 9 as well. However, if we wanted to "capture" variable 8, we would incur two false positives (5 and 7). This tradeoff between false positives and false negatives is discussed in more detail in Section 5.7.2.1.

The above example illustrates the "gold standard" for variable selection: the problem was sufficiently small (only 10 variables) that we were able to compute the full posterior exactly. Of course, variable selection is most useful in the cases where the number of dimensions is large. Since there are 2^D possible models (bit vectors), it will be impossible to compute the full posterior in general, and even finding summaries, such as the MAP estimate or marginal inclusion probabilities, will be intractable. We will therefore spend most of this chapter focussing on algorithmic speedups. But before we do that, we will explain how we computed $p(\boldsymbol{\gamma}|\mathcal{D})$ in the above example.

13.2.1 The spike and slab model

The posterior is given by

$$p(\boldsymbol{\gamma}|\mathcal{D}) \propto p(\boldsymbol{\gamma})p(\mathcal{D}|\boldsymbol{\gamma}) \tag{13.5}$$

We first consider the prior, then the likelihood.

It is common to use the following prior on the bit vector:

$$p(\boldsymbol{\gamma}) = \prod_{j=1}^{D} \text{Ber}(\gamma_j|\pi_0) = \pi_0^{||\boldsymbol{\gamma}||_0}(1-\pi_0)^{D-||\boldsymbol{\gamma}||_0} \tag{13.6}$$

where π_0 is the probability a feature is relevant, and $||\boldsymbol{\gamma}||_0 = \sum_{j=1}^{D} \gamma_j$ is the ℓ_0 **pseudo-norm**, that is, the number of non-zero elements of the vector. For comparison with later models, it is useful to write the log prior as follows:

$$\log p(\boldsymbol{\gamma}|\pi_0) = ||\boldsymbol{\gamma}||_0 \log \pi_0 + (D - ||\boldsymbol{\gamma}||_0)\log(1-\pi_0) \tag{13.7}$$

$$= ||\boldsymbol{\gamma}||_0 (\log \pi_0 - \log(1-\pi_0)) + \text{const} \tag{13.8}$$

$$= -\lambda ||\boldsymbol{\gamma}||_0 + \text{const} \tag{13.9}$$

where $\lambda \triangleq \log \frac{1-\pi_0}{\pi_0}$ controls the sparsity of the model.

We can write the likelihood as follows:

$$p(\mathcal{D}|\boldsymbol{\gamma}) = p(\mathbf{y}|\mathbf{X}, \boldsymbol{\gamma}) = \int \int p(\mathbf{y}|\mathbf{X}, \mathbf{w}, \boldsymbol{\gamma})p(\mathbf{w}|\boldsymbol{\gamma}, \sigma^2)p(\sigma^2)d\mathbf{w}d\sigma^2 \tag{13.10}$$

For notational simplicity, we have assumed the response is centered, (i.e., $\bar{y} = 0$), so we can ignore any offset term μ.

We now discuss the prior $p(\mathbf{w}|\boldsymbol{\gamma}, \sigma^2)$. If $\gamma_j = 0$, feature j is irrelevant, so we expect $w_j = 0$. If $\gamma_j = 1$, we expect w_j to be non-zero. If we standardize the inputs, a reasonable prior is $\mathcal{N}(0, \sigma^2\sigma_w^2)$, where σ_w^2 controls how big we expect the coefficients associated with the relevant variables to be (which is scaled by the overall noise level σ^2). We can summarize this prior as follows:

$$p(w_j|\sigma^2, \gamma_j) = \begin{cases} \delta_0(w_j) & \text{if } \gamma_j = 0 \\ \mathcal{N}(w_j|0, \sigma^2\sigma_w^2) & \text{if } \gamma_j = 1 \end{cases} \tag{13.11}$$

The first term is a "spike" at the origin. As $\sigma_w^2 \to \infty$, the distribution $p(w_j|\gamma_j = 1)$ approaches a uniform distribution, which can be thought of as a "slab" of constant height. Hence this is called the **spike and slab** model (Mitchell and Beauchamp 1988).

We can drop the coefficients w_j for which $w_j = 0$ from the model, since they are clamped to zero under the prior. Hence Equation 13.10 becomes the following (assuming a Gaussian likelihood):

$$p(\mathcal{D}|\boldsymbol{\gamma}) = \int \int \mathcal{N}(\mathbf{y}|\mathbf{X}_{\boldsymbol{\gamma}}\mathbf{w}_{\boldsymbol{\gamma}}, \sigma^2\mathbf{I}_N)\mathcal{N}(\mathbf{w}_{\boldsymbol{\gamma}}|\mathbf{0}_{\boldsymbol{\gamma}}, \sigma^2\sigma_w^2\mathbf{I}_{\boldsymbol{\gamma}})p(\sigma^2)d\mathbf{w}_{\boldsymbol{\gamma}}d\sigma^2 \tag{13.12}$$

where $\mathbf{X}_{\gamma}$ is the design matrix where we select only the columns of $\mathbf{X}$ where $\gamma_j = 1$, $\mathbf{0}_{\gamma} = \mathbf{0}_{||\gamma||_0}$ and $\mathbf{I}_{\gamma} = \mathbf{I}_{||\gamma||_0}$. Similarly, we write $\mathbf{w}_{\gamma}$ to remind the reader that this is a coefficient vector only for the dimensions where $\gamma_j = 1$. In what follows, we will generalize the prior slightly by defining $p(\mathbf{w}|\gamma, \sigma^2) = \mathcal{N}(\mathbf{w}_{\gamma}|\mathbf{0}_{\gamma}, \sigma^2 \boldsymbol{\Sigma}_{\gamma})$ for any positive definite matrix $\boldsymbol{\Sigma}_{\gamma}$, rather than requiring $\boldsymbol{\Sigma}_{\gamma} = \sigma_w^2 \mathbf{I}_{\gamma}$.[1]

Given these priors, we can now compute the marginal likelihood. If the noise variance is known, we can write down the marginal likelihood (using Equation 13.151) as follows:

$$p(\mathcal{D}|\gamma, \sigma^2) = \int \mathcal{N}(\mathbf{y}|\mathbf{X}_{\gamma}\mathbf{w}_{\gamma}, \sigma^2 \mathbf{I}) \mathcal{N}(\mathbf{w}_{\gamma}|\mathbf{0}, \sigma^2 \boldsymbol{\Sigma}_{\gamma}) d\mathbf{w}_{\gamma} = \mathcal{N}(\mathbf{y}|\mathbf{0}, \mathbf{C}_{\gamma}) \qquad (13.13)$$

$$\mathbf{C}_{\gamma} \triangleq \sigma^2 \mathbf{X}_{\gamma} \boldsymbol{\Sigma}_{\gamma} \mathbf{X}_{\gamma}^T + \sigma^2 \mathbf{I}_N \qquad (13.14)$$

If the noise is unknown, we can put a prior on it and integrate it out. It is common to use $p(\sigma^2) = \mathrm{IG}(\sigma^2|a_{\sigma}, b_{\sigma})$. Some guidelines on setting a, b can be found in (Kohn et al. 2001). If we use $a = b = 0$, we recover the Jeffrey's prior, $p(\sigma^2) \propto \sigma^{-2}$. When we integrate out the noise, we get the following more complicated expression for the marginal likelihood (Brown et al. 1998):

$$p(\mathcal{D}|\gamma) = \int \int p(\mathbf{y}|\gamma, \mathbf{w}_{\gamma}, \sigma^2) p(\mathbf{w}_{\gamma}|\gamma, \sigma^2) p(\sigma^2) d\mathbf{w}_{\gamma} d\sigma^2 \qquad (13.15)$$

$$\propto |\mathbf{X}_{\gamma}^T \mathbf{X}_{\gamma} + \boldsymbol{\Sigma}_{\gamma}^{-1}|^{-\frac{1}{2}} |\boldsymbol{\Sigma}_{\gamma}|^{-\frac{1}{2}} (2b_{\sigma} + S(\gamma))^{-(2a_{\sigma}+N-1)/2} \qquad (13.16)$$

where $S(\gamma)$ is the RSS:

$$S(\gamma) \triangleq \mathbf{y}^T \mathbf{y} - \mathbf{y}^T \mathbf{X}_{\gamma} (\mathbf{X}_{\gamma}^T \mathbf{X}_{\gamma} + \boldsymbol{\Sigma}_{\gamma}^{-1})^{-1} \mathbf{X}_{\gamma}^T \mathbf{y} \qquad (13.17)$$

See also Exercise 13.4.

When the marginal likelihood cannot be computed in closed form (e.g., if we are using logistic regression or a nonlinear model), we can approximate it using BIC, which has the form

$$\log p(\mathcal{D}|\gamma) \approx \log p(\mathbf{y}|\mathbf{X}, \hat{\mathbf{w}}_{\gamma}, \hat{\sigma}^2) - \frac{||\gamma||_0}{2} \log N \qquad (13.18)$$

where $\hat{\mathbf{w}}_{\gamma}$ is the ML or MAP estimate based on $\mathbf{X}_{\gamma}$, and $||\gamma||_0$ is the "degrees of freedom" of the model (Zou et al. 2007). Adding the log prior, the overall objective becomes

$$\log p(\gamma|\mathcal{D}) \approx \log p(\mathbf{y}|\mathbf{X}, \hat{\mathbf{w}}_{\gamma}, \hat{\sigma}^2) - \frac{||\gamma||_0}{2} \log N - \lambda ||\gamma||_0 + \mathrm{const} \qquad (13.19)$$

We see that there are two complexity penalties: one arising from the BIC approximation to the marginal likelihood, and the other arising from the prior on $p(\gamma)$.

---

1. It is common to use a g-prior of the form $\boldsymbol{\Sigma}_{\gamma} = g(\mathbf{X}_{\gamma}^T \mathbf{X}_{\gamma})^{-1}$ for reasons explained in Section 7.6.3.1 (see also Exercise 13.4). Various approaches have been proposed for setting g, including cross validation, empirical Bayes (Minka 2000b; George and Foster 2000), hierarchical Bayes (Liang et al. 2008), etc.

13.2.2   From the Bernoulli-Gaussian model to ℓ_0 regularization

Another model that is sometimes used (e.g., (Kuo and Mallick 1998; Zhou et al. 2009; Soussen et al. 2010)) is the following:

$$y_i|\mathbf{x}_i, \mathbf{w}, \boldsymbol{\gamma}, \sigma^2 \sim \mathcal{N}(\sum_j \gamma_j w_j x_{ij}, \sigma^2) \tag{13.20}$$

$$\gamma_j \sim \text{Ber}(\pi_0) \tag{13.21}$$

$$w_j \sim \mathcal{N}(0, \sigma_w^2) \tag{13.22}$$

In the signal processing literature (e.g., (Soussen et al. 2010)), this is called the **Bernoulli-Gaussian** model, although we could also call it the **binary mask** model, since we can think of the γ_j variables as "masking out" the weights w_j.

Unlike the spike and slab model, we do not integrate out the "irrelevant" coefficients; they always exist. In addition, the binary mask model has the form $\gamma_j \to \mathbf{y} \leftarrow w_j$, whereas the spike and slab model has the form $\gamma_j \to w_j \to \mathbf{y}$. In the binary mask model, only the product $\gamma_j w_j$ can be identified from the likelihood.

One interesting aspect of this model is that it can be used to derive an objective function that is widely used in the (non-Bayesian) subset selection literature. First, note that the joint prior has the form

$$p(\boldsymbol{\gamma}, \mathbf{w}) \propto \mathcal{N}(\mathbf{w}|\mathbf{0}, \sigma_w^2\mathbf{I})\pi_0^{||\boldsymbol{\gamma}||_0}(1 - \pi_0)^{D-||\boldsymbol{\gamma}||_0} \tag{13.23}$$

Hence the scaled unnormalized negative log posterior has the form

$$\begin{aligned} f(\boldsymbol{\gamma}, \mathbf{w}) &\triangleq -2\sigma^2 \log p(\boldsymbol{\gamma}, \mathbf{w}, \mathbf{y}|\mathbf{X}) = ||\mathbf{y} - \mathbf{X}(\boldsymbol{\gamma}. * \mathbf{w})||^2 \\ &+ \frac{\sigma^2}{\sigma_w^2}||\mathbf{w}||^2 + \lambda||\boldsymbol{\gamma}||_0 + \text{const} \end{aligned} \tag{13.24}$$

where

$$\lambda \triangleq 2\sigma^2 \log(\frac{1 - \pi_0}{\pi_0}) \tag{13.25}$$

Let us split $\mathbf{w}$ into two subvectors, $\mathbf{w}_{-\boldsymbol{\gamma}}$ and $\mathbf{w}_{\boldsymbol{\gamma}}$, indexed by the zero and non-zero entries of $\boldsymbol{\gamma}$ respectively. Since $\mathbf{X}(\boldsymbol{\gamma}. * \mathbf{w}) = \mathbf{X}_{\boldsymbol{\gamma}}\mathbf{w}_{\boldsymbol{\gamma}}$, we can just set $\mathbf{w}_{-\boldsymbol{\gamma}} = \mathbf{0}$.

Now consider the case where $\sigma_w^2 \to \infty$, so we do not regularize the non-zero weights (so there is no complexity penalty coming from the marginal likelihood or its BIC approximation). In this case, the objective becomes

$$f(\boldsymbol{\gamma}, \mathbf{w}) = ||\mathbf{y} - \mathbf{X}_{\boldsymbol{\gamma}}\mathbf{w}_{\boldsymbol{\gamma}}||_2^2 + \lambda||\boldsymbol{\gamma}||_0 \tag{13.26}$$

This is similar to the BIC objective above.

Instead of keeping track of the bit vector $\boldsymbol{\gamma}$, we can define the set of relevant variables to be the **support**, or set of non-zero entries, of $\mathbf{w}$. Then we can rewrite the above equation as follows:

$$f(\mathbf{w}) = ||\mathbf{y} - \mathbf{X}\mathbf{w}||_2^2 + \lambda||\mathbf{w}||_0 \tag{13.27}$$

This is called ℓ_0 **regularization**. We have converted the discrete optimization problem (over $\boldsymbol{\gamma} \in \{0, 1\}^D$) into a continuous one (over $\mathbf{w} \in \mathbb{R}^D$); however, the ℓ_0 pseudo-norm makes the objective very non smooth, so this is still hard to optimize. We will discuss different solutions to this in the rest of this chapter.

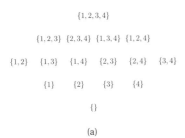
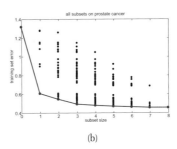

(a) (b)

Figure 13.2 (a) A lattice of subsets of $\{1, 2, 3, 4\}$. (b) Residual sum of squares versus subset size, on the prostate cancer data set. The lower envelope is the best RSS achievable for any set of a given size. Based on Figure 3.5 of (Hastie et al. 2001). Figure generated by `prostateSubsets`.

13.2.3 Algorithms

Since there are 2^D models, we cannot explore the full posterior, or find the globally optimal model. Instead we will have to resort to heuristics of one form or another. All of the methods we will discuss involve searching through the space of models, and evaluating the cost $f(\gamma)$ at each point. This requires fitting the model (i.e., computing $\arg\max p(\mathcal{D}|\mathbf{w})$), or evaluating its marginal likelihood (i.e., computing $\int p(\mathcal{D}|\mathbf{w})p(\mathbf{w})d\mathbf{w}$) at each step. This is sometimes called the **wrapper method**, since we "wrap" our search for the best model (or set of good models) around a generic model-fitting procedure.

In order to make wrapper methods efficient, it is important that we can quickly evaluate the score function for some new model, γ', given the score of a previous model, γ. This can be done provided we can efficiently update the sufficient statistics needed to compute $f(\gamma)$. This is possible provided γ' only differs from γ in one bit (corresponding to adding or removing a single variable), and provided $f(\gamma)$ only depends on the data via $\mathbf{X}_\gamma$. In this case, we can use rank-one matrix updates/ downdates to efficiently compute $\mathbf{X}_{\gamma'}^T \mathbf{X}_{\gamma'}$ from $\mathbf{X}_\gamma^T \mathbf{X}_\gamma$. These updates are usually applied to the QR decomposition of $\mathbf{X}$. See e.g., (Miller 2002; Schniter et al. 2008) for details.

13.2.3.1 Greedy search

Suppose we want to find the MAP model. If we use the ℓ_0-regularized objective in Equation 13.27, we can exploit properties of least squares to derive various efficient greedy forwards search methods, some of which we summarize below. For further details, see (Miller 2002; Soussen et al. 2010).

- **Single best replacement** The simplest method is to use greedy hill climbing, where at each step, we define the neighborhood of the current model to be all models than can be reached by flipping a single bit of γ, i.e., for each variable, if it is currently out of the model, we consider adding it, and if it is currently in the model, we consider removing it. In (Soussen et al. 2010), they call this the **single best replacement** (SBR). Since we are expecting a sparse solution, we can start with the empty set, $\gamma = \mathbf{0}$. We are essentially moving through the lattice of subsets, shown in Figure 13.2(a). We continue adding or removing until no improvement is possible.

- **Orthogonal least squares** If we set $\lambda = 0$ in Equation 13.27, so there is no complexity penalty, there will be no reason to perform deletion steps. In this case, the SBR algorithm is equivalent to **orthogonal least squares** (Chen and Wigger 1995), which in turn is equivalent to greedy **forwards selection**. In this algorithm, we start with the empty set and add the best feature at each step. The error will go down monotonically with $||\gamma||_0$, as shown in Figure 13.2(b). We can pick the next best feature j^* to add to the current set γ_t by solving

$$j^* = \arg \min_{j \notin \gamma_t} \min_{\mathbf{w}} ||\mathbf{y} - (\mathbf{X}_{\gamma_t \cup j})\mathbf{w}||^2 \tag{13.28}$$

We then update the active set by setting $\gamma_{t+1} = \gamma_t \cup \{j^*\}$. To choose the next feature to add at step t, we need to solve $D - D_t$ least squares problems at step t, where $D_t = ||\gamma_t||_0$ is the cardinality of the current active set. Having chosen the best feature to add, we need to solve an additional least squares problem to compute $\mathbf{w}_{t+1}$.

- **Orthogonal matching pursuits** Orthogonal least squares is somewhat expensive. A simplification is to "freeze" the current weights at their current value, and then to pick the next feature to add by solving

$$j^* = \arg \min_{j \notin \gamma_t} \min_{\beta} ||\mathbf{y} - \mathbf{X}\mathbf{w}_t - \beta \mathbf{x}_{:,j}||^2 \tag{13.29}$$

This inner optimization is easy to solve: we simply set $\beta = \mathbf{x}_{:,j}^T \mathbf{r}_t / ||\mathbf{x}_{:,j}||^2$, where $\mathbf{r}_t = \mathbf{y} - \mathbf{X}\mathbf{w}_t$ is the current residual vector. If the columns are unit norm, we have

$$j^* = \arg \max \mathbf{x}_{:,j}^T \mathbf{r}_t \tag{13.30}$$

so we are just looking for the column that is most correlated with the current residual. We then update the active set, and compute the new least squares estimate $\mathbf{w}_{t+1}$ using $\mathbf{X}_{\gamma_{t+1}}$. This method is called **orthogonal matching pursuits** or **OMP** (Mallat et al. 1994). This only requires one least squares calculation per iteration and so is faster than orthogonal least squares, but is not quite as accurate (Blumensath and Davies 2007).

- **Matching pursuits** An even more aggressive approximation is to just greedily add the feature that is most correlated with the current residual. This is called **matching pursuits** (Mallat and Zhang 1993). This is also equivalent to a method known as least squares boosting (Section 16.4.6).

- **Backwards selection** If we start with all variables in the model (the so-called **saturated model**), and then delete the worst one at each step, we get a method known as **backwards selection**. This is equivalent to performing a greedy search from the top of the lattice downwards. This can give better results than a bottom-up search, since the decision about whether to keep a variable or not is made in the context of all the other variables that might depend on it. However, this method is typically infeasible for large problems, since the saturated model will be too expensive to fit.

- **FoBa** The **forwards-backwards algorithm** of (Zhang 2008) is similar to the single best replacement algorithm presented above, except it uses an OMP-like approximation when choosing the next move to make. A similar "dual-pass" algorithm was described in (Moghaddam et al. 2008).

- **Bayesian Matching pursuit** The algorithm of (Schniter et al. 2008) is similiar to OMP except it uses a Bayesian marginal likelihood scoring criterion (under a spike and slab model) instead of a least squares objective. In addition, it uses a form of beam search to explore multiple paths through the lattice at once.

13.2.3.2 Stochastic search

If we want to approximate the posterior, rather than just computing a mode (e.g. because we want to compute marginal inclusion probabilities), one option is to use MCMC. The standard approach is to use Metropolis Hastings, where the proposal distribution just flips single bits. This enables us to efficiently compute $p(\boldsymbol{\gamma}'|\mathcal{D})$ given $p(\boldsymbol{\gamma}|\mathcal{D})$. The probability of a state (bit configuration) is estimated by counting how many times the random walk visits this state. See (O'Hara and Sillanpaa 2009) for a review of such methods, and (Bottolo and Richardson 2010) for a very recent method based on evolutionary MCMC.

However, in a discrete state space, MCMC is needlessly inefficient, since we can compute the (unnormalized) probability of a state directly using $p(\boldsymbol{\gamma}, \mathcal{D}) = \exp(-f(\boldsymbol{\gamma}))$; thus *there is no need to ever revisit a state*. A much more efficient alternative is to use some kind of stochastic search algorithm, to generate a set $\mathcal{S}$ of high scoring models, and then to make the following approximation

$$p(\boldsymbol{\gamma}|\mathcal{D}) \approx \frac{e^{-f(\boldsymbol{\gamma})}}{\sum_{\boldsymbol{\gamma}' \in \mathcal{S}} e^{-f(\boldsymbol{\gamma}')}} \qquad (13.31)$$

See (Heaton and Scott 2009) for a review of recent methods of this kind.

13.2.3.3 EM and variational inference *

It is tempting to apply EM to the spike and slab model, which has the form $\gamma_j \to w_j \to \mathbf{y}$. We can compute $p(\gamma_j = 1|w_j)$ in the E step, and optimize $\mathbf{w}$ in the M step. However, this will not work, because when we compute $p(\gamma_j = 1|w_j)$, we are comparing a delta-function, $\delta_0(w_j)$, with a Gaussian pdf, $\mathcal{N}(w_j|0, \sigma_w^2)$. We can replace the delta function with a narrow Gaussian, and then the E step amounts to classifying w_j under the two possible Gaussian models. However, this is likely to suffer from severe local minima.

An alternative is to apply EM to the Bernoulli-Gaussian model, which has the form $\gamma_j \to \mathbf{y} \leftarrow w_j$. In this case, the posterior $p(\boldsymbol{\gamma}|\mathcal{D}, \mathbf{w})$ is intractable to compute because all the bits become correlated due to explaining away. However, it is possible to derive a mean field approximation of the form $\prod_j q(\gamma_j)q(w_j)$ (Huang et al. 2007; Rattray et al. 2009; Carbonetto and Stephens 2012).

13.3 ℓ_1 regularization: basics

When we have many variables, it is computationally difficult to find the posterior mode of $p(\boldsymbol{\gamma}|\mathcal{D})$. And although greedy algorithms often work well (see e.g., (Zhang 2008) for a theoretical analysis), they can of course get stuck in local optima.

Part of the problem is due to the fact that the γ_j variables are discrete, $\gamma_j \in \{0, 1\}$. In the optimization community, it is common to relax hard constraints of this form by replacing discrete variables with continuous variables. We can do this by replacing the spike-and-slab style prior, that assigns finite probability mass to the event that $w_j = 0$, to continuous priors that "encourage" $w_j = 0$ by putting a lot of probability density near the origin, such as a zero-mean Laplace distribution. This was first introduced in Section 7.4 in the context of robust linear regression. There we exploited the fact that the Laplace has heavy tails. Here we exploit the fact that it has a spike near $\mu = 0$. More precisely, consider a prior of the form

$$p(\mathbf{w}|\lambda) = \prod_{j=1}^{D} \text{Lap}(w_j|0, 1/\lambda) \propto \prod_{j=1}^{D} e^{-\lambda|w_j|} \tag{13.32}$$

We will use a uniform prior on the offset term, $p(w_0) \propto 1$. Let us perform MAP estimation with this prior. The penalized negative log likelihood has the form

$$f(\mathbf{w}) = -\log p(\mathcal{D}|\mathbf{w}) - \log p(\mathbf{w}|\lambda) = \text{NLL}(\mathbf{w}) + \lambda||\mathbf{w}||_1 \tag{13.33}$$

where $||\mathbf{w}||_1 = \sum_{j=1}^{D} |w_j|$ is the ℓ_1 norm of $\mathbf{w}$. For suitably large λ, the estimate $\hat{\mathbf{w}}$ will be sparse, for reasons we explain below. Indeed, this can be thought of as a convex approximation to the non-convex ℓ_0 objective

$$\operatorname*{argmin}_{\mathbf{w}} \text{NLL}(\mathbf{w}) + \lambda||\mathbf{w}||_0 \tag{13.34}$$

In the case of linear regression, the ℓ_1 objective becomes

$$f(\mathbf{w}) \;\;=\;\; \sum_{i=1}^{N} -\frac{1}{2\sigma^2}(y_i - (w_0 + \mathbf{w}^T\mathbf{x}_i))^2 + \lambda||\mathbf{w}||_1 \tag{13.35}$$

$$=\;\; \text{RSS}(\mathbf{w}) + \lambda'||\mathbf{w}||_1 \tag{13.36}$$

where $\lambda' = 2\lambda\sigma^2$. This method is known as **basis pursuit denoising** or **BPDN** (Chen et al. 1998). The reason for this term will become clear later. In general, the technique of putting a zero-mean Laplace prior on the parameters and performing MAP estimation is called ℓ_1 **regularization**. It can be combined with any convex or non-convex NLL term. Many different algorithms have been devised for solving such problems, some of which we review in Section 13.4.

13.3.1 Why does ℓ_1 regularization yield sparse solutions?

We now explain why ℓ_1 regularization results in sparse solutions, whereas ℓ_2 regularization does not. We focus on the case of linear regression, although similar arguments hold for logistic regression and other GLMs.

The BPDN objective is the following non-smooth objective function:

$$\min_{\mathbf{w}} \text{RSS}(\mathbf{w}) + \lambda||\mathbf{w}||_1 \tag{13.37}$$

We can rewrite this as a constrained but smooth objective as follows:

$$\min_{\mathbf{w}} \text{RSS}(\mathbf{w}) \quad \text{s.t.} \quad ||\mathbf{w}||_1 \leq B \tag{13.38}$$

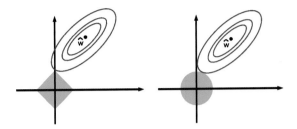

Figure 13.3 Illustration of ℓ_1 (left) vs ℓ_2 (right) regularization of a least squares problem. Based on Figure 3.12 of (Hastie et al. 2001).

where B is an upper bound on the ℓ_1-norm of the weights: a small (tight) bound B corresponds to a large penalty λ, and vice versa.[2] Equation 13.38 is known as **lasso**, which stands for "least absolute shrinkage and selection operator" (Tibshirani 1996). We will see why it has this name later.

Similarly, we can write ridge regression

$$\min_{\mathbf{w}} \text{RSS}(\mathbf{w}) + \lambda ||\mathbf{w}||_2^2 \tag{13.39}$$

or as a bound constrained form:

$$\min_{\mathbf{w}} \text{RSS}(\mathbf{w}) \quad \text{s.t.} \quad ||\mathbf{w}||_2^2 \leq B \tag{13.40}$$

In Figure 13.3, we plot the contours of the RSS objective function, as well as the contours of the ℓ_2 and ℓ_1 constraint surfaces. From the theory of constrained optimization, we know that the optimal solution occurs at the point where the lowest level set of the objective function intersects the constraint surface (assuming the constraint is active). It should be geometrically clear that as we relax the constraint B, we "grow" the ℓ_1 "ball" until it meets the objective; the corners of the ball are more likely to intersect the ellipse than one of the sides, especially in high dimensions, because the corners "stick out" more. The corners correspond to sparse solutions, which lie on the coordinate axes. By contrast, when we grow the ℓ_2 ball, it can intersect the objective at any point; there are no "corners", so there is no preference for sparsity.

To see this another away, notice that, with ridge regression, the prior cost of a sparse solution, such as $\mathbf{w} = (1, 0)$, is the same as the cost of a dense solution, such as $\mathbf{w} = (1/\sqrt{2}, 1/\sqrt{2})$, as long as they have the same ℓ_2 norm:

$$||(1,0)||_2 = ||(1/\sqrt{2}, 1/\sqrt{2}||_2 = 1 \tag{13.41}$$

However, for lasso, setting $\mathbf{w} = (1, 0)$ is cheaper than setting $\mathbf{w} = (1/\sqrt{2}, 1/\sqrt{2})$, since

$$||(1,0)||_1 = 1 < ||(1/\sqrt{2}, 1/\sqrt{2}||_1 = \sqrt{2} \tag{13.42}$$

The most rigorous way to see that ℓ_1 regularization results in sparse solutions is to examine conditions that hold at the optimum. We do this in Section 13.3.2.

---

2. Equation 13.38 is an example of a **quadratic program** or **QP**, since we have a quadratic objective subject to linear inequality constraints (see Exercise 13.11 for a way to rewrite the ℓ_1-norm constraint as a set of linear inequality constraints using an enlarged set of variables). Its Lagrangian is given by Equation 13.37.

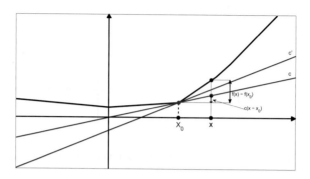

Figure 13.4   Illustration of some sub-derivatives of a function at point x_0. Based on a figure at `http://en.wikipedia.org/wiki/Subderivative`. Figure generated by `subgradientPlot`.

13.3.2   Optimality conditions for lasso

The lasso objective has the form

$$f(\boldsymbol{\theta}) = \text{RSS}(\boldsymbol{\theta}) + \lambda ||\mathbf{w}||_1 \tag{13.43}$$

Unfortunately, the $||\mathbf{w}||_1$ term is not differentiable whenever $w_j = 0$. This is an example of a **non-smooth** optimization problem.

To handle non-smooth functions, we need to extend the notion of a derivative. We define a **subderivative** or **subgradient** of a (convex) function $f : \mathcal{I} \to \mathbb{R}$ at a point θ_0 to be a scalar g such that

$$f(\theta) - f(\theta_0) \geq g(\theta - \theta_0) \quad \forall \theta \in \mathcal{I} \tag{13.44}$$

where $\mathcal{I}$ is some interval containing θ_0. See Figure 13.4 for an illustration.[3] We define the *set* of subderivatives as the interval $[a, b]$ where a and b are the one-sided limits

$$a = \lim_{\theta \to \theta_0^-} \frac{f(\theta) - f(\theta_0)}{\theta - \theta_0}, \quad b = \lim_{\theta \to \theta_0^+} \frac{f(\theta) - f(\theta_0)}{\theta - \theta_0} \tag{13.46}$$

The set $[a, b]$ of all subderivatives is called the **subdifferential** of the function f at θ_0 and is denoted $\partial f(\theta)|_{\theta_0}$. For example, in the case of the absolute value function $f(\theta) = |\theta|$, the subderivative is given by

$$\partial f(\theta) = \begin{cases} \{-1\} & \text{if } \theta < 0 \\ [-1, 1] & \text{if } \theta = 0 \\ \{+1\} & \text{if } \theta > 0 \end{cases} \tag{13.47}$$

---

3. In general, for a vector valued function, we say that $\mathbf{g}$ is a subgradient of f at $\boldsymbol{\theta}_0$ if for all vectors $\boldsymbol{\theta}$,

$$f(\boldsymbol{\theta}) - f(\boldsymbol{\theta}_0) \geq (\boldsymbol{\theta} - \boldsymbol{\theta}_0)^T \mathbf{g} \tag{13.45}$$

so $\mathbf{g}$ is a linear lower bound to the function at $\boldsymbol{\theta}_0$.

If the function is everywhere differentiable, then $\partial f(\theta) = \{\frac{df(\theta)}{d\theta}\}$. By analogy to the standard calculus result, one can show that the point $\hat{\theta}$ is a local minimum of f iff $0 \in \partial f(\theta)|_{\hat{\theta}}$.

Let us apply these concepts to the lasso problem. Let us initially ignore the non-smooth penalty term. One can show (Exercise 13.1) that

$$\frac{\partial}{\partial w_j}\mathrm{RSS}(\mathbf{w}) = a_j w_j - c_j \qquad (13.48)$$

$$a_j = 2\sum_{i=1}^{n} x_{ij}^2 \qquad (13.49)$$

$$c_j = 2\sum_{i=1}^{n} x_{ij}(y_i - \mathbf{w}_{-j}^T \mathbf{x}_{i,-j}) \qquad (13.50)$$

where $\mathbf{w}_{-j}$ is $\mathbf{w}$ without component j, and similarly for $\mathbf{x}_{i,-j}$. We see that c_j is (proportional to) the correlation between the j'th feature $\mathbf{x}_{:,j}$ and the residual due to the other features, $\mathbf{r}_{-j} = \mathbf{y} - \mathbf{X}_{:,-j}\mathbf{w}_{-j}$. Hence the magnitude of c_j is an indication of how relevant feature j is for predicting $\mathbf{y}$ (relative to the other features and the current parameters).

Adding in the penalty term, we find that the subderivative is given by

$$\partial_{w_j} f(\mathbf{w}) = (a_j w_j - c_j) + \lambda \partial_{w_j}||\mathbf{w}||_1 \qquad (13.51)$$

$$= \begin{cases} \{a_j w_j - c_j - \lambda\} & \text{if } w_j < 0 \\ [-c_j - \lambda, -c_j + \lambda] & \text{if } w_j = 0 \\ \{a_j w_j - c_j + \lambda\} & \text{if } w_j > 0 \end{cases} \qquad (13.52)$$

Consequently $\mathbf{w}$ is a local minimum iff the following conditions hold:

$$\mathbf{X}^T(\mathbf{X}\mathbf{w} - \mathbf{y})_j \in \begin{cases} \{-\lambda\} & \text{if } w_j < 0 \\ [-\lambda, \lambda] & \text{if } w_j = 0 \\ \{\lambda\} & \text{if } w_j > 0 \end{cases} \qquad (13.53)$$

Depending on the value of c_j, the solution to $\partial_{w_j} f(\mathbf{w}) = 0$ can occur at 3 different values of w_j, as follows:

1. If $c_j < -\lambda$, so the feature is strongly negatively correlated with the residual, then the subgradient is zero at $\hat{w}_j = \frac{c_j + \lambda}{a_j} < 0$.

2. If $c_j \in [-\lambda, \lambda]$, so the feature is only weakly correlated with the residual, then the subgradient is zero at $\hat{w}_j = 0$.

3. If $c_j > \lambda$, so the feature is strongly positively correlated with the residual, then the subgradient is zero at $\hat{w}_j = \frac{c_j - \lambda}{a_j} > 0$.

In summary, we have

$$\hat{w}_j(c_j) = \begin{cases} (c_j + \lambda)/a_j & \text{if } c_j < -\lambda \\ 0 & \text{if } c_j \in [-\lambda, \lambda] \\ (c_j - \lambda)/a_j & \text{if } c_j > \lambda \end{cases} \qquad (13.54)$$

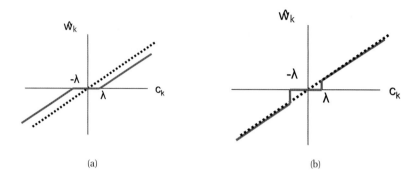

Figure 13.5 Left: soft thresholding. Right: hard thresholding. In both cases, the horizontal axis is the residual error incurred by making predictions using all the coefficients except for w_k, and the vertical axis is the estimated coefficient $\hat{w}_k$ that minimizes this penalized residual. The flat region in the middle is the interval $[-\lambda, +\lambda]$.

We can write this as follows:

$$\hat{w}_j = \text{soft}(\frac{c_j}{a_j}; \frac{\lambda}{a_j}) \tag{13.55}$$

where

$$\text{soft}(a; \delta) \triangleq \text{sign}(a)(|a| - \delta)_+ \tag{13.56}$$

and $x_+ = \max(x, 0)$ is the positive part of x. This is called **soft thresholding**. This is illustrated in Figure 13.5(a), where we plot $\hat{w}_j$ vs c_j. The dotted black line is the line $w_j = c_j/a_j$ corresponding to the least squares fit. The solid red line, which represents the regularized estimate $\hat{w}_j(c_j)$, shifts the dotted line down (or up) by λ, except when $-\lambda \leq c_j \leq \lambda$, in which case it sets $w_j = 0$.

By contrast, in Figure 13.5(b), we illustrate **hard thresholding**. This sets values of w_j to 0 if $-\lambda \leq c_j \leq \lambda$, but it does not shrink the values of w_j outside of this interval. The slope of the soft thresholding line does not coincide with the diagonal, which means that even large coefficients are shrunk towards zero; consequently lasso is a biased estimator. This is undesirable, since if the likelihood indicates (via c_j) that the coefficient w_j should be large, we do not want to shrink it. We will discuss this issue in more detail in Section 13.6.2.

Now we finally can understand why Tibshirani invented the term "lasso" in (Tibshirani 1996): it stands for "least absolute selection and shrinkage operator", since it selects a subset of the variables, and shrinks all the coefficients by penalizing the absolute values. If $\lambda = 0$, we get the OLS solution (of minimal ℓ_1 norm). If $\lambda \geq \lambda_{max}$, we get $\hat{\mathbf{w}} = \mathbf{0}$, where

$$\lambda_{max} = ||\mathbf{X}^T \mathbf{y}||_\infty = \max_j |\mathbf{y}^T \mathbf{x}_{:,j}| \tag{13.57}$$

This value is computed using the fact that $\mathbf{0}$ is optimal if $(\mathbf{X}^T \mathbf{y})_j \in [-\lambda, \lambda]$ for all j. In general, the maximum penalty for an ℓ_1 regularized objective is

$$\lambda_{max} = \max_j |\nabla_j NLL(\mathbf{0})| \tag{13.58}$$

13.3.3 Comparison of least squares, lasso, ridge and subset selection

We can gain further insight into ℓ_1 regularization by comparing it to least squares, and ℓ_2 and ℓ_0 regularized least squares. For simplicity, assume all the features of $\mathbf{X}$ are orthonormal, so $\mathbf{X}^T\mathbf{X} = \mathbf{I}$. In this case, the RSS is given by

$$\text{RSS}(\mathbf{w}) = ||\mathbf{y} - \mathbf{X}\mathbf{w}||^2 = \mathbf{y}^T\mathbf{y} + \mathbf{w}^T\mathbf{X}^T\mathbf{X}\mathbf{w} - 2\mathbf{w}^T\mathbf{X}^T\mathbf{y} \tag{13.59}$$

$$= \text{const} + \sum_k w_k^2 - 2\sum_k\sum_i w_k x_{ik} y_i \tag{13.60}$$

so we see this factorizes into a sum of terms, one per dimension. Hence we can write down the MAP and ML estimates analytically, as follows:

- **MLE** The OLS solution is given by

$$\hat{w}_k^{OLS} = \mathbf{x}_{:k}^T\mathbf{y} \tag{13.61}$$

where $\mathbf{x}_{:k}$ is the k'th column of $\mathbf{X}$. This follows trivially from Equation 13.60. We see that $\hat{w}_k^{OLS}$ is just the orthogonal projection of feature k onto the response vector (see Section 7.3.2).

- **Ridge** One can show that the ridge estimate is given by

$$\hat{w}_k^{ridge} = \frac{\hat{w}_k^{OLS}}{1 + \lambda} \tag{13.62}$$

- **Lasso** From Equation 13.55, and using the fact that $a_k = 2$ and $\hat{w}_k^{OLS} = c_k/2$, we have

$$\hat{w}_k^{lasso} = \text{sign}(\hat{w}_k^{OLS})\left(|\hat{w}_k^{OLS}| - \frac{\lambda}{2}\right)_+ \tag{13.63}$$

This corresponds to soft thresholding, shown in Figure 13.5(a).

- **Subset selection** If we pick the best K features using subset selection, the parameter estimate is as follows

$$\hat{w}_k^{SS} = \begin{cases} \hat{w}_k^{OLS} & \text{if rank}(|w_k^{OLS}|) \leq K \\ 0 & \text{otherwise} \end{cases} \tag{13.64}$$

where rank refers to the location in the sorted list of weight magnitudes. This corresponds to hard thresholding, shown in Figure 13.5(b).

Figure 13.6(a) plots the MSE vs λ for lasso for a degree 14 polynomial, and Figure 13.6(b) plots the MSE vs polynomial order. We see that lasso gives similar results to the subset selection method.

As another example, consider a data set concerning prostate cancer. We have $D = 8$ features and $N = 67$ training cases; the goal is to predict the log prostate-specific antigen levels (see (Hastie et al. 2009, p4) for more biological details). Table 13.1 shows that lasso gives better prediction accuracy (at least on this particular data set) than least squares, ridge, and best subset regression. (In each case, the strength of the regularizer was chosen by cross validation.) Lasso also gives rise to a sparse solution. Of course, for other problems, ridge may give better predictive accuracy. In practice, a combination of lasso and ridge, known as the elastic net, often performs best, since it provides a good combination of sparsity and regularization (see Section 13.5.3).

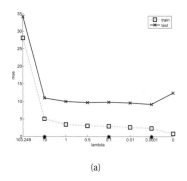
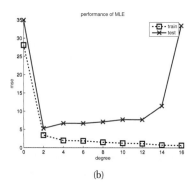

(a)   (b)

Figure 13.6   (a) MSE vs λ for lasso for a degree 14 polynomial. Note that λ decreases as we move to the right. Figure generated by `linregPolyLassoDemo`. (b) MSE versus polynomial degree. Note that the model order increases as we move to the right. See Figure 1.18 for a plot of some of these polynomial regression models. Figure generated by `linregPolyVsDegree`.

Term	LS	Best Subset	Ridge	Lasso
Intercept	2.452	2.481	2.479	2.480
lcavol	0.716	0.651	0.656	0.653
lweight	0.293	0.380	0.300	0.297
age	-0.143	-0.000	-0.129	-0.119
lbph	0.212	-0.000	0.208	0.200
svi	0.310	-0.000	0.301	0.289
lcp	-0.289	-0.000	-0.260	-0.236
gleason	-0.021	-0.000	-0.019	0.000
pgg45	0.277	0.178	0.256	0.226
Test Error	0.586	0.572	0.580	0.564

Table 13.1   Results of different methods on the prostate cancer data, which has 8 features and 67 training cases. Methods are: LS = least squares, Subset = best subset regression, Ridge, Lasso. Rows represent the coefficients; we see that subset regression and lasso give sparse solutions. Bottom row is the mean squared error on the test set (30 cases). Based on Table 3.3. of (Hastie et al. 2009). Figure generated by `prostateComparison`.

13.3.4   Regularization path

As we increase λ, the solution vector $\hat{\mathbf{w}}(\lambda)$ will tend to get sparser, although not necessarily monotonically. We can plot the values $\hat{w}_j(\lambda)$ vs λ for each feature j; this is known as the **regularization path**.

This is illustrated for ridge regression in Figure 13.7(a), where we plot $\hat{w}_j(\lambda)$ as the regularizer λ decreases. We see that when $\lambda = \infty$, all the coefficients are zero. But for any finite value of λ, all coefficients are non-zero; furthermore, they increase in magnitude as λ is decreased.

In Figure 13.7(b), we plot the analogous result for lasso. As we move to the right, the upper bound on the ℓ_1 penalty, B, increases. When $B = 0$, all the coefficients are zero. As we increase

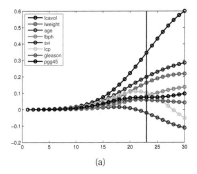 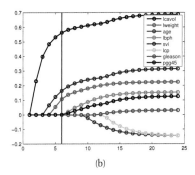

(a)   (b)

Figure 13.7 (a) Profiles of ridge coefficients for the prostate cancer example vs bound on ℓ_2 norm of $\mathbf{w}$, so small t (large λ) is on the left. The vertical line is the value chosen by 5-fold CV using the 1SE rule. Based on Figure 3.8 of (Hastie et al. 2009). Figure generated by `ridgePathProstate`. (b) Profiles of lasso coefficients for the prostate cancer example vs bound on ℓ_1 norm of $\mathbf{w}$, so small t (large λ) is on the left. Based on Figure 3.10 of (Hastie et al. 2009). Figure generated by `lassoPathProstate`.

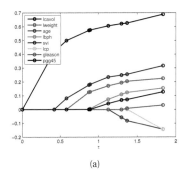 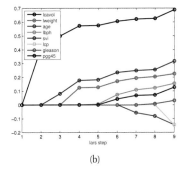

(a)   (b)

Figure 13.8 Illustration of piecewise linearity of regularization path for lasso on the prostate cancer example. (a) We plot $\hat{w}_j(B)$ vs B for the critical values of B. (b) We plot vs steps of the LARS algorithm. Figure generated by `lassoPathProstate`.

B, the coefficients gradually "turn on". But for any value between 0 and $B_{max} = ||\hat{\mathbf{w}}_{OLS}||_1$, the solution is sparse.[4]

Remarkably, it can be shown that the solution path is a piecewise linear function of B (Efron et al. 2004). That is, there are a set of critical values of B where the active set of non-zero coefficients changes. For values of B between these critical values, each non-zero coefficient increases or decreases in a linear fashion. This is illustrated in Figure 13.8(a). Furthermore, one can solve for these critical values analytically. This is the basis of the **LARS** algorithm (Efron et al. 2004), which stands for "least angle regression and shrinkage" (see Section 13.4.2 for details). Remarkably, LARS can compute the entire regularization path for roughly the same

---

4. It is common to plot the solution versus the **shrinkage factor**, defined as $s(B) = B/B_{max}$, rather than against B. This merely affects the scale of the horizontal axis, not the shape of the curves.

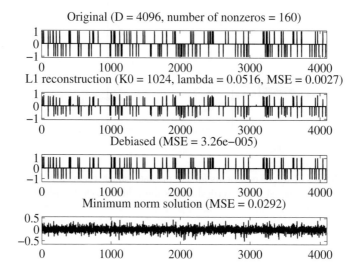

Figure 13.9 Example of recovering a sparse signal using lasso. See text for details. Based on Figure 1 of (Figueiredo et al. 2007). Figure generated by `sparseSensingDemo`, written by Mario Figueiredo.

computational cost as a single least squares fit (namely $O(\min(ND^2, DN^2))$).

In Figure 13.8(b), we plot the coefficients computed at each critical value of B. Now the piecewise linearity is more evident. Below we display the actual coefficient values at each step along the regularization path (the last line is the least squares solution):

Listing 13.1   Output of `lassoPathProstate`

0	0	0	0	0	0	0	0
0.4279	0	0	0	0	0	0	0
0.5015	0.0735	0	0	0	0	0	0
0.5610	0.1878	0	0	0.0930	0	0	0
0.5622	0.1890	0	0.0036	0.0963	0	0	0
0.5797	0.2456	0	0.1435	0.2003	0	0	0.0901
0.5864	0.2572	-0.0321	0.1639	0.2082	0	0	0.1066
0.6994	0.2910	-0.1337	0.2062	0.3003	-0.2565	0	0.2452
0.7164	0.2926	-0.1425	0.2120	0.3096	-0.2890	-0.0209	0.2773

By changing B from 0 to B_{max}, we can go from a solution in which all the weights are zero to a solution in which all weights are non-zero. Unfortunately, not all subset sizes are achievable using lasso. One can show that, if $D > N$, the optimal solution can have at most N variables in it, before reaching the complete set corresponding to the OLS solution of minimal ℓ_1 norm. In Section 13.5.3, we will see that by using an ℓ_2 regularizer as well as an ℓ_1 regularizer (a method known as the elastic net), we can achieve sparse solutions which contain more variables than training cases. This lets us explore model sizes between N and D.

13.3.5 Model selection

It is tempting to use ℓ_1 regularization to estimate the set of relevant variables. In some cases, we can recover the true sparsity pattern of $\mathbf{w}^*$, the parameter vector that generated the data. A method that can recover the true model in the $N \to \infty$ limit is called **model selection consistent**. The details on which methods enjoy this property, and when, are beyond the scope of this book; see e.g., (Buhlmann and van de Geer 2011) for details.

Instead of going into a theoretical discussion, we will just show a small example. We first generate a sparse signal $\mathbf{w}^*$ of size $D = 4096$, consisting of 160 randomly placed ± 1 spikes. Next we generate a random design matrix $\mathbf{X}$ of size $N \times D$, where $N = 1024$. Finally we generate a noisy observation $\mathbf{y} = \mathbf{X}\mathbf{w}^* + \boldsymbol{\epsilon}$, where $\epsilon_i \sim \mathcal{N}(0, 0.01^2)$. We then estimate $\mathbf{w}$ from $\mathbf{y}$ and $\mathbf{X}$.

The original $\mathbf{w}^*$ is shown in the first row of Figure 13.9. The second row is the ℓ_1 estimate $\hat{\mathbf{w}}_{L1}$ using $\lambda = 0.1\lambda_{max}$. We see that this has "spikes" in the right places, but they are too small. The third row is the least squares estimate of the coefficients which are estimated to be non-zero based on $\text{supp}(\hat{\mathbf{w}}_{L1})$. This is called **debiasing**, and is necessary because lasso shrinks the relevant coefficients as well as the irrelevant ones. The last row is the least squares estimate for all the coefficients jointly, ignoring sparsity. We see that the (debiased) sparse estimate is an excellent estimate of the original signal. By contrast, least squares without the sparsity assumption performs very poorly.

Of course, to perform model selection, we have to pick λ. It is common to use cross validation. However, it is important to note that cross validation is picking a value of λ that results in good predictive accuracy. This is not usually the same value as the one that is likely to recover the "true" model. To see why, recall that ℓ_1 regularization performs selection *and* shrinkage, that is, the chosen coefficients are brought closer to 0. In order to prevent relevant coefficients from being shrunk in this way, cross validation will tend to pick a value of λ that is not too large. Of course, this will result in a less sparse model which contains irrelevant variables (false positives). Indeed, it was proved in (Meinshausen and Buhlmann 2006) that the prediction-optimal value of λ does not result in model selection consistency. In Section 13.6.2, we will discuss some adaptive mechanisms for automatically tuning λ on a per-dimension basis that does result in model selection consistency.

A downside of using ℓ_1 regularization to select variables is that it can give quite different results if the data is perturbed slightly. The Bayesian approach, which estimates posterior marginal inclusion probabilities, $p(\gamma_j = 1 | \mathcal{D})$, is much more robust. A frequentist solution to this is to use bootstrap resampling (see Section 6.2.1), and to rerun the estimator on different versions of the data. By computing how often each variable is selected across different trials, we can approximate the posterior inclusion probabilities. This method is known as **stability selection** (Meinshausen and Buehlmann 2010).

We can threshold the stability selection (bootstrap) inclusion probabilities at some level, say 90%, and thus derive a sparse estimator. This is known as **bootstrap lasso** or **bolasso** (Bach 2008). It will include a variable if it occurs in at least 90% of sets returned by lasso (for a fixed λ). This process of intersecting the sets is a way of eliminating the false positives that vanilla lasso produces. The theoretical results in (Bach 2008) prove that bolasso is model selection consistent under a wider range of conditions than vanilla lasso.

As an illustration, we reproduced the experiments in (Bach 2008). In particular, we created

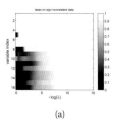
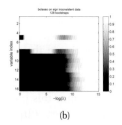
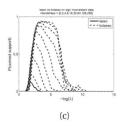

(a)   (b)   (c)

Figure 13.10   (a) Probability of selection of each variable (white = large probabilities, black = small probabilities) vs. regularization parameter for Lasso. As we move from left to right, we decrease the amount of regularization, and therefore select more variables. (b) Same as (a) but for bolasso. (c) Probability of correct sign estimation vs. regularization parameter. Bolasso (red, dashed) and Lasso (black, plain): The number of bootstrap replications is in $\{2, 4, 8, 16, 32, 64, 128, 256\}$. Based on Figures 1-3 of (Bach 2008). Figure generated by `bolassoDemo`.

256 datasets of size $N = 1000$ with $D = 16$ variables, of which 8 are relevant. See (Bach 2008) for more detail on the experimental setup. For dataset n, variable j, and sparsity level k, define $S(j, k, n) = \mathbb{I}(\hat{w}_j(\lambda_k, \mathcal{D}_n) \neq 0)$. Now define $P(j, k)$ be the average of $S(j, k, n)$ over the 256 datasets. In Figure 13.10(a-b), we plot P vs $-\log(\lambda)$ for lasso and bolasso. We see that for bolasso, there is a large range of λ where the true variables are selected, but this is not the case for lasso. This is emphasized in Figure 13.10(c), where we plot the empirical probability that the correct set of variables is recovered, for lasso and for bolasso with an increasing number of bootstrap samples. Of course, using more samples takes longer. In practice, 32 bootstraps seems to be a good compromise between speed and accuracy.

With bolasso, there is the usual issue of picking λ. Obviously we could use cross validation, but plots such as Figure 13.10(b) suggest another heuristic: shuffle the rows to create a large black block, and then pick λ to be in the middle of this region. Of course, operationalizing this intuition may be tricky, and will require various ad-hoc thresholds (it is reminiscent of the "find the knee in the curve" heuristic discussed in Section 11.5.2 when discussing how to pick K for mixture models). A Bayesian approach provides a more principled method for selecting λ.

13.3.6   Bayesian inference for linear models with Laplace priors

We have been focusing on MAP estimation in sparse linear models. It is also possible to perform Bayesian inference (see e.g., (Park and Casella 2008; Seeger 2008)). However, the posterior mean and median, as well as samples from the posterior, are not sparse; only the mode is sparse. This is another example of the phenomenon discussed in Section 5.2.1, where we said that the MAP estimate is often untypical of the bulk of the posterior.

Another argument in favor of using the posterior mean comes from Equation 5.108, which showed that plugging in the posterior mean, rather than the posterior mode, is the optimal thing to do if we want to minimize squared prediction error. (Schniter et al. 2008) shows experimentally, and (Elad and Yavnch 2009) shows theoretically, that using the posterior mean with a spike-and-slab prior results in better prediction accuracy than using the posterior mode with a Laplace prior, albeit at slightly higher computational cost.

13.4 ℓ_1 regularization: algorithms

In this section, we give a brief review of some algorithms that can be used to solve ℓ_1 regularized estimation problems. We focus on the lasso case, where we have a quadratic loss. However, most of the algorithms can be extended to more general settings, such as logistic regression (see (Yaun et al. 2010) for a comprehensive review of ℓ_1 regularized logistic regression). Note that this area of machine learning is advancing very rapidly, so the methods below may not be state of the art by the time you read this chapter. (See (Schmidt et al. 2009; Yaun et al. 2010; Yang et al. 2010) for some recent surveys.)

13.4.1 Coordinate descent

Sometimes it is hard to optimize all the variables simultaneously, but it easy to optimize them one by one. In particular, we can solve for the j'th coefficient with all the others held fixed:

$$w_j^* = \operatorname*{argmin}_z f(\mathbf{w} + z\mathbf{e}_j) - f(\mathbf{w}) \tag{13.65}$$

where $\mathbf{e}_j$ is the j'th unit vector. We can either cycle through the coordinates in a deterministic fashion, or we can sample them at random, or we can choose to update the coordinate for which the gradient is steepest.

The coordinate descent method is particularly appealing if each one-dimensional optimization problem can be solved analytically For example, the **shooting** algorithm (Fu 1998; Wu and Lange 2008) for lasso uses Equation 13.54 to compute the optimal value of w_j given all the other coefficients. See Algorithm 13.1 for the pseudo code (and `LassoShooting` for some Matlab code).

See (Yaun et al. 2010) for some extensions of this method to the logistic regression case. The resulting algorithm was the fastest method in their experimental comparison, which concerned document classification with large sparse feature vectors (representing bags of words). Other types of data (e.g., dense features and/or regression problems) might call for different algorithms.

Algorithm 13.1: Coordinate descent for lasso (aka shooting algorithm)

1 Initialize $\mathbf{w} = (\mathbf{X}^T\mathbf{X} + \lambda\mathbf{I})^{-1}\mathbf{X}^T\mathbf{y}$;
2 **repeat**
3   **for** $j = 1, \dots, D$ **do**
4         $a_j = 2\sum_{i=1}^n x_{ij}^2$;
5         $c_j = 2\sum_{i=1}^n x_{ij}(y_i - \mathbf{w}^T\mathbf{x}_i + w_j x_{ij})$;
6         $w_j = \operatorname{soft}(\frac{c_j}{a_j}, \frac{\lambda}{a_j})$;
7 **until** *converged*;

13.4.2 LARS and other homotopy methods

The problem with coordinate descent is that it only updates one variable at a time, so can be slow to converge. **Active set** methods update many variables at a time. Unfortunately, they are more complicated, because of the need to identify which variables are constrained to be zero, and which are free to be updated.

Active set methods typically only add or remove a few variables at a time, so they can take a long if they are started far from the solution. But they are ideally suited for generating a set of solutions for different values of λ, starting with the empty set, i.e., for generating the regularization path. These algorithms exploit the fact that one can quickly compute $\hat{\mathbf{w}}(\lambda_k)$ from $\hat{\mathbf{w}}(\lambda_{k-1})$ if $\lambda_k \approx \lambda_{k-1}$; this is known as **warm starting**. In fact, even if we only want the solution for a single value of λ, call it λ_*, it can sometimes be computationally more efficient to compute a set of solutions, from λ_{max} down to λ_*, using warm-starting; this is called a **continuation method** or **homotopy** method. This is often much faster than directly "cold-starting" at λ_*; this is particularly true if λ_* is small.

Perhaps the most well-known example of a homotopy method in machine learning is the **LARS** algorithm, which stands for "least angle regression and shrinkage" (Efron et al. 2004) (a similar algorithm was independently invented in (Osborne et al. 2000b,a)). This can compute $\hat{\mathbf{w}}(\lambda)$ for all possible values of λ in an efficient manner.

LARS works as follows. It starts with a large value of λ, such that only the variable that is most correlated with the response vector $\mathbf{y}$ is chosen. Then λ is decreased until a second variable is found which has the same correlation (in terms of magnitude) with the current residual as the first variable, where the residual at step k is defined as $\mathbf{r}_k = \mathbf{y} - \mathbf{X}_{:,F_k}\mathbf{w}_k$, where F_k is the current **active set** (cf., Equation 13.50). Remarkably, one can solve for this new value of λ analytically, by using a geometric argument (hence the term "least angle"). This allows the algorithm to quickly "jump" to the next point on the regularization path where the active set changes. This repeats until all the variables are added.

It is necessary to allow variables to be removed from the active set if we want the sequence of solutions to correspond to the regularization path of lasso. If we disallow variable removal, we get a slightly different algorithm called least angle regression or LAR. Remarkably, LAR costs the same as a single ordinary least squares fit, namely $O(ND\min(N,D))$, which is $O(ND^2)$ if $N > D$, and $O(N^2 D)$ if $D > N$. LAR is very similar to greedy forward selection, and a method known as least squares boosting (see Section 16.4.6).

There have been many attempts to extend the LARS algorithm to compute the full regularization path for ℓ_1 regularized GLMs, such as logistic regression. In general, one cannot analytically solve for the critical values of λ. Instead, the standard approach is to start at $\lambda_{\max}$, and then slowly decrease λ, tracking the solution as we go (i.e., to use a homotopy method). The method described in (Friedman et al. 2010) combines coordinate descent with this warm-starting strategy, and computes the full regularization path for any ℓ_1 regularized GLM. This has been implemented in the glmnet package, which is bundled with PMTK.

13.4.3 Proximal and gradient projection methods

In this section, we consider some methods that are suitable for very large scale problems, where homotopy methods made be too slow. These methods will also be easy to extend to other kinds of regularizers, beyond ℓ_1, as we will see later. Our presentation in this section is based on (Vandenberghe 2011; Yang et al. 2010).

Consider a convex objective of the form

$$f(\boldsymbol{\theta}) = L(\boldsymbol{\theta}) + R(\boldsymbol{\theta}) \tag{13.66}$$

where $L(\boldsymbol{\theta})$ (representing the loss) is convex and differentiable, and $R(\boldsymbol{\theta})$ (representing the regularizer) is convex but not necessarily differentiable. For example, $L(\boldsymbol{\theta}) = \text{RSS}(\boldsymbol{\theta})$ and $R(\boldsymbol{\theta}) = \lambda||\boldsymbol{\theta}||_1$ corresponds to the BPDN problem. As another example, the lasso problem can be formulated as follows: $L(\boldsymbol{\theta}) = \text{RSS}(\boldsymbol{\theta})$ and $R(\boldsymbol{\theta}) = I_C(\boldsymbol{\theta})$, where $C = \{\boldsymbol{\theta} : ||\boldsymbol{\theta}||_1 \leq B\}$, and $I_C(\boldsymbol{\theta})$ is the indicator function of a convex set C, defined as

$$I_C(\boldsymbol{\theta}) \triangleq \begin{cases} 0 & \boldsymbol{\theta} \in C \\ +\infty & \text{otherwise} \end{cases} \tag{13.67}$$

In some cases, it is easy to optimize functions of the form in Equation 13.66. For example, suppose $L(\boldsymbol{\theta}) = \text{RSS}(\boldsymbol{\theta})$, and the design matrix is simply $\mathbf{X} = \mathbf{I}$. Then the objective becomes $f(\boldsymbol{\theta}) = R(\boldsymbol{\theta}) + \frac{1}{2}||\boldsymbol{\theta} - \mathbf{y}||_2^2$. The minimizer of this is given by $\text{prox}_R(\mathbf{y})$, which is the **proximal operator** for the convex function R, defined by

$$\text{prox}_R(\mathbf{y}) = \underset{\mathbf{z}}{\operatorname{argmin}} \left(R(\mathbf{z}) + \frac{1}{2}||\mathbf{z} - \mathbf{y}||_2^2 \right) \tag{13.68}$$

Intuitively, we are returning a point that minimizes R but which is also close (proximal) to $\mathbf{y}$. In general, we will use this operator inside an iterative optimizer, in which case we want to stay close to the previous iterate. In this case, we use

$$\text{prox}_R(\boldsymbol{\theta}_k) = \underset{\mathbf{z}}{\operatorname{argmin}} \left(R(\mathbf{z}) + \frac{1}{2}||\mathbf{z} - \boldsymbol{\theta}_k||_2^2 \right) \tag{13.69}$$

The key issues are: how do we efficiently compute the proximal operator for different regularizers R, and how do we extend this technique to more general loss functions L? We discuss these issues below.

13.4.3.1 Proximal operators

If $R(\boldsymbol{\theta}) = \lambda||\boldsymbol{\theta}||_1$, the proximal operator is given by componentwise soft-thresholding:

$$\text{prox}_R(\boldsymbol{\theta}) = \text{soft}(\boldsymbol{\theta}, \lambda) \tag{13.70}$$

as we showed in Section 13.3.2. If $R(\boldsymbol{\theta}) = \lambda||\boldsymbol{\theta}||_0$, the proximal operator is given by componentwise hard-thresholding:

$$\text{prox}_R(\boldsymbol{\theta}) = \text{hard}(\boldsymbol{\theta}, \sqrt{2\lambda}) \tag{13.71}$$

where $\text{hard}(u, a) \triangleq u\mathbb{I}(|u| > a)$.

If $R(\boldsymbol{\theta}) = I_C(\boldsymbol{\theta})$, the proximal operator is given by the projection onto the set C:

$$\text{prox}_R(\boldsymbol{\theta}) = \underset{\mathbf{z} \in C}{\operatorname{argmin}} ||\mathbf{z} - \boldsymbol{\theta}||_2^2 = \text{proj}_C(\boldsymbol{\theta}) \tag{13.72}$$

For some convex sets, it is easy to compute the projection operator. For example, to project onto the rectangular set defined by the box constraints $C = \{\boldsymbol{\theta} : \ell_j \leq \theta_j \leq u_j\}$ we can use

$$\text{proj}_C(\boldsymbol{\theta})_j = \begin{cases} \ell_j & \theta_j \leq \ell_j \\ \theta_j & \ell_j \leq \theta_j \leq u_j \\ u_j & \theta_j \geq u_j \end{cases} \tag{13.73}$$

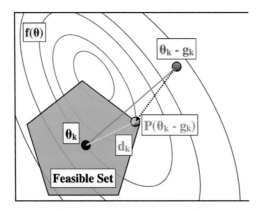

Figure 13.11   Illustration of projected gradient descent. The step along the negative gradient, to $\boldsymbol{\theta}_k - \mathbf{g}_k$, takes us outside the feasible set. If we project that point onto the closest point in the set we get $\boldsymbol{\theta}_{k+1} = \text{proj}_\Theta(\boldsymbol{\theta}_k - \mathbf{g}_k)$. We can then derive the implicit update direction using $\mathbf{d}_k = \boldsymbol{\theta}_{k+1} - \boldsymbol{\theta}_k$. Used with kind permission of Mark Schmidt.

To project onto the Euclidean ball $C = \{\boldsymbol{\theta} : ||\boldsymbol{\theta}||_2 \le 1\}$ we can use

$$\text{proj}_C(\boldsymbol{\theta}) = \begin{cases} \frac{\boldsymbol{\theta}}{||\boldsymbol{\theta}||_2} & ||\boldsymbol{\theta}||_2 > 1 \\ \boldsymbol{\theta} & ||\boldsymbol{\theta}||_2 \le 1 \end{cases} \tag{13.74}$$

To project onto the 1-norm ball $C = \{\boldsymbol{\theta} : ||\boldsymbol{\theta}||_1 \le 1\}$ we can use

$$\text{proj}_C(\boldsymbol{\theta}) = \text{soft}(\boldsymbol{\theta}, \lambda) \tag{13.75}$$

where $\lambda = 0$ if $||\boldsymbol{\theta}||_1 \le 1$, and otherwise λ is the solution to the equation

$$\sum_{j=1}^{D} \max(|\theta_j| - \lambda, 0) = 1 \tag{13.76}$$

We can implement the whole procedure in $O(D)$ time, as explained in (Duchi et al. 2008).

We will see an application of these different projection methods in Section 13.5.1.2.

13.4.3.2   Proximal gradient method

We now discuss how to use the proximal operator inside of a gradient descent routine. The basic idea is to minimize a simple quadratic approximation to the loss function[5], centered on the $\boldsymbol{\theta}_k$:

$$\boldsymbol{\theta}_{k+1} = \underset{\mathbf{z}}{\text{argmin}} \, R(\mathbf{z}) + L(\boldsymbol{\theta}_k) + \mathbf{g}_k^T(\mathbf{z} - \boldsymbol{\theta}_k) + \frac{1}{2t_k}||\mathbf{z} - \boldsymbol{\theta}_k||_2^2 \tag{13.77}$$

---

5. If we replace the Euclidean distance $||\mathbf{z} - \boldsymbol{\theta}_k||_2^2$ with a Bregman divergence, we get a method known as mirror descent (Beck and Teoulle 2003). When the parameters live on the probability simplex, a natural Bregman divergence to use is the KL divergence. In this case, mirror descent is known as the exponentiated gradient algorithm (Kivinen and Warmuth 1997).

where $\mathbf{g}_k = \nabla L(\boldsymbol{\theta}_k)$ is the gradient of the loss, t_k is a constant discussed below, and the last term arises from a simple approximation to the Hessian of the loss of the form $\nabla^2 L(\boldsymbol{\theta}_k) \approx \frac{1}{t_k}\mathbf{I}$.

Dropping terms that are independent of $\mathbf{z}$, and multiplying by t_k, we can rewrite the above expression in terms of a proximal operator as follows:

$$\boldsymbol{\theta}_{k+1} = \underset{\mathbf{z}}{\text{argmin}} \left[t_k R(\mathbf{z}) + \frac{1}{2}||\mathbf{z} - \mathbf{u}_k||_2^2 \right] = \text{prox}_{t_k R}(\mathbf{u}_k) \tag{13.78}$$

$$\mathbf{u}_k = \boldsymbol{\theta}_k - t_k \mathbf{g}_k \tag{13.79}$$

$$\mathbf{g}_k = \nabla L(\boldsymbol{\theta}_k) \tag{13.80}$$

If $R(\boldsymbol{\theta}) = 0$, this is equivalent to gradient descent. If $R(\boldsymbol{\theta}) = I_C(\boldsymbol{\theta})$, the method is equivalent to **projected gradient descent**, sketched in Figure 13.11. If $R(\boldsymbol{\theta}) = \lambda||\boldsymbol{\theta}||_1$, the method is known as **iterative soft thresholding**.

There are several ways to pick t_k, or equivalently, $\alpha_k = 1/t_k$. Given that $\alpha_k \mathbf{I}$ is an approximation to the Hessian $\nabla^2 L$, we require that

$$\alpha_k(\boldsymbol{\theta}_k - \boldsymbol{\theta}_{k-1}) \approx \mathbf{g}_k - \mathbf{g}_{k-1} \tag{13.81}$$

in the least squares sense. Hence

$$\alpha_k = \underset{\alpha}{\text{argmin}} \, ||\alpha(\boldsymbol{\theta}_k - \boldsymbol{\theta}_{k-1}) - (\mathbf{g}_k - \mathbf{g}_{k-1})||_2^2 = \frac{(\boldsymbol{\theta}_k - \boldsymbol{\theta}_{k-1})^T (\mathbf{g}_k - \mathbf{g}_{k-1})}{(\boldsymbol{\theta}_k - \boldsymbol{\theta}_{k-1})^T (\boldsymbol{\theta}_k - \boldsymbol{\theta}_{k-1})} \tag{13.82}$$

This is known as the **Barzilai-Borwein** (BB) or **spectral** stepsize (Barzilai and Borwein 1988; Fletcher 2005; Raydan 1997). This stepsize can be used with any gradient method, whether proximal or not. It does not lead to monotonic decrease of the objective, but it is much faster than standard line search techniques. (To ensure convergence, we require that the objective decrease "on average", where the average is computed over a sliding window of size $M + 1$.)

When we combine the BB stepsize with the iterative soft thresholding technique (for $R(\boldsymbol{\theta}) = \lambda||\boldsymbol{\theta}||_1$), plus a continuation method that gradually reduces λ, we get a fast method for the BPDN problem known as the SpaRSA algorithm, which stands for "sparse reconstruction by separable approximation" (Wright et al. 2009). However, we will call it the iterative shrinkage and thresholding algorithm. See Algorithm 12 for some pseudocode, and SpaRSA for some Matlab code. See also Exercise 13.11 for a related approach based on projected gradient descent.

13.4.3.3 Nesterov's method

A faster version of proximal gradient descent can be obtained by expanding the quadratic approximation around a point other than the most recent parameter value. In particular, consider performing updates of the form

$$\boldsymbol{\theta}_{k+1} = \text{prox}_{t_k R}(\boldsymbol{\phi}_k - t_k \mathbf{g}_k) \tag{13.83}$$

$$\mathbf{g}_k = \nabla L(\boldsymbol{\phi}_k) \tag{13.84}$$

$$\boldsymbol{\phi}_k = \boldsymbol{\theta}_k + \frac{k-1}{k+2}(\boldsymbol{\theta}_k - \boldsymbol{\theta}_{k-1}) \tag{13.85}$$

This is known as **Nesterov's method** (Nesterov 2004; Tseng 2008). As before, there are a variety of ways of setting t_k; typically one uses line search.

Algorithm 13.2: Iterative Shrinkage and Thresholding Algorithm (ISTA) for BPDN

---

1 Input: $\mathbf{X} \in \mathbb{R}^{N \times D}$, $\mathbf{y} \in \mathbb{R}^N$, parameters $\lambda \geq 0$, $M \geq 1$, $0 < s < 1$;
2 Initialize $\boldsymbol{\theta}_0 = \mathbf{0}$, $\alpha = 1$, $\mathbf{r} = \mathbf{y}$, $\lambda_0 = \infty$;
3 **repeat**
4 $\quad$ $\lambda_t = \max(s||\mathbf{X}^T\mathbf{r}||_\infty, \lambda)$ // Adapt the regularizer ;
5 $\quad$ **repeat**
6 $\quad\quad$ $\mathbf{g} = \nabla L(\boldsymbol{\theta})$;
7 $\quad\quad$ $\mathbf{u} = \boldsymbol{\theta} - \frac{1}{\alpha}\mathbf{g}$;
8 $\quad\quad$ $\boldsymbol{\theta} = \text{soft}(\mathbf{u}, \frac{\lambda_t}{\alpha})$;
9 $\quad\quad$ Update α using BB stepsize in Equation 13.82 ;
10 $\quad$ **until** $f(\boldsymbol{\theta})$ *increased too much within the past M steps*;
11 $\quad$ $\mathbf{r} = \mathbf{y} - \mathbf{X}\boldsymbol{\theta}$ // Update residual ;
12 **until** $\lambda_t = \lambda$;

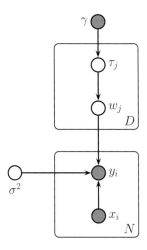

Figure 13.12   Representing lasso using a Gaussian scale mixture prior.

When this method is combined with the iterative soft thresholding technique (for $R(\boldsymbol{\theta}) = \lambda||\boldsymbol{\theta}||_1$), plus a continuation method that gradually reduces λ, we get a fast method for the BPDN problem known as the **fast iterative shrinkage thesholding algorithm** or **FISTA** (Beck and Teboulle 2009).

13.4.4   EM for lasso

In this section, we show how to solve the lasso problem using lasso. At first sight, this might seem odd, since there are no hidden variables. The key insight is that we can represent the Laplace distribution as a **Gaussian scale mixture** (GSM) (Andrews and Mallows 1974; West 1987)

as follows:

$$\text{Lap}(w_j|0, 1/\gamma) = \frac{\gamma}{2}e^{-\gamma|w_j|} = \int \mathcal{N}(w_j|0, \tau_j^2)\text{Ga}(\tau_j^2|1, \frac{\gamma^2}{2})d\tau_j^2 \tag{13.86}$$

Thus the Laplace is a GSM where the mixing distibution on the variances is the exponential distribution, $\text{Expon}(\tau_j^2|\frac{\gamma^2}{2}) = \text{Ga}(\tau_j^2|1, \frac{\gamma^2}{2})$. Using this decomposition, we can represent the lasso model as shown in Figure 13.12. The corresponding joint distribution has the form

$$p(\mathbf{y}, \mathbf{w}, \boldsymbol{\tau}, \sigma^2|\mathbf{X})$$

$$= \mathcal{N}(\mathbf{y}|\mathbf{Xw}, \sigma^2\mathbf{I}_N)\,\mathcal{N}(\mathbf{w}|\mathbf{0}, \boldsymbol{\Lambda}_\tau)\,\left[\prod_j \text{Ga}(\tau_j^2|1, \gamma^2/2)\right]\,\text{IG}(\sigma^2|a_\sigma, b_\sigma) \tag{13.87}$$

where $\boldsymbol{\Lambda}_\tau = \text{diag}(1/\tau_j^2)$, and where we have assumed for notational simplicity that $\mathbf{X}$ is standardized and that $\mathbf{y}$ is centered (so we can ignore the offset term μ). Expanding out, we get

$$p(\mathbf{y}, \mathbf{w}, \boldsymbol{\tau}, \sigma^2|\mathbf{X}) \quad \propto \quad (\sigma^2)^{-N/2}\exp\left(-\frac{1}{2\sigma^2}||\mathbf{y} - \mathbf{Xw}||_2^2\right)$$

$$|\boldsymbol{\Lambda}_\tau|^{-\frac{1}{2}}\,\exp\left(-\frac{1}{2}\mathbf{w}^T\boldsymbol{\Lambda}_\tau^{-1}\mathbf{w}\right)$$

$$\left[\prod_j \exp(-\frac{\gamma^2}{2}\tau_j^2)\right]\,\exp(-b_\sigma/\sigma^2)\,(\sigma^2)^{-(a_\sigma+1)} \tag{13.88}$$

Below we describe how to apply the EM algorithm to optimize this equation.[6] In brief, in the E step we infer τ_j^2 and σ^2, and in the M step we estimate $\mathbf{w}$. The resulting estimate $\hat{\mathbf{w}}$ is the same as the lasso estimator. This approach was first proposed in (Figueiredo 2003) (see also (Griffin and Brown 2007; Caron and Doucet 2008; Ding and Harrison 2010) for some extensions).

13.4.4.1 Why EM?

Before going into the details of EM, it is worthwhile asking why we are presenting this approach at all, given that there are a variety of other (often much faster) algorithms that directly solve the ℓ_1 MAP estimation problem (see `linregFitL1Test` for an empirical comparison). The reason is that the latent variable perspective brings several advantages, such as the following:

- It provides an easy way to derive an algorithm to find ℓ_1-regularized parameter estimates for a variety of other models, such as robust linear regression (Exercise 11.12) or probit regression (Exercise 13.9).

- It suggests trying other priors on the variances besides $\text{Ga}(\tau_j^2|1, \gamma^2/2)$. We will consider various extensions below.

- It makes it clear how we can compute the full posterior, $p(\mathbf{w}|\mathcal{D})$, rather than just a MAP estimate. This technique is known as the **Bayesian lasso** (Park and Casella 2008; Hans 2009).

---

6. To ensure the posterior is unimodal, one can follow (Park and Casella 2008) and slightly modify the model by making the prior variance for the weights depend on the observation noise: $p(w_j|\tau_j^2, \sigma^2) = \mathcal{N}(w_j|0, \sigma^2\tau_j^2)$. The EM algorithm is easy to modify.

13.4.4.2   The objective function

From Equation 13.88, the complete data penalized log likelihood is as follows (dropping terms that do not depend on $\mathbf{w}$):

$$\ell_c(\mathbf{w}) \quad = \quad -\frac{1}{2\sigma^2}||\mathbf{y} - \mathbf{Xw}||_2^2 - \frac{1}{2}\mathbf{w}^T\boldsymbol{\Lambda}_\tau\mathbf{w} + \text{const} \tag{13.89}$$

13.4.4.3   The E step

We can rewrite the prior term as the scalar $\text{tr}(\boldsymbol{\Lambda}_\tau\mathbf{w}\mathbf{w}^T)$, which is linear in $\boldsymbol{\Lambda}_\tau$. Hence we can substitute the expectation of $\boldsymbol{\Lambda}_\tau$ into the complete log likelihood, and then solve for $\mathbf{w}$ in the M step. Thus the key is to compute $\mathbb{E}\left[1/\tau_j^2|\mathbf{w}, \mathcal{D}\right]$, which we shall denote by $\mathbb{E}\left[1/\tau_j^2\right]$ for brevity (dropping the explicit conditioning). We can derive this directly, as shown in Exercise 13.8. However, to facilitate a Bayesian treatment of this model, it is useful to derive the full posterior for $1/\tau_j^2$. This is given by the following (Park and Casella 2008):

$$p(1/\tau_j^2|\mathbf{w}, \mathcal{D}) \quad = \quad \text{InverseGaussian}\left(\sqrt{\frac{\gamma^2}{w_j^2}}, \gamma^2\right) \tag{13.90}$$

(Note that the **inverse Gaussian** distribution is also known as the Wald distribution.) Hence

$$\mathbb{E}\left[\frac{1}{\tau_j^2}\right] = \frac{\gamma}{|w_j|} \tag{13.91}$$

Let $\overline{\boldsymbol{\Lambda}} = \text{diag}(\mathbb{E}\left[1/\tau_1^2\right], \ldots, \mathbb{E}\left[1/\tau_D^2\right])$ denote the result of this E step. Since many w_j elements may be zero, the expression $\gamma/|w_j|$ may not be defined. Fortunately, we can express the M step in terms of $\overline{\boldsymbol{\Lambda}}^{-1}$, as we will see below, which avoids these numerical problems.

We also need to infer σ^2. It is easy to show that that the posterior is

$$p(\sigma^2|\mathcal{D}, \mathbf{w}) = \text{IG}(a_\sigma + N/2, \ b_\sigma + \frac{1}{2}(\mathbf{y} - \mathbf{X}\hat{\mathbf{w}})^T(\mathbf{y} - \mathbf{X}\hat{\mathbf{w}})) = \text{IG}(a_N, b_N) \tag{13.92}$$

Hence

$$\mathbb{E}\left[1/\sigma^2\right] = \frac{a_N}{b_N} \triangleq \overline{\omega} \tag{13.93}$$

13.4.4.4   The M step

The M step consists of computing

$$\hat{\mathbf{w}} = \underset{\mathbf{w}}{\text{argmax}} -\frac{1}{2}\overline{\omega}||\mathbf{y} - \mathbf{Xw}||_2^2 - \frac{1}{2}\mathbf{w}^T\overline{\boldsymbol{\Lambda}}\mathbf{w} \tag{13.94}$$

This is just MAP estimation under a Gaussian prior:

$$\hat{\mathbf{w}} = (\frac{1}{\overline{\omega}}\overline{\boldsymbol{\Lambda}} + \mathbf{X}^T\mathbf{X})^{-1}\mathbf{X}^T\mathbf{y} \tag{13.95}$$

However, as we mentioned above, we expect many $w_j = 0$, making inverting $\overline{\Lambda}$ numerically unstable. Fortunately, we can use the SVD of $\mathbf{X}$, given by $\mathbf{X} = \mathbf{UDV}^T$, and then solve for $\mathbf{w}$ as follows:

$$\hat{\mathbf{w}} \;=\; \overline{\Lambda}^{-1}\mathbf{V}(\mathbf{V}^T\overline{\Lambda}^{-1}\mathbf{V} + \frac{1}{\omega}\mathbf{D}^{-2})^{-1}\mathbf{D}^{-1}\mathbf{U}^T\mathbf{y} \tag{13.96}$$

where

$$\overline{\Lambda}^{-1} = \mathrm{diag}\left(\frac{1}{\mathbb{E}\left[1/\tau_j^2\right]}\right) = \mathrm{diag}\left(\frac{|w_j|}{\gamma}\right) \tag{13.97}$$

13.4.4.5   Caveat

Since the lasso objective is convex, this method should always find the global optimum. Unfortunately, this sometimes does not happen, for numerical reasons. In particular, suppose that in the true solution, $w_j^* \neq 0$. Further, suppose that we set $\hat{w}_j = 0$ in an M step. In the following E step we infer that $\tau_j^2 = 0$, so then we set $\hat{w}_j = 0$ again; thus we can never "undo" our mistake. Fortunately, in practice, this situation seems to be rare. See (Hunter and Li 2005) for further discussion.

13.5 ℓ_1 regularization: extensions

In this section, we discuss various extensions of "vanilla" ℓ_1 regularization.

13.5.1   Group lasso

In standard ℓ_1 regularization, we assume that there is a 1:1 correspondence between parameters and variables, so that if $\hat{w}_j = 0$, we interpret this to mean that variable j is excluded. But in more complex models, there may be many parameters associated with a given variable. In particular, we may have a vector of weights for each input, $\mathbf{w}_j$. Here are some examples:

- **Multinomial logistic regression** Each feature is associated with C different weights, one per class.
- **Linear regression with categorical inputs** Each scalar input is one-hot encoded into a vector of length C.
- **Multi-task learning** In multi-task learning, we have multiple related prediction problems. For example, we might have C separate regression or binary classification problems. Thus each feature is associated with C different weights. We may want to use a feature for all of the tasks or none of the tasks, and thus select weights at the group level (Obozinski et al. 2007).

If we use an ℓ_1 regularizer of the form $||\mathbf{w}|| = \sum_j \sum_c |w_{jc}|$, we may end up with with some elements of $\mathbf{w}_{j,:}$ being zero and some not. To prevent this kind of situation, we partition the parameter vector into G groups. We now minimize the following objective

$$J(\mathbf{w}) = \mathrm{NLL}(\mathbf{w}) + \sum_{g=1}^{G} \lambda_g ||\mathbf{w}_g||_2 \tag{13.98}$$

where

$$\|\mathbf{w}_g\|_2 = \sqrt{\sum_{j \in g} w_j^2} \tag{13.99}$$

is the 2-norm of the group weight vector. If the NLL is least squares, this method is called **group lasso** (Yuan and Lin 2006).

We often use a larger penalty for larger groups, by setting $\lambda_g = \lambda \sqrt{d_g}$, where d_g is the number of elements in group g. For example, if we have groups $\{1, 2\}$ and $\{3, 4, 5\}$, the objective becomes

$$J(\mathbf{w}) = \text{NLL}(\mathbf{w}) + \lambda \left[\sqrt{2}\sqrt{(w_1^2 + w_2^2|)} + \sqrt{3}\sqrt{(w_3^2 + w_4^2 + w_5^2)} \right] \tag{13.100}$$

Note that if we had used the square of the 2-norms, the model would become equivalent to ridge regression, since

$$\sum_{g=1}^{G} \|\mathbf{w}_g\|_2^2 = \sum_{g} \sum_{j \in g} w_j^2 = \|\mathbf{w}\|_2^2 \tag{13.101}$$

By using the square root, we are penalizing the radius of a ball containing the group's weight vector: the only way for the radius to be small is if all elements are small. Thus the square root results in group sparsity.

A variant of this technique replaces the 2-norm with the infinity-norm (Turlach et al. 2005; Zhao et al. 2005):

$$\|\mathbf{w}_g\|_\infty = \max_{j \in g} |w_j| \tag{13.102}$$

It is clear that this will also result in group sparsity.

An illustration of the difference is shown in Figures 13.13 and 13.14. In both cases, we have a true signal $\mathbf{w}$ of size $D = 2^{12} = 4096$, divided into 64 groups each of size 64. We randomly choose 8 groups of $\mathbf{w}$ and assign them non-zero values. In the first example, the values are drawn from a $\mathcal{N}(0, 1)$. In the second example, the values are all set to 1. We then pick a random design matrix $\mathbf{X}$ of size $N \times D$, where $N = 2^{10} = 1024$. Finally, we generate $\mathbf{y} = \mathbf{X}\mathbf{w} + \boldsymbol{\epsilon}$, where $\boldsymbol{\epsilon} \sim \mathcal{N}(\mathbf{0}, 10^{-4}\mathbf{I}_N)$. Given this data, we estimate the support of $\mathbf{w}$ using ℓ_1 or group ℓ_1, and then estimate the non-zero values using least squares. We see that group lasso does a much better job than vanilla lasso, since it respects the known group structure.[7] We also see that the ℓ_∞ norm has a tendency to make all the elements within a block to have similar magnitude. This is appropriate in the second example, but not the first. (The value of λ was the same in all examples, and was chosen by hand.)

13.5.1.1   GSM interpretation of group lasso

Group lasso is equivalent to MAP estimation using the following prior

$$p(\mathbf{w}|\gamma, \sigma^2) \propto \exp\left(-\frac{\gamma}{\sigma} \sum_{g=1}^{G} \|\mathbf{w}_g\|_2 \right) \tag{13.103}$$

---

7. The slight non-zero "noise" in the ℓ_∞ group lasso results is presumably due to numerical errors.

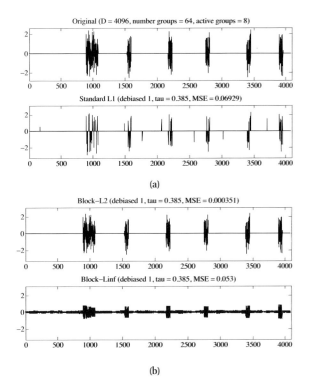

Figure 13.13 Illustration of group lasso where the original signal is piecewise Gaussian. First row: original signal. Second row: vanilla lasso estimate. Third row: group lasso estimate using a ℓ_2 norm on the blocks. Bottom row: group lasso estimate using an ℓ_∞ norm on the blocks. Based on Figures 3-4 of (Wright et al. 2009). Figure generated by `groupLassoDemo`, based on code by Mario Figueiredo.

Now one can show (Exercise 13.10) that this prior can be written as a GSM, as follows:

$$\mathbf{w}_g | \sigma^2, \tau_g^2 \quad \sim \quad \mathcal{N}(\mathbf{0}, \sigma^2 \tau_g^2 \mathbf{I}_{d_g}) \tag{13.104}$$

$$\tau_g^2 | \gamma \quad \sim \quad \text{Ga}\left(\frac{d_g + 1}{2}, \frac{\gamma}{2}\right) \tag{13.105}$$

where d_g is the size of group g. So we see that there is one variance term per group, each of which comes from a Gamma prior, whose shape parameter depends on the group size, and whose rate parameter is controlled by γ. Figure 13.15 gives an example, where we have 2 groups, one of size 2 and one of size 3.

This picture also makes it clearer why there should be a grouping effect. Suppose $w_{1,1}$ is small; then τ_1^2 will be estimated to be small, which will force $w_{1,2}$ to be small. Converseley, suppose $w_{1,1}$ is large; then τ_1^2 will be estimated to be large, which will allow $w_{1,2}$ to be become large as well.

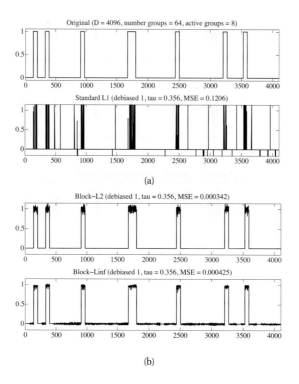

Figure 13.14   Same as Figure 13.13, except the original signal is piecewise constant.

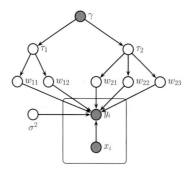

Figure 13.15   Graphical model for group lasso with 2 groups, the first has size $G_1 = 2$, the second has size $G_2 = 3$.

13.5.1.2 Algorithms for group lasso

There are a variety of algorithms for group lasso. Here we briefly mention two. The first approach is based on proximal gradient descent, discussed in Section 13.4.3. Since the regularizer is separable, $R(\mathbf{w}) = \sum_g ||\mathbf{w}_g||_p$, the proximal operator decomposes into G separate operators of the form

$$\text{prox}_R(\mathbf{b}) = \underset{\mathbf{z} \in \mathbb{R}^{D_g}}{\text{argmin}} ||\mathbf{z} - \mathbf{b}||_2^2 + \lambda ||\mathbf{z}||_p \tag{13.106}$$

where $\mathbf{b} = \boldsymbol{\theta}_{kg} - t_k \mathbf{g}_{kg}$. If $p = 2$, one can show (Combettes and Wajs 2005) that this can be implemented as follows

$$\text{prox}_R(\mathbf{b}) = \mathbf{b} - \text{proj}_{\lambda C}(\mathbf{b}) \tag{13.107}$$

where $C = \{\mathbf{z} : ||\mathbf{z}||_2 \le 1\}$ is the ℓ_2 ball. Using Equation 13.74, if $||\mathbf{b}||_2 < \lambda$, we have

$$\text{prox}_R(\mathbf{b}) = \mathbf{b} - \mathbf{b} = \mathbf{0} \tag{13.108}$$

otherwise we have

$$\text{prox}_R(\mathbf{b}) = \mathbf{b} - \lambda \frac{\mathbf{b}}{||\mathbf{b}||_2} = \mathbf{b} \frac{||\mathbf{b}||_2 - \lambda}{||\mathbf{b}||_2} \tag{13.109}$$

We can combine these into a vectorial soft-threshold function as follows (Wright et al. 2009):

$$\text{prox}_R(\mathbf{b}) = \mathbf{b} \frac{\max(||\mathbf{b}||_2 - \lambda, 0)}{\max(||\mathbf{b}||_2 - \lambda, 0) + \lambda} \tag{13.110}$$

If $p = \infty$, we use $C = \{\mathbf{z} : ||\mathbf{z}||_1 \le 1\}$, which is the ℓ_1 ball. We can project onto this in $O(d_g)$ time using an algorithm described in (Duchi et al. 2008).

Another approach is to modify the EM algorithm. The method is almost the same as for vanilla lasso. If we define $\tau_j^2 = \tau_{g(j)}^2$, where $g(j)$ is the group to which dimension j belongs, we can use the same full conditionals for σ^2 and $\mathbf{w}$ as before. The only changes are as follows:

- We must modify the full conditional for the weight precisions, which are estimated based on a shared set of weights:

$$\frac{1}{\tau_g^2} | \gamma, \mathbf{w}, \sigma^2, \mathbf{y}, \mathbf{X} \sim \text{InverseGaussian}\left(\sqrt{\frac{\gamma^2 \sigma^2}{||\mathbf{w}_g||_2^2}}, \gamma^2\right) \tag{13.111}$$

  where $||\mathbf{w}_g||_2^2 = \sum_{j \in g} w_{jg}^2$. For the E step, we can use

$$\mathbb{E}\left[\frac{1}{\tau_g^2}\right] = \frac{\gamma \sigma}{||\mathbf{w}_g||_2} \tag{13.112}$$

- We must modify the full conditional for the tuning parameter, which is now only estimated based on G values of τ_g^2:

$$p(\gamma^2 | \boldsymbol{\tau}) = \text{Ga}(a_\gamma + G/2, b_\gamma + \frac{1}{2} \sum_g^G \tau_g^2) \tag{13.113}$$

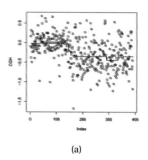  

(a) (b) (c)

Figure 13.16 (a) Example of the fused lasso. The vertical axis represents array CGH (chromosomal genome hybridization) intensity, and the horizontal axis represents location along a genome. Source: Figure 1 of (Hoefling 2010). (b) Noisy image. (c) Fused lasso estimate using 2d lattice prior. Source: Figure 2 of (Hoefling 2010). Used with kind permission of Holger Hoefling.

13.5.2 Fused lasso

In some problem settings (e.g., functional data analysis), we want neighboring coefficients to be similar to each other, in addition to being sparse. An example is given in Figure 13.16(a), where we want to fit a signal that is mostly "off", but in addition has the property that neighboring locations are typically similar in value. We can model this by using a prior of the form

$$p(\mathbf{w}|\sigma^2) \propto \exp\left(-\frac{\lambda_1}{\sigma}\sum_{j=1}^{D}|w_j| - \frac{\lambda_2}{\sigma}\sum_{j=1}^{D-1}|w_{j+1} - w_j|\right) \qquad (13.114)$$

This is known as the **fused lasso** penalty. In the context of functional data analysis, we often use $\mathbf{X} = \mathbf{I}$, so there is one coefficient for each location in the signal (see Section 4.4.2.3). In this case, the overall objective has the form

$$J(\mathbf{w}, \lambda_1, \lambda_2) = \sum_{i=1}^{N}(y_i - w_i)^2 + \lambda_1 \sum_{i=1}^{N}|w_i| + \lambda_2 \sum_{i=1}^{N-1}|w_{i+1} - w_i| \qquad (13.115)$$

This is a sparse version of Equation 4.148.

It is possible to generalize this idea beyond chains, and to consider other graph structures, using a penalty of the form

$$J(\mathbf{w}, \lambda_1, \lambda_2) = \sum_{s \in V}(y_s - w_s)^2 + \lambda_1 \sum_{s \in V}|w_s| + \lambda_2 \sum_{(s,t) \in E}|w_s - w_t| \qquad (13.116)$$

This is called **graph-guided fused lasso** (see e.g., (Chen et al. 2010)). The graph might come from some prior knowledge, e.g., from a database of known biological pathways. Another example is shown in Figure 13.16(b-c), where the graph structure is a 2d lattice.

13.5.2.1 GSM interpretation of fused lasso

One can show (Kyung et al. 2010) that the fused lasso model is equivalent to the following hierarchical model

$$\mathbf{w}|\sigma^2, \boldsymbol{\tau}, \boldsymbol{\omega} \quad \sim \quad \mathcal{N}(\mathbf{0}, \sigma^2 \boldsymbol{\Sigma}(\boldsymbol{\tau}, \boldsymbol{\omega})) \tag{13.117}$$

$$\tau_j^2|\gamma_1 \quad \sim \quad \text{Expon}(\frac{\gamma_1^2}{2}), \quad j = 1 : D \tag{13.118}$$

$$\omega_j^2|\gamma_2 \quad \sim \quad \text{Expon}(\frac{\gamma_2^2}{2}), \quad j = 1 : D - 1 \tag{13.119}$$

where $\boldsymbol{\Sigma} = \boldsymbol{\Omega}^{-1}$, and $\boldsymbol{\Omega}$ is a tridiagonal precision matrix with

$$\text{main diagonal} \quad = \quad \{\frac{1}{\tau_j^2} + \frac{1}{\omega_{j-1}^2} + \frac{1}{\omega_j^2}\} \tag{13.120}$$

$$\text{off diagonal} \quad = \quad \{-\frac{1}{\omega_j^2}\} \tag{13.121}$$

where we have defined $\omega_0^{-2} = \omega_D^{-2} = 0$. This is very similar to the model in Section 4.4.2.3, where we used a chain-structured Gaussian Markov random field as the prior, with fixed variance. Here we just let the variance be random. In the case of graph-guided lasso, the structure of the graph is reflected in the zero pattern of the Gaussian precision matrix (see Section 19.4.4).

13.5.2.2 Algorithms for fused lasso

It is possible to generalize the EM algorithm to fit the fused lasso model, by exploiting the Markov structure of the Gaussian prior for efficiency. Direct solvers (which don't use the latent variable trick) can also be derived (see e.g., (Hoefling 2010)). However, this model is undeniably more expensive to fit than the other variants we have considered.

13.5.3 Elastic net (ridge and lasso combined)

Although lasso has proved to be effective as a variable selection technique, it has several problems (Zou and Hastie 2005), such as the following:

- If there is a group of variables that are highly correlated (e.g., genes that are in the same pathway), then the lasso tends to select only one of them, chosen rather arbitrarily. (This is evident from the LARS algorithm: once one member of the group has been chosen, the remaining members of the group will not be very correlated with the new residual and hence will not be chosen.) It is usually better to select all the relevant variables in a group. If we know the grouping structure, we can use group lasso, but often we don't know the grouping structure.

- In the $D > N$ case, lasso can select at most N variables before it saturates.

- If $N > D$, but the variables are correlated, it has been empirically observed that the prediction performance of ridge is better than that of lasso.

Zou and Hastie (Zou and Hastie 2005) proposed an approach called the **elastic net**, which is a hybrid between lasso and ridge regression, which solves all of these problems. It is apparently called the "elastic net" because it is "like a stretchable fishing net that retains 'all the big fish'" (Zou and Hastie 2005).

13.5.3.1 Vanilla version

The vanilla version of the model defines the following objective function:

$$J(\mathbf{w}, \lambda_1, \lambda_2) = ||\mathbf{y} - \mathbf{X}\mathbf{w}||^2 + \lambda_2||\mathbf{w}||_2^2 + \lambda_1||\mathbf{w}||_1 \tag{13.122}$$

Notice that this penalty function is *strictly convex* (assuming $\lambda_2 > 0$) so there is a unique global minimum, even if $\mathbf{X}$ is not full rank.

It can be shown (Zou and Hastie 2005) that any strictly convex penalty on $\mathbf{w}$ will exhibit a **grouping effect**, which means that the regression coefficients of highly correlated variables tend to be equal (up to a change of sign if they are negatively correlated). For example, if two features are equal, so $\mathbf{X}_{:j} = \mathbf{X}_{:k}$, one can show that their estimates are also equal, $\hat{w}_j = \hat{w}_k$. By contrast, with lasso, we may have that $\hat{w}_j = 0$ and $\hat{w}_k \neq 0$ or vice versa.

13.5.3.2 Algorithms for vanilla elastic net

It is simple to show (Exercise 13.5) that the elastic net problem can be reduced to a lasso problem on modified data. In particular, define

$$\tilde{\mathbf{X}} = c \begin{pmatrix} \mathbf{X} \\ \sqrt{\lambda_2}\mathbf{I}_D \end{pmatrix}, \quad \tilde{\mathbf{y}} = \begin{pmatrix} \mathbf{y} \\ \mathbf{0}_{D \times 1} \end{pmatrix} \tag{13.123}$$

where $c = (1 + \lambda_2)^{-\frac{1}{2}}$. Then we solve

$$\tilde{\mathbf{w}} = \arg\min_{\tilde{\mathbf{w}}} ||\tilde{\mathbf{y}} - \tilde{\mathbf{X}}\tilde{\mathbf{w}}||^2 + c\lambda_1||\tilde{\mathbf{w}}||_1 \tag{13.124}$$

and set $\mathbf{w} = c\tilde{\mathbf{w}}$.

We can use LARS to solve this subproblem; this is known as the LARS-EN algorithm. If we stop the algorithm after m variables have been included, the cost is $O(m^3 + Dm^2)$. Note that we can use $m = D$ if we wish, since $\tilde{\mathbf{X}}$ has rank D. This is in contrast to lasso, which cannot select more than N variables (before jumping to the OLS solution) if $N < D$.

When using LARS-EN (or other ℓ_1 solvers), one typically uses cross-validation to select λ_1 and λ_2.

13.5.3.3 Improved version

Unfortunately it turns out that the "vanilla" elastic net does not produce functions that predict very accurately, unless it is very close to either pure ridge or pure lasso. Intuitively the reason is that it performs shrinkage twice: once due to the ℓ_2 penalty and again due to the ℓ_1 penalty. The solution is simple: undo the ℓ_2 shrinkage by scaling up the estimates from the vanilla version. In other words, if $\mathbf{w}^*$ is the solution of Equation 13.124, then a better estimate is

$$\hat{\mathbf{w}} = \sqrt{1 + \lambda_2}\tilde{\mathbf{w}} \tag{13.125}$$

We will call this a corrected estimate.

One can show that the corrected estimates are given by

$$\hat{\mathbf{w}} = \arg\min_{\mathbf{w}} \mathbf{w}^T \left(\frac{\mathbf{X}^T \mathbf{X} + \lambda_2 \mathbf{I}}{1 + \lambda_2} \right) \mathbf{w} - 2\mathbf{y}^T \mathbf{X} \mathbf{w} + \lambda_1 ||\mathbf{w}||_1 \qquad (13.126)$$

Now

$$\frac{\mathbf{X}^T \mathbf{X} + \lambda_2 \mathbf{I}}{1 + \lambda_2} = (1 - \rho)\hat{\boldsymbol{\Sigma}} + \rho \mathbf{I} \qquad (13.127)$$

where $\rho = \lambda_2/(1 + \lambda_2)$. So the the elastic net is like lasso but where we use a version of $\hat{\boldsymbol{\Sigma}}$ that is shrunk towards $\mathbf{I}$. (See Section 4.2.6 for more discussion of regularized estimates of covariance matrices.)

13.5.3.4   GSM interpretation of elastic net

The implicit prior being used by the elastic net obviously has the form

$$p(\mathbf{w}|\sigma^2) \propto \exp \left(-\frac{\gamma_1}{\sigma} \sum_{j=1}^{D} |w_j| - \frac{\gamma_2}{2\sigma^2} \sum_{j=1}^{D} w_j^2 \right) \qquad (13.128)$$

which is just a product of Gaussian and Laplace distributions.

This can be written as a hierarchical prior as follows (Kyung et al. 2010; Chen et al. 2011):

$$w_j|\sigma^2, \tau_j^2 \quad \sim \quad \mathcal{N}(0, \sigma^2(\tau_j^{-2} + \gamma_2)^{-1}) \qquad (13.129)$$

$$\tau_j^2|\gamma_1 \quad \sim \quad \text{Expon}(\frac{\gamma_1^2}{2}) \qquad (13.130)$$

Clearly if $\gamma_2 = 0$, this reduces to the regular lasso.

It is possible to perform MAP estimation in this model using EM, or Bayesian inference using MCMC (Kyung et al. 2010) or variational Bayes (Chen et al. 2011).

13.6   Non-convex regularizers

Although the Laplace prior results in a convex optimization problem, from a statistical point of view this prior is not ideal. There are two main problems with it. First, it does not put enough probability mass near 0, so it does not sufficiently suppress noise. Second, it does not put enough probability mass on large values, so it causes shrinkage of relevant coefficients, corresponding to "signal". (This can be seen in Figure 13.5(a): we see that ℓ_1 estimates of large coefficients are significantly smaller than their ML estimates, a phenomenon known as bias.)

Both problems can be solved by going to more flexible kinds of priors which have a larger spike at 0 and heavier tails. Even though we cannot find the global optimum anymore, these non-convex methods often outperform ℓ_1 regularization, both in terms of predictive accuracy and in detecting relevant variables (Fan and Li 2001; Schniter et al. 2008). We give some examples below.

13.6.1   Bridge regression

A natural generalization of ℓ_1 regularization, known as **bridge regression** (Frank and Friedman 1993), has the form

$$\hat{\mathbf{w}} = \text{NLL}(\mathbf{w}) + \lambda \sum_j |w_j|^b \tag{13.131}$$

for $b \geq 0$. This corresponds to MAP estimation using a **exponential power distribution** given by

$$\text{ExpPower}(w|\mu, a, b) \triangleq \frac{b}{2a\Gamma(1 + 1/b)} \exp\left(-\frac{|x - \mu|^b}{a}\right) \tag{13.132}$$

If $b = 2$, we get the Gaussian distribution (with $a = \sigma\sqrt{2}$), corresponding to ridge regression; if we set $b = 1$, we get the Laplace distribution, corresponding to lasso; if we set $b = 0$, we get ℓ_0 regression, which is equivalent to best subset selection. Unfortunately, the objective is not convex for $b < 1$, and is not sparsity promoting for $b > 1$. So the ℓ_1 norm is the tightest convex approximation to the ℓ_0 norm.

The effect of changing b is illustrated in Figure 13.17, where we plot the prior for $b = 2$, $b = 1$ and $b = 0.4$; we assume $p(\mathbf{w}) = p(w_1)p(w_2)$. We also plot the posterior after seeing a single observation, $(\mathbf{x}, y)$, which imposes a single linear constraint of the form, $y = \mathbf{w}^T\mathbf{x}$, with a certain tolerance controlled by the observation noise (compare to Figure 7.11). We see see that the mode of the Laplace is on the vertical axis, corresponding to $w_1 = 0$. By contrast, there are two modes when using $b = 0.4$, corresponding to two different sparse solutions. When using the Gaussian, the MAP estimate is not sparse (the mode does not lie on either of the coordinate axes).

13.6.2   Hierarchical adaptive lasso

Recall that one of the principal problems with lasso is that it results in biased estimates. This is because it needs to use a large value of λ to "squash" the irrelevant parameters, but this then over-penalizes the relevant parameters. It would be better if we could associate a different penalty parameter with each parameter. Of course, it is completely infeasible to tune D parameters by cross validation, but this poses no problem to the Bayesian: we simply make each τ_j^2 have its own private tuning parameter, γ_j, which are now treated as random variables coming from the conjugate prior $\gamma_j \sim \text{IG}(a, b)$. The full model is as follows:

$$\gamma_j \;\sim\; \text{IG}(a, b) \tag{13.133}$$
$$\tau_j^2|\gamma_j \;\sim\; \text{Ga}(1, \gamma_j^2/2) \tag{13.134}$$
$$w_j|\tau_j^2 \;\sim\; \mathcal{N}(0, \tau_j^2) \tag{13.135}$$

See Figure 13.18(a). This has been called the **hierarchical adaptive lasso** (HAL) (Lee et al. 2010, 2011; Cevher 2009; Armagan et al. 2011). We can integrate out τ_j^2, which induces a $\text{Lap}(w_j|0, 1/\gamma_j)$ distribution on w_j as before. The result is that $p(w_j)$ is now a scaled mixture of Laplacians. It turns out that we can fit this model (i.e., compute a *local* posterior mode) using EM, as we explain below. The resulting estimate, $\hat{\mathbf{w}}_{HAL}$, often works much better than the

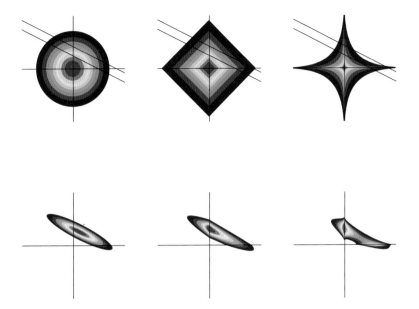

Figure 13.17 Top: plot of log *prior* for three different distributions with unit variance: Gaussian, Laplace and exponential power. Bottom: plot of log *posterior* after observing a single observation, corresponding to a single linear constraint. The precision of this observation is shown by the diagonal lines in the top figure. In the case of the Gaussian prior, the posterior is unimodal and symmetric. In the case of the Laplace prior, the posterior is unimodal and asymmetric (skewed). In the case of the exponential prior, the posterior is bimodal. Based on Figure 1 of (Seeger 2008). Figure generated by `sparsePostPlot`, written by Florian Steinke.

estimate returned by lasso, $\hat{\mathbf{w}}_{L1}$, in the sense that it is more likely to contain zeros in the right places (model selection consistency) and more likely to result in good predictions (prediction consistency) (Lee et al. 2010). We give an explanation for this behavior in Section 13.6.2.2.

13.6.2.1 EM for HAL

Since the inverse Gamma is conjugate to the Laplace, we find that the E step for γ_j is given by

$$p(\gamma_j|w_j) = \text{IG}(a+1, b+|w_j|) \tag{13.136}$$

The E step for σ^2 is the same as for vanilla lasso.

The prior for $\mathbf{w}$ has the following form:

$$p(\mathbf{w}|\boldsymbol{\gamma}) = \prod_j \frac{1}{2\gamma_j} \exp(-|w_j|/\gamma_j) \tag{13.137}$$

Hence the M step must optimize

$$\hat{\mathbf{w}}^{(t+1)} = \underset{\mathbf{w}}{\operatorname{argmax}} \log \mathcal{N}(\mathbf{y}|\mathbf{X}\mathbf{w}, \sigma^2) - \sum_j |w_j| \mathbb{E}\left[1/\gamma_j\right] \tag{13.138}$$

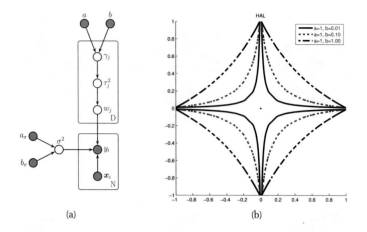

Figure 13.18   (a) DGM for hierarchical adaptive lasso. (b) Contours of Hierarchical adpative Laplace. Based on Figure 1 of (Lee et al. 2010). Figure generated by `normalGammaPenaltyPlotDemo`.

The expectation is given by

$$\mathbb{E}\left[1/\gamma_j\right] = \frac{a+1}{b+|w_j^{(t)}|} \triangleq s_j^{(t)} \tag{13.139}$$

Thus the M step becomes a weighted lasso problem:

$$\hat{\mathbf{w}}^{(t+1)} = \underset{\mathbf{w}}{\operatorname{argmin}} \, ||\mathbf{y} - \mathbf{X}\mathbf{w}||_2^2 + \sum_j s_j^{(t)} |w_j| \tag{13.140}$$

This is easily solved using standard methods (e.g., LARS). Note that if the coefficient was estimated to be large in the previous iteration (so $w_j^{(t)}$ is large), then the scaling factor $s_j^{(t)}$ will be small, so large coefficients are not penalized heavily. Conversely, small coefficients *do* get penalized heavily. This is the way that the algorithm adapts the penalization strength of each coefficient. The result is an estimate that is often much sparser than returned by lasso, but also less biased.

Note that if we set $a = b = 0$, and we only perform 1 iteration of EM, we get a method that is closely related to the **adaptive lasso** of (Zou 2006; Zou and Li 2008). This EM algorithm is also closely related to some iteratively reweighted ℓ_1 methods proposed in the signal processing community (Chartrand and Yin 2008; Candes et al. 2008).

13.6.2.2   Understanding the behavior of HAL

We can get a better understanding of HAL by integrating out γ_j to get the following marginal distribution,

$$p(w_j|a, b) = \frac{a}{2b} \left(\frac{|w_j|}{b} + 1\right)^{-(a+1)} \tag{13.141}$$

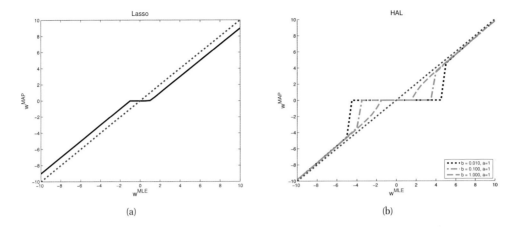

Figure 13.19 Thresholding behavior of two penalty functions (negative log priors). (a) Laplace. (b) Hierarchical adaptive Laplace. Based on Figure 2 of (Lee et al. 2010). Figure generated by `normalGammaThresholdPlotDemo`.

This is an instance of the **generalized t distribution** (McDonald and Newey 1988) (in (Cevher 2009; Armagan et al. 2011), this is called the double Pareto distribution) defined as

$$\text{GT}(w|\mu, a, c, q) \triangleq \frac{q}{2ca^{1/q}B(1/q, a)} \left(1 + \frac{|w - \mu|^q}{ac^q}\right)^{-(a+1/q)} \tag{13.142}$$

where c is the scale parameter (which controls the degree of sparsity), and a is related to the degrees of freedom. When $q = 2$ and $c = \sqrt{2}$ we recover the standard t distribution; when $a \to \infty$, we recover the exponential power distribution; and when $q = 1$ and $a = \infty$ we get the Laplace distribution. In the context of the current model, we see that $p(w_j|a, b) = \text{GT}(w_j|0, a, b/a, 1)$.

The resulting penalty term has the form

$$\pi_{\boldsymbol{\lambda}}(w_j) \triangleq -\log p(w_j) = (a + 1)\log(1 + \frac{|w_j|}{b}) + \text{const} \tag{13.143}$$

where $\boldsymbol{\lambda} = (a, b)$ are the tuning parameters. We plot this penalty in 2d (i.e., we plot $\pi_{\boldsymbol{\lambda}}(w_1) + \pi_{\boldsymbol{\lambda}}(w_2)$) in Figure 13.18(b) for various values of b. Compared to the diamond-shaped Laplace penalty, shown in Figure 13.3(a), we see that the HAL penalty looks more like a "star fish": it puts much more density along the "spines", thus enforcing sparsity more aggressively. Note that this penalty is clearly not convex.

We can gain further understanding into the behavior of this penalty function by considering applying it to the problem of linear regression with an orthogonal design matrix. In this case,

$p(\tau_j^2)$	$p(\gamma_j)$	$p(w_j)$	Ref
$\mathrm{Ga}(1, \frac{\gamma^2}{2})$	$\gamma_j = \gamma$	$\mathrm{Lap}(0, 1/\gamma)$	Sec. 13.4.4
$\mathrm{Ga}(1, \frac{\gamma^2}{2})$	$\mathrm{IG}(a, b)$	$\mathrm{GT}(0, a, b/a, 1)$	Sec. 13.6.2
$\mathrm{Ga}(1, \frac{\gamma^2}{2})$	$\mathrm{Ga}(a, b)$	$\mathrm{NEG}(a, b)$	(Griffin and Brown 2007, 2010; Chen et al. 2011)
$\mathrm{Ga}(\delta, \frac{\gamma^2}{2})$	$\gamma_j = \gamma$	$\mathrm{NG}(\delta, \gamma)$	(Griffin and Brown 2007, 2010)
$\mathrm{Ga}(\tau_j^2|0, 0)$	N/A	$\mathrm{NJ}(w_j)$	(Figueiredo 2003)
$\mathrm{IG}(\frac{\delta}{2}, \frac{\delta\gamma^2}{2})$	$\gamma_j = \gamma$	$\mathcal{T}(0, \delta, \gamma)$	(Andrews and Mallows 1974; West 1987)
$C^+(0, \gamma)$	$C^+(0, b)$	horseshoe(b)	(Carvahlo et al. 2010)

Table 13.2 Some scale mixtures of Gaussians. Abbreviations: C^+ = half-rectified Cauchy; Ga = Gamma (shape and rate parameterization); GT = generalized t; IG = inverse Gamma; NEG = Normal-Exponential-Gamma; NG = Normal-Gamma; NJ = Normal-Jeffreys. The horseshoe distribution is the name we give to the distribution induced on w_j by the prior described in (Carvahlo et al. 2010); this has no simple analytic form. The definitions of the NEG and NG densities are a bit complicated, but can be found in the references. The other distributions are defined in the text.

one can show that the objective becomes

$$J(\mathbf{w}) = \frac{1}{2}\|\mathbf{y} - \mathbf{X}\mathbf{w}\|_2^2 + \sum_{j=1}^{D} \pi_\lambda(|w_j|) \tag{13.144}$$

$$= \frac{1}{2}\|\mathbf{y} - \hat{\mathbf{y}}\|^2 + \frac{1}{2}\sum_{j=1}^{D}(\hat{w}_j^{mle} - w_j)^2 + \sum_{j=1}^{D} \pi_\lambda(|w_j|) \tag{13.145}$$

where $\hat{\mathbf{w}}^{mle} = \mathbf{X}^T\mathbf{y}$ is the MLE and $\hat{\mathbf{y}} = \mathbf{X}\hat{\mathbf{w}}^{mle}$. Thus we can compute the MAP estimate one dimension at a time by solving the following 1d optimization problem:

$$\hat{w}_j = \underset{w_j}{\mathrm{argmin}}\, \frac{1}{2}(\hat{w}_j^{mle} - w_j)^2 + \pi_\lambda(w_j) \tag{13.146}$$

In Figure 13.19(a) we plot the lasso estimate, $\hat{w}^{L1}$, vs the ML estimate, $\hat{w}^{mle}$. We see that the ℓ_1 estimator has the usual soft-thresholding behavior seen earlier in Figure 13.5(a). However, this behavior is undesirable since the large magnitude coefficients are also shrunk towards 0, whereas we would like them to be equal to their unshrunken ML estimates.

In Figure 13.19(b) we plot the HAL estimate, $\hat{w}^{HAL}$, vs the ML estimate $\hat{w}^{mle}$. We see that this approximates the more desirable hard thresholding behavior seen earlier in Figure 13.5(b) much more closely.

13.6.3   Other hierarchical priors

Many other hierarchical sparsity-promoting priors have been proposed; see Table 13.2 for a brief summary. In some cases, we can analytically derive the form of the marginal prior for w_j. Generally speaking, this prior is not concave.

A particularly interesting prior is the improper Normal-Jeffreys prior, which has been used in (Figueiredo 2003). This puts a non-informative Jeffreys prior on the variance, $\mathrm{Ga}(\tau_j^2|0, 0) \propto$

$1/\tau_j^2$; the resulting marginal has the form $p(w_j) = \mathrm{NJ}(w_j) \propto 1/|w_j|$. This gives rise to a thresholding rule that looks very similar to HAL in Figure 13.19(b), which in turn is very similar to hard thresholding. However, this prior has no free parameters, which is both a good thing (nothing to tune) and a bad thing (no ability to adapt the level of sparsity).

13.7   Automatic relevance determination (ARD)/sparse Bayesian learning (SBL)

All the methods we have considered so far (except for the spike-and-slab methods in Section 13.2.1) have used a **factorial prior** of the form $p(\mathbf{w}) = \prod_j p(w_j)$. We have seen how these priors can be represented in terms of Gaussian scale mixtures of the form $w_j \sim \mathcal{N}(0, \tau_j^2)$, where τ_j^2 has one of the priors listed in Table 13.2. Using these latent variances, we can represent the model in the form $\tau_j^2 \to w_j \to \mathbf{y} \leftarrow \mathbf{X}$. We can then use EM to perform MAP estimation, where in the E step we infer $p(\tau_j^2|w_j)$, and in the M step we estimate $\mathbf{w}$ from $\mathbf{y}$, $\mathbf{X}$ and $\boldsymbol{\tau}$. This M step either involves a closed-form weighted ℓ_2 optimization (in the case of Gaussian scale mixtures), or a weighted ℓ_1 optimization (in the case of Laplacian scale mixtures). We also discussed how to perform Bayesian inference in such models, rather than just computing MAP estimates.

In this section, we discuss an alternative approach based on type II ML estimation (empirical Bayes), whereby we integrate out $\mathbf{w}$ and maximize the marginal likelihood wrt $\boldsymbol{\tau}$. This EB procedure can be implemented via EM, or via a reweighted ℓ_1 scheme, as we will explain below. Having estimated the variances, we plug them in to compute the posterior mean of the weights, $\mathbb{E}[\mathbf{w}|\hat{\boldsymbol{\tau}}, \mathcal{D}]$; rather surprisingly (in view of the Gaussian prior), the result is an (approximately) sparse estimate, for reasons we explain below.

In the context of neural networks, this method is called **automatic relevance determination** or **ARD** (MacKay 1995b; Neal 1996): see Section 16.5.7.5. In the context of the linear models we are considering in this chapter, this method is called **sparse Bayesian learning** or **SBL** (Tipping 2001). Combining ARD/SBL with basis function expansion in a linear model gives rise to a technique called the relevance vector machine (RVM), which we will discuss in Section 14.3.2.

13.7.1   ARD for linear regression

We will explain the procedure in the context of linear regression; ARD for GLMs requires the use of the Laplace (or some other) approximation. It is conventional, when discussing ARD / SBL, to denote the weight precisions by $\alpha_j = 1/\tau_j^2$, and the measurement precision by $\beta = 1/\sigma^2$ (do not confuse this with the use of $\boldsymbol{\beta}$ in statistics to represent the regression coefficients!). In particular, we will assume the following model:

$$p(y|\mathbf{x}, \mathbf{w}, \beta) = \mathcal{N}(y|\mathbf{w}^T\mathbf{x}, 1/\beta) \tag{13.147}$$

$$p(\mathbf{w}) = \mathcal{N}(\mathbf{w}|\mathbf{0}, \mathbf{A}^{-1}) \tag{13.148}$$

where $\mathbf{A} = \mathrm{diag}(\boldsymbol{\alpha})$. The marginal likelihood can be computed analytically as follows:

$$
\begin{aligned}
p(\mathbf{y}|\mathbf{X}, \boldsymbol{\alpha}, \beta) &= \int \mathcal{N}(\mathbf{y}|\mathbf{X}\mathbf{w}, (1/\beta)\mathbf{I}_N)\mathcal{N}(\mathbf{w}|\mathbf{0}, \mathbf{A}^{-1})d\mathbf{w} & (13.149) \\
&= \mathcal{N}(\mathbf{y}|\mathbf{0}, \beta^{-1}\mathbf{I}_N + \mathbf{X}\mathbf{A}^{-1}\mathbf{X}^T) & (13.150) \\
&= (2\pi)^{-N/2}|\mathbf{C}_{\boldsymbol{\alpha}}|^{-\frac{1}{2}}\exp(-\frac{1}{2}\mathbf{y}^T\mathbf{C}_{\boldsymbol{\alpha}}^{-1}\mathbf{y}) & (13.151)
\end{aligned}
$$

where

$$
\mathbf{C}_{\boldsymbol{\alpha}} \triangleq \beta^{-1}\mathbf{I}_N + \mathbf{X}\mathbf{A}^{-1}\mathbf{X}^T \tag{13.152}
$$

Compare this to the marginal likelihood in Equation 13.13 in the spike and slab model; modulo the $\beta = 1/\sigma^2$ factor missing from the second term, the equations are the same, except we have replaced the binary $\gamma_j \in \{0, 1\}$ with continuous $\alpha_j \in \mathbb{R}^+$. In log form, the objective becomes

$$
\ell(\boldsymbol{\alpha}, \beta) \triangleq -\frac{1}{2}\log p(\mathbf{y}|\mathbf{X}, \boldsymbol{\alpha}, \beta) = \log|\mathbf{C}_{\boldsymbol{\alpha}}| + \mathbf{y}^T\mathbf{C}_{\boldsymbol{\alpha}}^{-1}\mathbf{y} \tag{13.153}
$$

To regularize the problem, we may put a conjugate prior on each precision, $\alpha_j \sim \mathrm{Ga}(a, b)$ and $\beta \sim \mathrm{Ga}(c, d)$. The modified objective becomes

$$
\begin{aligned}
\ell(\boldsymbol{\alpha}, \beta) &\triangleq -\frac{1}{2}\log p(\mathbf{y}|\mathbf{X}, \boldsymbol{\alpha}, \beta) + \sum_j \log \mathrm{Ga}(\alpha_j|a, b) + \log \mathrm{Ga}(\beta|c, d) & (13.154) \\
&= \log|\mathbf{C}_{\boldsymbol{\alpha}}| + \mathbf{y}^T\mathbf{C}_{\boldsymbol{\alpha}}^{-1}\mathbf{y} + \sum_j (a\log\alpha_j - b\alpha_j) + c\log\beta - d\beta & (13.155)
\end{aligned}
$$

This is useful when performing Bayesian inference for $\boldsymbol{\alpha}$ and β (Bishop and Tipping 2000). However, when performing (type II) point estimation, we will use the improper prior $a = b = c = d = 0$, which results in maximal sparsity.

Below we describe how to optimize $\ell(\boldsymbol{\alpha}, \beta)$ wrt the precision terms $\boldsymbol{\alpha}$ and β.[8] This is a proxy for finding the most probable model setting of $\boldsymbol{\gamma}$ in the spike and slab model, which in turn is closely related to ℓ_0 regularization. In particular, it can be shown (Wipf et al. 2010) that the objective in Equation 13.153 has many fewer local optima than the ℓ_0 objective, and hence is much easier to optimize.

Once we have estimated $\boldsymbol{\alpha}$ and β, we can compute the posterior over the parameters using

$$
p(\mathbf{w}|\mathcal{D}, \hat{\boldsymbol{\alpha}}, \hat{\beta}) = \mathcal{N}(\boldsymbol{\mu}, \boldsymbol{\Sigma}) \tag{13.156}
$$

$$
\boldsymbol{\Sigma}^{-1} = \hat{\beta}\mathbf{X}^T\mathbf{X} + \mathbf{A} \tag{13.157}
$$

$$
\boldsymbol{\mu} = \hat{\beta}\boldsymbol{\Sigma}\mathbf{X}^T\mathbf{y} \tag{13.158}
$$

The fact that we compute a posterior over $\mathbf{w}$, while simultaneously encouraging sparsity, is why the method is called "sparse Bayesian learning". Nevertheless, since there are many ways to be sparse and Bayesian, we will use the "ARD" term instead, even in the linear model context. (In addition, SBL is only "being Bayesian" about the values of the coefficients, rather than reflecting uncertainty about the set of relevant variables, which is typically of more interest.)

---

8. An alternative approach to optimizing β is to put a Gamma prior on β and to integrate it out to get a Student posterior for $\mathbf{w}$ (Buntine and Weigend 1991). However, it turns out that this results in a less accurate estimate for $\boldsymbol{\alpha}$ (MacKay 1999). In addition, working with Gaussians is easier than working with the Student distribution, and the Gaussian case generalizes more easily to other cases such as logistic regression.

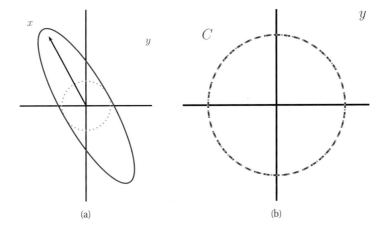

Figure 13.20 Illustration of why ARD results in sparsity. The vector of inputs $\mathbf{x}$ does not point towards the vector of outputs $\mathbf{y}$, so the feature should be removed. (a) For finite α, the probability density is spread in directions away from $\mathbf{y}$. (b) When $\alpha = \infty$, the probability density at $\mathbf{y}$ is maximized. Based on Figure 8 of (Tipping 2001).

13.7.2 Whence sparsity?

If $\hat{\alpha}_j \approx 0$, we find $\hat{w}_j \approx \hat{w}_j^{mle}$, since the Gaussian prior shrinking w_j towards 0 has zero precision. However, if we find that $\hat{\alpha}_j \approx \infty$, then the prior is very confident that $w_j = 0$, and hence that feature j is "irrelevant". Hence the posterior mean will have $\hat{w}_j \approx 0$. Thus irrelevant features automatically have their weights "turned off" or "pruned out".

We now give an intuitive argument, based on (Tipping 2001), about why ML-II should encourage $\alpha_j \to \infty$ for irrelevant features. Consider a 1d linear regression with 2 training examples, so $\mathbf{X} = \mathbf{x} = (x_1, x_2)$, and $\mathbf{y} = (y_1, y_2)$. We can plot $\mathbf{x}$ and $\mathbf{y}$ as vectors in the plane, as shown in Figure 13.20. Suppose the feature is irrelevant for predicting the response, so $\mathbf{x}$ points in a nearly orthogonal direction to $\mathbf{y}$. Let us see what happens to the marginal likelihood as we change α. The marginal likelihood is given by $p(\mathbf{y}|\mathbf{x}, \alpha, \beta) = \mathcal{N}(\mathbf{y}|\mathbf{0}, \mathbf{C})$, where

$$\mathbf{C} = \frac{1}{\beta}\mathbf{I} + \frac{1}{\alpha}\mathbf{x}\mathbf{x}^T \tag{13.159}$$

If α is finite, the posterior will be elongated along the direction of $\mathbf{x}$, as in Figure 13.20(a). However, if $\alpha = \infty$, we find $\mathbf{C} = \frac{1}{\beta}\mathbf{I}$, so $\mathbf{C}$ is spherical, as in Figure 13.20(b). If $|\mathbf{C}|$ is held constant, the latter assigns higher probability density to the observed response vector $\mathbf{y}$, so this is the preferred solution. In other words, the marginal likelihood "punishes" solutions where α_j is small but $\mathbf{X}_{:,j}$ is irrelevant, since these waste probability mass. It is more parsimonious (from the point of view of Bayesian Occam's razor) to eliminate redundant dimensions.

13.7.3 Connection to MAP estimation

ARD seems quite different from the MAP estimation methods we have been considering earlier in this chapter. In particular, in ARD, we are not integrating out α and optimizing $\mathbf{w}$, but vice versa. Because the parameters w_j become correlated in the posterior (due to explaining away), when we estimate α_j we are borrowing information from all the features, not just feature j. Consequently, the effective prior $p(\mathbf{w}|\hat{\boldsymbol{\alpha}})$ is **non-factorial**, and furthermore it depends on the data $\mathcal{D}$ (and σ^2). However, in (Wipf and Nagarajan 2007), it was shown that ARD can be viewed as the following MAP estimation problem:

$$\hat{\mathbf{w}}^{ARD} = \arg\min_{\mathbf{w}} \beta \|\mathbf{y} - \mathbf{X}\mathbf{w}\|_2^2 + g_{ARD}(\mathbf{w}) \tag{13.160}$$

$$g_{ARD}(\mathbf{w}) \triangleq \min_{\boldsymbol{\alpha} \geq 0} \sum_j \alpha_j w_j^2 + \log|\mathbf{C}_{\boldsymbol{\alpha}}| \tag{13.161}$$

The proof, which is based on convex analysis, is a little complicated and hence is omitted.

Furthermore, (Wipf and Nagarajan 2007; Wipf et al. 2010) prove that MAP estimation with non-factorial priors is strictly better than MAP estimation with any possible factorial prior in the following sense: the non-factorial objective always has fewer local minima than factorial objectives, while still satisfying the property that the global optimum of the non-factorial objective corresponds to the global optimum of the ℓ_0 objective — a property that ℓ_1 regularization, which has no local minima, does not enjoy.

13.7.4 Algorithms for ARD *

In this section, we review several different algorithms for implementing ARD.

13.7.4.1 EM algorithm

The easiest way to implement SBL/ARD is to use EM. The expected complete data log likelihood is given by

$$Q(\boldsymbol{\alpha}, \beta) = \mathbb{E}\left[\log \mathcal{N}(\mathbf{y}|\mathbf{X}\mathbf{w}, \sigma^2 \mathbf{I}) + \log \mathcal{N}(\mathbf{w}|\mathbf{0}, \mathbf{A}^{-1})\right] \tag{13.162}$$

$$= \frac{1}{2}\mathbb{E}\left[N\log\beta - \beta\|\mathbf{y} - \mathbf{X}\mathbf{w}\|^2 + \sum_j \log\alpha_j - \mathrm{tr}(\mathbf{A}\mathbf{w}\mathbf{w}^T)\right] + \mathrm{const} \tag{13.163}$$

$$= \frac{1}{2}N\log\beta - \frac{\beta}{2}\left(\|\mathbf{y} - \mathbf{X}\boldsymbol{\mu}\|^2 + \mathrm{tr}(\mathbf{X}^T\mathbf{X}\boldsymbol{\Sigma})\right)$$
$$+ \frac{1}{2}\sum_j \log\alpha_j - \frac{1}{2}\mathrm{tr}[\mathbf{A}(\boldsymbol{\mu}\boldsymbol{\mu}^T + \boldsymbol{\Sigma})] + \mathrm{const} \tag{13.164}$$

where $\boldsymbol{\mu}$ and $\boldsymbol{\Sigma}$ are computed in the E step using Equation 13.158.

Suppose we put a $\mathrm{Ga}(a, b)$ prior on α_j and a $\mathrm{Ga}(c, d)$ prior on β. The penalized objective becomes

$$Q'(\boldsymbol{\alpha}, \beta) = Q(\boldsymbol{\alpha}, \beta) + \sum_j (a\log\alpha_j - b\alpha_j) + c\log\beta - d\beta \tag{13.165}$$

Setting $\frac{dQ'}{d\alpha_j} = 0$ we get the following M step:

$$\alpha_j = \frac{1 + 2a}{\mathbb{E}\left[w_j^2\right] + 2b} = \frac{1 + 2a}{m_j^2 + \Sigma_{jj} + 2b} \tag{13.166}$$

If $\alpha_j = \alpha$, and $a = b = 0$, the update becomes

$$\alpha = \frac{D}{\mathbb{E}\left[\mathbf{w}^T\mathbf{w}\right]} = \frac{D}{\boldsymbol{\mu}^T\boldsymbol{\mu} + \text{tr}(\boldsymbol{\Sigma})} \tag{13.167}$$

The update for β is given by

$$\beta_{new}^{-1} = \frac{||\mathbf{y} - \mathbf{X}\boldsymbol{\mu}||^2 + \beta^{-1}\sum_j(1 - \alpha_j\Sigma_{jj}) + 2d}{N + 2c} \tag{13.168}$$

(Deriving this is Exercise 13.2.)

13.7.4.2 Fixed-point algorithm

A faster and more direct approach is to directly optimize the objective in Equation 13.155. One can show (Exercise 13.3) that the equations $\frac{d\ell}{d\alpha_j} = 0$ and $\frac{d\ell}{d\beta} = 0$ lead to the following fixed point updates:

$$\alpha_j \leftarrow \frac{\gamma_j + 2a}{m_j^2 + 2b} \tag{13.169}$$

$$\beta^{-1} \leftarrow \frac{||\mathbf{y} - \mathbf{X}\boldsymbol{\mu}||^2 + 2d}{N - \sum_j\gamma_j + 2c} \tag{13.170}$$

$$\gamma_j \triangleq 1 - \alpha_j\Sigma_{jj} \tag{13.171}$$

The quantity γ_j is a measure of how well-determined w_j is by the data (MacKay 1992). Hence $\gamma = \sum_j\gamma_j$ is the effective degrees of freedom of the model. See Section 7.5.3 for further discussion.

Since $\boldsymbol{\alpha}$ and β both depend on $\boldsymbol{\mu}$ and $\boldsymbol{\Sigma}$ (which can be computed using Equation 13.158 or the Laplace approximation), we need to re-estimate these equations until convergence. (Convergence properties of this algorithm have been studied in (Wipf and Nagarajan 2007).) At convergence, the results are formally identical to those obtained by EM, but since the objective is non-convex, the results can depend on the initial values.

13.7.4.3 Iteratively reweighted ℓ_1 algorithm

Another approach to solving the ARD problem is based on the view that it is a MAP estimation problem. Although the log prior $g(\mathbf{w})$ is rather complex in form, it can be shown to be a non-decreasing, concave function of $|w_j|$. This means that it can be solved by an iteratively reweighted ℓ_1 problem of the form

$$\mathbf{w}^{t+1} = \arg\min_{\mathbf{w}} \text{NLL}(\mathbf{w}) + \sum_j\lambda_j^{(t)}|w_j| \tag{13.172}$$

In (Wipf and Nagarajan 2010), the following procedure for setting the penalty terms is suggested (based on a convex bound to the penalty function). We initialize with $\lambda_j^{(0)} = 1$, and then at iteration $t + 1$, compute $\lambda_j^{(t+1)}$ by iterating the following equation a few times:[9]

$$\lambda_j \leftarrow \left[\mathbf{X}_{:,j} \left(\sigma^2 \mathbf{I} + \mathbf{X} \mathrm{diag}(1/\lambda_j) \mathrm{diag}(|w_j^{(t+1)}|) \right)^{-1} \mathbf{X}^T)^{-1} \mathbf{X}_{:,j} \right]^{\frac{1}{2}} \tag{13.173}$$

We see that the new penalty λ_j depends on *all* the old weights. This is quite different from the adaptive lasso method of Section 13.6.2.

To understand this difference, consider the noiseless case where $\sigma^2 = 0$, and assume $D \gg N$. In this case, there are $\binom{D}{N}$ solutions which perfectly reconstruct the data, $\mathbf{Xw} = \mathbf{y}$, and which have sparsity $||\mathbf{w}||_0 = N$; these are called basic feasible solutions or BFS. What we want are solutions that satsify $\mathbf{Xw} = \mathbf{y}$ but which are much sparser than this. Suppose the method has found a BFS. We do not want to increase the penalty on a weight just because it is small (as in adaptive lasso), since that will just reinforce our current local optimum. Instead, we want to increase the penalty on a weight if it is small and if we have $||\mathbf{w}^{(t+1)}|| < N$. The covariance term $(\mathbf{X} \mathrm{diag}(1/\lambda_j) \mathrm{diag}(|w_j^{(t+1)}|))^{-1}$ has this effect: if $\mathbf{w}$ is a BFS, this matrix will be full rank, so the penalty will not increase much, but if $\mathbf{w}$ is sparser than N, the matrix will not be full rank, so the penalties associated with zero-valued coefficients will increase, thus reinforcing this solution (Wipf and Nagarajan 2010).

13.7.5   ARD for logistic regression

Now consider binary logistic regression, $p(y|\mathbf{x}, \mathbf{w}) = \mathrm{Ber}(y|\mathrm{sigm}(\mathbf{w}^T \mathbf{x}))$, using the same Gaussian prior, $p(\mathbf{w}) = \mathcal{N}(\mathbf{w}|\mathbf{0}, \mathbf{A}^{-1})$. We can no longer use EM to estimate $\boldsymbol{\alpha}$, since the Gaussian prior is not conjugate to the logistic likelihood, so the E step cannot be done exactly. One approach is to use a variational approximation to the E step, as discussed in Section 21.8.1.1. A simpler approach is to use a Laplace approximation (see Section 8.4.1) in the E step. We can then use this approximation inside the same EM procedure as before, except we no longer need to update β. Note, however, that this is not guaranteed to converge.

An alternative is to use the techniques from Section 13.7.4.3. In this case, we can use exact methods to compute the inner weighted ℓ_1 regularized logistic regression problem, and no approximations are required.

13.8   Sparse coding *

So far, we have been concentrating on sparse priors for supervised learning. In this section, we discuss how to use them for unsupervised learning.

In Section 12.6, we discussed ICA, which is like PCA except it uses a non-Gaussian prior for the latent factors $\mathbf{z}_i$. If we make the non-Gaussian prior be sparsity promoting, such as a Laplace distribution, we will be approximating each observed vector $\mathbf{x}_i$ as a sparse combination of basis vectors (columns of $\mathbf{W}$); note that the sparsity pattern (controlled by $\mathbf{z}_i$) changes from data case to data case. If we relax the constraint that $\mathbf{W}$ is orthogonal, we get a method called

---

9. The algorithm in (Wipf and Nagarajan 2007) is equivalent to a single iteration of Equation 13.173. However, since the equation is cheap to compute (only $O(ND||\mathbf{w}^{(t+1)}||_0)$ time), it is worth iterating a few times before solving the more expensive ℓ_1 problem.

Method	$p(\mathbf{z}_i)$	$p(\mathbf{W})$	$\mathbf{W}$ orthogonal
PCA	Gauss	-	yes
FA	Gauss	-	no
ICA	Non-Gauss	-	yes
Sparse coding	Laplace	-	no
Sparse PCA	Gauss	Laplace	maybe
Sparse MF	Laplace	Laplace	no

Table 13.3 Summary of various latent factor models. A dash "-" in the $p(\mathbf{W})$ column means we are performing ML parameter estimation rather than MAP parameter estimation. Summary of abbreviations: PCA = principal components analysis; FA = factor analysis; ICA = independent components analysis; MF = matrix factorization.

sparse coding. In this context, we call the factor loading matrix $\mathbf{W}$ a **dictionary**; each column is referred to as an **atom**.[10] In view of the sparse representation, it is common for $L > D$, in which case we call the representation **overcomplete**.

In sparse coding, the dictionary can be fixed or learned. If it is fixed, it is common to use a wavelet or DCT basis, since many natural signals can be well approximated by a small number of such basis functions. However, it is also possible to learn the dictionary, by maximizing the likelihood

$$\log p(\mathcal{D}|\mathbf{W}) = \sum_{i=1}^{N} \log \int_{\mathbf{z}_i} \mathcal{N}(\mathbf{x}_i|\mathbf{W}\mathbf{z}_i, \sigma^2 \mathbf{I}) p(\mathbf{z}_i) d\mathbf{z}_i \tag{13.174}$$

We discuss ways to optimize this below, and then we present several interesting applications.

Do not confuse sparse coding with **sparse PCA** (see e.g., (Witten et al. 2009; Journee et al. 2010)): this puts a sparsity promoting prior on the regression weights $\mathbf{W}$, whereas in sparse coding, we put a sparsity promoting prior on the latent factors $\mathbf{z}_i$. Of course, the two techniques can be combined; we call the result **sparse matrix factorization**, although this term is non-standard. See Table 13.3 for a summary of our terminology.

13.8.1 Learning a sparse coding dictionary

Since Equation 13.174 is a hard objective to maximize, it is common to make the following approximation:

$$\log p(\mathcal{D}|\mathbf{W}) \approx \sum_{i=1}^{N} \max_{\mathbf{z}_i} \left[\log \mathcal{N}(\mathbf{x}_i|\mathbf{W}\mathbf{z}_i, \sigma^2 \mathbf{I}) + \log p(\mathbf{z}_i) \right] \tag{13.175}$$

If $p(\mathbf{z}_i)$ is Laplace, we can rewrite the NLL as

$$\text{NLL}(\mathbf{W}, \mathbf{Z}) = \sum_{i=1}^{N} \frac{1}{2} ||\mathbf{x}_i - \mathbf{W}\mathbf{z}_i||_2^2 + \lambda ||\mathbf{z}_i||_1 \tag{13.176}$$

---

10. It is common to denote the dictionary by $\mathbf{D}$, and to denote the latent factors by $\boldsymbol{\alpha}_i$. However, we will stick with the $\mathbf{W}$ and $\mathbf{z}_i$ notation.

To prevent $\mathbf{W}$ from becoming arbitrarily large, it is common to constrain the ℓ_2 norm of its columns to be less than or equal to 1. Let us denote this constraint set by

$$\mathcal{C} = \{\mathbf{W} \in \mathbb{R}^{D \times L} \quad \text{s.t.} \quad \mathbf{w}_j^T \mathbf{w}_j \leq 1\} \tag{13.177}$$

Then we want to solve $\min_{\mathbf{W} \in \mathcal{C}, \mathbf{Z} \in \mathbb{R}^{N \times L}} \text{NLL}(\mathbf{W}, \mathbf{Z})$. For a fixed $\mathbf{z}_i$, the optimization over $\mathbf{W}$ is a simple least squares problem. And for a fixed dictionary $\mathbf{W}$, the optimization problem over $\mathbf{Z}$ is identical to the lasso problem, for which many fast algorithms exist. This suggests an obvious iterative optimization scheme, in which we alternate between optimizing $\mathbf{W}$ and $\mathbf{Z}$. (Mumford 1994) called this kind of approach an **analysis-synthesis** loop, where estimating the basis $\mathbf{W}$ is the analysis phase, and estimating the coefficients $\mathbf{Z}$ is the synthesis phase. In cases where this is too slow, more sophisticated algorithms can be used, see e.g., (Mairal et al. 2010).

A variety of other models result in an optimization problem that is similar to Equation 13.176. For example, **non-negative matrix factorization** or **NMF** (Paatero and Tapper 1994; Lee and Seung 2001) requires solving an objective of the form

$$\min_{\mathbf{W} \in \mathcal{C}, \mathbf{Z} \in \mathbb{R}^{L \times N}} \frac{1}{2} \sum_{i=1}^{N} ||\mathbf{x}_i - \mathbf{W}\mathbf{z}_i||_2^2 \quad \text{s.t.} \quad \mathbf{W} \geq 0, \mathbf{z}_i \geq 0 \tag{13.178}$$

(Note that this has no hyper-parameters to tune.) The intuition behind this constraint is that the learned dictionary may be more interpretable if it is a positive sum of positive "parts", rather than a sparse sum of atoms that may be positive or negative. Of course, we can combine NMF with a sparsity promoting prior on the latent factors. This is called **non-negative sparse coding** (Hoyer 2004).

Alternatively, we can drop the positivity constraint, but impose a sparsity constraint on both the factors $\mathbf{z}_i$ and the dictionary $\mathbf{W}$. We call this **sparse matrix factorization**. To ensure strict convexity, we can use an elastic net type penalty on the weights (Mairal et al. 2010) resulting in

$$\min_{\mathbf{W}, \mathbf{Z}} \frac{1}{2} \sum_{i=1}^{N} ||\mathbf{x}_i - \mathbf{W}\mathbf{z}_i||_2^2 + \lambda ||\mathbf{z}_i||_1 \quad \text{s.t.} \quad ||\mathbf{w}_j||_2^2 + \gamma ||\mathbf{w}_j||_1 \leq 1 \tag{13.179}$$

There are several related objectives one can write down. For example, we can replace the lasso NLL with group lasso or fused lasso (Witten et al. 2009).

We can also use other sparsity-promoting priors besides the Laplace. For example, (Zhou et al. 2009) propose a model in which the latent factors $\mathbf{z}_i$ are made sparse using the binary mask model of Section 13.2.2. Each bit of the mask can be generated from a Bernoulli distribution with parameter π, which can be drawn from a beta distribution. Alternatively, we can use a non-parametric prior, such as the beta process. This allows the model to use dictionaries of unbounded size, rather than having to specify L in advance. One can perform Bayesian inference in this model using e.g., Gibbs sampling or variational Bayes. One finds that the effective size of the dictionary goes down as the noise level goes up, due to the Bayesian Occam's razor. This can prevent overfitting. See (Zhou et al. 2009) for details.

13.8.2   Results of dictionary learning from image patches

One reason that sparse coding has generated so much interest recently is because it explains an interesting phenomenon in neuroscience. In particular, the dictionary that is learned by applying

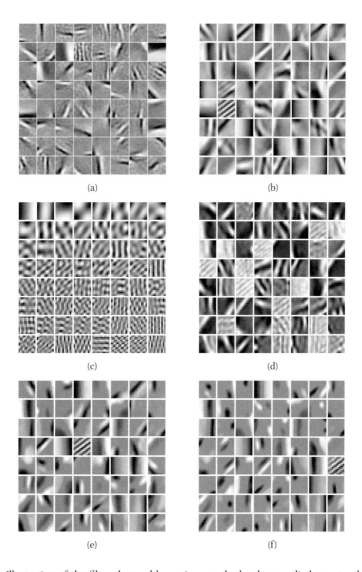

(a)

(b)

(c)

(d)

(e)

(f)

Figure 13.21 Illustration of the filters learned by various methods when applied to natural image patches. (Each patch is first centered and normalized to unit norm.) (a) ICA. Figure generated by `icaBasisDemo`, kindly provided by Aapo Hyvarinen. (b) sparse coding. (c) PCA. (d) non-negative matrix factorization. (e) sparse PCA with low sparsity on weight matrix. (f) sparse PCA with high sparsity on weight matrix. Figure generated by `sparseDictDemo`, written by Julien Mairal.

sparse coding to patches of natural images consists of basis vectors that look like the filters that are found in simple cells in the primary visual cortex of the mammalian brain (Olshausen and Field 1996). In particular, the filters look like bar and edge detectors, as shown in Figure 13.21(b). (In this example, the parameter λ was chosen so that the number of active basis functions (non-zero components of $\mathbf{z}_i$) is about 10.) Interestingly, using ICA gives visually similar results, as shown in Figure 13.21(a). By contrast, applying PCA to the same data results in sinusoidal gratings, as shown in Figure 13.21(c); these do not look like cortical cell response patterns.[11] It has therefore been conjectured that parts of the cortex may be performing sparse coding of the sensory input; the resulting latent representation is then further processed by higher levels of the brain.

Figure 13.21(d) shows the result of using NMF, and Figure 13.21(e-f) show the results of sparse PCA, as we increase the sparsity of the basis vectors.

13.8.3 Compressed sensing

Although it is interesting to look at the dictionaries learned by sparse coding, it is not necessarily very useful. However, there are some practical applications of sparse coding, which we discuss below.

Imagine that, instead of observing the data $\mathbf{x} \in \mathbb{R}^D$, we observe a low-dimensional projection of it, $\mathbf{y} = \mathbf{R}\mathbf{x} + \boldsymbol{\epsilon}$ where $\mathbf{y} \in \mathbb{R}^M$, $\mathbf{R}$ is a $M \times D$ matrix, $M \ll D$, and $\boldsymbol{\epsilon}$ is a noise term (usually Gaussian). We assume $\mathbf{R}$ is a known sensing matrix, corresponding to different linear projections of $\mathbf{x}$. For example, consider an MRI scanner: each beam direction corresponds to a vector, encoded as a row in $\mathbf{R}$. Figure 13.22 illustrates the modeling assumptions.

Our goal is to infer $p(\mathbf{x}|\mathbf{y}, \mathbf{R})$. How can we hope to recover all of $\mathbf{x}$ if we do not measure all of $\mathbf{x}$? The answer is: we can use Bayesian inference with an appropriate prior, that exploits the fact that natural signals can be expressed as a weighted combination of a small number of suitably chosen basis functions. That is, we assume $\mathbf{x} = \mathbf{W}\mathbf{z}$, where $\mathbf{z}$ has a sparse prior, and $\mathbf{W}$ is suitable dictionary. This is called **compressed sensing** or **compressive sensing** (Candes et al. 2006; Baruniak 2007; Candes and Wakin 2008; Bruckstein et al. 2009).

For CS to work, it is important to represent the signal in the right basis, otherwise it will not be sparse. In traditional CS applications, the dictionary is fixed to be a standard form, such as wavelets. However, one can get much better performance by learning a domain-specific dictionary using sparse coding (Zhou et al. 2009). As for the sensing matrix $\mathbf{R}$, it is often chosen to be a random matrix, for reasons explained in (Candes and Wakin 2008). However, one can get better performance by adapting the projection matrix to the dictionary (Seeger and Nickish 2008; Chang et al. 2009).

13.8.4 Image inpainting and denoising

Suppose we have an image which is corrupted in some way, e.g., by having text or scratches sparsely superimposed on top of it, as in Figure 13.23. We might want to estimate the underlying

---

11. The reason PCA discovers sinusoidal grating patterns is because it is trying to model the covariance of the data, which, in the case of image patches, is translation invariant. This means $\text{cov}\left[I(x, y), I(x', y')\right] = f\left[(x - x')^2 + (y - y')^2\right]$ for some function f, where $I(x, y)$ is the image intensity at location (x, y). One can show (Hyvarinen et al. 2009, p125) that the eigenvectors of a matrix of this kind are always sinusoids of different phases, i.e., PCA discovers a Fourier basis.

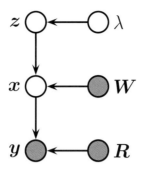

Figure 13.22 Schematic DGM for compressed sensing. We observe a low dimensional measurement $\mathbf{y}$ generated by passing $\mathbf{x}$ through a measurement matrix $\mathbf{R}$, and possibly subject to observation noise with variance σ^2. We assume that $\mathbf{x}$ has a sparse decomposition in terms of the dictionary $\mathbf{W}$ and the latent variables $\mathbf{z}$. the parameter λ controls the sparsity level.

(a)   (b)

Figure 13.23 An example of image inpainting using sparse coding. Left: original image. Right: reconstruction. Source: Figure 13 of (Mairal et al. 2008). Used with kind permission of Julien Mairal.

"clean" image. This is called **image inpainting**. One can use similar techniques for **image denoising**.

We can model this as a special kind of compressed sensing problem. The basic idea is as follows. We partition the image into overlapping patches, $\mathbf{y}_i$, and concatenate them to form $\mathbf{y}$. We define $\mathbf{R}$ so that the i'th row selects out patch i. Now define $\mathcal{V}$ to be the visible (uncorrupted) components of $\mathbf{y}$, and $\mathcal{H}$ to be the hidden components. To perform image inpainting, we just compute $p(\mathbf{y}_{\mathcal{H}}|\mathbf{y}_{\mathcal{V}}, \boldsymbol{\theta})$, where $\boldsymbol{\theta}$ are the model parameters, which specify the dictionary $\mathbf{W}$ and the sparsity level λ of $\mathbf{z}$. We can either learn a dictionary offline from a database of images, or we can learn a dictionary just for this image, based on the non-corrupted patches.

Figure 13.23 shows this technique in action. The dictionary (of size 256 atoms) was learned from 7×10^6 undamaged 12×12 color patches in the 12 mega-pixel image.

An alternative approach is to use a graphical model (e.g., the **fields of experts** model (Roth and Black 2009)) which directly encodes correlations between neighboring image patches, rather than using a latent variable model. Unfortunately such models tend to be computationally more expensive.

Exercises

Exercise 13.1 Partial derivative of the RSS

Define

$$RSS(\mathbf{w}) = ||\mathbf{X}\mathbf{w} - \mathbf{y}||_2^2 \tag{13.180}$$

a. Show that

$$\frac{\partial}{\partial w_k} RSS(\mathbf{w}) = a_k w_k - c_k \tag{13.181}$$

$$a_k = 2 \sum_{i=1}^{n} x_{ik}^2 = 2||\mathbf{x}_{:,k}||^2 \tag{13.182}$$

$$c_k = 2 \sum_{i=1}^{n} x_{ik}(y_i - \mathbf{w}_{-k}^T \mathbf{x}_{i,-k}) = 2\mathbf{x}_{:,k}^T \mathbf{r}_k \tag{13.183}$$

where $\mathbf{w}_{-k} = \mathbf{w}$ without component k, $\mathbf{x}_{i,-k}$ is $\mathbf{x}_i$ without component k, and $\mathbf{r}_k = \mathbf{y} - \mathbf{w}_{-k}^T \mathbf{x}_{:,-k}$ is the residual due to using all the features except feature k. Hint: Partition the weights into those involving k and those not involving k.

b. Show that if $\frac{\partial}{\partial w_k} RSS(\mathbf{w}) = 0$, then

$$\hat{w}_k = \frac{\mathbf{x}_{:,k}^T \mathbf{r}_k}{||\mathbf{x}_{:,k}||^2} \tag{13.184}$$

Hence when we sequentially add features, the optimal weight for feature k is computed by computing orthogonally projecting $\mathbf{x}_{:,k}$ onto the current residual.

Exercise 13.2 Derivation of M step for EB for linear regression

Derive Equations 13.166 and 13.168. Hint: the following identity should be useful

$$\mathbf{\Sigma}\mathbf{X}^T\mathbf{X} = \mathbf{\Sigma}\mathbf{X}^T\mathbf{X} + \beta^{-1}\mathbf{\Sigma}\mathbf{A} - \beta^{-1}\mathbf{\Sigma}\mathbf{A} \tag{13.185}$$

$$= \mathbf{\Sigma}(\mathbf{X}^T\mathbf{X}\beta + \mathbf{A})\beta^{-1} - \beta^{-1}\mathbf{\Sigma}\mathbf{A} \tag{13.186}$$

$$= (\mathbf{A} + \beta\mathbf{X}^T\mathbf{X})^{-1}(\mathbf{X}^T\mathbf{X}\beta + \mathbf{A})\beta^{-1} - \beta^{-1}\mathbf{\Sigma}\mathbf{A} \tag{13.187}$$

$$= (\mathbf{I} - \mathbf{A}\mathbf{\Sigma})\beta^{-1} \tag{13.188}$$

Exercise 13.3 Derivation of fixed point updates for EB for linear regression

Derive Equations 13.169 and 13.170. Hint: The easiest way to derive this result is to rewrite $\log p(\mathcal{D}|\boldsymbol{\alpha}, \beta)$ as in Equation 8.54. This is exactly equivalent, since in the case of a Gaussian prior and likelihood, the posterior is also Gaussian, so the Laplace "approximation" is exact. In this case, we get

$$\log p(\mathcal{D}|\boldsymbol{\alpha}, \beta) = \frac{N}{2}\log\beta - \frac{\beta}{2}||\mathbf{y} - \mathbf{X}\mathbf{w}||^2$$

$$+ \frac{1}{2}\sum_j \log\alpha_j - \frac{1}{2}\mathbf{m}^T\mathbf{A}\mathbf{m} + \frac{1}{2}\log|\mathbf{\Sigma}| - \frac{D}{2}\log(2\pi) \tag{13.189}$$

The rest is straightforward algebra.

Exercise 13.4 Marginal likelihood for linear regression

Suppose we use a g-prior of the form $\mathbf{\Sigma}_\gamma = g(\mathbf{X}_\gamma^T \mathbf{X}_\gamma)^{-1}$. Show that Equation 13.16 simplifies to

$$p(\mathcal{D}|\boldsymbol{\gamma}) \quad \propto \quad (1+g)^{-D_\gamma/2}(2b_\sigma + S(\boldsymbol{\gamma}))^{-(2a_\sigma+N-1)/2} \tag{13.190}$$

$$S(\boldsymbol{\gamma}) \quad = \quad \mathbf{y}^T\mathbf{y} - \frac{g}{1+g}\mathbf{y}^T\mathbf{X}_\gamma(\mathbf{X}_\gamma^T\mathbf{X}_\gamma)^{-1}\mathbf{X}_\gamma^T\mathbf{y} \tag{13.191}$$

Exercise 13.5 Reducing elastic net to lasso

Define

$$J_1(\mathbf{w}) = |\mathbf{y} - \mathbf{X}\mathbf{w}|^2 + \lambda_2|\mathbf{w}|^2 + \lambda_1|\mathbf{w}|_1 \tag{13.192}$$

and

$$J_2(\mathbf{w}) = |\tilde{\mathbf{y}} - \tilde{\mathbf{X}}\tilde{\mathbf{w}}|^2 + c\lambda_1|\mathbf{w}|_1 \tag{13.193}$$

where $c = (1+\lambda_2)^{-\frac{1}{2}}$ and

$$\tilde{\mathbf{X}} = c\begin{pmatrix} \mathbf{X} \\ \sqrt{\lambda_2}\mathbf{I}_d \end{pmatrix}, \quad \tilde{\mathbf{y}} = \begin{pmatrix} \mathbf{y} \\ \mathbf{0}_{d\times 1} \end{pmatrix} \tag{13.194}$$

Show

$$\arg\min J_1(\mathbf{w}) = c(\arg\min J_2(\mathbf{w})) \tag{13.195}$$

i.e.

$$J_1(c\mathbf{w}) = J_2(\mathbf{w}) \tag{13.196}$$

and hence that one can solve an elastic net problem using a lasso solver on modified data.

Exercise 13.6 Shrinkage in linear regression

(Source: Jaakkola.) Consider performing linear regression with an orthonormal design matrix, so $||\mathbf{x}_{:,k}||_2^2 = 1$ for each column (feature) k, and $\mathbf{x}_{:,k}^T\mathbf{x}_{:,j} = 0$, so we can estimate each parameter w_k separately.

Figure 13.24 plots $\hat{w}_k$ vs $c_k = 2\mathbf{y}^T\mathbf{x}_{:,k}$, the correlation of feature k with the response, for 3 different esimation methods: ordinary least squares (OLS), ridge regression with parameter λ_2, and lasso with parameter λ_1.

a. Unfortunately we forgot to label the plots. Which method does the solid (1), dotted (2) and dashed (3) line correspond to? Hint: see Section 13.3.3.

b. What is the value of λ_1?

c. What is the value of λ_2?

Exercise 13.7 Prior for the Bernoulli rate parameter in the spike and slab model

Consider the model in Section 13.2.1. Suppose we put a prior on the sparsity rates, $\pi_j \sim \text{Beta}(\alpha_1, \alpha_2)$. Derive an expression for $p(\boldsymbol{\gamma}|\boldsymbol{\alpha})$ after integrating out the π_j's. Discuss some advantages and disadvantages of this approach compared to assuming $\pi_j = \pi_0$ for fixed π_0.

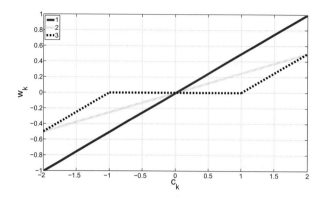

Figure 13.24   Plot of $\hat{w}_k$ vs amount of correlation c_k for three different estimators.

Exercise 13.8 Deriving E step for GSM prior

Show that

$$\mathbb{E}\left[\frac{1}{\tau_j^2}\big|w_j\right] \;=\; \frac{\pi'(w_j)}{|w_j|} \tag{13.197}$$

where $\pi(w_j) = -\log p(w_j)$ and $p(w_j) = \int \mathcal{N}(w_j|0,\tau_j^2)p(\tau_j^2)d\tau_j^2$. Hint 1:

$$\frac{1}{\tau_j^2}\mathcal{N}(w_j|0,\tau_j^2) \;\propto\; \frac{1}{\tau_j^2}\exp(-\frac{w_j^2}{2\tau_j^2}) \tag{13.198}$$

$$= \frac{-1}{|w_j|}\frac{-2w_j}{2\tau_j^2}\exp(-\frac{w_j^2}{2\tau_j^2}) \tag{13.199}$$

$$= \frac{-1}{|w_j|}\frac{d}{d|w_j|}\mathcal{N}(w_j|0,\tau_j^2) \tag{13.200}$$

Hint 2:

$$\frac{d}{d|w_j|}p(w_j) = \frac{1}{p(w_j)}\frac{d}{d|w_j|}\log p(w_j) \tag{13.201}$$

Exercise 13.9 EM for sparse probit regression with Laplace prior

Derive an EM algorithm for fitting a binary probit classifier (Section 9.4) using a Laplace prior on the weights. (If you get stuck, see (Figueiredo 2003; Ding and Harrison 2010).)

Exercise 13.10 GSM representation of group lasso

Consider the prior $\tau_j^2 \sim \mathrm{Ga}(\delta,\rho^2/2)$, ignoring the grouping issue for now. The marginal distribution induced on the weights by a Gamma mixing distribution is called the **normal Gamma** distribution and is given by

$$
\begin{aligned}
\mathrm{NG}(w_j|\delta,\rho) &= \int \mathcal{N}(w_j|0,\tau_j^2)\mathrm{Ga}(\tau_j^2|\delta,\rho^2/2)d\tau_j^2 && (13.202)\\
&= \frac{1}{Z}|w_j|^{\delta-1/2}\,\mathcal{K}_{\delta-\frac{1}{2}}(\rho|w_j|) && (13.203)\\
1/Z &= \frac{\rho^{\delta+\frac{1}{2}}}{\sqrt{\pi}\,2^{\delta-1/2}\,\rho(\delta)} && (13.204)
\end{aligned}
$$

where $\mathcal{K}_\alpha(x)$ is the modified Bessel function of the second kind (the `besselk` function in Matlab).

Now suppose we have the following prior on the variances

$$
p(\boldsymbol{\sigma}_{1:D}^2) = \prod_{g=1}^{G} p(\boldsymbol{\sigma}_{1:d_g}^2),\; p(\boldsymbol{\sigma}_{1:d_g}^2) = \prod_{j\in g}\mathrm{Ga}(\tau_j^2|\delta_g,\rho^2/2) \tag{13.205}
$$

The corresponding marginal for each group of weights has the form

$$
p(\mathbf{w}_g) \propto |u_g|^{\delta_g-d_g/2}\,\mathcal{K}_{\delta_g-d_g/2}(\rho u_g) \tag{13.206}
$$

where

$$
u_g \triangleq \sqrt{\sum_{j\in g} w_{g,j}^2} = ||\mathbf{w}_g||_2 \tag{13.207}
$$

Now suppose $\delta_g = (d_g+1)/2$, so $\delta_g - d_g/2 = \frac{1}{2}$. Conveniently, we have $\mathcal{K}_{\frac{1}{2}}(z) = \sqrt{\frac{\pi}{2z}}\exp(-z)$. Show that the resulting MAP estimate is equivalent to group lasso.

Exercise 13.11 Projected gradient descent for ℓ_1 regularized least squares

Consider the BPDN problem $\mathrm{argmin}_{\boldsymbol{\theta}}\,\mathrm{RSS}(\boldsymbol{\theta}) + \lambda||\boldsymbol{\theta}||_1$. By using the split variable trick introduced in Section 7.4 (i.e., by defining $\boldsymbol{\theta} = \boldsymbol{\theta}_+ - \boldsymbol{\theta}_-$), rewrite this as a quadratic program with a simple bound constraint. Then sketch how to use projected gradient descent to solve this problem. (If you get stuck, consult (Figueiredo et al. 2007).)

Exercise 13.12 Subderivative of the hinge loss function

Let $f(x) = (1-x)_+$ be the hinge loss function, where $(z)_+ = \max(0,z)$. What are $\partial f(0)$, $\partial f(1)$, and $\partial f(2)$?

Exercise 13.13 Lower bounds to convex functions

Let f be a convex function. Explain how to find a global affine lower bound to f at an arbitrary point $\mathbf{x} \in \mathrm{dom}(f)$.

14 *Kernels*

14.1 Introduction

So far in this book, we have been assuming that each object that we wish to classify or cluster or process in anyway can be represented as a fixed-size feature vector, typically of the form $\mathbf{x}_i \in \mathbb{R}^D$. However, for certain kinds of objects, it is not clear how to best represent them as fixed-sized feature vectors. For example, how do we represent a text document or protein sequence, which can be of variable length? or a molecular structure, which has complex 3d geometry? or an evolutionary tree, which has variable size and shape?

One approach to such problems is to define a generative model for the data, and use the inferred latent representation and/or the parameters of the model as features, and then to plug these features in to standard methods. For example, in Chapter 28, we discuss deep learning, which is essentially an unsupervised way to learn good feature representations.

Another approach is to assume that we have some way of measuring the similarity between objects, that doesn't require preprocessing them into feature vector format. For example, when comparing strings, we can compute the edit distance between them. Let $\kappa(\mathbf{x}, \mathbf{x}') \geq 0$ be some measure of similarity between objects $\mathbf{x}, \mathbf{x}' \in \mathcal{X}$, where $\mathcal{X}$ is some abstract space; we will call κ a **kernel function**. Note that the word "kernel" has several meanings; we will discuss a different interpretation in Section 14.7.1.

In this chapter, we will discuss several kinds of kernel functions. We then describe some algorithms that can be written purely in terms of kernel function computations. Such methods can be used when we don't have access to (or choose not to look at) the "inside" of the objects $\mathbf{x}$ that we are processing.

14.2 Kernel functions

We define a **kernel function** to be a real-valued function of two arguments, $\kappa(\mathbf{x}, \mathbf{x}') \in \mathbb{R}$, for $\mathbf{x}, \mathbf{x}' \in \mathcal{X}$. Typically the function is symmetric (i.e., $\kappa(\mathbf{x}, \mathbf{x}') = \kappa(\mathbf{x}', \mathbf{x})$), and non-negative (i.e., $\kappa(\mathbf{x}, \mathbf{x}') \geq 0$), so it can be interpreted as a measure of similarity, but this is not required. We give several examples below.

14.2.1 RBF kernels

The **squared exponential kernel** (SE kernel) or **Gaussian kernel** is defined by

$$\kappa(\mathbf{x}, \mathbf{x}') = \exp\left(-\frac{1}{2}(\mathbf{x} - \mathbf{x}')^T \Sigma^{-1} (\mathbf{x} - \mathbf{x}')\right) \tag{14.1}$$

If Σ is diagonal, this can be written as

$$\kappa(\mathbf{x}, \mathbf{x}') = \exp\left(-\frac{1}{2}\sum_{j=1}^{D}\frac{1}{\sigma_j^2}(x_j - x_j')^2\right) \tag{14.2}$$

We can interpret the σ_j as defining the **characteristic length scale** of dimension j. If $\sigma_j = \infty$, the corresponding dimension is ignored; hence this is known as the **ARD kernel**. If Σ is spherical, we get the isotropic kernel

$$\kappa(\mathbf{x}, \mathbf{x}') = \exp\left(-\frac{||\mathbf{x} - \mathbf{x}'||^2}{2\sigma^2}\right) \tag{14.3}$$

Here σ^2 is known as the **bandwidth**. Equation 14.3 is an example of a **radial basis function** or **RBF** kernel, since it is only a function of $||\mathbf{x} - \mathbf{x}'||$.

14.2.2 Kernels for comparing documents

When performing document classification or retrieval, it is useful to have a way of comparing two documents, $\mathbf{x}_i$ and $\mathbf{x}_{i'}$. If we use a bag of words representation, where x_{ij} is the number of times words j occurs in document i, we can use the **cosine similarity**, which is defined by

$$\kappa(\mathbf{x}_i, \mathbf{x}_{i'}) = \frac{\mathbf{x}_i^T \mathbf{x}_{i'}}{||\mathbf{x}_i||_2 ||\mathbf{x}_{i'}||_2} \tag{14.4}$$

This quantity measures the cosine of the angle between $\mathbf{x}_i$ and $\mathbf{x}_{i'}$ when interpreted as vectors. Since $\mathbf{x}_i$ is a count vector (and hence non-negative), the cosine similarity is between 0 and 1, where 0 means the vectors are orthogonal and therefore have no words in common.

Unfortunately, this simple method does not work very well, for two main reasons. First, if $\mathbf{x}_i$ has any word in common with $\mathbf{x}_{i'}$, it is deemed similar, even though some popular words, such as "the" or "and" occur in many documents, and are therefore not discriminative. (These are known as **stop words**.) Second, if a discriminative word occurs many times in a document, the similarity is artificially boosted, even though word usage tends to be bursty, meaning that once a word is used in a document it is very likely to be used again (see Section 3.5.5).

Fortunately, we can significantly improve performance using some simple preprocessing. The idea is to replace the word count vector with a new feature vector called the **TF-IDF** representation, which stands for "term frequency inverse document frequency". We define this as follows. First, the term frequency is defined as a log-transform of the count:

$$\text{tf}(x_{ij}) \triangleq \log(1 + x_{ij}) \tag{14.5}$$

This reduces the impact of words that occur many times within one document. Second, the inverse document frequency is defined as

$$\text{idf}(j) \triangleq \log \frac{N}{1 + \sum_{i=1}^{N} \mathbb{I}(x_{ij} > 0)} \tag{14.6}$$

where N is the total number of documents, and the denominator counts how many documents contain term j. Finally, we define

$$\text{tf-idf}(\mathbf{x}_i) \triangleq [\text{tf}(x_{ij}) \times \text{idf}(j)]_{j=1}^{V} \tag{14.7}$$

(There are several other ways to define the tf and idf terms, see (Manning et al. 2008) for details.) We then use this inside the cosine similarity measure. That is, our new kernel has the form

$$\kappa(\mathbf{x}_i, \mathbf{x}_{i'}) = \frac{\phi(\mathbf{x}_i)^T \phi(\mathbf{x}_{i'})}{||\phi(\mathbf{x}_i)||_2 ||\phi(\mathbf{x}_{i'})||_2} \tag{14.8}$$

where $\phi(\mathbf{x}) = \text{tf-idf}(\mathbf{x})$. This gives good results for information retrieval (Manning et al. 2008).

A probabilistic interpretation of the tf-idf kernel is given in (Elkan 2005).

14.2.3 Mercer (positive definite) kernels

Some methods that we will study require that the kernel function satisfy the requirement that the **Gram matrix**, defined by

$$\mathbf{K} = \begin{pmatrix} \kappa(\mathbf{x}_1, \mathbf{x}_1) & \cdots & \kappa(\mathbf{x}_1, \mathbf{x}_N) \\ & \vdots & \\ \kappa(\mathbf{x}_N, \mathbf{x}_1) & \cdots & \kappa(\mathbf{x}_N, \mathbf{x}_N) \end{pmatrix} \tag{14.9}$$

be positive definite for any set of inputs $\{\mathbf{x}_i\}_{i=1}^{N}$. We call such a kernel a **Mercer kernel**, or **positive definite kernel**. It can be shown (Schoelkopf and Smola 2002) that the Gaussian kernel is a Mercer kernel as is the cosine similarity kernel (Sahami and Heilman 2006).

The importance of Mercer kernels is the following result, known as **Mercer's theorem**. If the Gram matrix is positive definite, we can compute an eigenvector decomposition of it as follows

$$\mathbf{K} = \mathbf{U}^T \mathbf{\Lambda} \mathbf{U} \tag{14.10}$$

where $\mathbf{\Lambda}$ is a diagonal matrix of eigenvalues $\lambda_i > 0$. Now consider an element of $\mathbf{K}$:

$$k_{ij} = (\mathbf{\Lambda}^{\frac{1}{2}} \mathbf{U}_{:,i})^T (\mathbf{\Lambda}^{\frac{1}{2}} \mathbf{U}_{:j}) \tag{14.11}$$

Let us define $\phi(\mathbf{x}_i) = \mathbf{\Lambda}^{\frac{1}{2}} \mathbf{U}_{:i}$. Then we can write

$$k_{ij} = \phi(\mathbf{x}_i)^T \phi(\mathbf{x}_j) \tag{14.12}$$

Thus we see that the entries in the kernel matrix can be computed by performing an inner product of some feature vectors that are implicitly defined by the eigenvectors $\mathbf{U}$. In general, if the kernel is Mercer, then there exists a function ϕ mapping $\mathbf{x} \in \mathcal{X}$ to $\mathbb{R}^D$ such that

$$\kappa(\mathbf{x}, \mathbf{x}') = \phi(\mathbf{x})^T \phi(\mathbf{x}') \tag{14.13}$$

where ϕ depends on the eigen *functions* of κ (so D is a potentially infinite dimensional space).

For example, consider the (non-stationary) **polynomial kernel** $\kappa(\mathbf{x}, \mathbf{x}') = (\gamma \mathbf{x}^T \mathbf{x}' + r)^M$, where $r > 0$. One can show that the corresponding feature vector $\phi(\mathbf{x})$ will contain all terms up to degree M. For example, if $M = 2$, $\gamma = r = 1$ and $\mathbf{x}, \mathbf{x}' \in \mathbb{R}^2$, we have

$$\begin{aligned} (1 + \mathbf{x}^T \mathbf{x}')^2 &= (1 + x_1 x_1' + x_2 x_2')^2 \tag{14.14} \\ &= 1 + 2x_1 x_1' + 2x_2 x_2' + (x_1 x_1)^2 + (x_2 x_2')^2 + 2x_1 x_1' x_2 x_2' \tag{14.15} \end{aligned}$$

This can be written as $\phi(\mathbf{x})^T\phi(\mathbf{x}')$, where

$$\phi(\mathbf{x}) = [1, \sqrt{2}x_1, \sqrt{2}x_2, x_1^2, x_2^2, \sqrt{2}x_1x_2]^T \tag{14.16}$$

So using this kernel is equivalent to working in a 6 dimensional feature space. In the case of a Gaussian kernel, the feature map lives in an infinite dimensional space. In such a case, it is clearly infeasible to explicitly represent the feature vectors.

An example of a kernel that is not a Mercer kernel is the so-called **sigmoid kernel**, defined by

$$\kappa(\mathbf{x}, \mathbf{x}') = \tanh(\gamma\mathbf{x}^T\mathbf{x}' + r) \tag{14.17}$$

(Note that this uses the tanh function even though it is called a sigmoid kernel.) This kernel was inspired by the multi-layer perceptron (see Section 16.5), but there is no real reason to use it. (For a true "neural net kernel", which is positive definite, see Section 15.4.5.)

In general, establishing that a kernel is a Mercer kernel is difficult, and requires techniques from functional analysis. However, one can show that it is possible to build up new Mercer kernels from simpler ones using a set of standard rules. For example, if κ_1 and κ_2 are both Mercer, so is $\kappa(\mathbf{x}, \mathbf{x}') = \kappa_1(\mathbf{x}, \mathbf{x}') + \kappa_2(\mathbf{x}, \mathbf{x}')$. See e.g., (Schoelkopf and Smola 2002) for details.

14.2.4 Linear kernels

Deriving the feature vector implied by a kernel is in general quite difficult, and only possible if the kernel is Mercer. However, deriving a kernel from a feature vector is easy: we just use

$$\kappa(\mathbf{x}, \mathbf{x}') = \phi(\mathbf{x})^T\phi(\mathbf{x}') = \langle\phi(\mathbf{x}), \phi(\mathbf{x}')\rangle \tag{14.18}$$

If $\phi(\mathbf{x}) = \mathbf{x}$, we get the **linear kernel**, defined by

$$\kappa(\mathbf{x}, \mathbf{x}') = \mathbf{x}^T\mathbf{x}' \tag{14.19}$$

This is useful if the original data is already high dimensional, and if the original features are individually informative, e.g., a bag of words representation where the vocabulary size is large, or the expression level of many genes. In such a case, the decision boundary is likely to be representable as a linear combination of the original features, so it is not necessary to work in some other feature space.

Of course, not all high dimensional problems are linearly separable. For example, images are high dimensional, but individual pixels are not very informative, so image classification typically requires non-linear kernels (see e.g., Section 14.2.7).

14.2.5 Matern kernels

The **Matern kernel**, which is commonly used in Gaussian process regression (see Section 15.2), has the following form

$$\kappa(r) = \frac{2^{1-\nu}}{\Gamma(\nu)}\left(\frac{\sqrt{2\nu}r}{\ell}\right)^\nu K_\nu\left(\frac{\sqrt{2\nu}r}{\ell}\right) \tag{14.20}$$

where $r = ||\mathbf{x} - \mathbf{x}'||$, $\nu > 0$, $\ell > 0$, and K_ν is a modified Bessel function. As $\nu \to \infty$, this approaches the SE kernel. If $\nu = \frac{1}{2}$, the kernel simplifies to

$$\kappa(r) = \exp(-r/\ell) \tag{14.21}$$

If $D = 1$, and we use this kernel to define a Gaussian process (see Chapter 15), we get the **Ornstein-Uhlenbeck process**, which describes the velocity of a particle undergoing Brownian motion (the corresponding function is continuous but not differentiable, and hence is very "jagged").

14.2.6 String kernels

The real power of kernels arises when the inputs are structured objects. As an example, we now describe one way of comparing two variable length strings using a **string kernel**. We follow the presentation of (Rasmussen and Williams 2006, p100) and (Hastie et al. 2009, p668).

Consider two strings $\mathbf{x}$, and $\mathbf{x}'$ of lengths D, D', each defined over the alphabet $\mathcal{A}$. For example, consider two amino acid sequences, defined over the 20 letter alphabet $\mathcal{A} = \{A, R, N, D, C, E, Q, G, H, I, L, K, M, F, P, S, T, W, Y, V\}$. Let $\mathbf{x}$ be the following sequence of length 110

```
IPTSALVKETLALLSTHRTLLIANETLRIPVPVHKNHQLCTEEIFQGIGTLESQTVQGGTV
ERLFKNLSLIKKYIDGQKKKCGEERRRVNQFLDYLQEFLGVMNTEWI
```

and let $\mathbf{x}'$ be the following sequence of length 153

```
PHRRDLCSRSIWLARKIRSDLTALTESYVKHQGLWSELTEAERLQENLQAYRTFHVLLA
RLLEDQQVHFTPTEGDFHQAIHTLLLQVAAFAYQIEELMILLEYKIPRNEADGMLFEKK
LWGLKVLQELSQWTVRSIHDLRFISSHQTGIP
```

These strings have the substring LQE in common. We can define the similarity of two strings to be the number of substrings they have in common.

More formally and more generally, let us say that s is a substring of x if we can write $x = usv$ for some (possibly empty) strings u, s and v. Now let $\phi_s(x)$ denote the number of times that substring s appears in string x. We define the kernel between two strings x and x' as

$$\kappa(x, x') = \sum_{s \in \mathcal{A}^*} w_s \phi_s(x) \phi_s(x') \tag{14.22}$$

where $w_s \geq 0$ and $\mathcal{A}^*$ is the set of all strings (of any length) from the alphabet $\mathcal{A}$ (this is known as the Kleene star operator). This is a Mercer kernel, and can be computed in $O(|x| + |x'|)$ time (for certain settings of the weights $\{w_s\}$) using suffix trees (Leslie et al. 2003; Vishwanathan and Smola 2003; Shawe-Taylor and Cristianini 2004).

There are various cases of interest. If we set $w_s = 0$ for $|s| > 1$ we get a bag-of-characters kernel. This defines $\phi(x)$ to be the number of times each character in $\mathcal{A}$ occurs in x. If we require s to be bordered by white-space, we get a bag-of-words kernel, where $\phi(x)$ counts how many times each possible word occurs. Note that this is a very sparse vector, since most words

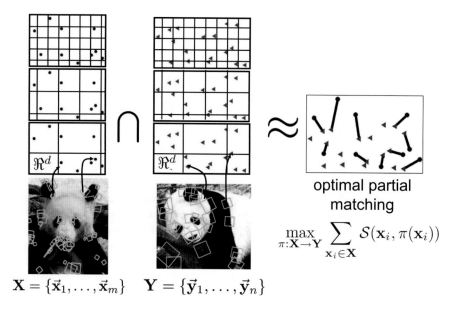

Figure 14.1 Illustration of a pyramid match kernel computed from two images. Used with kind permission of Kristen Grauman.

will not be present. If we only consider strings of a fixed length k, we get the **k-spectrum kernel**. This has been used to classify proteins into SCOP superfamilies (Leslie et al. 2003). For example if $k = 3$, we have $\phi_{LQE}(\mathbf{x}) = 1$ and $\phi_{LQE}(\mathbf{x}') = 2$ for the two strings above.

Various extensions are possible. For example, we can allow character mismatches (Leslie et al. 2003). And we can generalize string kernels to compare trees, as described in (Collins and Duffy 2002). This is useful for classifying (or ranking) parse trees, evolutionary trees, etc.

14.2.7 Pyramid match kernels

In computer vision, it is common to create a bag-of-words representation of an image by computing a feature vector (often using SIFT (Lowe 1999)) from a variety of points in the image, commonly chosen by an interest point detector. The feature vectors at the chosen places are then vector-quantized to create a bag of discrete symbols.

One way to compare two variable-sized bags of this kind is to use a **pyramid match kernel** (Grauman and Darrell 2007). The basic idea is illustrated in Figure 14.1. Each feature set is mapped to a multi-resolution histogram. These are then compared using weighted histogram intersection. It turns out that this provides a good approximation to the similarity measure one would obtain by performing an optimal bipartite match at the finest spatial resolution, and then summing up pairwise similarities between matched points. However, the histogram method is faster and is more robust to missing and unequal numbers of points. This is a Mercer kernel.

14.2.8 Kernels derived from probabilistic generative models

Suppose we have a probabilistic generative model of feature vectors, $p(\mathbf{x}|\boldsymbol{\theta})$. Then there are several ways we can use this model to define kernel functions, and thereby make the model suitable for discriminative tasks. We sketch two approaches below.

14.2.8.1 Probability product kernels

One approach is to define a kernel as follows:

$$\kappa(\mathbf{x}_i, \mathbf{x}_j) = \int p(\mathbf{x}|\mathbf{x}_i)^\rho p(\mathbf{x}|\mathbf{x}_j)^\rho d\mathbf{x} \tag{14.23}$$

where $\rho > 0$, and $p(\mathbf{x}|\mathbf{x}_i)$ is often approximated by $p(\mathbf{x}|\hat{\boldsymbol{\theta}}(\mathbf{x}_i))$, where $\hat{\boldsymbol{\theta}}(\mathbf{x}_i)$ is a parameter estimate computed using a single data vector. This is called a **probability product kernel** (Jebara et al. 2004).

Although it seems strange to fit a model to a single data point, it is important to bear in mind that the fitted model is only being used to see how similar two objects are. In particular, if we fit the model to $\mathbf{x}_i$ and then the model thinks $\mathbf{x}_j$ is likely, this means that $\mathbf{x}_i$ and $\mathbf{x}_j$ are similar. For example, suppose $p(\mathbf{x}|\boldsymbol{\theta}) = \mathcal{N}(\boldsymbol{\mu}, \sigma^2 \mathbf{I})$, where σ^2 is fixed. If $\rho = 1$, and we use $\hat{\boldsymbol{\mu}}(\mathbf{x}_i) = \mathbf{x}_i$ and $\hat{\boldsymbol{\mu}}(\mathbf{x}_j) = \mathbf{x}_j$, we find (Jebara et al. 2004, p825) that

$$\kappa(\mathbf{x}_i, \mathbf{x}_j) = \frac{1}{(4\pi\sigma^2)^{D/2}} \exp\left(-\frac{1}{4\sigma^2}||\mathbf{x}_i - \mathbf{x}_j||^2\right) \tag{14.24}$$

which is (up to a constant factor) the RBF kernel.

It turns out that one can compute Equation 14.23 for a variety of generative models, including ones with latent variables, such as HMMs. This provides one way to define kernels on variable length sequences. Furthermore, this technique works even if the sequences are of real-valued vectors, unlike the string kernel in Section 14.2.6. See (Jebara et al. 2004) for further details.

14.2.8.2 Fisher kernels

A more efficient way to use generative models to define kernels is to use a **Fisher kernel** (Jaakkola and Haussler 1998) which is defined as follows:

$$\kappa(\mathbf{x}, \mathbf{x}') = \mathbf{s}(\mathbf{x})^T \mathbf{I}^{-1} \mathbf{s}(\mathbf{x}') \tag{14.25}$$

where $\mathbf{s}$ is the gradient of the log likelihood, or score function, evaluated at the MLE $\hat{\boldsymbol{\theta}}$

$$\mathbf{s}(\mathbf{x}) \triangleq \nabla_{\boldsymbol{\theta}} \log p(\mathbf{x}|\boldsymbol{\theta})\big|_{\hat{\boldsymbol{\theta}}} \tag{14.26}$$

and $\mathbf{I}$ is the Fisher information matrix (see Section 6.2.2):

$$\mathbf{I} = -\nabla^2 \log p(\mathbf{x}|\boldsymbol{\theta})\big|_{\hat{\boldsymbol{\theta}}} \tag{14.27}$$

Note that $\hat{\boldsymbol{\theta}}$ is a function of all the data, so the similarity of $\mathbf{x}$ and $\mathbf{x}'$ is computed in the context of all the data as well. Also, note that we only have to fit one model.

The intuition behind the Fisher kernel is the following: let $\mathbf{g}(\mathbf{x})$ be the direction (in parameter space) in which $\mathbf{x}$ would like the parameters to move (from $\hat{\boldsymbol{\theta}}$) so as to maximize its own

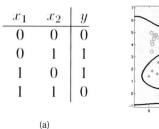

x_1	x_2	y
0	0	0
0	1	1
1	0	1
1	1	0

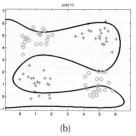

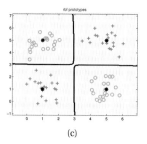

(a)   (b)   (c)

Figure 14.2   (a) xor truth table. (b) Fitting a linear logistic regression classifier using degree 10 polynomial expansion. (c) Same model, but using an RBF kernel with centroids specified by the 4 black crosses. Figure generated by `logregXorDemo`.

likelihood; call this the directional gradient. Then we say that two vectors $\mathbf{x}$ and $\mathbf{x}'$ are similar if their directional gradients are similar wrt the geometry encoded by the curvature of the likelihood function (see Section 7.5.3).

Interestingly, (Saunders et al. 2003) have shown that the string kernel of Section 14.2.6 is equivalent to the Fisher kernel derived from an L'th order Markov chain (see Section 17.2). Also, (Elkan 2005) showed that a kernel defined by the inner product of TF-IDF vectors (Section 14.2.2) is approximately equal to the Fisher kernel for a certain generative model of text based on the compound Dirichlet multinomial model (Section 3.5.5).

14.3   Using kernels inside GLMs

In this section, we discuss one simple way to use kernels for classification and regression. We will see other approaches later.

14.3.1   Kernel machines

We define a **kernel machine** to be a GLM where the input feature vector has the form

$$\boldsymbol{\phi}(\mathbf{x}) = [\kappa(\mathbf{x}, \boldsymbol{\mu}_1), \ldots, \kappa(\mathbf{x}, \boldsymbol{\mu}_K)] \tag{14.28}$$

where $\boldsymbol{\mu}_k \in \mathcal{X}$ are a set of K **centroids**. If κ is an RBF kernel, this is called an **RBF network**. We discuss ways to choose the $\boldsymbol{\mu}_k$ parameters below. We will call Equation 14.28 a **kernelised feature vector**. Note that in this approach, the kernel need not be a Mercer kernel.

We can use the kernelized feature vector for logistic regression by defining $p(y|\mathbf{x}, \boldsymbol{\theta}) = \text{Ber}(\mathbf{w}^T \boldsymbol{\phi}(\mathbf{x}))$. This provides a simple way to define a non-linear decision boundary. As an example, consider the data coming from the **exclusive or** or **xor** function. This is a binary-valued function of two binary inputs. Its truth table is shown in Figure 14.2(a). In Figure 14.2(b), we have shown some data labeled by the xor function, but we have **jittered** the points to make the picture clearer.[1] We see we cannot separate the data even using a degree 10 polynomial.

---

1. Jittering is a common visualization trick in statistics, wherein points in a plot/display that would otherwise land on top of each other are dispersed with uniform additive noise.

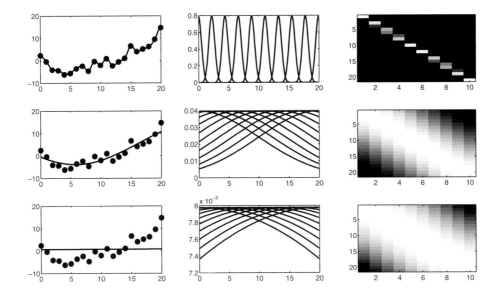

Figure 14.3 RBF basis in 1d. Left column: fitted function. Middle column: basis functions evaluated on a grid. Right column: design matrix. Top to bottom we show different bandwidths: $\tau = 0.1$, $\tau = 0.5$, $\tau = 50$. Figure generated by `linregRbfDemo`.

However, using an RBF kernel and just 4 prototypes easily solves the problem as shown in Figure 14.2(c).

We can also use the kernelized feature vector inside a linear regression model by defining $p(y|\mathbf{x}, \boldsymbol{\theta}) = \mathcal{N}(\mathbf{w}^T \phi(\mathbf{x}), \sigma^2)$. For example, Figure 14.3 shows a 1d data set fit with $K = 10$ uniformly spaced RBF prototypes, but with the bandwidth ranging from small to large. Small values lead to very wiggly functions, since the predicted function value will only be non-zero for points $\mathbf{x}$ that are close to one of the prototypes $\boldsymbol{\mu}_k$. If the bandwidth is very large, the design matrix reduces to a constant matrix of 1's, since each point is equally close to every prototype; hence the corresponding function is just a straight line.

14.3.2 L1VMs, RVMs, and other sparse vector machines

The main issue with kernel machines is: how do we choose the centroids $\boldsymbol{\mu}_k$? If the input is low-dimensional Euclidean space, we can uniformly tile the space occupied by the data with prototypes, as we did in Figure 14.2(c). However, this approach breaks down in higher numbers of dimensions because of the curse of dimensionality. If $\boldsymbol{\mu}_k \in \mathbb{R}^D$, we can try to perform numerical optimization of these parameters (see e.g., (Haykin 1998)), or we can use MCMC inference, (see e.g., (Andrieu et al. 2001; Kohn et al. 2001)), but the resulting objective function / posterior is highly multimodal. Furthermore, these techniques are hard to extend to structured input spaces, where kernels are most useful.

Another approach is to find clusters in the data and then to assign one prototype per cluster center (many clustering algorithms just need a similarity metric as input). However, the regions of space that have high density are not necessarily the ones where the prototypes are most useful for representing the output, that is, clustering is an unsupervised task that may not yield a representation that is useful for prediction. Furthermore, there is the need to pick the number of clusters.

A simpler approach is to make each example $\mathbf{x}_i$ be a prototype, so we get

$$\phi(\mathbf{x}) = [\kappa(\mathbf{x}, \mathbf{x}_1), \dots, \kappa(\mathbf{x}, \mathbf{x}_N)] \tag{14.29}$$

Now we see $D = N$, so we have as many parameters as data points. However, we can use any of the sparsity-promoting priors for $\mathbf{w}$ discussed in Chapter 13 to efficiently select a subset of the training exemplars. We call this a **sparse vector machine**.

The most natural choice is to use ℓ_1 regularization (Krishnapuram et al. 2005). (Note that in the multi-class case, it is necessary to use group lasso, since each exemplar is associated with C weights, one per class.) We call this **L1VM**, which stands for "ℓ_1-regularized vector machine". By analogy, we define the use of an ℓ_2 regularizer to be a **L2VM** or "ℓ_2-regularized vector machine"; this of course will not be sparse.

We can get even greater sparsity by using ARD/SBL, resulting in a method called the **relevance vector machine** or **RVM** (Tipping 2001). One can fit this model using generic ARD/SBL algorithms, although in practice the most common method is the greedy algorithm in (Tipping and Faul 2003) (this is the algorithm implemented in Mike Tipping's code, which is bundled with PMTK).

Another very popular approach to creating a sparse kernel machine is to use a **support vector machine** or **SVM**. This will be discussed in detail in Section 14.5. Rather than using a sparsity-promoting prior, it essentially modifies the likelihood term, which is rather unnatural from a Bayesian point of view. Nevertheless, the effect is similar, as we will see.

In Figure 14.4, we compare L2VM, L1VM, RVM and an SVM using the same RBF kernel on a binary classification problem in 2d. For simplicity, λ was chosen by hand for L2VM and L1VM; for RVMs, the parameters are estimated using empirical Bayes; and for the SVM, we use CV to pick $C = 1/\lambda$, since SVM performance is very sensitive to this parameter (see Section 14.5.3). We see that all the methods give similar performance. However, RVM is the sparsest (and hence fastest at test time), then L1VM, and then SVM. RVM is also the fastest to train. This is despite the fact that the RVM code is in Matlab and the SVM code is in the C language. The reason is that CV (needed for the SVM) requires fitting multiple models, which is slow, whereas empirical Bayes only has to fit a single model.

In Figure 14.5, we compare L2VM, L1VM, RVM and an SVM using an RBF kernel on a 1d regression problem. Again, we see that predictions are quite similar, but RVM is the sparsest, then L1VM, then SVM. This is further illustrated in Figure 14.6.

14.4   The kernel trick

Rather than defining our feature vector in terms of kernels, $\phi(\mathbf{x}) = [\kappa(\mathbf{x}, \mathbf{x}_1), \dots, \kappa(\mathbf{x}, \mathbf{x}_N)]$, we can instead work with the original feature vectors $\mathbf{x}$, but modify the algorithm so that it replaces all inner products of the form $\langle \mathbf{x}, \mathbf{x}' \rangle$ with a call to the kernel function, $\kappa(\mathbf{x}, \mathbf{x}')$. This is called the **kernel trick**. It turns out that many algorithms can be kernelized in this way. We

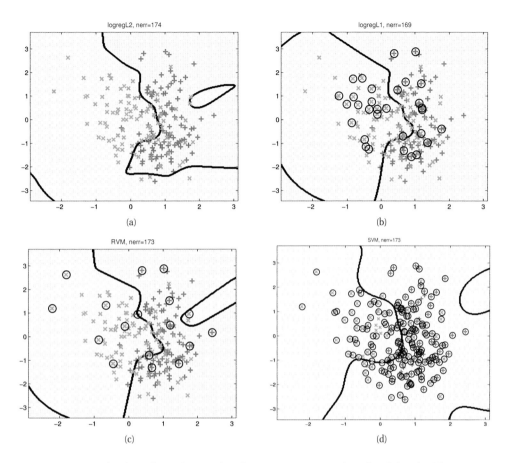

Figure 14.4   Example of non-linear binary classification using an RBF kernel with bandwidth $\sigma = 0.3$. (a) L2VM with $\lambda = 5$. (b) L1VM with $\lambda = 1$. (c) RVM. (d) SVM with $C = 1/\lambda$ chosen by cross validation. Black circles denote the support vectors. 178 out of the 200 points are chosen as SVs. Figure generated by `kernelBinaryClassifDemo`.

give some examples below. Note that we require that the kernel be a Mercer kernel for this trick to work.

14.4.1   Kernelized nearest neighbor classification

Recall that in a 1NN classifier (Section 1.4.2), we just need to compute the Euclidean distance of a test vector to all the training points, find the closest one, and look up its label. This can be kernelized by observing that

$$||\mathbf{x}_i - \mathbf{x}_{i'}||_2^2 = \langle \mathbf{x}_i, \mathbf{x}_i \rangle + \langle \mathbf{x}_{i'}, \mathbf{x}_{i'} \rangle - 2\langle \mathbf{x}_i, \mathbf{x}_{i'} \rangle \tag{14.30}$$

This allows us to apply the nearest neighbor classifier to structured data objects.

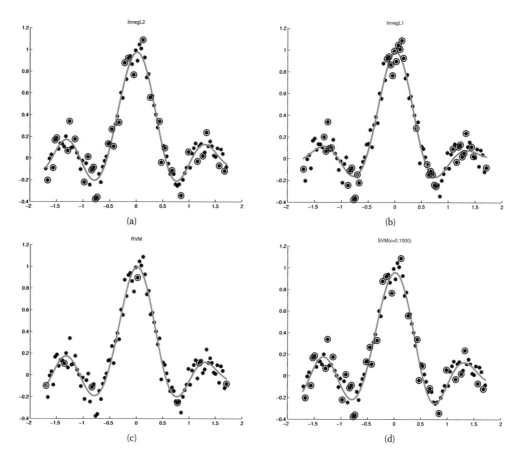

Figure 14.5 Example of kernel based regression on the noisy sinc function using an RBF kernel with bandwidth $\sigma = 0.3$. (a) L2VM with $\lambda = 0.5$. (b) L1VM with $\lambda = 0.5$. (c) RVM. (d) SVM regression with $C = 1/\lambda$. and $\epsilon = 0.1$ (the default for SVMlight). Red circles denote the retained training exemplars. Figure generated by kernelRegrDemo.

14.4.2 Kernelized K-medoids clustering

K-means clustering (Section 11.4.2.5) uses Euclidean distance to measure dissimilarity, which is not always appropriate for structured objects. We now describe how to develop a kernelized version of the algorithm.

The first step is to replace the K-means algorithm with the **K-medoids algorithm**. This is similar to K-means, but instead of representing each cluster's centroid by the mean of all data vectors assigned to this cluster, we make each centroid be one of the data vectors themselves. Thus we always deal with integer indexes, rather than data objects. We assign objects to their closest centroids as before. When we update the centroids, we look at each object that belongs to the cluster, and measure the sum of its distances to all the others in the same cluster; we

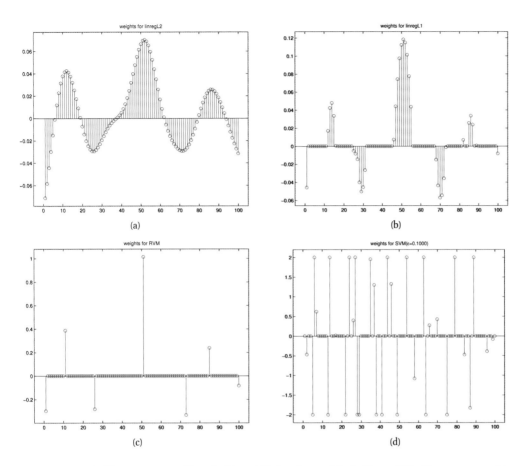

Figure 14.6 Coefficient vectors of length $N = 100$ for the models in Figure 14.5. Figure generated by `kernelRegrDemo`.

then pick the one which has the smallest such sum:

$$m_k = \underset{i:z_i=k}{\operatorname{argmin}} \sum_{i':z_{i'}=k} d(i, i') \tag{14.31}$$

where

$$d(i, i') \triangleq ||\mathbf{x}_i - \mathbf{x}_{i'}||_2^2 \tag{14.32}$$

This takes $O(n_k^2)$ work per cluster, whereas K-means takes $O(n_k D)$ to update each cluster. The pseudo-code is given in Algorithm 14.1. This method can be modified to derive a classifier, by computing the nearest medoid for each class. This is known as **nearest medoid classification** (Hastie et al. 2009, p671).

This algorithm can be kernelized by using Equation 14.30 to replace the distance computation, $d(i, i')$.

Algorithm 14.1: K-medoids algorithm

---

1 *initialize* $m_{1:K}$ as a random subset of size K from $\{1, \ldots, N\}$;
2 **repeat**
3     $z_i = \text{argmin}_k\, d(i, m_k)$ for $i = 1 : N$;
4     $m_k \leftarrow \text{argmin}_{i:z_i=k} \sum_{i':z_{i'}=k} d(i, i')$ for $k = 1 : K$;
5 **until** *converged*;

14.4.3 Kernelized ridge regression

Applying the kernel trick to distance-based methods was straightforward. It is not so obvious how to apply it to parametric models such as ridge regression. However, it can be done, as we now explain. This will serve as a good "warm up" for studying SVMs.

14.4.3.1 The primal problem

Let $\mathbf{x} \in \mathbb{R}^D$ be some feature vector, and $\mathbf{X}$ be the corresponding $N \times D$ design matrix. We want to minimize

$$J(\mathbf{w}) = (\mathbf{y} - \mathbf{X}\mathbf{w})^T (\mathbf{y} - \mathbf{X}\mathbf{w}) + \lambda ||\mathbf{w}||^2 \tag{14.33}$$

The optimal solution is given by

$$\mathbf{w} = (\mathbf{X}^T\mathbf{X} + \lambda \mathbf{I}_D)^{-1}\mathbf{X}^T\mathbf{y} = \left(\sum_i \mathbf{x}_i\mathbf{x}_i^T + \lambda \mathbf{I}_D\right)^{-1}\mathbf{X}^T\mathbf{y} \tag{14.34}$$

14.4.3.2 The dual problem

Equation 14.34 is not yet in the form of inner products. However, using the matrix inversion lemma (Equation 4.107) we rewrite the ridge estimate as follows

$$\mathbf{w} = \mathbf{X}^T(\mathbf{X}\mathbf{X}^T + \lambda \mathbf{I}_N)^{-1}\mathbf{y} \tag{14.35}$$

which takes $O(N^3 + N^2D)$ time to compute. This can be advantageous if D is large. Furthermore, we see that we can partially kernelize this, by replacing $\mathbf{X}\mathbf{X}^T$ with the Gram matrix $\mathbf{K}$. But what about the leading $\mathbf{X}^T$ term?

Let us define the following **dual variables**:

$$\boldsymbol{\alpha} \triangleq (\mathbf{K} + \lambda \mathbf{I}_N)^{-1}\mathbf{y} \tag{14.36}$$

Then we can rewrite the **primal variables** as follows

$$\mathbf{w} = \mathbf{X}^T\boldsymbol{\alpha} = \sum_{i=1}^{N} \alpha_i \mathbf{x}_i \tag{14.37}$$

This tells us that the solution vector is just a linear sum of the N training vectors. When we plug this in at test time to compute the predictive mean, we get

$$\hat{f}(\mathbf{x}) = \mathbf{w}^T\mathbf{x} = \sum_{i=1}^{N} \alpha_i \mathbf{x}_i^T\mathbf{x} = \sum_{i=1}^{N} \alpha_i \kappa(\mathbf{x}, \mathbf{x}_i) \tag{14.38}$$

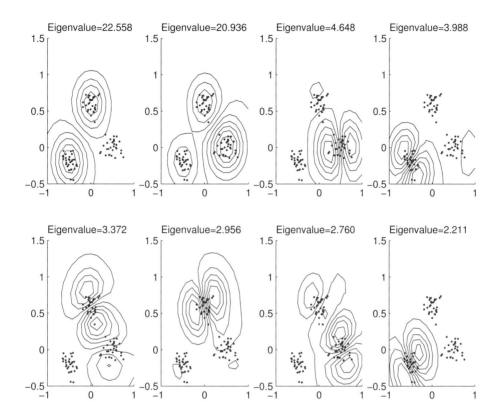

Figure 14.7 Visualization of the first 8 kernel principal component basis functions derived from some 2d data. We use an RBF kernel with $\sigma^2 = 0.1$. Figure generated by `kpcaScholkopf`, written by Bernhard Scholkopf.

So we have successfully kernelized ridge regression by changing from primal to dual variables. This technique can be applied to many other linear models, such as logistic regression.

14.4.3.3 Computational cost

The cost of computing the dual variables $\boldsymbol{\alpha}$ is $O(N^3)$, whereas the cost of computing the primal variables $\mathbf{w}$ is $O(D^3)$. Hence the kernel method can be useful in high dimensional settings, even if we only use a linear kernel (c.f., the SVD trick in Equation 7.44). However, prediction using the dual variables takes $O(ND)$ time, while prediction using the primal variables only takes $O(D)$ time. We can speedup prediction by making $\boldsymbol{\alpha}$ sparse, as we discuss in Section 14.5.

14.4.4 Kernel PCA

In Section 12.2, we saw how we could compute a low-dimensional linear embedding of some data using PCA. This required finding the eigenvectors of the sample covariance matrix $\mathbf{S} =$

$\frac{1}{N} \sum_{i=1}^{N} \mathbf{x}_i \mathbf{x}_i^T = (1/N) \mathbf{X}^T \mathbf{X}$. However, we can also compute PCA by finding the eigenvectors of the inner product matrix $\mathbf{X}\mathbf{X}^T$, as we show below. This will allow us to produce a nonlinear embedding, using the kernel trick, a method known as **kernel PCA** (Schoelkopf et al. 1998).

First, let $\mathbf{U}$ be an orthogonal matrix containing the eigenvectors of $\mathbf{X}\mathbf{X}^T$ with corresponding eigenvalues in $\mathbf{\Lambda}$. By definition we have $(\mathbf{X}\mathbf{X}^T)\mathbf{U} = \mathbf{U}\mathbf{\Lambda}$. Pre-multiplying by $\mathbf{X}^T$ gives

$$(\mathbf{X}^T\mathbf{X})(\mathbf{X}^T\mathbf{U}) = (\mathbf{X}^T\mathbf{U})\mathbf{\Lambda} \tag{14.39}$$

from which we see that the eigenvectors of $\mathbf{X}^T\mathbf{X}$ (and hence of $\mathbf{S}$) are $\mathbf{V} = \mathbf{X}^T\mathbf{U}$, with eigenvalues given by $\mathbf{\Lambda}$ as before. However, these eigenvectors are not normalized, since $||\mathbf{v}_j||^2 = \mathbf{u}_j^T \mathbf{X}\mathbf{X}^T \mathbf{u}_j = \lambda_j \mathbf{u}_j^T \mathbf{u}_j = \lambda_j$. So the normalized eigenvectors are given by $\mathbf{V}_{pca} = \mathbf{X}^T\mathbf{U}\mathbf{\Lambda}^{-\frac{1}{2}}$. This is a useful trick for regular PCA if $D > N$, since $\mathbf{X}^T\mathbf{X}$ has size $D \times D$, whereas $\mathbf{X}\mathbf{X}^T$ has size $N \times N$. It will also allow us to use the kernel trick, as we now show.

Now let $\mathbf{K} = \mathbf{X}\mathbf{X}^T$ be the Gram matrix. Recall from Mercer's theorem that the use of a kernel implies some underlying feature space, so we are implicitly replacing $\mathbf{x}_i$ with $\phi(\mathbf{x}_i) = \phi_i$. Let $\mathbf{\Phi}$ be the corresponding (notional) design matrix, and $\mathbf{S}_\phi = \frac{1}{N} \sum_i \phi_i \phi_i^T$ be the corresponding (notional) covariance matrix in feature space. The eigenvectors are given by $\mathbf{V}_{kpca} = \mathbf{\Phi}^T\mathbf{U}\mathbf{\Lambda}^{-\frac{1}{2}}$, where $\mathbf{U}$ and $\mathbf{\Lambda}$ contain the eigenvectors and eigenvalues of $\mathbf{K}$. Of course, we can't actually compute $\mathbf{V}_{kpca}$, since ϕ_i is potentially infinite dimensional. However, we can compute the projection of a test vector $\mathbf{x}_*$ onto the feature space as follows:

$$\phi_*^T \mathbf{V}_{kpca} = \phi_*^T \mathbf{\Phi}\mathbf{U}\mathbf{\Lambda}^{-\frac{1}{2}} = \mathbf{k}_*^T \mathbf{U}\mathbf{\Lambda}^{-\frac{1}{2}} \tag{14.40}$$

where $\mathbf{k}_* = [\kappa(\mathbf{x}_*, \mathbf{x}_1), \ldots, \kappa(\mathbf{x}_*, \mathbf{x}_N)]$.

There is one final detail to worry about. So far, we have assumed the projected data has zero mean, which is not the case in general. We cannot simply subtract off the mean in feature space. However, there is a trick we can use. Define the centered feature vector as $\tilde{\phi}_i = \phi(\mathbf{x}_i) - \frac{1}{N} \sum_{j=1}^{N} \phi(\mathbf{x}_j)$. The Gram matrix of the centered feature vectors is given by

$$
\begin{aligned}
\tilde{K}_{ij} &= \tilde{\phi}_i^T \tilde{\phi}_j & (14.41) \\
&= \phi_i^T \phi_j - \frac{1}{N} \sum_{k=1}^{N} \phi_i^T \phi_k - \frac{1}{N} \sum_{k=1}^{N} \phi_j^T \phi_k + \frac{1}{N^2} \sum_{k=1}^{N} \sum_{l=1}^{M} \phi_k^T \phi_l & (14.42) \\
&= \kappa(\mathbf{x}_i, \mathbf{x}_j) - \frac{1}{N} \sum_{k=1}^{N} \kappa(\mathbf{x}_i, \mathbf{x}_k) - \frac{1}{N} \sum_{k=1}^{N} \kappa(\mathbf{x}_j, \mathbf{x}_k) + \frac{1}{N^2} \sum_{k=1}^{N} \sum_{l=1}^{M} \kappa(\mathbf{x}_k, \mathbf{x}_l) & (14.43)
\end{aligned}
$$

This can be expressed in matrix notation as follows:

$$\tilde{\mathbf{K}} = \mathbf{H}\mathbf{K}\mathbf{H} \tag{14.44}$$

where $\mathbf{H} \triangleq \mathbf{I} - \frac{1}{N}\mathbf{1}_N\mathbf{1}_N^T$. is the **centering matrix**. We can convert all this algebra into the pseudocode shown in Algorithm 14.2.

Whereas linear PCA is limited to using $L \leq D$ components, in kPCA, we can use up to N components, since the rank of $\mathbf{\Phi}$ is $N \times D^*$, where D^* is the (potentially infinite) dimensionality of embedded feature vectors. Figure 14.7 gives an example of the method applied to some $D = 2$ dimensional data using an RBF kernel. We project points in the unit grid onto the first

---

Algorithm 14.2: Kernel PCA

1 Input: $\mathbf{K}$ of size $N \times N$, $\mathbf{K}_*$ of size $N_* \times N$, num. latent dimensions L;
2 $\mathbf{O} = \mathbf{1}_N \mathbf{1}_N^T / N$;
3 $\tilde{\mathbf{K}} = \mathbf{K} - \mathbf{OK} - \mathbf{KO} + \mathbf{OKO}$;
4 $[\mathbf{U}, \mathbf{\Lambda}] = \mathrm{eig}(\tilde{\mathbf{K}})$;
5 **for** $i = 1 : N$ **do**
6     $\mathbf{v}_i = \mathbf{u}_i / \sqrt{\lambda_i}$
7 $\mathbf{O}_* = \mathbf{1}_{N_*} \mathbf{1}_N^T / N$;
8 $\tilde{\mathbf{K}}_* = \mathbf{K}_* - \mathbf{O}_* \mathbf{K}_* - \mathbf{K}_* \mathbf{O}_* + \mathbf{O}_* \mathbf{K}_* \mathbf{O}_*$;
9 $\mathbf{Z} = \tilde{\mathbf{K}}_* \mathbf{V}(:, 1 : L)$

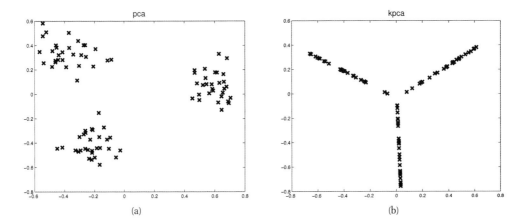

Figure 14.8 2d visualization of some 2d data. (a) PCA projection. (b) Kernel PCA projection. Figure generated by `kpcaDemo2`, based on code by L.J.P. van der Maaten.

8 components and visualize the corresponding surfaces using a contour plot. We see that the first two component separate the three clusters, and following components split the clusters.

Although the features learned by kPCA can be useful for classification (Schoelkopf et al. 1998), they are not necessarily so useful for data visualization. For example, Figure 14.8 shows the projection of the data from Figure 14.7 onto the first 2 principal bases computed using PCA and kPCA. Obviously PCA perfectly represents the data. kPCA represents each cluster by a different line.

Of course, there is no need to project 2d data back into 2d. So let us consider a different data set. We will use a 12 dimensional data set representing the three known phases of flow in an oil pipeline. (This data, which is widely used to compare data visualization methods, is synthetic, and comes from (Bishop and James 1993).) We project this into 2d using PCA and kPCA (with an RBF kernel). The results are shown in Figure 14.9. If we perform nearest neighbor classification in the low-dimensional space, kPCA makes 13 errors and PCA makes 20 (Lawrence

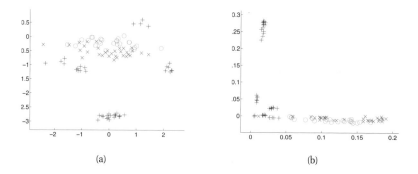

Figure 14.9   2d representation of 12 dimensional oil flow data. The different colors/symbols represent the 3 phases of oil flow. (a) PCA. (b) Kernel PCA with Gaussian kernel. Compare to Figure 15.10(b). From Figure 1 of (Lawrence 2005). Used with kind permission of Neil Lawrence.

2005). Nevertheless, the kPCA projection is rather unnatural. In Section 15.5, we will discuss how to make kernelized versions of *probabilistic* PCA.

Note that there is a close connection between kernel PCA and a technique known as multidimensional scaling or MDS. This methods finds a low-dimensional embedding such that Euclidean distance in the embedding space approximates the original dissimilarity matrix. See e.g., (Williams 2002) for details.

14.5   Support vector machines (SVMs)

In Section 14.3.2, we saw one way to derive a sparse kernel machine, namely by using a GLM with kernel basis functions, plus a sparsity-promoting prior such as ℓ_1 or ARD. An alternative approach is to change the objective function from negative log likelihood to some other loss function, as we discussed in Section 6.5.5. In particular, consider the ℓ_2 regularized empirical risk function

$$J(\mathbf{w}, \lambda) = \sum_{i=1}^{N} L(y_i, \hat{y}_i) + \lambda ||\mathbf{w}||^2 \tag{14.45}$$

where $\hat{y}_i = \mathbf{w}^T \mathbf{x}_i + w_0$. (So far this is in the original feature space; we introduce kernels in a moment.) If L is quadratic loss, this is equivalent to ridge regression, and if L is the log-loss defined in Equation 6.73, this is equivalent to logistic regression.

In the ridge regression case, we know that the solution to this has the form $\hat{\mathbf{w}} = (\mathbf{X}^T\mathbf{X} + \lambda\mathbf{I})^{-1}\mathbf{X}^T\mathbf{y}$, and plug-in predictions take the form $\hat{w}_0 + \hat{\mathbf{w}}^T\mathbf{x}$. As we saw in Section 14.4.3, we can rewrite these equations in a way that only involves inner products of the form $\mathbf{x}^T\mathbf{x}'$, which we can replace by calls to a kernel function, $\kappa(\mathbf{x}, \mathbf{x}')$. This is kernelized, but not sparse. However, if we replace the quadratic/ log-loss with some other loss function, to be explained below, we can ensure that the solution is sparse, so that predictions only depend on a subset of the training data, known as **support vectors**. This combination of the kernel trick plus a modified loss function is known as a **support vector machine** or **SVM**. This technique was

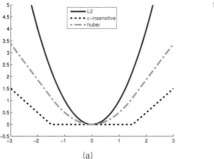
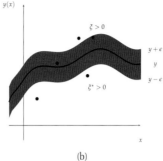

(a)   (b)

Figure 14.10 (a) Illustration of ℓ_2, Huber and ϵ-insensitive loss functions, where $\epsilon = 1.5$. Figure generated by `huberLossDemo`. (b) Illustration of the ϵ-tube used in SVM regression. Points above the tube have $\xi_i > 0$ and $\xi_i^* = 0$. Points below the tube have $\xi_i = 0$ and $\xi_i^* > 0$. Points inside the tube have $\xi_i = \xi_i^* = 0$. Based on Figure 7.7 of (Bishop 2006).

originally designed for binary classification, but can be extended to regression and multi-class classification as we explain below.

Note that SVMs are very unnatural from a probabilistic point of view. First, they encode sparsity in the loss function rather than the prior. Second, they encode kernels by using an algorithmic trick, rather than being an explicit part of the model. Finally, SVMs do not result in probabilistic outputs, which causes various difficulties, especially in the multi-class classification setting (see Section 14.5.2.4 for details).

It is possible to obtain sparse, probabilistic, multi-class kernel-based classifiers, which work as well or better than SVMs, using techniques such as the L1VM or RVM, discussed in Section 14.3.2. However, we include a discussion of SVMs, despite their non-probabilistic nature, for two main reasons. First, they are very popular and widely used, so all students of machine learning should know about them. Second, they have some computational advantages over probabilistic methods in the structured output case; see Section 19.7.

14.5.1 SVMs for regression

The problem with kernelized ridge regression is that the solution vector $\mathbf{w}$ depends on all the training inputs. We now seek a method to produce a sparse estimate.

Vapnik (Vapnik et al. 1997) proposed a variant of the Huber loss function (Section 7.4) called the **epsilon insensitive loss function**, defined by

$$L_\epsilon(y, \hat{y}) \triangleq \begin{cases} 0 & \text{if } |y - \hat{y}| < \epsilon \\ |y - \hat{y}| - \epsilon & \text{otherwise} \end{cases} \tag{14.46}$$

This means that any point lying inside an ϵ-**tube** around the prediction is not penalized, as in Figure 14.10.

The corresponding objective function is usually written in the following form

$$J = C \sum_{i=1}^{N} L_\epsilon(y_i, \hat{y}_i) + \frac{1}{2}||\mathbf{w}||^2 \tag{14.47}$$

where $\hat{y}_i = f(\mathbf{x}_i) = \mathbf{w}^T\mathbf{x}_i + w_0$ and $C = 1/\lambda$ is a regularization constant. This objective is convex and unconstrained, but not differentiable, because of the absolute value function in the loss term. As in Section 13.4, where we discussed the lasso problem, there are several possible algorithms we could use. One popular approach is to formulate the problem as a constrained optimization problem. In particular, we introduce **slack variables** to represent the degree to which each point lies outside the tube:

$$y_i \leq f(\mathbf{x}_i) + \epsilon + \xi_i^+ \tag{14.48}$$
$$y_i \geq f(\mathbf{x}_i) - \epsilon - \xi_i^- \tag{14.49}$$

Given this, we can rewrite the objective as follows:

$$J = C\sum_{i=1}^{N}(\xi_i^+ + \xi_i^-) + \frac{1}{2}||\mathbf{w}||^2 \tag{14.50}$$

This is a quadratic function of $\mathbf{w}$, and must be minimized subject to the linear constraints in Equations 14.48-14.49, as well as the positivity constraints $\xi_i^+ \geq 0$ and $\xi_i^- \geq 0$. This is a standard quadratic program in $2N + D + 1$ variables.

One can show (see e.g., (Schoelkopf and Smola 2002)) that the optimal solution has the form

$$\hat{\mathbf{w}} = \sum_i \alpha_i \mathbf{x}_i \tag{14.51}$$

where $\alpha_i \geq 0$. Furthermore, it turns out that the α vector is sparse, because we don't care about errors which are smaller than ϵ. The $\mathbf{x}_i$ for which $\alpha_i > 0$ are called the **support vectors**; these are points for which the errors lie on or outside the ϵ tube.

Once the model is trained, we can then make predictions using

$$\hat{y}(\mathbf{x}) = \hat{w}_0 + \hat{\mathbf{w}}^T\mathbf{x} \tag{14.52}$$

Plugging in the definition of $\hat{\mathbf{w}}$ we get

$$\hat{y}(\mathbf{x}) = \hat{w}_0 + \sum_i \alpha_i \mathbf{x}_i^T \mathbf{x} \tag{14.53}$$

Finally, we can replace $\mathbf{x}_i^T\mathbf{x}$ with $\kappa(\mathbf{x}_i, \mathbf{x})$ to get a kernelized solution:

$$\hat{y}(\mathbf{x}) = \hat{w}_0 + \sum_i \alpha_i \kappa(\mathbf{x}_i, \mathbf{x}) \tag{14.54}$$

14.5.2 SVMs for classification

We now discuss how to apply SVMs to classification. We first focus on the binary case, and then discuss the multi-class case in Section 14.5.2.4.

14.5.2.1 Hinge loss

In Section 6.5.5, we showed that the negative log likelihood of a logistic regression model,

$$L_{\text{nll}}(y, \eta) = -\log p(y|\mathbf{x}, \mathbf{w}) = \log(1 + e^{-y\eta}) \tag{14.55}$$

was a convex upper bound on the 0-1 risk of a binary classifier, where $\eta = f(\mathbf{x}) = \mathbf{w}^T \mathbf{x} + w_0$ is the log odds ratio, and we have assumed the labels are $y \in \{1, -1\}$ rather than $\{0, 1\}$. In this section, we replace the NLL loss with the **hinge loss**, defined as

$$L_{\text{hinge}}(y, \eta) = \max(0, 1 - y\eta) = (1 - y\eta)_+ \tag{14.56}$$

Here $\eta = f(\mathbf{x})$ is our "confidence" in choosing label $y = 1$; however, it need not have any probabilistic semantics. See Figure 6.7 for a plot. We see that the function looks like a door hinge, hence its name. The overall objective has the form

$$\min_{\mathbf{w}, w_0} \frac{1}{2} ||\mathbf{w}||^2 + C \sum_{i=1}^N (1 - y_i f(\mathbf{x}_i))_+ \tag{14.57}$$

Once again, this is non-differentiable, because of the max term. However, by introducing slack variables ξ_i, one can show that this is equivalent to solving

$$\min_{\mathbf{w}, w_0, \boldsymbol{\xi}} \frac{1}{2} ||\mathbf{w}||^2 + C \sum_{i=1}^N \xi_i \quad \text{s.t.} \quad \xi_i \geq 0, \ y_i(\mathbf{x}_i^T \mathbf{w} + w_0) \geq 1 - \xi_i, i = 1 : N \tag{14.58}$$

This is a quadratic program in $N + D + 1$ variables, subject to $O(N)$ constraints. We can eliminate the primal variables $\mathbf{w}$, w_0 and ξ_i, and just solve the N dual variables, which correspond to the Lagrange multipliers for the constraints. Standard solvers take $O(N^3)$ time. However, specialized algorithms, which avoid the use of generic QP solvers, have been developed for this problem, such as the **sequential minimal optimization** or **SMO** algorithm (Platt 1998). In practice this can take $O(N^2)$. However, even this can be too slow if N is large. In such settings, it is common to use linear SVMs, which take $O(N)$ time to train (Joachims 2006; Bottou et al. 2007).

One can show that the solution has the form

$$\hat{\mathbf{w}} = \sum_i \alpha_i \mathbf{x}_i \tag{14.59}$$

where $\alpha_i = \lambda_i y_i$ and where $\boldsymbol{\alpha}$ is sparse (because of the hinge loss). The $\mathbf{x}_i$ for which $\alpha_i > 0$ are called support vectors; these are points which are either incorrectly classified, or are classified correctly but are on or inside the margin (we disuss margins below). See Figure 14.12(b) for an illustration.

At test time, prediction is done using

$$\hat{y}(\mathbf{x}) = \text{sgn}(f(\mathbf{x})) = \text{sgn}\left(\hat{w}_0 + \hat{\mathbf{w}}^T \mathbf{x}\right) \tag{14.60}$$

Using Equation 14.59 and the kernel trick we have

$$\hat{y}(\mathbf{x}) = \text{sgn}\left(\hat{w}_0 + \sum_{i=1}^N \alpha_i \kappa(\mathbf{x}_i, \mathbf{x})\right) \tag{14.61}$$

This takes $O(sD)$ time to compute, where $s \leq N$ is the number of support vectors. This depends on the sparsity level, and hence on the regularizer C.

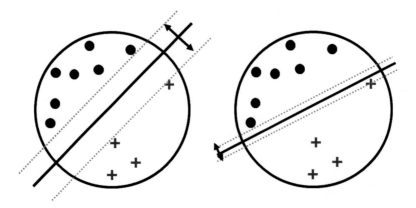

Figure 14.11   Illustration of the large margin principle. Left: a separating hyper-plane with large margin. Right: a separating hyper-plane with small margin.

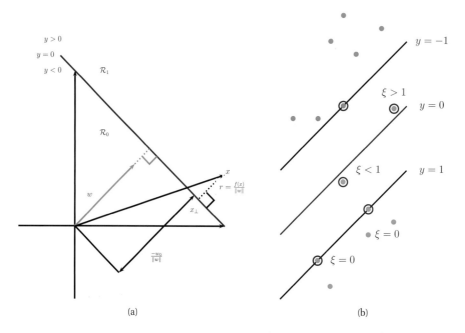

Figure 14.12   (a) Illustration of the geometry of a linear decision boundary in 2d. A point $\mathbf{x}$ is classified as belonging in decision region $\mathcal{R}_1$ if $f(\mathbf{x}) > 0$, otherwise it belongs in decision region $\mathcal{R}_2$; here $f(\mathbf{x})$ is known as a **discriminant function**. The decision boundary is the set of points such that $f(\mathbf{x}) = 0$. $\mathbf{w}$ is a vector which is perpendicular to the decision boundary. The term w_0 controls the distance of the decision boundary from the origin. The signed distance of $\mathbf{x}$ from its orthogonal projection onto the decision boundary, $\mathbf{x}_\perp$, is given by $f(\mathbf{x})/||\mathbf{w}||$. Based on Figure 4.1 of (Bishop 2006). (b) Illustration of the soft margin principle. Points with circles around them are support vectors. We also indicate the value of the corresponding slack variables. Based on Figure 7.3 of (Bishop 2006).

14.5.2.2 The large margin principle

In this section, we derive Equation 14.58 form a completely different perspective. Recall that our goal is to derive a discriminant function $f(\mathbf{x})$ which will be linear in the feature space implied by the choice of kernel. Consider a point $\mathbf{x}$ in this induced space. Referring to Figure 14.12(a), we see that

$$\mathbf{x} = \mathbf{x}_\perp + r \frac{\mathbf{w}}{||\mathbf{w}||} \tag{14.62}$$

where r is the distance of $\mathbf{x}$ from the decision boundary whose normal vector is $\mathbf{w}$, and $\mathbf{x}_\perp$ is the orthogonal projection of $\mathbf{x}$ onto this boundary. Hence

$$f(\mathbf{x}) = \mathbf{w}^T\mathbf{x} + w_0 = (\mathbf{w}^T\mathbf{x}_\perp + w_0) + r\frac{\mathbf{w}^T\mathbf{w}}{||\mathbf{w}||} = (\mathbf{w}^T\mathbf{x}_\perp + w_0) + r||\mathbf{w}|| \tag{14.63}$$

Now $0 = f(\mathbf{x}_\perp) = \mathbf{w}^T\mathbf{x}_\perp + w_0$, so $f(\mathbf{x}) = r||\mathbf{w}||$ and $r = \frac{f(\mathbf{x})}{||\mathbf{w}||}$.

We would like to make this distance $r = f(\mathbf{x})/||\mathbf{w}||$ as large as possible, for reasons illustrated in Figure 14.11. In particular, there might be many lines that perfectly separate the training data (especially if we work in a high dimensional feature space), but intuitively, the best one to pick is the one that maximizes the margin, i.e., the perpendicular distance to the closest point. In addition, we want to ensure each point is on the correct side of the boundary, hence we want $f(\mathbf{x}_i)y_i > 0$. So our objective becomes

$$\max_{\mathbf{w}, w_0} \min_{i=1}^{N} \frac{y_i(\mathbf{w}^T\mathbf{x}_i + w_0)}{||\mathbf{w}||} \tag{14.64}$$

Note that by rescaling the parameters using $\mathbf{w} \to k\mathbf{w}$ and $w_0 \to kw_0$, we do not change the distance of any point to the boundary, since the k factor cancels out when we divide by $||\mathbf{w}||$. Therefore let us define the scale factor such that $y_i f_i = 1$ for the point that is closest to the decision boundary. Hence we require $y_i f_i \geq 1$ for all i. Finally, note that maximizing $1/||\mathbf{w}||$ is equivalent to minimizing $||\mathbf{w}||^2$. Thus we get the new objective

$$\min_{\mathbf{w}, w_0} \frac{1}{2}||\mathbf{w}||^2 \quad \text{s.t.} \quad y_i(\mathbf{w}^T\mathbf{x}_i + w_0) \geq 1, i = 1 : N \tag{14.65}$$

(The fact of $\frac{1}{2}$ is added for convenience and doesn't affect the optimal parameters.) The constraint says that we want all points to be on the correct side of the decision boundary with a margin of at least 1. For this reason, we say that an SVM is an example of a **large margin classifier**.

If the data is not linearly separable (even after using the kernel trick), there will be no feasible solution in which $y_i f_i \geq 1$ for all i. We therefore introduce slack variables $\xi_i \geq 0$ such that $\xi_i = 0$ if the point is on or inside the correct margin boundary, and $\xi_i = |y_i - f_i|$ otherwise. If $0 < \xi_i \leq 1$ the point lies inside the margin, but on the correct side of the decision boundary. If $\xi_i > 1$, the point lies on the wrong side of the decision boundary. See Figure 14.12(b).

We replace the hard constraints that $y_i f_i \geq 0$ with the **soft margin constraints** that $y_i f_i \geq 1 - \xi_i$. The new objective becomes

$$\min_{\mathbf{w}, w_0, \boldsymbol{\xi}} \frac{1}{2}||\mathbf{w}||^2 + C\sum_{i=1}^{N}\xi_i \quad \text{s.t.} \quad \xi_i \geq 0, \ y_i(\mathbf{x}_i^T\mathbf{w} + w_0) \geq 1 - \xi_i \tag{14.66}$$

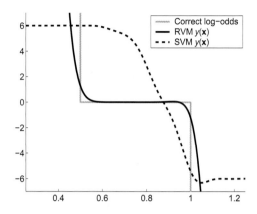

Figure 14.13 Log-odds vs x for 3 different methods. Based on Figure 10 of (Tipping 2001). Used with kind permission of Mike Tipping.

which is the same as Equation 14.58. Since $\xi_i > 1$ means point i is misclassified, we can interpret $\sum_i \xi_i$ as an upper bound on the number of misclassified points.

The parameter C is a regularization parameter that controls the number of errors we are willing to tolerate on the training set. It is common to define this using $C = 1/(\nu N)$, where $0 < \nu \le 1$ controls the fraction of misclassified points that we allow during the training phase. This is called a ν-**SVM classifier**. ν is usually set using cross-validation (see Section 14.5.3).

14.5.2.3   Probabilistic output

An SVM classifier produces a hard-labeling, $\hat{y}(\mathbf{x}) = \text{sign}(f(\mathbf{x}))$. However, we often want a measure of confidence in our prediction. One heuristic approach is to interpret $f(\mathbf{x})$ as the log-odds ratio, $\log \frac{p(y=1|\mathbf{x})}{p(y=0|\mathbf{x})}$. We can then convert the output of an SVM to a probability using

$$p(y = 1|\mathbf{x}, \boldsymbol{\theta}) = \sigma(af(\mathbf{x}) + b) \tag{14.67}$$

where a, b can be estimated by maximum likelihood on a separate validation set. (Using the training set to estimate a and b leads to severe overfitting.) This technique was first proposed in (Platt 2000).

However, the resulting probabilities are not particularly well calibrated, since there is nothing in the SVM training procedure that justifies interpreting $f(\mathbf{x})$ as a log-odds ratio. To illustrate this, consider an example from (Tipping 2001). Suppose we have 1d data where $p(x|y = 0) = \text{Unif}(0, 1)$ and $p(x|y = 1) = \text{Unif}(0.5, 1.5)$. Since the class-conditional distributions overlap in the $[0.5, 1]$ range, the log-odds of class 1 over class 0 should be zero in this region, and infinite outside this region. We sampled 1000 points from the model, and then fit an RVM and an SVM with a Gaussian kenel of width 0.1. Both models can perfectly capture the decision boundary, and achieve a generalizaton error of 25%, which is Bayes optimal in this problem. The probabilistic output from the RVM is a good approximation to the true log-odds, but this is not the case for the SVM, as shown in Figure 14.13.

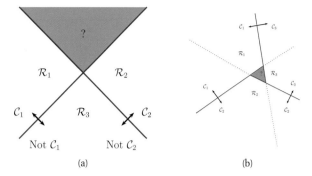

(a)   (b)

Figure 14.14 (a) The one-versus-rest approach. The green region is predicted to be both class 1 and class 2. (b) The one-versus-one approach. The label of the green region is ambiguous. Based on Figure 4.2 of (Bishop 2006).

14.5.2.4 SVMs for multi-class classification

In Section 8.3.7, we saw how we could "upgrade" a binary logistic regression model to the multi-class case, by replacing the sigmoid function with the softmax, and the Bernoulli distribution with the multinomial. Upgrading an SVM to the multi-class case is not so easy, since the outputs are not on a calibrated scale and hence are hard to compare to each other.

The obvious approach is to use a **one-versus-the-rest** approach (also called **one-vs-all**), in which we train C binary classifiers, $f_c(\mathbf{x})$, where the data from class c is treated as positive, and the data from all the other classes is treated as negative. However, this can result in regions of input space which are ambiguously labeled, as shown in Figure 14.14(a).

A common alternative is to pick $\hat{y}(\mathbf{x}) = \arg\max_c f_c(\mathbf{x})$. However, this technique may not work either, since there is no guarantee that the different f_c functions have comparable magnitudes. In addition, each binary subproblem is likely to suffer from the **class imbalance** problem. To see this, suppose we have 10 equally represented classes. When training f_1, we will have 10% positive examples and 90% negative examples, which can hurt performance. It is possible to devise ways to train all C classifiers simultaneously (Weston and Watkins 1999), but the resulting method takes $O(C^2 N^2)$ time, instead of the usual $O(CN^2)$ time.

Another approach is to use the **one-versus-one** or OVO approach, also called **all pairs**, in which we train $C(C-1)/2$ classifiers to discriminate all pairs $f_{c,c'}$. We then classify a point into the class which has the highest number of votes. However, this can also result in ambiguities, as shown in Figure 14.14(b). Also, it takes $O(C^2 N^2)$ time to train and $O(C^2 N_{sv})$ to test each data point, where N_{sv} is the number of support vectors.[2] See also (Allwein et al. 2000) for an approach based on error-correcting output codes.

It is worth remembering that all of these difficulties, and the plethora of heuristics that have been proposed to fix them, fundamentally arise because SVMs do not model uncertainty using probabilities, so their output scores are not comparable across classes. A simple alternative is to use kernels inside probabilistic multiclass classifiers, such as the RVM or L1VM.

---

2. We can reduce the test time by structuring the classes into a DAG (directed acyclic graph), and performing $O(C)$ pairwise comparisons (Platt et al. 2000). However, the $O(C^2)$ factor in the training time is unavoidable.

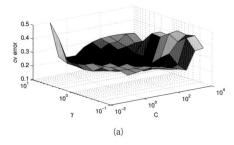 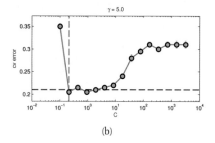

(a)   (b)

Figure 14.15 (a) A cross validation estimate of the 0-1 error for an SVM classifier with RBF kernel with different precisions $\gamma = 1/(2\sigma^2)$ and different regularizer $\lambda = 1/C$, applied to a synthetic data set drawn from a mixture of 2 Gaussians. (b) A slice through this surface for $\gamma = 5$ The red dotted line is the Bayes optimal error, computed using Bayes rule applied to the model used to generate the data. Based on Figure 12.6 of (Hastie et al. 2009). Figure generated by `svmCgammaDemo`.

14.5.3   Choosing C

SVMs for both classification and regression require that you specify the kernel function and the parameter C. Typically C is chosen by cross-validation. Note, however, that C interacts quite strongly with the kernel parameters. For example, suppose we are using an RBF kernel with precision $\gamma = \frac{1}{2\sigma^2}$. If $\gamma = 5$, corresponding to narrow kernels, we need heavy regularization, and hence small C (so $\lambda = 1/C$ is big). If $\gamma = 1$, a larger value of C should be used. So we see that γ and C are tightly coupled. This is illustrated in Figure 14.15, which shows the CV estimate of the 0-1 risk as a function of C and γ.

The authors of libsvm (Hsu et al. 2003) recommend using CV over a 2d grid with values $C \in \{2^{-5}, 2^{-3}, \dots, 2^{15}\}$ and $\gamma \in \{2^{-15}, 2^{-13}, \dots, 2^3\}$. In addition, it is important to standardize the data first, for a spherical Gaussian kernel to make sense.

To choose C efficiently, one can develop a path following algorithm in the spirit of lars (Section 13.3.4). The basic idea is to start with λ large, so that the margin $1/||\mathbf{w}(\lambda)||$ is wide, and hence all points are inside of it and have $\alpha_i = 1$. By slowly decreasing λ, a small set of points will move from inside the margin to outside, and their α_i values will change from 1 to 0, as they cease to be support vectors. When λ is maximal, the function is completely smoothed, and no support vectors remain. See (Hastie et al. 2004) for the details.

14.5.4   Summary of key points

Summarizing the above discussion, we recognize that SVM classifiers involve three key ingredients: the kernel trick, sparsity, and the large margin principle. The kernel trick is necessary to prevent underfitting, i.e., to ensure that the feature vector is sufficiently rich that a linear classifier can separate the data. (Recall from Section 14.2.3 that any Mercer kernel can be viewed as implicitly defining a potentially high dimensional feature vector.) If the original features are already high dimensional (as in many gene expression and text classification problems), it suffices to use a linear kernel, $\kappa(\mathbf{x}, \mathbf{x}') = \mathbf{x}^T \mathbf{x}'$, which is equivalent to working with the original features.

Method	Opt. $\mathbf{w}$	Opt. kernel	Sparse	Prob.	Multiclass	Non-Mercer	Section
L2VM	Convex	EB	No	Yes	Yes	Yes	14.3.2
L1VM	Convex	CV	Yes	Yes	Yes	Yes	14.3.2
RVM	Not convex	EB	Yes	Yes	Yes	Yes	14.3.2
SVM	Convex	CV	Yes	No	Indirectly	No	14.5
GP	N/A	EB	No	Yes	Yes	No	15

Table 14.1 Comparison of various kernel based classifiers. EB = empirical Bayes, CV = cross validation. See text for details.

The sparsity and large margin principles are necessary to prevent overfitting, i.e., to ensure that we do not use all the basis functions. These two ideas are closely related to each other, and both arise (in this case) from the use of the hinge loss function. However, there are other methods of achieving sparsity (such as ℓ_1), and also other methods of maximizing the margin (such as boosting). A deeper discussion of this point takes us outside of the scope of this book. See e.g., (Hastie et al. 2009) for more information.

14.5.5 A probabilistic interpretation of SVMs

In Section 14.3, we saw how to use kernels inside GLMs to derive probabilistic classifiers, such as the L1VM and RVM. And in Section 15.3, we will discuss Gaussian process classifiers, which is another way to build classifiers using kernels. However, all of these approaches use a logistic or probit likelihood, as opposed to the hinge loss used by SVMs. It is natural to wonder if one can interpret the SVM more directly as a probabilistic model. To do so, we must interpret $Cg(m)$ as a negative log likelihood, where $g(m) = (1 - m)_+$, where $m = yf(\mathbf{x})$ is the margin. Hence $p(y = 1|f) = \exp(-Cg(f))$ and $p(y = -1|f) = \exp(-Cg(-f))$. By summing over both values of y, we require that $\exp(-Cg(f)) + \exp(-Cg(-f))$ be a constant independent of f. But it turns out this is not possible for any $C > 0$ (Sollich 2002).

However, if we are willing to relax the sum-to-one condition, and work with a pseudo-likelihood, we *can* derive a probabilistic interpretation of the hinge loss (Polson and Scott 2011). In particular, one can show that

$$\exp(-2(1 - y_i\mathbf{x}_i^T\mathbf{w})_+) = \int_0^\infty \frac{1}{\sqrt{2\pi\lambda_i}} \exp\left(-\frac{1}{2}\frac{(1 + \lambda_i - y_i\mathbf{x}_i^T\mathbf{w})^2}{\lambda_i}\right) d\lambda_i \qquad (14.68)$$

Thus the exponential of the negative hinge loss can be represented as a Gaussian scale mixture. This allows one to fit an SVM using EM or Gibbs sampling, where λ_i are the latent variables. This in turn opens the door to Bayesian methods for setting the hyper-parameters for the prior on $\mathbf{w}$. See (Polson and Scott 2011) for details. (See also (Franc et al. 2011) for a different probabilistic interpretation of SVMs.)

14.6 Comparison of discriminative kernel methods

We have mentioned several different methods for classification and regression based on kernels, which we summarize in Table 14.1. (GP stands for "Gaussian process", which we discuss in Chapter 15.) The columns have the following meaning:

- Optimize **w**: a key question is whether the objective $J(\mathbf{w}) = -\log p(\mathcal{D}|\mathbf{w}) - \log p(\mathbf{w})$ is convex or not. L2VM, L1VM and SVMs have convex objectives. RVMs do not. GPs are Bayesian methods that do not perform parameter estimation.

- Optimize kernel: all the methods require that one "tune" the kernel parameters, such as the bandwidth of the RBF kernel, as well as the level of regularization. For methods based on Gaussian priors, including L2VM, RVMs and GPs, we can use efficient gradient based optimizers to maximize the marginal likelihood. For SVMs, and L1VM, we must use cross validation, which is slower (see Section 14.5.3).

- Sparse: L1VM, RVMs and SVMs are sparse kernel methods, in that they only use a subset of the training examples. GPs and L2VM are not sparse: they use all the training examples. The principle advantage of sparsity is that prediction at test time is usually faster. In addition, one can sometimes get improved accuracy.

- Probabilistic: All the methods except for SVMs produce probabilistic output of the form $p(y|\mathbf{x})$. SVMs produce a "confidence" value that can be converted to a probability, but such probabilities are usually very poorly calibrated (see Section 14.5.2.3).

- Multiclass: All the methods except for SVMs naturally work in the multiclass setting, by using a multinoulli output instead of Bernoulli. The SVM can be made into a multiclass classifier, but there are various difficulties with this approach, as discussed in Section 14.5.2.4.

- Mercer kernel: SVMs and GPs require that the kernel is positive definite; the other techniques do not.

Apart from these differences, there is the natural question: which method works best? In a small experiment[3], we found that all of these methods had similar accuracy when averaged over a range of problems, provided they have the same kernel, and provided the regularization constants are chosen appropriately.

Given that the statistical performance is roughly the same, what about the computational performance? GPs and L2VM are generally the slowest, taking $O(N^3)$ time, since they don't exploit sparsity (although various speedups are possible, see Section 15.6). SVMs also take $O(N^3)$ time to train (unless we use a linear kernel, in which case we only need $O(N)$ time (Joachims 2006)). However, the need to use cross validation can make SVMs slower than RVMs. L1VM should be faster than an RVM, since an RVM requires multiple rounds of ℓ_1 minimization (see Section 13.7.4.3). However, in practice it is common to use a greedy method to train RVMs, which is faster than ℓ_1 minimization. This is reflected in our empirical results.

The conclusion of all this is as follows: if speed matters, use an RVM, but if well-calibrated probabilistic output matters (e.g., for active learning or control problems), use a GP. The only circumstances under which using an SVM seems sensible is the structured output case, where likelihood-based methods can be slow (see Section 19.7). Section 16.7.1 gives a more extensive experimental comparison of supervised learning methods, including SVMs and various non kernel methods.

---

3. See `http://pmtk3.googlecode.com/svn/trunk/docs/tutorial/html/tutKernelClassif.html`.

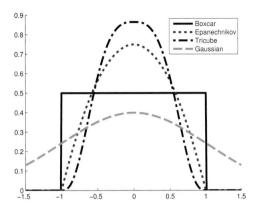

Figure 14.16 A comparison of some popular smoothing kernels. The boxcar kernel has compact support but is not smooth. The Epanechnikov kernel has compact support but is not differentiable at its boundary. The tri-cube has compact support and two continuous derivatives at the boundary of its support. The Gaussian is differentiable, but does not have compact support. Based on Figure 6.2 of (Hastie et al. 2009). Figure generated by `smoothingKernelPlot`.

14.7 Kernels for building generative models

There is a different kind of kernel known as a smoothing kernel which can be used to create non-parametric density estimates. This can be used for unsupervised density estimation, $p(\mathbf{x})$, as well as for creating generative models for classification and regression by making models of the form $p(y, \mathbf{x})$.

14.7.1 Smoothing kernels

A **smoothing kernel** is a function of one argument which satisfies the following properties:

$$\int \kappa(x)dx = 1, \quad \int x\kappa(x)dx = 0, \quad \int x^2\kappa(x)dx > 0 \qquad (14.69)$$

A simple example is the **Gaussian kernel**,

$$\kappa(x) \triangleq \frac{1}{(2\pi)^{\frac{1}{2}}} e^{-x^2/2} \qquad (14.70)$$

We can control the width of the kernel by introducing a **bandwidth** parameter h:

$$\kappa_h(x) \triangleq \frac{1}{h}\kappa(\frac{x}{h}) \qquad (14.71)$$

We can generalize to vector valued inputs by defining an RBF kernel:

$$\kappa_h(\mathbf{x}) = \kappa_h(||\mathbf{x}||) \qquad (14.72)$$

In the case of the Gaussian kernel, this becomes

$$\kappa_h(\mathbf{x}) = \frac{1}{h^D(2\pi)^{D/2}} \prod_{j=1}^{D} \exp(-\frac{1}{2h^2}x_j^2) \qquad (14.73)$$

Although Gaussian kernels are popular, they have unbounded support. An alternative kernel, with compact support, is the **Epanechnikov kernel**, defined by

$$\kappa(x) \triangleq \frac{3}{4}(1 - x^2)\mathbb{I}(|x| \le 1) \tag{14.74}$$

This is plotted in Figure 14.16. Compact support can be useful for efficiency reasons, since one can use fast nearest neighbor methods to evaluate the density.

Unfortunately, the Epanechnikov kernel is not differentiable at the boundary of its support. An alterative is the **tri-cube kernel**, defined as follows:

$$\kappa(x) \triangleq \frac{70}{81}(1 - |x|^3)^3\mathbb{I}(|x| \le 1) \tag{14.75}$$

This has compact support and has two continuous derivatives at the boundary of its support. See Figure 14.16.

The **boxcar kernel** is simply the uniform distribution:

$$\kappa(x) \triangleq \mathbb{I}(|x| \le 1) \tag{14.76}$$

We will use this kernel below.

14.7.2 Kernel density estimation (KDE)

Recall the Gaussian mixture model from Section 11.2.1. This is a parametric density estimator for data in $\mathbb{R}^D$. However, it requires specifying the number K and locations $\boldsymbol{\mu}_k$ of the clusters. An alternative to estimating the $\boldsymbol{\mu}_k$ is to allocate one cluster center per data point, so $\boldsymbol{\mu}_i = \mathbf{x}_i$. In this case, the model becomes

$$p(\mathbf{x}|\mathcal{D}) = \frac{1}{N}\sum_{i=1}^{N}\mathcal{N}(\mathbf{x}|\mathbf{x}_i, \sigma^2\mathbf{I}) \tag{14.77}$$

We can generalize the approach by writing

$$\hat{p}(\mathbf{x}) = \frac{1}{N}\sum_{i=1}^{N}\kappa_h(\mathbf{x} - \mathbf{x}_i) \tag{14.78}$$

This is called a **Parzen window density estimator**, or **kernel density estimator** (KDE), and is a simple non-parametric density model. The advantage over a parametric model is that no model fitting is required (except for tuning the bandwidth, usually done by cross-validation), and there is no need to pick K. The disadvantage is that the model takes a lot of memory to store, and a lot of time to evaluate. It is also of no use for clustering tasks.

Figure 14.17 illustrates KDE in 1d for two kinds of kernel. On the top, we use a boxcar kernel, $\kappa(x) = \mathbb{I}(-1 \le z \le 1)$. The result is equivalent to a **histogram** estimate of the density, since we just count how many data points land within an interval of size h around x_i. On the bottom, we use a Gaussian kernel, which results in a smoother fit.

The usual way to pick h is to minimize an estimate (such as cross validation) of the frequentist risk (see e.g., (Bowman and Azzalini 1997)). In Section 25.2, we discuss a Bayesian approach to non-parametric density estimation, based on Dirichlet process mixture models, which allows us

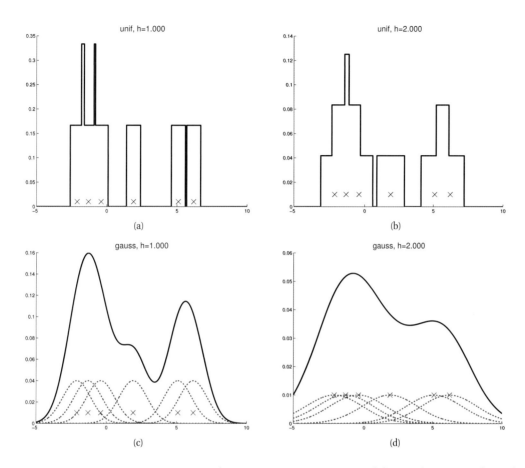

Figure 14.17 A nonparametric (Parzen) density estimator in 1D estimated from 6 data points, denoted by x. Top row: uniform kernel. Bottom row: Gaussian kernel. Rows represent increasingly large bandwidth parameters. Based on `http://en.wikipedia.org/wiki/Kernel_density_estimation`. Figure generated by `parzenWindowDemo2`.

to infer h. DP mixtures can also be more efficient than KDE, since they do not need to store all the data. See also Section 15.2.4 where we discuss an empirical Bayes approach to estimating kernel parameters in a Gaussian process model for classification/ regression.

14.7.3 From KDE to KNN

We can use KDE to define the class conditional densities in a generative classifier. This turns out to provide an alternative derivation of the nearest neighbors classifier, which we introduced in Section 1.4.2. To show this, we follow the presentation of (Bishop 2006, p125). In KDE with a boxcar kernel, we fixed the bandwidth and count how many data points fall within the hypercube centered on a datapoint. Suppose that, instead of fixing the bandwidth h, we allow the

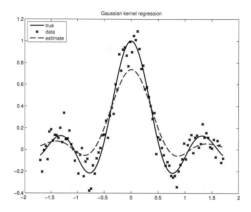

Figure 14.18 An example of kernel regression in 1d using a Gaussian kernel. Figure generated by `kernelRegressionDemo`, based on code by Yi Cao.

bandwidth or volume to be different for each data point. Specifically, we will "grow" a volume around $\mathbf{x}$ until we encounter K data points, regardless of their class label. Let the resulting volume have size $V(\mathbf{x})$ (this was previously h^D), and let there be $N_c(\mathbf{x})$ examples from class c in this volume. Then we can estimate the class conditional density as follows:

$$p(\mathbf{x}|y = c, \mathcal{D}) = \frac{N_c(\mathbf{x})}{N_c V(\mathbf{x})} \tag{14.79}$$

where N_c is the total number of examples in class c in the whole data set. The class prior can be estimated by

$$p(y = c|\mathcal{D}) = \frac{N_c}{N} \tag{14.80}$$

Hence the class posterior is given by

$$p(y = c|\mathbf{x}, \mathcal{D}) \quad = \quad \frac{\frac{N_c(\mathbf{x})}{N_c V(\mathbf{x})} \frac{N_c}{N}}{\sum_{c'} \frac{N_{c'}(\mathbf{x})}{N_{c'} V(\mathbf{x})} \frac{N_{c'}}{N}} = \frac{N_c(\mathbf{x})}{\sum_{c'} N_{c'}(\mathbf{x})} = \frac{N_c(\mathbf{x})}{K} \tag{14.81}$$

where we used the fact that $\sum_c N_c(\mathbf{x}) = K$, since we choose a total of K points (regardless of class) around every point. This is equivalent to Equation 1.2, since $N_c(\mathbf{x}) = \sum_{i \in N_K(\mathbf{x}, \mathcal{D})} \mathbb{I}(y_i = c)$.

14.7.4   Kernel regression

In Section 14.7.2, we discussed the use of kernel density estimation or KDE for unsupervised learning. We can also use KDE for regression. The goal is to compute the conditional expectation

$$f(\mathbf{x}) = \mathbb{E}[y|\mathbf{x}] = \int y \, p(y|\mathbf{x}) dy = \frac{\int y \, p(\mathbf{x}, y) dy}{\int p(\mathbf{x}, y) dy} \tag{14.82}$$

We can use KDE to approximate the joint density $p(\mathbf{x}, y)$ as follows:

$$p(\mathbf{x}, y) \approx \frac{1}{N} \sum_{i=1}^{N} \kappa_h(\mathbf{x} - \mathbf{x}_i) \kappa_h(y - y_i) \tag{14.83}$$

Hence

$$f(\mathbf{x}) = \frac{\frac{1}{N} \sum_{i=1}^{N} \kappa_h(\mathbf{x} - \mathbf{x}_i) \int y \kappa_h(y - y_i) dy}{\frac{1}{N} \sum_{i=1}^{N} \kappa_h(\mathbf{x} - \mathbf{x}_i) \int \kappa_h(y - y_i) dy} \tag{14.84}$$

$$= \frac{\sum_{i=1}^{N} \kappa_h(\mathbf{x} - \mathbf{x}_i) y_i}{\sum_{i=1}^{N} \kappa_h(\mathbf{x} - \mathbf{x}_i)} \tag{14.85}$$

To derive this result, we used two properties of smoothing kernels. First, that they integrate to one, i.e., $\int \kappa_h(y - y_i) dy = 1$. And second, the fact that $\int y \kappa_h(y - y_i) dy = y_i$. This follows by defining $y' = y - y_i$ and using the zero mean property of smoothing kernels:

$$\int (y' + y_i) \kappa_h(y) dy' = \int y' \kappa_h(y') dy' + y_i \int \kappa_h(y') dy' = 0 + y_i = y_i \tag{14.86}$$

We can rewrite the above result as follows:

$$f(\mathbf{x}) = \sum_{i=1}^{N} w_i(\mathbf{x}) y_i \tag{14.87}$$

$$w_i(\mathbf{x}) \triangleq \frac{\kappa_h(\mathbf{x} - \mathbf{x}_i)}{\sum_{i'=1}^{N} \kappa_h(\mathbf{x} - \mathbf{x}_{i'})} \tag{14.88}$$

We see that the prediction is just a weighted sum of the outputs at the training points, where the weights depend on how similar $\mathbf{x}$ is to the stored training points. This method is called **kernel regression**, **kernel smoothing**, or the **Nadaraya-Watson** model. See Figure 14.18 for an example, where we use a Gaussian kernel.

Note that this method only has one free parameter, namely h. One can show (Bowman and Azzalini 1997) that for 1d data, if the true density is Gaussian and we are using Gaussian kernels, the optimal bandwidth h is given by

$$h = \left(\frac{4}{3N} \right)^{1/5} \hat{\sigma} \tag{14.89}$$

We can compute a robust approximation to the standard deviation by first computing the **mean absolute deviation**

$$\text{MAD} = \text{median}(|\mathbf{x} - \text{median}(\mathbf{x})|) \tag{14.90}$$

and then using

$$\hat{\sigma} = 1.4826 \text{ MAD} = \frac{1}{0.6745} \text{ MAD} \tag{14.91}$$

The code used to produce Figure 14.18 estimated h_x and h_y separately, and then set $h = \sqrt{h_x h_y}$.

Although these heuristics seem to work well, their derivation rests on some rather dubious assumptions (such as Gaussianity of the true density). Furthermore, these heuristics are limited to tuning just a single parameter. In Section 15.2.4 we discuss an empirical Bayes approach to estimating multiple kernel parameters in a Gaussian process model for classification/ regression, which can handle many tuning parameters, and which is based on much more transparent principles (maximizing the marginal likelihood).

14.7.5 Locally weighted regression

If we define $\kappa_h(\mathbf{x} - \mathbf{x}_i) = \kappa(\mathbf{x}, \mathbf{x}_i)$, we can rewrite the prediction made by kernel regression as follows

$$\hat{f}(\mathbf{x}_*) = \sum_{i=1}^{N} y_i \frac{\kappa(\mathbf{x}_*, \mathbf{x}_i)}{\sum_{i'=1}^{N} \kappa(\mathbf{x}_*, \mathbf{x}_{i'})} \tag{14.92}$$

Note that $\kappa(\mathbf{x}, \mathbf{x}_i)$ need not be a smoothing kernel. If it is not, we no longer need the normalization term, so we can just write

$$\hat{f}(\mathbf{x}_*) = \sum_{i=1}^{N} y_i \kappa(\mathbf{x}_*, \mathbf{x}_i) \tag{14.93}$$

This model is essentially fitting a constant function locally. We can improve on this by fitting a linear regression model for each point $\mathbf{x}_*$ by solving

$$\min_{\boldsymbol{\beta}(\mathbf{x}_*)} \sum_{i=1}^{N} \kappa(\mathbf{x}_*, \mathbf{x}_i)[y_i - \boldsymbol{\beta}(\mathbf{x}_*)^T \phi(\mathbf{x}_i)]^2 \tag{14.94}$$

where $\phi(\mathbf{x}) = [1, \mathbf{x}]$. This is called **locally weighted regression**. An example of such a method is **LOESS**, aka **LOWESS**, which stands for "locally-weighted scatterplot smoothing" (Cleveland and Devlin 1988). See also (Edakunni et al. 2010) for a Bayesian version of this model.

We can compute the parameters $\boldsymbol{\beta}(\mathbf{x}_*)$ for each test case by solving the following weighted least squares problem:

$$\boldsymbol{\beta}(\mathbf{x}_*) = (\boldsymbol{\Phi}^T \mathbf{D}(\mathbf{x}_*)\boldsymbol{\Phi})^{-1} \boldsymbol{\Phi}^T \mathbf{D}(\mathbf{x}_*)\mathbf{y} \tag{14.95}$$

where $\boldsymbol{\Phi}$ is an $N \times (D + 1)$ design matrix and $\mathbf{D} = \mathrm{diag}(\kappa(\mathbf{x}_*, \mathbf{x}_i))$. The corresponding prediction has the form

$$\hat{f}(\mathbf{x}_*) = \phi(x_*)^T \boldsymbol{\beta}(\mathbf{x}_*) = (\boldsymbol{\Phi}^T \mathbf{D}(\mathbf{x}_*)\boldsymbol{\Phi})^{-1} \boldsymbol{\Phi}^T \mathbf{D}(\mathbf{x}_*)\mathbf{y} = \sum_{i=1}^{N} w_i(\mathbf{x}_*) y_i \tag{14.96}$$

The term $w_i(\mathbf{x}_*)$, which combines the local smoothing kernel with the effect of linear regression, is called the **equivalent kernel**. See also Section 15.4.2.

Exercises

Exercise 14.1 Fitting an SVM classifier by hand

(Source: Jaakkola.) Consider a dataset with 2 points in 1d: $(x_1 = 0, y_1 = -1)$ and $(x_2 = \sqrt{2}, y_2 = 1)$. Consider mapping each point to 3d using the feature vector $\phi(x) = [1, \sqrt{2}x, x^2]^T$. (This is equivalent to using a second order polynomial kernel.) The max margin classifier has the form

$$\min ||\mathbf{w}||^2 \quad \text{s.t.} \tag{14.97}$$

$$y_1(\mathbf{w}^T \phi(\mathbf{x}_1) + w_0) \geq 1 \tag{14.98}$$

$$y_2(\mathbf{w}^T \phi(\mathbf{x}_2) + w_0) \geq 1 \tag{14.99}$$

a. Write down a vector that is parallel to the optimal vector $\mathbf{w}$. Hint: recall from Figure 7.8 (12Apr10 version) that $\mathbf{w}$ is perpendicular to the decision boundary between the two points in the 3d feature space.

b. What is the value of the margin that is achieved by this $\mathbf{w}$? Hint: recall that the margin is the distance from each support vector to the decision boundary. Hint 2: think about the geometry of 2 points in space, with a line separating one from the other.

c. Solve for $\mathbf{w}$, using the fact the margin is equal to $1/||\mathbf{w}||$.

d. Solve for w_0 using your value for $\mathbf{w}$ and Equations 14.97 to 14.99. Hint: the points will be on the decision boundary, so the inequalities will be tight.

e. Write down the form of the discriminant function $f(x) = w_0 + \mathbf{w}^T \phi(x)$ as an explicit function of x.

Exercise 14.2 Linear separability

(Source: Koller.) Consider fitting an SVM with $C > 0$ to a dataset that is linearly separable. Is the resulting decision boundary guaranteed to separate the classes?

15 *Gaussian processes*

15.1 Introduction

In supervised learning, we observe some inputs $\mathbf{x}_i$ and some outputs y_i. We assume that $y_i = f(\mathbf{x}_i)$, for some unknown function f, possibly corrupted by noise. The optimal approach is to infer a *distribution over functions* given the data, $p(f|\mathbf{X}, \mathbf{y})$, and then to use this to make predictions given new inputs, i.e., to compute

$$p(y_*|\mathbf{x}_*, \mathbf{X}, \mathbf{y}) = \int p(y_*|f, \mathbf{x}_*)p(f|\mathbf{X}, \mathbf{y})df \tag{15.1}$$

Up until now, we have focussed on parametric representations for the function f, so that instead of inferring $p(f|\mathcal{D})$, we infer $p(\boldsymbol{\theta}|\mathcal{D})$. In this chapter, we discuss a way to perform Bayesian inference over functions themselves.

Our approach will be based on **Gaussian processes** or **GPs**. A GP defines a prior over functions, which can be converted into a posterior over functions once we have seen some data. Although it might seem difficult to represent a distribution over a function, it turns out that we only need to be able to define a distribution over the function's values at a finite, but arbitrary, set of points, say $\mathbf{x}_1, \ldots, \mathbf{x}_N$. A GP assumes that $p(f(\mathbf{x}_1), \ldots, f(\mathbf{x}_N))$ is jointly Gaussian, with some mean $\boldsymbol{\mu}(\mathbf{x})$ and covariance $\boldsymbol{\Sigma}(\mathbf{x})$ given by $\Sigma_{ij} = \kappa(\mathbf{x}_i, \mathbf{x}_j)$, where κ is a positive definite kernel function (see Section 14.2 information on kernels). The key idea is that if $\mathbf{x}_i$ and $\mathbf{x}_j$ are deemed by the kernel to be similar, then we expect the output of the function at those points to be similar, too. See Figure 15.1 for an illustration.

It turns out that, in the regression setting, all these computations can be done in closed form, in $O(N^3)$ time. (We discuss faster approximations in Section 15.6.) In the classification setting, we must use approximations, such as the Gaussian approximation, since the posterior is no longer exactly Gaussian.

GPs can be thought of as a Bayesian alternative to the kernel methods we discussed in Chapter 14, including L1VM, RVM and SVM. Although those methods are sparser and therefore faster, they do not give well-calibrated probabilistic outputs (see Section 15.4.4 for further discussion). Having properly tuned probabilistic output is important in certain applications, such as online tracking for vision and robotics (Ko and Fox 2009), reinforcement learning and optimal control (Engel et al. 2005; Deisenroth et al. 2009), global optimization of non-convex functions (Mockus et al. 1996; Lizotte 2008; Brochu et al. 2009), experiment design (Santner et al. 2003), etc.

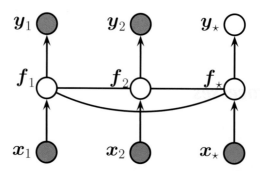

Figure 15.1 A Gaussian process for 2 training points and 1 testing point, represented as a mixed directed and undirected graphical model representing $p(\mathbf{y}, \mathbf{f}|\mathbf{x}) = \mathcal{N}(\mathbf{f}|\mathbf{0}, \mathbf{K}(\mathbf{x})) \prod_i p(y_i|f_i)$. The hidden nodes $f_i = f(\mathbf{x}_i)$ represent the value of the function at each of the data points. These hidden nodes are fully interconnected by undirected edges, forming a Gaussian graphical model; the edge strengths represent the covariance terms $\Sigma_{ij} = \kappa(\mathbf{x}_i, \mathbf{x}_j)$. If the test point $\mathbf{x}_*$ is similar to the training points $\mathbf{x}_1$ and $\mathbf{x}_2$, then the predicted output y_* will be similar to y_1 and y_2.

Our presentation is closely based on (Rasmussen and Williams 2006), which should be consulted for futher details. See also (Diggle and Ribeiro 2007), which discusses the related approach known as **kriging**, which is widely used in the spatial statistics literature.

15.2   GPs for regression

In this section, we discuss GPs for regression. Let the prior on the regression function be a GP, denoted by

$$f(\mathbf{x}) \sim GP(m(\mathbf{x}), \kappa(\mathbf{x}, \mathbf{x}')) \tag{15.2}$$

where $m(\mathbf{x})$ is the mean function and $\kappa(\mathbf{x}, \mathbf{x}')$ is the kernel or covariance function, i.e.,

$$m(\mathbf{x}) = \mathbb{E}\left[f(\mathbf{x})\right] \tag{15.3}$$

$$\kappa(\mathbf{x}, \mathbf{x}') = \mathbb{E}\left[(f(\mathbf{x}) - m(\mathbf{x}))(f(\mathbf{x}') - m(\mathbf{x}'))^T\right] \tag{15.4}$$

We obviously require that $\kappa()$ be a positive definite kernel. For any finite set of points, this process defines a joint Gaussian:

$$p(\mathbf{f}|\mathbf{X}) = \mathcal{N}(\mathbf{f}|\boldsymbol{\mu}, \mathbf{K}) \tag{15.5}$$

where $K_{ij} = \kappa(\mathbf{x}_i, \mathbf{x}_j)$ and $\boldsymbol{\mu} = (m(\mathbf{x}_1), \dots, m(\mathbf{x}_N))$.

Note that it is common to use a mean function of $m(\mathbf{x}) = 0$, since the GP is flexible enough to model the mean arbitrarily well, as we will see below. However, in Section 15.2.6 we will consider parametric models for the mean function, so the GP just has to model the residual errors. This semi-parametric approach combines the interpretability of parametric models with the accuracy of non-parametric models.

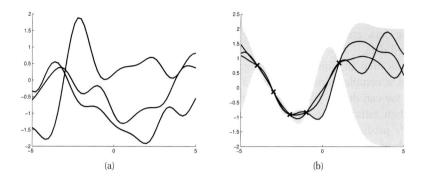

Figure 15.2 Left: some functions sampled from a GP prior with SE kernel. Right: some samples from a GP posterior, after conditioning on 5 noise-free observations. The shaded area represents $\mathbb{E}\left[f(\mathbf{x})\right]\pm 2\mathrm{std}(f(\mathbf{x}))$. Based on Figure 2.2 of (Rasmussen and Williams 2006). Figure generated by `gprDemoNoiseFree`.

15.2.1 Predictions using noise-free observations

Suppose we observe a training set $\mathcal{D} = \{(\mathbf{x}_i, f_i), i = 1 : N\}$, where $f_i = f(\mathbf{x}_i)$ is the noise-free observation of the function evaluated at $\mathbf{x}_i$. Given a test set $\mathbf{X}_*$ of size $N_* \times D$, we want to predict the function outputs $\mathbf{f}_*$.

If we ask the GP to predict $f(\mathbf{x})$ for a value of $\mathbf{x}$ that it has already seen, we want the GP to return the answer $f(\mathbf{x})$ with no uncertainty. In other words, it should act as an **interpolator** of the training data. This will only happen if we assume the observations are noiseless. We will consider the case of noisy observations below.

Now we return to the prediction problem. By definition of the GP, the joint distribution has the following form

$$\begin{pmatrix} \mathbf{f} \\ \mathbf{f}_* \end{pmatrix} \sim \mathcal{N}\left(\begin{pmatrix} \boldsymbol{\mu} \\ \boldsymbol{\mu}_* \end{pmatrix}, \begin{pmatrix} \mathbf{K} & \mathbf{K}_* \\ \mathbf{K}_*^T & \mathbf{K}_{**} \end{pmatrix}\right) \tag{15.6}$$

where $\mathbf{K} = \kappa(\mathbf{X}, \mathbf{X})$ is $N \times N$, $\mathbf{K}_* = \kappa(\mathbf{X}, \mathbf{X}_*)$ is $N \times N_*$, and $\mathbf{K}_{**} = \kappa(\mathbf{X}_*, \mathbf{X}_*)$ is $N_* \times N_*$. By the standard rules for conditioning Gaussians (Section 4.3), the posterior has the following form

$$\begin{aligned} p(\mathbf{f}_* | \mathbf{X}_*, \mathbf{X}, \mathbf{f}) &= \mathcal{N}(\mathbf{f}_* | \boldsymbol{\mu}_*, \boldsymbol{\Sigma}_*) & (15.7) \\ \boldsymbol{\mu}_* &= \boldsymbol{\mu}(\mathbf{X}_*) + \mathbf{K}_*^T \mathbf{K}^{-1}(\mathbf{f} - \boldsymbol{\mu}(\mathbf{X})) & (15.8) \\ \boldsymbol{\Sigma}_* &= \mathbf{K}_{**} - \mathbf{K}_*^T \mathbf{K}^{-1} \mathbf{K}_* & (15.9) \end{aligned}$$

This process is illustrated in Figure 15.2. On the left we show some samples from the prior, $p(\mathbf{f}|\mathbf{X})$, where we use a **squared exponential kernel**, aka Gaussian kernel or RBF kernel. In 1d, this is given by

$$\kappa(x, x') = \sigma_f^2 \exp(-\frac{1}{2\ell^2}(x - x')^2) \tag{15.10}$$

Here ℓ controls the horizontal length scale over which the function varies, and σ_f^2 controls the vertical variation. (We discuss how to estimate such kernel parameters below.) On the right we show samples from the posterior, $p(\mathbf{f}_*|\mathbf{X}_*, \mathbf{X}, \mathbf{f})$. We see that the model perfectly interpolates the training data, and that the predictive uncertainty increases as we move further away from the observed data.

One application of noise-free GP regression is as a computationally cheap proxy for the behavior of a complex simulator, such as a weather forecasting program. (If the simulator is stochastic, we can define f to be its mean output; note that there is still no observation noise.) One can then estimate the effect of changing simulator parameters by examining their effect on the GP's predictions, rather than having to run the simulator many times, which may be prohibitively slow. This strategy is known as DACE, which stands for design and analysis of computer experiments (Santner et al. 2003).

15.2.2 Predictions using noisy observations

Now let us consider the case where what we observe is a noisy version of the underlying function, $y = f(\mathbf{x}) + \epsilon$, where $\epsilon \sim \mathcal{N}(0, \sigma_y^2)$. In this case, the model is not required to interpolate the data, but it must come "close" to the observed data. The covariance of the observed noisy responses is

$$\operatorname{cov}[y_p, y_q] = \kappa(\mathbf{x}_p, \mathbf{x}_q) + \sigma_y^2 \delta_{pq} \tag{15.11}$$

where $\delta_{pq} = \mathbb{I}(p = q)$. In other words

$$\operatorname{cov}[\mathbf{y}|\mathbf{X}] = \mathbf{K} + \sigma_y^2 \mathbf{I}_N \triangleq \mathbf{K}_y \tag{15.12}$$

The second matrix is diagonal because we assumed the noise terms were independently added to each observation.

The joint density of the observed data and the latent, noise-free function on the test points is given by

$$\begin{pmatrix} \mathbf{y} \\ \mathbf{f}_* \end{pmatrix} \sim \mathcal{N}\left(\mathbf{0}, \begin{pmatrix} \mathbf{K}_y & \mathbf{K}_* \\ \mathbf{K}_*^T & \mathbf{K}_{**} \end{pmatrix}\right) \tag{15.13}$$

where we are assuming the mean is zero, for notational simplicity. Hence the posterior predictive density is

$$p(\mathbf{f}_*|\mathbf{X}_*, \mathbf{X}, \mathbf{y}) = \mathcal{N}(\mathbf{f}_*|\boldsymbol{\mu}_*, \boldsymbol{\Sigma}_*) \tag{15.14}$$

$$\boldsymbol{\mu}_* = \mathbf{K}_*^T \mathbf{K}_y^{-1} \mathbf{y} \tag{15.15}$$

$$\boldsymbol{\Sigma}_* = \mathbf{K}_{**} - \mathbf{K}_*^T \mathbf{K}_y^{-1} \mathbf{K}_* \tag{15.16}$$

In the case of a single test input, this simplifies as follows

$$p(f_*|\mathbf{x}_*, \mathbf{X}, \mathbf{y}) = \mathcal{N}(f_*|\mathbf{k}_*^T \mathbf{K}_y^{-1} \mathbf{y}, k_{**} - \mathbf{k}_*^T \mathbf{K}_y^{-1} \mathbf{k}_*) \tag{15.17}$$

where $\mathbf{k}_* = [\kappa(\mathbf{x}_*, \mathbf{x}_1), \ldots, \kappa(\mathbf{x}_*, \mathbf{x}_N)]$ and $k_{**} = \kappa(\mathbf{x}_*, \mathbf{x}_*)$. Another way to write the posterior mean is as follows:

$$\overline{f}_* = \mathbf{k}_*^T \mathbf{K}_y^{-1} \mathbf{y} = \sum_{i=1}^N \alpha_i \kappa(\mathbf{x}_i, \mathbf{x}_*) \tag{15.18}$$

where $\boldsymbol{\alpha} = \mathbf{K}_y^{-1} \mathbf{y}$. We will revisit this expression later.

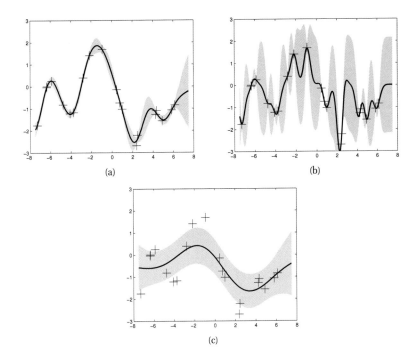

(a)

(b)

(c)

Figure 15.3 Some 1d GPs with SE kernels but different hyper-parameters fit to 20 noisy observations. The kernel has the form in Equation 15.19. The hyper-parameters $(\ell, \sigma_f, \sigma_y)$ are as follows: (a) (1,1,0.1) (b) (0.3, 1.08, 0.00005), (c) (3.0, 1.16, 0.89). Based on Figure 2.5 of (Rasmussen and Williams 2006). Figure generated by `gprDemoChangeHparams`, written by Carl Rasmussen.

15.2.3 Effect of the kernel parameters

The predictive performance of GPs depends exclusively on the suitability of the chosen kernel. Suppose we choose the following squared-exponential (SE) kernel for the noisy observations

$$\kappa_y(x_p, x_q) = \sigma_f^2 \exp\left(-\frac{1}{2\ell^2}(x_p - x_q)^2\right) + \sigma_y^2 \delta_{pq} \tag{15.19}$$

Here ℓ is the horizontal scale over which the function changes, σ_f^2 controls the vertical scale of the function, and σ_y^2 is the noise variance. Figure 15.3 illustrates the effects of changing these parameters. We sampled 20 noisy data points from the SE kernel using $(\ell, \sigma_f, \sigma_y) = (1, 1, 0.1)$, and then made predictions various parameters, conditional on the data. In Figure 15.3(a), we use $(\ell, \sigma_f, \sigma_y) = (1, 1, 0.1)$, and the result is a good fit. In Figure 15.3(b), we reduce the length scale to $\ell = 0.3$ (the other parameters were optimized by maximum (marginal) likelihood, a technique we discuss below); now the function looks more "wiggly". Also, the uncertainty goes up faster, since the effective distance from the training points increases more rapidly. In Figure 15.3(c), we increase the length scale to $\ell = 3$; now the function looks smoother.

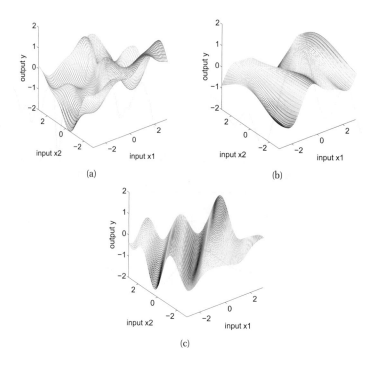

Figure 15.4 Some 2d functions sampled from a GP with an SE kernel but different hyper-parameters. The kernel has the form in Equation 15.20 where (a) $\mathbf{M} = \mathbf{I}$, (b) $\mathbf{M} = \text{diag}(1, 3)^{-2}$, (c) $\mathbf{M} = (1, -1; -1, 1) + \text{diag}(6, 6)^{-2}$. Based on Figure 5.1 of (Rasmussen and Williams 2006). Figure generated by `gprDemoArd`, written by Carl Rasmussen.

We can extend the SE kernel to multiple dimensions as follows:

$$\kappa_y(\mathbf{x}_p, \mathbf{x}_q) = \sigma_f^2 \exp(-\frac{1}{2}(\mathbf{x}_p - \mathbf{x}_q)^T \mathbf{M}(\mathbf{x}_p - \mathbf{x}_q)) + \sigma_y^2 \delta_{pq} \tag{15.20}$$

We can define the matrix $\mathbf{M}$ in several ways. The simplest is to use an isotropic matrix, $\mathbf{M}_1 = \ell^{-2}\mathbf{I}$. See Figure 15.4(a) for an example. We can also endow each dimension with its own characteristic length scale, $\mathbf{M}_2 = \text{diag}(\boldsymbol{\ell})^{-2}$. If any of these length scales become large, the corresponding feature dimension is deemed "irrelevant", just as in ARD (Section 13.7). In Figure 15.4(b), we use $\mathbf{M} = \mathbf{M}_2$ with $\boldsymbol{\ell} = (1, 3)$, so the function changes faster along the x_1 direction than the x_2 direction.

We can also create a matrix of the form $\mathbf{M}_3 = \boldsymbol{\Lambda}\boldsymbol{\Lambda}^T + \text{diag}(\boldsymbol{\ell})^{-2}$, where $\boldsymbol{\Lambda}$ is a $D \times K$ matrix, where $K < D$. (Rasmussen and Williams 2006, p107) calls this the **factor analysis distance** function, by analogy to the fact that factor analysis (Section 12.1) approximates a covariance matrix as a low rank matrix plus a diagonal matrix. The columns of $\boldsymbol{\Lambda}$ correspond to relevant directions in input space. In Figure 15.4(c), we use $\boldsymbol{\ell} = (6; 6)$ and $\boldsymbol{\Lambda} = (1; -1)$, so the function changes mostly rapidly in the direction which is perpendicular to (1,1).

15.2.4 Estimating the kernel parameters

To estimate the kernel parameters, we could use exhaustive search over a discrete grid of values, with validation loss as an objective, but this can be quite slow. (This is the approach used to tune kernels used by SVMs.) Here we consider an empirical Bayes approach, which will allow us to use continuous optimization methods, which are much faster. In particular, we will maximize the marginal likelihood[1]

$$p(\mathbf{y}|\mathbf{X}) = \int p(\mathbf{y}|\mathbf{f}, \mathbf{X})p(\mathbf{f}|\mathbf{X})d\mathbf{f} \tag{15.21}$$

Since $p(\mathbf{f}|\mathbf{X}) = \mathcal{N}(\mathbf{f}|\mathbf{0}, \mathbf{K})$, and $p(\mathbf{y}|\mathbf{f}) = \prod_i \mathcal{N}(y_i|f_i, \sigma_y^2)$, the marginal likelihood is given by

$$\log p(\mathbf{y}|\mathbf{X}) = \log \mathcal{N}(\mathbf{y}|\mathbf{0}, \mathbf{K}_y) = -\frac{1}{2}\mathbf{y}\mathbf{K}_y^{-1}\mathbf{y} - \frac{1}{2}\log|\mathbf{K}_y| - \frac{N}{2}\log(2\pi) \tag{15.22}$$

The first term is a data fit term, the second term is a model complexity term, and the third term is just a constant. To understand the tradeoff between the first two terms, consider a SE kernel in 1D, as we vary the length scale ℓ and hold σ_y^2 fixed. Let $J(\ell) = -\log p(\mathbf{y}|\mathbf{X}, \ell)$. For short length scales, the fit will be good, so $\mathbf{y}^T\mathbf{K}_y^{-1}\mathbf{y}$ will be small. However, the model complexity will be high: $\mathbf{K}$ will be almost diagonal (as in Figure 14.3, top right), since most points will not be considered "near" any others, so the $\log|\mathbf{K}_y|$ will be large. For long length scales, the fit will be poor but the model complexity will be low: $\mathbf{K}$ will be almost all 1's (as in Figure 14.3, bottom right), so $\log|\mathbf{K}_y|$ will be small.

We now discuss how to maximize the marginal likelihood. Let the kernel parameters (also called hyper-parameters) be denoted by $\boldsymbol{\theta}$. One can show that

$$\frac{\partial}{\partial\theta_j}\log p(\mathbf{y}|\mathbf{X}) = \frac{1}{2}\mathbf{y}^T\mathbf{K}_y^{-1}\frac{\partial\mathbf{K}_y}{\partial\theta_j}\mathbf{K}_y^{-1}\mathbf{y} - \frac{1}{2}\mathrm{tr}(\mathbf{K}_y^{-1}\frac{\partial\mathbf{K}_y}{\partial\theta_j}) \tag{15.23}$$

$$= \frac{1}{2}\mathrm{tr}\left((\boldsymbol{\alpha}\boldsymbol{\alpha}^T - \mathbf{K}_y^{-1})\frac{\partial\mathbf{K}_y}{\partial\theta_j}\right) \tag{15.24}$$

where $\boldsymbol{\alpha} = \mathbf{K}_y^{-1}\mathbf{y}$. It takes $O(N^3)$ time to compute $\mathbf{K}_y^{-1}$, and then $O(N^2)$ time per hyper-parameter to compute the gradient.

The form of $\frac{\partial\mathbf{K}_y}{\partial\theta_j}$ depends on the form of the kernel, and which parameter we are taking derivatives with respect to. Often we have constraints on the hyper-parameters, such as $\sigma_y^2 \geq 0$. In this case, we can define $\theta = \log(\sigma_y^2)$, and then use the chain rule.

Given an expression for the log marginal likelihood and its derivative, we can estimate the kernel parameters using any standard gradient-based optimizer. However, since the objective is not convex, local minima can be a problem, as we illustrate below.

15.2.4.1 Example

Consider Figure 15.5. We use the SE kernel in Equation 15.19 with $\sigma_f^2 = 1$, and plot $\log p(\mathbf{y}|\mathbf{X}, \ell, \sigma_y^2)$ (where $\mathbf{X}$ and $\mathbf{y}$ are the 7 data points shown in panels b and c) as we vary ℓ and σ_y^2. The two

---

1. The reason it is called the marginal likelihood, rather than just likelihood, is because we have marginalized out the latent Gaussian vector $\mathbf{f}$. This moves us up one level of the Bayesian hierarchy, and reduces the chances of overfitting (the number of kernel parameters is usually fairly small compared to a standard parametric model).

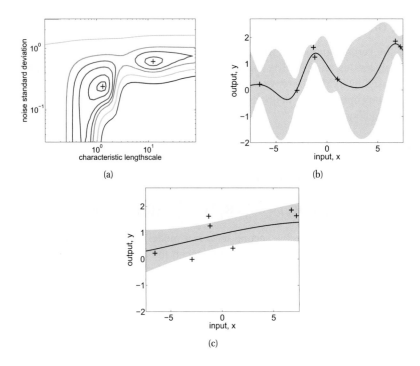

(a)   (b)

(c)

Figure 15.5 Illustration of local minima in the marginal likelihood surface. (a) We plot the log marginal likelihood vs σ_y^2 and ℓ, for fixed $\sigma_f^2 = 1$, using the 7 data points shown in panels b and c. (b) The function corresponding to the lower left local minimum, $(\ell, \sigma_n^2) \approx (1, 0.2)$. This is quite "wiggly" and has low noise. (c) The function corresponding to the top right local minimum, $(\ell, \sigma_n^2) \approx (10, 0.8)$. This is quite smooth and has high noise. The data was generated using $(\ell, \sigma_n^2) = (1, 0.1)$.   Source: Figure 5.5 of (Rasmussen and Williams 2006). Figure generated by `gprDemoMarglik`, written by Carl Rasmussen.

local optima are indicated by +. The bottom left optimum corresponds to a low-noise, short-length scale solution (shown in panel b). The top right optimum corresponds to a high-noise, long-length scale solution (shown in panel c). With only 7 data points, there is not enough evidence to confidently decide which is more reasonable, although the more complex model (panel b) has a marginal likelihood that is about 60% higher than the simpler model (panel c). With more data, the MAP estimate should come to dominate.

Figure 15.5 illustrates some other interesting (and typical) features. The region where $\sigma_y^2 \approx 1$ (top of panel a) corresponds to the case where the noise is very high; in this regime, the marginal likelihood is insensitive to the length scale (indicated by the horizontal contours), since all the data is explained as noise. The region where $\ell \approx 0.5$ (left hand side of panel a) corresponds to the case where the length scale is very short; in this regime, the marginal likelihood is insensitive to the noise level, since the data is perfectly interpolated. Neither of these regions would be chosen by a good optimizer.

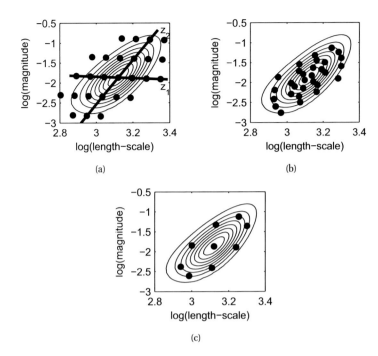

(a)   (b)

(c)

Figure 15.6 Three different approximations to the posterior over hyper-parameters: grid-based, Monte Carlo, and central composite design. Source: Figure 3.2 of (Vanhatalo 2010). Used with kind permission of Jarno Vanhatalo.

15.2.4.2 Bayesian inference for the hyper-parameters

An alternative to computing a point estimate of the hyper-parameters is to compute their posterior. Let $\boldsymbol{\theta}$ represent all the kernel parameters, as well as σ_y^2. If the dimensionality of $\boldsymbol{\theta}$ is small, we can compute a discrete grid of possible values, centered on the MAP estimate $\hat{\boldsymbol{\theta}}$ (computed as above). We can then approximate the posterior over the latent variables using

$$p(\mathbf{f}|\mathcal{D}) \propto \sum_{s=1}^{S} p(\mathbf{f}|\mathcal{D}, \boldsymbol{\theta}_s) p(\boldsymbol{\theta}_s|\mathcal{D}) \delta_s \tag{15.25}$$

where δ_s denotes the weight for grid point s.

In higher dimensions, a regular grid suffers from the curse of dimensionality. An obvious alternative is Monte Carlo, but this can be slow. Another approach is to use a form of quasi-Monte Carlo, whereby we place grid points at the mode, and at a distance ± 1sd from the mode along each dimension, for a total of $2|\boldsymbol{\theta}| + 1$ points. This is called a **central composite design** (Rue et al. 2009). (This is also used in the unscented Kalman filter, see Section 18.5.2.) To make this Gaussian-like approximation more reasonable, we often log-transform the hyper-parameters. See Figure 15.6 for an illustration.

15.2.4.3 Multiple kernel learning

A quite different approach to optimizing kernel parameters known as **multiple kernel learning**. The idea is to define the kernel as a weighted sum of base kernels, $\kappa(\mathbf{x}, \mathbf{x}') = \sum_j w_j \kappa_j(\mathbf{x}, \mathbf{x}')$, and then to optimize the weights w_j instead of the kernel parameters themselves. This is particularly useful if we have different kinds of data which we wish to fuse together. See e.g., (Rakotomamonjy et al. 2008) for an approach based on risk-minimization and convex optimization, and (Girolami and Rogers 2005) for an approach based on variational Bayes.

15.2.5 Computational and numerical issues *

The predictive mean is given by $\overline{f_*} = \mathbf{k}_*^T \mathbf{K}_y^{-1} \mathbf{y}$. For reasons of numerical stability, it is unwise to directly invert $\mathbf{K}_y$. A more robust alternative is to compute a Cholesky decomposition, $\mathbf{K}_y = \mathbf{L}\mathbf{L}^T$. We can then compute the predictive mean and variance, and the log marginal likelihood, as shown in the pseudo-code in Algorithm 15.1 (based on (Rasmussen and Williams 2006, p19)). It takes $O(N^3)$ time to compute the Cholesky decomposition, and $O(N^2)$ time to solve for $\boldsymbol{\alpha} = \mathbf{K}_y^{-1}\mathbf{y} = \mathbf{L}^{-T}\mathbf{L}^{-1}\mathbf{y}$. We can then compute the mean using $\mathbf{k}_*^T \boldsymbol{\alpha}$ in $O(N)$ time and the variance using $k_{**} - \mathbf{k}_*^T \mathbf{L}^{-T}\mathbf{L}^{-1}\mathbf{k}_*$ in $O(N^2)$ time for each test case.

An alternative to Cholesky decomposition is to solve the linear system $\mathbf{K}_y \boldsymbol{\alpha} = \mathbf{y}$ using conjugate gradients (CG). If we terminate this algorithm after k iterations, it takes $O(kN^2)$ time. If we run for $k = N$, it gives the exact solution in $O(N^3)$ time. Another approach is to approximate the matrix-vector multiplies needed by CG using the fast Gauss transform. (Yang et al. 2005); however, this doesn't scale to high-dimensional inputs. See also Section 15.6 for a discussion of other speedup techniques.

Algorithm 15.1: GP regression

1 $\mathbf{L} = \text{cholesky}(\mathbf{K} + \sigma_y^2 \mathbf{I})$;
2 $\boldsymbol{\alpha} = \mathbf{L}^T \setminus (\mathbf{L} \setminus \mathbf{y})$;
3 $\mathbb{E}[f_*] = \mathbf{k}_*^T \boldsymbol{\alpha}$;
4 $\mathbf{v} = \mathbf{L} \setminus \mathbf{k}_*$;
5 $\text{var}[f_*] = \kappa(\mathbf{x}_*, \mathbf{x}_*) - \mathbf{v}^T \mathbf{v}$;
6 $\log p(\mathbf{y}|\mathbf{X}) = -\frac{1}{2}\mathbf{y}^T \boldsymbol{\alpha} - \sum_i \log L_{ii} - \frac{N}{2}\log(2\pi)$

15.2.6 Semi-parametric GPs *

Sometimes it is useful to use a linear model for the mean of the process, as follows:

$$f(\mathbf{x}) = \boldsymbol{\beta}^T \boldsymbol{\phi}(\mathbf{x}) + r(\mathbf{x}) \tag{15.26}$$

where $r(\mathbf{x}) \sim \text{GP}(0, \kappa(\mathbf{x}, \mathbf{x}'))$ models the residuals. This combines a parametric and a non-parametric model, and is known as a **semi-parametric model**.

If we assume $\boldsymbol{\beta} \sim \mathcal{N}(\mathbf{b}, \mathbf{B})$, we can integrate these parameters out to get a new GP (O'Hagan 1978):

$$f(\mathbf{x}) \sim \text{GP}\left(\boldsymbol{\phi}(\mathbf{x})^T \mathbf{b}, \ \kappa(\mathbf{x}, \mathbf{x}') + \boldsymbol{\phi}(\mathbf{x})^T \mathbf{B} \boldsymbol{\phi}(\mathbf{x}')\right) \tag{15.27}$$

$\log p(y_i\|f_i)$	$\frac{\partial}{\partial f_i} \log p(y_i\|f_i)$	$\frac{\partial^2}{\partial f_i^2} \log p(y_i\|f_i)$
$\log \text{sigm}(y_i f_i)$	$t_i - \pi_i$	$-\pi_i(1 - \pi_i)$
$\log \Phi(y_i f_i)$	$\frac{y_i \phi(f_i)}{\Phi(y_i f_i)}$	$-\frac{\phi_i^2}{\Phi(y_i f_i)^2} - \frac{y_i f_i \phi(f_i)}{\Phi(y_i f_i)}$

Table 15.1 Likelihood, gradient and Hessian for binary logistic/ probit GP regression. We assume $y_i \in \{-1, +1\}$ and define $t_i = (y_i + 1)/2 \in \{0, 1\}$ and $\pi_i = \text{sigm}(f_i)$ for logistic regression, and $\pi_i = \Phi(f_i)$ for probit regression. Also, ϕ and Φ are the pdf and cdf of $\mathcal{N}(0, 1)$. From (Rasmussen and Williams 2006, p43).

Integrating out $\boldsymbol{\beta}$, the corresponding predictive distribution for test inputs $\mathbf{X}_*$ has the following form (Rasmussen and Williams 2006, p28):

$$p(\mathbf{f}_* | \mathbf{X}_*, \mathbf{X}, \mathbf{y}) = \mathcal{N}(\overline{\mathbf{f}_*}, \text{cov}[f_*]) \tag{15.28}$$

$$\overline{\mathbf{f}_*} = \boldsymbol{\Phi}_*^T \overline{\boldsymbol{\beta}} + \mathbf{K}_*^T \mathbf{K}_y^{-1}(\mathbf{y} - \boldsymbol{\Phi}\overline{\boldsymbol{\beta}}) \tag{15.29}$$

$$\overline{\boldsymbol{\beta}} = (\boldsymbol{\Phi}^T \mathbf{K}_y^{-1} \boldsymbol{\Phi} + \mathbf{B}^{-1})^{-1}(\boldsymbol{\Phi}\mathbf{K}_y^{-1}\mathbf{y} + \mathbf{B}^{-1}\mathbf{b}) \tag{15.30}$$

$$\text{cov}[\mathbf{f}_*] = \mathbf{K}_{**} - \mathbf{K}_*^T \mathbf{K}_y^{-1}\mathbf{K}_* + \mathbf{R}^T(\mathbf{B}^{-1} + \boldsymbol{\Phi}\mathbf{K}_y^{-1}\boldsymbol{\Phi}^T)^{-1}\mathbf{R} \tag{15.31}$$

$$\mathbf{R} = \boldsymbol{\Phi}_* - \boldsymbol{\Phi}\mathbf{K}_y^{-1}\boldsymbol{\Phi}_* \tag{15.32}$$

The predictive mean is the output of the linear model plus a correction term due to the GP, and the predictive covariance is the usual GP covariance plus an extra term due to the uncertainty in $\boldsymbol{\beta}$.

15.3 GPs meet GLMs

In this section, we extend GPs to the GLM setting (cf. (Chan and Dong 2011)), focussing on the classification case. As with Bayesian logistic regression, the main difficulty is that the Gaussian prior is not conjugate to the bernoulli/ multinoulli likelihood. There are several approximations one can adopt: Gaussian approximation (Section 8.4.3), expectation propagation (Kuss and Rasmussen 2005; Nickisch and Rasmussen 2008), variational (Girolami and Rogers 2006; Opper and Archambeau 2009), MCMC (Neal 1997; Christensen et al. 2006), etc. Here we focus on the Gaussian approximation, since it is the simplest and fastest.

15.3.1 Binary classification

In the binary case, we define the model as $p(y_i|\mathbf{x}_i) = \sigma(y_i f(\mathbf{x}_i))$, where, following (Rasmussen and Williams 2006), we assume $y_i \in \{-1, +1\}$, and we let $\sigma(z) = \text{sigm}(z)$ (logistic regression) or $\sigma(z) = \Phi(z)$ (probit regression). As for GP regression, we assume $f \sim \text{GP}(0, \kappa)$.

15.3.1.1 Computing the posterior

Define the log of the unnormalized posterior as follows:

$$\ell(\mathbf{f}) = \log p(\mathbf{y}|\mathbf{f}) + \log p(\mathbf{f}|\mathbf{X}) = \log p(\mathbf{y}|\mathbf{f}) - \frac{1}{2}\mathbf{f}^T\mathbf{K}^{-1}\mathbf{f} - \frac{1}{2}\log|\mathbf{K}| - \frac{N}{2}\log 2\pi \tag{15.33}$$

Let $J(f) \triangleq -\ell(f)$ be the function we want to minimize. The gradient and Hessian of this are given by

$$
\begin{aligned}
\mathbf{g} &= -\nabla \log p(\mathbf{y}|\mathbf{f}) + \mathbf{K}^{-1}\mathbf{f} & (15.34) \\
\mathbf{H} &= -\nabla\nabla \log p(\mathbf{y}|\mathbf{f}) + \mathbf{K}^{-1} = \mathbf{W} + \mathbf{K}^{-1} & (15.35)
\end{aligned}
$$

Note that $\mathbf{W} \triangleq -\nabla\nabla \log p(\mathbf{y}|\mathbf{f})$ is a diagonal matrix because the data are iid (conditional on $\mathbf{f}$). Expressions for the gradient and Hessian of the log likelihood for the logit and probit case are given in Sections 8.3.1 and 9.4.1, and summarized in Table 15.1.

We can use IRLS to find the MAP estimate. The update has the form

$$
\begin{aligned}
\mathbf{f}^{new} &= \mathbf{f} - \mathbf{H}^{-1}\mathbf{g} = \mathbf{f} + (\mathbf{K}^{-1} + \mathbf{W})^{-1}(\nabla \log p(\mathbf{y}|\mathbf{f}) - \mathbf{K}^{-1}\mathbf{f}) & (15.36) \\
&= (\mathbf{K}^{-1} + \mathbf{W})^{-1}(\mathbf{W}\mathbf{f} + \nabla \log p(\mathbf{y}|\mathbf{f})) & (15.37)
\end{aligned}
$$

At convergence, the Gaussian approximation of the posterior takes the following form:

$$
p(\mathbf{f}|\mathbf{X}, \mathbf{y}) \approx \mathcal{N}(\hat{\mathbf{f}}, (\mathbf{K}^{-1} + \mathbf{W})^{-1}) \tag{15.38}
$$

15.3.1.2 Computing the posterior predictive

We now compute the posterior predictive. First we predict the latent function at the test case $\mathbf{x}_*$. For the mean we have

$$
\begin{aligned}
\mathbb{E}\left[f_*|\mathbf{x}_*, \mathbf{X}, \mathbf{y}\right] &= \int \mathbb{E}\left[f_*|\mathbf{f}, \mathbf{x}_*, \mathbf{X}, \mathbf{y}\right] p(\mathbf{f}|\mathbf{X}, \mathbf{y}) d\mathbf{f} & (15.39) \\
&= \int \mathbf{k}_*^T \mathbf{K}^{-1}\mathbf{f}\, p(\mathbf{f}|\mathbf{X}, \mathbf{y}) d\mathbf{f} & (15.40) \\
&= \mathbf{k}_*^T \mathbf{K}^{-1}\mathbb{E}\left[\mathbf{f}|\mathbf{X}, \mathbf{y}\right] \approx \mathbf{k}_*^T \mathbf{K}^{-1}\hat{\mathbf{f}} & (15.41)
\end{aligned}
$$

where we used Equation 15.8 to get the mean of f_* given noise-free $\mathbf{f}$.

To compute the predictive variance, we use the rule of iterated variance:

$$
\text{var}\left[f_*\right] = \mathbb{E}\left[\text{var}\left[f_*|\mathbf{f}\right]\right] + \text{var}\left[\mathbb{E}\left[f_*|\mathbf{f}\right]\right] \tag{15.42}
$$

where all probabilities are conditioned on $\mathbf{x}_*, \mathbf{X}, \mathbf{y}$. From Equation 15.9 we have

$$
\mathbb{E}\left[\text{var}\left[f_*|\mathbf{f}\right]\right] = \mathbb{E}\left[k_{**} - \mathbf{k}_*^T\mathbf{K}^{-1}\mathbf{k}_*\right] = k_{**} - \mathbf{k}_*^T\mathbf{K}^{-1}\mathbf{k}_* \tag{15.43}
$$

From Equation 15.9 we have

$$
\text{var}\left[\mathbb{E}\left[f_*|\mathbf{f}\right]\right] = \text{var}\left[\mathbf{k}_*\mathbf{K}^{-1}\mathbf{f}\right] = \mathbf{k}_*^T\mathbf{K}^{-1}\text{cov}\left[\mathbf{f}\right]\mathbf{K}^{-1}\mathbf{k}_* \tag{15.44}
$$

Combining these we get

$$
\text{var}\left[f_*\right] = k_{**} - \mathbf{k}_*^T(\mathbf{K}^{-1} - \mathbf{K}^{-1}\text{cov}\left[\mathbf{f}\right]\mathbf{K}^{-1})\mathbf{k}_* \tag{15.45}
$$

From Equation 15.38 we have $\text{cov}\left[f\right] \approx (\mathbf{K}^{-1} + \mathbf{W})^{-1}$. Using the matrix inversion lemma we get

$$
\begin{aligned}
\text{var}\left[f_*\right] &\approx k_{**} - \mathbf{k}_*^T\mathbf{K}^{-1}\mathbf{k}_* + \mathbf{k}_*^T\mathbf{K}^{-1}(\mathbf{K}^{-1} + \mathbf{W})^{-1}\mathbf{K}^{-1}\mathbf{k}_* & (15.46) \\
&= k_{**} - \mathbf{k}_*^T(\mathbf{K} + \mathbf{W}^{-1})^{-1}\mathbf{k}_* & (15.47)
\end{aligned}
$$

So in summary we have

$$p(f_*|\mathbf{x}_*, \mathbf{X}, \mathbf{y}) = \mathcal{N}(\mathbb{E}[f_*], \mathrm{var}[f_*]) \tag{15.48}$$

To convert this in to a predictive distribution for binary responses, we use

$$\pi_* = p(y_* = 1|\mathbf{x}_*, \mathbf{X}, \mathbf{y}) \approx \int \sigma(f_*)p(f_*|\mathbf{x}_*, \mathbf{X}, \mathbf{y})df_* \tag{15.49}$$

This can be approximated using any of the methods discussed in Section 8.4.4, where we discussed Bayesian logistic regression. For example, using the probit approximation of Section 8.4.4.2, we have $\pi_* \approx \mathrm{sigm}(\kappa(v)\mathbb{E}[f_*])$, where $v = \mathrm{var}[f_*]$ and $\kappa^2(v) = (1 + \pi v/8)^{-1}$.

15.3.1.3 Computing the marginal likelihood

We need the marginal likelihood in order to optimize the kernel parameters. Using the Laplace approximation in Equation 8.54 we have

$$\log p(\mathbf{y}|\mathbf{X}) \approx \ell(\hat{\mathbf{f}}) - \frac{1}{2}\log|\mathbf{H}| + \mathrm{const} \tag{15.50}$$

Hence

$$\log p(\mathbf{y}|\mathbf{X}) \quad \approx \quad \log p(\mathbf{y}|\hat{\mathbf{f}}) - \frac{1}{2}\hat{\mathbf{f}}^T\mathbf{K}^{-1}\hat{\mathbf{f}} - \frac{1}{2}\log|\mathbf{K}| - \frac{1}{2}\log|\mathbf{K}^{-1} + \mathbf{W}| \tag{15.51}$$

Computing the derivatives $\frac{\partial \log p(\mathbf{y}|\mathbf{X}, \boldsymbol{\theta})}{\partial \theta_j}$ is more complex than in the regression case, since $\hat{\mathbf{f}}$ and $\mathbf{W}$, as well as $\mathbf{K}$, depend on $\boldsymbol{\theta}$. Details can be found in (Rasmussen and Williams 2006, p125).

15.3.1.4 Numerically stable computation *

To implement the above equations in a numerically stable way, it is best to avoid inverting $\mathbf{K}$ or $\mathbf{W}$. (Rasmussen and Williams 2006, p45) suggest defining

$$\mathbf{B} = \mathbf{I}_N + \mathbf{W}^{\frac{1}{2}}\mathbf{K}\mathbf{W}^{\frac{1}{2}} \tag{15.52}$$

which has eigenvalues bounded below by 1 (because of the $\mathbf{I}$) and above by $1 + \frac{N}{4}\max_{ij} K_{ij}$ (because $w_{ii} = \pi_i(1 - \pi) \leq 0.25$), and hence can be safely inverted.

One can use the matrix inversion lemma to show

$$(\mathbf{K}^{-1} + \mathbf{W})^{-1} = \mathbf{K} - \mathbf{K}\mathbf{W}^{\frac{1}{2}}\mathbf{B}^{-1}\mathbf{W}^{\frac{1}{2}}\mathbf{K} \tag{15.53}$$

Hence the IRLS update becomes

$$\mathbf{f}^{new} \quad = \quad (\mathbf{K}^{-1} + \mathbf{W})^{-1}\underbrace{(\mathbf{W}\mathbf{f} + \nabla \log p(\mathbf{y}|\mathbf{f}))}_{\mathbf{b}} \tag{15.54}$$

$$= \quad \mathbf{K}(\mathbf{I} - \mathbf{W}^{\frac{1}{2}}\mathbf{B}^{-1}\mathbf{W}^{\frac{1}{2}}\mathbf{K})\mathbf{b} \tag{15.55}$$

$$= \quad \mathbf{K}\underbrace{(\mathbf{b} - \mathbf{W}^{\frac{1}{2}}\mathbf{L}^T \backslash (\mathbf{L} \backslash (\mathbf{W}^{\frac{1}{2}}\mathbf{K}\mathbf{b})))}_{\mathbf{a}} \tag{15.56}$$

where $\mathbf{B} = \mathbf{LL}^T$ is a Cholesky decomposition of $\mathbf{B}$. The fitting algorithm takes in $O(TN^3)$ time and $O(N^2)$ space, where T is the number of Newton iterations.

At convergence we have $\mathbf{a} = \mathbf{K}^{-1}\hat{\mathbf{f}}$, so we can evaluate the log marginal likelihood (Equation 15.51) using

$$\log p(\mathbf{y}|\mathbf{X}) = \log p(\mathbf{y}|\hat{\mathbf{f}}) - \frac{1}{2}\mathbf{a}^T\hat{\mathbf{f}} - \sum_i \log L_{ii} \tag{15.57}$$

where we exploited the fact that

$$|\mathbf{B}| = |\mathbf{K}||\mathbf{K}^{-1} + \mathbf{W}| = |\mathbf{I}_N + \mathbf{W}^{\frac{1}{2}}\mathbf{K}\mathbf{W}^{\frac{1}{2}}| \tag{15.58}$$

We now compute the predictive distribution. Rather than using $\mathbb{E}[f_*] = \mathbf{k}_*^T\mathbf{K}^{-1}\hat{\mathbf{f}}$, we exploit the fact that at the mode, $\nabla\ell = 0$, so $\hat{\mathbf{f}} = \mathbf{K}(\nabla \log p(\mathbf{y}|\hat{\mathbf{f}}))$. Hence we can rewrite the predictive mean as follows:[2]

$$\mathbb{E}[f_*] = \mathbf{k}_*^T\nabla \log p(\mathbf{y}|\hat{\mathbf{f}}) \tag{15.59}$$

To compute the predictive variance, we exploit the fact that

$$(\mathbf{K} + \mathbf{W}^{-1})^{-1} = \mathbf{W}^{\frac{1}{2}}\mathbf{W}^{-\frac{1}{2}}(\mathbf{K} + \mathbf{W}^{-1})^{-1}\mathbf{W}^{-\frac{1}{2}}\mathbf{W}^{\frac{1}{2}} = \mathbf{W}^{\frac{1}{2}}\mathbf{B}^{-1}\mathbf{W}^{\frac{1}{2}} \tag{15.60}$$

to get

$$\text{var}[f_*] = k_{**} - \mathbf{k}_*^T\mathbf{W}^{\frac{1}{2}}(\mathbf{LL}^T)^{-1}\mathbf{W}^{\frac{1}{2}}\mathbf{k}_* = k_{**} - \mathbf{v}^T\mathbf{v} \tag{15.61}$$

where $\mathbf{v} = \mathbf{L} \setminus (\mathbf{W}^{\frac{1}{2}}\mathbf{k}_*)$. We can then compute π_*.

The whole algorithm is summarized in Algorithm 15.2, based on (Rasmussen and Williams 2006, p46). Fitting takes $O(N^3)$ time, and prediction takes $O(N^2N_*)$ time, where N_* is the number of test cases.

15.3.1.5   Example

In Figure 15.7, we show a synthetic binary classification problem in 2d. We use an SE kernel. On the left, we show predictions using hyper-parameters set by hand; we use a short length scale, hence the very sharp turns in the decision boundary. On the right, we show the predictions using the learned hyper-parameters; the model favors a more parsimonious explanation of the data.

15.3.2   Multi-class classification

In this section, we consider a model of the form $p(y_i|\mathbf{x}_i) = \text{Cat}(y_i|\mathcal{S}(\mathbf{f}_i))$, where $\mathbf{f}_i = (f_{i1}, \ldots, f_{iC})$, and we assume $f_{.c} \sim \text{GP}(0, \kappa_c)$. Thus we have one latent function per class, which are a priori independent, and which may use different kernels. As before, we will use a Gaussian approximation to the posterior. (A similar model, but using the multinomial probit function instead of the multinomial logit, is described in (Girolami and Rogers 2006).)

---

2. We see that training points that are well-predicted by the model, for which $\nabla_i \log p(y_i|f_i) \approx 0$, do not contribute strongly to the prediction at test points; this is similar to the behavior of support vectors in an SVM (see Section 14.5).

Algorithm 15.2: GP binary classification using Gaussian approximation

1 // First compute MAP estimate using IRLS;
2 $\mathbf{f} = \mathbf{0}$;
3 **repeat**
4      $\mathbf{W} = -\nabla\nabla \log p(\mathbf{y}|\mathbf{f})$;
5      $\mathbf{B} = \mathbf{I}_N + \mathbf{W}^{\frac{1}{2}}\mathbf{K}\mathbf{W}^{\frac{1}{2}}$;
6      $\mathbf{L} = \mathrm{cholesky}(\mathbf{B})$;
7      $\mathbf{b} = \mathbf{W}\mathbf{f} + \nabla \log p(\mathbf{y}|\mathbf{f})$;
8      $\mathbf{a} = \mathbf{b} - \mathbf{W}^{\frac{1}{2}}\mathbf{L}^T \backslash (\mathbf{L} \backslash (\mathbf{W}^{\frac{1}{2}}\mathbf{K}\mathbf{b}))$;
9      $\mathbf{f} = \mathbf{K}\mathbf{a}$;
10 **until** *converged*;
11 $\log p(\mathbf{y}|\mathbf{X}) = \log p(\mathbf{y}|\mathbf{f}) - \frac{1}{2}\mathbf{a}^T\mathbf{f} - \sum_i \log L_{ii}$;
12 // Now perform prediction ;
13 $\mathbb{E}\left[f_*\right] = \mathbf{k}_*^T \nabla \log p(\mathbf{y}|\mathbf{f})$;
14 $\mathbf{v} = \mathbf{L} \backslash (\mathbf{W}^{\frac{1}{2}}\mathbf{k}_*)$;
15 $\mathrm{var}\left[f_*\right] = k_{**} - \mathbf{v}^T\mathbf{v}$;
16 $p(y_* = 1) = \int \mathrm{sigm}(z)\mathcal{N}(z|\mathbb{E}\left[f_*\right], \mathrm{var}\left[f_*\right])dz$;

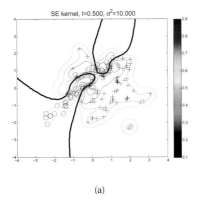
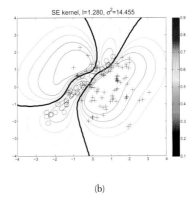

(a)   (b)

Figure 15.7 Contours of the posterior predictive probability for the red circle class generated by a GP with an SE kernel. Thick black line is the decision boundary if we threshold at a probability of 0.5. (a) Manual parameters, short length scale. (b) Learned parameters, long length scale. Figure generated by `gpcDemo2d`, based on code by Carl Rasmussen.

15.3.2.1 Computing the posterior

The unnormalized log posterior is given by

$$\ell(\mathbf{f}) = -\frac{1}{2}\mathbf{f}^T\mathbf{K}^{-1}\mathbf{f} + \mathbf{y}^T\mathbf{f} - \sum_{i=1}^{N} \log\left(\sum_{c=1}^{C} \exp f_{ic}\right) - \frac{1}{2}\log|\mathbf{K}| - \frac{CN}{2}\log 2\pi \tag{15.62}$$

where

$$\mathbf{f} = (f_{11}, \ldots, f_{N1}, f_{12}, \ldots, f_{N2}, \cdots, f_{1C}, \ldots, f_{NC})^T \tag{15.63}$$

and $\mathbf{y}$ is a dummy encoding of the y_i's which has the same layout as $\mathbf{f}$. Also, $\mathbf{K}$ is a block diagonal matrix containing $\mathbf{K}_c$, where $\mathbf{K}_c = [\kappa_c(\mathbf{x}_i, \mathbf{x}_j)]$ models the correlation of the c'th latent function.

The gradient and Hessian are given by

$$\nabla\ell = -\mathbf{K}^{-1}\mathbf{f} + \mathbf{y} - \boldsymbol{\pi} \tag{15.64}$$
$$\nabla\nabla\ell = -\mathbf{K}^{-1} - \mathbf{W} \tag{15.65}$$

where $\mathbf{W} \triangleq \mathrm{diag}(\boldsymbol{\pi}) - \boldsymbol{\Pi}\boldsymbol{\Pi}^T$, where $\boldsymbol{\Pi}$ is a $CN \times N$ matrix obtained by stacking $\mathrm{diag}(\boldsymbol{\pi}_{:c})$ vertically. (Compare these expressions to standard logistic regression in Section 8.3.7.)

We can use IRLS to compute the mode. The Newton step has the form

$$\mathbf{f}^{new} = (\mathbf{K}^{-1} + \mathbf{W})^{-1}(\mathbf{W}\mathbf{f} + \mathbf{y} - \boldsymbol{\pi}) \tag{15.66}$$

Naively implementing this would take $O(C^3 N^3)$ time. However, we can reduce this to $O(CN^3)$, as shown in (Rasmussen and Williams 2006, p52).

15.3.2.2 Computing the posterior predictive

We can compute the posterior predictive in a manner analogous to Section 15.3.1.2. For the mean of the latent response we have

$$\mathbb{E}[f_{*c}] = \mathbf{k}_c(\mathbf{x}_*)^T\mathbf{K}_c^{-1}\hat{\mathbf{f}}_c = \mathbf{k}_c(\mathbf{x}_*)^T(\mathbf{y}_c - \hat{\boldsymbol{\pi}}_c) \tag{15.67}$$

We can put this in vector form by writing

$$\mathbb{E}[\mathbf{f}_*] = \mathbf{Q}_*T(\mathbf{y} - \hat{\boldsymbol{\pi}}) \tag{15.68}$$

where

$$\mathbf{Q}_* = \begin{pmatrix} \mathbf{k}_1(\mathbf{x}_*) & \cdots & \mathbf{0} \\ & \ddots & \\ \mathbf{0} & \cdots & \mathbf{k}_C(\mathbf{x}_*) \end{pmatrix} \tag{15.69}$$

Using a similar argument to Equation 15.47, we can show that the covariance of the latent response is given by

$$\mathrm{cov}[\mathbf{f}_*] = \boldsymbol{\Sigma} + \mathbf{Q}_*^T\mathbf{K}^{-1}(\mathbf{K}^{-1} + \mathbf{W})^{-1}\mathbf{K}^{-1}\mathbf{Q}_* \tag{15.70}$$
$$= \mathrm{diag}(\mathbf{k}(\mathbf{x}_*, \mathbf{x}_*)) - \mathbf{Q}_*^T(\mathbf{K} + \mathbf{W}^{-1})^{-1}\mathbf{Q}_* \tag{15.71}$$

where $\boldsymbol{\Sigma}$ is a $C \times C$ diagonal matrix with $\Sigma_{cc} = \kappa_c(\mathbf{x}_*, \mathbf{x}_*) - \mathbf{k}_c^T(\mathbf{x}_*)\mathbf{K}_c^{-1}\mathbf{k}_c(\mathbf{x}_*)$, and $\mathbf{k}(\mathbf{x}_*, \mathbf{x}_*) = [\kappa_c(\mathbf{x}_*, \mathbf{x}_*)]$.

To compute the posterior predictive for the visible response, we need to use

$$p(y|\mathbf{x}_*, \mathbf{X}, \mathbf{y}) \approx \int \mathrm{Cat}(y|\mathcal{S}(\mathbf{f}_*))\mathcal{N}(\mathbf{f}_*|\mathbb{E}\,[\mathbf{f}_*]\,, \mathrm{cov}\,[\mathbf{f}_*])d\mathbf{f}_* \tag{15.72}$$

We can use any of deterministic approximations to the softmax function discussed in Section 21.8.1.1 to compute this. Alternatively, we can just use Monte Carlo.

15.3.2.3 Computing the marginal likelihood

Using arguments similar to the binary case, we can show that

$$\log p(\mathbf{y}|\mathbf{X}) \approx -\frac{1}{2}\hat{\mathbf{f}}^T\mathbf{K}^{-1}\hat{\mathbf{f}} + \mathbf{y}^T\hat{\mathbf{f}} - \sum_{i=1}^{N}\log\left(\sum_{c=1}^{C}\exp\hat{f}_{ic}\right) - \frac{1}{2}\log|\mathbf{I}_{C_N} + \mathbf{W}^{\frac{1}{2}}\mathbf{K}\mathbf{W}^{\frac{1}{2}}| \tag{15.73}$$

This can be optimized numerically in the usual way.

15.3.2.4 Numerical and computational issues

One can implement model fitting in $O(TCN^3)$ time and $O(CN^2)$ space, where T is the number of Newton iterations, using the techniques described in (Rasmussen and Williams 2006, p50). Prediction takes $O(CN^3 + CN^2N_*)$ time, where N_* is the number of test cases.

15.3.3 GPs for Poisson regression

In this section, we illustrate GPs for Poisson regression. An interesting application of this is to spatial **disease mapping**. For example, (Vanhatalo et al. 2010) discuss the problem of modeling the relative risk of heart attack in different regions in Finland. The data consists of the heart attacks in Finland from 1996-2000 aggregated into 20km x 20km lattice cells. The model has the following form:

$$y_i \sim \mathrm{Poi}(e_i r_i) \tag{15.74}$$

where e_i is the known expected number of deaths (related to the population of cell i and the overall death rate), and r_i is the **relative risk** of cell i which we want to infer. Since the data counts are small, we regularize the problem by sharing information with spatial neighbors. Hence we assume $f \triangleq \log(r) \sim \mathrm{GP}(0, \kappa)$, where we use a Matern kernel with $\nu = 3/2$, and a length scale and magnitude that are estimated from data.

Figure 15.8 gives an example of the kind of output one can obtain from this method, based on data from 911 locations. On the left we plot the posterior mean relative risk (RR), and on the right, the posterior variance. We see that the RR is higher in Eastern Finland, which is consistent with other studies. We also see that the variance in the North is higher, since there are fewer people living there.

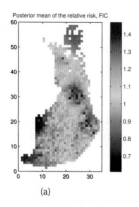
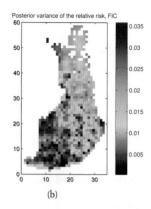

(a)   (b)

Figure 15.8 We show the relative risk of heart disease in Finland using a Poisson GP. Left: posterior mean. Right: posterior variance. Figure generated by `gpSpatialDemoLaplace`, written by Jarno Vanhatalo.

15.4   Connection with other methods

There are variety of other methods in statistics and machine learning that are closely related to GP regression/ classification. We give a brief review of some of these below.

15.4.1   Linear models compared to GPs

Consider Bayesian linear regression for D-dimensional features, where the prior on the weights is $p(\mathbf{w}) = \mathcal{N}(\mathbf{0}, \mathbf{\Sigma})$. The posterior predictive distribution is given by the following;

$$p(f_* | \mathbf{x}_*, \mathbf{X}, \mathbf{y}) = \mathcal{N}(\mu, \sigma^2) \tag{15.75}$$

$$\mu = \frac{1}{\sigma_y^2} \mathbf{x}_*^T \mathbf{A}^{-1} \mathbf{X}^T \mathbf{y} \tag{15.76}$$

$$\sigma^2 = \mathbf{x}_*^T \mathbf{A}^{-1} \mathbf{x}_* \tag{15.77}$$

where $\mathbf{A} = \sigma_y^{-2} \mathbf{X}^T \mathbf{X} + \mathbf{\Sigma}^{-1}$. One can show that we can rewrite the above distribution as follows

$$\mu = \mathbf{x}_*^T \mathbf{\Sigma} \mathbf{X}^T (\mathbf{K} + \sigma_y^2 \mathbf{I})^{-1} \mathbf{y} \tag{15.78}$$

$$\sigma^2 = \mathbf{x}_*^T \mathbf{\Sigma} \mathbf{x}_* - \mathbf{x}_*^T \mathbf{\Sigma} \mathbf{X}^T (\mathbf{K} + \sigma^2 \mathbf{I})^{-1} \mathbf{X} \mathbf{\Sigma} \mathbf{x}_* \tag{15.79}$$

where we have defined $\mathbf{K} = \mathbf{X} \mathbf{\Sigma} \mathbf{X}^T$, which is of size $N \times N$. Since the features only ever appear in the form $\mathbf{X} \mathbf{\Sigma} \mathbf{X}^T$, $\mathbf{x}_*^T \mathbf{\Sigma} \mathbf{X}^T$ or $\mathbf{x}_*^T \mathbf{\Sigma} \mathbf{x}_*$, we can kernelize the above expression by defining $\kappa(\mathbf{x}, \mathbf{x}') = \mathbf{x}^T \mathbf{\Sigma} \mathbf{x}'$.

Thus we see that Bayesian linear regression is equivalent to a GP with covariance function $\kappa(\mathbf{x}, \mathbf{x}') = \mathbf{x}^T \mathbf{\Sigma} \mathbf{x}'$. Note, however, that this is a **degenerate** covariance function, since it has at most D non-zero eigenvalues. Intuitively this reflects the fact that the model can only represent a limited number of functions. This can result in underfitting, since the model is not flexible enough to capture the data. What is perhaps worse, it can result in overconfidence, since the model's prior is so impoverished that its posterior will become too concentrated. So not only is the model wrong, it thinks it's right!

15.4.2 Linear smoothers compared to GPs

A **linear smoother** is a regression function which is a linear function of the training outputs:

$$\hat{f}(\mathbf{x}_*) = \sum_i w_i(\mathbf{x}_*) \, y_i \tag{15.80}$$

where $w_i(\mathbf{x}_*)$ is called the **weight function** (Silverman 1984). (Do not confuse this with a linear model, where the output is a linear function of the input vector.)

There are a variety of linear smoothers, such as kernel regression (Section 14.7.4), locally weighted regression (Section 14.7.5), smoothing splines (Section 15.4.6), and GP regression. To see that GP regession is a linear smoother, note that the mean of the posterior predictive distribution of a GP is given by

$$\overline{f}(\mathbf{x}_*) = \mathbf{k}_*^T (\mathbf{K} + \sigma_y^2 \mathbf{I}_N)^{-1} \mathbf{y} = \sum_{i=1}^{N} y_i w_i(\mathbf{x}_*) \tag{15.81}$$

where $w_i(\mathbf{x}_*) = [(\mathbf{K} + \sigma_y^2 \mathbf{I}_N)^{-1} \mathbf{k}_*]_i$.

In kernel regression, we derive the weight function from a smoothing kernel rather than a Mercer kernel, so it is clear that the weight function will then have local support. In the case of a GP, things are not as clear, since the weight function depends on the inverse of $\mathbf{K}$. For certain GP kernel functions, we can analytically derive the form of $w_i(\mathbf{x})$; this is known as the **equivalent kernel** (Silverman 1984). One can show that $\sum_{i=1}^{N} w_i(\mathbf{x}_*) = 1$, although we may have $w_i(\mathbf{x}_*) < 0$, so we are computing a linear combination but not a convex combination of the y_i's. More interestingly, $w_i(\mathbf{x}_*)$ is a local function, even if the original kernel used by the GP is not local. Futhermore the effective bandwidth of the equivalent kernel of a GP automatically decreases as the sample size N increases, whereas in kernel smoothing, the bandwidth h needs to be set by hand to adapt to N. See e.g., (Rasmussen and Williams 2006, Sec 2.6,Sec 7.1) for details.

15.4.2.1 Degrees of freedom of linear smoothers

It is clear why this method is called "linear", but why is it called a "smoother"? This is best explained in terms of GPs. Consider the prediction on the training set:

$$\overline{\mathbf{f}} = \mathbf{K}(\mathbf{K} + \sigma_y^2)^{-1} \mathbf{y} \tag{15.82}$$

Now let $\mathbf{K}$ have the eigendecomposition $\mathbf{K} = \sum_{i=1}^{N} \lambda_i \mathbf{u}_i \mathbf{u}_i^T$. Since $\mathbf{K}$ is real and symmetric positive definite, the eigenvalues λ_i are real and non-negative, and the eigenvectors $\mathbf{u}_i$ are orthonormal. Now let $\mathbf{y} = \sum_{i=1}^{N} \gamma_i \mathbf{u}_i$, where $\gamma_i = \mathbf{u}_i^T \mathbf{y}$. Then we can rewrite the above equation as follows:

$$\overline{\mathbf{f}} = \sum_{i=1}^{N} \frac{\gamma_i \lambda_i}{\lambda_i + \sigma_y^2} \mathbf{u}_i \tag{15.83}$$

This is the same as Equation 7.47, except we are working with the eigenvectors of the Gram matrix $\mathbf{K}$ instead of the data matrix $\mathbf{X}$. In any case, the interpretation is similar: if $\frac{\lambda_i}{\lambda_i + \sigma_y^2} \ll 1$, then the corresponding basis function $\mathbf{u}_i$ will not have much influence. Consequently the high-frequency components in $\mathbf{y}$ are smoothed out. The effective **degrees of freedom** of the linear smoother is defined as

$$
\text{dof} \triangleq \text{tr}(\mathbf{K}(\mathbf{K} + \sigma_y^2 \mathbf{I})^{-1}) = \sum_{i=1}^{N} \frac{\lambda_i}{\lambda_i + \sigma_y^2} \tag{15.84}
$$

This specifies how "wiggly" the curve is.

15.4.3   SVMs compared to GPs

We saw in Section 14.5.2 that the SVM objective for binary classification is given by Equation 14.57

$$
J(\mathbf{w}) = \frac{1}{2}||\mathbf{w}||^2 + C \sum_{i=1}^{N} (1 - y_i f_i)_+ \tag{15.85}
$$

We also know from Equation 14.59 that the optimal solution has the form $\mathbf{w} = \sum_i \alpha_i \mathbf{x}_i$, so $||\mathbf{w}||^2 = \sum_{i,j} \alpha_i \alpha_j \mathbf{x}_i^T \mathbf{x}_j$. Kernelizing we get $||\mathbf{w}||^2 = \boldsymbol{\alpha} \mathbf{K} \boldsymbol{\alpha}$. From Equation 14.61, and absorbing the $\hat{w}_0$ term into one of the kernels, we have $\mathbf{f} = \mathbf{K}\boldsymbol{\alpha}$, so $||\mathbf{w}||^2 = \mathbf{f}^T \mathbf{K}^{-1} \mathbf{f}$. Hence the SVM objective can be rewritten as

$$
J(\mathbf{f}) = \frac{1}{2}\mathbf{f}^T \mathbf{f} + C \sum_{i=1}^{N} (1 - y_i f_i)_+ \tag{15.86}
$$

Compare this to MAP estimation for GP classifier:

$$
J(\mathbf{f}) = \frac{1}{2}\mathbf{f}^T \mathbf{f} - \sum_{i=1}^{N} \log p(y_i|f_i) \tag{15.87}
$$

It is tempting to think that we can "convert" an SVM into a GP by figuring out what likelihood would be equivalent to the hinge loss. However, it turns out there is no such likelihood (Sollich 2002), although there is a pseudo-likelihood that matches the SVM (see Section 14.5.5).

From Figure 6.7 we saw that the hinge loss and the logistic loss (as well as the probit loss) are quite similar to each other. The main difference is that the hinge loss is strictly 0 for errors larger than 1. This gives rise to a sparse solution. In Section 14.3.2, we discussed other ways to derive sparse kernel machines. We discuss the connection between these methods and GPs below.

15.4.4   L1VM and RVMs compared to GPs

Sparse kernel machines are just linear models with basis function expansion of the form $\phi(\mathbf{x}) = [\kappa(\mathbf{x}, \mathbf{x}_1), \dots, \kappa(\mathbf{x}, \mathbf{x}_N)]$. From Section 15.4.1, we know that this is equivalent to a GP with the following kernel:

$$
\kappa(\mathbf{x}, \mathbf{x}') = \sum_{j=1}^{D} \frac{1}{\alpha_j} \phi_j(\mathbf{x})\phi_j(\mathbf{x}') \tag{15.88}
$$

where $p(\mathbf{w}) = \mathcal{N}(\mathbf{0}, \text{diag}(\alpha_j^{-1}))$. This kernel function has two interesting properties. First, it is degenerate, meaning it has at most N non-zero eigenvalues, so the joint distribution $p(\mathbf{f}, \mathbf{f}_*)$ will be highly constrained. Second, the kernel depends on the training data. This can cause the model to be overconfident when extrapolating beyond the training data. To see this, consider a point $\mathbf{x}_*$ far outside the convex hull of the data. All the basis functions will have values close to 0, so the prediction will back off to the mean of the GP. More worryingly, the variance will back off to the noise variance. By contrast, when using a non-degenerate kernel function, the predictive variance increases as we move away from the training data, as desired. See (Rasmussen and Quiñonero-Candela 2005) for further discussion.

15.4.5   Neural networks compared to GPs

In Section 16.5, we will discuss neural networks, which are a nonlinear generalization of GLMs. In the binary classification case, a neural network is defined by a logistic regression model applied to a logistic regression model:

$$p(y|\mathbf{x}, \boldsymbol{\theta}) = \text{Ber}\left(y|\text{sigm}\left(\mathbf{w}^T \text{sigm}(\mathbf{V}\mathbf{x})\right)\right) \tag{15.89}$$

It turns out there is an interesting connection between neural networks and Gaussian processes, as first pointed out by (Neal 1996).

To explain the connection, we follow the presentation of (Rasmussen and Williams 2006, p91). Consider a neural network for regression with one hidden layer. This has the form

$$p(y|\mathbf{x}, \boldsymbol{\theta}) = \mathcal{N}(y|f(\mathbf{x}; \boldsymbol{\theta}), \sigma^2) \tag{15.90}$$

where

$$f(\mathbf{x}) = b + \sum_{j=1}^{H} v_j g(\mathbf{x}; \mathbf{u}_j) \tag{15.91}$$

where b is the offset of bias term, v_j is the output weight from hidden unit j to the response y, $\mathbf{u}_j$ are the inputs weights to unit j from the input $\mathbf{x}$, and $g()$ is the hidden unit activation function. This is typically the sigmoid or tanh function, but can be any smooth function.

Let us use the following priors on the weights: where $b \sim \mathcal{N}(0, \sigma_b^2)$ $\mathbf{v} \sim \prod_j \mathcal{N}(v_j|0, \sigma_w^2)$, $\mathbf{u} \sim \prod_j p(\mathbf{u}_j)$ for some unspecified $p(\mathbf{u}_j)$. Denoting all the weights by $\boldsymbol{\theta}$ we have

$$\mathbb{E}_{\boldsymbol{\theta}}\left[f(\mathbf{x})\right] = 0 \tag{15.92}$$

$$\mathbb{E}_{\boldsymbol{\theta}}\left[f(\mathbf{x})f(\mathbf{x}')\right] = \sigma_b^2 + \sum_j \sigma_v^2 \mathbb{E}_{\mathbf{v}}\left[g(\mathbf{x}; \mathbf{u}_j)g(\mathbf{x}'; \mathbf{u}_j)\right] \tag{15.93}$$

$$= \sigma_b^2 + H\sigma_v^2 \mathbb{E}_{\mathbf{u}}\left[g(\mathbf{x}; \mathbf{u})g(\mathbf{x}'; \mathbf{u})\right] \tag{15.94}$$

where the last equality follows since the H hidden units are iid. If we let σ_v^2 scale as ω^2/H (since more hidden units will increase the input to the final node, so we should scale down the magnitude of the weights), then the last term becomes $\omega^2 \mathbb{E}_{\mathbf{u}}\left[g(\mathbf{x}; \mathbf{u})g(\mathbf{x}'; \mathbf{u})\right]$. This is a sum over H iid random variables. Assuming that g is bounded, we can apply the central limit theorem. The result is that as $H \to \infty$, we get a Gaussian process.

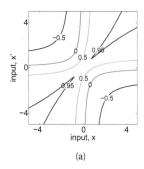
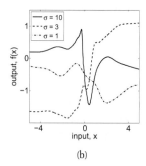

(a)

(b)

Figure 15.9 (a) Covariance function $\kappa_{NN}(x, x')$ for $\sigma_0 = 10$, $\sigma = 10$. (b) Samples from from a GP with this kernel, using various values of σ. Figure generated by `gpnnDemo`, written by Chris Williams.

If we use as activation / transfer function $g(\mathbf{x}; \mathbf{u}) = \text{erf}(u_0 + \sum_{j=1}^{D} u_j x_j)$, where $\text{erf}(z) = 2/\sqrt{\pi} \int_0^z e^{-t^2} dt$, and we choose $\mathbf{u} \sim \mathcal{N}(\mathbf{0}, \mathbf{\Sigma})$, then (Williams 1998) showed that the covariance kernel has the form

$$\kappa_{NN}(\mathbf{x}, \mathbf{x}') = \frac{2}{\pi} \sin^{-1} \left(\frac{2\tilde{\mathbf{x}}^T \mathbf{\Sigma} \tilde{\mathbf{x}}'}{\sqrt{(1 + 2\tilde{\mathbf{x}}^T \mathbf{\Sigma} \tilde{\mathbf{x}})(1 + 2(\tilde{\mathbf{x}}')^T \mathbf{\Sigma} \tilde{\mathbf{x}}')}} \right) \quad (15.95)$$

where $\tilde{\mathbf{x}} = (1, x_1, \ldots, x_D)$. This is a true "neural network" kernel, unlike the "sigmoid" kernel $\kappa(\mathbf{x}, \mathbf{x}') = \tanh(a + b\mathbf{x}^T \mathbf{x}')$, which is not positive definite.

Figure 15.9(a) illustrates this kernel when $D = 2$ and $\mathbf{\Sigma} = \text{diag}(\sigma_0^2, \sigma^2)$. Figure 15.9(b) shows some functions sampled from the corresponding GP. These are equivalent to functions which are superpositions of $\text{erf}(u_0 + ux)$ where u_0 and u are random. As σ^2 increases, the variance of u increases, so the function varies more quickly. Unlike the RBF kernel, functions sampled from this kernel do not tend to 0 away from the data, but rather they tend to remain at the same value they had at the "edge" of the data.

Now suppose we use an RBF network, which is equivalent to a hidden unit activation function of the form $g(\mathbf{x}; \mathbf{u}) = \exp(-|\mathbf{x} - \mathbf{u}|^2/(2\sigma_g^2))$. If $\mathbf{u} \sim \mathcal{N}(\mathbf{0}, \sigma_u^2 \mathbf{I})$, one can show that the coresponding kernel is equivalent to the RBF or SE kernel.

15.4.6 Smoothing splines compared to GPs *

Smoothing splines are a widely used non-parametric method for smoothly interpolating data (Green and Silverman 1994). They are are a special case of GPs, as we will see. They are usually used when the input is 1 or 2 dimensional.

15.4.6.1 Univariate splines

The basic idea is to fit a function f by minimizing the discrepancy to the data plus a smoothing term that penalizes functions that are "too wiggly". If we penalize the m'th derivative of the function, the objective becomes

$$J(f) = \sum_{i=1}^{N} (f(x_i) - y_i)^2 + \lambda \int \left(\frac{d^m}{dx^m} f(x)\right)^2 dx \tag{15.96}$$

One can show (Green and Silverman 1994) that the solution is a **piecewise polynomial** where the polynomials have order $2m - 1$ in the interior bins $[x_{i-1}, x_i]$ (denoted $\mathcal{I}$), and order $m - 1$ in the two outermost intervals $(-\infty, x_1]$ and $[x_N, \infty)$:

$$f(x) = \sum_{j=0}^{m-1} \beta_j x^j + \mathbb{I}(x \in \mathcal{I}) \left(\sum_{i=1}^{N} \alpha_i (x - x_i)_+^{2m-1}\right) + \mathbb{I}(x \notin \mathcal{I}) \left(\sum_{i=1}^{N} \alpha_i (x - x_i)_+^{m-1}\right) \tag{15.97}$$

For example, if $m = 2$, we get the (natural) **cubic spline**

$$f(x) = \beta_0 + \beta_1 x + \mathbb{I}(x \in \mathcal{I}) \left(\sum_{i=1}^{N} \alpha_i (x - x_i)_+^3\right) + \mathbb{I}(x \notin \mathcal{I}) \left(\sum_{i=1}^{N} \alpha_i (x - x_i)_+\right) \tag{15.98}$$

which is a series of truncated cubic polynomials, whose left hand sides are located at each of the N training points. (The fact that the model is linear on the edges prevents it from extrapolating too wildly beyond the range of the data; if we drop this requirement, we get an "unrestricted" spline.)

We can clearly fit this model using ridge regression: $\hat{\mathbf{w}} = (\mathbf{\Phi}^T \mathbf{\Phi} + \lambda \mathbf{I}_N)^{-1} \mathbf{\Phi}^T \mathbf{y}$, where the columns of $\mathbf{\Phi}$ are 1, x_i and $(x - x_i)_+^3$ for $i = 2 : N - 1$ and $(x - x_i)_+$ for $i = 1$ or $i = N$. However, we can also derive an $O(N)$ time method (Green and Silverman 1994, Sec 2.3.3).

15.4.6.2 Regression splines

In general, we can place the polynomials at a fixed set of K locations known as **knots**, denoted ξ_k. The result is called a **regression spline**. This is a parametric model, which uses basis function expansion of the following form (where we drop the interior/ exterior distinction for simplicity):

$$f(x) = \beta_0 + \beta_1 x + \sum_{k=1}^{K} \alpha_j (x - \xi_k)_+^3 \tag{15.99}$$

Choosing the number and locations of the knots is just like choosing the number and values of the support vectors in Section 14.3.2. If we impose an ℓ_2 regularizer on the regression coefficients α_j, the method is known as **penalized splines**. See Section 9.6.1 for a practical example of penalized splines.

15.4.6.3 The connection with GPs

One can show (Rasmussen and Williams 2006, p139) that the cubic spline is the MAP estimate of the following function

$$f(x) = \beta_0 + \beta_1 x + r(x) \tag{15.100}$$

where $p(\beta_j) \propto 1$ (so that we don't penalize the zero'th and first derivatives of f), and $r(x) \sim$ GP$(0, \sigma_f^2 \kappa_{sp}(x, x'))$, where

$$\kappa_{sp}(x, x') \triangleq \int_0^1 (x - u)_+ (x' - u)_+ du \tag{15.101}$$

Note that the kernel in Equation 15.101 is rather unnatural, and indeed posterior samples from the resulting GP are rather unsmooth. However, the posterior mode/mean is smooth. This shows that regularizers don't always make good priors.

15.4.6.4   2d input (thin-plate splines)

One can generalize cubic splines to 2d input by defining a regularizer of the following form:

$$\int \int \left[\left(\frac{\partial^2 f(x)}{\partial x_1^2} \right)^2 + 2 \left(\frac{\partial^2 f(x)}{\partial x_1 \partial x_2} \right)^2 + \left(\frac{\partial^2 f(x)}{\partial x_2^2} \right)^2 \right] dx_1 dx_2 \tag{15.102}$$

One can show that the solution has the form

$$f(x) = \beta_0 + \boldsymbol{\beta}_1^T \mathbf{x} + \sum_{i=1}^N \alpha_i \phi_i(\mathbf{x}) \tag{15.103}$$

where $\phi_i(\mathbf{x}) = \eta(||\mathbf{x} - \mathbf{x}_i||)$, and $\eta(z) = z^2 \log z^2$. This is known as a **thin plate spline**. This is equivalent to MAP estimation with a GP whose kernel is defined in (Williams and Fitzgibbon 2006).

15.4.6.5   Higher-dimensional inputs

It is hard to analytically solve for the form of the optimal solution when using higher-order inputs. However, in the parametric regression spline setting, where we forego the regularizer on f, we have more freedom in defining our basis functions. One way to handle multiple inputs is to use a **tensor product basis**, defined as the cross product of 1d basis functions. For example, for 2d input, we can define

$$
\begin{aligned}
f(x_1, x_2) \quad = \quad & \beta_0 + \sum_m \beta_{1m}(x_1 - \xi_{1m})_+ + \sum_m \beta_{2m}(x_2 - \xi_{2m})_+ & (15.104) \\
+ \quad & \sum_m \beta_{12m}(x_1 - \xi_{1m})_+(x_2 - \xi_{2m})_+ & (15.105)
\end{aligned}
$$

It is clear that for high-dimensional data, we cannot allow higher-order interactions, because there will be too many parameters to fit. One approach to this problem is to use a search procedure to look for useful interaction terms. This is known as MARS, which stands for "multivariate adaptive regression splines". See Section 16.3.3 for details.

15.4.7   RKHS methods compared to GPs *

We can generalize the idea of penalizing derivatives of functions, as used in smoothing splines, to fit functions with a more general notion of smoothness. Recall from Section 14.2.3 that

Mercer's theorem says that any positive definite kernel function can be represented in terms of eigenfunctions:

$$\kappa(\mathbf{x}, \mathbf{x}') = \sum_{i=1}^{\infty} \lambda_i \phi_i(\mathbf{x}) \phi_i(\mathbf{x}') \tag{15.106}$$

The ϕ_i form an orthormal basis for a function space:

$$\mathcal{H}_k = \{f : f(\mathbf{x}) = \sum_{i=1}^{\infty} f_i \phi_i(\mathbf{x}), \; \sum_{i=1}^{\infty} f_i^2 / \lambda_i < \infty\} \tag{15.107}$$

Now define the inner product between two functions $f(\mathbf{x}) = \sum_{i=1}^{\infty} f_i \phi_i(\mathbf{x})$ and $g(\mathbf{x}) = \sum_{i=1}^{\infty} g_i \phi_i(\mathbf{x})$ in this space as follows:

$$\langle f, g \rangle_{\mathcal{H}} \triangleq \sum_{i=1}^{\infty} \frac{f_i g_i}{\lambda_i} \tag{15.108}$$

In Exercise 15.1, we show that this definition implies that

$$\langle \kappa(\mathbf{x}_1, \cdot), \kappa(\mathbf{x}_2, \cdot) \rangle_{\mathcal{H}} = \kappa(\mathbf{x}_1, \mathbf{x}_2) \tag{15.109}$$

This is called the **reproducing property**, and the space of functions $\mathcal{H}_k$ is called a **reproducing kernel Hilbert space** or **RKHS**.

Now consider an optimization problem of the form

$$J(f) = \frac{1}{2\sigma_y^2} \sum_{i=1}^{N} (y_i - f(\mathbf{x}_i))^2 + \frac{1}{2}||f||_H^2 \tag{15.110}$$

where $||f||_J$ is the **norm of a function**:

$$||f||_H = \langle f, f \rangle_{\mathcal{H}} = \sum_{i=1}^{\infty} \frac{f_i^2}{\lambda_i} \tag{15.111}$$

The intuition is that functions that are complex wrt the kernel will have large norms, because they will need many eigenfunctions to represent them. We want to pick a simple function that provides a good fit to the data.

One can show (see e.g., (Schoelkopf and Smola 2002)) that the solution must have the form

$$f(\mathbf{x}) = \sum_{i=1}^{N} \alpha_i \kappa(\mathbf{x}, \mathbf{x}_i) \tag{15.112}$$

This is known as the **representer theorem**, and holds for other convex loss functions besides squared error.

We can solve for the $\boldsymbol{\alpha}$ by substituting in $f(\mathbf{x}) = \sum_{i=1}^{N} \alpha_i \kappa(\mathbf{x}, \mathbf{x}_i)$ and using the reproducing property to get

$$J(\boldsymbol{\alpha}) = \frac{1}{2\sigma_y^2} |\mathbf{y} - \mathbf{K}\boldsymbol{\alpha}|^2 + \frac{1}{2} \boldsymbol{\alpha}^T \mathbf{K} \boldsymbol{\alpha} \tag{15.113}$$

Minimizing wrt $\boldsymbol{\alpha}$ we find

$$\hat{\boldsymbol{\alpha}} = (\mathbf{K} + \sigma_y^2 \mathbf{I})^{-1} \tag{15.114}$$

and hence

$$\hat{f}(\mathbf{x}_*) = \sum_i \hat{\alpha}_i \kappa(\mathbf{x}_*, \mathbf{x}_i) = \mathbf{k}_*^T (\mathbf{K} + \sigma_y^2 \mathbf{I})^{-1} \mathbf{y} \tag{15.115}$$

This is identical to Equation 15.18, the posterior mean of a GP predictive distribution. Indeed, since the mean and mode of a Gaussian are the same, we can see that linear regresson with an RKHS regularizer is equivalent to MAP estimation with a GP. An analogous statement holds for the GP logistic regression case, which also uses a convex likelihood / loss function.

15.5   GP latent variable model

In Section 14.4.4, we discussed kernel PCA, which applies the kernel trick to regular PCA. In this section, we discuss a different way to combine kernels with probabilistic PCA. The resulting method is known as the **GP-LVM**, which stands for "Gaussian process latent variable model" (Lawrence 2005).

To explain the method, we start with PPCA. Recall from Section 12.2.4 that the PPCA model is as follows:

$$p(\mathbf{z}_i) = \mathcal{N}(\mathbf{z}_i | \mathbf{0}, \mathbf{I}) \tag{15.116}$$

$$p(\mathbf{y}_i | \mathbf{z}_i, \boldsymbol{\theta}) = \mathcal{N}(\mathbf{y}_i | \mathbf{W} \mathbf{z}_i, \sigma^2 \mathbf{I}) \tag{15.117}$$

We can fit this model by maximum likelihood, by integrating out the $\mathbf{z}_i$ and maximizing $\mathbf{W}$ (and σ^2). The objective is given by

$$p(\mathbf{Y} | \mathbf{W}, \sigma^2) = (2\pi)^{-DN/2} |\mathbf{C}|^{-N/2} \exp\left(-\frac{1}{2} \mathrm{tr}(\mathbf{C}^{-1} \mathbf{Y}^T \mathbf{Y})\right) \tag{15.118}$$

where $\mathbf{C} = \mathbf{W}\mathbf{W}^T + \sigma^2 \mathbf{I}$. As we showed in Theorem 12.2.2, the MLE for this can be computed in terms of the eigenvectors of $\mathbf{Y}^T \mathbf{Y}$.

Now we consider the dual problem, whereby we maximize $\mathbf{Z}$ and integrate out $\mathbf{W}$. We will use a prior of the form $p(\mathbf{W}) = \prod_j \mathcal{N}(\mathbf{w}_j | \mathbf{0}, \mathbf{I})$. The corresponding likelihood becomes

$$p(\mathbf{Y} | \mathbf{Z}, \sigma^2) = \prod_{d=1}^{D} \mathcal{N}(\mathbf{y}_{:,d} | \mathbf{0}, \mathbf{Z}\mathbf{Z}^T + \sigma^2 \mathbf{I}) \tag{15.119}$$

$$= (2\pi)^{-DN/2} |\mathbf{K}_z|^{-D/2} \exp\left(-\frac{1}{2} \mathrm{tr}(\mathbf{K}_z^{-1} \mathbf{Y} \mathbf{Y}^T)\right) \tag{15.120}$$

where $\mathbf{K}_z = \mathbf{Z}\mathbf{Z}^T + \sigma^2 \mathbf{I}$. Based on our discussion of the connection between the eigenvalues of $\mathbf{Y}\mathbf{Y}^T$ and of $\mathbf{Y}^T \mathbf{Y}$ in Section 14.4.4, it should come as no surprise that we can also solve the dual problem using eigenvalue methods (see (Lawrence 2005) for the details).

If we use a linear kernel, we recover PCA. But we can also use a more general kernel: $\mathbf{K}_z = \mathbf{K} + \sigma^2 \mathbf{I}$, where $\mathbf{K}$ is the Gram matrix for $\mathbf{Z}$. The MLE for $\hat{\mathbf{Z}}$ will no longer be available

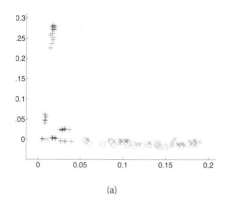
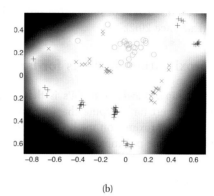

| (a) | (b) |

Figure 15.10 2d representation of 12 dimensional oil flow data. The different colors/symbols represent the 3 phases of oil flow. (a) Kernel PCA with Gaussian kernel. (b) GP-LVM with Gaussian kernel. The shading represents the precision of the posterior, where lighter pixels have higher precision. From Figure 1 of (Lawrence 2005). Used with kind permission of Neil Lawrence.

via eigenvalue methods; instead we must use gradient-based optimization. The objective is given by

$$\ell = -\frac{D}{2} \log |\mathbf{K}_z| - \frac{1}{2} \text{tr}(\mathbf{K}_z^{-1} \mathbf{Y} \mathbf{Y}^T) \qquad (15.121)$$

and the gradient is given by

$$\frac{\partial \ell}{\partial Z_{ij}} = \frac{\partial \ell}{\partial \mathbf{K}_z} \frac{\partial \mathbf{K}_z}{\partial Z_{ij}} \qquad (15.122)$$

where

$$\frac{\partial \ell}{\partial \mathbf{K}_z} = \mathbf{K}_z^{-1} \mathbf{Y} \mathbf{Y}^T \mathbf{K}_z^{-1} - D \mathbf{K}_z^{-1} \qquad (15.123)$$

The form of $\frac{\partial \mathbf{K}_z}{\partial Z_{ij}}$ will of course depend on the kernel used. (For example, with a linear kernel, where $\mathbf{K}_z = \mathbf{Z}\mathbf{Z}^T + \sigma^2 \mathbf{I}$, we have $\frac{\partial \mathbf{K}_z}{\partial \mathbf{Z}} = \mathbf{Z}$.) We can then pass this gradient to any standard optimizer, such as conjugate gradient descent.

Let us now compare GP-LVM to kernel PCA. In kPCA, we learn a kernelized mapping from the observed space to the latent space, whereas in GP-LVM, we learn a kernelized mapping from the latent space to the observed space. Figure 15.10 illustrates the results of applying kPCA and GP-LVM to visualize the 12 dimensional oil flow data shown in In Figure 14.9(a). We see that the embedding produced by GP-LVM is far better. If we perform nearest neighbor classification in the latent space, GP-LVM makes 4 errors, while kernel PCA (with the same kernel but separately optimized hyper-parameters) makes 13 errors, and regular PCA makes 20 errors.

GP-LVM inherits the usual advantages of probabilistic generative models, such as the ability to handle missing data and data of different types, the ability to use gradient-based methods (instead of grid search) to tune the kernel parameters, the ability to handle prior information, etc. For a discussion of some other probabilistic methods for (spectral) dimensionality reduction, see (Lawrence 2012).

15.6 Approximation methods for large datasets

The principal drawback of GPs is that they take $O(N^3)$ time to use. This is because of the need to invert (or compute the Cholesky decomposition of) the $N \times N$ kernel matrix $\mathbf{K}$. A variety of approximation methods have been devised which take $O(M^2 N)$ time, where M is a user-specifiable parameter. For details, see (Quinonero-Candela et al. 2007).

Exercises

Exercise 15.1 Reproducing property

Prove Equation 15.109.

16 *Adaptive basis function models*

16.1 Introduction

In Chapters 14 and 15, we discussed kernel methods, which provide a powerful way to create non-linear models for regression and classification. The prediction takes the form $f(\mathbf{x}) = \mathbf{w}^T \boldsymbol{\phi}(\mathbf{x})$, where we define

$$\boldsymbol{\phi}(\mathbf{x}) = [\kappa(\mathbf{x}, \boldsymbol{\mu}_1), \dots, \kappa(\mathbf{x}, \boldsymbol{\mu}_N)] \tag{16.1}$$

and where $\boldsymbol{\mu}_k$ are either all the training data or some subset. Models of this form essentially perform a form of **template matching**, whereby they compare the input $\mathbf{x}$ to the stored prototypes $\boldsymbol{\mu}_k$.

Although this can work well, it relies on having a good kernel function to measure the similarity between data vectors. Often coming up with a good kernel function is quite difficult. For example, how do we define the similarity between two images? Pixel-wise comparison of intensities (which is what a Gaussian kernel corresponds to) does not work well. Although it is possible (and indeed common) to hand-engineer kernels for specific tasks (see e.g., the pyramid match kernel in Section 14.2.7), it would be more interesting if we could learn the kernel.

In Section 15.2.4, we discussed a way to learn the parameters of a kernel function, by maximizing the marginal likelihood. For example, if we use the ARD kernel,

$$\kappa(\mathbf{x}, \mathbf{x}') = \theta_0 \exp\left(-\frac{1}{2} \sum_{j=1}^{D} \theta_j (x_j - x_j')^2\right) \tag{16.2}$$

we can can estimate the θ_j, and thus perform a form of nonlinear feature selection. However, such methods can be computationally expensive. Another approach, known as multiple kernel learning (see e.g., (Rakotomamonjy et al. 2008)) uses a convex combination of base kernels, $\kappa(\mathbf{x}, \mathbf{x}') = \sum_j w_j \kappa_j(\mathbf{x}, \mathbf{x}')$, and then estimates the mixing weights w_j. But this relies on having good base kernels (and is also computationally expensive).

An alternative approach is to dispense with kernels altogether, and try to learn useful features $\boldsymbol{\phi}(\mathbf{x})$ directly from the input data. That is, we will create what we call an **adaptive basis-function model** (ABM), which is a model of the form

$$f(\mathbf{x}) = w_0 + \sum_{m=1}^{M} w_m \phi_m(\mathbf{x}) \tag{16.3}$$

where $\phi_m(\mathbf{x})$ is the m'th basis function, which is learned from data. This framework covers all of the models we will discuss in this chapter.

Typically the basis functions are parametric, so we can write $\phi_m(\mathbf{x}) = \phi(\mathbf{x}; \mathbf{v}_m)$, where $\mathbf{v}_m$ are the parameters of the basis function itself. We will use $\boldsymbol{\theta} = (w_0, \mathbf{w}_{1:M}, \{\mathbf{v}_m\}_{m=1}^M)$ to denote the entire parameter set. The resulting model is not linear-in-the-parameters anymore, so we will only be able to compute a locally optimal MLE or MAP estimate of $\boldsymbol{\theta}$. Nevertheless, such models often significantly outperform linear models, as we will see.

16.2 Classification and regression trees (CART)

Classification and regression trees or **CART** models, also called **decision trees** (not to be confused with the decision trees used in decision theory) are defined by recursively partitioning the input space, and defining a local model in each resulting region of input space. This can be represented by a tree, with one leaf per region, as we explain below.

16.2.1 Basics

To explain the CART approach, consider the tree in Figure 16.1(a). The first node asks if x_1 is less than some threshold t_1. If yes, we then ask if x_2 is less than some other threshold t_2. If yes, we are in the bottom left quadrant of space, R_1. If no, we ask if x_1 is less than t_3. And so on. The result of these **axis parallel splits** is to partition 2d space into 5 regions, as shown in Figure 16.1(b). We can now associate a mean response with each of these regions, resulting in the piecewise constant surface shown in Figure 16.1(c).

We can write the model in the following form

$$f(\mathbf{x}) = \mathbb{E}\left[y|\mathbf{x}\right] = \sum_{m=1}^M w_m \mathbb{I}(\mathbf{x} \in R_m) = \sum_{m=1}^M w_m \phi(\mathbf{x}; \mathbf{v}_m) \tag{16.4}$$

where R_m is the m'th region, w_m is the mean response in this region, and $\mathbf{v}_m$ encodes the choice of variable to split on, and the threshold value, on the path from the root to the m'th leaf. This makes it clear that a CART model is just a an adaptive basis-function model, where the basis functions define the regions, and the weights specify the response value in each region. We discuss how to find these basis functions below.

We can generalize this to the classification setting by storing the distribution over class labels in each leaf, instead of the mean response. This is illustrated in Figure 16.2. This model can be used to classify the data in Figure 1.1. For example, we first check the color of the object. If it is blue, we follow the left branch and end up in a leaf labeled "4,0", which means we have 4 positive examples and 0 negative examples which match this criterion. Hence we predict $p(y = 1|\mathbf{x}) = 4/4$ if $\mathbf{x}$ is blue. If it is red, we then check the shape: if it is an ellipse, we end up in a leaf labeled "1,1", so we predict $p(y = 1|\mathbf{x}) = 1/2$. If it is red but not an ellipse, we predict $p(y = 1|\mathbf{x}) = 0/2$; If it is some other colour, we check the size: if less than 10, we predict $p(y = 1|\mathbf{x}) = 4/4$, otherwise $p(y = 1|\mathbf{x}) = 0/5$. These probabilities are just the empirical fraction of positive examples that satisfy each conjunction of feature values, which defines a path from the root to a leaf.

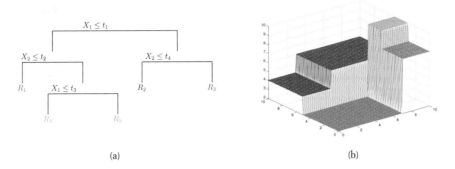

Figure 16.1 A simple regression tree on two inputs. Based on Figure 9.2 of (Hastie et al. 2009). Figure generated by `regtreeSurfaceDemo`.

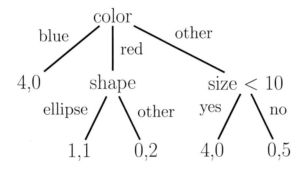

Figure 16.2 A simple decision tree for the data in Figure 1.1. A leaf labeled as (n_1, n_0) means that there are n_1 positive examples that match this path, and n_0 negative examples. In this tree, most of the leaves are "pure", meaning they only have examples of one class or the other; the only exception is leaf representing red ellipses, which has a label distribution of $(1, 1)$. We could distinguish positive from negative red ellipses by adding a further test based on size. However, it is not always desirable to construct trees that perfectly model the training data, due to overfitting.

16.2.2 Growing a tree

Finding the optimal partitioning of the data is NP-complete (Hyafil and Rivest 1976), so it is common to use the greedy procedure shown in Algorithm 16.1 to compute a locally optimal MLE. This method is used by CART, (Breiman et al. 1984) **C4.5**(Quinlan 1993), and **ID3** (Quinlan 1986), which are three popular implementations of the method. (See `dtfit` for a simple Matlab implementation.)

The split function chooses the best feature, and the best value for that feature, as follows:

$$(j^*, t^*) = \arg \min_{j \in \{1, \ldots, D\}} \min_{t \in \mathcal{T}_j} \text{cost}(\{\mathbf{x}_i, y_i : x_{ij} \leq t\}) + \text{cost}(\{\mathbf{x}_i, y_i : x_{ij} > t\}) \tag{16.5}$$

Algorithm 16.1: Recursive procedure to grow a classification/ regression tree

1 function fitTree(node, $\mathcal{D}$, depth) ;
2 node.prediction = mean($y_i : i \in \mathcal{D}$) // or class label distribution ;
3 $(j^*, t^*, \mathcal{D}_L, \mathcal{D}_R) = \text{split}(\mathcal{D})$;
4 **if** *not worthSplitting(depth, cost, $\mathcal{D}_L$, $\mathcal{D}_R$)* **then**
5 | return node
6 **else**
7 | node.test = $\lambda\mathbf{x}.x_{j^*} < t^*$ // anonymous function;
8 | node.left = fitTree(node, $\mathcal{D}_L$, depth+1);
9 | node.right = fitTree(node, $\mathcal{D}_R$, depth+1);
10 | return node;

---

where the cost function for a given dataset will be defined below. For notational simplicity, we have assumed all inputs are real-valued or ordinal, so it makes sense to compare a feature x_{ij} to a numeric value t. The set of possible thresholds $\mathcal{T}_j$ for feature j can be obtained by sorting the unique values of x_{ij}. For example, if feature 1 has the values $\{4.5, -12, 72, -12\}$, then we set $\mathcal{T}_1 = \{-12, 4.5, 72\}$. In the case of categorical inputs, the most common approach is to consider splits of the form $x_{ij} = c_k$ and $x_{ij} \neq c_k$, for each possible class label c_k. Although we could allow for multi-way splits (resulting in non-binary trees), this would result in **data fragmentation**, meaning too little data might "fall" into each subtree, resulting in overfitting.

The function that checks if a node is worth splitting can use several stopping heuristics, such as the following:

- is the reduction in cost too small? Typically we define the gain of using a feature to be a normalized measure of the reduction in cost:

$$\Delta \triangleq \text{cost}(\mathcal{D}) - \left(\frac{|\mathcal{D}_L|}{|\mathcal{D}|} \text{cost}(\mathcal{D}_L) + \frac{|\mathcal{D}_R|}{|\mathcal{D}|} \text{cost}(\mathcal{D}_R) \right) \tag{16.6}$$

- has the tree exceeded the maximum desired depth?
- is the distribution of the response in either $\mathcal{D}_L$ or $\mathcal{D}_R$ sufficiently homogeneous (e.g., all labels are the same, so the distribution is **pure**)?
- is the number of examples in either $\mathcal{D}_L$ or $\mathcal{D}_R$ too small?

All that remains is to specify the cost measure used to evaluate the quality of a proposed split. This depends on whether our goal is regression or classification. We discuss both cases below.

16.2.2.1 Regression cost

In the regression setting, we define the cost as follows:

$$\text{cost}(\mathcal{D}) = \sum_{i \in \mathcal{D}} (y_i - \overline{y})^2 \tag{16.7}$$

where $\bar{y} = \frac{1}{|\mathcal{D}|} \sum_{i \in \mathcal{D}} y_i$ is the mean of the response variable in the specified set of data. Alternatively, we can fit a linear regression model for each leaf, using as inputs the features that were chosen on the path from the root, and then measure the residual error.

16.2.2.2 Classification cost

In the classification setting, there are several ways to measure the quality of a split. First, we fit a multinoulli model to the data in the leaf satisfying the test $X_j < t$ by estimating the class-conditional probabilities as follows:

$$\hat{\pi}_c = \frac{1}{|\mathcal{D}|} \sum_{i \in \mathcal{D}} \mathbb{I}(y_i = c) \tag{16.8}$$

where $\mathcal{D}$ is the data in the leaf. Given this, there are several common error measures for evaluating a proposed partition:

- **Misclassification rate**. We define the most probable class label as $\hat{y}_c = \mathrm{argmax}_c \hat{\pi}_c$. The corresponding error rate is then

$$\frac{1}{|\mathcal{D}|} \sum_{i \in \mathcal{D}} \mathbb{I}(y_i \neq \hat{y}) = 1 - \hat{\pi}_{\hat{y}} \tag{16.9}$$

- **Entropy**, or **deviance**:

$$\mathbb{H}(\hat{\boldsymbol{\pi}}) = -\sum_{c=1}^{C} \hat{\pi}_c \log \hat{\pi}_c \tag{16.10}$$

Note that minimizing the entropy is equivalent to maximizing the **information gain** (Quinlan 1986) between test $X_j < t$ and the class label Y, defined by

$$\mathrm{infoGain}(X_j < t, Y) \triangleq \mathbb{H}(Y) - \mathbb{H}(Y|X_j < t) \tag{16.11}$$

$$= \left(-\sum_c p(y = c) \log p(y = c) \right) \tag{16.12}$$

$$+ \left(\sum_c p(y = c|X_j < t) \log p(c|X_j < t) \right) \tag{16.13}$$

since $\hat{\pi}_c$ is an MLE for the distribution $p(c|X_j < t)$.[1]

---

1. If X_j is categorical, and we use tests of the form $X_j = k$, then taking expectations over values of X_j gives the mutual information between X_j and Y: $\mathbb{E}[\mathrm{infoGain}(X_j, Y)] = \sum_k p(X_j = k)\mathrm{infoGain}(X_j = k, Y) = \mathbb{H}(Y) - \mathbb{H}(Y|X_j) = \mathbb{I}(Y; X_j)$.

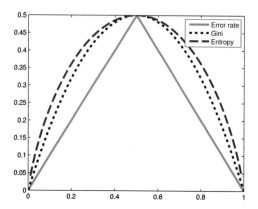

Figure 16.3 Node impurity measures for binary classification. The horizontal axis corresponds to p, the probability of class 1. The entropy measure has been rescaled to pass through (0.5,0.5). Based on Figure 9.3 of (Hastie et al. 2009). Figure generated by `giniDemo`.

- **Gini index**

$$\sum_{c=1}^{C} \hat{\pi}_c(1 - \hat{\pi}_c) = \sum_c \hat{\pi}_c - \sum_c \hat{\pi}_c^2 = 1 - \sum_c \hat{\pi}_c^2 \tag{16.14}$$

This is the expected error rate. To see this, note that $\hat{\pi}_c$ is the probability a random entry in the leaf belongs to class c, and $(1 - \hat{\pi}_c$ is the probability it would be misclassified.

In the two-class case, where $p = \pi_m(1)$, the misclassification rate is $1 - \max(p, 1 - p)$, the entropy is $\mathbb{H}_2(p)$, and the Gini index is $2p(1 - p)$. These are plotted in Figure 16.3. We see that the cross-entropy and Gini measures are very similar, and are more sensitive to changes in class probability than is the misclassification rate. For example, consider a two-class problem with 400 cases in each class. Suppose one split created the nodes (300,100) and (100,300), while the other created the nodes (200,400) and (200,0). Both splits produce a misclassification rate of 0.25. However, the latter seems preferable, since one of the nodes is **pure**, i.e., it only contains one class. The cross-entropy and Gini measures will favor this latter choice.

16.2.2.3   Example

As an example, consider two of the four features from the 3-class iris dataset, shown in Figure 16.4(a). The resulting tree is shown in Figure 16.5(a), and the decision boundaries are shown in Figure 16.4(b). We see that the tree is quite complex, as are the resulting decision boundaries. In Figure 16.5(b), we show that the CV estimate of the error is much higher than the training set error, indicating overfitting. Below we discuss how to perform a tree-pruning stage to simplify the tree.

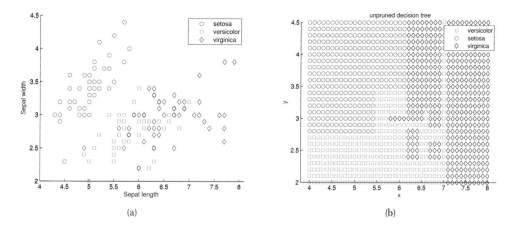

Figure 16.4 (a) Iris data. We only show the first two features, sepal length and sepal width, and ignore petal length and petal width. (b) Decision boundaries induced by the decision tree in Figure 16.5(a).

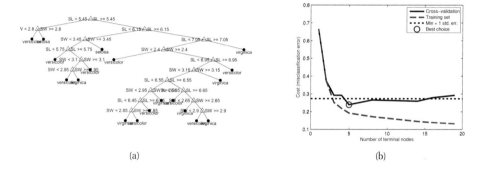

Figure 16.5 (a) Unpruned decision tree for Iris data. (b) Plot of misclassification error rate vs depth of tree. Figure generated by dtreeDemoIris.

16.2.3 Pruning a tree

To prevent overfitting, we can stop growing the tree if the decrease in the error is not sufficient to justify the extra complexity of adding an extra subtree. However, this tends to be too myopic. For example, on the xor data in Figure 14.2(c), it would might never make any splits, since each feature on its own has little predictive power.

The standard approach is therefore to grow a "full" tree, and then to perform **pruning**. This can be done using a scheme that prunes the branches giving the least increase in the error. See (Breiman et al. 1984) for details.

To determine how far to prune back, we can evaluate the cross-validated error on each such subtree, and then pick the tree whose CV error is within 1 standard error of the minimum. This is illustrated in Figure 16.4(b). The point with the minimum CV error corresponds to the simple tree in Figure 16.6(a).

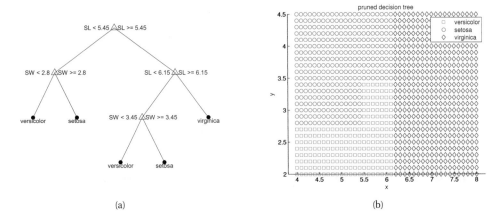

(a)   (b)

Figure 16.6   Pruned decision tree for Iris data. Figure generated by `dtreeDemoIris`.

16.2.4   Pros and cons of trees

CART models are popular for several reasons: they are easy to interpret[2], they can easily handle mixed discrete and continuous inputs, they are insensitive to monotone transformations of the inputs (because the split points are based on ranking the data points), they perform automatic variable selection, they are relatively robust to outliers, they scale well to large data sets, and they can be modified to handle missing inputs.[3]

However, CART models also have some disadvantages. The primary one is that they do not predict very accurately compared to other kinds of model. This is in part due to the greedy nature of the tree construction algorithm. A related problem is that trees are **unstable**: small changes to the input data can have large effects on the structure of the tree, due to the hierarchical nature of the tree-growing process, causing errors at the top to affect the rest of the tree. In frequentist terminology, we say that trees are high variance estimators. We discuss a solution to this below.

16.2.5   Random forests

One way to reduce the variance of an estimate is to average together many estimates. For example, we can train M different trees on different subsets of the data, chosen randomly with

---

2. We can postprocess the tree to derive a series of logical **rules** such as "If $x_1 < 5.45$ then ..." (Quinlan 1990).
3. The standard heuristic for handling missing inputs in decision trees is to look for a series of "backup" variables, which can induce a similar partition to the chosen variable at any given split; these can be used in case the chosen variable is unobserved at test time. These are called **surrogate splits**. This method finds highly correlated features, and can be thought of as learning a local joint model of the input. This has the advantage over a generative model of not modeling the entire joint distribution of inputs, but it has the disadvantage of being entirely ad hoc. A simpler approach, applicable to categorical variables, is to code "missing" as a new value, and then to treat the data as fully observed.

replacement, and then compute the ensemble

$$f(\mathbf{x}) = \sum_{m=1}^{M} \frac{1}{M} f_m(\mathbf{x}) \qquad (16.15)$$

where f_m is the m'th tree. This technique is called **bagging** (Breiman 1996), which stands for "bootstrap aggregating".

Unfortunately, simply re-running the same learning algorithm on different subsets of the data can result in highly correlated predictors, which limits the amount of variance reduction that is possible. The technique known as **random forests** (Breiman 2001a) tries to decorrelate the base learners by learning trees based on a randomly chosen subset of input variables, as well as a randomly chosen subset of data cases. Such models often have very good predictive accuracy (Caruana and Niculescu-Mizil 2006), and have been widely used in many applications (e.g., for body pose recognition using Microsoft's popular kinect sensor (Shotton et al. 2011)).

Bagging is a frequentist concept. It is also possible to adopt a Bayesian approach to learning trees. In particular, (Chipman et al. 1998; Denison et al. 1998; Wu et al. 2007) perform approximate inference over the space of trees (structure and parameters) using MCMC. This reduces the variance of the predictions. We can also perform Bayesian inference over the space of ensembles of trees, which tends to work much better. This is known as **Bayesian adaptive regression trees** or **BART** (Chipman et al. 2010). Note that the cost of these sampling-based Bayesian methods is comparable to the sampling-based random forest method. That is, both approaches are fairly slow to train, but produce high quality classifiers.

Unfortunately, methods that use multiple trees (whether derived from a Bayesian or frequentist standpoint) lose their nice interpretability properties. Fortunately, various post-processing measures can be applied, as discussed in Section 16.8.

16.2.6 CART compared to hierarchical mixture of experts *

An interesting alternative to a decision tree is known as the hierarchical mixture of experts. Figure 11.7(b) gives an illustration where we have two levels of experts. This can be thought of as a probabilistic decision tree of depth 2, since we recursively partition the space, and apply a different expert to each partition. Hastie et al. (Hastie et al. 2009, p331) write that "The HME approach is a promising competitor to CART trees". Some of the advantages include the following:

- The model can partition the input space using any set of nested linear decision boundaries. By contrast, standard decision trees are constrained to use axis-parallel splits.

- The model makes predictions by averaging over all experts. By contrast, in a standard decision tree, predictions are made only based on the model in the corresponding leaf. Since leaves often contain few training examples, this can result in overfitting.

- Fitting an HME involves solving a smooth continuous optimization problem (usually using EM), which is likely to be less prone to local optima than the standard greedy discrete optimization methods used to fit decision trees. For similar reasons, it is computationally easier to "be Bayesian" about the parameters of an HME (see e.g., (Peng et al. 1996; Bishop and Svensén 2003)) than about the structure and parameters of a decision tree (see e.g., (Wu et al. 2007)).

16.3   Generalized additive models

A simple way to create a nonlinear model with multiple inputs is to use a **generalized additive model** (Hastie and Tibshirani 1990), which is a model of the form

$$f(\mathbf{x}) = \alpha + f_1(x_1) + \cdots + f_D(x_D) \tag{16.16}$$

Here each f_j can be modeled by some scatterplot smoother, and $f(\mathbf{x})$ can be mapped to $p(y|\mathbf{x})$ using a link function, as in a GLM (hence the term *generalized* additive model).

If we use regression splines (or some other fixed basis function expansion approach) for the f_j, then each $f_j(x_j)$ can be written as $\boldsymbol{\beta}_j^T \boldsymbol{\phi}_j(x_j)$, so the whole model can be written as $f(\mathbf{x}) = \boldsymbol{\beta}^T \boldsymbol{\phi}(\mathbf{x})$, where $\boldsymbol{\phi}(\mathbf{x}) = [1, \boldsymbol{\phi}_1(x_1), \dots, \boldsymbol{\phi}_D(x_D)]$. However, it is more common to use smoothing splines (Section 15.4.6) for the f_j. In this case, the objective (in the regression setting) becomes

$$J(\alpha, f_1, \dots, f_D) = \sum_{i=1}^{N} \left(y_i - \alpha - \sum_{j=1}^{D} f_j(x_{ij}) \right)^2 + \sum_{j=1}^{D} \lambda_j \int f_j''(t_j)^2 dt_j \tag{16.17}$$

where λ_j is the strength of the regularizer for f_j.

16.3.1   Backfitting

We now discuss how to fit the model using MLE. The constant α is not uniquely identifiable, since we can always add or subtract constants to any of the f_j functions. The convention is to assume $\sum_{i=1}^{N} f_j(x_{ij}) = 0$ for all j. In this case, the MLE for α is just $\hat{\alpha} = \frac{1}{N} \sum_{i=1}^{N} y_i$.

To fit the rest of the model, we can center the responses (by subtracting $\hat{\alpha}$), and then iteratively update each f_j in turn, using as a target vector the residuals obtained by omitting term f_j:

$$\hat{f}_j := \text{smoother}(\{y_i - \sum_{k \neq j} \hat{f}_k(x_{ik})\}_{i=1}^{N}) \tag{16.18}$$

We should then ensure the output is zero mean using

$$\hat{f}_j := \hat{f}_j - \frac{1}{N} \sum_{i=1}^{N} \hat{f}_j(x_{ij}) \tag{16.19}$$

This is called the **backfitting** algorithm (Hastie and Tibshirani 1990). If $\mathbf{X}$ has full column rank, then the above objective is convex (since each smoothing spline is a linear operator, as shown in Section 15.4.2), so this procedure is guaranteed to converge to the global optimum.

In the GLM case, we need to modify the method somewhat. The basic idea is to replace the weighted least squares step of IRLS (see Section 8.3.4) with a weighted backfitting algorithm. In the logistic regression case, each response has weight $s_i = \mu_i(1 - \mu_i)$ associated with it, where $\mu_i = \text{sigm}(\hat{\alpha} + \sum_{j=1}^{D} \hat{f}_j(x_{ij}))$.)

16.3.2 Computational efficiency

Each call to the smoother takes $O(N)$ time, so the total cost is $O(NDT)$, where T is the number of iterations. If we have high-dimensional inputs, fitting a GAM is expensive. One approach is to combine it with a sparsity penalty, see e.g., the **SpAM** (sparse additive model) approach of (Ravikumar et al. 2009). Alternatively, we can use a greedy approach, such as boosting (see Section 16.4.6)

16.3.3 Multivariate adaptive regression splines (MARS)

We can extend GAMs by allowing for interaction effects. In general, we can create an ANOVA-like decomposition:

$$f(\mathbf{x}) = \beta_0 + \sum_{j=1}^{D} f_j(x_j) + \sum_{j,k} f_{jk}(x_j, x_k) + \sum_{j,k,l} f_{jkl}(x_j, x_k, x_l) + \cdots \quad (16.20)$$

Of course, we cannot allow for too many higher-order interactions, because there will be too many parameters to fit.

It is common to use greedy search to decide which variables to add. The **multivariate adaptive regression splines** or **MARS** algorithm is one example of this (Hastie et al. 2009, Sec9.4). It fits models of the form in Equation 16.20, where it uses a tensor product basis of regression splines to represent the multidimensional regression functions. For example, for 2d input, we might use

$$\begin{aligned}
f(x_1, x_2) &= \beta_0 + \sum_m \beta_{1m}(x_1 - t_{1m})_+ \\
&+ \sum_m \beta_{2m}(t_{2m} - x_2)_+ + \sum_m \beta_{12m}(x_1 - t_{1m})_+(t_{2m} - x_2)_+
\end{aligned} \quad (16.21)$$

To create such a function, we start with a set of candidate basis functions of the form

$$\mathcal{C} = \{(x_j - t)_+, (t - x_j)_+ : t \in \{x_{1j}, \dots, x_{Nj}\}, j = 1, \dots, D\} \quad (16.22)$$

These are 1d linear splines where the knots are at all the observed values for that variable. We consider splines sloping up in both directions; this is called a **reflecting pair**. See Figure 16.7(a).

Let $\mathcal{M}$ represent the current set of basis functions. We initialize by using $\mathcal{M} = \{1\}$. We consider creating a new basis function pair by multplying an $h_m \in \mathcal{M}$ with one of the reflecting pairs in $\mathcal{C}$. For example, we might initially get

$$f(\mathbf{x}) = 25 - 4(x_1 - 5)_+ + 20(5 - x_1)_+ \quad (16.23)$$

obtained by multiplying $h_0(\mathbf{x}) = 1$ with a reflecting pair involving x_1 with knot $t = 5$. This pair is added to $\mathcal{M}$. See Figure 16.7(b). At the next step, we might create a model such as

$$\begin{aligned}
f(\mathbf{x}) &= = 2 - 2(x_1 - 5)_+ + 3(5 - x_1)_+ \\
&- (x_2 - 10)_+ \times (5 - x_1) + -1.2(10 - x_2)_+ \times (5 - x_1)_+
\end{aligned} \quad (16.24)$$

obtained by multiplying $(5 - x_1)_+$ from $\mathcal{M}$ by the new reflecting pair $(x_2 - 10)_+$ and $(10 - x_2)_+$. This new function is shown in Figure 16.7(c).

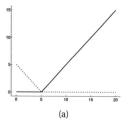
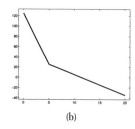
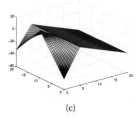

(a)   (b)   (c)

Figure 16.7 (a) Linear spline function with a knot at 5. Solid blue: $(x-5)_+$. Dotted red: $(5-x)_+$. (b) A MARS model in 1d given by Equation 16.23. (c) A simple MARS model in 2d given by Equation 16.24. Figure generated by `marsDemo`.

We proceed in this way until the model becomes very large. (We may impose an upper bound on the order of interactions.) Then we prune backwards, at each step eliminating the basis function that causes the smallest increase in the residual error, until the CV error stops improving.

The whole procedure is closely related to CART. To see this, suppose we replace the piecewise linear basis functions by step functions $\mathbb{I}(x_j > t)$ and $\mathbb{I}(x_j < t)$. Multiplying by a pair of reflected step functions is equivalent to splitting a node. Now suppose we impose the constraint that once a variable is involved in a multiplication by a candidate term, that variable gets replaced by the interaction, so the original variable is no longer available. This ensures that a variable can not be split more than once, thus guaranteeing that the resulting model can be represented as a tree. In this case, the MARS growing strategy is the same as the CART growing strategy.

16.4   Boosting

Boosting (Schapire and Freund 2012) is a greedy algorithm for fitting adaptive basis-function models of the form in Equation 16.3, where the ϕ_m are generated by an algorithm called a **weak learner** or a **base learner**. The algorithm works by applying the weak learner sequentially to weighted versions of the data, where more weight is given to examples that were misclassified by earlier rounds.

This weak learner can be any classification or regression algorithm, but it is common to use a CART model. In 1998, the late Leo Breiman called boosting, where the weak learner is a shallow decision tree, the "best off-the-shelf classifier in the world" (Hastie et al. 2009, p340). This is supported by an extensive empirical comparison of 10 different classifiers in (Caruana and Niculescu-Mizil 2006), who showed that boosted decision trees were the best both in terms of misclassification error and in terms of producing well-calibrated probabilities, as judged by ROC curves. (The second best method was random forests, invented by Breiman; see Section 16.2.5.) By contrast, single decision trees performed very poorly.

Boosting was originally derived in the computational learning theory literature (Schapire 1990; Freund and Schapire 1996), where the focus is binary classification. In these papers, it was proved that one could boost the performance (on the training set) of any weak learner arbitrarily

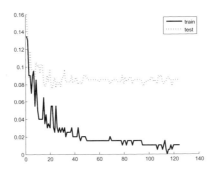

Figure 16.8 Performance of adaboost using a decision stump as a weak learner on the data in Figure 16.10. Training (solid blue) and test (dotted red) error vs number of iterations. Figure generated by `boostingDemo`, written by Richard Stapenhurst.

high, provided the weak learner could always perform slightly better than chance. For example, in Figure 16.8, we plot the training and test error for boosted decision stumps on a 2d dataset shown in Figure 16.10. We see that the training set error rapidly goes to near zero. What is more surprising is that the test set error continues to decline even after the training set error has reached zero (although the test set error will eventually go up). Thus boosting is very resistant to overfitting. (Boosted decision stumps form the basis of a very successful face detector (Viola and Jones 2001), which was used to generate the results in Figure 1.6, and which is used in many digital cameras.)

In view of its stunning empirical success, statisticians started to become interested in this method. Breiman (Breiman 1998) showed that boosting can be interpreted as a form of *gradient descent in function space*. This view was then extended in (Friedman et al. 2000), who showed how boosting could be extended to handle a variety of loss functions, including for regression, robust regression, Poisson regression, etc. In this section, we shall present this statistical interpretation of boosting, drawing on the reviews in (Buhlmann and Hothorn 2007) and (Hastie et al. 2009, ch10), which should be consulted for further details.

16.4.1 Forward stagewise additive modeling

The goal of boosting is to solve the following optimization problem:

$$\min_{f} \sum_{i=1}^{N} L(y_i, f(\mathbf{x}_i)) \tag{16.25}$$

and $L(y, \hat{y})$ is some loss function, and f is assumed to be an ABM model as in Equation 16.3. Common choices for the loss function are listed in Table 16.1.

If we use squared error loss, the optimal estimate is given by

$$f^*(\mathbf{x}) = \underset{f(\mathbf{x})}{\operatorname{argmin}} \, \mathbb{E}_{y|\mathbf{x}} \left[(Y - f(\mathbf{x}))^2 \right] = \mathbb{E}\left[Y|\mathbf{x}\right] \tag{16.26}$$

Name	Loss	Derivative	f^*	Algorithm
Squared error	$\frac{1}{2}(y_i - f(\mathbf{x}_i))^2$	$y_i - f(\mathbf{x}_i)$	$\mathbb{E}\left[y|\mathbf{x}_i\right]$	L2Boosting
Absolute error	$|y_i - f(\mathbf{x}_i)|$	$\text{sgn}(y_i - f(\mathbf{x}_i))$	$\text{median}(y|\mathbf{x}_i)$	Gradient boosting
Exponential loss	$\exp(-\tilde{y}_i f(\mathbf{x}_i))$	$-\tilde{y}_i \exp(-\tilde{y}_i f(\mathbf{x}_i))$	$\frac{1}{2}\log\frac{\pi_i}{1-\pi_i}$	AdaBoost
Logloss	$\log(1 + e^{-\tilde{y}_i f_i})$	$y_i - \pi_i$	$\frac{1}{2}\log\frac{\pi_i}{1-\pi_i}$	LogitBoost

Table 16.1 Some commonly used loss functions, their gradients, their population minimizers f^*, and some algorithms to minimize the loss. For binary classification problems, we assume $\tilde{y}_i \in \{-1, +1\}$, $y_i \in \{0, 1\}$ and $\pi_i = \text{sigm}(2f(\mathbf{x}_i))$. For regression problems, we assume $y_i \in \mathbb{R}$. Adapted from (Hastie et al. 2009, p360) and (Buhlmann and Hothorn 2007, p483).

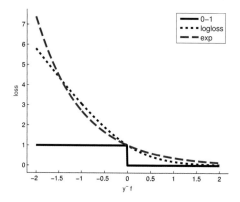

Figure 16.9 Illustration of various loss functions for binary classification. The horizontal axis is the margin $y\eta$, the vertical axis is the loss. The log loss uses log base 2. Figure generated by `hingeLossPlot`.

as we showed in Section 5.7.1.3. Of course, this cannot be computed in practice since it requires knowing the true conditional distribution $p(y|\mathbf{x})$. Hence this is sometimes called the **population minimizer**, where the expectation is interpreted in a frequentist sense. Below we will see that boosting will try to approximate this conditional expectation.

For binary classification, the obvious loss is 0-1 loss, but this is not differentiable. Instead it is common to use logloss, which is a convex upper bound on 0-1 loss, as we showed in Section 6.5.5. In this case, one can show that the optimal estimate is given by

$$f^*(\mathbf{x}) = \frac{1}{2}\log\frac{p(\tilde{y} = 1|\mathbf{x})}{p(\tilde{y} = -1|\mathbf{x})} \tag{16.27}$$

where $\tilde{y} \in \{-1, +1\}$. One can generalize this framework to the multiclass case, but we will not discuss that here.

An alternative convex upper bound is **exponential loss**, defined by

$$L(\tilde{y}, f) = \exp(-\tilde{y}f) \tag{16.28}$$

See Figure 16.9 for a plot. This will have some computational advantages over the logloss, to be discussed below. It turns out that the optimal estimate for this loss is also $f^*(\mathbf{x}) =$

$\frac{1}{2} \log \frac{p(\tilde{y}=1|\mathbf{x})}{p(\tilde{y}=-1|\mathbf{x})}$. To see this, we can just set the derivative of the expected loss (for each $\mathbf{x}$) to zero:

$$\frac{\partial}{\partial f(\mathbf{x})} \mathbb{E} \left[e^{-\tilde{y}f(\mathbf{x})} | \mathbf{x} \right] = \frac{\partial}{\partial f(\mathbf{x})} [p(\tilde{y}=1|\mathbf{x})e^{-f(\mathbf{x})} + p(\tilde{y}=-1|\mathbf{x})e^{f(\mathbf{x})}] \tag{16.29}$$

$$= -p(\tilde{y}=1|\mathbf{x})e^{-f(\mathbf{x})} + p(\tilde{y}=-1|\mathbf{x})e^{f(\mathbf{x})} \tag{16.30}$$

$$= 0 \Rightarrow \frac{p(\tilde{y}=1|\mathbf{x})}{p(\tilde{y}=-1|\mathbf{x})} = e^{2f(\mathbf{x})} \tag{16.31}$$

So in both cases, we can see that boosting should try to approximate (half) the log-odds ratio.

Since finding the optimal f is hard, we shall tackle it sequentially. We initialise by defining

$$f_0(\mathbf{x}) = \arg\min_{\boldsymbol{\gamma}} \sum_{i=1}^{N} L(y_i, f(\mathbf{x}_i; \boldsymbol{\gamma})) \tag{16.32}$$

For example, if we use squared error, we can set $f_0(\mathbf{x}) = \overline{y}$, and if we use log-loss or exponential loss , we can set $f_0(\mathbf{x}) = \frac{1}{2} \log \frac{\hat{\pi}}{1-\hat{\pi}}$, where $\hat{\pi} = \frac{1}{N} \sum_{i=1}^{N} \mathbb{I}(y_i = 1)$. We could also use a more powerful model for our baseline, such as a GLM.

Then at iteration m, we compute

$$(\beta_m, \boldsymbol{\gamma}_m) = \operatorname*{argmin}_{\beta, \boldsymbol{\gamma}} \sum_{i=1}^{N} L(y_i, f_{m-1}(\mathbf{x}_i) + \beta\phi(\mathbf{x}_i; \boldsymbol{\gamma})) \tag{16.33}$$

and then we set

$$f_m(\mathbf{x}) = f_{m-1}(\mathbf{x}) + \beta_m\phi(\mathbf{x}; \boldsymbol{\gamma}_m) \tag{16.34}$$

The key point is that we do not go back and adjust earlier parameters. This is why the method is called **forward stagewise additive modeling**.

We continue this for a fixed number of iterations M. In fact M is the main tuning parameter of the method. Often we pick it by monitoring the performance on a separate validation set, and then stopping once performance starts to decrease; this is called **early stopping**. Alternatively, we can use model selection criteria such as AIC or BIC (see e.g., (Buhlmann and Hothorn 2007) for details).

In practice, better (test set) performance can be obtained by performing "partial updates" of the form

$$f_m(\mathbf{x}) = f_{m-1}(\mathbf{x}) + \nu\beta_m\phi(\mathbf{x}; \boldsymbol{\gamma}_m) \tag{16.35}$$

Here $0 < \nu \leq 1$ is a step-size parameter. In practice it is common to use a small value such as $\nu = 0.1$. This is called **shrinkage**.

Below we discuss how to solve the suproblem in Equation 16.33. This will depend on the form of loss function. However, it is independent of the form of weak learner.

16.4.2 L2boosting

Suppose we used squared error loss. Then at step m the loss has the form

$$L(y_i, f_{m-1}(\mathbf{x}_i) + \beta\phi(\mathbf{x}_i; \boldsymbol{\gamma})) = (r_{im} - \phi(\mathbf{x}_i; \boldsymbol{\gamma}))^2 \tag{16.36}$$

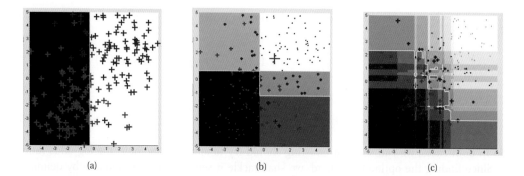

(a)   (b)   (c)

Figure 16.10 Example of adaboost using a decision stump as a weak learner. The degree of blackness represents the confidence in the red class. The degree of whiteness represents the confidence in the blue class. The size of the datapoints represents their weight. Decision boundary is in yellow. (a) After 1 round. (b) After 3 rounds. (c) After 120 rounds. Figure generated by `boostingDemo`, written by Richard Stapenhurst.

where $r_{im} \triangleq y_i - f_{m-1}(\mathbf{x}_i)$ is the current residual, and we have set $\beta = 1$ without loss of generality. Hence we can find the new basis function by using the weak learner to predict $\mathbf{r}_m$. This is called **L2boosting**, or **least squares boosting** (Buhlmann and Yu 2003). In Section 16.4.6, we will see that this method, with a suitable choice of weak learner, can be made to give the same results as LARS, which can be used to perform variable selection (see Section 13.4.2).

16.4.3   AdaBoost

Consider a binary classification problem with exponential loss. At step m we have to minimize

$$L_m(\phi) = \sum_{i=1}^{N} \exp[-\tilde{y}_i(f_{m-1}(\mathbf{x}_i) + \beta\phi(\mathbf{x}_i))] = \sum_{i=1}^{N} w_{i,m} \exp(-\beta \tilde{y}_i \phi(\mathbf{x}_i)) \tag{16.37}$$

where $w_{i,m} \triangleq \exp(-\tilde{y}_i f_{m-1}(\mathbf{x}_i))$ is a weight applied to datacase i, and $\tilde{y}_i \in \{-1, +1\}$. We can rewrite this objective as follows:

$$L_m = e^{-\beta} \sum_{\tilde{y}_i = \phi(\mathbf{x}_i)} w_{i,m} + e^{\beta} \sum_{\tilde{y}_i \neq \phi(\mathbf{x}_i)} w_{i,m} \tag{16.38}$$

$$= (e^{\beta} - e^{-\beta}) \sum_{i=1}^{N} w_{i,m} \mathbb{I}(\tilde{y}_i \neq \phi(\mathbf{x}_i)) + e^{-\beta} \sum_{i=1}^{N} w_{i,m} \tag{16.39}$$

Consequently the optimal function to add is

$$\phi_m = \underset{\phi}{\operatorname{argmin}} \, w_{i,m} \mathbb{I}(\tilde{y}_i \neq \phi(\mathbf{x}_i)) \tag{16.40}$$

This can be found by applying the weak learner to a weighted version of the dataset, with weights $w_{i,m}$. Subsituting ϕ_m into L_m and solving for β we find

$$\beta_m = \frac{1}{2} \log \frac{1 - \operatorname{err}_m}{\operatorname{err}_m} \tag{16.41}$$

where

$$\text{err}_m = \frac{\sum_{i=1}^{N} w_i \mathbb{I}(\tilde{y}_i \neq \phi_m(\mathbf{x}_i))}{\sum_{i=1}^{N} w_{i,m}} \tag{16.42}$$

The overall update becomes

$$f_m(\mathbf{x}) = f_{m-1}(\mathbf{x}) + \beta_m \phi_m(\mathbf{x}) \tag{16.43}$$

With this, the weights at the next iteration become

$$
\begin{align}
w_{i,m+1} &= w_{i,m} e^{-\beta_m \tilde{y}_i \phi_m(\mathbf{x}_i)} \tag{16.44} \\
&= w_{i,m} e^{\beta_m (2\mathbb{I}(\tilde{y}_i \neq \phi_m(\mathbf{x}_i))-1)} \tag{16.45} \\
&= w_{i,m} e^{2\beta_m \mathbb{I}(\tilde{y}_i \neq \phi_m(\mathbf{x}_i))} e^{-\beta_m} \tag{16.46}
\end{align}
$$

where we exploited the fact that $-\tilde{y}_i \phi_m(\mathbf{x}_i) = -1$ if $\tilde{y}_i = \phi_m(\mathbf{x}_i)$ and $-\tilde{y}_i \phi_m(\mathbf{x}_i) = +1$ otherwise. Since $e^{-\beta_m}$ will cancel out in the normalization step, we can drop it. The result is the algorithm shown in Algorithm 16.2, known **Adaboost.M1**.[4]

An example of this algorithm in action, using decision stumps as the weak learner, is given in Figure 16.10. We see that after many iterations, we can "carve out" a complex decision boundary. What is rather surprising is that AdaBoost is very slow to overfit, as is apparent in Figure 16.8. See Section 16.4.8 for a discussion of this point.

Algorithm 16.2: Adaboost.M1, for binary classification with exponential loss

---

1 $w_i = 1/N$;
2 **for** $m = 1 : M$ **do**
3 $\quad$ Fit a classifier $\phi_m(\mathbf{x})$ to the training set using weights $\mathbf{w}$;
4 $\quad$ Compute $\text{err}_m = \frac{\sum_{i=1}^{N} w_{i,m} \mathbb{I}(\tilde{y}_i \neq \phi_m(\mathbf{x}_i))}{\sum_{i=1}^{N} w_{i,m}}$;
5 $\quad$ Compute $\alpha_m = \log[(1 - \text{err}_m)/\text{err}_m]$;
6 $\quad$ Set $w_i \leftarrow w_i \exp[\alpha_m \mathbb{I}(\tilde{y}_i \neq \phi_m(\mathbf{x}_i))]$;
7 Return $f(\mathbf{x}) = \text{sgn}\left[\sum_{m=1}^{M} \alpha_m \phi_m(\mathbf{x})\right]$;

16.4.4 LogitBoost

The trouble with exponential loss is that it puts a lot of weight on misclassified examples, as is apparent from the exponential blowup on the left hand side of Figure 16.9. This makes the method very sensitive to outliers (mislabeled examples). In addition, $e^{-\tilde{y}f}$ is not the logarithm of any pmf for binary variables $\tilde{y} \in \{-1, +1\}$; consequently we cannot recover probability estimates from $f(\mathbf{x})$.

---

4. In (Friedman et al. 2000), this is called **discrete AdaBoost**, since it assumes that the base classifier ϕ_m returns a binary class label. If ϕ_m returns a probability instead, a modified algorithm, known as **real AdaBoost**, can be used. See (Friedman et al. 2000) for details.

A natural alternative is to use logloss instead. This only punishes mistakes linearly, as is clear from Figure 16.9. Furthermore, it means that we will be able to extract probabilities from the final learned function, using

$$p(y = 1|\mathbf{x}) = \frac{e^{f(\mathbf{x})}}{e^{-f(\mathbf{x})} + e^{f(\mathbf{x})}} = \frac{1}{1 + e^{-2f(\mathbf{x})}} \tag{16.47}$$

The goal is to minimze the expected log-loss, given by

$$L_m(\phi) \quad = \quad \sum_{i=1}^{N} \log\left[1 + \exp\left(-2\tilde{y}_i(f_{m-1}(\mathbf{x}) + \phi(\mathbf{x}_i))\right)\right] \tag{16.48}$$

By performing a Newton upate on this objective (similar to IRLS), one can derive the algorithm shown in Algorithm 16.3. This is known as **logitBoost** (Friedman et al. 2000). It can be generalized to the multi-class setting, as explained in (Friedman et al. 2000).

Algorithm 16.3: LogitBoost, for binary classification with log-loss

---

1 $w_i = 1/N$, $\pi_i = 1/2$;
2 **for** $m = 1 : M$ **do**
3   Compute the working response $z_i = \frac{y_i^* - \pi_i}{\pi_i(1 - \pi_i)}$;
4   Compute the weights $w_i = \pi_i(1 - \pi_i)$;
5      $\phi_m = \operatorname{argmin}_\phi \sum_{i=1}^{N} w_i(z_i - \phi(\mathbf{x}_i))^2$;
6   Update $f(\mathbf{x}) \leftarrow f(\mathbf{x}) + \frac{1}{2}\phi_m(\mathbf{x})$;
7   Compute $\pi_i = 1/(1 + \exp(-2f(\mathbf{x}_i)))$;
8 Return $f(\mathbf{x}) = \operatorname{sgn}\left[\sum_{m=1}^{M} \phi_m(\mathbf{x})\right]$;

16.4.5 Boosting as functional gradient descent

Rather than deriving new versions of boosting for every different loss function, it is possible to derive a generic version, known as **gradient boosting** (Friedman 2001; Mason et al. 2000). To explain this, imagine minimizing

$$\hat{\mathbf{f}} = \operatorname*{argmin}_{\mathbf{f}} L(\mathbf{f}) \tag{16.49}$$

where $\mathbf{f} = (f(\mathbf{x}_1), \ldots, f(\mathbf{x}_N))$ are the "parameters". We will solve this stagewise, using gradient descent. At step m, let $\mathbf{g}_m$ be the gradient of $L(\mathbf{f})$ evaluated at $\mathbf{f} = \mathbf{f}_{m-1}$:

$$g_{im} = \left[\frac{\partial L(y_i, f(\mathbf{x}_i))}{\partial f(\mathbf{x}_i)}\right]_{f = f_{m-1}} \tag{16.50}$$

Gradients of some common loss functions are given in Table 16.1. We then make the update

$$\mathbf{f}_m = \mathbf{f}_{m-1} - \rho_m \mathbf{g}_m \tag{16.51}$$

where ρ_m is the step length, chosen by

$$\rho_m = \underset{\rho}{\mathrm{argmin}}\, L(\mathbf{f}_{m-1} - \rho \mathbf{g}_m) \tag{16.52}$$

This is called **functional gradient descent**.

In its current form, this is not much use, since it only optimizes f at a fixed set of N points, so we do not learn a function that can generalize. However, we can modify the algorithm by fitting a weak learner to approximate the negative gradient signal. That is, we use this update

$$\boldsymbol{\gamma}_m = \underset{\boldsymbol{\gamma}}{\mathrm{argmin}} \sum_{i=1}^N (-g_{im} - \phi(\mathbf{x}_i; \boldsymbol{\gamma}))^2 \tag{16.53}$$

The overall algorithm is summarized in Algorithm 16.4. (We have omitted the line search step, which is not strictly necessary, as argued in (Buhlmann and Hothorn 2007).)

Algorithm 16.4: Gradient boosting

1 Initialize $f_0(\mathbf{x}) = \mathrm{argmin}_{\boldsymbol{\gamma}} \sum_{i=1}^N L(y_i, \phi(\mathbf{x}_i; \boldsymbol{\gamma}))$;
2 **for** $m = 1 : M$ **do**
3   Compute the gradient residual using $r_{im} = -\left[\frac{\partial L(y_i, f(\mathbf{x}_i))}{\partial f(\mathbf{x}_i)}\right]_{f(\mathbf{x}_i)=f_{m-1}(\mathbf{x}_i)}$;
4   Use the weak learner to compute $\boldsymbol{\gamma}_m$ which minimizes $\sum_{i=1}^N (r_{im} - \phi(\mathbf{x}_i; \boldsymbol{\gamma}_m))^2$;
5   Update $f_m(\mathbf{x}) = f_{m-1}(\mathbf{x}) + \nu \phi(\mathbf{x}; \boldsymbol{\gamma}_m)$;
6 Return $f(\mathbf{x}) = f_M(\mathbf{x})$

If we apply this algorithm using squared loss, we recover L2Boosting. If we apply this algorithm to log-loss, we get an algorithm known as **BinomialBoost** (Buhlmann and Hothorn 2007). The advantage of this over LogitBoost is that it does not need to be able to do weighted fitting: it just applies any black-box regression model to the gradient vector. Also, it is relatively easy to extend to the multi-class case (see (Hastie et al. 2009, p387)). We can also apply this algorithm to other loss functions, such as the Huber loss (Section 7.4), which is more robust to outliers than squared error loss.

16.4.6 Sparse boosting

Suppose we use as our weak learner the following algorithm: search over all possible variables $j = 1 : D$, and pick the one $j(m)$ that best predicts the residual vector:

$$j(m) = \underset{j}{\mathrm{argmin}} \sum_{i=1}^N (r_{im} - \hat{\beta}_{jm} x_{ij})^2 \tag{16.54}$$

$$\hat{\beta}_{jm} = \frac{\sum_{i=1}^N x_{ij} r_{im}}{\sum_{i=1}^N x_{ij}^2} \tag{16.55}$$

$$\phi_m(\mathbf{x}) = \hat{\beta}_{j(m),m}\, x_{j(m)} \tag{16.56}$$

This method, which is known as **sparse boosting** (Buhlmann and Yu 2006), is identical to the matching pursuit algorithm discussed in Section 13.2.3.1.

It is clear that this will result in a sparse estimate, at least if M is small. To see this, let us rewrite the update as follows:

$$\boldsymbol{\beta}_m := \boldsymbol{\beta}_{m-1} + \nu(0, \ldots, 0, \hat{\beta}_{j(m),m}, 0, \ldots, 0) \tag{16.57}$$

where the non-zero entry occurs in location $j(m)$. This is known as **forward stagewise linear regression** (Hastie et al. 2009, p608), which becomes equivalent to the LAR algorithm discussed in Section 13.4.2 as $\nu \to 0$. Increasing the number of steps m in boosting is analogous to decreasing the regularization penalty λ. If we modify boosting to allow some variable deletion steps (Zhao and Yu 2007), we can make it equivalent to the LARS algorithm, which computes the full regularization path for the lasso problem. The same algorithm can be used for sparse logistic regression, by simply modifying the residual to be the appropriate negative gradient.

Now consider a weak learner that is similar to the above, except it uses a smoothing spline instead of linear regression when mapping from x_j to the residual. The result is a sparse generalized additive model (see Section 16.3). It can obviously be extended to pick pairs of variables at a time. The resulting method often works much better than MARS (Buhlmann and Yu 2006).

16.4.7 Multivariate adaptive regression trees (MART)

It is quite common to use CART models as weak learners. It is usually advisable to use a shallow tree, so that the variance is low. Even though the bias will be high (since a shallow tree is likely to be far from the "truth"), this will compensated for in subsequent rounds of boosting.

The height of the tree is an additional tuning parameter (in addition to M, the number of rounds of boosting, and ν, the shrinkage factor). Suppose we restrict to trees with J leaves. If $J = 2$, we get a stump, which can only split on a single variable. If $J = 3$, we allow for two-variable interactions, etc. In general, it is recommended (e.g., in (Hastie et al. 2009, p363) and (Caruana and Niculescu-Mizil 2006)) to use $J \approx 6$.

If we combine the gradient boosting algorithm with (shallow) regression trees, we get a model known as **MART**, which stands for "multivariate adaptive regression trees". This actually includes a slight refinement to the basic gradient boosting algorithm: after fitting a regression tree to the residual (negative gradient), we re-estimate the parameters at the leaves of the tree to minimize the loss:

$$\gamma_{jm} = \underset{\gamma}{\operatorname{argmin}} \sum_{x_i \in R_{jm}} L(y_i, f_{m-1}(\mathbf{x}_i) + \gamma) \tag{16.58}$$

where R_{jm} is the region for leaf j in the m'th tree, and γ_{jm} is the corresponding parameter (the mean response of y for regression problems, or the most probable class label for classification problems).

16.4.8 Why does boosting work so well?

We have seen that boosting works very well, especially for classifiers. There are two main reasons for this. First, it can be seen as a form of ℓ_1 regularization, which is known to help prevent overfitting by eliminating "irrelevant" features. To see this, imagine pre-computing all possible weak-learners, and defining a feature vector of the form $\phi(\mathbf{x}) = [\phi_1(\mathbf{x}), \ldots, \phi_K(\mathbf{x})]$. We could use ℓ_1 regularization to select a subset of these. Alternatively we can use boosting, where at each step, the weak learner creates a new ϕ_k on the fly. It is possible to combine boosting and ℓ_1 regularization, to get an algorithm known as **L1-Adaboost** (Duchi and Singer 2009). Essentially this method greedily adds the best features (weak learners) using boosting, and then prunes off irrelevant ones using ℓ_1 regularization.

Another explanation has to do with the concept of margin, which we introduced in Section 14.5.2.2. (Schapire et al. 1998; Ratsch et al. 2001) proved that AdaBoost maximizes the margin on the training data. (Rosset et al. 2004) generalized this to other loss functions, such as log-loss.

16.4.9 A Bayesian view

So far, our presentation of boosting has been very frequentist, since it has focussed on greedily minimizing loss functions. A likelihood interpretation of the algorithm was given in (Neal and MacKay 1998; Meek et al. 2002). The idea is to consider a mixture of experts model of the form

$$p(y|\mathbf{x}, \boldsymbol{\theta}) = \sum_{m=1}^{M} \pi_m p(y|\mathbf{x}, \boldsymbol{\gamma}_m) \tag{16.59}$$

where each expert $p(y|\mathbf{x}, \boldsymbol{\gamma}_m)$ is like a weak learner. We usually fit all M experts at once using EM, but we can imagine a sequential scheme, whereby we only update the parameters for one expert at a time. In the E step, the posterior responsibilities will reflect how well the existing experts explain a given data point; if this is a poor fit, these data points will have more influence on the next expert that is fitted. (This view naturally suggest a way to use a boosting-like algorithm for unsupervised learning: we simply sequentially fit mixture models, instead of mixtures of experts.)

Notice that this is a rather "broken" MLE procedure, since it never goes back to update the parameters of an old expert. Similarly, if boosting ever wants to change the weight assigned to a weak learner, the only way to do this is to add the weak learner again with a new weight. This can result in unnecessarily large models. By contrast, the BART model (Chipman et al. 2006, 2010) uses a Bayesian version of backfitting to fit a small sum of weak learners (typically trees).

16.5 Feedforward neural networks (multilayer perceptrons)

A (feedforward) **neural network**, aka **multi-layer perceptron** (**MLP**), is a series of logistic regression models stacked on top of each other, with the final layer being either another logistic regression or a linear regression model, depending on whether we are solving a classification or regression problem. For example, if we have two layers, and we are solving a regression problem, the model has the form

$$
\begin{align}
p(y|\mathbf{x}, \boldsymbol{\theta}) &= \mathcal{N}(y|\mathbf{w}^T \mathbf{z}(\mathbf{x}), \sigma^2) \tag{16.60} \\
\mathbf{z}(\mathbf{x}) &= g(\mathbf{V}\mathbf{x}) = [g(\mathbf{v}_1^T \mathbf{x}), \ldots, g(\mathbf{v}_H^T \mathbf{x})] \tag{16.61}
\end{align}
$$

where $\mathbf{v}_j = \mathbf{V}_{j,:}$ is the j'th row of $\mathbf{V}$. Here $\mathbf{z}(\mathbf{x}) = \phi(\mathbf{x}, \mathbf{V})$ is called the **hidden layer** of size H; this is a deterministic function of the input, created by applying a linear mapping to each

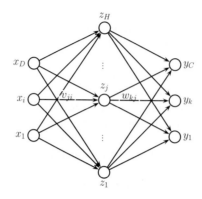

Figure 16.11   A neural network with one hidden layer.

input, and passing the result through a non-linear **activation** or **transfer function** g. This is commonly the logistic function, $g(u) = \text{sigm}(u)$ or the tanh function; more recently people have got better results using **rectified linear units** or ReLU (Glorot et al. 2011), which have the form $\max(0, u)$. It is important that g be non-linear, otherwise the whole model collapses into a large linear regression model of the form $y = \mathbf{w}^T(\mathbf{V}\mathbf{x})$.

To handle binary classification, we pass the output through a sigmoid, as in a GLM:

$$p(y|\mathbf{x}, \boldsymbol{\theta}) = \text{Ber}(y|\text{sigm}(\mathbf{w}^T\mathbf{z}(\mathbf{x}))) \tag{16.62}$$

We can easily extend the MLP to predict multiple outputs. For example, in the regression case, we have

$$p(\mathbf{y}|\mathbf{x}, \boldsymbol{\theta}) = \mathcal{N}(\mathbf{y}|\mathbf{W}\ \phi(\mathbf{x}, \mathbf{V}), \sigma^2\mathbf{I}) \tag{16.63}$$

See Figure 16.11 for an illustration. If we add **mutual inhibition** arcs between the output units, ensuring that only one of them turns on, we can enforce a sum-to-one constraint, which can be used for multi-class classification. The resulting model has the form

$$p(y|\mathbf{x}, \boldsymbol{\theta}) \quad = \quad \text{Cat}(y|\mathcal{S}(\mathbf{W}\mathbf{z}(\mathbf{x}))) \tag{16.64}$$

where $\mathbf{W}_{k,:}$ is the weight vector for class k. Obviously we could also make neural networks for Poisson regression, etc., just as in a GLM.

One can show that an MLP is a **universal approximator**, meaning it can model any suitably smooth function, given enough hidden units, to any desired level of accuracy (Hornik 1991). One can either make the model be "wide", or "deep"; the latter has some advantages, as discussed in Section 28.3.

16.5.1   Convolutional neural networks

The purpose of the hidden units is to learn non-linear combinations of the original inputs; this is called **feature extraction** or **feature construction**. These hidden features are then passed as

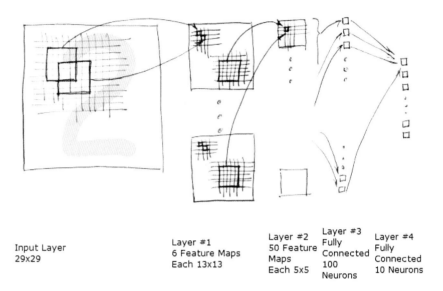

Input Layer
29x29

Layer #1
6 Feature Maps
Each 13x13

Layer #2
50 Feature
Maps
Each 5x5

Layer #3
Fully
Connected
100
Neurons

Layer #4
Fully
Connected
10 Neurons

Figure 16.12   The convolutional neural network from (Simard et al. 2003). Source: `http://www.codep roject.com/KB/library/NeuralNetRecognition.aspx` . Used with kind permission of Mike O'Neill.

input to the final GLM. This approach is particularly useful for problems where the original input features are not very individually informative. For example, each pixel in an image is not very informative; it is the combination of pixels that tells us what objects are present. Conversely, for a task such as document classification using a bag of words representation, each feature (word count) *is* informative on its own, so extracting "higher order" features is less important. Not suprisingly, then, much of the work in neural networks has been motivated by visual pattern recognition (e.g., (LeCun et al. 1989)), although they have also been applied to other types of data, including text (e.g., (Collobert and Weston 2008)).

A form of MLP which is particularly well suited to 1d signals like speech or text, or 2d signals like images, is the **convolutional neural network**. This is an MLP in which the hidden units have local **receptive fields** (as in the primary visual cortex), and in which the weights are **tied** or shared across the image, in order to reduce the number of parameters. Intuitively, the effect of such spatial parameter tying is that any useful features that are "discovered" in some portion of the image can be re-used everywhere else without having to be independently learned. The resulting network then exhibits **translation invariance**, meaning it can classify patterns no matter where they occur inside the input image.

Figure 16.12 gives an example of a convolutional network, designed by Simard and colleagues (Simard et al. 2003), with 5 layers (4 layers of adjustable parameters) designed to classify 29×29 gray-scale images of handwritten digits from the MNIST dataset (see Section 1.2.1.3). In layer 1, we have 6 **feature maps** each of which has size 13×13. Each hidden node in one of these feature maps is computed by convolving the image with a 5×5 weight matrix (sometimes called a kernel), adding a bias, and then passing the result through some form of nonlinearity. There are therefore $13 \times 13 \times 6 = 1014$ neurons in Layer 1, and $(5 \times 5 + 1) \times 6 = 156$ weights. (The "+1" is for the bias.) If we did not share these parameters, there would be $1014 \times 26 = 26,364$ weights at the first layer. In layer 2, we have 50 feature maps, each of which is obtained by convolving each feature map in layer 1 with a 5×5 weight matrix, adding them up, adding a bias, and passing through a nonlinearity. There are therefore $5 \times 5 \times 50 = 1250$ neurons in Layer 2, $(5 \times 5 + 1) \times 6 \times 50 = 7800$ adjustable weights (one kernel for each pair of feature maps in layers 1 and 2), and $1250 \times 26 = 32,500$ connections. Layer 3 is fully connected to layer 2, and has 100 neurons and $100 \times (1250 + 1) = 125,100$ weights. Finally, layer 4 is also fully connected, and has 10 neurons, and $10 \times (100 + 1) = 1010$ weights. Adding the above numbers, there are a total of 3,215 neurons, 134,066 adjustable weights, and 184,974 connections.

This model is usually trained using stochastic gradient descent (see Section 16.5.4 for details). A single pass over the data set is called an epoch. When Mike O'Neill did these experiments in 2006, he found that a single epoch took about 40 minutes (recall that there are 60,000 training examples in MNIST). Since it took about 30 epochs for the error rate to converge, the total training time was about 20 hours.[5] Using this technique, he obtained a misclassification rate on the 10,000 test cases of about 1.40%.

To further reduce the error rate, a standard trick is to expand the training set by including **distorted** versions of the original data, to encourage the network to be invariant to small changes that don't affect the identity of the digit. These can be created by applying a random flow field to shift pixels around. See Figure 16.13 for some examples. (If we use online training, such as stochastic gradient descent, we can create these distortions on the fly, rather than having to store them.) Using this technique, Mike O'Neill obtained a misclassification rate on the 10,000 test cases of about 0.74%, which is close to the current state of the art.[6]

Yann Le Cun and colleagues (LeCun et al. 1998) obtained similar performance using a slightly more complicated architecture shown in Figure 16.14. This model is known as **LeNet5**, and historically it came before the model in Figure 16.12. There are two main differences. First, LeNet5 has a **subsampling** layer between each convolutional layer, which either averages or computes the max over each small window in the previous layer, in order to reduce the size, and to obtain a small amount of shift invariance. The convolution and sub-sampling combination was inspired by Hubel and Wiesel's model of simple and complex cells in the visual cortex (Hubel and Wiesel 1962), and it continues to be popular in neurally-inspired models of visual object recognition (Riesenhuber and Poggio 1999). A similar idea first appeared in Fukushima's **neocognitron** (Fukushima 1975), though no globally supervised training algorithm was available at that time.

The second difference between LeNet5 and the Simard architecture is that the final layer is actually an RBF network rather than a more standard sigmoidal or softmax layer. This model gets a test error rate of about 0.95% when trained with no distortions, and 0.8% when trained with distortions. Figure 16.15 shows all 82 errors made by the system. Some are genuinely ambiguous, but several are errors that a person would never make. A web-based demo of the LeNet5 can be found at `http://yann.lecun.com/exdb/lenet/index.html`.

---

5. Implementation details: Mike used C++ code and a variety of speedup tricks. He was using standard 2006 era hardware (an Intel Pentium 4 hyperthreaded processor running at 2.8GHz). See `http://www.codeproject.com/KB/library/NeuralNetRecognition.aspx` for details.

6. A list of various methods, along with their misclassification rates on the MNIST test set, is available from `http://yann.lecun.com/exdb/mnist/`. Error rates within 0.1–0.2% of each other are not statistically significantly different.

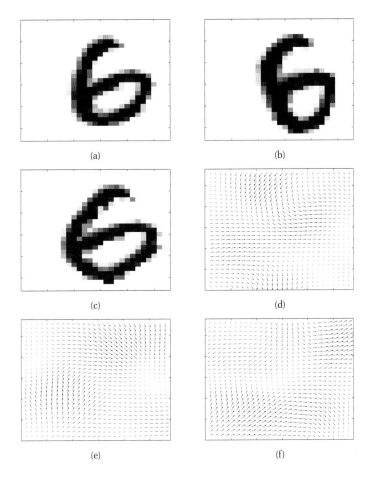

Figure 16.13 Several synthetic warpings of a handwritten digit. Based on Figure 5.14 of (Bishop 2006). Figure generated by elasticDistortionsDemo, written by Kevin Swersky.

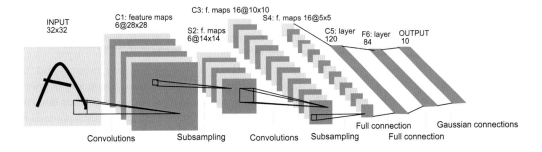

Figure 16.14 LeNet5, a convolutional neural net for classifying handwritten digits. Source: Figure 2 from (LeCun et al. 1998) . Used with kind permission of Yann LeCun.

Figure 16.15 These are the 82 errors made by LeNet5 on the 10,000 test cases of MNIST. Below each image is a label of the form correct-label → estimated-label.   Source: Figure 8 of (LeCun et al. 1998). Used with kind permission of Yann LeCun. (Compare to Figure 28.4(b) which shows the results of a deep generative model.)

Of course, classifying isolated digits is of limited applicability: in the real world, people usually write strings of digits or other letters. This requires both segmentation and classification. Le Cun and colleagues devised a way to combine convolutional neural networks with a model similar to a conditional random field (described in Section 19.6) to solve this problem. The system was eventually deployed by the US postal service. (See (LeCun et al. 1998) for a more detailed account of the system, which remains one of the best performing systems for this task.)

16.5.2 Other kinds of neural networks

Other network topologies are possible besides the ones discussed above. For example, we can have **skip arcs** that go directly from the input to the output, skipping the hidden layer; we can have sparse connections between the layers; etc. However, the MLP always requires that the weights form a directed acyclic graph. If we allow feedback connections, the model is known as a **recurrent neural network** (RNN); this defines a nonlinear dynamical system, but does not have a simple probabilistic interpretation. Such RNN models are currently the best approach for language modeling (i.e., performing word prediction in natural language) (Mikolov et al. 2011), significantly outperforming the standard n-gram-based methods discussed in Section 17.2.2. One difficulty with RNNs is that they are hard to train, since they suffer from the vanishing gradient problem. One solution is to use a model known as long short-term memory or LSTM (Hochreiter and Schmidhuber 1997), which is an RNN variant which has proved to be succesful at various tasks such as handwriting recognition. A simpler approach is to use a standard RNN, but with carefully initialized weights (Mikolov et al. 2011).

If we allow symmetric connections between the hidden units, the model is known as a **Hopfield network** or **associative memory**; its probabilistic counterpart is known as a Boltzmann machine (see Section 27.7) and can be used for unsupervised learning.

16.5.3 A brief history of the field

Neural networks have been the subject of great interest for many decades, due to the desire to understand the brain, and to build learning machines. It is not possible to review the entire history here. Instead, we just give a few "edited highlights".

The field is generally viewed as starting with McCulloch and Pitts (McCulloch and Pitts 1943), who devised a simple mathematical model of the neuron in 1943, in which they approximated the output as a weighted sum of inputs passed through a threshold function, $y = \mathbb{I}(\sum_i w_i x_i > \theta)$, for some threshold θ. This is similar to a sigmoidal activation function. Frank Rosenblatt invented the perceptron learning algorithm in 1957, which is a way to estimate the parameters of a McCulloch-Pitts neuron (see Section 8.5.4 for details). A very similar model called the **adaline** (for adaptive linear element) was invented in 1960 by Widrow and Hoff.

In 1969, Minsky and Papert (Minsky and Papert 1969) published a famous book called "Perceptrons" in which they showed that such linear models, with no hidden layers, were very limited in their power, since they cannot classify data that is not linearly separable. This considerably reduced interest in the field.

In 1986, Rumelhart, Hinton and Williams (Rumelhart et al. 1986) discovered the backpropagation algorithm (see Section 16.5.4), which allows one to fit models with hidden layers. (The backpropagation algorithm was originally discovered in (Bryson and Ho 1969), and independently in (Werbos 1974); however, it was (Rumelhart et al. 1986) that brought the algorithm to people's attention.) This spawned a decade of intense interest in these models.

In 1987, Sejnowski and Rosenberg (Sejnowski and Rosenberg 1987) created the famous **NETtalk** system, that learned a mapping from English words to phonetic symbols which could be fed into a speech synthesizer. An audio demo of the system as it learns over time can be found at `http://www.cnl.salk.edu/ParallelNetsPronounce/nettalk.mp3`. The systems starts by "babbling" and then gradually learns to pronounce English words. NETtalk learned a **distributed representation** (via its hidden layer) of various sounds, and its success spawned a big debate in psychology between **connectionism**, based on neural networks, and **computationalism**, based on syntactic rules. This debate lives on to some extent in the machine learning community, where there are still arguments about whether learning is best performed using low-level, "neural-like" representations, or using more structured models.

In 1989, Yann Le Cun and others (LeCun et al. 1989) created the famous LeNet system described in Section 16.5.1.

In 1992, the support vector machine (see Section 14.5) was invented (Boser et al. 1992). SVMs provide similar prediction accuracy to neural networks while being considerably easier to train (since they use a convex objective function). This spawned a decade of interest in kernel methods in general.[7] Note, however, that SVMs do not use adaptive basis functions, so they require a fair

---

7. It became part of the folklore during the 1990s that to get published in the top machine learning conference known as

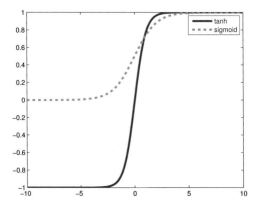

Figure 16.16 Two possible activation functions. $\tanh$ maps $\mathbb{R}$ to $[-1, +1]$ and is the preferred nonlinearity for the hidden nodes. sigm maps $\mathbb{R}$ to $[0, 1]$ and is the preferred nonlinearity for binary nodes at the output layer. Figure generated by `tanhPlot`.

amount of human expertise to design the right kernel function.

In 2002, Geoff Hinton invented the contrastive divergence training procedure (Hinton 2002), which provided a way, for the first time, to learn deep networks, by training one layer at a time in an unsupervised fashion (see Section 27.7.2.4 for details). This in turn has spawned renewed interest in neural networks over the last few years (see Chapter 28).

16.5.4 The backpropagation algorithm

Unlike a GLM, the NLL of an MLP is a non-convex function of its parameters. Nevertheless, we can find a locally optimal ML or MAP estimate using standard gradient-based optimization methods. Since MLPs have lots of parameters, they are often trained on very large data sets. Consequently it is common to use first-order online methods, such as stochastic gradient descent (Section 8.5.2), whereas GLMs are usually fit with IRLS, which is a second-order offline method.

We now discuss how to compute the gradient vector of the NLL by applying the chain rule of calculus. The resulting algorithm is known as **backpropagation**, for reasons that will become apparent.

For notational simplicity, we shall assume a model with just one hidden layer. It is helpful to distinguish the pre- and post-synaptic values of a neuron, that is, before and after we apply the nonlinearity. Let $\mathbf{x}_n$ be the n'th input, $\mathbf{a}_n = \mathbf{V}\mathbf{x}_n$ be the pre-synaptic hidden layer, and $\mathbf{z}_n = g(\mathbf{a}_n)$ be the post-synaptic hidden layer, where g is some **transfer function**. We typically use $g(a) = \text{sigm}(a)$, but we may also use $g(a) = \tanh(a)$: see Figure 16.16 for a comparison. (When the input to sigm or $\tanh$ is a vector, we assume it is applied component-wise.)

We now convert this hidden layer to the output layer as follows. Let $\mathbf{b}_n = \mathbf{W}\mathbf{z}_n$ be the pre-synaptic output layer, and $\hat{\mathbf{y}}_n = h(\mathbf{b}_n)$ be the post-synaptic output layer, where h is another nonlinearity, corresponding to the canonical link for the GLM. (We reserve the notation

---

NIPS, which stands for "neural information processing systems", it was important to ensure your paper did not contain the word "neural network"!

$\mathbf{y}_n$, without the hat, for the output corresponding to the n'th training case.) For a regression model, we use $h(\mathbf{b}) = \mathbf{b}$; for binary classifcation, we use $h(\mathbf{b}) = [\mathrm{sigm}(b_1), \ldots, \mathrm{sigm}(b_c)]$; for multi-class classification, we use $h(\mathbf{b}) = \mathcal{S}(\mathbf{b})$.

We can write the overall model as follows:

$$\mathbf{x}_n \xrightarrow{\mathbf{V}} \mathbf{a}_n \xrightarrow{g} \mathbf{z}_n \xrightarrow{\mathbf{W}} \mathbf{b}_n \xrightarrow{h} \hat{\mathbf{y}}_n \tag{16.65}$$

The parameters of the model are $\boldsymbol{\theta} = (\mathbf{V}, \mathbf{W})$, the first and second layer weight matrices. Offset or bias terms can be accomodated by clamping an element of $\mathbf{x}_n$ and $\mathbf{z}_n$ to 1.[8]

In the regression case, with K outputs, the NLL is given by the squared error:

$$J(\boldsymbol{\theta}) = -\sum_n \sum_k (\hat{y}_{nk}(\boldsymbol{\theta}) - y_{nk})^2 \tag{16.66}$$

In the classification case, with K classes, the NLL is given by the cross entropy

$$J(\boldsymbol{\theta}) = -\sum_n \sum_k y_{nk} \log \hat{y}_{nk}(\boldsymbol{\theta}) \tag{16.67}$$

Our task is to compute $\nabla_{\boldsymbol{\theta}} J$. We will derive this for each n separately; the overall gradient is obtained by summing over n, although often we just use a mini-batch (see Section 8.5.2).

Let us start by considering the output layer weights. We have

$$\nabla_{\mathbf{w}_k} J_n = \frac{\partial J_n}{\partial b_{nk}} \nabla_{\mathbf{w}_k} b_{nk} = \frac{\partial J_n}{\partial b_{nk}} \mathbf{z}_n \tag{16.68}$$

since $b_{nk} = \mathbf{w}_k^T \mathbf{z}_n$. Assuming h is the canonical link function for the output GLM, then Equation 9.91 tells us that

$$\frac{\partial J_n}{\partial b_{nk}} \triangleq \delta_{nk}^w = (\hat{y}_{nk} - y_{nk}) \tag{16.69}$$

which is the error signal. So the overall gradient is

$$\nabla_{\mathbf{w}_k} J_n = \delta_{nk}^w \mathbf{z}_n \tag{16.70}$$

which is the pre-synaptic input to the output layer, namely $\mathbf{z}_n$, times the error signal, namely δ_{nk}^w.

For the input layer weights, we have

$$\nabla_{\mathbf{v}_j} J_n = \frac{\partial J_n}{\partial a_{nj}} \nabla_{\mathbf{v}_j} a_{nj} \triangleq \delta_{nj}^v \mathbf{x}_n \tag{16.71}$$

where we exploited the fact that $a_{nj} = \mathbf{v}_j^T \mathbf{x}_n$. All that remains is to compute the first level error signal δ_{nj}^v. We have

$$\delta_{nj}^v = \frac{\partial J_n}{\partial a_{nj}} = \sum_{k=1}^K \frac{\partial J_n}{\partial b_{nk}} \frac{\partial b_{nk}}{\partial a_{nj}} = \sum_{k=1}^K \delta_{nk}^w \frac{\partial b_{nk}}{\partial a_{nj}} \tag{16.72}$$

---

8. In the regression setting, we can easily estimate the variance of the output noise using the empirical variance of the residual errors, $\hat{\sigma}^2 = \frac{1}{N} ||\hat{\mathbf{y}}(\hat{\boldsymbol{\theta}}) - \mathbf{y}||^2$, after training is complete. There will be one value of σ^2 for each output node, if we are performing multi-target regression, as we usually assume.

Now

$$b_{nk} = \sum_j w_{kj} g(a_{nj}) \tag{16.73}$$

so

$$\frac{\partial b_{nk}}{\partial a_{nj}} = w_{kj} g'(a_{nj}) \tag{16.74}$$

where $g'(a) = \frac{d}{da} g(a)$. For tanh units, $g'(a) = \frac{d}{da} \tanh(a) = 1 - \tanh^2(a) = \text{sech}^2(a)$, and for sigmoid units, $g'(a) = \frac{d}{da} \sigma(a) = \sigma(a)(1 - \sigma(a))$. Hence

$$\delta_{nj}^v = \sum_{k=1}^{K} \delta_{nk}^w w_{kj} g'(a_{nj}) \tag{16.75}$$

Thus the layer 1 errors can be computed by passing the layer 2 errors back through the $\mathbf{W}$ matrix; hence the term "backpropagation". The key property is that we can compute the gradients locally: each node only needs to know about its immediate neighbors. This is supposed to make the algorithm "neurally plausible", although this interpretation is somewhat controversial.

Putting it all together, we can compute all the gradients as follows: we first perform a forwards pass to compute $\mathbf{a}_n$, $\mathbf{z}_n$, $\mathbf{b}_n$ and $\hat{\mathbf{y}}_n$. We then compute the error for the output layer, $\boldsymbol{\delta}_n^{(2)} = \hat{\mathbf{y}}_n - \mathbf{y}_n$, which we pass backwards through $\mathbf{W}$ using Equation 16.75 to compute the error for the hidden layer, $\boldsymbol{\delta}_n^{(1)}$. We then compute the overall gradient vector by stacking the two component vectors as follows:

$$\nabla_{\boldsymbol{\theta}} J(\boldsymbol{\theta}) = [\sum_n \boldsymbol{\delta}_n^v \mathbf{x}_n; \sum_n \boldsymbol{\delta}_n^w \mathbf{z}_n] \tag{16.76}$$

16.5.5 Identifiability

It is easy to see that the parameters of a neural network are not identifiable. For example, we can change the sign of the weights going into one of the hidden units, so long as we change the sign of all the weights going out of it; these effects cancel, since tanh is an odd function, so $\tanh(-a) = -\tanh(a)$. There will be H such sign flip symmetries, leading to 2^H equivalent settings of the parameters. Similarly, we can change the identity of the hidden units without affecting the likelihood. There are $H!$ such permutations. The total number of equivalent parameter settings (with the same likelihood) is therefore $H!2^H$.

In addition, there may be local minima due to the non-convexity of the NLL. This can be a more serious problem, although with enough data, these local optima are often quite "shallow", and simple stochastic optimization methods can avoid them. In addition, it is common to perform multiple restarts, and to pick the best solution, or to average over the resulting predictions. (It does not make sense to average the parameters themselves, since they are not identifiable.)

16.5.6 Regularization

As usual, the MLE can overfit, especially if the number of nodes is large. A simple way to prevent this is called **early stopping**, which means stopping the training procedure when the error on the validation set first starts to increase. This method works because we usually initialize from small random weights, so the model is initially simple (since the tanh and sigm functions are nearly linear near the origin). As training progresses, the weights become larger, and the model becomes nonlinear. Eventually it will overfit.

Another way to prevent overfitting, that is more in keeping with the approaches used elsewhere in this book, is to impose a prior on the parameters, and then use MAP estimation. It is standard to use a $\mathcal{N}(0, \alpha^{-1}\mathbf{I})$ prior (equivalent to ℓ_2 regularization), where α is the precision (strength) of the prior. In the neural networks literature, this is called **weight decay**, since it encourages small weights, and hence simpler models. The penalized NLL objective becomes

$$J(\boldsymbol{\theta}) = -\sum_{n=1}^{N} \log p(y_n|\mathbf{x}_n, \boldsymbol{\theta}) + \frac{\alpha}{2}[\sum_{ij} v_{ij}^2 + \sum_{jk} w_{jk}^2] \tag{16.77}$$

(Note that we don't penalize the bias terms.) The gradient of the modified objective becomes

$$\nabla_{\boldsymbol{\theta}} J(\boldsymbol{\theta}) = [\sum_n \boldsymbol{\delta}_n^v \mathbf{x}_n + \alpha\mathbf{v}, \ \sum_n \boldsymbol{\delta}_n^w \mathbf{z}_n + \alpha\mathbf{w}] \tag{16.78}$$

as in Section 8.3.6.

If the regularization is sufficiently strong, it does not matter if we have too many hidden units (apart from wasted computation). Hence it is advisable to set H to be as large as you can afford (say 10–100), and then to choose an appropriate regularizer. We can set the α parameter by cross validation or empirical Bayes (see Section 16.5.7.5).

As with ridge regression, it is good practice to standardize the inputs to zero mean and unit variance, so that the spherical Gaussian prior makes sense.

16.5.6.1 Consistent Gaussian priors *

One can show (MacKay 1992) that using the same regularization parameter for both the first and second layer weights results in the lack of a certain desirable invariance property. In particular, suppose we linearly scale and shift the inputs and/or outputs to a neural network regression model. Then we would like the model to learn to predict the same function, by suitably scaling its internal weights and bias terms. However, the amount of scaling needed by the first and second layer weights to compensate for a change in the inputs and/or outputs is not the same. Therefore we need to use a different regularization strength for the first and second layer. Fortunately, this is easy to do — we just use the following prior:

$$p(\boldsymbol{\theta}) = \mathcal{N}(\mathbf{W}|\mathbf{0}, \frac{1}{\alpha_w}\mathbf{I})\mathcal{N}(\mathbf{V}|\mathbf{0}, \frac{1}{\alpha_v}\mathbf{I})\mathcal{N}(\mathbf{b}|\mathbf{0}, \frac{1}{\alpha_b}\mathbf{I})\mathcal{N}(\mathbf{c}|\mathbf{0}, \frac{1}{\alpha_c}\mathbf{I}) \tag{16.79}$$

where $\mathbf{b}$ and $\mathbf{c}$ are the bias terms.[9]

To get a feeling for the effect of these hyper-parameters, we can sample MLP parameters from this prior and plot the resulting random functions. Figure 16.17 shows some examples. Decreasing α_v allows the first layer weights to get bigger, making the sigmoid-like shape of

---

9. Since we are regularizing the output bias terms, it is helpful, in the case of regression, to normalize the target responses in the training set to zero mean, to be consistent with the fact that the prior on the output bias has zero mean.

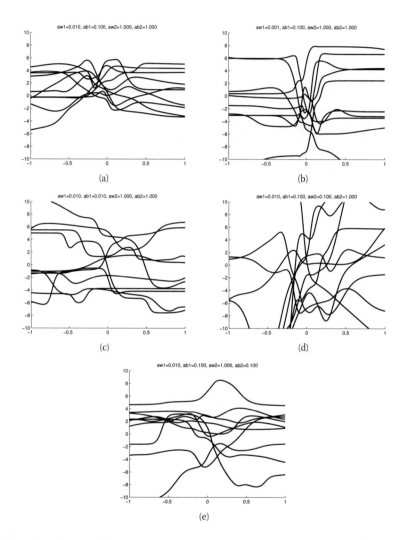

Figure 16.17 The effects of changing the hyper-parameters on an MLP. (a) Default parameter values $\alpha_v = 0.01$, $\alpha_b = 0.1$, $\alpha_w = 1$, $\alpha_c = 1$. (b) Decreasing α_v by factor of 10. (c) Decreasing α_b by factor of 10. (d) Decreasing α_w by factor of 10. (e) Decreasing α_c by factor of 10. Figure generated by `mlpPriorsDemo`.

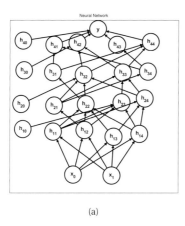
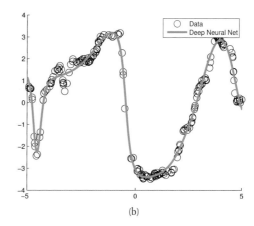

(a)   (b)

Figure 16.18 (a) A deep but sparse neural network. The connections are pruned using ℓ_1 regularization. At each level, nodes numbered 0 are clamped to 1, so their outgoing weights correspond to the offset/bias terms. (b) Predictions made by the model on the training set. Figure generated by `sparseNnetDemo`, written by Mark Schmidt.

the functions steeper. Decreasing α_b allows the first layer biases to get bigger, which allows the center of the sigmoid to shift left and right more. Decreasing α_w allows the second layer weights to get bigger, making the functions more "wiggly" (greater sensitivity to change in the input, and hence larger dynamic range). And decreasing α_c allows the second layer biases to get bigger, allowing the mean level of the function to move up and down more. (In Chapter 15, we will see an easier way to define priors over functions.)

16.5.6.2 Weight pruning

Since there are many weights in a neural network, it is often helpful to encourage sparsity. Various ad-hoc methods for doing this, with names such as "optimal brain damage", were devised in the 1990s; see e.g., (Bishop 1995) for details.

However, we can also use the more principled sparsity-promoting techniques we discussed in Chapter 13. One approach is to use an ℓ_1 regularizer. See Figure 16.18 for an example. Another approach is to use ARD; this is discussed in more detail in Section 16.5.7.5.

16.5.6.3 Soft weight sharing*

Another way to regularize the parameters is to encourage similar weights to share statistical strength. But how do we know which parameters to group together? We can learn this, by using a mixture model. That is, we model $p(\boldsymbol{\theta})$ as a mixture of (diagonal) Gaussians. Parameters that are assigned to the same cluster will share the same mean and variance and thus will have similar values (assuming the variance for that cluster is low). This is called **soft weight sharing** (Nowlan and Hinton 1992). In practice, this technique is not widely used. See e.g., (Bishop 2006, p271) if you want to know the details.

16.5.6.4   Semi-supervised embedding *

An interesting way to regularize "deep" feedforward neural networks is to encourage the hidden layers to assign similar objects to similar representations. This is useful because it is often easy to obtain "side" information consisting of sets of pairs of similar and dissimilar objects. For example, in a video classification task, neighboring frames can be deemed similar, but frames that are distant in time can be deemed dis-similar (Mobahi et al. 2009). Note that this can be done without collecting any labels.

Let $S_{ij} = 1$ if examples i and j are similar, and $S_{ij} = 0$ otherwise. Let $f(\mathbf{x}_i)$ be some embedding of item $\mathbf{x}_i$, e.g., $f(\mathbf{x}_i) = \mathbf{z}(\mathbf{x}_i, \boldsymbol{\theta})$, where $\mathbf{z}$ is the hidden layer of a neural network. Now define a loss function $L(f(\mathbf{x}_i), f(\mathbf{x}_j), S_{ij})$ that depends on the embedding of two objects, and the observed similarity measure. For example, we might want to force similar objects to have similar embeddings, and to force the embeddings of dissimilar objects to be a minimal distance apart:

$$L(\mathbf{f}_i, \mathbf{f}_j, S_{ij}) = \begin{cases} ||\mathbf{f}_i - \mathbf{f}_j||^2 & \text{if } S_{ij} = 1 \\ \max(0, m - ||\mathbf{f}_i - \mathbf{f}_j||^2) & \text{if } S_{ij} = 0 \end{cases} \tag{16.80}$$

where m is some minimal margin. We can now define an augmented loss function for training the neural network:

$$\sum_{i \in \mathcal{L}} \text{NLL}(f(\mathbf{x}_i), y_i) + \lambda \sum_{i,j \in \mathcal{U}} L(f(\mathbf{x}_i), f(\mathbf{x}_j), S_{ij}) \tag{16.81}$$

where $\mathcal{L}$ is the labeled training set, $\mathcal{U}$ is the unlabeled training set, and $\lambda \geq 0$ is some tradeoff parameter. This is called **semi-supervised embedding** (Weston et al. 2008).

Such an objective can be easily optimized by stochastic gradient descent. At each iteration, pick a random labeled training example, $(\mathbf{x}_n, y_n)$, and take a gradient step to optimize $\text{NLL}(f(\mathbf{x}_i), y_i)$. Then pick a random pair of similar unlabeled examples $\mathbf{x}_i$, $\mathbf{x}_j$ (these can sometimes be generated on the fly rather than stored in advance), and make a gradient step to optimize $\lambda L(f(\mathbf{x}_i), f(\mathbf{x}_j), 1)$, Finally, pick a random unlabeled example $\mathbf{x}_k$, which with high probability is dissimilar to $\mathbf{x}_i$, and make a gradient step to optimize $\lambda L(f(\mathbf{x}_i), f(\mathbf{x}_k), 0)$.

Note that this technique is effective because it can leverage massive amounts of data. In a related approach, (Collobert and Weston 2008) trained a neural network to distinguish valid English sentences from invalid ones. This was done by taking all 631 million words from English Wikipedia (en.wikipedia.org), and then creating windows of length 11 containing neighboring words. This constitutes the positive examples. To create negative examples, the middle word of each window was replaced by a random English word (this is likely to be an "invalid" sentence — either grammatically and/or semantically — with high probability). This neural network was then trained over the course of 1 week, and its latent representation was then used as the input to a supervised semantic role labeling task, for which very little labeled training data is available. (See also (Ando and Zhang 2005) for related work.)

16.5.7   Bayesian inference *

Although MAP estimation is a succesful way to reduce overfitting, there are still some good reasons to want to adopt a fully Bayesian approach to "fitting" neural networks:

- Integrating out the parameters instead of optimizing them is a much stronger form of regularization than MAP estimation.

- We can use Bayesian model selection to determine things like the hyper-parameter settings and the number of hidden units. This is likely to be much faster than cross validation, especially if we have many hyper-parameters (e.g., as in ARD).

- Modelling uncertainty in the parameters will induce uncertainty in our predictive distributions, which is important for certain problems such as active learning and risk-averse decision making.

- We can use online inference methods, such as the extended Kalman filter, to do online learning (Haykin 2001).

One can adopt a variety of approximate Bayesian inference techniques in this context. In this section, we discuss the Laplace approximation, first suggested in (MacKay 1992, 1995b). One can also use hybrid Monte Carlo (Neal 1996), or variational Bayes (Hinton and Camp 1993; Barber and Bishop 1998).

16.5.7.1 Parameter posterior for regression

We start by considering regression, following the presentation of (Bishop 2006, sec 5.7), which summarizes the work of (MacKay 1992, 1995b). We will use a prior of the form $p(\mathbf{w}) = \mathcal{N}(\mathbf{w}|\mathbf{0}, (1/\alpha)\mathbf{I})$, where $\mathbf{w}$ represents all the weights combined. We will denote the precision of the noise by $\beta = 1/\sigma^2$.

The posterior can be approximated as follows:

$$p(\mathbf{w}|\mathcal{D}, \alpha, \beta) \quad \propto \quad \exp(-E(\mathbf{w})) \tag{16.82}$$

$$E(\mathbf{w}) \quad \triangleq \quad \beta E_D(\mathbf{w}) + \alpha E_W(\mathbf{w}) \tag{16.83}$$

$$E_D(\mathbf{w}) \quad \triangleq \quad \frac{1}{2}\sum_{n=1}^{N}(y_n - f(\mathbf{x}_n, \mathbf{w}))^2 \tag{16.84}$$

$$E_W(\mathbf{w}) \quad \triangleq \quad \frac{1}{2}\mathbf{w}^T\mathbf{w} \tag{16.85}$$

where E_D is the data error, E_W is the prior error, and E is the overall error (negative log prior plus log likelihood). Now let us make a second-order Taylor series approximation of $E(\mathbf{w})$ around its minimum (the MAP estimate)

$$E(\mathbf{w}) \approx E(\mathbf{w}_{MP}) + \frac{1}{2}(\mathbf{w} - \mathbf{w}_{MP})^T\mathbf{A}(\mathbf{w} - \mathbf{w}_{MP}) \tag{16.86}$$

where $\mathbf{A}$ is the Hessian of E:

$$\mathbf{A} = \nabla\nabla E(\mathbf{w}_{MP}) = \beta\mathbf{H} + \alpha\mathbf{I} \tag{16.87}$$

where $\mathbf{H} = \nabla\nabla E_D(\mathbf{w}_{MP})$ is the Hessian of the data error. This can be computed exactly in $O(d^2)$ time, where d is the number of parameters, using a variant of backpropagation (see (Bishop 2006, sec 5.4) for details). Alternatively, if we use a quasi-Newton method to find the mode, we can use its internally computed (low-rank) approximation to $\mathbf{H}$. (Note that diagonal approximations of $\mathbf{H}$ are usually very inaccurate.) In either case, using this quadratic approximation, the posterior becomes Gaussian:

$$p(\mathbf{w}|\alpha, \beta, \mathcal{D}) \quad \approx \quad \mathcal{N}(\mathbf{w}|\mathbf{w}_{MP}, \mathbf{A}^{-1}) \tag{16.88}$$

16.5.7.2 Parameter posterior for classification

The classification case is the same as the regression case, except $\beta = 1$ and E_D is a cross-entropy error of the form

$$E_D(\mathbf{w}) \quad \triangleq \quad \sum_{n=1}^{N} [y_n \ln f(\mathbf{x}_n, \mathbf{w}) + (1 - y_n) \ln f(\mathbf{x}_n, \mathbf{w})] \tag{16.89}$$

$$\tag{16.90}$$

16.5.7.3 Predictive posterior for regression

The posterior predictive density is given by

$$p(y|\mathbf{x}, \mathcal{D}, \alpha, \beta) = \int \mathcal{N}(y|f(\mathbf{x}, \mathbf{w}), 1/\beta) \mathcal{N}(\mathbf{w}|\mathbf{w}_{MP}, \mathbf{A}^{-1}) d\mathbf{w} \tag{16.91}$$

This is not analytically tractable because of the nonlinearity of $f(\mathbf{x}, \mathbf{w})$. Let us therefore construct a first-order Taylor series approximation around the mode:

$$f(\mathbf{x}, \mathbf{w}) \approx f(\mathbf{x}, \mathbf{w}_{MP}) + \mathbf{g}^T(\mathbf{w} - \mathbf{w}_{MP}) \tag{16.92}$$

where

$$\mathbf{g} = \nabla_{\mathbf{w}} f(\mathbf{x}, \mathbf{w})|_{\mathbf{w}=\mathbf{w}_{MP}} \tag{16.93}$$

We now have a linear-Gaussian model with a Gaussian prior on the weights. From Equation 4.126 we have

$$p(y|\mathbf{x}, \mathcal{D}, \alpha, \beta) \approx \mathcal{N}(y|f(\mathbf{x}, \mathbf{w}_{MP}), \sigma^2(\mathbf{x})) \tag{16.94}$$

where the predictive variance depends on the input $\mathbf{x}$ as follows:

$$\sigma^2(\mathbf{x}) = \beta^{-1} + \mathbf{g}^T \mathbf{A}^{-1} \mathbf{g} \tag{16.95}$$

The error bars will be larger in regions of input space where we have little training data. See Figure 16.19 for an example.

16.5.7.4 Predictive posterior for classification

In this section, we discuss how to approximate $p(y|\mathbf{x}, \mathcal{D})$ in the case of binary classification. The situation is similar to the case of logistic regression, discussed in Section 8.4.4, except in addition the posterior predictive mean is a non-linear function of $\mathbf{w}$. Specifically, we have

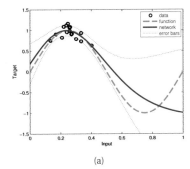 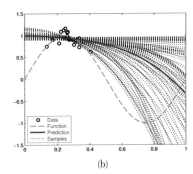

(a)   (b)

Figure 16.19 The posterior predictive density for an MLP with 3 hidden nodes, trained on 16 data points. The dashed green line is the true function. (a) Result of using a Laplace approximation, after performing empirical Bayes to optimize the hyperparameters. The solid red line is the posterior mean prediction, and the dotted blue lines are 1 standard deviation above and below the mean. Figure generated by `mlpRegEvidenceDemo`. (b) Result of using hybrid Monte Carlo, using the same trained hyperparameters as in (a). The solid red line is the posterior mean prediction, and the dotted blue lines are samples from the posterior predictive. Figure generated by `mlpRegHmcDemo`, written by Ian Nabney.

$\mu = \mathbb{E}\left[y|\mathbf{x}, \mathbf{w}\right] = \text{sigm}(a(\mathbf{x}, \mathbf{w}))$, where $a(\mathbf{x}, \mathbf{w})$ is the pre-synaptic output of the final layer. Let us make a linear approximation to this:

$$a(\mathbf{x}, \mathbf{w}) \approx a_{MP}(\mathbf{x}) + \mathbf{g}^T(\mathbf{w} - \mathbf{w}_{MP}) \tag{16.96}$$

where $a_{MP}(\mathbf{x}) = a(\mathbf{x}, \mathbf{w}_{MP})$ and $\mathbf{g} = \nabla_{\mathbf{x}} a(\mathbf{x}, \mathbf{w}_{MP})$ can be found by a modified version of backpropagation. Clearly

$$p(a|\mathbf{x}, \mathcal{D}) \approx \mathcal{N}(a(\mathbf{x}, \mathbf{w}_{MP}), \mathbf{g}(\mathbf{x})^T \mathbf{A}^{-1} \mathbf{g}(\mathbf{x})) \tag{16.97}$$

Hence the posterior predictive for the output is

$$p(y = 1|\mathbf{x}, \mathcal{D}) = \int \text{sigm}(a) p(a|\mathbf{x}, \mathcal{D}) da \approx \text{sigm}(\kappa(\sigma_a^2) \mathbf{b}^T \mathbf{w}_{MP}) \tag{16.98}$$

where κ is defined by Equation 8.70, which we repeat here for convenience:

$$\kappa(\sigma^2) \triangleq (1 + \pi\sigma^2/8)^{-\frac{1}{2}} \tag{16.99}$$

Of course, a simpler (and potentially more accurate) alternative to this is to draw a few samples from the Gaussian posterior and to approximate the posterior predictive using Monte Carlo.

In either case, the effect of taking uncertainty of the parameters into account, as in Section 8.4.4, is to "moderate" the confidence of the output; the decision boundary itself is unaffected, however.

16.5.7.5   ARD for neural networks

Once we have made the Laplace approximation to the posterior, we can optimize the marginal likelihood wrt the hyper-parameters $\boldsymbol{\alpha}$ using the same fixed-point equations as in Section 13.7.4.2.

Typically we use one hyper-parameter for the weight vector leaving each node, to achieve an effect similar to group lasso (Section 13.5.1). That is, the prior has the form

$$p(\boldsymbol{\theta}) = \prod_{i=1}^{D} \mathcal{N}(\mathbf{v}_{:,i}|\mathbf{0}, \frac{1}{\alpha_{v,i}}\mathbf{I}) \prod_{j=1}^{H} \mathcal{N}(\mathbf{w}_{:,j}|\mathbf{0}, \frac{1}{\alpha_{w,j}}\mathbf{I}) \tag{16.100}$$

If we find $\alpha_{v,i} = \infty$, then input feature i is irrelevant, and its weight vector $\mathbf{v}_{:,i}$ is pruned out. Similarly, if we find $\alpha_{w,j} = \infty$, then hidden feature j is irrelevant. This is known as automatic relevancy determination or ARD, which was discussed in detail in Section 13.7. Applying this to neural networks gives us an efficient means of variable selection in non-linear models.

The software package NETLAB contains a simple example of ARD applied to a neural network, called demard. This demo creates some data according to a nonlinear regression function $f(x_1, x_2, x_3) = \sin(2\pi x_1) + \epsilon$, where x_2 is a noisy copy of x_1. We see that x_2 and x_3 are irrelevant for predicting the target. However, x_2 is correlated with x_1, which is relevant. Using ARD, the final hyper-parameters are as follows:

$$\boldsymbol{\alpha} = [0.2, \quad 21.4, \quad 249001.8] \tag{16.101}$$

This clearly indicates that feature 3 is irrelevant, feature 2 is only weakly relevant, and feature 1 is very relevant.

16.6 Ensemble learning

Ensemble learning refers to learning a weighted combination of base models of the form

$$f(y|\mathbf{x}, \boldsymbol{\pi}) = \sum_{m \in \mathcal{M}} w_m f_m(y|\mathbf{x}) \tag{16.102}$$

where the w_m are tunable parameters. Ensemble learning is sometimes called a **committee method**, since each base model f_m gets a weighted "vote".

Clearly ensemble learning is closely related to learning adaptive-basis function models. In fact, one can argue that a neural net is an ensemble method, where f_m represents the m'th hidden unit, and w_m are the output layer weights. Also, we can think of boosting as kind of ensemble learning, where the weights on the base models are determined sequentially. Below we describe some other forms of ensemble learning.

16.6.1 Stacking

An obvious way to estimate the weights in Equation 16.102 is to use

$$\hat{\mathbf{w}} = \underset{\mathbf{w}}{\operatorname{argmin}} \sum_{i=1}^{N} L(y_i, \sum_{m=1}^{M} w_m f_m(\mathbf{x})) \tag{16.103}$$

However, this will result in overfitting, with w_m being large for the most complex model. A simple solution to this is to use cross-validation. In particular, we can use the LOOCV estimate

$$\hat{\mathbf{w}} = \underset{\mathbf{w}}{\operatorname{argmin}} \sum_{i=1}^{N} L(y_i, \sum_{m=1}^{M} w_m \hat{f}_m^{-i}(\mathbf{x})) \tag{16.104}$$

Class	C_1	C_2	C_3	C_4	C_5	C_6	$\cdots$	C_{15}
0	1	1	0	0	0	0	$\cdots$	1
1	0	0	1	1	1	1	$\cdots$	0
				$\vdots$				
9	0	1	1	1	0	0	$\cdots$	0

Table 16.2 Part of a 15-bit error-correcting output code for a 10-class problem. Each row defines a two-class problem. Based on Table 16.1 of (Hastie et al. 2009).

where $\hat{f}_m^{-i}(\mathbf{x})$ is the predictor obtained by training on data excluding $(\mathbf{x}_i, y_i)$. This is known as **stacking**, which stands for "stacked generalization" (Wolpert 1992). This technique is more robust to the case where the "true" model is not in the model class than standard BMA (Clarke 2003). This approach was used by the Netflix team known as "The Ensemble", which tied the submission of the winning team (BellKor's Pragmatic Chaos) in terms of accuracy (Sill et al. 2009). Stacking has also been used for problems such as image segmentation and labeling.

16.6.2 Error-correcting output codes

An interesting form of ensemble learning is known as **error-correcting output codes** or **ECOC** (Dietterich and Bakiri 1995), which can be used in the context of multi-class classification. The idea is that we are trying to decode a symbol (namely the class label) which has C possible states. We could use a bit vector of length $B = \lceil \log_2 C \rceil$ to encode the class label, and train B separate binary classifiers to predict each bit. However, by using more bits, and by designing the codewords to have maximal Hamming distance from each other, we get a method that is more resistant to individual bit-flipping errors (misclassification). For example, in Table 16.2, we use $B = 15$ bits to encode a $C = 10$ class problem. The minimum Hamming distance between any pair of rows is 7. The decoding rule is

$$\hat{c}(\mathbf{x}) = \min_c \sum_{b=1}^{B} |C_{cb} - \hat{p}_b(\mathbf{x})| \tag{16.105}$$

where C_{cb} is the b'th bit of the codeword for class c, and $\hat{p}_b(\mathbf{x})$ is the probability that output bit b turns on given input featres $\mathbf{x}$. (James and Hastie 1998) showed that a random code worked just as well as the optimal code: both methods work by averaging the results of multiple classifiers, thereby reducing variance.

16.6.3 Ensemble learning is not equivalent to Bayes model averaging

In Section 5.3, we discussed Bayesian model selection. An alternative to picking the best model, and then using this to make predictions, is to make a weighted average of the predictions made by each model, i.e., we compute

$$p(y|\mathbf{x}, \mathcal{D}) = \sum_{m \in \mathcal{M}} p(y|\mathbf{x}, m, \mathcal{D}) p(m|\mathcal{D}) \tag{16.106}$$

This is called **Bayes model averaging** (BMA), and can sometimes give better performance than using any single model (Hoeting et al. 1999). Of course, averaging over all models is typically computationally infeasible (analytical integration is obviously not possible in a discrete space, although one can sometimes use dynamic programming to perform the computation exactly, e.g., (Meila and Jaakkola 2006)). A simple approximation is to sample a few models from the posterior. An even simpler approximation (and the one most widely used in practice) is to just use the MAP model.

It is important to note that BMA is not equivalent to ensemble learning (Minka 2000c). This latter technique corresponds to enlarging the model space, by defining a single new model which is a convex combination of base models, as follows:

$$p(y|\mathbf{x}, \boldsymbol{\pi}) = \sum_{m \in \mathcal{M}} \pi_m p(y|\mathbf{x}, m) \tag{16.107}$$

In principle, we can now perform Bayesian inference to compute $p(\boldsymbol{\pi}|\mathcal{D})$; we then make predictions using $p(y|\mathbf{x}, \mathcal{D}) = \int p(y|\mathbf{x}, \boldsymbol{\pi}) p(\boldsymbol{\pi}|\mathcal{D}) d\boldsymbol{\pi}$. However, it is much more common to use point estimation methods for $\boldsymbol{\pi}$, as we saw above.

16.7 Experimental comparison

We have described many different methods for classification and regression. Which one should you use? That depends on which inductive bias you think is most appropriate for your domain. Usually this is hard to assess, so it is common to just try several different methods, and see how they perform empirically. Below we summarize two such comparisons that were carefully conducted (although the data sets that were used are relatively small). See the website mlcomp.org for a distributed way to perform large scale comparisons of this kind. Of course, we must always remember the no free lunch theorem (Section 1.4.9), which tells us that there is no universally best learning method.

16.7.1 Low-dimensional features

In 2006, Rich Caruana and Alex Niculescu-Mizil (Caruana and Niculescu-Mizil 2006) conducted a very extensive experimental comparison of 10 different binary classification methods, on 11 different data sets. The 11 data sets all had 5000 training cases, and had test sets containing $\sim 10,000$ examples on average. The number of features ranged from 9 to 200 (see Section 16.7.2 for a study using high dimensional features). 5-fold cross validation was used to assess average test error. (This is separate from any internal CV a method may need to use for model selection.)

The methods they compared are as follows (listed in roughly decreasing order of performance, as assessed by Table 16.3):

- BST-DT: boosted decision trees
- RF: random forest
- BAG-DT: bagged decision trees
- SVM: support vector machine
- ANN: artificial neural network
- KNN: K-nearest neighbors

MODEL	1ST	2ND	3RD	4TH	5TH	6TH	7TH	8TH	9TH	10TH
BST-DT	0.580	0.228	0.160	0.023	0.009	0.000	0.000	0.000	0.000	0.000
RF	0.390	0.525	0.084	0.001	0.000	0.000	0.000	0.000	0.000	0.000
BAG-DT	0.030	0.232	0.571	0.150	0.017	0.000	0.000	0.000	0.000	0.000
SVM	0.000	0.008	0.148	0.574	0.240	0.029	0.001	0.000	0.000	0.000
ANN	0.000	0.007	0.035	0.230	0.606	0.122	0.000	0.000	0.000	0.000
KNN	0.000	0.000	0.000	0.009	0.114	0.592	0.245	0.038	0.002	0.000
BST-STMP	0.000	0.000	0.002	0.013	0.014	0.257	0.710	0.004	0.000	0.000
DT	0.000	0.000	0.000	0.000	0.000	0.000	0.004	0.616	0.291	0.089
LOGREG	0.000	0.000	0.000	0.000	0.000	0.000	0.040	0.312	0.423	0.225
NB	0.000	0.000	0.000	0.000	0.000	0.000	0.000	0.030	0.284	0.686

Table 16.3 Fraction of time each method achieved a specified rank, when sorting by mean performance across 11 datasets and 8 metrics. Based on Table 4 of (Caruana and Niculescu-Mizil 2006). Used with kind permission of Alexandru Niculescu-Mizil.

- BST-STMP: boosted stumps
- DT: decision tree
- LOGREG: logistic regression
- NB: naive Bayes

They used 8 different performance measures, which can be divided into three groups. Threshold metrics just require a point estimate as output. These include accuracy, F-score (Section 5.7.2.3), etc. Ordering/ ranking metrics measure how well positive cases are ordered before the negative cases. These include area under the ROC curve (Section 5.7.2.1), average precision, and the precision/recall break even point. Finally, the probability metrics included cross-entropy (log-loss) and squared error, $(y - \hat{p})^2$. Methods such as SVMs that do not produce calibrated probabilities were post-processed using Platt's logistic regression trick (Section 14.5.2.3), or using isotonic regression. Performance measures were standardized to a 0:1 scale so they could be compared.

Obviously the results vary by dataset and by metric. Therefore just averaging the performance does not necessarily give reliable conclusions. However, one can perform a bootstrap analysis, which shows how robust the conclusions are to such changes. The results are shown in Table 16.3. We see that most of the time, boosted decision trees are the best method, followed by random forests, bagged decision trees, SVMs and neural networks. However, the following methods all did relatively poorly: KNN, stumps, single decision trees, logistic regression and naive Bayes.

These results are generally consistent with conventional wisdom of practioners in the field. Of course, the conclusions may change if there the features are high dimensional and/ or there are lots of irrelevant features (as in Section 16.7.2), or if there is lots of noise, etc.

16.7.2 High-dimensional features

In 2003, the NIPS conference ran a competition where the goal was to solve binary classification problems with large numbers of (mostly irrelevant) features, given small training sets. (This

Dataset	Domain	Type	D	% probes	N_{train}	N_{val}	N_{test}
Aracene	Mass spectrometry	Dense	10,000	30	100	100	700
Dexter	Text classification	Sparse	20,000	50	300	300	2000
Dorothea	Drug discovery	Sparse	100,000	50	800	350	800
Gisette	Digit recognition	Dense	5000	30	6000	1000	6500
Madelon	Artificial	Dense	500	96	2000	600	1800

Table 16.4 Summary of the data used in the NIPS 2003 "feature selection" challenge. For the Dorothea datasets, the features are binary. For the others, the features are real-valued.

Method	Screened features		ARD	
	Avg rank	Avg time	Avg rank	Avg time
HMC MLP	1.5	384 (138)	1.6	600 (186)
Boosted MLP	3.8	9.4 (8.6)	2.2	35.6 (33.5)
Bagged MLP	3.6	3.5 (1.1)	4.0	6.4 (4.4)
Boosted trees	3.4	3.03 (2.5)	4.0	34.1 (32.4)
Random forests	2.7	1.9 (1.7)	3.2	11.2 (9.3)

Table 16.5 Performance of different methods on the NIPS 2003 "feature selection" challenge. (HMC stands for hybrid Monte Carlo; see Section 24.5.4.) We report the average rank (lower is better) across the 5 datasets. We also report the average training time in minutes (standard error in brackets). The MCMC and bagged MLPs use two hidden layers of 20 and 8 units. The boosted MLPs use one hidden layer with 2 or 4 hidden units. The boosted trees used depths between 2 and 9, and shrinkage between 0.001 and 0.1. Each tree was trained on 80% of the data chosen at random at each step (so-called **stochastic gradient boosting**). From Table 11.3 of (Hastie et al. 2009).

was called a "feature selection" challenge, but performance was measured in terms of predictive accuracy, not in terms of the ability to select features.) The five datasets that were used are summarized in Table 16.4. The term **probe** refers to artifical variables that were added to the problem to make it harder. These have no predictive power, but are correlated with the original features.

Results of the competition are discussed in (Guyon et al. 2006). The overall winner was an approach based on Bayesian neural networks (Neal and Zhang 2006). In a follow-up study (Johnson 2009), Bayesian neural nets (MLPs with 2 hidden layers) were compared to several other methods based on bagging and boosting. Note that all of these methods are quite similar: in each case, the prediction has the form

$$\hat{f}(\mathbf{x}_*) = \sum_{m=1}^{M} w_m \mathbb{E}\left[y|\mathbf{x}_*, \boldsymbol{\theta}_m\right] \tag{16.108}$$

The Bayesian MLP was fit by MCMC (hybrid Monte Carlo), so we set $w_m = 1/M$ and set $\boldsymbol{\theta}_m$ to a draw from the posterior. In bagging, we set $w_m = 1/M$ and $\boldsymbol{\theta}_m$ is estimated by fitting the model to a bootstrap sample from the data. In boosting, we set $w_m = 1$ and the $\boldsymbol{\theta}_m$ are estimated sequentially.

To improve computational and statistical performance, some feature selection was performed. Two methods were considered: simple uni-variate screening using T-tests, and a method based on MLP+ARD. Results of this follow-up study are shown in Table 16.5. We see that Bayesian MLPs are again the winner. In second place are either random forests or boosted MLPs, depending on the preprocessing. However, it is not clear how statistically significant these differences are, since the test sets are relatively small.

In terms of training time, we see that MCMC is much slower than the other methods. It would be interesting to see how well deterministic Bayesian inference (e.g., Laplace approximation) would perform. (Obviously it will be much faster, but the question is: how much would one lose in statistical performance?)

16.8 Interpreting black-box models

Linear models are popular in part because they are easy to interpet. However, they often are poor predictors, which makes them a poor proxy for "nature's mechanism". Thus any conclusions about the importance of particular variables should only be based on models that have good predictive accuracy (Breiman 2001b). (Interestingly, many standard statistical tests of "goodness of fit" do not test the predictive accuracy of a model.)

In this chapter, we studied **black-box** models, which do have good predictive accuracy. Unfortunately, they are hard to interpret directly. Fortunately, there are various heuristics we can use to "probe" such models, in order to assess which input variables are the most important.

As a simple example, consider the following non-linear function, first proposed (Friedman 1991) to illustrate the power of MARS:

$$f(\mathbf{x}) = 10\sin(\pi x_1 x_2) + 20(x_3 - 0.5)^2 + 10x_4 + 5x_5 + \epsilon \tag{16.109}$$

where $\epsilon \sim \mathcal{N}(0, 1)$. We see that the output is a complex function of the inputs. By augmenting the $\mathbf{x}$ vector with additional irrelevant random variables, all drawn uniform on $[0, 1]$, we can create a challenging feature selection problem. In the experiments below, we add 5 extra dummy variables.

One useful way to measure the effect of a set s of variables on the output is to compute a **partial dependence plot** (Friedman 2001). This is a plot of $f(\mathbf{x}_s)$ vs $\mathbf{x}_s$, where $f(\mathbf{x}_s)$ is defined as the response to $\mathbf{x}_s$ with the other predictors averaged out:

$$f(\mathbf{x}_s) = \frac{1}{N} \sum_{i=1}^{N} f(\mathbf{x}_s, \mathbf{x}_{i,-s}) \tag{16.110}$$

Figure 16.20 shows an example where we use sets corresponding to each single variable. The data was generated from Equation 16.109, with 5 irrelevant variables added. We then fit a BART model (Section 16.2.5) and computed the partial dependence plots. We see that the predicted response is invariant for $s \in \{6, \dots, 10\}$, indicating that these variables are (marginally) irrelevant. The response is roughly linear in x_4 and x_5, and roughly quadratic in x_3. (The error bars are obtained by computing empirical quantiles of $f(\mathbf{x}, \boldsymbol{\theta})$ based on posterior samples of $\boldsymbol{\theta}$; alternatively, we can use bootstrap.)

Another very useful summary computes the **relative importance of predictor variables**. This can be thought of as a nonlinear, or even "model free", way of performing variable selection, although the technique is restricted to ensembles of trees. The basic idea, originally proposed in (Breiman et al. 1984), is to count how often variable j is used as a node in any of the trees.

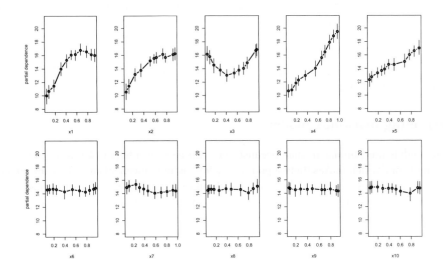

Figure 16.20 Partial dependence plots for the 10 predictors in Friedman's synthetic 5-dimensional regression problem.   Source: Figure 4 of (Chipman et al. 2010) . Used with kind permission of Hugh Chipman.

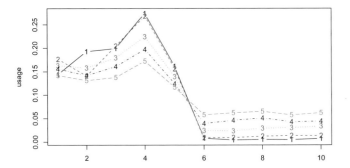

Figure 16.21 Average usage of each variable in a BART model fit to data where only the first 5 features are relevant. The different coloured lines correspond to different numbers of trees in the ensemble.   Source: Figure 5 of (Chipman et al. 2010) . Used with kind permission of Hugh Chipman.

In particular, let $v_j = \frac{1}{M} \sum_{m=1}^{M} \mathbb{I}(j \in T_m)$ be the proportion of all splitting rules that use x_j, where T_m is the m'th tree. If we can sample the posterior of trees, $p(T_{1:M}|\mathcal{D})$, we can easily compute the posterior for v_j. Alternatively, we can use bootstrap.

Figure 16.21 gives an example, using BART. We see that the five relevant variables are chosen much more than the five irrelevant variables. As we increase the number M of trees, all the variables are more likely to be chosen, reducing the sensitivity of this method, but for small M, the method is farily diagnostic.

See also (Cortez and Embrechts 2012) for a more recent approach for interpreting black box models.

Exercises

Exercise 16.1 Nonlinear regression for inverse dynamics

In this question, we fit a model which can predict what torques a robot needs to apply in order to make its arm reach a desired point in space. The data was collected from a SARCOS robot arm with 7 degrees of freedom. The input vector $\mathbf{x} \in \mathbb{R}^{21}$ encodes the desired position, velocity and accelaration of the 7 joints. The output vector $\mathbf{y} \in \mathbb{R}^7$ encodes the torques that should be applied to the joints to reach that point. The mapping from $\mathbf{x}$ to $\mathbf{y}$ is highly nonlinear.

We have $N = 48,933$ training points and $N_{test} = 4,449$ testing points. For simplicity, we following standard practice and focus on just predicting a scalar output, namely the torque for the first joint.

Download the data from `http://www.gaussianprocess.org/gpml`. Standardize the inputs so they have zero mean and unit variance on the training set, and center the outputs so they have zero mean on the training set. Apply the corresponding transformations to the test data. Below we will describe various models which you should fit to this transformed data. Then make predictions and compute the standardized mean squared error on the test set as follows:

$$SMSE = \frac{\frac{1}{N_{test}} \sum_{i=1}^{N_{test}} (y_i - \hat{y}_i)^2}{\sigma^2} \tag{16.111}$$

where $\sigma^2 = \frac{1}{N_{train}} \sum_{i=1}^{N_{train}} (y_i - \overline{y})^2$ is the variance of the output computed on the training set.

a. The first method you should try is standard linear regression. Turn in your numbers and code. (According to (Rasmussen and Williams 2006, p24), you should be able to achieve a SMSE of 0.075 using this method.)

b. Now try running K-means clustering (using cross validation to pick K). Then fit an RBF network to the data, using the $\boldsymbol{\mu}_k$ estimated by K-means. Use CV to estimate the RBF bandwidth. What SMSE do you get? Turn in your numbers and code. (According to (Rasmussen and Williams 2006, p24), Gaussian process regression can get an SMSE of 0.011, so the goal is to get close to that.)

c. Now try fitting a feedforward neural network. Use CV to pick the number of hidden units and the strength of the ℓ_2 regularizer. What SMSE do you get? Turn in your numbers and code.

17 *Markov and hidden Markov models*

17.1 Introduction

In this chapter, we discuss probabilistic models for sequences of observations, $X_1, \ldots, X_T$, of arbitrary length T. Such models have applications in computational biology, natural language processing, time series forecasting, etc. We focus on the case where we the observations occur at discrete "time steps", although "time" may also refer to locations within a sequence.

17.2 Markov models

Recall from Section 10.2.2 that the basic idea behind a Markov chain is to assume that X_t captures all the relevant information for predicting the future (i.e., we assume it is a sufficient statistic). If we assume discrete time steps, we can write the joint distribution as follows:

$$p(X_{1:T}) = p(X_1)p(X_2|X_1)p(X_3|X_2)\ldots = p(X_1) \prod_{t=2}^{T} p(X_t|X_{t-1}) \tag{17.1}$$

This is called a **Markov chain** or **Markov model**.

If we assume the transition function $p(X_t|X_{t-1})$ is independent of time, then the chain is called **homogeneous**, **stationary**, or **time-invariant**. This is an example of **parameter tying**, since the same parameter is shared by multiple variables. This assumption allows us to model an arbitrary number of variables using a fixed number of parameters; such models are called **stochastic processes**.

If we assume that the observed variables are discrete, so $X_t \in \{1, \ldots, K\}$, this is called a discrete-state or finite-state Markov chain. We will make this assumption throughout the rest of this section.

17.2.1 Transition matrix

When X_t is discrete, so $X_t \in \{1, \ldots, K\}$, the conditional distribution $p(X_t|X_{t-1})$ can be written as a $K \times K$ matrix, known as the **transition matrix** $\mathbf{A}$, where $A_{ij} = p(X_t = j|X_{t-1} = i)$ is the probability of going from state i to state j. Each row of the matrix sums to one, $\sum_j A_{ij} = 1$, so this is called a **stochastic matrix**.

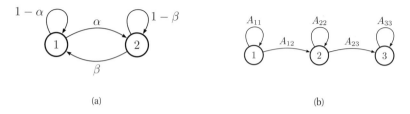

(a)   (b)

Figure 17.1   State transition diagrams for some simple Markov chains. Left: a 2-state chain. Right: a 3-state left-to-right chain.

A stationary, finite-state Markov chain is equivalent to a **stochastic automaton**. It is common to visualize such automata by drawing a directed graph, where nodes represent states and arrows represent legal transitions, i.e., non-zero elements of $\mathbf{A}$. This is known as a **state transition diagram**. The weights associated with the arcs are the probabilities. For example, the following 2-state chain

$$\mathbf{A} = \begin{pmatrix} 1 - \alpha & \alpha \\ \beta & 1 - \beta \end{pmatrix} \tag{17.2}$$

is illustrated in Figure 17.1(left). The following 3-state chain

$$\mathbf{A} = \begin{pmatrix} A_{11} & A_{12} & 0 \\ 0 & A_{22} & A_{23} \\ 0 & 0 & 1 \end{pmatrix} \tag{17.3}$$

is illustrated in Figure 17.1(right). This is called a **left-to-right transition matrix**, and is commonly used in speech recognition (Section 17.6.2).

The A_{ij} element of the transition matrix specifies the probability of getting from i to j in one step. The n-step transition matrix $\mathbf{A}(n)$ is defined as

$$A_{ij}(n) \triangleq p(X_{t+n} = j | X_t = i) \tag{17.4}$$

which is the probability of getting from i to j in exactly n steps. Obviously $\mathbf{A}(1) = \mathbf{A}$. The **Chapman-Kolmogorov** equations state that

$$A_{ij}(m + n) = \sum_{k=1}^{K} A_{ik}(m) A_{kj}(n) \tag{17.5}$$

In words, the probability of getting from i to j in $m + n$ steps is just the probability of getting from i to k in m steps, and then from k to j in n steps, summed up over all k. We can write the above as a matrix multiplication

$$\mathbf{A}(m + n) = \mathbf{A}(m)\mathbf{A}(n) \tag{17.6}$$

Hence

$$\mathbf{A}(n) = \mathbf{A}\,\mathbf{A}(n - 1) = \mathbf{A}\,\mathbf{A}\,\mathbf{A}(n - 2) = \cdots = \mathbf{A}^n \tag{17.7}$$

Thus we can simulate multiple steps of a Markov chain by "powering up" the transition matrix.

```
SAYS IT'S NOT IN THE CARDS LEGENDARY RECONNAISSANCE BY ROLLIE
DEMOCRACIES UNSUSTAINABLE COULD STRIKE REDLINING VISITS TO PROFIT
BOOKING WAIT HERE AT MADISON SQUARE GARDEN COUNTY COURTHOUSE WHERE HE
HAD BEEN DONE IN THREE ALREADY IN ANY WAY IN WHICH A TEACHER
```

Table 17.1 Example output from an 4-gram word model, trained using backoff smoothing on the Broadcast News corpus. The first 4 words are specified by hand, the model generates the 5th word, and then the results are fed back into the model. Source: `http://www.fit.vutbr.cz/~imikolov/rnnlm/gen-4gram.txt` .

17.2.2 Application: Language modeling

One important application of Markov models is to make statistical **language models**, which are probability distributions over sequences of words. We define the state space to be all the words in English (or some other language). The marginal probabilities $p(X_t = k)$ are called **unigram statistics**. If we use a first-order Markov model, then $p(X_t = k | X_{t-1} = j)$ is called a **bigram model**. If we use a second-order Markov model, then $p(X_t = k | X_{t-1} = j, X_{t-2} = i)$ is called a **trigram model**. And so on. In general these are called **n-gram models**. For example, Figure 17.2 shows 1-gram and 2-grams counts for the *letters* $\{a, \ldots, z, -\}$ (where - represents space) estimated from Darwin's *On The Origin Of Species*.

Language models can be used for several things, such as the following:

- **Sentence completion** A language model can predict the next word given the previous words in a sentence. This can be used to reduce the amount of typing required, which is particularly important for disabled users (see e.g., David Mackay's Dasher system[1]), or uses of mobile devices.
- **Data compression** Any density model can be used to define an encoding scheme, by assigning short codewords to more probable strings. The more accurate the predictive model, the fewer the number of bits it requires to store the data.
- **Text classification** Any density model can be used as a class-conditional density and hence turned into a (generative) classifier. Note that using a 0-gram class-conditional density (i.e., only unigram statistics) would be equivalent to a naive Bayes classifier (see Section 3.5).
- **Automatic essay writing** One can sample from $p(x_{1:t})$ to generate artificial text. This is one way of assessing the quality of the model. In Table 17.1, we give an example of text generated from a 4-gram model, trained on a corpus with 400 million words. ((Mikolov et al. 2011) describes a much better language model, based on a recurrent neural network, which generates much more semantically plausible text.)

---

1. `http://www.inference.phy.cam.ac.uk/dasher/`

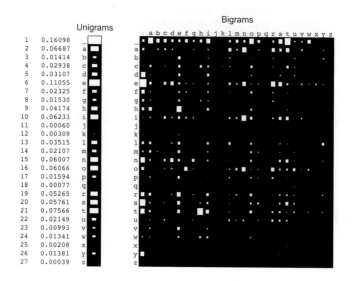

Figure 17.2 Unigram and bigram counts from Darwin's *On The Origin Of Species*. The 2D picture on the right is a Hinton diagram of the joint distribution. The size of the white squares is proportional to the value of the entry in the corresponding vector/ matrix. Based on (MacKay 2003, p22). Figure generated by `ngramPlot`.

17.2.2.1 MLE for Markov language models

We now discuss a simple way to estimate the transition matrix from training data. The probability of any particular sequence of length T is given by

$$p(x_{1:T}|\boldsymbol{\theta}) = \pi(x_1)A(x_1,x_2)\ldots A(x_{T-1},x_T) \tag{17.8}$$

$$= \prod_{j=1}^{K}(\pi_j)^{\mathbb{I}(x_1=j)}\prod_{t=2}^{T}\prod_{j=1}^{K}\prod_{k=1}^{K}(A_{jk})^{\mathbb{I}(x_t=k,x_{t-1}=j)} \tag{17.9}$$

Hence the log-likelihood of a set of sequences $\mathcal{D} = (\mathbf{x}_1,\ldots,\mathbf{x}_N)$, where $\mathbf{x}_i = (x_{i1},\ldots,x_{i,T_i})$ is a sequence of length T_i, is given by

$$\log p(\mathcal{D}|\boldsymbol{\theta}) = \sum_{i=1}^{N}\log p(\mathbf{x}_i|\boldsymbol{\theta}) = \sum_j N_j^1 \log \pi_j + \sum_j\sum_k N_{jk}\log A_{jk} \tag{17.10}$$

where we define the following counts:

$$N_j^1 \triangleq \sum_{i=1}^{N}\mathbb{I}(x_{i1}=j), \quad N_{jk} \triangleq \sum_{i=1}^{N}\sum_{t=1}^{T_i-1}\mathbb{I}(x_{i,t}=j,x_{i,t+1}=k) \tag{17.11}$$

Hence we can write the MLE as the normalized counts:

$$\hat{\pi}_j = \frac{N_j^1}{\sum_j N_j^1}, \quad \hat{A}_{jk} = \frac{N_{jk}}{\sum_k N_{jk}} \tag{17.12}$$

These results can be extended in a straightforward way to higher order Markov models. However, the problem of zero-counts becomes very acute whenever the number of states K, and/or the order of the chain, n, is large. An n-gram models has $O(K^n)$ parameters. If we have $K \sim 50,000$ words in our vocabulary, then a bi-gram model will have about 2.5 billion free parameters, corresponding to all possible word pairs. It is very unlikely we will see all of these in our training data. However, we do not want to predict that a particular word string is totally impossible just because we happen not to have seen it in our training text — that would be a severe form of overfitting.[2]

A simple solution to this is to use add-one smoothing, where we simply add one to all the empirical counts before normalizing. The Bayesian justification for this is given in Section 3.3.4.1. However add-one smoothing assumes all n-grams are equally likely, which is not very realistic. A more sophisticated Bayesian approach is discussed in Section 17.2.2.2.

An alternative to using smart priors is to gather lots and lots of data. For example, Google has fit n-gram models (for $n = 1 : 5$) based on one trillion words extracted from the web. Their data, which is over 100GB when uncompressed, is publically available.[3] An example of their data, for a set of 4-grams, is shown below.

```
serve as the incoming 92
serve as the incubator 99
serve as the independent 794
serve as the index 223
serve as the indication 72
serve as the indicator 120
serve as the indicators 45
serve as the indispensable 111
serve as the indispensible 40
serve as the individual 234
...
```

Although such an approach, based on "brute force and ignorance", can be successful, it is rather unsatisfying, since it is clear that this is not how humans learn (see e.g., (Tenenbaum and Xu 2000)). A more refined Bayesian approach, that needs much less data, is described in Section 17.2.2.2.

17.2.2.2 Empirical Bayes version of deleted interpolation

A common heuristic used to fix the sparse data problem is called **deleted interpolation** (Chen and Goodman 1996). This defines the transition matrix as a convex combination of the bigram

---

2. A famous example of an improbable, but syntactically valid, English word string, due to Noam Chomsky, is "colourless green ideas sleep furiously". We would not want our model to predict that this string is impossible. Even ungrammatical constructs should be allowed by our model with a certain probability, since people frequently violate grammatical rules, especially in spoken language.

3. See `http://googleresearch.blogspot.com/2006/08/all-our-n-gram-are-belong-to-you.html` for details.

frequencies $f_{jk} = N_{jk}/N_j$ and the unigram frequencies $f_k = N_k/N$:

$$A_{jk} = (1 - \lambda)f_{jk} + \lambda f_k \tag{17.13}$$

The term λ is usually set by cross validation. There is also a closely related technique called **backoff smoothing**; the idea is that if f_{jk} is too small, we "back off" to a more reliable estimate, namely f_k.

We will now show that the deleted interpolation heuristic is an approximation to the predictions made by a simple hierarchical Bayesian model. Our presentation follows (MacKay and Peto 1995). First, let us use an independent Dirichlet prior on each row of the transition matrix:

$$\mathbf{A}_j \sim \mathrm{Dir}(\alpha_0 m_1, \ldots, \alpha_0 m_K) = \mathrm{Dir}(\alpha_0 \mathbf{m}) = \mathrm{Dir}(\boldsymbol{\alpha}) \tag{17.14}$$

where $\mathbf{A}_j$ is row j of the transition matrix, $\mathbf{m}$ is the prior mean (satisfying $\sum_k m_k = 1$) and α_0 is the prior strength. We will use the same prior for each row: see Figure 17.3.

The posterior is given by $\mathbf{A}_j \sim \mathrm{Dir}(\boldsymbol{\alpha} + \mathbf{N}_j)$, where $\mathbf{N}_j = (N_{j1}, \ldots, N_{jK})$ is the vector that records the number of times we have transitioned out of state j to each of the other states. From Equation 3.51, the posterior predictive density is

$$p(X_{t+1} = k | X_t = j, \mathcal{D}) = \overline{A}_{jk} = \frac{N_{jk} + \alpha m_k}{N_j + \alpha_0} = \frac{f_{jk}N_j + \alpha m_k}{N_j + \alpha_0} = (1 - \lambda_j)f_{jk} + \lambda_j m_k \tag{17.15}$$

where $\overline{A}_{jk} = \mathbb{E}[A_{jk} | \mathcal{D}, \boldsymbol{\alpha}]$ and

$$\lambda_j = \frac{\alpha_j}{N_j + \alpha_0} \tag{17.16}$$

This is very similar to Equation 17.13 but not identical. The main difference is that the Bayesian model uses a context-dependent weight λ_j to combine m_k with the empirical frequency f_{jk}, rather than a fixed weight λ. This is like *adaptive* deleted interpolation. Furthermore, rather than backing off to the empirical marginal frequencies f_k, we back off to the model parameter m_k.

The only remaining question is: what values should we use for α and $\mathbf{m}$? Let's use empirical Bayes. Since we assume each row of the transition matrix is a priori independent given $\boldsymbol{\alpha}$, the marginal likelihood for our Markov model is found by applying Equation 5.24 to each row:

$$p(\mathcal{D}|\boldsymbol{\alpha}) = \prod_j \frac{B(\mathbf{N}_j + \boldsymbol{\alpha})}{B(\boldsymbol{\alpha})} \tag{17.17}$$

where $\mathbf{N}_j = (N_{j1}, \ldots, N_{jK})$ are the counts for leaving state j and $B(\boldsymbol{\alpha})$ is the generalized beta function.

We can fit this using the methods discussed in (Minka 2000e). However, we can also use the following approximation (MacKay and Peto 1995, p12):

$$m_k \propto |\{j : N_{jk} > 0\}| \tag{17.18}$$

This says that the prior probability of word k is given by the number of different contexts in which it occurs, rather than the number of times it occurs. To justify the reasonableness of this result, MacKay and Peto (1995) give the following example.

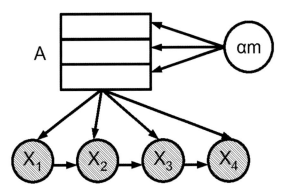

Figure 17.3 A Markov chain in which we put a different Dirichlet prior on every row of the transition matrix **A**, but the hyperparameters of the Dirichlet are shared.

```
Imagine, you see, that the language, you see, has, you see, a
frequently occuring couplet 'you see', you see, in which the second
word of the couplet, see, follows the first word, you, with very high
probability, you see. Then the marginal statistics, you see, are going
to become hugely dominated, you see, by the words you and see, with
equal frequency, you see.
```

If we use the standard smoothing formula, Equation 17.13, then P(you|novel) and P(see|novel), for some novel context word not seen before, would turn out to be the same, since the marginal frequencies of 'you' and 'see' are the same (11 times each). However, this seems unreasonable. 'You' appears in many contexts, so P(you|novel) should be high, but 'see' only follows 'you', so P(see|novel) should be low. If we use the Bayesian formula Equation 17.15, we will get this effect for free, since we back off to m_k not f_k, and m_k will be large for 'you' and small for 'see' by Equation 17.18.

Unfortunately, although elegant, this Bayesian model does not beat the state-of-the-art language model, known as **interpolated Kneser-Ney** (Kneser and Ney 1995; Chen and Goodman 1998). However, in (Teh 2006), it was shown how one can build a non-parametric Bayesian model which outperforms interpolated Kneser-Ney, by using variable-length contexts. In (Wood et al. 2009), this method was extended to create the "sequence memoizer", which is currently (2010) the best-performing language model.[4]

17.2.2.3 Handling out-of-vocabulary words

While the above smoothing methods handle the case where the counts are small or even zero, none of them deal with the case where the test set may contain a completely novel word. In particular, they all assume that the words in the vocabulary (i.e., the state space of X_t) is fixed and known (typically it is the set of unique words in the training data, or in some dictionary).

---

4. Interestingly, these non-parametric methods are based on posterior inference using MCMC (Section 24.1) and/or particle filtering (Section 23.5), rather than optimization methods such as EB. Despite this, they are quite efficient.

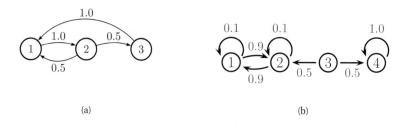

(a)   (b)

Figure 17.4   Some Markov chains. (a) A 3-state aperiodic chain. (b) A reducible 4-state chain.

Even if all $\overline{A}_{jk}$'s are non-zero, none of these models will predict a novel word outside of this set, and hence will assign zero probability to a test sentence with an unfamiliar word. (Unfamiliar words are bound to occur, because the set of words is an open class. For example, the set of proper nouns (names of people and places) is unbounded.)

A standard heuristic to solve this problem is to replace all novel words with the special symbol **unk**, which stands for "unknown". A certain amount of probability mass is held aside for this event.

A more principled solution would be to use a Dirichlet process, which can generate a countably infinite state space, as the amount of data increases (see Section 25.2.2). If all novel words are "accepted" as genuine words, then the system has no predictive power, since any misspelling will be considered a new word. So the novel word has to be seen frequently enough to warrant being added to the vocabulary. See e.g., (Friedman and Singer 1999; Griffiths and Tenenbaum 2001) for details.

17.2.3   Stationary distribution of a Markov chain *

We have been focussing on Markov models as a way of defining joint probability distributions over sequences. However, we can also interpret them as stochastic dynamical systems, where we "hop" from one state to another at each time step. In this case, we are often interested in the long term distribution over states, which is known as the **stationary distribution** of the chain. In this section, we discuss some of the relevant theory. Later we will consider two important applications: Google's PageRank algorithm for ranking web pages (Section 17.2.4), and the MCMC algorithm for generating samples from hard-to-normalize probability distributions (Chapter 24).

17.2.3.1   What is a stationary distribution?

Let $A_{ij} = p(X_t = j | X_{t-1} = i)$ be the one-step transition matrix, and let $\pi_t(j) = p(X_t = j)$ be the probability of being in state j at time t. It is conventional in this context to assume that $\boldsymbol{\pi}$ is a *row* vector. If we have an initial distribution over states of $\boldsymbol{\pi}_0$, then at time 1 we have

$$\pi_1(j) = \sum_i \pi_0(i) A_{ij} \tag{17.19}$$

or, in matrix notation,

$$\boldsymbol{\pi}_1 = \boldsymbol{\pi}_0 \mathbf{A} \tag{17.20}$$

We can imagine iterating these equations. If we ever reach a stage where

$$\boldsymbol{\pi} = \boldsymbol{\pi}\mathbf{A} \tag{17.21}$$

then we say we have reached the **stationary distribution** (also called the **invariant distribution** or **equilibrium distribution**). Once we enter the stationary distribution, we will never leave.

For example, consider the chain in Figure 17.4(a). To find its stationary distribution, we write

$$\begin{pmatrix} \pi_1 & \pi_2 & \pi_3 \end{pmatrix} = \begin{pmatrix} \pi_1 & \pi_2 & \pi_3 \end{pmatrix} \begin{pmatrix} 1 - A_{12} - A_{13} & A_{12} & A_{13} \\ A_{21} & 1 - A_{21} - A_{23} & A_{23} \\ A_{31} & A_{32} & 1 - A_{31} - A_{32} \end{pmatrix} \tag{17.22}$$

so

$$\pi_1 = \pi_1(1 - A_{12} - A_{12}) + \pi_2 A_{21} + \pi_3 A_{31} \tag{17.23}$$

or

$$\pi_1(A_{12} + A_{13}) = \pi_2 A_{21} + \pi_3 A_{31} \tag{17.24}$$

In general, we have

$$\pi_i \sum_{j \neq i} A_{ij} = \sum_{j \neq i} \pi_j A_{ji} \tag{17.25}$$

In other words, the probability of being in state i times the net flow out of state i must equal the probability of being in each other state j times the net flow from that state into i. These are called the **global balance equations**. We can then solve these equations, subject to the constraint that $\sum_j \pi_j = 1$.

17.2.3.2 Computing the stationary distribution

To find the stationary distribution, we can just solve the eigenvector equation $\mathbf{A}^T \mathbf{v} = \mathbf{v}$, and then to set $\boldsymbol{\pi} = \mathbf{v}^T$, where $\mathbf{v}$ is an eigenvector with eigenvalue 1. (We can be sure such an eigenvector exists, since $\mathbf{A}$ is a row-stochastic matrix, so $\mathbf{A}\mathbf{1} = \mathbf{1}$; also recall that the eigenvalues of $\mathbf{A}$ and $\mathbf{A}^T$ are the same.) Of course, since eigenvectors are unique only up to constants of proportionality, we must normalize $\mathbf{v}$ at the end to ensure it sums to one.

Note, however, that the eigenvectors are only guaranteed to be real-valued if the matrix is positive, $A_{ij} > 0$ (and hence $A_{ij} < 1$, due to the sum-to-one constraint). A more general approach, which can handle chains where some transition probabilities are 0 or 1 (such as Figure 17.4(a)), is as follows (Resnick 1992, p138). We have K constraints from $\boldsymbol{\pi}(\mathbf{I} - \mathbf{A}) = \mathbf{0}_{K \times 1}$ and 1 constraint from $\boldsymbol{\pi}\mathbf{1}_{K \times 1} = 0$. Since we only have K unknowns, this is overconstrained. So let us replace any column (e.g., the last) of $\mathbf{I} - \mathbf{A}$ with $\mathbf{1}$, to get a new matrix, call it $\mathbf{M}$. Next we define $\mathbf{r} = [0, 0, \ldots, 1]$, where the 1 in the last position corresponds to the column of all 1s in $\mathbf{M}$. We then solve $\boldsymbol{\pi}\mathbf{M} = \mathbf{r}$. For example, for a 3 state chain we have to solve this linear system:

$$\begin{pmatrix} \pi_1 & \pi_2 & \pi_3 \end{pmatrix} \begin{pmatrix} 1 - A_{11} & -A_{12} & 1 \\ -A_{21} & 1 - A_{22} & 1 \\ -A_{31} & -A_{32} & 1 \end{pmatrix} = \begin{pmatrix} 0 & 0 & 1 \end{pmatrix} \tag{17.26}$$

For the chain in Figure 17.4(a) we find $\boldsymbol{\pi} = [0.4, 0.4, 0.2]$. We can easily verify this is correct, since $\boldsymbol{\pi} = \boldsymbol{\pi}\mathbf{A}$. See `mcStatDist` for some Matlab code.

Unfortunately, not all chains have a stationary distribution. as we explain below.

17.2.3.3   When does a stationary distribution exist? *

Consider the 4-state chain in Figure 17.4(b). If we start in state 4, we will stay there forever, since 4 is an **absorbing state**. Thus $\boldsymbol{\pi} = (0, 0, 0, 1)$ is one possible stationary distribution. However, if we start in 1 or 2, we will oscillate between those two states for ever. So $\boldsymbol{\pi} = (0.5, 0.5, 0, 0)$ is another possible stationary distribution. If we start in state 3, we could end up in either of the above stationary distributions.

We see from this example that a necessary condition to have a unique stationary distribution is that the state transition diagram be a singly connected component, i.e., we can get from any state to any other state. Such chains are called **irreducible**.

Now consider the 2-state chain in Figure 17.1(a). This is irreducible provided $\alpha, \beta > 0$. Suppose $\alpha = \beta = 0.9$. It is clear by symmetry that this chain will spend 50% of its time in each state. Thus $\boldsymbol{\pi} = (0.5, 0.5)$. But now suppose $\alpha = \beta = 1$. In this case, the chain will oscillate between the two states, but the long-term distribution on states depends on where you start from. If we start in state 1, then on every odd time step (1,3,5,...) we will be in state 1; but if we start in state 2, then on every odd time step we will be in state 2.

This example motivates the following definition. Let us say that a chain has a **limiting distribution** if $\pi_j = \lim_{n\to\infty} A_{ij}^n$ exists and is independent of i, for all j. If this holds, then the long-run distribution over states will be independent of the starting state:

$$P(X_t = j) = \sum_i P(X_0 = i) A_{ij}(t) \to \pi_j \text{ as } t \to \infty \tag{17.27}$$

Let us now characterize when a limiting distribution exists. Define the **period** of state i to be

$$d(i) = \gcd\{t : A_{ii}(t) > 0\} \tag{17.28}$$

where gcd stands for **greatest common divisor**, i.e., the largest integer that divides all the members of the set. For example, in Figure 17.4(a), we have $d(1) = d(2) = gcd(2, 3, 4, 6, ...) = 1$ and $d(3) = gcd(3, 5, 6, ...) = 1$. We say a state i is **aperiodic** if $d(i) = 1$. (A sufficient condition to ensure this is if state i has a self-loop, but this is not a necessary condition.) We say a chain is aperiodic if all its states are aperiodic. One can show the following important result:

Theorem 17.2.1. *Every irreducible (singly connected), aperiodic finite state Markov chain has a limiting distribution, which is equal to $\boldsymbol{\pi}$, its unique stationary distribution.*

A special case of this result says that every regular finite state chain has a unique stationary distribution, where a **regular** chain is one whose transition matrix satisfies $A_{ij}^n > 0$ for some integer n and all i, j, i.e., it is possible to get from any state to any other state in n steps. Consequently, after n steps, the chain could be in any state, no matter where it started. One can show that sufficient conditions to ensure regularity are that the chain be irreducible (singly connected) and that every state have a self-transition.

To handle the case of Markov chains whose state-space is not finite (e.g, the countable set of all integers, or all the uncountable set of all reals), we need to generalize some of the earlier definitions. Since the details are rather technical, we just briefly state the main results without proof. See e.g., (Grimmett and Stirzaker 1992) for details.

For a stationary distribution to exist, we require irreducibility (singly connected) and aperiodicity, as before. But we also require that each state is **recurrent**. (A chain in which all states are recurrent is called a recurrent chain.) Recurrent means that you will return to that state with probability 1. As a simple example of a non-recurrent state (i.e., a **transient** state), consider Figure 17.4(b): states 3 is transient because one immediately leaves it and either spins around state 4 forever, or oscillates between states 1 and 2 forever. There is no way to return to state 3.

It is clear that any finite-state irreducible chain is recurrent, since you can always get back to where you started from. But now consider an example with an infinite state space. Suppose we perform a random walk on the integers, $\mathcal{X} = \{\ldots, -2, -1, 0, 1, 2, \ldots\}$. Let $A_{i,i+1} = p$ be the probability of moving right, and $A_{i,i-1} = 1 - p$ be the probability of moving left. Suppose we start at $X_1 = 0$. If $p > 0.5$, we will shoot off to $+\infty$; we are not guaranteed to return. Similarly, if $p < 0.5$, we will shoot off to $-\infty$. So in both cases, the chain is not recurrent, even though it is irreducible.

It should be intuitively obvious that we require all states to be recurrent for a stationary distribution to exist. However, this is not sufficient. To see this, consider the random walk on the integers again, and suppose $p = 0.5$. In this case, we can return to the origin an infinite number of times, so the chain is recurrent. However, it takes infinitely long to do so. This prohibits it from having a stationary distribution. The intuitive reason is that the distribution keeps spreading out over a larger and larger set of the integers, and never converges to a stationary distribution. More formally, we define a state to be **non-null recurrent** if the expected time to return to this state is finite. A chain in which all states are non-null is called a non-null chain.

For brevity, we we say that a state is **ergodic** if it is aperiodic, recurrent and non-null, and we say a chain is ergodic if all its states are ergodic.

We can now state our main theorem:

Theorem 17.2.2. *Every irreducible (singly connected), ergodic Markov chain has a limiting distribution, which is equal to $\boldsymbol{\pi}$, its unique stationary distribution.*

This generalizes Theorem 17.2.1, since for irreducible finite-state chains, all states are recurrent and non-null.

17.2.3.4 Detailed balance

Establishing ergodicity can be difficult. We now give an alternative condition that is easier to verify.

We say that a Markov chain $\mathbf{A}$ is **time reversible** if there exists a distribution $\boldsymbol{\pi}$ such that

$$\pi_i A_{ij} = \pi_j A_{ji} \tag{17.29}$$

These are called the **detailed balance equations**. This says that the flow from i to j must equal the flow from j to i, weighted by the appropriate source probabilities.

We have the following important result.

Theorem 17.2.3. *If a Markov chain with transition matrix $\mathbf{A}$ is regular and satisfies detailed balance wrt distribution $\boldsymbol{\pi}$, then $\boldsymbol{\pi}$ is a stationary distribution of the chain.*

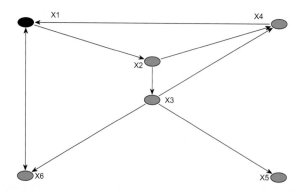

Figure 17.5   A very small world wide web. Figure generated by `pagerankDemo`, written by Tim Davis.

Proof. To see this, note that

$$\sum_i \pi_i A_{ij} = \sum_i \pi_j A_{ji} = \pi_j \sum_i A_{ji} = \pi_j \tag{17.30}$$

and hence $\pi = \mathbf{A}\pi$.   □

Note that this condition is sufficient but not necessary (see Figure 17.4(a) for an example of a chain with a stationary distribution which does not satisfy detailed balance).

In Section 24.1, we will discuss Markov chain Monte Carlo or MCMC methods. These take as input a desired distribution π and construct a transition matrix (or in general, a transition **kernel**) $\mathbf{A}$ which satisfies detailed balance wrt π. Thus by sampling states from such a chain, we will eventually enter the stationary distribution, and will visit states with probabilities given by π.

17.2.4   Application: Google's PageRank algorithm for web page ranking *

The results in Section 17.2.3 form the theoretical underpinnings to Google's **PageRank** algorithm, which is used for information retrieval on the world-wide web. We sketch the basic idea below; see (Bryan and Leise 2006) for a more detailed explanation.

We will treat the web as a giant directed graph, where nodes represent web pages (documents) and edges represent hyper-links.[5] We then perform a process called **web crawling**. We start at a few designated root nodes, such as `dmoz.org`, the home of the Open Directory Project, and then follows the links, storing all the pages that we encounter, until we run out of time.

Next, all of the words in each web page are entered into a data structure called an **inverted index**. That is, for each word, we store a list of the documents where this word occurs. (In practice, we store a list of hash codes representing the URLs.) At test time, when a user enters

---

5. In 2008, Google said it had indexed 1 trillion (10^{12}) unique URLs. If we assume there are about 10 URLs per page (on average), this means there were about 100 billion unique web pages. Estimates for 2010 are about 121 billion unique web pages. Source: `thenextweb.com/shareables/2011/01/11/infographic-how-big-is-the-internet`.

a query, we can just look up all the documents containing each word, and intersect these lists (since queries are defined by a conjunction of search terms). We can get a refined search by storing the location of each word in each document. We can then test if the words in a document occur in the same order as in the query.

Let us give an example, from `http://en.wikipedia.org/wiki/Inverted_index`. We have 3 documents, T_0 = "it is what it is", T_1 = "what is it" and T_2 = "it is a banana". Then we can create the following inverted index, where each pair represents a document and word location:

```
"a":      {(2, 2)}
"banana": {(2, 3)}
"is":     {(0, 1), (0, 4), (1, 1), (2, 1)}
"it":     {(0, 0), (0, 3), (1, 2), (2, 0)}
"what":   {(0, 2), (1, 0)}
```

For example, we see that the word "what" occurs at location 2 (counting from 0) in document 0, and location 0 in document 1. Suppose we search for "what is it". If we ignore word order, we retrieve the following documents:

$$\{T_0, T_1\} \cap \{T_0, T_1, T_2\} \cap \{T_0, T_1, T_2\} = \{T_0, T_1\} \tag{17.31}$$

If we require that the word order matches, only document T_1 would be returned. More generally, we can allow out-of-order matches, but can give "bonus points" to documents whose word order matches the query's word order, or to other features, such as if the words occur in the title of a document. We can then return the matching documents in decreasing order of their score/relevance. This is called document **ranking**.

So far, we have described the standard process of information retrieval. But the link structure of the web provides an additional source of information. The basic idea is that some web pages are more authoritative than others, so these should be ranked higher (assuming they match the query). A web page is an authority if it is linked to by many other pages. But to protect against the effect of so-called **link farms**, which are dummy pages which just link to a given site to boost its apparent relevance, we will weight each incoming link by the source's authority. Thus we get the following recursive definition for the authoritativeness of page j, also called its **PageRank**:

$$\pi_j = \sum_i A_{ij} \pi_i \tag{17.32}$$

where A_{ij} is the probability of following a link from i to j. We recognize Equation 17.32 as the stationary distribution of a Markov chain.

In the simplest setting, we define $A_{i\cdot}$ as a uniform distribution over all states that i is connected to. However, to ensure the distribution is unique, we need to make the chain into a regular chain. This can be done by allowing each state i to jump to any other state (including itself) with some small probability. This effectively makes the transition matrix aperiodic and fully connected (although the adjacency matrix G_{ij} of the web itself is highly sparse). [6]

---

6. There is a variant of this method called **personalized pagerank** (aka **random walks with restart**), which returns with some probability to a specific starting node u rather than a random node. We can compute a different stationary distribution for every node, $\boldsymbol{\pi}^u$; this gives us a measure of how "important" each node is relative to u.

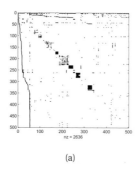   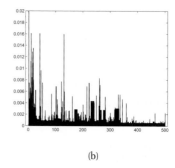

(a)   (b)

Figure 17.6   (a) Web graph of 500 sites rooted at `www.harvard.edu`. (b) Corresponding page rank vector. Figure generated by `pagerankDemoPmtk`, Based on code by Cleve Moler (Moler 2004).

We discuss efficient methods for computing the leading eigenvector of this giant matrix below. But first, let us give an example of the PageRank algorithm. Consider the small web in Figure 17.5. We find that the stationary distribution is

$$\boldsymbol{\pi} = (0.3209, 0.1706, 0.1065, 0.1368, 0.0643, 0.2008) \tag{17.33}$$

So a random surfer will visit site 1 about 32% of the time. We see that node 1 has a higher PageRank than nodes 4 or 6, even though they all have the same number of in-links. This is because being linked to from an influential nodehelps increase your PageRank score more than being linked to by a less influential node.

As a slightly larger example, Figure 17.6(a) shows a web graph, derived from the root of `harvard.edu`. Figure 17.6(b) shows the corresponding PageRank vector.

17.2.4.1   Efficiently computing the PageRank vector

Let $G_{ij} = 1$ iff there is a link from j to i. Now imagine performing a random walk on this graph, where at every time step, with probability $p = 0.85$ you follow one of the outlinks uniformly at random, and with probability $1 - p$ you jump to a random node, again chosen uniformly at random. If there are no outlinks, you just jump to a random page. (These random jumps, including self-transitions, ensure the chain is irreducible (singly connected) and regular. Hence we can solve for its unique stationary distribution using eigenvector methods.) This defines the following transition matrix:

$$M_{ij} = \begin{cases} pG_{ij}/c_j + \delta & \text{if } c_j \neq 0 \\ 1/n & \text{if } c_j = 0 \end{cases} \tag{17.34}$$

where n is the number of nodes, $\delta = (1 - p)/n$ is the probability of jumping from one page to another without following a link and $c_j = \sum_i G_{ij}$ represents the out-degree of page j. (If $n = 4 \cdot 10^9$ and $p = 0.85$, then $\delta = 3.75 \cdot 10^{-11}$.) Here $\mathbf{M}$ is a stochastic matrix in which *columns* sum to one. Note that $\mathbf{M} = \mathbf{A}^T$ in our earlier notation.

We can represent the transition matrix compactly as follows. Define the diagonal matrix $\mathbf{D}$

with entries

$$d_{jj} = \begin{cases} 1/c_j & \text{if } c_j \neq 0 \\ 0 & \text{if } c_j = 0 \end{cases} \tag{17.35}$$

Define the vector $\mathbf{z}$ with components

$$z_j = \begin{cases} \delta & \text{if } c_j \neq 0 \\ 1/n & \text{if } c_j = 0 \end{cases} \tag{17.36}$$

Then we can rewrite Equation 17.34 as follows:

$$\mathbf{M} = p\mathbf{GD} + \mathbf{1}\mathbf{z}^T \tag{17.37}$$

The matrix $\mathbf{M}$ is not sparse, but it is a rank one modification of a sparse matrix. Most of the elements of $\mathbf{M}$ are equal to the small constant δ. Obviously these do not need to be stored explicitly.

Our goal is to solve $\mathbf{v} = \mathbf{Mv}$, where $\mathbf{v} = \boldsymbol{\pi}^T$. One efficient method to find the leading eigenvector of a large matrix is known as the **power method**. This simply consists of repeated matrix-vector multiplication, followed by normalization:

$$\mathbf{v} \propto \mathbf{Mv} = p\mathbf{GDv} + \mathbf{1}\mathbf{z}^T\mathbf{v} \tag{17.38}$$

It is possible to implement the power method without using any matrix multiplications, by simply sampling from the transition matrix and counting how often you visit each state. This is essentially a Monte Carlo approximation to the sum implied by $\mathbf{v} = \mathbf{Mv}$. Applying this to the data in Figure 17.6(a) yields the stationary distribution in Figure 17.6(b). This took 13 iterations to converge, starting from a uniform distribution. (See also the function `pagerankDemo`, by Tim Davis, for an animation of the algorithm in action, applied to the small web example.) To handle changing web structure, we can re-run this algorithm every day or every week, starting $\mathbf{v}$ off at the old distribution (Langville and Meyer 2006).

For details on how to perform this Monte Carlo power method in a parallel distributed computing environment, see e.g., (Rajaraman and Ullman 2010).

17.2.4.2 Web spam

PageRank is not foolproof. For example, consider the strategy adopted by JC Penney, a department store in the USA. During the Christmas season of 2010, it planted many links to its home page on 1000s of irrelevant web pages, thus increasing its ranking on Google's search engine (Segal 2011). Even though each of these source pages has low PageRank, there were so many of them that their effect added up. Businesses call this **search engine optimization**; Google calls it **web spam**. When Google was notified of this scam (by the *New York Times*), it manually downweighted JC Penney, since such behavior violates Google's code of conduct. The result was that JC Penney dropped from rank 1 to rank 65, essentially making it disappear from view. Automatically detecting such scams relies on various techniques which are beyond the scope of this chapter.

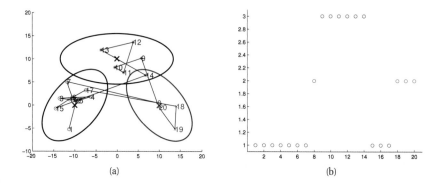

Figure 17.7 (a) Some 2d data sampled from a 3 state HMM. Each state emits from a 2d Gaussian. (b) The hidden state sequence. Based on Figure 13.8 of (Bishop 2006). Figure generated by `hmmLillypadDemo`.

17.3 Hidden Markov models

As we mentioned in Section 10.2.2, a **hidden Markov model** or **HMM** consists of a discrete-time, discrete-state Markov chain, with hidden states $z_t \in \{1, \ldots, K\}$, plus an **observation** model $p(\mathbf{x}_t | z_t)$. The corresponding joint distribution has the form

$$p(\mathbf{z}_{1:T}, \mathbf{x}_{1:T}) = p(\mathbf{z}_{1:T})p(\mathbf{x}_{1:T} | \mathbf{z}_{1:T}) = \left[p(z_1) \prod_{t=2}^{T} p(z_t | z_{t-1}) \right] \left[\prod_{t=1}^{T} p(\mathbf{x}_t | z_t) \right] \qquad (17.39)$$

The observations in an HMM can be discrete or continuous. If they are discrete, it is common for the observation model to be an observation matrix:

$$p(\mathbf{x}_t = l | z_t = k, \boldsymbol{\theta}) = B(k, l) \qquad (17.40)$$

If the observations are continuous, it is common for the observation model to be a conditional Gaussian:

$$p(\mathbf{x}_t | z_t = k, \boldsymbol{\theta}) = \mathcal{N}(\mathbf{x}_t | \boldsymbol{\mu}_k, \boldsymbol{\Sigma}_k) \qquad (17.41)$$

Figure 17.7 shows an example where we have 3 states, each of which emits a different Gaussian. The resulting model is similar to a Gaussian mixture model, except the cluster membership has Markovian dynamics. (Indeed, HMMs are sometimes called **Markov switching models** (Fruhwirth-Schnatter 2007).) We see that we tend to get multiple observations in the same location, and then a sudden jump to a new cluster.

17.3.1 Applications of HMMs

HMMs can be used as black-box density models on sequences. They have the advantage over Markov models in that they can represent long-range dependencies between observations, mediated via the latent variables. In particular, note that they do not assume the Markov property holds for the observations themselves. Such black-box models are useful for time-series prediction (Fraser 2008). They can also be used to define class-conditional densities inside a generative classifier.

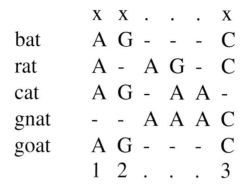

(a)

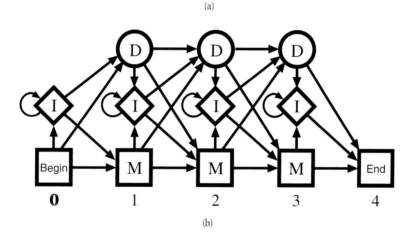

(b)

Figure 17.8 (a) Some DNA sequences. (b) State transition diagram for a profile HMM. Source: Figure 5.7 of (Durbin et al. 1998). Used with kind permission of Richard Durbin.

However, it is more common to imbue the hidden states with some desired meaning, and to then try to estimate the hidden states from the observations, i.e., to compute $p(z_t|\mathbf{x}_{1:t})$ if we are in an online scenario, or $p(z_t|\mathbf{x}_{1:T})$ if we are in an offline scenario (see Section 17.4.1 for further discussion of the differences between these two approaches). Below we give some examples of applications which use HMMs in this way:

- **Automatic speech recognition**. Here $\mathbf{x}_t$ represents features extracted from the speech signal, and z_t represents the word that is being spoken. The transition model $p(z_t|z_{t-1})$ represents the language model, and the observation model $p(\mathbf{x}_t|z_t)$ represents the acoustic model. See e.g., (Jelinek 1997; Jurafsky and Martin 2008) for details.
- **Activity recognition**. Here $\mathbf{x}_t$ represents features extracted from a video frame, and z_t is the class of activity the person is engaged in (e.g., running, walking, sitting, etc.) See e.g., (Szeliski 2010) for details.

- **Part of speech tagging**. Here x_t represents a word, and z_t represents its **part of speech** (noun, verb, adjective, etc.) See Section 19.6.2.1 for more information on POS tagging and related tasks.

- **Gene finding**. Here x_t represents the DNA nucleotides (A,C,G,T), and z_t represents whether we are inside a gene-coding region or not. See e.g., (Schweikerta et al. 2009) for details.

- **Protein sequence alignment**. Here x_t represents an amino acid, and z_t represents whether this matches the latent **consensus sequence** at this location. This model is called a **profile HMM** and is illustrated in Figure 17.8. The HMM has 3 states, called match, insert and delete. If z_t is a match state, then x_t is equal to the t'th value of the consensus. If z_t is an insert state, then x_t is generated from a uniform distribution that is unrelated to the consensus sequence. If z_t is a delete state, then $x_t = -$. In this way, we can generate noisy copies of the consensus sequence of different lengths. In Figure 17.8(a), the consensus is "AGC", and we see various versions of this below. A path through the state transition diagram, shown in Figure 17.8(b), specifies how to align a sequence to the consensus, e.g., for the gnat, the most probable path is D, D, I, I, I, M. This means we delete the A and G parts of the consensus sequence, we insert 3 A's, and then we match the final C. We can estimate the model parameters by counting the number of such transitions, and the number of emissions from each kind of state, as shown in Figure 17.8(c). See Section 17.5 for more information on training an HMM, and (Durbin et al. 1998) for details on profile HMMs.

Note that for some of these tasks, conditional random fields, which are essentially discriminative versions of HMMs, may be more suitable; see Chapter 19 for details.

17.4　Inference in HMMs

We now discuss how to infer the hidden state sequence of an HMM, assuming the parameters are known. Exactly the same algorithms apply to other chain-structured graphical models, such as chain CRFs (see Section 19.6.1). In Chapter 20, we generalize these methods to arbitrary graphs. And in Section 17.5.2, we show how we can use the output of inference in the context of parameter estimation.

17.4.1　Types of inference problems for temporal models

There are several different kinds of inferential tasks for an HMM (and SSM in general). To illustrate the differences, we will consider an example called the **occasionally dishonest casino**, from (Durbin et al. 1998). In this model, $x_t \in \{1, 2, \dots, 6\}$ represents which dice face shows up, and z_t represents the identity of the dice that is being used. Most of the time the casino uses a fair dice, $z = 1$, but occasionally it switches to a loaded dice, $z = 2$, for a short period. If $z = 1$ the observation distribution is a uniform multinoulli over the symbols $\{1, \dots, 6\}$. If $z = 2$, the observation distribution is skewed towards face 6 (see Figure 17.9). If we sample from this model, we may observe data such as the following:

Listing 17.1 Example output of `casinoDemo`

```
Rolls:    6641532161621152346532143566342616552342323151424641566663246
Die:      LLLLLLLLLLLLLLFFFFFFLLLLLLLLLLLLLLFFFFFFFFFFFFFFFFFFFFLLLLLLLL
```

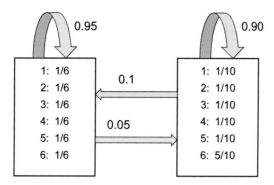

Figure 17.9 An HMM for the occasionally dishonest casino. The blue arrows visualize the state transition diagram **A**. Based on (Durbin et al. 1998, p54).

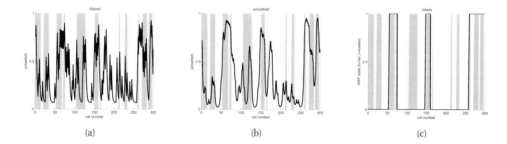

Figure 17.10 Inference in the dishonest casino. Vertical gray bars denote the samples that we generated using a loaded die. (a) Filtered estimate of probability of using a loaded dice. (b) Smoothed estimates. (c) MAP trajectory. Figure generated by `casinoDemo`.

Here "rolls" refers to the observed symbol and "die" refers to the hidden state (L is loaded and F is fair). Thus we see that the model generates a sequence of symbols, but the statistics of the distribution changes abruptly every now and then. In a typical application, we just see the rolls and want to infer which dice is being used. But there are different kinds of inference, which we summarize below.

- **Filtering** means to compute the **belief state** $p(z_t|\mathbf{x}_{1:t})$ online, or recursively, as the data streams in. This is called "filtering" because it reduces the noise more than simply estimating the hidden state using just the current estimate, $p(z_t|\mathbf{x}_t)$. We will see below that we can perform filtering by simply applying Bayes rule in a sequential fashion. See Figure 17.10(a) for an example.

- **Smoothing** means to compute $p(z_t|\mathbf{x}_{1:T})$ offline, given all the evidence. See Figure 17.10(b) for an example. By conditioning on past and future data, our uncertainty will be significantly reduced. To understand this intuitively, consider a detective trying to figure out who committed a crime. As he moves through the crime scene, his uncertainty is high until he finds

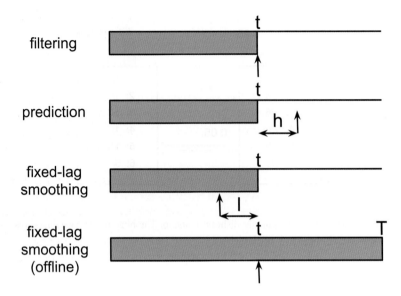

Figure 17.11 The main kinds of inference for state-space models. The shaded region is the interval for which we have data. The arrow represents the time step at which we want to perform inference. t is the current time, T is the sequence length, ℓ is the lag and h is the prediction horizon. See text for details.

the key clue; then he has an "aha" moment, his uncertainty is reduced, and all the previously confusing observations are, in **hindsight**, easy to explain.

- **Fixed lag smoothing** is an interesting compromise between online and offline estimation; it involves computing $p(z_{t-\ell}|\mathbf{x}_{1:t})$, where $\ell > 0$ is called the lag. This gives better performance than filtering, but incurs a slight delay. By changing the size of the lag, one can trade off accuracy vs delay.

- **Prediction** Instead of predicting the past given the future, as in fixed lag smoothing, we might want to predict the future given the past, i.e., to compute $p(z_{t+h}|\mathbf{x}_{1:t})$, where $h > 0$ is called the prediction **horizon**. For example, suppose $h = 2$; then we have

$$p(z_{t+2}|\mathbf{x}_{1:t}) \quad = \quad \sum_{z_{t+1}} \sum_{z_t} p(z_{t+2}|z_{t+1})p(z_{t+1}|z_t)p(z_t|\mathbf{x}_{1:t}) \tag{17.42}$$

It is straightforward to perform this computation: we just power up the transition matrix and apply it to the current belief state. The quantity $p(z_{t+h}|\mathbf{x}_{1:t})$ is a prediction about future hidden states; it can be converted into a prediction about future observations using

$$p(\mathbf{x}_{t+h}|\mathbf{x}_{1:t}) = \sum_{z_{t+h}} p(\mathbf{x}_{t+h}|z_{t+h})p(z_{t+h}|\mathbf{x}_{1:t}) \tag{17.43}$$

This is the posterior predictive density, and can be used for time-series forecasting (see (Fraser 2008) for details). See Figure 17.11 for a sketch of the relationship between filtering, smoothing, and prediction.

- **MAP estimation** This means computing $\arg\max_{\mathbf{z}_{1:T}} p(\mathbf{z}_{1:T}|\mathbf{x}_{1:T})$, which is a most probable state sequence. In the context of HMMs, this is known as **Viterbi decoding** (see Section 17.4.4). Figure 17.10 illustrates the difference between filtering, smoothing and MAP decoding for the occasionally dishonest casino HMM. We see that the smoothed (offline) estimate is indeed smoother than the filtered (online) estimate. If we threshold the estimates at 0.5 and compare to the true sequence, we find that the filtered method makes 71 errors out of 300, and the smoothed method makes 49/300; the MAP path makes 60/300 errors. It is not surprising that smoothing makes fewer errors than Viterbi, since the optimal way to minimize bit-error rate is to threshold the posterior marginals (see Section 5.7.1.1). Nevertheless, for some applications, we may prefer the Viterbi decoding, as we discuss in Section 17.4.4.

- **Posterior samples** If there is more than one plausible interpretation of the data, it can be useful to sample from the posterior, $\mathbf{z}_{1:T} \sim p(\mathbf{z}_{1:T}|\mathbf{x}_{1:T})$. These sample paths contain much more information than the sequence of marginals computed by smoothing.

- **Probability of the evidence** We can compute the probability of the evidence, $p(\mathbf{x}_{1:T})$, by summing up over all hidden paths, $p(\mathbf{x}_{1:T}) = \sum_{\mathbf{z}_{1:T}} p(\mathbf{z}_{1:T}, \mathbf{x}_{1:T})$. This can be used to classify sequences (e.g., if the HMM is used as a class conditional density), for model-based clustering, for anomaly detection, etc.

17.4.2 The forwards algorithm

We now describe how to recursively compute the filtered marginals, $p(z_t|\mathbf{x}_{1:t})$ in an HMM.

The algorithm has two steps. First comes the prediction step, in which we compute the **one-step-ahead predictive density**; this acts as the new prior for time t:

$$p(z_t = j|\mathbf{x}_{1:t-1}) = \sum_i p(z_t = j|z_{t-1} = i)p(z_{t-1} = i|\mathbf{x}_{1:t-1}) \tag{17.44}$$

Next comes the update step, in which we absorb the observed data from time t using Bayes rule:

$$\alpha_t(j) \triangleq p(z_t = j|\mathbf{x}_{1:t}) = p(z_t = j|\mathbf{x}_t, \mathbf{x}_{1:t-1}) \tag{17.45}$$

$$= \frac{1}{Z_t} p(\mathbf{x}_t|z_t = j, \cancel{\mathbf{x}_{1:t-1}})p(z_t = j|\mathbf{x}_{1:t-1}) \tag{17.46}$$

where the normalization constant is given by

$$Z_t \triangleq p(\mathbf{x}_t|\mathbf{x}_{1:t-1}) = \sum_j p(z_t = j|\mathbf{x}_{1:t-1})p(\mathbf{x}_t|z_t = j) \tag{17.47}$$

This process is known as the **predict-update cycle**. The distribution $p(z_t|\mathbf{x}_{1:t})$ is called the (filtered) **belief state** at time t, and is a vector of K numbers, often denoted by $\boldsymbol{\alpha}_t$.[7] In

---

7. In most other publications, $\alpha_t(j)$ is defined as the *joint probability* $p(z_t = j, \mathbf{x}_{1:t})$, rather than defining it as the conditional probability $p(z_t = j|\mathbf{x}_{1:t})$ as we do here. Defining it as $p(z_t = j, \mathbf{x}_{1:t})$, has two problems. First it rapidly suffers from numerical underflow, since the probability of observing any particular sequence of evidence $\mathbf{x}_{1:t}$ is very small. Second, the joint probability $p(z_t = j, \mathbf{x}_{1:t})$ is not as meaningful as the posterior distribution over states, $p(z_t = j|\mathbf{x}_{1:t})$. Note, however, that these two definitions only differ by a multiplicative constant, so the *algorithmic* difference is just one line of code. (In fact, all good implementations of the forwards algorithm normalize the belief state after each step, to avoid underflow, and hence they are actually computing $p(z_t = j|\mathbf{x}_{1:t})$!)

matrix-vector notation, we can write the update in the following simple form:

$$\boldsymbol{\alpha}_t \propto \boldsymbol{\psi}_t \odot (\boldsymbol{\Psi}^T \boldsymbol{\alpha}_{t-1}) \tag{17.48}$$

where $\psi_t(j) = p(\mathbf{x}_t|z_t = j)$ is the local evidence at time t, $\Psi(i,j) = p(z_t = j|z_{t-1} = i)$ is the transition matrix, and $\mathbf{u} \odot \mathbf{v}$ is the **Hadamard product**, representing elementwise vector multiplication. See Algorithm 17.1 for the pseudo-code, and hmmFilter for some Matlab code.

In addition to computing the hidden states, we can use this algorithm to compute the log probability of the evidence:

$$\log p(\mathbf{x}_{1:T}|\boldsymbol{\theta}) = \sum_{t=1}^{T} \log p(\mathbf{x}_t|\mathbf{x}_{1:t-1}) = \sum_{t=1}^{T} \log Z_t \tag{17.49}$$

(We need to work in the log domain to avoid numerical underflow.)

Algorithm 17.1: Forwards algorithm

---

1 Input: Transition matrices $\Psi(i,j) = p(z_t = j|z_{t-1} = i)$, local evidence vectors
  $\psi_t(j) = p(\mathbf{x}_t|z_t = j)$, initial state distribution $\pi(j) = p(z_1 = j)$;
2 $[\boldsymbol{\alpha}_1, Z_1] = \text{normalize}(\boldsymbol{\psi}_1 \odot \boldsymbol{\pi})$;
3 **for** $t = 2 : T$ **do**
4 $\quad \lfloor \; [\boldsymbol{\alpha}_t, Z_t] = \text{normalize}(\boldsymbol{\psi}_t \odot (\boldsymbol{\Psi}^T \boldsymbol{\alpha}_{t-1}))$;
5 Return $\boldsymbol{\alpha}_{1:T}$ and $\log p(\mathbf{x}_{1:T}) = \sum_t \log Z_t$;

6 Subroutine: $[\mathbf{v}, Z] = \text{normalize}(\mathbf{u}) : Z = \sum_j u_j; \quad v_j = u_j/Z$;

17.4.3   The forwards-backwards algorithm

In Section 17.4.2, we explained how to compute the filtered marginals $p(z_t = j|\mathbf{x}_{1:t})$ using online inference. We now discuss how to compute the smoothed marginals, $p(z_t = j|\mathbf{x}_{1:T})$, using offline inference.

17.4.3.1   Basic idea

The key decomposition relies on the fact that we can break the chain into two parts, the past and the future, by conditioning on z_t:

$$p(z_t = j|\mathbf{x}_{1:T}) \propto p(z_t = j, \mathbf{x}_{t+1:T}|\mathbf{x}_{1:t}) \propto p(z_t = j|\mathbf{x}_{1:t})p(\mathbf{x}_{t+1:T}|z_t = j, \mathbf{x}_{1:t}) \tag{17.50}$$

Let $\alpha_t(j) \triangleq p(z_t = j|\mathbf{x}_{1:t})$ be the filtered belief state as before. Also, define

$$\beta_t(j) \triangleq p(\mathbf{x}_{t+1:T}|z_t = j) \tag{17.51}$$

as the conditional likelihood of future evidence given that the hidden state at time t is j. (Note that this is not a probability distribution over states, since it does not need to satisfy $\sum_j \beta_t(j) = 1$.) Finally, define

$$\gamma_t(j) \triangleq p(z_t = j|\mathbf{x}_{1:T}) \tag{17.52}$$

as the desired smoothed posterior marginal. From Equation 17.50, we have

$$\gamma_t(j) \propto \alpha_t(j)\beta_t(j) \tag{17.53}$$

We have already described how to recursively compute the α's in a left-to-right fashion in Section 17.4.2. We now describe how to recursively compute the β's in a right-to-left fashion. If we have already computed β_t, we can compute β_{t-1} as follows:

$$
\begin{aligned}
\beta_{t-1}(i) &= p(\mathbf{x}_{t:T}|z_{t-1} = i) \tag{17.54} \\
&= \sum_j p(z_t = j, \mathbf{x}_t, \mathbf{x}_{t+1:T}|z_{t-1} = i) \tag{17.55} \\
&= \sum_j p(\mathbf{x}_{t+1:T}|z_t = j, \cancel{\mathbf{x}_t, z_{t-1} = i})p(z_t = j, \mathbf{x}_t|z_{t-1} = i) \tag{17.56} \\
&= \sum_j p(\mathbf{x}_{t+1:T}|z_t = j)p(\mathbf{x}_t|z_t = j, \cancel{z_{t-1} = i})p(z_t = j|z_{t-1} = i) \tag{17.57} \\
&= \sum_j \beta_t(j)\psi_t(j)\psi(i, j) \tag{17.58}
\end{aligned}
$$

We can write the resulting equation in matrix-vector form as

$$\boldsymbol{\beta}_{t-1} = \boldsymbol{\Psi}(\boldsymbol{\psi}_t \odot \boldsymbol{\beta}_t) \tag{17.59}$$

The base case is

$$\beta_T(i) = p(\mathbf{x}_{T+1:T}|z_T = i) = p(\emptyset|z_T = i) = 1 \tag{17.60}$$

which is the probability of a non-event.

Having computed the forwards and backwards messages, we can combine them to compute $\gamma_t(j) \propto \alpha_t(j)\beta_t(j)$. The overall algorithm is known as the **forwards-backwards algorithm**. The pseudo code is very similar to the forwards case; see `hmmFwdBack` for an implementation.

We can think of this algorithm as passing "messages" from left to right, and then from right to left, and then combining them at each node. We will generalize this intuition in Section 20.2, when we discuss belief propagation.

17.4.3.2 Two-slice smoothed marginals

When we estimate the parameters of the transition matrix using EM (see Section 17.5), we will need to compute the expected number of transitions from state i to state j:

$$N_{ij} = \sum_{t=1}^{T-1} \mathbb{E}\left[\mathbb{I}(z_t = i, z_{t+1} = j)|\mathbf{x}_{1:T}\right] = \sum_{t=1}^{T-1} p(z_t = i, z_{t+1} = j|\mathbf{x}_{1:T}) \tag{17.61}$$

The term $p(z_t = i, z_{t+1} = j | \mathbf{x}_{1:T})$ is called a (smoothed) **two-slice marginal**, and can be computed as follows

$$
\begin{align}
\xi_{t,t+1}(i,j) &\triangleq p(z_t = i, z_{t+1} = j | \mathbf{x}_{1:T}) \tag{17.62}\\
&\propto p(z_t | \mathbf{x}_{1:t}) p(z_{t+1} | z_t, \mathbf{x}_{t+1:T}) \tag{17.63}\\
&\propto p(z_t | \mathbf{x}_{1:t}) p(\mathbf{x}_{t+1:T} | z_t, z_{t+1}) p(z_{t+1} | z_t) \tag{17.64}\\
&\propto p(z_t | \mathbf{x}_{1:t}) p(\mathbf{x}_{t+1} | z_{t+1}) p(\mathbf{x}_{t+2:T} | z_{t+1}) p(z_{t+1} | z_t) \tag{17.65}\\
&= \alpha_t(i) \phi_{t+1}(j) \beta_{t+1}(j) \psi(i,j) \tag{17.66}
\end{align}
$$

In matrix-vector form, we have

$$
\boldsymbol{\xi}_{t,t+1} \propto \boldsymbol{\Psi} \odot \left(\boldsymbol{\alpha}_t (\boldsymbol{\phi}_{t+1} \odot \boldsymbol{\beta}_{t+1})^T \right) \tag{17.67}
$$

For another interpretation of these equations, see Section 20.2.4.3.

17.4.3.3   Time and space complexity

It is clear that a straightforward implementation of FB takes $O(K^2 T)$ time, since we must perform a $K \times K$ matrix multiplication at each step. For some applications, such as speech recognition, K is very large, so the $O(K^2)$ term becomes prohibitive. Fortunately, if the transition matrix is sparse, we can reduce this substantially. For example, in a left-to-right transition matrix, the algorithm takes $O(TK)$ time.

In some cases, we can exploit special properties of the state space, even if the transition matrix is not sparse. In particular, suppose the states represent a discretization of an underlying continuous state-space, and the transition matrix has the form $\psi(i,j) \propto \exp(-\sigma^2 |\mathbf{z}_i - \mathbf{z}_j|)$, where $\mathbf{z}_i$ is the continuous vector represented by state i. Then one can implement the forwards-backwards algorithm in $O(TK \log K)$ time. This is very useful for models with large state spaces. See Section 22.2.6.1 for details.

In some cases, the bottleneck is memory, not time. The expected sufficient statistics needed by EM are $\sum_t \xi_{t-1,t}(i,j)$; this takes constant space (independent of T); however, to compute them, we need $O(KT)$ working space, since we must store α_t for $t = 1, \ldots, T$ until we do the backwards pass. It is possible to devise a simple divide-and-conquer algorithm that reduces the space complexity from $O(KT)$ to $O(K \log T)$ at the cost of increasing the running time from $O(K^2 T)$ to $O(K^2 T \log T)$: see (Binder et al. 1997; Zweig and Padmanabhan 2000) for details.

17.4.4   The Viterbi algorithm

The **Viterbi** algorithm (Viterbi 1967) can be used to compute the most probable sequence of states in a chain-structured graphical model, i.e., it can compute

$$
\mathbf{z}^* = \arg \max_{\mathbf{z}_{1:T}} p(\mathbf{z}_{1:T} | \mathbf{x}_{1:T}) \tag{17.68}
$$

This is equivalent to computing a shortest path through the **trellis diagram** in Figure 17.12, where the nodes are possible states at each time step, and the node and edge weights are log probabilities. That is, the weight of a path $z_1, z_2, \ldots, z_T$ is given by

$$
\log \pi_1(z_1) + \log \phi_1(z_1) + \sum_{t=2}^{T} \left[\log \psi(z_{t-1}, z_t) + \log \phi_t(z_t) \right] \tag{17.69}
$$

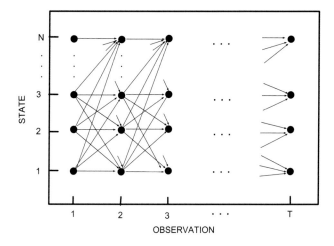

Figure 17.12 The trellis of states vs time for a Markov chain. Based on (Rabiner 1989).

17.4.4.1 MAP vs MPE

Before discussing how the algorithm works, let us make one important remark: the *(jointly) most probable sequence of states is not necessarily the same as the sequence of (marginally) most probable states.* The former is given by Equation 17.68, and is what Viterbi computes, whereas the latter is given by the maximizer of the posterior marginals or **MPM**:

$$\hat{\mathbf{z}} = (\arg\max_{z_1} p(z_1|\mathbf{x}_{1:T}), \dots, \arg\max_{z_T} p(z_T|\mathbf{x}_{1:T})) \tag{17.70}$$

As a simple example of the difference, consider a chain with two time steps, defining the following joint:

	$X_1 = 0$	$X_1 = 1$	
$X_2 = 0$	0.04	0.3	0.34
$X_2 = 1$	0.36	0.3	0.66
	0.4	0.6	

The joint MAP estimate is $(0, 1)$, whereas the sequence of marginal MPMs is $(1, 1)$.

The advantage of the joint MAP estimate is that it is always globally consistent. For example, suppose we are performing speech recognition and someones says "recognize speech". This could be mis-heard as "wreck a nice beach". Locally it may appear that "beach" is the most probable interpretation of that particular window of sound, but when we add the requirement that the data be explained by a single linguistically plausible path, this interpretation becomes less likely.

On the other hand, the MPM estimates can be more robust (Marroquin et al. 1987). To see why, note that in Viterbi, when we estimate z_t, we "max out" the other variables:

$$z_t^* = \arg\max_{z_t} \max_{\mathbf{z}_{1:t-1}, \mathbf{z}_{t+1:T}} p(\mathbf{z}_{1:t-1}, z_t, \mathbf{z}_{t+1:T}|\mathbf{x}_{1:T}) \tag{17.71}$$

whereas we when we use forwards-backwards, we sum out the other variables:

$$p(z_t|\mathbf{x}_{1:T}) = \sum_{\mathbf{z}_{1:t-1},\mathbf{z}_{t+1:T}} p(\mathbf{z}_{1:t-1}, z_t, \mathbf{z}_{t+1:T}|\mathbf{x}_{1:T}) \qquad (17.72)$$

This makes the MPM in Equation 17.70 more robust, since we estimate each node averaging over its neighbors, rather than conditioning on a specific value of its neighbors.[8]

17.4.4.2 Details of the algorithm

It is tempting to think that we can implement Viterbi by just replacing the sum-operator in forwards-backwards with a max-operator. The former is called the **sum-product**, and the latter the **max-product** algorithm. If there is a unique mode, running max-product and then computing using Equation 17.70 will give the same result as using Equation 17.68 (Weiss and Freeman 2001b), but in general, it can lead to incorrect results if there are multiple equally probably joint assignments. The reasons is that each node breaks ties independently and hence may do so in a manner that is inconsistent with its neighbors. The Viterbi algorithm is therefore not quite as simple as replacing sum with max. In particular, the forwards pass does use max-product, but the backwards pass uses a **traceback** procedure to recover the most probable path through the trellis of states. Essentially, once z_t picks its most probable state, the previous nodes condition on this event, and therefore they will break ties consistently.

In more detail, define

$$\delta_t(j) \triangleq \max_{z_1,\ldots,z_{t-1}} p(\mathbf{z}_{1:t-1}, z_t = j|\mathbf{x}_{1:t}) \qquad (17.73)$$

This is the probability of ending up in state j at time t, given that we take the most probable path. The key insight is that the most probable path to state j at time t must consist of the most probable path to some other state i at time $t-1$, followed by a transition from i to j. Hence

$$\delta_t(j) = \max_i \delta_{t-1}(i)\psi(i,j)\phi_t(j) \qquad (17.74)$$

We also keep track of the most likely previous state, for each possible state that we end up in:

$$a_t(j) = \operatorname*{argmax}_i \delta_{t-1}(i)\psi(i,j)\phi_t(j) \qquad (17.75)$$

That is, $a_t(j)$ tells us the most likely previous state on the most probable path to $z_t = j$. We initialize by setting

$$\delta_1(j) = \pi_j\phi_1(j) \qquad (17.76)$$

---

8. In general, we may want to mix max and sum. For example, consider a joint distribution where we observe v and we want to query q; let n be the remaining nuisance variables. We define the MAP estimate as $\mathbf{x}_q^* = \arg\max_{\mathbf{x}_q} \sum_{\mathbf{x}_n} p(\mathbf{x}_q, \mathbf{x}_n|\mathbf{x}_v)$, where we max over $\mathbf{x}_q$ and sum over $\mathbf{x}_n$. By contrast, we define the **MPE** or most probable explanation as $(\mathbf{x}_q^*, \mathbf{x}_n^*) = \arg\max_{\mathbf{x}_q,\mathbf{x}_n} p(\mathbf{x}_q, \mathbf{x}_n|\mathbf{x}_v)$, where we max over both $\mathbf{x}_q$ and $\mathbf{x}_n$. This terminology is due to (Pearl 1988), although it is not widely used outside the Bayes net literatire. Obviously MAP=MPE if $n = \emptyset$. However, if $n \neq \emptyset$, then summing out the nuisance variables can give different results than maxing them out. Summing out nuisance variables is more sensible, but computationally harder, because of the need to combine max and sum operations (Lerner and Parr 2001).

and we terminate by computing the most probable final state z_T^*:

$$z_T^* = \arg\max_i \delta_T(i) \tag{17.77}$$

We can then compute the most probable sequence of states using **traceback**:

$$z_t^* = a_{t+1}(z_{t+1}^*) \tag{17.78}$$

As usual, we have to worry about numerical underflow. We are free to normalize the $\boldsymbol{\delta}_t$ terms at each step; this will not affect the maximum. However, unlike the forwards-backwards case, we can also easily work in the log domain. The key difference is that $\log\max = \max\log$, whereas $\log\sum \neq \sum\log$. Hence we can use

$$\log \delta_t(j) \triangleq \max_{\mathbf{z}_{1:t-1}} \log p(\mathbf{z}_{1:t-1}, z_t = j | \mathbf{x}_{1:t}) \tag{17.79}$$

$$= \max_i \log \delta_{t-1}(i) + \log \psi(i, j) + \log \phi_t(j) \tag{17.80}$$

In the case of Gaussian observation models, this can result in a significant (constant factor) speedup, since computing $\log p(\mathbf{x}_t | z_t)$ can be much faster than computing $p(\mathbf{x}_t | z_t)$ for a high-dimensional Gaussian. This is one reason why the Viterbi algorithm is widely used in the E step of EM (Section 17.5.2) when training large speech recognition systems based on HMMs.

17.4.4.3 Example

Figure 17.13 gives a worked example of the Viterbi algorithm, based on (Russell et al. 1995). Suppose we observe the discrete sequence of observations $\mathbf{x}_{1:4} = (C_1, C_3, C_4, C_6)$, representing codebook entries in a vector-quantized version of a speech signal. The model starts in state z_1. The probability of generating C_1 in z_1 is 0.5, so we have $\delta_1(1) = 0.5$, and $\delta_1(i) = 0$ for all other states. Next we can self-transition to z_1 with probability 0.3, or transition to z_2 with proability 0.7. If we end up in z_1, the probability of generating C_3 is 0.3; if we end up in z_2, the probability of generating C_3 is 0.2. Hence we have

$$\delta_2(1) = \delta_1(1)\psi(1, 1)\phi_2(1) = 0.5 \cdot 0.3 \cdot 0.3 = 0.045 \tag{17.81}$$
$$\delta_2(2) = \delta_1(1)\psi(1, 2)\phi_2(2) = 0.5 \cdot 0.7 \cdot 0.2 = 0.07 \tag{17.82}$$

Thus state 2 is more probable at $t = 2$; see the second column of Figure 17.13(b). In time step 3, we see that there are two paths into z_2, from z_1 and from z_2. The bold arrow indicates that the latter is more probable. Hence this is the only one we have to remember. The algorithm continues in this way until we have reached the end of the sequence. One we have reached the end, we can follow the black arrows back to recover the MAP path (which is 1,2,2,3).

17.4.4.4 Time and space complexity

The time complexity of Viterbi is clearly $O(K^2T)$ in general, and the space complexity is $O(KT)$, both the same as forwards-backwards. If the transition matrix has the form $\psi(i, j) \propto \exp(-\sigma^2 ||\mathbf{z}_i - \mathbf{z}_j||^2)$, where $\mathbf{z}_i$ is the continuous vector represented by state i, we can implement Viterbi in $O(TK)$ time, instead of $O(TK \log K)$ needed by forwards-backwards. See Section 22.2.6.1 for details.

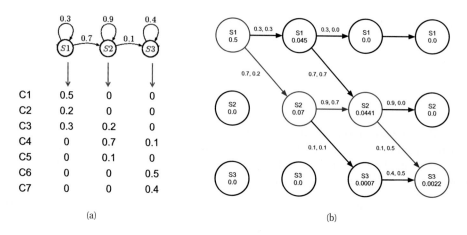

(a)   (b)

Figure 17.13 Illustration of Viterbi decoding in a simple HMM for speech recognition. (a) A 3-state HMM for a single phone. We are visualizing the state transition diagram. We assume the observations have been vector quantized into 7 possible symbols, $C_1, \ldots, C_7$. Each state z_1, z_2, z_3 has a different distribution over these symbols. Based on Figure 15.20 of (Russell and Norvig 2002). (b) Illustration of the Viterbi algorithm applied to this model, with data sequence $C1, C3, C4, C6$. The columns represent time, and the rows represent states. The numbers inside the circles represent the $\delta_t(j)$ value for that state. An arrow from state i at $t-1$ to state j at t is annotated with two numbers: the first is the probability of the $i \to j$ transition, and the second is the probability of generating observation $\mathbf{x}_t$ from state j. The red lines/ circles represent the most probable sequence of states. Based on Figure 24.27 of (Russell and Norvig 1995).

17.4.4.5   N-best list

The Viterbi algorithm returns one of the most probable paths. It can be extended to return the top N paths (Schwarz and Chow 1990; Nilsson and Goldberger 2001). This is called the **N-best list**. One can then use a discriminative method to rerank the paths based on global features derived from the fully observed state sequence (as well as the visible features). This technique is widely used in speech recognition. For example, consider the sentence "recognize speech". It is possible that the most probable interpretation by the system of this acoustic signal is "wreck a nice speech", or maybe "wreck a nice beach". Maybe the correct interpretation is much lower down on the list. However, by using a re-ranking system, we may be able to improve the score of the correct interpretation based on a more global context.

One problem with the N-best list is that often the top N paths are very similar to each other, rather than representing qualitatively different interpretations of the data. Instead we might want to generate a more diverse set of paths to more accurately represent posterior uncertainty. One way to do this is to sample paths from the posterior, as we discuss below. Another way is to use a determinantal point process (Kulesza and Taskar 2011; Zou and Adams 2012), which encourages points to be diverse (see also (Yadollahpour et al. 2011)).

17.4.5 Forwards filtering, backwards sampling

It is often useful to sample paths from the posterior:

$$\mathbf{z}_{1:T}^{s} \sim p(\mathbf{z}_{1:T}|\mathbf{x}_{1:T}) \tag{17.83}$$

We can do this is as follow: run forwards backwards, to compute the two-slice smoothed posteriors, $p(z_{t-1,t}|\mathbf{x}_{1:T})$; next compute the conditionals $p(z_t|z_{t-1}, \mathbf{x}_{1:T})$ by normalizing; sample from the initial pair of states, $z_{1,2}^* \sim p(z_{1,2}|\mathbf{x}_{1:T})$; finally, recursively sample $z_t^* \sim p(z_t|z_{t-1}^*, \mathbf{x}_{1:T})$.

Note that the above solution requires a forwards-backwards pass, and then an additional forwards sampling pass. An alternative is to do the forwards pass, and then perform sampling in the backwards pass. The key insight into how to do this is that we can write the joint from right to left using

$$p(\mathbf{z}_{1:T}|\mathbf{x}_{1:T}) = p(z_T|\mathbf{x}_{1:T}) \prod_{t=T-1}^{1} p(z_t|z_{t+1}, \mathbf{x}_{1:T}) \tag{17.84}$$

We can then sample z_t given future sampled states using

$$z_t^s \sim p(z_t|z_{t+1:T}, \mathbf{x}_{1:T}) = p(z_t|z_{t+1}, \cancel{\mathbf{z}_{t+2:T}}, \mathbf{x}_{1:t}, \cancel{\mathbf{x}_{t+1:T}}) = p(z_t|z_{t+1}^s, \mathbf{x}_{1:t}) \tag{17.85}$$

The sampling distribution is given by

$$
\begin{align}
p(z_t = i|z_{t+1} = j, \mathbf{x}_{1:t}) &= p(z_t|z_{t+1}, \mathbf{x}_{1:t}, \cancel{\mathbf{x}_{t+1}}) \tag{17.86} \\
&= \frac{p(z_{t+1}, z_t|\mathbf{x}_{1:t+1})}{p(z_{t+1}|\mathbf{x}_{1:t+1})} \tag{17.87} \\
&\propto \frac{p(\mathbf{x}_{t+1}|z_{t+1}, \cancel{z_t}, \cancel{\mathbf{x}_{1:t}})p(z_{t+1}, z_t|\mathbf{x}_{1:t})}{p(z_{t+1}|\mathbf{x}_{1:t+1})} \tag{17.88} \\
&= \frac{p(\mathbf{x}_{t+1}|z_{t+1})p(z_{t+1}|z_t, \cancel{\mathbf{x}_{1:t}})p(z_t|\mathbf{x}_{1:t})}{p(z_{t+1}|\mathbf{x}_{1:t+1})} \tag{17.89} \\
&= \frac{\phi_{t+1}(j)\psi(i, j)\alpha_t(i)}{\alpha_{t+1}(j)} \tag{17.90}
\end{align}
$$

The base case is

$$z_T^s \sim p(z_T = i|\mathbf{x}_{1:T}) = \alpha_T(i) \tag{17.91}$$

This algorithm forms the basis of blocked-Gibbs sampling methods for parameter inference, as we will see below.

17.5 Learning for HMMs

We now discuss how to estimate the parameters $\boldsymbol{\theta} = (\boldsymbol{\pi}, \mathbf{A}, \mathbf{B})$, where $\pi(i) = p(z_1 = i)$ is the initial state distribution, $A(i, j) = p(z_t = j|z_{t-1} = i)$ is the transition matrix, and $\mathbf{B}$ are the parameters of the class-conditional densities $p(\mathbf{x}_t|z_t = j)$. We first consider the case where $\mathbf{z}_{1:T}$ is observed in the training set, and then the harder case where $\mathbf{z}_{1:T}$ is hidden.

17.5.1 Training with fully observed data

If we observe the hidden state sequences, we can compute the MLEs for $\mathbf{A}$ and $\boldsymbol{\pi}$ exactly as in Section 17.2.2.1. If we use a conjugate prior, we can also easily compute the posterior.

The details on how to estimate $\mathbf{B}$ depend on the form of the observation model. The situation is identical to fitting a generative classifier. For example, if each state has a multinoulli distribution associated with it, with parameters $B_{jl} = p(X_t = l | z_t = j)$, where $l \in \{1, \ldots, L\}$ represents the observed symbol, the MLE is given by

$$\hat{B}_{jl} = \frac{N_{jl}^X}{N_j}, \quad N_{jl}^X \triangleq \sum_{i=1}^{N} \sum_{t=1}^{T_i} \mathbb{I}(z_{i,t} = j, x_{i,t} = l) \tag{17.92}$$

This result is quite intuitive: we simply add up the number of times we are in state j and we see a symbol l, and divide by the number of times we are in state j.

Similarly, if each state has a Gaussian distribution associated with it, we have (from Section 4.2.4) the following MLEs:

$$\hat{\boldsymbol{\mu}}_k = \frac{\overline{\mathbf{x}}_k}{N_k}, \quad \hat{\boldsymbol{\Sigma}}_k = \frac{(\overline{\mathbf{x}\mathbf{x}})_k^T - N_k \hat{\boldsymbol{\mu}}_k \hat{\boldsymbol{\mu}}_k^T}{N_k} \tag{17.93}$$

where the sufficient statistics are given by

$$\overline{\mathbf{x}}_k \triangleq \sum_{i=1}^{N} \sum_{t=1}^{T_i} \mathbb{I}(z_{i,t} = k) \mathbf{x}_{i,t} \tag{17.94}$$

$$(\overline{\mathbf{x}\mathbf{x}})_k^T \triangleq \sum_{i=1}^{N} \sum_{t=1}^{T_i} \mathbb{I}(z_{i,t} = k) \mathbf{x}_{i,t} \mathbf{x}_{i,t}^T \tag{17.95}$$

Analogous results can be derived for other kinds of distributions. One can also easily extend all of these results to compute MAP estimates, or even full posteriors over the parameters.

17.5.2 EM for HMMs (the Baum-Welch algorithm)

If the z_t variables are not observed, we are in a situation analogous to fitting a mixture model. The most common approach is to use the EM algorithm to find the MLE or MAP parameters, although of course one could use other gradient-based methods (see e.g., (Baldi and Chauvin 1994)). In this Section, we derive the EM algorithm. When applied to HMMs, this is also known as the **Baum-Welch** algorithm (Baum et al. 1970).

17.5.2.1 E step

It is straightforward to show that the expected complete data log likelihood is given by

$$Q(\boldsymbol{\theta}, \boldsymbol{\theta}^{old}) = \sum_{k=1}^{K} \mathbb{E}\left[N_k^1\right] \log \pi_k + \sum_{j=1}^{K} \sum_{k=1}^{K} \mathbb{E}\left[N_{jk}\right] \log A_{jk} \tag{17.96}$$

$$+ \sum_{i=1}^{N} \sum_{t=1}^{T_i} \sum_{k=1}^{K} p(z_t = k | \mathbf{x}_i, \boldsymbol{\theta}^{old}) \log p(\mathbf{x}_{i,t} | \boldsymbol{\phi}_k) \tag{17.97}$$

where the expected counts are given by

$$\mathbb{E}\left[N_k^1\right] = \sum_{i=1}^{N} p(z_{i1} = k | \mathbf{x}_i, \boldsymbol{\theta}^{old}) \tag{17.98}$$

$$\mathbb{E}\left[N_{jk}\right] = \sum_{i=1}^{N} \sum_{t=2}^{T_i} p(z_{i,t-1} = j, z_{i,t} = k | \mathbf{x}_i, \boldsymbol{\theta}^{old}) \tag{17.99}$$

$$\mathbb{E}\left[N_j\right] = \sum_{i=1}^{N} \sum_{t=1}^{T_i} p(z_{i,t} = j | \mathbf{x}_i, \boldsymbol{\theta}^{old}) \tag{17.100}$$

These expected sufficient statistics can be computed by running the forwards-backwards algorithm on each sequence. In particular, this algorithm computes the following smoothed node and edge marginals:

$$\gamma_{i,t}(j) \triangleq p(z_t = j | \mathbf{x}_{i,1:T_i}, \boldsymbol{\theta}) \tag{17.101}$$
$$\xi_{i,t}(j,k) \triangleq p(z_{t-1} = j, z_t = k | \mathbf{x}_{i,1:T_i}, \boldsymbol{\theta}) \tag{17.102}$$

17.5.2.2 M step

Based on Section 11.3, we have that the M step for $\mathbf{A}$ and $\boldsymbol{\pi}$ is to just normalize the expected counts:

$$\hat{A}_{jk} = \frac{\mathbb{E}\left[N_{jk}\right]}{\sum_{k'} \mathbb{E}\left[N_{jk'}\right]}, \quad \hat{\pi}_k = \frac{\mathbb{E}\left[N_k^1\right]}{N} \tag{17.103}$$

This result is quite intuitive: we simply add up the expected number of transitions from j to k, and divide by the expected number of times we transition from j to anything else.

For a multinoulli observation model, the expected sufficient statistics are

$$\mathbb{E}\left[M_{jl}\right] = \sum_{i=1}^{N} \sum_{t=1}^{T_i} \gamma_{i,t}(j) \mathbb{I}(x_{i,t} = l) = \sum_{i=1}^{N} \sum_{t:x_{i,t}=l} \gamma_{i,t}(j) \tag{17.104}$$

The M step has the form

$$\hat{B}_{jl} = \frac{\mathbb{E}\left[M_{jl}\right]}{\mathbb{E}\left[N_j\right]} \tag{17.105}$$

This result is quite intuitive: we simply add up the expected number of times we are in state j and we see a symbol l, and divide by the expected number of times we are in state j.

For a Gaussian observation model, the expected sufficient statistics are given by

$$\mathbb{E}\left[\bar{\mathbf{x}}_k\right] = \sum_{i=1}^{N} \sum_{t=1}^{T_i} \gamma_{i,t}(k) \mathbf{x}_{i,t} \tag{17.106}$$

$$\mathbb{E}\left[(\overline{\mathbf{x}\mathbf{x}})_k^T\right] = \sum_{i=1}^{N} \sum_{t=1}^{T_i} \gamma_{i,t}(k) \mathbf{x}_{i,t} \mathbf{x}_{i,t}^T \tag{17.107}$$

The M step becomes

$$\hat{\boldsymbol{\mu}}_k = \frac{\mathbb{E}\left[\overline{\mathbf{x}}_k\right]}{\mathbb{E}\left[N_k\right]}, \quad \hat{\boldsymbol{\Sigma}}_k = \frac{\mathbb{E}\left[(\overline{\mathbf{x}\mathbf{x}})_k^T\right] - \mathbb{E}\left[N_k\right]\hat{\boldsymbol{\mu}}_k\hat{\boldsymbol{\mu}}_k^T}{\mathbb{E}\left[N_k\right]} \tag{17.108}$$

This can (and should) be regularized in the same way we regularize GMMs.

17.5.2.3   Initialization

As usual with EM, we must take care to ensure that we initialize the parameters carefully, to minimize the chance of getting stuck in poor local optima. There are several ways to do this, such as

- Use some fully labeled data to initialize the parameters.

- Initially ignore the Markov dependencies, and estimate the observation parameters using the standard mixture model estimation methods, such as K-means or EM.

- Randomly initialize the parameters, use multiple restarts, and pick the best solution.

Techniques such as deterministic annealing (Ueda and Nakano 1998; Rao and Rose 2001) can help mitigate the effect of local minima. Also, just as K-means is often used to initialize EM for GMMs, so it is common to initialize EM for HMMs using **Viterbi training**, which means approximating the posterior over paths with the single most probable path. (This is not necessarily a good idea, since initially the parameters are often poorly estimated, so the Viterbi path will be fairly arbitrary. A safer option is to start training using forwards-backwards, and to switch to Viterbi near convergence.)

17.5.3   Bayesian methods for "fitting" HMMs *

EM returns a MAP estimate of the parameters. In this section, we briefly discuss some methods for Bayesian parameter estimation in HMMs. (These methods rely on material that we will cover later in the book.)

One approach is to use variational Bayes EM (VBEM), which we discuss in general terms in Section 21.6. The details for the HMM case can be found in (MacKay 1997; Beal 2003), but the basic idea is this: The E step uses forwards-backwards, but where (roughly speaking) we plug in the posterior mean parameters instead of the MAP estimates. The M step updates the parameters of the conjugate posteriors, instead of updating the parameters themselves.

An alternative to VBEM is to use MCMC. A particularly appealing algorithm is block Gibbs sampling, which we discuss in general terms in Section 24.2.8. The details for the HMM case can be found in (Fruhwirth-Schnatter 2007), but the basic idea is this: we sample $\mathbf{z}_{1:T}$ given the data and parameters using forwards-filtering, backwards-sampling, and we then sample the parameters from their posteriors, conditional on the sampled latent paths. This is simple to implement, but one does need to take care of unidentifiability (label switching), just as with mixture models (see Section 11.3.1).

17.5.4 Discriminative training

Sometimes HMMs are used as the class conditional density inside a generative classifier. In this case, $p(\mathbf{x}|y = c, \boldsymbol{\theta})$ can be computed using the forwards algorithm. We can easily maximize the joint likelihood $\prod_{i=1}^{N} p(\mathbf{x}_i, y_i | \boldsymbol{\theta})$ by using EM (or some other method) to fit the HMM for each class-conditional density separately.

However, we might like to find the parameters that maximize the conditional likelihood

$$\prod_{i=1}^{N} p(y_i | \mathbf{x}_i, \boldsymbol{\theta}) = \prod_i \frac{p(y_i | \boldsymbol{\theta}) p(\mathbf{x}_i | y_i, \boldsymbol{\theta})}{\sum_c p(y_i = c | \boldsymbol{\theta}) p(\mathbf{x}_i | c, \boldsymbol{\theta})} \qquad (17.109)$$

This is more expensive than maximizing the joint likelihood, since the denominator couples all C class-conditional HMMs together. Furthermore, EM can no longer be used, and one must resort to generic gradient based methods. Nevertheless, discriminative training can result in improved accuracies. The standard practice in speech recognition is to initially train the generative models separately using EM, and then to fine tune them discriminatively (Jelinek 1997).

17.5.5 Model selection

In HMMs, the two main model selection issues are: how many states, and what topology to use for the state transition diagram. We discuss both of these issues below.

17.5.5.1 Choosing the number of hidden states

Choosing the number of hidden states K in an HMM is analogous to the problem of choosing the number of mixture components. Here are some possible solutions:

- Use grid-search over a range of K's, using as an objective function cross-validated likelihood, the BIC score, or a variational lower bound to the log-marginal likelihood.
- Use reversible jump MCMC. See (Fruhwirth-Schnatter 2007) for details. Note that this is very slow and is not widely used.
- Use variational Bayes to "extinguish" unwanted components, by analogy to the GMM case discussed in Section 21.6.1.6. See (MacKay 1997; Beal 2003) for details.
- Use an "infinite HMM", which is based on the hierarchical Dirichlet process. See e.g., (Beal et al. 2002; Teh et al. 2006) for details.

17.5.5.2 Structure learning

The term **structure learning** in the context of HMMs refers to learning a sparse transition matrix. That is, we want to learn the structure of the state transition diagram, not the structure of the graphical model (which is fixed). A large number of heuristic methods have been proposed. Most alternate between parameter estimation and some kind of heuristic **split merge** method (see e.g., (Stolcke and Omohundro 1992)).

Alternatively, one can pose the problem as MAP estimation using a **minimum entropy prior**, of the form

$$p(\mathbf{A}_{i,:}) \propto \exp(-\mathbb{H}(\mathbf{A}_{i,:})) \qquad (17.110)$$

This prior prefers states whose outgoing distribution is nearly deterministic, and hence has low entropy (Brand 1999). The corresponding M step cannot be solved in closed form, but numerical methods can be used. The trouble with this is that we might prune out all incoming transitions to a state, creating isolated "islands" in state-space. The infinite HMM presents an interesting alternative to these methods. See e.g., (Beal et al. 2002; Teh et al. 2006) for details.

17.6   Generalizations of HMMs

Many variants of the basic HMM model have been proposed. We briefly discuss some of them below.

17.6.1   Variable duration (semi-Markov) HMMs

In a standard HMM, the probability we remain in state i for exactly d steps is

$$p(t_i = d) = (1 - A_{ii})A_{ii}^d \propto \exp(d \log A_{ii}) \tag{17.111}$$

where A_{ii} is the self-loop probability. This is called the **geometric distribution**. However, this kind of exponentially decaying function of d is sometimes unrealistic.

To allow for more general durations, one can use a **semi-Markov model**. It is called semi-Markov because to predict the next state, it is not sufficient to condition on the past state: we also need to know how long we've been in that state. When the state space is not observed directly, the result is called a **hidden semi-Markov model** (**HSMM**), a **variable duration HMM**, or an **explicit duration HMM**.

HSMMs are widely used in many gene finding programs, since the length distribution of exons and introns is not geometric (see e.g., (Schweikerta et al. 2009)), and in some chip-Seq data analysis programs (see e.g., (Kuan et al. 2009)).

HSMMs are useful not only because they can model the waiting time of each state more accurately, but also because they can model the distribution of a whole batch of observations at once, instead of assuming all observations are conditionally iid. That is, they can use likelihood models of the form $p(\mathbf{x}_{t:t+l}|z_t = k, d_t = l)$, which generate l correlated observations if the duration in state k is for l time steps. This is useful for modeling data that is piecewise linear, or shows other local trends (Ostendorf et al. 1996).

17.6.1.1   HSMM as augmented HMMs

One way to represent a HSMM is to use the graphical model shown in Figure 17.14. (In this figure, we have assumed the observations are iid within each state, but this is not required, as mentioned above.) The $D_t \in \{0, 1, \ldots, D\}$ node is a state duration counter, where D is the maximum duration of any state. When we first enter state j, we sample D_t from the duration distribution for that state, $D_t \sim p_j(\cdot)$. Thereafter, D_t deterministically counts down until $D_t = 0$. While $D_t > 0$, the state z_t is not allowed to change. When $D_t = 0$, we make a stochastic transition to a new state.

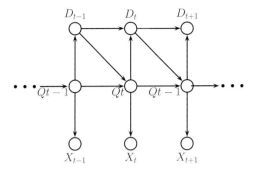

Figure 17.14 Encoding a hidden semi-Markov model as a DGM. D_t are deterministic duration counters.

More precisely, we define the CPDs as follows:

$$p(D_t = d' | D_{t-1} = d, z_t = j) = \begin{cases} p_j(d') & \text{if } d = 0 \\ 1 & \text{if } d' = d - 1 \text{ and } d \geq 1 \\ 0 & \text{otherwise} \end{cases} \tag{17.112}$$

$$p(z_t = k | z_{t-1} = j, D_{t-1} = d) = \begin{cases} 1 & \text{if } d > 0 \text{ and } j = k \\ A_{jk} & \text{if } d = 0 \\ 0 & \text{otherwise} \end{cases} \tag{17.113}$$

Note that $p_j(d)$ could be represented as a table (a non-parametric approach) or as some kind of parametric distribution, such as a Gamma distribution. If $p_j(d)$ is a geometric distribution, this emulates a standard HMM.

One can perform inference in this model by defining a mega-variable $Y_t = (D_t, z_t)$. However, this is rather inefficient, since D_t is deterministic. It is possible to marginalize D_t out, and derive special purpose inference procedures. See (Guedon 2003; Yu and Kobayashi 2006) for details. Unfortunately, all these methods take $O(TK^2D)$ time, where T is the sequence length, K is the number of states, and D is the maximum duration of any state.

17.6.1.2 Approximations to semi-Markov models

A more efficient, but less flexible, way to model non-geometric waiting times is to replace each state with n new states, each with the same emission probabilities as the original state. For example, consider the model in Figure 17.15(a). Obviously the smallest sequence this can generate is of length $n = 4$. Any path of length d through the model has probability $p^{d-n}(1-p)^n$; multiplying by the number of possible paths we find that the total probability of a path of length d is

$$p(d) = \binom{d-1}{n-1} p^{d-n}(1-p)^n \tag{17.114}$$

This is equivalent to the negative binomial distribution. By adjusting n and the self-loop probabilities p of each state, we can model a wide range of waiting times: see Figure 17.15(b).

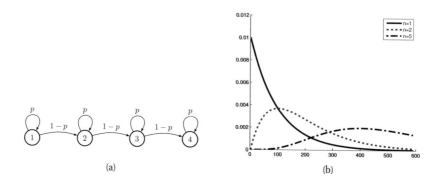

Figure 17.15   (a) A Markov chain with $n = 4$ repeated states and self loops. (b) The resulting distribution over sequence lengths, for $p = 0.99$ and various n. Figure generated by hmmSelfLoopDist.

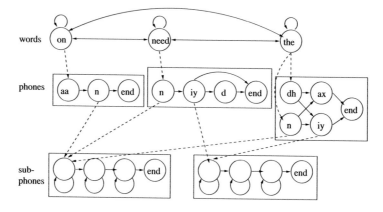

Figure 17.16   An example of an HHMM for an ASR system which can recognize 3 words. The top level represents bigram word probabilities. The middle level represents the phonetic spelling of each word. The bottom level represents the subphones of each phone. (It is traditional to represent a phone as a 3 state HMM, representing the beginning, middle and end.) Based on Figure 7.5 of (Jurafsky and Martin 2000).

Let E be the number of expansions of each state needed to approximate $p_j(d)$. Forwards-backwards on this model takes $O(T(KE)F_{in})$ time, where F_{in} is the average number of predecessor states, compared to $O(TK(F_{in}+D))$ for the HSMM. For typical speech recognition applications, $F_{in} \sim 3$, $D \sim 50$, $K \sim 10^6$, $T \sim 10^5$. (Similar figures apply to problems such as gene finding, which also often uses HSMMs.) Since $F_{in} + D \gg EF_{in}$, the expanded state method is much faster than an HSMM. See (Johnson 2005) for details.

17.6.2   Hierarchical HMMs

A **hierarchical HMM** (HHMM) (Fine et al. 1998) is an extension of the HMM that is designed to model domains with hierarchical structure. Figure 17.16 gives an example of an HHMM used in automatic speech recognition. The phone and subphone models can be "called" from different

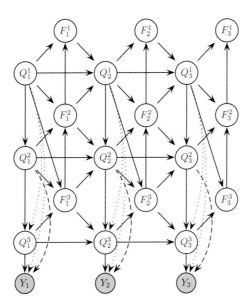

Figure 17.17 An HHMM represented as a DGM. Q_t^ℓ is the state at time t, level ℓ; $F_t^\ell = 1$ if the HMM at level ℓ has finished (entered its exit state), otherwise $F_t^\ell = 0$. Shaded nodes are observed; the remaining nodes are hidden. We may optionally clamp $F_T^\ell = 1$, where T is the length of the observation sequence, to ensure all models have finished by the end of the sequence. Source: Figure 2 of (Murphy and Paskin 2001).

higher level contexts. We can always "flatten" an HHMM to a regular HMM, but a factored representation is often easier to interpret, and allows for more efficient inference and model fitting.

HHMMs have been used in many application domains, e.g., speech recognition (Bilmes 2001), gene finding (Hu et al. 2000), plan recognition (Bui et al. 2002), monitoring transportation patterns (Liao et al. 2007), indoor robot localization (Theocharous et al. 2004), etc. HHMMs are less expressive than stochastic context free grammars (SCFGs), since they only allow hierarchies of bounded depth, but they support more efficient inference. In particular, inference in SCFGs (using the inside outside algorithm, (Jurafsky and Martin 2008)) takes $O(T^3)$ whereas inference in an HHMM takes $O(T)$ time (Murphy and Paskin 2001).

We can represent an HHMM as a directed graphical model as shown in Figure 17.17. Q_t^ℓ represents the state at time t and level ℓ. A state transition at level ℓ is only "allowed" if the chain at the level below has "finished", as determined by the $F_t^{\ell-1}$ node. (The chain below finishes when it chooses to enter its end state.) This mechanism ensures that higher level chains evolve more slowly than lower level chains, i.e., lower levels are nested within higher levels.

A variable duration HMM can be thought of as a special case of an HHMM, where the top level is a deterministic counter, and the bottom level is a regular HMM, which can only change states once the counter has "timed out". See (Murphy and Paskin 2001) for further details.

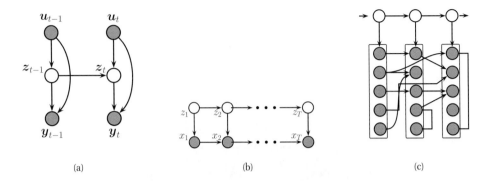

Figure 17.18   (a) Input-output HMM. (b) First-order auto-regressive HMM. (c) A second-order buried Markov model. Depending on the value of the hidden variables, the effective graph structure between the components of the observed variables (i.e., the non-zero elements of the regression matrix and the precision matrix) can change, although this is not shown.

17.6.3   Input-output HMMs

It is straightforward to extend an HMM to handle inputs, as shown in Figure 17.18(a). This defines a conditional density model for sequences of the form

$$p(\mathbf{y}_{1:T}, \mathbf{z}_{1:T} | \mathbf{u}_{1:T}, \boldsymbol{\theta}) \tag{17.115}$$

where $\mathbf{u}_t$ is the input at time t; this is sometimes called a control signal. If the inputs and outputs are continuous, a typical parameterization would be

$$p(z_t | \mathbf{x}_t, z_{t-1} = i, \boldsymbol{\theta}) = \text{Cat}(z_t | \mathcal{S}(\mathbf{W}_i \mathbf{u}_t)) \tag{17.116}$$

$$p(\mathbf{y}_t | \mathbf{x}_t, z_t = j, \boldsymbol{\theta}) = \mathcal{N}(\mathbf{y}_t | \mathbf{V}_j \mathbf{u}_t, \boldsymbol{\Sigma}_j) \tag{17.117}$$

Thus the transition matrix is a logistic regression model whose parameters depend on the previous state. The observation model is a Gaussian whose parameters depend on the current state. The whole model can be thought of as a hidden version of a maximum entropy Markov model (Section 19.6.1).

Conditional on the inputs $\mathbf{u}_{1:T}$ and the parameters $\boldsymbol{\theta}$, one can apply the standard forwards-backwards algorithm to estimate the hidden states. It is also straightforward to derive an EM algorithm to estimate the parameters (see (Bengio and Frasconi 1996) for details).

17.6.4   Auto-regressive and buried HMMs

The standard HMM assumes the observations are conditionally independent given the hidden state. In practice this is often not the case. However, it is straightforward to have direct arcs from $\mathbf{x}_{t-1}$ to $\mathbf{x}_t$ as well as from z_t to $\mathbf{x}_t$, as in Figure 17.18(b). This is known as an **auto-regressive HMM**, or a **regime switching Markov model**. For continuous data, the observation model becomes

$$p(\mathbf{x}_t | \mathbf{x}_{t-1}, z_t = j, \boldsymbol{\theta}) = \mathcal{N}(\mathbf{x}_t | \mathbf{W}_j \mathbf{x}_{t-1} + \boldsymbol{\mu}_j, \boldsymbol{\Sigma}_j) \tag{17.118}$$

This is a linear regression model, where the parameters are chosen according to the current hidden state. We can also consider higher-order extensions, where we condition on the last L observations:

$$p(\mathbf{x}_t|\mathbf{x}_{t-L:t-1}, z_t = j, \boldsymbol{\theta}) = \mathcal{N}(\mathbf{x}_t| \sum_{\ell=1}^{L} \mathbf{W}_{j,\ell}\mathbf{x}_{t-\ell} + \boldsymbol{\mu}_j, \boldsymbol{\Sigma}_j) \tag{17.119}$$

Such models are widely used in econometrics (Hamilton 1990). Similar models can be defined for discrete observations.

The AR-HMM essentially combines two Markov chains, one on the hidden variables, to capture long range dependencies, and one on the observed variables, to capture short range dependencies (Berchtold 1999). Since the X nodes are observed, the connections between them only change the computation of the local evidence; inference can still be performed using the standard forwards-backwards algorithm. Parameter estimation using EM is also straightforward: the E step is unchanged, as is the M step for the transition matrix. If we assume scalar observations for notational simplicty, the M step involves minimizing

$$\sum_t \mathbb{E} \left[\frac{1}{\sigma^2(s_t)}(y_t - \mathbf{y}_{t-L:t-1}^T \mathbf{w}(s_t))^2 + \log \sigma^2(s_t) \right] \tag{17.120}$$

Focussing on the $\mathbf{w}$ terms, we see that this requires solving K weighted least squares problems:

$$J(\mathbf{w}_{1:K}) = \sum_j \sum_t \frac{\gamma_t(j)}{\sigma^2(j)}(y_t - \mathbf{y}_{t-L:t-1}^T \mathbf{w}_j)^2 \tag{17.121}$$

where $\gamma_t(j) = p(z_t = k|\mathbf{x}_{1:T})$ is the smoothed posterior marginal. This is a weighted linear regression problem, where the design matrix has a Toeplitz form. This subproblem can be solved efficiently using the Levinson-Durbin method (Durbin and Koopman 2001).

Buried Markov models generalize AR-HMMs by allowing the dependency structure between the observable nodes to change based on the hidden state, as in Figure 17.18(c). Such a model is called a dynamic Bayesian **multi net**, since it is a mixture of different networks. In the linear-Gaussian setting, we can change the structure of the of $\mathbf{x}_{t-1} \to \mathbf{x}_t$ arcs by using sparse regression matrices, $\mathbf{W}_j$, and we can change the structure of the connections within the components of $\mathbf{x}_t$ by using sparse Gaussian graphical models, either directed or undirected. See (Bilmes 2000) for details.

17.6.5 Factorial HMM

An HMM represents the hidden state using a single discrete random variable $z_t \in \{1, \ldots, K\}$. To represent 10 bits of information would require $K = 2^{10} = 1024$ states. By contrast, consider a **distributed representation** of the hidden state, where each $z_{c,t} \in \{0, 1\}$ represents the c'th bit of the t'th hidden state. Now we can represent 10 bits using just 10 binary variables, as illustrated in Figure 17.19(a). This model is called a **factorial HMM** (Ghahramani and Jordan 1997). The hope is that this kind of model could capture different aspects of a signal, e.g., one chain would represent speaking style, another the words that are being spoken.

Unfortunately, conditioned on $\mathbf{x}_t$, all the hidden variables are correlated (due to explaining away the common observed child $\mathbf{x}_t$). This make exact state estimation intractable. However, we can derive efficient approximate inference algorithms, as we discuss in Section 21.4.1.

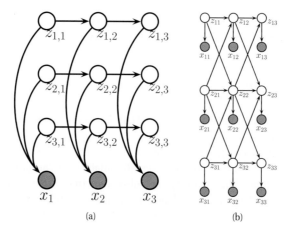

Figure 17.19   (a) A factorial HMM with 3 chains. (b) A coupled HMM with 3 chains.

17.6.6   Coupled HMM and the influence model

If we have multiple related data streams, we can use a **coupled HMM** (Brand 1996), as illustrated in Figure 17.19(b). This is a series of HMMs where the state transitions depend on the states of neighboring chains. That is, we represent the joint conditional distribution as

$$p(\mathbf{z}_t|\mathbf{z}_{t-1}) = \prod_c p(z_{ct}|\mathbf{z}_{t-1}) \tag{17.122}$$

$$p(z_{ct}|\mathbf{z}_{t-1}) = p(z_{ct}|z_{c,t-1},\ z_{c-1,t-1},\ z_{c+1,t-1}) \tag{17.123}$$

This has been used for various tasks, such as **audio-visual speech recognition** (Nefian et al. 2002) and modeling freeway traffic flows (Kwon and Murphy 2000).

The trouble with the above model is that it requires $O(CK^4)$ parameters to specify, if there are C chains with K states per chain, because each state depends on its own past plus the past of its two neighbors. There is a closely related model, known as the **influence model** (Asavathiratham 2000), which uses fewer parameters. It models the joint conditional distribution as

$$p(z_{ct}|\mathbf{z}_{t-1}) = \sum_{c'=1}^{C} \alpha_{c,c'} p(z_{ct}|z_{c',t-1}) \tag{17.124}$$

where $\sum_{c'} \alpha_{c,c'} = 1$ for each c. That is, we use a convex combination of pairwise transition matrices. The $\alpha_{c,c'}$ parameter specifies how much influence chain c has on chain c'. This model only takes $O(C^2 + CK^2)$ parameters to specify. Furthermore, it allows each chain to be influenced by all the other chains, not just its nearest neighbors. (Hence the corresponding graphical model is similar to Figure 17.19(b), except that each node has incoming edges from all the previous nodes.) This has been used for various tasks, such as modeling conversational interactions between people (Basu et al. 2001).

Unfortunately, inference in both of these models takes $O(T(K^C)^2)$ time, since all the chains

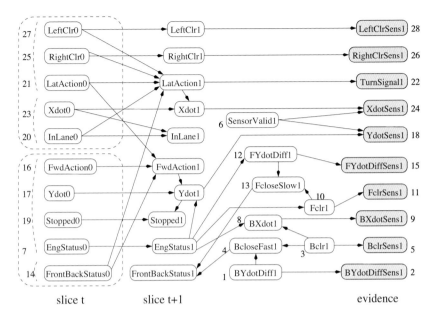

Figure 17.20 The BATnet DBN. The transient nodes are only shown for the second slice, to minimize clutter. The dotted lines can be ignored. Used with kind permission of Daphne Koller.

become fully correlated even if the interaction graph is sparse. Various approximate inference methods can be applied, as we discuss later.

17.6.7 Dynamic Bayesian networks (DBNs)

A **dynamic Bayesian network** is just a way to represent a stochastic process using a directed graphical model.[9] Note that the network is not dynamic (the structure and parameters are fixed), rather it is a network representation of a dynamical system. All of the HMM variants we have seen above could be considered to be DBNs. However, we prefer to reserve the term "DBN" for graph structures that are more "irregular" and problem-specific. An example is shown in Figure 17.20, which is a DBN designed to monitor the state of a simulated autonomous car known as the "Bayesian Automated Taxi", or "BATmobile" (Forbes et al. 1995).

Defining DBNs is straightforward: you just need to specify the structure of the first time-slice, the structure between two time-slices, and the form of the CPDs. Learning is also easy. The main problem is that exact inference can be computationally expensive, because all the hidden variables become correlated over time (this is known as **entanglement** — see e.g., (Koller and Friedman 2009, Sec. 15.2.4) for details). Thus a sparse graph does not necessarily result in tractable exact inference. However, later we will see algorithms that can exploit the graph structure for efficient approximate inference.

---

9. The acronym **DBN** can stand for either "dynamic Bayesian network" or "deep belief network" (Section 28.1) depending on the context. Geoff Hinton (who invented the term "deep belief network") has suggested the acronyms **DyBN** and **DeeBN** to avoid this ambiguity.

Exercises

Exercise 17.1 Derivation of Q function for HMM

Derive Equation 17.97.

Exercise 17.2 Two filter approach to smoothing in HMMs

Assuming that $\Pi_t(i) = p(S_t = i) > 0$ for all i and t, derive a recursive algorithm for updating $r_t(i) = p(S_t = i|\mathbf{x}_{t+1:T})$. Hint: it should be very similar to the standard forwards algorithm, but using a time-reversed transition matrix. Then show how to compute the posterior marginals $\gamma_t(i) = p(S_t = i|\mathbf{x}_{1:T})$ from the backwards filtered messages $r_t(i)$, the forwards filtered messages $\alpha_t(i)$, and the stationary distribution $\Pi_t(i)$.

Exercise 17.3 EM for for HMMs with mixture of Gaussian observations

Consider an HMM where the observation model has the form

$$p(\mathbf{x}_t|z_t = j, \boldsymbol{\theta}) = \sum_k w_{jk}\mathcal{N}(\mathbf{x}_t|\mu_{jk}, \boldsymbol{\Sigma}_{jk}) \tag{17.125}$$

- Draw the DGM.
- Derive the E step.
- Derive the M step.

Exercise 17.4 EM for for HMMs with tied mixtures

In many applications, it is common that the observations are high-dimensional vectors (e.g., in speech recognition, $\mathbf{x}_t$ is often a vector of cepstral coefficients and their derivatives, so $\mathbf{x}_t \in \mathbb{R}^{39}$), so estimating a full covariance matrix for KM values (where M is the number of mixture components per hidden state), as in Exercise 17.3, requires a lot of data. An alternative is to use just M Gaussians, rather than MK Gaussians, and to let the state influence the mixing weights but not the means and covariances. This is called a **semi-continuous HMM** or **tied-mixture HMM**.

- Draw the corresponding graphical model.
- Derive the E step.
- Derive the M step.

18 *State space models*

18.1   Introduction

A **state space model** or **SSM** is just like an HMM, except the hidden states are continuous. The model can be written in the following generic form:

$$
\begin{aligned}
\mathbf{z}_t &= g(\mathbf{u}_t, \mathbf{z}_{t-1}, \boldsymbol{\epsilon}_t) & (18.1) \\
\mathbf{y}_t &= h(\mathbf{z}_t, \mathbf{u}_t, \boldsymbol{\delta}_t) & (18.2)
\end{aligned}
$$

where $\mathbf{z}_t$ is the hidden state, $\mathbf{u}_t$ is an optional input or control signal, $\mathbf{y}_t$ is the observation, g is the **transition model**, h is the **observation model**, $\boldsymbol{\epsilon}_t$ is the system noise at time t, and $\boldsymbol{\delta}_t$ is the observation noise at time t. We assume that all parameters of the model, $\boldsymbol{\theta}$, are known; if not, they can be included into the hidden state, as we discuss below.

One of the primary goals in using SSMs is to recursively estimate the belief state, $p(\mathbf{z}_t|\mathbf{y}_{1:t}, \mathbf{u}_{1:t}, \boldsymbol{\theta})$. (Note: we will often drop the conditioning on $\mathbf{u}$ and $\boldsymbol{\theta}$ for brevity.) We will discuss algorithms for this later in this chapter. We will also discuss how to convert our beliefs about the hidden state into predictions about future observables by computing the posterior predictive $p(\mathbf{y}_{t+1}|\mathbf{y}_{1:t})$.

An important special case of an SSM is where all the CPDs are linear-Gaussian. In other words, we assume

- The transition model is a linear function

$$
\mathbf{z}_t = \mathbf{A}_t \mathbf{z}_{t-1} + \mathbf{B}_t \mathbf{u}_t + \boldsymbol{\epsilon}_t \tag{18.3}
$$

- The observation model is a linear function

$$
\mathbf{y}_t = \mathbf{C}_t \mathbf{z}_t + \mathbf{D}_t \mathbf{u}_t + \boldsymbol{\delta}_t \tag{18.4}
$$

- The system noise is Gaussian

$$
\boldsymbol{\epsilon}_t \sim \mathcal{N}(\mathbf{0}, \mathbf{Q}_t) \tag{18.5}
$$

- The observation noise is Gaussian

$$
\boldsymbol{\delta}_t \sim \mathcal{N}(\mathbf{0}, \mathbf{R}_t) \tag{18.6}
$$

This model is called a **linear-Gaussian SSM** (**LG-SSM**) or a **linear dynamical system** (**LDS**). If the parameters $\boldsymbol{\theta}_t = (\mathbf{A}_t, \mathbf{B}_t, \mathbf{C}_t, \mathbf{D}_t, \mathbf{Q}_t, \mathbf{R}_t)$ are independent of time, the model is called **stationary**.

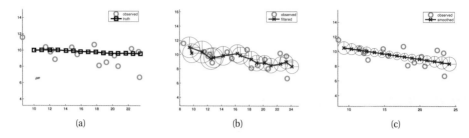

Figure 18.1 Illustration of Kalman filtering and smoothing. (a) Observations (green cirles) are generated by an object moving to the right (true location denoted by black squares). (b) Filtered estimated is shown by dotted red line. Red cross is the posterior mean, blue circles are 95% confidence ellipses derived from the posterior covariance. For clarity, we only plot the ellipses every other time step. (c) Same as (b), but using offline Kalman smoothing. Figure generated by `kalmanTrackingDemo`.

The LG-SSM is important because it supports exact inference, as we will see. In particular, if the initial belief state is Gaussian, $p(\mathbf{z}_1) = \mathcal{N}(\boldsymbol{\mu}_{1|0}, \boldsymbol{\Sigma}_{1|0})$, then all subsequent belief states will also be Gaussian; we will denote them by $p(\mathbf{z}_t|\mathbf{y}_{1:t}) = \mathcal{N}(\boldsymbol{\mu}_{t|t}, \boldsymbol{\Sigma}_{t|t})$. (The notation $\boldsymbol{\mu}_{t|\tau}$ denotes $\mathbb{E}[\mathbf{z}_t|\mathbf{y}_{1:\tau}]$, and similarly for $\boldsymbol{\Sigma}_{t|t}$; thus $\boldsymbol{\mu}_{t|0}$ denotes the prior for $\mathbf{z}_1$ before we have seen any data. For brevity we will denote the posterior belief states using $\boldsymbol{\mu}_{t|t} = \boldsymbol{\mu}_t$ and $\boldsymbol{\Sigma}_{t|t} = \boldsymbol{\Sigma}_t$.) We can compute these quantities efficiently using the celebrated Kalman filter, as we show in Section 18.3.1. But before discussing algorithms, we discuss some important applications.

18.2 Applications of SSMs

SSMs have many applications, some of which we discuss in the sections below. We mostly focus on LG-SSMs, for simplicity, although non-linear and/or non-Gaussian SSMs are even more widely used.

18.2.1 SSMs for object tracking

One of the earliest applications of Kalman filtering was for tracking objects, such as airplanes and missiles, from noisy measurements, such as radar. Here we give a simplified example to illustrate the key ideas. Consider an object moving in a 2D plane. Let z_{1t} and z_{2t} be the horizontal and vertical locations of the object, and $\dot{z}_{1t}$ and $\dot{z}_{2t}$ be the corresponding velocity. We can represent this as a state vector $\mathbf{z}_t \in \mathbb{R}^4$ as follows:

$$\mathbf{z}_t^T = \begin{pmatrix} z_{1t} & z_{2t} & \dot{z}_{1t} & \dot{z}_{2t} \end{pmatrix}. \tag{18.7}$$

Let us assume that the object is moving at constant velocity, but is "perturbed" by random Gaussian noise (e.g., due to the wind). Thus we can model the system dynamics as follows:

$$\mathbf{z}_t = \mathbf{A}_t \mathbf{z}_{t-1} + \boldsymbol{\epsilon}_t \tag{18.8}$$

$$\begin{pmatrix} z_{1t} \\ z_{2t} \\ \dot{z}_{1t} \\ \dot{z}_{2t} \end{pmatrix} = \begin{pmatrix} 1 & 0 & \Delta & 0 \\ 0 & 1 & 0 & \Delta \\ 0 & 0 & 1 & 0 \\ 0 & 0 & 0 & 1 \end{pmatrix} \begin{pmatrix} z_{1,t-1} \\ z_{2,t-1} \\ \dot{z}_{1,t-1} \\ \dot{z}_{2,t-1} \end{pmatrix} + \begin{pmatrix} \epsilon_{1t} \\ \epsilon_{2t} \\ \epsilon_{3t} \\ \epsilon_{4t} \end{pmatrix} \tag{18.9}$$

where $\boldsymbol{\epsilon}_t \sim \mathcal{N}(\mathbf{0}, \mathbf{Q})$ is the system noise, and Δ is the **sampling period**. This says that the new location $z_{j,t}$ is the old location $z_{j,t-1}$ plus Δ times the old velocity $\dot{z}_{j,t-1}$, plus random noise, ϵ_{jt}, for $j = 1 : 2$. Also, the new velocity $\dot{z}_{j,t}$ is the old velocity $\dot{z}_{j,t-1}$ plus random noise, ϵ_{jt}, for $j = 3 : 4$. This is called a **random accelerations model**, since the object moves according to Newton's laws, but is subject to random changes in velocity.

Now suppose that we can observe the location of the object but not its velocity. Let $\mathbf{y}_t \in \mathbb{R}^2$ represent our observation, which we assume is subject to Gaussian noise. We can model this as follows:

$$\mathbf{y}_t = \mathbf{C}_t \mathbf{z}_t + \boldsymbol{\delta}_t \tag{18.10}$$

$$\begin{pmatrix} y_{1t} \\ y_{2t} \end{pmatrix} = \begin{pmatrix} 1 & 0 & 0 & 0 \\ 0 & 1 & 0 & 0 \end{pmatrix} \begin{pmatrix} z_{1t} \\ z_{2t} \\ \dot{z}_{1t} \\ \dot{z}_{2t} \end{pmatrix} + \begin{pmatrix} \delta_{1t} \\ \delta_{2t} \\ \delta_{3t} \\ \delta_{4t} \end{pmatrix} \tag{18.11}$$

where $\boldsymbol{\delta}_t \sim \mathcal{N}(\mathbf{0}, \mathbf{R})$ is the measurement noise.

Finally, we need to specify our initial (prior) beliefs about the state of the object, $p(\mathbf{z}_1)$. We will assume this is a Gaussian, $p(\mathbf{z}_1) = \mathcal{N}(\mathbf{z}_1 | \boldsymbol{\mu}_{1|0}, \boldsymbol{\Sigma}_{1|0})$. We can represent prior ignorance by making $\boldsymbol{\Sigma}_{1|0}$ suitably "broad", e.g., $\boldsymbol{\Sigma}_{1|0} = \infty \mathbf{I}$. We have now fully specified the model and can perform sequential Bayesian updating to compute $p(\mathbf{z}_t | \mathbf{y}_{1:t})$ using an algorithm known as the Kalman filter, to be described in Section 18.3.1.

Figure 18.1(a) gives an example. The object moves to the right and generates an observation at each time step (think of "blips" on a radar screen). We observe these blips and filter out the noise by using the Kalman filter. At every step, we have $p(\mathbf{z}_t | \mathbf{y}_{1:t})$, from which we can compute $p(z_{1t}, z_{2t} | \mathbf{y}_{1:t})$ by marginalizing out the dimensions corresponding to the velocities. (This is easy to do since the posterior is Gaussian.) Our "best guess" about the location of the object is the posterior mean, $E[\mathbf{z}_t | \mathbf{y}_{1:t}]$, denoted as a red cross in Figure 18.1(b). Our uncertainty associated with this is represented as an ellipse, which contains 95% of the probability mass. We see that our uncertainty goes down over time, as the effects of the initial uncertainty get "washed out". We also see that the estimated trajectory has "filtered out" some of the noise. To obtain the much smoother plot in Figure 18.1(c), we need to use the Kalman smoother, which computes $p(\mathbf{z}_t | \mathbf{y}_{1:T})$; this depends on "future" as well as "past" data, as discussed in Section 18.3.2.

18.2.2 Robotic SLAM

Consider a robot moving around an unknown 2d world. It needs to learn a map and keep track of its location within that map. This problem is known as **simultaneous localization and**

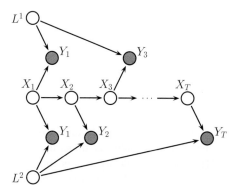

Figure 18.2 Illustration of graphical model underlying SLAM. L^i is the fixed location of landmark i, $\mathbf{x}_t$ is the location of the robot, and $\mathbf{y}_t$ is the observation. In this trace, the robot sees landmarks 1 and 2 at time step 1, then just landmark 2, then just landmark 1, etc. Based on Figure 15.A.3 of (Koller and Friedman 2009).

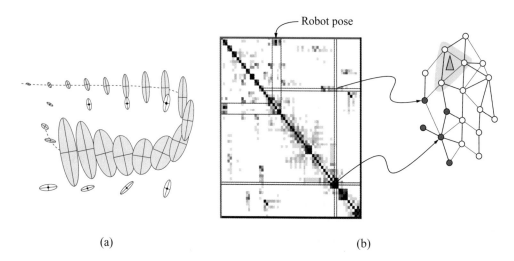

(a)   (b)

Figure 18.3 Illustration of the SLAM problem. (a) A robot starts at the top left and moves clockwise in a circle back to where it started. We see how the posterior uncertainty about the robot's location increases and then decreases as it returns to a familar location, closing the loop. If we performed smoothing, this new information would propagate backwards in time to disambiguate the entire trajectory. (b) We show the precision matrix, representing sparse correlations between the landmarks, and between the landmarks and the robot's position (pose). This sparse precision matrix can be visualized as a Gaussian graphical model, as shown.   Source: Figure 15.A.3 of (Koller and Friedman 2009) . Used with kind permission of Daphne Koller.

mapping, or **SLAM** for short, and is widely used in mobile robotics, as well as other applications such as indoor navigation using cellphones (since GPS does not work inside buildings).

Let us assume we can represent the map as the 2d locations of a fixed set of K landmarks, denote them by $L^1, \ldots, L^K$ (each is a vector in $\mathbb{R}^2$). For simplicity, we will assume these are uniquely identifiable. Let $\mathbf{x}_t$ represent the unknown location of the robot at time t. We define the state space to be $\mathbf{z}_t = (\mathbf{x}_t, \mathbf{L}^{1:K})$; we assume the landmarks are static, so their motion model is a constant, and they have no system noise. If $\mathbf{y}_t$ measures the distance from $\mathbf{x}_t$ to the set of closest landmarks, then the robot can update its estimate of the landmark locations based on what it sees. Figure 18.2 shows the corresponding graphical model for the case where $K = 2$, and where on the first step it sees landmarks 1 and 2, then just landmark 2, then just landmark 1, etc.

If we assume the observation model $p(\mathbf{y}_t | \mathbf{z}_t, \mathbf{L})$ is linear-Gaussian, and we use a Gaussian motion model for $p(\mathbf{x}_t | \mathbf{x}_{t-1}, \mathbf{u}_t)$, we can use a Kalman filter to maintain our belief state about the location of the robot and the location of the landmarks (Smith and Cheeseman 1986; Choset and Nagatani 2001).

Over time, the uncertainty in the robot's location will increase, due to wheel slippage etc., but when the robot returns to a familiar location, its uncertainty will decrease again. This is called **closing the loop**, and is illustrated in Figure 18.3(a), where we see the uncertainty ellipses, representing $\text{cov}\,[\mathbf{x}_t | \mathbf{y}_{1:t}, \mathbf{u}_{1:t}]$, grow and then shrink. (Note that in this section, we assume that a human is joysticking the robot through the environment, so $\mathbf{u}_{1:t}$ is given as input, i.e., we do not address the decision-theoretic issue of choosing where to explore.)

Since the belief state is Gaussian, we can visualize the posterior covariance matrix $\mathbf{\Sigma}_t$. Actually, it is more interesting to visualize the posterior precision matrix, $\mathbf{\Lambda}_t = \mathbf{\Sigma}_t^{-1}$, since that is fairly sparse, as shown in Figure 18.3(b). The reason for this is that zeros in the precision matrix correspond to absent edges in the corresponding undirected Gaussian graphical model (see Section 19.4.4). Initially all the landmarks are uncorrelated (assuming we have a diagonal prior on $\mathbf{L}$), so the GGM is a disconnected graph, and $\mathbf{\Lambda}_t$ is diagonal. However, as the robot moves about, it will induce correlation between nearby landmarks. Intuitively this is because the robot is estimating its position based on distance to the landmarks, but the landmarks' locations are being estimated based on the robot's position, so they all become inter-dependent. This can be seen more clearly from the graphical model in Figure 18.2: it is clear that L^1 and L^2 are not d-separated by $\mathbf{y}_{1:t}$, because there is a path between them via the unknown sequence of $\mathbf{x}_{1:t}$ nodes. As a consequence of the precision matrix becoming denser, exact inference takes $O(K^3)$ time. (This is an example of the entanglement problem for inference in DBNs.) This prevents the method from being applied to large maps.

There are two main solutions to this problem. The first is to notice that the correlation pattern moves along with the location of the robot (see Figure 18.3(b)). The remaining correlations become weaker over time. Consequently we can dynamically "prune out" weak edges from the GGM using a technique called the thin junction tree filter (Paskin 2003) (junction trees are explained in Section 20.4).

A second approach is to notice that, conditional on knowing the robot's path, $\mathbf{x}_{1:t}$, the landmark locations are independent. That is, $p(\mathbf{L} | \mathbf{x}_{1:t}, \mathbf{y}_{1:t}) = \prod_{k=1}^K p(\mathbf{L}^k | \mathbf{x}_{1:t}, \mathbf{y}_{1:t})$. This forms the basis of a method known as FastSLAM, which combines Kalman filtering and particle filtering, as discussed in Section 23.6.3.

(Thrun et al. 2006) provides a more detailed account of SLAM and mobile robotics.

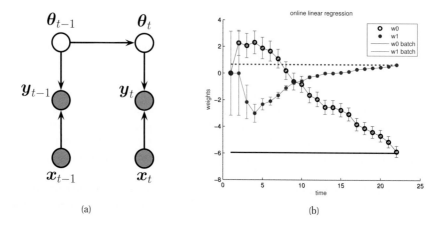

(a)   (b)

Figure 18.4   (a) A dynamic generalization of linear regression. (b) Illustration of the recursive least squares algorithm applied to the model $p(y|\mathbf{x}, \boldsymbol{\theta}) = \mathcal{N}(y|w_0 + w_1x, \sigma^2)$. We plot the marginal posterior of w_0 and w_1 vs number of data points. (Error bars represent $\mathbb{E}\left[w_j|y_{1:t}\right] \pm \sqrt{\mathrm{var}\left[w_j|y_{1:t}\right]}$.) After seeing all the data, we converge to the offline ML (least squares) solution, represented by the horizontal lines. Figure generated by `linregOnlineDemoKalman`.

18.2.3   Online parameter learning using recursive least squares

We can perform online Bayesian inference for the parameters of various statistical models using SSMs. In this section, we focus on linear regression; in Section 18.5.3.2, we discuss logistic regression.

The basic idea is to let the hidden state represent the regression parameters, and to let the (time-varying) observation model represent the current data vector. In more detail, define the prior to be $p(\boldsymbol{\theta}) = \mathcal{N}(\boldsymbol{\theta}|\boldsymbol{\theta}_0, \boldsymbol{\Sigma}_0)$. (If we want to do online ML estimation, we can just set $\boldsymbol{\Sigma}_0 = \infty\mathbf{I}$.) Let the hidden state be $\mathbf{z}_t = \boldsymbol{\theta}$; if we assume the regression parameters do not change, we can set $\mathbf{A}_t = \mathbf{I}$ and $\mathbf{Q}_t = 0\mathbf{I}$, so

$$p(\boldsymbol{\theta}_t|\boldsymbol{\theta}_{t-1}) = \mathcal{N}(\boldsymbol{\theta}_t|\boldsymbol{\theta}_{t-1}, 0\mathbf{I}) = \delta_{\boldsymbol{\theta}_{t-1}}(\boldsymbol{\theta}_t) \tag{18.12}$$

(If we do let the parameters change over time, we get a so-called **dynamic linear model** (Harvey 1990; West and Harrison 1997; Petris et al. 2009).) Let $\mathbf{C}_t = \mathbf{x}_t^T$, and $\mathbf{R}_t = \sigma^2$, so the (non-stationary) observation model has the form

$$\mathcal{N}(\mathbf{y}_t|\mathbf{C}_t\mathbf{z}_t, \mathbf{R}_t) = \mathcal{N}(\mathbf{y}_t|\mathbf{x}_t^T\boldsymbol{\theta}_t, \sigma^2) \tag{18.13}$$

Applying the Kalman filter to this model provides a way to update our posterior beliefs about the parameters as the data streams in. This is known as the **recursive least squares** or **RLS** algorithm.

We can derive an explicit form for the updates as follows. In Section 18.3.1, we show that the Kalman update for the posterior mean has the form

$$\boldsymbol{\mu}_t = \mathbf{A}_t\boldsymbol{\mu}_{t-1} + \mathbf{K}_t(\mathbf{y}_t - \mathbf{C}_t\mathbf{A}_t\boldsymbol{\mu}_{t-1}) \tag{18.14}$$

where $\mathbf{K}_t$ is known as the Kalman gain matrix. Based on Equation 18.39, one can show that $\mathbf{K}_t = \mathbf{\Sigma}_t \mathbf{C}_t^T \mathbf{R}_t^{-1}$. In this context, we have $\mathbf{K}_t = \mathbf{\Sigma}_t \mathbf{x}_t / \sigma^2$. Hence the update for the parameters becomes

$$\hat{\boldsymbol{\theta}}_t = \hat{\boldsymbol{\theta}}_{t-1} + \frac{1}{\sigma^2} \mathbf{\Sigma}_{t|t}(y_t - \mathbf{x}_t^T \hat{\boldsymbol{\theta}}_{t-1}) \mathbf{x}_t \tag{18.15}$$

If we approximate $\frac{1}{\sigma^2} \mathbf{\Sigma}_{t|t-1}$ with $\eta_t \mathbf{I}$, we recover the **least mean squares** or **LMS** algorithm, discussed in Section 8.5.3. In LMS, we need to specify how to adapt the update parameter η_t to ensure convergence to the MLE. Furthermore, the algorithm may take multiple passes through the data. By contrast, the RLS algorithm automatically performs step-size adaptation, and converges to the optimal posterior in one pass over the data. See Figure 18.4 for an example.

18.2.4 SSM for time series forecasting *

SSMs are very well suited for time-series forecasting, as we explain below. We focus on the case of scalar (one dimensional) time series, for simplicity. Our presentation is based on (Varian 2011). See also (Aoki 1987; Harvey 1990; West and Harrison 1997; Durbin and Koopman 2001; Petris et al. 2009; Prado and West 2010) for good books on this topic.

At first sight, it might not be apparent why SSMs are useful, since the goal in forecasting is to predict future visible variables, not to estimate hidden states of some system. Indeed, most classical methods for time series forecasting are just functions of the form $\hat{y}_{t+1} = f(\mathbf{y}_{1:t}, \boldsymbol{\theta})$, where hidden variables play no role (see Section 18.2.4.4). The idea in the state-space approach to time series is to create a generative model of the data in terms of latent processes, which capture different aspects of the signal. We can then integrate out the hidden variables to compute the posterior predictive of the visibles.

Since the model is linear-Gaussian, we can just add these processes together to explain the observed data. This is called a **structural time series** model. Below we explain some of the basic building blocks.

18.2.4.1 Local level model

The simplest latent process is known as the **local level model**, which has the form

$$y_t = a_t + \epsilon_t^y, \quad \epsilon_t^y \sim \mathcal{N}(0, R) \tag{18.16}$$
$$a_t = a_{t-1} + \epsilon_t^a, \quad \epsilon_t^a \sim \mathcal{N}(0, Q) \tag{18.17}$$

where the hidden state is just $\mathbf{z}_t = a_t$. This model asserts that the observed data $y_t \in \mathbb{R}$ is equal to some unknown level term $a_t \in \mathbb{R}$, plus observation noise with variance R. In addition, the level a_t evolves over time subject to system noise with variance Q. See Figure 18.5 for some examples.

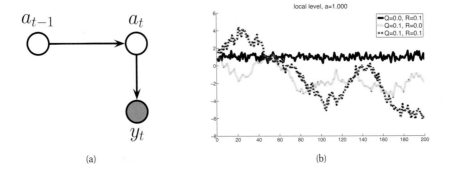

(a)   (b)

Figure 18.5   (a) Local level model. (b) Sample output, for $a_0 = 10$. Black solid line: $Q = 0$, $R = 1$ (deterministic system, noisy observations). Red dotted line: $Q = 0.1$, $R = 0$ (noisy system, deterministic observation). Blue dot-dash line: $Q = 0.1$, $R = 1$ (noisy system and observations). Figure generated by `ssmTimeSeriesSimple`.

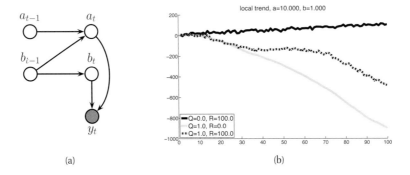

(a)   (b)

Figure 18.6   (a) Local Trend. (b) Sample output, for $a_0 = 10$, $b_0 = 1$. Color code as in Figure 18.5. Figure generated by `ssmTimeSeriesSimple`.

18.2.4.2   Local linear trend

Many time series exhibit linear trends upwards or downwards, at least locally. We can model this by letting the level a_t change by an amount b_t at each step as follows:

$$y_t = a_t + \epsilon_t^y, \quad \epsilon_t^y \sim \mathcal{N}(0, R) \tag{18.18}$$

$$a_t = a_{t-1} + b_{t-1} + \epsilon_t^a, \quad \epsilon_t^a \sim \mathcal{N}(0, Q_a) \tag{18.19}$$

$$b_t = b_{t-1} + \epsilon_t^b, \quad \epsilon_t^b \sim \mathcal{N}(0, Q_b) \tag{18.20}$$

See Figure 18.6(a). We can write this in standard form by defining $\mathbf{z}_t = (a_t, b_t)$ and

$$\mathbf{A} = \begin{pmatrix} 1 & 1 \\ 0 & 1 \end{pmatrix}, \quad \mathbf{C} = \begin{pmatrix} 1 & 0 \end{pmatrix}, \quad \mathbf{Q} = \begin{pmatrix} Q_a & 0 \\ 0 & Q_b \end{pmatrix} \tag{18.21}$$

When $Q_b = 0$, we have $b_t = b_0$, which is some constant defining the slope of the line. If in addition we have $Q_a = 0$, we have $a_t = a_{t-1} + b_0 t$. Unrolling this, we have $a_t = a_0 + b_0 t$, and

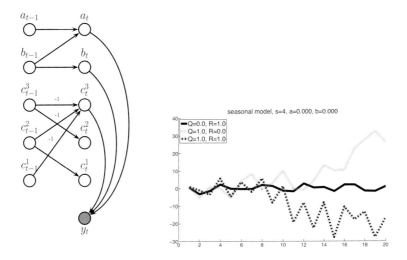

Figure 18.7 (a) Seasonal model. (b) Sample output, for $a_0 = b_0 = 0$, $\mathbf{c}_0 = (1, 1, 1)$, with a period of 4. Color code as in Figure 18.5. Figure generated by `ssmTimeSeriesSimple`.

hence $\mathbb{E}\left[y_t | \mathbf{y}_{1:t-1}\right] = a_0 + t b_0$. This is thus a generalization of the classic constant linear trend model, an example of which is shown in the black line of Figure 18.6(b).

18.2.4.3 Seasonality

Many time series fluctuate periodically, as illustrated in Figure 18.7(b). This can be modeled by adding a latent process consisting of a series offset terms, c_t, which sum to zero (on average) over a complete cycle of S steps:

$$c_t = -\sum_{s=1}^{S-1} c_{t-s} + \epsilon_t^c, \ \epsilon_t^c \sim \mathcal{N}(0, Q_c) \tag{18.22}$$

See Figure 18.7(a) for the graphical model for the case $S = 4$ (we only need 3 seasonal variable because of the sum-to-zero constraint). Writing this in standard LG-SSM form is left to Exercise 18.2.

18.2.4.4 ARMA models *

The classical approach to time-series forecasting is based on **ARMA** models, which stands for auto-regressive moving-average. An ARMA(p, q) model has the form

$$x_t = \sum_{i=1}^{p} \alpha_i x_{t-i} + \sum_{j=1}^{q} \beta_j w_{t-j} + v_t \tag{18.23}$$

where $v_t, w_t \sim \mathcal{N}(0, 1)$ are independent Gaussian noise terms. If $q = 0$, we have a pure AR model, where $x_t \perp x_i | x_{t-1:t-p}$, for $i < t - p$. For example, Figure 18.8(a) illustrates an AR(1)

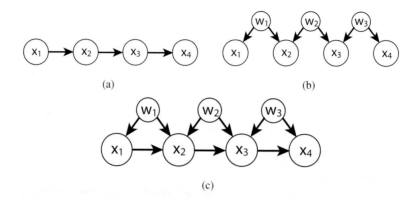

Figure 18.8 (a) An AR(1) model. (b) An MA(1) model. (c) An ARMA(1,1) model. Source: Figure 5.14 of (Choi 2011). Used with kind permission of Myung Choi.

model. (The v_t noise variables are implicit in the Gaussian CPD for x_t.)

If $p = 0$, we have a pure MA model, where $x_t \perp x_i$, for $i < t - q$. For example, Figure 18.8(b) illustrates an MA(1) model. Here the w_t noise variables are hidden common causes, which induces dependencies between adjacent time steps. This models short-range correlation.

Figure 18.8(c) shows an ARMA(1,1) model. This can captures correlation at short and long time scales, using the MA and AR components.

It turns out that ARMA models can be represented as SSMs, as explained in (Aoki 1987; Harvey 1990; West and Harrison 1997; Durbin and Koopman 2001; Petris et al. 2009; Prado and West 2010). However, the structural approach to time series is often easier to understand than the ARMA approach. In addition, the structural approach allows the parameters to evolve over time, which makes the models more adaptive to non-stationarity.

18.3 Inference in LG-SSM

In this section, we discuss exact inference in LG-SSM models. We first consider the online case, which is analogous to the forwards algorithm for HMMs. We then consider the offline case, which is analogous to the forwards-backwards algorithm for HMMs.

18.3.1 The Kalman filtering algorithm

The **Kalman filter** is an algorithm for exact Bayesian filtering for linear-Gaussian state space models. We will represent the marginal posterior at time t by

$$p(\mathbf{z}_t|\mathbf{y}_{1:t}, \mathbf{u}_{1:t}) = \mathcal{N}(\mathbf{z}_t|\boldsymbol{\mu}_t, \boldsymbol{\Sigma}_t) \tag{18.24}$$

Since everything is Gaussian, we can perform the prediction and update steps in closed form, as we explain below. The resulting algorithm is the Gaussian analog of the HMM filter in Section 17.4.2.

18.3.1.1 Prediction step

The prediction step is straightforward to derive:

$$
p(\mathbf{z}_t|\mathbf{y}_{1:t-1}, \mathbf{u}_{1:t}) = \int \mathcal{N}(\mathbf{z}_t|\mathbf{A}_t\mathbf{z}_{t-1} + \mathbf{B}_t\mathbf{u}_t, \mathbf{Q}_t)\mathcal{N}(\mathbf{z}_{t-1}|\boldsymbol{\mu}_{t-1}, \boldsymbol{\Sigma}_{t-1})d\mathbf{z}_{t-1} \tag{18.25}
$$

$$
= \mathcal{N}(\mathbf{z}_t|\boldsymbol{\mu}_{t|t-1}, \boldsymbol{\Sigma}_{t|t-1}) \tag{18.26}
$$

$$
\boldsymbol{\mu}_{t|t-1} \triangleq \mathbf{A}_t\boldsymbol{\mu}_{t-1} + \mathbf{B}_t\mathbf{u}_t \tag{18.27}
$$

$$
\boldsymbol{\Sigma}_{t|t-1} \triangleq \mathbf{A}_t\boldsymbol{\Sigma}_{t-1}\mathbf{A}_t^T + \mathbf{Q}_t \tag{18.28}
$$

18.3.1.2 Measurement step

The measurement step can be computed using Bayes rule, as follows

$$
p(\mathbf{z}_t|\mathbf{y}_t, \mathbf{y}_{1:t-1}, \mathbf{u}_{1:t}) \propto p(\mathbf{y}_t|\mathbf{z}_t, \mathbf{u}_t)p(\mathbf{z}_t|\mathbf{y}_{1:t-1}, \mathbf{u}_{1:t}) \tag{18.29}
$$

In Section 18.3.1.6, we show that this is given by

$$
p(\mathbf{z}_t|\mathbf{y}_{1:t}, \mathbf{u}_t) = \mathcal{N}(\mathbf{z}_t|\boldsymbol{\mu}_t, \boldsymbol{\Sigma}_t) \tag{18.30}
$$

$$
\boldsymbol{\mu}_t = \boldsymbol{\mu}_{t|t-1} + \mathbf{K}_t\mathbf{r}_t \tag{18.31}
$$

$$
\boldsymbol{\Sigma}_t = (\mathbf{I} - \mathbf{K}_t\mathbf{C}_t)\boldsymbol{\Sigma}_{t|t-1} \tag{18.32}
$$

where $\mathbf{r}_t$ is the **residual** or **innovation**, given by the difference between our predicted observation and the actual observation:

$$
\mathbf{r}_t \triangleq \mathbf{y}_t - \hat{\mathbf{y}}_t \tag{18.33}
$$

$$
\hat{\mathbf{y}}_t \triangleq \mathbb{E}[\mathbf{y}_t|\mathbf{y}_{1:t-1}, \mathbf{u}_{1:t}] = \mathbf{C}_t\boldsymbol{\mu}_{t|t-1} + \mathbf{D}_t\mathbf{u}_t \tag{18.34}
$$

and $\mathbf{K}_t$ is the **Kalman gain matrix**, given by

$$
\mathbf{K}_t \triangleq \boldsymbol{\Sigma}_{t|t-1}\mathbf{C}_t^T\mathbf{S}_t^{-1} \tag{18.35}
$$

where

$$
\mathbf{S}_t \triangleq \mathrm{cov}[\mathbf{r}_t|\mathbf{y}_{1:t-1}, \mathbf{u}_{1:t}] \tag{18.36}
$$

$$
= \mathbb{E}[(\mathbf{C}_t\mathbf{z}_t + \boldsymbol{\delta}_t - \hat{\mathbf{y}}_t)(\mathbf{C}_t\mathbf{z}_t + \boldsymbol{\delta}_t - \hat{\mathbf{y}}_t)^T|\mathbf{y}_{1:t-1}, \mathbf{u}_{1:t}] \tag{18.37}
$$

$$
= \mathbf{C}_t\boldsymbol{\Sigma}_{t|t-1}\mathbf{C}_t^T + \mathbf{R}_t \tag{18.38}
$$

where $\boldsymbol{\delta}_t \sim \mathcal{N}(\mathbf{0}, \mathbf{R}_t)$ is an observation noise term which is independent of all other noise sources. Note that by using the matrix inversion lemma, the Kalman gain matrix can also be written as

$$
\mathbf{K}_t = \boldsymbol{\Sigma}_{t|t-1}\mathbf{C}^T(\mathbf{C}\boldsymbol{\Sigma}_{t|t-1}\mathbf{C}^T + \mathbf{R})^{-1} = (\boldsymbol{\Sigma}_{t|t-1}^{-1} + \mathbf{C}^T\mathbf{R}\mathbf{C})^{-1}\mathbf{C}^T\mathbf{R}^{-1} \tag{18.39}
$$

We now have all the quantities we need to implement the algorithm; see `kalmanFilter` for some Matlab code.

Let us try to make sense of these equations. In particular, consider the equation for the mean update: $\boldsymbol{\mu}_t = \boldsymbol{\mu}_{t|t-1} + \mathbf{K}_t\mathbf{r}_t$. This says that the new mean is the old mean plus a correction factor, which is $\mathbf{K}_t$ times the error signal $\mathbf{r}_t$. The amount of weight placed on the error signal depends on the Kalman gain matrix. If $\mathbf{C}_t = \mathbf{I}$, then $\mathbf{K}_t = \boldsymbol{\Sigma}_{t|t-1}\mathbf{S}_t^{-1}$, which is the ratio between the covariance of the prior (from the dynamic model) and the covariance of the measurement error. If we have a strong prior and/or very noisy sensors, $|\mathbf{K}_t|$ will be small, and we will place little weight on the correction term. Conversely, if we have a weak prior and/or high precision sensors, then $|\mathbf{K}_t|$ will be large, and we will place a lot of weight on the correction term.

18.3.1.3   Marginal likelihood

As a byproduct of the algorithm, we can also compute the log-likelihood of the sequence using

$$\log p(\mathbf{y}_{1:T}|\mathbf{u}_{1:T}) = \sum_t \log p(\mathbf{y}_t|\mathbf{y}_{1:t-1}, \mathbf{u}_{1:t}) \tag{18.40}$$

where

$$p(\mathbf{y}_t|\mathbf{y}_{1:t-1}, \mathbf{u}_{1:t}) = \mathcal{N}(\mathbf{y}_t|\mathbf{C}_t\boldsymbol{\mu}_{t|t-1}, \mathbf{S}_t) \tag{18.41}$$

18.3.1.4   Posterior predictive

The one-step-ahead posterior predictive density for the observations can be computed as follows

$$\begin{aligned} p(\mathbf{y}_t|\mathbf{y}_{1:t-1}, \mathbf{u}_{1:t}) &= \int \mathcal{N}(\mathbf{y}_t|\mathbf{C}\mathbf{z}_t, \mathbf{R})\mathcal{N}(\mathbf{z}_t|\boldsymbol{\mu}_{t|t-1}, \boldsymbol{\Sigma}_{t|t-1})d\mathbf{z}_t \tag{18.42} \\ &= \mathcal{N}(\mathbf{y}_t|\mathbf{C}\boldsymbol{\mu}_{t|t-1}, \mathbf{C}\boldsymbol{\Sigma}_{t|t-1}\mathbf{C}^T + \mathbf{R}) \tag{18.43} \end{aligned}$$

This is useful for time series forecasting.

18.3.1.5   Computational issues

There are two dominant costs in the Kalman filter: the matrix inversion to compute the Kalman gain matrix, $\mathbf{K}_t$, which takes $O(|\mathbf{y}_t|^3)$ time; and the matrix-matrix multiply to compute $\boldsymbol{\Sigma}_t$, which takes $O(|\mathbf{z}_t|^2)$ time. In some applications (e.g., robotic mapping), we have $|\mathbf{z}_t| \gg |\mathbf{y}_t|$, so the latter cost dominates. However, in such cases, we can sometimes use sparse approximations (see (Thrun et al. 2006)).

In cases where $|\mathbf{y}_t| \gg |\mathbf{z}_t|$, we can precompute $\mathbf{K}_t$, since, suprisingly, it does not depend on the actual observations $\mathbf{y}_{1:t}$ (an unusual property that is specific to linear Gaussian systems). The iterative equations for updating $\boldsymbol{\Sigma}_t$ are called the **Ricatti equations**, and for time invariant systems (i.e., where $\boldsymbol{\theta}_t = \boldsymbol{\theta}$), they converge to a fixed point. This steady state solution can then be used instead of using a time-specific gain matrix.

In practice, more sophisticated implementations of the Kalman filter should be used, for reasons of numerical stability. One approach is the **information filter**, which recursively updates the canonical parameters of the Gaussian, $\boldsymbol{\Lambda}_t = \boldsymbol{\Sigma}_t^{-1}$ and $\boldsymbol{\eta}_t = \boldsymbol{\Lambda}_t\boldsymbol{\mu}_t$, instead of the moment parameters. Another approach is the **square root filter**, which works with the Cholesky decomposition or the $\mathbf{U}_t\mathbf{D}_t\mathbf{U}_t$ decomposition of $\boldsymbol{\Sigma}_t$. This is much more numerically stable than directly updating $\boldsymbol{\Sigma}_t$. Further details can be found at http://www.cs.unc.edu/~welch/kalman and in various books, such as (Simon 2006).

18.3.1.6   Derivation *

We now derive the Kalman filter equations. For notational simplicity, we will ignore the input terms $\mathbf{u}_{1:t}$. From Bayes rule for Gaussians (Equation 4.125), we have that the posterior precision is given by

$$\boldsymbol{\Sigma}_t^{-1} = \boldsymbol{\Sigma}_{t|t-1}^{-1} + \mathbf{C}_t^T \mathbf{R}_t^{-1} \mathbf{C}_t \tag{18.44}$$

From the matrix inversion lemma (Equation 4.106) we can rewrite this as

$$\boldsymbol{\Sigma}_t = \boldsymbol{\Sigma}_{t|t-1} - \boldsymbol{\Sigma}_{t|t-1} \mathbf{C}_t^T (\mathbf{R}_t + \mathbf{C}_t \boldsymbol{\Sigma}_{t|t-1} \mathbf{C}_t^T)^{-1} \mathbf{C}_t \boldsymbol{\Sigma}_{t|t-1} \tag{18.45}$$

$$= (\mathbf{I} - \mathbf{K}_t \mathbf{C}_t) \boldsymbol{\Sigma}_{t|t-1} \tag{18.46}$$

From Bayes rule for Gaussians (Equation 4.125), the posterior mean is given by

$$\boldsymbol{\mu}_t = \boldsymbol{\Sigma}_t \mathbf{C}_t \mathbf{R}_t^{-1} \mathbf{y}_t + \boldsymbol{\Sigma}_t \boldsymbol{\Sigma}_{t|t-1}^{-1} \boldsymbol{\mu}_{t|t-1} \tag{18.47}$$

We will now massage this into the form stated earlier. Applying the second matrix inversion lemma (Equation 4.107) to the first term of Equation 18.47 we have

$$\boldsymbol{\Sigma}_t \mathbf{C}_t \mathbf{R}_t^{-1} \mathbf{y}_t = (\boldsymbol{\Sigma}_{t|t-1}^{-1} + \mathbf{C}_t^T \mathbf{R}_t^{-1} \mathbf{C}_t)^{-1} \mathbf{C}_t \mathbf{R}_t^{-1} \mathbf{y}_t \tag{18.48}$$

$$= \boldsymbol{\Sigma}_{t|t-1} \mathbf{C}_t^T (\mathbf{R}_t + \mathbf{C}_t \boldsymbol{\Sigma}_{t|t-1} \mathbf{C}_t^T)^{-1} \mathbf{y}_t = \mathbf{K}_t \mathbf{y}_t \tag{18.49}$$

Now applying the matrix inversion lemma (Equation 4.106) to the second term of Equation 18.47 we have

$$\boldsymbol{\Sigma}_t \boldsymbol{\Sigma}_{t|t-1}^{-1} \boldsymbol{\mu}_{t|t-1} \tag{18.50}$$

$$= (\boldsymbol{\Sigma}_{t|t-1}^{-1} + \mathbf{C}_t^T \mathbf{R}_t^{-1} \mathbf{C}_t)^{-1} \boldsymbol{\Sigma}_{t|t-1}^{-1} \boldsymbol{\mu}_{t|t-1} \tag{18.51}$$

$$= \left[\boldsymbol{\Sigma}_{t|t-1} - \boldsymbol{\Sigma}_{t|t-1} \mathbf{C}^T (\mathbf{R}_t + \mathbf{C}_t^T \boldsymbol{\Sigma}_{t|t-1} \mathbf{C}_t^T) \mathbf{C}_t \boldsymbol{\Sigma}_{t|t-1} \right] \boldsymbol{\Sigma}_{t|t-1}^{-1} \boldsymbol{\mu}_{t|t-1} \tag{18.52}$$

$$= (\boldsymbol{\Sigma}_{t|t-1} - \mathbf{K}_t \mathbf{C}_t^T \boldsymbol{\Sigma}_{t|t-1}) \boldsymbol{\Sigma}_{t|t-1}^{-1} \boldsymbol{\mu}_{t|t-1} \tag{18.53}$$

$$= \boldsymbol{\mu}_{t|t-1} - \mathbf{K}_t \mathbf{C}_t^T \boldsymbol{\mu}_{t|t-1} \tag{18.54}$$

Putting the two together we get

$$\boldsymbol{\mu}_t = \boldsymbol{\mu}_{t|t-1} + \mathbf{K}_t (\mathbf{y}_t - \mathbf{C}_t \boldsymbol{\mu}_{t|t-1}) \tag{18.55}$$

18.3.2   The Kalman smoothing algorithm

In Section 18.3.1, we described the Kalman filter, which sequentially computes $p(\mathbf{z}_t|\mathbf{y}_{1:t})$ for each t. This is useful for online inference problems, such as tracking. However, in an offline setting, we can wait until all the data has arrived, and then compute $p(\mathbf{z}_t|\mathbf{y}_{1:T})$. By conditioning on past and future data, our uncertainty will be significantly reduced. This is illustrated in Figure 18.1(c), where we see that the posterior covariance ellipsoids are smaller for the smoothed trajectory than for the filtered trajectory. (The ellipsoids are larger at the beginning and end of the trajectory, since states near the boundary do not have as many useful neighbors from which to borrow information.)

We now explain how to compute the smoothed estimates, using an algorithm called the **RTS smoother**, named after its inventors, Rauch, Tung and Striebel (Rauch et al. 1965). It is also known as the **Kalman smoothing** algorithm. The algorithm is analogous to the forwards-backwards algorithm for HMMs, although there are some small differences which we discuss below.

18.3.2.1   Algorithm

Kalman filtering can be regarded as message passing on a graph, from left to right. When the messages have reached the end of the graph, we have successfully computed $p(\mathbf{z}_T|\mathbf{y}_{1:T})$. Now we work backwards, from right to left, sending information from the future back to the past, and them combining the two information sources. The question is: how do we compute these backwards equations? We first give the equations, then the derivation.

We have

$$p(\mathbf{z}_t|\mathbf{y}_{1:T}) = \mathcal{N}(\boldsymbol{\mu}_{t|T}, \boldsymbol{\Sigma}_{t|T}) \tag{18.56}$$

$$\boldsymbol{\mu}_{t|T} = \boldsymbol{\mu}_{t|t} + \mathbf{J}_t(\boldsymbol{\mu}_{t+1|T} - \boldsymbol{\mu}_{t+1|t}) \tag{18.57}$$

$$\boldsymbol{\Sigma}_{t|T} = \boldsymbol{\Sigma}_{t|t} + \mathbf{J}_t(\boldsymbol{\Sigma}_{t+1|T} - \boldsymbol{\Sigma}_{t+1|t})\mathbf{J}_t^T \tag{18.58}$$

$$\mathbf{J}_t \triangleq \boldsymbol{\Sigma}_{t|t}\mathbf{A}_{t+1}^T\boldsymbol{\Sigma}_{t+1|t}^{-1} \tag{18.59}$$

where $\mathbf{J}_t$ is the backwards Kalman gain matrix. The algorithm can be initialized from $\boldsymbol{\mu}_{T|T}$ and $\boldsymbol{\Sigma}_{T|T}$ from the Kalman filter. Note that this backwards pass does not need access to the data, that is, it does not need $\mathbf{y}_{1:T}$. This allows us to "throw away" potentially high dimensional observation vectors, and just keep the filtered belief states, which usually requires less memory.

18.3.2.2   Derivation *

We now derive the Kalman smoother, following the presentation of (Jordan 2007, sec 15.7).

The key idea is to leverage the Markov property, which says that $\mathbf{z}_t$ is independent of future data, $\mathbf{y}_{t+1:T}$, as long as $\mathbf{z}_{t+1}$ is known. Of course, $\mathbf{z}_{t+1}$ is not known, but we have a distribution over it. So we condition on $\mathbf{z}_{t+1}$ and then integrate it out, as follows.

$$p(\mathbf{z}_t|\mathbf{y}_{1:T}) = \int p(\mathbf{z}_t|\mathbf{y}_{1:T}, \mathbf{z}_{t+1})p(\mathbf{z}_{t+1}|\mathbf{y}_{1:T})d\mathbf{z}_{t+1} \tag{18.60}$$

$$= \int p(\mathbf{z}_t|\mathbf{y}_{1:t}, \cancel{\mathbf{y}_{t+1:T}}, \mathbf{z}_{t+1})p(\mathbf{z}_{t+1}|\mathbf{y}_{1:T})d\mathbf{z}_{t+1} \tag{18.61}$$

By induction, assume we have already computed the smoothed distribution for $t + 1$:

$$p(\mathbf{z}_{t+1}|\mathbf{y}_{1:T}) = \mathcal{N}(\mathbf{z}_{t+1}|\boldsymbol{\mu}_{t+1|T}, \boldsymbol{\Sigma}_{t+1|T}) \tag{18.62}$$

The question is: how do we perform the integration?

First, we compute the filtered two-slice distribution $p(\mathbf{z}_t, \mathbf{z}_{t+1}|\mathbf{y}_{1:t})$ as follows:

$$p(\mathbf{z}_t, \mathbf{z}_{t+1}|\mathbf{y}_{1:t}) = \mathcal{N}\left(\begin{pmatrix}\mathbf{z}_t \\ \mathbf{z}_{t+1}\end{pmatrix} \middle| \begin{pmatrix}\boldsymbol{\mu}_{t|t} \\ \boldsymbol{\mu}_{t+1|t}\end{pmatrix}, \begin{pmatrix}\boldsymbol{\Sigma}_{t|t} & \boldsymbol{\Sigma}_{t|t}\mathbf{A}_{t+1}^T \\ \mathbf{A}_{t+1}\boldsymbol{\Sigma}_{t|t} & \boldsymbol{\Sigma}_{t+1|t}\end{pmatrix}\right) \tag{18.63}$$

Now we use Gaussian conditioning to compute $p(\mathbf{z}_t|\mathbf{z}_{t+1}, \mathbf{y}_{1:t})$ as follows:

$$p(\mathbf{z}_t|\mathbf{z}_{t+1}, \mathbf{y}_{1:t}) \quad = \quad \mathcal{N}(\mathbf{z}_t|\boldsymbol{\mu}_{t|t} + \mathbf{J}_t(\mathbf{z}_{t+1} - \boldsymbol{\mu}_{t+1|t}), \boldsymbol{\Sigma}_{t|t} - \mathbf{J}_t\boldsymbol{\Sigma}_{t+1|t}\mathbf{J}_t^T) \tag{18.64}$$

We can compute the smoothed distribution for t using the rules of iterated expectation and iterated covariance. First, the mean:

$$\boldsymbol{\mu}_{t|T} \quad = \quad \mathbb{E}\left[\mathbb{E}\left[\mathbf{z}_t|\mathbf{z}_{t+1}, \mathbf{y}_{1:T}\right]|\mathbf{y}_{1:T}\right] \tag{18.65}$$

$$= \quad \mathbb{E}\left[\mathbb{E}\left[\mathbf{z}_t|\mathbf{z}_{t+1}, \mathbf{y}_{1:t}\right]|\mathbf{y}_{1:T}\right] \tag{18.66}$$

$$= \quad \mathbb{E}\left[\boldsymbol{\mu}_{t|t} + \mathbf{J}_t(\mathbf{z}_{t+1} - \boldsymbol{\mu}_{t+1|t})|\mathbf{y}_{1:T}\right] \tag{18.67}$$

$$= \quad \boldsymbol{\mu}_{t|t} + \mathbf{J}_t(\boldsymbol{\mu}_{t+1|T} - \boldsymbol{\mu}_{t+1|t}) \tag{18.68}$$

Now the covariance:

$$\boldsymbol{\Sigma}_{t|T} \quad = \quad \text{cov}\left[\mathbb{E}\left[\mathbf{z}_t|\mathbf{z}_{t+1}, \mathbf{y}_{1:T}\right]|\mathbf{y}_{1:T}\right] + \mathbb{E}\left[\text{cov}\left[\mathbf{z}_t|\mathbf{z}_{t+1}, \mathbf{y}_{1:T}\right]|\mathbf{y}_{1:T}\right] \tag{18.69}$$

$$= \quad \text{cov}\left[\mathbb{E}\left[\mathbf{z}_t|\mathbf{z}_{t+1}, \mathbf{y}_{1:t}\right]|\mathbf{y}_{1:T}\right] + \mathbb{E}\left[\text{cov}\left[\mathbf{z}_t|\mathbf{z}_{t+1}, \mathbf{y}_{1:t}\right]|\mathbf{y}_{1:T}\right] \tag{18.70}$$

$$= \quad \text{cov}\left[\boldsymbol{\mu}_{t|t} + \mathbf{J}_t(\mathbf{z}_{t+1} - \boldsymbol{\mu}_{t+1|t})|\mathbf{y}_{1:T}\right] + \mathbb{E}\left[\boldsymbol{\Sigma}_{t|t} - \mathbf{J}_t\boldsymbol{\Sigma}_{t+1|t}\mathbf{J}_t^T|\mathbf{y}_{1:T}\right] \tag{18.71}$$

$$= \quad \mathbf{J}_t\text{cov}\left[\mathbf{z}_{t+1} - \boldsymbol{\mu}_{t+1|t}|\mathbf{y}_{1:T}\right]\mathbf{J}_t^T + \boldsymbol{\Sigma}_{t|t} - \mathbf{J}_t\boldsymbol{\Sigma}_{t+1|t}\mathbf{J}_t^T \tag{18.72}$$

$$= \quad \mathbf{J}_t\boldsymbol{\Sigma}_{t+1|T}\mathbf{J}_t^T + \boldsymbol{\Sigma}_{t|t} - \mathbf{J}_t\boldsymbol{\Sigma}_{t+1|t}\mathbf{J}_t^T \tag{18.73}$$

$$= \quad \boldsymbol{\Sigma}_{t|t} + \mathbf{J}_t(\boldsymbol{\Sigma}_{t+1|T} - \boldsymbol{\Sigma}_{t+1|t})\mathbf{J}_t^T \tag{18.74}$$

The algorithm can be initialized from $\boldsymbol{\mu}_{T|T}$ and $\boldsymbol{\Sigma}_{T|T}$ from the last step of the filtering algorithm.

18.3.2.3 Comparison to the forwards-backwards algorithm for HMMs *

Note that in both the forwards and backwards passes for LDS, we always worked with normalized distributions, either conditioned on the past data or conditioned on all the data. Furthermore, the backwards pass depends on the results of the forwards pass. This is different from the usual presentation of forwards-backwards for HMMs, where the backwards pass can be computed independently of the forwards pass (see Section 17.4.3).

It turns out that we can rewrite the Kalman smoother in a modified form which makes it more similar to forwards-backwards for HMMs. In particular, we have

$$p(\mathbf{z}_t|\mathbf{y}_{1:T}) \quad = \quad \int p(\mathbf{z}_t|\mathbf{y}_{1:t}, \mathbf{z}_{t+1})p(\mathbf{z}_{t+1}|\mathbf{y}_{1:T})d\mathbf{z}_{t+1} \tag{18.75}$$

$$= \quad \int p(\mathbf{z}_t, \mathbf{z}_{t+1}|\mathbf{y}_{1:t})\frac{p(\mathbf{z}_{t+1}|\mathbf{y}_{1:T})}{p(\mathbf{z}_{t+1}|\mathbf{y}_{1:t})}d\mathbf{z}_{t+1} \tag{18.76}$$

Now

$$p(\mathbf{z}_{t+1}|\mathbf{y}_{1:T}) = \frac{p(\mathbf{y}_{t+1:T}|\mathbf{z}_{t+1}, \mathbf{y}_{1:t})p(\mathbf{z}_{t+1}|\mathbf{y}_{1:t})}{p(\mathbf{y}_{t+1:T}|\mathbf{y}_{1:t})} \tag{18.77}$$

so

$$\frac{p(\mathbf{z}_{t+1}|\mathbf{y}_{1:T})}{p(\mathbf{z}_{t+1}|\mathbf{y}_{1:t})} = \frac{p(\mathbf{z}_{t+1}|\mathbf{y}_{1:t})p(\mathbf{y}_{t+1:T}|\mathbf{z}_{t+1})}{p(\mathbf{z}_{t+1}|\mathbf{y}_{1:t})p(\mathbf{y}_{t+1:T}|\mathbf{y}_{1:t})} \propto p(\mathbf{y}_{t+1:T}|\mathbf{z}_{t+1}) \tag{18.78}$$

which is the conditional likelihood of the future data. This backwards message can be computed independently of the forwards message, similar to the HMM case. However, this approach has several disadvantages: (1) it needs access to the original observation sequence; (2) the backwards message is a likelihood, not a posterior, so it need not to integrate to 1 over $\mathbf{z}_t$ – in fact, it may not always be possible to represent $p(\mathbf{y}_{t+1:T}|\mathbf{z}_{t+1})$ as a Gaussian with positive definite covariance (this problem does not arise in discrete state-spaces, as used in HMMs); (3) when exact inference is not possible, it makes more sense to try to approximate the smoothed distribution rather than the backwards likelihood term (see Section 22.5).

There is yet another variant, known as **two-filter smoothing**, whereby we compute $p(\mathbf{z}_t|\mathbf{y}_{1:t})$ in the forwards pass as usual, and the filtered posterior $p(\mathbf{z}_t|\mathbf{y}_{t+1:T})$ in the backwards pass. These can then be combined to compute $p(\mathbf{z}_t|\mathbf{y}_{1:T})$. See (Kitagawa 2004; Briers et al. 2010) for details.

18.4   Learning for LG-SSM

In this section, we briefly discuss how to estimate the parameters of an LG-SSM. In the control theory community, this is known as **systems identification** (Ljung 1987).

When using SSMs for time series forecasting, and also in some physical state estimation problems, the observation matrix $\mathbf{C}$ and the transition matrix $\mathbf{A}$ are both known and fixed, by definition of the model. In such cases, all that needs to be learned are the noise covariances $\mathbf{Q}$ and $\mathbf{R}$. (The initial state estimate $\boldsymbol{\mu}_0$ is often less important, since it will get "washed away" by the data after a few time steps. This can be encouraged by setting the initial state covariance to be large, representing a weak prior.) Although we can estimate $\mathbf{Q}$ and $\mathbf{R}$ offline, using the methods described below, it is also possible to derive a recursive procedure to exactly compute the posterior $p(\mathbf{z}_t, \mathbf{R}, \mathbf{Q}|\mathbf{y}_{1:t})$, which has the form of a Normal-inverse-Wishart; see (West and Harrison 1997; Prado and West 2010) for details.

18.4.1   Identifiability and numerical stability

In the more general setting, where the hidden states have no pre-specified meaning, we need to learn $\mathbf{A}$ and $\mathbf{C}$. However, in this case we can set $\mathbf{Q} = \mathbf{I}$ without loss of generality, since an arbitrary noise covariance can be modeled by appropriately modifying $\mathbf{A}$. Also, by analogy with factor analysis, we can require $\mathbf{R}$ to be diagonal without loss of generality. Doing this reduces the number of free parameters and improves numerical stability.

Another constraint that is useful to impose is on the eigenvalues of the dynamics matrix $\mathbf{A}$. To see why this is important, consider the case of no system noise. In this case, the hidden state at time t is given by

$$\mathbf{z}_t = \mathbf{A}^t \mathbf{z}_1 = \mathbf{U} \mathbf{\Lambda}^t \mathbf{U}^{-1} \mathbf{z}_1 \tag{18.79}$$

where $\mathbf{U}$ is the matrix of eigenvectors for $\mathbf{A}$, and $\mathbf{\Lambda} = \text{diag}(\lambda_i)$ contains the eigenvalues. If any $\lambda_i > 1$, then for large t, $\mathbf{z}_t$ will blow up in magnitude. Consequently, to ensure stability, it is useful to require that all the eigenvalues are less than 1 (Siddiqi et al. 2007). Of course, if all the eigenvalues are less than 1, then $\mathbb{E}[\mathbf{z}_t] = \mathbf{0}$ for large t, so the state will return to the origin. Fortunately, when we add noise, the state become non-zero, so the model does not degenerate.

Below we discuss how to estimate the parameters. However, for simplicity of presentation, we do not impose any of the constraints mentioned above.

18.4.2 Training with fully observed data

If we observe the hidden state sequences, we can fit the model by computing the MLEs (or even the full posteriors) for the parameters by solving a multivariate linear regression problem for $\mathbf{z}_{t-1} \to \mathbf{z}_t$ and for $\mathbf{z}_t \to \mathbf{y}_t$. That is, we can estimate $\mathbf{A}$ by solving the least squares problem $J(\mathbf{A}) = \sum_{t=1}^{2} (\mathbf{z}_t - \mathbf{A}\mathbf{z}_{t-1})^2$, and similarly for $\mathbf{C}$. We can estimate the system noise covariance $\mathbf{Q}$ from the residuals in predicting $\mathbf{z}_t$ from $\mathbf{z}_{t-1}$, and estimate the observation noise covariance $\mathbf{R}$ from the residuals in predicting $\mathbf{y}_t$ from $\mathbf{z}_t$.

18.4.3 EM for LG-SSM

If we only observe the output sequence, we can compute ML or MAP estimates of the parameters using EM. The method is conceptually quite similar to the Baum-Welch algorithm for HMMs (Section 17.5), except we use Kalman smoothing instead of forwards-backwards in the E step, and use different calculations in the M step. We leave the details to Exercise 18.1.

18.4.4 Subspace methods

EM does not always give satisfactory results, because it is sensitive to the initial parameter estimates. One way to avoid this is to use a different approach known as a **subspace method** (Overschee and Moor 1996; Katayama 2005).

To understand this approach, let us initially assume there is no observation noise and no system noise. In this case, we have $\mathbf{z}_t = \mathbf{A}\mathbf{z}_{t-1}$ and $\mathbf{y}_t = \mathbf{C}\mathbf{z}_t$, and hence $\mathbf{y}_t = \mathbf{C}\mathbf{A}^{t-1}\mathbf{z}_1$. Consequently all the observations must be generated from a $\dim(\mathbf{z}_t)$-dimensional linear manifold or subspace. We can identify this subspace using PCA (see the above references for details). Once we have an estimate of the $\mathbf{z}_t$'s, we can fit the model as if it were fully observed. We can either use these estimates in their own right, or use them to initialize EM.

18.4.5 Bayesian methods for "fitting" LG-SSMs

There are various offline Bayesian alternatives to the EM algorithm, including variational Bayes EM (Beal 2003; Barber and Chiappa 2007) and blocked Gibbs sampling (Carter and Kohn 1994; Cappe et al. 2005; Fruhwirth-Schnatter 2007). The Bayesian approach can also be used to perform online learning, as we discussed in Section 18.2.3. Unfortunately, once we add the SSM parameters to the state space, the model is generally no longer linear Gaussian. Consequently we must use some of the approximate online inference methods to be discussed below.

18.5 Approximate online inference for non-linear, non-Gaussian SSMs

In Section 18.3.1, we discussed how to perform exact online inference for LG-SSMs. However, many models are non linear. For example, most moving objects do not move in straight lines. And even if they did, if we assume the parameters of the model are unknown and add them to the state space, the model becomes nonlinear. Furthermore, non-Gaussian noise is also very common, e.g., due to outliers, or when inferring parameters for GLMs instead of just linear regression. For these more general models, we need to use approximate inference.

The approximate inference algorithms we discuss below approximate the posterior by a Gaussian. In general, if $Y = f(X)$, where X has a Gaussian distribution and f is a non-linear function, there are two main ways to approximate $p(Y)$ by a Gaussian. The first is to use a first-order approximation of f. The second is to use the exact f, but to project $f(X)$ onto the space of Gaussians by moment matching. We discuss each of these methods in turn. (See also Section 23.5, where we discuss particle filtering, which is a stochastic algorithm for approximate online inference, which uses a non-parametric approximation to the posterior, which is often more accurate but slower to compute.)

18.5.1 Extended Kalman filter (EKF)

In this section, we focus on non-linear models, but we assume the noise is Gaussian. That is, we consider models of the form

$$
\begin{align}
\mathbf{z}_t &= g(\mathbf{u}_t, \mathbf{z}_{t-1}) + \mathcal{N}(\mathbf{0}, \mathbf{Q}_t) \tag{18.80} \\
\mathbf{y}_t &= h(\mathbf{z}_t) + \mathcal{N}(\mathbf{0}, \mathbf{R}_t) \tag{18.81}
\end{align}
$$

where the transition model g and the observation model h are nonlinear but differentiable functions. Furthermore, we focus on the case where we approximate the posterior by a single Gaussian. (The simplest way to handle more general posteriors (e.g., multi-modal, discrete, etc). is to use particle filtering, which we discuss in Section 23.5.)

The **extended Kalman filter** or **EKF** can be applied to nonlinear Gaussian dynamical systems of this form. The basic idea is to linearize g and h about the previous state estimate using a first order Taylor series expansion, and then to apply the standard Kalman filter equations. (The noise variance in the equations ($\mathbf{Q}$ and $\mathbf{R}$) is not changed, i.e., the additional error due to linearization is not modeled.) Thus we approximate the stationary non-linear dynamical system with a non-stationary linear dynamical system.

The intuition behind the approach is shown in Figure 18.9, which shows what happens when we pass a Gaussian distribution $p(x)$, shown on the bottom right, through a nonlinear function $y = g(x)$, shown on the top right. The resulting distribution (approximated by Monte Carlo) is shown in the shaded gray area in the top left corner. The best Gaussian approximation to this, computed from $\mathbb{E}[g(x)]$ and $\text{var}[g(x)]$ by Monte Carlo, is shown by the solid black line. The EKF approximates this Gaussian as follows: it linearizes the g function at the current mode, μ, and then passes the Gaussian distribution $p(x)$ through this linearized function. In this example, the result is quite a good approximation to the first and second moments of $p(y)$, for much less cost than an MC approximation.

In more detail, the method works as follows. We approximate the measurement model using

$$
p(\mathbf{y}_t | \mathbf{z}_t) \approx \mathcal{N}(\mathbf{y}_t | \mathbf{h}(\boldsymbol{\mu}_{t|t-1}) + \mathbf{H}_t(\mathbf{y}_t - \boldsymbol{\mu}_{t|t-1}), \mathbf{R}_t) \tag{18.82}
$$

where $\mathbf{H}_t$ is the Jacobian matrix of $\mathbf{h}$ evaluated at the prior mode:

$$
\begin{align}
H_{ij} &\triangleq \frac{\partial h_i(\mathbf{z})}{\partial z_j} \tag{18.83} \\
\mathbf{H}_t &\triangleq \mathbf{H}|_{\mathbf{z}=\boldsymbol{\mu}_{t|t-1}} \tag{18.84}
\end{align}
$$

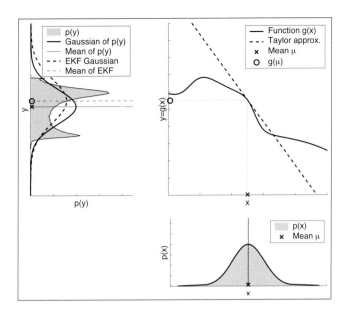

Figure 18.9 Nonlinear transformation of a Gaussian random variable. The prior $p(x)$ is shown on the bottom right. The function $y = g(x)$ is shown on the top right. The transformed distribution $p(y)$ is shown in the top left. A linear function induces a Gaussian distribution, but a non-linear function induces a complex distribution. The solid line is the best Gaussian approximation to this; the dotted line is the EKF approximation to this. Source: Figure 3.4 of (Thrun et al. 2006). Used with kind permission of Sebastian Thrun.

Similarly, we approximate the system model using

$$p(\mathbf{z}_t|\mathbf{z}_{t-1}, \mathbf{u}_t) \approx \mathcal{N}(\mathbf{z}_t|\mathbf{g}(\mathbf{u}_t, \boldsymbol{\mu}_{t-1}) + \mathbf{G}_t(\mathbf{z}_{t-1} - \boldsymbol{\mu}_{t-1}), \mathbf{Q}_t) \tag{18.85}$$

where

$$G_{ij}(\mathbf{u}) \triangleq \frac{\partial g_i(\mathbf{u}, \mathbf{z})}{\partial z_j} \tag{18.86}$$

$$\mathbf{G}_t \triangleq \mathbf{G}(\mathbf{u}_t)|_{\mathbf{z}=\boldsymbol{\mu}_{t-1}} \tag{18.87}$$

so $\mathbf{G}$ is the Jacobian matrix of $\mathbf{g}$ evaluated at the prior mode.

Given this, we can then apply the Kalman filter to compute the posterior as follows:

$$\boldsymbol{\mu}_{t|t-1} = \mathbf{g}(\mathbf{u}_t, \boldsymbol{\mu}_{t-1}) \tag{18.88}$$

$$\mathbf{V}_{t|t-1} = \mathbf{G}_t \mathbf{V}_{t-1} \mathbf{G}_t^T + \mathbf{Q}_t \tag{18.89}$$

$$\mathbf{K}_t = \mathbf{V}_{t|t-1} \mathbf{H}_t^T (\mathbf{H}_t \mathbf{V}_{t|t-1} \mathbf{H}_t^T + \mathbf{R}_t)^{-1} \tag{18.90}$$

$$\boldsymbol{\mu}_t = \boldsymbol{\mu}_{t|t-1} + \mathbf{K}_t(\mathbf{y}_t - \mathbf{h}(\boldsymbol{\mu}_{t|t-1})) \tag{18.91}$$

$$\mathbf{V}_t = (\mathbf{I} - \mathbf{K}_t \mathbf{H}_t) \mathbf{V}_{t|t-1} \tag{18.92}$$

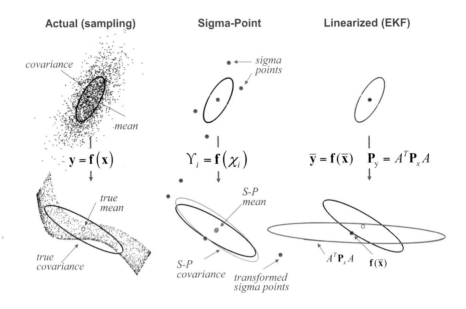

Figure 18.10   An example of the unscented transform in two dimensions.   Source: (Wan and der Merwe 2001).   Used with kind permission of Eric Wan.

We see that the only difference from the regular Kalman filter is that, when we compute the state prediction, we use $\mathbf{g}(\mathbf{u}_t, \boldsymbol{\mu}_{t-1})$ instead of $\mathbf{A}_t\boldsymbol{\mu}_{t-1} + \mathbf{B}_t\mathbf{u}_t$, and when we compute the measurement update we use $\mathbf{h}(\boldsymbol{\mu}_{t|t-1})$ instead of $\mathbf{C}_t\boldsymbol{\mu}_{t|t-1}$.

It is possible to improve performance by repeatedly re-linearizing the equations around $\boldsymbol{\mu}_t$ instead of $\boldsymbol{\mu}_{t|t-1}$; this is called the **iterated EKF**, and yields better results, although it is of course slower.

There are two cases when the EKF works poorly. The first is when the prior covariance is large. In this case, the prior distribution is broad, so we end up sending a lot of probability mass through different parts of the function that are far from the mean, where the function has been linearized. The other setting where the EKF works poorly is when the function is highly nonlinear near the current mean. In Section 18.5.2, we will discuss an algorithm called the UKF which works better than the EKF in both of these settings.

18.5.2   Unscented Kalman filter (UKF)

The **unscented Kalman filter** (**UKF**) is a better version of the EKF (Julier and Uhlmann 1997). (Apparently it is so-called because it "doesn't stink"!) The key intuition is this: it is easier to approximate a Gaussian than to approximate a function. So instead of performing a linear approximation to the function, and passing a Gaussian through it, instead pass a deterministically chosen set of points, known as **sigma points**, through the function, and fit a Gaussian to the resulting transformed points. This is known as the **unscented transform**, and is sketched in Figure 18.10. (We explain this figure in detail below.)

The UKF basically uses the unscented transform twice, once to approximate passing through the system model **g**, and once to approximate passing through the measurement model **h**. We give the details below. Note that the UKF and EKF both perform $O(d^3)$ operations per time step where d is the size of the latent state-space. However, the UKF is accurate to at least second order, whereas the EKF is only a first order approximation (although both the EKF and UKF can be extended to capture higher order terms). Furthermore, the unscented transform does not require the analytic evaluation of any derivatives or Jacobians (a so-called **derivative free filter**), making it simpler to implement and more widely applicable.

18.5.2.1 The unscented transform

Before explaining the UKF, we first explain the unscented transform. Assume $p(\mathbf{x}) = \mathcal{N}(\mathbf{x}|\boldsymbol{\mu}, \boldsymbol{\Sigma})$, and consider estimating $p(\mathbf{y})$, where $\mathbf{y} = \mathbf{f}(\mathbf{x})$ for some nonlinear function $\mathbf{f}$. The unscented transform does this as follows. First we create a set of $2d + 1$ sigma points $\mathbf{x}_i$, given by

$$\mathbf{x} = \left(\boldsymbol{\mu}, \{\boldsymbol{\mu} + (\sqrt{(d + \lambda)\boldsymbol{\Sigma}})_{:i}\}_{i=1}^d, \{\boldsymbol{\mu} - (\sqrt{(d + \lambda)\boldsymbol{\Sigma}})_{:i}\}_{i=1}^d \right) \tag{18.93}$$

where $\lambda = \alpha^2(d + \kappa) - d$ is a scaling parameter to be specified below, and the notation $\mathbf{M}_{:i}$ means the i'th column of matrix $\mathbf{M}$.

These sigma points are propagated through the nonlinear function to yield $\mathbf{y}_i = f(\mathbf{x}_i)$, and the mean and covariance for $\mathbf{y}$ is computed as follows:

$$\boldsymbol{\mu}_y = \sum_{i=0}^{2d} w_m^i \mathbf{y}_i \tag{18.94}$$

$$\boldsymbol{\Sigma}_y = \sum_{i=0}^{2d} w_c^i (\mathbf{y}_i - \boldsymbol{\mu}_y)(\mathbf{y}_i - \boldsymbol{\mu}_y)^T \tag{18.95}$$

where the w's are weighting terms, given by

$$w_m^0 = \frac{\lambda}{d + \lambda} \tag{18.96}$$

$$w_c^0 = \frac{\lambda}{d + \lambda} + (1 - \alpha^2 + \beta) \tag{18.97}$$

$$w_m^i = w_c^i = \frac{1}{2(d + \lambda)} \tag{18.98}$$

See Figure 18.10 for an illustration.

In general, the optimal values of α, β and κ are problem dependent, but when $d = 1$, they are $\alpha = 1$, $\beta = 0$, $\kappa = 2$. Thus in the 1d case, $\lambda = 2$, so the 3 sigma points are μ, $\mu + \sqrt{3}\sigma$ and $\mu - \sqrt{3}\sigma$.

18.5.2.2 The UKF algorithm

The UKF algorithm is simply two applications of the unscented tranform, one to compute $p(\mathbf{z}_t|\mathbf{y}_{1:t-1}, \mathbf{u}_{1:t})$ and the other to compute $p(\mathbf{z}_t|\mathbf{y}_{1:t}, \mathbf{u}_{1:t})$. We give the details below.

The first step is to approximate the predictive density $p(\mathbf{z}_t|\mathbf{y}_{1:t-1}, \mathbf{u}_{1:t}) \approx \mathcal{N}(\mathbf{z}_t|\overline{\boldsymbol{\mu}}_t, \overline{\boldsymbol{\Sigma}}_t)$ by passing the old belief state $\mathcal{N}(\mathbf{z}_{t-1}|\boldsymbol{\mu}_{t-1}, \boldsymbol{\Sigma}_{t-1})$ through the system model $\mathbf{g}$ as follows:

$$\mathbf{z}_{t-1}^0 = \left(\boldsymbol{\mu}_{t-1}, \{\boldsymbol{\mu}_{t-1} + \gamma(\sqrt{\boldsymbol{\Sigma}_{t-1}})_{:i}\}_{i=1}^d, \{\boldsymbol{\mu}_{t-1} - \gamma(\sqrt{\boldsymbol{\Sigma}_{t-1}})_{:i}\}_{i=1}^d\right) \tag{18.99}$$

$$\overline{\mathbf{z}}_t^{*i} = \mathbf{g}(\mathbf{u}_t, \mathbf{z}_{t-1}^{0i}) \tag{18.100}$$

$$\overline{\boldsymbol{\mu}}_t = \sum_{i=0}^{2d} w_m^i \overline{\mathbf{z}}_t^{*i} \tag{18.101}$$

$$\overline{\boldsymbol{\Sigma}}_t = \sum_{i=0}^{2d} w_c^i (\overline{\mathbf{z}}_t^{*i} - \overline{\boldsymbol{\mu}}_t)(\overline{\mathbf{z}}_t^{*i} - \overline{\boldsymbol{\mu}}_t)^T + \mathbf{Q}_t \tag{18.102}$$

where $\gamma = \sqrt{d + \lambda}$.

The second step is to approximate the likelihood $p(\mathbf{y}_t|\mathbf{z}_t) \approx \mathcal{N}(\mathbf{y}_t|\hat{\mathbf{y}}_t, \mathbf{S}_t)$ by passing the prior $\mathcal{N}(\mathbf{z}_t|\overline{\boldsymbol{\mu}}_t, \overline{\boldsymbol{\Sigma}}_t)$ through the observation model $\mathbf{h}$:

$$\overline{\mathbf{z}}_t^0 = \left(\overline{\boldsymbol{\mu}}_t, \{\overline{\boldsymbol{\mu}}_t + \gamma(\sqrt{\overline{\boldsymbol{\Sigma}}_t})_{:i}\}_{i=1}^d, \{\overline{\boldsymbol{\mu}}_t - \gamma(\sqrt{\overline{\boldsymbol{\Sigma}}_t})_{:i}\}_{i=1}^d\right) \tag{18.103}$$

$$\overline{\mathbf{y}}_t^{*i} = \mathbf{h}(\overline{\mathbf{z}}_t^{0i}) \tag{18.104}$$

$$\hat{\mathbf{y}}_t = \sum_{i=0}^{2d} w_m^i \overline{\mathbf{y}}_t^{*i} \tag{18.105}$$

$$\mathbf{S}_t = \sum_{i=0}^{2d} w_c^i (\overline{\mathbf{y}}_t^{*i} - \hat{\mathbf{y}}_t)(\overline{\mathbf{y}}_t^{*i} - \hat{\mathbf{y}}_t)^T + \mathbf{R}_t \tag{18.106}$$

Finally, we use Bayes rule for Gaussians to get the posterior $p(\mathbf{z}_t|\mathbf{y}_{1:t}, \mathbf{u}_{1:t}) \approx \mathcal{N}(\mathbf{z}_t|\boldsymbol{\mu}_t, \boldsymbol{\Sigma}_t)$:

$$\overline{\boldsymbol{\Sigma}}_t^{z,y} = \sum_{i=0}^{2d} w_c^i (\overline{\mathbf{z}}_t^{*i} - \overline{\boldsymbol{\mu}}_t)(\overline{\mathbf{y}}_t^{*i} - \hat{\mathbf{y}}_t)^T \tag{18.107}$$

$$\mathbf{K}_t = \overline{\boldsymbol{\Sigma}}_t^{z,y} \mathbf{S}_t^{-1} \tag{18.108}$$

$$\boldsymbol{\mu}_t = \overline{\boldsymbol{\mu}}_t + \mathbf{K}_t(\mathbf{y}_t - \hat{\mathbf{y}}_t) \tag{18.109}$$

$$\boldsymbol{\Sigma}_t = \overline{\boldsymbol{\Sigma}}_t - \mathbf{K}_t \mathbf{S}_t \mathbf{K}_t^T \tag{18.110}$$

18.5.3   Assumed density filtering (ADF)

In this section, we discuss inference where we perform an exact update step, but then approximate the posterior by a distribution of a certain convenient form, such as a Gaussian. More precisely, let the unknowns that we want to infer be denoted by $\boldsymbol{\theta}_t$. Suppose that $\mathcal{Q}$ is a set of tractable distributions, e.g., Gaussians with a diagonal covariance matrix, or a product of discrete distributions. Suppose that we have an approximate prior $q_{t-1}(\boldsymbol{\theta}_{t-1}) \approx p(\boldsymbol{\theta}_{t-1}|\mathbf{y}_{1:t-1})$, where $q_{t-1} \in \mathcal{Q}$. We can update this with the new measurement to get the approximate posterior

$$\hat{p}(\boldsymbol{\theta}_t) = \frac{1}{Z_t} p(\mathbf{y}_t|\boldsymbol{\theta}_t) q_{t|t-1}(\boldsymbol{\theta}_t) \tag{18.111}$$

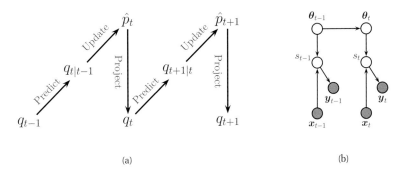

(a)   (b)

Figure 18.11 (a) Illustration of the predict-update-project cycle of assumed density filtering. (b) A dynamical logistic regression model. Compare to Figure 18.4(a).

where

$$Z_t = \int p(\mathbf{y}_t|\boldsymbol{\theta}_t)q_{t|t-1}(\boldsymbol{\theta}_t)d\boldsymbol{\theta}_t \tag{18.112}$$

is the normalization constant and

$$q_{t|t-1}(\boldsymbol{\theta}_t) = \int p(\boldsymbol{\theta}_t|\boldsymbol{\theta}_{t-1})q_{t-1}(\boldsymbol{\theta}_{t-1})d\boldsymbol{\theta}_{t-1} \tag{18.113}$$

is the one step ahead predictive distribution. If the prior is from a suitably restricted family, this one-step update process is usually tractable. However, we often find that the resulting posterior is no longer in our tractable family, $\hat{p}(\boldsymbol{\theta}_t) \notin \mathcal{Q}$. So after updating we seek the best tractable approximation by computing

$$q(\boldsymbol{\theta}_t) = \underset{q \in \mathcal{Q}}{\operatorname{argmin}} \, \mathbb{KL}\left(\hat{p}(\boldsymbol{\theta}_t)||q(\boldsymbol{\theta}_t)\right) \tag{18.114}$$

This minimizes the the Kullback-Leibler divergence (Section 2.8.2) from the approximation $q(\boldsymbol{\theta}_t)$ to the "exact" posterior $\hat{p}(\boldsymbol{\theta}_t)$, and can be thought of as projecting $\hat{p}$ onto the space of tractable distributions. The whole algorithm consists of **predict-update-project** cycles. This is known as **assumed density filtering** or **ADF** (Maybeck 1979). See Figure 18.11(a) for a sketch.

If q is in the exponential family, one can show that this KL minimization can be done by **moment matching**. We give some examples of this below.

18.5.3.1   Boyen-Koller algorithm for online inference in DBNs

If we are performing inference in a discrete-state dynamic Bayes net (Section 17.6.7), where θ_{tj} is the j'th hidden variable at time t, then the exact posterior $p(\boldsymbol{\theta}_t)$ becomes intractable to compute because of the entanglement problem. Suppose we use a fully factored approximation of the form $q(\boldsymbol{\theta}_t) = \prod_{j=1}^{D} \operatorname{Cat}(\theta_{t,j}|\boldsymbol{\pi}_{t,j})$, where $\pi_{tjk} = q(\theta_{t,j} = k)$ is the probability variable j is in state k, and D is the number of variables. In this case, the moment matching operation becomes

$$\pi_{tjk} = \hat{p}(\theta_{t,j} = k) \tag{18.115}$$

This can be computed by performing a predict-update step using the factored prior, and then computing the posterior marginals. This is known as the **Boyen-Koller** algorithm, named after the authors of (Boyen and Koller 1998), who demonstrated that the error incurred by this series of repeated approximations remains bounded (under certain assumptions about the stochasticity of the system).

18.5.3.2   Gaussian approximation for online inference in GLMs

Now suppose $q(\boldsymbol{\theta}_t) = \prod_{j=1}^{D} \mathcal{N}(\theta_{t,j} | \mu_{t,j}, \tau_{t,j})$, where $\tau_{t,j}$ is the variance. Then the optimal parameters of the tractable approximation to the posterior are

$$\mu_{t,j} = \mathbb{E}_{\hat{p}} [\theta_{t,j}] , \; \tau_{t,j} = \text{var}_{\hat{p}} [\theta_{t,j}] \tag{18.116}$$

This method can be used to do online inference for the parameters of many statistical models. For example, theTrueSkill system, used in Microsoft's Xbox to rank players over time, uses this form of approximation (Herbrich et al. 2007). We can also apply this method to simpler models, such as GLM, which have the advantage that the posterior is log-concave. Below we explain how to do this for binary logistic regression, following the presentation of (Zoeter 2007).

The model has the form

$$p(y_t | \mathbf{x}_t, \boldsymbol{\theta}_t) = \text{Ber}(y_t | \text{sigm}(\mathbf{x}_t^T \boldsymbol{\theta}_t)) \tag{18.117}$$

$$p(\boldsymbol{\theta}_t | \boldsymbol{\theta}_{t-1}) = \mathcal{N}(\boldsymbol{\theta}_t | \boldsymbol{\theta}_{t-1}, \sigma^2 \mathbf{I}) \tag{18.118}$$

where σ^2 is some process noise which allows the parameters to change slowly over time. (This can be set to 0, as in the recursive least squares method (Section 18.2.3), if desired.) We will assume $q_{t-1}(\boldsymbol{\theta}_{t-1}) = \prod_j \mathcal{N}(\theta_{t-1,j} | \mu_{t-1,j}, \tau_{t-1,j})$ is the tractable prior. We can compute the one-step-ahead predictive density $q_{t|t-1}(\boldsymbol{\theta}_t)$ using the standard linear-Gaussian update. So now we concentrate on the measurement update step.

Define the deterministic quantity $s_t = \boldsymbol{\theta}_t^T \mathbf{x}_t$, as shown in Figure 18.11(b). If $q_{t|t-1}(\boldsymbol{\theta}_t) = \prod_j \mathcal{N}(\theta_{t,j} | \mu_{t|t-1,j}, \tau_{t|t-1,j})$, then we can compute the predictive distribution for s_t as follows:

$$q_{t|t-1}(s_t) = \mathcal{N}(s_t | m_{t|t-1}, v_{t|t-1}) \tag{18.119}$$

$$m_{t|t-1} = \sum_j x_{t,j} \mu_{t|t-1,j} \tag{18.120}$$

$$v_{t|t-1} = \sum_j x_{t,j}^2 \tau_{t|t-1,j} \tag{18.121}$$

The posterior for s_t is given by

$$q_t(s_t) = \mathcal{N}(s_t | m_t, v_t) \tag{18.122}$$

$$m_t = \int s_t \frac{1}{Z_t} p(y_t | s_t) q_{t|t-1}(s_t) ds_t \tag{18.123}$$

$$v_t = \int s_t^2 \frac{1}{Z_t} p(y_t | s_t) q_{t|t-1}(s_t) ds_t - m_t^2 \tag{18.124}$$

$$Z_t = \int p(y_t | s_t) q_{t|t-1}(s_t) ds_t \tag{18.125}$$

where $p(y_t|s_t) = \text{Ber}(y_t|s_t)$. These integrals are one dimensional, and so can be computed using Gaussian quadrature (see (Zoeter 2007) for details). This is the same as one step of the UKF algorithm.

Having inferred $q(s_t)$, we need to compute $q(\boldsymbol{\theta}|s_t)$. This can be done as follows. Define δ_m as the change in the mean of s_t and δ_v as the change in the variance:

$$m_t = m_{t|t-1} + \delta_m, \ v_t = v_{t|t-1} + \delta_v \tag{18.126}$$

Then one can show that the new factored posterior over the model parameters is given by

$$
\begin{align}
q(\theta_{t,j}) &= \mathcal{N}(\theta_{t,j}|\mu_{t,j}, \tau_{t,j}) \tag{18.127}\\
\mu_{t,j} &= \mu_{t|t-1,j} + a_j\delta_m \tag{18.128}\\
\tau_{t,j} &= \tau_{t|t-1,j} + a_j^2\delta_v \tag{18.129}\\
a_j &\triangleq \frac{x_{t,j}\tau_{t|t-1,j}}{\sum_{j'} x_{t,j'}^2 \tau_{t|t-1,j}^2} \tag{18.130}
\end{align}
$$

Thus we see that the parameters which correspond to inputs with larger magnitude (big $|x_{t,j}|$) or larger uncertainty (big $\tau_{t|t-1,j}$) get updated most, which makes intuitive sense.

In (Opper 1998) a version of this algorithm is derived using a probit likelihood (see Section 9.4). In this case, the measurement update can be done in closed form, without the need for numerical integration. In either case, the algorithm only takes $O(D)$ operations per time step, so it can be applied to models with large numbers of parameters. And since it is an online algorithm, it can also handle massive datasets. For example (Zhang et al. 2010) use a version of this algorithm to fit a multi-class classifier online to very large datasets. They beat alternative (non Bayesian) online learning algorithms, and sometimes even outperform state of the art batch (offline) learning methods such as SVMs (described in Section 14.5).

18.6 Hybrid discrete/continuous SSMs

Many systems contain both discrete and continuous hidden variables; these are known as **hybrid systems**. For example, the discrete variables may indicate whether a measurement sensor is faulty or not, or which "regime" the system is in. We will see some other examples below.

A special case of a hybrid system is when we combine an HMM and an LG-SSM. This is called a **switching linear dynamical system** (SLDS), a **jump Markov linear system** (JMLS), or a **switching state space model** (SSSM). More precisely, we have a discrete latent variable, $q_t \in \{1, \ldots, K\}$, a continuous latent variable, $\mathbf{z}_t \in \mathbb{R}^L$, an continuous observed response $\mathbf{y}_t \in \mathbb{R}^D$ and an optional continuous observed input or control $\mathbf{u}_t \in \mathbb{R}^U$. We then assume that the continuous variables have linear Gaussian CPDs, conditional on the discrete states:

$$
\begin{align}
p(q_t = k|q_{t-1} = j, \boldsymbol{\theta}) &= A_{ij} \tag{18.131}\\
p(\mathbf{z}_t|\mathbf{z}_{t-1}, q_t = k, \mathbf{u}_t, \boldsymbol{\theta}) &= \mathcal{N}(\mathbf{z}_t|\mathbf{A}_k\mathbf{z}_{t-1} + \mathbf{B}_k\mathbf{u}_t, \mathbf{Q}_k) \tag{18.132}\\
p(\mathbf{y}_t|\mathbf{z}_t, q_t = k, \mathbf{u}_t, \boldsymbol{\theta}) &= \mathcal{N}(\mathbf{y}_t|\mathbf{C}_k\mathbf{z}_t + \mathbf{D}_k\mathbf{u}_t, \mathbf{R}_k) \tag{18.133}
\end{align}
$$

See Figure 18.12(a) for the DGM representation.

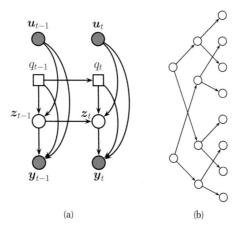

(a)   (b)

Figure 18.12   A switching linear dynamical system. (a) Squares represent discrete nodes, circles represent continuous nodes. (b) Illustration of how the number of modes in the belief state grows exponentially over time. We assume there are two binary states.

18.6.1   Inference

Unfortunately inference (i.e., state estimation) in hybrid models, including the switching LG-SSM model, is intractable. To see why, suppose q_t is binary, but that only the dynamics **A** depend on q_t, not the observation matrix. Our initial belief state will be a mixture of 2 Gaussians, corresponding to $p(\mathbf{z}_1|\mathbf{y}_1, q_1 = 1)$ and $p(\mathbf{z}_1|\mathbf{y}_1, q_1 = 2)$. The one-step-ahead predictive density will be a mixture of 4 Gaussians $p(\mathbf{z}_2|\mathbf{y}_1, q_1 = 1, q_2 = 1)$, $p(\mathbf{z}_2|\mathbf{y}_1, q_1 = 1, q_2 = 2)$, $p(\mathbf{z}_2|\mathbf{y}_1, q_1 = 2, q_2 = 1)$, and $p(\mathbf{z}_2|\mathbf{y}_1, q_1 = 2, q_2 = 2)$, obtained by passing each of the prior modes through the 2 possible transition models. The belief state at step 2 will also be a mixture of 4 Gaussians, obtained by updating each of the above distributions with $\mathbf{y}_2$. At step 3, the belief state will be a mixture of 8 Gaussians. And so on. So we see there is an exponential explosion in the number of modes (see Figure 18.12(b)).

Various approximate inference methods have been proposed for this model, such as the following:

- Prune off low probability trajectories in the discrete tree; this is the basis of **multiple hypothesis tracking** (Bar-Shalom and Fortmann 1988; Bar-Shalom and Li 1993).

- Use Monte Carlo. Essentially we just sample discrete trajectories, and apply an analytical filter to the continuous variables conditional on a trajectory. See Section 23.6 for details.

- Use ADF, where we approximate the exponentially large mixture of Gaussians with a smaller mixture of Gaussians. See Section 18.6.1.1 for details.

18.6.1.1   A Gaussian sum filter for switching SSMs

A **Gaussian sum filter** (Sorenson and Alspach 1971) approximates the belief state at each step by a mixture of K Gaussians. This can be implemented by running K Kalman filters in

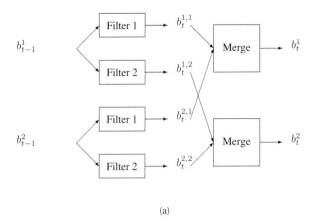

(a)

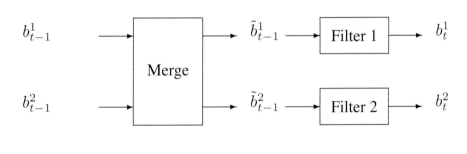

(b)

Figure 18.13 ADF for a switching linear dynamical system. (a) GPB2 method. (b) IMM method. See text for details.

parallel. This is particularly well suited to switching SSMs. We now describe one version of this algorithm, known as the "second order **generalized pseudo Bayes filter**" (GPB2) (Bar-Shalom and Fortmann 1988). We assume that the prior belief state b_{t-1} is a mixture of K Gaussians, one per discrete state:

$$b_{t-1}^i \triangleq p(\mathbf{z}_{t-1}, q_{t-1} = i | \mathbf{y}_{1:t-1}) = \pi_{t-1,i} \mathcal{N}(\mathbf{z}_{t-1} | \boldsymbol{\mu}_{t-1,i}, \boldsymbol{\Sigma}_{t-1,i}) \tag{18.134}$$

We then pass this through the K different linear models to get

$$b_t^{ij} \triangleq p(\mathbf{z}_t, q_{t-1} = i, q_t = j | \mathbf{y}_{1:t}) = \pi_{tij} \mathcal{N}(\mathbf{z}_t | \boldsymbol{\mu}_{t,ij}, \boldsymbol{\Sigma}_{t,ij}) \tag{18.135}$$

where $\pi_{tij} = \pi_{t-1,i} p(q_t = j | q_{t-1} = i)$. Finally, for each value of j, we collapse the K Gaussian mixtures down to a single mixture to give

$$b_t^j \triangleq p(\mathbf{z}_t, q_t = j | \mathbf{y}_{1:t}) = \pi_{tj} \mathcal{N}(\mathbf{z}_t | \boldsymbol{\mu}_{t,j}, \boldsymbol{\Sigma}_{t,j}) \tag{18.136}$$

See Figure 18.13(a) for a sketch.

The optimal way to approximate a mixture of Gaussians with a single Gaussian is given by $q = \arg\min_q \mathbb{KL}\left(q||p\right)$, where $p(\mathbf{z}) = \sum_k \pi_k \mathcal{N}(\mathbf{z}|\boldsymbol{\mu}_k, \boldsymbol{\Sigma}_k)$ and $q(\mathbf{z}) = \mathcal{N}(\mathbf{z}|\boldsymbol{\mu}, \boldsymbol{\Sigma})$. This can be solved by moment matching, that is,

$$\boldsymbol{\mu} = \mathbb{E}\left[\mathbf{z}\right] = \sum_k \pi_k \boldsymbol{\mu}_k \tag{18.137}$$

$$\boldsymbol{\Sigma} = \text{cov}\left[\mathbf{z}\right] = \sum_k \pi_k \left(\boldsymbol{\Sigma}_k + (\boldsymbol{\mu}_k - \boldsymbol{\mu})(\boldsymbol{\mu}_k - \boldsymbol{\mu})^T\right) \tag{18.138}$$

In the graphical model literature, this is called **weak marginalization** (Lauritzen 1992), since it preserves the first two moments. Applying these equations to our model, we can go from b_t^{ij} to b_t^j as follows (where we drop the t subscript for brevity):

$$\pi_j = \sum_i \pi_{ij} \tag{18.139}$$

$$\pi_{j|i} = \frac{\pi_{ij}}{\sum_{j'} \pi_{ij'}} \tag{18.140}$$

$$\boldsymbol{\mu}_j = \sum_i \pi_{j|i} \boldsymbol{\mu}_{ij} \tag{18.141}$$

$$\boldsymbol{\Sigma}_j = \sum_i \pi_{j|i} \left(\boldsymbol{\Sigma}_{ij} + (\boldsymbol{\mu}_{ij} - \boldsymbol{\mu}_j)(\boldsymbol{\mu}_{ij} - \boldsymbol{\mu}_j)^T\right) \tag{18.142}$$

This algorithm requires running K^2 filters at each step. A cheaper alternative is to represent the belief state by a single Gaussian, marginalizing over the discrete switch at each step. This is a straightforward application of ADF. An offline extension to this method, called **expectation correction**, is described in (Barber 2006; Mesot and Barber 2009).

Another heuristic approach, known as **interactive multiple models** or **IMM** (Bar-Shalom and Fortmann 1988), can be obtained by first collapsing the prior to a single Gaussian (by moment matching), and then updating it using K different Kalman filters, one per value of q_t. See Figure 18.13(b) for a sketch.

18.6.2 Application: data association and multi-target tracking

Suppose we are tracking K objects, such as airplanes, and at time t, we observe K' detection events, e.g., "blips" on a radar screen. We can have $K' < K$ due to occlusion or missed detections. We can have $K' > K$ due to clutter or false alarms. Or we can have $K' = K$. In any case, we need to figure out the **correspondence** between the K' detections $\mathbf{y}_{tk}$ and the K objects $\mathbf{z}_{tj}$. This is called the problem of **data association**, and it arises in many application domains.

Figure 18.14 gives an example in which we are tracking $K = 2$ objects. At each time step, q_t is the unknown mapping which specifies which objects caused which observations. It specifies the "wiring diagram" for time slice t. The standard way to solve this problem is to compute a weight which measures the "compatibility" between object j and measurement k, typically based on how close k is to where the model thinks j should be (the so-called **nearest neighbor data association** heuristic). This gives us a $K \times K'$ weight matrix. We can make this into a

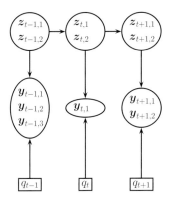

Figure 18.14 A model for tracking two objects in the presence of data-assocation ambiguity. We observe 3, 1 and 2 detections at time steps $t-1$, t and $t+1$. The q_t hidden variable encodes the association between the observations and the hidden causes.

square matrix of size $N \times N$, where $N = \max(K, K')$, by adding dummy background objects, which can explain all the false alarms, and adding dummy observations, which can explain all the missed detections. We can then compute the maximal weight bipartite matching using the **Hungarian algorithm**, which takes $O(N^3)$ time (see e.g., (Burkard et al. 2009)). Conditional on this, we can perform a Kalman filter update, where objects that are assigned to dummy observations do not perform a measurement update.

An extension of this method, to handle a variable and/or unknown number of objects, is known as **multi-target tracking**. This requires dealing with a variable-sized state space. There are many ways to do this, but perhaps the simplest and most robust methods are based on sequential Monte Carlo (e.g., (Ristic et al. 2004)) or MCMC (e.g., (Khan et al. 2006; Oh et al. 2009)). We will discuss these algorithms later.

18.6.3 Application: fault diagnosis

Consider the model in Figure 18.15(a). This represents an industrial plant consisting of various tanks of liquid, interconnected by pipes. In this example, we just have two tanks, for simplicity. We want to estimate the pressure inside each tank, based on a noisy measurement of the flow into and out of each tank. However, the measurement devices can sometimes fail. Furthermore, pipes can burst or get blocked; we call this a "resistance failure". This model is widely used as a benchmark in the **fault diagnosis** community (Mosterman and Biswas 1999).

We can create a probabilistic model of the system as shown in Figure 18.15(b). The square nodes represent discrete variables, such as measurement failures and resistance failures. The remaining variables are continuous. A variety of approximate inference algorithms can be applied to this model. See (Koller and Lerner 2001) for one approach, based on Rao-Blackwellized particle filtering (which is explained in Section 23.6).

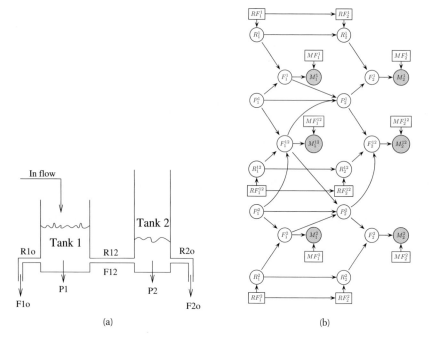

Figure 18.15 (a) The two-tank system. The goal is to infer when pipes are blocked or have burst, or sensors have broken, from (noisy) observations of the flow out of tank 1, $F1o$, out of tank 2, $F2o$, or between tanks 1 and 2, $F12$. $R1o$ is a hidden variable representing the resistance of the pipe out of tank 1, $P1$ is a hidden variable representing the pressure in tank 1, etc. Source: Figure 11 of (Koller and Lerner 2001) . Used with kind permission of Daphne Koller. (b) Dynamic Bayes net representation of the two-tank system. Discrete nodes are squares, continuous nodes are circles. Abbreviations: R = resistance, P = pressure, F = flow, M = measurement, RF = resistance failure, MF = measurement failure. Based on Figure 12 of (Koller and Lerner 2001).

18.6.4 Application: econometric forecasting

The switching LG-SSM model is widely used in **econometric forecasting**, where it is called a **regime switching** model. For example, we can combine two linear trend models (see Section 18.2.4.2), one in which $b_t > 0$ reflects a growing economy, and one in which $b_t < 0$ reflects a shrinking economy. See (West and Harrison 1997) for further details.

Exercises

Exercise 18.1 Derivation of EM for LG-SSM

Derive the E and M steps for computing a (locally optimal) MLE for an LG-SSM model. Hint: the results are in (Ghahramani and Hinton 1996b); your task is to derive these results.

Exercise 18.2 Seasonal LG-SSM model in standard form

Write the seasonal model in Figure 18.7(a) as an LG-SSM. Define the matrices $\mathbf{A}$, $\mathbf{C}$, $\mathbf{Q}$ and $\mathbf{R}$.

19 Undirected graphical models (Markov random fields)

19.1 Introduction

In Chapter 10, we discussed directed graphical models (DGMs), commonly known as Bayes nets. However, for some domains, being forced to choose a direction for the edges, as required by a DGM, is rather awkward. For example, consider modeling an image. We might suppose that the intensity values of neighboring pixels are correlated. We can create a DAG model with a 2d lattice topology as shown in Figure 19.1(a). This is known as a **causal MRF** or a **Markov mesh** (Abend et al. 1965). However, its conditional independence properties are rather unnatural. In particular, the Markov blanket (defined in Section 10.5) of the node X_8 in the middle is the other colored nodes (3, 4, 7, 9, 12 and 13) rather than just its 4 nearest neighbors as one might expect.

An alternative is to use an **undirected graphical model** (UGM), also called a **Markov random field** (MRF) or **Markov network**. These do not require us to specify edge orientations, and are much more natural for some problems such as image analysis and spatial statistics. For example, an undirected 2d lattice is shown in Figure 19.1(b); now the Markov blanket of each node is just its nearest neighbors, as we show in Section 19.2.

Roughly speaking, the main advantages of UGMs over DGMs are: (1) they are symmetric and therefore more "natural" for certain domains, such as spatial or relational data; and (2) discriminative UGMs (aka conditional random fields, or CRFs), which define conditional densities of the form $p(\mathbf{y}|\mathbf{x})$, work better than discriminative DGMs, for reasons we explain in Section 19.6.1. The main disadvantages of UGMs compared to DGMs are: (1) the parameters are less interpretable and less modular, for reasons we explain in Section 19.3; and (2) parameter estimation is computationally more expensive, for reasons we explain in Section 19.5. See (Domke et al. 2008) for an empirical comparison of the two approaches for an image processing task.

19.2 Conditional independence properties of UGMs

19.2.1 Key properties

UGMs define CI relationships via simple graph separation as follows: for sets of nodes A, B, and C, we say $\mathbf{x}_A \perp_G \mathbf{x}_B|\mathbf{x}_C$ iff C separates A from B in the graph G. This means that, when we remove all the nodes in C, if there are no paths connecting any node in A to any node in B, then the CI property holds. This is called the **global Markov property** for UGMs. For example, in Figure 19.2(b), we have that $\{1,2\} \perp \{6,7\}|\{3,4,5\}$.

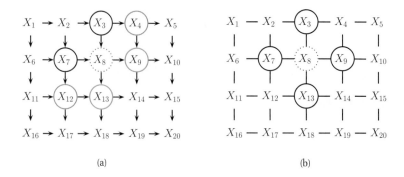

(a)   (b)

Figure 19.1   (a) A 2d lattice represented as a DAG. The dotted red node X_8 is independent of all other nodes (black) given its Markov blanket, which include its parents (blue), children (green) and co-parents (orange). (b) The same model represented as a UGM. The red node X_8 is independent of the other black nodes given its neighbors (blue nodes).

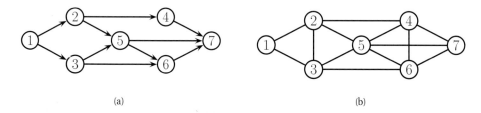

(a)   (b)

Figure 19.2   (a) A DGM. (b) Its moralized version, represented as a UGM.

The smallest set of nodes that renders a node t conditionally independent of all the other nodes in the graph is called t's **Markov blanket**; we will denote this by $\mathrm{mb}(t)$. Formally, the Markov blanket satisfies the following property:

$$t \perp \mathcal{V} \setminus \mathrm{cl}(t) | \mathrm{mb}(t) \tag{19.1}$$

where $\mathrm{cl}(t) \triangleq \mathrm{mb}(t) \cup \{t\}$ is the **closure** of node t. One can show that, in a UGM, a node's Markov blanket is its set of immediate neighbors. This is called the **undirected local Markov property**. For example, in Figure 19.2(b), we have $\mathrm{mb}(5) = \{2, 3, 4, 6, 7\}$.

From the local Markov property, we can easily see that two nodes are conditionally independent given the rest if there is no direct edge between them. This is called the **pairwise Markov property**. In symbols, this is written as

$$s \perp t | \mathcal{V} \setminus \{s, t\} \iff G_{st} = 0 \tag{19.2}$$

Using the three Markov properties we have discussed, we can derive the following CI properties (amongst others) from the UGM in Figure 19.2(b):

- **Pairwise** $1 \perp 7 | \text{rest}$
- **Local** $1 \perp \text{rest} | 2, 3$

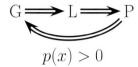

Figure 19.3 Relationship between Markov properties of UGMs.

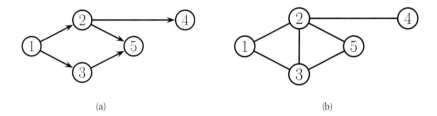

(a)   (b)

Figure 19.4 (a) The ancestral graph induced by the DAG in Figure 19.2(a) wrt $U = \{2, 4, 5\}$. (b) The moralized version of (a).

- **Global** $1, 2 \perp 6, 7|3, 4, 5$

It is obvious that global Markov implies local Markov which implies pairwise Markov. What is less obvious, but nevertheless true (assuming $p(\mathbf{x}) > 0$ for all $\mathbf{x}$, i.e., that p is a positive density), is that pairwise implies global, and hence that all these Markov properties are the same, as illustrated in Figure 19.3 (see e.g., (Koller and Friedman 2009, p119) for a proof).[1] The importance of this result is that it is usually easier to empirically assess pairwise conditional independence; such pairwise CI statements can be used to construct a graph from which global CI statements can be extracted.

19.2.2 An undirected alternative to d-separation

We have seen that determinining CI relationships in UGMs is much easier than in DGMs, because we do not have to worry about the directionality of the edges. In this section, we show how to determine CI relationships for a DGM using a UGM.

It is tempting to simply convert the DGM to a UGM by dropping the orientation of the edges, but this is clearly incorrect, since a v-structure $A \to B \leftarrow C$ has quite different CI properties than the corresponding undirected chain $A - B - C$. The latter graph incorrectly states that $A \perp C|B$. To avoid such incorrect CI statements, we can add edges between the "unmarried" parents A and C, and then drop the arrows from the edges, forming (in this case) a fully connected undirected graph. This process is called **moralization**. Figure 19.2(b) gives a larger

---

1. The restriction to positive densities arises because deterministic constraints can result in independencies present in the distribution that are not explicitly represented in the graph. See e.g., (Koller and Friedman 2009, p120) for some examples. Distributions with non-graphical CI properties are said to be **unfaithful** to the graph, so $I(p) \neq I(G)$.

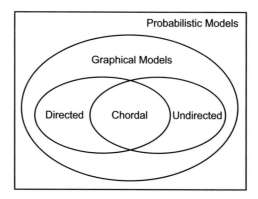

Figure 19.5 DGMs and UGMs can perfectly represent different sets of distributions. Some distributions can be perfectly represented by either DGMs or UGMs; the corresponding graph must be chordal.

example of moralization: we interconnect 2 and 3, since they have a common child 5, and we interconnect 4, 5 and 6, since they have a common child 7.

Unfortunately, moralization loses some CI information, and therefore we cannot use the moralized UGM to determine CI properties of the DGM. For example, in Figure 19.2(a), using d-separation, we see that $4 \perp 5|2$. Adding a moralization arc $4 - 5$ would lose this fact (see Figure 19.2(b)). However, notice that the 4-5 moralization edge, due to the common child 7, is not needed if we do not observe 7 or any of its descendants. This suggests the following approach to determining if $A \perp B|C$. First we form the **ancestral graph** of DAG G with respect to $U = A \cup B \cup C$. This means we remove all nodes from G that are not in U or are not ancestors of U. We then moralize this ancestral graph, and apply the simple graph separation rules for UGMs. For example, in Figure 19.4(a), we show the ancestral graph for Figure 19.2(a) using $U = \{2, 4, 5\}$. In Figure 19.4(b), we show the moralized version of this graph. It is clear that we now correctly conclude that $4 \perp 5|2$.

19.2.3   Comparing directed and undirected graphical models

Which model has more "expressive power", a DGM or a UGM? To formalize this question, recall that we say that G is an I-map of a distribution p if $I(G) \subseteq I(p)$. Now define G to be **perfect map** of p if $I(G) = I(p)$, in other words, the graph can represent all (and only) the CI properties of the distribution. It turns out that DGMs and UGMs are perfect maps for different sets of distributions (see Figure 19.5). In this sense, neither is more powerful than the other as a representation language.

As an example of some CI relationships that can be perfectly modeled by a DGM but not a UGM, consider a v-structure $A \to C \leftarrow B$. This asserts that $A \perp B$, and $A \not\perp B|C$. If we drop the arrows, we get $A - C - B$, which asserts $A \perp B|C$ and $A \not\perp B$, which is incorrect. In fact, there is no UGM that can precisely represent all and only the two CI statements encoded by a v-structure. In general, CI properties in UGMs are monotonic, in the following sense: if $A \perp B|C$, then $A \perp B|(C \cup D)$. But in DGMs, CI properties can be non-monotonic, since conditioning

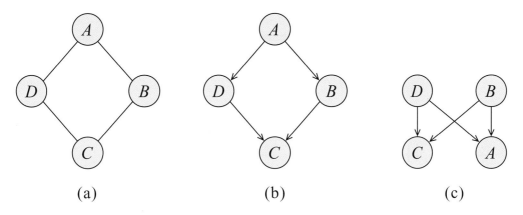

Figure 19.6 A UGM and two failed attempts to represent it as a DGM. Source: Figure 3.10 of (Koller and Friedman 2009). Used with kind permission of Daphne Koller.

on extra variables can eliminate conditional independencies due to explaining away.

As an example of some CI relationships that can be perfectly modeled by a UGM but not a DGM, consider the 4-cycle shown in Figure 19.6(a). One attempt to model this with a DGM is shown in Figure 19.6(b). This correctly asserts that $A \perp C|B, D$. However, it incorrectly asserts that $B \perp D|A$. Figure 19.6(c) is another incorrect DGM: it correctly encodes $A \perp C|B, D$, but incorrectly encodes $B \perp D$. In fact there is no DGM that can precisely represent all and only the CI statements encoded by this UGM.

Some distributions can be perfectly modeled by either a DGM or a UGM; the resulting graphs are called **decomposable** or **chordal**. Roughly speaking, this means the following: if we collapse together all the variables in each maximal clique, to make "mega-variables", the resulting graph will be a tree. Of course, if the graph is already a tree (which includes chains as a special case), it will be chordal. See Section 20.4.1 for further details.

19.3 Parameterization of MRFs

Although the CI properties of UGM are simpler and more natural than for DGMs, representing the joint distribution for a UGM is less natural than for a DGM, as we see below.

19.3.1 The Hammersley-Clifford theorem

Since there is no topological ordering associated with an undirected graph, we can't use the chain rule to represent $p(\mathbf{y})$. So instead of associating CPDs with each node, we associate **potential functions** or **factors** with each maximal clique in the graph. We will denote the potential function for clique c by $\psi_c(\mathbf{y}_c|\boldsymbol{\theta}_c)$. A potential function can be any non-negative function of its arguments. The joint distribution is then defined to be proportional to the product of clique potentials. Rather surprisingly, one can show that any positive distribution whose CI properties can be represented by a UGM can be represented in this way. We state this result more formally below.

Theorem 19.3.1 (Hammersley-Clifford). *A positive distribution $p(\mathbf{y}) > 0$ satisfies the CI properties of an undirected graph G iff p can be represented as a product of factors, one per maximal clique, i.e.,*

$$p(\mathbf{y}|\boldsymbol{\theta}) = \frac{1}{Z(\boldsymbol{\theta})} \prod_{c \in \mathcal{C}} \psi_c(\mathbf{y}_c|\boldsymbol{\theta}_c) \tag{19.3}$$

where $\mathcal{C}$ is the set of all the (maximal) cliques of G, and $Z(\boldsymbol{\theta})$ is the **partition function** *given by*

$$Z(\boldsymbol{\theta}) \triangleq \sum_{\mathbf{y}} \prod_{c \in \mathcal{C}} \psi_c(\mathbf{y}_c|\boldsymbol{\theta}_c) \tag{19.4}$$

Note that the partition function is what ensures the overall distribution sums to 1.[2]

The proof was never published, but can be found in e.g., (Koller and Friedman 2009).

For example, consider the MRF in Figure 10.1(b). If p satisfies the CI properties of this graph then we can write p as follows:

$$p(\mathbf{y}|\boldsymbol{\theta}) = \frac{1}{Z(\boldsymbol{\theta})} \psi_{123}(y_1, y_2, y_3) \psi_{234}(y_2, y_3, y_4) \psi_{35}(y_3, y_5) \tag{19.5}$$

where

$$Z = \sum_{\mathbf{y}} \psi_{123}(y_1, y_2, y_3) \psi_{234}(y_2, y_3, y_4) \psi_{35}(y_3, y_5) \tag{19.6}$$

There is a deep connection between UGMs and statistical physics. In particular, there is a model known as the **Gibbs distribution**, which can be written as follows:

$$p(\mathbf{y}|\boldsymbol{\theta}) = \frac{1}{Z(\boldsymbol{\theta})} \exp(-\sum_c E(\mathbf{y}_c|\boldsymbol{\theta}_c)) \tag{19.7}$$

where $E(\mathbf{y}_c) > 0$ is the energy associated with the variables in clique c. We can convert this to a UGM by defining

$$\psi_c(\mathbf{y}_c|\boldsymbol{\theta}_c) = \exp(-E(\mathbf{y}_c|\boldsymbol{\theta}_c)) \tag{19.8}$$

We see that high probability states correspond to low energy configurations. Models of this form are known as **energy based models**, and are commonly used in physics and biochemistry, as well as some branches of machine learning (LeCun et al. 2006).

Note that we are free to restrict the parameterization to the edges of the graph, rather than the maximal cliques. This is called a **pairwise MRF**. In Figure 10.1(b), we get

$$p(\mathbf{y}|\boldsymbol{\theta}) \propto \psi_{12}(y_1, y_2)\psi_{13}(y_1, y_3)\psi_{23}(y_2, y_3)\psi_{24}(y_2, y_4)\psi_{34}(y_3, y_4)\psi_{35}(y_3, y_5) \tag{19.9}$$

$$\propto \prod_{s \sim t} \psi_{st}(y_s, y_t) \tag{19.10}$$

This form is widely used due to its simplicity, although it is not as general.

---

2. The partition function is denoted by Z because of the German word *Zustandssumme*, which means "sum over states". This reflects the fact that a lot of pioneering working in statistical physics was done by German speakers.

19.3.2 Representing potential functions

If the variables are discrete, we can represent the potential or energy functions as tables of (non-negative) numbers, just as we did with CPTs. However, the potentials are not probabilities. Rather, they represent the relative "compatibility" between the different assignments to the potential. We will see some examples of this below.

A more general approach is to define the log potentials as a linear function of the parameters:

$$\log \psi_c(\mathbf{y}_c) \triangleq \phi_c(\mathbf{y}_c)^T \boldsymbol{\theta}_c \tag{19.11}$$

where $\phi_c(\mathbf{y}_c)$ is a feature vector derived from the values of the variables $\mathbf{y}_c$. The resulting log probability has the form

$$\log p(\mathbf{y}|\boldsymbol{\theta}) = \sum_c \phi_c(\mathbf{y}_c)^T \boldsymbol{\theta}_c - \log Z(\boldsymbol{\theta}) \tag{19.12}$$

This is also known as a **maximum entropy** or a **log-linear** model.

For example, consider a pairwise MRF, where for each edge, we associate a feature vector of length K^2 as follows:

$$\phi_{st}(y_s, y_t) = [\dots, \mathbb{I}(y_s = j, y_t = k), \dots] \tag{19.13}$$

If we have a weight for each feature, we can convert this into a $K \times K$ potential function as follows:

$$\psi_{st}(y_s = j, y_t = k) = \exp([\boldsymbol{\theta}_{st}^T \phi_{st}]_{jk}) = \exp(\theta_{st}(j, k)) \tag{19.14}$$

So we see that we can easily represent tabular potentials using a log-linear form. But the log-linear form is more general.

To see why this is useful, suppose we are interested in making a probabilistic model of English spelling. Since certain letter combinations occur together quite frequently (e.g., "ing"), we will need higher order factors to capture this. Suppose we limit ourselves to letter trigrams. A tabular potential still has $26^3 = 17,576$ parameters in it. However, most of these triples will never occur.

An alternative approach is to define indicator functions that look for certain "special" triples, such as "ing", "qu-", etc. Then we can define the potential on each trigram as follows:

$$\psi(y_{t-1}, y_t, y_{t+1}) = \exp(\sum_k \theta_k \phi_k(y_{t-1}, y_t, y_{t+1})) \tag{19.15}$$

where k indexes the different features, corresponding to "ing", "qu-", etc., and ϕ_k is the corresponding binary **feature function**. By tying the parameters across locations, we can define the probability of a word of any length using

$$p(\mathbf{y}|\boldsymbol{\theta}) \propto \exp(\sum_t \sum_k \theta_k \phi_k(y_{t-1}, y_t, y_{t+1})) \tag{19.16}$$

This raises the question of where these feature functions come from. In many applications, they are created by hand to reflect domain knowledge (we will see examples later), but it is also possible to learn them from data, as we discuss in Section 19.5.6.

19.4 Examples of MRFs

In this section, we show how several popular probability models can be conveniently expressed as UGMs.

19.4.1 Ising model

The **Ising model** is an example of an MRF that arose from statistical physics.[3] It was originally used for modeling the behavior of magnets. In particular, let $y_s \in \{-1, +1\}$ represent the spin of an atom, which can either be spin down or up. In some magnets, called **ferro-magnets**, neighboring spins tend to line up in the same direction, whereas in other kinds of magnets, called **anti-ferromagnets**, the spins "want" to be different from their neighbors.

We can model this as an MRF as follows. We create a graph in the form of a 2D or 3D lattice, and connect neighboring variables, as in Figure 19.1(b). We then define the following pairwise clique potential:

$$\psi_{st}(y_s, y_t) = \begin{pmatrix} e^{w_{st}} & e^{-w_{st}} \\ e^{-w_{st}} & e^{w_{st}} \end{pmatrix} \tag{19.17}$$

Here w_{st} is the coupling strength between nodes s and t. If two nodes are not connected in the graph, we set $w_{st} = 0$. We assume that the weight matrix $\mathbf{W}$ is symmetric, so $w_{st} = w_{ts}$. Often we assume all edges have the same strength, so $w_{st} = J$ (assuming $w_{st} \neq 0$).

If all the weights are positive, $J > 0$, then neighboring spins are likely to be in the same state; this can be used to model ferromagnets, and is an example of an **associative Markov network**. If the weights are sufficiently strong, the corresponding probability distribution will have two modes, corresponding to the all +1's state and the all -1's state. These are called the **ground states** of the system.

If all of the weights are negative, $J < 0$, then the spins want to be different from their neighbors; this can be used to model an anti-ferromagnet, and results in a **frustrated system**, in which not all the constraints can be satisfied at the same time. The corresponding probability distribution will have multiple modes. Interestingly, computing the partition function $Z(J)$ can be done in polynomial time for associative Markov networks, but is NP-hard in general (Cipra 2000).

There is an interesting analogy between Ising models and Gaussian graphical models. First, assuming $y_t \in \{-1, +1\}$, we can write the unnormalized log probability of an Ising model as follows:

$$\log \tilde{p}(\mathbf{y}) = -\sum_{s \sim t} y_s w_{st} y_t = -\frac{1}{2} \mathbf{y}^T \mathbf{W} \mathbf{y} \tag{19.18}$$

(The factor of $\frac{1}{2}$ arises because we sum each edge twice.) If $w_{st} = J > 0$, we get a low energy (and hence high probability) if neighboring states agree.

Sometimes there is an **external field**, which is an energy term which is added to each spin. This can be modelled using a local energy term of the form $-\mathbf{b}^T \mathbf{y}$, where $\mathbf{b}$ is sometimes called

---

3. Ernst Ising was a German-American physicist, 1900–1998.

a **bias term**. The modified distribution is given by

$$\log \tilde{p}(\mathbf{y}) \quad = \quad -\sum_{s \sim t} w_{st} y_s y_t + \sum_s b_s y_s = -\frac{1}{2} \mathbf{y}^T \mathbf{W} \mathbf{y} + \mathbf{b}^T \mathbf{y} \tag{19.19}$$

where $\boldsymbol{\theta} = (\mathbf{W}, \mathbf{b})$.

If we define $\boldsymbol{\Sigma}^{-1} = \mathbf{W}$, $\boldsymbol{\mu} \triangleq \boldsymbol{\Sigma} \mathbf{b}$, and $c \triangleq \frac{1}{2} \boldsymbol{\mu}^T \boldsymbol{\Sigma}^{-1} \boldsymbol{\mu}$, we can rewrite this in a form that looks similar to a Gaussian:

$$\tilde{p}(\mathbf{y}) \propto \exp(-\frac{1}{2}(\mathbf{y} - \boldsymbol{\mu})^T \boldsymbol{\Sigma}^{-1}(\mathbf{y} - \boldsymbol{\mu}) + c) \tag{19.20}$$

One very important difference is that, in the case of Gaussians, the normalization constant, $Z = |2\pi\boldsymbol{\Sigma}|$, requires the computation of a matrix determinant, which can be computed in $O(D^3)$ time, whereas in the case of the Ising model, the normalization constant requires summing over all 2^D bit vectors; this is equivalent to computing the matrix permanent, which is NP-hard in general (Jerrum et al. 2004).

19.4.2 Hopfield networks

A **Hopfield network** (Hopfield 1982) is a fully connected Ising model with a symmetric weight matrix, $\mathbf{W} = \mathbf{W}^T$. These weights, plus the bias terms $\mathbf{b}$, can be learned from training data using (approximate) maximum likelihood, as described in Section 19.5.[4]

The main application of Hopfield networks is as an **associative memory** or **content addressable memory**. The idea is this: suppose we train on a set of fully observed bit vectors, corresponding to patterns we want to memorize. Then, at test time, we present a partial pattern to the network. We would like to estimate the missing variables; this is called **pattern completion**. See Figure 19.7 for an example. This can be thought of as retrieving an example from memory based on a piece of the example itself, hence the term "associative memory".

Since exact inference is intractable in this model, it is standard to use a coordinate descent algorithm known as **iterative conditional modes** (ICM), which just sets each node to its most likely (lowest energy) state, given all its neighbors. The full conditional can be shown to be

$$p(y_s = 1 | \mathbf{y}_{-s}, \boldsymbol{\theta}) = \mathrm{sigm}(\mathbf{w}_{s,:}^T \mathbf{y}_{-s} + b_s) \tag{19.21}$$

Picking the most probable state amounts to using the rule $y_s^* = 1$ if $\sum_t w_{st} y_t > -b_s$ and using $y_s^* = 0$ otherwise. (Much better inference algorithms will be discussed later in this book.)

Since inference is deterministic, it is also possible to interpret this model as a **recurrent neural network**. (This is quite different from the feedforward neural nets studied in Section 16.5; they are univariate conditional density models of the form $p(y|\mathbf{x}, \boldsymbol{\theta})$ which can only be used for supervised learning.) See (Hertz et al. 1991) for further details on Hopfield networks.

A **Boltzmann machine** generalizes the Hopfield / Ising model by including some hidden nodes, which makes the model representationally more powerful. Inference in such models often uses Gibbs sampling, which is a stochastic version of ICM (see Section 24.2 for details).

---

4. Computing the parameter MLE works much better than the outer product rule proposed in (Hopfield 1982), because it not only lowers the energy of the observed patterns, but it also raises the energy of the non-observed patterns, in order to make the distribution sum to one (Hillar et al. 2012).

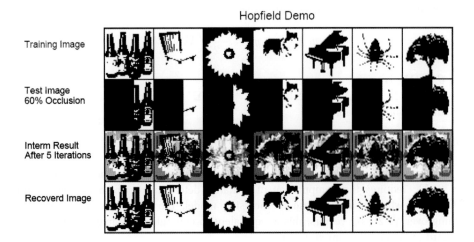

Figure 19.7 Examples of how an associative memory can reconstruct images. These are binary images of size 50×50 pixels. Top: training images. Row 2: partially visible test images. Row 3: estimate after 5 iterations. Bottom: final state estimate. Based on Figure 2.1 of Hertz et al. (1991). Figure generated by `hopfieldDemo`.

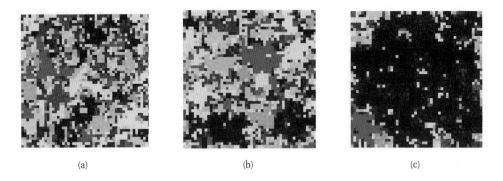

Figure 19.8 Visualizing a sample from a 10-state Potts model of size 128×128 for different association strengths: (a) $J = 1.42$, (b) $J = 1.44$ (crirical value), (c) $J = 1.46$. The regions are labeled according to size: blue is largest, red is smallest. Used with kind permission of Erik Sudderth. See `gibbsDemoIsing` for Matlab code to produce a similar plot for the Ising model.

However, we could equally well apply Gibbs to a Hopfield net and ICM to a Boltzmann machine: the inference algorithm is not part of the model definition. See Section 27.7 for further details on Boltzmann machines.

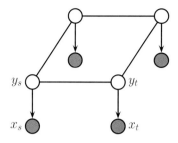

Figure 19.9 A grid-structured MRF with local evidence nodes.

19.4.3 Potts model

It is easy to generalize the Ising model to multiple discrete states, $y_t \in \{1, 2, \ldots, K\}$. For example, if $K = 3$, we use a potential function of the following form:

$$\psi_{st}(y_s, y_t) = \begin{pmatrix} e^{w_{st}} & e^0 & e^0 \\ e^0 & e^{w_{st}} & e^0 \\ e^0 & e^0 & e^{w_{st}} \end{pmatrix} \tag{19.22}$$

This is called the **Potts model**.[5] As before, we often assume tied weights of the form $w_{st} = J$. If $J > 0$, then neighboring nodes are encouraged to have the same label. Some samples from this model are shown in Figure 19.8. We see that for $J > 1.44$, large clusters occur, for $J < 1.44$, many small clusters occur, and at the **critical value** of $J^* = 1.44$, there is a mix of small and large clusters. This rapid change in behavior as we vary a parameter of the system is called a **phase transition**, and has been widely studied in the physics community. An analogous phenomenon occurs in the Ising model; see (MacKay 2003, ch 31) for details.

The Potts model can be used as a prior for **image segmentation**, since it says that neighboring pixels are likely to have the same discrete label and hence belong to the same segment. We can combine this prior with a likelihood term as follows:

$$p(\mathbf{y}, \mathbf{x}|\boldsymbol{\theta}) = p(\mathbf{y}|J) \prod_t p(x_t|y_t, \boldsymbol{\theta}) = \left[\frac{1}{Z(J)} \prod_{s \sim t} \psi(y_s, y_t; J) \right] \prod_t p(x_t|y_t, \boldsymbol{\theta}) \tag{19.23}$$

where $p(x_t|y_t = k, \boldsymbol{\theta})$ is the probability of observing pixel x_t given that the corresponding segment belongs to class k. This observation model can be modeled using a Gaussian or a non-parametric density. (Note that we label the hidden nodes y_t and the observed nodes x_t, to be compatible with Section 19.6.)

The corresponding graphical model is a mix of undirected and directed edges, as shown in Figure 19.9. The undirected 2d lattice represents the prior $p(\mathbf{y})$; in addition, there are directed edges from each y_t to its corresponding x_t, representing the **local evidence**. Technically speaking, this combination of an undirected and directed graph is called a **chain graph**. However,

---

5. Renfrey Potts was an Australian mathematician, 1925–2005.

since the x_t nodes are observed, they can be "absorbed" into the model, thus leaving behind an undirected "backbone".

This model is a 2d analog of an HMM, and could be called a **partially observed MRF**. As in an HMM, the goal is to perform posterior inference, i.e., to compute (some function of) $p(\mathbf{y}|\mathbf{x}, \boldsymbol{\theta})$. Unfortunately, the 2d case is provably much harder than the 1d case, and we must resort to approximate methods, as we discuss in later chapters.

Although the Potts prior is adequate for regularizing supervised learning problems, it is not sufficiently accurate to perform image segmentation in an unsupervised way, since the segments produced by this model do not accurately represent the kinds of segments one sees in natural images (Morris et al. 1996).[6] For the unsupervised case, one needs to use more sophisticated priors, such as the truncated Gaussian process prior of (Sudderth and Jordan 2008).

19.4.4 Gaussian MRFs

An undirected GGM, also called a **Gaussian MRF** (see e.g., (Rue and Held 2005)), is a pairwise MRF of the following form:

$$p(\mathbf{y}|\boldsymbol{\theta}) \quad \propto \quad \prod_{s \sim t} \psi_{st}(y_s, y_t) \prod_t \psi_t(y_t) \tag{19.24}$$

$$\psi_{st}(y_s, y_t) \quad = \quad \exp(-\frac{1}{2} y_s \Lambda_{st} y_t) \tag{19.25}$$

$$\psi_t(y_t) \quad = \quad \exp(-\frac{1}{2} \Lambda_{tt} y_t^2 + \eta_t y_t) \tag{19.26}$$

(Note that we could easily absorb the node potentials ψ_t into the edge potentials, but we have kept them separate for clarity.) The joint distribution can be written as follows:

$$p(\mathbf{y}|\boldsymbol{\theta}) \quad \propto \quad \exp[\boldsymbol{\eta}^T \mathbf{y} - \frac{1}{2} \mathbf{y}^T \boldsymbol{\Lambda} \mathbf{y}] \tag{19.27}$$

We recognize this as a multivariate Gaussian written in information form where $\boldsymbol{\Lambda} = \boldsymbol{\Sigma}^{-1}$ and $\boldsymbol{\eta} = \boldsymbol{\Lambda}\boldsymbol{\mu}$ (see Section 4.3.3).

If $\Lambda_{st} = 0$, then there is no pairwise term connecting s and t, so by the factorization theorem (Theorem 2.2.1), we conclude that

$$y_s \perp y_t | \mathbf{y}_{-(st)} \iff \Lambda_{st} = 0 \tag{19.28}$$

The zero entries in $\boldsymbol{\Lambda}$ are called **structural zeros**, since they represent the absent edges in the graph. Thus undirected GGMs correspond to sparse precision matrices, a fact which we exploit in Section 26.7.2 to efficiently learn the structure of the graph.

19.4.4.1 Comparing Gaussian DGMs and UGMs *

In Section 10.2.5, we saw that directed GGMs correspond to sparse regression matrices, whereas undirected GGMs correspond to sparse precision matrices. The advantage of the DAG formulation is that we can make the regression weights $\mathbf{W}$, and hence $\boldsymbol{\Sigma}$, be conditional on covariate

---

6. An influential paper (Geman and Geman 1984), which introduced the idea of a Gibbs sampler (Section 24.2), proposed using the Potts model as a prior for image segmentation, but the results in their paper are misleading because they did not run their Gibbs sampler for long enough. See Figure 24.10 for a vivid illustration of this point.

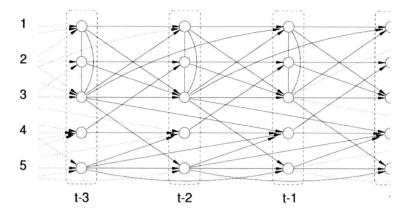

Figure 19.10 A VAR(2) process represented as a dynamic chain graph. Source: (Dahlhaus and Eichler 2000). Used with kind permission of Rainer Dahlhaus and Oxford University Press.

information (Pourahmadi 2004), without worrying about positive definite constraints. The disadavantage of the DAG formulation is its dependence on the order, although in certain domains, such as time series, there is already a natural ordering of the variables.

It is actually possible to combine both representations, resulting in a Gaussian chain graph. For example, consider a a discrete-time, second-order Markov chain in which the states are continuous, $\mathbf{y}_t \in \mathbb{R}^D$. The transition function can be represented as a (vector-valued) linear-Gaussian CPD:

$$p(\mathbf{y}_t|\mathbf{y}_{t-1}, \mathbf{y}_{t-2}, \boldsymbol{\theta}) = \mathcal{N}(\mathbf{y}_t|\mathbf{A}_1\mathbf{y}_{t-1} + \mathbf{A}_2\mathbf{y}_{t-2}, \boldsymbol{\Sigma}) \tag{19.29}$$

This is called **vector auto-regressive** or **VAR** process of order 2. Such models are widely used in econometrics for time-series forecasting.

The time series aspect is most naturally modeled using a DGM. However, if $\boldsymbol{\Sigma}^{-1}$ is sparse, then the correlation amongst the components within a time slice is most naturally modeled using a UGM. For example, suppose we have

$$\mathbf{A}_1 = \begin{pmatrix} \frac{3}{5} & 0 & \frac{1}{5} & 0 & 0 \\ 0 & \frac{3}{5} & 0 & -\frac{1}{5} & 0 \\ \frac{2}{5} & \frac{1}{3} & \frac{3}{5} & 0 & 0 \\ 0 & 0 & 0 & -\frac{1}{2} & \frac{1}{5} \\ 0 & 0 & \frac{1}{5} & 0 & \frac{2}{5} \end{pmatrix}, \ \mathbf{A}_2 = \begin{pmatrix} 0 & 0 & -\frac{1}{5} & 0 & 0 \\ 0 & 0 & 0 & 0 & 0 \\ 0 & 0 & 0 & 0 & 0 \\ 0 & 0 & \frac{1}{5} & 0 & \frac{1}{3} \\ 0 & 0 & 0 & 0 & -\frac{1}{5} \end{pmatrix} \tag{19.30}$$

and

$$\boldsymbol{\Sigma} = \begin{pmatrix} 1 & \frac{1}{2} & \frac{1}{3} & 0 & 0 \\ \frac{1}{2} & 1 & -\frac{1}{3} & 0 & 0 \\ \frac{1}{3} & -\frac{1}{3} & 1 & 0 & 0 \\ 0 & 0 & 0 & 1 & 0 \\ 0 & 0 & 0 & 0 & 1 \end{pmatrix}, \ \boldsymbol{\Sigma}^{-1} = \begin{pmatrix} 2.13 & -1.47 & -1.2 & 0 & 0 \\ -1.47 & 2.13 & 1.2 & 0 & 0 \\ -1.2 & 1.2 & 1.8 & 0 & 0 \\ 0 & 0 & 0 & 1 & 0 \\ 0 & 0 & 0 & 0 & 1 \end{pmatrix} \tag{19.31}$$

The resulting graphical model is illustrated in Figure 19.10. Zeros in the transition matrices $\mathbf{A}_1$

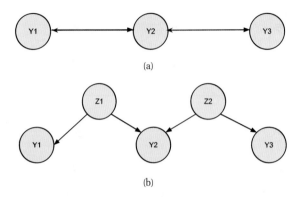

(a)

(b)

Figure 19.11   (a) A bi-directed graph. (b) The equivalent DAG. Here the z nodes are latent confounders. Based on Figures 5.12-5.13 of (Choi 2011).

and $\mathbf{A}_2$ correspond to absent directed arcs from $\mathbf{y}_{t-1}$ and $\mathbf{y}_{t-2}$ into $\mathbf{y}_t$. Zeros in the precision matrix $\mathbf{\Sigma}^{-1}$ correspond to absent undirected arcs between nodes in $\mathbf{y}_t$.

Sometimes we have a sparse covariance matrix rather than a sparse precision matrix. This can be represented using a **bi-directed graph**, where each edge has arrows in both directions, as in Figure 19.11(a). Here nodes that are not connected are unconditionally independent. For example in Figure 19.11(a) we see that $Y_1 \perp Y_3$. In the Gaussian case, this means $\Sigma_{1,3} = \Sigma_{3,1} = 0$. (A graph representing a sparse covariance matrix is called a **covariance graph**, see e.g., (Pena 2013)). By contrast, if this were an undirected model, we would have that $Y_1 \perp Y_3 | Y_2$, and $\Lambda_{1,3} = \Lambda_{3,1} = 0$, where $\mathbf{\Lambda} = \mathbf{\Sigma}^{-1}$.

A bidirected graph can be converted to a DAG with latent variables, where each bidirected edge is replaced with a hidden variable representing a hidden common cause, or **confounder**, as illustrated in Figure 19.11(b). The relevant CI properties can then be determined using d-separation.

We can combine bidirected and directed edges to get a **directed mixed graphical model**. This is useful for representing a variety of models, such as ARMA models (Section 18.2.4.4), structural equation models (Section 26.5.5), etc.

19.4.5   Markov logic networks *

In Section 10.2.2, we saw how we could "unroll" Markov models and HMMs for an arbitrary number of time steps in order to model variable-length sequences. Similarly, in Section 19.4.1, we saw how we could expand a lattice UGM to model images of any size. What about more complex domains, where we have a variable number of objects and relationships between them? Creating models for such scenarios is often done using **first-order logic** (see e.g., (Russell and Norvig 2010)). For example, consider the sentences "Smoking causes cancer" and "If two people are friends, and one smokes, then so does the other". We can write these sentences in first-order

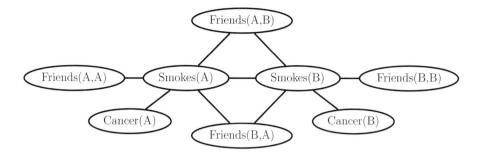

Figure 19.12 An example of a ground Markov logic network represented as a pairwise MRF for 2 people. Based on Figure 2.1 from (Domingos and Lowd 2009). Used with kind permission of Pedro Domingos.

logic as follows:

$$\forall x. Sm(x) \implies Ca(x) \tag{19.32}$$

$$\forall x. \forall y. Fr(x,y) \wedge Sm(x) \implies Sm(y) \tag{19.33}$$

where Sm and Ca are predicates, and Fr is a relation.[7]

Of course, such rules are not always true. Indeed, this brittleness is the main reason why logical approaches to AI are no longer widely used, at least not in their pure form. There have been a variety of attempts to combine first order logic with probability theory, an area known as **statistical relational AI** or **probabilistic relational modeling** (Kersting et al. 2011). One simple approach is to take logical rules and attach weights (known as **certainty factors**) to them, and then to interpret them as conditional probability distributions. For example, we might say $p(Ca(x) = 1|Sm(x) = 1) = 0.9$. Unfortunately, the rule does not say what to predict if $Sm(x) = 0$. Furthermore, combining CPDs in this way is not guaranteed to define a consistent joint distribution, because the resulting graph may not be a DAG.

An alternative approach is to treat these rules as a way of defining potential functions in an unrolled UGM. The result is known as a **Markov logic network** (Richardson and Domingos 2006; Domingos et al. 2006; Domingos and Lowd 2009). To specify the network, we first rewrite all the rules in **conjunctive normal form** (CNF), also known as **clausal form**. In this case, we get

$$\neg Sm(x) \vee Ca(x) \tag{19.34}$$

$$\neg Fr(x,y) \vee \neg Sm(x) \vee Sm(y) \tag{19.35}$$

The first clause can be read as "Either x does not smoke or he has cancer", which is logically equivalent to Equation 19.32. (Note that in a clause, any unbound variable, such as x, is assumed to be universally quantified.)

---

7. A predicate is just a function of one argument, known as an object, that evaluates to true or false, depending on whether the property holds or not of that object. A (logical) relation is just a function of two or more arguments (objects) that evaluates to true or false, depending on whether the relationship holds between that set of objects or not.

Inference in first-order logic is only semi-decidable, so it is common to use a restricted subset. A common approach (as used in Prolog) is to restrict the language to **Horn clauses**, which are clauses that contain at most one positive literal. Essentially this means the model is a series of if-then rules, where the right hand side of the rules (the "then" part, or consequence) has only a single term.

Once we have encoded our **knowledge base** as a set of clauses, we can attach weights to each one; these weights are the parameter of the model, and they define the clique potentials as follows:

$$\psi_c(\mathbf{x}_c) = \exp(w_c \phi_c(\mathbf{x}_c)) \tag{19.36}$$

where $\phi_c(\mathbf{x}_c)$ is a logical expression which evaluates clause c applied to the variables $\mathbf{x}_c$, and w_c is the weight we attach to this clause. Roughly speaking, the weight of a clause specifies the probability of a world in which this clause is satisfied relative to a world in which it is not satisfied.

Now suppose there are two objects (people) in the world, Anna and Bob, which we will denote by **constant symbols** A and B. We can make a **ground network** from the above clauses by creating binary random variables S_x, C_x, and $F_{x,y}$ for $x, y \in \{A, B\}$, and then "wiring these up" according to the clauses above. The result is the UGM in Figure 19.12 with 8 binary nodes. Note that we have not encoded the fact that Fr is a symmetric relation, so $Fr(A, B)$ and $Fr(B, A)$ might have different values. Similarly, we have the "degenerate" nodes $Fr(A, A)$ and $Fr(B, B)$, since we did not enforce $x \neq y$ in Equation 19.33. (If we add such constraints, then the model compiler, which generates the ground network, could avoid creating redundant nodes.)

In summary, we can think of MLNs as a convenient way of specifying a UGM **template**, that can get unrolled to handle data of arbitrary size. There are several other ways to define relational probabilistic models; see e.g., (Koller and Friedman 2009; Kersting et al. 2011) for details. In some cases, there is uncertainty about the number or existence of objects or relations (the so-called **open universe** problem). Section 18.6.2 gives a concrete example in the context of multi-object tracking. See e.g., (Russell and Norvig 2010; Kersting et al. 2011) and references therein for further details.

19.5 Learning

In this section, we discuss how to perform ML and MAP parameter estimation for MRFs. We will see that this is quite computationally expensive. For this reason, it is rare to perform Bayesian inference for the parameters of MRFs (although see (Qi et al. 2005)).

19.5.1 Training maxent models using gradient methods

Consider an MRF in log-linear form:

$$p(\mathbf{y}|\boldsymbol{\theta}) = \frac{1}{Z(\boldsymbol{\theta})} \exp \left(\sum_c \boldsymbol{\theta}_c^T \phi_c(\mathbf{y}) \right) \tag{19.37}$$

where c indexes the cliques. The scaled log-likelihood is given by

$$\ell(\boldsymbol{\theta}) \triangleq \frac{1}{N} \sum_i \log p(\mathbf{y}_i|\boldsymbol{\theta}) = \frac{1}{N} \sum_i \left[\sum_c \boldsymbol{\theta}_c^T \boldsymbol{\phi}_c(\mathbf{y}_i) - \log Z(\boldsymbol{\theta}) \right] \tag{19.38}$$

Since MRFs are in the exponential family, we know that this function is convex in $\boldsymbol{\theta}$ (see Section 9.2.3), so it has a unique global maximum which we can find using gradient-based optimizers. In particular, the derivative for the weights of a particular clique, c, is given by

$$\frac{\partial \ell}{\partial \boldsymbol{\theta}_c} = \frac{1}{N} \sum_i \left[\boldsymbol{\phi}_c(\mathbf{y}_i) - \frac{\partial}{\partial \boldsymbol{\theta}_c} \log Z(\boldsymbol{\theta}) \right] \tag{19.39}$$

Exercise 19.1 asks you to show that the derivative of the log partition function wrt $\boldsymbol{\theta}_c$ is the expectation of the c'th feature under the model, i.e.,

$$\frac{\partial \log Z(\boldsymbol{\theta})}{\partial \boldsymbol{\theta}_c} = \mathbb{E}\left[\boldsymbol{\phi}_c(\mathbf{y})|\boldsymbol{\theta}\right] = \sum_{\mathbf{y}} \boldsymbol{\phi}_c(\mathbf{y}) p(\mathbf{y}|\boldsymbol{\theta}) \tag{19.40}$$

Hence the gradient of the log likelihood is

$$\frac{\partial \ell}{\partial \boldsymbol{\theta}_c} = \left[\frac{1}{N} \sum_i \boldsymbol{\phi}_c(\mathbf{y}_i) \right] - \mathbb{E}\left[\boldsymbol{\phi}_c(\mathbf{y})\right] \tag{19.41}$$

In the first term, we fix $\mathbf{y}$ to its observed values; this is sometimes called the **clamped term**. In the second term, $\mathbf{y}$ is free; this is sometimes called the **unclamped term** or **contrastive term**. Note that computing the unclamped term requires inference in the model, and this must be done once per gradient step. This makes UGM training much slower than DGM training.

The gradient of the log likelihood can be rewritten as the expected feature vector according to the empirical distribution minus the model's expectation of the feature vector:

$$\frac{\partial \ell}{\partial \boldsymbol{\theta}_c} = \mathbb{E}_{p_{\text{emp}}}\left[\boldsymbol{\phi}_c(\mathbf{y})\right] - \mathbb{E}_{p(\cdot|\boldsymbol{\theta})}\left[\boldsymbol{\phi}_c(\mathbf{y})\right] \tag{19.42}$$

At the optimum, the gradient will be zero, so the empirical distribution of the features will match the model's predictions:

$$\mathbb{E}_{p_{\text{emp}}}\left[\boldsymbol{\phi}_c(\mathbf{y})\right] = \mathbb{E}_{p(\cdot|\boldsymbol{\theta})}\left[\boldsymbol{\phi}_c(\mathbf{y})\right] \tag{19.43}$$

This is called **moment matching**. This observation motivates a different optimization algorithm which we discuss in Section 19.5.7.

19.5.2 Training partially observed maxent models

Suppose we have missing data and/or hidden variables in our model. In general, we can represent such models as follows:

$$p(\mathbf{y}, \mathbf{h}|\boldsymbol{\theta}) = \frac{1}{Z(\boldsymbol{\theta})} \exp(\sum_c \boldsymbol{\theta}_c^T \boldsymbol{\phi}_c(\mathbf{h}, \mathbf{y})) \tag{19.44}$$

The log likelihood has the form

$$\ell(\boldsymbol{\theta}) = \frac{1}{N} \sum_{i=1}^{N} \log \left(\sum_{\mathbf{h}} p(\mathbf{y}_i, \mathbf{h}|\boldsymbol{\theta}) \right) = \frac{1}{N} \sum_{i=1}^{N} \log \left(\frac{1}{Z(\boldsymbol{\theta})} \sum_{\mathbf{h}} \tilde{p}(\mathbf{y}_i, \mathbf{h}|\boldsymbol{\theta}) \right) \tag{19.45}$$

where

$$\tilde{p}(\mathbf{y}, \mathbf{h}|\boldsymbol{\theta}) \triangleq \exp \left(\sum_{c} \boldsymbol{\theta}_c^T \boldsymbol{\phi}_c(\mathbf{h}, \mathbf{y}) \right) \tag{19.46}$$

is the unnormalized distribution. The term $\sum_{\mathbf{h}_i} \tilde{p}(\mathbf{y}_i, \mathbf{h}_i|\boldsymbol{\theta})$ is the same as the partition function for the whole model, except that $\mathbf{y}$ is fixed at $\mathbf{y}_i$. Hence the gradient is just the expected features where we clamp $\mathbf{y}_i$, but average over $\mathbf{h}$:

$$\frac{\partial}{\partial \boldsymbol{\theta}_c} \log \left(\sum_{\mathbf{h}} \tilde{p}(\mathbf{y}_i, \mathbf{h}|\boldsymbol{\theta}) \right) = \mathbb{E}\left[\boldsymbol{\phi}_c(\mathbf{h}, \mathbf{y}_i)|\boldsymbol{\theta} \right] \tag{19.47}$$

So the overall gradient is given by

$$\frac{\partial \ell}{\partial \boldsymbol{\theta}_c} = \frac{1}{N} \sum_{i} \left\{ \mathbb{E}\left[\boldsymbol{\phi}_c(\mathbf{h}, \mathbf{y}_i)|\boldsymbol{\theta} \right] - \mathbb{E}\left[\boldsymbol{\phi}_c(\mathbf{h}, \mathbf{y})|\boldsymbol{\theta} \right] \right\} \tag{19.48}$$

The first set of expectations are computed by "clamping" the visible nodes to their observed values, and the second set are computed by letting the visible nodes be free. In both cases, we marginalize over $\mathbf{h}_i$.

An alternative approach is to use generalized EM, where we use gradient methods in the M step. See (Koller and Friedman 2009, p956) for details.

In cases where exact inference is impossible, we can use approximate inference, as we discuss below. Section 27.7.2 discusses how to fit a special class of latent variable MRFs known as restricted Boltzmann machines by combining MCMC inference with stochastic gradient methods.

19.5.3   Approximate methods for computing the MLEs of MRFs

When fitting a UGM there is (in general) no closed form solution for the ML or the MAP estimate of the parameters, so we need to use gradient-based optimizers. This gradient requires inference. In models where inference is intractable, learning also becomes intractable. We can combine approximate inference with (gradient based) learning, but results are not always guaranteed to be good depending on which inference algorithm is used (Kulesza and Pereira 2007). This has motivated various computationally faster alternatives to ML/MAP estimation, which we list in Table 19.1. We dicsuss some of these alternatives below, and defer others to later sections.

19.5.4   Pseudo likelihood

One alternative to MLE is to maximize the **pseudo likelihood** (Besag 1975), defined as follows:

$$\ell_{PL}(\boldsymbol{\theta}) \triangleq \sum_{\mathbf{y}} \sum_{d=1}^{D} p_{\text{emp}}(\mathbf{y}) \log p(y_d|\mathbf{y}_{-d}) = \frac{1}{N} \sum_{i=1}^{N} \sum_{d=1}^{D} \log p(y_{id}|\mathbf{y}_{i,-d}, \boldsymbol{\theta}) \tag{19.49}$$

Method	Restriction	Exact MLE?	Section
Closed form	Only Chordal MRF	Exact	Section 19.5.7.4
IPF	Only Tabular / Gaussian MRF	Exact	Section 19.5.7
Gradient-based optimization	Low tree width	Exact	Section 19.5.1
Max-margin training	Only CRFs	N/A	Section 19.7
Pseudo-likelihood	No hidden variables	Approximate	Section 19.5.4
Stochastic ML	-	Exact (up to MC error)	Section 19.5.5
Contrastive divergence	-	Approximate	Section 27.7.2.4
Minimum probability flow	Can integrate out the hiddens	Approximate	Sohl-Dickstein et al. (2011)

Table 19.1 Some methods that can be used to compute approximate ML/ MAP parameter estimates for MRFs/ CRFs. Low tree-width means that, in order for the method to be efficient, the graph must be "tree-like"; see Section 20.5 for details.

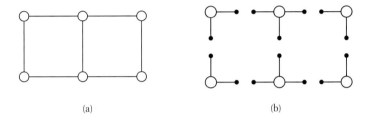

(a) (b)

Figure 19.13 (a) A small 2d lattice. (b) The representation used by pseudo likelihood. Solid nodes are observed neighbors. Based on Figure 2.2 of (Carbonetto 2003).

That is, we optimize the product of the full conditionals, also known as the **composite likelihood** (Lindsay 1988), Compare this to the objective for maximum likelihood:

$$\ell_{ML}(\boldsymbol{\theta}) = \sum_{\mathbf{y}} p_{\text{emp}}(\mathbf{y}) \log p(\mathbf{y}|\boldsymbol{\theta}) = \frac{1}{N} \sum_{i=1}^{N} \log p(\mathbf{y}_i|\boldsymbol{\theta}) \quad (19.50)$$

In the case of Gaussian MRFs, PL is equivalent to ML (Besag 1975), but this is not true in general (Liang and Jordan 2008).

The PL approach is illustrated in Figure 19.13 for a 2d grid. We learn to predict each node, given all of its neighbors. This objective is generally fast to compute since each full conditional $p(y_{id}|\mathbf{y}_{i,-d}, \boldsymbol{\theta})$ only requires summing over the states of a single node, y_{id}, in order to compute the local normalization constant. The PL approach is similar to fitting each full conditional separately (which is the method used to train dependency networks, discussed in Section 26.2.2), except that, in PL, the parameters are tied between adjacent nodes.

One problem with PL is that it is hard to apply to models with hidden variables (Parise and Welling 2005). Another more subtle problem is that each node assumes that its neighbors have known values. If node $t \in \text{nbr}(s)$ is a perfect predictor for node s, then s will learn to rely completely on node t, even at the expense of ignoring other potentially useful information, such as its local evidence.

However, experiments in (Parise and Welling 2005; Hoefling and Tibshirani 2009) suggest that PL works as well as exact ML for fully observed Ising models, and of course PL is *much* faster.

19.5.5 Stochastic maximum likelihood

Recall that the gradient of the log-likelihood for a fully observed MRF is given by

$$\nabla_{\boldsymbol{\theta}} \ell(\boldsymbol{\theta}) = \frac{1}{N} \sum_i \left[\phi(\mathbf{y}_i) - \mathbb{E}\left[\phi(\mathbf{y}) \right] \right] \tag{19.51}$$

The gradient for a partially observed MRF is similar. In both cases, we can approximate the model expectations using Monte Carlo sampling. We can combine this with stochastic gradient descent (Section 8.5.2), which takes samples from the empirical distribution. Pseudocode for the resulting method is shown in Algorithm 19.1.

Algorithm 19.1: Stochastic maximum likelihood for fitting an MRF

1 Initialize weights $\boldsymbol{\theta}$ randomly;
2 $k = 0$, $\eta = 1$;
3 **for** *each epoch* **do**
4   **for** *each minibatch of size B* **do**
5   **for** *each sample $s = 1 : S$* **do**
6   Sample $\mathbf{y}^{s,k} \sim p(\mathbf{y}|\boldsymbol{\theta}_k)$;
7         $\hat{E}(\phi(\mathbf{y})) = \frac{1}{S} \sum_{s=1}^{S} \phi(\mathbf{y}^{s,k})$;
8   **for** *each training case i in minibatch* **do**
9             $\mathbf{g}_{ik} = \phi(\mathbf{y}_i) - \hat{E}(\phi(\mathbf{y}))$;
10         $\mathbf{g}_k = \frac{1}{B} \sum_{i \in B} \mathbf{g}_{ik}$;
11         $\boldsymbol{\theta}_{k+1} = \boldsymbol{\theta}_k - \eta \mathbf{g}_k$;
12         $k = k + 1$;
13   Decrease step size η;

Typically we use MCMC to generate the samples. Of course, running MCMC to convergence at each step of the inner loop would be extremely slow. Fortunately, it was shown in (Younes 1989) that we can start the MCMC chain at its previous value, and just take a few steps. In otherwords, we sample $\mathbf{y}^{s,k}$ by initializing the MCMC chain at $\mathbf{y}^{s,k-1}$, and then run for a few iterations. This is valid since $p(\mathbf{y}|\boldsymbol{\theta}^k)$ is likely to be close to $p(\mathbf{y}|\boldsymbol{\theta}^{k-1})$, since we only changed the parameters by a small amount. We call this algorithm **stochastic maximum likelihood** or **SML**. (There is a closely related algorithm called persistent contrastive divergence which we discuss in Section 27.7.2.5.)

19.5.6 Feature induction for maxent models *

MRFs require a good set of features. One unsupervised way to learn such features, known as **feature induction**, is to start with a base set of features, and then to continually create new feature combinations out of old ones, greedily adding the best ones to the model. This approach was first proposed in (Pietra et al. 1997; Zhu et al. 1997), and was later extended to the CRF case in (McCallum 2003).

To illustrate the basic idea, we present an example from (Pietra et al. 1997), which described how to build unconditional probabilistic models to represent English spelling. Initially the model has no features, which represents the uniform distribution. The algorithm starts by choosing to add the feature

$$\phi_1(\mathbf{y}) = \sum_t \mathbb{I}(y_t \in \{a, \ldots, z\}) \tag{19.52}$$

which checks if any letter is lower case or not. After the feature is added, the parameters are (re)-fit by maximum likelihood. For this feature, it turns out that $\hat{\theta}_1 = 1.944$, which means that a word with a lowercase letter in any position is about $e^{1.944} \approx 7$ times more likely than the same word without a lowercase letter in that position. Some samples from this model, generated using (annealed) Gibbs sampling (Section 24.2), are shown below.[8]

```
m, r, xevo, ijjiir, b, to, jz, gsr, wq, vf, x, ga, msmGh, pcp, d, oziVlal,
hzagh, yzop, io, advzmxnv, ijv_bolft, x, emx, kayerf, mlj, rawzyb, jp, ag,
ctdnnnbg, wgdw, t, kguv, cy, spxcq, uzflbbf, dxtkkn, cxwx, jpd, ztzh, lv,
zhpkvnu, l^, r, qee, nynrx, atze4n, ik, se, w, lrh, hp+, yrqyka'h, zcngotcnx,
igcump, zjcjs, lqpWiqu, cefmfhc, o, lb, fdcY, tzby, yopxmvk, by, fz,, t, govyccm,
ijyiduwfzo, 6xr, duh, ejv, pk, pjw, l, fl, w
```

The second feature added by the algorithm checks if two adjacent characters are lower case:

$$\phi_2(\mathbf{y}) = \sum_{s \sim t} \mathbb{I}(y_s \in \{a, \ldots, z\}, y_t \in \{a, \ldots, z\}) \tag{19.53}$$

Now the model has the form

$$p(\mathbf{y}) = \frac{1}{Z} \exp(\theta_1 \phi_1(\mathbf{y}) + \theta_2 \phi_2(\mathbf{y})) \tag{19.54}$$

Continuing in this way, the algorithm adds features for the strings s> and ing>, where > represents the end of word, and for various regular expressions such as [0-9], etc. Some samples from the model with 1000 features, generated using (annealed) Gibbs sampling, are shown below.

```
was, reaser, in, there, to, will, ,, was, by, homes, thing, be, reloverated,
ther, which, conists, at, fores, anditing, with, Mr., proveral, the, ,, ***,
on't, prolling, prothere, ,, mento, at, yaou, 1, chestraing, for, have, to,
intrally, of, qut, ., best, compers, ***, cluseliment, uster, of, is, deveral,
this, thise, of, offect, inatever, thifer, constranded, stater, vill, in, thase,
in, youse, menttering, and, ., of, in, verate, of, to
```

This approach of feature learning can be thought of as a form of graphical model structure learning (Chapter 26), except it is more fine-grained: we add features that are useful, regardless of the resulting graph structure. However, the resulting graphs can become densely connected, which makes inference (and hence parameter estimation) intractable.

---

8. We thank John Lafferty for sharing this example.

19.5.7   Iterative proportional fitting (IPF) *

Consider a pairwise MRF where the potentials are represented as tables, with one parameter per variable setting. We can represent this in log-linear form using

$$\psi_{st}(y_s, y_t) = \exp\left(\boldsymbol{\theta}_{st}^T[\mathbb{I}(y_s = 1, y_t = 1), \ldots, \mathbb{I}(y_s = K, y_t = K)]\right) \tag{19.55}$$

and similarly for $\psi_t(y_t)$. Thus the feature vectors are just indicator functions.

From Equation 19.43, we have that, at the maximum of the likelihood, the empirical expectation of the features equals the model's expectation:

$$\mathbb{E}_{p_{\mathrm{emp}}}[\mathbb{I}(y_s = j, y_t = k)] \ = \ \mathbb{E}_{p(\cdot|\boldsymbol{\theta})}[\mathbb{I}(y_s = j, y_t = k)] \tag{19.56}$$

$$p_{\mathrm{emp}}(y_s = j, y_t = k) \ = \ p(y_s = j, y_t = k|\boldsymbol{\theta}) \tag{19.57}$$

where p_{emp} is the empirical probability:

$$p_{\mathrm{emp}}(y_s = j, y_t = k) = \frac{N_{st,jk}}{N} = \frac{\sum_{n=1}^N \mathbb{I}(y_{ns} = j, y_{nt} = k)}{N} \tag{19.58}$$

For a general graph, the condition that must hold at the optimum is

$$p_{\mathrm{emp}}(\mathbf{y}_c) = p(\mathbf{y}_c|\boldsymbol{\theta}) \tag{19.59}$$

For a special family of graphs known as decomposable graphs (defined in Section 20.4.1), one can show that $p(\mathbf{y}_c|\boldsymbol{\theta}) = \psi_c(\mathbf{y}_c)$. However, even if the graph is not decomposable, we can imagine trying to enforce this condition. This suggests an iterative coordinate ascent scheme where at each step we compute

$$\psi_c^{t+1}(\mathbf{y}_c) = \psi_c^t(\mathbf{y}_c) \times \frac{p_{\mathrm{emp}}(\mathbf{y}_c)}{p(\mathbf{y}_c|\psi^t)} \tag{19.60}$$

where the multiplication and division is elementwise. This is known as **iterative proportional fitting** or **IPF** (Fienberg 1970; Bishop et al. 1975). See Algorithm 19.2 for the pseudocode.

Algorithm 19.2: Iterative Proportional Fitting algorithm for tabular MRFs

---

1 Initialize $\psi_c = 1$ for $c = 1 : C$;
2 **repeat**
3   **for** $c = 1 : C$ **do**
4          $p_c = p(\mathbf{y}_c|\psi)$;
5          $\hat{p}_c = p_{\mathrm{emp}}(\mathbf{y}_c)$;
6          $\psi_c = \psi_c. * \frac{\hat{p}_c}{p_c}$;
7 **until** *converged*;

19.5.7.1   Example

Let us consider a simple example from http://en.wikipedia.org/wiki/Iterative_proportional_fitting. We have two binary variables, where Y_1 indicates if the person is male or female, and Y_2 indicates whether they are right or left handed. We can summarize the empirican count data using the following 2×2 contingency table:

	right-handed	left-handed	Total
male	43	9	52
female	44	4	48
Total	87	13	100

Suppose we want to fit a disconnected graphical model containing nodes Y_1 and Y_2 but with no edge between them. That is, we are making the approximation $p(Y_1, Y_2) = 1/Z\psi_1(Y_1)\psi_2(Y_2)$. We want to find the factors (2d vectors) $\boldsymbol{\psi}_1$ and $\boldsymbol{\psi}_2$. One possible solution is to use $\boldsymbol{\psi}_1 = [0.5200, 0.4800]$ and $\boldsymbol{\psi}_2 = [87, 13]$. Below we show the model's predictions, $\mathbf{M} = \boldsymbol{\psi}_1\boldsymbol{\psi}_2^T$.

	right-handed	left-handed	Total
male	45.24	6.76	52
female	41.76	6.24	48
Total	87	13	100

It is easy to see that the row and column sums of the model's predictions $\mathbf{M}$ match the row and column sums of the data, $\mathbf{C}$, which we means we have succesfully matched the moments of Y_1 and Y_2. See `IPFdemo2x2` for some Matlab code that computes these numbers.

19.5.7.2 Speed of IPF

IPF is a fixed point algorithm for enforcing the moment matching constraints and is guaranteed to converge to the global optimum (Bishop et al. 1975). The number of iterations depends on the form of the model. If the graph is decomposable, then IPF converges in a single iteration, but in general, IPF may require many iterations.

It is clear that the dominant cost of IPF is computing the required marginals under the model. Efficient methods, such as the junction tree algorithm (Section 20.4), can be used, resulting in something called **efficient IPF** (Jirousek and Preucil 1995).

Nevertheless, coordinate descent can be slow. An alternative method is to update all the parameters at once, by simply following the gradient of the likelihood. This gradient approach has the further significant advantage that it works for models in which the clique potentials may not be fully parameterized, i.e., the features may not consist of all possible indicators for each clique, but instead can be arbitrary. Although it is possible to adapt IPF to this setting of general features, resulting in a method known as **iterative scaling**, in practice the gradient method is much faster (Malouf 2002; Minka 2003).

19.5.7.3 Generalizations of IPF

We can use IPF to fit Gaussian graphical models: instead of working with empirical counts, we work with empirical means and covariances (Speed and Kiiveri 1986). It is also possible to create a Bayesian IPF algorithm for sampling from the posterior of the model's parameters (see e.g., (Dobra and Massam 2010)).

19.5.7.4 IPF for decomposable graphical models

There is a special family of undirected graphical models known as decomposable graphical models. This is formally defined in Section 20.4.1, but the basic idea is that it contains graphs which are "tree-like". Such graphs can be represented by UGMs or DGMs without any loss of information.

In the case of decomposable graphical models, IPF converges in one iteration. In fact, the MLE has a closed form solution (Lauritzen 1996). In particular, for tabular potentials we have

$$\hat{\psi}_c(\mathbf{y}_c = k) = \frac{\sum_{i=1}^N \mathbb{I}(\mathbf{y}_{i,c} = k)}{N} \tag{19.61}$$

and for Gaussian potentials, we have

$$\hat{\boldsymbol{\mu}}_c = \frac{\sum_{i=1}^N \mathbf{y}_{ic}}{N}, \ \hat{\boldsymbol{\Sigma}}_c = \frac{\sum_i (\mathbf{y}_{ic} - \hat{\boldsymbol{\mu}}_c)(\mathbf{x}_{ic} - \hat{\boldsymbol{\mu}}_c)^T}{N} \tag{19.62}$$

By using conjugate priors, we can also easily compute the full posterior over the model parameters in the decomposable case, just as we did in the DGM case. See (Lauritzen 1996) for details.

19.6   Conditional random fields (CRFs)

A **conditional random field** or **CRF** (Lafferty et al. 2001), sometimes a **discriminative random field** (Kumar and Hebert 2003), is just a version of an MRF where all the clique potentials are conditioned on input features:

$$p(\mathbf{y}|\mathbf{x}, \mathbf{w}) = \frac{1}{Z(\mathbf{x}, \mathbf{w})} \prod_c \psi_c(\mathbf{y}_c|\mathbf{x}, \mathbf{w}) \tag{19.63}$$

A CRF can be thought of as a **structured output** extension of logistic regression, where we model the correlation amongst the output labels conditioned on the input features.[9] We will usually assume a log-linear representation of the potentials:

$$\psi_c(\mathbf{y}_c|\mathbf{x}, \mathbf{w}) = \exp(\mathbf{w}_c^T \boldsymbol{\phi}(\mathbf{x}, \mathbf{y}_c)) \tag{19.64}$$

where $\boldsymbol{\phi}(\mathbf{x}, \mathbf{y}_c)$ is a feature vector derived from the global inputs $\mathbf{x}$ and the local set of labels $\mathbf{y}_c$. We will give some examples below which will make this notation clearer.

The advantage of a CRF over an MRF is analogous to the advantage of a discriminative classifier over a generative classifier (see Section 8.6), namely, we don't need to "waste resources" modeling things that we always observe. Instead we can focus our attention on modeling what we care about, namely the distribution of labels given the data.

Another important advantage of CRFs is that we can make the potentials (or factors) of the model be data-dependent. For example, in image processing applications, we may "turn off" the label smoothing between two neighboring nodes s and t if there is an observed discontinuity in the image intensity between pixels s and t. Similarly, in natural language processing problems, we can make the latent labels depend on global properties of the sentence, such as which language it is written in. It is hard to incorporate global features into generative models.

The disadvantage of CRFs over MRFs is that they require labeled training data, and they are slower to train, as we explain in Section 19.6.3. This is analogous to the strengths and weaknesses of logistic regression vs naive Bayes, discussed in Section 8.6.

---

9. An alternative approach is to predict the label for a node i given i's features and given the features of i's neighbors, as opposed to conditioning on the hidden label of i's neighbors. In general, however, this aproach needs much more data to work well (Jensen et al. 2004).

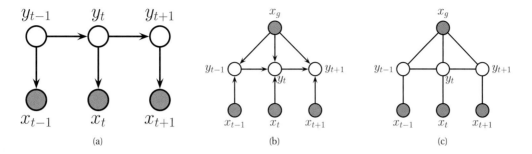

Figure 19.14 Various models for sequential data. (a) A generative directed HMM. (b) A discriminative directed MEMM. (c) A discriminative undirected CRF.

19.6.1 Chain-structured CRFs, MEMMs and the label-bias problem

The most widely used kind of CRF uses a chain-structured graph to model correlation amongst neighboring labels. Such models are useful for a variety of sequence labeling tasks (see Section 19.6.2).

Traditionally, HMMs (discussed in detail in Chapter 17) have been used for such tasks. These are joint density models of the form

$$p(\mathbf{x}, \mathbf{y}|\mathbf{w}) = \prod_{t=1}^{T} p(y_t|y_{t-1}, \mathbf{w}) p(\mathbf{x}_t|y_t, \mathbf{w}) \tag{19.65}$$

where we have dropped the initial $p(y_1)$ term for simplicity. See Figure 19.14(a). If we observe both $\mathbf{x}_t$ and y_t for all t, it is very easy to train such models, using techniques described in Section 17.5.1.

An HMM requires specifying a generative observation model, $p(\mathbf{x}_t|y_t, \mathbf{w})$, which can be difficult. Furthermore, each $\mathbf{x}_t$ is required to be local, since it is hard to define a generative model for the whole stream of observations, $\mathbf{x} = \mathbf{x}_{1:T}$.

An obvious way to make a discriminative version of an HMM is to "reverse the arrows" from y_t to $\mathbf{x}_t$, as in Figure 19.14(b). This defines a directed discriminative model of the form

$$p(\mathbf{y}|\mathbf{x}, \mathbf{w}) = \prod_{t} p(y_t|y_{t-1}, \mathbf{x}, \mathbf{w}) \tag{19.66}$$

where $\mathbf{x} = (\mathbf{x}_{1:T}, \mathbf{x}_g)$, $\mathbf{x}_g$ are global features, and $\mathbf{x}_t$ are features specific to node t. (This partition into local and global is not necessary, but helps when comparing to HMMs.) This is called a **maximum entropy Markov model** or **MEMM** (McCallum et al. 2000; Kakade et al. 2002).

An MEMM is simply a Markov chain in which the state transition probabilities are conditioned on the input features. (It is therefore a special case of an input-output HMM, discussed in Section 17.6.3.) This seems like the natural generalization of logistic regression to the structured-output setting, but it suffers from a subtle problem known (rather obscurely) as the **label bias** problem (Lafferty et al. 2001). The problem is that local features at time t do not influence states prior to time t. This follows by examining the DAG, which shows that $\mathbf{x}_t$ is d-separated from

y_{t-1} (and all earlier time points) by the v-structure at y_t, which is a hidden child, thus blocking the information flow.

To understand what this means in practice, consider the part of speech (POS) tagging task. Suppose we see the word "banks"; this could be a verb (as in "he banks at BoA"), or a noun (as in "the river banks were overflowing"). Locally the POS tag for the word is ambiguous. However, suppose that later in the sentence, we see the word "fishing"; this gives us enough context to infer that the sense of "banks" is "river banks". However, in an MEMM (unlike in an HMM and CRF), the "fishing" evidence will not flow backwards, so we will not be able to disambiguate "banks".

Now consider a chain-structured CRF. This model has the form

$$p(\mathbf{y}|\mathbf{x}, \mathbf{w}) = \frac{1}{Z(\mathbf{x}, \mathbf{w})} \prod_{t=1}^{T} \psi(y_t|\mathbf{x}, \mathbf{w}) \prod_{t=1}^{T-1} \psi(y_t, y_{t+1}|\mathbf{x}, \mathbf{w}) \tag{19.67}$$

From the graph in Figure 19.14(c), we see that the label bias problem no longer exists, since y_t does not block the information from $\mathbf{x}_t$ from reaching other $y_{t'}$ nodes.

The label bias problem in MEMMs occurs because directed models are **locally normalized**, meaning each CPD sums to 1. By contrast, MRFs and CRFs are **globally normalized**, which means that local factors do not need to sum to 1, since the partition function Z, which sums over all joint configurations, will ensure the model defines a valid distribution. However, this solution comes at a price: we do not get a valid probability distribution over $\mathbf{y}$ until we have seen the whole sentence, since only then can we normalize over all configurations. Consequently, CRFs are not as useful as DGMs (whether discriminative or generative) for online or real-time inference. Furthermore, the fact that Z depends on all the nodes, and hence all their parameters, makes CRFs much slower to train than DGMs, as we will see in Section 19.6.3.

19.6.2   Applications of CRFs

CRFs have been applied to many interesting problems; we give a representative sample below. These applications illustrate several useful modeling tricks, and will also provide motivation for some of the inference techniques we will discuss in Chapter 20.

19.6.2.1   Handwriting recognition

A natural application of CRFs is to classify hand-written digit strings, as illustrated in Figure 19.15. The key observation is that locally a letter may be ambiguous, but by depending on the (unknown) labels of one's neighbors, it is possible to use context to reduce the error rate. Note that the node potential, $\psi_t(y_t|\mathbf{x}_t)$, is often taken to be a probabilistic discriminative classifier, such as a neural network or RVM, that is trained on isolated letters, and the edge potentials, $\psi_{st}(y_s, y_t)$, are often taken to be a language bigram model. Later we will discuss how to train all the potentials jointly.

19.6.2.2   Noun phrase chunking

One common NLP task is **noun phrase chunking**, which refers to the task of segmenting a sentence into its distinct noun phrases (NPs). This is a simple example of a technique known as **shallow parsing**.

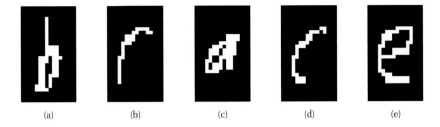

Figure 19.15 Example of handwritten letter recognition. In the word 'brace', the 'r' and the 'c' look very similar, but can be disambiguated using context. Source: (Taskar et al. 2003) . Used with kind permission of Ben Taskar.

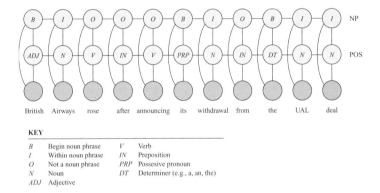

Figure 19.16 A CRF for joint POS tagging and NP segmentation. Source: Figure 4.E.1 of (Koller and Friedman 2009). Used with kind permission of Daphne Koller.

In more detail, we tag each word in the sentence with B (meaning beginning of a new NP), I (meaning inside a NP), or O (meaning outside an NP). This is called **BIO** notation. For example, in the following sentence, the NPs are marked with brackets:

```
    B    I       O    O    O       B    I       O    B    I    I
(British Airways) rose after announcing (its withdrawl) from (the UAI deal)
```

(We need the B symbol so that we can distinguish I I, meaning two words within a single NP, from B B, meaning two separate NPs.)

A standard approach to this problem would first convert the string of words into a string of POS tags, and then convert the POS tags to a string of BIOs. However, such a **pipeline** method can propagate errors. A more robust approach is to build a joint probabilistic model of the form $p(\text{NP}_{1:T}, \text{POS}_{1:T}|\text{words}_{1:T})$. One way to do this is to use the CRF in Figure 19.16. The connections between adjacent labels encode the probability of transitioning between the B, I and O states, and can enforce constraints such as the fact that B must preceed I. The features are usually hand engineered and include things like: does this word begin with a capital letter, is this word followed by a full stop, is this word a noun, etc. Typically there are $\sim 1,000 - 10,000$

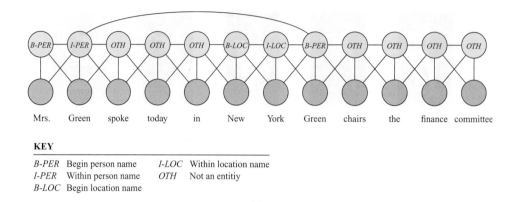

KEY

B-PER	Begin person name	*I-LOC*	Within location name
I-PER	Within person name	*OTH*	Not an entitiy
B-LOC	Begin location name		

Figure 19.17   A skip-chain CRF for named entity recognition.   Source: Figure 4.E.1 of (Koller and Friedman 2009).   Used with kind permission of Daphne Koller.

features per node.

The number of features has minimal impact on the inference time, since the features are observed and do not need to be summed over. (There is a small increase in the cost of evaluating potential functions with many features, but this is usually negligible; if not, one can use ℓ_1 regularization to prune out irrelevant features.) However, the graph structure can have a dramatic effect on inference time. The model in Figure 19.16 is tractable, since it is essentially a "fat chain", so we can use the forwards-backwards algorithm (Section 17.4.3) for exact inference in $O(T|\text{POS}|^2|\text{NP}|^2)$ time, where $|\text{POS}|$ is the number of POS tags, and $|\text{NP}|$ is the number of NP tags. However, the seemingly similar graph in Figure 19.17, to be explained below, is computationally intractable.

19.6.2.3   Named entity recognition

A task that is related to NP chunking is **named entity extraction**. Instead of just segmenting out noun phrases, we can segment out phrases to do with people and locations. Similar techniques are used to automatically populate your calendar from your email messages; this is called **information extraction**.

A simple approach to this is to use a chain-structured CRF, but to expand the state space from BIO to B-Per, I-Per, B-Loc, I-Loc, and Other. However, sometimes it is ambiguous whether a word is a person, location, or something else. (Proper nouns are particularly difficult to deal with because they belong to an **open class**, that is, there is an unbounded number of possible names, unlike the set of nouns and verbs, which is large but essentially fixed.) We can get better performance by considering long-range correlations between words. For example, we might add a link between all occurrences of the same word, and force the word to have the same tag in each occurence. (The same technique can also be helpful for resolving the identity of pronouns.) This is known as a **skip-chain CRF**. See Figure 19.17 for an illustration.

We see that the graph structure itself changes depending on the input, which is an additional advantage of CRFs over generative models. Unfortunately, inference in this model is gener-

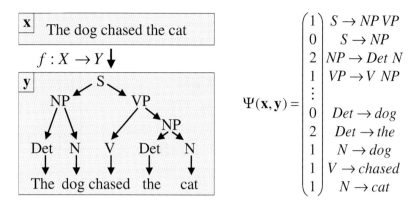

Figure 19.18 Illustration of a simple parse tree based on a context free grammar in Chomsky normal form. The feature vector $\phi(\mathbf{x}, \mathbf{y}) = \Psi(\mathbf{x}, \mathbf{y})$ counts the number of times each production rule was used. Source: Figure 5.2 of (Altun et al. 2007). Used with kind permission of Yasemin Altun.

ally more expensive than in a simple chain with local connections, for reasons explained in Section 20.5.

19.6.2.4 Natural language parsing

A generalization of chain-structured models for language is to use probabilistic grammars. In particular, a probabilistic **context free grammar** or **PCFG** is a set of re-write or production rules of the form $\sigma \to \sigma'\sigma''$ or $\sigma \to x$, where $\sigma, \sigma', \sigma'' \in \Sigma$ are non-terminals (analogous to parts of speech), and $x \in \mathcal{X}$ are terminals, i.e., words. See Figure 19.18 for an example. Each such rule has an associated probability. The resulting model defines a probability distribution over sequences of words. We can compute the probability of observing a particular sequence $\mathbf{x} = x_1 \ldots x_T$ by summing over all trees that generate it. This can be done in $O(T^3)$ time using the **inside-outside algorithm**; see e.g., (Jurafsky and Martin 2008; Manning and Schuetze 1999) for details.

PCFGs are generative models. It is possible to make discriminative versions which encode the probability of a labeled tree, $\mathbf{y}$, given a sequence of words, $\mathbf{x}$, by using a CRF of the form $p(\mathbf{y}|\mathbf{x}) \propto \exp(\mathbf{w}^T \phi(\mathbf{x}, \mathbf{y}))$. For example, we might define $\phi(\mathbf{x}, \mathbf{y})$ to count the number of times each production rule was used (which is analogous to the number of state transitions in a chain-structured model). See e.g., (Taskar et al. 2004) for details.

19.6.2.5 Hierarchical classification

Suppose we are performing multi-class classification, where we have a **label taxonomy**, which groups the classes into a hierarchy. We can encode the position of y within this hierarchy by defining a binary vector $\phi(y)$, where we turn on the bit for component y and for all its children. This can be combined with input features $\phi(\mathbf{x})$ using a tensor product, $\phi(\mathbf{x}, y) = \phi(\mathbf{x}) \otimes \phi(y)$. See Figure 19.19 for an example.

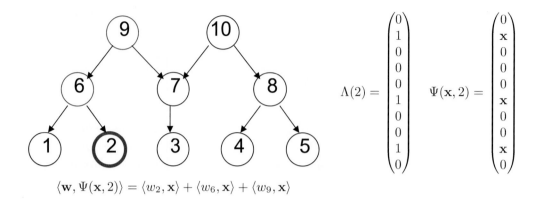

$$\langle \mathbf{w}, \Psi(\mathbf{x}, 2) \rangle = \langle w_2, \mathbf{x} \rangle + \langle w_6, \mathbf{x} \rangle + \langle w_9, \mathbf{x} \rangle$$

Figure 19.19 Illustration of a simple label taxonomy, and how it can be used to compute a distributed representation for the label for class 2. In this figure, $\phi(\mathbf{x}) = \mathbf{x}$, $\phi(y = 2) = \Lambda(2)$, $\phi(\mathbf{x}, y)$ is denoted by $\Psi(\mathbf{x}, 2)$, and $\mathbf{w}^T \phi(\mathbf{x}, y)$ is denoted by $\langle \mathbf{w}, \Psi(\mathbf{x}, 2) \rangle$. Source: Figure 5.1 of (Altun et al. 2007) . Used with kind permission of Yasemin Altun.

This method is widely used for text classification, where manually constructed taxnomies (such as the Open Directory Project at `www.dmoz.org`) are quite common. The benefit is that information can be shared between the parameters for nearby categories, enabling generalization across classes.

19.6.2.6   Protein side-chain prediction

An interesting analog to the skip-chain model arises in the problem of predicting the structure of protein side chains. Each residue in the side chain has 4 dihedral angles, which are usually discretized into 3 values called rotamers. The goal is to predict this discrete sequence of angles, $\mathbf{y}$, from the discrete sequence of amino acids, $\mathbf{x}$.

We can define an energy function $E(\mathbf{x}, \mathbf{y})$, where we include various pairwise interaction terms between nearby residues (elements of the $\mathbf{y}$ vector). This energy is usually defined as a weighted sum of individual energy terms, $E(\mathbf{x}, \mathbf{y}|\mathbf{w}) = \sum_{j=1}^{D} \theta_j E_j(\mathbf{x}, \mathbf{y})$, where the E_j are energy contribution due to various electrostatic charges, hydrogen bonding potentials, etc, and $\mathbf{w}$ are the parameters of the model. See (Yanover et al. 2007) for details.

Given the model, we can compute the most probable side chain configuration using $\mathbf{y}^* = \operatorname{argmin} E(\mathbf{x}, \mathbf{y}|\mathbf{w})$. In general, this problem is NP-hard, depending on the nature of the graph induced by the E_j terms, due to long-range connections between the variables. Nevertheless, some special cases can be efficiently handled, using methods discussed in Section 22.6.

19.6.2.7   Stereo vision

Low-level vision problems are problems where the input is an image (or set of images), and the output is a processed version of the image. In such cases, it is common to use 2d lattice-structured models; the models are similar to Figure 19.9, except that the features can be global, and are not generated by the model. We will assume a pairwise CRF.

A classic low-level vision problem is **dense stereo reconstruction**, where the goal is to estimate the depth of every pixel given two images taken from slightly different angles. In this section (based on (Sudderth and Freeman 2008)), we give a sketch of how a simple CRF can be used to solve this task. See e.g., (Sun et al. 2003) for a more sophisticated model.

By using some standard preprocessing techniques, one can convert depth estimation into a problem of estimating the **disparity** y_s between the pixel at location (i_s, j_s) in the left image and the corresponding pixel at location $(i_s + y_s, j_s)$ in the right image. We typically assume that corresponding pixels have similar intensity, so we define a local node potential of the form

$$\psi_s(y_s|\mathbf{x}) \propto \exp\left\{-\frac{1}{2\sigma^2}\left(x_L(i_s, j_s) - x_R(i_s + y_s, j_s)\right)^2\right\} \tag{19.68}$$

where x_L is the left image and x_R is the right image. This equation can be generalized to model the intensity of small windows around each location. In highly textured regions, it is usually possible to find the corresponding patch using cross correlation, but in regions of low texture, there will be considerable ambiguity about the correct value of y_s.

We can easily add a Gaussian prior on the edges of the MRF that encodes the assumption that neighboring disparities y_s, y_t should be similar, as follows:

$$\psi_{st}(y_s, y_t) \propto \exp\left(-\frac{1}{2\gamma^2}(y_s - y_t)^2\right) \tag{19.69}$$

The resulting model is a Gaussian CRF.

However, using Gaussian edge-potentials will oversmooth the estimate, since this prior fails to account for the occasional large changes in disparity that occur between neighboring pixels which are on different sides of an occlusion boundary. One gets much better results using a **truncated Gaussian potential** of the form

$$\psi_{st}(y_s, y_t) \propto \exp\left\{-\frac{1}{2\gamma^2}\min\left((y_s - y_t)^2, \delta_0^2\right)\right\} \tag{19.70}$$

where γ encodes the expected smoothness, and δ_0 encodes the maximum penalty that will be imposed if disparities are significantly different. This is called a **discontinuity preserving** potential; note that such penalties are not convex. The local evidence potential can be made robust in a similar way, in order to handle outliers due to specularities, occlusions, etc.

Figure 19.20 illustrates the difference between these two forms of prior. On the top left is an image from the standard Middlebury stereo benchmark dataset (Scharstein and Szeliski 2002). On the bottom left is the corresponding true disparity values. The remaining columns represent the estimated disparity after 0, 1 and an "infinite" number of rounds of loopy belief propagation (see Section 22.2), where by "infinite" we mean the results at convergence. The top row shows the results using a Gaussian edge potential, and the bottom row shows the results using the truncated potential. The latter is clearly better.

Unfortunately, performing inference with real-valued variables is computationally difficult, unless the model is jointly Gaussian. Consequently, it is common to discretize the variables. (For example, Figure 19.20(bottom) used 50 states.) The edge potentials still have the form given in Equation 19.69. The resulting model is called a **metric CRF**, since the potentials form a

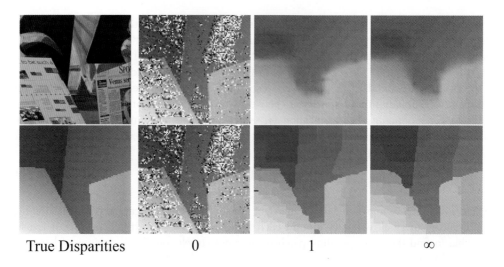

True Disparities   0   1   ∞

Figure 19.20   Illustration of belief propagation for stereo depth estimation. Left column: image and true disparities. Remaining columns: initial estimate, estimate after 1 iteration, and estimate at convergence. Top row: Gaussian edge potentials. Bottom row: robust edge potentials. Source: Figure 4 of (Sudderth and Freeman 2008). Used with kind permission of Erik Sudderth.

metric. [10] Inference in metric CRFs is more efficient than in CRFs where the discrete labels have no natural ordering, as we explain in Section 22.6.3.3. See Section 22.6.4 for a comparison of various approximate inference methods applied to low-level CRFs, and see (Blake et al. 2011; Prince 2012) for more details on probabilistic models for computer vision.

19.6.3   CRF training

We can modify the gradient based optimization of MRFs described in Section 19.5.1 to the CRF case in a straightforward way. In particular, the scaled log-likelihood becomes

$$\ell(\mathbf{w}) \triangleq \frac{1}{N} \sum_i \log p(\mathbf{y}_i | \mathbf{x}_i, \mathbf{w}) = \frac{1}{N} \sum_i \left[\sum_c \mathbf{w}_c^T \phi_c(\mathbf{y}_i, \mathbf{x}_i) - \log Z(\mathbf{w}, \mathbf{x}_i) \right] \tag{19.71}$$

and the gradient becomes

$$\frac{\partial \ell}{\partial \mathbf{w}_c} = \frac{1}{N} \sum_i \left[\phi_c(\mathbf{y}_i, \mathbf{x}_i) - \frac{\partial}{\partial \mathbf{w}_c} \log Z(\mathbf{w}, \mathbf{x}_i) \right] \tag{19.72}$$

$$= \frac{1}{N} \sum_i \left[\phi_c(\mathbf{y}_i, \mathbf{x}_i) - \mathbb{E}\left[\phi_c(\mathbf{y}, \mathbf{x}_i) \right] \right] \tag{19.73}$$

---

10. A function f is said to be a **metric** if it satisfies the following three properties: Reflexivity: $f(a, b) = 0$ iff $a = b$; Symmetry: $f(a, b) = f(b, a)$; and Triangle inequality: $f(a, b) + f(b, c) \geq f(a, c)$. If f satisfies only the first two properties, it is called a **semi-metric**.

Note that we now have to perform inference for every single training case inside each gradient step, which is $O(N)$ times slower than the MRF case. This is because the partition function depends on the inputs $\mathbf{x}_i$.

In most applications of CRFs (and some applications of MRFs), the size of the graph structure can vary. Hence we need to use parameter tying to ensure we can define a distribution of arbitrary size. In the pairwise case, we can write the model as follows:

$$p(\mathbf{y}|\mathbf{x}, \mathbf{w}) = \frac{1}{Z(\mathbf{w}, \mathbf{x})} \exp\left(\mathbf{w}^T \phi(\mathbf{y}, \mathbf{x})\right) \tag{19.74}$$

where $\mathbf{w} = [\mathbf{w}_n, \mathbf{w}_e]$ are the node and edge parameters, and

$$\phi(\mathbf{y}, \mathbf{x}) \triangleq [\sum_t \phi_t(y_t, \mathbf{x}), \sum_{s \sim t} \phi_{st}(y_s, y_t, \mathbf{x})] \tag{19.75}$$

are the summed node and edge features (these are the sufficient statistics). The gradient expression is easily modified to handle this case.

In practice, it is important to use a prior/ regularization to prevent overfitting. If we use a Gaussian prior, the new objective becomes

$$\ell'(\mathbf{w}) \triangleq \frac{1}{N} \sum_i \log p(\mathbf{y}_i|\mathbf{x}_i, \mathbf{w}) - \lambda ||\mathbf{w}||_2^2 \tag{19.76}$$

It is simple to modify the gradient expression.

Alternatively, we can use ℓ_1 regularization. For example, we could use ℓ_1 for the edge weights $\mathbf{w}_e$ to learn a sparse graph structure, and ℓ_2 for the node weights $\mathbf{w}_n$, as in (Schmidt et al. 2008). In other words, the objective becomes

$$\ell'(\mathbf{w}) \triangleq \frac{1}{N} \sum_i \log p(\mathbf{y}_i|\mathbf{x}_i, \mathbf{w}) - \lambda_1 ||\mathbf{w}_e||_1 - \lambda_2 ||\mathbf{w}_n||_2^2 \tag{19.77}$$

Unfortunately, the optimization algorithms are more complicated when we use ℓ_1 (see Section 13.4), although the problem is still convex.

To handle large datasets, we can use stochastic gradient descent (SGD), as described in Section 8.5.2. To handle cases where exact inference is intractable, we can use stochastic maximum likelihood (SML), as described in Section 19.5.5. This combines MCMC inference with SGD parameter learning.

Recently, an extension of SML has been proposed that estimates parameters to minimize a loss function, not just to maximize likelihood. This is known as **SampleRank** (Wick et al. 2011). The idea is as follows. Let $L(\mathbf{y}^i)$ be some loss function on the label vector sampled at iteration i, and let $\boldsymbol{\Delta} = \phi(\mathbf{y}^{i-1}) - \phi(\mathbf{y}^i)$. Then we increase $\boldsymbol{\theta}$ by $\alpha\boldsymbol{\Delta}$ (where α is a step size) if $p(\mathbf{y}^{i-1}) < p(\mathbf{y}^i)$ and $L(\mathbf{y}^{i-1}) < L(\mathbf{y}^i)$, i.e., if the previous label vector was less probable but had less loss; we decrease $\boldsymbol{\theta}$ by $\alpha\boldsymbol{\Delta}$ if $p(\mathbf{y}^{i-1}) > p(\mathbf{y}^i)$ and $L(\mathbf{y}^{i-1} > L(\mathbf{y}^i)$; otherwise we make no change. This forces the learning to focus on parameter values that have an affect on the loss function, and not just the likelihood.

It is possible (and useful) to define CRFs with hidden variables, for example to allow for an unknown alignment between the visible features and the hidden labels (see e.g., (Schnitzspan et al. 2010)). In this case, the objective function is no longer convex. Nevertheless, we can find a locally optimal ML or MAP parameter estimate using EM and/ or gradient methods, as discussed in Section 19.5.2.

19.7 Structural SVMs

We have seen that training a CRF requires inference, in order to compute the expected sufficient statistics needed to evaluate the gradient. For certain models, computing a joint MAP estimate of the states is provably simpler than computing marginals, as we discuss in Section 22.6. In this section, we discuss a way to train structured output classifiers that leverages the existence of fast MAP solvers. (To avoid confusion with MAP estimation of parameters, we will often refer to MAP estimation of states as **decoding**.) These methods are known as **structural SVMs** (support vector machines) or SSVMs for short (Tsochantaridis et al. 2005). (There is also a very similar class of methods known as **max margin Markov networks** or **M3nets** (Taskar et al. 2003); see Section 19.7.2 for a discussion of the differences.)

19.7.1 SSVMs: a probabilistic view

In this book, we have mostly concentrated on fitting models using MAP parameter estimation, i.e., by minimizing functions of the form

$$R_{MAP}(\mathbf{w}) = -\log p(\mathbf{w}) - \sum_{i=1}^{N} \log p(\mathbf{y}_i | \mathbf{x}_i, \mathbf{w}) \tag{19.78}$$

However, at test time, we pick the label so as to minimize the posterior expected loss (defined in Section 5.7):

$$\hat{\mathbf{y}}(\mathbf{x} | \mathbf{w}) = \operatorname*{argmin}_{\hat{\mathbf{y}}} \sum_{\mathbf{y}} \mathcal{L}(\mathbf{y}, \hat{\mathbf{y}}) p(\mathbf{y} | \mathbf{x}, \mathbf{w}) \tag{19.79}$$

where $\mathcal{L}(\mathbf{y}^*, \hat{\mathbf{y}})$ is the loss we incur when we estimate $\hat{\mathbf{y}}$ but the truth is $\mathbf{y}^*$. It therefore seems reasonable to take the loss function into account when performing parameter estimation.[11] So, following (Yuille and He 2012), let us instead minimize the posterior expected loss on the training set:

$$R_{EL}(\mathbf{w}) \triangleq -\log p(\mathbf{w}) + \sum_{i=1}^{N} \log \left[\sum_{\mathbf{y}} \mathcal{L}(\mathbf{y}_i, \mathbf{y}) p(\mathbf{y} | \mathbf{x}_i, \mathbf{w}) \right] \tag{19.80}$$

In the special case of 0-1 loss, $\mathcal{L}(\mathbf{y}_i, \mathbf{y}) = 1 - \delta_{\mathbf{y}, \mathbf{y}_i}$, this reduces to R_{MAP}.

We will assume that we can write our model in the following form:

$$p(\mathbf{y} | \mathbf{x}, \mathbf{w}) = \frac{\exp(\mathbf{w}^T \phi(\mathbf{x}, \mathbf{y}))}{Z(\mathbf{x}, \mathbf{w})} \tag{19.81}$$

$$p(\mathbf{w}) = \frac{\exp(-E(\mathbf{w}))}{Z} \tag{19.82}$$

---

11. Note that this violates the fundamental Bayesian distinction between inference and decision making. However, performing these tasks separately will only result in an optimal decision if we can compute the exact posterior. In most cases, this is intractable, so we need to perform **loss-calibrated inference** (Lacoste-Julien et al. 2011). In this section, we just perform loss-calibrated MAP parameter estimation, which is computationally simpler. (See also (Stoyanov et al. 2011).)

where $Z(\mathbf{x}, \mathbf{w}) = \sum_{\mathbf{y}} \exp(\mathbf{w}^T \phi(\mathbf{x}, \mathbf{y}))$. Also, let us define $L(\mathbf{y}_i, \mathbf{y}) = \log \mathcal{L}(\mathbf{y}_i, \mathbf{y})$. With this, we can rewrite our objective as follows:

$$R_{EL}(\mathbf{w}) = -\log p(\mathbf{w}) + \sum_i \log \left[\sum_{\mathbf{y}} \exp L(\mathbf{y}_i, \mathbf{y}) \frac{\exp(\mathbf{w}^T \phi(\mathbf{x}, \mathbf{y}))}{Z(\mathbf{x}_i, \mathbf{w})} \right] \quad (19.83)$$

$$= E(\mathbf{w}) +$$
$$\sum_i \left[-\log Z(\mathbf{x}_i, \mathbf{w}) + \log \sum_{\mathbf{y}} \exp \left(L(\mathbf{y}_i, \mathbf{y}) + \mathbf{w}^T \phi(\mathbf{x}_i, \mathbf{y}) \right) \right] \quad (19.84)$$

We will now consider various bounds in order to simplify this objective. First note that for any function $f(\mathbf{y})$ we have

$$\max_{\mathbf{y} \in \mathcal{Y}} f(\mathbf{y}) \le \log \sum_{\mathbf{y} \in \mathcal{Y}} \exp[f(\mathbf{y})] \le \log \left[|\mathcal{Y}| \exp \left(\max_{\mathbf{y}} f(\mathbf{y}) \right) \right] = \log |\mathcal{Y}| + \max_{\mathbf{y}} f(\mathbf{y}) \,(19.85)$$

For example, suppose $\mathcal{Y} = \{0, 1, 2\}$ and $f(y) = y$. Then we have

$$2 = \log[\exp(2)] \le \log[\exp(0) + \exp(1) + \exp(2)] \le \log[3 \times \exp(2)] = \log(3) + 2 \quad (19.86)$$

We can ignore the $\log |\mathcal{Y}|$ term, which is independent of $\mathbf{y}$, and treat $\max_{\mathbf{y} \in \mathcal{Y}} f(\mathbf{y})$ as both a lower and upper bound. Hence we see that

$$R_{EL}(\mathbf{w}) \sim E(\mathbf{w}) + \sum_{i=1}^{N} \left[\max_{\mathbf{y}} \left\{ L(\mathbf{y}_i, \mathbf{y}) + \mathbf{w}^T \phi(\mathbf{x}_i, \mathbf{y}) \right\} - \max_{\mathbf{y}} \mathbf{w}^T \phi(\mathbf{x}_i, \mathbf{y}) \right] \quad (19.87)$$

where $x \sim y$ means $c_1 + x \le y + c_2$ for some constants c_1, c_2. Unfortunately, this objective is not convex in $\mathbf{w}$. However, we can devise a convex upper bound by exploiting the following looser lower bound on the log-sum-exp function:

$$f(\mathbf{y}') \le \log \sum_{\mathbf{y}} \exp[f(\mathbf{y})] \quad (19.88)$$

for any $\mathbf{y}' \in \mathcal{Y}$. Applying this equation to our earlier example, for $f(y) = y$ and $y' = 1$, we get $1 = \log[\exp(1)] \le \log[\exp(0) + \exp(1) + \exp(2)]$. And applying this bound to R_{EL} we get

$$R_{EL}(\mathbf{w}) \le E(\mathbf{w}) + \sum_{i=1}^{N} \left[\max_{\mathbf{y}} \left\{ L(\mathbf{y}_i, \mathbf{y}) + \mathbf{w}^T \phi(\mathbf{x}_i, \mathbf{y}) \right\} - \mathbf{w}^T \phi(\mathbf{x}_i, \mathbf{y}_i) \right] \quad (19.89)$$

If we set $E(\mathbf{w}) = \frac{1}{2C} ||\mathbf{w}||_2^2$ (corresponding to a spherical Gaussian prior), we get

$$R_{SSVM}(\mathbf{w}) \triangleq \frac{1}{2} ||\mathbf{w}||^2 + C \sum_{i=1}^{N} \left[\max_{\mathbf{y}} \left\{ L(\mathbf{y}_i, \mathbf{y}) + \mathbf{w}^T \phi(\mathbf{x}_i, \mathbf{y}) \right\} - \mathbf{w}^T \phi(\mathbf{x}_i, \mathbf{y}_i) \right] (19.90)$$

This is the same objective as used in the SSVM approach of (Tsochantaridis et al. 2005).

In the special case that $\mathcal{Y} = \{-1, +1\}$ $L(y^*, y) = 1 - \delta_{y, y^*}$, and $\phi(\mathbf{x}, y) = \frac{1}{2} y \mathbf{x}$, this criterion reduces to the following (by considering the two cases that $y = y_i$ and $y \ne y_i$):

$$R_{SVM}(\mathbf{w}) \triangleq \frac{1}{2} ||\mathbf{w}||^2 + C \sum_{i=1}^{N} \left[\max\{0, 1 - y_i \mathbf{w}^T \mathbf{x}_i\} \right] \quad (19.91)$$

which is the standard binary SVM objective (see Equation 14.57).

So we see that the SSVM criterion can be seen as optimizing an upper bound on the Bayesian objective, a result first shown in (Yuille and He 2012). This bound will be tight (and hence the approximation will be a good one) when $||\mathbf{w}||$ is large, since in that case, $p(\mathbf{y}|\mathbf{x}, \mathbf{w})$ will concentrate its mass on $\text{argmax}_{\mathbf{y}}\, p(\mathbf{y}|\mathbf{x}, \mathbf{w})$. Unfortunately, a large $||\mathbf{w}||$ corresponds to a model that is likely to overfit, so it is unlikely that we will be working in this regime (because we will tune the strength of the regularizer to avoid this situation). An alternative justification for the SVM criterion is that it focusses effort on fitting parameters that affect the decision boundary. This is a better use of computational resources than fitting the full distribution, especially when the model is wrong.[12]

19.7.2 SSVMs: a non-probabilistic view

We now present SSVMs in a more traditional (non-probabilistic) way, following (Tsochantaridis et al. 2005). The resulting objective will be the same as the one above. However, this derivation will set the stage for the algorithms we discuss below.

Let $f(\mathbf{x}; \mathbf{w}) = \text{argmax}_{\mathbf{y} \in \mathcal{Y}}\, \mathbf{w}^T \phi(\mathbf{x}, \mathbf{y})$ be the prediction function. We can obtain zero loss on the training set using this predictor if

$$\forall i. \quad \max_{\mathbf{y} \in \mathcal{Y} \setminus \mathbf{y}_i} \mathbf{w}^T \phi(\mathbf{x}_i, \mathbf{y}) \leq \mathbf{w}^T \phi(\mathbf{x}_i, \mathbf{y}_i) \tag{19.92}$$

Each one of these nonlinear inequalities can be equivalently replaced by $|\mathcal{Y}| - 1$ linear inequalities, resulting in a total of $N|\mathcal{Y}| - N$ linear constraints of the following form:

$$\forall i. \forall \mathbf{y} \in \mathcal{Y} \setminus \mathbf{y}_i. \quad \mathbf{w}^T \phi(\mathbf{x}_i, \mathbf{y}_i) - \mathbf{w}^T \phi(\mathbf{x}_i, \mathbf{y}) \geq 0 \tag{19.93}$$

For brevity, we introduce the notation

$$\boldsymbol{\delta}_i(\mathbf{y}) \triangleq \phi(\mathbf{x}_i, \mathbf{y}_i) - \phi(\mathbf{x}_i, \mathbf{y}) \tag{19.94}$$

so we can rewrite these constraints as $\mathbf{w}^T \boldsymbol{\delta}_i(\mathbf{y}) \geq 0$.

If we can achieve zero loss, there will typically be multiple solution vectors $\mathbf{w}$. We pick the one that maximizes the margin, defined as

$$\gamma \triangleq \min_i \mathbf{w}^T \phi(\mathbf{x}_i, \mathbf{y}_i) - \max_{\mathbf{y}' \in \mathcal{Y} \setminus \mathbf{y}_i} \mathbf{w}^T \phi(\mathbf{x}_i, \mathbf{y}') \tag{19.95}$$

Since the margin can be made arbitrarily large by rescaling $\mathbf{w}$, we fix its norm to be 1, resulting in the optimization problem

$$\max_{\mathbf{w}: ||\mathbf{w}||=1} \gamma \quad \text{s.t.} \quad \forall i. \forall \mathbf{y} \in \mathcal{Y} \setminus \mathbf{y}_i. \quad \mathbf{w}^T \boldsymbol{\delta}_i(\mathbf{y}) \geq \gamma \tag{19.96}$$

Equivalently, we can write

$$\min_{\mathbf{w}} \frac{1}{2} ||\mathbf{w}||^2 \quad \text{s.t.} \quad \forall i. \forall \mathbf{y} \in \mathcal{Y} \setminus \mathbf{y}_i. \quad \mathbf{w}^T \boldsymbol{\delta}_i(\mathbf{y}) \geq 1 \tag{19.97}$$

---

12. See e.g., (Keerthi and Sundararajan 2007) for an experiemental comparison of SSVMs and CRFs for the task of sequence labeling; in this setting, their performance is quite similar, suggesting that the main benefit of SSVMs is when exact inference is intractable.

To allow for the case where zero loss cannot be achieved (equivalent to the data being inseparable in the case of binary classification), we relax the constraints by introducing slack terms ξ_i, one per data case. This yields

$$\min_{\mathbf{w},\boldsymbol{\xi}} \frac{1}{2}||\mathbf{w}||^2 + C \sum_{i=1}^{N} \xi_i \quad \text{s.t.} \quad \forall i. \forall \mathbf{y} \in \mathcal{Y} \setminus \mathbf{y}_i. \; \mathbf{w}^T \boldsymbol{\delta}_i(\mathbf{y}) \geq 1 - \xi_i, \xi_i \geq 0 \qquad (19.98)$$

In the case of structured outputs, we don't want to treat all constraint violations equally. For example, in a segmentation problem, getting one position wrong should be punished less than getting many positions wrong. One way to achieve this is to divide the slack variable by the size of the loss (this is called **slack re-scaling**). This yields

$$\min_{\mathbf{w},\boldsymbol{\xi}} \frac{1}{2}||\mathbf{w}||^2 + C \sum_{i=1}^{N} \xi_i \quad \text{s.t.} \quad \forall i. \forall \mathbf{y} \in \mathcal{Y} \setminus \mathbf{y}_i. \; \mathbf{w}^T \boldsymbol{\delta}_i(\mathbf{y}) \geq 1 - \frac{\xi_i}{L(\mathbf{y}_i, \mathbf{y})}, \xi_i \geq 0 \quad (19.99)$$

Alternatively, we can define the margin to be proportional to the loss (this is called **margin re-rescaling**). This yields

$$\min_{\mathbf{w},\boldsymbol{\xi}} \frac{1}{2}||\mathbf{w}||^2 + C \sum_{i=1}^{N} \xi_i \quad \text{s.t.} \quad \forall i. \forall \mathbf{y} \in \mathcal{Y} \setminus \mathbf{y}_i. \; \mathbf{w}^T \boldsymbol{\delta}_i(\mathbf{y}) \geq L(\mathbf{y}_i, \mathbf{y}) - \xi_i, \; \xi_i \geq 0 (19.100)$$

(In fact, we can write $\forall \mathbf{y} \in \mathcal{Y}$ instead of $\forall \mathbf{y} \in \mathcal{Y} \setminus \mathbf{y}_i$, since if $\mathbf{y} = \mathbf{y}_i$, then $\mathbf{w}^T \boldsymbol{\delta}_i(\mathbf{y}) = 0$ and $\xi_i = 0$. By using the simpler notation, which doesn't exclude $\mathbf{y}_i$, we add an extra but redundant constraint.) This latter approach is used in M3nets.

For future reference, note that we can solve for the ξ_i^* terms as follows:

$$\xi_i^*(\mathbf{w}) = \max\{0, \max_{\mathbf{y}}(L(\mathbf{y}_i, \mathbf{y}) - \mathbf{w}^T \boldsymbol{\delta}_i(\mathbf{y})))\} = \max_{\mathbf{y}}(L(\mathbf{y}_i, \mathbf{y}) - \mathbf{w}^T \boldsymbol{\delta}_i(\mathbf{y}))) \qquad (19.101)$$

Substituting in, and dropping the constraints, we get the following equivalent problem:

$$\min_{\mathbf{w}} \frac{1}{2}||\mathbf{w}||^2 + C \sum_{i} \max_{\mathbf{y}} \left\{ L(\mathbf{y}_i, \mathbf{y}) + \mathbf{w}^T \boldsymbol{\phi}(\mathbf{x}_i, \mathbf{y}) \right\} - \mathbf{w}^T \boldsymbol{\phi}(\mathbf{x}_i, \mathbf{y}_i) \qquad (19.102)$$

19.7.2.1 Empirical risk minimization

Let us pause and consider whether the above objective is reasonable. Recall that in the frequentist approach to machine learning (Section 6.5), the goal is to minimize the regularized empirical risk, defined by

$$\mathcal{R}(\mathbf{w}) + \frac{C}{N} \sum_{i=1}^{N} L(\mathbf{y}_i, f(\mathbf{x}_i, \mathbf{w})) \qquad (19.103)$$

where $\mathcal{R}(\mathbf{w})$ is the regularizer, and $f(\mathbf{x}_i, \mathbf{w}) = \operatorname{argmax}_{\mathbf{y}} \mathbf{w}^T \boldsymbol{\phi}(\mathbf{x}_i, \mathbf{y}) = \hat{\mathbf{y}}_i$ is the prediction. Since this objective is hard to optimize, because the loss is not differentiable, we will construct a convex upper bound instead.

We can show that

$$\mathcal{R}(\mathbf{w}) + \frac{C}{N} \sum_{i} \max_{\mathbf{y}}(L(\mathbf{y}_i, \mathbf{y}) - \mathbf{w}^T \boldsymbol{\delta}_i(\mathbf{y}))) \qquad (19.104)$$

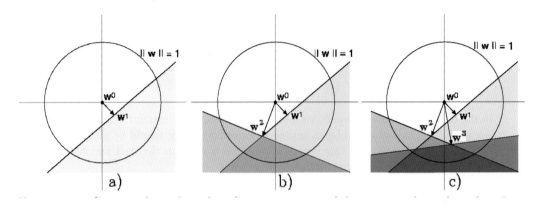

Figure 19.21 Illustration of the cutting plane algorithm in 2d. We start with the estimate $\mathbf{w} = \mathbf{w}_0 = \mathbf{0}$. (a) We add the first constraint; the shaded region is the new feasible set. The new minimum norm solution is $\mathbf{w}_1$. (b) We add another constraint; the dark shaded region is the new feasible set. (c) We add a third constraint. Source: Figure 5.3 of (Altun et al. 2007) . Used with kind permission of Yasemin Altun.

is such a convex upper bound. To see this, note that

$$L(\mathbf{y}_i, f(\mathbf{x}_i, \mathbf{w})) \quad \leq \quad L(\mathbf{y}_i, f(\mathbf{x}_i, \mathbf{w})) - \mathbf{w}^T \phi(\mathbf{x}_i, \mathbf{y}_i) + \mathbf{w}^T \phi(\mathbf{x}_i, \hat{\mathbf{y}}_i) \tag{19.105}$$

$$\leq \quad \max_{\mathbf{y}} L(\mathbf{y}_i, \mathbf{y}) - \mathbf{w}^T \phi(\mathbf{x}_i, \mathbf{y}_i) + \mathbf{w}^T \phi(\mathbf{x}_i, \mathbf{y}) \tag{19.106}$$

Using this bound and $\mathcal{R}(\mathbf{w}) = \frac{1}{2N} ||\mathbf{w}||^2$ yields Equation 19.102.

19.7.2.2   Computational issues

Although the above objectives are simple quadratic programs (QP), they have $O(N|\mathcal{Y}|)$ constraints. This is intractable, since $\mathcal{Y}$ is usually exponentially large. In the case of the margin rescaling formulation, it is possible to reduce the exponential number of constraints to a polynomial number, provided the loss function and the feature vector decompose according to a graphical model. This is the approach used in M3nets (Taskar et al. 2003).

An alternative approach is to work directly with the exponentially sized QP. This allows for the use of more general loss functions. There are several possible methods to make this feasible. One is to use cutting plane methods. Another is to use stochastic subgradient methods. We discuss both of these below.

19.7.3   Cutting plane methods for fitting SSVMs

In this section, we discuss an efficient algorithm for fitting SSVMs due to (Joachims et al. 2009). This method can handle general loss functions, and is implemented in the popular **SVMstruct** package[13]. The method is based on the **cutting plane** method from convex optimization (Kelley 1960).

---

13. `http://svmlight.joachims.org/svm_struct.html`

The basic idea is as follows. We start with an initial guess $\mathbf{w}$ and no constraints. At each iteration, we then do the following: for each example i, we find the "most violated" constraint involving $\mathbf{x}_i$ and $\hat{\mathbf{y}}_i$. If the loss-augmented margin violation exceeds the current value of ξ_i by more than ϵ, we add $\hat{\mathbf{y}}_i$ to the working set of constraints for this training case, $\mathcal{W}_i$, and then solve the resulting new QP to find the new $\mathbf{w}, \boldsymbol{\xi}$. See Figure 19.21 for a sketch, and Algorithm 19.3 for the pseudo code. (Since at each step we only add one new constraint, we can warm-start the QP solver.) We can can easily modify the algorithm to optimize the slack rescaling version by replacing the expression $L(\mathbf{y}_i, \mathbf{y}) - \mathbf{w}^T \boldsymbol{\delta}_i(\mathbf{y}_i)$ with $L(\mathbf{y}_i, \mathbf{y})(1 - \mathbf{w}^T \boldsymbol{\delta}_i(\mathbf{y}_i))$.

Algorithm 19.3: Cutting plane algorithm for SSVMs (margin rescaling, N-slack version)

1 Input $\mathcal{D} = \{(\mathbf{x}_1, \mathbf{y}_1), \ldots, (\mathbf{x}_N, \mathbf{y}_n)\}$, C, ϵ ;
2 $\mathcal{W}_i = \emptyset$, $\xi_i = 0$ for $i = 1 : N$;
3 **repeat**
4 $\quad$ **for** $i = 1 : N$ **do**
5 $\quad\quad$ $\hat{\mathbf{y}}_i = \mathrm{argmax}_{\hat{\mathbf{y}}_i \in \mathcal{Y}} L(\mathbf{y}_i, \hat{\mathbf{y}}_i) - \mathbf{w}^T \boldsymbol{\delta}_i(\hat{\mathbf{y}}_i)$;
6 $\quad\quad$ **if** $L(\mathbf{y}_i, \mathbf{y}) - \mathbf{w}^T \boldsymbol{\delta}_i(\hat{\mathbf{y}}_i) > \xi_i + \epsilon$ **then**
7 $\quad\quad\quad$ $\mathcal{W}_i = \mathcal{W}_i \cup \{\hat{\mathbf{y}}_i\}$;
8 $\quad\quad\quad$ $(\mathbf{w}, \boldsymbol{\xi}) = \mathrm{argmin}_{\mathbf{w}, \boldsymbol{\xi} \geq 0} \frac{1}{2}||\mathbf{w}||_2^2 + C \sum_{i=1}^N \xi_i$;
9 $\quad\quad\quad$ s.t. $\quad \forall i = 1 : N, \forall \mathbf{y} \in \mathcal{W}_i : \mathbf{w}^T \boldsymbol{\delta}_i(\mathbf{y}) \geq L(\mathbf{y}_i, \mathbf{y}) - \xi_i$;
10 **until** *no $\mathcal{W}_i$ has changed*;
11 Return $(\mathbf{w}, \boldsymbol{\xi})$

The key to the efficiency of this method is that only polynomially many constraints need to be added, and as soon as they are, the exponential number of other constraints are guaranteed to also be satisfied to within a tolerance of ϵ (see (Tsochantaridis et al. 2005) for the proof). The overall running time is $O(1/\epsilon^2)$ (Nowozin and Lampert 2011, p158).

19.7.3.1 Loss-augmented decoding

The other key to efficiency is the ability to find the most violated constraint in line 5 of the algorithm, i.e., to compute

$$\mathrm{argmax}_{\mathbf{y} \in \mathcal{Y}} L(\mathbf{y}_i, \mathbf{y}) - \mathbf{w}^T \boldsymbol{\delta}_i(\mathbf{y}) = \mathrm{argmax}_{\mathbf{y} \in \mathcal{Y}} L(\mathbf{y}_i, \mathbf{y}) + \mathbf{w}^T \boldsymbol{\phi}(\mathbf{x}_i, \mathbf{y}) \qquad (19.107)$$

We call this process **loss-augmented decoding**. (In (Joachims et al. 2009), this procedure is called the **separation oracle**.) If the loss function has an additive decomposition of the same form as the features, then we can fold the loss into the weight vector, i.e., we can find a new set of parameters $\mathbf{w}'$ such that $(\mathbf{w}')^T \boldsymbol{\delta}_i(\mathbf{y}) = -L(\mathbf{y}_i, \mathbf{y}) + \mathbf{w}^T \boldsymbol{\delta}_i(\mathbf{y})$. We can then use a standard decoding algorithm, such as Viterbi, on the model $p(\mathbf{y}|\mathbf{x}, \mathbf{w}')$.

In the special case of 0-1 loss, the optimum will either be the best solution, $\mathrm{argmax}_{\mathbf{y}} \mathbf{w}^T \boldsymbol{\phi}(\mathbf{x}_i, \mathbf{y})$, with a value of $0 - \mathbf{w}^T \boldsymbol{\delta}_i(\hat{\mathbf{y}})$, or it will be the second best solution, i.e.,

$$\tilde{\mathbf{y}} = \mathrm{argmax}_{\mathbf{y} \neq \hat{\mathbf{y}}} \mathbf{w}^T \boldsymbol{\phi}(\mathbf{x}_i, \mathbf{y}) \qquad (19.108)$$

which achieves an overall value of $1 - \mathbf{w}^T \delta_i(\tilde{\mathbf{y}})$. For chain structured CRFs, we can use the Viterbi algorithm to do decoding; the second best path will differ from the best path in a single position, which can be obtained by changing the variable whose max marginal is closest to its decision boundary to its second best value. We can generalize this (with a bit more work) to find the N-best list (Schwarz and Chow 1990; Nilsson and Goldberger 2001).

For Hamming loss, $L(\mathbf{y}^*, \mathbf{y}) = \sum_t \mathbb{I}(y_t^* \neq y_t)$, and for the F1 score (defined in Section 5.7.2.3), we can devise a dynamic programming algorithm to compute Equation 19.107. See (Altun et al. 2007) for details. Other models and loss function combinations will require different methods.

19.7.3.2   A linear time algorithm

Although the above algorithm takes polynomial time, we can do better, and devise an algorithm that runs in *linear* time, assuming we use a linear kernel (i.e., we work with the original features $\phi(\mathbf{x}, \mathbf{y})$ and do not apply the kernel trick). The basic idea, as explained in (Joachims et al. 2009), is to have a single slack variable, ξ, instead of N, but to use $|\mathcal{Y}|^N$ constraints, instead of just $N|\mathcal{Y}|$. Specifically, we optimize the following (assuming the margin rescaling formulation):

$$\min_{\mathbf{w}, \xi \geq 0} \quad \frac{1}{2}||\mathbf{w}||_2^2 + C\xi$$

$$\text{s.t.} \quad \forall (\overline{\mathbf{y}}_1, \ldots, \overline{\mathbf{y}}_N) \in \mathcal{Y}^N : \frac{1}{N}\mathbf{w}^T \sum_{i=1}^{N} \delta_i(\overline{\mathbf{y}}_i) \geq \frac{1}{N}\sum_{i=1}^{N} L(\mathbf{y}_i, \overline{\mathbf{y}}_i) - \xi \tag{19.109}$$

Compare this to the original version, which was

$$\min_{\mathbf{w}, \xi \geq 0} \frac{1}{2}||\mathbf{w}||_2^2 + \frac{C}{N}\sum_{i=1}^{N} \xi_i \quad \text{s.t.} \quad \forall i = 1 : N, \forall \mathbf{y} \in \mathcal{Y} : \mathbf{w}^T \delta_i(\overline{\mathbf{y}}_i) \geq L(\mathbf{y}_i, \overline{\mathbf{y}}_i) - \xi_i \tag{19.110}$$

One can show that any solution $\mathbf{w}^*$ of Equation 19.109 is also a solution of Equation 19.110 and vice versa, with $\xi^* = \frac{1}{N}\sum_{i=1}^{N} \xi_i^*$.

Algorithm 19.4: Cutting plane algorithm for SSVMs (margin rescaling, 1-slack version)

---

1 Input $\mathcal{D} = \{(\mathbf{x}_1, \mathbf{y}_1), \ldots, (\mathbf{x}_N, \mathbf{y}_n)\}$, C, ϵ ;
2 $\mathcal{W} = \emptyset$;
3 **repeat**
4 $\quad (\mathbf{w}, \xi) = \text{argmin}_{\mathbf{w}, \xi \geq 0} \frac{1}{2}||\mathbf{w}||_2^2 + C\xi$;
5 $\qquad \text{s.t.} \quad \forall (\overline{\mathbf{y}}_1, \ldots, \overline{\mathbf{y}}_N) \in \mathcal{W} : \frac{1}{N}\mathbf{w}^T \sum_{i=1}^{N} \delta_i(\overline{\mathbf{y}}_i) \geq \frac{1}{N}\sum_{i=1}^{N} L(\mathbf{y}_i, \overline{\mathbf{y}}_i) - \xi$;
6 $\quad$ **for** $i = 1 : N$ **do**
7 $\qquad \lfloor \; \hat{\mathbf{y}}_i = \text{argmax}_{\hat{\mathbf{y}}_i \in \mathcal{Y}} L(\mathbf{y}_i, \hat{\mathbf{y}}_i) + \mathbf{w}^T \phi(\mathbf{x}_i, \hat{\mathbf{y}}_i)$
8 $\quad \mathcal{W} = \mathcal{W} \cup \{(\hat{\mathbf{y}}_1, \ldots, \hat{\mathbf{y}}_N)\}$;
9 **until** $\frac{1}{N}\sum_{i=1}^{N} L(\mathbf{y}_i, \hat{\mathbf{y}}_i) - \frac{1}{N}\mathbf{w}^T \sum_{i=1}^{N} \delta_i(\hat{\mathbf{y}}_i) \leq \xi + \epsilon$;
10 Return $(\mathbf{w}, \xi)$

We can optimize Equation 19.109 using the cutting plane algorithm in Algorithm 19.4. (This is what is implemented in SVMstruct.) The inner QP in line 4 can be solved in $O(N)$ time using the method of (Joachims 2006). In line 7 we make N calls to the loss-augmented decoder. Finally, it can be shown that the number of iterations is independent of N. Thus the overall running time is linear.

19.7.4 Online algorithms for fitting SSVMs

Although the cutting plane algorithm can be made to run in time linear in the number of data points, that can still be slow if we have a large dataset. In such cases, it is preferable to use online learning. We briefly mention a few possible algorithms below.

19.7.4.1 The structured perceptron algorithm

A very simple algorithm for fitting SSVMs is the **structured perceptron algorithm** (Collins 2002). This method is an extension of the regular perceptron algorithm of Section 8.5.4. At each step, we compute $\hat{\mathbf{y}} = \operatorname{argmax} p(\mathbf{y}|\mathbf{x})$ (e.g., using the Viterbi algorithm) for the current training sample $\mathbf{x}$. If $\hat{\mathbf{y}} = \mathbf{y}$, we do nothing, otherwise we update the weight vector using

$$\mathbf{w}_{k+1} = \mathbf{w}_k + \phi(\mathbf{y}, \mathbf{x}) - \phi(\hat{\mathbf{y}}, \mathbf{x}) \tag{19.111}$$

To get good performance, it is necessary to average the parameters over the last few updates (see Section 8.5.2 for details), rather than using the most recent value.

19.7.4.2 Stochastic subgradient descent

The disadvantage of the structured perceptron algorithm is that it implicitly assumes 0-1 loss, and it does not enforce any kind of margin. An alternative approach is to perform stochastic subgradient descent. A specific instance of this is the **Pegasos** algorithm (Shalev-Shwartz et al. 2007), which stands for "primal estimated sub-gradient solver for SVM". Pegasos was designed for binary SVMs, but can be extended to SSVMS.

Let us start by considering the objective function:

$$f(\mathbf{w}) = \sum_{i=1}^{N} \max_{\hat{\mathbf{y}}_i} \left[L(\mathbf{y}_i, \hat{\mathbf{y}}_i) + \mathbf{w}^T \phi(\mathbf{x}_i, \hat{\mathbf{y}}_i) \right] - \mathbf{w}^T \phi(\mathbf{x}_i, \mathbf{y}_i) + \lambda ||\mathbf{w}||^2 \tag{19.112}$$

Letting $\hat{\mathbf{y}}_i$ be the argmax of this max. Then the subgradient of this objective function is

$$g(\mathbf{w}) \;=\; \sum_{i=1}^{N} \phi(\mathbf{x}_i, \hat{\mathbf{y}}_i) - \phi(\mathbf{x}_i, \mathbf{y}_i) + 2\lambda \mathbf{w} \tag{19.113}$$

In stochastic subgradient descent, we approximate this gradient with a single term, i, and then perform an update:

$$\mathbf{w}_{k+1} = \mathbf{w}_k - \eta_k g_i(\mathbf{w}_k) = \mathbf{w}_k - \eta_k [\phi(\mathbf{x}_i, \hat{\mathbf{y}}_i) - \phi(\mathbf{x}_i, \mathbf{y}_i) + (2/N)\lambda \mathbf{w}] \tag{19.114}$$

where η_k is the step size parameter, which should satisfy the Robbins-Monro conditions (Section 8.5.2.1). (Notice that the perceptron algorithm is just a special case where $\lambda = 0$ and $\eta_k = 1$.) To ensure that $\mathbf{w}$ has unit norm, we can project it onto the ℓ_2 ball after each update.

19.7.5   Latent structural SVMs

In many applications of interest, we have latent or hidden variables $\mathbf{h}$. For example, in object detections problems, we may know that an object is present in an image patch, but not know where the parts are located (see e.g., (Felzenszwalb et al. 2010)). Or in machine translation, we may know the source text $\mathbf{x}$ (say English) and the target text $\mathbf{y}$ (say French), but we typically do not know the alignment between the words.

We will extend our model as follows, to get a **latent CRF**:

$$p(\mathbf{y}, \mathbf{h}|\mathbf{x}, \mathbf{w}) \;\; = \;\; \frac{\exp(\mathbf{w}^T \phi(\mathbf{x}, \mathbf{y}, \mathbf{h}))}{Z(\mathbf{x}, \mathbf{w})} \tag{19.115}$$

$$Z(\mathbf{x}, \mathbf{w}) \;\; = \;\; \sum_{\mathbf{y}, \mathbf{h}} \exp(\mathbf{w}^T \phi(\mathbf{x}, \mathbf{y}, \mathbf{h})) \tag{19.116}$$

In addition, we introduce the loss function $\mathcal{L}(\mathbf{y}^*, \mathbf{y}, \mathbf{h})$; this measures the loss when the "action" that we take is to predict $\mathbf{y}$ using latent variables $\mathbf{h}$. (As before, we define $L(\mathbf{y}^*, \mathbf{y}, \mathbf{h}) = \log \mathcal{L}(\mathbf{y}^*, \mathbf{y}, \mathbf{h})$.) We could just use $\mathcal{L}(\mathbf{y}^*, \mathbf{y})$ as before, since $\mathbf{h}$ is usually a nuisance variable and not of direct interest. However, $\mathbf{h}$ can sometimes play a useful role in defining a loss function.[14]

Given the loss function, we define our objective as

$$R_{EL}(\mathbf{w}) \;\; = \;\; -\log p(\mathbf{w}) + \sum_i \log \left[\sum_{\mathbf{y}, \mathbf{h}} \exp L(\mathbf{y}_i, \mathbf{y}, \mathbf{h}) \frac{\exp(\mathbf{w}^T \phi(\mathbf{x}, \mathbf{y}, \mathbf{h}))}{Z(\mathbf{x}, \mathbf{w})} \right] \tag{19.117}$$

Using the same loose lower bound as before, we get

$$
\begin{aligned}
R_{EL}(\mathbf{w}) \;\; \leq \;\; & E(\mathbf{w}) + \sum_{i=1}^{N} \max_{\mathbf{y}, \mathbf{h}} \left\{ L(\mathbf{y}_i, \mathbf{y}, \mathbf{h}) + \mathbf{w}^T \phi(\mathbf{x}_i, \mathbf{y}, \mathbf{h}) \right\} \\
& - \sum_{i=1}^{N} \max_{\mathbf{h}} \mathbf{w}^T \phi(\mathbf{x}_i, \mathbf{y}_i, \mathbf{h})
\end{aligned}
\tag{19.118}
$$

If we set $E(\mathbf{w}) = -\frac{1}{2C}||\mathbf{w}||_2^2$, we get the same objective as is optimized in **latent SVMs** (Yu and Joachims 2009).

Unfortunately, this objective is no longer convex. However, it is a difference of convex functions, and hence can be solved efficiently using the **CCCP** or **concave-convex procedure** (Yuille and Rangarajan 2003). This is a method for minimizing functions of the form $f(\mathbf{w}) - g(\mathbf{w})$, where f and g are convex. The method alternates between finding a linear upper bound

---

14. For example, consider the problem of learning to classify a set of documents as relevant or not to a query. That is, given n documents $\mathbf{x}_1, \ldots, \mathbf{x}_n$ for a single query q, we want to produce a labeling $y_j \in \{-1, +1\}$, representing whether document j is relevant to q or not. Suppose our goal is to maximize the precision at k, which is a metric widely used in ranking (see Section 9.7.4). We will introduce a latent variable for each document h_j representing its degree of relevance. This corresponds to a latent total ordering, that has to be consistent with the observed partial ordering $\mathbf{y}$. Given this, we can define the following loss function: $L(\mathbf{y}, \hat{\mathbf{y}}, \hat{\mathbf{h}}) = \min\{1, \frac{n(\mathbf{y})}{k}\} - \frac{1}{k} \sum_{j=1}^{k} \mathbb{I}(y_{h_j} = 1)$, where $n(\mathbf{y})$ is the total number of relevant documents. This loss is essentially just 1 minus the precision@k, except we replace 1 with $n(\mathbf{y})/k$ so that the loss will have a minimum of zero. See (Yu and Joachims 2009) for details.

u on $-g$, and then minimizing the convex function $f(\mathbf{w}) + u(\mathbf{w})$; see Algorithm 19.5 for the pseudocode. CCCP is guaranteed to decrease the objective at every iteration, and to converge to a local minimum or a saddle point.

Algorithm 19.5: Concave-Convex Procedure (CCCP)

1 Set $t = 0$ and initialize $\mathbf{w}_0$;
2 **repeat**
3   Find hyperplane $\mathbf{v}_t$ such that $-g(\mathbf{w}) \leq -g(\mathbf{w}_t) + (\mathbf{w} - \mathbf{w}_t)^T \mathbf{v}_t$ for all $\mathbf{w}$;
4   Solve $\mathbf{w}_{t+1} = \operatorname{argmin}_{\mathbf{w}} f(\mathbf{w}) + \mathbf{w}^T \mathbf{v}_t$;
5   Set $t = t + 1$
6 **until** *converged*;

When applied to latent SSVMs, CCCP is very similar to (hard) EM. In the "E step", we compute the linear upper bound by setting $\mathbf{v}_t = -C \sum_{i=1}^{N} \phi(\mathbf{x}_i, \mathbf{y}_i, \mathbf{h}_i^*)$, where

$$\mathbf{h}_i^* = \operatorname*{argmax}_{\mathbf{h}} \mathbf{w}_t^T \phi(\mathbf{x}_i, \mathbf{y}_i, \mathbf{h}) \tag{19.119}$$

In the "M step", we estimate $\mathbf{w}$ using techniques for solving fully visible SSVMs. Specifically, we minimize

$$\frac{1}{2}||\mathbf{w}||_2 + C \sum_{i=1}^{N} \max_{\mathbf{y},\mathbf{h}} \left\{ \tilde{L}(\mathbf{y}_i, \mathbf{y}, \mathbf{h}) + \mathbf{w}^T \phi(\mathbf{x}_i, \mathbf{y}, \mathbf{h}) \right\} - C \sum_{i=1}^{N} \mathbf{w}^T \phi(\mathbf{x}_i, \mathbf{y}_i, \mathbf{h}_i^*) \tag{19.120}$$

Exercises

Exercise 19.1 Derivative of the log partition function

Derive Equation 19.40.

Exercise 19.2 CI properties of Gaussian graphical models

(Source: Jordan.)

In this question, we study the relationship between sparse matrices and sparse graphs for Gaussian graphical models. Consider a multivariate Gaussian $\mathcal{N}(x|\mu, \Sigma)$ in 3 dimensions. Suppose $\mu = (0, 0, 0)^T$ throughout.

Recall that for jointly Gaussian random variables, we know that X_i and X_j are independent iff they are uncorrelated, ie. $\Sigma_{ij} = 0$. (This is not true in general, or even if X_i and X_j are Gaussian but not jointly Gaussian.) Also, X_i is conditionally independent of X_j given all the other variables iff $\Sigma_{ij}^{-1} = 0$.

a. Suppose

$$\Sigma = \begin{pmatrix} 0.75 & 0.5 & 0.25 \\ 0.5 & 1.0 & 0.5 \\ 0.25 & 0.5 & 0.75 \end{pmatrix}$$

Are there any marginal independencies amongst X_1, X_2 and X_3? What about conditional independencies? Hint: compute Σ^{-1} and expand out $x^T \Sigma^{-1} x$: which pairwise terms $x_i x_j$ are missing? Draw an undirected graphical model that captures as many of these independence statements (marginal and conditional) as possible, but does not make any false independence assertions.

b. Suppose

$$\Sigma = \begin{pmatrix} 2 & 1 & 0 \\ 1 & 2 & 1 \\ 0 & 1 & 2 \end{pmatrix}$$

Are there any marginal independencies amongst X_1, X_2 and X_3? Are there any conditional independencies amongst X_1, X_2 and X_3? Draw an undirected graphical model that captures as many of these independence statements (marginal and conditional) as possible, but does not make any false independence assertions.

c. Now suppose the distribution on X can be represented by the following DAG:

$$X_1 \to X_2 \to X_3$$

Let the CPDs be as follows:

$$P(X_1) = \mathcal{N}(X_1; 0, 1), \ P(X_2|x_1) = \mathcal{N}(X_2; x_1, 1), \ P(X_3|x_2) = \mathcal{N}(X_3; x_2, 1) \qquad (19.121)$$

Multiply these 3 CPDs together and complete the square to find the corresponding joint distribution $\mathcal{N}(X_{1:3}|\mu, \Sigma)$. (You may find it easier to solve for Σ^{-1} rather than Σ.)

d. For the DAG model in the previous question: Are there any marginal independencies amongst X_1, X_2 and X_3? What about conditional independencies? Draw an undirected graphical model that captures as many of these independence statements as possible, but does not make any false independence assertions (either marginal or conditional).

Exercise 19.3 Independencies in Gaussian graphical models

(Source: MacKay.)

a. Consider the DAG $X1 \leftarrow X2 \to X3$. Assume that all the CPDs are linear-Gaussian. Which of the following matrices *could* be the covariance matrix?

$$A = \begin{pmatrix} 9 & 3 & 1 \\ 3 & 9 & 3 \\ 1 & 3 & 9 \end{pmatrix}, B = \begin{pmatrix} 8 & -3 & 1 \\ -3 & 9 & -3 \\ 1 & -3 & 8 \end{pmatrix},$$

$$C = \begin{pmatrix} 9 & 3 & 0 \\ 3 & 9 & 3 \\ 0 & 3 & 9 \end{pmatrix}, D = \begin{pmatrix} 9 & -3 & 0 \\ -3 & 10 & -3 \\ 0 & -3 & 9 \end{pmatrix} \qquad (19.122)$$

b. Which of the above matrices could be inverse covariance matrix?

c. Consider the DAG $X1 \to X2 \leftarrow X3$. Assume that all the CPDs are linear-Gaussian. Which of the above matrices could be the covariance matrix?

d. Which of the above matrices could be the inverse covariance matrix?

e. Let three variables x_1, x_2, x_4 have covariance matrix $\Sigma_{(1:3)}$ and precision matrix $\Omega_{(1:3)} = \Sigma_{(1:3)}^{-1}$ as follows

$$\Sigma_{(1:3)} = \begin{pmatrix} 1 & 0.5 & 0 \\ 0.5 & 1 & 0.5 \\ 0 & 0.5 & 1 \end{pmatrix}, \Omega_{(1:3)} = \begin{pmatrix} 1.5 & -1 & 0.5 \\ -1 & 2 & -1 \\ 0.5 & -1 & 1.5 \end{pmatrix} \qquad (19.123)$$

Now focus on x_1 and x_2. Which of the following statements about their covariance matrix $\Sigma_{(1:2)}$ and precision matrix $\Omega_{(1:2)}$ are true?

$$A : \Sigma_{(1:2)} = \begin{pmatrix} 1 & 0.5 \\ 0.5 & 1 \end{pmatrix}, \ B : \Omega_{(1:2)} = \begin{pmatrix} 1.5 & -1 \\ -1 & 2 \end{pmatrix} \qquad (19.124)$$

Exercise 19.4 Cost of training MRFs and CRFs

(Source: Koller.) Consider the process of gradient-ascent training for a log-linear model with k features, given a data set with N training instances. Assume for simplicity that the cost of computing a single feature over a single instance in our data set is constant, as is the cost of computing the expected value of each feature once we compute a marginal over the variables in its scope. Assume that it takes c time to compute all the marginals for each data case. Also, assume that we need r iterations for the gradient process to converge.

- Using this notation, what is the time required to train an MRF in big-O notation?

- Using this notation, what is the time required to train a CRF in big-O notation?

Exercise 19.5 Full conditional in an Ising model

Consider an Ising model

$$p(x_1, \ldots, x_n | \boldsymbol{\theta}) = \frac{1}{Z(\boldsymbol{\theta})} \prod_{<ij>} \exp(J_{ij} x_i x_j) \prod_{i=1}^{n} \exp(h_i x_i) \tag{19.125}$$

where $< ij >$ denotes all unique pairs (i.e., all edges), $J_{ij} \in \mathbb{R}$ is the coupling strength (weight) on edge $i - j$, $h_i \in \mathbb{R}$ is the local evidence (bias term), and $\boldsymbol{\theta} = (\mathbf{J}, \mathbf{h})$ are all the parameters.

If $x_i \in \{0, 1\}$, derive an expression for the full conditional

$$p(x_i = 1 | \mathbf{x}_{-i}, \boldsymbol{\theta}) = p(x_i = 1 | \mathbf{x}_{nb_i}, \boldsymbol{\theta}) \tag{19.126}$$

where $\mathbf{x}_{-i}$ are all nodes except i, and nb_i are the neighbors of i in the graph. Hint: you answer should use the sigmoid/ logistic function $\sigma(z) = 1/(1 + e^{-z})$. Now suppose $x_i \in \{-1, +1\}$. Derive a related expression for $p(x_i | \mathbf{x}_{-i}, \boldsymbol{\theta})$ in this case. (This result can be used when applying Gibbs sampling to the model.)

20 Exact inference for graphical models

20.1 Introduction

In Section 17.4.3, we discussed the forwards-backwards algorithm, which can exactly compute the posterior marginals $p(x_t|\mathbf{v}, \boldsymbol{\theta})$ in any chain-structured graphical model, where $\mathbf{x}$ are the hidden variables (assumed discrete) and $\mathbf{v}$ are the visible variables. This algorithm can be modified to compute the posterior mode and posterior samples. A similar algorithm for linear-Gaussian chains, known as the Kalman smoother, was discussed in Section 18.3.2. Our goal in this chapter is to generalize these exact inference algorithms to arbitrary graphs. The resulting methods apply to both directed and undirected graphical models. We will describe a variety of algorithms, but we omit their derivations for brevity. See e.g., (Cowell et al. 1999; Darwiche 2009; Koller and Friedman 2009) for a detailed exposition of exact inference techniques for graphical models.

20.2 Belief propagation for trees

In this section, we generalize the forwards-backwards algorithm from chains to trees. The resulting algorithm is known as **belief propagation** (**BP**) (Pearl 1988), or the **sum-product algorithm**.

20.2.1 Serial protocol

We initially assume (for notational simplicity) that the model is a pairwise MRF (or CRF), i.e.,

$$p(\mathbf{x}|\mathbf{v}) = \frac{1}{Z(\mathbf{v})} \prod_{s \in \mathcal{V}} \psi_s(x_s) \prod_{(s,t) \in \mathcal{E}} \psi_{s,t}(x_s, x_t) \tag{20.1}$$

where ψ_s is the local evidence for node s, and ψ_{st} is the potential for edge $s - t$. We will consider the case of models with higher order cliques (such as directed trees) later on.

One way to implement BP for undirected trees is as follows. Pick an arbitrary node and call it the root, r. Now orient all edges away from r (intuitively, we can imagine "picking up the graph" at node r and letting all the edges "dangle" down). This gives us a well-defined notion of parent and child. Now we send messages up from the leaves to the root (the **collect evidence** phase) and then back down from the root (the **distribute evidence** phase), in a manner analogous to forwards-backwards on chains.

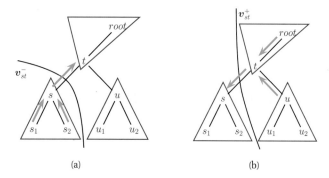

Figure 20.1   Message passing on a tree. (a) Collect-to-root phase. (b) Distribute-from-root phase.

To explain the process in more detail, consider the example in Figure 20.1. Suppose we want to compute the belief state at node t. We will initially condition the belief only on evidence that is at or below t in the graph, i.e., we want to compute $\text{bel}_t^-(x_t) \triangleq p(x_t|\mathbf{v}_t^-)$, where $\mathbf{v}_t^-$ is all the evidence at or below node t in the tree. We will call this a "bottom-up belief state". Suppose, by induction, that we have computed "messages" from t's two children, summarizing what they think t should know about the evidence in their subtrees, i.e., we have computed $m_{s \to t}^-(x_t) = p(x_t|\mathbf{v}_{st}^-)$, where $\mathbf{v}_{st}^-$ is all the evidence on the downstream side of the $s - t$ edge (see Figure 20.1(a)), and similarly we have computed $m_{u \to t}(x_t)$. Then we can compute the bottom-up belief state at t as follows:

$$\text{bel}_t^-(x_t) \triangleq p(x_t|\mathbf{v}_t^-) = \frac{1}{Z_t} \psi_t(x_t) \prod_{c \in \text{ch}(t)} m_{c \to t}^-(x_t) \tag{20.2}$$

where $\psi_t(x_t) \propto p(x_t|\mathbf{v}_t)$ is the local evidence for node t, and Z_t is the local normalization constant. In words, we multiply all the incoming messages from our children, as well as the incoming message from our local evidence, and then normalize.

We have explained how to compute the bottom-up belief states from the bottom-up messages. How do we compute the messages themselves? Consider computing $m_{s \to t}^-(x_t)$, where s is one of t's children. Assume, by recursion, that we have computed $\text{bel}_s^-(x_s) = p(x_s|\mathbf{v}_s^-)$. Then we can compute the message as follows:

$$m_{s \to t}^-(x_t) = \sum_{x_s} \psi_{st}(x_s, x_t) \text{bel}_s^-(x_s) \tag{20.3}$$

Essentially we convert beliefs about x_s into beliefs about x_t by using the edge potential ψ_{st}.

We continue in this way up the tree until we reach the root. Once at the root, we have "seen" all the evidence in the tree, so we can compute our local belief state at the root using

$$\text{bel}_r(x_r) \triangleq p(x_r|\mathbf{v}) = p(x_t|\mathbf{v}_r^-) \propto \psi_r(x_r) \prod_{c \in \text{ch}(r)} m_{c \to r}^-(x_r) \tag{20.4}$$

This completes the end of the upwards pass, which is analogous to the forwards pass in an HMM. As a "side effect", we can compute the probability of the evidence by collecting the normalization constants:

$$p(\mathbf{v}) = \prod_t Z_t \tag{20.5}$$

We can now pass messages down from the root. For example, consider node s, with parent t, as shown in Figure 20.1(b). To compute the belief state for s, we need to combine the bottom-up belief for s together with a top-down message from t, which summarizes all the information in the rest of the graph, $m_{t \to s}^+(x_s) \triangleq p(x_t | \mathbf{v}_{st}^+)$, where $\mathbf{v}_{st}^+$ is all the evidence on the upstream (root) side of the $s - t$ edge, as shown in Figure 20.1(b). We then have

$$\text{bel}_s(x_s) \triangleq p(x_s | \mathbf{v}) \propto \text{bel}_s^-(x_s) \prod_{t \in \text{pa}(s)} m_{t \to s}^+(x_s) \tag{20.6}$$

How do we compute these downward messages? For example, consider the message from t to s. Suppose t's parent is r, and t's children are s and u, as shown in Figure 20.1(b). We want to include in $m_{t \to s}^+$ all the information that t has received, except for the information that s sent it:

$$m_{t \to s}^+(x_s) \triangleq p(x_s | \mathbf{v}_{st}^+) = \sum_{x_t} \psi_{st}(x_s, x_t) \frac{\text{bel}_t(x_t)}{m_{s \to t}^-(x_t)} \tag{20.7}$$

Rather than dividing out the message sent up to t, we can plug in the equation of bel_t to get

$$m_{t \to s}^+(x_s) = \sum_{x_t} \psi_{st}(x_s, x_t) \psi_t(x_t) \prod_{c \in \text{ch}(t), c \neq s} m_{c \to t}^-(x_t) \prod_{p \in \text{pa}(t)} m_{p \to t}^+(x_t) \tag{20.8}$$

In other words, we multiply together all the messages coming into t from all nodes except for the recipient s, combine together, and then pass through the edge potential ψ_{st}. In the case of a chain, t only has one child s and one parent p, so the above simplifies to

$$m_{t \to s}^+(x_s) = \sum_{x_t} \psi_{st}(x_s, x_t) \psi_t(x_t) m_{p \to t}^+(x_t) \tag{20.9}$$

The version of BP in which we use division is called **belief updating**, and the version in which we multiply all-but-one of the messages is called **sum-product**. The belief updating version is analogous to how we formulated the Kalman smoother in Section 18.3.2: the top-down messages m^+ depend on the bottom-up messages m^- as well as the filtered belief states bel_t. The sum-product version is analogous to how we formulated the backwards algorithm in Section 17.4.3: in the case of a chain, the top-down (backward) messages m^+ are completely independent of the bottom-up (forward) messages m^-, and do not depend on the filtered belief states. See Section 18.3.2.3 for a more detailed discussion of this subtle difference.

20.2.2 Parallel protocol

So far, we have presented a serial version of the algorithm, in which we send messages up to the root and back. This is the optimal approach for a tree, and is a natural extension of forwards-backwards on chains. However, as a prelude to handling general graphs with loops, we now consider a parallel version of BP. This gives equivalent results to the serial version but is less efficient when implemented on a serial machine.

The basic idea is that all nodes receive messages from their neighbors in parallel, they then updates their belief states, and finally they send new messages back out to their neighbors. This process repeats until convergence. This kind of computing architecture is called a **systolic array**, due to its resemblance to a beating heart.

More precisely, we initialize all messages to the all 1's vector. Then, in parallel, each node absorbs messages from all its neighbors using

$$\text{bel}_s(x_s) \propto \psi_s(x_s) \prod_{t \in \text{nbr}_s} m_{t \to s}(x_s) \tag{20.10}$$

Then, in parallel, each node sends messages to each of its neighbors:

$$m_{s \to t}(x_t) = \sum_{x_s} \left(\psi_s(x_s) \psi_{st}(x_s, x_t) \prod_{u \in \text{nbr}_s \setminus t} m_{u \to s}(x_s) \right) \tag{20.11}$$

The $m_{s \to t}$ message is computed by multiplying together all incoming messages, except the one sent by the recipient, and then passing through the ψ_{st} potential.

At iteration T of the algorithm, $\text{bel}_s(x_s)$ represents the posterior belief of x_s conditioned on the evidence that is T steps away in the graph. After $D(G)$ steps, where $D(G)$ is the **diameter** of the graph (the largest distance between any two pairs of nodes), every node has obtained information from all the other nodes. Its local belief state is then the correct posterior marginal. Since the diameter of a tree is at most $|\mathcal{V}| - 1$, the algorithm converges in a linear number of steps.

We can actually derive the up-down version of the algorithm by imposing the condition that a node can only send a message once it has received messages from all its other neighbors. This means we must start with the leaf nodes, which only have one neighbor. The messages then propagate up to the root and back. We can also update the nodes in a random order. The only requirement is that each node get updated $D(G)$ times. This is just enough time for information to spread throughout the whole tree.

Similar parallel, distributed algorithms for solving linear systems of equations are discussed in (Bertsekas 1997). In particular, the Gauss-Seidel algorithm is analogous to the serial up-down version of BP, and the Jacobi algorithm is analogous to the parallel version of BP.

20.2.3 Gaussian belief propagation *

Now consider the case where $p(\mathbf{x}|\mathbf{v})$ is jointly Gaussian, so it can be represented as a Gaussian pairwise MRF, as in Section 19.4.4. We now present Gaussian belief propagation, following the presentation of (Bickson 2009) (see also (Malioutov et al. 2006)). We will assume the following node and edge potentials:

$$\psi_t(x_t) = \exp(-\frac{1}{2} A_{tt} x_t^2 + b_t x_t) \tag{20.12}$$

$$\psi_{st}(x_s, x_t) = \exp(-\frac{1}{2} x_s A_{st} x_t) \tag{20.13}$$

so the overall model has the form

$$p(\mathbf{x}|\mathbf{v}) \propto \exp(-\frac{1}{2} \mathbf{x}^T \mathbf{A} \mathbf{x} + \mathbf{b}^T \mathbf{x}) \tag{20.14}$$

This is the information form of the MVN (see Exercise 9.2), where $\mathbf{A}$ is the precision matrix. Note that by completing the square, the local evidence can be rewritten as a Gaussian:

$$\psi_t(x_t) \propto \mathcal{N}(b_t/A_{tt}, A_{tt}^{-1}) \triangleq \mathcal{N}(m_t, \ell_t^{-1}) \tag{20.15}$$

Below we describe how to use BP to compute the posterior node marginals,

$$p(x_t|\mathbf{v}) = \mathcal{N}(\mu_t, \lambda_t^{-1}) \tag{20.16}$$

If the graph is a tree, the method is exact. If the graph is loopy, the posterior means may still be exact, but the posterior variances are often too small (Weiss and Freeman 1999).

Although the precision matrix $\mathbf{A}$ is often sparse, computing the posterior mean requires inverting it, since $\boldsymbol{\mu} = \mathbf{A}^{-1}\mathbf{b}$. BP provides a way to exploit graph structure to perform this computation in $O(D)$ time instead of $O(D^3)$. This is related to various methods from linear algebra, as discussed in (Bickson 2009).

Since the model is jointly Gaussian, all marginals and all messages will be Gaussian. The key operations we need are to multiply together two Gaussian factors, and to marginalize out a variable from a joint Gaussian factor.

For multiplication, we can use the fact that the product of two Gaussians is Gaussian:

$$\mathcal{N}(x|\mu_1, \lambda_1^{-1}) \times \mathcal{N}(x|\mu_2, \lambda_2^{-1}) = C\mathcal{N}(x|\mu, \lambda^{-1}) \tag{20.17}$$
$$\lambda = \lambda_1 + \lambda_2 \tag{20.18}$$
$$\mu = \lambda^{-1}(\mu_1\lambda_1 + \mu_2\lambda_2) \tag{20.19}$$

where

$$C = \sqrt{\frac{\lambda}{\lambda_1\lambda_2}} \exp\left(\frac{1}{2}(\lambda_1\mu_1^2(\lambda^{-1}\lambda_1 - 1) + \lambda_2\mu_2^2(\lambda^{-1}\lambda_2 - 1) + 2\lambda^{-1}\lambda_1\lambda_2\mu_1\mu_2)\right) \tag{20.20}$$

See Exercise 20.2 for the proof.

For marginalization, we have the following result:

$$\int \exp(-ax^2 + bx)dx = \sqrt{\pi/a}\exp(b^2/4a) \tag{20.21}$$

which follows from the normalization constant of a Gaussian (Exercise 2.11).

We now have all the pieces we need. In particular, let the message $m_{s\to t}(x_t)$ be a Gaussian with mean μ_{st} and precision λ_{st}. From Equation 20.10, the belief at node s is given by the product of incoming messages times the local evidence (Equation 20.15) and hence

$$\text{bel}_s(x_s) = \psi_s(x_s) \prod_{t\in\text{nbr}(s)} m_{ts}(x_s) = \mathcal{N}(x_s|\mu_s, \lambda_s^{-1}) \tag{20.22}$$
$$\lambda_s = \ell_s + \sum_{t\in\text{nbr}(s)} \lambda_{ts} \tag{20.23}$$
$$\mu_s = \lambda_s^{-1}\left(\ell_s m_s + \sum_{t\in\text{nbr}(s)} \lambda_{ts}\mu_{ts}\right) \tag{20.24}$$

To compute the messages themselves, we use Equation 20.11, which is given by

$$m_{s \to t}(x_t) = \int_{x_s} \left(\psi_{st}(x_s, x_t) \psi_s(x_s) \prod_{u \in \mathrm{nbr}_s \backslash t} m_{u \to s}(x_s) \right) dx_s \qquad (20.25)$$

$$= \int_{x_s} \psi_{st}(x_s, x_t) f_{s \backslash t}(x_s) dx_s \qquad (20.26)$$

where $f_{s \backslash t}(x_s)$ is the product of the local evidence and all incoming messages excluding the message from t:

$$f_{s \backslash t}(x_s) \triangleq \psi_s(x_s) \prod_{u \in \mathrm{nbr}_s \backslash t} m_{u \to s}(x_s) \qquad (20.27)$$

$$= \mathcal{N}(x_s | \mu_{s \backslash t}, \lambda_{s \backslash t}^{-1}) \qquad (20.28)$$

$$\lambda_{s \backslash t} \triangleq \ell_s + \sum_{u \in \mathrm{nbr}(s) \backslash t} \lambda_{us} \qquad (20.29)$$

$$\mu_{s \backslash t} \triangleq \lambda_{s \backslash t}^{-1} \left(\ell_s m_s + \sum_{u \in \mathrm{nbr}(s) \backslash t} \lambda_{us} \mu_{us} \right) \qquad (20.30)$$

Returning to Equation 20.26 we have

$$m_{s \to t}(x_t) = \int_{x_s} \underbrace{\exp(-x_s A_{st} x_t)}_{\psi_{st}(x_s, x_t)} \underbrace{\exp(-\lambda_{s \backslash t}/2(x_s - \mu_{s \backslash t})^2)}_{f_{s \backslash t}(x_s)} dx_s \qquad (20.31)$$

$$= \int_{x_s} \exp \left((-\lambda_{s \backslash t} x_s^2 / 2) + (\lambda_{s \backslash t} \mu_{s \backslash t} - A_{st} x_t) x_s \right) dx_s + \mathrm{const} \qquad (20.32)$$

$$\propto \exp \left((\lambda_{s \backslash t} \mu_{s \backslash t} - A_{st} x_t)^2 / (2\lambda_{s \backslash t}) \right) \qquad (20.33)$$

$$\propto \mathcal{N}(\mu_{st}, \lambda_{st}^{-1}) \qquad (20.34)$$

$$\lambda_{st} = A_{st}^2 / \lambda_{s \backslash t} \qquad (20.35)$$

$$\mu_{st} = A_{st} \mu_{s \backslash t} / \lambda_{st} \qquad (20.36)$$

One can generalize these equations to the case where each node is a vector, and the messages become small MVNs instead of scalar Gaussians (Alag and Agogino 1996). If we apply the resulting algorithm to a linear dynamical system, we recover the Kalman smoothing algorithm of Section 18.3.2.

To perform message passing in models with non-Gaussian potentials, one can use sampling methods to approximate the relevant integrals. This is called **non-parametric BP** (Sudderth et al. 2003; Isard 2003; Sudderth et al. 2010).

20.2.4   Other BP variants *

In this section, we briefly discuss several variants of the main algorithm.

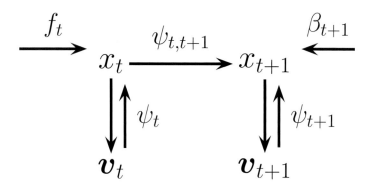

Figure 20.2 Illustration of how to compute the two-slice distribution for an HMM. The ψ_t and ψ_{t+1} terms are the local evidence messages from the visible nodes $\mathbf{v}_t$, $\mathbf{v}_{t+1}$ to the hidde nodes x_t, x_{t+1} respectively; f_t is the forwards message from x_{t-1} and β_{t+1} is the backwards message from x_{t+2}.

20.2.4.1 Max-product algorithm

It is possible to devise a **max-product** version of the BP algorithm, by replacing the $\sum$ operator with the max operator. We can then compute the local MAP marginal of each node. However, if there are ties, this might not be globally consistent, as discussed in Section 17.4.4. Fortunately, we can generalize the Viterbi algorithm to trees, where we use max and argmax in the collect-to-root phase, and perform traceback in the distribute-from-root phase. See (Dawid 1992) for details.

20.2.4.2 Sampling from a tree

It is possible to draw samples from a tree structured model by generalizing the forwards filtering / backwards sampling algorithm discussed in Section 17.4.5. See (Dawid 1992) for details.

20.2.4.3 Computing posteriors on sets of variables

In Section 17.4.3.2, we explained how to compute the "two-slice" distribution $\xi_{t,t+1}(i,j) = p(x_t = i, x_{t+1} = j|\mathbf{v})$ in an HMM, namely by using

$$\xi_{t,t+1}(i,j) \quad = \quad \alpha_t(i)\psi_{t+1}(j)\beta_{t+1}(j)\psi_{t,t+1}(i,j) \tag{20.37}$$

Since $\alpha_t(i) \propto \psi_t(i)f_t(i)$, where $f_t = p(x_t|\mathbf{v}_{1:t-1})$ is the forwards message, we can think of this as sending messages f_t and ψ_t into x_t, β_{t+1} and ϕ_{t+1} into x_{t+1}, and then combining them with the $\mathbf{\Psi}$ matrix, as shown in Figure 20.2. This is like treating x_t and x_{t+1} as a single "mega node", and then multiplying all the incoming messages as well as all the local factors (here, $\psi_{t,t+1}$).

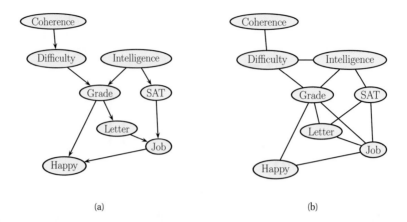

Figure 20.3 Left: The "student" DGM. Right: the equivalent UGM. We add moralization arcs D-I, G-J and L-S. Based on Figure 9.8 of (Koller and Friedman 2009).

20.3 The variable elimination algorithm

We have seen how to use BP to compute exact marginals on chains and trees. In this section, we discuss an algorithm to compute $p(\mathbf{x}_q|\mathbf{x}_v)$ for any kind of graph.

We will explain the algorithm by example. Consider the DGM in Figure 20.3(a). This model, from (Koller and Friedman 2009), is a hypothetical model relating various variables pertaining to a typical student. The corresponding joint has the following form:

$$P(C, D, I, G, S, L, J, H) \tag{20.38}$$
$$= P(C)P(D|C)P(I)P(G|I, D)P(S|I)P(L|G)P(J|L, S)P(H|G, J) \tag{20.39}$$

Note that the forms of the CPDs do not matter, since all our calculations will be symbolic. However, for illustration purposes, we will assume all variables are binary.

Before proceeding, we convert our model to undirected form. This is not required, but it makes for a more unified presentation, since the resulting method can then be applied to both DGMs and UGMs (and, as we will see in Section 20.3.1, to a variety of other problems that have nothing to do with graphical models). Since the computational complexity of inference in DGMs and UGMs is, generally speaking, the same, nothing is lost in this transformation from a computational point of view.[1]

To convert the DGM to a UGM, we simply define a potential or factor for every CPD, yielding

$$p(C, D, I, G, S, L, J, H) = \psi_C(C)\psi_D(D, C)\psi_I(I)\psi_G(G, I, D) \tag{20.40}$$
$$\psi_S(S, I)\psi_L(L, G)\psi_J(J, L, S)\psi_H(H, G, J) \tag{20.41}$$

---

1. There are a few "tricks" one can exploit in the directed case that cannot easily be exploited in the undirected case. One important example is **barren node removal**. To explain this, consider a naive Bayes classifier, as in Figure 10.2. Suppose we want to infer y and we observe x_1 and x_2, but not x_3 and x_4. It is clear that we can safely remove x_3 and x_4, since $\sum_{x_3} p(x_3|y) = 1$, and similarly for x_4. In general, once we have removed hidden leaves, we can apply this process recursively. Since potential functions do not necessary sum to one, we cannot use this trick in the undirected case. See (Koller and Friedman 2009) for a variety of other speedup tricks.

Since all the potentials are **locally normalized**, since they are CPDs, there is no need for a global normalization constant, so $Z = 1$. The corresponding undirected graph is shown in Figure 20.3(b). Note that it has more edges than the DAG. In particular, any "unmarried" nodes that share a child must get "married", by adding an edge between them; this process is known as **moralization**. Only then can the arrows be dropped. In this example, we added D-I, G-J, and L-S moralization arcs. The reason this operation is required is to ensure that the CI properties of the UGM match those of the DGM, as explained in Section 19.2.2. It also ensures there is a clique that can "store" the CPDs of each family.

Now suppose we want to compute $p(J = 1)$, the marginal probability that a person will get a job. Since we have 8 binary variables, we could simply enumerate over all possible assignments to all the variables (except for J), adding up the probability of each joint instantiation:

$$p(J) = \sum_L \sum_S \sum_G \sum_H \sum_I \sum_D \sum_C p(C, D, I, G, S, L, J, H) \tag{20.42}$$

However, this would take $O(2^7)$ time. We can be smarter by **pushing sums inside products**. This is the key idea behind the **variable elimination** algorithm (Zhang and Poole 1996), also called **bucket elimination** (Dechter 1996), or, in the context of genetic pedigree trees, the **peeling algorithm** (Cannings et al. 1978). In our example, we get

$$
\begin{aligned}
p(J) &= \sum_{L,S,G,H,I,D,C} p(C, D, I, G, S, L, J, H) \\
&= \sum_{L,S,G,H,I,D,C} \psi_C(C)\psi_D(D, C)\psi_I(I)\psi_G(G, I, D)\psi_S(S, I)\psi_L(L, G) \\
&\quad \times \psi_J(J, L, S)\psi_H(H, G, J) \\
&= \sum_{L,S} \psi_J(J, L, S) \sum_G \psi_L(L, G) \sum_H \psi_H(H, G, J) \sum_I \psi_S(S, I)\psi_I(I) \\
&\quad \times \sum_D \psi_G(G, I, D) \sum_C \psi_C(C)\psi_D(D, C)
\end{aligned}
$$

We now evaluate this expression, working right to left as shown in Table 20.1. First we multiply together all the terms in the scope of the $\sum_C$ operator to create the temporary factor

$$\tau_1'(C, D) = \psi_C(C)\psi_D(D, C) \tag{20.43}$$

Then we marginalize out C to get the new factor

$$\tau_1(D) = \sum_C \tau_1'(C, D) \tag{20.44}$$

Next we multiply together all the terms in the scope of the $\sum_D$ operator and then marginalize out to create

$$
\begin{aligned}
\tau_2'(G, I, D) &= \psi_G(G, I, D)\tau_1(D) \tag{20.45} \\
\tau_2(G, I) &= \sum_D \tau_2'(G, I, D) \tag{20.46}
\end{aligned}
$$

$$\sum_L \sum_S \psi_J(J, L, S) \sum_G \psi_L(L, G) \sum_H \psi_H(H, G, J) \sum_I \psi_S(S, I) \psi_I(I) \sum_D \psi_G(G, I, D) \underbrace{\sum_C \psi_C(C) \psi_D(D, C)}_{\tau_1(D)}$$

$$\sum_L \sum_S \psi_J(J, L, S) \sum_G \psi_L(L, G) \sum_H \psi_H(H, G, J) \sum_I \psi_S(S, I) \psi_I(I) \underbrace{\sum_D \psi_G(G, I, D) \tau_1(D)}_{\tau_2(G, I)}$$

$$\sum_L \sum_S \psi_J(J, L, S) \sum_G \psi_L(L, G) \sum_H \psi_H(H, G, J) \underbrace{\sum_I \psi_S(S, I) \psi_I(I) \tau_2(G, I)}_{\tau_3(G, S)}$$

$$\sum_L \sum_S \psi_J(J, L, S) \sum_G \psi_L(L, G) \underbrace{\sum_H \psi_H(H, G, J) \, \tau_3(G, S)}_{\tau_4(G, J)}$$

$$\sum_L \sum_S \psi_J(J, L, S) \underbrace{\sum_G \psi_L(L, G) \tau_4(G, J) \tau_3(G, S)}_{\tau_5(J, L, S)}$$

$$\sum_L \sum_S \underbrace{\psi_J(J, L, S) \tau_5(J, L, S)}_{\tau_6(J, L)}$$

$$\underbrace{\sum_L \tau_6(J, L)}_{\tau_7(J)}$$

Table 20.1   Eliminating variables from Figure 20.3 in the order C, D, I, H, G, S, L to compute $P(J)$.

Next we multiply together all the terms in the scope of the $\sum_I$ operator and then marginalize out to create

$$\tau_3'(G, I, S) \quad = \quad \psi_S(S, I) \psi_I(I) \tau_2(G, I) \tag{20.47}$$

$$\tau_3(G, S) \quad = \quad \sum_I \tau_3'(G, I, S) \tag{20.48}$$

And so on.

The above technique can be used to compute any marginal of interest, such as $p(J)$ or $p(J, H)$. To compute a conditional, we can take a ratio of two marginals, where the visible variables have been clamped to their known values (and hence don't need to be summed over). For example,

$$p(J = j | I = 1, H = 0) = \frac{p(J = j, I = 1, H = 0)}{\sum_{j'} p(J = j', I = 1, H = 0)} \tag{20.49}$$

In general, we can write

$$p(\mathbf{x}_q | \mathbf{x}_v) \quad = \quad \frac{p(\mathbf{x}_q, \mathbf{x}_v)}{p(\mathbf{x}_v)} = \frac{\sum_{\mathbf{x}_h} p(\mathbf{x}_h, \mathbf{x}_q, \mathbf{x}_v)}{\sum_{\mathbf{x}_h} \sum_{\mathbf{x}_q'} p(\mathbf{x}_h, \mathbf{x}_q', \mathbf{x}_v)} \tag{20.50}$$

The normalization constant in the denominator, $p(\mathbf{x}_v)$, is called the **probability of the evidence**.

See `variableElimination` for a simple Matlab implementation of this algorithm, which works for arbitrary graphs, and arbitrary discrete factors. But before you go too crazy, please read Section 20.3.2, which points out that VE can be exponentially slow in the worst case.

20.3.1 The generalized distributive law *

Abstractly, VE can be thought of as computing the following expression:

$$p(\mathbf{x}_q|\mathbf{x}_v) \propto \sum_{\mathbf{x}} \prod_c \psi_c(\mathbf{x}_c) \tag{20.51}$$

It is understood that the visible variables $\mathbf{x}_v$ are clamped, and not summed over. VE uses **non-serial dynamic programming** (Bertele and Brioschi 1972), caching intermediate results to avoid redundant computation.

However, there are other tasks we might like to solve for any given graphical model. For example, we might want the MAP estimate:

$$\mathbf{x}^* = \operatorname*{argmax}_{\mathbf{x}} \prod_c \psi_c(\mathbf{x}_c) \tag{20.52}$$

Fortunately, essentially the same algorithm can also be used to solve this task: we just replace sum with max. (We also need a **traceback** step, which actually recovers the argmax, as opposed to just the value of max; these details are explained in Section 17.4.4.)

In general, VE can be applied to any **commutative semi-ring**. This is a set K, together with two binary operations called "+" and "×", which satisfy the following three axioms:

1. The operation "+" is associative and commutative, and there is an additive identity element called "0" such that $k + 0 = k$ for all $k \in K$.

2. The operation "×" is associative and commutative, and there is a multiplicative identity element called "1" such that $k \times 1 = k$ for all $k \in K$.

3. The **distributive law** holds, i.e.,

$$(a \times b) + (a \times c) = a \times (b + c) \tag{20.53}$$

   for all triples (a, b, c) from K.

This framework covers an extremely wide range of important applications, including constraint satisfaction problems (Bistarelli et al. 1997; Dechter 2003), the fast Fourier transform (Aji and McEliece 2000), etc. See Table 20.2 for some examples.

20.3.2 Computational complexity of VE

The running time of VE is clearly exponential in the size of the largest factor, since we have sum over all of the corresponding variables. Some of the factors come from the original model (and are thus unavoidable), but new factors are created in the process of summing out. For example,

Domain	+	×	Name
$[0, \infty)$	$(+, 0)$	$(\times, 1)$	sum-product
$[0, \infty)$	$(\max, 0)$	$(\times, 1)$	max-product
$(-\infty, \infty]$	$(\min, \infty)$	$(+, 0)$	min-sum
$\{T, F\}$	$(\vee, F)$	$(\wedge, T)$	Boolean satisfiability

Table 20.2 Some commutative semirings.

$$\sum_D \sum_C \psi_D(D,C) \sum_H \sum_L \sum_S \psi_J(J,L,S) \sum_I \psi_I(I)\psi_S(S,I) \underbrace{\sum_G \psi_G(G,I,D)\psi_L(L,)\psi_H(H,G,J)}_{\tau_1(I,D,L,J,H)}$$

$$\sum_D \sum_C \psi_D(D,C) \sum_H \sum_L \sum_S \psi_J(J,L,S) \underbrace{\sum_I \psi_I(I)\psi_S(S,I)\tau_1(I,D,L,J,H)}_{\tau_2(D,L,S,J,H)}$$

$$\sum_D \sum_C \psi_D(D,C) \sum_H \sum_L \underbrace{\sum_S \psi_J(J,L,S)\tau_2(D,L,S,J,H)}_{\tau_3(D,L,J,H)}$$

$$\sum_D \sum_C \psi_D(D,C) \sum_H \underbrace{\sum_L \tau_3(D,L,J,H)}_{\tau_4(D,J,H)}$$

$$\sum_D \sum_C \psi_D(D,C) \underbrace{\sum_H \tau_4(D,J,H)}_{\tau_5(D,J)}$$

$$\sum_D \underbrace{\sum_C \psi_D(D,C)\tau_5(D,J)}_{\tau_6(D,J)}$$

$$\underbrace{\sum_D \tau_6(D,J)}_{\tau_7(J)}$$

Table 20.3 Eliminating variables from Figure 20.3 in the order G, I, S, L, H, C, D.

in Equation 20.47, we created a factor involving G, I and S; but these nodes were not originally present together in any factor.

The order in which we perform the summation is known as the **elimination order**. This can have a large impact on the size of the intermediate factors that are created. For example, consider the ordering in Table 20.1: the largest created factor (beyond the original ones in the model) has size 3, corresponding to $\tau_5(J, L, S)$. Now consider the ordering in Table 20.3: now the largest factors are $\tau_1(I, D, L, J, H)$ and $\tau_2(D, L, S, J, H)$, which are much bigger.

We can determine the size of the largest factor graphically, without worrying about the actual numerical values of the factors. When we eliminate a variable X_t, we connect it to all variables

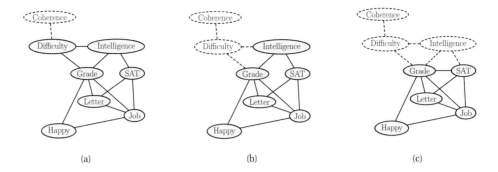

Figure 20.4 Example of the elimination process, in the order C, D, I, etc. When we eliminate I (figure c), we add a fill-in edge between G and S, since they are not connected. Based on Figure 9.10 of (Koller and Friedman 2009).

that share a factor with X_t (to reflect the new temporary factor τ'_t). The edges created by this process are called **fill-in edges**. For example, Figure 20.4 shows the fill-in edges introduced when we eliminate in the order $C, D, I, \ldots$. The first two steps do not introduce any fill-ins, but when we eliminate I, we connect G and S, since they co-occur in Equation 20.48.

Let $G(\prec)$ be the (undirected) graph induced by applying variable elimination to G using elimination ordering $\prec$. The temporary factors generated by VE correspond to maximal cliques in the graph $G(\prec)$. For example, with ordering (C, D, I, H, G, S, L), the maximal cliques are as follows:

$$\{C, D\}, \{D, I, G\}, \{G, L, S, J\}, \{G, J, H\}, \{G, I, S\} \tag{20.54}$$

It is clear that the time complexity of VE is

$$\sum_{c \in \mathcal{C}(G(\prec))} K^{|c|} \tag{20.55}$$

where $\mathcal{C}$ are the cliques that are created, $|c|$ is the size of the clique c, and we assume for notational simplicity that all the variables have K states each.

Let us define the **induced width** of a graph given elimination ordering $\prec$, denoted $w(\prec)$, as the size of the largest factor (i.e., the largest clique in the induced graph) minus 1. Then it is easy to see that the complexity of VE with ordering $\prec$ is $O(K^{w(\prec)+1})$.

Obviously we would like to minimize the running time, and hence the induced width. Let us define the **treewidth** of a graph as the minimal induced width.

$$w \triangleq \min_{\prec} \max_{c \in G(\prec)} |c| - 1 \tag{20.56}$$

Then clearly the best possible running time for VE is $O(DK^{w+1})$. Unfortunately, one can show that for arbitrary graphs, finding an elimination ordering $\prec$ that minimizes $w(\prec)$ is NP-hard (Arnborg et al. 1987). In practice greedy search techniques are used to find reasonable orderings (Kjaerulff 1990), although people have tried other heuristic methods for discrete optimization, such as genetic algorithms (Larranaga et al. 1997). It is also possible to derive approximate algorithms with provable performance guarantees (Amir 2010).

In some cases, the optimal elimination ordering is clear. For example, for chains, we should work forwards or backwards in time. For trees, we should work from the leaves to the root. These orderings do not introduce any fill-in edges, so $w = 1$. Consequently, inference in chains and trees takes $O(VK^2)$ time. This is one reason why Markov chains and Markov trees are so widely used.

Unfortunately, for other graphs, the treewidth is large. For example, for an $m \times n$ 2d lattice, the treewidth is $O(\min\{m, n\})$ (Lipton and Tarjan 1979). So VE on a 100×100 Ising model would take $O(2^{100})$ time.

Of course, just because VE is slow doesn't mean that there isn't some smarter algorithm out there. We discuss this issue in Section 20.5.

20.3.3 A weakness of VE

The main disadvantage of the variable elimination algorithm (apart from its exponential dependence on treewidth) is that it is inefficient if we want to compute multiple queries conditioned on the same evidence. For example, consider computing all the marginals in a chain-structured graphical model such as an HMM. We can easily compute the final marginal $p(x_T|\mathbf{v})$ by eliminating all the nodes x_1 to x_{T-1} in order. This is equivalent to the forwards algorithm, and takes $O(K^2T)$ time. But now suppose we want to compute $p(x_{T-1}|\mathbf{v})$. We have to run VE again, at a cost of $O(K^2T)$ time. So the total cost to compute all the marginals is $O(K^2T^2)$. However, we know that we can solve this problem in $O(K^2T)$ using forwards-backwards. The difference is that FB caches the messages computed on the forwards pass, so it can reuse them later.

The same argument holds for BP on trees. For example, consider the 4-node tree in Figure 20.5. We can compute $p(x_1|\mathbf{v})$ by eliminating $\mathbf{x}_{2:4}$; this is equivalent to sending messages up to x_1 (the messages correspond to the τ factors created by VE). Similarly we can compute $p(x_2|\mathbf{v})$, $p(x_3|\mathbf{v})$ and then $p(x_4|\mathbf{v})$. We see that some of the messages used to compute the marginal on one node can be re-used to compute the marginals on the other nodes. By storing the messages for later re-use, we can compute all the marginals in $O(DK^2)$ time. This is what the up-down (collect-distribute) algorithm on trees does.

The question is: how can we combine the efficiency of BP on trees with the generality of VE? The answer is given in Section 20.4.

20.4 The junction tree algorithm *

The **junction tree algorithm** or **JTA** generalizes BP from trees to arbitrary graphs. We sketch the basic idea below; for details, see e.g., (Koller and Friedman 2009).

20.4.1 Creating a junction tree

The basic idea behind the JTA is this. We first run the VE algorithm "symbolically", adding fill-in edges as we go, according to a given elimination ordering. The resulting graph will be a **chordal graph**, which means that every undirected cycle $X_1 - X_2 \cdots X_k - X_1$ of length $k \geq 4$ has a

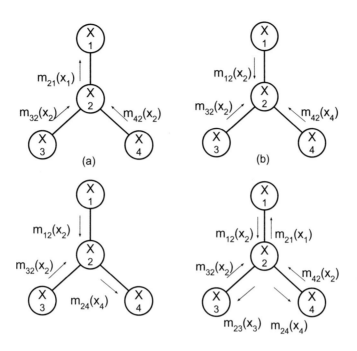

Figure 20.5 Sending multiple messages along a tree. (a) X_1 is root. (b) X_2 is root. (c) X_4 is root. (d) All of the messages needed to compute all singleton marginals. Based on Figure 4.3 of (Jordan 2007).

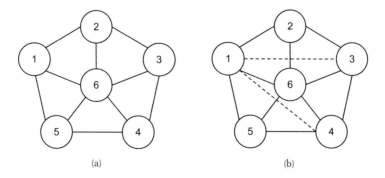

Figure 20.6 Left: this graph is not triangulated, despite appearances, since it contains a chordless 5-cycle 1-2-3-4-5-1. Right: one possible triangulation, by adding the 1-3 and 1-4 fill-in edges. Based on (Armstrong 2005, p46)

chord, i.e., an edge connects X_i, X_j for all non-adjacent nodes *i,j* in the cycle.[2]

Having created a chordal graph, we can extract its maximal cliques. In general, finding max cliques is computationally hard, but it turns out that it can be done efficiently from this special kind of graph. Figure 20.7(b) gives an example, where the max cliques are as follows:

$$\{C, D\}, \{G, I, D\}, \{G, S, I\}, \{G, J, S, L\}, \{H, G, J\} \tag{20.57}$$

Note that if the original graphical model was already chordal, the elimination process would not add any extra fill-in edges (assuming the optimal elimination ordering was used). We call such models **decomposable**, since they break into little pieces defined by the cliques.

It turns out that the cliques of a chordal graph can be arranged into a special kind of tree known as a **junction tree**. This enjoys the **running intersection property** (RIP), which means that any subset of nodes containing a given variable forms a connected component. Figure 20.7(c) gives an example of such a tree. We see that the node *I* occurs in two adjacent tree nodes, so they can share information about this variable. A similar situation holds for all the other variables.

One can show that if a tree that satisfies the running intersection property, then applying BP to this tree (as we explain below) will return the exact values of $p(\mathbf{x}_c|\mathbf{v})$ for each node *c* in the tree (i.e., clique in the induced graph). From this, we can easily extract the node and edge marginals, $p(x_t|\mathbf{v})$ and $p(x_s, x_t|\mathbf{v})$ from the original model, by marginalizing the clique distributions.[3]

20.4.2 Message passing on a junction tree

Having constructed a junction tree, we can use it for inference. The process is very similar to belief propagation on a tree. As in Section 20.2, there are two versions: the sum-product form, also known as the **Shafer-Shenoy** algorithm, named after (Shafer and Shenoy 1990); and the belief updating form (which involves division), also known as the **Hugin** (named after a company) or the **Lauritzen-Spiegelhalter** algorithm (named after (Lauritzen and Spiegelhalter 1988)). See (Lepar and Shenoy 1998) for a detailed comparison of these methods. Below we sketch how the Hugin algorithm works.

We assume the original model has the following form:

$$p(\mathbf{x}) = \frac{1}{Z} \prod_{c \in \mathcal{C}(G)} \psi_c(\mathbf{x}_c) \tag{20.58}$$

where $\mathcal{C}(G)$ are the cliques of the original graph. On the other hand, the tree defines a distribution of the following form:

$$p(\mathbf{x}) = \frac{\prod_{c \in \mathcal{C}(T)} \psi_c(\mathbf{x}_c)}{\prod_{s \in \mathcal{S}(T)} \psi_s(\mathbf{x}_s)} \tag{20.59}$$

---

2. The largest loop in a chordal graph is length 3. Consequently chordal graphs are sometimes called **triangulated**. However, it is not enough for the graph to look like it is made of little triangles. For example, Figure 20.6(a) is not chordal, even though it is made of little triangles, since it contains the chordless 5-cycle 1-2-3-4-5-1.

3. If we want the joint distribution of some variables that are not in the same clique — a so-called **out-of-clique query** — we can adapt the technique described in Section 20.2.4.3 as follows: create a mega node containing the query variables and any other nuisance variables that lie on the path between them, multiply in messages onto the boundary of the mega node, and then marginalize out the internal nuisance variables. This internal marginalization may require the use of BP or VE. See (Koller and Friedman 2009) for details.

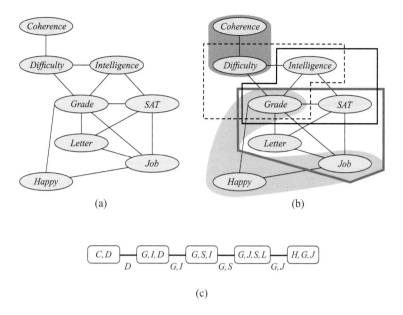

Figure 20.7   (a) The student graph with fill-in edges added. (b) The maximal cliques. (c) The junction tree. An edge between nodes s and t is labeled by the intersection of the sets on nodes s and t; this is called the **separating set**. From Figure 9.11 of (Koller and Friedman 2009). Used with kind permission of Daphne Koller.

where $\mathcal{C}(T)$ are the nodes of the junction tree (which are the cliques of the chordal graph), and $\mathcal{S}(T)$ are the separators of the tree. To make these equal, we initialize by defining $\psi_s = 1$ for all separators and $\psi_c = 1$ for all cliques. Then, for each clique in the original model, $c \in \mathcal{C}(G)$, we find a clique in the tree $c' \in \mathcal{C}(T)$ which contains it, $c' \supseteq c$. We then multiply ψ_c onto $\psi_{c'}$ by computing $\psi_{c'} = \psi_{c'} \, \psi_c$. After doing this for all the cliques in the original graph, we have

$$\prod_{c \in \mathcal{C}(T)} \psi_c(\mathbf{x}_c) = \prod_{c \in \mathcal{C}(G)} \psi_c(\mathbf{x}_c) \tag{20.60}$$

As in Section 20.2.1, we now send messages from the leaves to the root and back, as sketched in Figure 20.1. In the upwards pass, also known as the **collect-to-root** phase, node i sends to its parent j the following message:

$$m_{i \to j}(S_{ij}) = \sum_{C_i \setminus S_{ij}} \psi_i(C_i) \tag{20.61}$$

That is, we marginalize out the variables that node i "knows about" which are irrelevant to j, and then we send what is left over. Once a node has received messages from all its children, it updates its belief state using

$$\psi_i(C_i) \propto \psi_i(C_i) \prod_{j \in \mathrm{ch}_i} m_{j \to i}(S_{ij}) \tag{20.62}$$

At the root, $\psi_r(C_r)$ represents $p(\mathbf{x}_{C_r}|\mathbf{v})$, which is the posterior over the nodes in clique C_r conditioned on all the evidence. Its normalization constant is $p(\mathbf{v})/Z_0$, where Z_0 is the normalization constant for the unconditional prior, $p(\mathbf{x})$. (We have $Z_0 = 1$ if the original model was a DGM.)

In the downwards pass, also known as the **distribute-from-root** phase, node i sends to its children j the following message:

$$m_{i \rightarrow j}(S_{ij}) = \frac{\sum_{C_i \backslash S_{ij}} \psi_i(C_i)}{m_{j \rightarrow i}(S_{ij})} \tag{20.63}$$

We divide out by what j sent to i to avoid double counting the evidence. This requires that we store the messages from the upwards pass. Once a node has received a top-down message from its parent, it can compute its final belief state using

$$\psi_j(C_j) \propto \psi_j(C_j) m_{i \rightarrow j}(S_{ij}) \tag{20.64}$$

An equivalent way to present this algorithm is based on storing the messages inside the separator potentials. So on the way up, sending from i to j we compute the separator potential

$$\psi_{ij}^*(S_{ij}) = \sum_{C_i \backslash S_{ij}} \psi_i(C_i) \tag{20.65}$$

and then update the recipient potential:

$$\psi_j^*(C_j) \propto \psi_j(C_j) \frac{\psi_{ij}^*(S_{ij})}{\psi_{ij}(S_{ij})} \tag{20.66}$$

(Recall that we initialize $\psi_{ij}(S_{ij}) = 1$.) This is sometimes called **passing a flow** from i to j. On the way down, from i to j, we compute the separator potential

$$\psi_{ij}^{**}(S_{ij}) = \sum_{C_i \backslash S_{ij}} \psi_i^*(C_i) \tag{20.67}$$

and then update the recipient potential:

$$\psi_j^{**}(C_j) \propto \psi_j^*(C_j) \frac{\psi_{ij}^{**}(S_{ij})}{\psi_{ij}^*(S_{ij})} \tag{20.68}$$

This process is known as junction tree **calibration**. See Figure 20.1 for an illustration. Its correctness follows from the fact that each edge partitions the evidence into two distinct groups, plus the fact that the tree satisfies RIP, which ensures that no information is lost by only performing local computations.

20.4.2.1 Example: jtree algorithm on a chain

It is interesting to see what happens if we apply this process to a chain structured graph such as an HMM. A detailed discussion can be found in (Smyth et al. 1997), but the basic idea is this. The cliques are the edges, and the separators are the nodes, as shown in Figure 20.8. We initialize the potentials as follows: we set $\psi_s = 1$ for all the separators, we set $\psi_c(x_{t-1}, x_t) = p(x_t|x_{t-1})$ for clique $c = (X_{t-1}, X_t)$, and we set $\psi_c(x_t, y_t) = p(y_t|x_t)$ for clique $c = (X_t, Y_t)$.

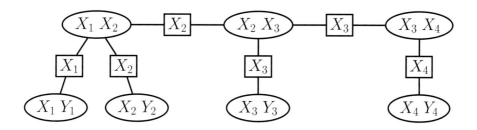

Figure 20.8   The junction tree derived from an HMM/SSM of length $T = 4$.

Next we send messages from left to right. Consider clique (X_{t-1}, X_t) with potential $p(X_t|X_{t-1})$. It receives a message from clique (X_{t-2}, X_{t-1}) via separator X_{t-1} of the form $\sum_{x_{t-2}} p(X_{t-2}, X_{t-1}|\mathbf{v}_{1:t-1}) = p(X_{t-1}|\mathbf{v}_{1:t-1})$. When combined with the clique potential, this becomes the two-slice predictive density

$$p(X_t|X_{t-1})p(X_{t-1}|\mathbf{v}_{1:t-1}) = p(X_{t-1}, X_t|\mathbf{v}_{1:t-1}) \tag{20.69}$$

The clique (X_{t-1}, X_t) also receives a message from (X_t, Y_t) via separator X_t of the form $p(y_t|X_t)$, which corresponds to its local evidence. When combined with the updated clique potential, this becomes the two-slice filtered posterior

$$p(X_{t-1}, X_t|\mathbf{v}_{1:t-1})p(\mathbf{v}_t|X_t) = p(X_{t-1}, X_t|\mathbf{v}_{1:t}) \tag{20.70}$$

Thus the messages in the forwards pass are the filtered belief states α_t, and the clique potentials are the two-slice distributions. In the backwards pass, the messages are the update factors $\frac{\gamma_t}{\alpha_t}$, where $\gamma_t(k) = p(x_t = k|\mathbf{v}_{1:T})$ and $\alpha_t(k) = p(x_t = k|\mathbf{v}_{1:t})$. By multiplying by this message, we "swap out" the old α_t message and "swap in" the new γ_t message. We see that the backwards pass involves working with posterior beliefs, not conditional likelihoods. See Section 18.3.2.3 for further discussion of this difference.

20.4.3   Computational complexity of JTA

If all nodes are discrete with K states each, it is clear that the JTA takes $O(|\mathcal{C}|K^{w+1})$ time and space, where $|\mathcal{C}|$ is the number of cliques and w is the treewidth of the graph, i.e., the size of the largest clique minus 1. Unfortunately, choosing a triangulation so as to minimize the treewidth is NP-hard, as explained in Section 20.3.2.

The JTA can be modified to handle the case of Gaussian graphical models. The graph-theoretic steps remain unchanged. Only the message computation differs. We just need to define how to multiply, divide, and marginalize Gaussian potential functions. This is most easily done in information form. See e.g., (Lauritzen 1992; Murphy 1998; Cemgil 2001) for the details. The algorithm takes $O(|\mathcal{C}|w^3)$ time and $O(|\mathcal{C}|w^2)$ space. When applied to a chain structured graph, the algorithm is equivalent to the Kalman smoother in Section 18.3.2.

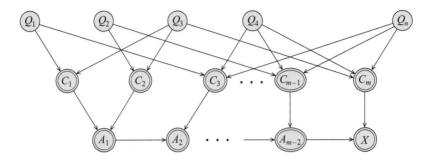

Figure 20.9 Encoding a 3-SAT problem on n variables and m clauses as a DGM. The Q_s variables are binary random variables. The C_t variables are deterministic functions of the Q_s's, and compute the truth value of each clause. The A_t nodes are a chain of AND gates, to ensure that the CPT for the final x node has bounded size. The double rings denote nodes with deterministic CPDs. Source: Figure 9.1 of (Koller and Friedman 2009). Used with kind permission of Daphne Koller.

20.4.4   JTA generalizations *

We have seen how to use the JTA algorithm to compute posterior marginals in a graphical model. There are several possible generalizations of this algorithm, some of which we mention below. All of these exploit graph decomposition in some form or other. They only differ in terms of how they define/ compute messages and "beliefs". The key requirement is that the operators which compute messages form a commutative semiring (see Section 20.3.1).

- Computing the MAP estimate. We just replace the sum-product with max-product in the collect phase, and use traceback in the distribute phase, as in the Viterbi algorithm (Section 17.4.4). See (Dawid 1992) for details.
- Computing the N-most probable configurations (Nilsson 1998).
- Computing posterior samples. The collect pass is the same as usual, but in the distribute pass, we sample variables given the values higher up in the tree, thus generalizing forwards-filtering backwards-sampling for HMMs described in Section 17.4.5. See (Dawid 1992) for details.
- Solving constraint satisfaction problems (Dechter 2003).
- Solving logical reasoning problems (Amir and McIlraith 2005).

20.5   Computational intractability of exact inference in the worst case

As we saw in Sections 20.3.2 and 20.4.3, VE and JTA take time that is exponential in the treewidth of a graph. Since the treewidth can be O(number of nodes) in the worst case, this means these algorithms can be exponential in the problem size.

Of course, just because VE and JTA are slow doesn't mean that there isn't some smarter algorithm out there. Unfortunately, this seems unlikely, since it is easy to show that exact inference is NP-hard (Dagum and Luby 1993). The proof is a simple reduction from the satisfiability prob-

Method	Restriction	Section
Forwards-backwards	Chains, D or LG	Section 17.4.3
Belief propagation	Trees, D or LG	Section 20.2
Variable elimination	Low treewidth, D or LG, single query	Section 20.3
Junction tree algorithm	Low treewidth, D or LG	Section 20.4
Mean field	Approximate, C-E	Section 21.3
Loopy belief propagation	Approximate, D or LG	Section 22.2
Importance sampling	Approximate	Section 23.4.3
Gibbs sampling	Approximate	Section 24.2

Table 20.4 Summary of some methods that can be used for inference in graphical models. Methods above the line are exact; those below the line are approximate. "D" means that all the hidden variables must be discrete. "L-G" means that all the factors must be linear-Gaussian. The term "single query" refers to the restriction that VE only computes one marginal $p(\mathbf{x}_q|\mathbf{x}_v)$ at a time. See Section 20.3.3 for a discussion of this point. "C-E" stands for "conjugate exponential"; this means that variational mean field only applies to models where the likelihood is in the exponential family, and the prior is conjugate. This includes the D and LG case, but many others as well, as we will see in Section 21.3.

lem. In particular, note that we can encode any 3-SAT problem[4] as a DGM with deterministic links, as shown in Figure 20.9. We clamp the final node, x, to be on, and we arrange the CPTs so that $p(x = 1) > 0$ iff there is a satisfying assignment. Computing any posterior marginal requires evaluating the normalization constant $p(x = 1)$, which represents the probability of the evidence, so inference in this model implicitly solves the SAT problem.

In fact, exact inference is #P-hard (Roth 1996), which is even harder than NP-hard. (See e.g., (Arora and Barak 2009) for definitions of these terms.) The intuitive reason for this is that to compute the normalizing constant Z, we have to *count* how many satisfying assignments there are. By contrast, MAP estimation is provably easier for some model classes (Greig et al. 1989), since, intuitively speaking, it only requires finding one satisfying assignment, not counting all of them.

20.5.1 Approximate inference

Many popular probabilistic models support efficient exact inference, since they are based on chains, trees or low treewidth graphs. But there are many other models for which exact inference is intractable. In fact, even simple two node models of the form $\boldsymbol{\theta} \rightarrow \mathbf{x}$ may not support exact inference if the prior on $\boldsymbol{\theta}$ is not conjugate to the likelihood $p(\mathbf{x}|\boldsymbol{\theta})$.[5]

Therefore we will need to turn to **approximate inference** methods. Unfortunately, even this is computationally hard in general (Dagum and Luby 1993; Roth 1996). Table 20.4 summarizes some of the key approximate methods which we will discuss in coming chapters. Most of these

---

4. A 3-SAT problem is a logical expression of the form $(Q_1 \wedge Q_2 \wedge \neg Q_3) \vee (Q_1 \wedge \neg Q_4 \wedge Q_5) \cdots$, where the Q_i are binary variables, and each clause consists of the conjunction of three variables (or their negation). The goal is to find a satisfying assignment, which is a set of values for the Q_i variables such that the expression evaluates to true.
5. For discrete random variables, conjugacy is not a concern, since discrete distributions are always closed under conditioning and marginalization. Consequently, graph-theoretic considerations are of more importance when discussing inference in models with discrete hidden states.

methods do not come with any guarantee as to their accuracy or running time. Theoretical computer scientists would therefore describe them as **heuristics** rather than approximation algorithms. Fortunately, we will see that for many of these heuristic methods often perform well in practice.

Exercises

Exercise 20.1 Variable elimination

Consider the MRF in Figure 20.10(a).

a. Suppose we want to compute the partition function using the elimination ordering $\prec = (1, 2, 3, 4, 5, 6)$, i.e.,

$$\sum_{x_6, x_5, x_4, x_3, x_2, x_1} \psi_{12}(x_1, x_2) \psi_{13}(x_1, x_3) \psi_{24}(x_2, x_4) \psi_{34}(x_3, x_4) \psi_{45}(x_4, x_5) \psi_{56}(x_5, x_6) \quad (20.71)$$

If we use the variable elimination algorithm, we will create new intermediate factors. What is the largest intermediate factor?

b. Add an edge to the original MRF between every pair of variables that end up in the same factor. (These are called fill in edges.) Draw the resulting MRF. What is the size of the largest maximal clique in this graph?

c. Now consider elimination ordering $\prec = (4, 1, 2, 3, 5, 6)$, i.e.,

$$\sum_{x_6, x_5, x_3, x_2, x_1, x_4} \psi_{12}(x_1, x_2) \psi_{13}(x_1, x_3) \psi_{24}(x_2, x_4) \psi_{34}(x_3, x_4) \psi_{45}(x_4, x_5) \psi_{56}(x_5, x_6) \quad (20.72)$$

If we use the variable elimination algorithm, we will create new intermediate factors. What is the largest intermediate factor?

d. Add an edge to the original MRF between every pair of variables that end up in the same factor. (These are called fill in edges.) Draw the resulting MRF. What is the size of the largest maximal clique in this graph?

Exercise 20.2 Gaussian times Gaussian is Gaussian

Prove Equation 20.17. Hint: use completing the square.

Exercise 20.3 Message passing on a tree

Consider the DGM in Figure 20.10(b) which represents the following fictitious biological model. Each G_i represents the genotype of a person: $G_i = 1$ if they have a healthy gene and $G_i = 2$ if they have an unhealthy gene. G_2 and G_3 may inherit the unhealthy gene from their parent G_1. $X_i \in \mathbb{R}$ is a continuous measure of blood pressure, which is low if you are healthy and high if you are unhealthy. We define the CPDs as follows

$$p(G_1) = [0.5, 0.5] \quad (20.73)$$

$$p(G_2|G_1) = \begin{pmatrix} 0.9 & 0.1 \\ 0.1 & 0.9 \end{pmatrix} \quad (20.74)$$

$$p(G_3|G_1) = \begin{pmatrix} 0.9 & 0.1 \\ 0.1 & 0.9 \end{pmatrix} \quad (20.75)$$

$$p(X_i|G_i = 1) = \mathcal{N}(X_i|\mu = 50, \sigma^2 = 10) \quad (20.76)$$

$$p(X_i|G_i = 2) = \mathcal{N}(X_i|\mu = 60, \sigma^2 = 10) \quad (20.77)$$

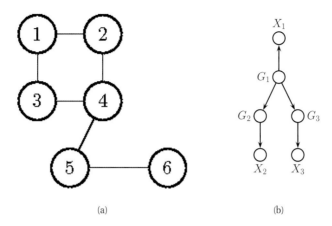

Figure 20.10 (a) An MRF on nodes. (b) A simple DAG representing inherited diseases.

The meaning of the matrix for $p(G_2|G_1)$ is that $p(G_2 = 1|G_1 = 1) = 0.9$, $p(G_2 = 1|G_1 = 2) = 0.1$, etc.

a. Suppose you observe $X_2 = 50$, and X_1 is unobserved. What is the posterior belief on G_1, i.e., $p(G_1|X_2 = 50)$?

b. Now suppose you observe $X_2 = 50$ amd $X_3 = 50$. What is $p(G_1|X_2, X_3)$? Explain your answer intuitively.

c. Now suppose $X_2 = 60$, $X_3 = 60$. What is $p(G_1|X_2, X_3)$? Explain your answer intuitively.

d. Now suppose $X_2 = 50$, $X_3 = 60$. What is $p(G_1|X_1, X_2)$? Explain your answer intuitively.

Exercise 20.4 Inference in 2D lattice MRFs

Consider an MRF with a 2D $m \times n$ lattice graph structure, so each hidden node, X_{ij}, is connected to its 4 nearest neighbors, as in an Ising model. In addition, each hidden node has its own local evidence, Y_{ij}. Assume all hidden nodes have $K > 2$ states. In general, exact inference in such models is intractable, because the maximum cliques of the corresponding triangulated graph have size $O(\max\{m, n\})$. Suppose $m \ll n$ i.e., the lattice is short and fat.

a. How can one *efficiently* perform exact inference (using a deterministic algorithm) in such models? (By exact inference, I mean computing marginal probabilities $P(X_{ij}|\vec{y})$ exactly, where $\vec{y}$ is all the evidence.) Give a *brief* description of your method.

b. What is the asymptotic complexity (running time) of your algorithm?

c. Now suppose the lattice is large and square, so $m = n$, but all hidden states are binary (ie $K = 2$). In this case, how can one efficiently exactly compute (using a deterministic algorithm) the MAP estimate $\arg\max_x P(x|y)$, where x is the joint assignment to all hidden nodes?

21 *Variational inference*

21.1 Introduction

We have now seen several algorithms for computing (functions of) a posterior distribution. For discrete graphical models, we can use the junction tree algorithm to perform exact inference, as explained in Section 20.4. However, this takes time exponential in the treewidth of the graph, rendering exact inference often impractical. Furthermore, the JTA does not work for continuous variables, except for jointly Gaussian distributions.

We can handle more general distributions providing the graph is simple, specifically if it has the form $\mathbf{x} \rightarrow \mathcal{D}$. If the prior $p(\mathbf{x})$ is conjugate to the likelihood $p(\mathcal{D}|\mathbf{x})$, we can compute the exact posterior $p(\mathbf{x}|\mathcal{D})$, as explained in Chapter 5. (Note that in this chapter, $\mathbf{x}$ represent the unknown variables, whereas in Chapter 5, we used $\boldsymbol{\theta}$ to represent the unknowns.) If the prior is not conjugate, we can use methods such as the Gaussian approximation, discussed In Section 8.4.1.

The Gaussian approximation is simple. However, some posteriors are not naturally modelled using Gaussians. For example, when inferring multinomial parameters, a Dirichlet distribution is a better choice, and when inferring discrete variables, a categorical distribution is a better choice. Furthermore, there is still the question of how to apply these approximations to arbitrary graphs.

In this chapter, we will study a general class of deterministic approximate inference algorithms based on **variational inference** (Jordan et al. 1998; Jaakkola and Jordan 2000; Jaakkola 2001; Wainwright and Jordan 2008). The basic idea is to pick an approximation $q(\mathbf{x})$ from some tractable family, and then to try to make this approximation as close as possible to the true posterior, $p^*(\mathbf{x}) \triangleq p(\mathbf{x}|\mathcal{D})$, usually by minimizing the KL divergence from p^* to q. This reduces inference to an optimization problem. By relaxing the constraints that q is a proper distribution, and/or by approximating the KL objective function, we can trade accuracy for speed.

21.2 Variational inference

Suppose $p^*(\mathbf{x})$ is our true but intractable distribution and $q(\mathbf{x})$ is some approximation, chosen from some tractable family, such as a multivariate Gaussian or a factored distribution. We assume q has some free parameters which we want to optimize so as to make q "similar to" p^*.

An obvious cost function to try to minimize is the KL divergence:

$$\mathbb{KL}\left(p^{*}||q\right)=\sum_{\mathbf{x}}p^{*}(\mathbf{x})\log\frac{p^{*}(\mathbf{x})}{q(\mathbf{x})} \tag{21.1}$$

However, this is hard to compute, since taking expectations wrt p^{*} is assumed to be intractable. A natural alternative is the reverse KL divergence:

$$\mathbb{KL}\left(q||p^{*}\right)=\sum_{\mathbf{x}}q(\mathbf{x})\log\frac{q(\mathbf{x})}{p^{*}(\mathbf{x})} \tag{21.2}$$

The main advantage of this objective is that computing expectations wrt q is tractable (by choosing a suitable form for q). We discuss the statistical differences between these two objectives in Section 21.2.2.

Unfortunately, Equation 21.2 is still not tractable as written, since even evaluating $p^{*}(\mathbf{x}) = p(\mathbf{x}|\mathcal{D})$ pointwise is hard, since it requires evaluating the intractable normalization constant $Z = p(\mathcal{D})$. However, usually the unnormalized distribution $\tilde{p}(\mathbf{x}) \triangleq p(\mathbf{x},\mathcal{D}) = p^{*}(\mathbf{x})Z$ is tractable to compute. We therefore define our new objective function as follows:

$$J(q)\quad\triangleq\quad\mathbb{KL}\left(q||\tilde{p}\right) \tag{21.3}$$

where we are slightly abusing notation, since $\tilde{p}$ is not a normalized distribution. Plugging in the definition of KL, we get

$$J(q)\quad=\quad\sum_{\mathbf{x}}q(\mathbf{x})\log\frac{q(\mathbf{x})}{\tilde{p}(\mathbf{x})} \tag{21.4}$$

$$=\quad\sum_{\mathbf{x}}q(\mathbf{x})\log\frac{q(\mathbf{x})}{Zp^{*}(\mathbf{x})} \tag{21.5}$$

$$=\quad\sum_{\mathbf{x}}q(\mathbf{x})\log\frac{q(\mathbf{x})}{p^{*}(\mathbf{x})}-\log Z \tag{21.6}$$

$$=\quad\mathbb{KL}\left(q||p^{*}\right)-\log Z \tag{21.7}$$

Since Z is a constant, by minimizing $J(q)$, we will force q to become close to p^{*}.

Since KL divergence is always non-negative, we see that $J(q)$ is an upper bound on the NLL (negative log likelihood), which we would like to minimize:

$$J(q)=\mathbb{KL}\left(q||p^{*}\right)-\log Z\geq-\log Z=-\log p(\mathcal{D}) \tag{21.8}$$

Equivalently, we can try to *maximize* the following quantity (in (Koller and Friedman 2009), this is referred to as the **energy functional**), which is a lower bound on the log likelihood of the data:

$$L(q)\triangleq-J(q)=-\mathbb{KL}\left(q||p^{*}\right)+\log Z\leq\log Z=\log p(\mathcal{D}) \tag{21.9}$$

Since this lower bound bound is tight when $q = p^{*}$, we see that variational inference is closely related to EM (see Section 11.4.7 for details).

21.2.1 Alternative interpretations of the variational objective

There are several equivalent ways of writing this objective that provide different insights. One formulation is as follows:

$$J(q) = \mathbb{E}_q \left[\log q(\mathbf{x}) \right] + \mathbb{E}_q \left[-\log \tilde{p}(\mathbf{x}) \right] = -\mathbb{H}(q) + \mathbb{E}_q \left[E(\mathbf{x}) \right] \tag{21.10}$$

which is the expected energy (recall $E(\mathbf{x}) = -\log \tilde{p}(\mathbf{x})$) minus the entropy of the system. In statistical physics, $J(q)$ is called the **variational free energy** or the **Helmholtz free energy**. (It is called "free" because the variables $\mathbf{x}$ are free to vary, rather than being fixed. The variational free energy is a function of the distribution q, whereas the regular energy is a function of the state vector $\mathbf{x}$.)

Another formulation of the objective is as follows:

$$
\begin{aligned}
J(q) &= \mathbb{E}_q \left[\log q(\mathbf{x}) - \log p(\mathbf{x}) p(\mathcal{D}|\mathbf{x}) \right] & (21.11) \\
&= \mathbb{E}_q \left[\log q(\mathbf{x}) - \log p(\mathbf{x}) - \log p(\mathcal{D}|\mathbf{x}) \right] & (21.12) \\
&= \mathbb{E}_q \left[-\log p(\mathcal{D}|\mathbf{x}) \right] + \mathbb{KL}\left(q(\mathbf{x}) || p(\mathbf{x}) \right) & (21.13)
\end{aligned}
$$

This is the expected NLL, plus a penalty term that measures how far the approximate posterior is from the exact prior.

We can also interpret the variational objective from the point of view of information theory (the so-called bits-back argument). See (Hinton and Camp 1993; Honkela and Valpola 2004), for details.

21.2.2 Forward or reverse KL? *

Since the KL divergence is not symmetric in its arguments, minimizing $\mathbb{KL}(q||p)$ wrt q will give different behavior than minimizing $\mathbb{KL}(p||q)$. Below we discuss these two different methods.

First, consider the reverse KL, $\mathbb{KL}(q||p)$, also known as an **I-projection** or **information projection**. By definition, we have

$$\mathbb{KL}(q||p) = \sum_{\mathbf{x}} q(\mathbf{x}) \ln \frac{q(\mathbf{x})}{p(\mathbf{x})} \tag{21.14}$$

This is infinite if $p(\mathbf{x}) = 0$ and $q(\mathbf{x}) > 0$. Thus if $p(\mathbf{x}) = 0$ we must ensure $q(\mathbf{x}) = 0$. We say that the reverse KL is **zero forcing** for q. Hence q will typically under-estimate the support of p.

Now consider the forwards KL, also known as an **M-projection** or **moment projection**:

$$\mathbb{KL}(p||q) = \sum_{\mathbf{x}} p(\mathbf{x}) \ln \frac{p(\mathbf{x})}{q(\mathbf{x})} \tag{21.15}$$

This is infinite if $q(\mathbf{x}) = 0$ and $p(\mathbf{x}) > 0$. So if $p(\mathbf{x}) > 0$ we must ensure $q(\mathbf{x}) > 0$. We say that the forwards KL is **zero avoiding** for q. Hence q will typically over-estimate the support of p.

The difference between these methods is illustrated in Figure 21.1. We see that when the true distribution is multimodal, using the forwards KL is a bad idea (assuming q is constrained to be unimodal), since the resulting posterior mode/mean will be in a region of low density, right

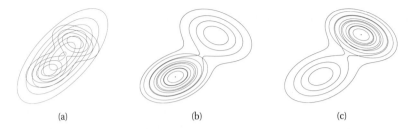

(a)   (b)   (c)

Figure 21.1   Illustrating forwards vs reverse KL on a bimodal distribution. The blue curves are the contours of the true distribution p. The red curves are the contours of the unimodal approximation q. (a) Minimizing forwards KL: q tends to "cover" p. (b-c) Minimizing reverse KL: q locks on to one of the two modes. Based on Figure 10.3 of (Bishop 2006). Figure generated by `KLfwdReverseMixGauss`.

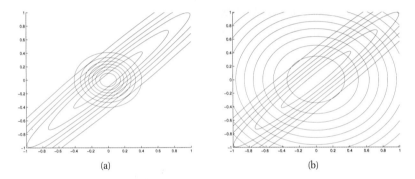

(a)   (b)

Figure 21.2   Illustrating forwards vs reverse KL on a symmetric Gaussian. The blue curves are the contours of the true distribution p. The red curves are the contours of a factorized approximation q. (a) Minimizing $\mathbb{KL}(q||p)$. (b) Minimizing $\mathbb{KL}(p||q)$. Based on Figure 10.2 of (Bishop 2006). Figure generated by `KLpqGauss`.

between the two peaks. In such contexts, the reverse KL is not only more tractable to compute, but also more sensible statistically.

Another example of the difference is shown in Figure 21.2, where the target distribution is an elongated 2d Gaussian and the approximating distribution is a product of two 1d Gaussians. That is, $p(\mathbf{x}) = \mathcal{N}(\mathbf{x}|\boldsymbol{\mu}, \boldsymbol{\Lambda}^{-1})$, where

$$\boldsymbol{\mu} = \begin{pmatrix} \mu_1 \\ \mu_2 \end{pmatrix}, \quad \boldsymbol{\Lambda} = \begin{pmatrix} \Lambda_{11} & \Lambda_{12} \\ \Lambda_{21} & \Lambda_{22} \end{pmatrix} \tag{21.16}$$

In Figure 21.2(a) we show the result of minimizing $\mathbb{KL}(q||p)$. In this simple example, one can show that the solution has the form

$$q(\mathbf{x}) = \mathcal{N}(x_1|m_1, \Lambda_{11}^{-1})\mathcal{N}(x_2|m_2, \Lambda_{22}^{-1}) \tag{21.17}$$

$$m_1 = \mu_1 - \Lambda_{11}^{-1}\Lambda_{12}(m_2 - \mu_2) \tag{21.18}$$

$$m_2 = \mu_2 - \Lambda_{22}^{-1}\Lambda_{21}(m_1 - \mu_1) \tag{21.19}$$

Figure 21.2(a) shows that we have correctly captured the mean, but the approximation is too compact: its variance is controlled by the direction of smallest variance of p. In fact, it is often the case (although not always (Turner et al. 2008)) that minimizing $\mathbb{KL}\left(q||p\right)$, where q is factorized, results in an approximation that is overconfident.

In Figure 21.2(b), we show the result of minimizing $\mathbb{KL}\left(p||q\right)$. As we show in Exercise 21.7, the optimal solution when minimizing the forward KL wrt a factored approximation is to set q to be the product of marginals. Thus the solution has the form

$$q(\mathbf{x}) = \mathcal{N}(x_1|\mu_1, \Sigma_{11})\mathcal{N}(x_2|\mu_2, \Sigma_{22}) \tag{21.20}$$

where $\boldsymbol{\Sigma} = \boldsymbol{\Lambda}^{-1}$. Figure 21.2(b) shows that this is too broad, since it is an over-estimate of the support of p.

For the rest of this chapter, and for most of the next, we will focus on minimizing $\mathbb{KL}\left(q||p\right)$. In Section 22.5, when we discuss expectation proagation, we will discuss ways to locally optimize $\mathbb{KL}\left(p||q\right)$.

One can create a family of divergence measures indexed by a parameter $\alpha \in \mathbb{R}$ by defining the **alpha divergence** as follows:

$$D_\alpha(p||q) \triangleq \frac{4}{1-\alpha^2} \left(1 - \int p(x)^{(1+\alpha)/2} q(x)^{(1-\alpha)/2} dx \right) \tag{21.21}$$

This measure satisfies $D_\alpha(p||q) = 0$ iff $p = q$, but is obviously not symmetric, and hence is not a metric. $\mathbb{KL}\left(p||q\right)$ corresponds to the limit $\alpha \to 1$, whereas $\mathbb{KL}\left(q||p\right)$ corresponds to the limit $\alpha \to -1$. When $\alpha = 0$, we get a symmetric divergence measure that is linearly related to the **Hellinger distance**, defined by

$$D_H(p||q) \triangleq \int \left(p(x)^{\frac{1}{2}} - q(x)^{\frac{1}{2}} \right)^2 dx \tag{21.22}$$

Note that $\sqrt{D_H(p||q)}$ is a valid distance metric, that is, it is symmetric, non-negative and satisfies the triangle inequality. See (Minka 2005) for details.

21.3 The mean field method

One of the most popular forms of variational inference is called the **mean field** approximation (Opper and Saad 2001). In this approach, we assume the posterior is a fully factorized approximation of the form

$$q(\mathbf{x}) = \prod_{i=1}^{D} q_i(\mathbf{x}_i) \tag{21.23}$$

Our goal is to solve this optimization problem:

$$\min_{q_1,\dots,q_D} \mathbb{KL}\left(q||p\right) \tag{21.24}$$

where we optimize over the parameters of each marginal distribution q_i. In Section 21.3.1, we derive a coordinate descent method for solving this problem. We show that at each step we make the following update:

$$\log q_j(\mathbf{x}_j) = \mathbb{E}_{-q_j}\left[\log \tilde{p}(\mathbf{x})\right] + \text{const} \tag{21.25}$$

Model	Section
Ising model	Section 21.3.2
Factorial HMM	Section 21.4.1
Univariate Gaussian	Section 21.5.1
Linear regression	Section 21.5.2
Logistic regression	Section 21.8.1.1
Mixtures of Gaussians	Section 21.6.1
Latent Dirichlet allocation	Section 27.3.6.3

Table 21.1 Some models in this book for which we provide detailed derivations of the mean field inference algorithm.

where $\tilde{p}(\mathbf{x}) = p(\mathbf{x}, \mathcal{D})$ is the unnormalized posterior and the notation $\mathbb{E}_{-q_j}[f(\mathbf{x})]$ means to take the expectation over $f(\mathbf{x})$ with respect to all the variables except for x_j. For example, if we have three variables, then

$$\mathbb{E}_{-q_2}[f(\mathbf{x})] = \sum_{x_1}\sum_{x_3} q(x_1)q_3(x_3)f(x_1, x_2, x_3) \tag{21.26}$$

where sums get replaced by integrals where necessary.

When updating q_j, we only need to reason about the variables which share a factor with x_j, i.e., the terms in j's Markov blanket (see Section 10.5.3); the other terms get absorbed into the constant term. Since we are replacing the neighboring values by their mean value, the method is known as mean field. This is very similar to Gibbs sampling (Section 24.2), except instead of sending sampled values between neighboring nodes, we send mean values between nodes. This tends to be more efficient, since the mean can be used as a proxy for a large number of samples. (On the other hand, mean field messages are dense, whereas samples are sparse; this can make sampling more scalable to very large models.)

Of course, updating one distribution at a time can be slow, since it is a form of coordinate descent. Several methods have been proposed to speed up this basic approach, including using pattern search (Honkela et al. 2003), and techniques based on parameter expansion (Qi and Jaakkola 2008). However, we will not consider these methods in this chapter.

It is important to note that the mean field method can be used to infer discrete or continuous latent quantities, using a variety of parametric forms for q_i, as we will see below. This is in contrast to some of the other variational methods we will encounter later, which are more restricted in their applicability. Table 21.1 lists some of the examples of mean field that we cover in this book.

21.3.1 Derivation of the mean field update equations

Recall that the goal of variational inference is to minimize the upper bound $J(q) \geq -\log p(\mathcal{D})$. Equivalently, we can try to maximize the lower bound

$$L(q) \triangleq -J(q) = \sum_{\mathbf{x}} q(\mathbf{x})\log\frac{\tilde{p}(\mathbf{x})}{q(\mathbf{x})} \leq \log p(\mathcal{D}) \tag{21.27}$$

We will do this one term at a time.

If we write the objective singling out the terms that involve q_j, and regarding all the other terms as constants, we get

$$
L(q_j) = \sum_{\mathbf{x}} \prod_i q_i(\mathbf{x}_i) \left[\log \tilde{p}(\mathbf{x}) - \sum_k \log q_k(\mathbf{x}_k) \right] \tag{21.28}
$$

$$
= \sum_{\mathbf{x}_j} \sum_{\mathbf{x}_{-j}} q_j(\mathbf{x}_j) \prod_{i \neq j} q_i(\mathbf{x}_i) \left[\log \tilde{p}(\mathbf{x}) - \sum_k \log q_k(\mathbf{x}_k) \right] \tag{21.29}
$$

$$
= \sum_{\mathbf{x}_j} q_j(\mathbf{x}_j) \sum_{\mathbf{x}_{-j}} \prod_{i \neq j} q_i(\mathbf{x}_i) \log \tilde{p}(\mathbf{x})
$$

$$
- \sum_{\mathbf{x}_j} q_j(\mathbf{x}_j) \sum_{\mathbf{x}_{-j}} \prod_{i \neq j} q_i(\mathbf{x}_i) \left[\sum_{k \neq j} \log q_k(\mathbf{x}_k) + \log q_j(\mathbf{x}_j) \right] \tag{21.30}
$$

$$
= \sum_{\mathbf{x}_j} q_j(\mathbf{x}_j) \log f_j(\mathbf{x}_j) - \sum_{\mathbf{x}_j} q_j(\mathbf{x}_j) \log q_j(\mathbf{x}_j) + \text{const} \tag{21.31}
$$

where

$$
\log f_j(\mathbf{x}_j) \triangleq \sum_{\mathbf{x}_{-j}} \prod_{i \neq j} q_i(\mathbf{x}_i) \log \tilde{p}(\mathbf{x}) = \mathbb{E}_{-q_j} \left[\log \tilde{p}(\mathbf{x}) \right] \tag{21.32}
$$

So we average out all the hidden variables except for $\mathbf{x}_j$. Thus we can rewrite $L(q_j)$ as follows:

$$
L(q_j) = -\mathbb{KL} \left(q_j || f_j \right) + \text{const} \tag{21.33}
$$

We can maximize L by minimizing this KL, which we can do by setting $q_j = f_j$, as follows:

$$
q_j(\mathbf{x}_j) = \frac{1}{Z_j} \exp \left(\mathbb{E}_{-q_j} \left[\log \tilde{p}(\mathbf{x}) \right] \right) \tag{21.34}
$$

We can usually ignore the local normalization constant Z_j, since we know q_j must be a normalized distribution. Hence we usually work with the form

$$
\log q_j(\mathbf{x}_j) = \mathbb{E}_{-q_j} \left[\log \tilde{p}(\mathbf{x}) \right] + \text{const} \tag{21.35}
$$

The functional form of the q_j distributions will be determined by the type of variables $\mathbf{x}_j$, as well as the form of the model. (This is sometimes called **free-form optimization**.) If x_j is a discrete random variable, then q_j will be a discrete distribution; if $\mathbf{x}_j$ is a continuous random variable, then q_j will be some kind of pdf. We will see examples of this below.

21.3.2 Example: mean field for the Ising model

Consider the image denoising example from Section 19.4.1, where $x_i \in \{-1, +1\}$ are the hidden pixel values of the "clean" image. We have a joint model of the form

$$
p(\mathbf{x}, \mathbf{y}) = p(\mathbf{x})p(\mathbf{y}|\mathbf{x}) \tag{21.36}
$$

where the prior has the form of an Ising model

$$p(\mathbf{x}) \quad = \quad \frac{1}{Z_0} \exp(-E_0(\mathbf{x})) \tag{21.37}$$

$$E_0(\mathbf{x}) \quad = \quad -\sum_{i=1}^{D} \sum_{j \in \text{nbr}_i} W_{ij} x_i x_j \tag{21.38}$$

and the likelihood has the form

$$p(\mathbf{y}|\mathbf{x}) = \prod_i p(\mathbf{y}_i|x_i) = \exp(\sum_i -L_i(x_i)) \tag{21.39}$$

Therefore the posterior has the form

$$p(\mathbf{x}|\mathbf{y}) \quad = \quad \frac{1}{Z} \exp(-E(\mathbf{x})) \tag{21.40}$$

$$E(\mathbf{x}) \quad = \quad E_0(\mathbf{x}) - \sum_i L_i(x_i) \tag{21.41}$$

We will now approximate this by a fully factored approximation

$$q(\mathbf{x}) = \prod_i q(x_i, \mu_i) \tag{21.42}$$

where μ_i is the mean value of node i. To derive the update for the variational parameter μ_i, we first write out $\log \tilde{p}(\mathbf{x}) = -E(\mathbf{x})$, dropping terms that do not involve x_i:

$$\log \tilde{p}(\mathbf{x}) = x_i \sum_{j \in \text{nbr}_i} W_{ij} x_j + L_i(x_i) + \text{const} \tag{21.43}$$

This only depends on the states of the neighboring nodes. Now we take expectations of this wrt $\prod_{j \neq i} q_j(x_j)$ to get

$$q_i(x_i) \propto \exp \left(x_i \sum_{j \in \text{nbr}_i} W_{ij} \mu_j + L_i(x_i) \right) \tag{21.44}$$

Thus we replace the states of the neighbors by their average values. Let

$$m_i = \sum_{j \in \text{nbr}_i} W_{ij} \mu_j \tag{21.45}$$

be the mean field influence on node i. Also, let $L_i^+ \triangleq L_i(+1)$ and $L_i^- \triangleq L_i(-1)$. The approximate marginal posterior is given by

$$q_i(x_i = 1) \quad = \quad \frac{e^{m_i + L_i^+}}{e^{m_i + L_i^+} + e^{-m_i + L_i^-}} = \frac{1}{1 + e^{-2m_i + L_i^- - L_i^+}} = \text{sigm}(2a_i) \tag{21.46}$$

$$a_i \quad \triangleq \quad m_i + 0.5(L_i^+ - L_i^-) \tag{21.47}$$

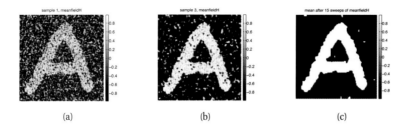

Figure 21.3 Example of image denoising using mean field (with parallel updates and a damping factor of 0.5). We use an Ising prior with $W_{ij} = 1$ and a Gaussian noise model with $\sigma = 2$. We show the results after 1, 3 and 15 iterations across the image. Compare to Figure 24.1. Figure generated by `isingImageDenoiseDemo`.

Similarly, we have $q_i(x_i = -1) = \text{sigm}(-2a_i)$. From this we can compute the new mean for site i:

$$\mu_i = \mathbb{E}_{q_i}[x_i] = q_i(x_i = +1) \cdot (+1) + q_i(x_i = -1) \cdot (-1) \tag{21.48}$$

$$= \frac{1}{1 + e^{-2a_i}} - \frac{1}{1 + e^{2a_i}} = \frac{e^{a_i}}{e^{a_i} + e^{-a_i}} - \frac{e^{-a_i}}{e^{-a_i} + e^{a_i}} = \tanh(a_i) \tag{21.49}$$

Hence the update equation becomes

$$\mu_i = \tanh\left(\sum_{j \in \text{nbr}_i} W_{ij}\mu_j + 0.5(L_i^+ - L_i^-)\right) \tag{21.50}$$

See also Exercise 21.6 for an alternative derivation of these equations.

We can turn the above equations in to a fixed point algorithm by writing

$$\mu_i^t = \tanh\left(\sum_{j \in \text{nbr}_i} W_{ij}\mu_j^{t-1} + 0.5(L_i^+ - L_i^-)\right) \tag{21.51}$$

It is usually better to use **damped updates** of the form

$$\mu_i^t = (1 - \lambda)\mu_i^{t-1} + \lambda \tanh\left(\sum_{j \in \text{nbr}_i} W_{ij}\mu_j^{t-1} + 0.5(L_i^+ - L_i^-)\right) \tag{21.52}$$

for $0 < \lambda < 1$. We can update all the nodes in parallel, or update them asynchronously.

Figure 21.3 shows the method in action, applied to a 2d Ising model with homogeneous attractive potentials, $W_{ij} = 1$. We use parallel updates with a damping factor of $\lambda = 0.5$. (If we don't use damping, we tend to get "checkerboard" artefacts.)

21.4 Structured mean field *

Assuming that all the variables are independent in the posterior is a very strong assumption that can lead to poor results. Sometimes we can exploit **tractable substructure** in our problem, so

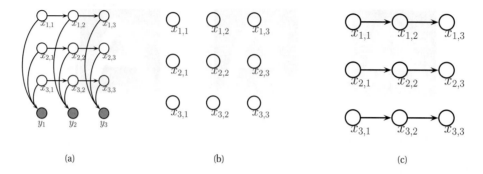

Figure 21.4   (a) A factorial HMM with 3 chains. (b) A fully factorized approximation. (c) A product-of-chains approximation. Based on Figure 2 of (Ghahramani and Jordan 1997).

that we can efficiently handle some kinds of dependencies. This is called the **structured mean field** approach (Saul and Jordan 1995). The approach is the same as before, except we group sets of variables together, and we update them simultaneously. (This follows by simply treating all the variables in the i'th group as a single "mega-variable", and then repeating the derivation in Section 21.3.1.) As long as we can perform efficient inference in each q_i, the method is tractable overall. We give an example below. See (Bouchard-Cote and Jordan 2009) for some more recent work in this area.

21.4.1   Example: factorial HMM

Consider the factorial HMM model (Ghahramani and Jordan 1997) introduced in Section 17.6.5. Suppose there are M chains, each of length T, and suppose each hidden node has K states. The model is defined as follows

$$p(\mathbf{x}, \mathbf{y}) = \prod_m \prod_t p(x_{tm} | x_{t-1,m}) p(\mathbf{y}_t | x_{tm}) \tag{21.53}$$

where $p(x_{tm} = k | x_{t-1,m} = j) = A_{mjk}$ is an entry in the transition matrix for chain m, $p(x_{1m} = k | x_{0m}) = p(x_{1m} = k) = \pi_{mk}$, is the initial state distribution for chain m, and

$$p(\mathbf{y}_t | \mathbf{x}_t) = \mathcal{N}\left(\mathbf{y}_t | \sum_{m=1}^{M} \mathbf{W}_m \mathbf{x}_{tm}, \boldsymbol{\Sigma}\right) \tag{21.54}$$

is the observation model, where $\mathbf{x}_{tm}$ is a 1-of-K encoding of x_{tm} and $\mathbf{W}_m$ is a $D \times K$ matrix (assuming $\mathbf{y}_t \in \mathbb{R}^D$). Figure 21.4(a) illustrates the model for the case where $M = 3$. Even though each chain is a priori independent, they become coupled in the posterior due to having an observed common child, $\mathbf{y}_t$. The junction tree algorithm applied to this graph takes $O(TMK^{M+1})$ time. Below we will derive a structured mean field algorithm that takes $O(TMK^2 I)$ time, where I is the number of mean field iterations (typically $I \sim 10$ suffices for good performance).

We can write the exact posterior in the following form:

$$p(\mathbf{x}|\mathbf{y}) \;=\; \frac{1}{Z}\exp(-E(\mathbf{x},\mathbf{y})) \tag{21.55}$$

$$E(\mathbf{x},\mathbf{y}) \;=\; \frac{1}{2}\sum_{t=1}^{T}\left(\mathbf{y}_t-\sum_m \mathbf{W}_m\mathbf{x}_{tm}\right)^{T}\boldsymbol{\Sigma}^{-1}\left(\mathbf{y}_t-\sum_m \mathbf{W}_m\mathbf{x}_{tm}\right)$$
$$-\sum_m \mathbf{x}_{1m}^{T}\tilde{\boldsymbol{\pi}}_m-\sum_{t=2}^{T}\sum_m \mathbf{x}_{tm}^{T}\tilde{\mathbf{A}}_m\mathbf{x}_{t-1,m} \tag{21.56}$$

where $\tilde{\mathbf{A}}_m \triangleq \log \mathbf{A}_m$ and $\tilde{\boldsymbol{\pi}}_m \triangleq \log \boldsymbol{\pi}_m$ (both interpreted elementwise).

We can approximate the posterior as a product of marginals, as in Figure 21.4(b), but a better approximation is to use a product of chains, as in Figure 21.4(c). Each chain can be tractably updated individually, using the forwards-backwards algorithm. More precisely, we assume

$$q(\mathbf{x}|\mathbf{y}) \;=\; \frac{1}{Z_q}\prod_{m=1}^{M} q(x_{1m}|\boldsymbol{\xi}_{1m})\prod_{t=2}^{T} q(x_{tm}|x_{t-1,m},\boldsymbol{\xi}_{tm}) \tag{21.57}$$

$$q(x_{1m}|\boldsymbol{\xi}_{1m}) \;=\; \prod_{k=1}^{K}(\xi_{1mk}\pi_{mk})^{x_{1mk}} \tag{21.58}$$

$$q(x_{tm}|x_{t-1,m},\boldsymbol{\xi}_{tm}) \;=\; \prod_{k=1}^{K}\left(\xi_{tmk}\prod_{j=1}^{K}(A_{mjk})^{x_{t-1,m,j}}\right)^{x_{tmk}} \tag{21.59}$$

We see that the ξ_{tmk} parameters play the role of an approximate local evidence, averaging out the effects of the other chains. This is contrast to the exact local evidence, which couples all the chains together.

We can rewrite the approximate posterior as $q(\mathbf{x})=\frac{1}{Z_q}\exp(-E_q(\mathbf{x}))$, where

$$E_q(\mathbf{x}) \;=\; -\sum_{t=1}^{T}\sum_{m=1}^{M}\mathbf{x}_{tm}^{T}\tilde{\boldsymbol{\xi}}_{tm}-\sum_{m=1}^{M}\mathbf{x}_{1m}^{T}\tilde{\boldsymbol{\pi}}_m-\sum_{t=2}^{T}\sum_{m=1}^{M}\mathbf{x}_{tm}^{T}\tilde{\mathbf{A}}_m\mathbf{x}_{t-1,m} \tag{21.60}$$

where $\tilde{\boldsymbol{\xi}}_{tm}=\log\boldsymbol{\xi}_{tm}$. We see that this has the same temporal factors as the exact posterior, but the local evidence term is different. The objective function is given by

$$\mathbb{KL}\,(q||p) = \mathbb{E}\,[E] - \mathbb{E}\,[E_q] - \log Z_q + \log Z \tag{21.61}$$

where the expectations are taken wrt q. One can show (Exercise 21.8) that the update has the form

$$\boldsymbol{\xi}_{tm} \;=\; \exp\left(\mathbf{W}_m^{T}\boldsymbol{\Sigma}^{-1}\tilde{\mathbf{y}}_{tm}-\frac{1}{2}\boldsymbol{\delta}_m\right) \tag{21.62}$$

$$\boldsymbol{\delta}_m \;\triangleq\; \mathrm{diag}(\mathbf{W}_m^{T}\boldsymbol{\Sigma}^{-1}\mathbf{W}_m) \tag{21.63}$$

$$\tilde{\mathbf{y}}_{tm} \;\triangleq\; \mathbf{y}_t-\sum_{\ell\neq m}^{M}\mathbf{W}_\ell\mathbb{E}\,[\mathbf{x}_{t,\ell}] \tag{21.64}$$

The $\boldsymbol{\xi}_{tm}$ parameter plays the role of the local evidence, averaging over the neighboring chains. Having computed this for each chain, we can perform forwards-backwards in parallel, using these approximate local evidence terms to compute $q(\mathbf{x}_{t,m}|\mathbf{y}_{1:T})$ for each m and t.

The update cost is $O(TMK^2)$ for a full "sweep" over all the variational parameters, since we have to run forwards-backwards M times, for each chain independently. This is the same cost as a fully factorized approximation, but is much more accurate.

21.5   Variational Bayes

So far we have been concentrating on inferring latent variables $\mathbf{z}_i$ assuming the parameters $\boldsymbol{\theta}$ of the model are known. Now suppose we want to infer the parameters themselves. If we make a fully factorized (i.e., mean field) approximation, $p(\boldsymbol{\theta}|\mathcal{D}) \approx \prod_k q(\boldsymbol{\theta}_k)$, we get a method known as **variational Bayes** or **VB** (Hinton and Camp 1993; MacKay 1995a; Attias 2000; Beal and Ghahramani 2006; Smidl and Quinn 2005).[1] We give some examples of VB below, assuming that there are no latent variables. If we want to infer both latent variables and parameters, and we make an approximation of the form $p(\boldsymbol{\theta}, \mathbf{z}_{1:N}|\mathcal{D}) \approx q(\boldsymbol{\theta}) \prod_i q_i(\mathbf{z}_i)$, we get a method known as variational Bayes EM, which we described in Section 21.6.

The algebra needed to derive VB algorithms can sometimes be a bit complicated. However, the effort is often worthwhile, since the resulting methods are usually as fast as MAP estimation but enjoy the statistical benefits of the Bayesian approach.

21.5.1   Example: VB for a univariate Gaussian

Following (MacKay 2003, p429), let us consider how to apply VB to infer the posterior over the parameters for a 1d Gaussian, $p(\mu, \lambda|\mathcal{D})$, where $\lambda = 1/\sigma^2$ is the precision. For convenience, we will use a conjugate prior of the form

$$p(\mu, \lambda) = \mathcal{N}(\mu|\mu_0, (\kappa_0\lambda)^{-1})\mathrm{Ga}(\lambda|a_0, b_0) \tag{21.65}$$

However, we will use an approximate factored posterior of the form

$$q(\mu, \lambda) = q_\mu(\mu)q_\lambda(\lambda) \tag{21.66}$$

We do not need to specify the forms for the distributions q_μ and q_λ; the optimal forms will "fall out" automatically during the derivation (and conveniently, they turn out to be Gaussian and Gamma respectively).

You might wonder why we would want to do this, since we know how to compute the exact posterior for this model (Section 4.6.3.7). There are two reasons. First, it is a useful pedagogical exercise, since we can compare the quality of our approximation to the exact posterior. Second, it is simple to modify the method to handle a semi-conjugate prior of the form $p(\mu, \lambda) = \mathcal{N}(\mu|\mu_0, \tau_0)\mathrm{Ga}(\lambda|a_0, b_0)$, for which exact inference is no longer possible.

---

1. This method was originally called **ensemble learning** (MacKay 1995a), since we are using an ensemble of parameters (a distribution) instead of a point estimate. However, the term "ensemble learning" is also used to describe methods such as boosting, so we prefer the term VB.

21.5.1.1 Target distribution

The unnormalized log posterior has the form

$$
\begin{aligned}
\log \tilde{p}(\mu, \lambda) &= \log p(\mu, \lambda, \mathcal{D}) = \log p(\mathcal{D}|\mu, \lambda) + \log p(\mu|\lambda) + \log p(\lambda) & (21.67) \\
&= \frac{N}{2} \log \lambda - \frac{\lambda}{2} \sum_{i=1}^{N} (x_i - \mu)^2 - \frac{\kappa_0 \lambda}{2} (\mu - \mu_0)^2 \\
&\quad + \frac{1}{2} \log(\kappa_0 \lambda) + (a_0 - 1) \log \lambda - b_0 \lambda + \text{const} & (21.68)
\end{aligned}
$$

21.5.1.2 Updating $q_\mu(\mu)$

The optimal form for $q_\mu(\mu)$ is obtained by averaging over λ:

$$
\begin{aligned}
\log q_\mu(\mu) &= \mathbb{E}_{q_\lambda} \left[\log p(\mathcal{D}|\mu, \lambda) + \log p(\mu|\lambda) \right] + \text{const} & (21.69) \\
&= -\frac{\mathbb{E}_{q_\lambda}[\lambda]}{2} \left\{ \kappa_0(\mu - \mu_0)^2 + \sum_{i=1}^{N}(x_i - \mu)^2 \right\} + \text{const} & (21.70)
\end{aligned}
$$

By completing the square one can show that $q_\mu(\mu) = \mathcal{N}(\mu|\mu_N, \kappa_N^{-1})$, where

$$
\mu_N = \frac{\kappa_0 \mu_0 + N\overline{x}}{\kappa_0 + N}, \quad \kappa_N = (\kappa_0 + N)\mathbb{E}_{q_\lambda}[\lambda] \tag{21.71}
$$

At this stage we don't know what $q_\lambda(\lambda)$ is, and hence we cannot compute $\mathbb{E}[\lambda]$, but we will derive this below.

21.5.1.3 Updating $q_\lambda(\lambda)$

The optimal form for $q_\lambda(\lambda)$ is given by

$$
\begin{aligned}
\log q_\lambda(\lambda) &= \mathbb{E}_{q_\mu} \left[\log p(\mathcal{D}|\mu, \lambda) + \log p(\mu|\lambda) + \log p(\lambda) \right] + \text{const} & (21.72) \\
&= (a_0 - 1) \log \lambda - b_0 \lambda + \frac{1}{2} \log \lambda + \frac{N}{2} \log \lambda \\
&\quad -\frac{\lambda}{2} \mathbb{E}_{q_\mu} \left[\kappa_0(\mu - \mu_0)^2 + \sum_{i=1}^{N}(x_i - \mu)^2 \right] + \text{const} & (21.73)
\end{aligned}
$$

We recognize this as the log of a Gamma distribution, hence $q_\lambda(\lambda) = \text{Ga}(\lambda|a_N, b_N)$, where

$$
a_N = a_0 + \frac{N+1}{2} \tag{21.74}
$$

$$
b_N = b_0 + \frac{1}{2} \mathbb{E}_{q_\mu} \left[\kappa_0(\mu - \mu_0)^2 + \sum_{i=1}^{N}(x_i - \mu)^2 \right] \tag{21.75}
$$

21.5.1.4 Computing the expectations

To implement the updates, we have to specify how to compute the various expectations. Since $q(\mu) = \mathcal{N}(\mu | \mu_N, \kappa_N^{-1})$, we have

$$
\mathbb{E}_{q(\mu)}[\mu] = \mu_N \tag{21.76}
$$

$$
\mathbb{E}_{q(\mu)}[\mu^2] = \frac{1}{\kappa_N} + \mu_N^2 \tag{21.77}
$$

Since $q(\lambda) = \text{Ga}(\lambda | a_N, b_N)$, we have

$$
\mathbb{E}_{q(\lambda)}[\lambda] = \frac{a_N}{b_N} \tag{21.78}
$$

We can now give explicit forms for the update equations. For $q(\mu)$ we have

$$
\mu_N = \frac{\kappa_0 \mu_0 + N\overline{x}}{\kappa_0 + N} \tag{21.79}
$$

$$
\kappa_N = (\kappa_0 + N)\frac{a_N}{b_N} \tag{21.80}
$$

and for $q(\lambda)$ we have

$$
a_N = a_0 + \frac{N+1}{2} \tag{21.81}
$$

$$
b_N = b_0 + \kappa_0(\mathbb{E}[\mu^2] + \mu_0^2 - 2\mathbb{E}[\mu]\mu_0) + \frac{1}{2}\sum_{i=1}^{N}\left(x_i^2 + \mathbb{E}[\mu^2] - 2\mathbb{E}[\mu]x_i\right) \tag{21.82}
$$

We see that μ_N and a_N are in fact fixed constants, and only κ_N and b_N need to be updated iteratively. (In fact, one can solve for the fixed points of κ_N and b_N analytically, but we don't do this here in order to illustrate the iterative updating scheme.)

21.5.1.5 Illustration

Figure 21.5 gives an example of this method in action. The green contours represent the exact posterior, which is Gaussian-Gamma. The dotted red contours represent the variational approximation over several iterations. We see that the final approximation is reasonably close to the exact solution. However, it is more "compact" than the true distribution. It is often the case that mean field inference underestimates the posterior uncertainty; See Section 21.2.2 for more discussion of this point.

21.5.1.6 Lower bound *

In VB, we are maximizing $L(q)$, which is a lower bound on the log marginal likelihood:

$$
L(q) \leq \log p(\mathcal{D}) = \log \int \int p(\mathcal{D}|\mu, \lambda)p(\mu, \lambda)d\mu d\lambda \tag{21.83}
$$

It is very useful to compute the lower bound itself, for three reasons. First, it can be used to assess convergence of the algorithm. Second, it can be used to assess the correctness of one's

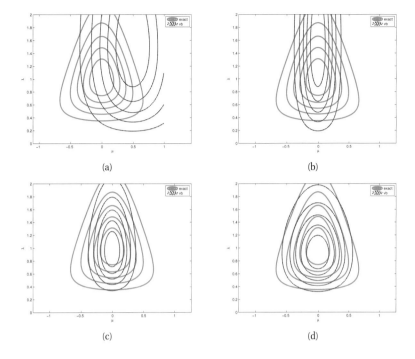

(a)

(b)

(c)

(d)

Figure 21.5 Factored variational approximation (red) to the Gaussian-Gamma distribution (green). (a) Initial guess. (b) After updating q_μ. (c) After updating q_λ. (d) At convergence (after 5 iterations). Based on 10.4 of (Bishop 2006). Figure generated by `unigaussVbDemo`.

code: as with EM, if the bound does not increase monotonically, there must be a bug. Third, the bound can be used as an approximation to the marginal likelihood, which can be used for Bayesian model selection.

Unfortunately, computing this lower bound involves a fair amount of tedious algebra. We work out the details for this example, but for other models, we will just state the results without proof, or even omit discussion of the bound altogether, for brevity.

For this model, $L(q)$ can be computed as follows:

$$
\begin{aligned}
L(q) &= \int \int q(\mu, \lambda) \log \frac{p(\mathcal{D}, \mu, \lambda)}{q(\mu, \lambda)} d\mu d\lambda && (21.84) \\
&= \mathbb{E}\left[\log p(\mathcal{D}|\mu, \lambda)\right] + \mathbb{E}\left[\log p(\mu|\lambda)\right] + \mathbb{E}\left[\log p(\lambda)\right] \\
&\quad - \mathbb{E}\left[\log q(\mu)\right] - \mathbb{E}\left[\log q(\lambda)\right] && (21.85)
\end{aligned}
$$

where all expectations are wrt $q(\mu, \lambda)$. We recognize the last two terms as the entropy of a Gaussian and the entropy of a Gamma distribution, which are given by

$$
\begin{aligned}
\mathbb{H}\left(\mathcal{N}(\mu_N, \kappa_N^{-1})\right) &= -\frac{1}{2}\log \kappa_N + \frac{1}{2}(1 + \log(2\pi)) && (21.86) \\
\mathbb{H}\left(\text{Ga}(a_N, b_N)\right) &= \log \Gamma(a_N) - (a_N - 1)\psi(a_N) - \log(b_N) + a_N && (21.87)
\end{aligned}
$$

where $\psi()$ is the digamma function.

To compute the other terms, we need the following facts:

$$\mathbb{E}\left[\log x | x \sim \text{Ga}(a, b)\right] \;=\; \psi(a) - \log(b) \tag{21.88}$$

$$\mathbb{E}\left[x | x \sim \text{Ga}(a, b)\right] \;=\; \frac{a}{b} \tag{21.89}$$

$$\mathbb{E}\left[x | x \sim \mathcal{N}(\mu, \sigma^2)\right] \;=\; \mu \tag{21.90}$$

$$\mathbb{E}\left[x^2 | x \sim \mathcal{N}(\mu, \sigma^2)\right] \;=\; \mu + \sigma^2 \tag{21.91}$$

For the expected log likelihood, one can show that

$$\mathbb{E}_{q(\mu,\lambda)}\left[\log p(\mathcal{D}|\mu, \lambda)\right] \tag{21.92}$$

$$= \;-\frac{N}{2}\log(2\pi) + \frac{N}{2}\mathbb{E}_{q(\lambda)}\left[\log \lambda\right] - \frac{\mathbb{E}\left[\lambda\right]_{q(\lambda)}}{2}\sum_{i=1}^{N}\mathbb{E}_{q(\mu)}\left[(x_i - \mu)^2\right]$$

$$= \;-\frac{N}{2}\log(2\pi) + \frac{N}{2}\left(\psi(a_N) - \log b_N\right) \tag{21.93}$$

$$-\frac{N a_N}{2 b_N}\left(\hat{\sigma}^2 + \bar{x}^2 - 2\mu_N \bar{x} + \mu_N^2 + \frac{1}{\kappa_N}\right) \tag{21.94}$$

where $\bar{x}$ and $\hat{\sigma}^2$ are the empirical mean and variance.

For the expected log prior of λ, we have

$$\mathbb{E}_{q(\lambda)}\left[\log p(\lambda)\right] \;=\; (a_0 - 1)\mathbb{E}\left[\log \lambda\right] - b_0 \mathbb{E}\left[\lambda\right] + a_0 \log b_0 - \log \Gamma(a_0) \tag{21.95}$$

$$= \;(a_0 - 1)(\psi(a_N) - \log b_N) - b_0 \frac{a_N}{b_N} + a_0 \log b_0 - \log \Gamma(a_0) \tag{21.96}$$

For the expected log prior of μ, one can show that

$$\mathbb{E}_{q(\mu,\lambda)}\left[\log p(\mu|\lambda)\right] \;=\; \frac{1}{2}\log\frac{\kappa_0}{2\pi} + \frac{1}{2}\mathbb{E}\left[\log \lambda\right]q(\lambda) - \frac{1}{2}\mathbb{E}_{q(\mu,\lambda)}\left[(\mu - \mu_0)^2 \kappa_0 \lambda\right]$$

$$= \;\frac{1}{2}\log\frac{\kappa_0}{2\pi} + \frac{1}{2}\left(\psi(a_N) - \log b_N\right)$$

$$-\frac{\kappa_0}{2}\frac{a_N}{b_N}\left[\frac{1}{\kappa_N} + (\mu_N - \mu_0)^2\right] \tag{21.97}$$

Putting it altogether, one can show that

$$L(q) \;=\; \frac{1}{2}\log\frac{1}{\kappa_N} + \log \Gamma(a_N) - a_N \log b_N + \text{const} \tag{21.98}$$

This quantity monotonically increases after each VB update.

21.5.2   Example: VB for linear regression

In Section 7.6.4, we discussed an empirical Bayes approach to setting the hyper-parameters for ridge regression known as the evidence procedure. In particular, we assumed a likelihood of the form $p(\mathbf{y}|\mathbf{X}, \boldsymbol{\theta}) = \mathcal{N}(\mathbf{Xw}, \lambda^{-1})$ and a prior of the form $p(\mathbf{w}) = \mathcal{N}(\mathbf{w}|\mathbf{0}, \alpha^{-1}\mathbf{I})$. We then computed a type II estimate of α and λ. The same approach was extended in Section 13.7 to handle a prior of the form $\mathcal{N}(\mathbf{w}|\mathbf{0}, \text{diag}(\boldsymbol{\alpha})^{-1})$, which allows one hyper-parameter per feature, a technique known as automatic relevancy determination.

In this section, we derive a VB algorithm for this model. We follow the presentation of (Drugowitsch 2008).[2] Initially we will use the following prior:

$$p(\mathbf{w}, \lambda, \alpha) \quad = \quad \mathcal{N}(\mathbf{w}|\mathbf{0}, (\lambda\alpha)^{-1}\mathbf{I})\text{Ga}(\lambda|a_0^\lambda, b_0^\lambda)\text{Ga}(\alpha|a_0^\alpha, b_0^\alpha) \tag{21.99}$$

We choose to use the following factorized approximation to the posterior:

$$q(\mathbf{w}, \alpha, \lambda) = q(\mathbf{w}, \lambda)q(\alpha) \tag{21.100}$$

Given these assumptions, one can show (see (Drugowitsch 2008)) that the optimal form for the posterior is

$$q(\mathbf{w}, \alpha, \lambda) \quad = \quad \mathcal{N}(\mathbf{w}|\mathbf{w}_N, \lambda^{-1}\mathbf{V}_N)\text{Ga}(\lambda|a_N^\lambda, b_N^\lambda)\text{Ga}(\alpha|a_N^\alpha, b_N^\alpha) \tag{21.101}$$

where

$$\mathbf{V}_N^{-1} \quad = \quad \overline{\mathbf{A}} + \mathbf{X}^{\mathbf{X}} \tag{21.102}$$

$$\mathbf{w}_N \quad = \quad \mathbf{V}_N\mathbf{X}^T\mathbf{y} \tag{21.103}$$

$$a_N^\lambda \quad = \quad a_0^\lambda + \frac{N}{2} \tag{21.104}$$

$$b_N^\lambda \quad = \quad b_0^\lambda + \frac{1}{2}(\|\mathbf{y} - \mathbf{Xw}\|^2 + \mathbf{w}_N^T\overline{\mathbf{A}}\mathbf{w}_N) \tag{21.105}$$

$$a_N^\alpha \quad = \quad a_0^\alpha + \frac{D}{2} \tag{21.106}$$

$$b_N^\alpha \quad = \quad b_0^\alpha + \frac{1}{2}\left(\frac{a_N^\lambda}{b_N^\lambda}\mathbf{w}_N^T\mathbf{w}_N + \text{tr}(\mathbf{V}_N)\right) \tag{21.107}$$

$$\overline{\mathbf{A}} \quad = \quad \langle\alpha\rangle\mathbf{I} = \frac{a_N^\alpha}{b_N^\alpha}\mathbf{I} \tag{21.108}$$

This method can be extended to the ARD case in a straightforward way, by using the following priors:

$$p(\mathbf{w}) \quad = \quad \mathcal{N}(\mathbf{0}, \text{diag}(\alpha)^{-1}) \tag{21.109}$$

$$p(\boldsymbol{\alpha}) \quad = \quad \prod_{j=1}^{D} \text{Ga}(\alpha_j|a_0^\alpha, b_0^\alpha) \tag{21.110}$$

The posterior for $\mathbf{w}$ and λ is computed as before, except we use $\overline{\mathbf{A}} = \text{diag}(a_N^\alpha/b_{N_j}^\alpha)$ instead of

---

2. Note that Drugowitsch uses a_0, b_0 as the hyper-parameters for $p(\lambda)$ and c_0, d_0 as the hyper-parameters for $p(\alpha)$, whereas (Bishop 2006, Sec 10.3) uses a_0, b_0 as the hyper-parameters for $p(\alpha)$ and treats λ as fixed. To (hopefully) avoid confusion, I use a_0^λ, b_0^λ as the hyper-parameters for $p(\lambda)$, and a_0^α, b_0^α as the hyper-parameters for $p(\alpha)$.

$a_N^\alpha / b_N^\alpha \mathbf{I}$. The posterior for $\boldsymbol{\alpha}$ has the form

$$q(\boldsymbol{\alpha}) = \prod_j \mathrm{Ga}(\alpha_j | a_N^\alpha, b_{N_j}^\alpha) \tag{21.111}$$

$$a_N^\alpha = a_0^\alpha + \frac{1}{2} \tag{21.112}$$

$$b_{N_j}^\alpha = b_0^\alpha + \frac{1}{2}\left(\frac{a_N^\lambda}{b_N^\lambda} w_{N,j}^2 + (\mathbf{V}_N)_{jj}\right) \tag{21.113}$$

The algorithm alternates between updating $q(\mathbf{w}, \lambda)$ and $q(\boldsymbol{\alpha})$. Once $\mathbf{w}$ and λ have been inferred, the posterior predictive is a Student distribution, as shown in Equation 7.76. Specifically, for a single data case, we have

$$p(y|\mathbf{x}, \mathcal{D}) = \mathcal{T}(y|\mathbf{w}_N^T \mathbf{x}, \frac{b_N^\lambda}{a_N^\lambda}(1 + \mathbf{x}^T \mathbf{V}_N \mathbf{x}), 2a_N^\lambda) \tag{21.114}$$

The exact marginal likelihood, which can be used for model selection, is given by

$$p(\mathcal{D}) = \int \int \int p(\mathbf{y}|\mathbf{X}, \mathbf{w}, \lambda) p(\mathbf{w}|\alpha) p(\lambda) d\mathbf{w}\, d\alpha\, d\lambda \tag{21.115}$$

We can compute a lower bound on $\log p(\mathcal{D})$ as follows:

$$
\begin{aligned}
L(q) = & -\frac{N}{2}\log(2\pi) - \frac{1}{2}\sum_{i=1}^N \left(\frac{a_N^\lambda}{b_N^\lambda}(y_i - \mathbf{w}_N^T \mathbf{x}_i)^2 + \mathbf{x}_i^T \mathbf{V}_N \mathbf{x}_i\right) \\
& + \frac{1}{2}\log|\mathbf{V}_N| + \frac{D}{2} \\
& - \log\Gamma(a_0^\lambda) + a_0^\lambda \log b_0^\lambda - b_0^\lambda \frac{a_N^\lambda}{b_N^\lambda} + \log\Gamma(a_N^\lambda) - a_N^\lambda \log b_N^\lambda + a_N^\lambda \\
& - \log\Gamma(a_0^\alpha) + a_0^\alpha \log b_0^\alpha + \log\Gamma(a_N^\alpha) - a_N^\alpha \log b_N^\alpha
\end{aligned}
\tag{21.116}
$$

In the ARD case, the last line becomes

$$\sum_{j=1}^D \left[-\log\Gamma(a_0^\alpha) + a_0^\alpha \log b_0^\alpha + \log\Gamma(a_N^\alpha) - a_N^\alpha \log b_{N_j}^\alpha \right] \tag{21.117}$$

Figure 21.6 compare VB and EB on a model selection problem for polynomial regression. We see that VB gives similar results to EB, but the precise behavior depends on the sample size. When $N = 5$, VB's estimate of the posterior over models is more diffuse than EB's, since VB models uncertainty in the hyper-parameters. When $N = 30$, the posterior estimate of the hyper-parameters becomes more well-determined. Indeed, if we compute $\mathbb{E}[\alpha|\mathcal{D}]$ when we have an uninformative prior, $a_0^\alpha = b_0^\alpha = 0$, we get

$$\overline{\alpha} = \frac{a_N^\alpha}{b_N^\alpha} = \frac{D/2}{\frac{1}{2}\left(\frac{a_N^\lambda}{b_N^\lambda}\mathbf{w}_N^T \mathbf{w}_N + \mathrm{tr}(\mathbf{V}_N)\right)} \tag{21.118}$$

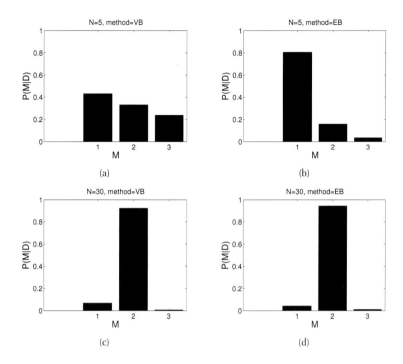

Figure 21.6 We plot the posterior over models (polynomials of degree 1, 2 and 3) assuming a uniform prior $p(m) \propto 1$. We approximate the marginal likelihood using (a,c) VB and (b,d) EB. In (a-b), we use $N = 5$ data points (shown in Figure 5.7). In (c-d), we use $N = 30$ data points (shown in Figure 5.8). Figure generated by `linregEbModelSelVsN`.

Compare this to Equation 13.167 for EB:

$$\hat{\alpha} = \frac{D}{\mathbb{E}\left[\mathbf{w}^T \mathbf{w}\right]} = \frac{D}{\mathbf{w}_N^T \mathbf{w}_N + \text{tr}(\mathbf{V}_N)} \tag{21.119}$$

Modulo the a_N^λ and b_N^λ terms, these are the same. In hindsight this is perhaps not that surprising, since EB is trying to maximize $\log p(\mathcal{D})$, and VB is trying to maximize a lower bound on $\log p(\mathcal{D})$.

21.6 Variational Bayes EM

Now consider latent variable models of the form $\mathbf{z}_i \rightarrow \mathbf{x}_i \leftarrow \boldsymbol{\theta}$. This includes mixtures models, PCA, HMMs, etc. There are now two kinds of unknowns: parameters, $\boldsymbol{\theta}$, and latent variables, $\mathbf{z}_i$. As we saw in Section 11.4, it is common to fit such models using EM, where in the E step we infer the posterior over the latent variables, $p(\mathbf{z}_i|\mathbf{x}_i, \boldsymbol{\theta})$, and in the M step, we compute a point estimate of the parameters, $\boldsymbol{\theta}$. The justification for this is two-fold. First, it results in simple algorithms. Second, the posterior uncertainty in $\boldsymbol{\theta}$ is usually less than in $\mathbf{z}_i$, since the $\boldsymbol{\theta}$ are informed by all N data cases, whereas $\mathbf{z}_i$ is only informed by $\mathbf{x}_i$; this makes a MAP estimate of

$\boldsymbol{\theta}$ more reasonable than a MAP estimate of $\mathbf{z}_i$.

However, VB provides a way to be "more Bayesian", by modeling uncertainty in the parameters $\boldsymbol{\theta}$ as well in the latent variables $\mathbf{z}_i$, at a computational cost that is essentially the same as EM. This method is known as **variational Bayes EM** or **VBEM**. The basic idea is to use mean field, where the approximate posterior has the form

$$p(\boldsymbol{\theta}, \mathbf{z}_{1:N}|\mathcal{D}) \approx q(\boldsymbol{\theta})q(\mathbf{z}) = q(\boldsymbol{\theta}) \prod_i q(\mathbf{z}_i) \qquad (21.120)$$

The first factorization, between $\boldsymbol{\theta}$ and $\mathbf{z}$, is a crucial assumption to make the algorithm tractable. The second factorization follows from the model, since the latent variables are iid conditional on $\boldsymbol{\theta}$.

In VBEM, we alternate between updating $q(\mathbf{z}_i|\mathcal{D})$ (the variational E step) and updating $q(\boldsymbol{\theta}|\mathcal{D})$ (the variational M step). We can recover standard EM from VBEM by approximating the parameter posterior using a delta function, $q(\boldsymbol{\theta}|\mathcal{D}) \approx \delta_{\hat{\boldsymbol{\theta}}}(\boldsymbol{\theta})$.

The variational E step is similar to a standard E step, except instead of plugging in a MAP estimate of the parameters and computing $p(\mathbf{z}_i|\mathcal{D}, \hat{\boldsymbol{\theta}})$, we need to average over the parameters. Roughly speaking, this can be computed by plugging in the posterior mean of the parameters instead of the MAP estimate, and then computing $p(\mathbf{z}_i|\mathcal{D}, \overline{\boldsymbol{\theta}})$ using standard algorithms, such as forwards-backwards. Unfortunately, things are not quite this simple, but this is the basic idea. The details depend on the form of the model; we give some examples below.

The variational M step is similar to a standard M step, except instead of computing a point estimate of the parameters, we update the hyper-parameters, using the expected sufficient statistics. This process is usually very similar to MAP estimation in regular EM. Again, the details on how to do this depend on the form of the model.

The principle advantage of VBEM over regular EM is that by marginalizing out the parameters, we can compute a lower bound on the marginal likelihood, which can be used for model selection. We will see an example of this in Section 21.6.1.6. VBEM is also "egalitarian", since it treats parameters as "first class citizens", just like any other unknown quantity, whereas EM makes an artificial distinction between parameters and latent variables.

21.6.1 Example: VBEM for mixtures of Gaussians *

Let us consider how to "fit" a mixture of Gaussians using VBEM. (We use scare quotes since we are not estimating the model parameters, but inferring a posterior over them.) We will follow the presentation of (Bishop 2006, Sec 10.2). Unfortunately, the details are rather complicated. Fortunately, as with EM, one gets used to it after a bit of practice. (As usual with math, simply reading the equations won't help much, you should really try deriving these results yourself (or try some of the exercises) if you want to learn this stuff in depth.)

21.6.1.1 The variational posterior

The likelihood function is the usual one for Gaussian mixture models:

$$p(\mathbf{z}, \mathbf{X}|\boldsymbol{\theta}) = \prod_i \prod_k \pi_k^{z_{ik}} \mathcal{N}(\mathbf{x}_i|\boldsymbol{\mu}_k, \boldsymbol{\Lambda}_k^{-1})^{z_{ik}} \qquad (21.121)$$

where $z_{ik} = 1$ if data point i belongs to cluster k, and $z_{ik} = 0$ otherwise.

We will assume the following factored conjugate prior

$$p(\boldsymbol{\theta}) = \mathrm{Dir}(\boldsymbol{\pi}|\boldsymbol{\alpha}_0) \prod_k \mathcal{N}(\boldsymbol{\mu}_k|\mathbf{m}_0, (\beta_0 \boldsymbol{\Lambda}_k)^{-1}) \mathrm{Wi}(\boldsymbol{\Lambda}_k|\mathbf{L}_0, \nu_0) \tag{21.122}$$

where $\boldsymbol{\Lambda}_k$ is the precision matrix for cluster k. The subscript 0 means these are parameters of the prior; we assume all the prior parameters are the same for all clusters. For the mixing weights, we usually use a symmetric prior, $\boldsymbol{\alpha}_0 = \alpha_0 \mathbf{1}$.

The exact posterior $p(\mathbf{z}, \boldsymbol{\theta}|\mathcal{D})$ is a mixture of K^N distributions, corresponding to all possible labelings $\mathbf{z}$. We will try to approximate the volume around one of these modes. We will use the standard VB approximation to the posterior:

$$p(\boldsymbol{\theta}, \mathbf{z}_{1:N}|\mathcal{D}) \approx q(\boldsymbol{\theta}) \prod_i q(\mathbf{z}_i) \tag{21.123}$$

At this stage we have not specified the forms of the q functions; these will be determined by the form of the likelihood and prior. Below we will show that the optimal form is as follows:

$$q(\mathbf{z}, \boldsymbol{\theta}) = q(\mathbf{z}|\boldsymbol{\theta})q(\boldsymbol{\theta}) = \left[\prod_i \mathrm{Cat}(\mathbf{z}_i|\mathbf{r}_i) \right] \tag{21.124}$$

$$\left[\mathrm{Dir}(\boldsymbol{\pi}|\boldsymbol{\alpha}) \prod_k \mathcal{N}(\boldsymbol{\mu}_k|\mathbf{m}_k, (\beta_k \boldsymbol{\Lambda}_k)^{-1}) \mathrm{Wi}(\boldsymbol{\Lambda}_k|\mathbf{L}_k, \nu_k) \right] \tag{21.125}$$

(The lack of 0 subscript means these are parameters of the posterior, not the prior.) Below we will derive the update equations for these variational parameters.

21.6.1.2 Derivation of $q(\mathbf{z})$ (variational E step)

The form for $q(\mathbf{z})$ can be obtained by looking at the complete data log joint, ignoring terms that do not involve $\mathbf{z}$, and taking expectations of what's left over wrt all the hidden variables except for $\mathbf{z}$. We have

$$\log q(\mathbf{z}) = \mathbb{E}_{q(\boldsymbol{\theta})}\left[\log p(\mathbf{x}, \mathbf{z}, \boldsymbol{\theta})\right] + \mathrm{const} \tag{21.126}$$

$$= \sum_i \sum_k z_{ik} \log \rho_{ik} + \mathrm{const} \tag{21.127}$$

where we define

$$\log \rho_{ik} \triangleq \mathbb{E}_{q(\boldsymbol{\theta})}\left[\log \pi_k\right] + \frac{1}{2}\mathbb{E}_{q(\boldsymbol{\theta})}\left[\log|\boldsymbol{\Lambda}_k|\right] - \frac{D}{2}\log(2\pi)$$
$$- \frac{1}{2}\mathbb{E}_{q(\boldsymbol{\theta})}\left[(\mathbf{x}_i - \boldsymbol{\mu}_k)^T \boldsymbol{\Lambda}_k (\mathbf{x}_i - \boldsymbol{\mu}_k)\right] \tag{21.128}$$

Using the fact that $q(\boldsymbol{\pi}) = \mathrm{Dir}(\boldsymbol{\pi})$, we have

$$\log \tilde{\pi}_k \triangleq \mathbb{E}\left[\log \pi_k\right] = \psi(\alpha_k) - \psi(\sum_{k'} \alpha_{k'}) \tag{21.129}$$

where $\psi()$ is the digamma function. (See Exercise 21.5 for the detailed derivation.) Next, we use the fact that

$$q(\boldsymbol{\mu}_k, \boldsymbol{\Lambda}_k) = \mathcal{N}(\boldsymbol{\mu}_k|\mathbf{m}_k, (\beta_k \boldsymbol{\Lambda}_k)^{-1})\text{Wi}(\boldsymbol{\Lambda}_k|\mathbf{L}_k, \nu_k) \tag{21.130}$$

to get

$$\log \tilde{\Lambda}_k \triangleq \mathbb{E}\left[\log |\boldsymbol{\Lambda}_k|\right] = \sum_{j=1}^{D} \psi\left(\frac{\nu_k + 1 - j}{2}\right) + D \log 2 + \log |\boldsymbol{\Lambda}_k| \tag{21.131}$$

Finally, for the expected value of the quadratic form, we get

$$\mathbb{E}\left[(\mathbf{x}_i - \boldsymbol{\mu}_k)^T \boldsymbol{\Lambda}_k (\mathbf{x}_i - \boldsymbol{\mu}_k)\right] = D\beta_k^{-1} + \nu_k (\mathbf{x}_i - \mathbf{m}_k)^T \boldsymbol{\Lambda}_k (\mathbf{x}_i - \mathbf{m}_k) \tag{21.132}$$

Putting it altogether, we get that the posterior responsibility of cluster k for datapoint i is

$$r_{ik} \quad \propto \quad \tilde{\pi}_k \tilde{\Lambda}_k^{\frac{1}{2}} \exp\left(-\frac{D}{2\beta_k} - \frac{\nu_k}{2}(\mathbf{x}_i - \mathbf{m}_k)^T \boldsymbol{\Lambda}_k (\mathbf{x}_i - \mathbf{m}_k)\right) \tag{21.133}$$

Compare this to the expression used in regular EM:

$$r_{ik}^{EM} \quad \propto \quad \hat{\pi}_k |\hat{\boldsymbol{\Lambda}}|_k^{\frac{1}{2}} \exp\left(-\frac{1}{2}(\mathbf{x}_i - \hat{\boldsymbol{\mu}}_k)^T \hat{\boldsymbol{\Lambda}}_k (\mathbf{x}_i - \hat{\boldsymbol{\mu}}_k)\right) \tag{21.134}$$

The significance of this difference is discussed further in Section 21.6.1.7.

21.6.1.3 Derivation of $q(\theta)$ (variational M step)

Using the mean field recipe, we have

$$\begin{aligned}
\log q(\boldsymbol{\theta}) \quad = \quad & \log p(\boldsymbol{\pi}) + \sum_k \log p(\mu_k, \boldsymbol{\Lambda}_k) + \sum_i \mathbb{E}_{q(\mathbf{z})}\left[\log p(\mathbf{z}_i|\boldsymbol{\pi})\right] \\
& + \sum_k \sum_i \mathbb{E}_{q(\mathbf{z})}\left[z_{ik}\right] \log \mathcal{N}(\mathbf{x}_i|\boldsymbol{\mu}_k, \boldsymbol{\Lambda}_k^{-1}) + \text{const}
\end{aligned} \tag{21.135}$$

We see this factorizes into the form

$$q(\boldsymbol{\theta}) \quad = \quad q(\boldsymbol{\pi}) \prod_k q(\boldsymbol{\mu}_k, \boldsymbol{\Lambda}_k) \tag{21.136}$$

For the $\boldsymbol{\pi}$ term, we have

$$\log q(\boldsymbol{\pi}) \quad = \quad (\alpha_0 - 1) \sum_k \log \pi_k + \sum_k \sum_i r_{ik} \log \pi_k + \text{const} \tag{21.137}$$

Exponentiating, we recognize this as a Dirichlet distribution:

$$\begin{aligned}
q(\boldsymbol{\pi}) \quad &= \quad \text{Dir}(\boldsymbol{\pi}|\boldsymbol{\alpha}) \tag{21.138} \\
\alpha_k \quad &= \quad \alpha_0 + N_k \tag{21.139} \\
N_k \quad &= \quad \sum_i r_{ik} \tag{21.140}
\end{aligned}$$

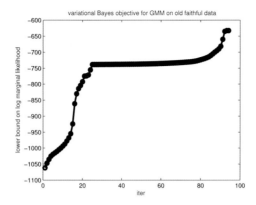

Figure 21.7 Lower bound vs iterations for the VB algorithm in Figure 21.8. The steep parts of the curve correspond to places where the algorithm figures out that it can increase the bound by "killing off" unnecessary mixture components, as described in Section 21.6.1.6. The plateaus correspond to slowly moving the clusters around. Figure generated by `mixGaussVbDemoFaithful`.

For the $\boldsymbol{\mu}_k$ and $\boldsymbol{\Lambda}_k$ terms, we have

$$q(\boldsymbol{\mu}_k, \boldsymbol{\Lambda}_k) = \mathcal{N}(\boldsymbol{\mu}_k | \mathbf{m}_k, (\beta_k \boldsymbol{\Lambda}_k)^{-1}) \mathrm{Wi}(\boldsymbol{\Lambda}_k | \mathbf{L}_k, \nu_k) \tag{21.141}$$

$$\beta_k = \beta_0 + N_k \tag{21.142}$$

$$\mathbf{m}_k = (\beta_0 \mathbf{m}_0 + N_k \overline{\mathbf{x}}_k)/\beta_k \tag{21.143}$$

$$\mathbf{L}_k^{-1} = \mathbf{L}_0^{-1} + N_k \mathbf{S}_k + \frac{\beta_0 N_k}{\beta_0 + N_k}(\overline{\mathbf{x}}_k - \mathbf{m}_0)(\overline{\mathbf{x}}_k - \mathbf{m}_0)^T \tag{21.144}$$

$$\nu_k = \nu_0 + N_k + 1 \tag{21.145}$$

$$\overline{\mathbf{x}}_k = \frac{1}{N_k} \sum_i r_{ik} \mathbf{x}_i \tag{21.146}$$

$$\mathbf{S}_k = \frac{1}{N_k} \sum_i r_{ik}(\mathbf{x}_i - \overline{\mathbf{x}}_k)(\mathbf{x}_i - \overline{\mathbf{x}}_k)^T \tag{21.147}$$

This is very similar to the M step for MAP estimation discussed in Section 11.4.2.8, except here we are computing the parameters of the posterior over $\boldsymbol{\theta}$, rather than MAP estimates of $\boldsymbol{\theta}$.

21.6.1.4 Lower bound on the marginal likelihood

The algorithm is trying to maximize the following lower bound

$$\mathcal{L} = \sum_{\mathbf{z}} \int q(\mathbf{z}, \boldsymbol{\theta}) \log \frac{p(\mathbf{x}, \mathbf{z}, \boldsymbol{\theta})}{q(\mathbf{z}, \boldsymbol{\theta})} d\boldsymbol{\theta} \leq \log p(\mathcal{D}) \tag{21.148}$$

This quantity should increase monotonically with each iteration, as shown in Figure 21.7. Unfortunately, deriving the bound is a bit messy, because we need to compute expectations of the unnormalized log posterior as well as entropies of the q distribution. We leave the details (which are similar to Section 21.5.1.6) to Exercise 21.4.

21.6.1.5 Posterior predictive distribution

We showed that the approximate posterior has the form

$$q(\boldsymbol{\theta}) = \text{Dir}(\boldsymbol{\pi}|\boldsymbol{\alpha}) \prod_k \mathcal{N}(\boldsymbol{\mu}_k|\mathbf{m}_k, (\beta_k \boldsymbol{\Lambda}_k)^{-1}) \text{Wi}(\boldsymbol{\Lambda}_k|\mathbf{L}_k, \nu_k) \tag{21.149}$$

Consequently the posterior predictive density can be approximated as follows, using the results from Section 4.6.3.6:

$$p(\mathbf{x}|\mathcal{D}) \approx \sum_z \int p(\mathbf{x}|z, \boldsymbol{\theta}) p(z|\boldsymbol{\theta}) q(\boldsymbol{\theta}) d\boldsymbol{\theta} \tag{21.150}$$

$$= \sum_k \int \pi_k \mathcal{N}(\mathbf{x}|\boldsymbol{\mu}_k, \boldsymbol{\Lambda}_k^{-1}) q(\boldsymbol{\theta}) d\boldsymbol{\theta} \tag{21.151}$$

$$= \sum_k \frac{\alpha_k}{\sum_{k'} \alpha_{k'}} \mathcal{T}(\mathbf{x}|\mathbf{m}_k, \mathbf{M}_k, \nu_k + 1 - D) \tag{21.152}$$

$$\mathbf{M}_k = \frac{(\nu_k + 1 - D)\beta_k}{1 + \beta_k} \mathbf{L}_k \tag{21.153}$$

This is just a weighted sum of Student distributions. If instead we used a plug-in approximation, we would get a weighted sum of Gaussian distributions.

21.6.1.6 Model selection using VBEM

The simplest way to select K when using VB is to fit several models, and then to use the variational lower bound to the log marginal likelihood, $\mathcal{L}(K) \leq \log p(\mathcal{D}|K)$, to approximate $p(K|\mathcal{D})$:

$$p(K|\mathcal{D}) = \frac{e^{\mathcal{L}(K)}}{\sum_{K'} e^{\mathcal{L}(K')}} \tag{21.154}$$

However, the lower bound needs to be modified somewhat to take into account the lack of identifiability of the parameters (Section 11.3.1). In particular, although VB will approximate the volume occupied by the parameter posterior, it will only do so around one of the local modes. With K components, there are $K!$ equivalent modes, which differ merely by permuting the labels. Therefore we should use $\log p(\mathcal{D}|K) \approx \mathcal{L}(K) + \log(K!)$.

21.6.1.7 Automatic sparsity inducing effects of VBEM

Although VB provides a reasonable approximation to the marginal likelihood (better than BIC (Beal and Ghahramani 2006)), this method still requires fitting multiple models, one for each value of K being considered. A faster alternative is to fit a single model, where K is set large, but where α_0 is set very small, $\alpha_0 \ll 1$. From Figure 2.14(d), we see that the resulting prior for the mixing weights $\boldsymbol{\pi}$ has "spikes" near the corners of the simplex, encouraging a sparse mixing weight vector.

In regular EM, the MAP estimate of the mixing weights will have the form $\hat{\pi}_k \propto (\alpha_k - 1)$, where $\alpha_k = \alpha_0 + N_k$. Unfortunately, this can be negative if $\alpha_0 = 0$ and $N_k = 0$ (Figueiredo

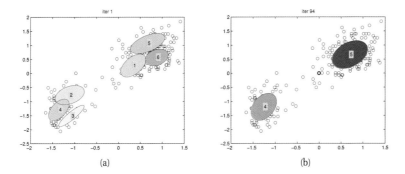

Figure 21.8 We visualize the posterior mean parameters at various stages of the VBEM algorithm applied to a mixture of Gaussians model on the Old Faithful data. Shading intensity is proportional to the mixing weight. We initialize with K-means and use $\alpha_0 = 0.001$ as the Dirichlet hyper-parameter. Based on Figure 10.6 of (Bishop 2006). Figure generated by `mixGaussVbDemoFaithful`, based on code by Emtiyaz Khan.

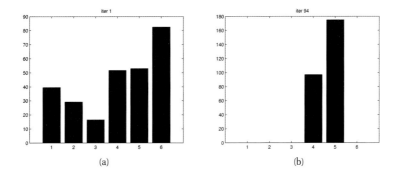

Figure 21.9 We visualize the posterior values of $\boldsymbol{\alpha}_k$ for the model in Figure 21.8. We see that unnecessary components get "killed off". Figure generated by `mixGaussVbDemoFaithful`.

and Jain 2002). However, in VBEM, we use

$$\tilde{\pi}_k \quad = \quad \frac{\exp[\Psi(\alpha_k)]}{\exp[\Psi(\sum_{k'} \alpha_{k'})]} \tag{21.155}$$

Now $\exp(\Psi(x)) \approx x - 0.5$ for $x > 1$. So if $\alpha_k = 0$, when we compute $\tilde{\pi}_k$, it's like we substract 0.5 from the posterior counts. This will hurt small clusters more than large clusters (like a regressive tax).[3] The effect is that clusters which have very few (weighted) members become more and more empty over successive iterations, whereas the popular clusters get more and more members. This is called the **rich get richer** phenomenon; we will encounter it again in Section 25.2, when we discuss Dirichlet process mixture models.

   This automatic pruning method is demonstrated in Figure 21.8. We fit a mixture of 6 Gaussians to the Old Faithful dataset, but the data only really "needs" 2 clusters, so the rest get "killed off".

---

3. For more details, see (Liang et al. 2007).

In this example, we used $\alpha_0 = 0.001$; if we use a larger α_0, we do not get a sparsity effect. In Figure 21.9, we plot $q(\boldsymbol{\alpha}|\mathcal{D})$ at various iterations; we see that the unwanted components get extinguished. This provides an efficient alternative to performing a discrete search over the number of clusters.

21.7   Variational message passing and VIBES

We have seen that mean field methods, at least of the fully-factorized variety, are all very similar: just compute each node's full conditional, and average out the neighbors. This is very similar to Gibbs sampling (Section 24.2), except the derivation of the equations is usually a bit more work. Fortunately it is possible to derive a general purpose set of update equations that work for any DGM for which all CPDs are in the exponential family, and for which all parent nodes have conjugate distributions (Ghahramani and Beal 2001). (See (Wand et al. 2011; Knowles and Minka 2011) for recent extensions to handle non-conjugate priors.) One can then sweep over the graph, updating nodes one at a time, in a manner similar to Gibbs sampling. This is known as **variational message passing** or **VMP** (Winn and Bishop 2005), and has been implemented in the open-source program **VIBES**[4]. This is a VB analog to BUGS, which is a popular generic program for Gibbs sampling discussed in Section 24.2.6.

VMP/ mean field is best-suited to inference where one or more of the hidden nodes are continuous (e.g., when performing "Bayesian learning"). For models where all the hidden nodes are discrete, more accurate approximate inference algorithms can be used, as we discuss in Chapter 22.

21.8   Local variational bounds *

So far, we have been focusing on mean field inference, which is a form of variational inference based on minimizing $\mathbb{KL}\left(q||\tilde{p}\right)$, where q is the approximate posterior, assumed to be factorized, and $\tilde{p}$ is the exact (but unnormalized) posterior. However, there is another kind of variational inference, where we replace a specific term in the joint distribution with a simpler function, to simplify computation of the posterior. Such an approach is sometimes called a **local variational approximation**, since we are only modifying one piece of the model, unlike mean field, which is a global approximation. In this section, we study several examples of this method.

21.8.1   Motivating applications

Before we explain how to derive local variational bounds, we give some examples of where this is useful.

21.8.1.1   Variational logistic regression

Consider the problem of how to approximate the parameter posterior for multiclass logistic regression model under a Gaussian prior. One approach is to use a Gaussian (Laplace) approximation, as discussed in Section 8.4.3. However, a variational approach can produce a more

---

4. Available at `http://vibes.sourceforge.net/`.

accurate approximation to the posterior, since it has tunable parameters. Another advantage is that the variational approach monotonically optimizes a lower bound on the likelihood of the data, as we will see.

To see why we need a bound, note that the likelihood can be written as follows:

$$p(\mathbf{y}|\mathbf{X}, \mathbf{w}) \quad = \quad \prod_{i=1}^{N} \exp\left[\mathbf{y}_i^T \boldsymbol{\eta}_i - \mathrm{lse}(\boldsymbol{\eta}_i)\right] \tag{21.156}$$

where $\boldsymbol{\eta}_i = \mathbf{X}_i \mathbf{w}_i = [\mathbf{x}_i^T \mathbf{w}_1, \ldots, \mathbf{x}_i^T \mathbf{w}_M]$, where $M = C - 1$ (since we set $\mathbf{w}_C = \mathbf{0}$ for identifiability), and where we define the **log-sum-exp** or **lse** function as follows:

$$\mathrm{lse}(\boldsymbol{\eta}_i) \triangleq \log\left(1 + \sum_{m=1}^{M} e^{\eta_{im}}\right) \tag{21.157}$$

The main problem is that this likelihood is not conjugate to the Gaussian prior. Below we discuss how to compute "Gaussian-like" lower bounds to this likelihood, which give rise to approximate Gaussian posteriors.

21.8.1.2 Multi-task learning

One important application of Bayesian inference for logistic regression is where we have multiple related classifiers we want to fit. In this case, we want to share information between the parameters for each classifier; this requires that we maintain a posterior distibution over the parameters, so we have a measure of confidence as well as an estimate of the value. We can embed the above variational method inside of a larger hierarchical model in order to perform such multi-task learning, as described in e.g., (Braun and McAuliffe 2010).

21.8.1.3 Discrete factor analysis

Another situation where variational bounds are useful arises when we fit a factor analysis model to discrete data. This model is just like multinomial logistic regression, except the input variables are hidden factors. We need to perform inference on the hidden variables as well as the regression weights. For simplicity, we might perform point estimation of the weights, and just integrate out the hidden variables. We can do this using variational EM, where we use the variational bound in the E step. See Section 12.4 for details.

21.8.1.4 Correlated topic model

A topic model is a latent variable model for text documents and other forms of discrete data; see Section 27.3 for details. Often we assume the distribution over topics has a Dirichlet prior, but a more powerful model, known as the correlated topic model, uses a Gaussian prior, which can model correlations more easily (see Section 27.4.1 for details). Unfortunately, this also involves the lse function. However, we can use our variational bounds in the context of a variational EM algorithm, as we will see later.

21.8.2 Bohning's quadratic bound to the log-sum-exp function

All of the above examples require dealing with multiplying a Gaussian prior by a multinomial likelihood; this is difficult because of the log-sum-exp (lse) term. In this section, we derive a way to derive a "Gaussian-like" lower bound on this likelihood.

Consider a Taylor series expansion of the lse function around $\boldsymbol{\psi}_i \in \mathbb{R}^M$:

$$\text{lse}(\boldsymbol{\eta}_i) = \text{lse}(\boldsymbol{\psi}_i) + (\boldsymbol{\eta}_i - \boldsymbol{\psi}_i)^T \mathbf{g}(\boldsymbol{\psi}_i) + \frac{1}{2}(\boldsymbol{\eta}_i - \boldsymbol{\psi}_i)^T \mathbf{H}(\boldsymbol{\psi}_i)(\boldsymbol{\eta}_i - \boldsymbol{\psi}_i) \tag{21.158}$$

$$\mathbf{g}(\boldsymbol{\psi}_i) = \exp[\boldsymbol{\psi}_i - \text{lse}(\boldsymbol{\psi}_i)] = \mathcal{S}(\boldsymbol{\psi}_i) \tag{21.159}$$

$$\mathbf{H}(\boldsymbol{\psi}_i) = \text{diag}(\mathbf{g}(\boldsymbol{\psi}_i)) - \mathbf{g}(\boldsymbol{\psi}_i)\mathbf{g}(\boldsymbol{\psi}_i)^T \tag{21.160}$$

where $\mathbf{g}$ and $\mathbf{H}$ are the gradient and Hessian of lse, and $\boldsymbol{\psi}_i \in \mathbb{R}^M$ is chosen such that equality holds. An upper bound to lse can be found by replacing the Hessian matrix $\mathbf{H}(\boldsymbol{\psi}_i)$ with a matrix $\mathbf{A}_i$ such that $\mathbf{A}_i \prec \mathbf{H}(\boldsymbol{\psi}_i)$. (Bohning 1992) showed that this can be achieved if we use the matrix $\mathbf{A}_i = \frac{1}{2}\left[\mathbf{I}_M - \frac{1}{M+1}\mathbf{1}_M\mathbf{1}_M^T\right]$. (Recall that $M + 1 = C$ is the number of classes.) Note that $\mathbf{A}_i$ is independent of $\boldsymbol{\psi}_i$; however, we still write it as $\mathbf{A}_i$ (rather than dropping the i subscript), since other bounds that we consider below will have a data-dependent curvature term. The upper bound on lse therefore becomes

$$\text{lse}(\boldsymbol{\eta}_i) \le \frac{1}{2}\boldsymbol{\eta}_i^T \mathbf{A}_i \boldsymbol{\eta}_i - \mathbf{b}_i^T \boldsymbol{\eta}_i + c_i \tag{21.161}$$

$$\mathbf{A}_i = \frac{1}{2}\left[\mathbf{I}_M - \frac{1}{M+1}\mathbf{1}_M\mathbf{1}_M^T\right] \tag{21.162}$$

$$\mathbf{b}_i = \mathbf{A}_i\boldsymbol{\psi}_i - \mathbf{g}(\boldsymbol{\psi}_i) \tag{21.163}$$

$$c_i = \frac{1}{2}\boldsymbol{\psi}_i^T \mathbf{A}_i \boldsymbol{\psi}_i - \mathbf{g}(\boldsymbol{\psi}_i)^T \boldsymbol{\psi}_i + \text{lse}(\boldsymbol{\psi}_i) \tag{21.164}$$

where $\boldsymbol{\psi}_i \in \mathbb{R}^M$ is a vector of variational parameters.

We can use the above result to get the following lower bound on the softmax likelihood:

$$\log p(y_i = c|\mathbf{x}_i, \mathbf{w}) \ge \left[\mathbf{y}_i^T \mathbf{X}_i \mathbf{w} - \frac{1}{2}\mathbf{w}^T \mathbf{X}_i \mathbf{A}_i \mathbf{X}_i \mathbf{w} + \mathbf{b}_i^T \mathbf{X}_i \mathbf{w} - c_i\right]_c \tag{21.165}$$

To simplify notation, define the pseudo-measurement

$$\tilde{\mathbf{y}}_i \triangleq \mathbf{A}_i^{-1}(\mathbf{b}_i + \mathbf{y}_i) \tag{21.166}$$

Then we can get a "Gaussianized" version of the observation model:

$$p(\mathbf{y}_i|\mathbf{x}_i, \mathbf{w}) \ge f(\mathbf{x}_i, \boldsymbol{\psi}_i)\,\mathcal{N}(\tilde{\mathbf{y}}_i|\mathbf{X}_i \mathbf{w}, \mathbf{A}_i^{-1}) \tag{21.167}$$

where $f(\mathbf{x}_i, \boldsymbol{\psi}_i)$ is some function that does not depend on $\mathbf{w}$. Given this, it is easy to compute the posterior $q(\mathbf{w}) = \mathcal{N}(\mathbf{m}_N, \mathbf{V}_N)$, using Bayes rule for Gaussians. Below we will explain how to update the variational parameters $\boldsymbol{\psi}_i$.

21.8.2.1 Applying Bohning's bound to multinomial logistic regression

Let us see how to apply this bound to multinomial logistic regression. From Equation 21.13, we can define the goal of variational inference as maximizing

$$
L(q) \triangleq -\mathbb{KL}\left(q(\mathbf{w})||p(\mathbf{w}|\mathcal{D})\right) + \mathbb{E}_q\left[\sum_{i=1}^{N} \log p(y_i|\mathbf{x}_i, \mathbf{w})\right] \tag{21.168}
$$

$$
= -\mathbb{KL}\left(q(\mathbf{w})||p(\mathbf{w}|\mathcal{D})\right) + \mathbb{E}_q\left[\sum_{i=1}^{N} \mathbf{y}_i^T \boldsymbol{\eta}_i - \mathrm{lse}(\boldsymbol{\eta}_i)\right] \tag{21.169}
$$

$$
= -\mathbb{KL}\left(q(\mathbf{w})||p(\mathbf{w}|\mathcal{D})\right) + \sum_{i=1}^{N} \mathbf{y}_i^T \mathbb{E}_q\left[\boldsymbol{\eta}_i\right] - \sum_{i=1}^{N} \mathbb{E}_q\left[\mathrm{lse}(\boldsymbol{\eta}_i)\right] \tag{21.170}
$$

where $q(\mathbf{w}) = \mathcal{N}(\mathbf{w}|\mathbf{m}_N, \mathbf{V}_N)$ is the approximate posterior. The first term is just the KL divergence between two Gaussians, which is given by

$$
\begin{aligned}
-\mathbb{KL}\left(\mathcal{N}(\mathbf{m}_0, \mathbf{V}_0)||\mathcal{N}(\mathbf{m}_N, \mathbf{V}_N)\right) = & -\frac{1}{2}\left[\mathrm{tr}(\mathbf{V}_N\mathbf{V}_0^{-1}) - \log|\mathbf{V}_N\mathbf{V}_0^{-1}|\right. \\
& \left. +(\mathbf{m}_N - \mathbf{m}_0)^T\mathbf{V}_0^{-1}(\mathbf{m}_N - \mathbf{m}_0) - DM\right]
\end{aligned} \tag{21.171}
$$

where DM is the dimensionality of the Gaussian, and we assume a prior of the form $p(\mathbf{w}) = \mathcal{N}(\mathbf{m}_0, \mathbf{V}_0)$, where typically $\boldsymbol{\mu}_0 = \mathbf{0}_{DM}$, and $\mathbf{V}_0$ is block diagonal. The second term is simply

$$
\sum_{i=1}^{N} \mathbf{y}_i^T \mathbb{E}_q\left[\boldsymbol{\eta}_i\right] = \sum_{i=1}^{N} \mathbf{y}_i^T \tilde{\mathbf{m}}_i \tag{21.172}
$$

where $\tilde{\mathbf{m}}_i \triangleq \mathbf{X}_i\mathbf{m}_N$. The final term can be lower bounded by taking expectations of our quadratic upper bound on lse as follows:

$$
-\sum_{i=1}^{N} \mathbb{E}_q\left[\mathrm{lse}(\boldsymbol{\eta}_i)\right] \geq -\frac{1}{2}\mathrm{tr}(\mathbf{A}_i\tilde{\mathbf{V}}_i) - \frac{1}{2}\tilde{\mathbf{m}}_i\mathbf{A}_i\tilde{\mathbf{m}}_i + \mathbf{b}_i^T\tilde{\mathbf{m}}_i - c_i \tag{21.173}
$$

where $\tilde{\mathbf{V}}_i \triangleq \mathbf{X}_i\mathbf{V}_N\mathbf{X}_i^T$. Putting it altogether, we have

$$
\begin{aligned}
L_{QJ}(q) \geq & -\frac{1}{2}\left[\mathrm{tr}(\mathbf{V}_N\mathbf{V}_0^{-1}) - \log|\mathbf{V}_N\mathbf{V}_0^{-1}| + (\mathbf{m}_N - \mathbf{m}_0)^T\mathbf{V}_0^{-1}(\mathbf{m}_N - \mathbf{m}_0)\right] \\
& -\frac{1}{2}DM + \sum_{i=1}^{N} \mathbf{y}_i^T \tilde{\mathbf{m}}_i - \frac{1}{2}\mathrm{tr}(\mathbf{A}_i\tilde{\mathbf{V}}_i) - \frac{1}{2}\tilde{\mathbf{m}}_i\mathbf{A}_i\tilde{\mathbf{m}}_i + \mathbf{b}_i^T\tilde{\mathbf{m}}_i - c_i
\end{aligned} \tag{21.174}
$$

This lower bound combines Jensen's inequality (as in mean field inference), plus the quadratic lower bound due to the lse term, so we write it as L_{QJ}.

We will use coordinate ascent to optimize this lower bound. That is, we update the variational posterior parameters $\mathbf{V}_N$ and $\mathbf{m}_N$, and then the variational likelihood parameters $\boldsymbol{\psi}_i$. We leave the detailed derivation as an exercise, and just state the results. We have

$$\mathbf{V}_N = \left(\mathbf{V}_0 + \sum_{i=1}^{N} \mathbf{X}_i^T \mathbf{A}_i \mathbf{X}_i \right)^{-1} \tag{21.175}$$

$$\mathbf{m}_N = \mathbf{V}_n \left(\mathbf{V}_0^{-1} \mathbf{m}_0 + \sum_{i=1}^{N} \mathbf{X}_i^T (\mathbf{y}_i + \mathbf{b}_i) \right) \tag{21.176}$$

$$\boldsymbol{\psi}_i = \tilde{\mathbf{m}}_i = \mathbf{X}_i \mathbf{m}_N \tag{21.177}$$

We can exploit the fact that $\mathbf{A}_i$ is a constant matrix, plus the fact that $\mathbf{X}_i$ has block structure, to simplify the first two terms as follows:

$$\mathbf{V}_N = \left(\mathbf{V}_0 + \mathbf{A} \otimes \sum_{i=1}^{N} \mathbf{x}_i \mathbf{x}_i^T \right)^{-1} \tag{21.178}$$

$$\mathbf{m}_N = \mathbf{V}_n \left(\mathbf{V}_0^{-1} \mathbf{m}_0 + \sum_{i=1}^{N} (\mathbf{y}_i + \mathbf{b}_i) \otimes \mathbf{x}_i \right) \tag{21.179}$$

where $\otimes$ denotes the kronecker product. See Algorithm 21.1 for some pseudocode, and `http://www.cs.ubc.ca/~emtiyaz/software/catLGM.html` for some Matlab code.

Algorithm 21.1: Variational inference for multi-class logistic regression using Bohning's bound

---

1   Input: $y_i \in \{1, \dots, C\}$, $\mathbf{x}_i \in \mathbb{R}^D$, $i = 1 : N$, prior $\mathbf{m}_0$, $\mathbf{V}_0$;
2   Define $M := C - 1$; dummy encode $\mathbf{y}_i \in \{0, 1\}^M$; define $\mathbf{X}_i = \text{blockdiag}(\mathbf{x}_i^T)$;
3   Define $\mathbf{y} := [\mathbf{y}_1; \dots; \mathbf{y}_N]$, $\mathbf{X} := [\mathbf{X}_1; \dots; \mathbf{X}_N]$ and $\mathbf{A} := \frac{1}{2} \left[\mathbf{I}_M - \frac{1}{M+1} \mathbf{1}_M \mathbf{1}_M^T \right]$;
4   $\mathbf{V}_N := \left(\mathbf{V}_0^{-1} + \sum_{i=1}^{n} \mathbf{X}_i^T \mathbf{A} \mathbf{X}_i \right)^{-1}$;
5   Initialize $\mathbf{m}_N := \mathbf{m}_0$;
6   **repeat**
7      $\boldsymbol{\psi} := \mathbf{X} \mathbf{m}_N$;
8      $\boldsymbol{\Psi} := \text{reshape}(\mathbf{m}, M, N)$;
9      $\mathbf{G} := \exp(\boldsymbol{\Psi} - \text{lse}(\boldsymbol{\Psi}))$;
10     $\mathbf{B} := \mathbf{A} \boldsymbol{\Psi} - \mathbf{G}$;
11     $\mathbf{b} := (\mathbf{B})$;
12     $\mathbf{m}_N := \mathbf{V}_N \left(\mathbf{V}_0^{-1} \mathbf{m}_0 + \mathbf{X}^T (\mathbf{y} + \mathbf{b}) \right)$;
13   Compute the lower bound L_{QJ} using Equation 21.174;
14   **until** *converged*;
15   Return $\mathbf{m}_N$ and $\mathbf{V}_N$;

21.8.3   Bounds for the sigmoid function

In many models, we just have binary data. In this case, we have $y_i \in \{0, 1\}$, $M = 1$ and $\eta_i = \mathbf{w}^T \mathbf{x}_i$ where $\mathbf{w} \in \mathbb{R}^D$ is a weight vector (not matrix). In this case, the Bohning bound

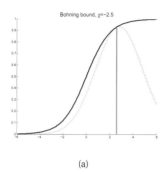
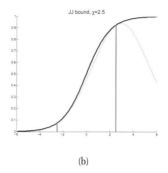

(a)   (b)

Figure 21.10   Quadratic lower bounds on the sigmoid (logistic) function. In solid red, we plot $\mathrm{sigm}(x)$ vs x. In dotted blue, we plot the lower bound $L(x, \xi)$ vs x for $\xi = 2.5$. (a) Bohning bound. This is tight at $-\xi = 2.5$. (b) JJ bound. This is tight at $\xi = \pm 2.5$. Figure generated by `sigmoidLowerBounds`.

becomes

$$\log(1 + e^{\eta}) \leq \frac{1}{2} a \eta^2 - b \eta + c \tag{21.180}$$

$$a = \frac{1}{4} \tag{21.181}$$

$$b = A \psi - (1 + e^{-\psi})^{-1} \tag{21.182}$$

$$c = \frac{1}{2} A \psi^2 - (1 + e^{-\psi})^{-1} \psi + \log(1 + e^{\psi}) \tag{21.183}$$

It is possible to derive an alternative quadratic bound for this case, as shown in (Jaakkola and Jordan 1996b, 2000). This has the following form

$$\log(1 + e^{\eta}) \leq \lambda(\xi)(\eta^2 - \xi^2) + \frac{1}{2}(\eta - \xi) + \log(1 + e^{\xi}) \tag{21.184}$$

$$\lambda(\xi) \triangleq \frac{1}{4\xi} \tanh(\xi/2) = \frac{1}{2\xi} \left[\mathrm{sigm}(\xi) - \frac{1}{2} \right] \tag{21.185}$$

We shall refer to this as the **JJ bound**, after its inventors, (Jaakkola and Jordan 1996b, 2000).

To facilitate comparison with Bohning's bound, let us rewrite the JJ bound as a quadratic form as follows

$$\log(1 + e^{\eta}) \leq \frac{1}{2} a(\xi) \eta^2 - b(\xi) \eta + c(\xi) \tag{21.186}$$

$$a(\xi) = 2\lambda(\xi) \tag{21.187}$$

$$b(\xi) = -\frac{1}{2} \tag{21.188}$$

$$c(\xi) = -\lambda(\xi)\xi^2 - \frac{1}{2}\xi + \log(1 + e^{\xi}) \tag{21.189}$$

The JJ bound has an adaptive curvature term, since a depends on ξ. In addition, it is tight at two points, as is evident from Figure 21.10(b). By contrast, the Bohning bound is a constant curvature bound, and is only tight at one point, as is evident from Figure 21.10(a).

If we wish to use the JJ bound for binary logistic regression, we can make some small modifications to Algorithm 21.1. First, we use the new definitions for a_i, b_i and c_i. The fact that a_i is not constant when using the JJ bound, unlike when using the Bohning bound, means we cannot compute $\mathbf{V}_N$ outside of the main loop, making the method a constant factor slower. Next we note that $\mathbf{X}_i = \mathbf{x}_i^T$, so the updates for the posterior become

$$\mathbf{V}_N^{-1} = \mathbf{V}_0^{-1} + 2\sum_{i=1}^{N} \lambda(\xi_i)\mathbf{x}_i\mathbf{x}_i^T \tag{21.190}$$

$$\mathbf{m}_N = \mathbf{V}_N\left(\mathbf{V}_0^{-1}\mathbf{m}_0 + \sum_{i=1}^{N}(y_i - \frac{1}{2})\mathbf{x}_i\right) \tag{21.191}$$

Finally, to compute the update for ξ_i, we isolate the terms in L_{QJ} that depend on ξ_i to get

$$L(\boldsymbol{\xi}) = \sum_{i=1}^{N}\left\{\ln \operatorname{sigm}(\xi_i) - \xi_i/2 - \lambda(\xi_i)(\mathbf{x}_i^T\mathbb{E}_q\left[\mathbf{w}\mathbf{w}^T\right]\mathbf{x}_i - \xi_i^2)\right\} + \text{const} \tag{21.192}$$

Optimizing this wrt ξ_i gives the equation

$$0 = \lambda'(\xi_i)(\mathbf{x}_i^T\mathbb{E}_q\left[\mathbf{w}\mathbf{w}^T\right]\mathbf{x}_i - \xi_i^2) \tag{21.193}$$

Now $\lambda'(\xi_i)$ is monotonic for $\xi_i \geq 0$, and we do not need to consider negative values of ξ_i by symmetry of the bound around $\xi_i = 0$ (see Figure 21.10). Hence the only way to make the above expression 0 is if we have $(\mathbf{x}_i^T\mathbb{E}\left[\mathbf{w}\mathbf{w}^T\right]\mathbf{x}_i - \xi_i^2) = 0$. Hence the update becomes

$$(\xi_i^{new})^2 = \mathbf{x}_i^T(\mathbf{V}_N + \mathbf{m}_N\mathbf{m}_N^T)\mathbf{x}_i \tag{21.194}$$

Although the JJ bound is tighter than the Bohning bound, sometimes it is not tight enough in order to estimate the posterior covariance accurately. A more accurate approach, which uses a piecewise quadratic upper bound to lse, is described in (Marlin et al. 2011). By increasing the number of pieces, the bound can be made arbitrarily tight.

21.8.4 Other bounds and approximations to the log-sum-exp function *

There are several other bounds and approximations to the multiclass lse function which we can use, which we briefly summarize below. Note, however, that all of these require numerical optimization methods to compute $\mathbf{m}_N$ and $\mathbf{V}_N$, making them more complicated to implement.

21.8.4.1 Product of sigmoids

The approach in (Bouchard 2007) exploits the fact that

$$\log\left(\sum_{k=1}^{K} e^{\eta_k}\right) \leq \alpha + \sum_{k=1}^{K}\log(1 + e^{\eta_k - \alpha}) \tag{21.195}$$

It then applies the JJ bound to the term on the right.

21.8.4.2 Jensen's inequality

The approach in (Blei and Lafferty 2006a, 2007) uses Jensen's inequality as follows:

$$
\mathbb{E}_q\left[\mathrm{lse}(\boldsymbol{\eta}_i)\right] \;=\; \mathbb{E}_q\left[\log\left(1+\sum_{c=1}^{M}\exp(\mathbf{x}_i^T\mathbf{w}_c)\right)\right] \tag{21.196}
$$

$$
\leq\; \log\left(1+\sum_{c=1}^{M}\mathbb{E}_q\left[\exp(\mathbf{x}_i^T\mathbf{w}_c)\right]\right) \tag{21.197}
$$

$$
\leq\; \log\left(1+\sum_{c=1}^{M}\exp(\mathbf{x}_i^T\mathbf{m}_{N,c}+\tfrac{1}{2}\mathbf{x}_i^T\mathbf{V}_{N,cc}\mathbf{x}_i)\right) \tag{21.198}
$$

where the last term follows from the mean of a log-normal distribution, which is $e^{\mu+\sigma^2/2}$.

21.8.4.3 Multivariate delta method

The approach in (Ahmed and Xing 2007; Braun and McAuliffe 2010) uses the **multivariate delta method**, which is a way to approximate moments of a function using a Taylor series expansion. In more detail, let $f(\mathbf{w})$ be the function of interest. Using a second-order approximation around $\mathbf{m}$ we have

$$
f(\mathbf{w}) \;\approx\; f(\mathbf{m})+(\mathbf{w}-\mathbf{m})^T\mathbf{g}(\mathbf{w}-\mathbf{m})+\frac{1}{2}(\mathbf{w}-\mathbf{m})^T\mathbf{H}(\mathbf{w}-\mathbf{m}) \tag{21.199}
$$

where $\mathbf{g}$ and $\mathbf{H}$ are the gradient and Hessian evaluated at $\mathbf{m}$. If $q(\mathbf{w})=\mathcal{N}(\mathbf{w}|\mathbf{m},\mathbf{V})$, we have

$$
\mathbb{E}_q\left[f(\mathbf{w})\right] \;\approx\; f(\mathbf{m})+\frac{1}{2}\mathrm{tr}[\mathbf{H}\mathbf{V}] \tag{21.200}
$$

If we use $f(\mathbf{w})=\mathrm{lse}(\mathbf{X}_i\mathbf{w})$, we get

$$
\mathbb{E}_q\left[\mathrm{lse}(\mathbf{X}_i\mathbf{w})\right]\approx \mathrm{lse}(\mathbf{X}_i\mathbf{m})+\frac{1}{2}\mathrm{tr}[\mathbf{X}_i\mathbf{H}\mathbf{X}_i^T\mathbf{V}] \tag{21.201}
$$

where $\mathbf{g}$ and $\mathbf{H}$ for the lse function are defined in Equations 21.159 and 21.160.

21.8.5 Variational inference based on upper bounds

So far, we have been concentrating on lower bounds. However, sometimes we need to use an upper bound. For example, (Saul et al. 1996) derives a mean field algorithm for sigmoid belief nets, which are DGMs in which each CPD is a logistic regression function (Neal 1992). Unlike the case of Ising models, the resulting MRF is not pairwise, but contains higher order interactions. This makes the standard mean field updates intractable. In particular, they turn out to involve computing an expression which requires evaluating

$$
\mathbb{E}\left[\log(1+e^{-\sum_{j\in\mathrm{pa}_i}w_{ij}x_j})\right]=\mathbb{E}\left[-\log\mathrm{sigm}(\mathbf{w}_i^T\mathbf{x}_{\mathrm{pa}(i)})\right] \tag{21.202}
$$

(Notice the minus sign in front.) (Saul et al. 1996) show how to derive an upper bound on the sigmoid function so as to make this update tractable, resulting in a monotonically convergent inference procedure.

Exercises

Exercise 21.1 Laplace approximation to $p(\mu, \log \sigma | \mathcal{D})$ for a univariate Gaussian.

In this exercise, we will compute a Laplace approximation of $p(\mu, \log \sigma | \mathcal{D})$ for a Gaussian, using an uninformative prior $p(\mu, \log \sigma) \propto 1$.

First, let $\ell = \log \sigma$ and $s^2 = \frac{1}{n} \sum_{i=1}^{n} (x_i - \overline{x})^2$. The log posterior is

$$\log p(\mu, \ell | \mathcal{D}) = -n \log \sigma - \frac{1}{2\sigma^2} [ns^2 + n(\overline{x} - \mu)^2] + \text{const} \tag{21.203}$$

where we ignore the dependence of $\log p(\mathcal{D})$ on the parameters.

a. Show that the first derivatives are

$$\frac{\partial}{\partial \mu} \log p(\mu, \ell | \mathcal{D}) = \frac{n(\overline{x} - \mu)}{\sigma^2} \tag{21.204}$$

$$\frac{\partial}{\partial \ell} \log p(\mu, \ell | \mathcal{D}) = -n + \frac{ns^2 + n(\overline{x} - \mu)^2}{\sigma^2} \tag{21.205}$$

b. Show that the Hessian matrix is given by

$$\mathbf{H} = \begin{pmatrix} \frac{\partial^2}{\partial \mu^2} \log p(\mu, \ell | \mathcal{D}) & \frac{\partial^2}{\partial \mu \partial \ell} \log p(\mu, \ell | \mathcal{D}) \\ \frac{\partial^2}{\partial \ell^2} \log p(\mu, \ell | \mathcal{D}) & \frac{\partial^2}{\partial \ell^2} \log p(\mu, \ell | \mathcal{D}) \end{pmatrix} \tag{21.206}$$

$$= \begin{pmatrix} -\frac{n}{\sigma^2} & -2n\frac{\overline{x} - \mu}{\sigma^2} \\ -2n\frac{\overline{x} - \mu}{\sigma^2} & -\frac{2}{\sigma^2}(ns^2 + n(\overline{x} - \mu)^2) \end{pmatrix} \tag{21.207}$$

c. Use this to derive a Laplace approximation to the posterior $p(\mu, \ell | \mathcal{D})$.

Exercise 21.2 Laplace approximation to $p(\boldsymbol{\mu}, \log \boldsymbol{\Sigma} | \mathcal{D})$ for a multivariate Gaussian

Extend Exercise 21.1 to the multivariate case.

Exercise 21.3 Variational lower bound for VB for univariate Gaussian

Fill in the details of the derivation in Section 21.5.1.6.

Exercise 21.4 Variational lower bound for VB for GMMs

Consider VBEM for GMMs as in Section 21.6.1.4. Show that the lower bound has the following form

$$\begin{aligned} \mathcal{L} = \ & \mathbb{E}\left[\ln p(\mathbf{x}|\mathbf{z}, \boldsymbol{\mu}, \boldsymbol{\Lambda})\right] + \mathbb{E}\left[\ln p(\mathbf{z}|\boldsymbol{\pi})\right] + \mathbb{E}\left[\ln p(\boldsymbol{\pi})\right] + \mathbb{E}\left[\ln p(\boldsymbol{\mu}, \boldsymbol{\Lambda})\right] \\ & - \mathbb{E}\left[\ln q(\mathbf{z})\right] - \mathbb{E}\left[\ln q(\boldsymbol{\pi})\right] - \mathbb{E}\left[\ln q(\boldsymbol{\mu}, \boldsymbol{\Lambda})\right] \end{aligned} \tag{21.208}$$

where

$$
\mathbb{E}\left[\ln p(\mathbf{x}|\mathbf{z}, \boldsymbol{\mu}, \boldsymbol{\Lambda})\right] = \frac{1}{2} \sum_k N_k \left\{ \ln \tilde{\Lambda}_k - D\beta_k^{-1} - \nu_k \mathrm{tr}(\mathbf{S}_k \mathbf{L}_k) \right.
$$
$$
\left. -\nu_k (\bar{\mathbf{x}}_k - \mathbf{m}_k)^T \mathbf{L}_k (\bar{\mathbf{x}}_k - \mathbf{m}_k) - D\ln(2\pi) \right\} \tag{21.209}
$$

$$
\mathbb{E}\left[\ln p(\mathbf{z}|\boldsymbol{\pi})\right] = \sum_i \sum_k r_{ik} \ln \tilde{\pi}_k \tag{21.210}
$$

$$
\mathbb{E}\left[\ln p(\boldsymbol{\pi})\right] = \ln C_{dir}(\boldsymbol{\alpha}_0) + (\alpha_0 - 1) \sum_k \ln \tilde{\pi}_k \tag{21.211}
$$

$$
\mathbb{E}\left[\ln p(\boldsymbol{\mu}, \boldsymbol{\Lambda})\right] = \frac{1}{2} \sum_k \left\{ D\ln(\beta_0/2\pi) + \ln \tilde{\Lambda}_k - \frac{D\beta_0}{\beta_k} \right.
$$
$$
-\beta_0 \nu_k (\mathbf{m}_k - \mathbf{m}_0)^T \mathbf{L}_k (\mathbf{m}_k - \mathbf{m}_0)
$$
$$
\left. + \ln C_{Wi}(\mathbf{L}_0, \nu_0) + \frac{\nu_0 - D - 1}{2} \ln \tilde{\Lambda}_k - \frac{1}{2} \nu_k \mathrm{tr}(\mathbf{L}_0^{-1} \mathbf{L}_k) \right\} \tag{21.212}
$$

$$
\mathbb{E}\left[\ln q(\mathbf{z})\right] = \sum_i \sum_k r_{ik} \ln r_{ik} \tag{21.213}
$$

$$
\mathbb{E}\left[\ln q(\boldsymbol{\pi})\right] = \sum_k (\alpha_k - 1) \ln \tilde{\pi}_k + \ln C_{dir}(\boldsymbol{\alpha}) \tag{21.214}
$$

$$
\mathbb{E}\left[\ln q(\boldsymbol{\mu}, \boldsymbol{\Lambda})\right] = \sum_k \left\{ \frac{1}{2} \ln \tilde{\Lambda}_k + \frac{D}{2} \ln\left(\frac{\beta_k}{2\pi}\right) - \frac{D}{2} - \mathbb{H}\left(q(\boldsymbol{\Lambda}_k)\right) \right\} \tag{21.215}
$$

where the normalization constant for the Dirichlet and Wishart is given by

$$
C_{dir}(\boldsymbol{\alpha}) \triangleq \frac{\Gamma(\sum_k \alpha_k)}{\prod_k \Gamma(\alpha_k)} \tag{21.216}
$$

$$
C_{Wi}(\mathbf{L}, \nu) \triangleq |\mathbf{L}|^{-\nu/2} \left(2^{\nu D/2} \Gamma_D(\nu/2) \right)^{-1} \tag{21.217}
$$

$$
\Gamma_D(\alpha) \triangleq \pi^{D(D-1)/4} \prod_{j=1}^{D} \Gamma\left(\alpha + (1-j)/2\right) \tag{21.218}
$$

where $\Gamma_D(\nu)$ is the multivariate Gamma function. Finally, the entropy of the Wishart is given by

$$
\mathbb{H}\left(\mathrm{Wi}(\mathbf{L}, \nu)\right) = -\ln C_{Wi}(\mathbf{L}, \nu) - \frac{\nu - D - 1}{2} \mathbb{E}\left[\ln |\boldsymbol{\Lambda}|\right] + \frac{\nu D}{2} \tag{21.219}
$$

where $\mathbb{E}\left[\ln |\boldsymbol{\Lambda}|\right]$ is given in Equation 21.131.

Exercise 21.5 Derivation of $\mathbb{E}\left[\log \pi_k\right]$ under a Dirichlet distribution
Show that

$$
\exp(\mathbb{E}\left[\log \pi_k\right]) = \frac{\exp(\Psi(\alpha_k))}{\exp(\Psi(\sum_{k'} \alpha_{k'}))} \tag{21.220}
$$

where $\boldsymbol{\pi} \sim \mathrm{Dir}(\boldsymbol{\alpha})$.

Exercise 21.6 Alternative derivation of the mean field updates for the Ising model
Derive Equation 21.50 by directly optimizing the variational free energy one term at a time.

Exercise 21.7 Forwards vs reverse KL divergence

(Source: Exercise 33.7 of (MacKay 2003).) Consider a factored approximation $q(x, y) = q(x)q(y)$ to a joint distribution $p(x, y)$. Show that to minimize the forwards KL $\mathbb{KL}(p||q)$ we should set $q(x) = p(x)$ and $q(y) = p(y)$, i.e., the optimal approximation is a product of marginals

Now consider the following joint distribution, where the rows represent y and the columns x.

	x			
	1	2	3	4
1	1/8	1/8	0	0
2	1/8	1/8	0	0
3	0	0	1/4	0
4	0	0	0	1/4

Show that the reverse KL $\mathbb{KL}(q||p)$ for this p has three distinct minima. Identify those minima and evaluate $\mathbb{KL}(q||p)$ at each of them. What is the value of $\mathbb{KL}(q||p)$ if we set $q(x, y) = p(x)p(y)$?

Exercise 21.8 Derivation of the structured mean field updates for FHMM

Derive the updates in Section 21.4.1.

Exercise 21.9 Variational EM for binary FA with sigmoid link

Consider the binary FA model:

$$p(\mathbf{x}_i|\mathbf{z}_i, \boldsymbol{\theta}) = \prod_{j=1}^{D} \text{Ber}(x_{ij}|\text{sigm}(\mathbf{w}_j^T\mathbf{z}_i + \beta_j)) = \prod_{j=1}^{D} \text{Ber}(x_{ij}|\text{sigm}(\eta_{ij})) \tag{21.221}$$

$$\eta_i = \tilde{\mathbf{W}}\tilde{\mathbf{z}}_i \tag{21.222}$$

$$\tilde{\mathbf{z}}_i \triangleq (\mathbf{z}_i; 1) \tag{21.223}$$

$$\tilde{\mathbf{W}} \triangleq (\mathbf{W}, \boldsymbol{\beta}) \tag{21.224}$$

$$p(\mathbf{z}_i) = \mathcal{N}(\mathbf{0}, \mathbf{I}) \tag{21.225}$$

Derive an EM algorithm to fit this model, using the Jaakkola-Jordan bound. Hint: the answer is in (Tipping 1998), but the exercise asks you to derive these equations.

Exercise 21.10 VB for binary FA with probit link

In Section 11.4.6, we showed how to use EM to fit probit regression, using a model of the form $p(y_i = 1|z_i) = \mathbb{I}(z_i > 0)$, where $z_i \sim \mathcal{N}(\mathbf{w}^T\mathbf{x}_i, 1)$ is latent. Now consider the case where the inputs $\mathbf{x}_i$ are also unknown, as in binary factor analysis. Show how to fit this model using variational Bayes, making an approximation to the posterior of the form $q(\mathbf{x}, \mathbf{z}, \mathbf{W}) = \prod_{i=1}^{N} q(\mathbf{x}_i)q(z_i) \prod_{l=1}^{L} q(\mathbf{w}_l)$. Hint: $q(\mathbf{x}_i)$ and $q(\mathbf{w}_i)$ will be Gaussian, and $q(z_i)$ will be a truncated univariate Gaussian.

22 *More variational inference*

22.1 Introduction

In Chapter 21, we discussed mean field inference, which approximates the posterior by a product of marginal distributions. This allows us to use different parametric forms for each variable, which is particularly useful when performing Bayesian inference for the parameters of statistical models (such as the mean and variance of a Gaussian or GMM, or the regression weights in a GLM), as we saw when we discussed variational Bayes and VB-EM.

In this chapter, we discuss a slightly different kind of variational inference. The basic idea is to minimize $J(q) = \mathbb{KL}\left(q||\tilde{p}\right)$, where $\tilde{p}$ is the exact but unnormalized posterior as before, but where we no longer require q to be factorized. In fact, we do not even require q to be a globally valid joint distribution. Instead, we only require that q is locally consistent, meaning that the joint distribution of two adjacent nodes agrees with the corresponding marginals (we will define this more precisely below).

In addition to this new kind of inference, we will discuss approximate methods for MAP state estimation in discrete graphical models. It turns out that algorithms for solving the MAP problem are very similar to some approximate methods for computing marginals, as we will see.

22.2 Loopy belief propagation: algorithmic issues

There is a very simple approximate inference algorithm for discrete (or Gaussian) graphical models known as **loopy belief propagation** or **LBP**. The basic idea is extremely simple: we apply the belief propagation algorithm of Section 20.2 to the graph, even if it has loops (i.e., even if it is not a tree). This method is simple and efficient, and often works well in practice, outperforming mean field (Weiss 2001). In this section, we discuss the algorithm in more detail. In the next section, we analyse this algorithm in terms of variational inference.

22.2.1 A brief history

When applied to loopy graphs, BP is not guaranteed to give correct results, and may not even converge. Indeed, Judea Pearl, who invented belief propagation for trees, wrote the following about loopy BP in 1988:

> When loops are present, the network is no longer singly connected and local propagation

schemes will invariably run into trouble ... If we ignore the existence of loops and permit the nodes to continue communicating with each other as if the network were singly connected, messages may circulate indefinitely around the loops and the process may not converge to a stable equilibrium ... Such oscillations do not normally occur in probabilistic networks ... which tend to bring all messages to some stable equilibrium as time goes on. However, this asymptotic equilibrium is not coherent, in the sense that it does not represent the posterior probabilities of all nodes of the network — (Pearl 1988, p.195)

Despite these reservations, Pearl advocated the use of belief propagation in loopy networks as an approximation scheme (J. Pearl, personal communication) and exercise 4.7 in (Pearl 1988) investigates the quality of the approximation when it is applied to a particular loopy belief network.

However, the main impetus behind the interest in BP arose when McEliece et al. (1998) showed that a popular algorithm for error correcting codes known as turbo codes (Berrou et al. 1993) could be viewed as an instance of BP applied to a certain kind of graph. This was an important observation since turbo codes have gotten very close to the theoretical lower bound on coding efficiency proved by Shannon. (Another approach, known as low density parity check or LDPC codes, has achieved comparable performance; it also uses LBP for decoding — see Figure 22.1 for an example.) In (Murphy et al. 1999), LBP was experimentally shown to also work well for inference in other kinds of graphical models beyond the error-correcting code context, and since then, the method has been widely used in many different applications.

22.2.2   LBP on pairwise models

We now discuss how to apply LBP to an undirected graphical model with pairwise factors (we discuss the directed case, which can involve higher order factors, in the next section). The method is simple: just continually apply Equations 20.11 and 20.10 until convergence. See Algorithm 22.1 for the pseudocode, and `beliefPropagation` for some Matlab code. We will discuss issues such as convergence and accuracy of this method shortly.

Algorithm 22.1: Loopy belief propagation for a pairwise MRF

---

1 Input: node potentials $\psi_s(x_s)$, edge potentials $\psi_{st}(x_s, x_t)$;
2 Initialize messages $m_{s \to t}(x_t) = 1$ for all edges $s - t$;
3 Initialize beliefs $\mathrm{bel}_s(x_s) = 1$ for all nodes s;
4 **repeat**
5 $\quad$ Send message on each edge
$$m_{s \to t}(x_t) = \sum_{x_s} \left(\psi_s(x_s) \psi_{st}(x_s, x_t) \prod_{u \in \mathrm{nbr}_s \setminus t} m_{u \to s}(x_s) \right);$$
6 $\quad$ Update belief of each node $\mathrm{bel}_s(x_s) \propto \psi_s(x_s) \prod_{t \in \mathrm{nbr}_s} m_{t \to s}(x_s)$;
7 **until** *beliefs don't change significantly*;
8 Return marginal beliefs $\mathrm{bel}_s(x_s)$;

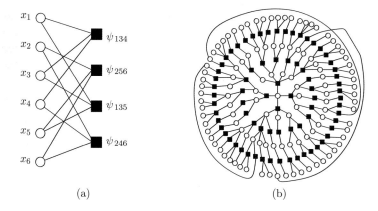

$$\begin{array}{cc} \text{(a)} & \text{(b)} \end{array}$$

Figure 22.1 (a) A simple factor graph representation of a (2,3) low-density parity check code (factor graphs are defined in Section 22.2.3.1). Each message bit (hollow round circle) is connected to two parity factors (solid black squares), and each parity factor is connected to three bits. Each parity factor has the form $\psi_{stu}(x_s, x_t, x_u) = \mathbb{I}(x_s \otimes x_t \otimes x_u = 1)$, where $\otimes$ is the xor operator. The local evidence factors for each hidden node are not shown. (b) A larger example of a random LDPC code. We see that this graph is "locally tree-like", meaning there are no short cycles; rather, each cycle has length $\sim \log m$, where m is the number of nodes. This gives us a hint as to why loopy BP works so well on such graphs. (Note, however, that some error correcting code graphs have short loops, so this is not the full explanation.)   Source: Figure 2.9 from (Wainwright and Jordan 2008).   Used with kind permission of Martin Wainwright.

22.2.3   LBP on a factor graph

To handle models with higher-order clique potentials (which includes directed models where some nodes have more than one parent), it is useful to use a representation known as a factor graph. We explain this representation below, and then describe how to apply LBP to such models.

22.2.3.1   Factor graphs

A **factor graph** (Kschischang et al. 2001; Frey 2003) is a graphical representation that unifies directed and undirected models, and which simplifies certain message passing algorithms. More precisely, a factor graph is an undirected bipartite graph with two kinds of nodes. Round nodes represent variables, square nodes represent factors, and there is an edge from each variable to every factor that mentions it. For example, consider the MRF in Figure 22.2(a). If we assume one potential per maximal clique, we get the factor graph in Figure 22.2(b), which represents the function

$$f(x_1, x_2, x_3, x_4) = f_{124}(x_1, x_2, x_4) f_{234}(x_2, x_3, x_4) \tag{22.1}$$

If we assume one potential per edge. we get the factor graph in Figure 22.2(c), which represents the function

$$f(x_1, x_2, x_3, x_4) = f_{14}(x_1, x_4) f_{12}(x_1, x_2) f_{34}(x_3, x_4) f_{23}(x_2, x_3) f_{24}(x_2, x_4) \tag{22.2}$$

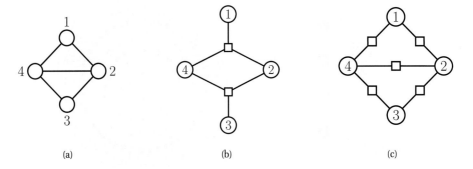

Figure 22.2   (a) A simple UGM. (b) A factor graph representation assuming one potential per maximal clique. (c) A factor graph representation assuming one potential per edge.

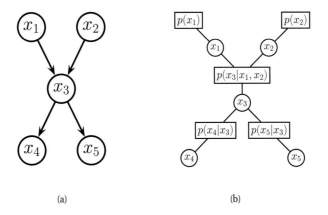

Figure 22.3   (a) A simple DGM. (b) Its corresponding factor graph. Based on Figure 5 of (Yedidia et al. 2001)..

We can also convert a DGM to a factor graph: just create one factor per CPD, and connect that factor to all the variables that use that CPD. For example, Figure 22.3 represents the following factorization:

$$f(x_1, x_2, x_3, x_4, x_5) = f_1(x_1) f_2(x_2) f_{123}(x_1, x_2, x_3) f_{34}(x_3, x_4) f_{35}(x_3, x_5) \qquad (22.3)$$

where we define $f_{123}(x_1, x_2, x_3) = p(x_3 | x_1, x_2)$, etc. If each node has at most one parent (and hence the graph is a chain or simple tree), then there will be one factor per edge (root nodes can have their prior CPDs absorbed into their children's factors). Such models are equivalent to pairwise MRFs.

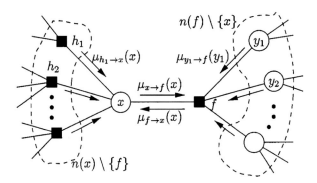

Figure 22.4 Message passing on a bipartite factor graph. Square nodes represent factors, and circles represent variables. Source: Figure 6 of (Kschischang et al. 2001). Used with kind permission of Brendan Frey.

22.2.3.2 BP on a factor graph

We now derive a version of BP that sends messages on a factor graph, as proposed in (Kschischang et al. 2001). Specifically, we now have two kinds of messages: variables to factors

$$m_{x \to f}(x) = \prod_{h \in \mathrm{nbr}(x) \setminus \{f\}} m_{h \to x}(x) \tag{22.4}$$

and factors to variables:

$$m_{f \to x}(x) = \sum_{\mathbf{y}} f(x, \mathbf{y}) \prod_{y \in \mathrm{nbr}(f) \setminus \{x\}} m_{y \to f}(y) \tag{22.5}$$

Here $\mathrm{nbr}(x)$ are all the factors that are connected to variable x, and $\mathrm{nbr}(f)$ are all the variables that are connected to factor f. These messages are illustrated in Figure 22.4. At convergence, we can compute the final beliefs as a product of incoming messages:

$$\mathrm{bel}(x) \propto \prod_{f \in \mathrm{nbr}(x)} m_{f \to x}(x) \tag{22.6}$$

For details on how to derive a parallel distributed implementation of this algorithm, see (Gonzalez et al. 2009).

In the following sections, we will focus on LBP for pairwise models, rather than for factor graphs, but this is just for notational simplicity.

22.2.4 Convergence

LBP does not always converge, and even when it does, it may converge to the wrong answers. This raises several questions: how can we predict when convergence will occur? what can we do to increase the probability of convergence? what can we do to increase the rate of convergence?

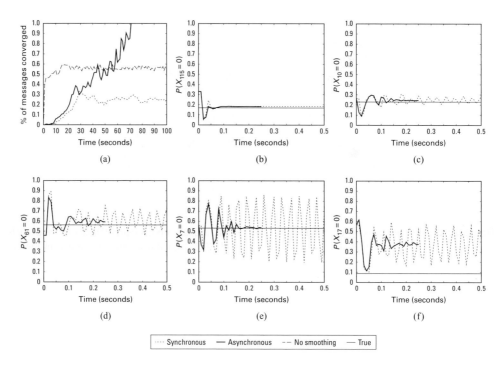

Figure 22.5 Illustration of the behavior of loopy belief propagation on an 11×11 Ising grid with random potentials, $w_{ij} \sim \text{Unif}(-C, C)$, where $C = 11$. For larger C, inference becomes harder. (a) Percentage of messasges that have converged vs time for 3 different update schedules: Dotted = damped sychronous (few nodes converge), dashed = undamped asychnronous (half the nodes converge), solid = damped asychnronous (all nodes converge). (b-f) Marginal beliefs of certain nodes vs time. Solid straight line = truth, dashed = sychronous, solid = damped asychronous.   Source: Figure 11.C.1 of (Koller and Friedman 2009). Used with kind permission of Daphne Koller.

We briefly discuss these issues below. We then discuss the issue of accuracy of the results at convergence.

22.2.4.1   When will LBP converge?

The details of the analysis of when LBP will converge are beyond the scope of this chapter, but we briefly sketch the basic idea. The key analysis tool is the **computation tree**, which visualizes the messages that are passed as the algorithm proceeds. Figure 22.6 gives a simple example. In the first iteration, node 1 receives messages from nodes 2 and 3. In the second iteration, it receives one message from node 3 (via node 2), one from node 2 (via node 3), and two messages from node 4 (via nodes 2 and 3). And so on.

The key insight is that T iterations of LBP is equivalent to exact computation in a computation tree of height $T + 1$. If the strengths of the connections on the edges is sufficiently weak, then the influence of the leaves on the root will diminish over time, and convergence will occur.

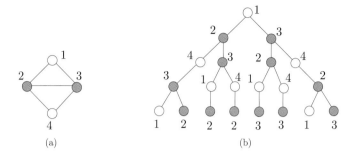

Figure 22.6 (a) A simple loopy graph. (b) The computation tree, rooted at node 1, after 4 rounds of message passing. Nodes 2 and 3 occur more often in the tree because they have higher degree than nodes 1 and 2. Source: Figure 8.2 of (Wainwright and Jordan 2008). Used with kind permission of Martin Wainwright.

See (Mooij and Kappen 2005; Wainwright and Jordan 2008) and references therein for more information.

22.2.4.2 Making LBP converge

Although the theoretical convergence analysis is very interesting, in practice, when faced with a model where LBP is not converging, what should we do?

One simple way to reduce the chance of oscillation is to use **damping**. That is, instead of sending the message M_{ts}^k, we send a damped message of the form

$$\tilde{M}_{ts}^k(x_s) = \lambda M_{ts}(x_s) + (1 - \lambda)\tilde{M}_{ts}^{k-1}(x_s) \tag{22.7}$$

where $0 \leq \lambda \leq 1$ is the damping factor Clearly if $\lambda = 1$ this reduces to the standard scheme, but for $\lambda < 1$, this partial updating scheme can help improve convergence. Using a value such as $\lambda \sim 0.5$ is standard practice. The benefits of this approach are shown in Figure 22.5, where we see that damped updating results in convergence much more often than undamped updating.

It is possible to devise methods, known as **double loop algorithms**, which are guaranteed to converge to a local minimum of the same objective that LBP is minimizing (Yuille 2001; Welling and Teh 2001). Unfortunately, these methods are rather slow and complicated, and the accuracy of the resulting marginals is usually not much greater than with standard LBP. (Indeed, oscillating marginals is sometimes a sign that the LBP approximation itself is a poor one.) Consequently, these techniques are not very widely used. In Section 22.4.2, we will see a different convergent version of BP that is widely used.

22.2.4.3 Increasing the convergence rate: message scheduling

Even if LBP converges, it may take a long time. The standard approach when implementing LBP is to perform **synchronous updates**, where all nodes absorb messages in parallel, and then send out messages in parallel. That is, the new messages at iteration $k + 1$ are computed in parallel using

$$\mathbf{m}^{k+1} = (f_1(\mathbf{m}^k), \dots, f_E(\mathbf{m}^k)) \tag{22.8}$$

where E is the number of edges, and $f_{st}(\mathbf{m})$ is the function that computes the message for edge $s \to t$ given all the old messages. This is analogous to the Jacobi method for solving linear systems of equations. It is well known (Bertsekas 1997) that the Gauss-Seidel method, which performs **asynchronous updates** in a fixed round-robin fashion, converges faster when solving linear systems of equations. We can apply the same idea to LBP, using updates of the form

$$\mathbf{m}_i^{k+1} = f_i \left(\{\mathbf{m}_j^{k+1} : j < i\}, \{\mathbf{m}_j^k : j > i\} \right) \tag{22.9}$$

where the message for edge i is computed using new messages (iteration $k+1$) from edges earlier in the ordering, and using old messages (iteration k) from edges later in the ordering.

This raises the question of what order to update the messages in. One simple idea is to use a fixed or random order. The benefits of this approach are shown in Figure 22.5, where we see that (damped) asynchronous updating results in convergence much more often than synchronous updating.

A smarter approach is to pick a set of spanning trees, and then to perform an up-down sweep on one tree at a time, keeping all the other messages fixed. This is known as **tree reparameterization** (TRP) (Wainwright et al. 2001), which should not be confused with the more sophisticated tree-reweighted BP (often abbreviated to TRW) to be discussed in Section 22.4.2.1.

However, we can do even better by using an adaptive ordering. The intuition is that we should focus our computational efforts on those variables that are most uncertain. (Elidan et al. 2006) proposed a technique known as **residual belief propagation**, in which messages are scheduled to be sent according to the norm of the difference from their previous value. That is, we define the residual of new message m_{st} at iteration k to be

$$r(s, t, k) = || \log \dot{m}_{st} - \log m_{st}^k ||_\infty = \max_i | \log \frac{m_{st}(i)}{m_{st}^k(i)} | \tag{22.10}$$

We can store messages in a priority queue, and always send the one with highest residual. When a message is sent from s to t, all of the other messages that depend on m_{st} (i.e., messages of the form m_{tu} where $u \in \text{nbr}(t) \setminus s$) need to be recomputed; their residual is recomputed, and they are added back to the queue. In (Elidan et al. 2006), they showed (experimentally) that this method converges more often, and much faster, than using synchronous updating, asynchronous updating with a fixed order, and the TRP approach.

A refinement of residual BP was presented in (Sutton and McCallum 2007). In this paper, they use an upper bound on the residual of a message instead of the actual residual. This means that messages are only computed if they are going to be sent; they are not just computed for the purposes of evaluating the residual. This was observed to be about five times faster than residual BP, although the quality of the final results is similar.

22.2.5 Accuracy of LBP

For a graph with a single loop, one can show that the max-product version of LBP will find the correct MAP estimate, if it converges (Weiss 2000). For more general graphs, one can bound the error in the approximate marginals computed by LBP, as shown in (Wainwright et al. 2003; Vinyals et al. 2010). Much stronger results are available in the case of Gaussian models (Weiss and Freeman 2001a; Johnson et al. 2006; Bickson 2009). In particular, in the Gaussian case, if the method converges, the means are exact, although the variances are not (typically the beliefs are over confident).

22.2.6 Other speedup tricks for LBP *

There are several tricks one can use to make BP run faster. We discuss some of them below.

22.2.6.1 Fast message computation for large state spaces

The cost of computing each message in BP (whether in a tree or a loopy graph) is $O(K^f)$, where K is the number of states, and f is the size of the largest factor ($f = 2$ for pairwise UGMs). In many vision problems (e.g., image denoising), K is quite large (say 256), because it represents the discretization of some underlying continuous space, so $O(K^2)$ per message is too expensive. Fortunately, for certain kinds of pairwise potential functions of the form $\psi_{st}(x_s, x_t) = \psi(x_s - x_t)$, one can compute the sum-product messages in $O(K \log K)$ time using the fast Fourier transform or FFT, as explained in (Felzenszwalb and Huttenlocher 2006). The key insight is that message computation is just convolution:

$$M_{st}^k(x_t) = \sum_{x_s} \psi(x_s - x_t) h(x_s) \qquad (22.11)$$

where $h(x_s) = \psi_s(x_s) \prod_{v \in \text{nbr}(s) \setminus t} M_{vs}^{k-1}(x_s)$. If the potential function $\psi(z)$ is a Gaussian-like potential, we can compute the convolution in $O(K)$ time by sequentially convolving with a small number of box filters (Felzenszwalb and Huttenlocher 2006).

For the max-product case, a technique called the **distance transform** can be used to compute messages in $O(K)$ time. However, this only works if $\psi(z) = \exp(-E(z))$ and where $E(z)$ has one the following forms: quadratic, $E(z) = z^2$; truncated linear, $E(z) = \min(c_1|z|, c_2)$; or Potts model, $E(z) = c \, \mathbb{I}(z \neq 0)$. See (Felzenszwalb and Huttenlocher 2006) for details.

22.2.6.2 Multi-scale methods

A method which is specific to 2d lattice structures, which commonly arise in computer vision, is based on multi-grid techniques. Such methods are widely used in numerical linear algebra, where one of the core problems is the fast solution of linear systems of equations; this is equivalent to MAP estimation in a Gaussian MRF. In the computer vision context, (Felzenszwalb and Huttenlocher 2006) suggested using the following heuristic to significantly speedup BP: construct a coarse-to-fine grid, compute messages at the coarse level, and use this to initialize messages at the level below; when we reach the bottom level, just a few iterations of standard BP are required, since long-range communication has already been achieved via the initialization process.

The beliefs at the coarse level are computed over a small number of large blocks. The local evidence is computed from the average log-probability each possible block label assigns to all the pixels in the block. The pairwise potential is based on the discrepancy between labels of neighboring blocks, taking into account their size. We can then run LBP at the coarse level, and then use this to initialize the messages one level down. Note that the *model* is still a flat grid; however, the *initialization process* exploits the multi-scale nature of the problem. See (Felzenszwalb and Huttenlocher 2006) for details.

22.2.6.3 Cascades

Another trick for handling high-dimensional state-spaces, that can also be used with exact inference (e.g., for chain-structured CRFs), is to prune out improbable states based on a computationally cheap filtering step. In fact, one can create a hierarchy of models which tradeoff speed and accuracy. This is called a computational **cascade**. In the case of chains, one can guarantee that the cascade will never filter out the true MAP solution (Weiss et al. 2010).

22.3 Loopy belief propagation: theoretical issues *

We now attempt to understand the LBP algorithm from a variational point of view. Our presentation is closely based on an excellent 300-page review article (Wainwright and Jordan 2008). This paper is sometimes called "the monster" (by its own authors!) in view of its length and technical difficulty. This section just sketches some of the main results.

To simplify the presentation, we focus on the special case of pairwise UGMs with discrete variables and tabular potentials. Many of the results generalize to UGMs with higher-order clique potentials (which includes DGMs), but this makes the notation more complex (see (Koller and Friedman 2009) for details of the general case).

22.3.1 UGMs represented in exponential family form

We assume the distribution has the following form:

$$p(\mathbf{x}|\boldsymbol{\theta}, G) = \frac{1}{Z(\boldsymbol{\theta})} \exp \left\{ \sum_{s \in \mathcal{V}} \theta_s(x_s) + \sum_{(s,t) \in \mathcal{E}} \theta_{st}(x_s, x_t) \right\} \tag{22.12}$$

where graph G has nodes $\mathcal{V}$ and edges $\mathcal{E}$. (Henceforth we will drop the explicit conditioning on $\boldsymbol{\theta}$ and G for brevity, since we assume both are known and fixed.) We can rewrite this in exponential family form as follows:

$$p(\mathbf{x}|\boldsymbol{\theta}) = \frac{1}{Z(\boldsymbol{\theta})} \exp(-E(\mathbf{x})) \tag{22.13}$$

$$E(\mathbf{x}) \triangleq -\boldsymbol{\theta}^T \boldsymbol{\phi}(\mathbf{x}) \tag{22.14}$$

where $\boldsymbol{\theta} = (\{\theta_{s;j}\}, \{\theta_{s,t;j,k}\})$ are all the node and edge parameters (the canonical parameters), and $\boldsymbol{\phi}(\mathbf{x}) = (\{\mathbb{I}(x_s = j)\}, \{\mathbb{I}(x_s = j, x_t = k)\})$ are all the node and edge indicator functions (the sufficient statistics). Note: we use $s, t \in \mathcal{V}$ to index nodes and $j, k \in \mathcal{X}$ to index states.

The mean of the sufficient statistics are known as the mean parameters of the model, and are given by

$$\boldsymbol{\mu} = \mathbb{E}\left[\boldsymbol{\phi}(\mathbf{x})\right] = (\{p(x_s = j)\}_s, \{p(x_s = j, x_t = k)\}_{s \neq t}) = (\{\mu_{s;j}\}_s, \{\mu_{st;jk}\}_{s \neq t}) \tag{22.15}$$

This is a vector of length $d = |\mathcal{X}||V| + |\mathcal{X}|^2|E|$, containing the node and edge marginals. It completely characterizes the distribution $p(\mathbf{x}|\boldsymbol{\theta})$, so we sometimes treat $\boldsymbol{\mu}$ as a distribution itself.

Equation 22.12 is called the **standard overcomplete representation**. It is called "overcomplete" because it ignores the sum-to-one constraints. In some cases, it is convenient to remove this redundancy. For example, consider an Ising model where $X_s \in \{0, 1\}$. The model can be written as

$$p(\mathbf{x}) = \frac{1}{Z(\boldsymbol{\theta})} \exp \left\{ \sum_{s \in V} \theta_s x_s + \sum_{(s,t) \in \mathcal{E}} \theta_{st} x_s x_t \right\} \tag{22.16}$$

Hence we can use the following minimal parameterization

$$\boldsymbol{\phi}(\mathbf{x}) = (x_s, s \in V; x_s x_t, (s, t) \in E) \in \mathbb{R}^d \tag{22.17}$$

where $d = |V| + |E|$. The corresponding mean parameters are $\mu_s = p(x_s = 1)$ and $\mu_{st} = p(x_s = 1, x_t = 1)$.

22.3.2   The marginal polytope

The space of allowable $\boldsymbol{\mu}$ vectors is called the **marginal polytope**, and is denoted $\mathbb{M}(G)$, where G is the structure of the graph defining the UGM. This is defined to be the set of all mean parameters for the given model that can be generated from a valid probability distribution:

$$\mathbb{M}(G) \triangleq \{\boldsymbol{\mu} \in \mathbb{R}^d : \exists p \quad \text{s.t.} \quad \boldsymbol{\mu} = \sum_{\mathbf{x}} \boldsymbol{\phi}(\mathbf{x}) p(\mathbf{x}) \text{ for some } p(\mathbf{x}) \geq 0, \sum_{\mathbf{x}} p(\mathbf{x}) = 1 \} \tag{22.18}$$

For example, consider an Ising model. If we have just two nodes connected as $X_1 - X_2$, one can show that we have the following minimal set of constraints: $0 \leq \mu_{12}$, $0 \leq \mu_{12} \leq \mu_1$, $0 \leq \mu_{12} \leq \mu_2$, and $1 + \mu_{12} - \mu_1 - \mu_2 \geq 0$. We can write these in matrix-vector form as

$$\begin{pmatrix} 0 & 0 & 1 \\ 1 & 0 & -1 \\ 0 & 1 & -1 \\ -1 & -1 & 1 \end{pmatrix} \begin{pmatrix} \mu_1 \\ \mu_2 \\ \mu_{12} \end{pmatrix} \geq \begin{pmatrix} 0 \\ 0 \\ 0 \\ -1 \end{pmatrix} \tag{22.19}$$

These four constraints define a series of half-planes, whose intersection defines a polytope, as shown in Figure 22.7(a).

Since $\mathbb{M}(G)$ is obtained by taking a convex combination of the $\boldsymbol{\phi}(\mathbf{x})$ vectors, it can also be written as the **convex hull** of the feature set:

$$\mathbb{M}(G) = \text{conv}\{\boldsymbol{\phi}_1(\mathbf{x}), \dots, \boldsymbol{\phi}_d(\mathbf{x})\} \tag{22.20}$$

For example, for a 2 node MRF $X_1 - X_2$ with binary states, we have

$$\mathbb{M}(G) = \text{conv}\{(0, 0, 0), (1, 0, 0), (0, 1, 0), (1, 1, 1)\} \tag{22.21}$$

These are the four black dots in Figure 22.7(a). We see that the convex hull defines the same volume as the intersection of half-spaces.

The marginal polytope will play a crucial role in the approximate inference algorithms we discuss in the rest of this chapter.

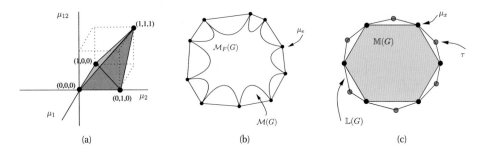

Figure 22.7 (a) Illustration of the marginal polytope for an Ising model with two variables. (b) Cartoon illustration of the set $\mathbb{M}_F(G)$, which is a nonconvex inner bound on the marginal polytope $\mathbb{M}(G)$. $\mathbb{M}_F(G)$ is used by mean field. (c) Cartoon illustration of the relationship between $\mathbb{M}(G)$ and $\mathbb{L}(G)$, which is used by loopy BP. The set $\mathbb{L}(G)$ is always an outer bound on $\mathbb{M}(G)$, and the inclusion $\mathbb{M}(G) \subset \mathbb{L}(G)$ is strict whenever G has loops. Both sets are polytopes, which can be defined as an intersection of half-planes (defined by facets), or as the convex hull of the vertices. $\mathbb{L}(G)$ actually has fewer facets than $\mathbb{M}(G)$, despite the picture. In fact, $\mathbb{L}(G)$ has $O(|\mathcal{X}||V| + |\mathcal{X}|^2|E|)$ facets, where $|\mathcal{X}|$ is the number of states per variable, $|V|$ is the number of variables, and $|E|$ is the number of edges. By contrast, $\mathbb{M}(G)$ has $O(|\mathcal{X}|^{|V|})$ facets. On the other hand, $\mathbb{L}(G)$ has more vertices than $\mathbb{M}(G)$, despite the picture, since $\mathbb{L}(G)$ contains all the binary vector extreme points $\mu \in \mathbb{M}(G)$, plus additional fractional extreme points. Source: Figures 3.6, 5.4 and 4.2 of (Wainwright and Jordan 2008). Used with kind permission of Martin Wainwright.

22.3.3   Exact inference as a variational optimization problem

Recall from Section 21.2 that the goal of variational inference is to find the distribution q that maximizes the **energy functional**

$$L(q) = -\mathbb{KL}\left(q||p\right) + \log Z = \mathbb{E}_q\left[\log \tilde{p}(\mathbf{x})\right] + \mathbb{H}\left(q\right) \leq \log Z \tag{22.22}$$

where $\tilde{p}(\mathbf{x}) = Zp(\mathbf{x})$ is the unnormalized posterior. If we write $\log \tilde{p}(\mathbf{x}) = \boldsymbol{\theta}^T \boldsymbol{\phi}(\mathbf{x})$, and we let $q = p$, then the exact energy functional becomes

$$\max_{\boldsymbol{\mu} \in \mathbb{M}(G)} \boldsymbol{\theta}^T \boldsymbol{\mu} + \mathbb{H}\left(\boldsymbol{\mu}\right) \tag{22.23}$$

where $\boldsymbol{\mu} = \mathbb{E}_p\left[\boldsymbol{\phi}(\mathbf{x})\right]$ is a joint distribution over all state configurations $\mathbf{x}$ (so it is valid to write $\mathbb{H}\left(\boldsymbol{\mu}\right)$). Since the KL divergence is zero when $p = q$, we know that

$$\max_{\boldsymbol{\mu} \in \mathbb{M}(G)} \boldsymbol{\theta}^T \boldsymbol{\mu} + \mathbb{H}\left(\boldsymbol{\mu}\right) = \log Z(\boldsymbol{\theta}) \tag{22.24}$$

This is a way to cast exact inference as a variational optimization problem.

Equation 22.24 seems easy to optimize: the objective is concave, since it is the sum of a linear function and a concave function (see Figure 2.21 to see why entropy is concave); furthermore, we are maximizing this over a convex set. However, the marginal polytope $\mathbb{M}(G)$ has exponentially many facets. In some cases, there is structure to this polytope that can be exploited by dynamic programming (as we saw in Chapter 20), but in general, exact inference takes exponential time. Most of the existing deterministic approximate inference schemes that have been proposed in the literature can be seen as different approximations to the marginal polytope, as we explain below.

22.3.4 Mean field as a variational optimization problem

We discussed mean field at length in Chapter 21. Let us re-interpret mean field inference in our new more abstract framework. This will help us compare it to other approximate methods which we discuss below.

First, let F be an edge subgraph of the original graph G, and let $\mathcal{I}(F) \subseteq \mathcal{I}$ be the subset of sufficient statistics associated with the cliques of F. Let Ω be the set of canonical parameters for the full model, and define the canonical parameter space for the submodel as follows:

$$\Omega(F) \triangleq \{\boldsymbol{\theta} \in \Omega : \boldsymbol{\theta}_\alpha = 0 \ \forall \alpha \in \mathcal{I} \setminus \mathcal{I}(F)\} \tag{22.25}$$

In other words, we require that the natural parameters associated with the sufficient statistics α outside of our chosen class to be zero. For example, in the case of a fully factorized approximation, F_0, we remove all edges from the graph, giving

$$\Omega(F_0) \triangleq \{\boldsymbol{\theta} \in \Omega : \boldsymbol{\theta}_{st} = 0 \ \forall (s,t) \in E\} \tag{22.26}$$

In the case of structured mean field (Section 21.4), we set $\theta_{st} = 0$ for edges which are not in our tractable subgraph.

Next, we define the mean parameter space of the restricted model as follows:

$$\mathbb{M}_F(G) \triangleq \{\boldsymbol{\mu} \in \mathbb{R}^d : \boldsymbol{\mu} = \mathbb{E}_{\boldsymbol{\theta}}[\boldsymbol{\phi}(\mathbf{x})] \ \text{ for some } \boldsymbol{\theta} \in \Omega(F)\} \tag{22.27}$$

This is called an **inner approximation** to the marginal polytope, since $\mathbb{M}_F(G) \subseteq \mathbb{M}(G)$. See Figure 22.7(b) for a sketch. Note that $\mathbb{M}_F(G)$ is a non-convex polytope, which results in multiple local optima. By contrast, some of the approximations we will consider later will be convex.

We define the entropy of our approximation $\mathbb{H}(\boldsymbol{\mu}(F))$ as the entropy of the distribution $\boldsymbol{\mu}$ defined on submodel F. Then we define the **mean field energy functional** optimization problem as follows:

$$\max_{\boldsymbol{\mu} \in \mathbb{M}_F(G)} \boldsymbol{\theta}^T \boldsymbol{\mu} + \mathbb{H}(\boldsymbol{\mu}) \leq \log Z(\boldsymbol{\theta}) \tag{22.28}$$

In the case of the fully factorized mean field approximation for pairwise UGMs, we can write this objective as follows:

$$\max_{\boldsymbol{\mu} \in \mathcal{P}^d} \sum_{s \in V} \sum_{x_s} \theta_s(x_s)\mu_s(x_s) + \sum_{(s,t) \in E} \sum_{x_s, x_t} \theta_{st}(x_s, x_t)\mu_s(x_s)\mu_t(x_t) + \sum_{s \in V} \mathbb{H}(\boldsymbol{\mu}_s) \tag{22.29}$$

where $\boldsymbol{\mu}_s \in \mathcal{P}$, and $\mathcal{P}$ is the probability simplex over $\mathcal{X}$.

Mean field involves a concave objective being maximized over a non-convex set. It is typically optimized using coordinate ascent, since it is easy to optimize a scalar concave function over $\mathcal{P}$ for each μ_s. For example, for a pairwise UGM we get

$$\mu_s(x_s) \propto \exp(\theta_s(x_s)) \exp\left(\sum_{t \in \text{nbr}(s)} \sum_{x_t} \mu_t(x_t)\theta_{st}(x_s, x_t)\right) \tag{22.30}$$

22.3.5 LBP as a variational optimization problem

In this section, we explain how LBP can be viewed as a variational inference problem.

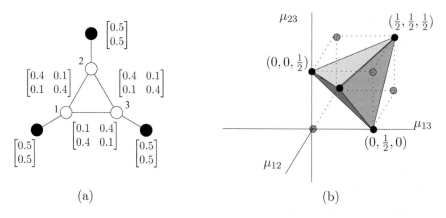

(a)   (b)

Figure 22.8   (a) Illustration of pairwise UGM on binary nodes, together with a set of pseudo marginals that are not globally consistent. (b) A slice of the marginal polytope illustrating the set of feasible edge marginals, assuming the node marginals are clamped at $\mu_1 = \mu_2 = \mu_3 = 0.5$.   Source: Figure 4.1 of (Wainwright and Jordan 2008).   Used with kind permission of Martin Wainwright.

22.3.5.1   An outer approximation to the marginal polytope

If we want to consider all possible probability distributions which are Markov wrt our model, we need to consider all vectors $\boldsymbol{\mu} \in \mathbb{M}(G)$. Since the set $\mathbb{M}(G)$ is exponentially large, it is usually infeasible to optimize over. A standard strategy in combinatorial optimization is to relax the constraints. In this case, instead of requiring probability vector $\boldsymbol{\mu}$ to live in $\mathbb{M}(G)$, we consider a vector $\boldsymbol{\tau}$ that only satisfies the following **local consistency** constraints:

$$\sum_{x_s} \tau_s(x_s) \;\; = \;\; 1 \tag{22.31}$$

$$\sum_{x_t} \tau_{st}(x_s, x_t) \;\; = \;\; \tau_s(x_s) \tag{22.32}$$

The first constraint is called the normalization constraint, and the second is called the marginalization constraint. We then define the set

$$\mathbb{L}(G) \triangleq \{ \boldsymbol{\tau} \geq 0 : (22.31) \text{ holds } \forall s \in V \text{ and } (22.32) \text{ holds } \forall (s,t) \in E \} \tag{22.33}$$

The set $\mathbb{L}(G)$ is also a polytope, but it only has $O(|V| + |E|)$ constraints. It is a convex **outer approximation** on $\mathbb{M}(G)$, as shown in Figure 22.7(c).

We call the terms $\tau_s, \tau_{st} \in \mathbb{L}(G)$ **pseudo marginals**, since they may not correspond to marginals of any valid probability distribution. As an example of this, consider Figure 22.8(a). The picture shows a set of pseudo node and edge marginals, which satisfy the local consistency requirements. However, they are not globally consistent. To see why, note that τ_{12} implies $p(X_1 = X_2) = 0.8$, τ_{23} implies $p(X_2 = X_3) = 0.8$, but τ_{13} implies $p(X_1 = X_3) = 0.2$, which is not possible (see (Wainwright and Jordan 2008, p81) for a formal proof). Indeed, Figure 22.8(b) shows that $\mathbb{L}(G)$ contains points that are not in $\mathbb{M}(G)$.

We claim that $\mathbb{M}(G) \subseteq \mathbb{L}(G)$, with equality iff G is a tree. To see this, first consider an element $\boldsymbol{\mu} \in \mathbb{M}(G)$. Any such vector must satisfy the normalization and marginalization constraints, hence $\mathbb{M}(G) \subseteq \mathbb{L}(G)$.

Now consider the converse. Suppose T is a tree, and let $\boldsymbol{\mu} \in \mathbb{L}(T)$. By definition, this satisfies the normalization and marginalization constraints. However, any tree can be represented in the form

$$p_{\boldsymbol{\mu}}(\mathbf{x}) = \prod_{s \in V} \mu_s(x_s) \prod_{(s,t) \in E} \frac{\mu_{st}(x_s, x_t)}{\mu_s(x_s)\mu_t(x_t)} \tag{22.34}$$

Hence satsifying normalization and local consistency is enough to define a valid distribution for any tree. Hence $\boldsymbol{\mu} \in \mathbb{M}(T)$ as well.

In contrast, if the graph has loops, we have that $\mathbb{M}(G) \neq \mathbb{L}(G)$. See Figure 22.8(b) for an example of this fact.

22.3.5.2 The entropy approximation

From Equation 22.34, we can write the exact entropy of any tree structured distribution $\boldsymbol{\mu} \in \mathbb{M}(T)$ as follows:

$$\mathbb{H}(\boldsymbol{\mu}) = \sum_{s \in V} H_s(\mu_s) - \sum_{(s,t) \in E} I_{st}(\mu_{st}) \tag{22.35}$$

$$H_s(\mu_s) = - \sum_{x_s \in \mathcal{X}_s} \mu_s(x_s) \log \mu_s(x_s) \tag{22.36}$$

$$I_{st}(\mu_{st}) = \sum_{(x_s, x_t) \in \mathcal{X}_s \times \mathcal{X}_t} \mu_{st}(x_s, x_t) \log \frac{\mu_{st}(x_s, x_t)}{\mu_s(x_s)\mu_t(x_t)} \tag{22.37}$$

Note that we can rewrite the mutual information term in the form $I_{st}(\mu_{st}) = H_s(\mu_s) + H_t(\mu_t) - H_{st}(\mu_{st})$, and hence we get the following alternative but equivalent expression:

$$\mathbb{H}(\boldsymbol{\mu}) = - \sum_{s \in V} (d_s - 1) H_s(\mu_s) + \sum_{(s,t) \in E} H_{st}(\mu_{st}) \tag{22.38}$$

where d_s is the degree (number of neighbors) for node s.

The **Bethe**[1] approximation to the entropy is simply the use of Equation 22.35 even when we don't have a tree:

$$\mathbb{H}_{\text{Bethe}}(\boldsymbol{\tau}) = \sum_{s \in V} H_s(\tau_s) - \sum_{(s,t) \in E} I_{st}(\tau_{st}) \tag{22.39}$$

We define the **Bethe free energy** as

$$F_{\text{Bethe}}(\boldsymbol{\tau}) \triangleq - \left[\boldsymbol{\theta}^T \boldsymbol{\tau} + \mathbb{H}_{\text{Bethe}}(\boldsymbol{\tau}) \right] \tag{22.40}$$

We define the **Bethe energy functional** as the negative of the Bethe free energy.

---

1. Hans Bethe was a German-American physicist, 1906–2005.

22.3.5.3   The LBP objective

Combining the outer approximation $\mathbb{L}(G)$ with the Bethe approximation to the entropy, we get the following Bethe variational problem (BVP):

$$\min_{\boldsymbol{\tau} \in \mathbb{L}(G)} F_{\text{Bethe}}(\boldsymbol{\tau}) = \max_{\boldsymbol{\tau} \in \mathbb{L}(G)} \boldsymbol{\theta}^T \boldsymbol{\tau} + \mathbb{H}_{\text{Bethe}}(\boldsymbol{\tau}) \tag{22.41}$$

The space we are optimizing over is a convex set, but the objective itself is not concave (since $\mathbb{H}_{\text{Bethe}}$ is not concave). Thus there can be multiple local optima of the BVP.

The value obtained by the BVP is an approximation to $\log Z(\boldsymbol{\theta})$. In the case of trees, the approximation is exact, and in the case of models with attractive potentials, the approximation turns out to be an upper bound (Sudderth et al. 2008).

22.3.5.4   Message passing and Lagrange multipliers

In this subsection, we will show that any fixed point of the LBP algorithm defines a stationary point of the above constrained objective. Let us define the normalization constraint at $C_{ss}(\boldsymbol{\tau}) \triangleq 1 - \sum_{x_s} \tau_s(x_s)$, and the marginalization constraint as $C_{ts}(x_s; \boldsymbol{\tau}) \triangleq \tau_s(x_s) - \sum_{x_t} \tau_{st}(x_s, x_t)$ for each edge $t \to s$. We can now write the Lagrangian as

$$\mathcal{L}(\boldsymbol{\tau}, \boldsymbol{\lambda}; \boldsymbol{\theta}) \triangleq \boldsymbol{\theta}^T \boldsymbol{\tau} + \mathbb{H}_{\text{Bethe}}(\boldsymbol{\tau}) + \sum_s \lambda_{ss} C_{ss}(\boldsymbol{\tau})$$

$$+ \sum_{s,t} \left[\sum_{x_s} \lambda_{ts}(x_s) C_{ts}(x_s; \boldsymbol{\tau}) + \sum_{x_t} \lambda_{st}(x_t) C_{st}(x_t; \boldsymbol{\tau}) \right] \tag{22.42}$$

(The constraint that $\boldsymbol{\tau} \geq 0$ is not explicitly enforced, but one can show that it will hold at the optimum since $\boldsymbol{\theta} > 0$.) Some simple algebra then shows that $\nabla_{\boldsymbol{\tau}} \mathcal{L} = \boldsymbol{0}$ yields

$$\log \tau_s(x_s) = \lambda_{ss} + \theta_s(x_s) + \sum_{t \in \text{nbr}(s)} \lambda_{ts}(x_s) \tag{22.43}$$

$$\log \frac{\tau_{st}(x_s, x_t)}{\tilde{\tau}_s(x_s)\tilde{\tau}_t(x_t)} = \theta_{st}(x_s, x_t) - \lambda_{ts}(x_s) - \lambda_{st}(x_t) \tag{22.44}$$

where we have defined $\tilde{\tau}_s(x_s) \triangleq \sum_{x_t} \tau(x_s, x_t)$. Using the fact that the marginalization constraint implies $\tilde{\tau}_s(x_s) = \tau_s(x_s)$, we get

$$\log \tau_{st}(x_s, x_t) = \lambda_{ss} + \lambda_{tt} + \theta_{st}(x_s, x_t) + \theta_s(x_s) + \theta_t(x_t)$$

$$+ \sum_{u \in \text{nbr}(s) \setminus t} \lambda_{us}(x_s) + \sum_{u \in \text{nbr}(t) \setminus s} \lambda_{ut}(x_t) \tag{22.45}$$

To make the connection to message passing, define $M_{ts}(x_s) = \exp(\lambda_{ts}(x_s))$. With this notation, we can rewrite the above equations (after taking exponents of both sides) as follows:

$$\tau_s(x_s) \propto \exp(\theta_s(x_s)) \prod_{t \in \text{nbr}(s)} M_{ts}(x_s) \tag{22.46}$$

$$\tau_{st}(x_s, x_t) \propto \exp\left(\theta_{st}(x_s, x_t) + \theta_s(x_s) + \theta_t(x_t)\right)$$

$$\times \prod_{u \in \text{nbr}(s) \setminus t} M_{us}(x_s) \prod_{u \in \text{nbr}(t) \setminus s} M_{ut}(x_t) \tag{22.47}$$

where the λ terms are absorbed into the constant of proportionality. We see that this is equivalent to the usual expression for the node and edge marginals in LBP.

To derive an equation for the messages in terms of other messages (rather than in terms of λ_{ts}), we enforce the marginalization condition $\sum_{x_t} \tau_{st}(x_s, x_t) = \tau_s(x_s)$. Then one can show that

$$M_{ts}(x_s) \propto \sum_{x_t} \left[\exp\{\theta_{st}(x_s, x_t) + \theta_t(x_t)\} \prod_{u \in \text{nbr}(t) \setminus s} M_{ut}(x_t) \right] \tag{22.48}$$

We see that this is equivalent to the usual expression for the messages in LBP.

22.3.6 Loopy BP vs mean field

It is interesting to compare the naive mean field (MF) and LBP approximations. There are several obvious differences. First, LBP is exact for trees whereas MF is not, suggesting LBP will in general be more accurate (see (Wainwright et al. 2003) for an analysis). Second, LBP optimizes over node and edge marginals, whereas MF only optimizes over node marginals, again suggesting LBP will be more accurate. Third, in the case that the true edge marginals factorize, so $\mu_{st} = \mu_s \mu_t$, the free energy approximations will be the same in both cases.

What is less obvious, but which nevertheless seems to be true, is that the MF objective has many more local optima than the LBP objective, so optimizing the MF objective seems to be harder. In particular, (Weiss 2001), shows empirically that optimizing MF starting from uniform or random initial conditions often leads to poor results, whereas optimizing BP from uniform initial messages often leads to good results. Furthermore, initializing MF with the BP marginals also leads to good results (although MF tends to be more overconfident than BP), indicating that the problem is caused not by the inaccuracy of the MF approximation, but rather by the severe non-convexity of the MF objective, and by the weakness of the standard coordinate descent optimization method used by MF.[2] However, the advantage of MF is that it gives a lower bound on the partition function, unlike BP, which is useful when using it as a subroutine inside a learning algorithm. Also, MF is easier to extend to other distributions besides discrete and Gaussian, as we saw in Chapter 21. Intuitively, this is because MF only works with marginal distributions, which have a single type, rather than needing to define pairwise distributions, which may need to have two different types.

22.4 Extensions of belief propagation *

In this section, we discuss various extensions of LBP.

22.4.1 Generalized belief propagation

We can improve the accuracy of loopy BP by clustering together nodes that form a tight loop. This is known as the **cluster variational method**. The result is a hyper-graph, which is a graph

---

2. (Honkela et al. 2003) discusses the use of the pattern search algorithm to speedup mean field inference in the case of continuous random variables. It is possible that similar ideas could be adapted to the discrete case, although there may be no reason to do this, given that LBP already works well in the discrete case.

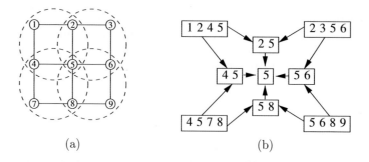

Figure 22.9 (a) Kikuchi clusters superimposed on a 3×3 lattice graph. (b) Corresponding hyper-graph. Source: Figure 4.5 of (Wainwright and Jordan 2008). Used with kind permission of Martin Wainwright.

where there are hyper-edges between sets of vertices instead of between single vertices. Note that a junction tree (Section 20.4.1) is a kind of hyper-graph. We can represent hyper-graph using a poset (partially ordered set) diagram, where each node represents a hyper-edge, and there is an arrow $e_1 \rightarrow e_2$ if $e_2 \subset e_1$. See Figure 22.9 for an example.

Let t be the size of the largest hyper-edge in the hyper-graph. If we allow t to be as large as the treewidth of the graph, then we can represent the hyper-graph as a tree, and the method will be exact, just as LBP is exact on regular trees (with treewidth 1). In this way, we can define a continuum of approximations, from LBP all the way to exact inference.

Define $\mathbb{L}_t(G)$ to be the set of all pseudo-marginals such that normalization and marginalization constraints hold on a hyper-graph whose largest hyper-edge is of size $t + 1$. For example, in Figure 22.9, we impose constraints of the form

$$\sum_{x_1, x_2} \tau_{1245}(x_1, x_2, x_4, x_5) = \tau_{45}(x_4, x_5), \quad \sum_{x_6} \tau_{56}(x_5, x_6) = \tau_5(x_5), \ldots \tag{22.49}$$

Furthermore, we approximate the entropy as follows:

$$\mathbb{H}_{\text{Kikuchi}}(\boldsymbol{\tau}) \triangleq \sum_{g \in E} c(g) H_g(\tau_g) \tag{22.50}$$

where $H_g(\tau_g)$ is the entropy of the joint (pseudo) distribution on the vertices in set g, and $c(g)$ is called the **overcounting number** of set g. These are related to **Mobious numbers** in set theory. Rather than giving a precise definition, we just give a simple example. For the graph in Figure 22.9, we have

$$\mathbb{H}_{\text{Kikuchi}}(\boldsymbol{\tau}) = [H_{1245} + H_{2356} + H_{4578} + H_{5689}]$$
$$- [H_{25} + H_{45} + H_{56} + H_{58}] + H_5 \tag{22.51}$$

Putting these two approximations together, we can define the **Kikuchi free energy**[3] as follows:

$$F_{\text{Kikuchi}}(\boldsymbol{\tau}) \triangleq - \left[\boldsymbol{\theta}^T \boldsymbol{\tau} + \mathbb{H}_{\text{Kikuchi}}(\boldsymbol{\tau}) \right] \tag{22.52}$$

---

3. Ryoichi Kikuchi is a Japanese physicist.

Our variational problem becomes

$$\min_{\boldsymbol{\tau} \in \mathbb{L}_t(G)} F_{\text{Kikuchi}}(\boldsymbol{\tau}) = \max_{\boldsymbol{\tau} \in \mathbb{L}_t(G)} \boldsymbol{\theta}^T \boldsymbol{\tau} + \mathbb{H}_{\text{Kikuchi}}(\boldsymbol{\tau}) \tag{22.53}$$

Just as with the Bethe free energy, this is not a concave objective. There are several possible algorithms for finding a local optimum of this objective, including a message passing algorithm known as **generalized belief propagation**. However, the details are beyond the scope of this chapter. See e.g., (Wainwright and Jordan 2008, Sec 4.2) or (Koller and Friedman 2009, Sec 11.3.2) for more information. Suffice it to say that the method gives more accurate results than LBP, but at increased computational cost (because of the need to handle clusters of nodes). This cost, plus the complexity of the approach, have precluded it from widespread use.

22.4.2 Convex belief propagation

The mean field energy functional is concave, but it is maximized over a non-convex inner approximation to the marginal polytope. The Bethe and Kikuchi energy functionals are not concave, but they are maximized over a convex outer approximation to the marginal polytope. Consequently, for both MF and LBP, the optimization problem has multiple optima, so the methods are sensitive to the initial conditions. Given that the exact formulation (Equation 22.24) is a concave objective maximized over a convex set, it is natural to try to come up with an appproximation of a similar form, without local optima.

We now describe one method, known as **convex belief propagation**. This involves working with a set of tractable submodels, $\mathcal{F}$, such as trees or planar graphs. For each model $F \subset G$, the entropy is higher, $\mathbb{H}(\boldsymbol{\mu}(F)) \geq \mathbb{H}(\boldsymbol{\mu}(G))$, since F has fewer constraints. Consequently, any convex combination of such subgraphs will have higher entropy, too:

$$\mathbb{H}(\boldsymbol{\mu}(G)) \leq \sum_{F \in \mathcal{F}} \rho(F) \mathbb{H}(\boldsymbol{\mu}(F)) \triangleq \mathbb{H}(\boldsymbol{\mu}, \rho) \tag{22.54}$$

where $\rho(F) \geq 0$ and $\sum_F \rho(F) = 1$. Furthermore, $\mathbb{H}(\boldsymbol{\mu}, \rho)$ is a concave function of $\boldsymbol{\mu}$. We now define the convex free energy as

$$F_{\text{Convex}}(\boldsymbol{\mu}, \rho) \triangleq - \left[\boldsymbol{\mu}^T \boldsymbol{\theta} + \mathbb{H}(\boldsymbol{\mu}, \rho) \right] \tag{22.55}$$

We define the concave energy functional as the negative of the convex free energy. We discuss how to optimize ρ below.

Having defined an upper bound on the entropy, we now consider a convex outerbound on the marginal polytope of mean parameters. We want to ensure we can evaluate the entropy of any vector $\boldsymbol{\tau}$ in this set, so we restrict it so that the projection of $\boldsymbol{\tau}$ onto the subgraph G lives in the projection of $\mathbb{M}$ onto F:

$$\mathbb{L}(G; \mathcal{F}) \triangleq \{ \boldsymbol{\tau} \in \mathbb{R}^d : \boldsymbol{\tau}(F) \in \mathbb{M}(F) \ \forall F \in \mathcal{F} \} \tag{22.56}$$

This is a convex set since each $\mathbb{M}(F)$ is a projection of a convex set. Hence we define our problem as

$$\min_{\boldsymbol{\tau} \in \mathbb{L}(G; \mathcal{F})} F_{\text{Convex}}(\boldsymbol{\tau}, \rho) = \max_{\boldsymbol{\tau} \in \mathbb{L}(G; \mathcal{F})} \boldsymbol{\tau}^T \boldsymbol{\theta} + \mathbb{H}(\boldsymbol{\tau}, \rho) \tag{22.57}$$

This is a concave objective being maximized over a convex set, and hence has a unique maximum. We give a specific example below. (See (Schwing et al. 2011) for details on how to derive a parallel distributed implementation of convex BP.)

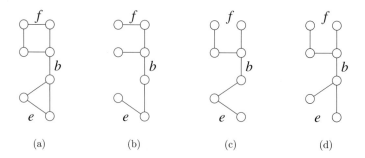

Figure 22.10   (a) A graph. (b-d) Some of its spanning trees.   Source: Figure 7.1 of (Wainwright and Jordan 2008).   Used with kind permission of Martin Wainwright.

22.4.2.1   Tree-reweighted belief propagation

Consider the specific case where $\mathcal{F}$ is all spanning trees of a graph. For any given tree, the entropy is given by Equation 22.35. To compute the upper bound, obtained by averaging over all trees, note that the terms $\sum_F \rho(F) H(\mu(F)_s)$ for single nodes will just be H_s, since node s appears in every tree, and $\sum_F \rho(F) = 1$. But the mutual information term I_{st} receives weight $\rho_{st} = \mathbb{E}_\rho \left[\mathbb{I}((s,t) \in E(T)) \right]$, known as the **edge appearance probability**. Hence we have the following upper bound on the entropy:

$$\mathbb{H}(\boldsymbol{\mu}) \leq \sum_{s \in V} H_s(\mu_s) - \sum_{(s,t) \in E} \rho_{st} I_{st}(\mu_{st}) \tag{22.58}$$

The edge appearance probabilities live in a space called the **spanning tree polytope**. This is because they are constrained to arise from a distribution over trees. Figure 22.10 gives an example of a graph and three of its spanning trees. Suppose each tree has equal weight under ρ. The edge f occurs in 1 of the 3 trees, so $\rho_f = 1/3$. The edge e occurs in 2 of the 3 trees, so $\rho_e = 2/3$. The edge b appears in all of the trees, so $\rho_b = 1$. And so on. Ideally we can find a distribution ρ, or equivalently edge probabilities in the spanning tree polytope, that make the above bound as tight as possible. An algorithm to do this is described in (Wainwright et al. 2005). (A simpler approach is to generate spanning trees of G at random until all edges are covered, or use all single edges with weight $\rho_e = 1/E$.)

What about the set we are optimizing over? We require $\boldsymbol{\mu}(T) \in \mathbb{M}(T)$ for each tree T, which means enforcing normalization and local consistency. Since we have to do this for every tree, we are enforcing normalization and local consistency on every edge. Hence $\mathbb{L}(G; \mathcal{F}) = \mathbb{L}(G)$. So our final optimization problem is as follows:

$$\max_{\boldsymbol{\tau} \in \mathbb{L}(G)} \left\{ \boldsymbol{\tau}^T \boldsymbol{\theta} + \sum_{s \in V} H_s(\tau_s) - \sum_{(s,t) \in E(G)} \rho_{st} I_{st}(\tau_{st}) \right\} \tag{22.59}$$

which is the same as the LBP objective except for the crucial ρ_{st} weights. So long as $\rho_{st} > 0$ for all edges (s,t), this problem is strictly concave with a unique maximum.

How can we find this global optimum? As for LBP, there are several algorithms, but perhaps the simplest is a modification of belief propagation known as **tree reweighted belief propagation**, also called **TRW** or **TRBP** for short. The message from t to s is now a function of all messages sent from other neighbors v to t, as before, but now it is also a function of the message sent from s to t. Specifically

$$M_{ts}(x_s) \quad \propto \quad \sum_{x_t} \exp\left(\frac{1}{\rho_{st}}\theta_{st}(x_s, x_t) + \theta_t(x_t)\right) \frac{\prod_{v \in \text{nbr}(t) \backslash s}[M_{vt}(x_t)]^{\rho_{vt}}}{[M_{st}(x_t)]^{1-\rho_{ts}}} \qquad (22.60)$$

At convergence, the node and edge pseudo marginals are given by

$$\tau_s(x_s) \quad \propto \quad \exp(\theta_s(x_s)) \prod_{v \in \text{nbr}(s)} [M_{vs}(x_s)]^{\rho_{vs}} \qquad (22.61)$$

$$\tau_{st}(x_s, x_t) \quad \propto \quad \varphi_{st}(x_s, x_t) \frac{\prod_{v \in \text{nbr}(s)\backslash t}[M_{vs}(x_s)]^{\rho_{vs}}}{[M_{ts}(x_s)]^{1-\rho_{st}}} \frac{\prod_{v \in \text{nbr}(t)\backslash s}[M_{vt}(x_t)]^{\rho_{vt}}}{[M_{st}(x_t)]^{1-\rho_{ts}}} \qquad (22.62)$$

$$\varphi_{st}(x_s, x_t) \quad \triangleq \quad \exp\left(\frac{1}{\rho_{st}}\theta_{st}(x_s, x_t) + \theta_s(x_s) + \theta_t(x_t)\right) \qquad (22.63)$$

This algorithm can be derived using a method similar to that described in Section 22.3.5.4.

If $\rho_{st} = 1$ for all edges $(s,t) \in E$, the algorithm reduces to the standard LBP algorithm. However, the condition $\rho_{st} = 1$ implies every edge is present in every spanning tree with probability 1, which is only possible if the original graph is a tree. Hence the method is only equivalent to standard LBP on trees, when the method is of course exact.

In general, this message passing scheme is not guaranteed to converge to the unique global optimum. One can devise double-loop methods that are guaranteed to converge (Hazan and Shashua 2008), but in practice, using damped updates as in Equation 22.7 is often sufficient to ensure convergence.

It is also possible to produce a convex version of the Kikuchi free energy, which one can optimize with a modified version of generalized belief propagation. See (Wainwright and Jordan 2008, Sec 7.2.2) for details.

From Equation 22.59, and using the fact that the TRBP entropy approximation is an upper bound on the true entropy, wee see that the TRBP objective is an upper bound on $\log Z$. Using the fact that $I_{st} = H_s + H_t - H_{st}$, we can rewrite the upper bound as follows:

$$\log \hat{Z}(\boldsymbol{\theta}) \triangleq \boldsymbol{\tau}^T \boldsymbol{\theta} + \sum_{st} \rho_{st} H_{st}(\tau_{st}) + \sum_s c_s H_s(\tau_s) \leq \log Z(\boldsymbol{\theta}) \qquad (22.64)$$

where $c_s \triangleq 1 - \sum_t \rho_{st}$.

22.5   Expectation propagation

Expectation propagation (EP) (Minka 2001c) is a form of belief propagation where the messages are approximated. It is a generalization of the assumed density filtering (ADF) algorithm, discussed in Section 18.5.3. In that method, we approximated the posterior at each step using an assumed functional form, such as a Gaussian. This posterior can be computed using moment matching, which locally optimizes $\mathbb{KL}(p||q)$ for a single term. From this, we derived the message to send to the next time step.

ADF works well for sequential Bayesian updating, but the answer it gives depends on the order in which the data is seen. EP essentially corrects this flaw by making multiple passes over the data (thus EP is an offline or batch inference algorithm).

22.5.1 EP as a variational inference problem

We now explain how to view EP in terms of variational inference. We follow the presentation of (Wainwright and Jordan 2008, Sec 4.3), which should be consulted for further details.

Suppose the joint distribution can be written in exponential family form as follows:

$$p(\mathbf{x}|\boldsymbol{\theta}, \tilde{\boldsymbol{\theta}}) \propto f_0(\mathbf{x}) \exp(\boldsymbol{\theta}^T \phi(\mathbf{x})) \prod_{i=1}^{d_I} \exp(\tilde{\boldsymbol{\theta}}_i^T \boldsymbol{\Phi}_i(\mathbf{x})) \tag{22.65}$$

where we have partitioned the parameters and the sufficient statistics into a tractable term $\boldsymbol{\theta}$ of size d_T and d_I intractable terms $\tilde{\boldsymbol{\theta}}_i$, each of size b.

For example, consider the problem of inferring an unknown vector $\mathbf{x}$, when the observation model is a mixture of two Gaussians, one centered at $\mathbf{x}$ and one centered at $\mathbf{0}$. (This can be used to represent outliers, for example.) Minka (who invented EP) calls this the **clutter problem**. More formally, we assume an observation model of the form

$$p(\mathbf{y}|\mathbf{x}) = (1 - w)\mathcal{N}(\mathbf{y}|\mathbf{x}, \mathbf{I}) + w\mathcal{N}(\mathbf{y}|\mathbf{0}, a\mathbf{I}) \tag{22.66}$$

where $0 < w < 1$ is the known mixing weight (fraction of outliers), and $a > 0$ is the variance of the background distribution. Assuming a fixed prior of the form $p(\mathbf{x}) = \mathcal{N}(\mathbf{x}|\mathbf{0}, \boldsymbol{\Sigma})$, we can write our model in the required form as follows:

$$p(\mathbf{x}|\mathbf{y}_{1:N}) \quad \propto \quad \mathcal{N}(\mathbf{x}|\mathbf{0}, \boldsymbol{\Sigma}) \prod_{i=1}^{N} p(\mathbf{y}_i|\mathbf{x}) \tag{22.67}$$

$$= \quad \exp\left(-\frac{1}{2}\mathbf{x}^T \boldsymbol{\Sigma}^{-1} \mathbf{x}\right) \exp\left(\sum_{i=1}^{N} \log p(\mathbf{y}_i|\mathbf{x})\right) \tag{22.68}$$

This matches our canonical form where $f_0(\mathbf{x}) \exp(\boldsymbol{\theta}^T \phi(\mathbf{x}))$ corresponds to $\exp\left(-\frac{1}{2}\mathbf{x}^T \boldsymbol{\Sigma}^{-1}\mathbf{x}\right)$, using $\phi(\mathbf{x}) = (\mathbf{x}, \mathbf{x}\mathbf{x}^T)$, and we set $\boldsymbol{\Phi}_i(\mathbf{x}) = \log p(\mathbf{y}_i|\mathbf{x})$, $\tilde{\boldsymbol{\theta}}_i = 1$, and $d_I = N$.

The exact inference problem corresponds to

$$\max_{(\boldsymbol{\tau}, \tilde{\boldsymbol{\tau}}) \in \mathcal{M}(\phi, \boldsymbol{\Phi})} \boldsymbol{\tau}^T \boldsymbol{\theta} + \tilde{\boldsymbol{\tau}}^T \tilde{\boldsymbol{\theta}} + \mathbb{H}\left((\boldsymbol{\tau}, \tilde{\boldsymbol{\tau}})\right) \tag{22.69}$$

where $\mathcal{M}(\phi, \boldsymbol{\Phi})$ is the set of mean parameters realizable by any probability distribution as seen through the eyes of the sufficient statistics:

$$\mathcal{M}(\phi, \boldsymbol{\Phi}) = \{(\boldsymbol{\mu}, \tilde{\boldsymbol{\mu}}) \in \mathbb{R}^{d_T} \times \mathbb{R}^{d_I b} : (\boldsymbol{\mu}, \tilde{\boldsymbol{\mu}}) = \mathbb{E}\left[(\phi(\mathbf{X}), \boldsymbol{\Phi}_1(\mathbf{X}), \ldots, \boldsymbol{\Phi}_{d_I}(\mathbf{X}))\right]\} \tag{22.70}$$

As it stands, it is intractable to perform inference in this distribution. For example, in our clutter example, the posterior contains 2^N modes. But suppose we incorporate just one of the intractable terms, say the i'th one; we will call this the $\boldsymbol{\Phi}_i$-augmented distribution:

$$p(\mathbf{x}|\boldsymbol{\theta}, \tilde{\boldsymbol{\theta}}_i) \quad \propto \quad f_0(\mathbf{x}) \exp(\boldsymbol{\theta}^T \phi(\mathbf{x})) \exp(\tilde{\boldsymbol{\theta}}_i^T \boldsymbol{\Phi}_i(\mathbf{x})) \tag{22.71}$$

In our clutter example, this becomes

$$p(\mathbf{x}|\boldsymbol{\theta}, \tilde{\boldsymbol{\theta}}_i) = \exp\left(-\frac{1}{2}\mathbf{x}^T \boldsymbol{\Sigma}^{-1}\mathbf{x}\right)[w\mathcal{N}(\mathbf{y}_i|\mathbf{0}, a\mathbf{I}) + (1 - w)\mathcal{N}(\mathbf{y}_i|\mathbf{x}, \mathbf{I})] \tag{22.72}$$

This *is* tractable to compute, since it is just a mixture of 2 Gaussians.

The key idea behind EP is to work with these the $\boldsymbol{\Phi}_i$-augmented distributions in an iterative fashion. First, we approximate the convex set $\mathcal{M}(\phi, \boldsymbol{\Phi})$ with another, larger convex set:

$$\mathcal{L}(\phi, \boldsymbol{\Phi}) \triangleq \{(\boldsymbol{\tau}, \tilde{\boldsymbol{\tau}}) : \boldsymbol{\tau} \in \mathcal{M}(\phi), (\boldsymbol{\tau}, \tilde{\boldsymbol{\tau}}_i) \in \mathcal{M}(\phi, \boldsymbol{\Phi}_i)\} \tag{22.73}$$

where $\mathcal{M}(\phi) = \{\boldsymbol{\mu} \in \mathbb{R}^{d_T} : \boldsymbol{\mu} = \mathbb{E}[\phi(\mathbf{X})]\}$ and $\mathcal{M}(\phi, \boldsymbol{\Phi}_i) = \{(\boldsymbol{\mu}, \tilde{\boldsymbol{\mu}}_i) \in \mathbb{R}^{d_T} \times \mathbb{R}^b :$ $(\boldsymbol{\mu}, \tilde{\boldsymbol{\mu}}_i) = \mathbb{E}[(\phi(\mathbf{X}), \boldsymbol{\Phi}_i(\mathbf{X}))]$. Next we approximate the entropy by the following term-by-term approximation:

$$\mathbb{H}_{\text{ep}}(\boldsymbol{\tau}, \tilde{\boldsymbol{\tau}}) \triangleq \mathbb{H}(\boldsymbol{\tau}) + \sum_{i=1}^{d_I}[\mathbb{H}(\boldsymbol{\tau}, \tilde{\boldsymbol{\tau}}_i) - \mathbb{H}(\boldsymbol{\tau})] \tag{22.74}$$

Then the EP problem becomes

$$\max_{(\boldsymbol{\tau}, \tilde{\boldsymbol{\tau}}) \in \mathcal{L}(\phi, \boldsymbol{\Phi})} \boldsymbol{\tau}^T \boldsymbol{\theta} + \tilde{\boldsymbol{\tau}}^T \tilde{\boldsymbol{\theta}} + \mathbb{H}_{\text{ep}}(\boldsymbol{\tau}, \tilde{\boldsymbol{\tau}}) \tag{22.75}$$

22.5.2 Optimizing the EP objective using moment matching

We now discuss how to maximize the EP objective in Equation 22.75. Let us duplicate $\boldsymbol{\tau}$ d_I times to yield $\boldsymbol{\eta}_i = \boldsymbol{\tau}$. The augmented set of parameters we need to optimize is now

$$(\boldsymbol{\tau}, (\boldsymbol{\eta}_i, \tilde{\boldsymbol{\tau}}_i)_{i=1}^{d_I}) \in \mathbb{R}^{d_T} \times (\mathbb{R}^{d_T} \times \mathbb{R}^b)^{d_I} \tag{22.76}$$

subject to the constraints that $\boldsymbol{\eta}_i = \boldsymbol{\tau}$ and $(\boldsymbol{\eta}_i, \tilde{\boldsymbol{\tau}}_i) \in \mathcal{M}(\phi; \boldsymbol{\Phi}_i)$. Let us associate a vector of Lagrange multipliers $\boldsymbol{\lambda}_i \in \mathbb{R}^{d_T}$ with the first set of constraints. Then the partial Lagrangian becomes

$$L(\boldsymbol{\tau}; \boldsymbol{\lambda}) = \boldsymbol{\tau}^T \boldsymbol{\theta} + \mathbb{H}(\boldsymbol{\tau}) + \sum_{i=1}^{d_i}\left[\tilde{\boldsymbol{\tau}}_i^T \tilde{\boldsymbol{\theta}}_i + \mathbb{H}((\boldsymbol{\eta}_i, \tilde{\boldsymbol{\tau}}_i)) - \mathbb{H}(\boldsymbol{\eta}_i) + \boldsymbol{\lambda}_i^T(\boldsymbol{\tau} - \boldsymbol{\eta}_i)\right] \tag{22.77}$$

By solving $\nabla_{\boldsymbol{\tau}}L(\boldsymbol{\tau}; \boldsymbol{\lambda}) = \mathbf{0}$, we can show that the corresponding distribution in $\mathcal{M}(\phi)$ has the form

$$q(\mathbf{x}|\boldsymbol{\theta}, \boldsymbol{\lambda}) \propto f_0(\mathbf{x})\exp\{(\boldsymbol{\theta} + \sum_{i=1}^{d_I}\boldsymbol{\lambda}_i)^T\phi(\mathbf{x})\} \tag{22.78}$$

The $\boldsymbol{\lambda}_i^T\phi(\mathbf{x})$ terms represents an approximation to the i'th intractable term using the sufficient statistics from the base distribution, as we will see below. Similarly, by solving $\nabla_{(\boldsymbol{\eta}_i, \tilde{\boldsymbol{\tau}}_i)}L(\boldsymbol{\tau}; \boldsymbol{\lambda}) = \mathbf{0}$, we find that the corresponding distribution in $\mathcal{M}(\phi, \boldsymbol{\Phi}_i)$ has the form

$$q_i(\mathbf{x}|\boldsymbol{\theta}, \tilde{\boldsymbol{\theta}}_i, \boldsymbol{\lambda}) \propto f_0(\mathbf{x})\exp\{(\boldsymbol{\theta} + \sum_{j \neq i}\boldsymbol{\lambda}_j)^T\phi(\mathbf{x}) + \tilde{\boldsymbol{\theta}}_i^T\boldsymbol{\Phi}_i(\mathbf{x})\} \tag{22.79}$$

This corresponds to removing the approximation to the i'th term, $\boldsymbol{\lambda}_i$, from the base distribution, and adding in the correct i'th term, $\boldsymbol{\Phi}_i$. Finally, $\nabla_\lambda L(\boldsymbol{\tau}; \boldsymbol{\lambda}) = \mathbf{0}$ just enforces the constraints that $\boldsymbol{\tau} = \mathbb{E}_q[\phi(\mathbf{X})]$ and $\boldsymbol{\eta}_i = \mathbb{E}_{q_i}[\phi(\mathbf{X})]$ are equal. In other words, we get the following moment matching constraints:

$$\int q(\mathbf{x}|\boldsymbol{\theta}, \boldsymbol{\lambda})\phi(\mathbf{x})d\mathbf{x} = \int q_i(\mathbf{x}|\boldsymbol{\theta}, \tilde{\boldsymbol{\theta}}_i, \boldsymbol{\lambda})\phi(\mathbf{x})d\mathbf{x} \tag{22.80}$$

Thus the overall algorithm is as follows. First we initialize the $\boldsymbol{\lambda}_i$. Then we iterate the following to convergence: pick a term i; compute q_i (corresponding to removing the old approximation to $\boldsymbol{\Phi}_i$ and adding in the new one); then update the $\boldsymbol{\lambda}_i$ term in q by solving the moment matching equation $\mathbb{E}_{q_i}[\phi(\mathbf{X})] = \mathbb{E}_q[\phi(\mathbf{X})]$. (Note that this particular optimization scheme is not guaranteed to converge to a fixed point.)

An equivalent way of stating the algorithm is as follows. Let us assume the true distribution is given by

$$p(\mathbf{x}|\mathcal{D}) = \frac{1}{Z}\prod_i f_i(\mathbf{x}) \tag{22.81}$$

We approximate each f_i by $\tilde{f}_i$ and set

$$q(\mathbf{x}) = \frac{1}{Z}\prod_i \tilde{f}_i(\mathbf{x}) \tag{22.82}$$

Now we repeat the following until convergence:

1. Choose a factor $\tilde{f}_i$ to refine.

2. Remove $\tilde{f}_i$ from the posterior by dividing it out:

$$q_{-i}(\mathbf{x}) = \frac{q(\mathbf{x})}{\tilde{f}_i(\mathbf{x})} \tag{22.83}$$

This can be implemented by substracting off the natural parameters of $\tilde{f}_i$ from q.

3. Compute the new posterior $q^{new}(\mathbf{x})$ by solving

$$\min_{q^{new}(\mathbf{x})} \mathbb{KL}\left(\frac{1}{Z_i}f_i(\mathbf{x})q_{-i}(\mathbf{x})||q^{new}(\mathbf{x})\right) \tag{22.84}$$

This can be done by equating the moments of $q^{new}(\mathbf{x})$ with those of $q_i(\mathbf{x}) \propto q_{-i}(\mathbf{x})f_i(\mathbf{x})$. The corresponding normalization constant has the form

$$Z_i = \int q_{-i}(\mathbf{x})f_i(\mathbf{x})d\mathbf{x} \tag{22.85}$$

4. Compute the new factor (message) that was implicitly used (so it can be later removed):

$$\tilde{f}_i(\mathbf{x}) = Z_i \frac{q^{new}(\mathbf{x})}{q_{-i}(\mathbf{x})} \tag{22.86}$$

After convergence, we can approximate the marginal likelihood using

$$p(\mathcal{D}) \approx \int \prod_i \tilde{f}_i(\mathbf{x}) d\mathbf{x} \tag{22.87}$$

We will give some examples of this below which will make things clearer.

22.5.3 EP for the clutter problem

Let us return to considering the clutter problem. Our presentation is based on (Bishop 2006).[4] For simplicity, we will assume that the prior is a spherical Gaussian, $p(\mathbf{x}) = \mathcal{N}(\mathbf{0}, b\mathbf{I})$. Also, we choose to approximate the posterior by a spherical Gaussian, $q(\mathbf{x}) = \mathcal{N}(\mathbf{m}, v\mathbf{I})$. We set $f_0(\mathbf{x})$ to be the prior; this can be held fixed. The factor approximations will be "Gaussian like" terms of the form

$$\tilde{f}_i(\mathbf{x}) = s_i \mathcal{N}(\mathbf{x}|\mathbf{m}_i, v_i\mathbf{I}) \tag{22.88}$$

Note, however, that in the EP updates, the variances may be negative! Thus these terms should be interpreted as functions, but not necessarily probability distributions. (If the variance is negative, it means the that $\tilde{f}_i$ curves upwards instead of downwards.)

First we remove $\tilde{f}_i(\mathbf{x})$ from $q(\mathbf{x})$ by division, which yields $q_{-i}(\mathbf{x}) = \mathcal{N}(\mathbf{m}_{-i}, v_{-i}\mathbf{I})$, where

$$v_{-i}^{-1} = v^{-1} - v_i^{-1} \tag{22.89}$$

$$\mathbf{m}_{-i} = \mathbf{m} + v_{-i}v_i^{-1}(\mathbf{m} - \mathbf{m}_i) \tag{22.90}$$

The normalization constant is given by

$$Z_i = (1-w)\mathcal{N}(\mathbf{y}_i|\mathbf{m}_{-i}, (v_{-i}+1)\mathbf{I}) + w\mathcal{N}(\mathbf{y}_i|\mathbf{0}, a\mathbf{I}) \tag{22.91}$$

Next we compute $q^{new}(\mathbf{x})$ by computing the mean and variance of $q_{-i}(\mathbf{x})f_i(\mathbf{x})$ as follows:

$$\mathbf{m} = \mathbf{m}_{-i} + \rho_i \frac{v_{-i}}{v_{-i}+1}(\mathbf{y}_i - \mathbf{m}_{-i}) \tag{22.92}$$

$$v = v_{-i} - \rho_i \frac{v_{-i}^2}{v_{-i}+1} + \rho_i(1-\rho_i)\frac{v_{-i}^2\|\mathbf{y}_i - \mathbf{m}_i\|^2}{D(v_{-i}+1)^2} \tag{22.93}$$

$$\rho_i = 1 - \frac{w}{Z_i}\mathcal{N}(\mathbf{y}_i|\mathbf{0}, a\mathbf{I}) \tag{22.94}$$

where D is the dimensionality of $\mathbf{x}$ and ρ_i can be interpreted as the probability that $\mathbf{y}_i$ is not clutter.

Finally, we compute the new factor $\tilde{f}_i$ whose parameters are given by

$$v_i^{-1} = v^{-1} - v_{-i}^{-1} \tag{22.95}$$

$$\mathbf{m}_i = \mathbf{m}_{-i} + (v_i + v_{-i})v_{-i}^{-1}(\mathbf{m} - \mathbf{m}_{-i}) \tag{22.96}$$

$$s_i = \frac{Z_i}{(2\pi v_i)^{D/2}\mathcal{N}(\mathbf{m}_i|\mathbf{m}_{-i}, (v_i + v_{-i})\mathbf{I})} \tag{22.97}$$

---

4. For a handy "crib sheet", containing many of the standard equations needed for deriving Gaussian EP algorithms, see
http://research.microsoft.com/en-us/um/people/minka/papers/ep/minka-ep-quickref.pdf.

At convergence, we can approximate the marginal likelihood as follows:

$$p(\mathcal{D}) \approx (2\pi v)^{D/2} \exp(c/2) \prod_{i=1}^{N} s_i (2\pi v_i)^{-D/2} \tag{22.98}$$

$$c \triangleq \frac{\mathbf{m}^T \mathbf{m}}{v} - \sum_{i=1}^{N} \frac{\mathbf{m}_i^T \mathbf{m}_i}{v_i} \tag{22.99}$$

In (Minka 2001d), it is shown that, at least on this example, EP gives better accuracy per unit of CPU time than VB and MCMC.

22.5.4 LBP is a special case of EP

We now show that loopy belief propagation is a special case of EP, where the base distribution contains the node marginals and the "intractable" terms correspond to the edge potentials. We assume the model has the pairwise form shown in Equation 22.12. If there are m nodes, the base distribution takes the form

$$p(\mathbf{x}|\boldsymbol{\theta}_1, \ldots, \boldsymbol{\theta}_m, \mathbf{0}) \propto \prod_{s \in V} \exp(\theta_s(x_s)) \tag{22.100}$$

The entropy of this distribution is simply

$$\mathbb{H}(\boldsymbol{\tau}_{1:m}) = \sum_s \mathbb{H}(\boldsymbol{\tau}_s) \tag{22.101}$$

If we add in the $u - v$ edge, the $\boldsymbol{\Phi}_{uv}$ augmented distribution has the form

$$p(\mathbf{x}|\boldsymbol{\theta}_{1:m}, \boldsymbol{\theta}_{uv}) \propto \left[\prod_{s \in V} \exp(\theta_s(x_s)) \right] \exp(\theta_{uv}(x_u, x_v)) \tag{22.102}$$

Since this graph is a tree, the exact entropy of this distribution is given by

$$\mathbb{H}(\boldsymbol{\tau}_{1:m}, \tilde{\boldsymbol{\tau}}_{uv}) = \sum_s \mathbb{H}(\boldsymbol{\tau}_s) - I(\tilde{\boldsymbol{\tau}}_{uv}) \tag{22.103}$$

where $I(\boldsymbol{\tau}_{uv}) = \mathbb{H}(\boldsymbol{\tau}_u) + \mathbb{H}(\boldsymbol{\tau}_v) - \mathbb{H}(\boldsymbol{\tau}_{uv})$ is the mutual information. Thus the EP approximation to the entropy of the full distribution is given by

$$\mathbb{H}_{\text{ep}}(\boldsymbol{\tau}, \tilde{\boldsymbol{\tau}}) = \mathbb{H}(\boldsymbol{\tau}) + \sum_{(u,v) \in E} [\mathbb{H}(\boldsymbol{\tau}_{1:m}, \tilde{\boldsymbol{\tau}}_{uv}) - \mathbb{H}(\boldsymbol{\tau})] \tag{22.104}$$

$$= \sum_s \mathbb{H}(\boldsymbol{\tau}_s) + \sum_{(u,v) \in E} \left[\sum_s \mathbb{H}(\boldsymbol{\tau}_s) - I(\tilde{\boldsymbol{\tau}}_{uv}) - \sum_s \mathbb{H}(\boldsymbol{\tau}_s) \right] \tag{22.105}$$

$$= \sum_s \mathbb{H}(\boldsymbol{\tau}_s) - \sum_{(u,v) \in E} I(\tilde{\boldsymbol{\tau}}_{uv}) \tag{22.106}$$

which is precisely the Bethe approximation to the entropy.

We now show that the convex set that EP is optimizing over, $\mathcal{L}(\phi, \mathbf{\Phi})$ given by Equation 22.73, is the same as the one that LBP is optimizing over, $\mathbb{L}(G)$ given in Equation 22.33. First, let us consider the set $\mathcal{M}(\phi)$. This consists of all marginal distributions $(\boldsymbol{\tau}_s, s \in V)$, realizable by a factored distribution. This is therefore equivalent to the set of all distributions which satisfy non-negativity $\tau_s(x_s) \geq 0$ and the local normalization constraint $\sum_{x_s} \tau(x_s) = 1$. Now consider the set $\mathcal{M}(\phi, \mathbf{\Phi}_{uv})$ for a single $u-v$ edge. This is equivalent to the marginal polytope $\mathbb{M}(G_{uv})$, where G_{uv} is the graph with the single $u - v$ edge added. Since this graph corresponds to a tree, this set also satisfies the marginalization conditions

$$\sum_{x_v} \tau_{uv}(x_u, x_v) = \tau_u(x_u), \ \sum_{x_u} \tau_{uv}(x_u, x_v) = \tau_v(x_v) \tag{22.107}$$

Since $\mathcal{L}(\phi, \mathbf{\Phi})$ is the union of such sets, as we sweep over all edges in the graph, we recover the same set as $\mathbb{L}(G)$.

We have shown that the Bethe approximation is equivalent to the EP approximation. We now show how the EP algorithm reduces to LBP. Associated with each intractable term $i = (u, v)$ will be a pair of Lagrange multipliers, $(\lambda_{uv}(x_v), \lambda_{vu}(x_u))$. Recalling that $\boldsymbol{\theta}^T \phi(\mathbf{x}) = [\theta_s(x_s)]_s$, the base distribution in Equation 22.78 has the form

$$q(\mathbf{x}|\boldsymbol{\theta}, \boldsymbol{\lambda}) \quad \propto \quad \prod_s \exp(\theta_s(x_s)) \prod_{(u,v) \in E} \exp(\lambda_{uv}(x_v) + \lambda_{vu}(x_u)) \tag{22.108}$$

$$= \quad \prod_s \exp\left(\theta_s(x_s) + \sum_{t \in \mathcal{N}(s)} \lambda_{ts}(x_s)\right) \tag{22.109}$$

Similarly, the augmented distribution in Equation 22.79 has the form

$$q_{uv}(\mathbf{x}|\boldsymbol{\theta}, \boldsymbol{\lambda}) \quad \propto \quad q(\mathbf{x}|\boldsymbol{\theta}, \boldsymbol{\lambda}) \exp\left(\theta_{uv}(x_u, x_v) - \lambda_{uv}(x_v) - \lambda_{vu}(x_u)\right) \tag{22.110}$$

We now need to update $\tau_u(x_u)$ and $\tau_v(x_v)$ to enforce the moment matching constraints:

$$(\mathbb{E}_q[x_s], \mathbb{E}_q[x_t]) = (\mathbb{E}_{q_{uv}}[x_s], \mathbb{E}_{q_{uv}}[x_t]) \tag{22.111}$$

It can be shown that this can be done by performing the usual sum-product message passing step along the $u - v$ edge (in both directions), where the messages are given by $M_{uv}(x_v) = \exp(\lambda_{uv}(x_v))$, and $M_{vu}(x_u) = \exp(\lambda_{vu}(x_u))$. Once we have updated q, we can derive the corresponding messages λ_{uv} and λ_{vu}.

The above analysis suggests a natural extension, where we make the base distribution be a tree structure instead of a fully factored distribution. We then add in one edge at a time, absorb its effect, and approximate the resulting distribution by a new tree. This is known as **tree EP** (Minka and Qi 2003), and is more accurate than LBP, and sometimes faster. By considering other kinds of structured base distributions, we can derive algorothms that outperform generalization belief propagation (Welling et al. 2005).

22.5.5 Ranking players using TrueSkill

We now present an interesting application of EP to the problem of ranking players who compete in games. Microsoft uses this method — known as **TrueSkill** (Herbrich et al. 2007) — to rank

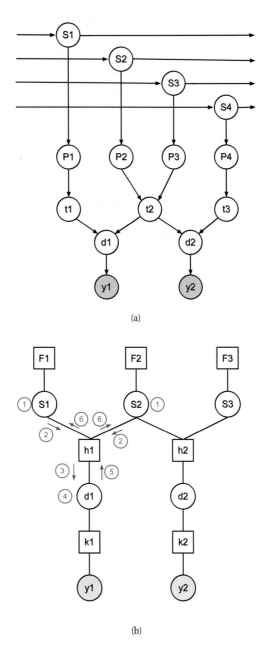

(a)

(b)

Figure 22.11   (a) A DGM representing the TrueSkill model for 4 players and 3 teams, where team 1 is player 1, team 2 is players 2 and 3, and team 3 is player 4. We assume there are two games, team 1 vs team 2, and team 2 vs team 3. Nodes with double circles are deterministic. (b) A factor graph representation of the model where we assume there are 3 players (and no teams). There are 2 games, player 1 vs player 2, and player 2 vs player 3. The numbers inside circles represent steps in the message passing algorithm.

players who use the Xbox 360 Live online gaming system; this system process over 10^5 games per day, making this one of the largest application of Bayesian statistics to date.[5] The same method can also be applied to other games, such as tennis or chess.[6]

The basic idea is shown in Figure 22.11(a). We assume each player i has a latent or true underlying skill level $s_i \in \mathbb{R}$. These skill levels can evolve over time according to a simple dynamical model, $p(s_i^t | s_i^{t-1}) = \mathcal{N}(s_i^t | s_i^{t-1}, \gamma^2)$. In any given game, we define the performance of player i to be p_i, which has the conditional distribution $p(p_i | s_i) = \mathcal{N}(p_i | s_i, \beta^2)$. We then define the performance of a team to be the sum of the performance of its constituent players. For example, in Figure 22.11(a), we assume team 2 is composed of players 2 and 3, so we define $t_2 = p_2 + p_3$. Finally, we assume that the outcome of a game depends on the difference in performance levels of the two teams. For example, in Figure 22.11(a), we assume $y_1 = \text{sign}(d_1)$, where $d_1 = t_1 - t_2$, and where $y_1 = +1$ means team 1 won, and $y_1 = -1$ means team 2 won. Thus the prior probability that team 1 wins is

$$p(y_1 = +1 | \mathbf{s}) = \int p(d_1 > 0 | t_1, t_2) p(t_1 | s_1) p(t_2 | s_2) dt_1 dt_2 \qquad (22.112)$$

where $t_1 \sim \mathcal{N}(s_1, \beta^2)$ and $t_2 \sim \mathcal{N}(s_2 + s_3, \beta^2)$.[7]

To simplify the presentation of the algorithm, we will ignore the dynamical model and assume a common static factored Gaussian prior, $\mathcal{N}(\mu_0, \sigma_0^2)$, on the skills. Also, we will assume that each team consists of 1 player, so $t_i = p_i$, and that there can be no ties. Finally, we will integrate out the performance variables p_i, and assume $\beta^2 = 1$, leading to a final model of the form

$$p(\mathbf{s}) = \prod_i \mathcal{N}(s_i | \mu_0, \sigma^2) \qquad (22.113)$$

$$p(d_g | \mathbf{s}) = \mathcal{N}(d_g | s_{i_g} - s_{j_g}, 1) \qquad (22.114)$$

$$p(y_g | d_g) = \mathbb{I}(y_g = \text{sign}(d_g)) \qquad (22.115)$$

where i_g is the first player of game g, and j_g is the second player. This is represented in factor graph form in in Figure 22.11(b). We have 3 kinds of factors: the prior factor, $f_i(s_i) = \mathcal{N}(s_i | \mu_0, \sigma_0^2)$, the game factor, $h_g(s_{i_g}, s_{j_g}, d_g) = \mathcal{N}(d_g | s_{i_g} - s_{j_g}, 1)$, and the outcome factor, $k_g(d_g, y_g) = \mathbb{I}(y_g = \text{sign}(d_g))$.

Since the likelihood term $(y_g | d_g)$ is not conjugate to the Gaussian priors, we will have to perform approximate inference. Thus even when the graph is a tree, we will need to iterate. (If there were an additional game, say between player 1 and player 3, then the graph would no longer be a tree.) We will represent all messages and marginal beliefs by 1d Gaussians. We will use the notation μ and v for the mean and variance (the moment parameters), and $\lambda = 1/v$ and $\eta = \lambda\mu$ for the precision and precision-adjusted mean (the natural parameters).

---

5. Naive Bayes classifiers, which are widely used in spam filters, are often described as the most common application of Bayesian methods. However, the parameters of such models are usually fit using non-Bayesian methods, such as penalized maximum likelihood.

6. Our presentation of this algorithm is based in part on lecture notes by Carl Rasmussen and Joaquin Quinonero-Candela, available at `http://mlg.eng.cam.ac.uk/teaching/4f13/1112/lect13.pdf`.

7. Note that this is very similar to probit regression, discussed in Section 9.4, except the inputs are (the differences of) latent 1 dimensional factors. If we assume a logistic noise model instead of a Gaussian noise model, we recover the **Bradley Terry model of ranking**.

We initialize by assuming that at iteration 0, the initial upward messages from factors h_g to variables s_i are uniform, i.e.,

$$m^0_{h_g \to s_{i_g}}(s_{i_g}) = 1, \quad v^0_{h_g \to s_{i_g}}(s_{i_g}) = \infty, \quad \lambda^0_{h_g \to s_{i_g}} = 0, \quad \eta^0_{h_g \to s_{i_g}} = 0 \tag{22.116}$$

and similarly $m^0_{h_g \to s_{j_g}}(s_{j_g}) = 1$. The messages passing algorithm consists of 6 steps per game, as illustrated in Figure 22.11(b). We give the details of these steps below.

1. Compute the posterior over the skills variables:

$$q^t(s_i) \quad = \quad f(s_i) \prod_g m^{t-1}_{h_g \to s_i}(s_i) = \mathcal{N}_c(s_i | \eta^t_i, \lambda^t_i) \tag{22.117}$$

$$\lambda^t_i \quad = \quad \lambda_0 + \sum_g \lambda^{t-1}_{h_g \to s_i}, \quad \eta^t_i = \eta_0 + \sum_g \eta^{t-1}_{h_g \to s_i} \tag{22.118}$$

2. Compute the message from the skills variables down to the game factor h_g:

$$m^t_{s_{i_g} \to h_g}(s_{i_g}) = \frac{q^t(s_{i_g})}{m^t_{h_g \to s_{i_g}}(s_{i_g})}, \quad m^t_{s_{j_g} \to h_g}(s_{j_g}) = \frac{q^t(s_{j_g})}{m^t_{h_g \to s_{j_g}}(s_{j_g})} \tag{22.119}$$

where the division is implemented by subtracting the natural parameters as follows:

$$\lambda^t_{s_{i_g} \to h_g} = \lambda^t_{s_{i_g}} - \lambda^t_{h_g \to s_{i_g}}, \quad \eta^t_{s_{i_g} \to h_g} = \eta^t_{s_{i_g}} - \eta^t_{h_g \to s_{i_g}} \tag{22.120}$$

and similarly for s_{j_g}.

3. Compute the message from the game factor h_g down to the difference variable d_g:

$$m^t_{h_g \to d_g}(d_g) \quad = \quad \int \int h_g(d_g, s_{i_g}, s_{j_g}) m^t_{s_{i_g} \to h_g}(s_{i_g}) m^t_{s_{j_g} \to h_g}(s_{j_g}) ds_{i_g} ds_{j_g} \tag{22.121}$$

$$= \quad \int \int \mathcal{N}(d_g | s_{i_g} - s_{j_g}, 1) \mathcal{N}(s_{i_g} | \mu^t_{s_{i_g} \to h_g}, v^t_{s_{i_g} \to h_g}) \tag{22.122}$$

$$\mathcal{N}(s_{j_g} | \mu^t_{s_{j_g} \to h_g}, v^t_{s_{j_g} \to h_g}) ds_{i_g} ds_{j_g} \tag{22.123}$$

$$= \quad \mathcal{N}(d_g | \mu^t_{h_g \to d_g}, v^t_{h_g \to d_g}) \tag{22.124}$$

$$v^t_{h_g \to d_g} \quad = \quad 1 + v^t_{s_{i_g} \to h_g} + v^t_{s_{j_g} \to h_g} \tag{22.125}$$

$$\mu^t_{h_g \to d_g} \quad = \quad \mu^t_{s_{i_g} \to h_g} - \mu^t_{s_{j_g} \to h_g} \tag{22.126}$$

4. Compute the posterior over the difference variables:

$$q^t(d_g) \quad \propto \quad m^t_{h_g \to d_g}(d_g) m_{k_g \to d_g}(d_g) \tag{22.127}$$

$$= \quad \mathcal{N}(d_g | \mu^t_{h_g \to d_g}, v^t_{h_g \to d_g}) \mathbb{I}(y_g = \text{sign}(d_g)) \tag{22.128}$$

$$\approx \quad \mathcal{N}(d_g | \mu^t_g, v^t_g) \tag{22.129}$$

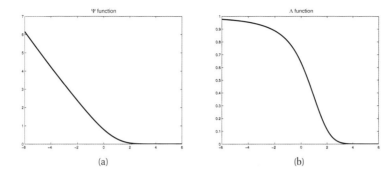

Figure 22.12 (a) Ψ function. (b) Λ function. Based on Figure 2 of (Herbrich et al. 2007). Figure generated by `trueskillPlot`.

(Note that the upward message from the k_g factor is constant.) We can find these parameters by moment matching as follows (see Exercise 22.5):

$$\mu_g^t = \mu_{h_g \to d_g}^t + y_g \sigma_{h_g \to d_g}^t \Psi \left(\frac{y_g \mu_{h_g \to d_g}^t}{\sigma_{h_g \to d_g}^t} \right) \tag{22.130}$$

$$v_g^t = v_{h_g \to d_g}^t \left[1 - \Lambda \left(\frac{y_g \mu_{h_g \to d_g}^t}{\sigma_{h_g \to d_g}^t} \right) \right] \tag{22.131}$$

$$\Psi(x) \triangleq \frac{\mathcal{N}(x|0,1)}{\Phi(x)} \tag{22.132}$$

$$\Lambda(x) \triangleq \Psi(x)(\Psi(x) + x) \tag{22.133}$$

where $\sigma_{h_g \to d_g}^t = \sqrt{v_{h_g \to d_g}^t}$. These functions are plotted in Figure 22.12. Let us try to understand these equations. Suppose $\mu_{h_g \to d_g}^t$ is a large positive number. That means we expect, based on the current estimate of the skills, that d_g will be large and positive. Consequently, if we observe $y_g = +1$, we will not be surprised that i_g is the winner, which is reflected in the fact that the update factor for the mean is small, $\Psi(y_g \mu_{h_g \to d_g}^t) \approx 0$. Similarly, the update factor for the variance is small, $\Lambda(y_g \mu_{h_g \to d_g}^t) \approx 0$. However, if we observe $y_g = -1$, then the update factor for the mean and variance becomes quite large.

5. Compute the upward message from the difference variable to the game factor h_g:

$$m_{d_g \to h_g}^t(d_g) = \frac{q^t(d_g)}{m_{h_g \to d_g}^t(d_g)} \tag{22.134}$$

$$\lambda_{d_g \to h_h}^t = \lambda_g^t - \lambda_{h_g \to d_g}^t, \quad \eta_{d_g \to h_h}^t = \eta_g^t - \eta_{h_g \to d_g}^t \tag{22.135}$$

6. Compute the upward messages from the game factor to the skill variables. Let us assume

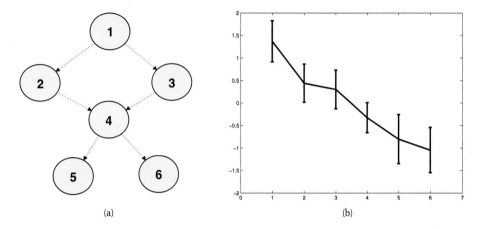

Figure 22.13 (a) A DAG representing a partial ordering of players. (b) Posterior mean plus/minus 1 standard deviation for the latent skills of each player based on 26 games. Figure generated by `trueskillDemo`.

that i_g is the winner, and j_g is the loser. Then we have

$$m^t_{h_g \to s_{i_g}}(s_{i_g}) = \int \int h_g(d_g, s_{i_g}, s_{j_g}) m^t_{d_g \to h_g}(d_g) m^t_{s_{j_g} \to h_g}(s_{j_g}) dd_g ds_{j_g} \tag{22.136}$$

$$= \mathcal{N}(s_{i_g} | \mu^t_{h_g \to s_{i_g}}, v^t_{h_g \to s_{i_g}}) \tag{22.137}$$

$$v^t_{h_g \to s_{i_g}} = 1 + v^t_{d_g \to h_g} + v^t_{s_{j_g} \to h_g} \tag{22.138}$$

$$\mu^t_{h_g \to s_{i_g}} = \mu^t_{d_g \to h_g} + \mu^t_{s_{j_g} \to h_g} \tag{22.139}$$

And similarly

$$m^t_{h_g \to s_{j_g}}(s_{j_g}) = \int \int h_g(d_g, s_{i_g}, s_{j_g}) m^t_{d_g \to h_g}(d_g) m^t_{s_{i_g} \to h_g}(s_{i_g}) dd_g ds_{i_g} \tag{22.140}$$

$$= \mathcal{N}(s_{j_g} | \mu^t_{h_g \to s_{j_g}}, v^t_{h_g \to s_{j_g}}) \tag{22.141}$$

$$v^t_{h_g \to s_{j_g}} = 1 + v^t_{d_g \to h_g} + v^t_{s_{i_g} \to h_g} \tag{22.142}$$

$$\mu^t_{h_g \to s_{j_g}} = \mu^t_{d_g \to h_g} - \mu^t_{s_{i_g} \to h_g} \tag{22.143}$$

When we compute $q^{t+1}(s_{i_g})$ at the next iteration, by combining $m^t_{h_g \to s_{i_g}}(s_{i_g})$ with the prior factor, we will see that the posterior mean of s_{i_g} goes up. Similarly, the posterior mean of s_{j_g} goes down.

It is straightforward to combine EP with ADF to perform online inference, which is necessary for most practical applications.

Let us consider a simple example of this method. We create a partial ordering of 5 players as shown in Figure 22.13(a). We then sample some game outcomes from this graph, where a parent always beats a child. We pass this data into (5 iterations of) the EP algorithm and infer the posterior mean and variance for each player's skill level. The results are shown in Figure 22.13(b). We see that the method has correctly inferred the rank ordering of the players.

22.5.6 Other applications of EP

The TrueSkill model was developed by researchers at Microsoft, although it has since been extended by others (see e.g., (Weng and Lin 2011)). The Microsoft group have extended the TrueSkill model to a variety of other interesting applications, including personalized ad recommendation (Stern et al. 2009), predicting click-through-rate on ads in the Bing search engine (Graepel et al. 2010), etc. They have also developed a general purpose Bayesian inference toolbox based on EP called **infer.net** (Minka et al. 2010).

EP has also been used for a variety of other models, such as Gaussian process classification (Nickisch and Rasmussen 2008). See `http://research.microsoft.com/en-us/um/people/minka/papers/ep/roadmap.html` for a list of other EP applications.

22.6 MAP state estimation

In this section, we consider the problem of finding the most probable configuration of variables in a discrete-state graphical model, i.e., our goal is to find a MAP assignment of the following form:

$$\mathbf{x}^* = \arg \max_{\mathbf{x} \in \mathcal{X}^m} p(\mathbf{x}|\boldsymbol{\theta}) = \arg \max_{\mathbf{x} \in \mathcal{X}^m} \sum_{i \in V} \theta_i(x_i) + \sum_{f \in F} \theta_f(\mathbf{x}_f) = \arg \max_{\mathbf{x} \in \mathcal{X}^m} \boldsymbol{\theta}^T \phi(\mathbf{x}) \quad (22.144)$$

where θ_i are the singleton node potentials, and θ_f are the factor potentials. (In this section, we follow the notation of (Sontag et al. 2011), which considers the case of general potentials, not just pairwise ones.) Note that the partition function $Z(\boldsymbol{\theta})$ plays no role in MAP estimation.

If the treewidth is low, we can solve this problem with the junction tree algorithm (Section 20.4), but in general this problem is intractable. In this section, we discuss various approximations, building on the material from Section 22.3.

22.6.1 Linear programming relaxation

We can rewrite the objective in terms of the variational parameters as follows:

$$\arg \max_{\mathbf{x} \in \mathcal{X}^m} \boldsymbol{\theta}^T \phi(\mathbf{x}) = \arg \max_{\boldsymbol{\mu} \in \mathbb{M}(G)} \boldsymbol{\theta}^T \boldsymbol{\mu} \quad (22.145)$$

where $\phi(\mathbf{x}) = [\{\mathbb{I}(x_s = j)\}, \{\mathbb{I}(\mathbf{x}_f = k)\}]$ and $\boldsymbol{\mu}$ is a probability vector in the marginal polytope. To see why this equation is true, note that we can just set $\boldsymbol{\mu}$ to be a degenerate distribution with $\mu(x_s) = \mathbb{I}(x_s = x_s^*)$, where x_s^* is the optimal assigment of node s. So instead of optimizing over discrete assignments, we now optimize over probability distributions $\boldsymbol{\mu}$.

It seems like we have an easy problem to solve, since the objective in Equation 22.145 is linear in $\boldsymbol{\mu}$, and the constraint set $\mathbb{M}(G)$ is convex. The trouble is, $\mathbb{M}(G)$ in general has a number of facets that is exponential in the number of nodes.

A standard strategy in combinatorial optimization is to relax the constraints. In this case, instead of requiring probability vector $\boldsymbol{\mu}$ to live in the marginal polytope $\mathbb{M}(G)$, we allow it to live inside a convex outer bound $\mathbb{L}(G)$. Having defined this relaxed constraint set, we have

$$\max_{\mathbf{x} \in \mathcal{X}^m} \boldsymbol{\theta}^T \boldsymbol{\phi}(\mathbf{x}) = \max_{\boldsymbol{\mu} \in \mathbb{M}(G)} \boldsymbol{\theta}^T \boldsymbol{\mu} \leq \max_{\boldsymbol{\tau} \in \mathbb{L}(G)} \boldsymbol{\theta}^T \boldsymbol{\tau} \tag{22.146}$$

If the solution is integral, it is exact; if it is fractional, it is an approximation. This is called a (first order) **linear programming relaxtion**. The reason it is called first-order is that the constraints that are enforced are those that correspond to consistency on a tree, which is a graph of treewidth 1. It is possible to enforce higher-order consistency, using graphs with larger treewidth (see (Wainwright and Jordan 2008, sec 8.5) for details).

How should we actually perform the optimization? We can use a generic linear programming package, but this is often very slow. Fortunately, in the case of graphical models, it is possible to devise specialised distributed message passing algorithms for solving this optimization problem, as we explain below.

22.6.2   Max-product belief propagation

The MAP objective in Equation 22.145, $\max_{\boldsymbol{\mu} \in \mathbb{M}(G)} \boldsymbol{\theta}^T \boldsymbol{\mu}$, is almost identical to the inference objective in Equation 22.23, $\max_{\boldsymbol{\mu} \in \mathbb{M}(G)} \boldsymbol{\theta}^T \boldsymbol{\mu} + \mathbb{H}(\boldsymbol{\mu})$, apart from the entropy term. One heuristic way to proceed would be to consider the **zero temperature limit** of the probability distribution $\boldsymbol{\mu}$, where the probability distribution has all its mass centered on its mode (see Section 4.2.2). In such a setting, the entropy term becomes zero. We can then modify the message passing methods used to solve the inference problem so that they solve the MAP estimation problem instead. In particular, in the zero temperature limit, the sum operator becomes the max operator, which results in a method called **max-product belief propagation**.

In more detail, let

$$A(\boldsymbol{\theta}) \triangleq \max_{\boldsymbol{\mu} \in \mathbb{M}(G)} \boldsymbol{\theta}^T \boldsymbol{\mu} + \mathbb{H}(\boldsymbol{\mu}) \tag{22.147}$$

Now consider an inverse temperature β going to infinity. We have

$$\lim_{\beta \to +\infty} \frac{A(\beta \boldsymbol{\theta})}{\beta} = \lim_{\beta \to +\infty} \frac{1}{\beta} \max_{\boldsymbol{\mu} \in \mathbb{M}(G)} \left\{ (\beta \boldsymbol{\theta})^T \boldsymbol{\mu} + \mathbb{H}(\boldsymbol{\mu}) \right\} \tag{22.148}$$

$$= \max_{\boldsymbol{\mu} \in \mathbb{M}(G)} \left\{ \boldsymbol{\theta}^T \boldsymbol{\mu} + \lim_{\beta \to +\infty} \frac{1}{\beta} \mathbb{H}(\boldsymbol{\mu}) \right\} \tag{22.149}$$

$$= \max_{\boldsymbol{\mu} \in \mathbb{M}(G)} \boldsymbol{\theta}^T \boldsymbol{\mu} \tag{22.150}$$

It is the concavity of the objective function that allows us to interchange the lim and max operators (see (Wainwright and Jordan 2008, p274) for details).

Now consider the Bethe approximation, which has the form $\max_{\boldsymbol{\tau} \in \mathbb{L}(G)} \boldsymbol{\theta}^T \boldsymbol{\tau} + \mathbb{H}_{\text{Bethe}}(\boldsymbol{\tau})$. We showed that loopy BP finds a local optimum of this objective. In the zero temperature limit, this objective is equivalent to the LP relaxation of the MAP problem. Unfortunately, max-product loopy BP does not solve this LP relaxation unless the graph is a tree (Wainwright and Jordan 2008, p211). The reason is that Bethe energy functional is not concave (except on trees), so we are not licensed to swap the limit and max operators in the above zero-temperature derivation. However, if we use tree-reweighted BP, or TRBP/ TRW, we have a concave objective. In this case, one can show (Kolmogorov and Wainwright 2005) that the max-product version of TRBP does solve the above LP relaxation.

A certain scheduling of this algorithm, known as **sequential TRBP**, **TRBP-S**, or **TRW-S**, can be shown to always converge (Kolmogorov 2006), and furthermore, it typically does so faster than the standard parallel updates. The idea is to pick an arbitrary node ordering $X_1, \ldots, X_N$. We then consider a set of trees which is a subsequence of this ordering. At each iteration, we perform max-product BP from X_1 towards X_N and back along one of these trees. It can be shown that this monotonically minimizes a lower bound on the energy, and thus is guaranteed to converge to the global optimum of the LP relaxation. (See (Schwing et al. 2011; Banerjee et al. 2012) for details of how to derive parallel distributed implementation of this LP relaxation.)

22.6.3 Graphcuts

In this section, we show how to find MAP state estimates, or equivalently, minimum energy configurations, by using the **max flow/min cut** algorithm for graphs.[8] This class of methods is known as **graphcuts** and is very widely used, especially in computer vision applications.

We will start by considering the case of MRFs with binary nodes and a restricted class of potentials; in this case, graphcuts will find the exact global optimum. We then consider the case of multiple states per node, which are assumed to have some underlying ordering; we can approximately solve this case by solving a series of binary subproblems, as we will see.

22.6.3.1 Graphcuts for the generalized Ising model

Let us start by considering a binary MRF where the edge energies have the following form:

$$E_{uv}(x_u, x_v) = \begin{cases} 0 & \text{if } x_u = x_v \\ \lambda_{st} & \text{if } x_u \neq x_v \end{cases} \tag{22.151}$$

where $\lambda_{st} \geq 0$ is the edge cost. This encourages neighboring nodes to have the same value (since we are trying to minimize energy). Since we are free to add any constant we like to the overall energy without affecting the MAP state estimate, let us rescale the local energy terms such that either $E_u(1) = 0$ or $E_u(0) = 0$.

Now let us construct a graph which has the same set of nodes as the MRF, plus two distinguished nodes: the source s and the sink t. If $E_u(1) = 0$, we add the edge $x_u \to t$ with cost $E_u(0)$. (This ensures that if u is not in partition $\mathcal{X}_t$, meaning u is assigned to state 0, we will pay a cost of $E_u(0)$ in the cut.) Similarly, If $E_u(0) = 0$, we add the edge $x_u \to s$ with cost $E_u(1)$. Finally, for every pair of variables that are connected in the MRF, we add edges $x_u \to x_v$ and $x_v \to x_u$, both with cost $\lambda_{u,v} \geq 0$. Figure 22.14 illustrates this construction for an MRF with 4 nodes, and with the following non-zero energy values:

$$E_1(0) = 7, E_2(1) = 2, E_3(1) = 1, E_4(1) = 6 \tag{22.152}$$
$$\lambda_{1,2} = 6, \lambda_{2,3} = 6, \lambda_{3,4} = 2, \lambda_{1,4} = 1 \tag{22.153}$$

---

8. There are a variety of ways to implement this algorithm, see e.g., (Sedgewick and Wayne 2011). The best take $O(EV \log V)$ or $O(V^3)$ time, where E is the number of edges and V is the number of nodes.

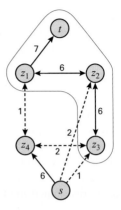

Figure 22.14 Illustration of graphcuts applied to an MRF with 4 nodes. Dashed lines are ones which contribute to the cost of the cut (for bidirected edges, we only count one of the costs). Here the min cut has cost 6.   Source: Figure 13.5 from (Koller and Friedman 2009).   Used with kind permission of Daphne Koller.

Having constructed the graph, we compute a minimal $s - t$ cut. This is a partition of the nodes into two sets, $\mathcal{X}_s$, which are nodes connected to s, and $\mathcal{X}_t$, which are nodes connected to t. We pick the partition which minimizes the sum of the cost of the edges between nodes on different sides of the partition:

$$\mathrm{cost}(\mathcal{X}_s, \mathcal{X}_t) = \sum_{x_u \in \mathcal{X}_s, x_v \in \mathcal{X}_t} \mathrm{cost}(x_u, s_v) \tag{22.154}$$

In Figure 22.14, we see that the min-cut has cost 6.

Minimizing the cost in this graph is equivalent to minimizing the energy in the MRF. Hence nodes that are assigned to s have an optimal state of 0, and the nodes that are assigned to t have an optimal state of 1. In Figure 22.14, we see that the optimal MAP estimate is $(1, 1, 1, 0)$.

22.6.3.2   Graphcuts for binary MRFs with submodular potentials

We now discuss how to extend the graphcuts construction to binary MRFs with more general kinds of potential functions. In particular, suppose each pairwise energy satisfies the following condition:

$$E_{uv}(1, 1) + E_{uv}(0, 0) \le E_{uv}(1, 0) + E_{uv}(0, 1) \tag{22.155}$$

In other words, the sum of the diagonal energies is less than the sum of the off-diagonal energies. In this case, we say the energies are **submodular** (Kolmogorov and Zabin 2004).[9] An example

---

9. Submodularity is the discrete analog of convexity. Intuitively, it corresponds to the "law of diminishing returns", that is, the extra value of adding one more element to a set is reduced if the set is already large. More formally, we say that $f : 2^S \to R$ is submodular if for any $A \subset B \subset S$ and $x \in S$, we have $f(A \cup \{x\}) - f(A) \ge f(B \cup \{x\}) - f(B)$. If $-f$ is submodular, then f is **supermodular**.

of a submodular energy is an Ising model where $\lambda_{uv} > 0$. This is also known as an **attractive MRF** or **associative MRF**, since the model "wants" neighboring states to be the same.

To apply graphcuts to a binary MRF with submodular potentials, we construct the pairwise edge weights as follows:

$$E'_{u,v}(0,1) = E_{u,v}(1,0) + E_{u,v}(0,1) - E_{u,v}(0,0) - E_{u,v}(1,1) \tag{22.156}$$

This is guaranteed to be non-negative by virtue of the submodularity assumption. In addition, we construct new local edge weights as follows: first we initialize $E'(u) = E(u)$, and then for each edge pair (u, v), we update these values as follows:

$$E'_u(1) = E'_u(1) + (E_{u,v}(1,0) - E_{u,v}(0,0)) \tag{22.157}$$
$$E'_v(1) = E'_v(1) + (E_{u,v}(1,1) - E_{u,v}(1,0)) \tag{22.158}$$

We now construct a graph in a similar way to before. Specifically, if $E'_u(1) > E'_u(0)$, we add the edge $u \to s$ with cost $E'_u(1) - E'_u(0)$, otherwise we add the edge $u \to t$ with cost $E'_u(0) - E'_u(1)$. Finally for every MRF edge for which $E'_{u,v}(0,1) > 0$, we add a graphcuts edge $x_u - x_v$ with cost $E'_{u,v}(0,1)$. (We don't need to add the edge in both directions.)

One can show (Exercise 22.1) that the min cut in this graph is the same as the minimum energy configuration. Thus we can use max flow/min cut to find the globally optimal MAP estimate (Greig et al. 1989).

22.6.3.3 Graphcuts for nonbinary metric MRFs

We now discuss how to use graphcuts for approximate MAP estimation in MRFs where each node can have multiple states (Boykov et al. 2001). However, we require that the pairwise energies form a metric. We call such a model a **metric MRF**. For example, suppose the states have a natural ordering, as commonly arises if they are a discretization of an underlying continuous space. In this case, we can define a metric of the form $E(x_s, x_t) = \min(\delta, ||x_s - x_t||)$ or a semi-metric of the form $E(x_s, x_t) = \min(\delta, (x_s - x_t)^2)$, for some constant $\delta > 0$. This energy encourages neighbors to have similar labels, but never "punishes" them by more than δ. This δ term prevents over-smoothing, which we illustrate in Figure 19.20.

One version of graphcuts is the **alpha expansion**. At each step, it picks one of the available labels or states and calls it α; then it solves a binary subproblem where each variable can choose to remain in its current state, or to become state α (see Figure 22.15(d) for an illustration). More precisely, we define a new MRF on binary nodes, and we define the energies of this new model, relative to the current assignment $\mathbf{x}$, as follows:

$$E'_u(0) = E_u(x_u), E'_u(1) = E_u(\alpha), E'_{u,v}(0,0) = E_{u,v}(x_u, x_v) \tag{22.159}$$
$$E'_{u,v}(0,1) = E_{u,v}(x_u, \alpha), E'_{u,v}(1,0) = E_{u,v}(\alpha, x_v), E'_{u,v}(1,1) = E_{u,v}(\alpha, \alpha) \tag{22.160}$$

To optimize E' using graph cuts (and thus figure out the optimal alpha expansion move), we require that the energies be submodular. Plugging in the definition we get the following constraint:

$$E_{u,v}(x_u, x_v) + E_{u,v}(\alpha, \alpha) \leq E_{u,v}(x_u, \alpha) + E_{u,v}(\alpha, x_v) \tag{22.161}$$

For any distance function, $E_{u,v}(\alpha, \alpha) = 0$, and the remaining inequality follows from the triangle inequality. Thus we can apply the alpha expansion move to any metric MRF.

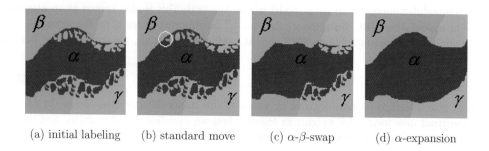

| (a) initial labeling | (b) standard move | (c) α-β-swap | (d) α-expansion |

Figure 22.15 (a) An image with 3 labels. (b) A standard local move (e.g., by iterative conditional modes) just flips the label of one pixel. (c) An $\alpha - \beta$ swap allows all nodes that are currently labeled as α to be relabeled as β if this decreases the energy. (d) An α expansion allows all nodes that are not currently labeled as α to be relabeled as α if this decreases the energy.   Source: Figure 2 of (Boykov et al. 2001). Used with kind permission of Ramin Zabih.

At each step of alpha expansion, we find the optimal move from amongst an exponentially large set; thus we reach a **strong local optimum**, of much lower energy than the local optima found by standard greedy label flipping methods such as iterative conditional modes. In fact, one can show that, once the algorithm has converged, the energy of the resulting solution is at most $2c$ times the optimal energy, where

$$c = \max_{(u,v) \in \mathcal{E}} \frac{\max_{\alpha \neq \beta} E_{uv}(\alpha, \beta)}{\min_{\alpha \neq \beta} E_{uv}(\alpha, \beta)} \tag{22.162}$$

See Exercise 22.3 for the proof. In the case of the Potts model, $c = 1$, so we have a 2-approximation.

Another version of graphcuts is the **alpha-beta swap**. At each step, two labels are chosen, call them α and β. All the nodes currently labeled α can change to β (and vice versa) if this reduces the energy (see Figure 22.15(c) for an illustration). The resulting binary subproblem can be solved exactly, even if the energies are only semi-metric (that is, the triangle inequality need not hold; see Exercise 22.2). Although the $\alpha - \beta$ swap version can be applied to a broader class of models than the α-expansion version, it is theoretically not as powerful. Indeed, in various low-level vision problems, (Szeliski et al. 2008) show empirically that the expansion version is usually better than the swap version (see Section 22.6.4).

22.6.4   Experimental comparison of graphcuts and BP

In Section 19.6.2.7, we described lattice-structured CRFs for various low-level vision problems. (Szeliski et al. 2008) performed an extensive comparison of different approximate optimization techniques for this class of problems. (See (Kappes et al. 2013) for a more recent comparison.) Some of the results, for the problem of stereo depth estimation, are shown in Figure 22.16. We see that the graphcut and tree-reweighted max-product BP (TRW) give the best results, with regular max-product BP being much worse. In terms of speed, graphcuts is the fastest, with TRW a close second. Other algorithms, such as ICM, simulated annealing or a standard domain-

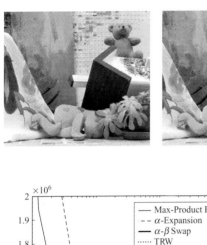

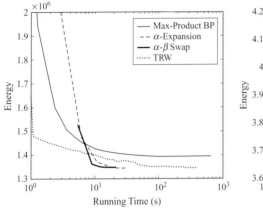
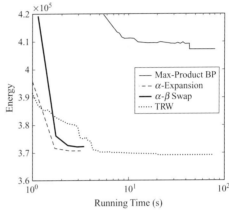

Figure 22.16 Energy minimization on a CRF for stereo depth estimation. Top row: two input images along with the ground truth depth values. Bottom row: energy vs time for 4 different optimization algorithms. Bottom left: results are for the Teddy image (shown in top row). Bottom right: results are for the Tsukuba image (shown in Figure 22.17(a)). Source: Figure 13.B.1 of (Koller and Friedman 2009). Used with kind permission of Daphne Koller.

specific heuristic known as normalize correlation, are even worse, as shown qualitatively in Figure 22.17.

Since TRW is optimizing the dual of the relaxed LP problem, we can use its value at convergence to evaluate the optimal energy. It turns out that for many of the images in the stereo benchmark dataset, the ground truth has higher energy (lower probability) than the globally optimal estimate (Meltzer et al. 2005). This indicates that we are optimizing the wrong model. This is not surprising, since the pairwise CRF ignores known long-range constraints. Unfortunately, if we add these constraints to the model, the graph either becomes too dense (making BP slow), and/or the potentials become non-submodular (making graphcuts inapplicable).

One way around this is to generate a diverse set of local modes, using repeated applications of graph cuts, as described in (Yadollahpour et al. 2011). We can then apply a more sophisticated model, which uses global features, to rerank the solutions.

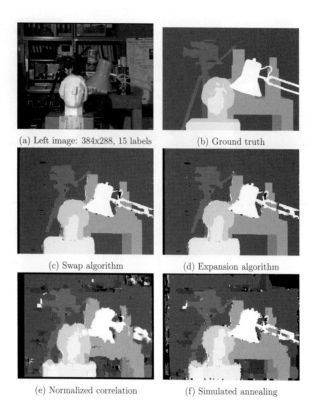

(a) Left image: 384x288, 15 labels   (b) Ground truth

(c) Swap algorithm   (d) Expansion algorithm

(e) Normalized correlation   (f) Simulated annealing

Figure 22.17 An example of stereo depth estimation using an MRF. (a) Left image, of size 384×288 pixels, from the University of Tsukuba. (The corresponding right image is similar, but not shown.) (b) Ground truth depth map, quantized to 15 levels. (c-f): MAP estimates using different methods: (c) $\alpha - \beta$ swap, (d) α expansion, (e) normalized cross correlation, (f) simulated annealing.   Source: Figure 10 of (Boykov et al. 2001).   Used with kind permission of Ramin Zabih.

22.6.5 Dual decomposition

We are interested in computing

$$p^* = \max_{\mathbf{x} \in \mathcal{X}^m} \sum_{i \in V} \theta_i(x_i) + \sum_{f \in F} \theta_f(\mathbf{x}_f) \tag{22.163}$$

where F represents a set of factors. We will assume that we can tractably optimize each local factor, but the combination of all of these factors makes the problem intractable. One way to proceed is to optimize each term independently, but then to introduce constraints that force all the local estimates of the variables' values to agree with each other. We explain this in more detail below, following the presentation of (Sontag et al. 2011).

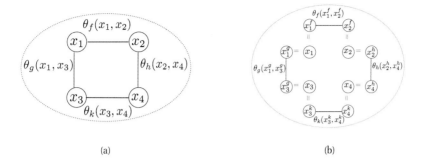

(a)   (b)

Figure 22.18 (a) A pairwise MRF with 4 different edge factors. (b) We have 4 separate variables, plus a copy of each variable for each factor it participates in. Source: Figure 1.2-1.3 of (Sontag et al. 2011). Used with kind permission of David Sontag.

22.6.5.1 Basic idea

Let us duplicate the variables x_i, once for each factor, and then force them to be equal. Specifically, let $\mathbf{x}_f^f = \{x_i^f\}_{i \in f}$ be the set of variables used by factor f. This construction is illustrated in Figure 22.18. We can reformulate the objective as follows:

$$p^* = \max_{\mathbf{x}, \mathbf{x}^f} \sum_{i \in V} \theta_i(x_i) + \sum_{f \in F} \theta_f(\mathbf{x}_f^f) \qquad \text{s.t.} \quad x_i^f = x_i \ \forall f, i \in f \tag{22.164}$$

Let us now introduce Lagrange multipliers, or dual variables, $\delta_{fi}(k)$, to enforce these constraints. The Lagrangian becomes

$$L(\boldsymbol{\delta}, \mathbf{x}, \mathbf{x}^f) = \sum_{i \in V} \theta_i(x_i) + \sum_{f \in F} \theta_f(\mathbf{x}_f^f) \tag{22.165}$$

$$+ \sum_{f \in F} \sum_{i \in f} \sum_{\hat{x}_i} \delta_{fi}(\hat{x}_i) \left(\mathbb{I}(x_i = \hat{x}_i) - \mathbb{I}(x_i^f = \hat{x}_i) \right) \tag{22.166}$$

This is equivalent to our original problem in the following sense: for any value of $\boldsymbol{\delta}$, we have

$$p^* = \max_{\mathbf{x}, \mathbf{x}^f} L(\boldsymbol{\delta}, \mathbf{x}, \mathbf{x}^f) \qquad \text{s.t.} \quad x_i^f = x_i \ \forall f, i \in f \tag{22.167}$$

since if the constraints hold, the last term is zero. We can get an upper bound by dropping the consistency constraints, and just optimizing the following upper bound:

$$L(\boldsymbol{\delta}) \triangleq \max_{\mathbf{x}, \mathbf{x}^f} L(\boldsymbol{\delta}, \mathbf{x}, \mathbf{x}^f) \tag{22.168}$$

$$= \sum_i \max_{x_i} \left(\theta_i(x_i) + \sum_{f: i \in f} \delta_{fi}(x_i) \right) + \sum_f \max_{\mathbf{x}_f} \left(\theta_f(\mathbf{x}_f) - \sum_{i \in f} \delta_{fi}(x_i) \right) \tag{22.169}$$

See Figure 22.19 for an illustration.

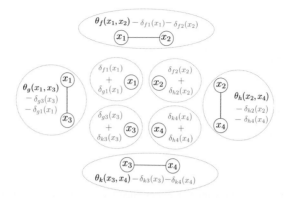

Figure 22.19   Illustration of dual decomposition.   Source: Figure 1.2 of (Sontag et al. 2011).   Used with kind permission of David Sontag.

This objective is tractable to optimize, since each $\mathbf{x}_f$ term is decoupled. Furthermore, we see that $L(\boldsymbol{\delta}) \geq p^*$, since by relaxing the consistency constraints, we are optimizing over a larger space. Furthermore, we have the property that

$$\min_{\boldsymbol{\delta}} L(\boldsymbol{\delta}) = p^* \tag{22.170}$$

so the upper bound is tight at the optimal value of $\boldsymbol{\delta}$, which enforces the original constraints.

Minimizing this upper bound is known as **dual decomposition** or **Lagrangian relaxation** (Komodakis et al. 2011; Sontag et al. 2011; Rush and Collins 2012). Furthemore, it can be shown that $L(\boldsymbol{\delta})$ is the dual to the same LP relaxation we saw before. We will discuss several possible optimization algorithms below.

The main advantage of dual decomposition from a practical point of view is that it allows one to mix and match different kinds of optimization algorithms in a convenient way. For example, we can combine a grid structured graph with local submodular factors to perform image segmentation, together with a tree structured model to perform pose estimation (see Exercise 22.4). Analogous methods can be used in natural language processing, where we often have a mix of local and global constraints (see e.g., (Koo et al. 2010; Rush and Collins 2012)).

22.6.5.2   Theoretical guarantees

What can we say about the quality of the solutions obtained in this way? To understand this, let us first introduce some more notation:

$$\overline{\theta}_i^{\boldsymbol{\delta}}(x_i) \triangleq \theta_i(x_i) + \sum_{f:i\in f} \delta_{fi}(x_i) \tag{22.171}$$

$$\overline{\theta}_f^{\boldsymbol{\delta}}(\mathbf{x}_f) \triangleq \theta_f(\mathbf{x}_f) - \sum_{i\in f} \delta_{fi}(x_i) \tag{22.172}$$

This represents a reparameterization of the original problem, in the sense that

$$\sum_i \theta_i(x_i) + \sum_f \theta_f(\mathbf{x}_f) = \sum_i \overline{\theta}_i^{\delta}(x_i) + \sum_f \overline{\theta}_f^{\delta}(\mathbf{x}_f) \tag{22.173}$$

and hence

$$L(\boldsymbol{\delta}) = \sum_i \max_{x_i} \overline{\theta}_i^{\delta}(x_i) + \sum_f \max_{\mathbf{x}_f} \overline{\theta}_f^{\delta}(\mathbf{x}_f) \tag{22.174}$$

Now suppose there is a set of dual variables $\boldsymbol{\delta}^*$ and an assignment $\mathbf{x}^*$ such that the maximizing assignments to the singleton terms agrees with the assignments to the factor terms, i.e., so that $x_i^* \in \operatorname{argmax}_{x_i} \overline{\theta}_i^{\delta^*}(x_i)$ and $\mathbf{x}_f^* \in \operatorname{argmax}_{\mathbf{x}_f} \overline{\theta}_f^{\delta^*}(\mathbf{x}_f)$. In this case, we have

$$L(\boldsymbol{\delta}^*) = \sum_i \overline{\theta}_i^{\delta^*}(x_i^*) + \sum_f \overline{\theta}_f^{\delta^*}(\mathbf{x}_f^*) = \sum_i \theta_i(x_i^*) + \sum_f \theta_f(\mathbf{x}_f^*) \tag{22.175}$$

Now since

$$\sum_i \theta_i(x_i^*) + \sum_f \theta_f(\mathbf{x}_f^*) \le p^* \le L(\boldsymbol{\delta}^*) \tag{22.176}$$

we conclude that $L(\boldsymbol{\delta}^*) = p^*$, so $\mathbf{x}^*$ is the MAP assignment.

So if we can find a solution where all the subproblems agree, we can be assured that it is the global optimum. This happens surprisingly often in practical problems.

22.6.5.3 Subgradient descent

$L(\boldsymbol{\delta})$ is a convex and continuous objective, but it is non-differentiable at points $\boldsymbol{\delta}$ where $\overline{\theta}_i^{\delta}(x_i)$ or $\overline{\theta}_f^{\delta}(\mathbf{x}_f)$ have multiple optima. One approach is to use subgradient descent. This updates all the elements of $\boldsymbol{\delta}$ at the same time, as follows:

$$\delta_{fi}^{t+1}(x_i) = \delta_{fi}^t(x_i) - \alpha_t g_{fi}^t(x_i) \tag{22.177}$$

where $\mathbf{g}^t$ the subgradient of $L(\boldsymbol{\delta})$ at $\boldsymbol{\delta}^t$. If the step sizes α_t are set appropriately (see Section 8.5.2.1), this method is guaranteed to converge to a global optimum of the dual. (See (Komodakis et al. 2011) for details.)

One can show that the gradient is given by the following sparse vector. First let $x_i^s \in \operatorname{argmax}_{x_i} \overline{\theta}_i^{\delta^t}(x_i)$ and $\mathbf{x}_f^f \in \operatorname{argmax}_{\mathbf{x}_f} \overline{\theta}_f^{\delta^t}(\mathbf{x}_f)$. Next let $g_{fi}(x_i) = 0$ for all elements. Finally, if $x_i^f \ne x_i^s$ (so factor f disagrees with the local term on how to set variable i), we set $g_{fi}(x_i^s) = +1$ and $g_{fi}(x_i^f) = -1$. This has the effect of decreasing $\overline{\theta}_i^{\delta^t}(x_i^s)$ and increasing $\overline{\theta}_i^{\delta^t}(x_i^f)$, bringing them closer to agreement. Similarly, the subgradient update will decrease the value of $\overline{\theta}_f^{\delta^t}(x_i^f, \mathbf{x}_{f\backslash i})$ and increasing the value of $\overline{\theta}_f^{\delta^t}(x_i^s, \mathbf{x}_{f\backslash i})$.

To compute the gradient, we need to be able to solve subproblems of the following form:

$$\operatorname{argmax}_{\mathbf{x}_f} \overline{\theta}_f^{\delta^t}(\mathbf{x}_f) = \operatorname{argmax}_{\mathbf{x}_f} \left[\theta_f(\mathbf{x}_f) - \sum_{i \in f} \delta_{fi}^t(x_i) \right] \tag{22.178}$$

(In (Komodakis et al. 2011), these subproblems are called slaves, whereas $L(\boldsymbol{\delta})$ is called the master.) Obviously if the scope of factor f is small, this is simple. For example, if each factor is pairwise, and each variable has K states, the cost is just K^2. However, there are some kinds of global factors that also support exact and efficient maximization, including the following:

- Graphical models with low tree width.

- Factors that correspond to bipartite graph matchings (see e.g., (Duchi et al. 2007)). This is useful for data association problems, where we must match up a sensor reading with an unknown source. We can find the maximal matching using the so-called Hungarian algorithm in $O(|f|^3)$ time (see e.g., (Padadimitriou and Steiglitz 1982)).

- Supermodular functions. We discuss this case in more detail in Section 22.6.3.2.

- Cardinality constraints. For example, we might have a factor over a large set of binary variables that enforces that a certain number of bits are turned on; this can be useful in problems such as image segmentation. In particular, suppose $\theta_f(\mathbf{x}_f) = 0$ if $\sum_{i \in f} x_i = L$ and $\theta_f(\mathbf{x}_f) = -\infty$ otherwise. We can find the maximizing assignment in $O(|f| \log |f|)$ time as follows: first define $e_i = \delta_{fi}(1) - \delta_{fi}(0)$; now sort the e_i; finally set $x_i = 1$ for the first L values, and $x_i = 0$ for the rest (Tarlow et al. 2010).

- Factors which are constant for all but a small set S of distinguished values of $\mathbf{x}_f$. Then we can optimize over the factor in $O(|S|)$ time (Rother et al. 2009).

22.6.5.4 Coordinate descent

An alternative to updating the entire $\boldsymbol{\delta}$ vector at once (albeit sparsely) is to update it using block coordinate descent. By choosing the size of the blocks, we can trade off convergence speed with ease of the local optimization problem.

One approach, which optimizes $\delta_{fi}(x_i)$ for all $i \in f$ and all x_i at the same time (for a fixed factor f), is known as **max product linear programming** (Globerson and Jaakkola 2008). Algorithmically, this is similar to belief propagation on a factor graph. In particular, we define $\delta_{f \to i}$ as messages sent from factor f to variable i, and we define $\delta_{i \to f}$ as messages sent from variable i to factor f. These messages can be computed as follows (see (Globerson and Jaakkola 2008) for the derivation):[10]

$$\delta_{i \to f}(x_i) = \theta_i(x_i) + \sum_{g \neq f} \delta_{g \to i}(x_i) \tag{22.179}$$

$$\delta_{f \to i}(x_i) = -\delta_{i \to f}(x_i) + \frac{1}{|f|} \max_{\mathbf{x}_{f \backslash i}} \left[\theta_f(\mathbf{x}_f) + \sum_{j \in f} \delta_{j \to f}(x_j) \right] \tag{22.180}$$

We then set the dual variables $\delta_{fi}(x_i)$ to be the messages $\delta_{f \to i}(x_i)$.

For example, consider a 2×2 grid MRF, with the following pairwise factors: $\theta_f(x_1, x_2)$, $\theta_g(x_1, x_3)$, $\theta_h(x_2, x_4)$, and $\theta_k(x_3, x_4)$. The outgoing message from factor f to variable 2 is a

---

10. Note that we denote their $\delta_i^{-f}(x_i)$ by $\delta_{i \to f}(x_i)$.

function of all messages coming into f, and f's local factor:

$$\delta_{f \to 2}(x_2) = -\delta_{2 \to f}(x_2) + \frac{1}{2} \max_{x_1} \left[\theta_f(x_1, x_2) + \delta_{1 \to f}(x_1) + \delta_{2 \to f}(x_2) \right] \tag{22.181}$$

Similarly, the outgoing message from variable 2 to factor f is a function of all the messages sent into variable 2 from other connected factors (in this example, just factor h) and the local potential:

$$\delta_{2 \to f}(x_2) = \theta_2(_2) + \delta_{h2}(x_2) \tag{22.182}$$

The key computational bottleneck is computing the max marginals of each factor, where we max out all the variables from $\mathbf{x}_f$ except for x_i, i.e., we need to be able to compute the following max marginals efficiently:

$$\max_{\mathbf{x}_{f \setminus i}} h(\mathbf{x}_{f \setminus i}, x_i), \quad h(\mathbf{x}_{f \setminus i}, x_i) \triangleq \theta_f(\mathbf{x}_f) + \sum_{j \in f} \delta_{jf}(x_j) \tag{22.183}$$

The difference from Equation 22.178 is that we are maxing over all but one of the variables. We can solve this efficiently for low treewidth graphical models using message passing; we can also solve this efficiently for factors corresponding to bipartite matchings (Duchi et al. 2007) or to cardinality constraints (Tarlow et al. 2010). However, there are cases where maximizing over all the variables in a factor's scope is computationally easier than maximizing over all-but-one (see (Sontag et al. 2011, Sec 1.5.4) for an example); in such cases, we may prefer to use a subgradient method.

Coordinate descent is a simple algorithm that is often much faster at minimizing the dual than gradient descent, especially in the early iterations. It also reduces the objective monotonically, and does not need any step size parameters. Unfortunately, it is not guaranteed to converge to the global optimum, since $L(\boldsymbol{\delta})$ is convex but not strictly convex (which implies there may be more than one globally optimizing value). One way to ensure convergence is to replace the max function in the definition of $L(\boldsymbol{\delta})$ with the soft-max function, which makes the objective strictly convex (see e.g., (Hazan and Shashua 2010) for details).

22.6.5.5 Recovering the MAP assignment

So far, we have been focussing on finding the optimal value of $\boldsymbol{\delta}^*$. But what we really want is the optimal value of $\mathbf{x}^*$. In general, computing $\mathbf{x}^*$ from $\boldsymbol{\delta}^*$ is NP-hard, even if the LP relaxation is tight and the MAP assignment is unique (Sontag et al. 2011, Theorem 1.4). (The troublesome cases arise when there are fractional assignments with the same optimal value as the MAP estimate.)

However, suppose that each $\overline{\theta}_i^{\boldsymbol{\delta}^*}$ has a unique maximum, x_i^*; in this case, we say that $\boldsymbol{\delta}^*$ is **locally decodable** to $\mathbf{x}^*$. One can show than in this case, the LP relaxation is unique and its solution is indeed $\mathbf{x}^*$. If many, but not all, of the nodes are uniquely decodable, we can "clamp" the uniquely decodable ones to their MAP value, and then use exact inference algorithms to figure out the optimal assignment to the remaining variables. Using this method, (Meltzer et al. 2005) was able to optimally solve various stereo vision CRF estimation problems, and (Yanover et al. 2007) was able to optimally solve various protein side-chain structure prediction problems.

Another approach is to use the upper bound provided by the dual in a branch and bound search procedure (Geoffrion 1974).

Exercises

Exercise 22.1 Graphcuts for MAP estimation in binary submodular MRFs

(Source: Ex. 13.14 of (Koller and Friedman 2009).). Show that using the graph construction described in Section 22.6.3.2, the cost of the cut is equal to the energy of the corresponding assignment, up to an irrelevant constant. (Warning: this exercise involves a lot of algebraic book-keeping.)

Exercise 22.2 Graphcuts for alpha-beta swap

(Source: Ex. 13.15 of (Koller and Friedman 2009).). Show how the optimal alpha-beta swap can be found by running min-cut on an appropriately constructed graph. More precisely,

a. Define a set of binary variables $t_1, \ldots, t_n$ such that $t_i = 0$ means $x_i' = \alpha$, $t_i = 1$ if $x_i' = \beta$, and $x_i' = x_i$ is unchanged f $x_i \neq \alpha$ and $x_i \neq \beta$.

b. Define an energy function over the new variables such that $E'(\mathbf{t}) = E(\mathbf{x}) + \text{const}$.

c. Show that E' is submodular if E is a semimetric.

Exercise 22.3 Constant factor optimality for alpha-expansion

Let $\mathcal{X}$ be a pairwise metric Markov random field over a graph $G = (V, E)$. Suppose that the variables are nonbinary and that the node potentials are nonnegative. Let $\mathcal{A}$ denote the set of labels for each $X \in \mathcal{X}$. Though it is not possible to (tractably) find the globally optimal assignment $x^\star$ in general, the α-expansion algorithm provides a method for finding assignments $\hat{x}$ that are locally optimal with respect to a large set of transformations, *i.e.*, the possible α-expansion moves.

Despite the fact that α-expansion only produces a locally optimal MAP assignment, it is possible to prove that the energy of this assignment is within a known factor of the energy of the globally optimal solution $x^\star$. In fact, this is a special case of a more general principle that applies to a wide variety of algorithms, including max-product belief propagation and more general move-making algorithms: If one can prove that the solutions obtained by the algorithm are 'strong local minima', *i.e.*, local minima with respect to a large set of potential moves, then it is possible to derive bounds on the (global) suboptimality of these solutions, and the quality of the bounds will depend on the nature of the moves considered. (There is a precise definition of 'large set of moves'.)

Consider the following approach to proving the suboptimality bound for α-expansion (details are in (Boykov et al. 2001) if you get stuck).

a. Let $\hat{x}$ be a local minimum with respect to expansion moves. For each $\alpha \in \mathcal{A}$, let $V^\alpha = \{s \in V \mid x_s^\star = \alpha\}$, *i.e.*, the set of nodes labelled α in the global minimum. Let x' be an assignment that is equal to $x^\star$ on V^α and equal to $\hat{x}$ elsewhere; this is an α-expansion of $\hat{x}$. Verify that $E(x^\star) \leq E(\hat{x}) \leq E(x')$.

b. Building on the previous part, show that $E(\hat{x}) \leq 2cE(x^\star)$, where $c = \max_{(s,t) \in E} \left(\frac{\max_{\alpha \neq \beta} \varepsilon_{st}(\alpha, \beta)}{\min_{\alpha \neq \beta} \varepsilon_{st}(\alpha, \beta)} \right)$ and E denotes the energy of an assignment.

Hint. Think about where x' agrees with $\hat{x}$ and where it agrees with $x^\star$.

Exercise 22.4 Dual decomposition for pose segmentation

(Source: Daphne Koller.). Two important problems in computer vision are that of parsing articulated objects (*e.g.*, the human body), called *pose estimation*, and segmenting the foreground and the background, called *segmentation*. Intuitively, these two problems are linked, in that solving either one would be easier if the solution to the other were available. We consider solving these problems simultaneously using a joint model over human poses and foreground/background labels and then using dual decomposition for MAP inference in this model.

We construct a two-level model, where the high level handles pose estimation and the low level handles pixel-level background segmentation. Let $G = (\mathcal{V}, \mathcal{E})$ be an undirected grid over the pixels. Each node $i \in \mathcal{V}$ represents a pixel. Suppose we have one binary variable x_i for each pixel, where $x_i = 1$ means that pixel i is in the foreground. Denote the full set of these variables by $\mathbf{x} = (x_i)$.

In addition, suppose we have an undirected tree structure $T = (\mathcal{V}', \mathcal{E}')$ on the parts. For each body part, we have a discrete set of candidate poses that the part can be in, where each pose is characterized by parameters specifying its position and orientation. (These candidates are generated by a procedure external to the algorithm described here.) Define y_{jk} to be a binary variable indicating whether body part $j \in \mathcal{V}'$ is in configuration k. Then the full set of part variables is given by $\mathbf{y} = (y_{jk})$, with $j \in \mathcal{V}'$ and $k = 1, \ldots, K$, where J is the total number of body parts and K is the number of candidate poses for each part. Note that in order to describe a valid configuration, $\mathbf{y}$ must satisfy the constraint that $\sum_{k=1}^{K} y_{jk} = 1$ for each j.

Suppose we have the following energy function on pixels:

$$E_1(\mathbf{x}) = \sum_{i \in \mathcal{V}} \mathbb{1}[x_i = 1] \cdot \theta_i + \sum_{(i,j) \in \mathcal{E}} \mathbb{1}[x_i \neq x_j] \cdot \theta_{ij}.$$

Assume that the θ_{ij} arises from a metric (*e.g.*, based on differences in pixel intensities), so this can be viewed as the energy for a pairwise metric MRF with respect to G.

We then have the following energy function for parts:

$$E_2(\mathbf{y}) = \sum_{p \in \mathcal{V}'} \theta_p(y_p) + \sum_{(p,q) \in \mathcal{E}'} \theta_{pq}(y_p, y_q).$$

Since each part candidate y_{jk} is assumed to come with a position and orientation, we can compute a binary mask in the image plane. The mask assigns a value to each pixel, denoted by $\{w_{jk}^i\}_{i \in \mathcal{V}}$, where $w_{jk}^i = 1$ if pixel i lies on the skeleton and decreases as we move away. We can use this to define an energy function relating the parts and the pixels:

$$E_3(\mathbf{x}, \mathbf{y}) = \sum_{i \in \mathcal{V}} \sum_{j \in \mathcal{V}'} \sum_{k=1}^{K} \mathbb{1}[x_i = 0, y_{jk} = 1] \cdot w_{jk}^i.$$

In other words, this energy term only penalizes the case where a part candidate is active but the pixel underneath is labeled as background.

Formulate the minimization of $E_1 + E_2 + E_3$ as an integer program and show how you can use dual decomposition to solve the dual of this integer program. Your solution should describe the decomposition into slaves, the method for solving each one, and the update rules for the overall algorithm. Briefly justify your design choices, particularly your choice of inference algorithms for the slaves.

Exercise 22.5 Derivation of the EP updates for trueskill

Derive Equation 22.130. Hint: if we define the rectified truncated Gaussian as follows

$$\mathbb{R}(x; \mu, \sigma^2, l, u) = \mathbb{I}_{x \in (l,u)} \frac{\mathcal{N}(x; \mu, \sigma^2)}{\Phi(u; \mu, \sigma^2) - \Phi(l; \mu, \sigma^2)} \tag{22.184}$$

then one can show (see Equations 4.2 and 4.4 of (Herbrich 2005)) that its mean is given by

$$\mu_{\mathbb{R}} = \mu + \sigma \frac{\mathcal{N}(\frac{l}{\sigma} - \frac{\mu}{\sigma}) - \mathcal{N}(\frac{u}{\sigma} - \frac{\mu}{\sigma})}{\Phi(\frac{u}{\sigma} - \frac{\mu}{\sigma}) - \Phi(\frac{l}{\sigma} - \frac{\mu}{\sigma})} \tag{22.185}$$

23 *Monte Carlo inference*

23.1 Introduction

So far, we discussed various deterministic algorithms for posterior inference. These methods enjoy many of the benefits of the Bayesian approach, while still being about as fast as optimization-based point-estimation methods. The trouble with these methods is that they can be rather complicated to derive, and they are somewhat limited in their domain of applicability (e.g., they usually assume conjugate priors and exponential family likelihoods, although see (Wand et al. 2011) for some recent extensions of mean field to more complex distributions). Furthermore, although they are fast, their accuracy is often limited by the form of the approximation which we choose.

In this chapter, we discuss an alternative class of algorithms based on the idea of Monte Carlo approximation, which we first introduced in Section 2.7. The idea is very simple: generate some (unweighted) samples from the posterior, $\mathbf{x}^s \sim p(\mathbf{x}|\mathcal{D})$, and then use these to compute any quantity of interest, such as a posterior marginal, $p(x_1|\mathcal{D})$, or the posterior of the difference of two quantities, $p(x_1 - x_2|\mathcal{D})$, or the posterior predictive, $p(y|\mathcal{D})$, etc. All of these quantities can be approximated by $\mathbb{E}[f|\mathcal{D}] \approx \frac{1}{S}\sum_{s=1}^{S} f(\mathbf{x}^s)$ for some suitable function f.

By generating enough samples, we can achieve any desired level of accuracy we like. The main issue is: how do we efficiently generate samples from a probability distribution, particularly in high dimensions? In this chapter, we discuss non-iterative methods for generating independent samples. In the next chapter, we discuss an iterative method known as Markov Chain Monte Carlo, or MCMC for short, which produces dependent samples but which works well in high dimensions. Note that sampling is a large topic. The reader should consult other books, such as (Liu 2001; Robert and Casella 2004), for more information.

23.2 Sampling from standard distributions

We briefly discuss some ways to sample from 1 or 2 dimensional distributions of standard form. These methods are often used as subroutines by more complex methods.

23.2.1 Using the cdf

The simplest method for sampling from a univariate distribution is based on the **inverse probability transform**. Let F be a cdf of some distribution we want to sample from, and let F^{-1}

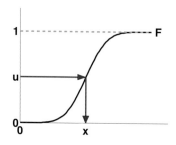

Figure 23.1   Sampling using an inverse CDF. Figure generated by `sampleCdf`.

be its inverse. Then we have the following result.

Theorem 23.2.1. *If $U \sim U(0,1)$ is a uniform rv, then $F^{-1}(U) \sim F$.*

Proof.

$$
\begin{aligned}
\Pr(F^{-1}(U) \leq x) &= \Pr(U \leq F(x)) \quad \text{(applying F to both sides)} & (23.1) \\
&= F(x) \quad \text{(because $\Pr(U \leq y) = y$)} & (23.2)
\end{aligned}
$$

where the first line follows since F is a monotonic function, and the second line follows since U is uniform on the unit interval.   □

Hence we can sample from any univariate distribution, for which we can evaluate its inverse cdf, as follows: generate a random number $u \sim U(0,1)$ using a **pseudo random number generator** (see e.g., (Press et al. 1988) for details). Let u represent the height up the y axis. Then "slide along" the x axis until you intersect the F curve, and then "drop down" and return the corresponding x value. This corresponds to computing $x = F^{-1}(u)$. See Figure 23.1 for an illustration.

For example, consider the exponential distribution

$$
\text{Expon}(x|\lambda) \triangleq \lambda e^{-\lambda x} \, \mathbb{I}(x \geq 0) \tag{23.3}
$$

The cdf is

$$
F(x) = 1 - e^{-\lambda x} \, \mathbb{I}(x \geq 0) \tag{23.4}
$$

whose inverse is the quantile function

$$
F^{-1}(p) = -\frac{\ln(1-p)}{\lambda} \tag{23.5}
$$

By the above theorem, if $U \sim \text{Unif}(0,1)$, we know that $F^{-1}(U) \sim \text{Expon}(\lambda)$. Furthermore, since $1 - U \sim \text{Unif}(0,1)$ as well, we can sample from the exponential distribution by first sampling from the uniform and then transforming the results using $-\ln(u)/\lambda$.

23.2.2 Sampling from a Gaussian (Box-Muller method)

We now describe a method to sample from a Gaussian. The idea is we sample uniformly from a unit radius circle, and then use the change of variables formula to derive samples from a spherical 2d Gaussian. This can be thought of as two samples from a 1d Gaussian.

In more detail, sample $z_1, z_2 \in (-1, 1)$ uniformly, and then discard pairs that do not satisfy $z_1^2 + z_2^2 \leq 1$. The result will be points uniformly distributed inside the unit circle, so $p(\mathbf{z}) = \frac{1}{\pi}\mathbb{I}(z \text{ inside circle})$. Now define

$$x_i = z_i \left(\frac{-2 \ln r^2}{r^2} \right)^{\frac{1}{2}} \tag{23.6}$$

for $i = 1 : 2$, where $r^2 = z_1^2 + z_2^2$. Using the multivariate change of variables formula, we have

$$p(x_1, x_2) = p(z_1, z_2) |\frac{\partial(z_1, z_2)}{\partial(x_1, x_2)}| = \left[\frac{1}{\sqrt{2\pi}} \exp(-\frac{1}{2}x_1^2) \right] \left[\frac{1}{\sqrt{2\pi}} \exp(-\frac{1}{2}x_2^2) \right] \tag{23.7}$$

Hence x_1 and x_2 are two independent samples from a univariate Gaussian. This is known as the **Box-Muller** method.

To sample from a multivariate Gaussian, we first compute the Cholesky decomposition of its covariance matrix, $\boldsymbol{\Sigma} = \mathbf{L}\mathbf{L}^T$, where $\mathbf{L}$ is lower triangular. Next we sample $\mathbf{x} \sim \mathcal{N}(\mathbf{0}, \mathbf{I})$ using the Box-Muller method. Finally we set $\mathbf{y} = \mathbf{L}\mathbf{x} + \boldsymbol{\mu}$. This is valid since

$$\text{cov}\left[\mathbf{y}\right] = \mathbf{L}\text{cov}\left[\mathbf{x}\right]\mathbf{L}^T = \mathbf{L}\,\mathbf{I}\,\mathbf{L}^T = \boldsymbol{\Sigma} \tag{23.8}$$

23.3 Rejection sampling

When the inverse cdf method cannot be used, one simple alternative is to use **rejection sampling**, which we now explain.

23.3.1 Basic idea

In rejection sampling, we create a **proposal distribution** $q(x)$ which satisifes $Mq(x) \geq \tilde{p}(x)$, for some constant M, where $\tilde{p}(x)$ is an unnormalized version of $p(x)$ (i.e., $p(x) = \tilde{p}(x)/Z_p$ for some possibly unknown constant Z_p). The function $Mq(x)$ provides an upper envelope for $\tilde{p}$. We then sample $x \sim q(x)$, which corresponds to picking a random x location, and then we sample $u \sim U(0, 1)$, which corresponds to picking a random height (y location) under the envelope. If $u > \frac{\tilde{p}(x)}{Mq(x)}$, we reject the sample, otherwise we accept it. See Figure 23.2(a). where the acceptance region is shown shaded, and the rejection region is the white region between the shaded zone and the upper envelope.

We now prove that this procedure is correct. Let

$$S = \{(x, u) : u \leq \tilde{p}(x)/Mq(x)\}, \ S_0 = \{(x, u) : x \leq x_0, u \leq \tilde{p}(x)/Mq(x)\} \tag{23.9}$$

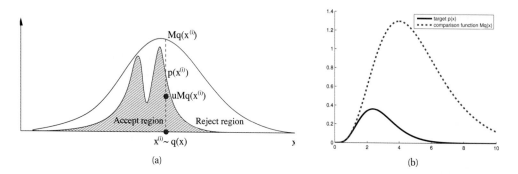

Figure 23.2 (a) Schematic illustration of rejection sampling. Source: Figure 2 of (Andrieu et al. 2003). Used with kind permission of Nando de Freitas. (b) Rejection sampling from a $\mathrm{Ga}(\alpha = 5.7, \lambda = 2)$ distribution (solid blue) using a proposal of the form $M\mathrm{Ga}(k, \lambda - 1)$ (dotted red), where $k = \lfloor 5.7 \rfloor = 5$. The curves touch at $\alpha - k = 0.7$. Figure generated by `rejectionSamplingDemo`.

Then the cdf of the accepted points is given by

$$P(x \leq x_0 | x \text{ accepted}) \quad = \quad \frac{P(x \leq x_0, x \text{ accepted})}{P(x \text{ accepted})} \tag{23.10}$$

$$= \quad \frac{\int \int \mathbb{I}((x, u) \in S_0) q(x) du dx}{\int \int \mathbb{I}((x, u) \in S) q(x) du dx} = \frac{\int_{-\infty}^{x_0} \tilde{p}(x) dx}{\int_{-\infty}^{\infty} \tilde{p}(x) dx} \tag{23.11}$$

which is the cdf of $p(x)$, as desired.

How efficient is this method? Since we generate with probability $q(x)$ and accept with probability $\frac{\tilde{p}(x)}{Mq(x)}$, the probability of acceptance is

$$p(\text{accept}) = \int \frac{\tilde{p}(x)}{Mq(x)} q(x) dx = \frac{1}{M} \int \tilde{p}(x) dx \tag{23.12}$$

Hence we want to choose M as small as possible while still satisfying $Mq(x) \geq \tilde{p}(x)$.

23.3.2 Example

For example, suppose we want to sample from a Gamma distribution:[1]

$$\mathrm{Ga}(x|\alpha, \lambda) = \frac{1}{\Gamma(\alpha)} x^{\alpha-1} \lambda^{\alpha} \exp(-\lambda x) \tag{23.13}$$

One can show that if $X_i \overset{iid}{\sim} \mathrm{Expon}(\lambda)$, and $Y = X_1 + \cdots + X_k$, then $Y \sim \mathrm{Ga}(k, \lambda)$. For non-integer shape parameters, we cannot use this trick. However, we can use rejection sampling

---

1. This section is based on notes by Ioana A. Cosma, available at `http://users.aims.ac.za/~ioana/cp2.pdf`.

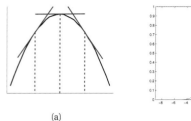
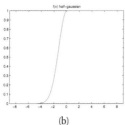
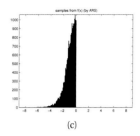

(a) (b) (c)

Figure 23.3 (a) Idea behind adaptive rejection sampling. We place piecewise linear upper (and lower) bounds on the log-concave density. Based on Figure 1 of (Gilks and Wild 1992). Figure generated by `arsEnvelope`. (b-c) Using ARS to sample from a half-Gaussian. Figure generated by `arsDemo`, written by Daniel Eaton.

using a $\text{Ga}(k, \lambda - 1)$ distribution as a proposal, where $k = \lfloor \alpha \rfloor$. The ratio has the form

$$\frac{p(x)}{q(x)} = \frac{\text{Ga}(x|\alpha, \lambda)}{\text{Ga}(x|k, \lambda - 1)} = \frac{x^{\alpha-1}\lambda^{\alpha}\exp(-\lambda x)/\Gamma(\alpha)}{x^{k-1}(\lambda - 1)^{k}\exp(-(\lambda - 1)x)/\Gamma(k)} \tag{23.14}$$

$$= \frac{\Gamma(k)\lambda^{\alpha}}{\Gamma(\alpha)(\lambda - 1)^{k}}x^{\alpha-k}\exp(-x) \tag{23.15}$$

This ratio attains its maximum when $x = \alpha - k$. Hence

$$M = \frac{\text{Ga}(\alpha - k|\alpha, \lambda)}{\text{Ga}(\alpha - k|k, \lambda - 1)} \tag{23.16}$$

See Figure 23.2(b) for a plot. (Exercise 23.2 asks you to devise a better proposal distribution based on the Cauchy distribution.)

23.3.3 Application to Bayesian statistics

Suppose we want to draw (unweighted) samples from the posterior, $p(\boldsymbol{\theta}|\mathcal{D}) = p(\mathcal{D}|\boldsymbol{\theta})p(\boldsymbol{\theta})/p(\mathcal{D})$. We can use rejection sampling with $\tilde{p}(\boldsymbol{\theta}) = p(\mathcal{D}|\boldsymbol{\theta})p(\boldsymbol{\theta})$ as the target distribution, $q(\boldsymbol{\theta}) = p(\boldsymbol{\theta})$ as our proposal, and $M = p(\mathcal{D}|\hat{\boldsymbol{\theta}})$, where $\hat{\boldsymbol{\theta}} = \arg\max p(\mathcal{D}|\boldsymbol{\theta})$ is the MLE; this was first suggested in (Smith and Gelfand 1992). We accept points with probability

$$\frac{\tilde{p}(\boldsymbol{\theta})}{Mq(\boldsymbol{\theta})} = \frac{p(\mathcal{D}|\boldsymbol{\theta})}{p(\mathcal{D}|\hat{\boldsymbol{\theta}})} \tag{23.17}$$

Thus samples from the prior that have high likelihood are more likely to be retained in the posterior. Of course, if there is a big mismatch between prior and posterior (which will be the case if the prior is vague and the likelihood is informative), this procedure is very inefficient. We discuss better algorithms later.

23.3.4 Adaptive rejection sampling

We now describe a method that can automatically come up with a tight upper envelope $q(x)$ to any log concave density $p(x)$. The idea is to upper bound the log density with a piecewise linear function, as illustrated in Figure 23.3(a). We choose the initial locations for the pieces based on a fixed grid over the support of the distribution. We then evaluate the gradient of the log density at these locations, and make the lines be tangent at these points.

Since the log of the envelope is piecewise linear, the envelope itself is piecewise exponential:

$$q(x) = M_i \lambda_i \exp(-\lambda_i(x - x_{i-1})), \quad x_{i-1} < x \leq x_i \tag{23.18}$$

where x_i are the grid points. It is relatively straightforward to sample from this distribution. If the sample x is rejected, we create a new grid point at x, and thereby refine the envelope. As the number of grid points is increased, the tightness of the envelope improves, and the rejection rate goes down. This is known as **adaptive rejection sampling** (ARS) (Gilks and Wild 1992). Figure 23.3(b-c) gives an example of the method in action. As with standard rejection sampling, it can be applied to unnormalized distributions.

23.3.5 Rejection sampling in high dimensions

It is clear that we want to make our proposal $q(x)$ as close as possible to the target distribution $p(x)$, while still being an upper bound. But this is quite hard to achieve, especially in high dimensions. To see this, consider sampling from $p(\mathbf{x}) = \mathcal{N}(\mathbf{0}, \sigma_p^2 \mathbf{I})$ using as a proposal $q(\mathbf{x}) = \mathcal{N}(\mathbf{0}, \sigma_q^2 \mathbf{I})$. Obviously we must have $\sigma_q^2 \geq \sigma_p^2$ in order to be an upper bound. In D dimensions, the optimum value is given by $M = (\sigma_q/\sigma_p)^D$. The acceptance rate is $1/M$ (since both p and q are normalized), which decreases exponentially fast with dimension. For example, if σ_q exceeds σ_p by just 1%, then in 1000 dimensions the acceptance ratio will be about 1/20,000. This is a fundamental weakness of rejection sampling.

In Chapter 24, we will describe MCMC sampling, which is a more efficient way to sample from high dimensional distributions. Sometimes this uses (adaptive) rejection sampling as a subroutine, which is known as **adaptive rejection Metropolis sampling** (Gilks et al. 1995).

23.4 Importance sampling

We now describe a Monte Carlo method known as **importance sampling** for approximating integrals of the form

$$I = \mathbb{E}[f] = \int f(\mathbf{x}) p(\mathbf{x}) d\mathbf{x} \tag{23.19}$$

23.4.1 Basic idea

The idea is to draw samples $\mathbf{x}$ in regions which have high probability, $p(\mathbf{x})$, but also where $|f(\mathbf{x})|$ is large. The result can be **super efficient**, meaning it needs less samples than if we were to sample from the exact distribution $p(\mathbf{x})$. The reason is that the samples are focussed on the important parts of space. For example, suppose we want to estimate the probability of a **rare event**. Define $f(\mathbf{x}) = \mathbb{I}(\mathbf{x} \in E)$, for some set E. Then it is better to sample from a proposal of the form $q(\mathbf{x}) \propto f(\mathbf{x}) p(\mathbf{x})$ than to sample from $p(\mathbf{x})$ itself.

Importance sampling samples from any proposal, $q(\mathbf{x})$. It then uses these samples to estimate the integral as follows:

$$\mathbb{E}\left[f\right] = \int f(\mathbf{x}) \frac{p(\mathbf{x})}{q(\mathbf{x})} q(\mathbf{x}) d\mathbf{x} \approx \frac{1}{S} \sum_{s=1}^{S} w_s f(\mathbf{x}^s) = \hat{I} \tag{23.20}$$

where $w_s \triangleq \frac{p(\mathbf{x}^s)}{q(\mathbf{x}^s)}$ are the **importance weights**. Note that, unlike rejection sampling, we use all the samples.

How should we choose the proposal? A natural criterion is to minimize the variance of the estimate $\hat{I} = \sum_s \frac{p(\mathbf{x}^s) f(\mathbf{x}^s)}{q(\mathbf{x}^s)}$. We now show how to minimize this. We have

$$\text{var}_q \left[\frac{p(\mathbf{x}) f(\mathbf{x})}{q(\mathbf{x})} \right] = \mathbb{E}_q \left[\frac{p^2(\mathbf{x}) f^2(\mathbf{x})}{q^2(\mathbf{x})} \right] - \left(\mathbb{E}_q \left[\frac{p(\mathbf{x}) f(\mathbf{x})}{q(\mathbf{x})} \right] \right)^2 \tag{23.21}$$

Since the last term is independent of q, we can ignore it. To ensure the first term is finite, we require $\mathbb{E}_q \left[\frac{p^2(\mathbf{x}) f^2(\mathbf{x})}{q^2(\mathbf{x})} \right] = \mathbb{E}_p \left[\frac{p(\mathbf{x}) f^2(\mathbf{x})}{q(\mathbf{x})} \right]$ to be finite. Sufficient conditions for this are discussed in (Robert and Casella 2004, p95). In practice we try to ensure q has heavier tails than p.

Amongst all the proposal distributions which obtain finite variance, we can find the optimal one as follows. By Jensen's inequality, we have $\mathbb{E}\left[u^2(\mathbf{x})\right] \geq (\mathbb{E}\left[u(\mathbf{x})\right])^2$ for $u(\mathbf{x}) \geq 0$. Setting $u(\mathbf{x}) = p(\mathbf{x})|f(\mathbf{x})|/q(\mathbf{x})$, we have the following lower bound:

$$\mathbb{E}_q \left[\frac{p^2(\mathbf{x}) f^2(\mathbf{x})}{q^2(\mathbf{x})} \right] \geq \left(\mathbb{E}_q \left[\frac{p(\mathbf{x})|f(\mathbf{x})|}{q(\mathbf{x})} \right] \right)^2 = \left(\int p(\mathbf{x})|f(\mathbf{x})| d\mathbf{x} \right)^2 \tag{23.22}$$

This lower bound holds for any q. The bound can be attained by using the following optimal importance distribution:

$$q^*(\mathbf{x}) = \frac{|f(\mathbf{x})| p(\mathbf{x})}{\int |f(\mathbf{x}')| p(\mathbf{x}') d\mathbf{x}'} \tag{23.23}$$

(This is easy to verify — just plug in $q = q^*$ into Equation 23.22 and notice that the left and right hand sides are equal.) Of course, computing this optimal proposal may be difficult.

When we don't have a particular target function $f(\mathbf{x})$ in mind, we often just try to make $q(\mathbf{x})$ as close as possible to $p(\mathbf{x})$. In general, this is difficult, especially in high dimensions, but it is possible to adapt the proposal distribution to improve the approximation. This is known as **adaptive importance sampling** (Oh and Berger 1992).

23.4.2 Handling unnormalized distributions

It is frequently the case that we can evaluate the unnormalized target distribution, $\tilde{p}(\mathbf{x})$, but not its normalization constant, Z_p. We may also want to use an unnormalized proposal, $\tilde{q}(\mathbf{x})$, with possibly unknown normalization constant Z_q. We can do this as follows. First we evaluate

$$\mathbb{E}\left[f\right] = \frac{Z_q}{Z_p} \int f(\mathbf{x}) \frac{\tilde{p}(\mathbf{x})}{\tilde{q}(\mathbf{x})} q(\mathbf{x}) d\mathbf{x} \approx \frac{Z_q}{Z_p} \frac{1}{S} \sum_{s=1}^{S} \tilde{w}_s f(\mathbf{x}^s) \tag{23.24}$$

where $\tilde{w}_s \triangleq \frac{\tilde{p}(\mathbf{x}^s)}{\tilde{q}(\mathbf{x}^s)}$ is the unnormalized importance weight. We can use the same set of samples to evaluate the ratio Z_p/Z_q as follows:

$$\frac{Z_p}{Z_q} = \frac{1}{Z_q} \int \tilde{p}(\mathbf{x}) d\mathbf{x} = \int \frac{\tilde{p}(\mathbf{x})}{\tilde{q}(\mathbf{x})} q(\mathbf{x}) d\mathbf{x} \approx \frac{1}{S} \sum_{s=1}^{S} \tilde{w}_s \tag{23.25}$$

Hence

$$\hat{I} = \frac{\frac{1}{S}\sum_s \tilde{w}_s f(\mathbf{x}^s)}{\frac{1}{S}\sum_s \tilde{w}_s} = \sum_{s=1}^{S} w_s f(\mathbf{x}^s) \tag{23.26}$$

where

$$w_s \triangleq \frac{\tilde{w}_s}{\sum_{s'} \tilde{w}_{s'}} \tag{23.27}$$

are the normalized importance weights. The resulting estimate is a ratio of two estimates, and hence is biased. However, as $S \to \infty$, we have that $\hat{I} \to I$, under weak assumptions (see e.g., (Robert and Casella 2004) for details).

23.4.3 Importance sampling for a DGM: likelihood weighting

We now describe a way to use importance sampling to generate samples from a distribution which can be represented as a directed graphical model (Chapter 10).

If we have no evidence, we can sample from the unconditional joint distribution of a DGM $p(\mathbf{x})$ as follows: first sample the root nodes, then sample their children, then sample their children, etc. This is known as **ancestral sampling**. It works because, in a DAG, we can always topologically order the nodes so that parents preceed children. (Note that there is no equivalent easy method for sampling from an unconditional *undirected* graphical model.)

Now suppose we have some evidence, so some nodes are "clamped" to observed values, and we want to sample from the posterior $p(\mathbf{x}|\mathcal{D})$. If all the variables are discrete, we can use the following simple procedure: perform ancestral sampling, but as soon as we sample a value that is inconsistent with an observed value, reject the whole sample and start again. This is known as **logic sampling** (Henrion 1988).

Needless to say, logic sampling is very inefficient, and it cannot be applied when we have real-valued evidence. However, it can be modified as follows. Sample unobserved variables as before, conditional on their parents. But don't sample observed variables; instead we just use their observed values. This is equivalent to using a proposal of the form

$$q(\mathbf{x}) = \prod_{t \notin E} p(x_t|\mathbf{x}_{\mathrm{pa}(t)}) \prod_{t \in E} \delta_{x_t^*}(x_t) \tag{23.28}$$

where E is the set of observed nodes, and x_t^* is the observed value for node t. We should therefore give the overall sample an importance weight as follows:

$$w(\mathbf{x}) = \frac{p(\mathbf{x})}{q(\mathbf{x})} = \prod_{t \notin E} \frac{p(x_t|\mathbf{x}_{\mathrm{pa}(t)})}{p(x_t|\mathbf{x}_{\mathrm{pa}(t)})} \prod_{t \in E} \frac{p(x_t|\mathbf{x}_{\mathrm{pa}(t)})}{1} = \prod_{t \in E} p(x_t|\mathbf{x}_{\mathrm{pa}(t)}) \tag{23.29}$$

This technique is known as **likelihood weighting** (Fung and Chang 1989; Shachter and Peot 1989).

23.4.4 Sampling importance resampling (SIR)

We can draw unweighted samples from $p(x)$ by first using importance sampling (with proposal q) to generate a distribution of the form

$$p(\mathbf{x}) \approx \sum_s w_s \delta_{\mathbf{x}^s}(\mathbf{x}) \tag{23.30}$$

where w_s are the normalized importance weights. We then sample with replacement from Equation 23.30, where the probability that we pick $\mathbf{x}^s$ is w_s. Let this procedure induce a distribution denoted by $\hat{p}$. To see that this is valid, note that

$$\hat{p}(x \leq x_0) \;=\; \sum_s \mathbb{I}(x^s \leq x_0) w_s = \frac{\sum_s \mathbb{I}(x^s \leq x_0) \tilde{p}(x^s)/q(x^s)}{\sum_s \tilde{p}(x^s)/q(x^s)} \tag{23.31}$$

$$\to \; \frac{\int \mathbb{I}(x \leq x_0) \frac{\tilde{p}(x)}{q(x)} q(x) dx}{\int \frac{\tilde{p}(x)}{q(x)} q(x) dx} \tag{23.32}$$

$$= \; \frac{\int \mathbb{I}(x \leq x_0) \tilde{p}(x) dx}{\int \tilde{p}(x) dx} = \int \mathbb{I}(x \leq x_0) p(x) dx = p(x \leq x_0) \tag{23.33}$$

This is known as **sampling importance resampling** (SIR) (Rubin 1998). The result is an unweighted approximation of the form

$$p(\mathbf{x}) \approx \frac{1}{S'} \sum_{s=1}^{S'} \delta_{\mathbf{x}^s}(\mathbf{x}) \tag{23.34}$$

Note that we typically take $S' \ll S$.

This algorithm can be used to perform Bayesian inference in low-dimensional settings (Smith and Gelfand 1992). That is, suppose we want to draw (unweighted) samples from the posterior, $p(\boldsymbol{\theta}|\mathcal{D}) = p(\mathcal{D}|\boldsymbol{\theta})p(\boldsymbol{\theta})/p(\mathcal{D})$. We can use importance sampling with $\tilde{p}(\boldsymbol{\theta}) = p(\mathcal{D}|\boldsymbol{\theta})p(\boldsymbol{\theta})$ as the unnormalized posterior, and $q(\boldsymbol{\theta}) = p(\boldsymbol{\theta})$ as our proposal. The normalized weights have the form

$$w_s = \frac{\tilde{p}(\boldsymbol{\theta}_s)/q(\boldsymbol{\theta}_s)}{\sum_{s'} \tilde{p}(\boldsymbol{\theta}_{s'})/q(\boldsymbol{\theta}_{s'})} = \frac{p(\mathcal{D}|\boldsymbol{\theta}_s)}{\sum_{s'} p(\mathcal{D}|\boldsymbol{\theta}_{s'})} \tag{23.35}$$

We can then use SIR to sample from $p(\boldsymbol{\theta}|\mathcal{D})$.

Of course, if there is a big discrepancy between our proposal (the prior) and the target (the posterior), we will need a huge number of importance samples for this technique to work reliably, since otherwise the variance of the importance weights will be very large, implying that most samples carry no useful information. (This issue will come up again in Section 23.5, when we discuss particle filtering.)

23.5 Particle filtering

Particle filtering (PF) is a Monte Carlo, or **simulation based**, algorithm for recursive Bayesian inference. That is, it approximates the predict-update cycle described in Section 18.3.1. It is very widely used in many areas, including tracking, time-series forecasting, online parameter learning, etc. We explain the basic algorithm below. For a book-length treatment, see (Doucet et al. 2001); for a good tutorial, see (Arulampalam et al. 2002), or just read on.

23.5.1   Sequential importance sampling

The basic idea is to appproximate the belief state (of the entire state trajectory) using a weighted set of particles:

$$p(\mathbf{z}_{1:t}|\mathbf{y}_{1:t}) \approx \sum_{s=1}^{S} \hat{w}_t^s \delta_{\mathbf{z}_{1:t}^s}(\mathbf{z}_{1:t}) \tag{23.36}$$

where $\hat{w}_t^s$ is the normalized weight of sample s at time t. From this representation, we can easily compute the marginal distribution over the most recent state, $p(\mathbf{z}_t|\mathbf{y}_{1:t})$, by simply ignoring the previous parts of the trajectory, $\mathbf{z}_{1:t-1}$. (The fact that PF samples in the space of entire trajectories has various implications which we will discuss later.)

We update this belief state using importance sampling. If the proposal has the form $q(\mathbf{z}_{1:t}^s|\mathbf{y}_{1:t})$, then the importance weights are given by

$$w_t^s \propto \frac{p(\mathbf{z}_{1:t}^s|\mathbf{y}_{1:t})}{q(\mathbf{z}_{1:t}^s|\mathbf{y}_{1:t})} \tag{23.37}$$

which can be normalized as follows:

$$\hat{w}_t^s = \frac{w_t^s}{\sum_{s'} w_t^{s'}} \tag{23.38}$$

We can rewrite the numerator recursively as follows:

$$p(\mathbf{z}_{1:t}|\mathbf{y}_{1:t}) = \frac{p(\mathbf{y}_t|\mathbf{z}_{1:t}, \mathbf{y}_{1:t-1})p(\mathbf{z}_{1:t}|\mathbf{y}_{1:t-1})}{p(\mathbf{y}_t|\mathbf{y}_{1:t-1})} \tag{23.39}$$

$$= \frac{p(\mathbf{y}_t|\mathbf{z}_t)p(\mathbf{z}_t|\mathbf{z}_{1:t-1}, \mathbf{y}_{1:t-1})p(\mathbf{z}_{1:t-1}|\mathbf{y}_{1:t-1})}{p(\mathbf{y}_t|\mathbf{y}_{1:t-1})} \tag{23.40}$$

$$\propto p(\mathbf{y}_t|\mathbf{z}_t)p(\mathbf{z}_t|\mathbf{z}_{t-1})p(\mathbf{z}_{1:t-1}|\mathbf{y}_{1:t-1}) \tag{23.41}$$

where we have made the usual Markov assumptions. We will restrict attention to proposal densities of the following form:

$$q(\mathbf{z}_{1:t}|\mathbf{y}_{1:t}) = q(\mathbf{z}_t|\mathbf{z}_{1:t-1}, \mathbf{y}_{1:t})q(\mathbf{z}_{1:t-1}|\mathbf{y}_{1:t-1}) \tag{23.42}$$

so that we can "grow" the trajectory by adding the new state $\mathbf{z}_t$ to the end. In this case, the importance weights simplify to

$$w_t^s \propto \frac{p(\mathbf{y}_t|\mathbf{z}_t^s)p(\mathbf{z}_t^s|\mathbf{z}_{t-1}^s)p(\mathbf{z}_{1:t-1}^s|\mathbf{y}_{1:t-1})}{q(\mathbf{z}_t^s|\mathbf{z}_{1:t-1}^s, \mathbf{y}_{1:t})q(\mathbf{z}_{1:t-1}^s|\mathbf{y}_{1:t-1})} \tag{23.43}$$

$$= w_{t-1}^s \frac{p(\mathbf{y}_t|\mathbf{z}_t^s)p(\mathbf{z}_t^s|\mathbf{z}_{t-1}^s)}{q(\mathbf{z}_t^s|\mathbf{z}_{1:t-1}^s, \mathbf{y}_{1:t})} \tag{23.44}$$

If we further assume that $q(\mathbf{z}_t|\mathbf{z}_{1:t-1}, \mathbf{y}_{1:t}) = q(\mathbf{z}_t|\mathbf{z}_{t-1}, \mathbf{y}_t)$, then we only need to keep the most recent part of the trajectory and observation sequence, rather than the whole history, in order to compute the new sample. In this case, the weight becomes

$$w_t^s \propto w_{t-1}^s \frac{p(\mathbf{y}_t|\mathbf{z}_t^s)p(\mathbf{z}_t^s|\mathbf{z}_{t-1}^s)}{q(\mathbf{z}_t^s|\mathbf{z}_{t-1}^s, \mathbf{y}_t)} \tag{23.45}$$

Hence we can approximate the posterior filtered density using

$$p(\mathbf{z}_t|\mathbf{y}_{1:t}) \approx \sum_{s=1}^{S} \hat{w}_t^s \delta_{\mathbf{z}_t^s}(\mathbf{z}_t) \tag{23.46}$$

As $S \to \infty$, one can show that this approaches the true posterior (Crisan et al. 1999).

The basic algorithm is now very simple: for each old sample s, propose an extension using $\mathbf{z}_t^s \sim q(\mathbf{z}_t|\mathbf{z}_{t-1}^s, \mathbf{y}_t)$, and give this new particle weight w_t^s using Equation 23.45. Unfortunately, this basic algorithm does not work very well, as we discuss below.

23.5.2 The degeneracy problem

The basic sequential importance sampling algorithm fails after a few steps because most of the particles will have negligible weight. This is called the **degeneracy problem**, and occurs because we are sampling in a high-dimensional space (in fact, the space is growing in size over time), using a myopic proposal distribution.

We can quantify the degree of degeneracy using the **effective sample size**, defined by

$$S_{\text{eff}} \triangleq \frac{S}{1 + \text{var}\left[w_t^{*s}\right]} \tag{23.47}$$

where $w_t^{*s} = p(\mathbf{z}_t^s|\mathbf{y}_{1:t})/q(\mathbf{z}_t^s|\mathbf{z}_{t-1}^s, \mathbf{y}_t)$ is the "true weight" of particle s. This quantity cannot be computed exactly, since we don't know the true posterior, but we can approximate it using

$$\hat{S}_{\text{eff}} = \frac{1}{\sum_{s=1}^{S}(w_t^s)^2} \tag{23.48}$$

If the variance of the weights is large, then we are wasting our resources updating particles with low weight, which do not contribute much to our posterior estimate.

There are two main solutions to the degeneracy problem: adding a resampling step, and using a good proposal distribution. We discuss both of these in turn.

23.5.3 The resampling step

The main improvement to the basic SIS algorithm is to monitor the effective sampling size, and whenever it drops below a threshold, to eliminate particles with low weight, and then to create replicates of the surviving particles. (Hence PF is sometimes called **survival of the fittest** (Kanazawa et al. 1995).) In particular, we generate a new set $\{\mathbf{z}_t^{s*}\}_{s=1}^{S}$ by sampling with replacement S times from the weighted distribution

$$p(\mathbf{z}_t|\mathbf{y}_{1:t}) \approx \sum_{s=1}^{S} \hat{w}_t^s \delta_{\mathbf{z}_t^s}(\mathbf{z}_t) \tag{23.49}$$

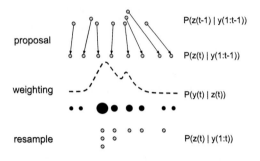

Figure 23.4 Illustration of particle filtering.

where the probability of choosing particle j for replication is w_t^j. (This is sometimes called **rejuvenation**.) The result is an iid *unweighted* sample from the discrete density Equation 23.49, so we set the new weights to $w_t^s = 1/S$. This scheme is illustrated in Figure 23.4.

There are a variety of algorithms for peforming the resampling step. The simplest is **multinomial resampling**, which computes

$$(K_1, \ldots, K_S) \sim \text{Mu}(S, (w_t^1, \ldots, w_t^S)) \tag{23.50}$$

We then make K_s copies of $\mathbf{z}_t^s$. Various improvements exist, such as **systematic resampling** **residual resampling**, and **stratified sampling**, which can reduce the variance of the weights. All these methods take $O(S)$ time. See (Doucet et al. 2001) for details.

The overall particle filtering algorithm is summarized in Algorithm 23.1. (Note that if an estimate of the state is required, it should be computed before the resampling step, since this will result in lower variance.)

Algorithm 23.1: One step of a generic particle filter

1 **for** $s = 1 : S$ **do**
2   Draw $\mathbf{z}_t^s \sim q(\mathbf{z}_t | \mathbf{z}_{t-1}^s, \mathbf{y}_t)$;
3   Compute weight $w_t^s \propto w_{t-1}^s \frac{p(\mathbf{y}_t | \mathbf{z}_t^s) p(\mathbf{z}_t^s | \mathbf{z}_{t-1}^s)}{q(\mathbf{z}_t^s | \mathbf{z}_{t-1}^s, \mathbf{y}_t)}$;
4 Normalize weights: $w_t^s = \frac{w_t^s}{\sum_{s'} w_t^{s'}}$;
5 Compute $\hat{S}_{\text{eff}} = \frac{1}{\sum_{s=1}^S (w_t^s)^2}$;
6 **if** $\hat{S}_{\text{eff}} < S_{min}$ **then**
7   Resample S indices $\boldsymbol{\pi} \sim \mathbf{w}_t$;
8      $\mathbf{z}_t^{\cdot} = \mathbf{z}_t^{\boldsymbol{\pi}}$;
9      $w_t^s = 1/S$;

Although the resampling step helps with the degeneracy problem, it introduces problems of its own. In particular, since the particles with high weight will be selected many times, there is a loss of diversity amongst the population. This is known as **sample impoverishment**. In the extreme case of no process noise (e.g., if we have static but unknown parameters as part of the state space), then all the particles will collapse to a single point within a few iterations.

To mitigate this problem, several solutions have been proposed. (1) Only resample when necessary, not at every time step. (The original **bootstrap filter** (Gordon 1993) resampled at every step, but this is suboptimal.) (2) After replicating old particles, sample new values using an MCMC step which leaves the posterior distribution invariant (see e.g., the **resample-move** algorithm in (Gilks and Berzuini 2001)). (3) Create a kernel density estimate on top of the particles,

$$p(\mathbf{z}_t|\mathbf{y}_{1:t}) \approx \sum_{s=1}^{S} w_t^s \kappa(\mathbf{z}_t - \mathbf{z}_t^s) \tag{23.51}$$

where κ is some smoothing kernel. We then sample from this smoothed distribution. This is known as a **regularized particle filter** (Musso et al. 2001). (4) When performing inference on static parameters, add some artificial process noise. (If this is undesirable, other algorithms must be used for online parameter estimation, e.g., (Andrieu et al. 2005)).

23.5.4 The proposal distribution

The simplest and most widely used proposal distribution is to sample from the prior:

$$q(\mathbf{z}_t|\mathbf{z}_{t-1}^s, \mathbf{y}_t) = p(\mathbf{z}_t|\mathbf{z}_{t-1}^s) \tag{23.52}$$

In this case, the weight update simplifies to

$$w_t^s \propto w_{t-1}^s p(\mathbf{y}_t|\mathbf{z}_t^s) \tag{23.53}$$

This can be thought of a "generate and test" approach: we sample values from the dynamic model, and then evaluate how good they are after we see the data (see Figure 23.4). This is the approach used in the **condensation** algorithm (which stands for "conditional density propagation") used for visual tracking (Isard and Blake 1998). However, if the likelihood is narrower than the dynamical prior (meaning the sensor is more informative than the motion model, which is often the case), this is a very inefficient approach, since most particles will be assigned very low weight.

It is much better to actually look at the data $\mathbf{y}_t$ when generating a proposal. In fact, the optimal proposal distribution has the following form:

$$q(\mathbf{z}_t|\mathbf{z}_{t-1}^s, \mathbf{y}_t) = p(\mathbf{z}_t|\mathbf{z}_{t-1}^s, \mathbf{y}_t) = \frac{p(\mathbf{y}_t|\mathbf{z}_t)p(\mathbf{z}_t|\mathbf{z}_{t-1}^s)}{p(\mathbf{y}_t|\mathbf{z}_{t-1}^s)} \tag{23.54}$$

If we use this proposal, the new weight is given by

$$w_t^s \propto w_{t-1}^s p(\mathbf{y}_t|\mathbf{z}_{t-1}^s) = w_{t-1}^s \int p(\mathbf{y}_t|\mathbf{z}_t')p(\mathbf{z}_t'|\mathbf{z}_{t-1}^s)d\mathbf{z}_t' \tag{23.55}$$

This proposal is optimal since, for any given $\mathbf{z}_{t-1}^s$, the new weight w_t^s takes the same value regardless of the value drawn for $\mathbf{z}_t^s$. Hence, conditional on the old values $\mathbf{z}_{t-1}$, the variance of true weights var $[w_t^{*s}]$, is zero.

In general, it is intractable to sample from $p(\mathbf{z}_t|\mathbf{z}_{t-1}^s, \mathbf{y}_t)$ and to evaluate the integral needed to compute the predictive density $p(\mathbf{y}_t|\mathbf{z}_{t-1}^s)$. However, there are two cases when the optimal proposal distribution can be used. The first setting is when $\mathbf{z}_t$ is discrete, so the integral becomes a sum. Of course, if the entire state space is discrete, we can use an HMM filter instead, but in some cases, some parts of the state are discrete, and some continuous. The second setting is when $p(\mathbf{z}_t|\mathbf{z}_{t-1}^s, \mathbf{y}_t)$ is Gaussian. This occurs when the dynamics are nonlinear but the observations are linear. See Exercise 23.3 for the details.

In cases where the model is not linear-Gaussian, we may still compute a Gaussian approximation to $p(\mathbf{z}_t|\mathbf{z}_{t-1}^s, \mathbf{y}_t)$ using the unscented transform (Section 18.5.2) and use this as a proposal. This is known as the **unscented particle filter** (van der Merwe et al. 2000). In more general settings, we can use other kinds of **data-driven proposals**, perhaps based on discriminative models. Unlike MCMC, we do not need to worry about the proposals being reversible.

23.5.5 Application: robot localization

Consider a mobile robot wandering around an office environment. We will assume that it already has a map of the world, represented in the form of an **occupancy grid**, which just specifies whether each grid cell is empty space or occupied by an something solid like a wall. The goal is for the robot to estimate its location. This can be solved optimally using an HMM filter, since we are assuming the state space is discrete. However, since the number of states, K, is often very large, the $O(K^2)$ time complexity per update is prohibitive. We can use a particle filter as a sparse approximation to the belief state. This is known as **Monte Carlo localization**, and is described in detail in (Thrun et al. 2006).

Figure 23.5 gives an example of the method in action. The robot uses a sonar range finder, so it can only sense distance to obstacles. It starts out with a uniform prior, reflecting the fact that the owner of the robot may have turned it on in an arbitrary location. (Figuring out where you are, starting from a uniform prior, is called **global localization**.) After the first scan, which indicates two walls on either side, the belief state is shown in (b). The posterior is still fairly broad, since the robot could be in any location where the walls are fairly close by, such as a corridor or any of the narrow rooms. After moving to location 2, the robot is pretty sure it must be in the corridor, as shown in (c). After moving to location 3, the sensor is able to detect the end of the corridor. However, due to symmetry, it is not sure if it is in location I (the true location) or location II. (This is an example of **perceptual aliasing**, which refers to the fact that different things may look the same.) After moving to locations 4 and 5, it is finally able to figure out precisely where it is. The whole process is analogous to someone getting lost in an office building, and wandering the corridors until they see a sign they recognize.

In Section 23.6.3, we discuss how to estimate location and the map at the same time.

23.5.6 Application: visual object tracking

Our next example is concerned with tracking an object (in this case, a remote-controlled helicopter) in a video sequence. The method uses a simple linear motion model for the centroid of the object, and a color histogram for the likelihood model, using **Bhattacharya distance** to compare histograms. The proposal distribution is obtained by sampling from the likelihood. See (Nummiaro et al. 2003) for further details.

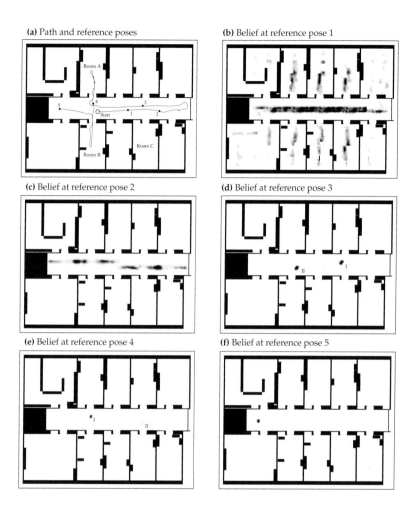

Figure 23.5 Illustration of Monte Carlo localization. Source: Figure 8.7 of (Thrun et al. 2006). Used with kind permission of Sebastian Thrun.

Figure 23.6 shows some example frames. The system uses $S = 250$ particles, with an effective sample size of $\hat{S}_{\mathrm{eff}} = 134$. (a) shows the belief state at frame 1. The system has had to resample 5 times to keep the effective sample size above the threshold of 150; (b) shows the belief state at frame 251; the red lines show the estimated location of the center of the object over the last 250 frames. (c) shows that the system can handle visual clutter, as long as it does not have the same color as the target object. (d) shows that the system is confused between the grey of the helicopter and the grey of the building. The posterior is bimodal. The green ellipse, representing the posterior mean and covariance, is in between the two modes. (e) shows that the probability mass has shifted to the wrong mode: the system has lost track. (f) shows the particles spread out over the gray building; recovery of the object is very unlikely from this state using this

Figure 23.6 Example of particle filtering applied to visual object tracking, based on color histograms. (a-c) succesful tracking: green ellipse is on top of the helicopter. (d-f): tracker gets distracted by gray clutter in the background. See text for details. Figure generated by `pfColorTrackerDemo`, written by Sebastien Paris.

proposal.

We see that the method is able to keep track for a fairly long time, despite the presence of clutter. However, eventually it loses track of the object. Note that since the algorithm is stochastic, simply re-running the demo may fix the problem. But in the real world, this is not an option. The simplest way to improve performance is to use more particles. An alternative is to perform **tracking by detection**, by running an object detector over the image every few frames. See (Forsyth and Ponce 2002; Szeliski 2010; Prince 2012) for details.

23.5.7 Application: time series forecasting

In Section 18.2.4, we discussed how to use the Kalman filter to perform time series forecasting. This assumes that the model is a linear-Gaussian state-space model. There are many models which are either non-linear and/or non-Gaussian. For example, **stochastic volatility** models, which are widely used in finance, assume that the variance of the system and/or observation noise changes over time. Particle filtering is widely used in such settings. See e.g., (Doucet et al. 2001) and references therein for details.

23.6 Rao-Blackwellised particle filtering (RBPF)

In some models, we can partition the hidden variables into two kinds, $\mathbf{q}_t$ and $\mathbf{z}_t$, such that we can analytically integrate out $\mathbf{z}_t$ provided we know the values of $\mathbf{q}_{1:t}$. This means we only have to sample $\mathbf{q}_{1:t}$, and can represent $p(\mathbf{z}_t|\mathbf{q}_{1:t})$ parametrically. Thus each particle s represents a value for $\mathbf{q}_{1:t}^s$ and a distribution of the form $p(\mathbf{z}_t|\mathbf{y}_{1:t}, \mathbf{q}_{1:t}^s)$. These hybrid particles are are sometimes called **distributional particles** or **collapsed particles** (Koller and Friedman 2009, Sec 12.4).

The advantage of this approach is that we reduce the dimensionality of the space in which we are sampling, which reduces the variance of our estimate. Hence this technique is known as **Rao-Blackwellised particle filtering** or **RBPF** for short, named after Theorem 24.20. The method is best explained using a specific example.

23.6.1 RBPF for switching LG-SSMs

A canonical example for which RBPF can be applied is the switching linear dynamical system (SLDS) model discussed in Section 18.6 (Chen and Liu 2000; Doucet et al. 2001). We can represent $p(\mathbf{z}_t|\mathbf{y}_{1:t}, \mathbf{q}_{1:t}^s)$ using a mean and covariance matrix for each particle s, where $q_t \in \{1, \dots, K\}$.

If we propose from the prior, $q(q_t = k|q_{t-1}^s)$, the weight update becomes

$$w_t^s \propto w_{t-1}^s p(\mathbf{y}_t|q_t = k, \mathbf{q}_{1:t-1}^s, \mathbf{y}_{1:t-1}) = w_{t-1}^s L_{t,k}^s \tag{23.56}$$

where

$$L_{tk}^s = \int p(\mathbf{y}_t|q_t = k, \mathbf{z}_t, \mathbf{y}_{1:t-1}, \mathbf{q}_{1:t-1}^s) p(\mathbf{z}_t|q_t = k, \mathbf{y}_{1:t-1}\mathbf{q}_{1:t-1}^s,) d\mathbf{z}_t \tag{23.57}$$

The quantity L_{tk}^s is the predictive density for the new observation $\mathbf{y}_t$ conditioned on $q_t = k$ and the history $\mathbf{q}_{1:t-1}^s$. In the case of SLDS models, this can be computed using the normalization constant of the Kalman filter, Equation 18.41.

We give some pseudo-code in Algorithm 23.2. (The step marked "KFupdate" refers to the Kalman filter update equations in Section 18.3.1.) This is known as a **mixture of Kalman filters**.

If K is small, we can compute the optimal proposal distribution, which is

$$p(q_t = k|\mathbf{y}_{1:t}, \mathbf{q}_{1:t-1}^s) = \hat{p}_{t-1}^s(q_t = k|\mathbf{y}_t) \tag{23.58}$$

$$= \frac{\hat{p}_{t-1}^s(\mathbf{y}_t|q_t = k)\hat{p}_{t-1}^s(q_t = k)}{\hat{p}_{t-1}^s(\mathbf{y}_t)} \tag{23.59}$$

$$= \frac{L_{tk}^s p(q_t = k|q_{t-1}^s)}{\sum_{k'} L_{tk'}^s p(q_t = k'|q_{t-1}^s)} \tag{23.60}$$

Algorithm 23.2: One step of RBPF for SLDS using prior as proposal

---

1 **for** $s = 1 : S$ **do**
2      $k \sim p(q_t | q_{t-1}^s)$;
3      $q_t^s := k$;
4      $(\boldsymbol{\mu}_t^s, \boldsymbol{\Sigma}_t^s, L_{tk}^s) = \text{KFupdate}(\boldsymbol{\mu}_{t-1}^s, \boldsymbol{\Sigma}_{t-1}^s, \mathbf{y}_t, \boldsymbol{\theta}_k)$;
5      $w_t^s = w_{t-1}^s L_{ts}^k$;
6 Normalize weights: $w_t^s = \frac{w_t^s}{\sum_{s'} w_t^{s'}}$;
7 Compute $\hat{S}_{\text{eff}} = \frac{1}{\sum_{s=1}^S (w_t^s)^2}$;
8 **if** $\hat{S}_{\text{eff}} < S_{min}$ **then**
9   Resample S indices $\boldsymbol{\pi} \sim \mathbf{w}_t$;
10      $\mathbf{q}_t^{\cdot} = \mathbf{q}_t^{\boldsymbol{\pi}}, \boldsymbol{\mu}_t^{\cdot} = \boldsymbol{\mu}_t^{\boldsymbol{\pi}}, \boldsymbol{\Sigma}_t^{\cdot} = \boldsymbol{\Sigma}_t^{\boldsymbol{\pi}}, $;
11      $w_t^s = 1/S$;

---

where we use the following shorthand:

$$\hat{p}_{t-1}^s(\cdot) = p(\cdot | \mathbf{y}_{1:t-1}, \mathbf{q}_{1:t-1}^s) \tag{23.61}$$

We then sample from $p(q_t | \mathbf{q}_{1:t-1}^s, \mathbf{y}_{1:t})$ and give the resulting particle weight

$$w_t^s \propto w_{t-1}^s p(\mathbf{y}_t | \mathbf{q}_{1:t-1}^s, \mathbf{y}_{1:t-1}) = w_{t-1}^s \sum_k \left[L_{tk}^s p(q_t = k | q_{t-1}^s) \right] \tag{23.62}$$

Since the weights of the particles in Equation 23.62 are independent of the new value that is actually sampled for q_t, we can compute these weights first, and use them to decide which particles to propagate. That is, we choose the fittest particles at time $t - 1$ using information from time t. This is called **look-ahead RBPF** (de Freitas et al. 2004).

In more detail, the idea is this. We pass each sample in the prior through all K models to get K posteriors, one per sample. The normalization constants of this process allow us to compute the optimal weights in Equation 23.62. We then resample S indices. Finally, for each old particle s that is chosen, we sample one new state $q_t^s = k$, and use the corresponding posterior from the K possible alternative that we have already computed. The pseudo-code is shown in Algorithm 23.3. This method needs $O(KS)$ storage, but has the advantage that each particle is chosen using the latest information, $\mathbf{y}_t$.

A further improvement can be obtained by exploiting the fact that the state space is discrete. Hence we can use the resampling method of (Fearnhead 2004) which avoids duplicating particles.

23.6.2 Application: tracking a maneuvering target

One application of SLDS is to track moving objects that have piecewise linear dynamics. For example, suppose we want to track an airplane or missile; q_t can specify if the object is flying normally or is taking evasive action. This is called **maneuvering target tracking**.

Figure 23.7 gives an example of an object moving in 2d. The setup is essentially the same as in Section 18.2.1, except that we add a three-state discrete Markov chain which controls the

---

Algorithm 23.3: One step of look-ahead RBPF for SLDS using optimal proposal

---

1 **for** $s = 1 : S$ **do**
2   **for** $k = 1 : K$ **do**
3      $(\boldsymbol{\mu}_{tk}^s, \boldsymbol{\Sigma}_{tk}^s, L_{ts}^k) = \text{KFupdate}(\boldsymbol{\mu}_{t-1}^s, \boldsymbol{\Sigma}_{t-1}^s, \mathbf{y}_t, \boldsymbol{\theta}_k);$
4    $w_t^s = w_{t-1}^s [\sum_k L_{ts}^k p(q_t = k | q_{t-1}^s)];$
5 Normalize weights: $w_t^s = \frac{w_t^s}{\sum_{s'} w_t^{s'}};$
6 Resample S indices $\boldsymbol{\pi} \sim \mathbf{w}_t;$
7 **for** $s \in \boldsymbol{\pi}$ **do**
8   Compute optimal proposal $p(k | \mathbf{q}_{1:t-1}^s, \mathbf{y}_{1:t}) = \frac{L_{tk}^s p(q_t = k | q_{t-1}^s)}{\sum_{k'} L_{tk}^s p(q_t = k | q_{t-1}^s)};$
9   Sample $k \sim p(k | \mathbf{q}_{1:t-1}^s, \mathbf{y}_{1:t});$
10    $q_t^s = k, \boldsymbol{\mu}_t^s = \boldsymbol{\mu}_{tk}^s, \boldsymbol{\Sigma}_t^s = \boldsymbol{\Sigma}_{tk}^s;$
11    $w_t^s = 1/S;$

Method	misclassification rate	MSE	Time (seconds)
PF	0.440	21.051	6.086
RBPF	0.340	18.168	10.986

Table 23.1 Comparison of PF an RBPF on the maneuvering target problem in Figure 23.7.

input to the system. We define $\mathbf{u}_t = 1$ and set

$$\mathbf{B}_1 = (0, 0, 0, 0)^T, \mathbf{B}_2 = (-1.225, -0.35, 1.225, 0.35)^T, \mathbf{B}_3 = (1.225, 0.35, -1.225, -0.35)^T$$

so the system will turn in different directions depending on the discrete state.

Figure 23.7(a) shows the true state of the system from a sample run, starting at $(0, 0)$: the colored symbols denote the discrete state, and the location of the symbol denotes the (x, y) location. The small dots represent noisy observations. Figure 23.7(b) shows the estimate of the state computed using particle filtering with 500 particles, where the proposal is to sample from the prior. The colored symbols denote the MAP estimate of the state, and the location of the symbol denotes the MMSE (minimum mean square error) estimate of the location, which is given by the posterior mean. Figure 23.7(c) shows the estimate computing using RBPF with 500 particles, using the optimal proposal distribution. A more quantitative comparison is shown in Table 23.1. We see that RBPF has slightly better performance, although it is also slightly slower.

Figure 23.8 visualizes the belief state of the system. In (a) we show the distribution over the discrete states. We see that the particle filter estimate of the belief state (second column) is not as accurate as the RBPF estimate (third column) in the beginning, although after the first few observations performance is similar for both methods. In (b), we plot the posterior over the x locations. For simplicity, we use the PF estimate, which is a set of weighted samples, but we could also have used the RBPF estimate, which is a set of weighted Gaussians.

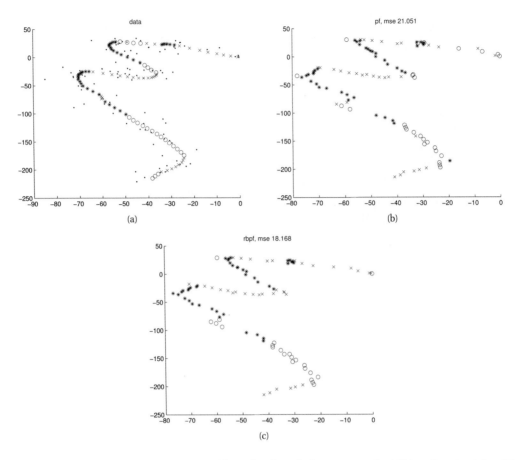

Figure 23.7 (a) A maneuvering target. The colored symbols represent the hidden discrete state. (b) Particle filter estimate. (c) RBPF estimate. Figure generated by `rbpfManeuverDemo`, based on code by Nando de Freitas.

23.6.3 Application: Fast SLAM

In Section 18.2.2, we introduced the problem of simultaneous localization and mapping or SLAM for mobile robotics. The main problem with the Kalman filter implementation is that it is cubic in the number of landmarks. However, by looking at the DGM in Figure 18.2, we see that, conditional on knowing the robot's path, $\mathbf{q}_{1:t}$, where $\mathbf{q}_t \in \mathbb{R}^2$, the landmark locations $\mathbf{z} \in \mathbb{R}^{2L}$ are independent. (We assume the landmarks don't move, so we drop the t subscript). That is, $p(\mathbf{z}|\mathbf{q}_{1:t}, \mathbf{y}_{1:t}) = \prod_{l=1}^{L} p(\mathbf{z}_l|\mathbf{q}_{1:t}, \mathbf{y}_{1:t})$. Consequently we can use RBPF, where we sample the robot's trajectory, $\mathbf{q}_{1:t}$, and we run L independent 2d Kalman filters inside each particle. This takes $O(L)$ time per particle. Fortunately, the number of particles needed for good performance is quite small (this partly depends on the control / exploration policy), so the algorithm is essentially linear in the number of particles. This technique has the additional advantage that

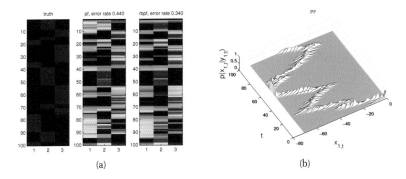

Figure 23.8 Belief states corresponding to Figure 23.7. (a) Discrete state. The system starts in state 2 (red x in Figure 23.7), then moves to state 3 (black * in Figure 23.7), returns briefly to state 2, then switches to state 1 (blue circle in Figure 23.7), etc. (b) Horizontal location (PF estimate). Figure generated by `rbpfManeuverDemo`, based on code by Nando de Freitas.

it is easy to use sampling to handle the data association ambiguity, and that it allows for other representations of the map, such as occupancy grids. This idea was first suggested in (Murphy 2000), and was subsequently extended and made practical in (Thrun et al. 2004), who christened the technique **FastSLAM**. See `rbpfSlamDemo` for a simple demo in a discrete grid world.

Exercises

Exercise 23.1 Sampling from a Cauchy

Show how to use inverse probability transform to sample from a standard Cauchy, $\mathcal{T}(x|0, 1, 1)$.

Exercise 23.2 Rejection sampling from a Gamma using a Cauchy proposal

Show how to use a Cauchy proposal to perform rejection sampling from a Gamma distribution. Derive the optimal constant M, and plot the density and its upper envelope.

Exercise 23.3 Optimal proposal for particle filtering with linear-Gaussian measurement model

Consider a state-space model of the following form:

$$\mathbf{z}_t = f_t(\mathbf{z}_{t-1}) + \mathcal{N}(\mathbf{0}, \mathbf{Q}_{t-1}) \tag{23.63}$$
$$\mathbf{y}_t = \mathbf{H}_t \mathbf{z}_t + \mathcal{N}(\mathbf{0}, \mathbf{R}_t) \tag{23.64}$$

Derive expressions for $p(\mathbf{z}_t|\mathbf{z}_{t-1}, \mathbf{y}_t)$ and $p(\mathbf{y}_t|\mathbf{z}_{t-1})$, which are needed to compute the optimal (minimum variance) proposal distribution. Hint: use Bayes rule for Gaussians.

24 *Markov chain Monte Carlo (MCMC) inference*

24.1 Introduction

In Chapter 23, we introduced some simple Monte Carlo methods, including rejection sampling and importance sampling. The trouble with these methods is that they do not work well in high dimensional spaces. The most popular method for sampling from high-dimensional distributions is **Markov chain Monte Carlo** or **MCMC**. In a survey by *SIAM News*[1], MCMC was placed in the top 10 most important algorithms of the 20th century.

The basic idea behind MCMC is to construct a Markov chain (Section 17.2) on the state space $\mathcal{X}$ whose stationary distribution is the target density $p^*(\mathbf{x})$ of interest (this may be a prior or a posterior). That is, we perform a random walk on the state space, in such a way that the fraction of time we spend in each state $\mathbf{x}$ is proportional to $p^*(\mathbf{x})$. By drawing (correlated!) samples $\mathbf{x}_0, \mathbf{x}_1, \mathbf{x}_2, \ldots$, from the chain, we can perform Monte Carlo integration wrt p^*. We give the details below.

The MCMC algorithm has an interesting history. It was discovered by physicists working on the atomic bomb at Los Alamos during World War II, and was first published in the open literature in (Metropolis et al. 1953) in a chemistry journal. An extension was published in the statistics literature in (Hastings 1970), but was largely unnoticed. A special case (Gibbs sampling, Section 24.2) was independently invented in 1984 in the context of Ising models and was published in (Geman and Geman 1984). But it was not until (Gelfand and Smith 1990) that the algorithm became well-known to the wider statistical community. Since then it has become wildly popular in Bayesian statistics, and is becoming increasingly popular in machine learning.

It is worth briefly comparing MCMC to variational inference (Chapter 21). The advantages of variational inference are (1) for small to medium problems, it is usually faster; (2) it is deterministic; (3) it is easy to determine when to stop; (4) it often provides a lower bound on the log likelihood. The advantages of sampling are: (1) it is often easier to implement; (2) it is applicable to a broader range of models, such as models whose size or structure changes depending on the values of certain variables (e.g., as happens in matching problems), or models without nice conjugate priors; (3) sampling can be faster than variational methods when applied to really huge models or datasets.[2]

---

1. Source: `http://www.siam.org/pdf/news/637.pdf`.
2. The reason is that sampling passes specific values of variables (or sets of variables), whereas in variational inference, we pass around distributions. Thus sampling passes sparse messages, whereas variational inference passes dense messages For comparisons of the two approaches, see e.g., (Yoshida and West 2010) and articles in (Bekkerman et al.

24.2   Gibbs sampling

In this section, we present one of the most popular MCMC algorithms, known as **Gibbs sampling**.[3] (In physics, this method is known as **Glauber dynamics** or the **heat bath** method.) This is the MCMC analog of coordinate descent.

24.2.1   Basic idea

The idea behind Gibbs sampling is that we sample each variable in turn, conditioned on the values of all the other variables in the distribution. That is, given a joint sample $\mathbf{x}^s$ of all the variables, we generate a new sample $\mathbf{x}^{s+1}$ by sampling each component in turn, based on the most recent values of the other variables. For example, if we have $D = 3$ variables, we use

- $x_1^{s+1} \sim p(x_1 | x_2^s, x_3^s)$

- $x_2^{s+1} \sim p(x_2 | x_1^{s+1}, x_3^s)$

- $x_3^{s+1} \sim p(x_3 | x_1^{s+1}, x_2^{s+1})$

This readily generalizes to D variables. If x_i is a visible variable, we do not sample it, since its value is already known.

The expression $p(x_i | \mathbf{x}_{-i})$ is called the **full conditional** for variable i. In general, x_i may only depend on some of the other variables. If we represent $p(\mathbf{x})$ as a graphical model, we can infer the dependencies by looking at i's Markov blanket, which are its neighbors in the graph. Thus to sample x_i, we only need to know the values of i's neighbors. In this sense, Gibbs sampling is a distributed algorithm. However, it is not a parallel algorithm, since the samples must be generated sequentially.

For reasons that we will explain in Section 24.4.1, it is necessary to discard some of the initial samples until the Markov chain has **burned in**, or entered its stationary distribution. We discuss how to estimate when burnin has occured in Section 24.4.1. In the examples below, we just discard the initial 25% of the samples, for simplicity.

24.2.2   Example: Gibbs sampling for the Ising model

In Section 21.3.2, we applied mean field to an Ising model. Here we apply Gibbs sampling. Gibbs sampling in pairwise MRF/CRF takes the form

$$p(x_t | \mathbf{x}_{-t}, \boldsymbol{\theta}) \propto \prod_{s \in \mathrm{nbr}(t)} \psi_{st}(x_s, x_t) \tag{24.1}$$

In the case of an Ising model with edge potentials $\psi(x_s, x_t) = \exp(Jx_sx_t)$, where $x_t \in$

2011)

3. Josiah Willard Gibbs, 1839-1903, was an American physicist.

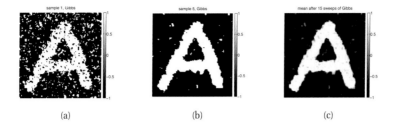

Figure 24.1 Example of image denoising. We use an Ising prior with $W_{ij} = J = 1$ and a Gaussian noise model with $\sigma = 2$. We use Gibbs sampling (Section 24.2) to perform approximate inference. (a) Sample from the posterior after one sweep over the image. (b) Sample after 5 sweeps. (c) Posterior mean, computed by averaging over 15 sweeps. Compare to Figure 21.3 which shows the results of using mean field inference. Figure generated by isingImageDenoiseDemo.

$\{-1, +1\}$, the full conditional becomes

$$p(x_t = +1 | \mathbf{x}_{-t}, \boldsymbol{\theta}) = \frac{\prod_{s \in \mathrm{nbr}(t)} \psi_{st}(x_t = +1, x_s)}{\prod_{s \in \mathrm{nbr}(t)} \psi(s_t = +1, x_s) + \prod_{s \in \mathrm{nbr}(t)} \psi(x_t = -1, x_s)} \qquad (24.2)$$

$$= \frac{\exp[J \sum_{s \in \mathrm{nbr}(t)} x_s]}{\exp[J \sum_{s \in \mathrm{nbr}(t)} x_s] + \exp[-J \sum_{s \in \mathrm{nbr}(t)} x_s]} \qquad (24.3)$$

$$= \frac{\exp[J \eta_t]}{\exp[J \eta_t] + \exp[-J \eta_t]} = \mathrm{sigm}(2 J \eta_t) \qquad (24.4)$$

where J is the coupling strength, $\eta_t \triangleq \sum_{s \in \mathrm{nbr}(t)} x_t$ and $\mathrm{sigm}(u) = 1/(1 + e^{-u})$ is the sigmoid function. It is easy to see that $\eta_t = x_t(a_t - d_t)$, where a_t is the number of neighbors that agree with (have the same sign as) t, and d_t is the number of neighbors who disagree. If this number is equal, the "forces" on x_t cancel out, so the full conditional is uniform.

We can combine an Ising prior with a local evidence term ψ_t. For example, with a Gaussian observation model, we have $\psi_t(x_t) = \mathcal{N}(y_t | x_t, \sigma^2)$. The full conditional becomes

$$p(x_t = +1 | \mathbf{x}_{-t}, \mathbf{y}, \boldsymbol{\theta}) = \frac{\exp[J \eta_t] \psi_t(+1)}{\exp[J \eta_t] \psi_t(+1) + \exp[-J \eta_t] \psi_t(-1)} \qquad (24.5)$$

$$= \mathrm{sigm}\left(2 J \eta_t - \log \frac{\psi_t(+1)}{\psi_t(-1)}\right) \qquad (24.6)$$

Now the probability of x_t entering each state is determined both by compatibility with its neighbors (the Ising prior) and compatibility with the data (the local likelihood term).

See Figure 24.1 for an example of this algorithm applied to a simple image denoising problem. The results are similar to mean field (Figure 21.3) except that the final estimate (based on averaging the samples) is somewhat "blurrier", due to the fact that mean field tends to be over-confident.

24.2.3   Example: Gibbs sampling for inferring the parameters of a GMM

It is straightforward to derive a Gibbs sampling algorithm to "fit" a mixture model, especially if we use conjugate priors. We will focus on the case of mixture of Gaussians, although the results are easily extended to other kinds of mixture models. (The derivation, which follows from the results of Section 4.6, is much easier than the corresponding variational Bayes algorithm in Section 21.6.1.)

Suppose we use a semi-conjugate prior. Then the full joint distribution is given by

$$p(\mathbf{x}, \mathbf{z}, \boldsymbol{\mu}, \boldsymbol{\Sigma}, \boldsymbol{\pi}) \quad = \quad p(\mathbf{x}|\mathbf{z}, \boldsymbol{\mu}, \boldsymbol{\Sigma})p(\mathbf{z}|\boldsymbol{\pi})p(\boldsymbol{\pi}) \prod_{k=1}^{K} p(\boldsymbol{\mu}_k)p(\boldsymbol{\Sigma}_k) \tag{24.7}$$

$$= \quad \left(\prod_{i=1}^{N} \prod_{k=1}^{K} (\pi_k \mathcal{N}(\mathbf{x}_i|\boldsymbol{\mu}_k, \boldsymbol{\Sigma}_k))^{\mathbb{I}(z_i=k)} \right) \times \tag{24.8}$$

$$\mathrm{Dir}(\boldsymbol{\pi}|\boldsymbol{\alpha}) \prod_{k=1}^{K} \mathcal{N}(\boldsymbol{\mu}_k|\mathbf{m}_0, \mathbf{V}_0)\mathrm{IW}(\boldsymbol{\Sigma}_k|\mathbf{S}_0, \nu_0) \tag{24.9}$$

We use the same prior for each mixture component. The full conditionals are as follows. For the discrete indicators, we have

$$p(z_i = k|\mathbf{x}_i, \boldsymbol{\mu}, \boldsymbol{\Sigma}, \boldsymbol{\pi}) \quad \propto \quad \pi_k \mathcal{N}(\mathbf{x}_i|\boldsymbol{\mu}_k, \boldsymbol{\Sigma}_k) \tag{24.10}$$

For the mixing weights, we have (using results from Section 3.4)

$$p(\boldsymbol{\pi}|\mathbf{z}) \quad = \quad \mathrm{Dir}(\{\alpha_k + \sum_{i=1}^{N} \mathbb{I}(z_i = k)\}_{k=1}^{K}) \tag{24.11}$$

For the means, we have (using results from Section 4.6.1)

$$p(\boldsymbol{\mu}_k|\boldsymbol{\Sigma}_k, \mathbf{z}, \mathbf{x}) \quad = \quad \mathcal{N}(\boldsymbol{\mu}_k|\mathbf{m}_k, \mathbf{V}_k) \tag{24.12}$$

$$\mathbf{V}_k^{-1} \quad = \quad \mathbf{V}_0^{-1} + N_k \boldsymbol{\Sigma}_k^{-1} \tag{24.13}$$

$$\mathbf{m}_k \quad = \quad \mathbf{V}_k(\boldsymbol{\Sigma}_k^{-1} N_k \overline{\mathbf{x}}_k + \mathbf{V}_0^{-1}\mathbf{m}_0) \tag{24.14}$$

$$N_k \quad \triangleq \quad \sum_{i=1}^{N} \mathbb{I}(z_i = k) \tag{24.15}$$

$$\overline{\mathbf{x}}_k \quad \triangleq \quad \frac{\sum_{i=1}^{N} \mathbb{I}(z_i = k)\mathbf{x}_i}{N_k} \tag{24.16}$$

For the covariances, we have (using results from Section 4.6.2)

$$p(\boldsymbol{\Sigma}_k|\boldsymbol{\mu}_k, \mathbf{z}, \mathbf{x}) \quad = \quad \mathrm{IW}(\boldsymbol{\Sigma}_k|\mathbf{S}_k, \nu_k) \tag{24.17}$$

$$\mathbf{S}_k \quad = \quad \mathbf{S}_0 + \sum_{i=1}^{N} \mathbb{I}(z_i = k)(\mathbf{x}_i - \boldsymbol{\mu}_k)(\mathbf{x}_i - \boldsymbol{\mu}_k)^{T} \tag{24.18}$$

$$\nu_k \quad = \quad \nu_0 + N_k \tag{24.19}$$

See gaussMissingFitGibbs for some Matlab code. (This code can also sample missing values for $\mathbf{x}$, if necessary.)

24.2.3.1 Label switching

Although it is simple to implement, Gibbs sampling for mixture models has a fundamental weakness. The problem is that the parameters of the model $\boldsymbol{\theta}$, and the indicator functions $\mathbf{z}$, are unidentifiable, since we can arbitrarily permute the hidden labels without affecting the likelihood (see Section 11.3.1). Consequently, we cannot just take a Monte Carlo average of the samples to compute posterior means, since what one sample considers the parameters for cluster 1 may be what another sample considers the parameters for cluster 2. Indeed, if we could average over all modes, we would find $\mathbb{E}\left[\boldsymbol{\mu}_k|\mathcal{D}\right]$ is the same for all k (assuming a symmetric prior). This is called the **label switching** problem.

This problem does not arise in EM or VBEM, which just "lock on" to a single mode. However, it arises in any method that visits multiple modes. In 1d problems, one can try to prevent this problem by introducing constraints on the parameters to ensure identifiability, e.g., $\mu_1 < \mu_2 < \mu_3$ (Richardson and Green 1997). However, this does not always work, since the likelihood might overwhelm the prior and cause label switching anyway. Furthermore, this technique does not scale to higher dimensions. Another approach is to post-process the samples by searching for a global label permutation to apply to each sample that minimizes some loss function (Stephens 2000); however, this can be slow.

Perhaps the best solution is simply to "not ask" questions that cannot be uniquely identified. For example, instead of asking for the probability that data point i belongs to cluster k, ask for the probability that data points i and j belong to the same cluster. The latter question is invariant to the labeling. Furthermore, it only refers to observable quantities (are i and j grouped together or not), rather than referring to unobservable quantities, such as latent clusters. This approach has the further advantage that it extends to infinite mixture models, discussed in Section 25.2, where K is unbounded; in such models, the notion of a hidden cluster is not well defined, but the notion of a partitioning of the data *is* well defined

24.2.4 Collapsed Gibbs sampling *

In some cases, we can analytically integrate out some of the unknown quantities, and just sample the rest. This is called a **collapsed Gibbs sampler**, and it tends to be much more efficient, since it is sampling in a lower dimensional space.

More precisely, suppose we sample $\mathbf{z}$ and integrate out $\boldsymbol{\theta}$. Thus the $\boldsymbol{\theta}$ parameters do not participate in the Markov chain; consequently we can draw conditionally independent samples $\boldsymbol{\theta}^s \sim p(\boldsymbol{\theta}|\mathbf{z}^s, \mathcal{D})$, which will have much lower variance than samples drawn from the joint state space (Liu et al. 1994). This process is called **Rao-Blackwellisation**, named after the following theorem:

Theorem 24.2.1 (Rao-Blackwell). *Let $\mathbf{z}$ and $\boldsymbol{\theta}$ be dependent random variables, and $f(\mathbf{z}, \boldsymbol{\theta})$ be some scalar function. Then*

$$\text{var}_{\mathbf{z},\boldsymbol{\theta}}\left[f(\mathbf{z}, \boldsymbol{\theta})\right] \geq \text{var}_{\mathbf{z}}\left[\mathbb{E}_{\boldsymbol{\theta}}\left[f(\mathbf{z}, \boldsymbol{\theta})|\mathbf{z}\right]\right] \tag{24.20}$$

This theorem guarantees that the variance of the estimate created by analytically integrating out $\boldsymbol{\theta}$ will always be lower (or rather, will never be higher) than the variance of a direct MC estimate. In collapsed Gibbs, we sample $\mathbf{z}$ with $\boldsymbol{\theta}$ integrated out; the above Rao-Blackwell theorem still applies in this case (Liu et al. 1994).

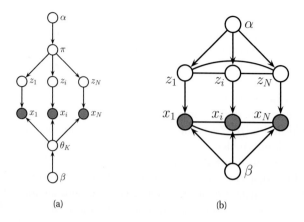

Figure 24.2 (a) A mixture model. (b) After integrating out the parameters.

We discuss Rao-Blackwellisation in Section 23.6. Although it can reduce statistical variance, it is only worth doing if the integrating out can be done quickly, otherwise we will not be able to produce as many samples per second as the naive method. We give an example of this below.

24.2.4.1 Example: collapsed Gibbs for fitting a GMM

Consider a GMM with a fully conjugate prior. In this case we can analytically integrate out the model parameters $\boldsymbol{\mu}_k$, $\boldsymbol{\Sigma}_k$ and $\boldsymbol{\pi}$, and just sample the indicators $\mathbf{z}$. Once we integrate out $\boldsymbol{\pi}$, all the z_i nodes become inter-dependent. Similarly, once we integrate out $\boldsymbol{\theta}_k = (\boldsymbol{\mu}_k, \boldsymbol{\Sigma}_k)$, all the $\mathbf{x}_i$ nodes become inter-dependent, as shown in Figure 24.2(b). Nevertheless, we can easily compute the full conditionals as follows:

$$
\begin{aligned}
p(z_i = k|\mathbf{z}_{-i}, \mathbf{x}, \boldsymbol{\alpha}, \boldsymbol{\beta}) &\propto p(z_i = k|\mathbf{z}_{-i}, \boldsymbol{\alpha}, \boldsymbol{\beta})p(\mathbf{x}|z_i = k, \mathbf{z}_{-i}, \boldsymbol{\alpha}, \boldsymbol{\beta}) & (24.21) \\
&\propto p(z_i = k|\mathbf{z}_{-i}, \boldsymbol{\alpha})p(\mathbf{x}_i|\mathbf{x}_{-i}, z_i = k, \mathbf{z}_{-i}, \boldsymbol{\beta}) \\
&\quad p(\mathbf{x}_{-i}|z_i = k, \mathbf{z}_{-i}, \boldsymbol{\beta}) & (24.22) \\
&\propto p(z_i = k|\mathbf{z}_{-i}, \boldsymbol{\alpha})p(\mathbf{x}_i|\mathbf{x}_{-i}, z_i = k, \mathbf{z}_{-i}, \boldsymbol{\beta}) & (24.23)
\end{aligned}
$$

where $\boldsymbol{\beta} = (\mathbf{m}_0, \mathbf{V}_0, \mathbf{S}_0, \nu_0)$ are the hyper-parameters for the class-conditional densities. The first term can be obtained by integrating out $\boldsymbol{\pi}$. Suppose we use a symmetric prior of the form $\boldsymbol{\pi} \sim \text{Dir}(\boldsymbol{\alpha})$, where $\alpha_k = \alpha/K$. From Equation 5.26 we have

$$
p(z_1, \ldots, z_N|\alpha) = \frac{\Gamma(\alpha)}{\Gamma(N + \alpha)} \prod_{k=1}^{K} \frac{\Gamma(N_k + \alpha/K)}{\Gamma(\alpha/K)} \tag{24.24}
$$

Hence

$$p(z_i = k | \mathbf{z}_{-i}, \alpha) = \frac{p(\mathbf{z}_{1:N} | \alpha)}{p(\mathbf{z}_{-i} | \alpha)} = \frac{\frac{1}{\Gamma(N+\alpha)}}{\frac{1}{\Gamma(N+\alpha-1)}} \times \frac{\Gamma(N_k + \alpha/K)}{\Gamma(N_{k,-i} + \alpha/K)} \qquad (24.25)$$

$$= \frac{\Gamma(N+\alpha-1)}{\Gamma(N+\alpha)} \frac{\Gamma(N_{k,-i} + 1 + \alpha/K)}{\Gamma(N_{k,-i} + \alpha/K)} = \frac{N_{k,-i} + \alpha/K}{N+\alpha-1} \qquad (24.26)$$

where $N_{k,-i} \triangleq \sum_{n \neq i} \mathbb{I}(z_n = k) = N_k - 1$, and where we exploited the fact that $\Gamma(x+1) = x\Gamma(x)$.

To obtain the second term in Equation 24.23, which is the posterior predictive distribution for $\mathbf{x}_i$ given all the other data and all the assignments, we use the fact that

$$p(\mathbf{x}_i | \mathbf{x}_{-i}, \mathbf{z}_{-i}, z_i = k, \boldsymbol{\beta}) = p(\mathbf{x}_i | \mathcal{D}_{-i,k}) \qquad (24.27)$$

where $\mathcal{D}_{-i,k} = \{\mathbf{x}_j : z_j = k, j \neq i\}$ is all the data assigned to cluster k except for $\mathbf{x}_i$. If we use a conjugate prior for $\boldsymbol{\theta}_k$, we can compute $p(\mathbf{x}_i | \mathcal{D}_{-i,k})$ in closed form. Furthermore, we can efficiently update these predictive likelihoods by caching the sufficient statistics for each cluster. To compute the above expression, we remove $\mathbf{x}_i$'s statistics from its current cluster (namely z_i), and then evaluate $\mathbf{x}_i$ under each cluster's posterior predictive distribution. Once we have picked a new cluster, we add $\mathbf{x}_i$'s statistics to this new cluster.

Some pseudo-code for one step of the algorithm is shown in Algorithm 24.1, based on (Sudderth 2006, p94). (We update the nodes in random order to improve the mixing time, as suggested in (Roberts and Sahu 1997).) We can initialize the sample by sequentially sampling from $p(z_i | \mathbf{z}_{1:i-1}, \mathbf{x}_{1:i})$. (See fmGibbs for some Matlab code, by Yee-Whye Teh.) In the case of GMMs, both the naive sampler and collapsed sampler take $O(NKD)$ time per step.

Algorithm 24.1: Collapsed Gibbs sampler for a mixture model

---

1 **for** *each* $i = 1 : N$ *in random order* **do**
2   Remove $\mathbf{x}_i$'s sufficient statistics from old cluster z_i ;
3   **for** *each* $k = 1 : K$ **do**
4   Compute $p_k(\mathbf{x}_i) \triangleq p(\mathbf{x}_i | \{\mathbf{x}_j : z_j = k, j \neq i\})$;
5   Compute $p(z_i = k | \mathbf{z}_{-i}, \mathcal{D}) \propto (N_{k,-i} + \alpha/K) p_k(\mathbf{x}_i)$;
6   Sample $z_i \sim p(z_i | \cdot)$;
7   Add $\mathbf{x}_i$'s sufficient statistics to new cluster z_i

---

A comparison of this method with the standard Gibbs sampler is shown in Figure 24.3. The vertical axis is the data log probability at each iteration, computed using

$$\log p(\mathcal{D} | \mathbf{z}, \boldsymbol{\theta}) = \sum_{i=1}^{N} \log \left[\pi_{z_i} p(\mathbf{x}_i | \boldsymbol{\theta}_{z_i}) \right] \qquad (24.28)$$

To compute this quantity using the collapsed sampler, we have to sample $\boldsymbol{\theta} = (\boldsymbol{\pi}, \boldsymbol{\theta}_{1:K})$ given the data and the current assignment $\mathbf{z}$.

In Figure 24.3 we see that the collapsed sampler does indeed generally work better than the vanilla sampler. Occasionally, however, both methods can get stuck in poor local modes. (Note

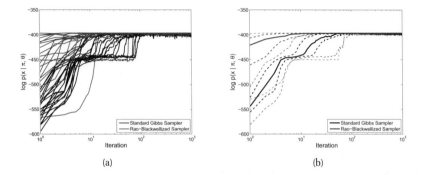

Figure 24.3 Comparison of collapsed (red) and vanilla (blue) Gibbs sampling for a mixture of $K = 4$ two-dimensional Gaussians applied to $N = 300$ data points (similar to Figure 25.7). We plot log probability of the data vs iteration. (a) 20 different random initializations. (b) logprob averaged over 100 different random initializations. Solid line is the median, thick dashed in the 0.25 and 0.75 quantiles, and thin dashed are the 0.05 and 0.95 quintiles. Source: Figure 2.20 of (Sudderth 2006). Used with kind permission of Erik Sudderth.

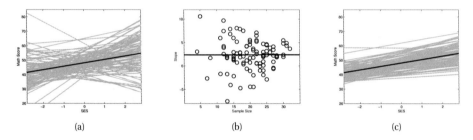

Figure 24.4 (a) Least squares regression lines for math scores vs socio-economic status for 100 schools. Population mean (pooled estimate) is in bold. (b) Plot of $\hat{w}_{2j}$ (the slope) vs N_j (sample size) for the 100 schools. The extreme slopes tend to correspond to schools with smaller sample sizes. (c) Predictions from the hierarchical model. Population mean is in bold. Based on Figure 11.1 of (Hoff 2009). Figure generated by `multilevelLinregDemo`, written by Emtiyaz Khan.

that the error bars in Figure 24.3(b) are averaged over starting values, whereas the theorem refers to MC samples in a single run.)

24.2.5 Gibbs sampling for hierarchical GLMs

Often we have data from multiple related sources. If some sources are more reliable and/or data-rich than others, it makes sense to model all the data simultaneously, so as to enable the borrowing of statistical strength. One of the most natural way to solve such problems is to use hierarchical Bayesian modeling, also called multi-level modeling. In Section 9.6, we discussed a way to perform approximate inference in such models using variational methods. Here we discuss how to use Gibbs sampling.

To explain the method, consider the following example. Suppose we have data on students

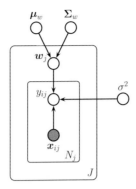

Figure 24.5 Multi-level model for linear regression.

in different schools. Such data is naturally modeled in a two-level hierarchy: we let y_{ij} be the response variable we want to predict for student i in school j. This prediction can be based on school and student specific covariates, $\mathbf{x}_{ij}$. Since the quality of schools varies, we want to use a separate parameter for each school. So our model becomes

$$y_{ij} = \mathbf{x}_{ij}^T \mathbf{w}_j + \epsilon_{ij} \tag{24.29}$$

We will illustrate this model below, using a dataset from (Hoff 2009, p197), where x_{ij} is the socio-economic status (SES) of student i in school y, and y_{ij} is their math score.

We could fit each $\mathbf{w}_j$ separately, but this can give poor results if the sample size of a given school is small. This is illustrated in Figure 24.4(a), which plots the least squares regression line estimated separately for each of the $J = 100$ schools. We see that most of the slopes are positive, but there are a few "errant" cases where the slope is negative. It turns out that the lines with extreme slopes tend to be in schools with small sample size, as shown in Figure 24.4(b). Thus we may not necessarily trust these fits.

We can get better results if we construct a hierarchical Bayesian model, in which the $\mathbf{w}_j$ are assumed to come from a common prior: $\mathbf{w}_j \sim \mathcal{N}(\boldsymbol{\mu}_w, \boldsymbol{\Sigma}_w)$. This is illustrated in Figure 24.5. In this model, the schools with small sample size borrow statistical strength from the schools with larger sample size, because the $\mathbf{w}_j$'s are correlated via the latent common parents $(\boldsymbol{\mu}_w, \boldsymbol{\Sigma}_w)$. (It is crucial that these hyper-parameters be inferrred from data; if they were fixed constants, the $\mathbf{w}_j$ would be conditionally independent, and there would be no information sharing between them.)

To complete the model specification, we must specify priors for the shared parameters. Following (Hoff 2009, p198), we will use the following semi-conjugate forms, for convenience:

$$\boldsymbol{\mu}_w \sim \mathcal{N}(\boldsymbol{\mu}_0, \mathbf{V}_0) \tag{24.30}$$

$$\boldsymbol{\Sigma}_w \sim \text{IW}(\eta_0, \mathbf{S}_0^{-1}) \tag{24.31}$$

$$\sigma^2 \sim \text{IG}(\nu_0/2, \nu_0 \sigma_0^2/2) \tag{24.32}$$

Given this, it is simple to show that the full conditionals needed for Gibbs sampling have the following forms. For the group-specific weights:

$$p(\mathbf{w}_j | \mathcal{D}_j, \boldsymbol{\theta}) = \mathcal{N}(\mathbf{w}_j | \boldsymbol{\mu}_j, \boldsymbol{\Sigma}_j) \qquad (24.33)$$

$$\boldsymbol{\Sigma}_j^{-1} = \boldsymbol{\Sigma}^{-1} + \mathbf{X}_j^T \mathbf{X}_j / \sigma^2 \qquad (24.34)$$

$$\boldsymbol{\mu}_j = \boldsymbol{\Sigma}_j (\boldsymbol{\Sigma}^{-1} \boldsymbol{\mu} + \mathbf{X}_j^T \mathbf{y}_j / \sigma^2) \qquad (24.35)$$

For the overall mean:

$$p(\boldsymbol{\mu}_w | \mathbf{w}_{1:J}, \boldsymbol{\Sigma}_w) = \mathcal{N}(\boldsymbol{\mu} | \boldsymbol{\mu}_N, \boldsymbol{\Sigma}_N) \qquad (24.36)$$

$$\boldsymbol{\Sigma}_N^{-1} = \mathbf{V}_0^{-1} + J\boldsymbol{\Sigma}^{-1} \qquad (24.37)$$

$$\boldsymbol{\mu}_N = \boldsymbol{\Sigma}_N (\mathbf{V}_0^{-1} \boldsymbol{\mu}_0 + J\boldsymbol{\Sigma}^{-1} \overline{\mathbf{w}}) \qquad (24.38)$$

where $\overline{\mathbf{w}} = \frac{1}{J} \sum_j \mathbf{w}_j$. For the overall covariance:

$$p(\boldsymbol{\Sigma}_w | \boldsymbol{\mu}_w, \mathbf{w}_{1:J}) = \text{IW}((\mathbf{S}_0 + \mathbf{S}_\mu)^{-1}, \eta_0 + J) \qquad (24.39)$$

$$\mathbf{S}_\mu = \sum_j (\mathbf{w}_j - \boldsymbol{\mu}_w)(\mathbf{w}_j - \boldsymbol{\mu}_w)^T \qquad (24.40)$$

For the noise variance:

$$p(\sigma^2 | \mathcal{D}, \mathbf{w}_{1:J}) = \text{IG}([\nu_0 + N]/2, [\nu_0 \sigma_0^2 + \text{SSR}(\mathbf{w}_{1:J})]/2) \qquad (24.41)$$

$$\text{SSR}(\mathbf{w}_{1:J}) = \sum_{j=1}^{J} \sum_{i=1}^{N_j} (y_{ij} - \mathbf{w}_j^T \mathbf{x}_{ij})^2 \qquad (24.42)$$

Applying Gibbs sampling to our hierarchical model, we get the results shown in Figure 24.4(c). The light gray lines plot the mean of the posterior predictive distribution for each school:

$$\mathbb{E}[y_j | \mathbf{x}_{ij}] = \mathbf{x}_{ij}^T \hat{\mathbf{w}}_j \qquad (24.43)$$

where

$$\hat{\mathbf{w}}_j = \mathbb{E}[\mathbf{w}_j | \mathcal{D}] \approx \frac{1}{S} \sum_{s=1}^{S} \mathbf{w}_j^{(s)} \qquad (24.44)$$

The dark gray line in the middle plots the prediction using the overall mean parameters, $\mathbf{x}_{ij}^T \hat{\boldsymbol{\mu}}_w$. We see that the method has regularized the fits quite nicely, without enforcing too much uniformity. (The amount of shrinkage is controlled by $\boldsymbol{\Sigma}_w$, which in turns depends on the hyper-parameters; in this example, we used vague values.)

24.2.6   BUGS and JAGS

One reason Gibbs sampling is so popular is that it is possible to design general purpose software that will work for almost any model. This software just needs a model specification, usually in the form a directed graphical model (specified in a file, or created with a graphical user interface), and a library of methods for sampling from different kinds of full conditionals. (This can often be done using adaptive rejection sampling, described in Section 23.3.4.) An example of such a package is **BUGS** (Lunn et al. 2000, 2012), which stands for "Bayesian updating using Gibbs Sampling". BUGS is very widely used in biostatistics and social science. Another more recent, but very similar, package is **JAGS** (Plummer 2003), which stands for "Just Another Gibbs Sampler". This uses a similar model specification language to BUGS.

For example, we can describe the model in Figure 24.5 as follows:

```
model {
 for (i in 1:N) {
    for (j in 1:J) {
        y[i,j] ~ dnorm(y.hat[i,j], tau.y)
        y.hat[i,j] <- inprod(W[j, ], X[i, j, ])
    }
}
tau.y <- pow(sigma.y, -2)
sigma.y ~ dunif(0,100)

for (j in 1:J) {
  W[j,] ~ dmnorm(mu, SigmaInv)
}
SigmaInv  ~ dwish(S0[,], eta0)
mu ~ dmnorm(mu0, V0inv)
}
```

We can then just pass this model to BUGS or JAGS, which will generate samples for us. See the webpages for details.

Although this approach is appealing, unfortunately it can be much slower than using hand-written code, especially for complex models. There has been some work on automatically deriving model-specific optimized inference code (Fischer and Schumann 2003), but fast code still typically requires human expertise.

24.2.7 The Imputation Posterior (IP) algorithm

The **Imputation Posterior** or IP algorithm (Tanner and Wong 1987) is a special case of Gibbs sampling in which we group the variables into two classes: hidden variables $\mathbf{z}$ and parameters $\boldsymbol{\theta}$. This should sound familiar: it is basically an MCMC version of EM, where the E step gets replaced by the I step, and the M step gets replaced the P step. This is an example of a more general strategy called **data augmentation**, whereby we introduce auxiliary variables in order to simplify the posterior computations (here the computation of $p(\boldsymbol{\theta}|\mathcal{D})$). See (Tanner 1996; van Dyk and Meng 2001) for more information.

24.2.8 Blocking Gibbs sampling

Gibbs sampling can be quite slow, since it only updates one variable at a time (so-called **single site updating**). If the variables are highly correlated, it will take a long time to move away from the current state. This is illustrated in Figure 24.6, where we illustrate sampling from a 2d Gaussian (see Exercise 24.1 for the details). If the variables are highly correlated, the algorithm

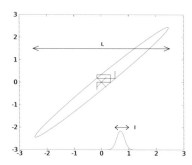

Figure 24.6   Illustration of potentially slow sampling when using Gibbs sampling for a skewed 2D Gaussian. Based on Figure 11.11 of (Bishop 2006). Figure generated by `gibbsGaussDemo`.

will move very slowly through the state space. In particular, the size of the moves is controlled by the variance of the conditional distributions. If this is ℓ in the x_1 direction, and the support of the distribution is L along this dimension, then we need $O((L/\ell)^2)$ steps to obtain an independent sample.

In some cases we can efficiently sample groups of variables at a time. This is called **blocking Gibbs sampling** or **blocked Gibbs sampling** (Jensen et al. 1995; Wilkinson and Yeung 2002), and can make much bigger moves through the state space.

24.3   Metropolis Hastings algorithm

Although Gibbs sampling is simple, it is somewhat restricted in the set of models to which it can be applied. For example, it is not much help in computing $p(\mathbf{w}|\mathcal{D})$ for a logistic regression model, since the corresponding graphical model has no useful Markov structure. In addition, Gibbs sampling can be quite slow, as we mentioned above.

Fortunately, there is a more general algorithm that can be used, known as the **Metropolis Hastings** or **MH** algorithm, which we describe below.

24.3.1   Basic idea

The basic idea in MH is that at each step, we propose to move from the current state $\mathbf{x}$ to a new state $\mathbf{x}'$ with probability $q(\mathbf{x}'|\mathbf{x})$, where q is called the **proposal distribution** (also called the **kernel**). The user is free to use any kind of proposal they want, subject to some conditions which we explain below. This makes MH quite a flexible method. A commonly used proposal is a symmetric Gaussian distribution centered on the current state, $q(\mathbf{x}'|\mathbf{x}) = \mathcal{N}(\mathbf{x}'|\mathbf{x}, \boldsymbol{\Sigma})$; this is called a **random walk Metropolis algorithm**. We discuss how to choose $\boldsymbol{\Sigma}$ in Section 24.3.3. If we use a proposal of the form $q(\mathbf{x}'|\mathbf{x}) = q(\mathbf{x}')$, where the new state is independent of the old state, we get a method known as the **independence sampler**, which is similar to importance sampling (Section 23.4).

Having proposed a move to $\mathbf{x}'$, we then decide whether to **accept** this proposal or not according to some formula, which ensures that the fraction of time spent in each state is proportional to $p^*(\mathbf{x})$. If the proposal is accepted, the new state is $\mathbf{x}'$, otherwise the new state is the same as the current state, $\mathbf{x}$ (i.e., we repeat the sample).

If the proposal is symmetric, so $q(\mathbf{x}'|\mathbf{x}) = q(\mathbf{x}|\mathbf{x}')$, the acceptance probability is given by the following formula:

$$r = \min(1, \frac{p^*(\mathbf{x}')}{p^*(\mathbf{x})}) \tag{24.45}$$

We see that if $\mathbf{x}'$ is more probable than $\mathbf{x}$, we definitely move there (since $\frac{p^*(\mathbf{x}')}{p^*(\mathbf{x})} > 1$), but if $\mathbf{x}'$ is less probable, we may still move there anyway, depending on the relative probabilities. So instead of greedily moving to only more probable states, we occasionally allow "downhill" moves to less probable states. In Section 24.3.6, we prove that this procedure ensures that the fraction of time we spend in each state $\mathbf{x}$ is proportional to $p^*(\mathbf{x})$.

If the proposal is asymmetric, so $q(\mathbf{x}'|\mathbf{x}) \neq q(\mathbf{x}|\mathbf{x}')$, we need the **Hastings correction**, given by the following:

$$r = \min(1, \alpha) \tag{24.46}$$

$$\alpha = \frac{p^*(\mathbf{x}')q(\mathbf{x}|\mathbf{x}')}{p^*(\mathbf{x})q(\mathbf{x}'|\mathbf{x})} = \frac{p^*(\mathbf{x}')/q(\mathbf{x}'|\mathbf{x})}{p^*(\mathbf{x})/q(\mathbf{x}|\mathbf{x}')} \tag{24.47}$$

This correction is needed to compensate for the fact that the proposal distribution itself (rather than just the target distribution) might favor certain states.

An important reason why MH is a useful algorithm is that, when evaluating α, we only need to know the target density up to a normalization constant. In particular, suppose $p^*(\mathbf{x}) = \frac{1}{Z}\tilde{p}(\mathbf{x})$, where $\tilde{p}(\mathbf{x})$ is an unnormalized distribution and Z is the normalization constant. Then

$$\alpha = \frac{(\tilde{p}(\mathbf{x}')/Z) \, q(\mathbf{x}|\mathbf{x}')}{(\tilde{p}(\mathbf{x})/Z) \, q(\mathbf{x}'|\mathbf{x})} \tag{24.48}$$

so the Z's cancel. Hence we can sample from p^* even if Z is unknown. In particular, all we have to do is evaluate $\tilde{p}$ pointwise, where $\tilde{p}(\mathbf{x}) = p^*(\mathbf{x})Z$.

The overall algorithm is summarized in Algorithm 24.2.

24.3.2 Gibbs sampling is a special case of MH

It turns out that Gibbs sampling, which we discussed in Section 24.2, is a special case of MH. In particular, it is equivalent to using MH with a sequence of proposals of the form

$$q(\mathbf{x}'|\mathbf{x}) = p(x_i'|\mathbf{x}_{-i})\mathbb{I}(\mathbf{x}_{-i}' = \mathbf{x}_{-i}) \tag{24.49}$$

That is, we move to a new state where x_i is sampled from its full conditional, but $\mathbf{x}_{-i}$ is left unchanged.

We now prove that the acceptance rate of each such proposal is 1, so the overall algorithm also has an acceptance rate of 100%. We have

$$\alpha = \frac{p(\mathbf{x}')q(\mathbf{x}|\mathbf{x}')}{p(\mathbf{x})q(\mathbf{x}'|\mathbf{x})} = \frac{p(x_i'|\mathbf{x}_{-i})p(\mathbf{x}_{-i}')p(x_i|\mathbf{x}_{-i}')}{p(x_i|\mathbf{x}_{-i})p(\mathbf{x}_{-i})p(x_i'|\mathbf{x}_{-i})} \tag{24.50}$$

$$= \frac{p(x_i'|\mathbf{x}_{-i})p(\mathbf{x}_{-i})p(x_i|\mathbf{x}_{-i})}{p(x_i|\mathbf{x}_{-i})p(\mathbf{x}_{-i})p(x_i'|\mathbf{x}_{-i})} = 1 \tag{24.51}$$

Algorithm 24.2: Metropolis Hastings algorithm

---

1 Initialize x^0 ;
2 **for** $s = 0, 1, 2, \ldots$ **do**
3   Define $x = x^s$;
4   Sample $x' \sim q(x'|x)$;
5   Compute acceptance probability

$$\alpha = \frac{\tilde{p}(x')q(x|x')}{\tilde{p}(x)q(x'|x)}$$

    Compute $r = \min(1, \alpha)$;
6   Sample $u \sim U(0, 1)$;
7   Set new sample to

$$x^{s+1} = \begin{cases} x' & \text{if } u < r \\ x^s & \text{if } u \geq r \end{cases}$$

---

where we exploited the fact that $\mathbf{x}'_{-i} = \mathbf{x}_{-i}$, and that $q(\mathbf{x}'|\mathbf{x}) = p(x'_i|\mathbf{x}_{-i})$.

The fact that the acceptance rate is 100% does not necessarily mean that Gibbs will converge rapidly, since it only updates one coordinate at a time (see Section 24.2.8). Fortunately, there are many other kinds of proposals we can use, as we discuss below.

24.3.3   Proposal distributions

For a given target distribution p^*, a proposal distribution q is valid or admissible if it gives a non-zero probability of moving to the states that have non-zero probability in the target. Formally, we can write this as

$$\text{supp}(p^*) \subseteq \cup_x \text{supp}(q(\cdot|x)) \tag{24.52}$$

For example, a Gaussian random walk proposal has non-zero probability density on the entire state space, and hence is a valid proposal for any continuous state space.

Of course, in practice, it is important that the proposal spread its probability mass in just the right way. Figure 24.7 shows an example where we use MH to sample from a mixture of two 1D Gaussians using a random walk proposal, $q(x'|x) = \mathcal{N}(x'|x, v)$. This is a somewhat tricky target distribution, since it consists of two well separated modes. It is very important to set the variance of the proposal v correctly: If the variance is too low, the chain will only explore one of the modes, as shown in Figure 24.7(a), but if the variance is too large, most of the moves will be rejected, and the chain will be very **sticky**, i.e., it will stay in the same state for a long time. This is evident from the long stretches of repeated values in Figure 24.7(b). If we set the proposal's variance just right, we get the trace in Figure 24.7(c), where the samples clearly explore the support of the target distribution. We discuss how to tune the proposal below.

One big advantage of Gibbs sampling is that one does not need to choose the proposal

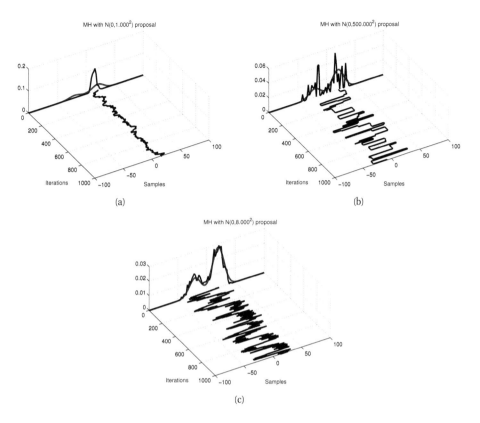

Figure 24.7 An example of the Metropolis Hastings algorithm for sampling from a mixture of two 1D Gaussians ($\boldsymbol{\mu} = (-20, 20)$, $\boldsymbol{\pi} = (0.3, 0.7)$, $\boldsymbol{\sigma} = (100, 100)$), using a Gaussian proposal with variances of $v \in \{1, 500, 8\}$. (a) When $v = 1$, the chain gets trapped near the starting state and fails to sample from the mode at $\mu = -20$. (b) When $v = 500$, the chain is very "sticky", so its effective sample size is low (as reflected by the rough histogram approximation at the end). (c) Using a variance of $v = 8$ is just right and leads to a good approximation of the true distribution (shown in red). Figure generated by `mcmcGmmDemo`. Based on code by Christophe Andrieu and Nando de Freitas.

distribution, and furthermore, the acceptance rate is 100%. Of course, a 100% acceptance can trivially be achieved by using a proposal with variance 0 (assuming we start at a mode), but this is obviously not exploring the posterior. So having a high acceptance is not the ultimate goal. We can increase the amount of exploration by increasing the variance of the Gaussian kernel. Often one experiments with different parameters until the acceptance rate is between 25% and 40%, which theory suggests is optimal, at least for Gaussian target distributions. These short initial runs, used to tune the proposal, are called **pilot runs**.

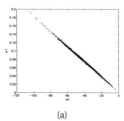 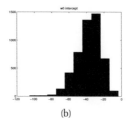 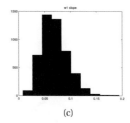

(a) (b) (c)

Figure 24.8 (a) Joint posterior of the parameters for 1d logistic regression when applied to some SAT data. (b) Marginal for the offset w_0. (c) Marginal for the slope w_1. We see that the marginals do not capture the fact that the parameters are highly correlated. Figure generated by `logregSatMhDemo`.

24.3.3.1 Gaussian proposals

If we have a continuous state space, the Hessian $\mathbf{H}$ at a local mode $\hat{\mathbf{w}}$ can be used to define the covariance of a Gaussian proposal distribution. This approach has the advantage that the Hessian models the local curvature and length scales of each dimension; this approach therefore avoids some of the slow mixing behavior of Gibbs sampling shown in Figure 24.6.

There are two obvious approaches: (1) an independence proposal, $q(\mathbf{w}'|\mathbf{w}) = \mathcal{N}(\mathbf{w}'|\hat{\mathbf{w}}, \mathbf{H}^{-1})$ or (2), a random walk proposal, $q(\mathbf{w}'|\mathbf{w}) = \mathcal{N}(\mathbf{w}'|\mathbf{w}, s^2\mathbf{H}^{-1})$, where s^2 is a scale factor chosen to facilitate rapid mixing. (Roberts and Rosenthal 2001) prove that, if the posterior is Gaussian, the asymptotically optimal value is to use $s^2 = 2.38^2/D$, where D is the dimensionality of $\mathbf{w}$; this results in an acceptance rate of 0.234.

For example, consider MH for binary logistic regression. From Equation 8.7, we have that the Hessian of the log-likelihood is $\mathbf{H}_l = \mathbf{X}^T\mathbf{D}\mathbf{X}$, where $\mathbf{D} = \text{diag}(\mu_i(1 - \mu_i))$ and $\mu_i = \text{sigm}(\hat{\mathbf{w}}^T\mathbf{x}_i)$. If we assume a Gaussian prior, $p(\mathbf{w}) = \mathcal{N}(\mathbf{0}, \mathbf{V}_0)$, we have $\mathbf{H} = \mathbf{V}_0^{-1} + \mathbf{H}_l$, so the asymptotically optimal Gaussian proposal has the form

$$q(\mathbf{w}'|\mathbf{w}) = \mathcal{N}\left(\mathbf{w}, \frac{2.38^2}{D}\left(\mathbf{V}_0^{-1} + \mathbf{X}^T\mathbf{D}\mathbf{X}\right)^{-1}\right) \tag{24.53}$$

See (Gamerman 1997; Rossi et al. 2006; Fruhwirth-Schnatter and Fruhwirth 2010) for further details. The approach is illustrated in Figure 24.8, where we sample parameters from a 1d logistic regression model fit to some SAT data. We initialize the chain at the mode, computed using IRLS, and then use the above random walk Metropolis sampler.

If you cannot afford to compute the mode or its Hessian $\mathbf{X}\mathbf{D}\mathbf{X}$, an alternative approach, suggested in (Scott 2009), is to approximate the above proposal as follows:

$$q(\mathbf{w}'|\mathbf{w}) = \mathcal{N}\left(\mathbf{w}, \left(\mathbf{V}_0^{-1} + \frac{6}{\pi^2}\mathbf{X}^T\mathbf{X}\right)^{-1}\right) \tag{24.54}$$

24.3.3.2 Mixture proposals

If there are several proposals that might be useful, one can combine them using a **mixture proposal**, which is a convex combination of base proposals:

$$q(\mathbf{x}'|\mathbf{x}) = \sum_{k=1}^{K} w_k q_k(\mathbf{x}'|\mathbf{x}) \tag{24.55}$$

where w_k are the mixing weights. As long as each q_k is an individually valid proposal, and each $w_k > 0$, then the overall mixture proposal will also be valid.

24.3.3.3 Data-driven MCMC

The most efficient proposals depend not just on the previous hidden state, but also the visible data, i.e., they have the form $q(\mathbf{x}'|\mathbf{x}, \mathcal{D})$. This is called **data-driven MCMC** (see e.g., (Tu and Zhu 2002)). To create such proposals, one can sample $(\mathbf{x}, \mathcal{D})$ pairs from the forwards model and then train a discriminative classifier to predict $p(\mathbf{x}|f(\mathcal{D}))$, where $f(\mathcal{D})$ are some features extracted from the visible data.

Typically $\mathbf{x}$ is a high-dimensional vector (e.g., position and orientation of all the limbs of a person in a visual object detector), so it is hard to predict the entire state vector, $p(\mathbf{x}|f(\mathcal{D}))$. Instead we might train a discriminative detector to predict parts of the state-space, $p(x_k|f_k(\mathcal{D}))$, such as the location of just the face of a person. We can then use a proposal of the form

$$q(\mathbf{x}'|\mathbf{x}, \mathcal{D}) = \pi_0 q_0(\mathbf{x}'|\mathbf{x}) + \sum_k \pi_k q_k(x'_k|f_k(\mathcal{D})) \tag{24.56}$$

where q_0 is a standard data-independent proposal (e.g., random walk), and q_k updates the k'th component of the state space. For added efficiency, the discriminative proposals should suggest joint changes to multiple variables, but this is often hard to do.

The overall procedure is a form of **generate and test**: the discriminative proposals $q(\mathbf{x}'|\mathbf{x})$ generate new hypotheses, which are then "tested" by computing the posterior ratio $\frac{p(\mathbf{x}'|\mathcal{D})}{p(\mathbf{x}|\mathcal{D})}$, to see if the new hypothesis is better or worse. By adding an annealing step, one can modify the algorithm to find posterior modes; this is called **simulated annealing**, and is described in Section 24.6.1. One advantage of using the mode-seeking version of the algorithm is that we do not need to ensure the proposal distribution is reversible.

24.3.4 Adaptive MCMC

One can change the parameters of the proposal as the algorithm is running to increase efficiency. This is called **adaptive MCMC**. This allows one to start with a broad covariance (say), allowing large moves through the space until a mode is found, followed by a narrowing of the covariance to ensure careful exploration of the region around the mode.

However, one must be careful not to violate the Markov property; thus the parameters of the proposal should not depend on the entire history of the chain. It turns out that a sufficient condition to ensure this is that the adaption is "faded out" gradually over time. See e.g., (Andrieu and Thoms 2008) for details.

24.3.5   Initialization and mode hopping

It is necessary to start MCMC in an initial state that has non-zero probability. If the model has deterministic constraints, finding such a legal configuration may be a hard problem in itself. It is therefore common to initialize MCMC methods at a local mode, found using an optimizer.

In some domains (especially with discrete state spaces), it is a more effective use of computation time to perform multiple restarts of an optimizer, and to average over these modes, rather than exploring similar points around a local mode. However, in continuous state spaces, the mode contains negligible volume (Section 5.2.1.3), so it is necessary to locally explore around each mode, in order to visit enough posterior probability mass.

24.3.6   Why MH works *

To prove that the MH procedure generates samples from p^*, we have to use a bit of Markov chain theory, so be sure to read Section 17.2.3 first.

The MH algorithm defines a Markov chain with the following transition matrix:

$$p(\mathbf{x}'|\mathbf{x}) = \begin{cases} q(\mathbf{x}'|\mathbf{x})r(\mathbf{x}'|\mathbf{x}) & \text{if } \mathbf{x}' \neq \mathbf{x} \\ q(\mathbf{x}|\mathbf{x}) + \sum_{\mathbf{x}' \neq \mathbf{x}} q(\mathbf{x}'|\mathbf{x})(1 - r(\mathbf{x}'|\mathbf{x})) & \text{otherwise} \end{cases} \tag{24.57}$$

This follows from a case analysis: if you move to $\mathbf{x}'$ from $\mathbf{x}$, you must have proposed it (with probability $q(\mathbf{x}'|\mathbf{x})$) and it must have been accepted (with probability $r(\mathbf{x}'|\mathbf{x})$); otherwise you stay in state $\mathbf{x}$, either because that is what you proposed (with probability $q(\mathbf{x}|\mathbf{x})$), or because you proposed something else (with probability $q(\mathbf{x}'|\mathbf{x})$) but it was rejected (with probability $1 - r(\mathbf{x}'|\mathbf{x})$).

Let us analyse this Markov chain. Recall from Section 17.2.3.4 that a chain satisfies **detailed balance** if

$$p(\mathbf{x}'|\mathbf{x})p^*(\mathbf{x}) = p(\mathbf{x}|\mathbf{x}')p^*(\mathbf{x}') \tag{24.58}$$

We also showed that if a chain satisfies detailed balance, then p^* is its stationary distribution. Our goal is to show that the MH algorithm defines a transition function that satisfies detailed balance and hence that p^* is its stationary distribution. (If Equation 24.58 holds, we say that p^* is an **invariant** distribution wrt the Markov transition kernel q.)

Theorem 24.3.1. *If the transition matrix defined by the MH algorithm (given by Equation 24.57) is ergodic and irreducible, then p^* is its unique limiting distribution.*

Proof. Consider two states $\mathbf{x}$ and $\mathbf{x}'$. Either

$$p^*(\mathbf{x})q(\mathbf{x}'|\mathbf{x}) < p^*(\mathbf{x}')q(\mathbf{x}|\mathbf{x}') \tag{24.59}$$

or

$$p^*(\mathbf{x})q(\mathbf{x}'|\mathbf{x}) > p^*(\mathbf{x}')q(\mathbf{x}|\mathbf{x}') \tag{24.60}$$

We will ignore ties (which occur with probability zero for continuous distributions). Without loss of generality, assume that $p^*(\mathbf{x})q(\mathbf{x}'|\mathbf{x}) > p^*(\mathbf{x}')q(\mathbf{x}|\mathbf{x}')$. Hence

$$\alpha(\mathbf{x}'|\mathbf{x}) = \frac{p^*(\mathbf{x}')q(\mathbf{x}|\mathbf{x}')}{p^*(\mathbf{x})q(\mathbf{x}'|\mathbf{x})} < 1 \tag{24.61}$$

Hence we have $r(\mathbf{x}'|\mathbf{x}) = \alpha(\mathbf{x}'|\mathbf{x})$ and $r(\mathbf{x}|\mathbf{x}') = 1$.

Now to move from $\mathbf{x}$ to $\mathbf{x}'$ we must first propose $\mathbf{x}'$ and then accept it. Hence

$$p(\mathbf{x}'|\mathbf{x}) = q(\mathbf{x}'|\mathbf{x})r(\mathbf{x}'|\mathbf{x}) = q(\mathbf{x}'|\mathbf{x})\frac{p^*(\mathbf{x}')q(\mathbf{x}|\mathbf{x}')}{p^*(\mathbf{x})q(\mathbf{x}'|\mathbf{x})} = \frac{p^*(\mathbf{x}')}{p^*(\mathbf{x})}q(\mathbf{x}|\mathbf{x}') \tag{24.62}$$

Hence

$$p^*(\mathbf{x})p(\mathbf{x}'|\mathbf{x}) = p^*(\mathbf{x}')q(\mathbf{x}|\mathbf{x}') \tag{24.63}$$

The backwards probability is

$$p(\mathbf{x}|\mathbf{x}') = q(\mathbf{x}|\mathbf{x}')r(\mathbf{x}|\mathbf{x}') = q(\mathbf{x}|\mathbf{x}') \tag{24.64}$$

since $r(\mathbf{x}|\mathbf{x}') = 1$. Inserting this into Equation 24.63 we get

$$p^*(\mathbf{x})p(\mathbf{x}'|\mathbf{x}) = p^*(\mathbf{x}')p(\mathbf{x}|\mathbf{x}') \tag{24.65}$$

so detailed balance holds wrt p^*. Hence, from Theorem 17.2.3, p^* is a stationary distribution. Furthermore, from Theorem 17.2.2, this distribution is unique, since the chain is ergodic and irreducible. $\square$

24.3.7 Reversible jump (trans-dimensional) MCMC *

Suppose we have a set of models with different numbers of parameters, e.g., mixture models in which the number of mixture components is unknown. Let the model be denoted by m, and let its unknowns (e.g., parameters) be denoted by $\mathbf{x}_m \in \mathcal{X}_m$ (e.g., $\mathcal{X}_m = \mathbb{R}^{n_m}$, where n_m is the dimensionality of model m). Sampling in spaces of differing dimensionality is called **trans-dimensional MCMC** (Green 2003). We could sample the model indicator $m \in \{1, \ldots, M\}$ and sample all the parameters from the product space $\prod_{m=1}^{M} \mathcal{X}_m$, but this is very inefficient. It is more parsimonious to sample in the union space $\mathcal{X} = \cup_{m=1}^{M}\{m\} \times \mathcal{X}_m$, where we only worry about parameters for the currently active model.

The difficulty with this approach arises when we move between models of different dimensionality. The trouble is that when we compute the MH acceptance ratio, we are comparing densities defined in different dimensionality spaces, which is meaningless. It is like trying to compare a sphere with a circle. The solution, proposed by (Green 1998) and known as **reversible jump MCMC** or **RJMCMC**, is to augment the low dimensional space with extra random variables so that the two spaces have a common measure.

Unfortunately, we do not have space to go into details here. Suffice it to say that the method can be made to work in theory, although it is a bit tricky in practice. If, however, the continuous parameters can be integrated out (resulting in a method called collapsed RJMCMC), much of the difficulty goes away, since we are just left with a discrete state space, where there is no need to worry about change of measure. For example, (Denison et al. 2002) includes many examples of applications of collapsed RJMCMC applied to Bayesian inference for adaptive basis-function models. They sample basis functions from a fixed set of candidates (e.g., centered on the data points), and integrate out the other parameters analytically. This provides a Bayesian alternative to using RVMs or SVMs.

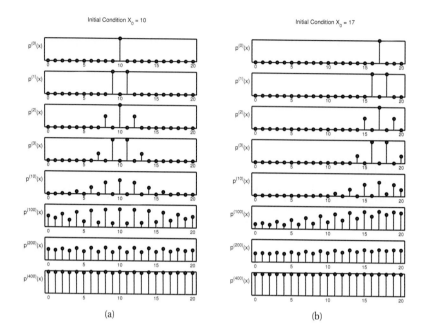

Figure 24.9 Illustration of convergence to the uniform distribution over $\{0, 1, \ldots, 20\}$ using a symmetric random walk starting from (left) state 10, and (right) state 17. Based on Figures 29.14 and 29.15 of (MacKay 2003). Figure generated by `randomWalk0to20Demo`.

24.4 Speed and accuracy of MCMC

In this section, we discuss a number of important theoretical and practical issues to do with MCMC.

24.4.1 The burn-in phase

We start MCMC from an arbitrary initial state. As we explained in Section 17.2.3, only when the chain has "forgotten" where it started from will the samples be coming from the chain's stationary distribution. Samples collected before the chain has reached its stationary distribution do not come from p^*, and are usually thrown away. The initial period, whose samples will be ignored, is called the **burn-in phase**.

For example, consider a uniform distribution on the integers $\{0, 1, \ldots, 20\}$. Suppose we sample from this using a symmetric random walk. In Figure 24.9, we show two runs of the algorithm. On the left, we start in state 10; on the right, we start in state 17. Even in this small problem it takes over 200 steps until the chain has "forgotten" where it started from.

It is difficult to diagnose when the chain has burned in, an issue we discuss in more detail below. (This is one of the fundamental weaknesses of MCMC.) As an interesting example of what can happen if you start collecting samples too early, consider the Potts model. Figure 24.10(a), shows a sample after 200 iterations of Gibbs sampling. This suggests that the model likes

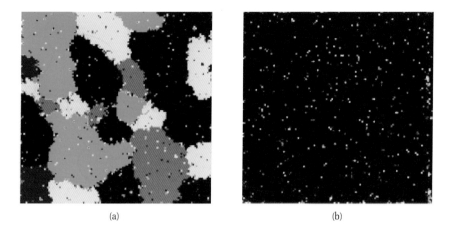

(a)   (b)

Figure 24.10 Illustration of problems caused by poor mixing. (a) One sample from a 5-state Potts model on a 128×128 grid with 8 nearest neighbor connectivity and $J = 2/3$ (as in (Geman and Geman 1984)), after 200 iterations. (b) One sample from the same model after 10,000 iterations. Used with kind permission of Erik Sudderth.

medium-sized regions where the label is the same, implying the model would make a good prior for image segmentation. Indeed, this was suggested in the original Gibbs sampling paper (Geman and Geman 1984).

However, it turns out that if you run the chain long enough, you get isolated speckles, as in Figure 24.10(b). The results depend on the coupling strength, but in general, it is very hard to find a setting which produces nice medium-sized blobs: most parameters result in a few super-clusters, or lots of small fragments. In fact, there is a rapid phase transition between these two regimes. This led to a paper called "The Ising/Potts model is not well suited to segmentation tasks" (Morris et al. 1996). It is possible to create priors more suited to image segmentation (e.g., (Sudderth and Jordan 2008)), but the main point here is that sampling before reaching convergence can lead to erroneous conclusions.

24.4.2 Mixing rates of Markov chains *

The amount of time it takes for a Markov chain to converge to the stationary distribution, and forget its initial state, is called the **mixing time**. More formally, we say that the mixing time from state x_0 is the minimal time such that, for any constant $\epsilon > 0$, we have that

$$\tau_\epsilon(x_0) \triangleq \min\{t : ||\delta_{x_0}(x)T^t - p^*||_1 \le \epsilon\} \tag{24.66}$$

where $\delta_{x_0}(x)$ is a distribution with all its mass in state x_0, T is the transition matrix of the chain (which depends on the target p^* and the proposal q), and $\delta_{x_0}(x)T^t$ is the distribution after t steps. The mixing time of the chain is defined as

$$\tau_\epsilon \triangleq \max_{x_0} \tau_\epsilon(x_0) \tag{24.67}$$

The mixing time is determined by the eigengap $\gamma = \lambda_1 - \lambda_2$, which is the difference of the

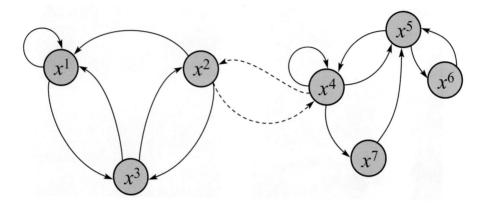

Figure 24.11 A Markov chain with low conductance. The dotted arcs represent transitions with very low probability. Source: Figure 12.6 of (Koller and Friedman 2009). Used with kind permission of Daphne Koller.

first and second eigenvalues of the transition matrix. In particular, one can show that

$$\tau_\epsilon \leq O(\frac{1}{\gamma} \log \frac{n}{\epsilon}) \tag{24.68}$$

where n is the number of states. Since computing the transition matrix can be hard to do, especially for high dimensional and/or continuous state spaces, it is useful to find other ways to estimate the mixing time.

An alternative approach is to examine the geometry of the state space. For example, consider the chain in Figure 24.11. We see that the state space consists of two "islands", each of which is connected via a narrow "bottleneck". (If they were completely disconnected, the chain would not be ergodic, and there would no longer be a unique stationary distribution.) We define the **conductance** ϕ of a chain as the minimum probability, over all subsets of states, of transitioning from that set to its complement:

$$\phi \triangleq \min_{S: 0 \leq p^*(S) \leq 0.5} \frac{\sum_{x \in S, x' \in S^c} T(x \to x')}{p^*(S)}, \tag{24.69}$$

One can show that

$$\tau_\epsilon \leq O(\frac{1}{\phi^2} \log \frac{n}{\epsilon}) \tag{24.70}$$

Hence chains with low conductance have high mixing time. For example, distributions with well-separated modes usually have high mixing time. Simple MCMC methods often do not work well in such cases, and more advanced algorithms, such as parallel tempering, are necessary (see e.g., (Liu 2001)).

24.4.3 Practical convergence diagnostics

Computing the mixing time of a chain is in general quite difficult, since the transition matrix is usually very hard to compute. In practice various heuristics have been proposed to diagnose convergence — see (Geyer 1992; Cowles and Carlin 1996; Brooks and Roberts 1998) for a review. Strictly speaking, these methods do not diagnose convergence, but rather non-convergence. That is, the method may claim the chain has converged when in fact it has not. This is a flaw common to all convergence diagnostics, since diagnosing convergence is computationally intractable in general (Bhatnagar et al. 2010).

One of the simplest approaches to assessing when the method has converged is to run multiple chains from very different **overdispersed** starting points, and to plot the samples of some variables of interest. This is called a **trace plot**. If the chain has mixed, it should have "forgotten" where it started from, so the trace plots should converge to the same distribution, and thus overlap with each other.

Figure 24.12 gives an example. We show the traceplot for x which was sampled from a mixture of two 1D Gaussians using four different methods: MH with a symmetric Gaussian proposal of variance $\sigma^2 \in \{1, 8, 500\}$, and Gibbs sampling. We see that $\sigma^2 = 1$ has not mixed, which is also evident from Figure 24.7(a), which shows that a single chain never leaves the area where it started. The results for the other methods indicate that the chains rapidly converge to the stationary distribution, no matter where they started. (The sticky nature of the $\sigma^2 = 500$ proposal is very evident. This reduces the computational efficiency, as we discuss below, but not the statistical validity.)

24.4.3.1   Estimated potential scale reduction (EPSR)

We can assess convergence more quantitatively as follows. The basic idea is to compare the variance of a quantity within each chain to its variance across chains. More precisely, suppose we collect S samples (after burn-in) from each of C chains of D variables, x_{isc}, $i = 1 : D$, $s = 1 : S$, $c = 1 : C$. Let y_{sc} be a scalar quantity of interest derived from $\mathbf{x}_{1:D,s,c}$ (e.g., $y_{sc} = x_{isc}$ for some chosen i). Define the within-sequence mean and overall mean as

$$\overline{y}_{\cdot c} \triangleq \frac{1}{S} \sum_{s=1}^{S} y_{sc}, \quad \overline{y}_{\cdot\cdot} \triangleq \frac{1}{C} \sum_{c=1}^{C} \overline{y}_{\cdot c} \tag{24.71}$$

Define the between-sequence and within-sequence variance as

$$B \triangleq \frac{S}{C-1} \sum_{c=1}^{C} (\overline{y}_{\cdot c} - \overline{y}_{\cdot\cdot})^2, \quad W \triangleq \frac{1}{C} \sum_{c=1}^{C} \left[\frac{1}{S-1} \sum_{s=1}^{S} (y_{sc} - \overline{y}_{\cdot c})^2 \right] \tag{24.72}$$

We can now construct two estimates of the variance of y. The first estimate is W: this should underestimate $\mathrm{var}[y]$ if the chains have not ranged over the full posterior. The second estimate is

$$\hat{V} = \frac{S-1}{S} W + \frac{1}{S} B \tag{24.73}$$

This is an estimate of $\mathrm{var}[y]$ that is unbiased under stationarity, but is an overestimate if the starting points were overdispersed (Gelman and Rubin 1992). From this, we can define the following convergence diagnostic statistic, known as the **estimated potential scale reduction** or **EPSR**:

$$\hat{R} \triangleq \sqrt{\frac{\hat{V}}{W}} \tag{24.74}$$

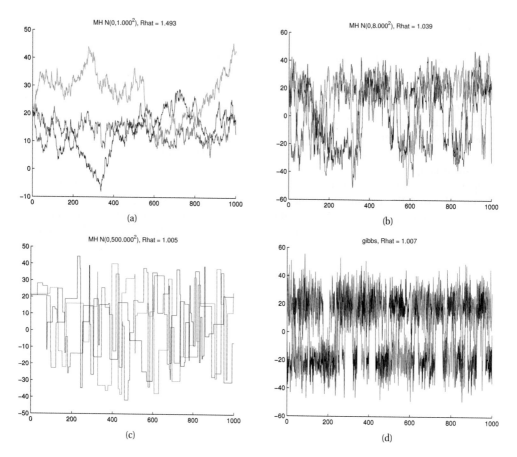

Figure 24.12 Traceplots for MCMC samplers. Each color represents the samples from a different starting point. (a-c) MH with proposal $\mathcal{N}(x'|x, \sigma^2)$ for $\sigma^2 \in \{1, 8, 500\}$, corresponding to Figure 24.7. (d) Gibbs sampling. Figure generated by `mcmcGmmDemo`.

This quantity, which was first proposed in (Gelman and Rubin 1992), measures the degree to which the posterior variance would decrease if we were to continue sampling in the $S \rightarrow \infty$ limit. If $\hat{R} \approx 1$ for any given quantity, then that estimate is reliable (or at least is not unreliable). The $\hat{R}$ values for the four samplers in Figure 24.12 are 1.493, 1.039, 1.005 and 1.007. So this diagnostic has correctly identified that the sampler using the first ($\sigma^2 = 1$) proposal is untrustworthy.

24.4.4 Accuracy of MCMC

The samples produced by MCMC are auto-correlated, and this reduces their information content relative to independent or "perfect" samples. We can quantify this as follows.[4] Suppose we want

---

4. This Section is based on (Hoff 2009, Sec 6.6).

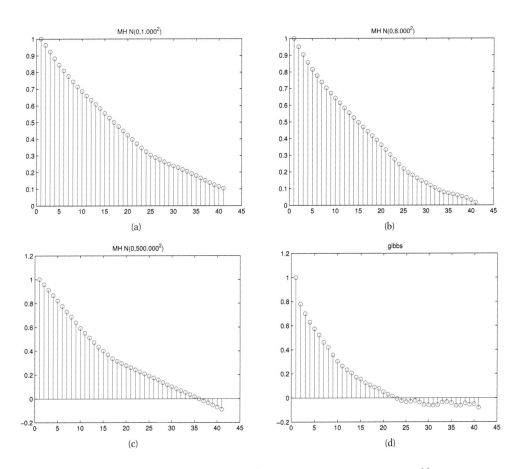

Figure 24.13 Autocorrelation functions corresponding to Figure 24.12. Figure generated by `mcmcGmmDemo`.

to estimate the mean of $f(X)$, for some function f, where $X \sim p()$. Denote the true mean by

$$f^* \triangleq \mathbb{E}\left[f(X)\right] \tag{24.75}$$

A Monte Carlo estimate is given by

$$\overline{f} = \frac{1}{S} \sum_{s=1}^{S} f_s \tag{24.76}$$

where $f_s \triangleq f(x_s)$ and $x_s \sim p(x)$. An MCMC estimate of the variance of this estimate is given by

$$
\text{Var}_{MCMC}[\overline{f}] \;=\; \mathbb{E}\left[(\overline{f} - f^*)^2\right] \tag{24.77}
$$

$$
= \; \mathbb{E}\left[\left\{\frac{1}{S}\sum_{s=1}^{S}(f_s - f^*)\right\}^2\right] \tag{24.78}
$$

$$
= \; \frac{1}{S^2}\mathbb{E}\left[\sum_{s=1}^{S}(f_s - f^*)^2\right] + \frac{1}{S^2}\sum_{s \neq t}\mathbb{E}\left[(f_s - f^*)(f_t - f^*)\right] \tag{24.79}
$$

$$
= \; \text{Var}_{MC}(\overline{f}) + \frac{1}{S^2}\sum_{s \neq t}\mathbb{E}\left[(f_s - f^*)(f_t - f^*)\right] \tag{24.80}
$$

where the first term is the Monte Carlo estimate of the variance if the samples weren't correlated, and the second term depends on the correlation of the samples. We can measure this as follows. Define the sample-based auto-correlation at lag t of a set of samples $f_1, \dots, f_S$ as follows:

$$
\rho_t \triangleq \frac{\frac{1}{S-t}\sum_{s=1}^{S-t}(f_s - \overline{f})(f_{s+t} - \overline{f})}{\frac{1}{S-1}\sum_{s=1}^{S}(f_s - \overline{f})^2} \tag{24.81}
$$

This is called the **autocorrelation function** (ACF). This is plotted in Figure 24.13 for our four samplers for the Gaussian mixture model. We see that the ACF of the Gibbs sampler (bottom right) dies off to 0 much more rapidly than the MH samplers, indicating that each Gibbs sample is "worth" more than each MH sample.

A simple method to reduce the autocorrelation is to use **thinning**, in which we keep every n'th sample. This does not increase the efficiency of the underlying sampler, but it does save space, since it avoids storing highly correlated samples.

We can estimate the information content of a set of samples by computing the **effective sample size** (ESS) S_{eff}, defined by

$$
S_{\text{eff}} \triangleq \frac{\text{Var}_{MC}(f)}{\text{Var}_{MCMC}(\overline{f})} \tag{24.82}
$$

From Figure 24.12, it is clear that the effective sample size of the Gibbs sampler is higher than that of the other samplers (in this example).

24.4.5   How many chains?

A natural question to ask is: how many chains should we run? We could either run one long chain to ensure convergence, and then collect samples spaced far apart, or we could run many short chains, but that wastes the burnin time. In practice it is common to run a medium number of chains (say 3) of medium length (say 100,000 steps), and to take samples from each after discarding the first half of the samples. If we initialize at a local mode, we may be able to use all the samples, and not wait for burn-in.

Model	Goal	Method	Reference
Probit	MAP	Gradient	Section 9.4.1
Probit	MAP	EM	Section 11.4.6
Probit	Post	EP	(Nickisch and Rasmussen 2008)
Probit	Post	Gibbs+	Exercise 24.6
Probit	Post	Gibbs with ARS	(Dellaportas and Smith 1993)
Probit	Post	MH using IRLS proposal	(Gamerman 1997)
Logit	MAP	Gradient	Section 8.3.4
Logit	Post	Gibbs+ with Student	(Fruhwirth-Schnatter and Fruhwirth 2010)
Logit	Post	Gibbs+ with KS	(Holmes and Held 2006)

Table 24.1 Summary of some possible algorithms for estimation and inference for binary classification problems using Gaussian priors. Abbreviations: ARS = adaptive rejection sampling, EP = expectation propagation, Gibbs+ = Gibbs sampling with auxiliary variables, IRLS = iterative reweighted least squares, KS = Kolmogorov Smirnov, MAP = maximum a posteriori, MH = Metropolis Hastings, Post = posterior.

24.5 Auxiliary variable MCMC *

Sometimes we can dramatically improve the efficiency of sampling by introducing dummy **auxiliary variables**, in order to reduce correlation between the original variables. If the original variables are denoted by $\mathbf{x}$, and the auxiliary variables by $\mathbf{z}$, we require that $\sum_{\mathbf{z}} p(\mathbf{x}, \mathbf{z}) = p(\mathbf{x})$, and that $p(\mathbf{x}, \mathbf{z})$ is easier to sample from than just $p(\mathbf{x})$. If we meet these two conditions, we can sample in the enlarged model, and then throw away the sampled $\mathbf{z}$ values, thereby recovering samples from $p(\mathbf{x})$. We give some examples below.

24.5.1 Auxiliary variable sampling for logistic regression

In Section 9.4.2, we discussed the latent variable interpretation of probit regression. Recall that this had the form

$$z_i \triangleq \mathbf{w}^T \mathbf{x}_i + \epsilon_i \tag{24.83}$$

$$\epsilon_i \sim \mathcal{N}(0, 1) \tag{24.84}$$

$$y_i = 1 = \mathbb{I}(z_i \geq 0) \tag{24.85}$$

We exploited this representation in Section 11.4.6, where we used EM to find an ML estimate. It is straightforward to convert this into an auxiliary variable Gibbs sampler (Exercise 24.6), since $p(\mathbf{w}|\mathcal{D})$ is Gaussian and $p(z_i|\mathbf{x}_i, y_i, \mathbf{w})$ is truncated Gaussian, both of which are easy to sample from.

Now let us discuss how to derive an auxiliary variable Gibbs sampler for logistic regression. Let ϵ_i follow a **logistic distribution**, with pdf

$$p_{\text{Logistic}}(\epsilon) = \frac{e^{-\epsilon}}{(1 + e^{-\epsilon})^2} \tag{24.86}$$

with mean $\mathbb{E}[\epsilon] = 0$ and variance $\text{var}[\epsilon] = \pi^2/3$. The cdf has the form $F(\epsilon) = \text{sigm}(\epsilon)$, which is the logistic function. Since $y_i = 1$ iff $\mathbf{w}^T\mathbf{x}_i + \epsilon > 0$, we have, by symmetry, that

$$p(y_i = 1|\mathbf{x}_i, \mathbf{w}) = \int_{-\mathbf{w}^T\mathbf{x}_i}^{\infty} f(\epsilon)d\epsilon = \int_{-\infty}^{\mathbf{w}^T\mathbf{x}_i} f(\epsilon)d\epsilon = F(\mathbf{w}^T\mathbf{x}_i) = \text{sigm}(\mathbf{w}^T\mathbf{x}_i) \quad (24.87)$$

as required.

We can derive an auxiliary variable Gibbs sampler by sampling from $p(\mathbf{z}|\mathbf{w}, \mathcal{D})$ and $p(\mathbf{w}|\mathbf{z}, \mathcal{D})$. Unfortunately, sampling directly from $p(\mathbf{w}|\mathbf{z}, \mathcal{D})$ is not possible. One approach is to define $\epsilon_i \sim \mathcal{N}(0, \lambda_i)$, where $\lambda_i = (2\psi_i)^2$ and $\psi_i \sim \text{KS}$, the Kolmogorov Smirnov distribution, and then to sample $\mathbf{w}$, $\mathbf{z}$, $\boldsymbol{\lambda}$ and $\boldsymbol{\psi}$ (Holmes and Held 2006).

A simpler approach is to approximate the logistic distribution by the Student distribution (Albert and Chib 1993). Specifically, we will make the approximation $\epsilon_i \sim \mathcal{T}(0, 1, \nu)$, where $\nu \approx 8$. We can now use the scale mixture of Gaussians representation of the Student to simplify inference. In particular, we write

$$\lambda_i \sim \text{Ga}(\nu/2, \nu/2) \quad (24.88)$$
$$\epsilon_i \sim \mathcal{N}(0, \lambda_i^{-1}) \quad (24.89)$$
$$z_i \triangleq \mathbf{w}^T\mathbf{x}_i + \epsilon_i \quad (24.90)$$
$$y_i = 1|z_i = \mathbb{I}(z_i \geq 0) \quad (24.91)$$

All of the full conditionals now have a simple form; see Exercise 24.7 for the details.

Note that if we set $\nu = 1$, then $z_i \sim \mathcal{N}(\mathbf{w}^T\mathbf{x}_i, 1)$, which is equivalent to probit regression (see Section 9.4). Rather than choosing between probit or logit regression, we can simply estimate the ν parameter. There is no convenient conjugate prior, but we can consider a finite range of possible values and evaluate the posterior as follows:

$$p(\nu|\boldsymbol{\lambda}) \propto p(\nu) \prod_{i=1}^{N} \frac{1}{\Gamma(\nu/2)(\nu/2)^{\nu/2}} \lambda_i^{\nu/2-1} e^{-\nu\lambda_i/2} \quad (24.92)$$

Furthermore, if we define $\mathbf{V}_0 = v_0\mathbf{I}$, we can sample v_0 as well. For example, suppose we use a $\text{IG}(\delta_1, \delta_2)$ prior for v_0. The posterior is given by $p(v_0|\mathbf{w}) = \text{IG}(\delta_1 + \frac{1}{2}D, \delta_2 + \frac{1}{2}\sum_{j=1}^{D} w_j^2)$. This can be interleaved with the other Gibbs sampling steps, and provides an appealing Bayesian alternative to cross validation for setting the strength of the regularizer.

See Table 24.1 for a summary of various algorithms for fitting probit and logit models. Many of these methods can also be extended to the multinomial logistic regression case. For details, see (Scott 2009; Fruhwirth-Schnatter and Fruhwirth 2010).

24.5.2   Slice sampling

Consider sampling from a univariate, but multimodal, distribution $\tilde{p}(x)$. We can sometimes improve the ability to make large moves by adding an auxiliary variable u. We define the joint distribution as follows:

$$\hat{p}(x, u) = \begin{cases} 1/Z_p & \text{if } 0 \leq u \leq \tilde{p}(x) \\ 0 & \text{otherwise} \end{cases} \quad (24.93)$$

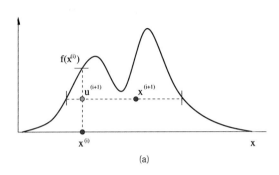
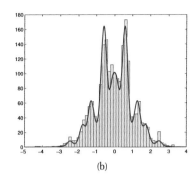

(a)   (b)

Figure 24.14 (a) Illustration of the principle behind slice sampling. Given a previous sample x^i, we sample u^{i+1} uniformly on $[0, f(x^i)]$, where f is the target density. We then sample x^{i+1} along the slice where $f(x) \geq u^{i+1}$. Source: Figure 15 of (Andrieu et al. 2003) . Used with kind permission of Nando de Freitas. (b) Slice sampling in action. Figure generated by `sliceSamplingDemo1d`.

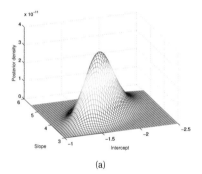
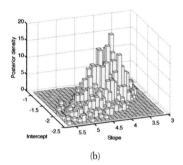

(a)   (b)

Figure 24.15 Binomial regression for 1d data. (a) Grid approximation to posterior. (b) Slice sampling approximation. Figure generated by `sliceSamplingDemo2d`.

where $Z_p = \int \tilde{p}(x)dx$. The marginal distribution over x is given by

$$\int \hat{p}(x,u)du = \int_0^{\tilde{p}(x)} \frac{1}{Z_p}du = \frac{\tilde{p}(x)}{Z_p} = p(x) \tag{24.94}$$

so we can sample from $p(x)$ by sampling from $\hat{p}(x,u)$ and then ignoring u. The full conditionals have the form

$$p(u|x) = U_{[0,\tilde{p}(x)]}(u) \tag{24.95}$$
$$p(x|u) = U_A(x) \tag{24.96}$$

where $A = \{x : \tilde{p}(x) \geq u\}$ is the set of points on or above the chosen height u. This corresponds to a slice through the distribution, hence the term **slice sampling** (Neal 2003a). See Figure 24.14(a).

In practice, it can be difficult to identify the set A. So we can use the following approach: construct an interval $x_{min} \leq x \leq x_{max}$ around the current point x^s of some width. We then test to see if each end point lies within the slice. If it does, we keep extending in that direction until it lies outside the slice. This is called **stepping out**. A candidate value x' is then chosen uniformly from this region. If it lies within the slice, it is kept, so $x^{s+1} = x'$. Otherwise we shrink the region such that x' forms one end and such that the region still contains x^s. Then another sample is drawn. We continue in this way until a sample is accepted.

To apply the method to multivariate distributions, we can sample one extra auxiliary variable for each dimension. The advantage of slice sampling over Gibbs is that it does not need a specification of the full-conditionals, just the unnormalized joint. The advantage of slice sampling over MH is that it does not need a user-specified proposal distribution (although it does require a specification of the width of the stepping out interval).

Figure 24.14(b) illustrates the algorithm in action on a synthetic 1d problem. Figure 24.15 illustrates its behavior on a slightly harder problem, namely binomial logistic regression. The model has the form

$$y_i \sim \text{Bin}(n_i, \text{logit}(\beta_1 + \beta_2 x_i)) \tag{24.97}$$

We use a vague Gaussian prior for the β_j's. Figure 24.15(a) shows a grid-based approximation to the posterior, and Figure 24.15(b) shows a sample-based approximation. In this example, the grid is faster to compute, but for any problem with more than 2 dimensions, the grid approach is infeasible.

24.5.3   Swendsen Wang

Consider an Ising model of the following form:

$$p(\mathbf{x}) = \frac{1}{Z} \prod_e f_e(\mathbf{x}_e) \tag{24.98}$$

where $\mathbf{x}_e = (x_i, x_j)$ for edge $e = (i, j)$, $x_i \in \{+1, -1\}$, and the edge factor f_e is defined by $\begin{pmatrix} e^J & e^{-J} \\ e^{-J} & e^J \end{pmatrix}$, where J is the edge strength. Gibbs sampling in such models can be slow when J is large in absolute value, because neighboring states can be highly correlated. The **Swendsen Wang** algorithm (Swendsen and Wang 1987) is a auxiliary variable MCMC sampler which mixes much faster, at least for the case of attractive or ferromagnetic models, with $J > 0$.

Suppose we introduce auxiliary binary variables, one per edge. [5] These are called **bond variables**, and will be denoted by $\mathbf{z}$. We then define an extended model $p(\mathbf{x}, \mathbf{z})$ of the form

$$p(\mathbf{x}, \mathbf{z}) = \frac{1}{Z'} \prod_e g_e(\mathbf{x}_e, z_e) \tag{24.99}$$

where $z_e \in \{0, 1\}$, and we define the new factor as follows: $g_e(\mathbf{x}_e, z_e = 0) = \begin{pmatrix} e^{-J} & e^{-J} \\ e^{-J} & e^{-J} \end{pmatrix}$, and $g_e(\mathbf{x}_e, z_e = 1) = \begin{pmatrix} e^J - e^{-J} & 0 \\ 0 & e^J - e^{-J} \end{pmatrix}$. It is clear that $\sum_{z_e=0}^{1} g_e(\mathbf{x}_e, z_e) = f_e(\mathbf{x}_e)$,

---

5. Our presentation of the method is based on some notes by David Mackay, available from http://www.inference .phy.cam.ac.uk/mackay/itila/swendsen.pdf.

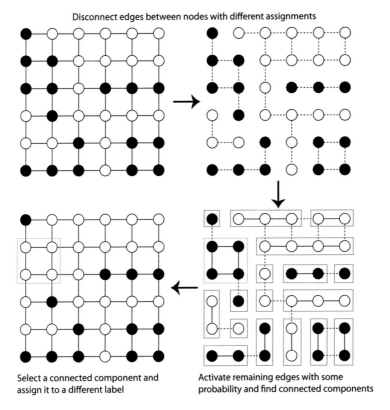

Figure 24.16 Illustration of the Swendsen Wang algorithm on a 2d grid. Used with kind permission of Kevin Tang.

and hence that $\sum_{\mathbf{z}} p(\mathbf{x}, \mathbf{z}) = p(\mathbf{x})$. So if we can sample from this extended model, we can just throw away the $\mathbf{z}$ samples and get valid $\mathbf{x}$ samples from the original distribution.

Fortunately, it is easy to apply Gibbs sampling to this extended model. The full conditional $p(\mathbf{z}|\mathbf{x})$ factorizes over the edges, since the bond variables are conditionally independent given the node variables. Furthermore, the full conditional $p(z_e|\mathbf{x}_e)$ is simple to compute: if the nodes on either end of the edge are in the same state ($x_i = x_j$), we set the bond z_e to 1 with probability $p = 1 - e^{-2J}$, otherwise we set it to 0. In Figure 24.16 (top right), the bonds that could be turned on (because their corresponding nodes are in the same state) are represented by dotted edges. In Figure 24.16 (bottom right), the bonds that are randomly turned on are represented by solid edges.

To sample $p(\mathbf{x}|\mathbf{z})$, we proceed as follows. Find the connected components defined by the graph induced by the bonds that are turned on. (Note that a connected component may consist of a singleton node.) Pick one of these components uniformly at random. All the nodes in each such component must have the same state, since the off-diagonal terms in the $g_e(\mathbf{x}_e, z_e = 1)$ factor are 0. Pick a state ± 1 uniformly at random, and force all the variables in this component to adopt this new state. This is illustrated in Figure 24.16 (bottom left), where the green square denotes the selected connected component, and we choose to force all nodes within in to enter the white state.

The validity of this algorithm is left as an exercise, as is the extension to handle local evidence and non-stationary potentials.

It should be intuitively clear that Swendsen Wang makes much larger moves through the state space than Gibbs sampling. In fact, SW mixes much faster than Gibbs sampling on 2d lattice Ising models for a variety of values of the coupling parameter, provided $J > 0$. More precisely, let the edge strength be parameterized by J/T, where $T > 0$ is a computational temperature. For large T, the nodes are roughly independent, so both methods work equally well. However, as T approaches a **critical temperature** T_c, the typical states of the system have very long correlation lengths, and Gibbs sampling takes a very long time to generate independent samples. As the temperature continues to drop, the typical states are either all on or all off. The frequency with which Gibbs sampling moves between these two modes is exponentiall small. By contrast, SW mixes rapidly at all temperatures.

Unfortunately, if any of the edge weights are negative, $J < 0$, the system is **frustrated**, and there are exponentially many modes, even at low temperature. SW does not work very well in this setting, since it tries to force many neighboring variables to have the same state. In fact, computation in this regime is provably hard for any algorithm (Jerrum and Sinclair 1993, 1996).

24.5.4 Hybrid/Hamiltonian MCMC *

In this section, we briefly mention a way to perform MCMC sampling for continuous state spaces, for which we can compute the gradient of the (unnormalized) log-posterior. This is the case in neural network models, for example.

The basic idea is to think of the parameters as a particle in space, and to create auxiliary variables which represent the "momentum" of this particle. We then update this parameter/ momentum pair according to certain rules (see e.g., (Duane et al. 1987; Neal 1993; MacKay 2003; Neal 2010) for details). The resulting method is called **hybrid MCMC** or **Hamiltonian MCMC**. The two main parameters that the user must specify are how many **leapfrog steps** to take when updating the position/ momentum, and how big to make these steps. Performance can be quite sensitive to these parameters (although see (Hoffman and Gelman 2011) for a recent way to set them automatically). This method can be combined with stochastic gradient descent (Section 8.5.2) in order to handle large datasets, as explained in (Ahn et al. 2012).

Recently, a more powerful extension of this method has been developed, that exploits second-order gradient information. See (Girolami et al. 2010) for details.

24.6 Annealing methods

Many distributions are multimodal and hence hard to sample from. However, by analogy to the way metals are heated up and then cooled down in order to make the molecules align, we can imagine using a computational temperature parameter to smooth out a distribution, gradually cooling it to recover the original "bumpy" distribution. We first explain this idea in more detail in the context of an algorithm for MAP estimation. We then discuss extensions to the sampling case.

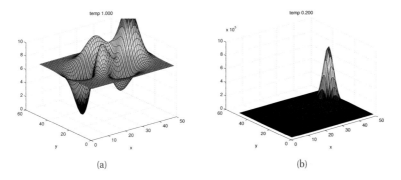

Figure 24.17 An energy surface at different temperatures. Note the different vertical scales. (a) $T = 1$. (b) $T = 0.5$. Figure generated by saDemoPeaks.

24.6.1 Simulated annealing

Simulated annealing (Kirkpatrick et al. 1983) is a stochastic algorithm that attempts to find the global optimum of a black-box function $f(\mathbf{x})$. It is closely related to the Metropolis-Hastings algorithm for generating samples from a probability distribution, which we discussed in Section 24.3. SA can be used for both discrete and continuous optimization.

The method is inspired by statistical physics. The key quantity is the **Boltzmann distribution**, which specifies that the probability of being in any particular state $\mathbf{x}$ is given by

$$p(\mathbf{x}) \propto \exp(-f(\mathbf{x})/T) \tag{24.100}$$

where $f(\mathbf{x})$ is the "energy" of the system and T is the computational temperature. As the temperature approaches 0 (so the system is cooled), the system spends more and more time in its minimum energy (most probable) state.

Figure 24.17 gives an example of a 2d function at different temperatures. At high temperatures, $T \gg 1$, the surface is approximately flat, and hence it is easy to move around (i.e., to avoid local optima). As the temperature cools, the largest peaks become larger, and the smallest peaks disappear. By cooling slowly enough, it is possible to "track" the largest peak, and thus find the global optimum. This is an example of a **continuation method**.

We can generate an algorithm from this as follows. At each step, sample a new state according to some proposal distribution $\mathbf{x}' \sim q(\cdot|\mathbf{x}_k)$. For real-valued parameters, this is often simply a random walk proposal, $\mathbf{x}' = \mathbf{x}_k + \boldsymbol{\epsilon}_k$, where $\boldsymbol{\epsilon}_k \sim \mathcal{N}(\mathbf{0}, \boldsymbol{\Sigma})$. For discrete optimization, other kinds of local moves must be defined.

Having proposed a new state, we compute

$$\alpha = \exp\left((f(\mathbf{x}) - f(\mathbf{x}'))/T\right) \tag{24.101}$$

We then accept the new state (i.e., set $\mathbf{x}_{k+1} = \mathbf{x}'$) with probability $\min(1, \alpha)$, otherwise we stay in the current state (i.e., set $\mathbf{x}_{k+1} = \mathbf{x}_k$). This means that if the new state has lower energy (is more probable), we will definitely accept it, but it it has higher energy (is less probable), we might still accept, depending on the current temperature. Thus the algorithm allows "down-hill" moves in probability space (up-hill in energy space), but less frequently as the temperature drops.

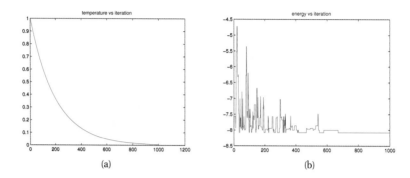

(a)   (b)

Figure 24.18 A run of simulated annealing on the energy surface in Figure 24.17. (a) Temperature vs iteration. (b) Energy vs iteration. Figure generated by `saDemoPeaks`.

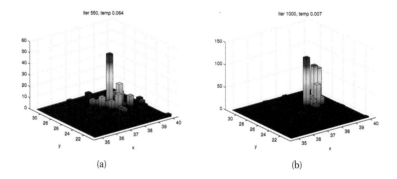

(a)   (b)

Figure 24.19 Histogram of samples from the annealed "posterior" at 2 different time points produced by simulated annealing on the energy surface shown in Figure 24.17. Note that at cold temperatures, most of the samples are concentrated near the peak at (38,25). Figure generated by `saDemoPeaks`.

The rate at which the temperature changes over time is called the **cooling schedule**. It has been shown (Kirkpatrick et al. 1983) that if one cools sufficiently slowly, the algorithm will provably find the global optimum. However, it is not clear what "sufficient slowly" means. In practice it is common to use an **exponential cooling schedule** of the following form: $T_k = T_0 C^k$, where T_0 is the initial temperature (often $T_0 \sim 1$) and C is the cooling rate (often $C \sim 0.8$). See Figure 24.18(a) for a plot of this cooling schedule. Cooling too quickly means one can get stuck in a local maximum, but cooling too slowly just wastes time. The best cooling schedule is difficult to determine; this is one of the main drawbacks of simulated annealing.

Figure 24.18(b) shows an example of simulated annealing applied to the function in Figure 24.17 using a random walk proposal. We see that the method stochastically reduces the energy over time. Figures 24.19 illustrate (a histogram of) samples drawn from the cooled probability distribution over time. We see that most of the samples are concentrated near the global maximum. When the algorithm has converged, we just return the largest value found.

24.6.2 Annealed importance sampling

We now describe a method known as **annealed importance sampling** (Neal 2001) that combines ideas from simulated annealing and importance sampling in order to draw independent samples from difficult (e.g., multimodal) distributions.

Suppose we want to sample from $p_0(\mathbf{x}) \propto f_0(\mathbf{x})$, but we cannot do so easily; for example, this might represent a multimodal posterior. Suppose however that there is an easier distribution which we can sample from, call it $p_n(\mathbf{x}) \propto f_n(\mathbf{x})$; for example, this might be the prior. We can now construct a sequence of intermediate distributions than move slowly from p_n to p_0 as follows:

$$f_j(\mathbf{x}) = f_0(\mathbf{x})^{\beta_j} f_n(\mathbf{x})^{1-\beta_j} \tag{24.102}$$

where $1 = \beta_0 > \beta_1 > \cdots > \beta_n = 0$, where β_j is an inverse temperature. (Contrast this to the scheme used by simulated annealing which has the form $f_j(\mathbf{x}) = f_0(\mathbf{x})^{\beta_j}$; this makes it hard to sample from p_n.) Furthermore, suppose we have a series of Markov chains $T_j(\mathbf{x}, \mathbf{x}')$ (from $\mathbf{x}$ to $\mathbf{x}'$) which leave each p_j invariant. Given this, we can sample $\mathbf{x}$ from p_0 by first sampling a sequence $\mathbf{z} = (\mathbf{z}_{n-1}, \ldots, \mathbf{z}_0)$ as follows: sample $\mathbf{z}_{n-1} \sim p_n$; sample $\mathbf{z}_{n-2} \sim T_{n-1}(\mathbf{z}_{n-1}, \cdot)$; ...; sample $\mathbf{z}_0 \sim T_1(\mathbf{z}_1, \cdot)$. Finally we set $\mathbf{x} = \mathbf{z}_0$ and give it weight

$$w = \frac{f_{n-1}(\mathbf{z}_{n-1})}{f_n(\mathbf{z}_{n-1})} \frac{f_{n-2}(\mathbf{z}_{n-2})}{f_{n-1}(\mathbf{z}_{n-2})} \cdots \frac{f_1(\mathbf{z}_1)}{f_2(\mathbf{z}_1)} \frac{f_0(\mathbf{z}_0)}{f_1(\mathbf{z}_0)} \tag{24.103}$$

This can be shown to be correct by viewing the algorithm as a form of importance sampling in an extended state space $\mathbf{z} = (\mathbf{z}_0, \ldots, \mathbf{z}_{n-1})$. Consider the following distribution on this state space:

$$p(\mathbf{z}) \propto f(\mathbf{z}) = f_0(\mathbf{z}_0)\tilde{T}_1(\mathbf{z}_0, \mathbf{z}_1)\tilde{T}_2(\mathbf{z}_1, \mathbf{z}_2) \cdots \tilde{T}_{n-1}(\mathbf{z}_{n-2}, \mathbf{z}_{n-1}) \tag{24.104}$$

where $\tilde{T}_j$ is the reversal of T_j:

$$\tilde{T}_j(\mathbf{z}, \mathbf{z}') = T_j(\mathbf{z}', \mathbf{z})p_j(\mathbf{z}')/p_j(\mathbf{z}) = T_j(\mathbf{z}', \mathbf{z})f_j(\mathbf{z}')/f_j(\mathbf{z}) \tag{24.105}$$

It is clear that $\sum_{\mathbf{z}_1, \ldots, \mathbf{z}_{n-1}} f(\mathbf{z}) = f_0(\mathbf{z}_0)$, so we can safely just use the $\mathbf{z}_0$ part of these sequences to recover the original ditribution.

Now consider the proposal distribution defined by the algorithm:

$$q(\mathbf{z}) \propto g(\mathbf{z}) = f_n(\mathbf{z}_{n-1})T_{n-1}(\mathbf{z}_{n-1}, \mathbf{z}_{n-2}) \cdots T_2(\mathbf{z}_2, \mathbf{z}_1)T_1(\mathbf{z}_1, \mathbf{z}_0) \tag{24.106}$$

One can show that the importance weights $w = \frac{f(\mathbf{z}_0, \ldots, \mathbf{z}_{n-1})}{g(\mathbf{z}_0, \ldots, \mathbf{z}_{n-1})}$ are given by Equation 24.103.

24.6.3 Parallel tempering

Another way to combine MCMC and annealing is to run multiple chains in parallel at different temperatures, and allow one chain to sample from another chain at a neighboring temperature. In this way, the high temperature chain can make long distance moves through the state space, and have this influence lower temperature chains. This is known as **parallel tempering**. See e.g., (Earl and Deem 2005) for details.

24.7   Approximating the marginal likelihood

The marginal likelihood $p(\mathcal{D}|M)$ is a key quantity for Bayesian model selection, and is given by

$$p(\mathcal{D}|M) = \int p(\mathcal{D}|\boldsymbol{\theta}, M)p(\boldsymbol{\theta}|M)d\boldsymbol{\theta} \tag{24.107}$$

Unfortunately, this integral is often intractable to compute, for example if we have non conjugate priors, and/or we have hidden variables. In this section, we briefly discuss some ways to approximate this expression using Monte Carlo. See (Gelman and Meng 1998) for a more extensive review.

24.7.1   The candidate method

There is a simple method for approximating the marginal likelihood known as the **Candidate method** (Chib 1995). This exploits the following identity:

$$p(\mathcal{D}|M) = \frac{p(\mathcal{D}|\boldsymbol{\theta}, M)p(\boldsymbol{\theta}|M)}{p(\boldsymbol{\theta}|\mathcal{D}, M)} \tag{24.108}$$

This holds for any value of $\boldsymbol{\theta}$. Once we have picked some value, we can evaluate $p(\mathcal{D}|\boldsymbol{\theta}, M)$ and $p(\boldsymbol{\theta}|M)$ quite easily. If we have some estimate of the posterior near $\boldsymbol{\theta}$, we can then evaluate the denominator as well. This posterior is often approximated using MCMC.

The flaw with this method is that it relies on the assumption that $p(\boldsymbol{\theta}|\mathcal{D}, M)$ has marginalized over all the modes of the posterior, which in practice is rarely possible. Consequently the method can give very inaccurate results in practice (Neal 1998).

24.7.2   Harmonic mean estimate

Newton and Raftery (1994) proposed a simple method for approximating $p(\mathcal{D})$ using the output of MCMC, as follows:

$$1/p(\mathcal{D}) \approx \frac{1}{S} \sum_{s=1}^{S} \frac{1}{p(\mathcal{D}|\boldsymbol{\theta}^s)} \tag{24.109}$$

where $\boldsymbol{\theta}^s \sim p(\boldsymbol{\theta}|\mathcal{D})$. This expression is the harmonic mean of the likelihood of the data under each sample. The theoretical correctness of this expression follows from the following identity:

$$\int \frac{1}{p(\mathcal{D}|\boldsymbol{\theta})}p(\boldsymbol{\theta}|\mathcal{D})d\boldsymbol{\theta} = \int \frac{1}{p(\mathcal{D}|\boldsymbol{\theta})}\frac{p(\mathcal{D}|\boldsymbol{\theta})p(\boldsymbol{\theta})}{p(\mathcal{D})}d\boldsymbol{\theta} = \frac{1}{p(\mathcal{D})}\int p(\boldsymbol{\theta})d\boldsymbol{\theta} = \frac{1}{p(\mathcal{D})} \tag{24.110}$$

Unfortunately, in practice this method works very poorly. Indeed, Radford Neal called this "the worst Monte Carlo method ever"[6]. The reason it is so bad is that it depends only on samples drawn from the posterior. But the posterior is often very insensitive to the prior, whereas the marginal likelihood is not. We present a better method below.

---

6. Source: radfordneal.wordpress.com/2008/08/17/the-harmonic-mean-of-the-likelihood-worst-mon te-carlo-method-ever.

24.7.3 Annealed importance sampling

We can use annealed importance sampling (Section 24.6.2) to evaluate a ratio of partition functions. Notice that $Z_0 = \int f_0(\mathbf{x})d\mathbf{x} = \int f(\mathbf{z})d\mathbf{z}$, and $Z_n = \int f_n(\mathbf{x})d\mathbf{x} = \int g(\mathbf{z})d\mathbf{z}$. Hence

$$\frac{Z_0}{Z_n} = \frac{\int f(\mathbf{z})d\mathbf{z}}{\int g(\mathbf{z})d\mathbf{z}} = \frac{\int \frac{f(\mathbf{z})}{g(\mathbf{z})}g(\mathbf{z})d\mathbf{z}}{\int g(\mathbf{z})d\mathbf{z}} = \mathbb{E}_q\left[\frac{f(\mathbf{z})}{g(\mathbf{z})}\right] \approx \frac{1}{S}\sum_{s=1}^{S} w_s \qquad (24.111)$$

where $w_s = f(\mathbf{z}_s)/g(\mathbf{z}_s)$. If f_n is a prior and f_0 is the posterior, we can estimate $Z_n = p(\mathcal{D})$ using the above equation, provided the prior has a known normalization constant Z_0. This is generally considered the method of choice for evaluating difficult partition functions.

Exercises

Exercise 24.1 Gibbs sampling from a 2D Gaussian

Suppose $\mathbf{x} \sim \mathcal{N}(\boldsymbol{\mu}, \boldsymbol{\Sigma})$, where $\boldsymbol{\mu} = (1, 1)$ and $\boldsymbol{\Sigma} = (1, -0.5; -0.5, 1)$. Derive the full conditionals $p(x_1|x_2)$ and $p(x_2|x_1)$. Implement the algorithm and plot the 1d marginals $p(x_1)$ and $p(x_2)$ as histograms. Superimpose a plot of the exact marginals.

Exercise 24.2 Gibbs sampling for a 1D Gaussian mixture model

Consider applying Gibbs sampling to a univariate mixture of Gaussians, as in Section 24.2.3. Derive the expressions for the full conditionals. Hint: if we know $z_n = j$ (say), then μ_j gets "connected" to x_n, but all other values of μ_i, for all $i \neq j$, are irrelevant. (This is an example of context-specific independence, where the structure of the graph simplifies once we have assigned values to some of the nodes.) Hence, given all the z_n values, the posteriors of the μ's should be independent, so the conditional of μ_j should be independent of $\boldsymbol{\mu}_{-j}$. (Similarly for σ_j.)

Exercise 24.3 Gibbs sampling from the Potts model

Modify the code in `gibbsDemoIsing` to draw samples from a Potts prior at different temperatures, as in Figure 19.8.

Exercise 24.4 Full conditionals for hierarchical model of Gaussian means

Let us reconsider the Gaussian-Gaussian model in Section 5.6.2 for modelling multiple related mean parameters θ_j. In this exercise we derive a Gibbs sampler instead of using EB. Suppose, following (Hoff 2009, p134)), that we use the following conjugate priors on the hyper-parameters:

$$\mu \;\sim\; \mathcal{N}(\mu_0, \gamma_0^2) \qquad (24.112)$$
$$\tau^2 \;\sim\; \mathrm{IG}(\eta_0/2, \eta_0\tau_0^2/2) \qquad (24.113)$$
$$\sigma^2 \;\sim\; \mathrm{IG}(\nu_0/2, \nu_0\sigma_0^2/2) \qquad (24.114)$$

We can set $\boldsymbol{\eta} = (\mu_0, \gamma_0, \eta_0, \tau_0, \nu_0, \sigma_0)$ to uninformative values. Given this model specification, show that the full conditionals for μ, τ, σ and the θ_j are as follows:

$$p(\mu|\theta_{1:D}, \tau^2) = \mathcal{N}(\mu|\frac{D\bar{\theta}/\tau^2 + \mu_0/\gamma_0^2}{D/\tau^2 + 1/\gamma_0^2}, [D/\tau^2 + 1/\gamma_0^2]^{-1}) \tag{24.115}$$

$$p(\theta_j|\mu, \tau^2, \mathcal{D}_j, \sigma^2) = \mathcal{N}(\theta_j|\frac{N_j\bar{x}_j/\sigma^2 + 1/\tau^2}{N_j/\sigma^2 + 1/\tau^2}, [N_j/\sigma^2 + 1/\tau^2]^{-1}) \tag{24.116}$$

$$p(\tau^2|\theta_{1:D}, \mu) = \text{IG}(\tau^2|\frac{\eta_0 + D}{2}, \frac{\eta_0\tau_0^2 + \sum_j(\theta_j - \mu)^2}{2}) \tag{24.117}$$

$$p(\sigma^2|\boldsymbol{\theta}_{1:D}, \mathcal{D}) = \text{IG}(\sigma^2|\frac{1}{2}[\nu_0 + \sum_{j=1}^{D} N_j], \frac{1}{2}[\nu_0\sigma_0^2 + \sum_{j=1}^{D}\sum_{i=1}^{N_j}(x_{ij} - \theta_j)^2]) \tag{24.118}$$

Exercise 24.5 *Gibbs sampling for robust linear regression with a Student t likelihood*

Modify the EM algorithm in Exercise 11.12 to perform Gibbs sampling for $p(\mathbf{w}, \sigma^2, \mathbf{z}|\mathcal{D}, \nu)$.

Exercise 24.6 *Gibbs sampling for probit regression*

Modify the EM algorithm in Section 11.4.6 to perform Gibbs sampling for $p(\mathbf{w}, \mathbf{z}|\mathcal{D})$. Hint: we can sample from a truncated Gaussian, $\mathcal{N}(z|\mu, \sigma)\mathbb{I}(a \leq z \leq b)$ in two steps: first sample $u \sim U(\Phi((a - \mu)/\sigma), \Phi((b - \mu)/\sigma))$, then set $z = \mu + \sigma\Phi^{-1}(u)$ (Robert 1995).

Exercise 24.7 *Gibbs sampling for logistic regression with the Student approximation*

Derive the full conditionals for the joint model defined by Equations 24.88 to 24.91.

25 *Clustering*

25.1 Introduction

Clustering is the process of grouping similar objects together. There are two kinds of inputs we might use. In **similarity-based clustering**, the input to the algorithm is an $N \times N$ **dissimilarity matrix** or **distance matrix** $\mathbf{D}$. In **feature-based clustering**, the input to the algorithm is an $N \times D$ feature matrix or design matrix $\mathbf{X}$. Similarity-based clustering has the advantage that it allows for easy inclusion of domain-specific similarity or kernel functions (Section 14.2). Feature-based clustering has the advantage that it is applicable to "raw", potentially noisy data. We will see examples of both below.

In addition to the two types of input, there are two possible types of output: **flat clustering**, also called **partitional clustering**, where we partition the objects into disjoint sets; and **hierarchical clustering**, where we create a nested tree of partitions. We will discuss both of these below. Not surprisingly, flat clusterings are usually faster to create ($O(ND)$ for flat vs $O(N^2 \log N)$ for hierarchical), but hierarchical clusterings are often more useful. Furthermore, most hierarchical clustering algorithms are deterministic and do not require the specification of K, the number of clusters, whereas most flat clustering algorithms are sensitive to the initial conditions and require some model selection method for K. (We will discuss how to choose K in more detail below.)

The final distinction we will make in this chapter is whether the method is based on a probabilistic model or not. One might wonder why we even bother discussing non-probabilistic methods for clustering. The reason is two-fold: first, they are widely used, so readers should know about them; second, they often contain good ideas, which can be used to speed up inference in a probabilistic models.

25.1.1 Measuring (dis)similarity

A dissimilarity matrix $\mathbf{D}$ is a matrix where $d_{i,i} = 0$ and $d_{i,j} \geq 0$ is a measure of "distance" between objects i and j. Subjectively judged dissimilarities are seldom distances in the strict sense, since the **triangle inequality**, $d_{i,j} \leq d_{i,k} + d_{j,k}$, often does not hold. Some algorithms require $\mathbf{D}$ to be a true distance matrix, but many do not. If we have a similarity matrix $\mathbf{S}$, we can convert it to a dissimilarity matrix by applying any monotonically decreasing function, e.g., $\mathbf{D} = \max(\mathbf{S}) - \mathbf{S}$.

The most common way to define dissimilarity between objects is in terms of the dissimilarity of their attributes:

$$\Delta(\mathbf{x}_i, \mathbf{x}_{i'}) = \sum_{j=1}^{D} \Delta_j(x_{ij}, x_{i'j}) \tag{25.1}$$

Some common attribute dissimilarity functions are as follows:

- Squared (Euclidean) distance:

$$\Delta_j(x_{ij}, x_{i'j}) = (x_{ij} - x_{i'j})^2 \tag{25.2}$$

Of course, this only makes sense if attribute j is real-valued.

- Squared distance strongly emphasizes large differences (because differences are squared). A more robust alternative is to use an ℓ_1 distance:

$$\Delta_j(x_{ij}, x_{i'j}) = |x_{ij} - x_{i'j}| \tag{25.3}$$

This is also called **city block distance**, since, in 2D, the distance can be computed by counting how many rows and columns we have to move horizontally and vertically to get from $\mathbf{x}_i$ to $\mathbf{x}_{i'}$.

- If $\mathbf{x}_i$ is a vector (e.g., a time-series of real-valued data), it is common to use the correlation coefficient (see Section 2.5.1). If the data is standardized, then $\text{corr}[\mathbf{x}_i, \mathbf{x}_{i'}] = \sum_j x_{ij} x_{i'j}$, and hence $\sum_j(x_{ij} - x_{i'j})^2 = 2(1 - \text{corr}[\mathbf{x}_i, \mathbf{x}_{i'}])$. So clustering based on correlation (similarity) is equivalent to clustering based on squared distance (dissimilarity).

- For ordinal variables, such as {low, medium, high}, it is standard to encode the values as real-valued numbers, say $1/3, 2/3, 3/3$ if there are 3 possible values. One can then apply any dissimilarity function for quantitative variables, such as squared distance.

- For categorical variables, such as {red, green, blue}, we usually assign a distance of 1 if the features are different, and a distance of 0 otherwise. Summing up over all the categorical features gives

$$\Delta(\mathbf{x}_i, \mathbf{x}_i) = \sum_{j=1}^{D} \mathbb{I}(x_{ij} \neq x_{i'j}) \tag{25.4}$$

This is called the **hamming distance**.

25.1.2   Evaluating the output of clustering methods *

The validation of clustering structures is the most difficult and frustrating part of cluster analysis. Without a strong effort in this direction, cluster analysis will remain a black art accessible only to those true believers who have experience and great courage. — Jain and Dubes (Jain and Dubes 1988)

Figure 25.1 Three clusters with labeled objects inside. Based on Figure 16.4 of (Manning et al. 2008).

Clustering is an unsupervised learning technique, so it is hard to evaluate the quality of the output of any given method. If we use probabilistic models, we can always evaluate the likelihood of a test set, but this has two drawbacks: first, it does not directly assess any clustering that is discovered by the model; and second, it does not apply to non-probabilistic methods. So now we discuss some performance measures not based on likelihood.

Intuitively, the goal of clustering is to assign points that are similar to the same cluster, and to ensure that points that are dissimilar are in different clusters. There are several ways of measuring these quantities e.g., see (Jain and Dubes 1988; Kaufman and Rousseeuw 1990). However, these internal criteria may be of limited use. An alternative is to rely on some external form of data with which to validate the method. For example, suppose we have labels for each object, as in Figure 25.1. (Equivalently, we can have a reference clustering; given a clustering, we can induce a set of labels and vice versa.) Then we can compare the clustering with the labels using various metrics which we describe below. We will use some of these metrics later, when we compare clustering methods.

25.1.2.1 Purity

Let N_{ij} be the number of objects in cluster i that belong to class j, and let $N_i = \sum_{j=1}^{C} N_{ij}$ be the total number of objects in cluster i. Define $p_{ij} = N_{ij}/N_i$; this is the empirical distribution over class labels for cluster i. We define the **purity** of a cluster as $p_i \triangleq \max_j p_{ij}$, and the overall purity of a clustering as

$$\text{purity} \triangleq \sum_i \frac{N_i}{N} p_i \tag{25.5}$$

For example, in Figure 25.1, we have that the purity is

$$\frac{6}{17}\frac{5}{6} + \frac{6}{17}\frac{4}{6} + \frac{5}{17}\frac{3}{5} = \frac{5+4+3}{17} = 0.71 \tag{25.6}$$

The purity ranges between 0 (bad) and 1 (good). However, we can trivially achieve a purity of 1 by putting each object into its own cluster, so this measure does not penalize for the number of clusters.

25.1.2.2 Rand index

Let $U = \{u_1, \ldots, u_R\}$ and $V = \{v_1, \ldots, V_C\}$ be two different partitions of the N data points, i.e., two different (flat) clusterings. For example, U might be the estimated clustering and V is reference clustering derived from the class labels. Now define a 2×2 contingency table, containing the following numbers: TP is the number of pairs that are in the same cluster in both U and V (true positives); TN is the number of pairs that are in the different clusters in both U and V (true negatives); FN is the number of pairs that are in the different clusters in U but the same cluster in V (false negatives); and FP is the number of pairs that are in the same cluster in U but different clusters in V (false positives). A common summary statistic is the **Rand index**:

$$R \triangleq \frac{TP + TN}{TP + FP + FN + TN} \tag{25.7}$$

This can be interpreted as the fraction of clustering decisions that are correct. Clearly $0 \leq R \leq 1$.

For example, consider Figure 25.1, The three clusters contain 6, 6 and 5 points, so the number of "positives" (i.e., pairs of objects put in the same cluster, regardless of label) is

$$TP + FP = \binom{6}{2} + \binom{6}{2} + \binom{5}{2} = 40 \tag{25.8}$$

Of these, the number of true positives is given by

$$TP = \binom{5}{2} + \binom{4}{2} + \binom{3}{2} + \binom{2}{2} = 20 \tag{25.9}$$

where the last two terms come from cluster 3: there are $\binom{3}{2}$ pairs labeled C and $\binom{2}{2}$ pairs labeled A. So $FP = 40 - 20 = 20$. Similarly, one can show $FN = 24$ and $TN = 72$. So the Rand index is $(20 + 72)/(20 + 20 + 24 + 72) = 0.68$.

The Rand index only achieves its lower bound of 0 if $TP = TN = 0$, which is a rare event. One can define an **adjusted Rand index** (Hubert and Arabie 1985) as follows:

$$AR \triangleq \frac{\text{index} - \text{expected index}}{\text{max index} - \text{expected index}} \tag{25.10}$$

Here the model of randomness is based on using the generalized hyper-geometric distribution, i.e., the two partitions are picked at random subject to having the original number of classes and objects in each, and then the expected value of $TP + TN$ is computed. This model can be used to compute the statistical significance of the Rand index.

The Rand index weights false positives and false negatives equally. Various other summary statistics for binary decision problems, such as the F-score (Section 5.7.2.2), can also be used. One can compute their frequentist sampling distribution, and hence their statistical significance, using methods such as bootstrap.

25.1.2.3   Mutual information

Another way to measure cluster quality is to compute the mutual information between U and V (Vaithyanathan and Dom 1999). To do this, let $p_{UV}(i, j) = \frac{|u_i \cap v_j|}{N}$ be the probability that a randomly chosen object belongs to cluster u_i in U and v_j in V. Also, let $p_U(i) = |u_i|/N$ be the be the probability that a randomly chosen object belongs to cluster u_i in U; define

$p_V(j) = |v_j|/N$ similarly. Then we have

$$\mathbb{I}(U, V) = \sum_{i=1}^{R} \sum_{j=1}^{C} p_{UV}(i, j) \log \frac{p_{UV}(i, j)}{p_U(i) p_V(j)} \qquad (25.11)$$

This lies between 0 and $\min\{\mathbb{H}(U), \mathbb{H}(V)\}$. Unfortunately, the maximum value can be achieved by using lots of small clusters, which have low entropy. To compensate for this, we can use the **normalized mutual information**,

$$NMI(U, V) \triangleq \frac{\mathbb{I}(U, V)}{(\mathbb{H}(U) + \mathbb{H}(V))/2} \qquad (25.12)$$

This lies between 0 and 1. A version of this that is adjusted for chance (under a particular random data model) is described in (Vinh et al. 2009). Another variant, called **variation of information**, is described in (Meila 2005).

25.2 Dirichlet process mixture models

The simplest approach to (flat) clustering is to use a finite mixture model, as we discussed in Section 11.2.3. This is sometimes called **model-based clustering**, since we define a probabilistic model of the data, and optimize a well-defined objective (the likelihood or posterior), as opposed to just using some heuristic algorithm.

The principle problem with finite mixture models is how to choose the number of components K. We discussed several techniques in Section 11.5. However, in many cases, there is no well-defined number of clusters. Even in the simple 2d height-weight data (Figure 1.8), it is not clear if the "correct" value of K should be 2, 3, or 4. It would be much better if we did not have to choose K at all.

In this section, we discuss **infinite mixture models**, in which we do not impose any a priori bound on K. To do this, we will use a **non-parametric prior** based on the **Dirichlet process** (DP). This allows the number of clusters to grow as the amount of data increases. It will also prove useful later when we discuss hiearchical clustering.

The topic of **non-parametric Bayes** is currently very active, and we do not have space to go into details (see (Hjort et al. 2010) for a recent book on the topic). Instead we just give a brief review of the DP and its application to mixture modeling, based on the presentation in (Sudderth 2006, sec 2.2).

25.2.1 From finite to infinite mixture models

Consider a finite mixture model, as shown in Figure 25.2(a). The usual representation is as follows:

$$p(\mathbf{x}_i | z_i = k, \boldsymbol{\theta}) = p(\mathbf{x}_i | \boldsymbol{\theta}_k) \qquad (25.13)$$
$$p(z_i = k | \boldsymbol{\pi}) = \pi_k \qquad (25.14)$$
$$p(\boldsymbol{\pi} | \alpha) = \text{Dir}(\boldsymbol{\pi} | (\alpha/K)\mathbf{1}_K) \qquad (25.15)$$

The form of $p(\boldsymbol{\theta}_k | \lambda)$ is chosen to be conjugate to $p(\mathbf{x}_i | \boldsymbol{\theta}_k)$. We can write $p(\mathbf{x}_i | \boldsymbol{\theta}_k)$ as $\mathbf{x}_i \sim F(\boldsymbol{\theta}_{z_i})$, where F is the observation distribution. Similarly, we can write $\boldsymbol{\theta}_k \sim H(\lambda)$, where H is the prior.

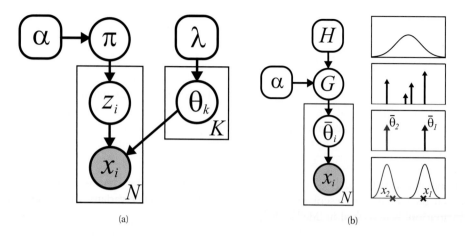

Figure 25.2 Two different representations of a finite mixture model. Left: traditional representation. Right: representation where parameters are samples from G, a discrete measure. The picture on the right illustrates the case where $K = 4$, and we sample 4 Gaussian means θ_k from a Gaussian prior $H(.|\lambda)$. The height of the spikes reflects the mixing weights π_k. This weighted sum of delta functions is G. We then generate two parameters, $\bar{\theta}_1$ and $\bar{\theta}_2$, from G, one per data point. Finally, we generate two data points, x_1 and x_2, from $\mathcal{N}(\bar{\theta}_1, \sigma^2)$ and $\mathcal{N}(\bar{\theta}_2, \sigma^2)$. Source: Figure 2.9 of (Sudderth 2006) . Used with kind permission of Erik Sudderth.

An equivalent representation for this model is shown in Figure 25.2(b). Here $\bar{\theta}_i$ is the parameter used to generate observation $\mathbf{x}_i$; these parameters are sampled from distribution G, which has the form

$$G(\boldsymbol{\theta}) = \sum_{k=1}^{K} \pi_k \delta_{\boldsymbol{\theta}_k}(\boldsymbol{\theta}) \tag{25.16}$$

where $\boldsymbol{\pi} \sim \text{Dir}(\frac{\alpha}{K}\mathbf{1})$, and $\boldsymbol{\theta}_k \sim H$. Thus we see that G is a finite mixture of delta functions, centered on the cluster parameters $\boldsymbol{\theta}_k$. The probability that $\bar{\boldsymbol{\theta}}_i$ is equal to $\boldsymbol{\theta}_k$ is exactly π_k, the prior probability for that cluster.

If we sample from this model, we will always (with probability one) get exactly K clusters, with data points scattered around the cluster centers. We would like a more flexible model, that can generate a variable number of clusters. Furthermore, the more data we generate, the more likely we should be to see a new cluster. The way to do this is to replace the discrete distribution G with a **random probability measure**. Below we will show that the Dirichlet process, denoted $G \sim \text{DP}(\alpha, H)$, is one way to do this.

Before we go into the details, we show some samples from this non-parametric model in Figure 25.3. We see that it has the desired properties of generating a variable number of clusters, with more clusters as the amount of data increases. The resulting samples look much more like real data than samples from a finite mixture model.

Of course, working with an "infinite" model sounds scary. Fortunately, as we show below, even though this model is potentially infinite, we can perform inference using an amount of computation that is not only tractable, but is often much less than that required to fit a set

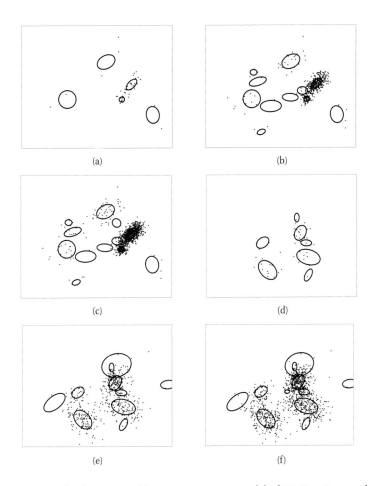

Figure 25.3 Some samples from a Dirichlet process mixture model of 2D Gaussians, with concentration parameter $\alpha = 1$. From left to right, we show $N = 50$, $N = 500$ and $N = 1000$ samples. Each row is a different run. We also show the model parameters as ellipses, which are sampled from a vague NIW base distribution. Based on Figure 2.25 of (Sudderth 2006). Figure generated by `dpmSampleDemo`, written by Yee-Whye Teh.

of finite mixture models for different K. The intuitive reason is that we can get evidence that certain values of K are appropriate (have high posterior support) long before we have been able to estimate the parameters, so we can focus our computational efforts on models of appropriate complexity. Thus going to the infinite limit can sometimes be faster. This is especially true when we have multiple model selection problems to solve.

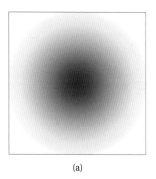
(a)

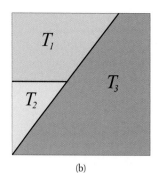
(b)

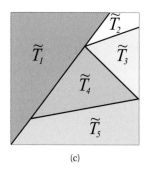
(c)

Figure 25.4 (a) A base measure H on a 2d space Θ. (b) One possible partition into $K = 3$ regions, where the shading of cell T_k is proportional to $\mathbb{E}\left[G(T_k)\right] = H(T_k)$. (c) A refined partition into $K = 5$ regions. Source: Figure 2.21 of (Sudderth 2006). Used with kind permission of Erik Sudderth.

25.2.2 The Dirichlet process

Recall from Chapter 15 that a Gaussian process is a distribution over functions of the form $f : \mathcal{X} \to R$. It is defined implicitly by the requirement that $p(f(\mathbf{x}_1), \ldots, f(\mathbf{x}_N))$ be jointly Gaussian, for any set of points $\mathbf{x}_i \in \mathcal{X}$. The parameters of this Gaussian can be computed using a mean function $\mu()$ and covariance (kernel) function $K()$. We write $f \sim \text{GP}(\mu(), K())$. Furthermore, the GP is consistently defined, so that $p(f(\mathbf{x}_1))$ can be derived from $p(f(\mathbf{x}_1), f(\mathbf{x}_2))$, etc.

A **Dirichlet process** is a distribution over probability measures $G : \Theta \to \mathbb{R}^+$, where we require $G(\theta) \geq 0$ and $\int_\Theta G(\theta) d\theta = 1$. The DP is defined implicitly by the requirement that $(G(T_1), \ldots, G(T_K))$ has a joint Dirichlet distribution

$$\text{Dir}(\alpha H(T_1), \ldots, \alpha H(T_K)) \tag{25.17}$$

for any finite partition $(T_1, \ldots, T_K)$ of Θ. If this is the case, we write $G \sim \text{DP}(\alpha, H)$, where α is called the **concentration parameter** and H is called the **base measure**.[1]

An example of a DP is shown in Figure 25.4, where the base measure is a 2d Gaussian. The distribution over all the cells, $p(G(T_1), \ldots, G(T_K))$, is Dirichlet, so the marginals in each cell are beta distributed:

$$\text{Beta}(\alpha H(T_i), \alpha \sum_{j \neq i} H(T_j)) \tag{25.18}$$

The DP is consistently defined in the sense that if T_1 and T_2 form a partition of $\tilde{T}_1$, then $G(T_1) + G(T_2)$ and $G(\tilde{T}_1)$ both follow the same beta distribution.

Recall that if $\boldsymbol{\pi} \sim \text{Dir}(\boldsymbol{\alpha})$, and $z | \boldsymbol{\pi} \sim \text{Cat}(\boldsymbol{\pi})$, then we can integrate out $\boldsymbol{\pi}$ to get the predictive distribution for the Dirichlet-multinoulli model:

$$z \sim \text{Cat}(\alpha_1/\alpha_0, \ldots, \alpha_K/\alpha_0) \tag{25.19}$$

---

1. Unlike a GP, knowing something about $G(T_k)$ does not tell us anything about $G(T_{k'})$, beyond the sum-to-one constraint; we say that the DP is a **neutral process**. Other stochastic processes can be defined that do not have this property, but they are not so computationally convenient.

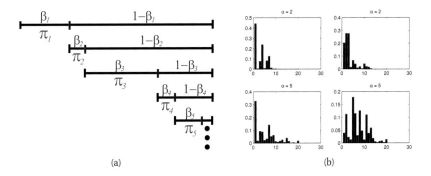

(a)   (b)

Figure 25.5 Illustration of the stick breaking construction. (a) We have a unit length stick, which we break at a random point β_1; the length of the piece we keep is called π_1; we then recursively break off pieces of the remaining stick, to generate $\pi_2, \pi_3, \ldots$. Source: Figure 2.22 of (Sudderth 2006). Used with kind permission of Erik Sudderth. (b) Samples of π_k from this process for $\alpha = 2$ (top row) and $\alpha = 5$ (bottom row). Figure generated by `stickBreakingDemo`, written by Yee-Whye Teh.

where $\alpha_0 = \sum_k \alpha_k$. In other words, $p(z = k | \boldsymbol{\alpha}) = \alpha_k / \alpha_0$. Also, the updated posterior for $\boldsymbol{\pi}$ given one observation is given by

$$\boldsymbol{\pi} | z \sim \text{Dir}(\alpha_1 + \mathbb{I}(z = 1), \ldots, \alpha_K + \mathbb{I}(z = K)) \tag{25.20}$$

The DP generalizes this to arbitrary partitions. If $G \sim \text{DP}(\alpha, H)$, then $p(\boldsymbol{\theta} \in T_i) = H(T_i)$ and the posterior is

$$p(G(T_1), \ldots, G(T_K) | \boldsymbol{\theta}, \alpha, H) = \text{Dir}(\alpha H(T_1) + \mathbb{I}(\boldsymbol{\theta} \in T_1), \ldots, \alpha H(T_K) + \mathbb{I}(\boldsymbol{\theta} \in T_K)) \tag{25.21}$$

This holds for any set of partitions. Hence if we observe multiple samples $\overline{\boldsymbol{\theta}}_i \sim G$, the new posterior is given by

$$G | \overline{\boldsymbol{\theta}}_1, \ldots, \overline{\boldsymbol{\theta}}_N, \alpha, H \sim \text{DP}\left(\alpha + N, \frac{1}{\alpha + N}\left(\alpha H + \sum_{i=1}^{N} \delta_{\boldsymbol{\theta}_i}\right)\right) \tag{25.22}$$

Thus we see that the DP effectively defines a conjugate prior for arbitrary measurable spaces. The concentration parameter α is like the effective sample size of the base measure H.

25.2.2.1 Stick breaking construction of the DP

Our discussion so far has been very abstract. We now give a constructive definition for the DP, known as the **stick-breaking construction**.

Let $\boldsymbol{\pi} = \{\pi_k\}_{k=1}^{\infty}$ be an infinite sequence of mixture weights derived from the following process:

$$\beta_k \quad \sim \quad \text{Beta}(1, \alpha) \tag{25.23}$$

$$\pi_k \quad = \quad \beta_k \prod_{l=1}^{k-1} (1 - \beta_l) = \beta_k (1 - \sum_{l=1}^{k-1} \pi_l) \tag{25.24}$$

This is often denoted by

$$\pi \sim \text{GEM}(\alpha) \tag{25.25}$$

where GEM stands for Griffiths, Engen and McCloskey (this term is due to (Ewens 1990)). Some samples from this process are shown in Figure 25.5. One can show that this process process will terminate with probability 1, although the number of elements it generates increases with α. Furthermore, the size of the π_k components decreases on average.

Now define

$$G(\boldsymbol{\theta}) = \sum_{k=1}^{\infty} \pi_k \delta_{\boldsymbol{\theta}_k}(\boldsymbol{\theta}) \tag{25.26}$$

where $\pi \sim \text{GEM}(\alpha)$ and $\boldsymbol{\theta}_k \sim H$. Then one can show that $G \sim \text{DP}(\alpha, H)$.

As a consequence of this construction, we see that samples from a DP are **discrete with probability one**. In other words, if you keep sampling it, you will get more and more repetitions of previously generated values. So if we sample $\overline{\boldsymbol{\theta}}_i \sim G$, we will see repeated values; let us number the unique values $\boldsymbol{\theta}_1$, $\boldsymbol{\theta}_2$, etc. Data sampled from $\overline{\boldsymbol{\theta}}_i$ will therefore cluster around the $\boldsymbol{\theta}_k$. This is evident in Figure 25.3, where most data comes from the Gaussians with large π_k values, represented by ellipses with thick borders. This is our first indication that the DP might be useful for clustering.

25.2.2.2 The Chinese restaurant process (CRP)

Working with infinite dimensional sticks is problematic. However, we can exploit the clustering property to draw samples form a GP, as we now show.

The key result is this: If $\overline{\boldsymbol{\theta}}_i \sim G$ are N observations from $G \sim \text{DP}(\alpha, H)$, taking on K distinct values $\boldsymbol{\theta}_k$, then the predictive distribution of the next observation is given by

$$p(\overline{\boldsymbol{\theta}}_{N+1} = \boldsymbol{\theta}|\overline{\boldsymbol{\theta}}_{1:N}, \alpha, H) = \frac{1}{\alpha + N}\left(\alpha H(\boldsymbol{\theta}) + \sum_{k=1}^{K} N_k \delta_{\overline{\boldsymbol{\theta}}_k}(\boldsymbol{\theta})\right) \tag{25.27}$$

where N_k is the number of previous observations equal to $\boldsymbol{\theta}_k$. This is called the **Polya urn** or **Blackwell-MacQueen** sampling scheme. This provides a constructive way to sample from a DP.

It is much more convenient to work with discrete variables z_i which specify which value of $\boldsymbol{\theta}_k$ to use. That is, we define $\overline{\boldsymbol{\theta}}_i = \boldsymbol{\theta}_{z_i}$. Based on the above expression, we have

$$p(z_{N+1} = z|\mathbf{z}_{1:N}, \alpha) = \frac{1}{\alpha + N}\left(\alpha \mathbb{I}(z = k^*) + \sum_{k=1}^{K} N_k \mathbb{I}(z = k)\right) \tag{25.28}$$

where k^* represents a new cluster index that has not yet been used. This is called the **Chinese restaurant process** or **CRP**, based on the seemingly infinite supply of tables at certain Chinese restaurants. The analogy is as follows: The tables are like clusters, and the customers are like observations. When a person enters the restaurant, he may choose to join an existing table with probability proportional to the number of people already sitting at this table (the N_k); otherwise, with a probability that diminishes as more people enter the room (due to the $1/(\alpha + N)$ term),

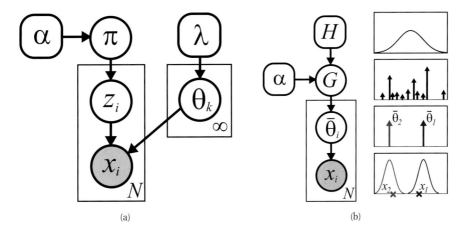

Figure 25.6 Two views of a DP mixture model. Left: infinite number of clusters parameters, $\boldsymbol{\theta}_k$, and $\pi \sim \text{GEM}(\alpha)$. Right: G is drawn from a DP. Compare to Figure 25.2. Source: Figure 2.24 of (Sudderth 2006). Used with kind permission of Erik Sudderth.

he may choose to sit at a new table k^*. The result is a distribution over **partitions of the integers**, which is like a distribution of customers to tables.

The fact that currently occupied tables are more likely to get new customers is sometimes called the **rich get richer** phenomenon. Indeed, one can derive an expression for the distribution of cluster sizes induced by this prior process; it is basically a power law. The number of occupied tables K almost surely approaches $\alpha \log(N)$ as $N \to \infty$, showing that the model complexity will indeed grow logarithmically with dataset size. More flexible priors over cluster sizes can also be defined, such as the two-parameter **Pitman-Yor process**.

25.2.3 Applying Dirichlet processes to mixture modeling

The DP is not particularly useful as a model for data directly, since data vectors rarely repeat exactly. However, it is useful as a prior for the parameters of a stochastic data generating mechanism, such as a mixture model. To create such a model, we follow exactly the same setup as Section 11.2, but we define $G \sim \text{DP}(\alpha, H)$. Equivalently, we can write the model as follows:

$$\boldsymbol{\pi} \quad \sim \quad \text{GEM}(\alpha) \tag{25.29}$$

$$z_i \quad \sim \quad \boldsymbol{\pi} \tag{25.30}$$

$$\boldsymbol{\theta}_k \quad \sim \quad H(\lambda) \tag{25.31}$$

$$\mathbf{x}_i \quad \sim \quad F(\boldsymbol{\theta}_{z_i}) \tag{25.32}$$

This is illustrated in Figure 25.6. We see that G is now a random draw of an unbounded number of parameters $\boldsymbol{\theta}_k$ from the base distribution H, each with weight π_k. Each data point $\mathbf{x}_i$ is generated by sampling its own "private" parameter $\bar{\boldsymbol{\theta}}_i$ from G. As we get more and more data, it becomes increasingly likely that $\bar{\boldsymbol{\theta}}_i$ will be equal to one of the $\boldsymbol{\theta}_k$'s we have seen before, and thus $\mathbf{x}_i$ will be generated close to an existing datapoint.

25.2.4 Fitting a DP mixture model

The simplest way to fit a DPMM is to modify the collapsed Gibbs sampler of Section 24.2.4. From Equation 24.23 we have

$$p(z_i = k|\mathbf{z}_{-i}, \mathbf{x}, \alpha, \boldsymbol{\lambda}) \quad \propto \quad p(z_i = k|\mathbf{z}_{-i}, \alpha)p(\mathbf{x}_i|\mathbf{x}_{-i}, z_i = k, \mathbf{z}_{-i}, \boldsymbol{\lambda}) \tag{25.33}$$

By exchangeability, we can assume that z_i is the last customer to enter the restaurant. Hence the first term is given by

$$p(z_i|\mathbf{z}_{-i}, \alpha) = \frac{1}{\alpha + N - 1} \left(\alpha \mathbb{I}(z_i = k^*) + \sum_{k=1}^{K} N_{k,-i} \mathbb{I}(z_i = k) \right) \tag{25.34}$$

where K is the number of clusters used by $\mathbf{z}_{-i}$, and k^* is a new cluster. Another way to write this is as follows:

$$p(z_i = k|\mathbf{z}_{-i}, \alpha) = \begin{cases} \frac{N_{k,-i}}{\alpha + N - 1} & \text{if } k \text{ has been seen before} \\ \frac{\alpha}{\alpha + N - 1} & \text{if } k \text{ is a new cluster} \end{cases} \tag{25.35}$$

Interestingly, this is equivalent to Equation 24.26, which has the form $p(z_i = k|\mathbf{z}_{-i}, \alpha) = \frac{N_{k,-i} + \alpha/K}{\alpha + N - 1}$, in the $K \to \infty$ limit (Rasmussen 2000; Neal 2000).

To compute the second term, $p(\mathbf{x}_i|\mathbf{x}_{-i}, z_i = k, \mathbf{z}_{-i}, \boldsymbol{\lambda})$, let us partition the data $\mathbf{x}_{-i}$ into clusters based on $\mathbf{z}_{-i}$. Let $\mathbf{x}_{-i,c} = \{\mathbf{x}_j : z_j = c, j \neq i\}$ be the data assigned to cluster c. If $z_i = k$, then $\mathbf{x}_i$ is conditionally independent of all the data points except those assigned to cluster k. Hence we have

$$p(\mathbf{x}_i|\mathbf{x}_{-i}, \mathbf{z}_{-i}, z_i = k, \boldsymbol{\lambda}) = p(\mathbf{x}_i|\mathbf{x}_{-i,k}, \boldsymbol{\lambda}) = \frac{p(\mathbf{x}_i, \mathbf{x}_{-i,k}|\boldsymbol{\lambda})}{p(\mathbf{x}_{-i,k}|\boldsymbol{\lambda})} \tag{25.36}$$

where

$$p(\mathbf{x}_i, \mathbf{x}_{-i,k}|\boldsymbol{\lambda}) = \int p(\mathbf{x}_i|\boldsymbol{\theta}_k) \left[\prod_{j \neq i: z_j = k} p(\mathbf{x}_j|\boldsymbol{\theta}_k) \right] H(\boldsymbol{\theta}_k|\boldsymbol{\lambda}) d\boldsymbol{\theta}_k \tag{25.37}$$

is the marginal likelihood of all the data assigned to cluster k, including i, and $p(\mathbf{x}_{-i,k}|\boldsymbol{\lambda})$ is an analogous expression excluding i. Thus we see that the term $p(\mathbf{x}_i|\mathbf{x}_{-i}, \mathbf{z}_{-i}, z_i = k, \boldsymbol{\lambda})$ is the posterior preditive distribution for cluster k evaluated at $\mathbf{x}_i$.

If $z_i = k^*$, corresponding to a new cluster, we have

$$p(\mathbf{x}_i|\mathbf{x}_{-i}, \mathbf{z}_{-i}, z_i = k^*, \boldsymbol{\lambda}) = p(\mathbf{x}_i|\boldsymbol{\lambda}) = \int p(\mathbf{x}_i|\boldsymbol{\theta}) H(\boldsymbol{\theta}|\boldsymbol{\lambda}) d\boldsymbol{\theta} \tag{25.38}$$

which is just the prior predictive distribution for a new cluster evaluated at $\mathbf{x}_i$.

See Algorithm 25.1 for the pseudocode. (This is called "Algorithm 3" in (Neal 2000).) This is very similar to collapsed Gibbs for finite mixtures except that we have to consider the case $z_i = k^*$.

An example of this procedure in action is shown in Figure 25.7. The sample clusterings, and the induced posterior over K, seems reasonable. The method tends to rapidly discover a good

---

Algorithm 25.1: Collapsed Gibbs sampler for DP mixtures

---

1 **for** *each* $i = 1 : N$ *in random order* **do**
2   Remove $\mathbf{x}_i$'s sufficient statistics from old cluster z_i ;
3   **for** *each* $k = 1 : K$ **do**
4   Compute $p_k(\mathbf{x}_i) = p(\mathbf{x}_i | \mathbf{x}_{-i}(k))$;
5   Set $N_{k,-i} = \dim(\mathbf{x}_{-i}(k))$;
6   Compute $p(z_i = k | \mathbf{z}_{-i}, \mathcal{D}) = \frac{N_{k,-i}}{\alpha + N - 1}$;
7   Compute $p_*(\mathbf{x}_i) = p(\mathbf{x}_i | \boldsymbol{\lambda})$;
8   Compute $p(z_i = * | \mathbf{z}_{-i}, \mathcal{D}) = \frac{\alpha}{\alpha + N - 1}$;
9   Normalize $p(z_i | \cdot)$;
10   Sample $z_i \sim p(z_i | \cdot)$;
11   Add $\mathbf{x}_i$'s sufficient statistics to new cluster z_i ;
12   If any cluster is empty, remove it and decrease K;

---

clustering. By contrast, Gibbs sampling (and EM) for a finite mixture model often gets stuck in poor local optima (not shown). This is because the DPMM is able to create extra redundant clusters early on, and to use them to escape local optima. Figure 25.8 shows that most of the time, the DPMM converges more rapidly than a finite mixture model.

A variety of other fitting methods have been proposed. (Daume 2007a) shows how one can use A^* search and beam search to quickly find an approximate MAP estimate. (Mansinghka et al. 2007) discusses how to fit a DPMM online using particle filtering, which is a like a stochastic version of beam search. This can be more efficient than Gibbs sampling, particularly for large datasets. (Kurihara et al. 2006) develops a variational approximation that is even faster (see also (Zobay 2009)). Extensions to the case of non-conjugate priors are discussed in (Neal 2000).

Another important issue is how to set the hyper-parameters. For the DP, the value of α does not have much impact on predictive accuracy, but it does affect the number of clusters. One approach is to put a $\text{Ga}(a, b)$ prior for α, and then to form its posterior, $p(\alpha | K, N, a, b)$, using auxiliary variable methods (Escobar and West 1995). Alternatively, one can use empirical Bayes (McAuliffe et al. 2006). Similarly, for the base distribution, we can either sample the hyper-parameters $\boldsymbol{\lambda}$ (Rasmussen 2000) or use empirical Bayes (McAuliffe et al. 2006).

25.3 Affinity propagation

Mixture models, whether finite or infinite, require access to the raw $N \times D$ data matrix, and need to specify a generative model of the data. An alternative approach takes as input an $N \times N$ similarity matrix, and then tries to identify **examplars**, which will act as cluster centers. The K-medoids or **K-centers** algorithm (Section 14.4.2) is one approach, but it can suffer from local minima. Here we describe an alternative approach called **affinity propagation** (Frey and Dueck 2007) that works substantially better in practice.

The idea is that each data point must choose another data point as its exemplar or centroid; some data points will choose themselves as centroids, and this will automatically determine the

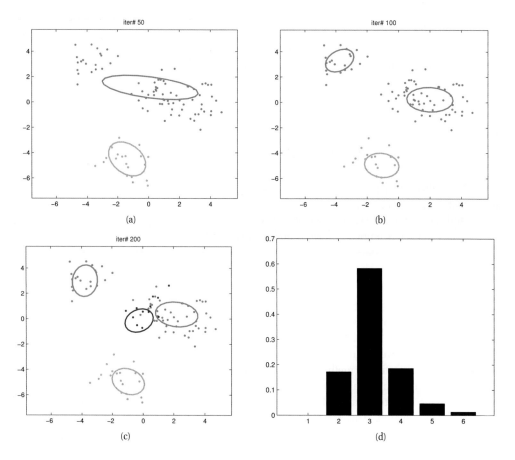

Figure 25.7 100 data points in 2d are clustered using a DP mixture fit with collapsed Gibbs sampling. We show samples from the posterior after 50,100, 200 samples. We also show the posterior over K, based on 200 samples, discarding the first 50 as burnin. Figure generated by `dpmGauss2dDemo`, written by Yee Whye Teh.

number of clusters. More precisely, let $c_i \in \{1, \ldots, N\}$ represent the centroid for datapoint i. The goal is to maximize the following function

$$S(\mathbf{c}) = \sum_{i=1}^{N} s(i, c_i) + \sum_{k=1}^{N} \delta_k(\mathbf{c}) \tag{25.39}$$

The first term measures the similarity of each point to its centroid. The second term is a penalty term that is $-\infty$ if some data point i has chosen k as its exemplar (i.e., $c_i = k$), but k has not chosen itself as an exemplar (i.e., we do not have $c_k = k$). More formally,

$$\delta_k(\mathbf{c}) = \begin{cases} -\infty & \text{if } c_k \neq k \text{ but } \exists i : c_i = k \\ 0 & \text{otherwise} \end{cases} \tag{25.40}$$

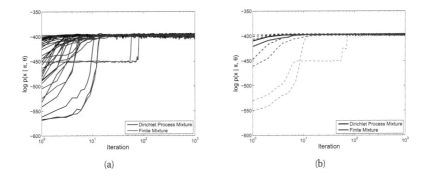

Figure 25.8 Comparison of collapsed Gibbs samplers for a DP mixture (dark blue) and a finite mixture (light red) with $K = 4$ applied to $N = 300$ data points (similar to Figure 25.7). Left: logprob vs iteration for 20 different starting values. Right: median (thick line) and quantiles (dashed lines) over 100 different starting values. Source: Figure 2.27 of (Sudderth 2006). Used with kind permission of Erik Sudderth.

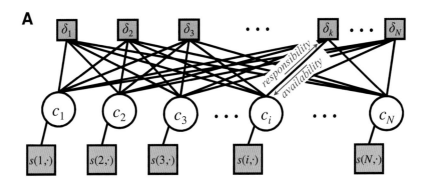

Figure 25.9 Factor graphs for affinity propagation. Circles are variables, squares are factors. Each c_i node has N possible states. From Figure S2 of (Frey and Dueck 2007). Used with kind permission of Brendan Frey.

The objective function can be represented as a factor graph. We can either use N nodes, each with N possible values, as shown in Figure 25.9, or we can use N^2 binary nodes (see (Givoni and Frey 2009) for the details). We will assume the former representation.

We can find a strong local maximum of the objective by using max-product loopy belief propagation (Section 22.2). Referring to the model in Figure 25.9, each variable nodes c_i sends a message to each factor node δ_k. It turns out that this vector of N numbers can be reduced to a scalar message, denote $r_{i \to k}$, known as the responsibility. This is a measure of how much i thinks k would make a good exemplar, compared to all the other exemplars i has looked at. In addition, each factor node δ_k sends a message to each variable node c_i. Again this can be reduced to a scalar message, $a_{i \leftarrow k}$, known as the availability. This is a measure of how strongly k believes it should an exemplar for i, based on all the other data points k has looked at.

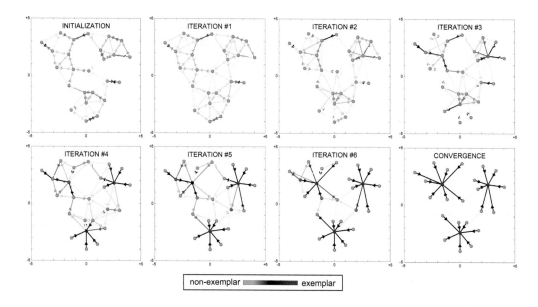

Figure 25.10 Example of affinity propagation. Each point is colored coded by how much it wants to be an exemplar (red is the most, green is the least). This can be computed by summing up all the incoming availability messages and the self-similarity term. The darkness of the $i \rightarrow k$ arrow reflects how much point i wants to belong to exemplar k. From Figure 1 of (Frey and Dueck 2007). Used with kind permission of Brendan Frey.

As usual with loopy BP, the method might oscillate, and convergence is not guaranteed. However, by using damping, the method is very reliable in practice. If the graph is densely connected, message passing takes $O(N^2)$ time, but with sparse similarity matrices, it only takes $O(E)$ time, where E is the number of edges or non-zero entries in $\mathbf{S}$.

The number of clusters can be controlled by scaling the diagonal terms $S(i, i)$, which reflect how much each data point wants to be an exemplar. Figure 25.10 gives a simple example of some 2d data, where the negative Euclidean distance was used to measured similarity. The $S(i, i)$ values were set to be the median of all the pairwise similarities. The result is 3 clusters. Many other results are reported in (Frey and Dueck 2007), who show that the method significantly outperforms K-medoids.

25.4 Spectral clustering

An alternative view of clustering is in terms of **graph cuts**. The idea is we create a weighted undirected graph $\mathbf{W}$ from the similarity matrix $\mathbf{S}$, typically by using the nearest neighbors of each point; this ensures the graph is sparse, which speeds computation. If we want to find a partition into K clusters, say $A_1, \ldots, A_K$, one natural criterion is to minimize

$$\text{cut}(A_1, \ldots, A_K) \triangleq \frac{1}{2} \sum_{k=1}^{K} W(A_k, \overline{A}_k) \tag{25.41}$$

where $\overline{A}_k = V \setminus A_k$ is the complement of A_k, and $W(A, B) \triangleq \sum_{i \in A, j \in B} w_{ij}$. For $K = 2$ this problem is easy to solve. Unfortunately the optimal solution often just partitions off a single data point from the rest. To ensure the sets are reasonably large, we can define the **normalized cut** to be

$$\text{Ncut}(A_1, \ldots, A_K) \triangleq \frac{1}{2} \sum_{k=1}^{K} \frac{\text{cut}(A_k, \overline{A}_k)}{\text{vol}(A_k)} \tag{25.42}$$

where $\text{vol}(A) \triangleq \sum_{i \in A} d_i$, and $d_i = \sum_{j=1}^{N} w_{ij}$ is the weighted degree of node i. This splits the graph into K clusters such that nodes within each cluster are similar to each other, but are different to nodes in other clusters.

We can formulate the Ncut problem in terms of searching for binary vectors $\mathbf{c}_i \in \{0, 1\}^N$, where $c_{ik} = 1$ if point i belongs to cluster k, that minimize the objective. Unfortunately this is NP-hard (Wagner and Wagner 1993). Affinity propagation is one way to solve the problem. Another is to relax the constraints that $\mathbf{c}_i$ be binary, and allow them to be real-valued. The result turns into an eigenvector problem known as **spectral clustering** (see e.g., (Shi and Malik 2000)). In general, the technique of performing eigenalysis of graphs is called **spectral graph theory** (Chung 1997).

Going into the details would take us too far afield, but below we give a very brief summary, based on (von Luxburg 2007), since we will encounter some of these ideas later on.

25.4.1 Graph Laplacian

Let $\mathbf{W}$ be a symmetric weight matrix for a graph, where $w_{ij} = w_{ji} \geq 0$. Let $\mathbf{D} = \text{diag}(d_i)$ be a diagonal matrix containing the weighted degree of each node. We define the **graph Laplacian** as follows:

$$\mathbf{L} \triangleq \mathbf{D} - \mathbf{W} \tag{25.43}$$

This matrix has various important properties. Because each row sums to zero, we have that $\mathbf{1}$ is an eigenvector with eigenvalue 0. Furthermore, the matrix is symmetric and positive semi-definite. To see this, note that

$$\mathbf{f}^T \mathbf{L} \mathbf{f} = \mathbf{f}^T \mathbf{D} \mathbf{f} - \mathbf{f}^T \mathbf{W} \mathbf{f} = \sum_i d_i f_i^2 - \sum_{i,j} f_i f_j w_{ij} \tag{25.44}$$

$$= \frac{1}{2} \left(\sum_i d_i f_i^2 - 2 \sum_{i,j} f_i f_j w_{ij} + \sum_j d_j f_j^2 \right) = \frac{1}{2} \sum_{i,j} w_{ij} (f_i - f_j)^2 \tag{25.45}$$

Hence $\mathbf{f}^T \mathbf{L} \mathbf{f} \geq 0$ for all $\mathbf{f} \in \mathbb{R}^N$. Consequently we see that $\mathbf{L}$ has N non-negative, real-valued eigenvalues, $0 \leq \lambda_1 \leq \lambda_2 \leq \ldots \leq \lambda_N$.

To get some intuition as to why $\mathbf{L}$ might be useful for graph-based clustering, we note the following result.

Theorem 25.4.1. *The set of eigenvectors of* $\mathbf{L}$ *with eigenvalue 0 is spanned by the indicator vectors* $\mathbf{1}_{A_1}, \ldots, \mathbf{1}_{A_K}$, *where* A_k *are the* K *connected components of the graph.*

Proof. Let us start with the case $K = 1$. If $\mathbf{f}$ is an eigenvector with eigenvalue 0, then $0 = \sum_{ij} w_{ij}(f_i - f_j)^2$. If two nodes are connected, so $w_{ij} > 0$, we must have that $f_i = f_j$. Hence $\mathbf{f}$ is constant for all vertices which are connected by a path in the graph. Now suppose $K > 1$. In this case, $\mathbf{L}$ will be block diagonal. A similar argument to the above shows that we will have K indicator functions, which "select out" the connected components. □

This suggests the following algorithm. Compute the first K eigenvectors $\mathbf{u}_k$ of $\mathbf{L}$. Let $\mathbf{U} = [\mathbf{u}_1, \ldots, \mathbf{u}_K]$ be an $N \times K$ matrix with the eigenvectors in its columns. Let $\mathbf{y}_i \in \mathbb{R}^K$ be the i'th row of $\mathbf{U}$. Since these $\mathbf{y}_i$ will be piecewise constant, we can apply K-means clustering to them to recover the connected components. Now assign point i to cluster k iff row i of $\mathbf{Y}$ was assigned to cluster k.

In reality, we do not expect a graph derived from a real similarity matrix to have isolated connected components — that would be too easy. But it is reasonable to suppose the graph is a small "perturbation" from such an ideal. In this case, one can use results from perturbation theory to show that the eigenvectors of the perturbed Laplacian will be close to these ideal indicator functions (Ng et al. 2001).

Note that this approach is related to kernel PCA (Section 14.4.4). In particular, KPCA uses the largest eigenvectors of $\mathbf{W}$; these are equivalent to the smallest eigenvectors of $\mathbf{I} - \mathbf{W}$. This is similar to the above method, which computes the smallest eigenvectors of $\mathbf{L} = \mathbf{D} - \mathbf{W}$. See (Bengio et al. 2004) for details. In practice, spectral clustering gives much better results than KPCA.

25.4.2 Normalized graph Laplacian

In practice, it is important to normalize the graph Laplacian, to account for the fact that some nodes are more highly connected than others. There are two common ways to do this. One method, used in e.g., (Shi and Malik 2000; Meila 2001), creates a random walk matrix as follows (the connection to random walks will be explained below):

$$\mathbf{L}_{rw} \triangleq \mathbf{D}^{-1}\mathbf{L} = \mathbf{I} - \mathbf{D}^{-1}\mathbf{W} \tag{25.46}$$

The eigenvalues and eigenvectors of $\mathbf{L}$ and $\mathbf{L}_{rw}$ are closely related to each other (see (von Luxburg 2007) for details). Furthermore, one can show that for $\mathbf{L}_{rw}$, the eigenspace of 0 is again spanned by the indicator vectors $\mathbf{1}_{A_k}$. This suggests the following algorithm: find the smallest K eigenvectors of $\mathbf{L}_{rw}$, create $\mathbf{U}$, cluster the rows of $\mathbf{U}$ using K-means, then infer the partitioning of the original points (Shi and Malik 2000). (Note that the eigenvectors/ values of $\mathbf{L}_{rw}$ are equivalent to the generalized eigenvectors/ values of $\mathbf{L}$, which solve $\mathbf{Lu} = \lambda \mathbf{DU}$.)

Another method, used in e.g., (Ng et al. 2001), creates a symmetric matrix

$$\mathbf{L}_{sym} \triangleq \mathbf{D}^{-\frac{1}{2}}\mathbf{L}\mathbf{D}^{-\frac{1}{2}} = \mathbf{I} - \mathbf{D}^{-\frac{1}{2}}\mathbf{W}\mathbf{D}^{-\frac{1}{2}} \tag{25.47}$$

This time the eigenspace of 0 is spanned by $\mathbf{D}^{\frac{1}{2}}\mathbf{1}_{A_k}$. This suggest the following algorithm: find the smallest K eigenvectors of $\mathbf{L}_{sym}$, create $\mathbf{U}$, normalize each row to unit norm by creating $t_{ij} = u_{ij}/\sqrt{(\sum_k u_{ik}^2)}$ to make the matrix $\mathbf{T}$, cluster the rows of $\mathbf{T}$ using K-means, then infer the partitioning of the original points.

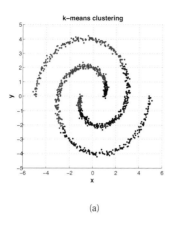
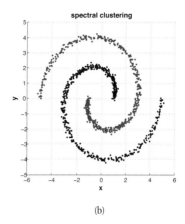

(a)   (b)

Figure 25.11 Clustering data consisting of 2 spirals. (a) K-means. (b) Spectral clustering. Figure generated by `spectralClusteringDemo`, written by Wei-Lwun Lu.

There is an interesting connection between Ncuts and random walks on a graph (Meila 2001). First note that $\mathbf{P} = \mathbf{D}^{-1}\mathbf{W} = \mathbf{I} - \mathbf{L}_{rw}$ is a stochastic matrix, where $p_{ij} = w_{ij}/d_i$ can be interpreted as the probability of going from i to j. If the graph is connected and non-bipartite, it possesses a unique stationary distribution $\boldsymbol{\pi} = (\pi_1, \ldots, \pi_N)$, where $\pi_i = d_i/\text{vol}(V)$. Furthermore, one can show that

$$\text{Ncut}(A, \overline{A}) = p(\overline{A}|A) + p(A|\overline{A}) \tag{25.48}$$

This means that we are looking for a cut such that a random walk rarely makes transitions from A to $\overline{A}$ or vice versa.

25.4.3 Example

Figure 25.11 illustrates the method in action. In Figure 25.11(a), we see that K-means does a poor job of clustering, since it implicitly assumes each cluster corresponds to a spherical Gaussian. Next we try spectral clustering. We define a similarity matrix using the Gaussian kernel. We compute the first two eigenvectors of the Laplacian. From this we can infer the clustering in Figure 25.11(b).

Since the method is based on finding the smallest K eigenvectors of a sparse matrix, it takes $O(N^3)$ time. However, a variety of methods can be used to scale it up for large datasets (see e.g., (Yan et al. 2009)).

25.5 Hierarchical clustering

Mixture models, whether finite or infinite, produce a "flat" clustering. Often we want to learn a **hierarchical clustering**, where clusters can be nested inside each other.

There are two main approaches to hierarchical clustering: bottom-up or **agglomerative clustering**, and top-down or **divisive clustering**. Both methods take as input a dissimilarity matrix

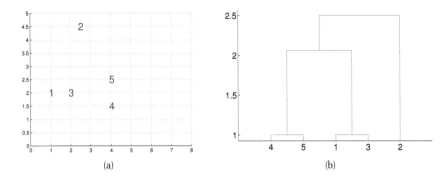

(a)   (b)

Figure 25.12   (a) An example of single link clustering using city block distance. Pairs (1,3) and (4,5) are both distance 1 apart, so get merged first. (b) The resulting dendrogram. Based on Figure 7.5 of (Alpaydin 2004). Figure generated by `agglomDemo`.

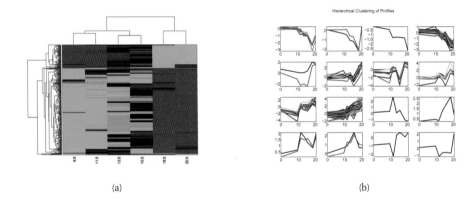

(a)   (b)

Figure 25.13   Hierarchical clustering applied to the yeast gene expression data. (a) The rows are permuted according to a hierarchical clustering scheme (average link agglomerative clustering), in order to bring similar rows close together. (b) 16 clusters induced by cutting the average linkage tree at a certain height. Figure generated by `hclustYeastDemo`.

between the objects. In the bottom-up approach, the most similar groups are merged at each step. In the top-down approach, groups are split using various different criteria. We give the details below.

Note that agglomerative and divisive clustering are both just heuristics, which do not optimize any well-defined objective function. Thus it is hard to assess the quality of the clustering they produce in any formal sense. Furthermore, they will always produce a clustering of the input data, even if the data has no structure at all (e.g., it is random noise). Later in this section we will discuss a probabilistic version of hierarchical clustering that solves both these problems.

Algorithm 25.2: Agglomerative clustering

---

1 *initialize* clusters as singletons: **for** $i \leftarrow 1$ **to** n **do** $C_i \leftarrow \{i\}$;
2 ;
3 *initialize* set of clusters available for merging: $S \leftarrow \{1, \ldots, n\}$;
4 **repeat**
5   Pick 2 most similar clusters to merge: $(j, k) \leftarrow \arg\min_{j,k \in S} d_{j,k}$;
6   Create new cluster $C_\ell \leftarrow C_j \cup C_k$;
7   Mark j and k as unavailable: $S \leftarrow S \setminus \{j, k\}$;
8   **if** $C_\ell \neq \{1, \ldots, n\}$ **then**
9   Mark ℓ as available, $S \leftarrow S \cup \{\ell\}$;
10   **foreach** $i \in S$ **do**
11   Update dissimilarity matrix $d(i, \ell)$;
12 **until** *no more clusters are available for merging*;

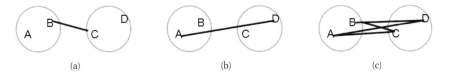

Figure 25.14 Illustration of (a) Single linkage. (b) Complete linkage. (c) Average linkage.

25.5.1 Agglomerative clustering

Agglomerative clustering starts with N groups, each initially containing one object, and then at each step it merges the two most similar groups until there is a single group, containing all the data. See Algorithm 25.2 for the pseudocode. Since picking the two most similar clusters to merge takes $O(N^2)$ time, and there are $O(N)$ steps in the algorithm, the total running time is $O(N^3)$. However, by using a priority queue, this can be reduced to $O(N^2 \log N)$ (see e.g., (Manning et al. 2008, ch. 17) for details). For large N, a common heuristic is to first run K-means, which takes $O(KND)$ time, and then apply hierarchical clustering to the estimated cluster centers.

The merging process can be represented by a **binary tree**, called a **dendrogram**, as shown in Figure 25.12(b). The initial groups (objects) are at the leaves (at the bottom of the figure), and every time two groups are merged, we join them in the tree. The height of the branches represents the dissimilarity between the groups that are being joined. The root of the tree (which is at the top) represents a group containing all the data. If we cut the tree at any given height, we induce a clustering of a given size. For example, if we cut the tree in Figure 25.12(b) at height 2, we get the clustering $\{\{\{4, 5\}, \{1, 3\}\}, \{2\}\}$. We discuss the issue of how to choose the height/ number of clusters below.

A more complex example is shown in Figure 25.13(a), where we show some gene expression data. If we cut the tree in Figure 25.13(a) at a certain height, we get the 16 clusters shown in Figure 25.13(b).

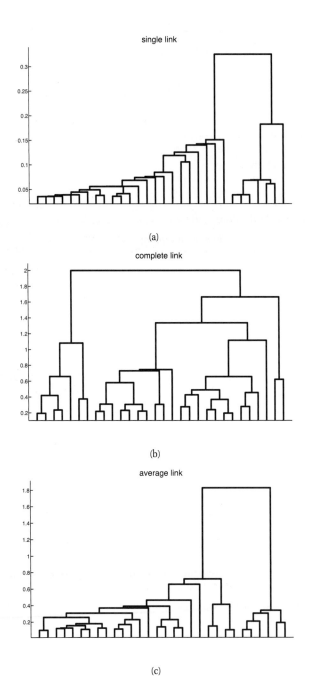

Figure 25.15 Hierarchical clustering of yeast gene expression data. (a) Single linkage. (b) Complete linkage. (c) Average linkage. Figure generated by `hclustYeastDemo`.

There are actually three variants of agglomerative clustering, depending on how we define the dissimilarity between groups of objects. These can give quite different results, as shown in Figure 25.15. We give the details below.

25.5.1.1 Single link

In **single link clustering**, also called **nearest neighbor clustering**, the distance between two groups G and H is defined as the distance between the two closest members of each group:

$$d_{SL}(G, H) = \min_{i \in G, i' \in H} d_{i,i'} \tag{25.49}$$

See Figure 25.14(a).

The tree built using single link clustering is a minimum spanning tree of the data, which is a tree that connects all the objects in a way that minimizes the sum of the edge weights (distances). To see this, note that when we merge two clusters, we connect together the two closest members of the clusters; this adds an edge between the corresponding nodes, and this is guaranteed to be the "lightest weight" edge joining these two clusters. And once two clusters have been merged, they will never be considered again, so we cannot create cycles. As a consequence of this, we can actually implement single link clustering in $O(N^2)$ time, whereas the other variants take $O(N^3)$ time.

25.5.1.2 Complete link

In **complete link clustering**, also called **furthest neighbor clustering**, the distance between two groups is defined as the distance between the two most distant pairs:

$$d_{CL}(G, H) = \max_{i \in G, i' \in H} d_{i,i'} \tag{25.50}$$

See Figure 25.14(b).

Single linkage only requires that a single pair of objects be close for the two groups to be considered close together, regardless of the similarity of the other members of the group. Thus clusters can be formed that violate the **compactness** property, which says that all the observations within a group should be similar to each other. In particular if we define the **diameter** of a group as the largest dissimilarity of its members, $d_G = \max_{i \in G, i' \in G} d_{i,i'}$, then we can see that single linkage can produce clusters with large diameters. Complete linkage represents the opposite extreme: two groups are considered close only if all of the observations in their union are relatively similar. This will tend to produce clusterings with small diameter, i.e., compact clusters.

25.5.1.3 Average link

In practice, the preferred method is **average link clustering**, which measures the average distance between all pairs:

$$d_{avg}(G, H) = \frac{1}{n_G n_H} \sum_{i \in G} \sum_{i' \in H} d_{i,i'} \tag{25.51}$$

where n_G and n_H are the number of elements in groups G and H. See Figure 25.14(c).

Average link clustering represents a compromise between single and complete link clustering. It tends to produce relatively compact clusters that are relatively far apart. However, since it involves averaging of the $d_{i,i'}$'s, any change to the measurement scale can change the result. In contrast, single linkage and complete linkage are invariant to monotonic transformations of $d_{i,i'}$, since they leave the relative ordering the same.

25.5.2   Divisive clustering

Divisive clustering starts with all the data in a single cluster, and then recursively divides each cluster into two daughter clusters, in a top-down fashion. Since there are $2^{N-1} - 1$ ways to split a group of N items into 2 groups, it is hard to compute the optimal split, so various heuristics are used. One approach is pick the cluster with the largest diameter, and split it in two using the K-means or K-medoids algorithm with $K = 2$. This is called the **bisecting K-means** algorithm (Steinbach et al. 2000). We can repeat this until we have any desired number of clusters. This can be used as an alternative to regular K-means, but it also induces a hierarchical clustering.

Another method is to build a minimum spanning tree from the dissimilarity graph, and then to make new clusters by breaking the link corresponding to the largest dissimilarity. (This actually gives the same results as single link agglomerative clustering.)

Yet another method, called **dissimilarity analysis** (Macnaughton-Smith et al. 1964), is as follows. We start with a single cluster containing all the data, $G = \{1, \ldots, N\}$. We then measure the average dissimilarity of $i \in G$ to all the other $i' \in G$:

$$d_i^G = \frac{1}{n_G} \sum_{i' \in G} d_{i,i'} \tag{25.52}$$

We remove the most dissimilar object and put it in its own cluster H:

$$i^* = \arg\max_{i \in G} d_i^G, \quad G = G \setminus \{i^*\}, \quad H = \{i^*\} \tag{25.53}$$

We now continue to move objects from G to H until some stopping criterion is met. Specifically, we pick a point i^* to move that maximizes the average dissimilarity to each $i' \in G$ but minimizes the average dissimilarity to each $i' \in H$:

$$d_i^H = \frac{1}{n_H} \sum_{i' \in H} d_{i,i'}, \quad i^* = \arg\max_{i \in G} d_i^G - d_i^H \tag{25.54}$$

We continue to do this until $d_i^G - d_i^H$ is negative. The final result is that we have split G into two daughter clusters, G and H. We can then recursively call the algorithm on G and/or H, or on any other node in the tree. For example, we might choose to split the node G whose average dissimilarity is highest, or whose maximum dissimilarity (i.e., diameter) is highest. We continue the process until the average dissimilarity within each cluster is below some threshold, and/or all clusters are singletons.

Divisive clustering is less popular than agglomerative clustering, but it has two advantages. First, it can be faster, since if we only split for a constant number of levels, it takes just $O(N)$ time. Second, the splitting decisions are made in the context of seeing all the data, whereas bottom-up methods make myopic merge decisions.

25.5.3 Choosing the number of clusters

It is difficult to choose the "right" number of clusters, since a hierarchical clustering algorithm will always create a hierarchy, even if the data is completely random. But, as with choosing K for K-means, there is the hope that there will be a visible "gap" in the lengths of the links in the dendrogram (represent the dissimilarity between merged groups) between natural clusters and unnatural clusters. Of course, on real data, this gap might be hard to detect. In Section 25.5.4, we will present a Bayesian approach to hierarchical clustering that nicely solves this problem.

25.5.4 Bayesian hierarchical clustering

There are several ways to make probabilistic models which produce results similar to hierarchical clustering, e.g., (Williams 2000; Neal 2003b; Castro et al. 2004; Lau and Green 2006). Here we present one particular approach called **Bayesian hierarchical clustering** (Heller and Ghahramani 2005). Algorithmically it is very similar to standard bottom-up agglomerative clustering, and takes comparable time, whereas several of the other techniques referenced above are much slower. However, it uses Bayesian hypothesis tests to decide which clusters to merge (if any), rather than computing the similarity between groups of points in some ad-hoc way. These hypothesis tests are closely related to the calculations required to do inference in a Dirichlet process mixture model, as we will see. Furthermore, the input to the model is a data matrix, not a dissimilarity matrix.

25.5.4.1 The algorithm

Let $\mathcal{D} = \{\mathbf{x}_1, \ldots, \mathbf{x}_N\}$ represent all the data, and let $\mathcal{D}_i$ be the set of datapoints at the leaves of the substree T_i. At each step, we compare two trees T_i and T_j to see if they should be merged into a new tree. Define $\mathcal{D}_{ij}$ as their merged data, and let $M_{ij} = 1$ if they should be merged, and $M_{ij} = 0$ otherwise.

The probability of a merge is given by

$$r_{ij} \triangleq \frac{p(\mathcal{D}_{ij}|M_{ij} = 1)p(M_{ij} = 1)}{p(\mathcal{D}_{ij}|T_{ij})} \tag{25.55}$$

$$p(\mathcal{D}_{ij}|T_{ij}) = p(\mathcal{D}_{ij}|M_{ij} = 1)p(M_{ij} = 1) + p(\mathcal{D}_{ij}|M_{ij} = 0)p(M_{ij} = 0) \tag{25.56}$$

Here $p(M_{ij} = 1)$ is the prior probability of a merge, which can be computed using a bottom-up algorithm described below. We now turn to the likelihood terms. If $M_{ij} = 1$, the data in $\mathcal{D}_{ij}$ is assumed to come from the same model, and hence

$$p(\mathcal{D}_{ij}|M_{ij} = 1) = \int \left[\prod_{\mathbf{x}_n \in \mathcal{D}_{ij}} p(\mathbf{x}_n|\boldsymbol{\theta}) \right] p(\boldsymbol{\theta}|\lambda)d\boldsymbol{\theta} \tag{25.57}$$

If $M_{ij} = 0$, the data in $\mathcal{D}_{ij}$ is assumed to have been generated by each tree independently, so

$$p(\mathcal{D}_{ij}|M_{ij} = 0) = p(\mathcal{D}_i|T_i)p(\mathcal{D}_j|T_j) \tag{25.58}$$

These two terms will have already been computed by the bottom-up process. Consequently we have all the quantities we need to decide which trees to merge. See Algorithm 25.3 for the pseudocode, assuming $p(M_{ij})$ is uniform. When finished, we can cut the tree at points where $r_{ij} < 0.5$.

Algorithm 25.3: Bayesian hierarchical clustering

---

1 Initialize $\mathcal{D}_i = \{\mathbf{x}_i\}$, $i = 1 : N$;
2 Compute $p(\mathcal{D}_i|T_i)$, $i = 1 : N$;
3 **repeat**
4   **for** *each pair of clusters i, j* **do**
5   Compute $p(\mathcal{D}_{ij}|T_{ij})$
6   Find the pair $\mathcal{D}_i$ and $\mathcal{D}_j$ with highest merge probability r_{ij};
7   Merge $\mathcal{D}_k := \mathcal{D}_i \cup \mathcal{D}_j$;
8   Delete $\mathcal{D}_i$, $\mathcal{D}_j$;
9 **until** *all clusters merged*;

25.5.4.2 The connection with Dirichlet process mixture models

In this section, we will establish the connection between BHC and DPMMs. This will in turn give us an algorithm to compute the prior probabilities $p(M_{ij} = 1)$.

Note that the marginal likelihood of a DPMM, summing over all $2^N - 1$ partitions, is given by

$$p(\mathcal{D}_k) = \sum_{v \in \mathcal{V}} p(v) p(\mathcal{D}_v) \tag{25.59}$$

$$p(v) = \frac{\alpha^{m_v} \prod_{l=1}^{m_v} \Gamma(n_l^v)}{\frac{\Gamma(n_k + \alpha)}{\Gamma(\alpha)}} \tag{25.60}$$

$$p(\mathcal{D}_v) = \prod_{l=1}^{m_v} p(\mathcal{D}_l^v) \tag{25.61}$$

where $\mathcal{V}$ is the set of all possible partitions of $\mathcal{D}_k$, $p(v)$ is the probability of partition v, m_v is the number of clusters in partition v, n_l^v is the number of points in cluster l of partition v, $\mathcal{D}_l^v$ are the points in cluster l of partition v, and n_k are the number of points in $\mathcal{D}_k$.

One can show (Heller and Ghahramani 2005) that $p(\mathcal{D}_k|T_k)$ computed by the BHC algorithm is similar to $p(\mathcal{D}_k)$ given above, except for the fact that it only sums over partitions which are consistent with tree T_k. (The number of tree-consistent partitions is exponential in the number of data points for balanced binary trees, but this is obviously a subset of all possible partitions.) In this way, we can use the BHC algorithm to compute a lower bound on the marginal likelihood of the data from a DPMM. Furthermore, we can interpret the algorithm as greedily searching through the exponentially large space of tree-consistent partitions to find the best ones of a given size at each step.

We are now in a position to compute $\pi_k = p(M_k = 1)$, for each node k with children i and j. This is equal to the probability of cluster $\mathcal{D}_k$ coming from the DPMM, relative to all other partitions of $\mathcal{D}_k$ consistent with the current tree. This can be computed as follows: initialize $d_i = \alpha$ and $\pi_i = 1$ for each leaf i; then as we build the tree, for each internal node k, compute $d_k = \alpha\Gamma(n_k) + d_i d_j$, and $\pi_k = \frac{\alpha\Gamma(n_k)}{d_k}$, where i and j are k's left and right children.

Data Set	Single Linkage	Complete Linkage	Average Linkage	BHC
Synthetic	0.599 ± 0.033	0.634 ± 0.024	0.668 ± 0.040	$\mathbf{0.828 \pm 0.025}$
Newsgroups	0.275 ± 0.001	0.315 ± 0.008	0.282 ± 0.002	$\mathbf{0.465 \pm 0.016}$
Spambase	0.598 ± 0.017	0.699 ± 0.017	0.668 ± 0.019	$\mathbf{0.728 \pm 0.029}$
Digits	0.224 ± 0.004	0.299 ± 0.006	0.342 ± 0.005	$\mathbf{0.393 \pm 0.015}$
Fglass	0.478 ± 0.009	0.476 ± 0.009	$\mathbf{0.491 \pm 0.009}$	0.467 ± 0.011

Table 25.1 Purity scores for various hierarchical clustering schemes applied to various data sets. The synthetic data has $N = 200, D = 2, C = 4$ and real features. Newsgroups is extracted from the 20 newsgroups dataset ($D = 500, N = 800, C = 4$, binary features). Spambase has $N = 100, C = 2, D = 57$, binary features. Digits is the CEDAR Buffalo digits ($N = 200, C = 10, D = 64$, binarized features). Fglass is forensic glass dataset ($N = 214, C = 6, D = 9$, real features). Source: Table 1 of (Heller and Ghahramani 2005). Used with kind permission of Katherine Heller.

25.5.4.3 Learning the hyper-parameters

The model has two free-parameters: α and λ, where λ are the hyper-parameters for the prior on the parameters $\boldsymbol{\theta}$. In (Heller and Ghahramani 2005), they show how one can back-propagate gradients of the form $\frac{\partial p(\mathcal{D}_k|T_k)}{\partial \lambda}$ through the tree, and thus perform an empirical Bayes estimate of the hyper-parameters.

25.5.4.4 Experimental results

(Heller and Ghahramani 2005) compared BHC with traditional agglomerative clustering algorithms on various data sets in terms of purity scores. The results are shown in Table 25.1. We see that BHC did much better than the other methods on all datasets except the forensic glass one.

Figure 25.16 visualizes the tree structure estimated by BHC and agglomerative hierarchical clustering (AHC) on the newsgroup data (using a beta-Bernoulli model). The BHC tree is clearly superior (look at the colors at the leaves, which represent class labels). Figure 25.17 is a zoom-in on the top few nodes of these two trees. BHC splits off clusters concerning sports from clusters concerning cars and space. AHC keeps sports and cars merged together. Although sports and cars both fall under the same "rec" newsgroup heading (as opposed to space, that comes under the "sci" newsgroup heading), the BHC clustering still seems more reasonable, and this is borne out by the quantitative purity scores.

BHC has also been applied to gene expression data, with good results (Savage et al. 2009).

25.6 Clustering datapoints and features

So far, we have been concentrating on clustering datapoints. But each datapoint is often described by multiple features, and we might be interested in clustering them as well. Below we describe some methods for doing this.

4 Newsgroups Average Linkage Clustering

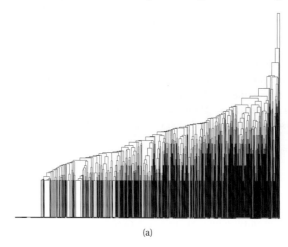

(a)

4 Newsgroups Bayesian Hierarchical Clustering

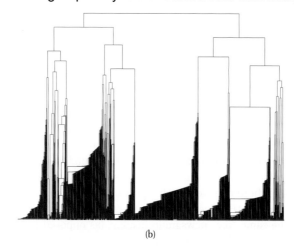

(b)

Figure 25.16 Hierarchical clustering applied to 800 documents from 4 newsgroups (red is rec.autos, blue is rec.sport.baseball, green is rec.sport.hockey, and magenta is sci.space). Top: average linkage hierarchical clustering. Bottom: Bayesian hierarchical clustering. Each of the leaves is labeled with a color, according to which newsgroup that document came from. We see that the Bayesian method results in a clustering that is more consistent with these labels (which were not used during model fitting). Source: Figure 7 of (Heller and Ghahramani 2005). Used with kind permission of Katherine Heller.

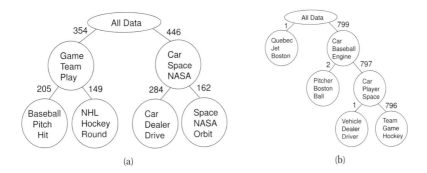

Figure 25.17 Zoom-in on the top nodes in the trees of Figure 25.16. (a) Bayesian method. (b) Average linkage. We show the 3 most probable words per cluster. The number of documents at each cluster is also given. Source: Figure 5 of (Heller and Ghahramani 2005). Used with kind permission of Katherine Heller.

25.6.1 Biclustering

Clustering the rows and columns is known as **biclustering** or **coclustering**. This is widely used in bioinformatics, where the rows often represent genes and the columns represent conditions. It can also be used for collaborative filtering, where the rows represent users and the columns represent movies.

A variety of ad hoc methods for biclustering have been proposed; see (Madeira and Oliveira 2004) for a review. Here we present a simple probabilistic generative model, based on (Kemp et al. 2006) (see also (Sheng et al. 2003) for a related approach). The idea is to associate each row and each column with a latent indicator, $r_i \in \{1, \ldots, K^r\}$, $c_j \in \{1, \ldots, K^c\}$. We then assume the data are iid across samples and across features within each block:

$$p(\mathbf{x}|\mathbf{r}, \mathbf{c}, \boldsymbol{\theta}) = \prod_i \prod_j p(x_{ij}|r_i, c_j, \boldsymbol{\theta}) = p(x_{ij}|\boldsymbol{\theta}_{r_i, c_j}) \tag{25.62}$$

where $\boldsymbol{\theta}_{a,b}$ are the parameters for row cluster a and column cluster b. Rather than using a finite number of clusters for the rows and columns, we can use a Dirchlet process, as in the infinite relational model which we discuss in Section 27.6.1. We can fit this model using e.g., (collapsed) Gibbs sampling.

The behavior of this model is illustrated in Figure 25.18. The data has the form $X(i, j) = 1$ iff animal i has feature j, where $i = 1 : 50$ and $j = 1 : 85$. The animals represent whales, bears, horses, etc. The features represent properties of the habitat (jungle, tree, coastal), or anatomical properties (has teeth, quadrapedal), or behavioral properties (swims, eats meat), etc. The model, using a Bernoulli likelihood, was fit to the data. It discovered 12 animal clusters and 33 feature clusters. For example, it discovered a bicluster that represents the fact that mammals tend to have aquatic features.

25.6.2 Multi-view clustering

The problem with biclustering is that each object (row) can only belong to one cluster. Intuitively, an object can have multiple roles, and can be assigned to different clusters depending on which

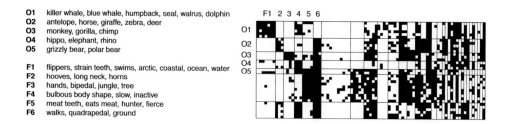

O1	killer whale, blue whale, humpback, seal, walrus, dolphin
O2	antelope, horse, giraffe, zebra, deer
O3	monkey, gorilla, chimp
O4	hippo, elephant, rhino
O5	grizzly bear, polar bear

F1	flippers, strain teeth, swims, arctic, coastal, ocean, water
F2	hooves, long neck, horns
F3	hands, bipedal, jungle, tree
F4	bulbous body shape, slow, inactive
F5	meat teeth, eats meat, hunter, fierce
F6	walks, quadrapedal, ground

Figure 25.18   Illustration of biclustering. We show 5 of the 12 animal clusters, and 6 of the 33 feature clusters. The original data matrix is shown, partitioned according to the discovered clusters. From Figure 3 of (Kemp et al. 2006). Used with kind permission of Charles Kemp.

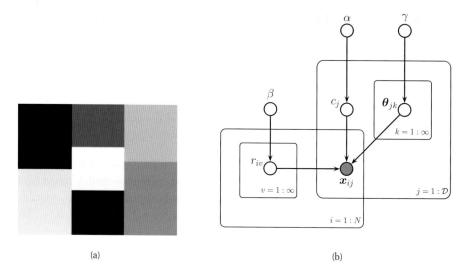

(a)   (b)

Figure 25.19   (a) Illustration of multi-view clustering. Here we have 3 views (column partitions). In the first view, we have 2 clusters (row partitions). In the second view, we have 3 clusters. In the third view, we have 2 clusters. The number of views and partitions are inferred from data. Rows within each colored block are assumed to generated iid; however, each column can have a different distributional form, which is useful for modeling discrete and continuous data. From Figure 1 of (Guan et al. 2010). Used with kind permission of Jennifer Dy. (b) Corresponding DGM.

subset of features you use. For example, in the animal dataset, we may want to group the animals on the basis of anatomical features (e.g., mammals are warm blooded, reptiles are not), or on the basis of behavioral features (e.g., predators vs prey).

We now present a model that can capture this phenomenon. This model was independently proposed in (Shafto et al. 2006; Mansinghka et al. 2011), who call it **crosscat** (for cross-categorization), and in (Guan et al. 2010; Cui et al. 2010), who call it (non-parametric) **multi-clust**. (See also (Rodriguez and Ghosh 2011) for a very similar model.) The idea is that we partition the columns (features) into V groups or **views**, so $c_j \in \{1, \dots, V\}$, where $j \in \{1, \dots, D\}$ indexes features. We will use a Dirichlet process prior for $p(c)$, which allows V to grow automatically. Then for each partition of the columns (i.e., each view), call it v, we partition the rows, again using a DP, as illustrated in Figure 25.19(a). Let $r_{iv} \in \{1, \ldots, K(v)\}$ be the cluster to which the i'th row belongs in view v. Finally, having partitioned the rows and columns, we generate the data: we assume all the rows and columns within a block are iid. We can define the model more precisely as follows:

$$
\begin{aligned}
p(\mathbf{c}, \mathbf{r}, \mathcal{D}) &= p(\mathbf{c})p(\mathbf{r}|\mathbf{c})p(\mathcal{D}|\mathbf{r}, \mathbf{c}) & (25.63) \\
p(\mathbf{c}) &= \mathrm{DP}(\mathbf{c}|\alpha) & (25.64) \\
p(\mathbf{r}|\mathbf{c}) &= \prod_{v=1}^{V(\mathbf{c})} \mathrm{DP}(\mathbf{r}_v|\beta) & (25.65) \\
p(\mathcal{D}|\mathbf{r}, \mathbf{c}) &= \prod_{v=1}^{V(\mathbf{c})} \prod_{j:c_j=v} \left[\prod_{k=1}^{K(\mathbf{r}_v)} \int \prod_{i:r_{iv}=k} p(x_{ij}|\boldsymbol{\theta}_{jk})p(\boldsymbol{\theta}_{jk})d\boldsymbol{\theta}_{jk} \right] & (25.66)
\end{aligned}
$$

See Figure 25.19(b) for the DGM.[2]

If the data is binary, and we use a $\mathrm{Beta}(\gamma, \gamma)$ prior for $\boldsymbol{\theta}_{jk}$, the likelihood reduces to

$$
p(\mathcal{D}|\mathbf{r}, \mathbf{c}, \gamma) = \prod_{v=1}^{V(\mathbf{c})} \prod_{j:c_j=v} \prod_{k=1}^{K(\mathbf{r}_v)} \frac{\mathrm{Beta}(n_{j,k,v} + \gamma, \overline{n}_{j,k,v} + \gamma)}{\mathrm{Beta}(\gamma, \gamma)} \tag{25.67}
$$

where $n_{j,k,v} = \sum_{i:r_{i,v}=k} \mathbb{I}(x_{ij} = 1)$ counts the number of features which are on in the j'th column for view v and for row cluster k. Similarly, $\overline{n}_{j,k,v}$ counts how many features are off. The model is easily extended to other kinds of data, by replacing the beta-Bernoulli with, say, the Gaussian-Gamma-Gaussian model, as discussed in (Guan et al. 2010; Mansinghka et al. 2011).

Approximate MAP estimation can be done using stochastic search (Shafto et al. 2006), and approximate inference can be done using variational Bayes (Guan et al. 2010) or Gibbs sampling (Mansinghka et al. 2011). The hyper-parameter γ for the likelihood can usually be set in a non-informative way. However, results are more sensitive to the parameters of the DP priors, since α controls the number of column partitions, and β controls the number of row partitions. Hence a more robust technique is to infer the hyper-parameters using MH. This also speeds up convergence (Mansinghka et al. 2011).

Figure 25.20 illustrates the model applied to some binary data containing 22 animals and 106 features. The figures shows the (approximate) MAP partition. The first partition of the columns contains taxonomic features, such as "has bones", "is warm-blooded", "lays eggs", etc. This divides the animals into birds, reptiles/ amphibians, mammals, and invertebrates. The second partition of the columns contains features that are treated as noise, with no apparent structure (except for the single row labeled "frog"). The third partition of the columns contains ecological features like "dangerous", "carnivorous", "lives in water", etc. This divides the animals into prey, land predators, sea predators and air predators. Thus each animal (row) can belong to a different

---

2. The dependence between $\mathbf{r}$ and $\mathbf{c}$ is not shown, since it is not a dependence between the values of r_{iv} and c_j, but between the cardinality of v and c_j. In other words, the number of row partitions we need to specify (the number of views, indexed by v) depends on the number of column partitions (clusters) that we have.

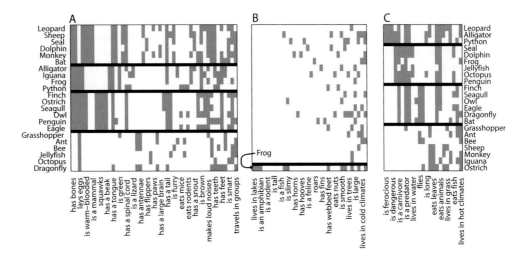

Figure 25.20 MAP estimate produced by the crosscat system when applied to a binary data matrix of animals (rows) by features (columns). See text for details. Source: Figure 7 of (Shafto et al. 2006) . Used with kind permission of Vikash Mansingkha.

cluster depending on what set of features are considered. Uncertainty about the partitions can be handled by sampling.

It is interesting to compare this model to a standard infinite mixture model. While the standard model can represent any density on fixed-sized vectors as $N \to \infty$, it cannot cope with $D \to \infty$, since it has no way to handle irrelevant, noisy or redundant features. By contrast, the crosscat/multi-clust system is robust to irrelevant features: it can just partition them off, and cluster the rows only using the relevant features. Note, however, that it does not need a separate "background" model, since everything is modelled using the same mechanism. This is useful, since one's person's noise is another person's signal. (Indeed, this symmetry may explain why multi-clust outperformed the sparse mixture model approach of (Law et al. 2004) in the experiments reported in (Guan et al. 2010).)

26

Graphical model structure learning

26.1   Introduction

We have seen how graphical models can be used to express conditional independence assumptions between variables. In this chapter, we discuss how to learn the structure of the graphical model itself. That is, we want to compute $p(G|\mathcal{D})$, where G is the graph structure, represented as an $V \times V$ adjacency matrix.

As we discussed in Section 1.3.3, there are two main applications of structure learning: knowledge discovery and density estimation. The former just requires a graph topology, whereas the latter requires a fully specified model.

The main obstacle in structure learning is that the number of possible graphs is exponential in the number of nodes: a simple upper bound is $O(2^{V(V-1)/2})$. Thus the full posterior $p(G|\mathcal{D})$ is prohibitively large: even if we could afford to compute it, we could not even store it. So we will seek appropriate summaries of the posterior. These summary statistics depend on our task.

If our goal is knowledge discovery, we may want to compute posterior edge marginals, $p(G_{st} = 1|\mathcal{D})$; we can then plot the corresponding graph, where the thickness of each edge represents our confidence in its presence. By setting a threshold, we can generate a sparse graph, which can be useful for visualization purposes (see Figure 1.11).

If our goal is density estimation, we may want to compute the MAP graph, $\hat{G} \in \text{argmax}_G\, p(G|\mathcal{D})$. In most cases, finding the globally optimal graph will take exponential time, so we will use discrete optimization methods such as heuristic search. However, in the case of trees, we can find the globally optimal graph structure quite efficiently using exact methods, as we discuss in Section 26.3.

If density estimation is our only goal, it is worth considering whether it would be more appropriate to learn a latent variable model, which can capture correlation between the visible variables via a set of latent common causes (see Chapters 12 and 27). Such models are often easier to learn and, perhaps more importantly, they can be applied (for prediction purposes) much more efficiently, since they do not require performing inference in a learned graph with potentially high treewidth. The downside with such models is that the latent factors are often unidentifiable, and hence hard to interpret. Of course, we can combine graphical model structure learning and latent variable learning, as we will show later in this chapter.

In some cases, we don't just want to model the observed correlation between variables; instead, we want to model the *causal* structure behind the data, so we can predict the effects of manipulating variables. This is a much more challenging task, which we briefly discuss in

Figure 26.1 Part of a relevance network constructed from the 20-news data shown in Figure 1.2. We show edges whose mutual information is greater than or equal to 20% of the maximum pairwise MI. For clarity, the graph has been cropped, so we only show a subset of the nodes and edges. Figure generated by `relevanceNetworkNewsgroupDemo`.

Section 26.6.

26.2   Structure learning for knowledge discovery

Since computing the MAP graph or the exact posterior edge marginals is in general computationally intractable (Chickering 1996), in this section we discuss some "quick and dirty" methods for learning graph structures which can be used to visualize one's data. The resulting models do not constitute consistent joint probability distributions, so they cannot be used for prediction, and they cannot even be formally evaluated in terms of goodness of fit. Nevertheless, these methods are a useful ad hoc tool to have in one's data visualization toolbox, in view of their speed and simplicity.

26.2.1   Relevance networks

A **relevance network** is a way of visualizing the pairwise mutual information between multiple random variables: we simply choose a threshold and draw an edge from node i to node j if $\mathbb{I}(X_i; X_j)$ is above this threshold. In the Gaussian case, $\mathbb{I}(X_i; X_j) = -\frac{1}{2}\log(1 - \rho_{ij}^2)$, where ρ_{ij} is the correlation coefficient (see Exercise 2.13), so we are essentially visualizing $\boldsymbol{\Sigma}$; this is known as the covariance graph (Section 19.4.4.1).

This method is quite popular in systems biology (Margolin et al. 2006), where it is used to visualize the interaction between genes. The trouble with biological examples is that they are hard for non-biologists to understand. So let us instead illustrate the idea using natural language text. Figure 26.1 gives an example, where we visualize the MI between words in the newsgroup dataset from Figure 1.2. The results seem intuitively reasonable.

However, relevance networks suffer from a major problem: the graphs are usually very dense, since most variables are dependent on most other variables, even after thresholding the MIs. For example, suppose X_1 directly influences X_2 which directly influences X_3 (e.g., these form components of a signalling cascade, $X_1 - X_2 - X_3$). Then X_1 has non-zero MI with X_3 (and vice versa), so there will be a $1 - 3$ edge in the relevance network. Indeed, most pairs will be

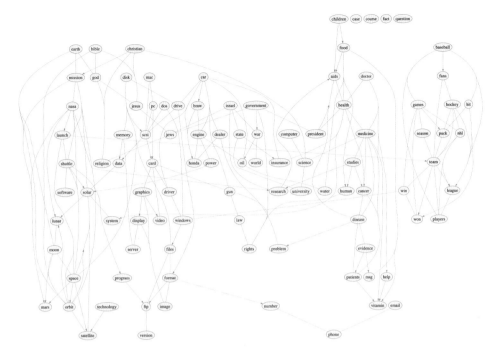

Figure 26.2 A dependency network constructed from the 20-news data. We show all edges with regression weight above 0.5 in the Markov blankets estimated by ℓ_1 penalized logistic regression. Undirected edges represent cases where a directed edge was found in both directions. From Figure 4.9 of (Schmidt 2010). Used with kind permission of Mark Schmidt.

connected.

A better approach is to use graphical models, which represent conditional *independence*, rather than *dependence*. In the above example, X_1 is conditionally independent of X_3 given X_2, so there will not be a $1-3$ edge. Consequently graphical models are usually much sparser than relevance networks, and hence are a more useful way of visualizing interactions between multiple variables.

26.2.2 Dependency networks

A simple and efficient way to learn a graphical model structure is to independently fit D sparse full-conditional distributions $p(x_t|\mathbf{x}_{-t})$; this is called a **dependency network** (Heckerman et al. 2000). The chosen variables constitute the inputs to the node, i.e., its Markov blanket. We can then visualize the resulting sparse graph. The advantage over relevance networks is that redundant variables will not be selected as inputs.

We can use any kind of sparse regression or classification method to fit each CPD. (Heckerman et al. 2000) uses classification/ regression trees, (Meinshausen and Buhlmann 2006) use ℓ_1-regularized linear regression, (Wainwright et al. 2006) use ℓ_1-regularized logistic regression (see depnetFit for some code), (Dobra 2009) uses Bayesian variable selection, etc. (Meinshausen and Buhlmann 2006) discuss theoretical conditions under which ℓ_1-regularized linear regression can recover the true graph structure, assuming the data was generated from a sparse Gaussian graphical model.

Figure 26.2 shows a dependency network that was learned from the 20-newsgroup data using ℓ_1 regularized logistic regression, where the penalty parameter λ was chosen by BIC. Many of the words present in these estimated Markov blankets represent fairly natural associations (aids:disease, baseball:fans, bible:god, bmw:car, cancer:patients, etc.). However, some of the estimated statistical dependencies seem less intuitive, such as baseball:windows and bmw:christian. We can gain more insight if we look not only at the sparsity pattern, but also the values of the regression weights. For example, here are the incoming weights for the first 5 words:

- **aids**: children (0.53), disease (0.84), fact (0.47), health (0.77), president (0.50), research (0.53)

- **baseball**: *christian* (-0.98), *drive* (-0.49), games (0.81), *god* (-0.46), *government* (-0.69), hit (0.62), *memory* (-1.29), players (1.16), season (0.31), *software* (-0.68), *windows* (-1.45)

- **bible**: *car* (-0.72), *card* (-0.88), christian (0.49), fact (0.21), god (1.01), jesus (0.68), orbit (0.83), *program* (-0.56), religion (0.24), version (0.49)

- **bmw**: car (0.60), *christian* (-11.54), engine (0.69), *god* (-0.74), *government* (-1.01), *help* (-0.50), *windows* (-1.43)

- **cancer**: disease (0.62), medicine (0.58), patients (0.90), research (0.49), studies (0.70)

Words in italic red have negative weights, which represents a dissociative relationship. For example, the model reflects that baseball:windows is an unlikely combination. It turns out that most of the weights are negative (1173 negative, 286 positive, 8541 zero) in this model.

In addition to visualizing the data, a dependency network can be used for inference. However, the only algorithm we can use is Gibbs sampling, where we repeatedly sample the nodes with missing values from their full conditionals. Unfortunately, a product of full conditionals does not, in general, constitute a representation of any valid joint distribution (Heckerman et al. 2000), so the output of the Gibbs sampler may not be meaningful. Nevertheless, the method can sometimes give reasonable results if there is not much missing data, and it is a useful method for data imputation (Gelman and Raghunathan 2001). In addition, the method can be used as an initialization technique for more complex structure learning methods that we discuss below.

26.3   Learning tree structures

For the rest of this chapter, we focus on learning fully specified joint probability models, which can be used for density estimation, prediction and knowledge discovery.

Since the problem of structure learning for general graphs is NP-hard (Chickering 1996), we start by considering the special case of trees. Trees are special because we can learn their structure efficiently, as we discuss below, and because, once we have learned the tree, we can use them for efficient exact inference, as discussed in Section 20.2.

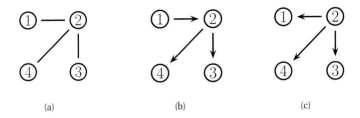

Figure 26.3 An undirected tree and two equivalent directed trees.

26.3.1 Directed or undirected tree?

Before continuing, we need to discuss the issue of whether we should use directed or undirected trees. A directed tree, with a single root node r, defines a joint distribution as follows:

$$p(\mathbf{x}|T) = \prod_{t \in V} p(x_t | x_{\mathrm{pa}(t)}) \tag{26.1}$$

where we define $\mathrm{pa}(r) = \emptyset$. For example, in Figure 26.3(b-c), we have

$$
\begin{aligned}
p(x_1, x_2, x_3, x_4 | T) &= p(x_1)p(x_2|x_1)p(x_3|x_2)p(x_4|x_2) \tag{26.2}\\
&= p(x_2)p(x_1|x_2)p(x_3|x_2)p(x_4|x_2) \tag{26.3}
\end{aligned}
$$

We see that the choice of root does not matter: both of these models are equivalent.

To make the model more symmetric, it is preferable to use an undirected tree. This can be represented as follows:

$$p(\mathbf{x}|T) = \prod_{t \in V} p(x_t) \prod_{(s,t) \in E} \frac{p(x_s, x_t)}{p(x_s)p(x_t)} \tag{26.4}$$

where $p(x_s, x_t)$ is an edge marginal and $p(x_t)$ is a node marginal. For example, in Figure 26.3(a) we have

$$p(x_1, x_2, x_3, x_4 | T) = p(x_1)p(x_2)p(x_3)p(x_4) \frac{p(x_1, x_2)p(x_2, x_3)p(x_2, x_4)}{p(x_1)p(x_2)p(x_2)p(x_3)p(x_2)p(x_4)} \tag{26.5}$$

To see the equivalence with the directed representation, let us cancel terms to get

$$
\begin{aligned}
p(x_1, x_2, x_3, x_4 | T) &= p(x_1, x_2) \frac{p(x_2, x_3)}{p(x_2)} \frac{p(x_2, x_4)}{p(x_2)} \tag{26.6}\\
&= p(x_1)p(x_2|x_1)p(x_3|x_2)p(x_4|x_2) \tag{26.7}\\
&= p(x_2)p(x_1|x_2)p(x_3|x_2)p(x_4|x_2) \tag{26.8}
\end{aligned}
$$

where $p(x_t|x_s) = p(x_s, x_t)/p(x_s)$.

Thus a tree can be represented as either an undirected or directed graph: the number of parameters is the same, and hence the complexity of learning is the same. And of course, inference is the same in both representations, too. The undirected representation, which is symmetric, is useful for structure learning, but the directed representation is more convenient for parameter learning.

26.3.2   Chow-Liu algorithm for finding the ML tree structure

Using Equation 26.4, we can write the log-likelihood for a tree as follows:

$$
\log p(\mathcal{D}|\boldsymbol{\theta}, T) \;=\; \sum_{t}\sum_{k} N_{tk}\log p(x_t = k|\boldsymbol{\theta})
$$

$$
+ \sum_{s,t}\sum_{j,k} N_{stjk}\log \frac{p(x_s = j, x_t = k|\boldsymbol{\theta})}{p(x_s = j|\boldsymbol{\theta})p(x_t = k|\boldsymbol{\theta})} \tag{26.9}
$$

where N_{stjk} is the number of times node s is in state j and node t is in state k, and N_{tk} is the number of times node t is in state k. We can rewrite these counts in terms of the empirical distribution: $N_{stjk} = N p_{\text{emp}}(x_s = j, x_t = k)$ and $N_{tk} = N p_{\text{emp}}(x_t = k)$. Setting $\boldsymbol{\theta}$ to the MLEs, this becomes

$$
\frac{\log p(\mathcal{D}|\boldsymbol{\theta}, T)}{N} \;=\; \sum_{t\in\mathcal{V}}\sum_{k} p_{\text{emp}}(x_t = k)\log p_{\text{emp}}(x_t = k) \tag{26.10}
$$

$$
+ \sum_{(s,t)\in\mathcal{E}(T)} \mathbb{I}(x_s, x_t|\hat{\boldsymbol{\theta}}_{st}) \tag{26.11}
$$

where $\mathbb{I}(x_s, x_t|\hat{\boldsymbol{\theta}}_{st}) \geq 0$ is the mutual information between x_s and x_t given the empirical distribution:

$$
\mathbb{I}(x_s, x_t|\hat{\boldsymbol{\theta}}_{st}) = \sum_{j}\sum_{k} p_{\text{emp}}(x_s = j, x_t = k)\log \frac{p_{\text{emp}}(x_s = j, x_t = k)}{p_{\text{emp}}(x_s = j)p_{\text{emp}}(x_t = k)} \tag{26.12}
$$

Since the first term in Equation 26.11 is independent of the topology T, we can ignore it when learning structure. Thus the tree topology that maximizes the likelihood can be found by computing the maximum weight spanning tree, where the edge weights are the pairwise mutual informations, $\mathbb{I}(y_s, y_t|\hat{\boldsymbol{\theta}}_{st})$. This is called the **Chow-Liu algorithm** (Chow and Liu 1968).

There are several algorithms for finding a max spanning tree (MST). The two best known are Prim's algorithm and Kruskal's algorithm. Both can be implemented to run in $O(E\log V)$ time, where $E = V^2$ is the number of edges and V is the number of nodes. See e.g., (Sedgewick and Wayne 2011, 4.3) for details. Thus the overall running time is $O(NV^2 + V^2\log V)$, where the first term is the cost of computing the sufficient statistics.

Figure 26.4 gives an example of the method in action, applied to the binary 20 newsgroups data shown in Figure 1.2. The tree has been arbitrarily rooted at the node representing "email". The connections that are learned seem intuitively reasonable.

26.3.3   Finding the MAP forest

Since all trees have the same number of parameters, we can safely use the maximum likelihood score as a model selection criterion without worrying about overfitting. However, sometimes we may want to fit a **forest** rather than a single tree, since inference in a forest is much faster than in a tree (we can run belief propagation in each tree in the forest in parallel). The MLE criterion will never choose to omit an edge. However, if we use the marginal likelihood or a penalized likelihood (such as BIC), the optimal solution may be a forest. Below we give the details for the marginal likelihood case.

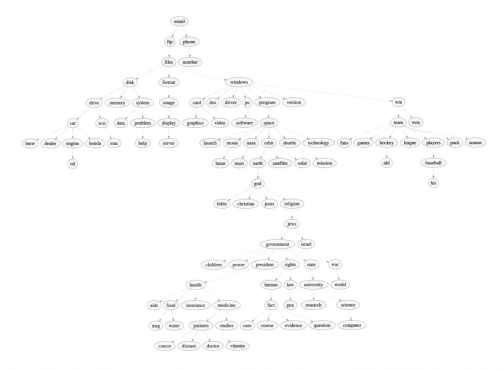

Figure 26.4 The MLE tree on the 20-newsgroup data. From Figure 4.11 of (Schmidt 2010). Used with kind permission of Mark Schmidt. (A topologically equivalent tree can be produced using `chowliuTreeDemo`.)

In Section 26.4.2.2, we explain how to compute the marginal likelihood of any DAG using a Dirichlet prior for the CPTs. The resulting expression can be written as follows:

$$\log p(\mathcal{D}|T) = \sum_{t \in \mathcal{V}} \log \int \prod_{i=1}^{N} p(x_{it}|\mathbf{x}_{i,\mathrm{pa}(t)}|\boldsymbol{\theta}_t)p(\boldsymbol{\theta}_t)d\boldsymbol{\theta}_t = \sum_{t} \mathrm{score}(\mathbf{N}_{t,\mathrm{pa}(t)}) \tag{26.13}$$

where $\mathbf{N}_{t,\mathrm{pa}(t)}$ are the counts (sufficient statistics) for node t and its parents, and score is defined in Equation 26.28.

Now suppose we only allow DAGs with at most one parent. Following (Heckerman et al. 1995, p227), let us associate a weight with each $s \to t$ edge, $w_{s,t} \triangleq \mathrm{score}(t|s) - \mathrm{score}(t|0)$, where $\mathrm{score}(t|0)$ is the score when t has no parents. Note that the weights might be negative (unlike the MLE case, where edge weights are aways non-negative because they correspond to mutual information). Then we can rewrite the objective as follows:

$$\log p(\mathcal{D}|T) = \sum_{t} \mathrm{score}(t|\mathrm{pa}(t)) = \sum_{t} w_{\mathrm{pa}(t),t} + \sum_{t} \mathrm{score}(t|0) \tag{26.14}$$

The last term is the same for all trees T, so we can ignore it. Thus finding the most probable tree amounts to finding a **maximal branching** in the corresponding weighted directed graph. This can be found using the algorithm in (Gabow et al. 1984).

If the scoring function is prior and likelihood equivalent (these terms are explained in Section 26.4.2.3), we have

$$\text{score}(s|t) + \text{score}(t|0) = \text{score}(t|s) + \text{score}(s|0) \qquad (26.15)$$

and hence the weight matrix is symmetric. In this case, the maximal branching is the same as the maximal weight forest. We can apply a slightly modified version of the MST algorithm to find this (Edwards et al. 2010). To see this, let $G = (V, E)$ be a graph with both positive and negative edge weights. Now let G' be a graph obtained by omitting all the negative edges from G. This cannot reduce the total weight, so we can find the maximum weight forest of G by finding the MST for each connected component of G'. We can do this by running Kruskal's algorithm directly on G': there is no need to find the connected components explicitly.

26.3.4 Mixtures of trees

A single tree is rather limited in its expressive power. Later in this chapter we discuss ways to learn more general graphs. However, the resulting graphs can be expensive to do inference in. An interesting alternative is to learn a **mixture of trees** (Meila and Jordan 2000), where each mixture component may have a different tree topology. This is like an unsupervised version of the TAN classifier discussed in Section 10.2.1. We can fit a mixture of trees by using EM: in the E step, we compute the responsibilities of each cluster for each data point, and in the M step, we use a weighted version of the Chow-Liu algorithm. See (Meila and Jordan 2000) for details.

In fact, it is possible to create an "infinite mixture of trees", by integrating out over all possible trees. Remarkably, this can be done in V^3 time using the matrix tree theorem. This allows us to perform exact Bayesian inference of posterior edge marginals etc. However, it is not tractable to use this infinite mixture for inference of hidden nodes. See (Meila and Jaakkola 2006) for details.

26.4 Learning DAG structures

In this section, we discuss how to compute (functions of) $p(G|\mathcal{D})$, where G is constrained to be a DAG. This is often called **Bayesian network structure learning**. In this section, we assume there is no missing data, and that there are no hidden variables. This is called the **complete data assumption**. For simplicity, we will focus on the case where all the variables are categorical and all the CPDs are tables, although the results generalize to real-valued data and other kinds of CPDs, such as linear-Gaussian CPDs.

Our presentation is based in part on (Heckerman et al. 1995), although we will follow the notation of Section 10.4.2. In particular, let $x_{it} \in \{1, \dots, K_t\}$ be the value of node t in case i, where K_t is the number of states for node t. Let $\theta_{tck} \triangleq p(x_t = k | \mathbf{x}_{\text{pa}(t)} = c)$, for $k = 1 : K_t$, and $c = 1 : C_t$, where C_t is the number of parent combinations (possible conditioning cases). For notational simplicity, we will often assume $K_t = K$, so all nodes have the same number of states. We will also let $d_t = \dim(\text{pa}(t))$ be the degree or fan-in of node t, so that $C_t = K^{d_t}$.

26.4.1 Markov equivalence

In this section, we discuss some fundamental limits to our ability to learn DAG structures from data.

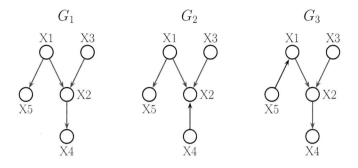

Figure 26.5 Three DAGs. G_1 and G_3 are Markov equivalent, G_2 is not.

Consider the following 3 DGMs: $X \to Y \to Z$, $X \leftarrow Y \leftarrow Z$ and $X \leftarrow Y \to Z$. These all represent the same set of CI statements, namely

$$X \perp Z | Y, \ \ X \not\perp Z \tag{26.16}$$

We say these graphs are **Markov equivalent**, since they encode the same set of CI assumptions. That is, they all belong to the same Markov **equivalence class**. However, the v-structure $X \to Y \leftarrow Z$ encodes $X \perp Z$ and $X \not\perp Z | Y$, which represents the opposite set of CI assumptions.

One can prove the following theorem.

Theorem 26.4.1. *(Verma and Pearl 1990). Two structures are Markov equivalent iff they have the same undirected skeleton and the same set of v-structures.*

For example, referring to Figure 26.5, we see that $G_1 \not\equiv G_2$, since reversing the $2 \to 4$ arc creates a new v-structure. However, $G_1 \equiv G_3$, since reversing the $1 \to 5$ arc does not create a new v-structure.

We can represent a Markov equivalence class using a single **partially directed acyclic graph** (PDAG), also called an **essential graph** or **pattern**, in which some edges are directed and some undirected. The undirected edges represent reversible edges; any combination is possible so long as no new v-structures are created. The directed edges are called **compelled edges**, since changing their orientation would change the v-structures and hence change the equivalence class. For example, the PDAG $X - Y - Z$ represents $\{X \to Y \to Z, X \leftarrow Y \leftarrow Z, X \leftarrow Y \to Z\}$ which encodes $X \not\perp Z$ and $X \perp Z | Y$. See Figure 26.6.

The significance of the above theorem is that, when we learn the DAG structure from data, we will not be able to uniquely identify all of the edge directions, even given an infinite amount of data. We say that we can learn DAG structure "up to Markov equivalence". This also cautions us not to read too much into the meaning of particular edge orientations, since we can often change them without changing the model in any observable way.

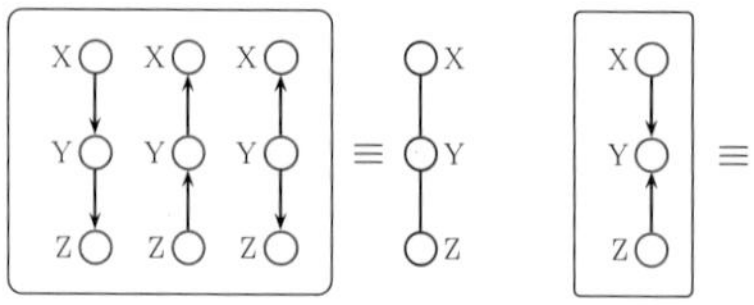

Figure 26.6   PDAG representation of Markov equivalent DAGs.

26.4.2   Exact structural inference

In this section, we discuss how to compute the exact posterior over graphs, $p(G|\mathcal{D})$, ignoring for now the issue of computational tractability.

26.4.2.1   Deriving the likelihood

Assuming there is no missing data, and that all CPDs are tabular, the likelihood can be written as follows:

$$p(\mathcal{D}|G,\boldsymbol{\theta}) = \prod_{i=1}^{N}\prod_{t=1}^{V} \mathrm{Cat}(x_{it}|\mathbf{x}_{i,\mathrm{pa}(t)},\boldsymbol{\theta}_t) \tag{26.17}$$

$$= \prod_{i=1}^{N}\prod_{t=1}^{V}\prod_{c=1}^{C_t} \mathrm{Cat}(x_{it}|\boldsymbol{\theta}_{tc})^{\mathbb{I}(\mathbf{x}_{i,\mathrm{pa}(t)}=c)} \tag{26.18}$$

$$= \prod_{i=1}^{N}\prod_{t=1}^{V}\prod_{c=1}^{C_t}\prod_{k=1}^{K_t} \theta_{tck}^{\mathbb{I}(x_{i,t}=k,\mathbf{x}_{i,\mathrm{pa}(t)}=c)} \tag{26.19}$$

$$= \prod_{t=1}^{V}\prod_{c=1}^{C_t}\prod_{k=1}^{K_t} \theta_{tck}^{N_{tck}} \tag{26.20}$$

where N_{tck} is the number of times node t is in state k and its parents are in state c. (Technically these counts depend on the graph structure G, but we drop this from the notation.)

26.4.2.2   Deriving the marginal likelihood

Of course, choosing the graph with the maximum likelihood will always pick a fully connected graph (subject to the acyclicity constraint), since this maximizes the number of parameters. To avoid such overfitting, we will choose the graph with the maximum marginal likelihood, $p(\mathcal{D}|G)$; the magic of the Bayesian Occam's razor will then penalize overly complex graphs.

To compute the marginal likelihood, we need to specify priors on the parameters. We will make two standard assumptions. First, we assume **global prior parameter independence**, which means

$$p(\boldsymbol{\theta}) = \prod_{t=1}^{V} p(\boldsymbol{\theta}_t) \tag{26.21}$$

Second, we assume **local prior parameter independence**, which means

$$p(\boldsymbol{\theta}_t) = \prod_{c=1}^{C_t} p(\boldsymbol{\theta}_{tc}) \tag{26.22}$$

for each t. It turns out that these assumtions imply that the prior for each row of each CPT must be a Dirichlet (Geiger and Heckerman 1997), that is,

$$p(\boldsymbol{\theta}_{tc}) = \text{Dir}(\boldsymbol{\theta}_{tc}|\boldsymbol{\alpha}_{tc}) \tag{26.23}$$

Given these assumptions, and using the results of Section 5.3.2.2, we can write down the marginal likelihood of any DAG as follows:

$$p(\mathcal{D}|G) = \prod_{t=1}^{V}\prod_{c=1}^{C_t} \int \left[\prod_{i:x_{i,\text{pa}(t)}=c} \text{Cat}(x_{it}|\boldsymbol{\theta}_{tc}) \right] \text{Dir}(\boldsymbol{\theta}_{tc})d\boldsymbol{\theta}_{tc} \tag{26.24}$$

$$= \prod_{t=1}^{V}\prod_{c=1}^{C_t} \frac{B(\mathbf{N}_{tc}+\boldsymbol{\alpha}_{tc})}{B(\boldsymbol{\alpha}_{tc})} \tag{26.25}$$

$$= \prod_{t=1}^{V}\prod_{c=1}^{C_t} \frac{\Gamma(N_{tc})}{\Gamma(N_{tc}+\alpha_{tc})} \prod_{k=1}^{K_t} \frac{\Gamma(N_{tck}+\alpha_{tck}^G)}{\Gamma(\alpha_{ijk}^G)} \tag{26.26}$$

$$= \prod_{t=1}^{V} \text{score}(\mathbf{N}_{t,\text{pa}(t)}) \tag{26.27}$$

where $N_{tc} = \sum_k N_{tck}$, $\alpha_{tc} = \sum_k \alpha_{tck}$, $\mathbf{N}_{t,\text{pa}(t)}$ is the vector of counts (sufficient statistics) for node t and its parents, and score() is a local scoring function defined by

$$\text{score}(\mathbf{N}_{t,\text{pa}(t)}) \triangleq \prod_{c=1}^{C_t} \frac{B(\mathbf{N}_{tc}+\boldsymbol{\alpha}_{tc})}{B(\boldsymbol{\alpha}_{tc})} \tag{26.28}$$

We say that the marginal likelihood **decomposes** or factorizes according to the graph structure.

26.4.2.3   Setting the prior

How should we set the hyper-parameters α_{tck}? It is tempting to use a Jeffreys prior of the form $\alpha_{tck} = \frac{1}{2}$ (Equation 5.62). However, it turns out that this violates a property called **likelihood equivalence**, which is sometimes considered desirable. This property says that if G_1 and G_2 are Markov equivalent (Section 26.4.1), they should have the same marginal likelihood, since they are essentially equivalent models. Geiger and Heckerman (1997) proved that, for complete graphs, the only prior that satisfies likelihood equivalence and parameter independence is the Dirichlet prior, where the pseudo counts have the form

$$\alpha_{tck} = \alpha\, p_0(x_t = k, \mathbf{x}_{\text{pa}(t)} = c) \tag{26.29}$$

where $\alpha > 0$ is called the **equivalent sample size**, and p_0 is some prior joint probability distribution. This is called the **BDe** prior, which stands for Bayesian Dirichlet likelihood equivalent.

To derive the hyper-parameters for other graph structures, Geiger and Heckerman (1997) invoked an additional assumption called **parameter modularity**, which says that if node X_t has the same parents in G_1 and G_2, then $p(\boldsymbol{\theta}_t|G_1) = p(\boldsymbol{\theta}_t|G_2)$. With this assumption, we can always derive $\boldsymbol{\alpha}_t$ for a node t in any other graph by marginalizing the pseudo counts in Equation 26.29.

Typically the prior distribution p_0 is assumed to be uniform over all possible joint configurations. In this case, we have

$$\alpha_{tck} = \frac{\alpha}{K_t C_t} \tag{26.30}$$

since $p_0(x_t = k, \mathbf{x}_{\mathrm{pa}(t)} = c) = \frac{1}{K_t C_t}$. Thus if we sum the pseudo counts over all $C_t \times K_t$ entries in the CPT, we get a total equivalent sample size of α. This is called the **BDeu** prior, where the "u" stands for uniform. This is the most widely used prior for learning Bayes net structures. For advice on setting the global tuning parameter α, see (Silander et al. 2007).

26.4.2.4   Simple worked example

We now give a very simple worked example from (Neapolitan 2003, p.438). Suppose we have just 2 binary nodes, and the following 8 data cases:

X_1	X_2
1	1
1	2
1	1
2	2
1	1
2	1
1	1
2	2

Suppose we are interested in two possible graphs: G_1 is $X_1 \rightarrow X_2$ and G_2 is the disconnected graph. The empirical counts for node 1 in G_1 are $\mathbf{N}_1 = (5, 3)$ and for node 2 are

	$X_2 = 1$	$X_2 = 2$
$X_1 = 1$	4	1
$X_1 = 2$	1	2

The BDeu prior for G_1 is $\boldsymbol{\alpha}_1 = (\alpha/2, \alpha/2)$, $\boldsymbol{\alpha}_{2|x_1=1} = (\alpha/4, \alpha/4)$ and $\boldsymbol{\alpha}_{2|x_1=2} = (\alpha/4, \alpha/4)$. For G_2, the prior for $\boldsymbol{\theta}_1$ is the same, and for $\boldsymbol{\theta}_2$ it is $\boldsymbol{\alpha}_{2|x_1=1} = (\alpha/2, \alpha/2)$ and $\boldsymbol{\alpha}_{2|x_1=2} = (\alpha/2, \alpha/2)$. If we set $\alpha = 4$, and use the BDeu prior, we find $p(\mathcal{D}|G_1) = 7.2150 \times 10^{-6}$ and $p(\mathcal{D}|G_2) = 6.7465 \times 10^{-6}$. Hence the posterior probabilites, under a uniform graph prior, are $p(G_1|\mathcal{D}) = 0.51678$ and $p(G_2|\mathcal{D}) = 0.48322$.

26.4.2.5   Example: analysis of the college plans dataset

We now consider a more interesting example from (Heckerman et al. 1997). Consider the data set collected in 1968 by Sewell and Shah which measured 5 variables that might influence the decision of high school students about whether to attend college. Specifically, the variables are as follows:

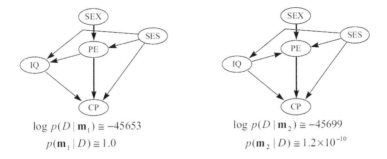

Figure 26.7 The two most probable DAGs learned from the Sewell-Shah data. Source: (Heckerman et al. 1997) . Used with kind permission of David Heckerman

- **Sex** Male or female
- **SES** Socio economic status: low, lower middle, upper middle or high.
- **IQ** Intelligence quotient: discretized into low, lower middle, upper middle or high.
- **PE** Parental encouragment: low or high
- **CP** College plans: yes or no.

These variables were measured for 10,318 Wisconsin high school seniors. There are $2 \times 4 \times 4 \times 2 \times = 128$ possible joint configurations.

Heckerman et al. computed the exact posterior over all 29,281 possible 5 node DAGs, except for ones in which SEX and/or SES have parents, and/or CP have children. (The prior probability of these graphs was set to 0, based on domain knowledge.) They used the BDeu score with $\alpha = 5$, although they said that the results were robust to any α in the range 3 to 40. The top two graphs are shown in Figure 26.7. We see that the most probable one has approximately all of the probability mass, so the posterior is extremely peaked.

It is tempting to interpret this graph in terms of causality (see Section 26.6). In particular, it seems that socio-economic status, IQ and parental encouragment all causally influence the decision about whether to go to college, which makes sense. Also, sex influences college plans only indirectly through parental encouragement, which also makes sense. However, the direct link from socio economic status to IQ seems surprising; this may be due to a hidden common cause. In Section 26.5.1.4 we will re-examine this dataset allowing for the presence of hidden variables.

26.4.2.6 The K2 algorithm

Suppose we know a total ordering of the nodes. Then we can compute the distribution over parents for each node independently, without the risk of introducing any directed cycles: we simply enumerate over all possible subsets of ancestors and compute their marginal likelihoods.[1]

---

1. We can make this method more efficient by using ℓ_1-regularization to select the parents (Schmidt et al. 2007). In this case, we need to approximate the marginal likelihood as we discuss below.

If we just return the best set of parents for each node, we get the the **K2 algorithm** (Cooper and Herskovits 1992).

26.4.2.7   Handling non-tabular CPDs

If all CPDs are linear Gaussian, we can replace the Dirichlet-multinomial model with the normal-gamma model, and thus derive a different exact expression for the marginal likelihood. See (Geiger and Heckerman 1994) for the details. In fact, we can easily combine discrete nodes and Gaussian nodes, as long as the discrete nodes always have discrete parents; this is called a **conditional Gaussian** DAG. Again, we can compute the marginal likelihood in closed form. See (Bottcher and Dethlefsen 2003) for the details.

In the general case (i.e., everything except Gaussians and CPTs), we need to approximate the marginal likelihood. The simplest approach is to use the BIC approximation, which has the form

$$\sum_t \log p(\mathcal{D}_t | \hat{\boldsymbol{\theta}}_t) - \frac{K_t C_t}{2} \log N \tag{26.31}$$

26.4.3   Scaling up to larger graphs

The main challenge in computing the posterior over DAGs is that there are so many possible graphs. More precisely, (Robinson 1973) showed that the number of DAGs on D nodes satisfies the following recurrence:

$$f(D) = \sum_{i=1}^{D} (-1)^{i+1} \binom{D}{i} 2^{i(D-i)} f(D-i) \tag{26.32}$$

for $D > 2$. The base case is $f(1) = 1$. Solving this recurrence yields the following sequence: 1, 3, 25, 543, 29281, 3781503, etc.[2] In view of the enormous size of the hypothesis space, we are generally forced to use approximate methods, some of which we review below.

26.4.3.1   Approximating the mode of the posterior

We can use dynamic programming to find the globally optimal MAP DAG (up to Markov equivalence) (Koivisto and Sood 2004; Silander and Myllymaki 2006). Unfortunately this method takes $V2^V$ time and space, making it intractable beyond about 16 nodes. Indeed, the general problem of finding the globally optimal MAP DAG is provably NP-complete (Chickering 1996),

Consequently, we must settle for finding a locally optimal MAP DAG. The most common method is greedy hill climbing: at each step, the algorithm proposes small changes to the current graph, such as adding, deleting or reversing a single edge; it then moves to the neighboring graph which most increases the posterior. The method stops when it reaches a local maximum. It is important that the method only proposes local changes to the graph, since this enables the change in marginal likelihood (and hence the posterior) to be computed in constant time (assuming we cache the sufficient statistics). This is because all but one

---

2. A longer list of values can be found at `http://www.research.att.com/~njas/sequences/A003024`. Interestingly, the number of DAGs is equal to the number of (0,1) matrices all of whose eigenvalues are positive real numbers (McKay et al. 2004).

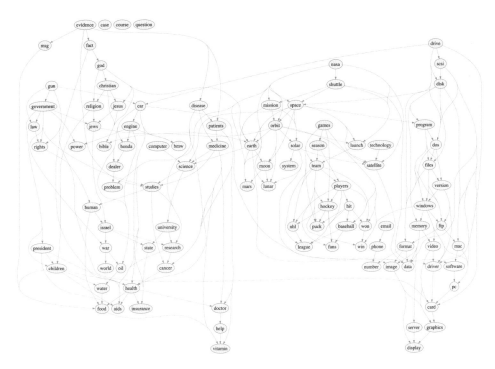

Figure 26.8 A locally optimal DAG learned from the 20-newsgroup data. From Figure 4.10 of (Schmidt 2010). Used with kind permission of Mark Schmidt.

or two of the terms in Equation 26.25 will cancel out when computing the log Bayes factor
$$\delta(G \to G') = \log p(G'|\mathcal{D}) - \log p(G|\mathcal{D}).$$

We can initialize the search from the best tree, which can be found using exact methods discussed in Section 26.3. For speed, we can restrict the search so it only adds edges which are part of the Markov blankets estimated from a dependency network (Schmidt 2010). Figure 26.8 gives an example of a DAG learned in this way from the 20-newsgroup data.

We can use techniques such as multiple random restarts to increase the chance of finding a good local maximum. We can also use more sophisticated local search methods, such as genetic algorithms or simulated annealing, for structure learning.

26.4.3.2 Approximating other functions of the posterior

If our goal is knowledge discovery, the MAP DAG can be misleading, for reasons we discussed in Section 5.2.1. A better approach is to compute the probability that each edge is present, $p(G_{st} = 1|\mathcal{D})$, of the probability there is a path from s to t. We can do this exactly using dynamic programming (Koivisto 2006; Parviainen and Koivisto 2011). Unfortunately these methods take $V2^V$ time in the general case, making them intractable for graphs with more than about 16 nodes.

An approximate method is to sample DAGs from the posterior, and then to compute the fraction of times there is an $s \to t$ edge or path for each (s, t) pair. The standard way to draw samples is to use the Metropolis Hastings algorithm (Section 24.3), where we use the same local proposal as we did in greedy search (Madigan and Raftery 1994).

A faster-mixing method is to use a collapsed MH sampler, as suggested in (Friedman and Koller 2003). This exploits the fact that, if a total ordering of the nodes is known, we can select the parents for each node independently, without worrying about cycles, as discussed in Section 26.4.2.6. By summing over all possible choice of parents, we can marginalize out this part of the problem, and just sample total orders. (Ellis and Wong 2008) also use order-space (collapsed) MCMC, but this time with a parallel tempering MCMC algorithm.

26.5   Learning DAG structure with latent variables

Sometimes the complete data assumption does not hold, either because we have missing data, and/ or because we have hidden variables. In this case, the marginal likelihood is given by

$$p(\mathcal{D}|G) = \int \sum_{\mathbf{h}} p(\mathcal{D}, \mathbf{h}|\boldsymbol{\theta}, G) p(\boldsymbol{\theta}|G) d\boldsymbol{\theta} = \sum_{\mathbf{h}} \int p(\mathcal{D}, \mathbf{h}|\boldsymbol{\theta}, G) p(\boldsymbol{\theta}|G) d\boldsymbol{\theta} \qquad (26.33)$$

where $\mathbf{h}$ represents the hidden or missing data.

In general this is intractable to compute. For example, consider a mixture model, where we don't observe the cluster label. In this case, there are K^N possible completions of the data (assuming we have K clusters); we can evaluate the inner integral for each one of these assignments to $\mathbf{h}$, but we cannot afford to evaluate all of the integrals. (Of course, most of these integrals will correspond to hypotheses with little posterior support, such as assigning single data points to isolated clusters, but we don't know ahead of time the relative weight of these assignments.)

In this section, we discuss some ways for learning DAG structure when we have latent variables and/or missing data.

26.5.1   Approximating the marginal likelihood when we have missing data

The simplest approach is to use standard structure learning methods for fully visible DAGs, but to approximate the marginal likelihood. In Section 24.7, we discussed some Monte Carlo methods for approximating the marginal likelihood. However, these are usually too slow to use inside of a search over models. Below we mention some faster deterministic approximations.

26.5.1.1   BIC approximation

A simple approximation is to use the BIC score, which is given by

$$\text{BIC}(G) \triangleq \log p(\mathcal{D}|\hat{\boldsymbol{\theta}}, G) - \frac{\log N}{2} \dim(G) \qquad (26.34)$$

where $\dim(G)$ is the number of degrees of freedom in the model and $\hat{\boldsymbol{\theta}}$ is the MAP or ML estimate. However, the BIC score often severely underestimates the true marginal likelihood (Chickering and Heckerman 1997), resulting in it selecting overly simple models.

26.5.1.2   Cheeseman-Stutz approximation

We now present a better method known as the **Cheeseman-Stutz approximation** (CS) (Cheeseman and Stutz 1996). We first compute a MAP estimate of the parameters $\hat{\boldsymbol{\theta}}$ (e.g., using EM). Denote the expected sufficient statistics of the data by $\overline{\mathcal{D}} = \overline{\mathcal{D}}(\hat{\boldsymbol{\theta}})$; in the case of discrete variables, we just "fill in" the hidden variables with their expectation. We then use the exact marginal likelihood equation on this filled-in data:

$$p(\mathcal{D}|G) \approx p(\overline{\mathcal{D}}|G) = \int p(\overline{\mathcal{D}}|\boldsymbol{\theta}, G)p(\boldsymbol{\theta}|G)d\boldsymbol{\theta} \tag{26.35}$$

However, comparing this to Equation 26.33, we can see that the value will be exponentially smaller, since it does not sum over all values of $\mathbf{h}$. To correct for this, we first write

$$\log p(\mathcal{D}|G) = \log p(\overline{\mathcal{D}}|G) + \log p(\mathcal{D}|G) - \log p(\overline{\mathcal{D}}|G) \tag{26.36}$$

and then we apply a BIC approximation to the last two terms:

$$\log p(\mathcal{D}|G) - \log p(\overline{\mathcal{D}}|G) \approx \left[\log p(\mathcal{D}|\hat{\boldsymbol{\theta}}, G) - \frac{N}{2}\dim(\hat{\boldsymbol{\theta}})\right] \tag{26.37}$$

$$- \left[\log p(\overline{\mathcal{D}}|\hat{\boldsymbol{\theta}}, G) - \frac{N}{2}\dim(\hat{\boldsymbol{\theta}})\right] \tag{26.38}$$

$$= \log p(\mathcal{D}|\hat{\boldsymbol{\theta}}, G) - \log p(\overline{\mathcal{D}}|\hat{\boldsymbol{\theta}}, G) \tag{26.39}$$

Putting it altogether we get

$$\log p(\mathcal{D}|G) \approx \log p(\overline{\mathcal{D}}|G) + \log p(\mathcal{D}|\hat{\boldsymbol{\theta}}, G) - \log p(\overline{\mathcal{D}}|\hat{\boldsymbol{\theta}}, G) \tag{26.40}$$

The first term $p(\overline{\mathcal{D}}|G)$ can be computed by plugging in the filled-in data into the exact marginal likelihood. The second term $p(\mathcal{D}|\hat{\boldsymbol{\theta}}, G)$, which involves an exponential sum (thus matching the "dimensionality" of the left hand side) can be computed using an inference algorithm. The final term $p(\overline{\mathcal{D}}|\hat{\boldsymbol{\theta}}, G)$ can be computed by plugging in the filled-in data into the regular likelihood.

26.5.1.3   Variational Bayes EM

An even more accurate approach is to use the variational Bayes EM algorithm. Recall from Section 21.6 that the key idea is to make the following factorization assumption:

$$p(\boldsymbol{\theta}, \mathbf{z}_{1:N}|\mathcal{D}) \approx q(\boldsymbol{\theta})q(\mathbf{z}) = q(\boldsymbol{\theta})\prod_i q(\mathbf{z}_i) \tag{26.41}$$

where $\mathbf{z}_i$ are the hidden variables in case i. In the E step, we update the $q(\mathbf{z}_i)$, and in the M step, we update $q(\boldsymbol{\theta})$. The corresponding variational free energy provides a lower bound on the log marginal likelihood. In (Beal and Ghahramani 2006), it is shown that this bound is a much better approximation to the true log marginal likelihood (as estimated by a slow annealed importance sampling procedure) than either BIC or CS. In fact, one can prove that the variational bound will always be more accurate than CS (which in turn is always more accurate than BIC).

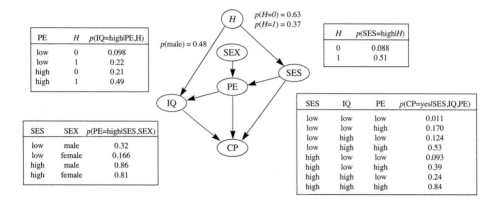

Figure 26.9 The most probable DAG with a single binary hidden variable learned from the Sewell-Shah data. MAP estimates of the CPT entries are shown for some of the nodes. Source: (Heckerman et al. 1997). Used with kind permission of David Heckerman.

26.5.1.4   Example: college plans revisited

Let us revisit the college plans dataset from Section 26.4.2.5. Recall that if we ignore the possibility of hidden variables there was a direct link from socio economic status to IQ in the MAP DAG. Heckerman et al. decided to see what would happen if they introduced a hidden variable H, which they made a parent of both SES and IQ, representing a hidden common cause. They also considered a variant in which H points to SES, IQ and PE. For both such cases, they considered dropping none, one, or both of the SES-PE and PE-IQ edges. They varied the number of states for the hidden node from 2 to 6. Thus they computed the approximate posterior over $8 \times 5 = 40$ different models, using the CS approximation.

The most probable model which they found is shown in Figure 26.9. This is $2 \cdot 10^{10}$ times more likely than the best model containing no hidden variable. It is also $5 \cdot 10^9$ times more likely than the second most probable model with a hidden variable. So again the posterior is very peaked.

These results suggests that there is indeed a hidden common cause underlying both the socio-economic status of the parents and the IQ of the children. By examining the CPT entries, we see that both SES and IQ are more likely to be high when H takes on the value 1. They interpret this to mean that the hidden variable represents "parent quality" (possibly a genetic factor). Note, however, that the arc between H and SES can be reversed without changing the v-structures in the graph, and thus without affecting the likelihood; this underscores the difficulty in interpreting hidden variables.

Interestingly, the hidden variable model has the same conditional independence assumptions amongst the visible variables as the most probable visible variable model. So it is not possible to distinguish between these hypotheses by merely looking at the empirical conditional independencies in the data (which is the basis of the **constraint-based approach** to structure learning (Pearl and Verma 1991; Spirtes et al. 2000)). Instead, by adopting a Bayesian approach, which takes parsimony into account (and not just conditional independence), we can discover the possible existence of hidden factors. This is the basis of much of scientific and everday human reasoning (see e.g. (Griffiths and Tenenbaum 2009) for a discussion).

26.5.2 Structural EM

One way to perform structural inference in the presence of missing data is to use a standard search procedure (deterministic or stochastic), and to use the methods from Section 26.5.1 to estimate the marginal likelihood. However, this approach is very efficient, because the marginal likelihood does not decompose when we have missing data, and nor do its approximations. For example, if we use the CS approximation or the VBEM approximation, we have to perform inference in every neighboring model, just to evaluate the quality of a single move!

(Friedman 1997b; Thiesson et al. 1998) presents a much more efficient approach called the **structural EM** algorithm. The basic idea is this: instead of fitting each candidate neighboring graph and then filling in its data, fill in the data once, and use this filled-in data to evaluate the score of all the neighbors. Although this might be a bad approximation to the marginal likelihood, it can be a good enough approximation of the difference in marginal likelihoods between different models, which is all we need in order to pick the best neighbor.

More precisely, define $\overline{\mathcal{D}}(G_0, \hat{\boldsymbol{\theta}}_0)$ to be the data filled in using model G_0 with MAP parameters $\hat{\boldsymbol{\theta}}_0$. Now define a modified BIC score as follows:

$$\text{score}_{\text{BIC}}(G, \mathcal{D}) \triangleq \log p(\mathcal{D}|\hat{\boldsymbol{\theta}}, G) - \frac{\log N}{2} \dim(G) + \log p(G) + \log p(\hat{\boldsymbol{\theta}}|G) \quad (26.42)$$

where we have included the log prior for the graph and parameters. One can show (Friedman 1997b) that if we pick a graph G which increases the BIC score relative to G_0 on the expected data, it will also increase the score on the actual data, i.e.,

$$\begin{aligned} \text{score}_{\text{BIC}}(G, \overline{\mathcal{D}}(G_0, \hat{\boldsymbol{\theta}}_0)) - \text{score}_{\text{BIC}}(G_0, \overline{\mathcal{D}}(G_0, \hat{\boldsymbol{\theta}}_0)) &\leq \\ \text{score}_{\text{BIC}}(G, \mathcal{D}) - \text{score}_{\text{BIC}}(G_0, \mathcal{D}) \end{aligned} \quad (26.43)$$

To convert this into an algorithm, we proceed as follows. First we initialize with some graph G_0 and some set of parameters $\boldsymbol{\theta}_0$. Then we fill-in the data using the current parameters — in practice, this means when we ask for the expected counts for any particular family, we perform inference using our current model. (If we know which counts we will need, we can precompute all of them, which is much faster.) We then evaluate the BIC score of all of our neighbors using the filled-in data, and we pick the best neighbor. We then refit the model parameters, fill-in the data again, and repeat. For increased speed, we may choose to only refit the model every few steps, since small changes to the structure hopefully won't invalidate the parameter estimates and the filled-in data too much.

One interesting application is to learn a phylogenetic tree structure. Here the observed leaves are the DNA or protein sequences of currently alive species, and the goal is to infer the topology of the tree and the values of the missing internal nodes. There are many classical algorithms for this task (see e.g., (Durbin et al. 1998)), but one that uses SEM is discussed in (Friedman et al. 2002).

Another interesting application of this method is to learn sparse mixture models (Barash and Friedman 2002). The idea is that we have one hidden variable C specifying the cluster, and we have to choose whether to add edges $C \rightarrow X_t$ for each possible feature X_t. Thus some features

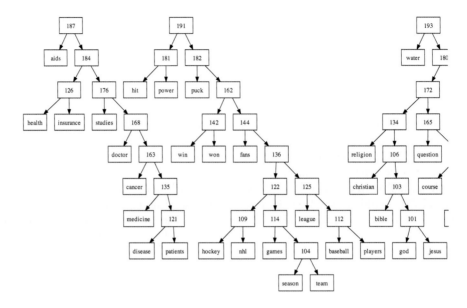

Figure 26.10 Part of a hierarchical latent tree learned from the 20-newsgroup data. From Figure 2 of (Harmeling and Williams 2011). Used with kind permission of Stefan Harmeling.

will be dependent on the cluster id, and some will be independent. (See also (Law et al. 2004) for a different way to perform this task, using regular EM and a set of bits, one per feature, that are free to change across data cases.)

26.5.3 Discovering hidden variables

In Section 26.5.1.4, we introduced a hidden variable "by hand", and then figured out the local topology by fitting a series of different models and computing the one with the best marginal likelihood. How can we automate this process?

Figure 11.1 provides one useful intuition: if there is a hidden variable in the "true model", then its children are likely to be densely connected. This suggest the following heuristic (Elidan et al. 2000): perform structure learning in the visible domain, and then look for **structural signatures**, such as sets of densely connected nodes (near-cliques); introduce a hidden variable and connect it to all nodes in this near-clique; and then let structural EM sort out the details. Unfortunately, this technique does not work too well, since structure learning algorithms are biased against fitting models with densely connected cliques.

Another useful intuition comes from clustering. In a flat mixture model, also called a **latent class model**, the discrete latent variable provides a compressed representation of its children. Thus we want to create hidden variables with high mutual information with their children.

One way to do this is to create a tree-structured hierarchy of latent variables, each of which only has to explain a small set of children. (Zhang 2004) calls this a **hierarchical latent class model**. They propose a greedy local search algorithm to learn such structures, based on adding

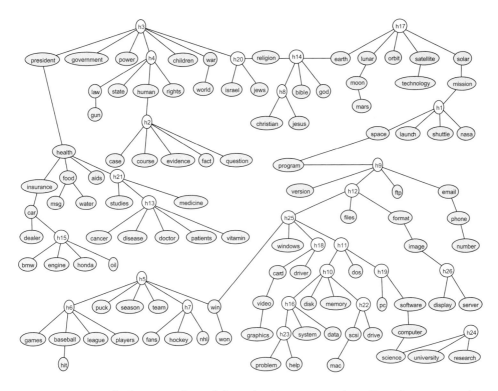

Figure 26.11 A partially latent tree learned from the 20-newsgroup data. Note that some words can have multiple meanings, and get connected to different latent variables, representing different "topics". For example, the word "win" can refer to a sports context (represented by h5) or the Microsoft Windows context (represented by h25). From Figure 12 of (Choi et al. 2011). Used with kind permission of Jin Choi.

or deleting hidden nodes, adding or deleting edges, etc. (Note that learning the optimal latent tree is NP-hard (Roch 2006).)

Recently (Harmeling and Williams 2011) proposed a faster greedy algorithm for learning such models based on agglomerative hierarchical clustering. Rather than go into details, we just give an example of what this system can learn. Figure 26.10 shows part of a latent forest learned from the 20-newsgroup data. The algorithm imposes the constraint that each latent node has exactly two children, for speed reasons. Nevertheless, we see interpretable clusters arising. For example, Figure 26.10 shows separate clusters concerning medicine, sports and religion. This provides an alternative to LDA and other topic models (Section 4.2.2), with the added advantage that inference in latent trees is exact and takes time linear in the number of nodes.

An alternative approach is proposed in (Choi et al. 2011), in which the observed data is not constrained to be at the leaves. This method starts with the Chow-Liu tree on the observed data, and then adds hidden variables to capture higher-order dependencies between internal nodes. This results in much more compact models, as shown in Figure 26.11. This model also has better predictive accuracy than other approaches, such as mixture models, or trees where all the observed data is forced to be at the leaves. Interestingly, one can show that this method

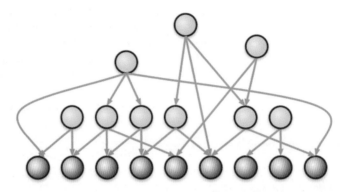

Figure 26.12 Google's rephil model. Leaves represent presence or absence of words. Internal nodes represent clusters of co-occuring words, or "concepts". All nodes are binary, and all CPDs are noisy-OR. The model contains 12 million word nodes, 1 million latent cluster nodes, and 350 million edges. Used with kind permission of Brian Milch.

can recover the exact latent tree structure, providing the data is generated from a tree. See (Choi et al. 2011) for details. Note, however, that this approach, unlike (Zhang 2004; Harmeling and Williams 2011), requires that the cardinality of all the variables, hidden and observed, be the same. Furthermore, if the observed variables are Gaussian, the hidden variables must be Gaussian also.

26.5.4   Case study: Google's Rephil

In this section, we describe a huge DGM called **Rephil**, which was automatically learned from data.[3] The model is widely used inside Google for various purposes, including their famous AdSense system.[4]

The model structure is shown in Figure 26.12. The leaves are binary nodes, and represent the presence or absence of words or compounds (such as "New York City") in a text document or query. The latent variables are also binary, and represent clusters of co-occuring words. All CPDs are noisy-OR, since some leaf nodes (representing words) can have many parents. This means each edge can be augmented with a hidden variable specifying if the link was activated or not; if the link is not active, then the parent cannot turn the child on. (A very similar model was proposed independently in (Singliar and Hauskrecht 2006).)

Parameter learning is based on EM, where the hidden activation status of each edge needs

---

3. The original system, called "Phil", was developed by Georges Harik and Noam Shazeer,. It has been published as US Patent #8024372, "Method and apparatus for learning a probabilistic generative model for text", filed in 2004. Rephil is a more probabilistically sound version of the method, developed by Uri Lerner et al. The summary below is based on notes by Brian Milch (who also works at Google).

4. AdSense is Google's system for matching web pages with content-appropriate ads in an automatic way, by extracting semantic keywords from web pages. These keywords play a role analogous to the words that users type in when searching; this latter form of information is used by Google's AdWords system. The details are secret, but (Levy 2011) gives an overview.

to be inferred (Meek and Heckerman 1997). Structure learning is based on the old neuroscience idea that "**nodes that fire together should wire together**". To implement this, we run inference and check for cluster-word and cluster-cluster pairs that frequently turn on together. We then add an edge from parent to child if the link can significantly increase the probability of the child. Links that are not activated very often are pruned out. We initialize with one cluster per "document" (corresponding to a set of semantically related phrases). We then merge clusters A and B if A explains B's top words and vice versa. We can also discard clusters that are used too rarely.

The model was trained on about 100 billion text snippets or search queries; this takes several weeks, even on a parallel distributed computing architecture. The resulting model contains 12 million word nodes and about 1 million latent cluster nodes. There are about 350 million links in the model, including many cluster-cluster dependencies. The longest path in the graph has length 555, so the model is quite deep.

Exact inference in this model is obviously infeasible. However note that most leaves will be off, since most words do not occur in a given query; such leaves can be analytically removed, as shown in Exercise 10.7. We an also prune out unlikely hidden nodes by following the strongest links from the words that are on up to their parents to get a candidate set of concepts. We then perform iterative conditional modes to find a good set of local maxima. At each step of ICM, each node sets its value to its most probable state given the values of its neighbors in its Markov blanket. This continues until it reaches a local maximum. We can repeat this process a few times from random starting configurations. At Google, this can be made to run in 15 milliseconds!

26.5.5 Structural equation models *

A **structural equation model** (Bollen 1989) is a special kind of directed mixed graph (Section 19.4.4.1), possibly cyclic, in which all CPDs are linear Gaussian, and in which all bidirected edges represent correlated Gaussian noise. Such models are also called **path diagrams**. SEMs are widely used, especially in economics and social science. It is common to interpret the edge directions in terms of causality, where directed cycles are interpreted is in terms of **feedback loops** (see e.g., (Pearl 2000, Ch.5)). However, the model is really just a way of specifying a joint Gaussian, as we show below. There is nothing inherently "causal" about it at all. (We discuss causality in Section 26.6.)

We can define an SEM as a series of full conditionals as follows:

$$x_i = \mu_i + \sum_{j \neq i} w_{ij} x_j + \epsilon_i \tag{26.44}$$

where $\epsilon \sim \mathcal{N}(\mathbf{0}, \mathbf{\Psi})$. We can rewrite the model in matrix form as follows:

$$\mathbf{x} = \mathbf{W}\mathbf{x} + \boldsymbol{\mu} + \boldsymbol{\epsilon} \Rightarrow \mathbf{x} = (\mathbf{I} - \mathbf{W})^{-1}(\boldsymbol{\epsilon} + \boldsymbol{\mu}) \tag{26.45}$$

Hence the joint distribution is given by $p(\mathbf{x}) = \mathcal{N}(\boldsymbol{\mu}, \boldsymbol{\Sigma})$ where

$$\mathbf{\Sigma} = (\mathbf{I} - \mathbf{W})^{-1} \mathbf{\Psi} (\mathbf{I} - \mathbf{W})^{-T} \tag{26.46}$$

We draw an arc $X_i \leftarrow X_j$ if $|w_{ij}| > 0$. If $\mathbf{W}$ is lower triangular then the graph is acyclic. If, in addition, $\mathbf{\Psi}$ is diagonal, then the model is equivalent to a Gaussian DGM, as discussed in

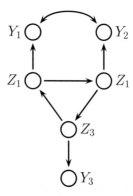

Figure 26.13 A cyclic directed mixed graphical model (non-recursive SEM). Note the $Z_1 \to Z_2 \to Z_3 \to Z_1$ feedback loop.

Section 10.2.5; such models are called **recursive**. If Ψ is not diagonal, then we draw a bidirected arc $X_i \leftrightarrow X_j$ for each non-zero off-diagonal term. Such edges represent correlation, possibly due to a hidden common cause.

When using structural equation models, it is common to partition the variables into latent variables, Z_t, and observed or **manifest** variables Y_t. For example, Figure 26.13 illustrates the following model:

$$\begin{pmatrix} X_1 \\ X_2 \\ X_3 \\ X_4 \\ X_5 \\ X_6 \end{pmatrix} = \begin{pmatrix} Z_1 \\ Z_2 \\ Z_3 \\ Y_1 \\ Y_2 \\ Y_3 \end{pmatrix} = \begin{pmatrix} 0 & 0 & w_{13} & 0 & 0 & 0 \\ w_{21} & 0 & 0 & 0 & 0 & 0 \\ 0 & w_{32} & 0 & 0 & 0 & 0 \\ w_{41} & 0 & 0 & 0 & 0 & 0 \\ 0 & w_{52} & 0 & 0 & 0 & 0 \\ 0 & 0 & w_{63} & 0 & 0 & 0 \end{pmatrix} \begin{pmatrix} Z_1 \\ Z_2 \\ Z_3 \\ Y_1 \\ Y_2 \\ Y_3 \end{pmatrix} + \begin{pmatrix} \epsilon_1 \\ \epsilon_2 \\ \epsilon_3 \\ \epsilon_4 \\ \epsilon_5 \\ \epsilon_6 \end{pmatrix}, \qquad (26.47)$$

where

$$\Psi = \begin{pmatrix} \Psi_{11} & 0 & 0 & 0 & 0 & 0 \\ 0 & \Psi_{22} & 0 & 0 & 0 & 0 \\ 0 & 0 & \Psi_{33} & 0 & 0 & 0 \\ 0 & 0 & 0 & \Psi_{44} & \Psi_{45} & 0 \\ 0 & 0 & 0 & \Psi_{54} & \Psi_{55} & 0 \\ 0 & 0 & 0 & 0 & 0 & \Psi_{66} \end{pmatrix} \qquad (26.48)$$

The presence of a feedback loop $Z_1 \to Z_2 \to Z_3$ is evident from the fact that $\mathbf{W}$ is not lower triangular. Also the presence of confounding between Y_1 and Y_2 is evident in the off-diagonal terms in Ψ.

Often we assume there are multiple observations for each latent variable. To ensure identifiability, we can set the mean of the latent variables Z_t to 0, and we can set the regression weights of $Z_t \to Y_t$ to 1. This essentially defines the scale of each latent variable. (In addition to the Z's, there are the extra hidden variables implied by the presence of the bidirected edges.)

The standard practice in the SEM community, as exemplified by the popular commercial software package called **LISREL** (available from `http://www.ssicentral.com/lisrel/`), is to build the structure by hand, to estimate the parameters by maximum likelihood, and then to test if any of the regression weights are significantly different from 0, using standard frequentist methods. However, one can also use Bayesian inference for the parameters (see e.g., (Dunson et al. 2005)). Structure learning in SEMs is rare, but since recursive SEMs are equivalent to Gaussian DAGs, many of the techniques we have been discussing in this section can be applied.

SEMs are closely related to factor analysis (FA) models (Chapter 12). The basic difference is that in an FA model, the latent Gaussian has a low-rank covariance matrix, and the observed noise has a diagonal covariance (hence no bidirected edges). In an SEM, the covariance of the latent Gaussian has a sparse Cholesky decomposition (at least if **W** is acyclic), and the observed noise might have a full covariance matrix.

Note that SEMs can be extended in many ways. For example, we can add covariates/ input variables (possibly noisily observed), we can make some of the observations be discrete (e.g., by using probit links), and so on.

26.6 Learning causal DAGs

Causal models are models which can predict the effects of interventions to, or manipulations of, a system. For example, an electronic circuit diagram implicitly provides a compact encoding of what will happen if one removes any given component, or cuts any wire. A causal medical model might predict that if I continue to smoke, I am likely to get lung cancer (and hence if I cease smoking, I am less likely to get lung cancer). Causal claims are inherently stronger, yet more useful, than purely **associative** claims, such as "people who smoke often have lung cancer".

Causal models are often represented by DAGs (Pearl 2000), although this is somewhat controversial (Dawid 2010). We explain this causal interpretation of DAGs below. We then show how to use a DAG to do causal reasoning. Finally, we briefly discuss how to learn the structure of causal DAGs. A more detailed description of this topic can be found in (Pearl 2000) and (Koller and Friedman 2009, Ch.21).

26.6.1 Causal interpretation of DAGs

In this section, we define a directed edge $A \to B$ in a DAG to mean that "A directly causes B", so if we manipulate A, then B will change. This is known as the **causal Markov assumption**. (Of course, we have not defined the word "causes", and we cannot do that by appealing to a DAG, lest we end up with a cyclic definition; see (Dawid 2010) for further disussion of this point.)

We will also assume that all relevant variables are included in the model, i.e., there are no unknown **confounders**, reflecting hidden common causes. This is called the **causal sufficiency** assumption. (If there are known to be confounders, they should be added to the model, although one can sometimes use directed mixed graphs (Section 19.4.4.1) as a way to avoid having to model confounders explicitly.)

Assuming we are willing to make the causal Markov and causal sufficiency assumptions, we can use DAGs to answer causal questions. The key abstraction is that of a **perfect intervention**; this represents the act of setting a variable to some known value, say setting X_i to x_i. A real

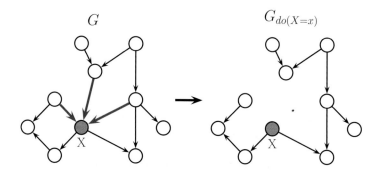

Figure 26.14   Surgical intervention on X. Based on (Pe'er 2005).

world example of such a perfect intervention is a gene knockout experiment, in which a gene is "silenced". We need some notational convention to distinguish this from observing that X_i happens to have value x_i. We use Pearl's **do calculus** notation (as in the verb "to do") and write $\mathrm{do}(X_i = x_i)$ to denote the event that we set X_i to x_i. A causal model can be used to make inferences of the form $p(\mathbf{x}|\mathrm{do}(X_i = x_i))$, which is different from making inferences of the form $p(\mathbf{x}|X_i = x_i)$.

To understand the difference between conditioning on interventions and conditioning on observations (i.e., the difference between doing and seeing), consider a 2 node DGM $S \to Y$, in which $S = 1$ if you smoke and $S = 0$ otherwise, and $Y = 1$ if you have yellow-stained fingers, and $Y = 0$ otherwise. If I observe you have yellow fingers, I am licensed to infer that you are probably a smoker (since nicotine causes yellow stains):

$$p(S = 1|Y = 1) > p(S = 1) \tag{26.49}$$

However, if I intervene and *paint* your fingers yellow, I am no longer licensed to infer this, since I have disrupted the normal causal mechanism. Thus

$$p(S = 1|\mathrm{do}(Y = 1)) = p(S = 1) \tag{26.50}$$

One way to model perfect interventions is to use **graph surgery**: represent the joint distribution by a DGM, and then cut the arcs coming into any nodes that were set by intervention. See Figure 26.14 for an example. This prevents any information flow from the nodes that were intervened on from being sent back up to their parents. Having perform this surgery, we can then perform probabilistic inference in the resulting "mutilated" graph in the usual way to reason about the effects of interventions. We state this formally as follows.

Theorem 26.6.1 (Manipulation theorem (Pearl 2000; Spirtes et al. 2000)). *Assuming the causal Markov condition, we can compute $p(X_i|\mathrm{do}(X_j))$ for sets of nodes i, j as follows: perform surgical intervention on the X_j nodes and then use standard probabilistic inference in the mutilated graph.*

We can generalize the notion of a perfect intervention by adding interventions as explicit action nodes to the graph. The result is like an influence diagram, except there are no utility

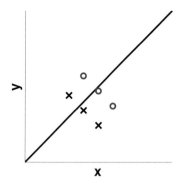

Figure 26.15 Illustration of Simpson's paradox. Figure generated by `simpsonsParadoxGraph`.

nodes (Lauritzen 2000; Dawid 2002). This has been called the **augmented DAG** (Pearl 2000). We can then define the CPD $p(X_i|\text{do}(X_i))$ to be anything we want. We can also allow an action to affect multiple nodes. This is called a **fat hand** intervention, a reference to someone trying to change a single component of some system (e.g., an electronic circuit), but accidently touching multiple components and thereby causing various side effects (see (Eaton and Murphy 2007) for a way to model this using augmented DAGs).

26.6.2 Using causal DAGs to resolve Simpson's paradox

In this section, we assume we know the causal DAG. We can then do causal reasoning by applying d-separation to the mutilated graph. In this section, we give an example of this, and show how causal reasoning can help resolve a famous paradox, known as **Simpon's paradox**.

Simpson's paradox says that any statistical relationship between two variables can be reversed by including additional factors in the analysis. For example, suppose some cause C (say, taking a drug) makes some effect E (say getting better) more likely

$$P(E|C) \;>\; P(E|\neg C)$$

and yet, when we condition on the gender of the patient, we find that taking the drug makes the effect less likely in both females (F) and males ($\neg F$):

$$P(E|C,F) \;<\; P(E|\neg C,F)$$
$$P(E|C,\neg F) \;<\; P(E|\neg C,\neg F)$$

This seems impossible, but by the rules of probability, this is perfectly possible, because the event space where we condition on $(\neg C, F)$ or $(\neg C, \neg F)$ can be completely different to the event space when we just condition on $\neg C$. The table of numbers below shows a concrete example (from (Pearl 2000, p175)):

	Combined				Male				Female			
	E	$\neg E$	Total	Rate	E	$\neg E$	Total	Rate	E	$\neg E$	Total	Rate
C	20	20	40	50%	18	12	30	60%	2	8	10	20%
$\neg C$	16	24	40	40%	7	3	10	70%	9	21	30	30%
Total	36	44	80		25	15	40		11	29	40	

From this table of numbers, we see that

$$p(E|C) = 20/40 = 0.5 \quad > \quad p(E|\neg C) = 16/40 = 0.4 \tag{26.51}$$

$$p(E|C, F) = 2/10 = 0.2 \quad < \quad p(E|\neg C, F) = 9/30 = 0.3 \tag{26.52}$$

$$p(E|C, \neg F) = 18/30 = 0.6 \quad < \quad p(E|\neg, \neg F) = 7/10 = 0.7 \tag{26.53}$$

A visual representation of the paradox is given in in Figure 26.15. The line which goes up and to the right shows that the effect (y-axis) increases as the cause (x-axis) increases. However, the dots represent the data for females, and the crosses represent the data for males. Within each subgroup, we see that the effect decreases as we increase the cause.

It is clear that the effect is real, but it is still very counter-intuitive. The reason the paradox arises is that we are interpreting the statements causally, but we are not using proper causal reasoning when performing our calculations. The statement that the drug C causes recovery E is

$$P(E|\mathrm{do}(C)) \quad > \quad P(E|\mathrm{do}(\neg C)) \tag{26.54}$$

whereas the data merely tell us

$$P(E|C) \quad > \quad P(E|\neg C) \tag{26.55}$$

This is not a contradiction. Observing C is positive evidence for E, since more males than females take the drug, and the male recovery rate is higher (regardless of the drug). Thus Equation 26.55 does not imply Equation 26.54.

Nevertheless, we are left with a practical question: should we use the drug or not? It seems like if we don't know the patient's gender, we should use the drug, but as soon as we discover if they are male or female, we should stop using it. Obviously this conclusion is ridiculous.

To answer the question, we need to make our assumptions more explicit. Suppose reality can be modeled by the causal DAG in Figure 26.16(a). To compute the causal effect of C on E, we need to **adjust for** (i.e., condition on) the **confounding variable** F. This is necessary because there is a **backdoor path** from C to E via F, so we need to check the $C \rightarrow E$ relationship for each value of F separately, to make sure the relationship between C and E is not affected by any value of F.

Suppose that for each value of F, taking the drug is harmful, that is,

$$p(E|\mathrm{do}(C), F) \quad < \quad p(E|\mathrm{do}(\neg C), F) \tag{26.56}$$

$$p(E|\mathrm{do}(C), \neg F) \quad < \quad p(E|\mathrm{do}(\neg C), \neg F) \tag{26.57}$$

Then we can show that taking the drug is harmful overall:

$$p(E|\mathrm{do}(C)) < p(E|\mathrm{do}(\neg C)) \tag{26.58}$$

The proof is as follows (Pearl 2000, p181). First, from our assumptions in Figure 26.16(a), we see that drugs have no effect on gender

$$p(F|\mathrm{do}(C)) = p(F|\mathrm{do}(\neg C)) = p(F) \tag{26.59}$$

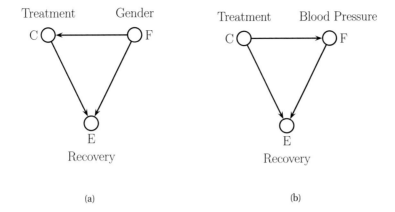

Figure 26.16 Two different models uses to illustrate Simpson's paradox. (a) F is gender and is a confounder for C and E. (b) F is blood pressure and is caused by C.

Now using the law of total probability,

$$p(E|\mathrm{do}(C)) = p(E|\mathrm{do}(C), F)p(F|\mathrm{do}(C)) + p(E|\mathrm{do}(C), \neg F)p(\neg F|\mathrm{do}(C)) \quad (26.60)$$

$$= p(E|\mathrm{do}(C), F)p(F) + p(E|\mathrm{do}(C), \neg F)p(\neg F) \quad (26.61)$$

Similarly,

$$p(E|\mathrm{do}(\neg C)) = p(E|\mathrm{do}(\neg C), F)p(F) + p(E|\mathrm{do}(\neg C), \neg F)p(\neg F) \quad (26.62)$$

Since every term in Equation 26.61 is less than the corresponding term in Equation 26.62, we conclude that

$$p(E|\mathrm{do}(C)) < p(E|\mathrm{do}(\neg C)) \quad (26.63)$$

So if the model in Figure 26.16(a) is correct, we should not administer the drug, since it reduces the probability of the effect.

Now consider a different version of this example. Suppose we keep the data the same but interpret F as something that is affected by C, such as blood pressure. See Figure 26.16(b). In this case, we can no longer assume

$$p(F|\mathrm{do}(C)) = p(F|\mathrm{do}(\neg C)) = p(F) \quad (26.64)$$

and the above proof breaks down. So $p(E|\mathrm{do}(C)) - p(E|\mathrm{do}(\neg C))$ may be positive or negaitve.

If the true model is Figure 26.16(b), then we should not condition on F when assessing the effect of C on E, since there is no backdoor path in this case, because of the v-structure at F. That is, conditioning on F might block one of the causal pathways. In other words, by comparing patients with the same post-treatment blood pressure (value of F), we may mask the effect of one of the two pathways by which the drug operates to bring about recovery.

Thus we see that different causal assumptions lead to different causal conclusions, and hence different courses of action. This raises the question on whether we can learn the causal model from data. We discuss this issue below.

26.6.3   Learning causal DAG structures

In this section, we discuss some ways to learn causal DAG structures.

26.6.3.1   Learning from observational data

In Section 26.4, we discussed various methods for learning DAG structures from observational data. It is natural to ask whether these methods can recover the "true" DAG structure that was used to generate the data. Clearly, even if we have infinite data, an optimal method can only identify the DAG up to Markov equivalence (Section 26.4.1). That is, it can identify the PDAG (partially directed acyclic graph), but not the complete DAG structure, because all DAGs which are Markov equivalent have the same likelihood.

There are several algorithms (e.g., the **greedy equivalence search** method of (Chickering 2002)) that are consistent estimators of PDAG structure, in the sense that they identify the true Markov equivalence class as the sample size goes to infinity, assuming we observe all the variables. However, we also have to assume that the generating distribution p is **faithful** to the generating DAG G. This means that all the conditional indepence (CI) properties of p are exactly captured by the graphical structure, so $I(p) = I(G)$; this means there cannot be any CI properties in p that are due to particular settings of the parameters (such as zeros in a regression matrix) that are not graphically explicit. For this reason, a faithful distribution is also called a **stable** distribution.

Suppose the assumptions hold and we learn a PDAG. What can we do with it? Instead of recovering the full graph, we can focus on the causal analog of edge marginals, by computing the magnitude of the causal effect of one node on another (say A on B). If we know the DAG, we can do this using techniques described in (Pearl 2000). If the DAG is unknown, we can compute a lower bound on the effect as follows (Maathuis et al. 2009): learn an equivalence class (PDAG) from data; enumerate all the DAGs in the equivalence class; apply Pearl's do-calculus to compute the magnitude of the causal effect of A on B in each DAG; finally, take the minimum of these effects as the lower bound. It is usually computationally infeasible to compute all DAGs in the equivalence class, but fortunately one only needs to be able to identify the local neighborhood of A and B, which can be esimated more efficiently, as described in (Maathuis et al. 2009). This technique is called **IDA**, which is short for "intervention-calculus when the DAG is absent".

In (Maathuis et al. 2010), this technique was applied to some yeast gene expression data. Gene knockout data was used to estimate the "ground truth" effect of each 234 single-gene deletions on the remaining 5,361 genes. Then the algorithm was applied to 63 unperturbed (wild-type) samples, and was used to rank order the likely targets of each of the 234 genes. The method had a precision of 66% when the recall was set to 10%; while low, this is substantially more than rival variable-selection methods, such as lasso and elastic net, which were only slightly above chance.

26.6.3.2   Learning from interventional data

If we want to distinguish between DAGs within the equivalence class, we need to use **interventional data**, where certain variables have been set, and the consequences have been measured. An example of this is the dataset in Figure 26.17(a), where proteins in a signalling pathway were perturbed, and their phosphorylation status was measured using a technique called flow

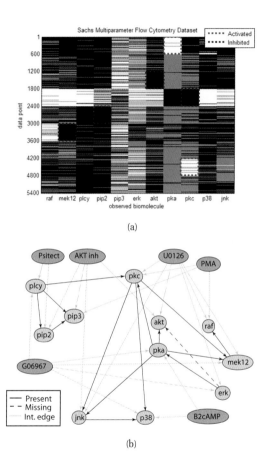

(a)

(b)

Figure 26.17 (a) A design matrix consisting of 5400 data points (rows) measuring the status (using flow cytometry) of 11 proteins (columns) under different experimental conditions. The data has been discretized into 3 states: low (black), medium (grey) and high (white). Some proteins were explicitly controlled using activating or inhibiting chemicals. (b) A directed graphical model representing dependencies between various proteins (blue circles) and various experimental interventions (pink ovals), which was inferred from this data. We plot all edges for which $p(G_{st} = 1|\mathcal{D}) > 0.5$. Dotted edges are believed to exist in nature but were not discovered by the algorithm (1 false negative). Solid edges are true positives. The light colored edges represent the effects of intervention. Source: Figure 6d of (Eaton and Murphy 2007) . This figure can be reproduced using the code at `http://www.cs.ubc.ca/~murphyk/Software/BDAGL/index.html`.

cytometry (Sachs et al. 2005).

It is straightforward to modify the standard Bayesian scoring criteria, such as the marginal likelihood or BIC score, to handle learning from mixed observational and experimental data: we just compute the sufficient statistics for a CPD's parameter by skipping over the cases where that node was set by intervention (Cooper and Yoo 1999). For example, when using tabular CPDs, we modify the counts as follows:

$$
N_{tck} \triangleq \sum_{i:x_{it} \text{ not set}} \mathbb{I}(x_{i,t} = k, \mathbf{x}_{i,\mathrm{pa}(t)} = c) \tag{26.65}
$$

The justification for this is that in cases where node t is set by force, it is not sampled from its usual mechanism, so such cases should be ignored when inferring the parameter $\boldsymbol{\theta}_t$. The modified scoring criterion can be combined with any of the standard structure learning algorithms. (He and Geng 2009) discusses some methods for choosing which interventions to perform, so as to reduce the posterior uncertainty as quickly as possible (a form of active learning).

The preceding method assumes the interventions are perfect. In reality, experimenters can rarely control the state of individual molecules. Instead, they inject various stimulant or inhibitor chemicals which are designed to target specific molecules, but which may have side effects. We can model this quite simply by adding the intervention nodes to the DAG, and then learning a larger augmented DAG structure, with the constraint that there are no edges between the intervention nodes, and no edges from the "regular" nodes back to the intervention nodes.

Figure 26.17(b) shows the augmented DAG that was learned from the interventional flow cytometry data depicted in Figure 26.17(a). In particular, we plot the median graph, which includes all edges for which $p(G_{ij} = 1|\mathcal{D}) > 0.5$. These were computed using the exact algorithm of (Koivisto 2006). It turns out that, in this example, the median model has exactly the same structure as the optimal MAP model, $\mathrm{argmax}_G \, p(G|\mathcal{D})$, which was computed using the algorithm of (Koivisto and Sood 2004; Silander and Myllymaki 2006).

26.7 Learning undirected Gaussian graphical models

Learning the structured of undirected graphical models is easier than learning DAG structure because we don't need to worry about acyclicity. On the other hand, it is harder than learning DAG structure since the likelihood does not decompose (see Section 19.5). This precludes the kind of local search methods (both greedy search and MCMC sampling) we used to learn DAG structures, because the cost of evaluating each neighboring graph is too high, since we have to refit each model from scratch (there is no way to incrementally update the score of a model).

In this section, we discuss several solutions to this problem, in the context of **Gaussian random fields** or undirected Gaussian graphical models (GGM)s. We consider structure learning for discrete undirected models in Section 26.8.

26.7.1 MLE for a GGM

Before discussing structure learning, we need to discuss parameter estimation. The task of computing the MLE for a (non-decomposable) GGM is called **covariance selection** (Dempster 1972).

From Equation 4.19, the log likelihood can be written as

$$\ell(\mathbf{\Omega}) = \log \det \mathbf{\Omega} - \text{tr}(\mathbf{S}\mathbf{\Omega}) \tag{26.66}$$

where $\mathbf{\Omega} = \mathbf{\Sigma}^{-1}$ is the precision matrix, and $\mathbf{S} = \frac{1}{N} \sum_{i=1}^{N} (\mathbf{x}_i - \bar{\mathbf{x}})(\mathbf{x}_i - \bar{\mathbf{x}})^T$ is the empirical covariance matrix. (For notational simplicity, we assume we have already estimated $\hat{\boldsymbol{\mu}} = \bar{\mathbf{x}}$.) One can show that the gradient of this is given by

$$\nabla \ell(\mathbf{\Omega}) = \mathbf{\Omega}^{-1} - \mathbf{S} \tag{26.67}$$

However, we have to enforce the constraints that $\Omega_{st} = 0$ if $G_{st} = 0$ (structural zeros), and that $\mathbf{\Omega}$ is positive definite. The former constraint is easy to enforce, but the latter is somewhat challenging (albeit still a convex constraint). One approach is to add a penalty term to the objective if $\mathbf{\Omega}$ leaves the positive definite cone; this is the approach used in `ggmFitMinfunc` (see also (Dahl et al. 2008)). Another approach is to use a coordinate descent method, described in (Hastie et al. 2009, p633), and implemented in `ggmFitHtf`. Yet another approach is to use iterative proportional fitting, described in Section 19.5.7. However, IPF requires identifying the cliques of the graph, which is NP-hard in general.

Interestingly, one can show that the MLE must satisfy the following property: $\Sigma_{st} = S_{st}$ if $G_{st} = 1$ or $s = t$, i.e., the covariance of a pair that are connected by an edge must match the empirical covariance. In addition, we have $\Omega_{st} = 0$ if $G_{st} = 0$, by definition of a GGM, i.e., the precision of a pair that are not connected must be 0. We say that $\mathbf{\Sigma}$ is a positive definite **matrix completion** of $\mathbf{S}$, since it retains as many of the entries in $\mathbf{S}$ as possible, corresponding to the edges in the graph, subject to the required sparsity pattern on $\mathbf{\Sigma}^{-1}$, corresponding to the absent edges; the remaining entries in $\mathbf{\Sigma}$ are filled in so as to maximize the likelihood.

Let us consider a worked example from (Hastie et al. 2009, p652). We will use the following adjacency matrix, representing the cyclic structure, $X_1 - X_2 - X_3 - X_4 - X_1$, and the following empirical covariance matrix:

$$\mathbf{G} = \begin{pmatrix} 0 & 1 & 0 & 1 \\ 1 & 0 & 1 & 0 \\ 0 & 1 & 0 & 1 \\ 1 & 0 & 1 & 0 \end{pmatrix}, \quad \mathbf{S} = \begin{pmatrix} 10 & 1 & 5 & 4 \\ 1 & 10 & 2 & 6 \\ 5 & 2 & 10 & 3 \\ 4 & 6 & 3 & 10 \end{pmatrix} \tag{26.68}$$

The MLE is given by

$$\mathbf{\Sigma} = \begin{pmatrix} 10.00 & 1.00 & \mathbf{1.31} & 4.00 \\ 1.00 & 10.00 & 2.00 & \mathbf{0.87} \\ \mathbf{1.31} & 2.00 & 10.00 & 3.00 \\ 4.00 & \mathbf{0.87} & 3.00 & 10.00 \end{pmatrix}, \quad \mathbf{\Omega} = \begin{pmatrix} 0.12 & -0.01 & \mathbf{0} & -0.05 \\ -0.01 & 0.11 & -0.02 & \mathbf{0} \\ \mathbf{0} & -0.02 & 0.11 & -0.03 \\ -0.05 & \mathbf{0} & -0.03 & 0.13 \end{pmatrix} \tag{26.69}$$

(See `ggmFitDemo` for the code to reproduce these numbers.) The constrained elements in $\mathbf{\Omega}$, and the free elements in $\mathbf{\Sigma}$, both of which correspond to absent edges, have been highlighted.

26.7.2 Graphical lasso

We now discuss one way to learn a sparse GRF structure, which exploits the fact that there is a 1:1 correspondence between zeros in the precision matrix and absent edges in the graph. This

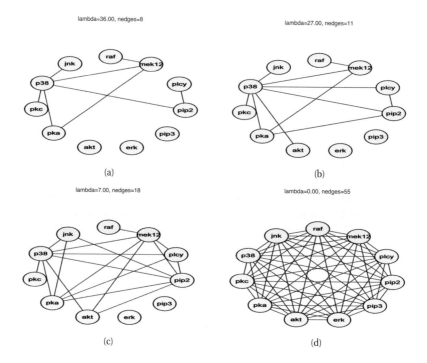

Figure 26.18   Sparse GGMs learned using graphical lasso applied to the flow cytometry data. (a) $\lambda = 36$. (b) $\lambda = 27$. (c) $\lambda = 7$. (d) $\lambda = 0$. Figure generated by `ggmLassoDemo`.

suggests that we can learn a sparse graph structure by using an objective that encourages zeros in the precision matrix. By analogy to lasso (see Section 13.3), one can define the following ℓ_1 penalized NLL:

$$J(\boldsymbol{\Omega}) = -\log\det\boldsymbol{\Omega} + \text{tr}(\mathbf{S}\boldsymbol{\Omega}) + \lambda||\boldsymbol{\Omega}||_1 \qquad (26.70)$$

where $||\boldsymbol{\Omega}||_1 = \sum_{j,k} |\omega_{jk}|$ is the 1-norm of the matrix. This is called the **graphical lasso** or **Glasso**.

Although the objective is convex, it is non-smooth (because of the non-differentiable ℓ_1 penalty) and is constrained (because $\boldsymbol{\Omega}$ must be a positive definite matrix). Several algorithms have been proposed for optimizing this objective (Yuan and Lin 2007; Banerjee et al. 2008; Duchi et al. 2008), although arguably the simplest is the one in (Friedman et al. 2008), which uses a coordinate descent algorithm similar to the shooting algorithm for lasso. See `ggmLassoHtf` for an implementation. (See also (Mazumder and Hastie 2012) for a more recent version of this algorithm.)

As an example, let us apply the method to the flow cytometry dataset from (Sachs et al. 2005). A discretized version of the data is shown in Figure 26.17(a). Here we use the original continuous data. However, we are ignoring the fact that the data was sampled under intervention. In Figure 26.18, we illustrate the graph structures that are learned as we sweep λ from 0 to a large value. These represent a range of plausible hypotheses about the connectivity of these proteins.

It is worth comparing this with the DAG that was learned in Figure 26.17(b). The DAG has the advantage that it can easily model the interventional nature of the data, but the disadvantage that it cannot model the feedback loops that are known to exist in this biological pathway (see the discussion in (Schmidt and Murphy 2009)). Note that the fact that we show many UGMs and only one DAG is incidental: we could easily use BIC to pick the "best" UGM, and conversely, we could easily display several DAG structures, sampled from the posterior.

26.7.3 Bayesian inference for GGM structure *

Although the graphical lasso is reasonably fast, it only gives a point estimate of the structure. Furthermore, it is not model-selection consistent (Meinshausen 2005), meaning it cannot recover the true graph even as $N \to \infty$. It would be preferable to integrate out the parameters, and perform posterior inference in the space of graphs, i.e., to compute $p(G|\mathcal{D})$. We can then extract summaries of the posterior, such as posterior edge marginals, $p(G_{ij} = 1|\mathcal{D})$, just as we did for DAGs. In this section, we discuss how to do this.

Note that the situation is analogous to Chapter 13, where we discussed variable selection. In Section 13.2, we discussed Bayesian variable selection, where we integrated out the regression weights and computed $p(\gamma|\mathcal{D})$ and the marginal inclusion probabilities $p(\gamma_j = 1|\mathcal{D})$. Then in Section 13.3, we discussed methods based on ℓ_1 regularization. Here we have the same dichotomy, but we are presenting them in the opposite order.

If the graph is decomposable, and if we use conjugate priors, we can compute the marginal likelihood in closed form (Dawid and Lauritzen 1993). Furthermore, we can efficiently identify the decomposable neighbors of a graph (Thomas and Green 2009), i.e., the set of legal edge additions and removals. This means that we can perform relatively efficient stochastic local search to approximate the posterior (see e.g. (Giudici and Green 1999; Armstrong et al. 2008; Scott and Carvalho 2008)).

However, the restriction to decomposable graphs is rather limiting if one's goal is knowledge discovery, since the number of decomposable graphs is much less than the number of general undirected graphs.[5]

A few authors have looked at Bayesian inference for GGM structure in the non-decomposable case (e.g., (Dellaportas et al. 2003; Wong et al. 2003; Jones et al. 2005)), but such methods cannot scale to large models because they use an expensive Monte Carlo approximation to the marginal likelihood (Atay-Kayis and Massam 2005). (Lenkoski and Dobra 2008) suggested using a Laplace approximation. This requires computing the MAP estimate of the parameters for Ω under a G-Wishart prior (Roverato 2002). In (Lenkoski and Dobra 2008), they used the iterative proportional scaling algorithm (Speed and Kiiveri 1986; Hara and Takimura 2008) to find the mode. However, this is very slow, since it requires knowing the maximal cliques of the graph, which is NP-hard in general.

In (Moghaddam et al. 2009), a much faster method is proposed. In particular, they modify the gradient-based methods from Section 26.7.1 to find the MAP estimate; these algorithms do not need to know the cliques of the graph. A further speedup is obtained by just using a

---

5. The number of decomposable graphs on V nodes, for $V = 2, \ldots, 8$, is as follows ((Armstrong 2005, p158)): 2; 8; 61; 822; 18,154; 61,7675; 30,888,596. If we divide these numbers by the number of undirected graphs, which is $2^{V(V-1)/2}$, we find the ratios are: 1, 1, 0.95, 0.8, 0.55, 0.29, 0.12. So we see that decomposable graphs form a vanishing fraction of the total hypothesis space.

diagonal Laplace approximation, which is more accurate than BIC, but has essentially the same cost. This, plus the lack of restriction to decomposable graphs, enables fairly fast stochastic search methods to be used to approximate $p(G|\mathcal{D})$ and its mode. This approach significantly outperfomed graphical lasso, both in terms of predictive accuracy and structural recovery, for a comparable computational cost.

26.7.4 Handling non-Gaussian data using copulas *

The graphical lasso and variants is inhertently limited to data that is jointly Gaussian, which is a rather severe restriction. Fortunately the method can be generalized to handle non-Gaussian, but still continuous, data in a fairly simple fashion. The basic idea is to estimate a set of D univariate monotonic transformations f_j, one per variable j, such that the resulting transformed data is jointly Gaussian. If this is possible, we say the data belongs to the nonparametric Normal distribution, or **nonparanormal** distribution (Liu et al. 2009). This is equivalent to the family of **Gaussian copulas** (Klaassen and Wellner 1997). Details on how to estimate the f_j transformations from the empirical cdf's of each variable can be found in (Liu et al. 2009). After transforming the data, we can compute the correlation matrix and then apply glasso in the usual way. One can show, under various assumptions, that this is a consistent estimator of the graph structure, representing the CI assumptions of the original distribution(Liu et al. 2009).

26.8 Learning undirected discrete graphical models

The problem of learning the structure for UGMs with discrete variables is harder than the Gaussian case, because computing the partition function $Z(\boldsymbol{\theta})$, which is needed for parameter estimation, has complexity comparable to computing the permanent of a matrix, which in general is intractable (Jerrum et al. 2004). By contrast, in the Gaussian case, computing Z only requires computing a matrix determinant, which is at most $O(V^3)$.

Since stochastic local search is not tractable for general discrete UGMs, below we mention some possible alternative approaches that have been tried.

26.8.1 Graphical lasso for MRFs/CRFs

It is possible to extend the graphical lasso idea to the discrete MRF and CRF case. However, now there is a set of parameters associated with each edge in the graph, so we have to use the graph analog of group lasso (see Section 13.5.1). For example, consider a pairwise CRF with ternary nodes, and node and edge potentials given by

$$\psi_t(y_t, \mathbf{x}) = \begin{pmatrix} \mathbf{v}_{t1}^T\mathbf{x} \\ \mathbf{v}_{t2}^T\mathbf{x} \\ \mathbf{v}_{t3}^T\mathbf{x} \end{pmatrix}, \; \psi_{st}(y_s, y_t, \mathbf{x}) = \begin{pmatrix} \mathbf{w}_{t11}^T\mathbf{x} & \mathbf{w}_{st12}^T\mathbf{x} & \mathbf{w}_{st13}^T\mathbf{x} \\ \mathbf{w}_{st21}^T\mathbf{x} & \mathbf{w}_{st22}^T\mathbf{x} & \mathbf{w}_{st23}^T\mathbf{x} \\ \mathbf{w}_{st31}^T\mathbf{x} & \mathbf{w}_{st32}^T\mathbf{x} & \mathbf{w}_{st33}^T\mathbf{x} \end{pmatrix} \tag{26.71}$$

where we assume $\mathbf{x}$ begins with a constant 1 term, to account for the offset. (If $\mathbf{x}$ only contains 1, the CRF reduces to an MRF.) Note that we may choose to set some of the $\mathbf{v}_{tk}$ and $\mathbf{w}_{stjk}$ weights to 0, to ensure identifiability, although this can also be taken care of by the prior, as shown in Exercise 8.5.

To learn sparse structure, we can minimize the following objective:

$$
\begin{aligned}
J \;=\; & -\sum_{i=1}^{N}\left[\sum_{t}\log\psi_t(y_{it},\mathbf{x}_i,\mathbf{v}_t)+\sum_{s=1}^{V}\sum_{t=s+1}^{V}\log\psi_{st}(y_{is},y_{it},\mathbf{x}_i,\mathbf{w}_{st})\right] \\
& +\lambda_1\sum_{s=1}^{V}\sum_{t=s+1}^{V}||\mathbf{w}_{st}||_p+\lambda_2\sum_{t=1}^{V}||\mathbf{v}_t||_2^2
\end{aligned}
\tag{26.72}
$$

where $||\mathbf{w}_{st}||_p$ is the p-norm; common choices are $p=2$ or $p=\infty$, as explained in Section 13.5.1. This method of CRF structure learning was first suggested in (Schmidt et al. 2008). (The use of ℓ_1 regularization for learning the structure of binary MRFs was proposed in (Lee et al. 2006).)

Although this objective is convex, it can be costly to evaluate, since we need to perform inference to compute its gradient, as explained in Section 19.6.3 (this is true also for MRFs). We should therefore use an optimizer that does not make too many calls to the objective function or its gradient, such as the projected quasi-Newton method in (Schmidt et al. 2009). In addition, we can use approximate inference, such as convex belief propagation (Section 22.4.2), to compute an approximate objective and gradient more quickly. Another approach is to apply the group lasso penalty to the pseudo-likelihood discussed in Section 19.5.4. This is much faster, since inference is no longer required (Hoefling and Tibshirani 2009). Figure 26.19 shows the result of applying this procedure to the 20-newsgroup data, where y_{it} indicates the presence of word t in document i, and $\mathbf{x}_i=1$ (so the model is an MRF).

26.8.2 Thin junction trees

So far, we have been concerned with learning "sparse" graphs, but these do not necessarily have low treewidth. For example, a $D\times D$ grid is sparse, but has treewidth $O(D)$. This means that the models we learn may be intractable to use for inference purposes, which defeats one of the two main reasons to learn graph structure in the first place (the other reason being "knowledge discovery"). There have been various attempts to learn graphical models with bounded treewidth (e.g., (Bach and Jordan 2001; Srebro 2001; Elidan and Gould 2008; Shahaf et al. 2009)), also known as **thin junction trees**, but the exact problem in general is hard.

An alternative approach is to learn a model with low **circuit complexity** (Gogate et al. 2010; Poon and Domingos 2011). Such models may have high treewidth, but they exploit context-specific independence and determinism to enable fast exact inference (see e.g., (Darwiche 2009)).

Exercises

Exercise 26.1 Causal reasoning in the sprinkler network

Consider the causal network in Figure 26.20. Let T represent true and F represent false.

a. Suppose I perform a perfect intervention and make the grass wet. What is the probability the sprinkler is on, $p(S=T|\mathrm{do}(W=T))$?

b. Suppose I perform a perfect intervention and make the grass dry. What is the probability the sprinkler is on, $p(S=T|\mathrm{do}(W=F))$?

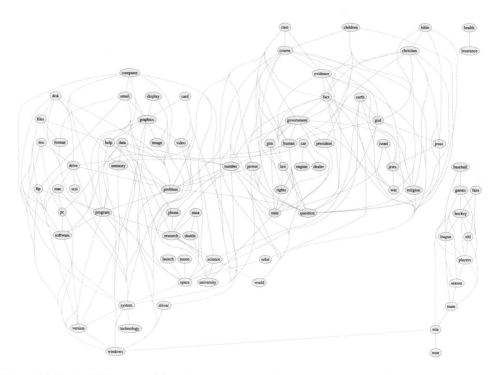

Figure 26.19   An MRF estimated from the 20-newsgroup data using group ℓ_1 regularization with $\lambda = 256$. Isolated nodes are not plotted. From Figure 5.9 of (Schmidt 2010). Used with kind permission of Mark Schmidt.

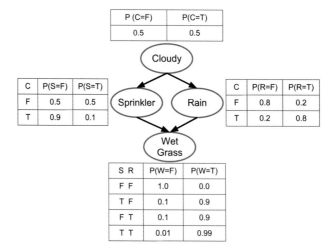

Figure 26.20   Water sprinkler DGM with corresponding binary CPTs. T and F stand for true and false.

c. Suppose I perform a perfect intervention and make the clouds "turn on" (e.g., by seeding them). What is the probability the sprinkler is on, $p(S = T|\mathrm{do}(C = T))$?

27 Latent variable models for discrete data

27.1 Introduction

In this chapter, we are concerned with latent variable models for discrete data, such as bit vectors, sequences of categorical variables, count vectors, graph structures, relational data, etc. These models can be used to analyze voting records, text and document collections, low-intensity images, movie ratings, etc. However, we will mostly focus on text analysis, and this will be reflected in our terminology.

Since we will be dealing with so many different kinds of data, we need some precise notation to keep things clear. When modeling variable-length sequences of categorical variables (i.e., symbols or **tokens**), such as words in a document, we will let $y_{il} \in \{1, \dots, V\}$ represent the identity of the l'th word in document i, where V is the number of possible words in the vocabulary. We assume $l = 1 : L_i$, where L_i is the (known) length of document i, and $i = 1 : N$, where N is the number of documents.

We will often ignore the word order, resulting in a **bag of words**. This can be reduced to a fixed length vector of counts (a histogram). We will use $n_{iv} \in \{0, 1, \dots, L_i\}$ to denote the number of times word v occurs in document i, for $v = 1 : V$. Note that the $N \times V$ count matrix $\mathbf{N}$ is often large but sparse, since we typically have many documents, but most words do not occur in any given document.

In some cases, we might have multiple different bags of words, e.g., bags of text words and bags of visual words. These correspond to different "channels" or types of features. We will denote these by y_{irl}, for $r = 1 : R$ (the number of responses) and $l = 1 : L_{ir}$. If $L_{ir} = 1$, it means we have a single token (a bag of length 1); in this case, we just write $y_{ir} \in \{1, \dots, V_r\}$ for brevity. If every channel is just a single token, we write the fixed-size response vector as $\mathbf{y}_{i,1:R}$; in this case, the $N \times R$ design matrix $\mathbf{Y}$ will not be sparse. For example, in social science surveys, y_{ir} could be the response of person i to the r'th multi-choice question.

Out goal is to build joint probability models of $p(\mathbf{y}_i)$ or $p(\mathbf{n}_i)$ using latent variables to capture the correlations. We will then try to interpret the latent variables, which provide a compressed representation of the data. We provide an overview of some approaches in Section 27.2, before going into more detail in later sections.

Towards the end of the chapter, we will consider modeling graphs and relations, which can also be represented as sparse discrete matrices. For example, we might want to model the graph of which papers may cite which other papers. We will denote these relations by $\mathbf{R}$, reserving the symbol $\mathbf{Y}$ for any categorical data (e.g., text) associated with the nodes.

27.2 Distributed state LVMs for discrete data

In this section, we summarize a variety of possible approaches for constructing models of the form $p(\mathbf{y}_{i,1:L_i})$, for bags of tokens; $p(\mathbf{y}_{1:R})$, for vectors of tokens; and $p(\mathbf{n}_i)$, for vectors of integer counts.

27.2.1 Mixture models

The simplest approach is to use a finite mixture model (Chapter 11). This associates a single discrete latent variable, $q_i \in \{1, \ldots, K\}$, with every document, where K is the number of clusters. We will use a discrete prior, $q_i \sim \mathrm{Cat}(\boldsymbol{\pi})$. For variable length documents, we can define $p(y_{il}|q_i = k) = b_{kv}$, where b_{kv} is the probability that cluster k generates word v. The value of q_i is known as a **topic**, and the vector $\mathbf{b}_k$ is the k'th topic's word distribution. That is, the likelihood has the form

$$p(\mathbf{y}_{i,1:L_i}|q_i = k) = \prod_{l=1}^{L_i} \mathrm{Cat}(y_{il}|\mathbf{b}_k) \qquad (27.1)$$

The induced distribution on the visible data is given by

$$p(\mathbf{y}_{i,1:L_i}) = \sum_k \pi_k \left[\prod_{l=1}^{L_i} \mathrm{Cat}(y_{il}|\mathbf{b}_k) \right] \qquad (27.2)$$

The "generative story" which this encodes is as follows: for document i, pick a topic q_i from $\boldsymbol{\pi}$, call it k, and then for each word $l = 1 : L_i$, pick a word from $\mathbf{b}_k$. We will consider more sophisticated generative models later in this chapter.

If we have a fixed set of categorical observations, we can use a different topic matrix for each output variable:

$$p(\mathbf{y}_{i,1:R}|q_i = k) = \prod_{r=1}^{R} \mathrm{Cat}(y_{il}|\mathbf{b}_k^{(r)}) \qquad (27.3)$$

This is an unsupervised analog of naive Bayes classification.

We can also model count vectors. If the sum $L_i = \sum_v n_{iv}$ is known, we can use a multinomial:

$$p(\mathbf{n}_i|L_i, q_i = k) = \mathrm{Mu}(\mathbf{n}_i|L_i, \mathbf{b}_k) \qquad (27.4)$$

If the sum is unknown, we can use a Poisson class-conditional density to give

$$p(\mathbf{n}_i|q_i = k) = \prod_{v=1}^{V} \mathrm{Poi}(n_{iv}|\lambda_{vk}) \qquad (27.5)$$

In this case, $L_i|q_i = k \sim \mathrm{Poi}(\sum_v \lambda_{vk})$.

27.2.2 Exponential family PCA

Unfortunately, finite mixture models are very limited in their expressive power. A more flexible model is to use a vector of real-valued continuous latent variables, similar to the factor analysis (FA) and PCA models in Chapter 12. In PCA, we use a Gaussian prior of the form $p(\mathbf{z}_i) = \mathcal{N}(\boldsymbol{\mu}, \boldsymbol{\Sigma})$, where $\mathbf{z}_i \in \mathbb{R}^K$, and a Gaussian likelihood of the form $p(\mathbf{y}_i|\mathbf{z}_i) = \mathcal{N}(\mathbf{W}\mathbf{z}_i, \sigma^2 \mathbf{I})$. This method can certainly be applied to discrete or count data. Indeed, the methods known as **latent semantic indexing** (**LSI**, (Deerwester et al. 1990)) and **latent semantic analysis** (**LSA**, (Dumais and Landauer 1997)) are exactly equivalent to applying PCA to a term by document count matrix. (LSI corresponds to treating the columns (documents) as samples, and measures document similarity; LSA corresponds to treating the rows (words) as samples, and measures word similarity.)

A better method for modeling categorical data is to use a multinoulli or multinomial distribution. We just have to change the likelihood to

$$p(\mathbf{y}_{i,1:L_i}|\mathbf{z}_i) = \prod_{l=1}^{L_i} \mathrm{Cat}(y_{il}|\mathcal{S}(\mathbf{W}\mathbf{z}_i)) \tag{27.6}$$

where $\mathbf{W} \in \mathbb{R}^{V \times K}$ is a weight matrix and $\mathcal{S}$ is the softmax function. If we have a fixed number of categorical responses, we can use

$$p(\mathbf{y}_{1:R}|\mathbf{z}_i) = \prod_{r=1}^{R} \mathrm{Cat}(y_{ir}|\mathcal{S}(\mathbf{W}_r\mathbf{z}_i)) \tag{27.7}$$

where $\mathbf{W}_r \in \mathbb{R}^{V \times K}$ is the weight matrix for the r'th response variable. This model is called **categorical PCA**, and is illustrated in Figure 27.1(a); see Section 12.4 for further discussion. If we have counts, we can use a multinomial model

$$p(\mathbf{n}_i|L_i, \mathbf{z}_i) = \mathrm{Mu}(\mathbf{n}_i|L_i, \mathcal{S}(\mathbf{W}\mathbf{z}_i)) \tag{27.8}$$

or a Poisson model

$$p(\mathbf{n}_i|\mathbf{z}_i) = \prod_{v=1}^{V} \mathrm{Poi}(n_{iv}|\exp(\mathbf{w}_{v,:}^T \mathbf{z}_i)) \tag{27.9}$$

All of these models are examples of **exponential family PCA** or **ePCA** (Collins et al. 2002; Mohamed et al. 2008), which is an unsupervised analog of GLMs. The corresponding induced distribution on the visible variables has the form

$$p(\mathbf{y}_{i,1:L_i}) = \int \left[\prod_{l=1}^{L_i} p(y_{il}|\mathbf{z}_i, \mathbf{W}) \right] \mathcal{N}(\mathbf{z}_i|\boldsymbol{\mu}, \boldsymbol{\Sigma}) d\mathbf{z}_i \tag{27.10}$$

Fitting this model is tricky, due to the lack of conjugacy. (Collins et al. 2002) proposed a coordinate ascent method that alternates between estimating the $\mathbf{z}_i$ and $\mathbf{W}$. This can be regarded as a degenerate version of EM, that computes a point estimate of $\mathbf{z}_i$ in the E step. The problem with the degenerate approach is that it is very prone to overfitting, since the number of latent variables is proportional to the number of datacases (Welling et al. 2008). A true EM algorithm would marginalize out the latent variables $\mathbf{z}_i$. A way to do this for categorical PCA, using variational EM, is discussed in Section 12.4. For more general models, one can use MCMC (Mohamed et al. 2008).

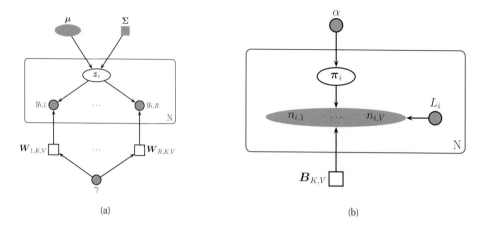

Figure 27.1 Two LVMs for discrete data. Circles are scalar nodes, ellipses are vector nodes, squares are matrix nodes. (a) Categorical PCA. (b) Multinomial PCA.

27.2.3 LDA and mPCA

In ePCA, the quantity $\mathbf{W}\mathbf{z}_i$ represents the natural parameters of the exponential family. Sometimes it is more convenient to use the dual parameters. For example, for the multinomial, the dual parameter is the probability vector, whereas the natural parameter is the vector of log odds.

If we want to use the dual parameters, we need to constrain the latent variables so they live in the appropriate parameter space. In the case of categorical data, we will need to ensure the latent vector lives in S_K, the K-dimensional probability simplex. To avoid confusion with ePCA, we will denote such a latent vector by $\boldsymbol{\pi}_i$. In this case, the natural prior for the latent variables is the Dirichlet, $\boldsymbol{\pi}_i \sim \text{Dir}(\boldsymbol{\alpha})$. Typically we set $\boldsymbol{\alpha} = \alpha \mathbf{1}_K$. If we set $\alpha \ll 1$, we encourage $\boldsymbol{\pi}_i$ to be sparse, as shown in Figure 2.14.

When we have a count vector whose total sum is known, the likelihood is given by

$$p(\mathbf{n}_i|L_i, \boldsymbol{\pi}_i) = \text{Mu}(\mathbf{n}_i|L_i, \mathbf{B}\boldsymbol{\pi}_i) \tag{27.11}$$

This model is called **multinomial PCA** or **mPCA** (Buntine 2002; Buntine and Jakulin 2004, 2006). See Figure 27.1(b). Since we are assuming $n_{iv} = \sum_k b_{vk}\pi_{iv}$, this can be seen as a form of matrix factorization for the count matrix. Note that we use $b_{v,k}$ to denote the parameter vector, rather than $w_{v,k}$, since we impose the constraints that $0 \le b_{v,k} \le 1$ and $\sum_v b_{v,k} = 1$. The corresponding marginal distribution has the form

$$p(\mathbf{n}_i|L_i) = \int \text{Mu}(\mathbf{n}_i|L_i, \mathbf{B}\boldsymbol{\pi}_i)\text{Dir}(\boldsymbol{\pi}_i|\boldsymbol{\alpha})d\boldsymbol{\pi}_i \tag{27.12}$$

Unfortunately, this integral cannot be computed analytically.

If we have a variable length sequence (of known length), we can use

$$p(\mathbf{y}_{i,1:L_i}|\boldsymbol{\pi}_i) = \prod_{l=1}^{L_i} \text{Cat}(y_{il}|\mathbf{B}\boldsymbol{\pi}_i) \tag{27.13}$$

This is called **latent Dirichlet allocation** or **LDA** (Blei et al. 2003), and will be described in much greater detail below. LDA can be thought of as a probabilistic extension of LSA, where the latent quantities π_{ik} are non-negative and sum to one. By contrast, in LSA, z_{ik} can be negative which makes interpetation difficult.

A predecessor to LDA, known as **probabilistic latent semantic indexing** or **PLSI** (Hofmann 1999), uses the same model but computes a point estimate of π_i for each document (similar to ePCA), rather than integrating it out. Thus in PLSI, there is no prior for π_i.

We can modify LDA to handle a fixed number of different categorical responses as follows:

$$p(\mathbf{y}_{i,1:R}|\boldsymbol{\pi}_i) = \prod_{r=1}^{R} \mathrm{Cat}(y_{il}|\mathbf{B}^{(r)}\boldsymbol{\pi}_i) \tag{27.14}$$

This has been called the **user rating profile** (URP) model (Marlin 2003), and the **simplex factor model** (Bhattacharya and Dunson 2011).

27.2.4 GaP model and non-negative matrix factorization

Now consider modeling count vectors where we do not constrain the sum to be observed. In this case, the latent variables just need to be non-negative, so we will denote them by $\mathbf{z}_i^+$. This can be ensured by using a prior of the form

$$p(\mathbf{z}_i^+) = \prod_{k=1}^{K} \mathrm{Ga}(z_{ik}^+|\alpha_k, \beta_k) \tag{27.15}$$

The likelihood is given by

$$p(\mathbf{n}_i|\mathbf{z}_i^+) = \prod_{v=1}^{V} \mathrm{Poi}(n_{iv}|\mathbf{b}_{v,:}^T\mathbf{z}_i^+) \tag{27.16}$$

This is called the **GaP** (Gamma-Poisson) model (Canny 2004). See Figure 27.2(a).

In (Buntine and Jakulin 2006), it is shown that the GaP model, when conditioned on a fixed L_i, reduces to the mPCA model. This follows since a set of Poisson random variables, when conditioned on their sum, becomes a multinomial distribution (see e.g., (Ross 1989)).

If we set $\alpha_k = \beta_k = 0$ in the GaP model, we recover a method known as **non-negative matrix factorization** or **NMF** (Lee and Seung 2001), as shown in (Buntine and Jakulin 2006). NMF is not a probabilistic generative model, since it does not specify a proper prior for $\mathbf{z}_i^+$. Furthermore, the algorithm proposed in (Lee and Seung 2001) is another degenerate EM algorithm, so suffers from overfitting. Some procedures to fit the GaP model, which overcome these problems, are given in (Buntine and Jakulin 2006).

To encourage $\mathbf{z}_i^+$ to be sparse, we can modify the prior to be a spike-and-Gamma type prior as follows:

$$p(z_{ik}^+) = \rho_k \mathbb{I}(z_{ik}^+ = 0) + (1 - \rho_k)\mathrm{Ga}(z_{ik}^+|\alpha_k, \beta_k) \tag{27.17}$$

where ρ_k is the probability of the spike at 0. This is called the conditional Gamma Poisson model (Buntine and Jakulin 2006). It is simple to modify Gibbs sampling to handle this kind of prior, although we will not go into detail here.

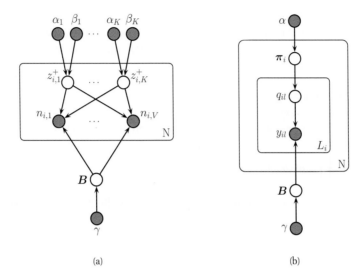

(a)   (b)

Figure 27.2   (a) Gaussian-Poisson (GAP) model. (b) Latent Dirichlet allocation (LDA) model.

27.3   Latent Dirichlet allocation (LDA)

In this section, we explain the **latent Dirichlet allocation** or **LDA** (Blei et al. 2003) model in detail.

27.3.1   Basics

In a mixture of multinoullis, every document is assigned to a single topic, $q_i \in \{1, \ldots, K\}$, drawn from a global distribution $\boldsymbol{\pi}$. In LDA, every word is assigned to its own topic, $q_{il} \in \{1, \ldots, K\}$, drawn from a document-specific distribution $\boldsymbol{\pi}_i$. Since a document belongs to a distribution over topics, rather than a single topic, the model is called an **admixture mixture** or **mixed membership model** (Erosheva et al. 2004). This model has many other applications beyond text analysis, e.g., genetics (Pritchard et al. 2000), health science (Erosheva et al. 2007), social network analysis (Airoldi et al. 2008), etc.

Adding conjugate priors to the parameters, the full model is as follows:[1]

$$\boldsymbol{\pi}_i | \alpha \quad \sim \quad \mathrm{Dir}(\alpha \mathbf{1}_K) \tag{27.18}$$

$$q_{il} | \boldsymbol{\pi}_i \quad \sim \quad \mathrm{Cat}(\boldsymbol{\pi}_i) \tag{27.19}$$

$$\mathbf{b}_k | \gamma \quad \sim \quad \mathrm{Dir}(\gamma \mathbf{1}_V) \tag{27.20}$$

$$y_{il} | q_{il} = k, \mathbf{B} \quad \sim \quad \mathrm{Cat}(\mathbf{b}_k) \tag{27.21}$$

This is illustrated in Figure 27.2(b). We can marginalize out the q_i variables, thereby creating a

---

1. Our notation is similar to the one we use elsewhere in this book, but is different from that used by most LDA papers. They typically use w_{nd} for the identity of word n in document d, z_{nd} to represent the discrete indicator, $\boldsymbol{\theta}_d$ as the continuous latent vector for document d, and $\boldsymbol{\beta}_k$ as the k'th topic vector.

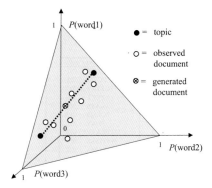

Figure 27.3 Geometric interpretation of LDA. We have $K = 2$ topics and $V = 3$ words. Each document (white dots), and each topic (black dots), is a point in the 3d simplex. Source: Figure 5 of (Steyvers and Griffiths 2007). Used with kind permission of Tom Griffiths.

direct arc from $\boldsymbol{\pi}_i$ to y_{il}, with the following CPD:

$$p(y_{il} = v|\boldsymbol{\pi}_i) = \sum_k p(y_{il} = v|q_{il} = k)p(q_{il} = k) = \sum_k \pi_{ik}b_{kv} \tag{27.22}$$

As we mentioned in the introduction, this is very similar to the multinomial PCA model proposed in (Buntine 2002), which in turn is closely related to categorical PCA, GaP, NMF, etc.

LDA has an interesting geometric interpretation. Each vector $\mathbf{b}_k$ defines a distribution over V words; each k is known as a **topic**. Each document vector $\boldsymbol{\pi}_i$ defines a distribution over K topics. So we model each document as an admixture over topics. Equivalently, we can think of LDA as a form of dimensionality reduction (assuming $K < V$, as is usually the case), where we project a point in the V-dimensional simplex (a normalized document count vector $\mathbf{x}_i$) onto the K-dimensional simplex. This is illustrated in Figure 27.3, where we have $V = 3$ words and $K = 2$ topics. The observed documents (which live in the 3d simplex) are approximated as living on a 2d simplex spanned by the 2 topic vectors, each of which lives in the 3d simplex.

One advantage of using the simplex as our latent space rather than Euclidean space is that the simplex can handle ambiguity. This is importance since in natural language, words can often have multiple meanings, a phenomomen known as **polysemy**. For example, "play" might refer to a verb (e.g., "to play ball" or "to play the coronet"), or to a noun (e.g., "Shakespeare's play"). In LDA, we can have multiple topics, each of which can generate the word "play", as shown in Figure 27.4, reflecting this ambiguity.

Given word l in document i, we can compute $p(q_{il} = k|\mathbf{y}_i, \boldsymbol{\theta})$, and thus infer its most likely topic. By looking at the word in isolation, it might be hard to know what sense of the word is meant, but we can disambiguate this by looking at other words in the document. In particular, given $\mathbf{x}_i$, we can infer the topic distribution $\boldsymbol{\pi}_i$ for the document; this acts as a prior for disambiguating q_{il}. This is illustrated in Figure 27.5, where we show three documents from the TASA corpus.[2] In the first document, there are a variety of music related words, which suggest

---

2. The TASA corpus is a collection of 37,000 high-school level English documents, comprising over 10 million words,

Topic 77

word	prob.
MUSIC	.090
DANCE	.034
SONG	.033
PLAY	.030
SING	.026
SINGING	.026
BAND	.026
PLAYED	.023
SANG	.022
SONGS	.021
DANCING	.020
PIANO	.017
PLAYING	.016
RHYTHM	.015
ALBERT	.013
MUSICAL	.013

Topic 82

word	prob.
LITERATURE	.031
POEM	.028
POETRY	.027
POET	.020
PLAYS	.019
POEMS	.019
PLAY	.015
LITERARY	.013
WRITERS	.013
DRAMA	.012
WROTE	.012
POETS	.011
WRITER	.011
SHAKESPEARE	.010
WRITTEN	.009
STAGE	.009

Topic 166

word	prob.
PLAY	.136
BALL	.129
GAME	.065
PLAYING	.042
HIT	.032
PLAYED	.031
BASEBALL	.027
GAMES	.025
BAT	.019
RUN	.019
THROW	.016
BALLS	.015
TENNIS	.011
HOME	.010
CATCH	.010
FIELD	.010

Figure 27.4 Three topics related to the word *play*. Source: Figure 9 of (Steyvers and Griffiths 2007). Used with kind permission of Tom Griffiths.

Document #29795

Bix beiderbecke, at age^{060} fifteen^{207}, sat^{174} on the slope^{071} of a bluff^{055} overlooking^{027} the mississippi^{137} river^{137}. He was listening^{077} to music^{077} coming^{009} from a passing^{043} riverboat. The music^{077} had already captured^{006} his heart^{157} as well as his ear^{119}. It was jazz^{077}. Bix beiderbecke had already had music^{077} lessons^{077}. He showed^{002} promise^{134} on the piano^{077}, and his parents^{035} hoped^{268} he might consider^{118} becoming a concert^{077} pianist^{077}. But bix was interested^{268} in another kind^{050} of music^{077}. He wanted^{268} to play^{077} the cornet. And he wanted^{268} to play^{077} jazz^{077}...

Document #1883

There is a simple^{050} reason^{106} why there are so few periods^{078} of really great theater^{082} in our whole western^{046} world. Too many things^{300} have to come right at the very same time. The dramatists must have the right actors^{082}, the actors^{082} must have the right playhouses, the playhouses must have the right audiences^{082}. We must remember^{288} that plays^{082} exist^{143} to be performed^{077}, not merely^{050} to be read^{254}. (even when you read^{254} a play^{082} to yourself, try^{288} to perform^{062} it, to put^{174} it on a stage^{078}, as you go along.) as soon^{028} as a play^{082} has to be performed^{082}, then some kind^{126} of theatrical^{082}...

Document #21359

Jim^{296} has a game^{166} book^{254}. Jim^{296} reads^{254} the book^{254}. Jim^{296} sees^{081} a game^{166} for one. Jim^{296} plays^{166} the game^{166}. Jim^{296} likes^{081} the game^{166} for one. The game^{166} book^{254} helps^{081} jim^{296}. Don^{180} comes^{040} into the house^{038}. Don^{180} and jim^{296} read^{254} the game^{166} book^{254}. The boys^{020} see a game^{166} for two. The two boys^{020} play^{166} the game^{166}. The boys^{020} play^{166} the game^{166} for two. The boys^{020} like the game^{166}. Meg^{282} comes^{040} into the house^{282}. Meg^{282} and don^{180} and jim^{296} read^{254} the book^{254}. They see a game^{166} for three. Meg^{282} and don^{180} and jim^{296} play^{166} the game^{166}. They play^{166}...

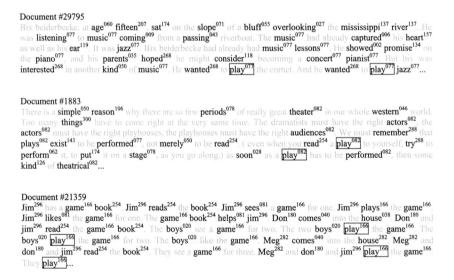

Figure 27.5 Three documents from the TASA corpus containing different senses of the word *play*. Grayed out words were ignored by the model, because they correspond to uninteresting stop words (such as "and", "the", etc.) or very low frequency words. Source: Figure 10 of (Steyvers and Griffiths 2007). Used with kind permission of Tom Griffiths.

π_i will put most of its mass on the music topic (number 77); this in turn makes the music interpretation of "play" the most likely, as shown by the superscript. The second document interprets play in the theatrical sense, and the third in the sports sense. Note that is crucial that π_i be a latent variable, so information can flow between the q_{il}'s, thus enabling local disambiguation to use the full set of words.

27.3.2 Unsupervised discovery of topics

One of the main purposes of LDA is discover topics in a large collection or **corpus** of documents (see Figure 27.12 for an example). Unfortunately, since the model is unidentifiable, the interpretation of the topics can be difficult (Chang et al. 2009).. One approach, known as labeled LDA (Ramage et al. 2009), exploits the existence of tags on documents as a way to ensure identifiability. In particular, it forces the topics to correspond to the tags, and then it learns a distribution over words for each tag. This can make the results easier to interpret.

27.3.3 Quantitatively evaluating LDA as a language model

In order to evaluate LDA quantitatively, we can treat it as a **language model**, i.e., a probability distribution over sequences of words. Of course, it is not a very good language model, since it ignores word order and just looks at single words (unigrams), but it is interesting to compare LDA to other unigram-based models, such as mixtures of multinoullis, and pLSI. Such simple language models are sometimes useful for information retrieval purposes. The standard way to measure the quality of a language model is to use perplexity, which we now define below.

27.3.3.1 Perplexity

The **perplexity** of language model q given a stochastic process[3] p is defined as

$$\text{perplexity}(p, q) \triangleq 2^{H(p,q)} \tag{27.23}$$

where $H(p, q)$ is the cross-entropy of the two stochastic processes, defined as

$$H(p, q) \triangleq \lim_{N \to \infty} -\frac{1}{N} \sum_{\mathbf{y}_{1:N}} p(\mathbf{y}_{1:N}) \log q(\mathbf{y}_{1:N}) \tag{27.24}$$

The cross entropy (and hence perplexity) is minimized if $q = p$; in this case, the model can predict as well as the "true" distribution.

We can approximate the stochastic process by using a single long test sequence (composed of multiple documents and multiple sentences, complete with end-of-sentence markers), call it $\mathbf{y}_{1:N}^*$. (This approximation becomes more and more accurate as the sequence gets longer, provided the process is stationary and ergodic (Cover and Thomas 2006).) Define the empirical distribution (an approximation to the stochastic process) as

$$p_{\text{emp}}(\mathbf{y}_{1:N}) = \delta_{\mathbf{y}_{1:N}^*}(\mathbf{y}_{1:N}) \tag{27.25}$$

---

collated by a company formerly known as Touchstone Applied Science Associates, but now known as Questar Assessment Inc www.questarai.com.

3. A stochastic process is one which can define a joint distribution over an arbitrary number of random variables. We can think of natural language as a stochastic process, since it can generate an infinite stream of words.

In this case, the cross-entropy becomes

$$H(p_{\text{emp}}, q) = -\frac{1}{N} \log q(\mathbf{y}_{1:N}^*) \tag{27.26}$$

and the perplexity becomes

$$\text{perplexity}(p_{\text{emp}}, q) = 2^{H(p_{\text{emp}}, q)} = q(\mathbf{y}_{1:N}^*)^{-1/N} = \sqrt[N]{\prod_{i=1}^{N} \frac{1}{q(y_i^* | \mathbf{y}_{1:i-1}^*)}} \tag{27.27}$$

We see that this is the geometric mean of the inverse predictive probabilities, which is the usual definition of perplexity (Jurafsky and Martin 2008, p96).

In the case of unigram models, the cross entropy term is given by

$$H = -\frac{1}{N} \sum_{i=1}^{N} \frac{1}{L_i} \sum_{l=1}^{L_i} \log q(y_{il}) \tag{27.28}$$

where N is the number of documents and L_i is the number of words in document i. Hence the perplexity of model q is given by

$$\text{perplexity}(p_{\text{emp}}, p) = \exp\left(-\frac{1}{N} \sum_{i=1}^{N} \frac{1}{L_i} \sum_{l=1}^{L_i} \log q(y_{il})\right) \tag{27.29}$$

Intuitively, perplexity mesures the weighted average **branching factor** of the model's predictive distribution. Suppose the model predicts that each symbol (letter, word, whatever) is equally likely, so $p(y_i | \mathbf{y}_{1:i-1}) = 1/K$. Then the perplexity is $((1/K)^N)^{-1/N} = K$. If some symbols are more likely than others, and the model correctly reflects this, its perplexity will be lower than K. Of course, $H(p, p) = H(p) \leq H(p, q)$, so we can never reduce the perplexity below the entropy of the underlying stochastic process.

27.3.3.2 Perplexity of LDA

The key quantity is $p(v)$, the predictive distribution of the model over possible words. (It is implicitly conditioned on the training set.) For LDA, this can be approximated by plugging in **B** (e.g., the posterior mean estimate) and approximately integrating out **q** using mean field inference (see (Wallach et al. 2009) for a more accurate way to approximate the predictive likelihood).

In Figure 27.6, we compare LDA to several other simple unigram models, namely MAP estimation of a multinoulli, MAP estimation of a mixture of multinoullis, and pLSI. (When performing MAP estimation, the same Dirichlet prior on **B** was used as in the LDA model.) The metric is perplexity, as in Equation 27.29, and the data is a subset of the TREC AP corpus containing 16,333 newswire articles with 23,075 unique terms. We see that LDA significantly outperforms these other methods.

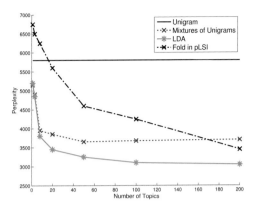

Figure 27.6 Perplexity vs number of topics on the TREC AP corpus for various language models. Based on Figure 9 of (Blei et al. 2003). Figure generated by `bleiLDAperplexityPlot`.

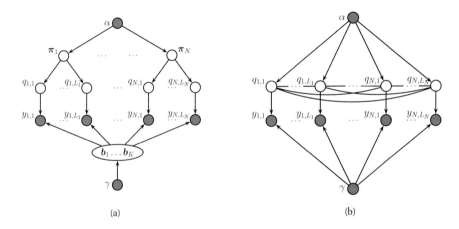

Figure 27.7 (a) LDA unrolled for N documents. (b) Collapsed LDA, where we integrate out the $\boldsymbol{\pi}_i$ and the $\mathbf{b}_k$.

27.3.4 Fitting using (collapsed) Gibbs sampling

It is straightforward to derive a Gibbs sampling algorithm for LDA. The full conditionals are as follows:

$$p(q_{il} = k|\cdot) \quad \propto \quad \exp[\log \pi_{ik} + \log b_{k,x_{il}}] \tag{27.30}$$

$$p(\boldsymbol{\pi}_i|\cdot) \quad = \quad \text{Dir}(\{\alpha_k + \sum_l \mathbb{I}(z_{il} = k)\}) \tag{27.31}$$

$$p(\mathbf{b}_k|\cdot) \quad = \quad \text{Dir}(\{\gamma_v + \sum_i \sum_l \mathbb{I}(x_{il} = v, z_{il} = k)\}) \tag{27.32}$$

However, one can get better performance by analytically integrating out the $\boldsymbol{\pi}_i$'s and the $\mathbf{b}_k$'s, both of which have a Dirichlet distribution, and just sampling the discrete q_{il}'s. This approach was first suggested in (Griffiths and Steyvers 2004), and is an example of **collapsed Gibbs sampling**. Figure 27.7(b) shows that now all the q_{il} variables are fully correlated. However, we can sample them one at a time, as we explain below.

First, we need some notation. Let $c_{ivk} = \sum_{l=1}^{L_i} \mathbb{I}(q_{il} = k, y_{il} = v)$ be the number of times word v is assigned to topic k in document i. Let $c_{ik} = \sum_v c_{ivk}$ be the number of times any word from document i has been assigned to topic k. Let $c_{vk} = \sum_i c_{ivk}$ be the number of times word v has been assigned to topic k in any document. Let $n_{iv} = \sum_k c_{ivk}$ be the number of times word v occurs in document i; this is observed. Let $c_k = \sum_v c_{vk}$ be the number of words assigned to topic k. Finally, let $L_i = \sum_k c_{ik}$ be the number of words in document i; this is observed.

We can now derive the marginal prior. By applying Equation 5.24, one can show that

$$p(\mathbf{q}|\alpha) = \prod_i \int \left[\prod_{l=1}^{L_i} \text{Cat}(q_{il}|\boldsymbol{\pi}_i) \right] \text{Dir}(\boldsymbol{\pi}_i|\alpha\mathbf{1}_K)d\boldsymbol{\pi}_i \tag{27.33}$$

$$= \left(\frac{\Gamma(K\alpha)}{\Gamma(\alpha)^K} \right)^N \prod_{i=1}^N \frac{\prod_{k=1}^K \Gamma(c_{ik} + \alpha)}{\Gamma(L_i + K\alpha)} \tag{27.34}$$

By similar reasoning, one can show

$$p(\mathbf{y}|\mathbf{q}, \gamma) = \prod_k \int \left[\prod_{il:q_{il}=k} \text{Cat}(y_{il}|\mathbf{b}_k) \right] \text{Dir}(\mathbf{b}_k|\gamma\mathbf{1}_V)d\mathbf{b}_k \tag{27.35}$$

$$= \left(\frac{\Gamma(V\beta)}{\Gamma(\beta)^V} \right)^K \prod_{k=1}^K \frac{\prod_{v=1}^V \Gamma(c_{vk} + \beta)}{\Gamma(c_k + V\beta)} \tag{27.36}$$

From the above equations, and using the fact that $\Gamma(x+1)/\Gamma(x) = x$, we can derive the full conditional for $p(q_{il}|\mathbf{q}_{-i,l})$. Define c_{ivk}^- to be the same as c_{ivk} except it is compute by summing over all locations in document i except for q_{il}. Also, let $y_{il} = v$. Then

$$p(q_{i,l} = k|\mathbf{q}_{-i,l}, \mathbf{y}, \alpha, \gamma) \propto \frac{c_{v,k}^- + \gamma}{c_k^- + V\gamma} \frac{c_{i,k}^- + \alpha}{L_i + K\alpha} \tag{27.37}$$

We see that a word in a document is assigned to a topic based both on how often that word is generated by the topic (first term), and also on how often that topic is used in that document (second term).

Given Equation 27.37, we can implement the collapsed Gibbs sampler as follows. We randomly assign a topic to each word, $q_{il} \in \{1, \ldots, K\}$. We can then sample a new topic as follows: for a given word in the corpus, decrement the relevant counts, based on the topic assigned to the current word; draw a new topic from Equation 27.37, update the count matrices; and repeat. This algorithm can be made efficient since the count matrices are very sparse.

27.3.5   Example

This process is illustrated in Figure 27.8 on a small example with two topics, and five words. The left part of the figure illustrates 16 documents that were sampled from the LDA model using

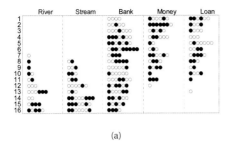
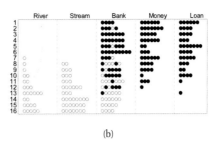

(a) (b)

Figure 27.8 Illustration of (collapsed) Gibbs sampling applied to a small LDA example. There are $N = 16$ documents, each containing a variable number of words drawn from a vocabulary of $V = 5$ words, There are two topics. A white dot means word the word is assigned to topic 1, a black dot means the word is assigned to topic 2. (a) The initial random assignment of states. (b) A sample from the posterior after 64 steps of Gibbs sampling. Source: Figure 7 of (Steyvers and Griffiths 2007). Used with kind permission of Tom Griffiths.

$p(\text{money}|k = 1) = p(\text{loan}|k = 1) = p(\text{bank}|k = 1) = 1/3$ and $p(\text{river}|k = 2) = p(\text{stream}|k = 2) = p(\text{bank}|k = 2) = 1/3$. For example, we see that the first document contains the word "bank" 4 times (indicated by the four dots in row 1 of the "bank" column), as well as various other financial terms. The right part of the figure shows the state of the Gibbs sampler after 64 iterations. The "correct" topic has been assigned to each token in most cases. For example, in document 1, we see that the word "bank" has been correctly assigned to the financial topic, based on the presence of the words "money" and "loan". The posterior mean estimate of the parameters is given by $\hat{p}(\text{money}|k = 1) = 0.32$, $\hat{p}(\text{loan}|k = 1) = 0.29$, $\hat{p}(\text{bank}|k = 1) = 0.39$, $\hat{p}(\text{river}|k = 2) = 0.25$, $\hat{p}(\text{stream}|k = 2) = 0.4$, and $\hat{p}(\text{bank}|k = 2) = 0.35$, which is impressively accurate, given that there are only 16 training examples.

27.3.6 Fitting using batch variational inference

A faster alternative to MCMC is to use variational EM. (We cannot use exact EM since exact inference of $\boldsymbol{\pi}_i$ and $\mathbf{q}_i$ is intractable.) We give the details below.

27.3.6.1 Sequence version

Following (Blei et al. 2003), we will use a fully factorized (mean field) approximation of the form

$$q(\boldsymbol{\pi}_i, \mathbf{q}_i) = \text{Dir}(\boldsymbol{\pi}_i|\tilde{\boldsymbol{\pi}}_i) \prod_l \text{Cat}(q_{il}|\tilde{\mathbf{q}}_{il}) \tag{27.38}$$

We will follow the usual mean field recipe. For $q(q_{il})$, we use Bayes' rule, but where we need to take expectations over the prior:

$$\tilde{q}_{ilk} \quad \propto \quad b_{y_{i,l},k} \exp(\mathbb{E}\left[\log \pi_{ik}\right]) \tag{27.39}$$

where

$$\mathbb{E}\left[\log \pi_{ik}\right] = \psi_k(\tilde{\boldsymbol{\pi}}_{i.}) \triangleq \Psi(\tilde{\pi}_{ik}) - \Psi(\sum_{k'} \tilde{\pi}_{ik'}) \tag{27.40}$$

where Ψ is the digamma function. The update for $q(\boldsymbol{\pi}_i)$ is obtained by adding up the expected counts:

$$\tilde{\pi}_{ik} \;=\; \alpha_k + \sum_l \tilde{q}_{ilk} \tag{27.41}$$

The M step is obtained by adding up the expected counts and normalizing:

$$\hat{b}_{vk} \;\propto\; \gamma_v + \sum_{i=1}^{N} \sum_{l=1}^{L_i} \tilde{q}_{ilk} \mathbb{I}(y_{il} = v) \tag{27.42}$$

27.3.6.2 Count version

Note that the E step takes $O((\sum_i L_i)VK)$ space to store the $\tilde{q}_{ilk}$. It is much more space efficient to perform inference in the mPCA version of the model, which works with counts; these only take $O(NVK)$ space, which is a big savings if documents are long. (By contrast, the collapsed Gibbs sampler must work explicitly with the q_{il} variables.)

We will focus on approximating $p(\boldsymbol{\pi}_i, \mathbf{c}_i | \mathbf{n}_i, L_i)$, where we write $\mathbf{c}_i$ as shorthand for $\mathbf{c}_{i..}$. We will again use a fully factorized (mean field) approximation of the form

$$q(\boldsymbol{\pi}_i, \mathbf{c}_i) = \text{Dir}(\boldsymbol{\pi}_i | \tilde{\boldsymbol{\pi}}_i) \prod_v \text{Mu}(\mathbf{c}_{iv.} | n_{iv}, \tilde{\mathbf{c}}_{iv.}) \tag{27.43}$$

The new E step becomes

$$\tilde{\pi}_{ik} \;=\; \alpha_k + \sum_v n_{iv} \tilde{c}_{ivk} \tag{27.44}$$

$$\tilde{c}_{ivk} \;\propto\; b_{vk} \exp(\mathbb{E}\left[\log \pi_{ik}\right]) \tag{27.45}$$

The new M step becomes

$$\hat{b}_{vk} \;\propto\; \gamma_v + \sum_i n_{iv} \tilde{c}_{ivk} \tag{27.46}$$

27.3.6.3 VB version

We now modify the algorithm to use VB instead of EM, so that we infer the parameters as well as the latent variables. There are two advantages to this. First, by setting $\gamma \ll 1$, VB will encourage $\mathbf{B}$ to be sparse (as in Section 21.6.1.6). Second, we will be able to generalize this to the online learning setting, as we discuss below.

Our new posterior approximation becomes

$$q(\boldsymbol{\pi}_i, \mathbf{c}_i, \mathbf{B}) = \text{Dir}(\boldsymbol{\pi}_i | \tilde{\boldsymbol{\pi}}_i) \prod_v \text{Mu}(\mathbf{c}_{iv.} | n_{iv}, \tilde{\mathbf{c}}_{iv.}) \prod_k \text{Dir}(\mathbf{b}_{.k} | \tilde{\mathbf{b}}_{.k}) \tag{27.47}$$

The update for $\tilde{c}_{ivk}$ changes, to the following:

$$\tilde{c}_{ivk} \;\propto\; \exp\left(\mathbb{E}\left[\log b_{vk}\right] + \mathbb{E}\left[\log \pi_{ik}\right]\right) \tag{27.48}$$

Algorithm 27.1: Batch VB for LDA

1 Input: n_{iv}, K, α_k, γ_v;
2 Estimate $\tilde{b}_{vk}$ using EM for multinomial mixtures;
3 Initialize counts n_{iv};
4 **while** *not converged* **do**
5   // E step ;
6     $s_{vk} = 0$ // expected sufficient statistics;
7   **for** *each document* $i = 1 : N$ **do**
8        $(\tilde{\boldsymbol{\pi}}_i, \tilde{\mathbf{c}}_i) = \text{Estep}(\mathbf{n}_i, \tilde{\mathbf{B}}, \boldsymbol{\alpha})$;
9        $s_{vk} += n_{iv}\tilde{c}_{ivk}$;
10   // M step ;
11   **for** *each topic* $k = 1 : K$ **do**
12       $\tilde{b}_{vk} = \gamma_v + s_{vk}$;

13 function $(\tilde{\boldsymbol{\pi}}_i, \tilde{\mathbf{c}}_i) = \text{Estep}(\mathbf{n}_i, \tilde{\mathbf{B}}, \boldsymbol{\alpha})$;
14 Initialize $\tilde{\pi}_{ik} = \alpha_k$;
15 **repeat**
16    $\tilde{\pi}_{i.}^{old} = \tilde{\pi}_{i.}$, $\tilde{\pi}_{ik} = \alpha_k$;
17   **for** *each word* $v = 1 : V$ **do**
18   **for** *each topic* $k = 1 : K$ **do**
19          $\tilde{c}_{ivk} = \exp\left(\psi_k(\tilde{\mathbf{b}}_{v.}) + \psi_k(\tilde{\boldsymbol{\pi}}_{i.}^{old})\right)$;
20       $\tilde{\mathbf{c}}_{iv.} = \text{normalize}(\tilde{\mathbf{c}}_{iv.})$;
21       $\tilde{\pi}_{ik} += n_{iv}\tilde{c}_{ivk}$
22 **until** $\frac{1}{K}\sum_k |\tilde{\pi}_{ik} - \tilde{\pi}_{ik}^{old}| < thresh$;

Also, the M step becomes

$$\tilde{b}_{vk} = \gamma_v + \sum_i \tilde{c}_{ivk} \tag{27.49}$$

No normalization is required, since we are just updating the pseudocounts. The overall algorithm is summarized in Algorithm 27.1.

27.3.7 Fitting using online variational inference

In the batch version, the E step clearly takes $O(NKVT)$ time, where T is the number of mean field updates (typically $T \sim 5$). This can be slow if we have many documents. This can be reduced by using stochastic gradient descent (Section 8.5.2) to perform online variational inference, as we now explain.

We can derive an online version, following (Hoffman et al. 2010). We perform an E step in the usual way. We then compute the variational parameters for $\mathbf{B}$ treating the expected sufficient statistics from the single data case as if the whole data set had those statistics. Finally, we make

---

Algorithm 27.2: Online variational Bayes for LDA

---

1 Input: n_{iv}, K, α_k, γ_v, τ_0, κ;

2 Initialize $\tilde{b}_{vk}$ randomly;

3 **for** $t = 1 : \infty$ **do**

4   Set step size $\rho_t = (\tau_0 + t)^{-\kappa}$;

5   Pick document $i = i(t)$; ;

6      $(\tilde{\boldsymbol{\pi}}_i, \tilde{\mathbf{c}}_i) = \mathrm{Estep}(\mathbf{n}_i, \tilde{\mathbf{B}}, \boldsymbol{\alpha})$;

7      $\tilde{b}_{vk}^{new} = \gamma_v + N n_{iv} \tilde{c}_{ivk}$;

8      $\tilde{b}_{vk} = (1 - \rho_t)\tilde{b}_{vk} + \rho_t \tilde{b}_{vk}^{new}$;

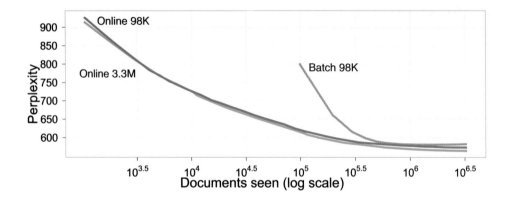

Figure 27.9 Test perplexity vs number of training documents for batch and online VB-LDA. From Figure 1 of (Hoffman et al. 2010). Used with kind permission of David Blei.

a partial update for the variational parameters for **B**, putting weight ρ_t on the new estimate and weight $1 - \rho_t$ on the old estimate. The step size ρ_t decays over time, as in Equation 8.83. The overall algorithm is summarized in Algorithm 27.2. In practice, we should use mini-batches, as explained in Section 8.5.2.3. In (Hoffman et al. 2010), they used a batch of size 256–4096.

Figure 27.9 plots the perplexity on a test set of size 1000 vs number of analyzed documents (E steps), where the data is drawn from (English) Wikipedia. The figure shows that online variational inference is much faster than offline inference, yet produces similar results.

27.3.8   Determining the number of topics

Choosing K, the number of topics, is a standard model selection problem. Here are some approaches that have been taken:

- Use annealed importance sampling (Section 24.6.2) to approximate the evidence (Wallach et al. 2009).
- Cross validation, using the log likelihood on a test set.

- Use the variational lower bound as a proxy for $\log p(\mathcal{D}|K)$.
- Use non-parametric Bayesian methods (Teh et al. 2006).

27.4 Extensions of LDA

Many extensions of LDA have been proposed since the first paper came out in 2003. We briefly discuss a few of these below.

27.4.1 Correlated topic model

One weakness of LDA is that it cannot capture correlation between topics. For example, if a document has the "business" topic, it is reasonable to expect the "finance" topic to co-occcur. The source of the problem is the use of a Dirichlet prior for $\boldsymbol{\pi}_i$. The problem with the Dirichlet it that it is characterized by just a mean vector and a strength parameter, but its covariance is fixed ($\Sigma_{ij} = -\alpha_i\alpha_j$), rather than being a free parameter.

One way around this is to replace the Dirichlet prior with the logistic normal distribution, as in categorical PCA (Section 27.2.2). The model becomes

$$\mathbf{b}_k|\gamma \quad \sim \quad \mathrm{Dir}(\gamma\mathbf{1}_V) \tag{27.50}$$

$$\mathbf{z}_i \quad \sim \quad \mathcal{N}(\boldsymbol{\mu}, \boldsymbol{\Sigma}) \tag{27.51}$$

$$\boldsymbol{\pi}_i|\mathbf{z}_i \quad = \quad \mathcal{S}(\mathbf{z}_i) \tag{27.52}$$

$$q_{il}|\boldsymbol{\pi}_i \quad \sim \quad \mathrm{Cat}(\boldsymbol{\pi}_i) \tag{27.53}$$

$$y_{il}|q_{il} = k, \mathbf{B} \quad \sim \quad \mathrm{Cat}(\mathbf{b}_k) \tag{27.54}$$

This is known as the **correlated topic model** (Blei and Lafferty 2007). This is very similar to categorical PCA, but slightly different. To see the difference, let us marginalize out the q_{il} and $\boldsymbol{\pi}_i$. Then in the CTM we have

$$y_{il} \sim \mathrm{Cat}(\mathbf{B}\mathcal{S}(\mathbf{z}_i)) \tag{27.55}$$

where $\mathbf{B}$ is a stochastic matrix. By contrast, in catPCA we have

$$y_{il} \sim \mathrm{Cat}(\mathcal{S}(\mathbf{W}\mathbf{z}_i)) \tag{27.56}$$

where $\mathbf{W}$ is an unconstrained matrix.

Fitting this model is tricky, since the prior for $\boldsymbol{\pi}_i$ is no longer conjugate to the multinomial likelihood for q_{il}. However, we can use any of the variational methods in Section 21.8.1.1, where we discussed Bayesian multiclass logistic regression. In the CTM case, things are even harder since the categorical response variables $\mathbf{q}_i$ are hidden, but we can handle this by using an additional mean field approximation. See (Blei and Lafferty 2007) for details.

Having fit the model, one can then convert $\hat{\boldsymbol{\Sigma}}$ to a sparse precision matrix $\hat{\boldsymbol{\Sigma}}^{-1}$ by pruning low-strength edges, to get a sparse Gaussian graphical model. This allows you to visualize the correlation between topics. Figure 27.10 shows the result of applying this procedure to articles from *Science* magazine, from 1990-1999. (This corpus contains 16,351 documents, and 5.7M words (19,088 of them unique), after stop-word and low-frequency removal.) Nodes represent topics, with the top 5 words per topic listed inside. The font size reflects the overall prevalence of the topic in the corpus. Edges represent significant elements of the precision matrix.

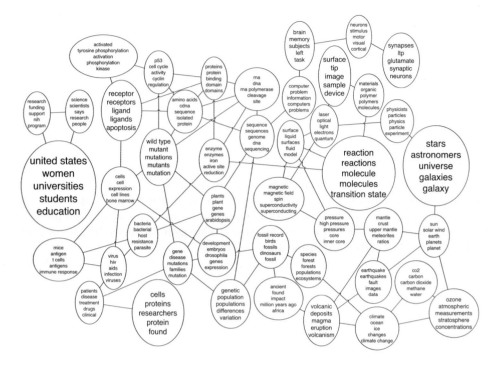

Figure 27.10 Output of the correlated topic model (with $K = 50$ topics) when applied to articles from *Science*. Nodes represent topics, with the 5 most probable phrases from each topic shown inside. Font size reflects overall prevalence of the topic. See http://www.cs.cmu.edu/~lemur/science/ for an interactive version of this model with 100 topics. Source: Figure 2 of (Blei and Lafferty 2007). Used with kind permission of David Blei.

27.4.2 Dynamic topic model

In LDA, the topics (distributions over words) are assumed to be static. In some cases, it makes sense to allow these distributions to evolve smoothly over time. For example, an article might use the topic "neuroscience", but if it was written in the 1900s, it is more likely to use words like "nerve", whereas if it was written in the 2000s, it is more likely to use words like "calcium receptor" (this reflects the general trend of neuroscience towards molecular biology).

One way to model this is use a dynamic logistic normal model, as illustrated in Figure 27.11. In particular, we assume the topic distributions evolve according to a Gaussian random walk, and then we map these Gaussian vectors to probabilities via the softmax function:

$$
\begin{aligned}
\mathbf{b}_{t,k}|\mathbf{b}_{t-1,k} &\sim \mathcal{N}(\mathbf{b}_{t-1,k}, \sigma^2 \mathbf{1}_V) & (27.57)\\
\boldsymbol{\pi}_i^t &\sim \mathrm{Dir}(\alpha \mathbf{1}_K) & (27.58)\\
q_{il}^t|\boldsymbol{\pi}_i^t &\sim \mathrm{Cat}(\boldsymbol{\pi}_i^t) & (27.59)\\
y_{il}^t|q_{il}^t = k, \mathbf{B}^t &\sim \mathrm{Cat}(\mathcal{S}(\mathbf{b}_k^t)) & (27.60)
\end{aligned}
$$

This is known as a **dynamic topic model** (Blei and Lafferty 2006b).

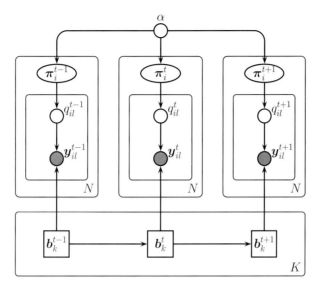

Figure 27.11 The dynamic topic model.

One can perform approximate inference in this model using a structured mean field method (Section 21.4), that exploits the Kalman smoothing algorithm (Section 18.3.1) to perform exact inference on the linear-Gaussian chain between the $\mathbf{b}_{t,k}$ nodes (see (Blei and Lafferty 2006b) for details).

Figure 27.12 illustrates a typical output of the system when applied to 100 years of articles from *Science*. On the top, we visualize the top 10 words from a specific topic (which seems to be related to neuroscience) after 10 year intervals. On the bottom left, we plot the probability of some specific words belonging to this topic. On the bottom right, we list the titles of some articles that contained this topic.

One interesting application of this model is to perform temporally-corrected document retrieval. That is, suppose we look for documents about the inheritance of disease. Modern articles will use words like "DNA", but older articles (before the discovery of DNA) may use other terms such as "heritable unit". But both articles are likely to use the same topics. Similar ideas can be used to perform cross-language information retrieval, see e.g., (Cimiano et al. 2009).

27.4.3 LDA-HMM

The LDA model assumes words are exchangeable, which is clearly not true. A simple way to model sequential dependence between words is to use a hidden Markov model or HMM. The trouble with HMMs is that they can only model short-range dependencies, so they cannot capture the overall gist of a document. Hence they can generate syntactically correct sentences (see e.g., Table 17.1). but not semantically plausible ones.

It is possible to combine LDA with HMM to create a model called **LDA-HMM** (Griffiths et al.

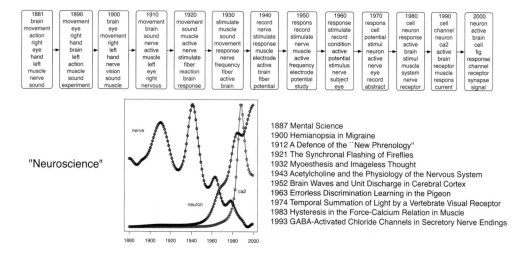

Figure 27.12 Part of the output of the dynamic topic model when applied to articles from *Science*. We show the top 10 words for the neuroscience topic over time. We also show the probability of three words within this topic over time, and some articles that contained this topic.   Source: Figure 4 of (Blei and Lafferty 2006b).   Used with kind permission of David Blei.

2004). This model uses the HMM states to model function or syntactic words, such as "and" or "however", and uses the LDA to model content or semantic words, which are harder to predict. There is a distinguished HMM state which specifies when the LDA model should be used to generate the word; the rest of the time, the HMM generates the word.

More formally, for each document i, the model defines an HMM with states $z_{il} \in \{0, \dots, C\}$. In addition, each document has an LDA model associated with it. If $z_{il} = 0$, we generate word y_{il} from the semantic LDA model, with topic specified by q_{il}; otherwise we generate word y_{il} from the syntactic HMM model. The DGM is shown in Figure 27.13. The CPDs are as follows:

$$p(\boldsymbol{\pi}_i) \;=\; \text{Dir}(\boldsymbol{\pi}_i | \alpha \mathbf{1}_K) \tag{27.61}$$

$$p(q_{il} = k | \boldsymbol{\pi}_i) \;=\; \pi_{ik} \tag{27.62}$$

$$p(z_{il} = c' | z_{i,l-1} = c) \;=\; A^{HMM}(c, c') \tag{27.63}$$

$$p(y_{il} = v | q_{il} = k, z_{il} = c) \;=\; \begin{cases} B^{LDA}(k, v) & \text{if } c = 0 \\ B^{HMM}(c, v) & \text{if } c > 0 \end{cases} \tag{27.64}$$

where $\mathbf{B}^{LDA}$ is the usual topic-word matrix, $\mathbf{B}^{HMM}$ is the state-word HMM emission matrix and $\mathbf{A}^{HMM}$ is the state-state HMM transition matrix.

Inference in this model can be done with collapsed Gibbs sampling, analytically integrating out all the continuous quantities. See (Griffiths et al. 2004) for the details.

The results of applying this model (with $K = 200$ LDA topics and $C = 20$ HMM states) to the combined Brown and TASA corpora[4] are shown in Table 27.1. We see that the HMM generally is

---

4. The Brown corpus consists of 500 documents and 1,137,466 word tokens, with part-of-speech tags for each token.

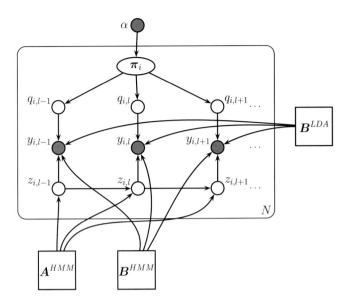

Figure 27.13 LDA-HMM model.

1. In contrast to this approach, we study here how the overall network activity can control single cell parameters such as input resistance, as well as time and space constants, parameters that are crucial for excitability and spariotemporal (sic) integration.

   The integrated architecture in this paper combines feed forward control and error feedback adaptive control using neural networks.

2. In other words, for our proof of convergence, we require the softassign algorithm to return a doubly stochastic matrix as *sinkhorn theorem guarantees that it will instead of a matrix which is merely close to being doubly stochastic based on some reasonable metric.

   The aim is to construct a portfolio with a maximal expected return for a given risk level and time horizon while simultaneously obeying *institutional or *legally required constraints.

3. The left graph is the standard experiment the right from a training with # samples.

   The graph G is called the *guest graph, and H is called the host graph.

Figure 27.14 Function and content words in the NIPS corpus, as distinguished by the LDA-HMM model. Graylevel indicates posterior probability of assignment to LDA component, with black being highest. The boxed word appears as a function word in one sentence, and as a content word in another sentence. Asterisked words had low frequency, and were treated as a single word type by the model. Source: Figure 4 of (Griffiths et al. 2004). Used with kind permission of Tom Griffiths.

the	the	the	the	the	a	the	the	the
blood	,	,	of	a	the	,	,	,
,	and	and	,	of	of	of	a	a
of	of	of	to	,	,	a	of	in
body	a	in	in	in	in	and	and	game
heart	in	land	and	to	water	in	drink	ball
and	trees	to	classes	picture	is	story	alcohol	and
in	tree	farmers	government	film	and	is	to	team
to	with	for	a	image	matter	to	bottle	to
is	on	farm	state	lens	are	as	in	play
blood	forest	farmers	government	light	water	story	drugs	ball
heart	trees	land	state	eye	matter	stories	drug	game
pressure	forests	crops	federal	lens	molecules	poem	alcohol	team
body	land	farm	public	image	liquid	characters	people	*
lungs	soil	food	local	mirror	particles	poetry	drinking	baseball
oxygen	areas	people	act	eyes	gas	character	person	players
vessels	park	farming	states	glass	solid	author	effects	football
arteries	wildlife	wheat	national	object	substance	poems	marijuana	player
*	area	farms	laws	objects	temperature	life	body	field
breathing	rain	corn	department	lenses	changes	poet	use	basketball
the	in	he	*	be	said	can	time	,
a	for	it	new	have	made	would	way	;
his	to	you	other	see	used	will	years	(
this	on	they	first	make	came	could	day	:
their	with	i	same	do	went	may	part	)
these	at	she	great	know	found	had	number	
your	by	we	good	get	called	must	kind	
her	from	there	small	go		do	place	
my	as	this	little	take		have		
some	into	who	old	find		did		

Table 27.1 Upper row: Topics extracted by the LDA model when trained on the combined Brown and TASA corpora. Middle row: topics extracted by LDA part of LDA-HMM model. Bottom row: topics extracted by HMM part of LDA-HMM model. Each column represents a single topic/class, and words appear in order of probability in that topic/class. Since some classes give almost all probability to only a few words, a list is terminated when the words account for 90% of the probability mass. Source: Figure 2 of (Griffiths et al. 2004). Used with kind permission of Tom Griffiths.

responsible for syntactic words, and the LDA for semantics words. If we did not have the HMM, the LDA topics would get "polluted" by function words (see top of figure), which is why such words are normally removed during preprocessing.

The model can also help disambiguate when the same word is being used syntactically or semantically. Figure 27.14 shows some examples when the model was applied to the NIPS corpus.[5] We see that the roles of words are distinguished, e.g., "we require the algorithm to *return* a matrix" (verb) vs "the maximal expected *return*" (noun). In principle, a part of speech tagger could disambiguate these two uses, but note that (1) the LDA-HMM method is fully unsupervised (no POS tags were used), and (2) sometimes a word can have the same POS tag, but different senses, e.g., "the left graph" (a synactic role) vs "the graph G" (a semantic role).

The topic of probabilistic models for syntax and semantics is a vast one, which we do not

---

The TASA corpus is an untagged collection of educational materials consisting of 37,651 documents and 12,190,931 word tokens. Words appearing in fewer than 5 documents were replaced with an asterisk, but punctuation was included. The combined vocabulary was of size 37,202 unique words.

5. NIPS stands for "Neural Information Processing Systems". It is one of the top machine learning conferences. The NIPS corpus volumes 1-12 contains 1713 documents.

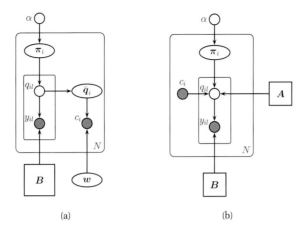

Figure 27.15 (a) Supervised LDA. (b) Discriminative LDA.

have space to delve into any more. See e.g., (Jurafsky and Martin 2008) for further information.

27.4.4 Supervised LDA

In this section, we discuss extensions of LDA to handle side information of various kinds beyond just words.

27.4.4.1 Generative supervised LDA

Suppose we have a variable length sequence of words $y_{il} \in \{1, \ldots, V\}$ as usual, but we also have a class label $c_i \in \{1, \ldots, C\}$. How can we predict c_i from $\mathbf{y}_i$? There are many possible approaches, but most are direct mappings from the words to the class. In some cases, such as **sentiment analysis**, we can get better performance by first performing inference, to try to disambiguate the meaning of words. For example, suppose the goal is to determine if a document is a favorable review of a movie or not. If we encounter the phrase "Brad Pitt was excellent until the middle of the movie", the word "excellent" may lead us to think the review is positive, but clearly the overall sentiment is negative.

One way to tackle such problems is to build a joint model of the form $p(c_i, \mathbf{y}_i | \boldsymbol{\theta})$. (Blei and McAuliffe 2010) proposes an approach, called **supervised LDA**, where the class label c_i is generated from the topics as follows:

$$p(c_i | \overline{\mathbf{q}}_i) = \text{Ber}(\text{sigm}(\mathbf{w}^T \overline{\mathbf{q}}_i)) \tag{27.65}$$

Here $\overline{\mathbf{q}}_i$ is the empirical topic distribution for document i:

$$\overline{q}_{ik} \triangleq \frac{1}{L_i} \sum_{i=1}^{L_i} q_{ilk} \tag{27.66}$$

See Figure 27.15(a) for an illustration.

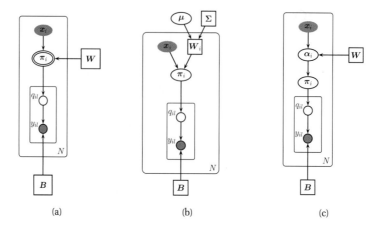

Figure 27.16 Discriminative variants of LDA. (a) Mixture of experts aka MR-LDA. The double ring denotes a node that $\boldsymbol{\pi}_i$ a deterministic function of its parents. (b) Mixture of experts with random effects. (c) DMR-LDA.

We can fit this model using Monte Carlo EM: run the collapsed Gibbs sampler in the E step, to compute $\mathbb{E}\left[\bar{q}_{ik}\right]$, and then use this as the input feature to a standard logistic regression package.

27.4.4.2 Discriminative supervised LDA

An alternative approach, known as **discriminative LDA** (Lacoste-Julien et al. 2009), is shown in Figure 27.15(b). This is a discriminative model of the form $p(\mathbf{y}_i|c_i, \boldsymbol{\theta})$. The only change from regular LDA is that the topic prior becomes input dependent, as follows:

$$p(q_{il}|\boldsymbol{\pi}_i, c_i = c, \boldsymbol{\theta}) = \mathrm{Cat}(\mathbf{A}_c\boldsymbol{\pi}) \tag{27.67}$$

where $\mathbf{A}_c$ is a $K \times K$ stochastic matrix.

So far, we have assumed the "side information" is a single categorical variable c_i. Often we have high dimensional covariates $\mathbf{x}_i \in \mathbb{R}^D$. For example, consider the task of **image tagging**. The idea is that y_{il} represent correlated tags or labels, which we want to predict given $\mathbf{x}_i$. We now discuss several attempts to extend LDA so that it can generate tags given the inputs.

The simplest approach is to use a mixture of experts (Section 11.2.4) with multiple outputs. This is just like LDA except we replace the Dirichlet prior on $\boldsymbol{\pi}_i$ with a deterministic function of the input:

$$\boldsymbol{\pi}_i = \mathcal{S}(\mathbf{W}\mathbf{x}_i) \tag{27.68}$$

In (Law et al. 2010), this is called **multinomial regression LDA**. See Figure 27.16(a). Eliminating the deterministic $\boldsymbol{\pi}_i$ we have

$$p(q_{il}|\mathbf{x}_i, \mathbf{W}) = \mathrm{Cat}(\mathcal{S}(\mathbf{W}\mathbf{x}_i)) \tag{27.69}$$

We can fit this with EM in the usual way. However, (Law et al. 2010) suggest an alternative. First fit an unsupervised LDA model based only on $\mathbf{y}_i$; then treat the inferred $\boldsymbol{\pi}_i$ as data, and fit a multinomial logistic regression model mapping $\mathbf{x}_i$ to $\boldsymbol{\pi}_i$. Although this is fast, fitting LDA in an unsupervised fashion does not necessarily result in a discriminative set of latent variables, as discussed in (Blei and McAuliffe 2010).

There is a more subtle problem with this model. Since $\boldsymbol{\pi}_i$ is a deterministic function of the inputs, it is effectively observed, rendering the q_{il} (and hence the tags y_{il}) independent. In other words,

$$p(\mathbf{y}_i|\mathbf{x}_i, \boldsymbol{\theta}) = \prod_{l=1}^{L_i} p(y_{il}|\mathbf{x}_i, \boldsymbol{\theta}) = \prod_{l=1}^{L_i} \sum_k p(y_{il}|q_{il} = k, \mathbf{B})p(q_{il} = k|\mathbf{x}_i, \mathbf{W}) \qquad (27.70)$$

This means that if we observe the value of one tag, it will have no influence on any of the others. This may explain why the results in (Law et al. 2010) only show negligible improvement over predicting each tag independently.

One way to induce correlations is to make $\mathbf{W}$ a random variable. The resulting model is shown in Figure 27.16(b). We call this a **random effects mixture of experts**. We typically assume a Gaussian prior on $\mathbf{W}_i$. If $\mathbf{x}_i = 1$, then $p(q_{il}|\mathbf{x}_i, \mathbf{w}_i) = \text{Cat}(\mathcal{S}(\mathbf{w}_i))$, so we recover the correlated topic model. It is possible to extend this model by adding Markovian dynamics to the q_{il} variables. This is called a **conditional topic random field** (Zhu and Xing 2010).

A closely related approach, known as **Dirichlet multinomial regression LDA** (Mimno and McCallum 2008), is shown in Figure 27.16(c). This is identical to standard LDA except we make $\boldsymbol{\alpha}$ a function of the input

$$\boldsymbol{\alpha}_i = \exp(\mathbf{W}\mathbf{x}_i) \qquad (27.71)$$

where $\mathbf{W}$ is a $K \times D$ matrix. Eliminating the deterministic $\boldsymbol{\alpha}_i$ we have

$$\boldsymbol{\pi}_i \sim \text{Dir}(\exp(\mathbf{W}\mathbf{x}_i)) \qquad (27.72)$$

Unlike (Law et al. 2010), this model allows information to flow between tags via the latent $\boldsymbol{\pi}_i$.

A variant of this model, where $\mathbf{x}_i$ corresponds to a bag of discrete labels and $\boldsymbol{\pi}_i \sim \text{Dir}(\boldsymbol{\alpha} \odot \mathbf{x}_i)$, is known as **labeled LDA** (Ramage et al. 2009). In this case, the labels $\mathbf{x}_i$ are in 1:1 correspondence with the latent topics, which makes the resulting topics much more interpretable. An extension, known as **partially labeled LDA** (Ramage et al. 2011), allows each label to have multiple latent sub-topics; this model includes LDA, labeled LDA and a multinomial mixture model as special cases.

27.4.4.3 Discriminative categorical PCA

An alternative to using LDA is to expand the categorical PCA model with inputs, as shown in Figure 27.17(a). Since the latent space is now real-valued, we can use simple linear regression for the input-hidden mapping. For the hidden-output mapping, we use traditional catPCA:

$$p(\mathbf{z}_i|\mathbf{x}_i, \mathbf{V}) = \mathcal{N}(\mathbf{V}\mathbf{x}_i, \boldsymbol{\Sigma}) \qquad (27.73)$$

$$p(\mathbf{y}_i|\mathbf{z}_i, \mathbf{W}) = \prod_l \text{Cat}(y_{il}|\mathcal{S}(\mathbf{W}\mathbf{z}_i)) \qquad (27.74)$$

This model is essentially a probabilistic neural network with one hidden layer, as shown in Figure 27.17(b), but with exchangeable output (e.g., to handle variable numbers of tags). The

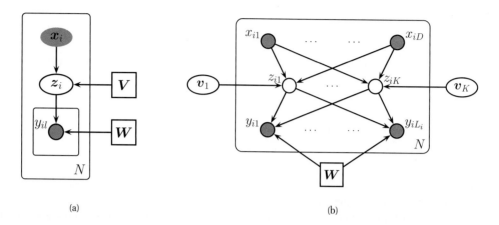

Figure 27.17   (a) Categorical PCA with inputs and exchangeable outputs. (b) Same as (a), but with the vector nodes expanded out.

key difference from a neural net is that information can flow between the y_{il}'s via the latent **bottleneck layer** z_i. This should work better than a conventional neural net when the output labels are highly correlated, even after conditioning on the features; this problem frequently arises in multi label classification. Note that we could allow a direct x_i to y_i arc, but this would require too many parameters if the number of labels is large.[6]

We can fit this model with a small modification of the variational EM algorithm in Section 12.4. If we use this model for regression, rather than classification, we can perform the E step exactly, by modifying the EM algorithm for factor analysis. (Ma et al. 1997) reports that this method converges faster than standard backpropagation.

We can also extend the model so that the prior on z_i is a mixture of Gaussians using input-dependent means. If the output is Gaussian, this corresponds to a mixture of discriminative factor analysers (Fokoue 2005; Zhou and Liu 2008). If the output is categorical, this would be an (as yet unpublished) model, which we could call "discriminative mixtures of categorical factor analyzers".

27.5   LVMs for graph-structured data

Another source of discrete data is when modeling graph or network structures. To see the connection, recall that any graph on D nodes can be represented as a $D \times D$ **adjacency matrix** $\mathbf{G}$, where $G(i,j) = 1$ iff there is an edge from node i to node j. Such matrices are binary, and often very sparse. See Figure 27.19 for an example.

Graphs arise in many application areas, such as modeling social networks, protein-protein interaction networks, or patterns of disease transmission between people or animals. There are usually two primary goals when analysing such data: first, try to discover some "interesting

---

6. A non-probabilistic version of this idea, using squared loss, was proposed in (Ji et al. 2010). This is similar to a linear feed-forward neural network with an additional edge from $\mathbf{x}_i$ directly to $\mathbf{y}_i$.

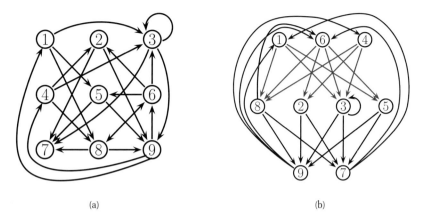

Figure 27.18   (a) A directed graph. (b) The same graph, with the nodes partitioned into 3 groups, making the block structure more apparent.

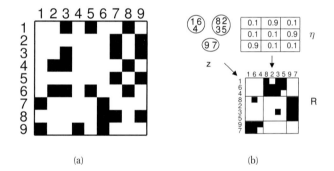

Figure 27.19   (a) Adjacency matrix for the graph in Figure 27.18(a). (b) Rows and columns are shown permuted to show the block structure. We also sketch of how the stochastic block model can generate this graph. From Figure 1 of (Kemp et al. 2006). Used with kind permission of Charles Kemp.

structure" in the graph, such as clusters or communities; second, try to predict which links might occur in the future (e.g., who will make friends with whom). Below we summarize some models that have been proposed for these tasks, some of which are related to LDA. Futher details on these and other approaches can be found in e.g., (Goldenberg et al. 2009) and the references therein.

27.5.1   Stochastic block model

In Figure 27.18(a) we show a directed graph on 9 nodes. There is no apparent structure. However, if we look more deeply, we see it is possible to partition the nodes into three groups or blocks, $B_1 = \{1, 4, 6\}$, $B_2 = \{2, 3, 5, 8\}$, and $B_3 = \{7, 9\}$, such that most of the connections go from nodes in B_1 to B_2, or from B_2 to B_3, or from B_3 to B_1. This is illustrated in Figure 27.18(b).

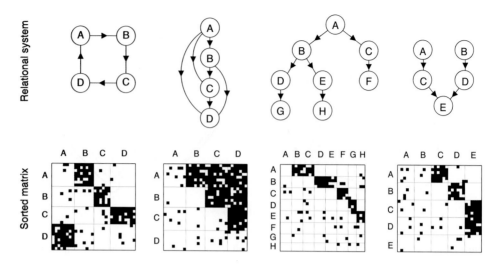

Figure 27.20 Some examples of graphs generated using the stochastic block model with different kinds of connectivity patterns between the blocks. The abstract graph (between blocks) represent a ring, a dominance hierarchy, a common-cause structure, and a common-effect structure. From Figure 4 of (Kemp et al. 2010). Used with kind permission of Charles Kemp.

The problem is easier to understand if we plot the adjacency matrices. Figure 27.19(a) shows the matrix for the graph with the nodes in their original ordering. Figure 27.19(b) shows the matrix for the graph with the nodes in their permtuted ordering. It is clear that there is block structure.

We can make a generative model of block structured graphs as follows. First, for every node, sample a latent block $q_i \sim \text{Cat}(\boldsymbol{\pi})$, where π_k is the probability of choosing block k, for $k = 1 : K$. Second, choose the probability of connecting group a to group b, for all pairs of groups; let us denote this probability by $\eta_{a,b}$. This can come from a beta prior. Finally, generate each edge R_{ij} using the following model:

$$p(R_{ij} = r | q_i = a, q_j = b, \boldsymbol{\eta}) = \text{Ber}(r | \eta_{a,b}) \tag{27.75}$$

This is called the **stochastic block model** (Nowicki and Snijders 2001). Figure 27.21(a) illustrates the model as a DGM, and Figure 27.19(c) illustrates how this model can be used to cluster the nodes in our example.

Note that this is quite different from a conventional clustering problem. For example, we see that all the nodes in block 3 are grouped together, even though there are no connections between them. What they share is the property that they "like to" connect to nodes in block 1, and to receive connections from nodes in block 2. Figure 27.20 illustrates the power of the model for generating many different kinds of graph structure. For example, some social networks have hierarchical structure, which can be modeled by clustering people into different social strata, whereas others consist of a set of cliques.

Unlike a standard mixture model, it is not possible to fit this model using exact EM, because all the latent q_i variables become correlated. However, one can use variational EM (Airoldi et al.

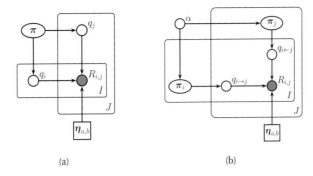

(a)   (b)

Figure 27.21   (a) Stochastic block model. (b) Mixed membership stochastic block model.

2008), collapsed Gibbs sampling (Kemp et al. 2006), etc. We omit the details (which are similar to the LDA case).

In (Kemp et al. 2006), they lifted the restriction that the number of blocks K be fixed, by replacing the Dirichlet prior on π by a Dirichlet process (see Section 25.2.2). This is known as the **infinite relational model**. See Section 27.6.1 for details.

If we have features associated with each node, we can make a discriminative version of this model, for example by defining

$$p(R_{ij} = r | q_i = a, q_j = b, \mathbf{x}_i, \mathbf{x}_j, \boldsymbol{\theta}) = \mathrm{Ber}(r | \mathbf{w}_{a,b}^T f(\mathbf{x}_i, \mathbf{x}_j)) \qquad (27.76)$$

where $f(\mathbf{x}_i, \mathbf{x}_j)$ is some way of combining the feature vectors. For example, we could use concatenation, $[\mathbf{x}_i, \mathbf{x}_j]$, or elementwise product $\mathbf{x}_i \otimes \mathbf{x}_j$ as in supervised LDA. The overall model is like a relational extension of the mixture of experts model.

27.5.2   Mixed membership stochastic block model

In (Airoldi et al. 2008), they lifted the restriction that each node only belong to one cluster. That is, they replaced $q_i \in \{1, \dots, K\}$ with $\pi_i \in S_K$. This is known as the **mixed membership stochastic block model**, and is similar in spirit to **fuzzy clustering** or **soft clustering**. Note that π_{ik} is not the same as $p(z_i = k | \mathcal{D})$; the former represents **ontological uncertainty** (to what degree does each object belong to a cluster) wheras the latter represents **epistemological uncertainty** (which cluster does an object belong to). If we want to combine epistemological and ontological uncertainty, we can compute $p(\pi_i | \mathcal{D})$.

In more detail, the generative process is as follows. First, each node picks a distribution over blocks, $\pi_i \sim \mathrm{Dir}(\boldsymbol{\alpha})$. Second, choose the probability of connecting group a to group b, for all pairs of groups, $\eta_{a,b} \sim \beta(\alpha, \beta)$. Third, for each edge, sample two discrete variables, one for each direction:

$$q_{i \to j} \sim \mathrm{Cat}(\pi_i), \; q_{i \leftarrow j} \sim \mathrm{Cat}(\pi_j) \qquad (27.77)$$

Finally, generate each edge R_{ij} using the following model:

$$p(R_{ij} = 1 | q_{i \to j} = a, q_{i \leftarrow j} = b, \boldsymbol{\eta}) = \eta_{a,b} \qquad (27.78)$$

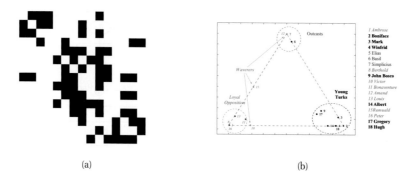

(a)   (b)

Figure 27.22   (a) Who-likes-whom graph for Sampson's monks. (b) Mixed membership of each monk in one of three groups. From Figures 2-3 of (Airoldi et al. 2008). Used with kind permission of Edo Airoldi.

See Figure 27.21(b) for the DGM.

Unlike the regular stochastic block model, each node can play a different role, depending on who it is connecting to. As an illustration of this, we will consider a data set that is widely used in the social networks analysis literature. The data concerns who-likes-whom amongst of group of 18 monks. It was collected by hand in 1968 by Sampson (Sampson 1968) over a period of months. (These days, in the era of social media such as Facebook, a social network with only 18 people is trivially small, but the methods we are discussing can be made to scale.) Figure 27.22(a) plots the raw data, and Figure 27.22(b) plots $\mathbb{E}\left[\pi\right]_i$ for each monk, where $K = 3$. We see that most of the monk belong to one of the three clusters, known as the "young turks", the "outcasts" and the "loyal opposition". However, some individuals, notably monk 15, belong to two clusters; Sampson called these monks the "waverers". It is interesting to see that the model can recover the same kinds of insights as Sampson derived by hand.

One prevalent problem in social network analysis is missing data. For example, if $R_{ij} = 0$, it may be due to the fact that person i and j have not had an opportunity to interact, or that data is not available for that interaction, as opposed to the fact that these people don't want to interact. In other words, *absence of evidence is not evidence of absence*. We can model this by modifying the observation model so that with probability ρ, we generate a 0 from the background model, and we only force the model to explain observed 0s with probability $1 - \rho$. In other words, we robustify the observation model to allow for outliers, as follows:

$$p(R_{ij} = r | q_{i \to j} = a, q_{i \leftarrow j} = b, \boldsymbol{\eta}) = \rho \delta_0(r) + (1 - \rho)\text{Ber}(r | \eta_{a,b}) \tag{27.79}$$

See (Airoldi et al. 2008) for details.

27.5.3   Relational topic model

In many cases, the nodes in our network have atttributes. For example, if the nodes represent academic papers, and the edges represent citations, then the attributes include the text of the document itself. It is therefore desirable to create a model that can explain the text and the link structure concurrently. Such a model can predict links given text, or even vice versa.

The **relational topic model** (RTM) (Chang and Blei 2010) is one way to do this. This is a

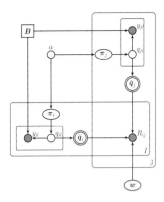

Figure 27.23 DGM for the relational topic model.

simple extension of supervised LDA (Section 27.4.4.1), where the response variable R_{ij} (which represents whether there is an edge between nodes i and j) is modeled as follows:

$$p(R_{ij} = 1|\overline{\mathbf{q}}_i, \overline{\mathbf{q}}_j, \boldsymbol{\theta}) = \mathrm{sigm}(\mathbf{w}^T(\overline{\mathbf{q}}_i \otimes \overline{\mathbf{q}}_j) + w_0) \tag{27.80}$$

Recall that $\overline{\mathbf{q}}_i$ is the empirical topic distribution for document i, $\overline{q}_{ik} \triangleq \frac{1}{L_i}\sum_{i=1}^{L_i} q_{ilk}$. See Figure 27.23

Note that it is important that R_{ij} depend on the actual topics chosen, $\overline{\mathbf{q}}_i$ and $\overline{\mathbf{q}}_j$, and not on the topic distributions, $\boldsymbol{\pi}_i$ and $\boldsymbol{\pi}_j$, otherwise predictive performance is not as good. The intuitive reason for this is as follows: if R_{ij} is a child of $\boldsymbol{\pi}_i$ and $\boldsymbol{\pi}_j$, it will be treated as just another word, similar to the q_{il}'s and y_{il}'s; but since there are many more words than edges, the graph structure information will get "washed out". By making R_{ij} a child of $\overline{\mathbf{q}}_i$ and $\overline{\mathbf{q}}_j$, the graph information can influence the choice of topics more directly.

One can fit this model in a manner similar to SLDA. See (Chang and Blei 2010) for details. The method does better at predicting missing links than the simpler approach of first fitting an LDA model, and then using the $\overline{\mathbf{q}}_i$'s as inputs to a logistic regression problem. The reason is analogous to the superiority of partial least squares (Section 12.5.2) to PCA+ linear regression, namely that the RTM learns a latent space that is forced to be predictive of the graph structure and words, whereas LDA might learn a latent space that is not useful for predicting the graph.

27.6 LVMs for relational data

Graphs can be used to represent data which represents the relation amongst variables of a certain type, e.g., friendship relationships between people. But often we have multiple types of objects, and multiple types of relations. For example, Figure 27.24 illustrates two relations, one between people and people, and one between people and movies.

In general, we define a k-ary **relation** R as a subset of k-tuples of the appropriate types:

$$R \subseteq T_1 \times T_2 \times \cdots \times T_k \tag{27.81}$$

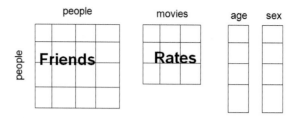

Figure 27.24   Example of relational data. There are two types of objects, *people* and *movies*; one 2-ary relation, *friends*: *people* × *people* → {0, 1} and one 2-ary function, *rates*: *people* × *movie* → ℝ. *Age* and *sex* are attributes (unary functions) of the *people* class.

where T_i are sets or types. A binary, pairwise or dyadic relation is a relation defined on pairs of objects. For example, the *seen* relation between people and movies might be represented as the set of movies that people have seen. We can either represent this explicitly as a set, such as

```
seen  = { (Bob, StarWars), (Bob, TombRaider), (Alice, Jaws)}
```

or implicitly, using an indicator function for the set:

```
seen(Bob, StarWars)=1, seen(Bob, TombRaider)=1, seen(Alice, Jaws)=1
```

A relation between two entities of types T^1 and T^2 can be represented as a binary function $R : T^1 \times T^2 \to \{0, 1\}$, and hence as a binary matrix. This can also be represented as a bipartite graph, in which we have nodes of two types. If $T^1 = T^2$, this becomes a regular directed graph, as in Section 27.5. However, there are some situations that are not so easily modelled by graphs, but which can still be modelled by relations. For example, we might have a ternary relation, $R : T^1 \times T^1 \times T^2 \to \{0, 1\}$, where, say, $R(i, j, k) = 1$ iff protein i interacts with protein j when chemical k is present. This can be modelled by a 3d binary matrix. We will give some examples of this in Section 27.6.1.

Making probabilistic models of relational data is called **statistical relational learning** (Getoor and Taskar 2007). One approach is to directly model the relationship between the variables using graphical models; this is known as **probabilistic relational modeling**. Another approach is to use latent variable models, as we discuss below.

27.6.1   Infinite relational model

It is straightforward to extend the stochastic block model to model relational data: we just associate a latent variable $q_i^t \in \{1, \dots, K_t\}$ with each entity i of each type t. We then define the probability of the relation holding between specific entities by looking up the probability of the relation holding between entities of that type. For example, if $R : T^1 \times T^1 \times T^2 \to \{0, 1\}$, we have

$$p(R(i, j, k)|q_i^1 = a, q_j^1 = b, q_k^2 = c, \boldsymbol{\eta}) = \text{Ber}(\eta_{a,b,c}) \qquad (27.82)$$

We can also have real-valued relations, where each edge has a weight. For example, we can write $p(R(i, j, k)|q_i^1 = a, q_j^1 = b, q_k^2 = c, \boldsymbol{\mu}) = \mathcal{N}(\mu_{a,b,c} + \mu_i + \mu_j + \mu_k, \sigma^2)$, where $\mu_{a,b,c}$

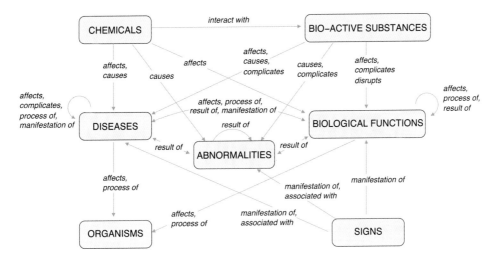

Figure 27.25 Illustration of an ontology learned by IRM applied to the Unified Medical Language System. The boxes represent 7 of the 14 concept clusters. Predicates that belong to the same cluster are grouped together, and associated with edges to which they pertain. All links with weight above 0.8 have been included. From Figure 9 of (Kemp et al. 2010). Used with kind permission of Charles Kemp.

captures the average response for that group of clusters, and μ_i, μ_j and μ_k are offsets for specific entities. (Allowing a different offset for every combination of i, j and k would require too many parameters.) This model was proposed in (Banerjee et al. 2007), who fit the model using an alternating minimization procedure.

If we allow the number of clusters K_t for each type of entity to be unbounded, by using a Dirichlet process, the model is called the **infinite relational model** (IRM) (Kemp et al. 2006), also known as an **infinite hidden relational model** (IHRM) (Xu et al. 2006). We can fit this model with variational Bayes (Xu et al. 2006, 2007) or collapsed Gibbs sampling (Kemp et al. 2006). Rather than go into algorithmic detail, we just sketch some interesting applications.

27.6.1.1 Learning ontologies

An **ontology** refers to an organisation of knowledge. In AI, ontologies are often built by hand (see e.g., (Russell and Norvig 2010)), but it is interesting to try and learn them from data. In (Kemp et al. 2006), they show how this can be done using the IRM.

The data comes from the Unified Medical Language System (McCray 2003), which defines a semantic network with 135 concepts (such as "disease or syndrome", "diagnostic procedure", "animal"), and 49 binary predicates (such as "affects", "prevents"). We can represent this as a ternary relation $R : T^1 \times T^1 \times T^2 \to \{0, 1\}$, where T^1 is the set of concepts and T^2 is the set of binary predicates. The result is a 3d cube. We can then apply the IRM to partition the cube into regions of roughly homogoneous response. The system found 14 concept clusters and 21 predicate clusters. Some of these are shown in Figure 27.25. The system learns, for example, that biological functions affect organisms (since $\eta_{a,b,c} \approx 1$ where a represents the biological function cluster, b represents the organism cluster, and c represents the affects cluster).

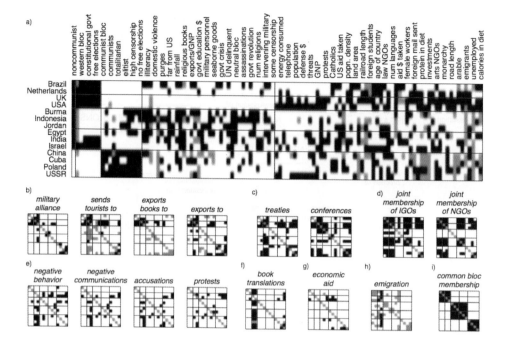

Figure 27.26 Illustration of IRM applied to some political data containing features and pairwise interactions. Top row (a): the partition of the countries into 5 clusters and the features into 5 clusters. Every second column is labelled with the name of the corresponding feature. Small squares at bottom (b-i): these are 8 of the 18 clusters of interaction types. From Figure 6 of (Kemp et al. 2006). Used with kind permission of Charles Kemp.

27.6.1.2 Clustering based on relations and features

We can also use IRM to cluster objects based on their relations and their features. For example, (Kemp et al. 2006) consider a political dataset (from 1965) consisting of 14 countries, 54 binary predicates representing interaction types between countries (e.g., "sends tourists to", "economic aid"), and 90 features (e.g., "communist", "monarchy"). To create a binary dataset, real-valued features were thresholded at their mean, and categorical variables were dummy-encoded. The data has 3 types: T^1 represents countries, T^2 represents interactions, and T^3 represents features. We have two relations: $R^1 : T^1 \times T^1 \times T^2 \to \{0, 1\}$, and $R^2 : T^1 \times T^3 \to \{0, 1\}$. (This problem therefore combines aspects of both the biclustering model and the ontology discovery model.) When given multiple relations, the IRM treats them as conditionally independent. In this case, we have

$$p(\mathbf{R}^1, \mathbf{R}^2, \mathbf{q}^1, \mathbf{q}^2, \mathbf{q}^3 | \boldsymbol{\theta}) = p(\mathbf{R}^1 | \mathbf{q}^1, \mathbf{q}^2, \boldsymbol{\theta}) p(\mathbf{R}^2 | \mathbf{q}^1, \mathbf{q}^3, \boldsymbol{\theta}) \tag{27.83}$$

The results are shown in Figure 27.26. The IRM divides the 90 features into 5 clusters, the first of which contains "noncommunist", which captures one of the most important aspects of this Cold-War era dataset. It also clusters the 14 countries into 5 clusters, reflecting natural geo-political groupings (e.g., US and UK, or the Communist Bloc), and the 54 predicates into 18 clusters, reflecting similar relationships (e.g., "negative behavior and "accusations").

27.6.2 Probabilistic matrix factorization for collaborative filtering

As discussed in Section 1.3.4.2, collaborative filtering (CF) requires predicting entries in a matrix $R : T^1 \times T^2 \to \mathbb{R}$, where for example $R(i, j)$ is the rating that user i gave to movie j. Thus we see that CF is a kind of relational learning problem (and one with particular commercial importance).

Much of the work in this area makes use of the data that Netflix made available in their competition. In particular, a large $17{,}770 \times 480{,}189$ movie x user ratings matrix is provided. The full matrix would have $\sim 8.6 \times 10^9$ entries, but only 100,480,507 (about 1%) of the entries are observed, so the matrix is extremely sparse. In addition the data is quite imbalanced, with many users rating fewer than 5 movies, and a few users rating over 10,000 movies. The validation set is 1,408,395 (movie,user) pairs. Finally, there is a separate test set with 2,817,131 (movie,user) pairs, for which the ranking is known but withheld from contestants. The performance measure is root mean square error:

$$
RMSE = \sqrt{\frac{1}{N} \sum_{i=1}^{N} (X(m_i, u_i) - \hat{X}(m_i, u_i))^2} \tag{27.84}
$$

where $X(m_i, u_i)$ is the true rating of user u_i on movie m_i, and $\hat{X}(m_i, u_i)$ is the prediction. The baseline system, known as Cinematch, had an RMSE on the training set of 0.9514, and on the test set of 0.9525. To qualify for the grand prize, teams needed to reduce the test RMSE by 10%, i.e., get a test RMSE of 0.8563 or less. We will discuss some of the basic methods used byt the winning team below.

Since the ratings are drawn from the set $\{0, 1, 2, 3, 4, 5\}$, it is tempting to use a categorical observation model. However, this does not capture the fact that the ratings are ordered. Although we could use an ordinal observation model, in practice people use a Gaussian observation model for simplicity. One way to make the model better match the data is to pass the model's predicted mean response through a sigmoid, and then to map the $[0, 1]$ interval to $[0, 5]$ (Salakhutdinov and Mnih 2008). Alternatively we can make the data a better match to the Gaussian model by transforming the data using $R_{ij} = \sqrt{6 - R_{ij}}$ (Aggarwal and Merugu 2007).

We could use the IRM for the CF task, by associating a discrete latent variable for each user q_i^u and for each movie or video q_j^v, and then defining

$$
p(R_{ij} = r | q_i^u = a, q_j^v = b, \boldsymbol{\theta}) = \mathcal{N}(r | \mu_{a,b}, \sigma^2) \tag{27.85}
$$

This is just another example of co-clustering. We can also extend the model to generate side information, such as attributes about each user and/or movie. See Figure 27.27 for an illustration.

Another possibility is to replace the discrete latent variables with continuous latent variables $\boldsymbol{\pi}_i^u \in S_{K_u}$ and $\boldsymbol{\pi}_j^v \in S_{K_v}$. However, it has been found (see e.g., (Shan and Banerjee 2010)) that one obtains much better results [7] by using unconstrained real-valued latent factors for each user

---

7. Good results with discrete latent variables have been obtained on some datasets that are smaller than Netflix, such as

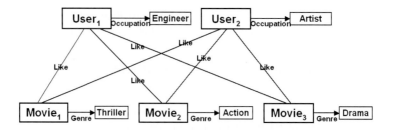

Figure 27.27   Visualization of a small relational dataset, where we have one relation, likes(user, movie), and features for movies (here, genre) and users (here, occupation). From Figure 5 of (Xu et al. 2008). Used with kind permission of Zhao Xu.

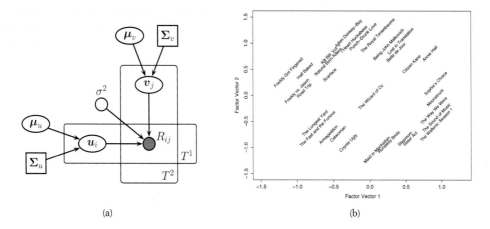

(a)   (b)

Figure 27.28   (a) A DGM for probabilistic matrix factorization. (b) Visualization of the first two factors in the PMF model estimated from the Netflix challenge data. Each movie j is plotted at the location specified $\hat{\mathbf{v}}_j$. On the left we have low-brow humor and horror movies (*Half Baked*, *Freddy vs Jason*), and on the right we have more serious dramas (*Sophie's Choice*, *Moonstruck*). On the top we have critically acclaimed independent movies (*Punch-Drunk Love*, *I Heart Huckabees*), and on the bottom we have mainstream Hollywood blockbusters (*Armageddon*, *Runway Bride*). The *Wizard of Oz* is right in the middle of these axes. From Figure 3 of (Koren et al. 2009). Used with kind permission of Yehuda Koren.

$\mathbf{u}_i \in \mathbb{R}^K$ and each movie $\mathbf{v}_j \in \mathbb{R}^K$. We then use a likelihood of the form

$$p(R_{ij} = r | \mathbf{u}_i, \mathbf{v}_j) = \mathcal{N}(r | \mathbf{u}_i^T \mathbf{v}_j, \sigma^2) \tag{27.86}$$

This has been called **probabilistic matrix factorization** (PMF) (Salakhutdinov and Mnih 2008). See Figure 27.28(a) for the DGM. The intuition behind this method is that each user and each movie get embedded into the same low-dimensional continuous space (see Figure 27.28(b)). If a

MovieLens and EachMovie. However, these datasets are much easier to predict, because there is less imbalance between the number of reviews performed by different users (in Netflix, some users have rated more than 10,000 movies, whereas others have rated less than 5).

user is close to a movie in that space, they are likely to rate it highly. All of the best entries in the Netflix competition used this approach in one form or another.[8]

PMF is closely related to the SVD. In particular, if there is no missing data, then computing the MLE for the $\mathbf{u}_i$'s and the $\mathbf{v}_j$'s is equivalent to finding a rank K approximation to $\mathbf{R}$. However, as soon as we have missing data, the problem becomes non-convex, as shown in (Srebro and Jaakkola 2003), and standard SVD methods cannot be applied. (Recall that in the Netflix challenge, only about 1% of the matrix is observed.)

The most straightforward way to fit the PMF model is to minimize the overall NLL:

$$ J(\mathbf{U}, \mathbf{V}) = -\log p(\mathbf{R}|\mathbf{U}, \mathbf{V}, \mathbf{O}) = -\log \left(\prod_{i=1}^{N} \prod_{j=1}^{M} \left[\mathcal{N}(R_{ij}|\mathbf{u}_i^T \mathbf{v}_j, \sigma^2) \right]^{\mathbb{I}(O_{ij}=1)} \right) \quad (27.87) $$

where $O_{ij} = 1$ if user i has seen movie j. Since this is non-convex, we can just find a locally optimal MLE. Since the Netflix data is so large (about 100 million observed entries), it is common to use stochastic gradient descent (Section 8.5.2) for this task. The gradient for $\mathbf{u}_i$ is given by

$$ \frac{dJ}{d\mathbf{u}_i} = \frac{d}{d\mathbf{u}_i} \frac{1}{2} \sum_{ij} \mathbb{I}(O_{ij} = 1)(R_{ij} - \mathbf{u}_i^T \mathbf{v}_j)^2 = -\sum_{j:O_{ij}=1} e_{ij} \mathbf{v}_j \quad (27.88) $$

where $e_{ij} = R_{ij} - \mathbf{u}_i^T \mathbf{v}_j$ is the error term. By stochastically sampling a single movie j that user i has watched, the update takes the following simple form:

$$ \mathbf{u}_i \quad = \quad \mathbf{u}_i + \eta e_{ij} \mathbf{v}_j \quad (27.89) $$

where η is the learning rate. The update for $\mathbf{v}_j$ is similar.

Of course, just maximizing the likelihood results in overfitting, as shown in Figure 27.29(a). We can regularize this by imposing Gaussian priors:

$$ p(\mathbf{U}, \mathbf{V}) = \prod_i \mathcal{N}(\mathbf{u}_i|\boldsymbol{\mu}_u, \boldsymbol{\Sigma}_u) \prod_j \mathcal{N}(\mathbf{v}_j|\boldsymbol{\mu}_v, \boldsymbol{\Sigma}_v) \quad (27.90) $$

If we use $\boldsymbol{\mu}_u = \boldsymbol{\mu}_v = \mathbf{0}$, $\boldsymbol{\Sigma}_u = \sigma_U^2 \mathbf{I}_K$, and $\boldsymbol{\Sigma}_v = \sigma_V^2 \mathbf{I}_K$, the new objective becomes

$$ \begin{aligned} J(\mathbf{U}, \mathbf{V}) \quad &= \quad -\log p(\mathbf{R}, \mathbf{U}, \mathbf{V}|\mathbf{O}, \boldsymbol{\theta}) & (27.91) \\ &= \quad \sum_i \sum_j \mathbb{I}(O_{ij} = 1)(R_{ij} - \mathbf{u}_i^T \mathbf{v}_j)^2 \\ &\quad + \lambda_U \sum_i ||\mathbf{u}_i||_2^2 + \lambda_V \sum_j ||\mathbf{v}_j||_2^2 + \text{const} & (27.92) \end{aligned} $$

where we have defined $\lambda_U = \sigma^2/\sigma_U^2$ and $\lambda_V = \sigma^2/\sigma_V^2$. By varying the regularizers, we can reduce the effect of overfitting, as shown in Figure 27.29(a). We can find MAP estimates using stochastic gradient descent. We can also compute approximate posteriors using variational Bayes (Ilin and Raiko 2010).

---

8. The winning entry was actually an ensemble of different methods, including PMF, nearest neighbor methods, etc. For details, see `http://www.netflixprize.com/community/viewtopic.php?id=1537`.

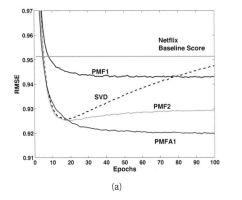
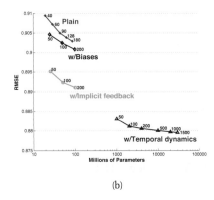

(a)   (b)

Figure 27.29 (a) RMSE on the validation set for different PMF variants vs number of passes through the data. "SVD" is the unregularized version, $\lambda_U = \lambda_V = 0$. "PMF" corresponds to $\lambda_U = 0.01$ and $\lambda_V = 0.001$, while "PMF2" corresponds to $\lambda_U = 0.001$ and $\lambda_V = 0.0001$. "PMFA1" corresponds to a version where the mean and diagonal covariance of the Gaussian prior were learned from data. From Figure 2a of (Salakhutdinov and Mnih 2008). Used with kind permission of Ruslan Salakhutdinov. (b) RMSE on the test set (quiz portion) vs number of parameters for several different models. "Plain" is the baseline PMF with suitably chosen λ_U, λ_V. "With biases" adds f_i and g_j offset terms. "With implicit feedback" means we exploited knowledge of which movies were observed in the test set (but did not use the test ratings, which are what we are trying to predict). "With temporal dynamics" allows the offset terms to change over time. The Netflix baseline system achieves an RMSE of 0.9514, and the grand prize's required accuracy is 0.8563 (which was obtained on 21 September 2009). Figure generated by `netflixResultsPlot`. From Figure 4 of (Koren et al. 2009). Used with kind permission of Yehuda Koren.

If we use diagonal covariances for the priors, we can penalize each latent dimension by a different amount. Also, if we use non-zero means for the priors, we can account for offset terms. Optimizing the prior parameters $(\boldsymbol{\mu}_u, \boldsymbol{\Sigma}_u, \boldsymbol{\mu}_v, \boldsymbol{\Sigma}_v)$ at the same time as the model parameters $(\mathbf{U}, \mathbf{V}, \sigma^2)$ is a way to create an adaptive prior. This avoids the need to search for the optimal values of λ_U and λ_V, and gives even better results, as shown in Figure 27.29(a).

It turns out that much of the variation in the data can be explained by movie-specific or user-specific effects. For example, some movies are popular for all types of users. And some users give low scores for all types of movies. We can model this by allowing for user and movie specific offset or bias terms as follows:

$$p(R_{ij} = r|\mathbf{u}_i, \mathbf{v}_j, \boldsymbol{\theta}) = \mathcal{N}(r|\mathbf{u}_i^T \mathbf{v}_j + \mu + f_i + g_j, \sigma^2) \tag{27.93}$$

where μ is the overall mean, f_i is the user bias, g_j is the movie bias, and $\mathbf{u}_i^T \mathbf{v}_j$ is the interaction term. This is equivalent to applying PMF just to the residual matrix, and gives much better results, as shown in Figure 27.29(b). We can estimate the f_i, g_j and μ terms using stochastic gradient descent, just as we estimated $\mathbf{U}$, $\mathbf{V}$ and $\boldsymbol{\theta}$.

We can also allow the bias terms to evolve over time, to reflect the changing preferences of users (Koren 2009b). This is important since in the Netflix competition, the test data was more recent than the training data. Figure 27.29(b) shows that allowing for temporal dynamics can help a lot.

Often we also have **side information** of various kinds. In the Netflix competition, entrants knew which movies the user had rated in the test set, even though they did not know the values of these ratings. That is, they knew the value of the (dense) **O** matrix even on the test set. If a user chooses to rate a movie, it is likely because they have seen it, which in turns means they thought they would like it. Thus the very act of rating reveals information. Conversely, if a user chooses not rate a movie, it suggests they knew they would not like it. So the data is not missing at random (see e.g., (Marlin and Zemel 2009)). Exploiting this can improve performance, as shown in Figure 27.29(b). In real problems, information on the test set is not available. However, we often know which movies the user has watched or declined to watch, even if they did not rate them (this is called **implicit feedback**), and this can be used as useful side information.

Another source of side information concerns the content of the movie, such as the movie genre, the list of the actors, or a synopsis of the plot. This can be denoted by $\mathbf{x}_v$, the features of the video. (In the case where we just have the id of the video, we can treat $\mathbf{x}_v$ as a $|\mathcal{V}|$-dimensional bit vector with just one bit turned on.) We may also know features about the user, which we can denote by $\mathbf{x}_u$. These can be used in addition to, or instead of, the latent features.

In some cases, we only know if the user clicked on the video or not, that is, we may not have a numerical rating. We can modify the model by replacing the Gaussian likelihood with a Bernoulli likelihood. For example, a model of binary responses with visible features for the user and movie is given by

$$p(R(u,v) = 1|\mathbf{x}_u, \mathbf{x}_v, \boldsymbol{\theta}) = \text{sigm}((\mathbf{U}\mathbf{x}_u)^T(\mathbf{V}\mathbf{x}_v)) \tag{27.94}$$

where $\mathbf{U}$ is a $|\mathcal{U}| \times K$ matrix, and $\mathbf{V}$ is a $|\mathcal{V}| \times K$ matrix (we can incorporate an offset term by appending a 1 to $\mathbf{x}_u$ and $\mathbf{x}_v$ in the usual way). A method for computing the approximate posterior $p(\mathbf{U}, \mathbf{V}|\mathcal{D})$ in an online fashion, using ADF and EP, was described in (Stern et al. 2009). This was implemented by Microsoft and has been deployed to predict click through rates on all the ads used by the Bing search engine.

Unfortunately, fitting this model just from positive binary data can result in an over prediction of links, since no negative examples are included. Better performance is obtained if one has access to the set of all videos shown to the user, of which at most one was picked; data of this form is known as an **impression log**. In this case, we can use a multinomial model instead of a binary model; in (Yang et al. 2011), this was shown to work much better than a binary model. To understand why, suppose some is presented with a choice of an action movie starring Arnold Schwarzenegger, an action movie starring Vin Diesel, and a comedy starring Hugh Grant. If the user picks Arnold Schwarzenegger, we learn not only that they like prefer action movies to comedies, but also that they prefer Schwarzenegger to Diesel. So we get information on the relative preferences of people wrt certain items.

27.7   Restricted Boltzmann machines (RBMs)

So far, all the models we have proposed in this chapter have been representable by directed graphical models. But some models are better represented using undirected graphs. For example, the **Boltzmann machine** (Ackley et al. 1985) is a pairwise MRF with hidden nodes **h** and visible nodes **v**, as shown in Figure 27.30(a). The main problem with the Boltzmann machine is that exact inference is intractable, and even approximate inference, using e.g., Gibbs sampling, can

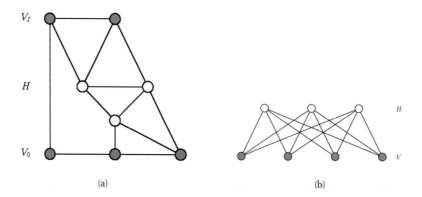

Figure 27.30 (a) A general Boltzmann machine, with an arbitrary graph structure. The shaded (visible) nodes are partitioned into input and output, although the model is actually symmetric and defines a joint density on all the nodes. (b) A restricted Boltzmann machine with a bipartite structure. Note the lack of intra-layer connections.

be slow. However, suppose we restrict the architecture so that the nodes are arranged in layers, and so that there are no connections between nodes within the same layer (see Figure 27.30(b)). Then the model has the form

$$p(\mathbf{h}, \mathbf{v}|\boldsymbol{\theta}) = \frac{1}{Z(\boldsymbol{\theta})} \prod_{r=1}^{R} \prod_{k=1}^{K} \psi_{rk}(v_r, h_k) \tag{27.95}$$

where R is the number of visible (response) variables, K is the number of hidden variables, and $\mathbf{v}$ plays the role of $\mathbf{y}$ earlier in this chapter. This model is known as a **restricted Boltzmann machine** (**RBM**) (Hinton 2002), or a **harmonium** (Smolensky 1986).

An RBM is a special case of a **product of experts** (PoE) (Hinton 1999), which is so-called because we are multiplying together a set of "experts" (here, potential functions on each edge) and then normalizing, whereas in a mixture of experts, we take a convex combination of normalized distributions. The intuitive reason why PoE models might work better than a mixture is that each expert can enforce a constraint (if the expert has a value which is $\gg 1$ or $\ll 1$) or a "don't care" condition (if the expert has value 1). By multiplying these experts together in different ways we can create "sharp" distributions which predict data which satisfies the specified constraints (Hinton and Teh 2001). For example, consider a distributed model of text. A given document might have the topics "government", "mafia" and "playboy". If we "multiply" the predictions of each topic together, the model may give very high probability to the word "Berlusconi"[9] (Salakhutdinov and Hinton 2010). By contrast, adding together experts can only make the distribution broader (see Figure 14.17).

Typically the hidden nodes in an RBM are binary, so $\mathbf{h}$ specifies which constraints are active. It is worth comparing this with the directed models we have discussed. In a mixture model, we have one hidden variable $q \in \{1, \ldots, K\}$. We can represent this using a set of K bits, with the

---

9. Silvio Berlusconi is the current (2011) prime minister of Italy.

Visible	Hidden	Name	Reference
Binary	Binary	Binary RBM	(Hinton 2002)
Gaussian	Binary	Gaussian RBM	(Welling and Sutton 2005)
Categorical	Binary	Categorical RBM	(Salakhutdinov et al. 2007)
Multiple categorical	Binary	Replicated softmax/ undirected LDA	(Salakhutdinov and Hinton 2010)
Gaussian	Gaussian	Undirected PCA	(Marks and Movellan 2001)
Binary	Gaussian	Undirected binary PCA	(Welling and Sutton 2005)

Table 27.2 Summary of different kinds of RBM.

restriction that exactly one bit is on at a time. This is called a **localist encoding**, since only one hidden unit is used to generate the response vector. This is analogous to the hypothetical notion of **grandmother cells** in the brain, that are able to recognize only one kind of object. By contrast, an RBM uses a **distributed encoding**, where many units are involved in generating each output. Models that used vector-valued hidden variables, such as $\boldsymbol{\pi} \in S_K$, as in mPCA/ LDA, or $\mathbf{z} \in \mathbb{R}^K$, as in ePCA also use distributed encodings.

The main difference between an RBM and directed two-layer models is that the hidden variables are conditionally independent given the visible variables, so the posterior factorizes:

$$p(\mathbf{h}|\mathbf{v}, \boldsymbol{\theta}) = \prod_k p(h_k|\mathbf{v}, \boldsymbol{\theta}) \tag{27.96}$$

This makes inference much simpler than in a directed model, since we can estimate each h_k independently and in parallel, as in a feedforward neural network. The disadvantage is that training undirected models is much harder, as we discuss below.

27.7.1 Varieties of RBMs

In this section, we describe various forms of RBMs, by defining different pairwise potential functions. See Table 27.2 for a summary. All of these are special cases of the **exponential family harmonium** (Welling et al. 2004).

27.7.1.1 Binary RBMs

The most common form of RBM has binary hidden nodes and binary visible nodes. The joint distribution then has the following form:

$$p(\mathbf{v}, \mathbf{h}|\boldsymbol{\theta}) = \frac{1}{Z(\boldsymbol{\theta})} \exp(-E(\mathbf{v}, \mathbf{h}; \boldsymbol{\theta})) \tag{27.97}$$

$$E(\mathbf{v}, \mathbf{h}; \boldsymbol{\theta}) \triangleq -\sum_{r=1}^{R}\sum_{k=1}^{K} v_r h_k W_{rk} - \sum_{r=1}^{R} v_r b_r - \sum_{k=1}^{K} h_k c_k \tag{27.98}$$

$$= -(\mathbf{v}^T \mathbf{W}\mathbf{h} + \mathbf{v}^T \mathbf{b} + \mathbf{h}^T \mathbf{c}) \tag{27.99}$$

$$Z(\boldsymbol{\theta}) = \sum_{\mathbf{v}}\sum_{\mathbf{h}} \exp(-E(\mathbf{v}, \mathbf{h}; \boldsymbol{\theta})) \tag{27.100}$$

where E is the energy function, $\mathbf{W}$ is a $R \times K$ weight matrix, $\mathbf{b}$ are the visible bias terms, $\mathbf{c}$ are the hidden bias terms, and $\boldsymbol{\theta} = (\mathbf{W}, \mathbf{b}, \mathbf{c})$ are all the parameters. For notational simplicity, we will absorb the bias terms into the weight matrix by clamping dummy units $v_0 = 1$ and $h_0 = 1$ and setting $\mathbf{w}_{0,:} = \mathbf{c}$ and $\mathbf{w}_{:,0} = \mathbf{b}$. Note that naively computing $Z(\boldsymbol{\theta})$ takes $O(2^R 2^K)$ time but we can reduce this to $O(\min\{R2^K, K2^R\})$ time (Exercise 27.1).

When using a binary RBM, the posterior can be computed as follows:

$$p(\mathbf{h}|\mathbf{v}, \boldsymbol{\theta}) = \prod_{k=1}^{K} p(h_k|\mathbf{v}, \boldsymbol{\theta}) = \prod_k \mathrm{Ber}(h_k|\mathrm{sigm}(\mathbf{w}_{:,k}^T \mathbf{v})) \tag{27.101}$$

By symmetry, one can show that we can generate data given the hidden variables as follows:

$$p(\mathbf{v}|\mathbf{h}, \boldsymbol{\theta}) = \prod_r p(v_r|\mathbf{h}, \boldsymbol{\theta}) = \prod_r \mathrm{Ber}(v_r|\mathrm{sigm}(\mathbf{w}_{r,:}^T \mathbf{h})) \tag{27.102}$$

We can write this in matrix-vector notation as follows:

$$\mathbb{E}[\mathbf{h}|\mathbf{v}, \boldsymbol{\theta}] = \mathrm{sigm}(\mathbf{W}^T \mathbf{v}) \tag{27.103}$$

$$\mathbb{E}[\mathbf{v}|\mathbf{h}, \boldsymbol{\theta}] = \mathrm{sigm}(\mathbf{W}\mathbf{h}) \tag{27.104}$$

The weights in $\mathbf{W}$ are called the **generative weights**, since they are used to generate the observations, and the weights in $\mathbf{W}^T$ are called the **recognition weights**, since they are used to recognize the input.

From Equation 27.101, we see that we activate hidden node k in proportion to how much the input vector $\mathbf{v}$ "looks like" the weight vector $\mathbf{w}_{:,k}$ (up to scaling factors). Thus each hidden node captures certain features of the input, as encoded in its weight vector, similar to a feedforward neural network.

27.7.1.2　Categorical RBM

We can extend the binary RBM to categorical visible variables by using a 1-of-C encoding, where C is the number of states for each v_r. We define a new energy function as follows (Salakhutdinov et al. 2007; Salakhutdinov and Hinton 2010):

$$E(\mathbf{v}, \mathbf{h}; \boldsymbol{\theta}) \triangleq -\sum_{r=1}^{R} \sum_{k=1}^{K} \sum_{c=1}^{C} v_r^c h_k w_{rk}^c - \sum_{r=1}^{R} \sum_{c=1}^{C} v_r^c b_r^c - \sum_{k=1}^{K} h_k c_k \tag{27.105}$$

The full conditionals are given by

$$p(v_r|\mathbf{h}, \boldsymbol{\theta}) = \mathrm{Cat}(\mathcal{S}(\{b_r^c + \sum_k h_k w_{rk}^c\}_{c=1}^C)) \tag{27.106}$$

$$p(h_k = 1|\mathbf{c}, \boldsymbol{\theta}) = \mathrm{sigm}(c_k + \sum_r \sum_c v_r^c w_{rk}^c) \tag{27.107}$$

where $\mathcal{S}()$ is the softmax function.

27.7.1.3　Gaussian RBM

We can generalize the model to handle real-valued data. In particular, a **Gaussian RBM** has the following energy function:

$$E(\mathbf{v}, \mathbf{h}|\boldsymbol{\theta}) = -\sum_{r=1}^{R} \sum_{k=1}^{K} w_{rk} h_k v_r - \frac{1}{2} \sum_{r=1}^{R} (v_r - b_r)^2 - \sum_{k=1}^{K} a_k h_k \tag{27.108}$$

The parameters of the model are $\boldsymbol{\theta} = (w_{rk}, a_k, b_r)$. (We have assumed the data is standardized, so we fix the variance to $\sigma^2 = 1$.) Compare this to a Gaussian in canonical or information form (see Section 4.3.3):

$$\mathcal{N}_c(\mathbf{v}|\boldsymbol{\eta}, \boldsymbol{\Lambda}) \propto \exp(\boldsymbol{\eta}^T \mathbf{v} - \frac{1}{2}\mathbf{v}^T \boldsymbol{\Lambda} \mathbf{v}) \tag{27.109}$$

where $\boldsymbol{\eta} = \boldsymbol{\Lambda}\boldsymbol{\mu}$. We see that we have set $\boldsymbol{\Lambda} = \mathbf{I}$, and $\boldsymbol{\eta} = \sum_k h_k \mathbf{w}_{:,k}$. Thus the mean is given by $\boldsymbol{\mu} = \boldsymbol{\Lambda}^{-1}\boldsymbol{\eta} = \sum_k h_k \mathbf{w}_{:,k}$. The full conditionals, which are needed for inference and learning, are given by

$$p(v_r|\mathbf{h}, \boldsymbol{\theta}) = \mathcal{N}(v_r|b_r + \sum_k w_{rk}h_k, 1) \tag{27.110}$$

$$p(h_k = 1|\mathbf{v}, \boldsymbol{\theta}) = \text{sigm}\left(c_k + \sum_r w_{rk}v_r\right) \tag{27.111}$$

We see that each visible unit has a Gaussian distribution whose mean is a function of the hidden bit vector. More powerful models, which make the (co)variance depend on the hidden state, can also be developed (Ranzato and Hinton 2010).

27.7.1.4 RBMs with Gaussian hidden units

If we use Gaussian latent variables and Gaussian visible variables, we get an undirected version of factor analysis. However, it turns out that it is identical to the standard directed version (Marks and Movellan 2001).

If we use Gaussian latent variables and categorical observed variables, we get an undirected version of categorical PCA (Section 27.2.2). In (Salakhutdinov et al. 2007), this was applied to the Netflix collaborative filtering problem, but was found to be significantly inferior to using binary latent variables, which have more expressive power.

27.7.2 Learning RBMs

In this section, we discuss some ways to compute ML parameter estimates of RBMs, using gradient-based optimizers. It is common to use stochastic gradient descent, since RBMs often have many parameters and therefore need to be trained on very large datasets. In addition, it is standard to use ℓ_2 regularization, a technique that is often called weight decay in this context. This requires a very small change to the objective and gradient, as discussed in Section 8.3.6.

27.7.2.1 Deriving the gradient using $p(\mathbf{h}, \mathbf{v}|\theta)$

To compute the gradient, we can modify the equations from Section 19.5.2, which show how to fit a generic latent variable maxent model. In the context of the Boltzmann machine, we have one feature per edge, so the gradient is given by

$$\frac{\partial \ell}{\partial w_{rk}} = \frac{1}{N}\sum_{i=1}^{N} \mathbb{E}\left[v_r h_k | \mathbf{v}_i, \boldsymbol{\theta}\right] - \mathbb{E}\left[v_r h_k | \boldsymbol{\theta}\right] \tag{27.112}$$

We can write this in matrix-vector form as follows:

$$\nabla_{\mathbf{w}} \ell = \mathbb{E}_{p_{\text{emp}}(\cdot|\boldsymbol{\theta})} \left[\mathbf{v}\mathbf{h}^T \right] - \mathbb{E}_{p(\cdot|\boldsymbol{\theta})} \left[\mathbf{v}\mathbf{h}^T \right] \tag{27.113}$$

where $p_{\text{emp}}(\mathbf{v}, \mathbf{h}|\boldsymbol{\theta}) \triangleq p(\mathbf{h}|\mathbf{v}, \boldsymbol{\theta})p_{\text{emp}}(\mathbf{v})$, and $p_{\text{emp}}(\mathbf{v}) = \frac{1}{N} \sum_{i=1}^{N} \delta_{\mathbf{v}_i}(\mathbf{v})$ is the empirical distribution. (We can derive a similar expression for the bias terms by setting $v_r = 1$ or $h_k = 1$.)

The first term on the gradient, when $\mathbf{v}$ is fixed to a data case, is sometimes called the **clamped phase**, and the second term, when $\mathbf{v}$ is free, is sometimes called the **unclamped phase**. When the model expectations match the empirical expectations, the two terms cancel out, the gradient becomes zero and learning stops. This algorithm was first proposed in (Ackley et al. 1985). The main problem is efficiently computing the expectations. We discuss some ways to do this below.

27.7.2.2 Deriving the gradient using $p(\mathbf{v}|\theta)$

We now present an alternative way to derive Equation 27.112, which also applies to other energy based models. First we marginalize out the hidden variables and write the RBM in the form $p(\mathbf{v}|\boldsymbol{\theta}) = \frac{1}{Z(\boldsymbol{\theta})} \exp(-F(\mathbf{v}; \boldsymbol{\theta}))$, where $F(\mathbf{v}; \boldsymbol{\theta})$ is the **free energy**:

$$F(\mathbf{v}) \triangleq -\log \tilde{p}(\mathbf{v}) \tag{27.114}$$

$$\tilde{p}(\mathbf{v}) = \sum_{\mathbf{h}} \exp(-E(\mathbf{v}, \mathbf{h})) = \sum_{\mathbf{h}} \exp\left(\sum_{r=1}^{R} \sum_{k=1}^{K} v_r h_k w_{rk} \right) \tag{27.115}$$

$$= \sum_{\mathbf{h}} \prod_{k=1}^{K} \exp\left(\sum_{r=1}^{R} v_r h_k w_{rk} \right) = \prod_{k=1}^{K} \sum_{h_k \in \{0,1\}} \exp\left(\sum_{r=1}^{R} v_r h_r w_{rk} \right) \tag{27.116}$$

$$= \prod_{k=1}^{K} \left(1 + \exp(\sum_{r=1}^{R} v_r w_{rk}) \right) \tag{27.117}$$

Next we write the (scaled) log-likelihood in the following form:

$$\ell(\boldsymbol{\theta}) = \frac{1}{N} \sum_{i=1}^{N} \log p(\mathbf{v}_i|\boldsymbol{\theta}) = -\frac{1}{N} \sum_{i=1}^{N} F(\mathbf{v}_i|\boldsymbol{\theta}) - \log Z(\boldsymbol{\theta}) \tag{27.118}$$

Using the fact that $Z(\boldsymbol{\theta}) = \sum_{\mathbf{v}} \exp(-F(\mathbf{v}; \boldsymbol{\theta}))$ we have

$$\nabla \ell(\boldsymbol{\theta}) = -\frac{1}{N} \sum_{i=1}^{N} \nabla F(\mathbf{v}_i) - \frac{\nabla Z}{Z} \tag{27.119}$$

$$= -\frac{1}{N} \sum_{i=1}^{N} \nabla F(\mathbf{v}_i) + \sum_{\mathbf{v}} \nabla F(\mathbf{v}) \frac{\exp(-F(\mathbf{v}))}{Z} \tag{27.120}$$

$$= -\frac{1}{N} \sum_{i=1}^{N} \nabla F(\mathbf{v}_i) + \mathbb{E}\left[\nabla F(\mathbf{v}) \right] \tag{27.121}$$

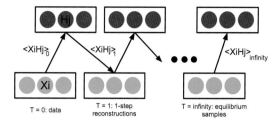

Figure 27.31 Illustration of Gibbs sampling in an RBM. The visible nodes are initialized at a datavector, then we sample a hidden vector, then another visible vector, etc. Eventually (at "infinity") we will be producing samples from the joint distribution $p(\mathbf{v}, \mathbf{h}|\boldsymbol{\theta})$.

Plugging in the free energy (Equation 27.117), one can show that

$$\frac{\partial}{\partial w_{rk}} F(\mathbf{v}) = -v_r \mathbb{E}\left[h_k | \mathbf{v}, \boldsymbol{\theta}\right] = -\mathbb{E}\left[v_r h_k | \mathbf{v}, \boldsymbol{\theta}\right] \tag{27.122}$$

Hence

$$\frac{\partial}{\partial w_{rk}} \ell(\boldsymbol{\theta}) = \frac{1}{N} \sum_{i=1}^{N} \mathbb{E}\left[v_r h_k | \mathbf{v}, \boldsymbol{\theta}\right] - \mathbb{E}\left[v_r h_k | \boldsymbol{\theta}\right] \tag{27.123}$$

which matches Equation 27.112.

27.7.2.3 Approximating the expectations

We can approximate the expectations needed to evaluate the gradient by performing block Gibbs sampling, using Equations 27.101 and 27.102. In more detail, we can sample from the joint distribution $p(\mathbf{v}, \mathbf{h}|\boldsymbol{\theta})$ as follows: initialize the chain at vv^1 (e.g. by setting $\mathbf{v}^1 = \mathbf{v}_i$ for some data vector), and then sample from $\mathbf{h}^1 \sim p(\mathbf{h}|\mathbf{v}^1)$, then from $\mathbf{v}^2 \sim p(\mathbf{v}|\mathbf{h}^1)$, then from $\mathbf{h}^2 \sim p(\mathbf{h}|\mathbf{v}^2)$, etc. See Figure 27.31 for an illustration. Note, however, that we have to wait until the Markov chain reaches equilibrium (i.e., until it has "burned in") before we can interpret the samples as coming from the joint distribution of interest, and this might take a long time.

A faster alternative is to use mean field, where we make the approximation $\mathbb{E}\left[v_r h_k\right] \approx \mathbb{E}\left[v_r\right] \mathbb{E}\left[h_k\right]$. However, since $p(\mathbf{v}, \mathbf{h})$ is typically multimodal, this is usually a very poor approximation, since it will average over different modes (see Section 21.2.2). Furthermore, there is a more subtle reason not to use mean field: since the gradient has the form $\mathbb{E}\left[v_r h_k | \mathbf{v}\right] - \mathbb{E}\left[v_r h_k\right]$, we see that the negative sign in front means that the method will try to make the variational bound as loose as possible (Salakhutdinov and Hinton 2009). This explains why earlier attempts to use mean field to learn Boltzmann machines (e.g., (Kappen and Rodriguez 1998)) did not work.

27.7.2.4 Contrastive divergence

The problem with using Gibbs sampling to compute the gradient is that it is slow. We now present a faster method known as **contrastive divergence** or **CD** (Hinton 2002). CD was originally derived by approximating an objective function defined as the difference of two KL divergences, rather than trying to maximize the likelihood itself. However, from an algorithmic point of view, it can be thought of as similar to stochastic gradient descent, except it approximates the "unclamped" expectations with "brief" Gibbs sampling where we initialize each Markov chain at the data vectors. That is, we approximate the gradient for one datavector as follows:

$$\nabla_{\mathbf{w}} \ell \approx \mathbb{E}\left[\mathbf{v}\mathbf{h}^T | \mathbf{v}_i\right] - \mathbb{E}_q\left[\mathbf{v}\mathbf{h}^T\right] \tag{27.124}$$

where q corresponds to the distribution generated by K up-down Gibbs sweeps, started at $\mathbf{v}_i$, as in Figure 27.31. This is known as CD-K. In more detail, the procedure (for $K = 1$) is as follows:

$$\mathbf{h}_i \sim p(\mathbf{h}|\mathbf{v}_i, \boldsymbol{\theta}) \tag{27.125}$$

$$\mathbf{v}'_i \sim p(\mathbf{v}|\mathbf{h}_i, \boldsymbol{\theta}) \tag{27.126}$$

$$\mathbf{h}'_i \sim p(\mathbf{h}|\mathbf{v}'_i, \boldsymbol{\theta}) \tag{27.127}$$

We then make the approximation

$$\mathbb{E}_q\left[\mathbf{v}\mathbf{h}^T\right] \approx \mathbf{v}_i (\mathbf{h}'_i)^T \tag{27.128}$$

Such samples are sometimes called **fantasy data**. We can think of $\mathbf{v}'_i$ as the model's best attempt at reconstructing $\mathbf{v}_i$ after being coded and then decoded by the model. This is similar to the way we train auto-encoders, which are models which try to "squeeze" the data through a restricted parametric "bottleneck" (see Section 28.3.2).

In practice, it is common to use $\mathbb{E}[\mathbf{h}|\mathbf{v}'_i]$ instead of a sampled value $\mathbf{h}'_i$ in the final upwards pass, since this reduces the variance. However, it is not valid to use $\mathbb{E}[\mathbf{h}|\mathbf{v}_i]$ instead of sampling $\mathbf{h}_i \sim p(\mathbf{h}|\mathbf{v}_i)$ in the earlier upwards passes, because then each hidden unit would be able to pass more than 1 bit of information, so it would not act as much of a bottleneck.

The whole procedure is summarized in Algorithm 27.3. (Note that we follow the positive gradient since we are maximizing likelihood.) Various tricks can be used to speed this algorithm up, such as using a momentum term (Section 8.3.2), using mini-batches, averaging the updates, etc. Such details can be found in (Hinton 2010; Swersky et al. 2010).

27.7.2.5 Persistent CD

In Section 19.5.5, we presented a technique called stochastic maximum likelihood (SML) for fitting maxent models. This avoids the need to run MCMC to convergence at each iteration, by exploiting the fact that the parameters are changing slowly, so the Markov chains will not be pushed too far from equilibrium after each update (Younes 1989). In other words, there are two dynamical processes running at different time scales: the states change quickly, and the parameters change slowly. This algorithm was independently rediscovered in (Tieleman 2008), who called it **persistent CD**. See Algorithm 27.4 for the pseudocode.

PCD often works better than CD (see e.g., (Tieleman 2008; Marlin et al. 2010; Swersky et al. 2010)), although CD can be faster in the early stages of learning. See also (Wick et al. 2011) for a more general algorithm, known as sample rank, that can optimize other loss functions beyond just likelihood.

Algorithm 27.3: CD-1 training for an RBM with binary hidden and visible units

1 Initialize weights $\mathbf{W} \in \mathbb{R}^{R \times K}$ randomly;
2 $t := 0$;
3 **for** *each epoch* **do**
4     $t := t + 1$;
5   **for** *each minibatch of size B* **do**
6   Set minibatch gradient to zero, $\mathbf{g} := \mathbf{0}$;
7   **for** *each case $\mathbf{v}_i$ in the minibatch* **do**
8   Compute $\boldsymbol{\mu}_i = \mathbb{E}\left[\mathbf{h}|\mathbf{v}_i, \mathbf{W}\right]$;
9   Sample $\mathbf{h}_i \sim p(\mathbf{h}|\mathbf{v}_i, \mathbf{W})$;
10   Sample $\mathbf{v}_i' \sim p(\mathbf{v}|\mathbf{h}_i, \mathbf{W})$;
11   Compute $\boldsymbol{\mu}_i' = \mathbb{E}\left[\mathbf{h}|\mathbf{v}_i', \mathbf{W}\right]$;
12   Compute gradient $\nabla_{\mathbf{W}} = (\mathbf{v}_i)(\boldsymbol{\mu}_i)^T - (\mathbf{v}_i')(\boldsymbol{\mu}_i')^T$;
13   Accumulate $\mathbf{g} := \mathbf{g} + \nabla_{\mathbf{W}}$;
14   Update parameters $\mathbf{W} := \mathbf{W} + (\alpha_t/B)\mathbf{g}$

Algorithm 27.4: Persistent CD for training an RBM with binary hidden and visible units

1 Initialize weights $\mathbf{W} \in \mathbb{R}^{D \times L}$ randomly;
2 Initialize chains $(\mathbf{v}_s, \mathbf{h}_s)_{s=1}^{S}$ randomly ;
3 **for** $t = 1, 2, \ldots$ **do**
4   // Mean field updates ;
5   **for** *each case $i = 1 : N$* **do**
6        $\mu_{ik} = \text{sigm}(\mathbf{v}_i^T \mathbf{w}_{:,k})$
7   // MCMC updates ;
8   **for** *each sample $s = 1 : S$* **do**
9   Generate $(\mathbf{v}_s, \mathbf{h}_s)$ by brief Gibbs sampling from old $(\mathbf{v}_s, \mathbf{h}_s)$
10   // Parameter updates ;
11     $\mathbf{g} = \frac{1}{N}\sum_{i=1}^{N} \mathbf{v}_i(\boldsymbol{\mu}_i)^T - \frac{1}{S}\sum_{s=1}^{S} \mathbf{v}_s(\mathbf{h}_s)^T$;
12     $\mathbf{W} := \mathbf{W} + \alpha_t \mathbf{g}$;
13   Decrease α_t

27.7.3 Applications of RBMs

The main application of RBMs is as a building block for deep generative models, which we discuss in Section 28.2. But they can also be used as substitutes for directed two-layer models. They are particularly useful in cases where inference of the hidden states at test time must be fast. We give some examples below.

Data set	Number of docs		K	$\bar{D}$	St. Dev.	Avg. Test perplexity per word (in nats)			
	Train	Test				LDA-50	LDA-200	R. Soft-50	Unigram
NIPS	1,690	50	13,649	98.0	245.3	3576	3391	3405	4385
20-news	11,314	7,531	2,000	51.8	70.8	1091	1058	953	1335
Reuters	794,414	10,000	10,000	94.6	69.3	1437	1142	988	2208

Figure 27.32 Comparison of RBM (replicated softmax) and LDA on three corpora. K is the number of words in the vocabulary, $\bar{D}$ is the average document length, and St. Dev. is the standard deviation of the document length. Source: (Salakhutdinov and Hinton 2010) .

27.7.3.1 Language modeling and document retrieval

We can use a categorical RBM to define a generative model for bag-of-words, as an alternative to LDA. One subtlety is that the partition function in an undirected models depends on how big the graph is, and therefore on how long the document is. A solution to this was proposed in (Salakhutdinov and Hinton 2010): use a categorical RBM with tied weights, but multiply the hidden activation bias terms c_k by the document length L to compensate form the fact that the observed word-count vector $\mathbf{v}$ is larger in magnitude:

$$E(\mathbf{v}, \mathbf{h}; \boldsymbol{\theta}) \triangleq -\sum_{k=1}^{K}\sum_{c=1}^{C} v^c h_k w_k^c - \sum_{c=1}^{C} v^c b_r^c - L\sum_{k=1}^{K} h_k c_k \tag{27.129}$$

where $v^c = \sum_{l=1}^{L} \mathbb{I}(y_{il} = c)$. This is like having a single multinomial node (so we have dropped the r subscript) with C states, where C is the number of words in the vocabulary. This is called the **replicated softmax model** (Salakhutdinov and Hinton 2010), and is an undirected alternative to mPCA/ LDA.

We can compare the modeling power of RBMs vs LDA by measuring the perplexity on a test set. This can be approximated using annealed importance sampling (Section 24.6.2). The results are shown in Figure 27.32. We see that the LDA is significantly better than a unigram model, but that an RBM is significantly better than LDA.

Another advantage of the RBM is that inference is fast and exact: just a single matrix-vector multiply followed by a sigmoid nonlinearity, as in Equation 27.107. In addition to being faster, the RBM is more accurate. This is illustrated in Figure 27.33, which shows precision-recall curves for RBMs and LDA on two different corpora. These curves were generated as follows: a query document from the test set is taken, its similarity to all the training documents is computed, where the similarity is defined as the cosine of the angle between the two topic vectors, and then the top M documents are returned for varying M. A retrieved document is considered relevant if it has the same class label as that of the query's (this is the only place where labels are used).

27.7.3.2 RBMs for collaborative filtering

RBMs have been applied to the Netflix collaborative filtering competition (Salakhutdinov et al. 2007). In fact, an RBM with binary hidden nodes and categorical visible nodes can slightly

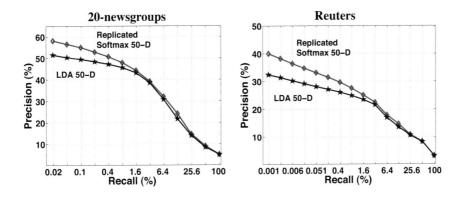

Figure 27.33 Precision-recall curves for RBM (replicated softmax) and LDA on two corpora. From Figure 3 of (Salakhutdinov and Hinton 2010). Used with kind permission of Ruslan Salakhutdinov.

outperform SVD. By combining the two methods, performance can be further improved. (The winning entry in the challenge was an ensemble of many different types of model (Koren 2009a).)

Exercises

Exercise 27.1 Partition function for an RBM

Show how to compute $Z(\boldsymbol{\theta})$ for an RBM with K binary hidden nodes and R binary observed nodes in $O(R2^K)$ time, assuming $K < R$.

28 *Deep learning*

28.1 Introduction

Many of the models we have looked at in this book have a simple two-layer architecture of the form $\mathbf{z} \to \mathbf{y}$ for unsupervised latent variable models, or $\mathbf{x} \to \mathbf{y}$ for supervised models. However, when we look at the brain, we seem many levels of processing. It is believed that each level is learning features or representations at increasing levels of abstraction. For example, the **standard model** of the visual cortex (Hubel and Wiesel 1962; Serre et al. 2005; Ranzato et al. 2007) suggests that (roughly speaking) the brain first extracts edges, then patches, then surfaces, then objects, etc. (See e.g., (Palmer 1999; Kandel et al. 2000) for more information about how the brain might perform vision.)

This observation has inspired a recent trend in machine learning known as **deep learning** (see e.g., (Bengio 2009), `deeplearning.net`, and the references therein), which attempts to replicate this kind of architecture in a computer. (Note the idea can be applied to non-vision problems as well, such as speech and language.)

In this chapter, we give a brief overview of this new field. However, we caution the reader that the topic of deep learning is currently evolving very quickly, so the material in this chapter may soon be outdated.

28.2 Deep generative models

Deep models often have millions of parameters. Acquiring enough labeled data to train such models is diffcult, despite crowd sourcing sites such as Mechanical Turk. In simple settings, such as hand-written character recognition, it is possible to generate lots of labeled data by making modified copies of a small manually labeled training set (see e.g., Figure 16.13), but it seems unlikely that this approach will scale to complex scenes.[1]

To overcome the problem of needing labeled training data, we will focus on unsupervised learning. The most natural way to perform this is to use generative models. In this section, we discuss three different kinds of deep generative models: directed, undirected, and mixed.

---

1. There have been some attempts to use computer graphics and video games to generate realistic-looking images of complex scenes, and then to use this as training data for computer vision systems. However, often graphics programs cut corners in order to make perceptually appealing images which are not reflective of the natural statistics of real-world images.

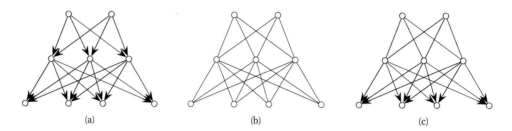

Figure 28.1 Some deep multi-layer graphical models. Observed variables are at the bottom. (a) A directed model. (b) An undirected model (deep Boltzmann machine). (c) A mixed directed-undirected model (deep belief net).

28.2.1 Deep directed networks

Perhaps the most natural way to build a deep generative model is to construct a deep directed graphical model, as shown in Figure 28.1(a). The bottom level contains the observed pixels (or whatever the data is), and the remaining layers are hidden. We have assumed just 3 layers for notational simplicity. The number and size of layers is usually chosen by hand, although one can also use non-parametric Bayesian methods (Adams et al. 2010) or boosting (Chen et al. 2010) to infer the model structure.

We shall call models of this form **deep directed networks** or DDNs. If all the nodes are binary, and all CPDs are logistic functions, this is called a **sigmoid belief net** (Neal 1992). In this case, the model defines the following joint distribution:

$$p(\mathbf{h}_1, \mathbf{h}_2, \mathbf{h}_3, \mathbf{v}|\boldsymbol{\theta}) = \prod_i \text{Ber}(v_i|\text{sigm}(\mathbf{h}_1^T \mathbf{w}_{0i})) \prod_j \text{Ber}(h_{1j}|\text{sigm}(\mathbf{h}_2^T \mathbf{w}_{1j})) \tag{28.1}$$

$$\prod_k \text{Ber}(h_{2k}|\text{sigm}(\mathbf{h}_3^T \mathbf{w}_{2k})) \prod_l \text{Ber}(h_{3l}|w_{3l}) \tag{28.2}$$

where $\boldsymbol{\theta} = (\mathbf{w}_0, \mathbf{w}_1, \mathbf{w}_2, \mathbf{w}_3)$ are all the parameters in the model.

Unfortunately, inference in directed models such as these is intractable because the posterior on the hidden nodes is correlated due to explaining away. One can use fast mean field approximations (Jaakkola and Jordan 1996a; Saul and Jordan 2000), but these may not be very accurate, since they approximate the correlated posterior with a factorial posterior. One can also use MCMC inference (Neal 1992; Adams et al. 2010), but this can be quite slow because the variables are highly correlated. Slow inference also results in slow learning.

28.2.2 Deep Boltzmann machines

A natural alternative to a directed model is to construct a deep undirected model. For example, we can stack a series of RBMs (Section 27.7) on top of each other, as shown in Figure 28.1(b). This is known as a **deep Boltzmann machine** or **DBM** (Salakhutdinov and Hinton 2009). If we have 3 hidden layers, the model is defined as follows:

$$p(\mathbf{h}_1, \mathbf{h}_2, \mathbf{h}_3, \mathbf{v} | \boldsymbol{\theta}) = \frac{1}{Z(\boldsymbol{\theta})} \exp \left(\sum_{ij} v_i h_{1j} W_{1ij} + \sum_{jk} h_{1j} h_{2j} W_{2jk} + \sum_{kl} h_{2k} h_{3l} W_{3kl} \right) \quad (28.3)$$

where we are ignoring constant offset or bias terms.

The main advantage over the directed model is that one can perform efficient block (layer-wise) Gibbs sampling, or block mean field, since all the nodes in each layer are conditionally independent of each other given the layers above and below (Salakhutdinov and Larochelle 2010). The main disadvantage is that training undirected models is more difficult, because of the partition function. However, below we will see a greedy layer-wise strategy for learning deep undirected models.

28.2.3 Deep belief networks

An interesting compromise is to use a model that is partially directed and partially undirected. In particular, suppose we construct a layered model which has directed arrows, except at the top, where there is an undirected bipartite graph, as shown in Figure 28.1(c). This model is known as a **deep belief network** (Hinton et al. 2006) or **DBN**.[2] If we have 3 hidden layers, the model is defined as follows:

$$p(\mathbf{h}_1, \mathbf{h}_2, \mathbf{h}_3, \mathbf{v} | \boldsymbol{\theta}) = \prod_i \text{Ber}(v_i | \text{sigm}(\mathbf{h}_1^T \mathbf{w}_{1i})) \prod_j \text{Ber}(h_{1j} | \text{sigm}(\mathbf{h}_2^T \mathbf{w}_{2j})) \quad (28.4)$$

$$\frac{1}{Z(\boldsymbol{\theta})} \exp \left(\sum_{kl} h_{2k} h_{3l} W_{3kl} \right) \quad (28.5)$$

Essentially the top two layers act as an associative memory, and the remaining layers then generate the output.

The advantage of this peculiar architecture is that we can infer the hidden states in a fast, bottom-up fashion. To see why, suppose we only have two hidden layers, and that $\mathbf{W}_2 = \mathbf{W}_1^T$, so the second level weights are tied to the first level weights (see Figure 28.2(a)). This defines a model of the form $p(\mathbf{h}_1, \mathbf{h}_2, \mathbf{v} | \mathbf{W}_1)$. One can show that the distribution $p(\mathbf{h}_1, \mathbf{v} | \mathbf{W}_1) = \sum_{\mathbf{h}_2} p(\mathbf{h}_1, \mathbf{h}_2, \mathbf{v} | \mathbf{W}_1)$ has the form $p(\mathbf{h}_1, \mathbf{v} | \mathbf{W}_1) = \frac{1}{Z(\mathbf{W}_1)} \exp(\mathbf{v}^T \mathbf{W}_1 \mathbf{h}_1)$, which is equivalent to an RBM. Since the DBN is equivalent to the RBM as far as $p(\mathbf{h}_1, \mathbf{v} | \mathbf{W}_1)$ is concerned, we can infer the posterior $p(\mathbf{h}_1 | \mathbf{v}, \mathbf{W}_1)$ in the DBN exactly as in the RBM. This posterior is exact, even though it is fully factorized.

Now the only way to get a factored posterior is if the prior $p(\mathbf{h}_1 | \mathbf{W}_1)$ is a **complementary prior**. This is a prior which, when multiplied by the likelihood $p(\mathbf{v} | \mathbf{h}_1)$, results in a perfectly factored posterior. Thus we see that the top level RBM in a DBN acts as a complementary prior for the bottom level directed sigmoidal likelihood function.

If we have multiple hidden levels, and/or if the weights are not tied, the correspondence between the DBN and the RBM does not hold exactly any more, but we can still use the factored

---

2. Unforunately the acronym "DBN" also stands for "dynamic Bayes net" (Section 17.6.7). Geoff Hinton, who invented deep belief networks, has suggested the acronyms **DeeBNs** and **DyBNs** for these two different meanings. However, this terminology is non-standard.

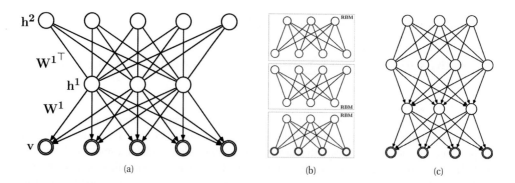

Figure 28.2 (a) A DBN with two hidden layers and tied weights that is equivalent to an RBM. Source: Figure 2.2 of (Salakhutdinov 2009). (b) A stack of RBMs trained greedily. (c) The corresponding DBN. Source: Figure 2.3 of (Salakhutdinov 2009). Used with kind permission of Ruslan Salakhutdinov.

inference rule as a form of approximate bottom-up inference. Below we show that this is a valid variational lower bound. This bound also suggests a layer-wise training strategy, that we will explain in more detail later. Note, however, that top-down inference in a DBN is not tractable, so DBNs are usually only used in a feedforward manner.

28.2.4 Greedy layer-wise learning of DBNs

The equivalence between DBNs and RBMs suggests the following strategy for learning a DBN.

*   Fit an RBM to learn $\mathbf{W}_1$ using methods described in Section 27.7.2.
*   Unroll the RBM into a DBN with 2 hidden layers, as in Figure 28.2(a). Now "freeze" the directed weights $\mathbf{W}_1$ and let $\mathbf{W}_2$ be "untied" so it is no longer forced to be equal to $\mathbf{W}_1^T$. We will now learn a better prior for $p(\mathbf{h}_1|\mathbf{W}_2)$ by fitting a second RBM. The input data to this new RBM is the activation of the hidden units $\mathbb{E}[\mathbf{h}_1|\mathbf{v},\mathbf{W}_1]$ which can be computed using a factorial approximation.
*   Continue to add more hidden layers until some stopping criterion is satisfied, e.g., you run out of time or memory, or you start to overfit the validation set. Construct the DBN from these RBMs, as illustrated in Figure 28.2(c).

One can show (Hinton et al. 2006) that this procedure always increases a lower bound the observed data likelihood. Of course this procedure might result in overfitting, but that is a different matter.

In practice, we want to be able to use any number of hidden units in each level. This means we will not be able to initialize the weights so that $\mathbf{W}_\ell = \mathbf{W}_{\ell-1}^T$. This voids the theoretical guarantee. Nevertheless the method works well in practice, as we will see. The method can also be extended to train DBMs in a greedy way (Salakhutdinov and Larochelle 2010).

After using the greedy layer-wise training strategy, it is standard to "fine tune" the weights, using a technique called **backfitting**. This works as follows. Perform an upwards sampling pass to the top. Then perform brief Gibbs sampling in the top level RBM, and perform a CD update of the RBM parameters (see Section 27.7.2.4). Finally, perform a downwards ancestral sampling pass (which is an approximate sample from the posterior), and update the logistic CPD parameters using a small gradient step. This is called the **up-down** procedure (Hinton et al. 2006). Unfortunately this procedure is very slow.

28.3 Deep neural networks

Given that DBNs are often only used in a feed-forward, or bottom-up, mode, they are effectively acting like neural networks. In view of this, it is natural to dispense with the generative story and try to fit deep neural networks directly, as we discuss below. The resulting training methods are often simpler to implement, and can be faster.

Note, however, that performance with **deep neural nets** (DNNs) is sometimes not as good as with probabilistic models (Bengio et al. 2007). One reason for this is that probabilistic models support top-down inference as well as bottom-up inference. (DBNs do not support efficient top-down inference, but DBMs do, and this has been shown to help (Salakhutdinov and Larochelle 2010).) Top-down inference is useful when there is a lot of ambiguity about the correct interpretation of the signal.

It is interesting to note that in the mammalian visual cortex, there are many more feedback connections than there are feedforward connections (see e.g., (Palmer 1999; Kandel et al. 2000)). The role of these feedback connections is not precisely understood, but they presumably provide contextual prior information (e.g., coming from the previous "frame" or retinal glance) which can be used to disambiguate the current bottom-up signals (Lee and Mumford 2003).

Of course, we can simulate the effect of top-down inference using a neural network. However the models we discuss below do not do this.

28.3.1 Deep multi-layer perceptrons

Many decision problems can be reduced to classification, e.g., predict which object (if any) is present in an image patch, or predict which phoneme is present in a given acoustic feature vector. We can solve such problems by creating a deep feedforward neural network or multi-layer perceptron (MLP), as in Section 16.5, and then fitting the parameters using gradient descent (aka back-propagation).

Unfortunately, this method does not work very well. One problem is that the gradient becomes weaker the further we move away from the data; this is known as the "**vanishing gradient**" problem (Bengio and Frasconi 1995). A related problem is that there can be large plateaus in the error surface, which cause simple first-order DNNs. gadient-based methods to get stuck (Glorot and Bengio 2010). Consequently early attempts to learn deep neural networks proved unsuccessful.

Recently there has been considerable progress in training DNNs, due to several factors: faster computers, in particular the adoption of GPUs (Ciresan et al. 2010); better optimization methods, such as Hessian-free Newton methods (Martens 2010); carefully choosing the variance of the distribution from which the initial random weights are drawn (Bengio 2012); and a method known as **generative pre-training**, which uses unsupervised learning to initialize the parameters (Erhan et al. 2010; Glorot and Bengio 2010).

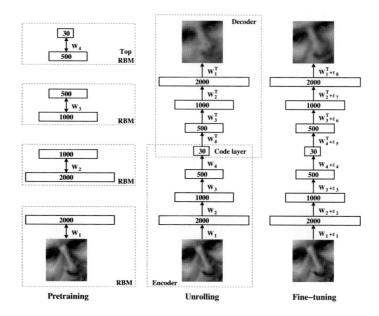

Figure 28.3 Training a deep autoencoder. (a) First we greedily train some RBMs. (b) Then we construct the auto-encoder by replicating the weights. (c) Finally we fine-tune the weights using back-propagation. From Figure 1 of (Hinton and Salakhutdinov 2006). Used with kind permission of Ruslan Salakhutdinov.

28.3.2   Deep auto-encoders

An **auto-encoder** is a kind of unsupervised neural network that is used for dimensionality reduction and feature discovery. More precisely, an auto-encoder is a feedforward neural network that is trained to predict the input itself. To prevent the system from learning the trivial identity mapping, the hidden layer in the middle is usually constrained to be a narrow **bottleneck**. The system can minimize the reconstruction error by ensuring the hidden units capture the most relevant aspects of the data.

Suppose the system has one hidden layer, so the model has the form $\mathbf{v} \to \mathbf{h} \to \mathbf{v}$. Further, suppose all the functions are linear. In this case, one can show that the weights to the K hidden units will span the same subspace as the first K principal components of the data (Karhunen and Joutsensalo 1995; Japkowicz et al. 2000). In other words, linear auto-encoders are equivalent to PCA. However, by using nonlinear activation functions, one can discover nonlinear representations of the data.

More powerful representations can be learned by using **deep auto-encoders**. Unfortunately training such models using back-propagation does not work well, because the gradient signal becomes too small as it passes back through multiple layers, and the learning algorithm often gets stuck in poor local minima.

One solution to this problem is to greedily train a series of RBMs and to use these to initialize an auto-encoder, as illustrated in Figure 28.3. The whole system can then be fine-tuned using backprop in the usual fashion. This approach, first suggested in (Hinton and Salakhutdinov

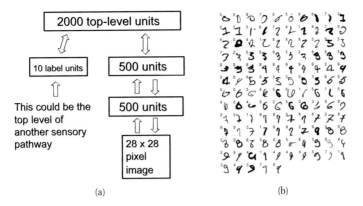

(a)   (b)

Figure 28.4 (a) A DBN architecture for classifying MNIST digits. Source: Figure 1 of (Hinton et al. 2006). Used with kind permission of Geoff Hinton. (b) These are the 125 errors made by the DBN on the 10,000 test cases of MNIST. Above each image is the estimated label. Source: Figure 6 of (Hinton et al. 2006). Used with kind permission of Geoff Hinton. Compare to Figure 16.15.

2006), works much better than trying to fit the deep auto-encoder directly starting with random weights.

28.3.3 Stacked denoising auto-encoders

A standard way to train an auto-encoder is to ensure that the hidden layer is narrower than the visible layer. This prevents the model from learning the identity function. But there are other ways to prevent this trivial solution, which allow for the use of an over-complete representation. One approach is to impose sparsity constraints on the activation of the hidden units (Ranzato et al. 2006). Another approach is to add noise to the inputs; this is called a **denoising auto-encoder** (Vincent et al. 2010). For example, we can corrupt some of the inputs, for example by setting them to zero, so the model has to learn to predict the missing entries. This can be shown to be equivalent to a certain approximate form of maximum likelihood training (known as score matching) applied to an RBM (Vincent 2011).

Of course, we can stack these models on top of each other to learn a deep stacked denoising auto-encoder, which can be discriminatively fine-tuned just like a feedforward neural network, if desired.

28.4 Applications of deep networks

In this section, we mention a few applications of the models we have been discussing.

28.4.1 Handwritten digit classification using DBNs

Figure 28.4(a) shows a DBN (from (Hinton et al. 2006)) consisting of 3 hidden layers. The visible layer corresponds to binary images of handwritten digits from the MNIST data set. In addition, the top RBM is connected to a softmax layer with 10 units, representing the class label.

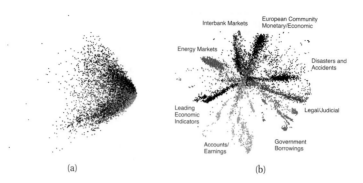

Figure 28.5 2d visualization of some bag of words data from the Reuters RCV1-v2 corpus. (a) Results of using LSA. (b) results of using a deep auto-encoder. Source: Figure 4 of (Hinton and Salakhutdinov 2006). Used with kind permission of Ruslan Salakhutdinov.

The first 2 hidden layers were trained in a greedy unsupervised fashion from 50,000 MNIST digits, using 30 epochs (passes over the data) and stochastic gradient descent, with the CD heuristic. This process took "a few hours per layer" (Hinton et al. 2006, p1540). Then the top layer was trained using as input the activations of the lower hidden layer, as well as the class labels. The corresponding generative model had a test error of about 2.5%. The network weights were then carefully fine-tuned on all 60,000 training images using the up-down procedure. This process took "about a week" (Hinton et al. 2006, p1540). The model can be used to classify by performing a deterministic bottom-up pass, and then computing the free energy for the top-level RBM for each possible class label. The final error on the test set was about 1.25%. The misclassified examples are shown in Figure 28.4(b).

This was the best error rate of any method on the permutation-invariant version of MNIST at that time. (By permutation-invariant, we mean a method that does not exploit the fact that the input is an image. Generic methods work just as well on permuted versions of the input (see Figure 1.5), and can therefore be applied to other kinds of datasets.) The only other method that comes close is an SVM with a degree 9 polynomial kernel, which has achieved an error rate of 1.4% (Decoste and Schoelkopf 2002). By way of comparison, 1-nearest neighbor (using all 60,000 examples) achieves 3.1% (see `mnist1NNdemo`). This is not as good, although 1-NN is much simpler.[3]

28.4.2 Data visualization and feature discovery using deep auto-encoders

Deep autoencoders can learn informative features from raw data. Such features are often used as input to standard supervised learning methods.

To illustrate this, consider fitting a deep auto-encoder with a 2d hidden bottleneck to some

---

3. One can get much improved performance on this task by exploiting the fact that the input is an image. One way to do this is to create distorted versions of the input, adding small shifts and translations (see Figure 16.13 for some examples). Applying this trick reduced the SVM error rate to 0.56%. Similar error rates can be achieved using convolutional neural networks (Section 16.5.1) trained on distorted images ((Simard et al. 2003) got 0.4%). However, the point of DBNs is that they offer a way to learn such prior knowledge, without it having to be hand-crafted.

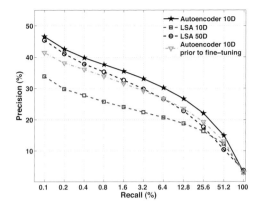

Figure 28.6 Precision-recall curves for document retrieval in the Reuters RCV1-v2 corpus. Source: Figure 3.9 of (Salakhutdinov 2009). Used with kind permission of Ruslan Salakhutdinov.

text data. The results are shown in Figure 28.5. On the left we show the 2d embedding produced by LSA (Section 27.2.2), and on the right, the 2d embedding produced by the auto-encoder. It is clear that the low-dimensional representation created by the auto-encoder has captured a lot of the meaning of the documents, even though class labels were not used.[4]

Note that various other ways of learning low-dimensional continuous embeddings of words have been proposed. See e.g., (Turian et al. 2010) for details.

28.4.3 Information retrieval using deep auto-encoders (semantic hashing)

In view of the sucess of RBMs for information retrieval discussed in Section 27.7.3.1, it is natural to wonder if deep models can do even better. In fact they can, as is shown in Figure 28.6.

More interestingly, we can use a binary low-dimensional representation in the middle layer of the deep auto-encoder, rather than a continuous representation as we used above. This enables very fast retrieval of related documents. For example, if we use a 20-bit code, we can precompute the binary representation for all the documents, and then create a hash-table mapping codewords to documents. This approach is known as **semantic hashing**, since the binary representation of semantically similar documents will be close in Hamming distance.

For the 402,207 test documents in Reuters RCV1-v2, this results in about 0.4 documents per entry in the table. At test time, we compute the codeword for the query, and then simply retrieve the relevant documents in *constant time* by looking up the contents of the relevant address in memory. To find other other related documents, we can compute all the codewords within a

---

4. Some details. Salakhutdinov and Hinton used the Reuters RCV1-v2 data set, which consists of 804,414 newswire articles, manually classified into 103 topics. They represent each document by counting how many times each of the top 2000 most frequent words occurs. They trained a deep auto-encoder with 2000-500-250-125-2 layers on half of the data. The 2000 visible units use a replicated softmax distribution, the 2 hidden units in the middle layer have a Gaussian distribution, and the remaining units have the usual Bernoulli-logistic distribution. When fine tuning the auto-encoder, a cross-entropy loss function (equivalent to maximum likelihood under a multinoulli distribution) was used. See (Hinton and Salakhutdinov 2006) for further details.

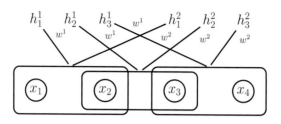

Figure 28.7 A small 1d convolutional RBM with two groups of hidden units, each associated with a filter of size 2. h_1^1 and h_1^2 are two different "views" of the data in the first window, (x_1, x_2). The first view is computed using the filter $\mathbf{w}^1$, the second view using filter $\mathbf{w}^2$. Similarly, h_2^1 and h_2^2 are the views of the data in the second window, (x_2, x_3), computed using $\mathbf{w}^1$ and $\mathbf{w}^2$ respectively.

Hamming distance of, say, 4. This results in retrieving about $6196 \times 0.4 \approx 2500$ documents[5]. The key point is that the total time is independent of the size of the corpus.

Of course, there are other techniques for fast document retrieval, such as inverted indices. These rely on the fact that individual words are quite informative, so we can simply intersect all the documents that contain each word. However, when performing image retrieval, it is clear that we do not want to work at the pixel level. Recently (Krizhevsky and Hinton 2010) showed that a deep autoencoder could learn a good semantic hashing function that outperformed previous techniques (Torralba et al. 2008; Weiss et al. 2008) on the 80 million tiny images dataset. It is hard to apply inverted indexing techniques to real-valued data (although one could imagine vector quantizing image patches).

28.4.4 Learning audio features using 1d convolutional DBNs

To apply DBNs to time series of unbounded length, it is necessary to use some form of parameter tying. One way to do this is to use **convolutional DBNs** (Lee et al. 2009; Desjardins and Bengio 2008), which use convolutional RBMs as their basic unit. These models are a generative version of convolutional neural nets discussed in Section 16.5.1. The basic idea is illustrated in Figure 28.7. The hidden activation vector for each group is computed by convolving the input vector with that group's filter (weight vector or matrix). In other words, each node within a hidden group is a weighted combination of a subset of the inputs. We compute the activaton of all the hidden nodes by "sliding" this weight vector over the input. This allows us to model **translation invariance**, since we use the same weights no matter where in the input vector the pattern occurs.[6] Each group has its own filter, corresponding to its own pattern detector.

---

5. Note that $6196 = \sum_{k=0}^{4} \binom{20}{k}$ is the number of bit vectors that are up to a Hamming distance of 4 away.
6. It is often said that the goal of deep learnng is to discover **invariant features**, e.g., a representation of an object that does not change even as nuisance variables, such as the lighting, do change. However, sometimes these so-called "nuisance variables" may be the variables of interest. For example if the task is to determine if a photograph was taken in the morning or the evening, then lighting is one of the more salient features, and object identity may be less relevant. As always, one task's "signal" is another task's "noise", so it is unwise to "throw away" apparently irrelevant information

More formally, for binary 1d signals, we can define the full conditionals in a convolutional RBM as follows (Lee et al. 2009):

$$p(h_t^k = 1|\mathbf{v}) \quad = \quad \text{sigm}((\mathbf{w}^k \otimes \mathbf{v})_t + b_t) \tag{28.6}$$

$$p(v_s = 1|\mathbf{h}) \quad = \quad \text{sigm}(\sum_k (\mathbf{w}^k \otimes \mathbf{h}^k)_s + c_s) \tag{28.7}$$

where $\mathbf{w}^k$ is the weight vector for group k, b_t and c_s are bias terms, and $\mathbf{a} \otimes \mathbf{b}$ represents the convolution of vectors $\mathbf{a}$ and $\mathbf{b}$.

It is common to add a **max pooling** layer as well as a convolutional layer, which computes a local maximum over the filtered response. This allows for a small amount of translation invariance. It also reduces the size of the higher levels, which speeds up computation considerably. Defining this for a neural network is simple, but defining this in a way which allows for information flow backwards as well as forwards is a bit more involved. The basic idea is similar to a noisy-OR CPD (Section 10.2.3), where we define a probabilistic relationship between the max node and the parts it is maxing over. See (Lee et al. 2009) for details. Note, however, that the top-down generative process will be difficult, since the max pooling operation throws away so much information.

(Lee et al. 2009) applies 1d convolutional DBNs of depth 2 to auditory data. When the input consists of speech signals, the method recovers a representation that is similar to phonemes. When applied to music classification and speaker identification, their method outperforms techniques using standard features such as MFCC. (All features were fed into the same discriminative classifier.)

In (Seide et al. 2011), a deep neural net was used in place of a GMM inside a conventional HMM. The use of DNNs significantly improved performance on conversational speech recognition. In an interview, the tech lead of this project said "historically, there have been very few individual technologies in speech recognition that have led to improvements of this magnitude".[7] See (Hinton et al. 2012) for more details.

28.4.5 Learning image features using 2d convolutional DBNs

We can extend a convolutional DBN from 1d to 2d in a straightforward way (Lee et al. 2009), as illustrated in Figure 28.8. The results of a 3 layer system trained on four classes of visual objects (cars, motorbikes, faces and airplanes) from the Caltech 101 dataset are shown in Figure 28.9. We only show the results for layers 2 and 3, because layer 1 learns Gabor-like filters that are very similar to those learned by sparse coding, shown in Figure 13.21(b). We see that layer 2 has learned some generic visual parts that are shared amongst object classes, and layer 3 seems to have learned filters that look like grandmother cells, that are specific to individual object classes, and in some cases, to individual objects.

---

too early.
7. Source: http://research.microsoft.com/en-us/news/features/speechrecognition-082911.aspx.

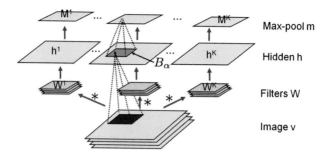

Figure 28.8 A 2d convolutional RBM with max-pooling layers. The input signal is a stack of 2d images (e.g., color planes). Each input layer is passed through a different set of filters. Each hidden unit is obtained by convolving with the appropriate filter, and then summing over the input planes. The final layer is obtained by computing the local maximum within a small window. Source: Figure 1 of (Chen et al. 2010) . Used with kind permission of Bo Chen.

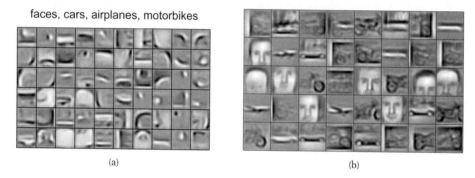

Figure 28.9 Visualization of the filters learned by a convolutional DBN in layers two and three. Source: Figure 3 of (Lee et al. 2009). Used with kind permission of Honglak Lee.

28.5 Discussion

So far, we have been discussing models inspired by low-level processing in the brain. These models have produced useful features for simple classification tasks. But can this pure bottom-up approach scale to more challenging problems, such as scene interpretation or natural language understanding?

To put the problem in perspective, consider the DBN for handwritten digit classification in Figure 28.4(a). This has about 1.6M free parameters ($28 \times 28 \times 500 + 500 \times 500 + 510 \times 2000 = 1,662,000$). Although this is a lot, it is tiny compared to the number of neurons in the human brain, which is about 100 billion. As Hinton says,

> [This DBN] has about as many parameters as 0.002 cubic millimetres of mouse cortex, and several hundred networks of this complexity could fit within a single voxel of a high-resolution fMRI scan. This suggests that much bigger networks may be required to compete with human shape recognition abilities. — (Hinton et al. 2006, p1547).

To scale up to more challenging problems, various groups are using GPUs (see e.g., (Raina et al. 2009)) and/or parallel computing (see e.g., (Dean et al. 2012)). But perhaps a more efficient approach is to work at a higher level of abstraction, where inference is done in the space of objects or their parts, rather than in the space of bits and pixels. That is, we want to bridge the **signal-to-symbol** divide, where by "symbol" we mean something atomic, that can be combined with other symbols in a compositional way.

The question of how to convert low level signals into a more structured/ "semantic" representation is known as the **symbol grounding** problem (Harnard 1990). Traditionally such symbols are associated with words in natural language, but it seems unlikely we can jump directly from low-level signals to high-level semantic concepts. Instead, what we need is an intermediate level of symbolic or atomic parts.

A very simple way to create such parts from real-valued signals, such as images, is to apply vector quantization. This generates a set of **visual words**. These can then be modelled using some of the techniques from Chapter 27 for modeling bags of words. Such models, however, are still quite "shallow".

It is possible to define, and learn, deep models which use discrete latent parts. Here we just mention a few recent approaches, to give a flavor of the possibilites. (Salakhutdinov et al. 2011) combine RBMs with hierarchical latent Dirichlet allocation methods, trained in an unsupervised way. (Zhu et al. 2010) use latent and-or graphs, trained in a manner similar to a latent structural SVM. A similar approach, based on grammars, is described in (Girshick et al. 2011). What is interesting about these techniques is that they apply data-driven machine learning methods to rich structured/symbolic "AI-style" models. This seems like a promising future direction for machine learning (see also (Tenenbaum et al. 2011)).

Notation

Introduction

It is very difficult to come up with a single, consistent notation to cover the wide variety of data, models and algorithms that we discuss. Furthermore, conventions differ between machine learning and statistics, and between different books and papers. Nevertheless, we have tried to be as consistent as possible. Below we summarize most of the notation used in this book, although individual sections may introduce new notation. Note also that the same symbol may have different meanings depending on the context, although we try to avoid this where possible.

General math notation

Symbol	Meaning
$\lfloor x \rfloor$	Floor of x, i.e., round down to nearest integer
$\lceil x \rceil$	Ceiling of x, i.e., round up to nearest integer
$\mathbf{x} \otimes \mathbf{y}$	Convolution of $\mathbf{x}$ and $\mathbf{y}$
$\mathbf{x} \odot \mathbf{y}$	Hadamard (elementwise) product of $\mathbf{x}$ and $\mathbf{y}$
$a \wedge b$	logical AND
$a \vee b$	logical OR
$\neg a$	logical NOT
$\mathbb{I}(x)$	Indicator function, $\mathbb{I}(x) = 1$ if x is true, else $\mathbb{I}(x) = 0$
∞	Infinity
$\rightarrow$	Tends towards, e.g., $n \rightarrow \infty$
$\propto$	Proportional to, so $y = ax$ can be written as $y \propto x$
$|x|$	Absolute value
$|\mathcal{S}|$	Size (cardinality) of a set
$n!$	Factorial function
∇	Vector of first derivatives
∇^2	Hessian matrix of second derivatives
$\triangleq$	Defined as
$O(\cdot)$	Big-O: roughly means order of magnitude
$\mathbb{R}$	The real numbers
$1 : n$	Range (Matlab convention): $1 : n = \{1, 2, \ldots, n\}$
$\approx$	Approximately equal to
$\operatorname{argmax}_x f(x)$	Argmax: the value x that maximizes f

$B(a,b)$ Beta function, $B(a,b) = \frac{\Gamma(a)\Gamma(b)}{\Gamma(a+b)}$

$B(\boldsymbol{\alpha})$ Multivariate beta function, $\frac{\prod_k \Gamma(\alpha_k)}{\Gamma(\sum_k \alpha_k)}$

$\binom{n}{k}$ n choose k, equal to $n!/(k!(n-k)!)$

$\delta(x)$ Dirac delta function, $\delta(x) = \infty$ if $x = 0$, else $\delta(x) = 0$

δ_{ij} Kronecker delta, equals 1 if $i = j$, otherwise equals 0

$\delta_x(y)$ Kronecker delta, equals 1 if $x = y$, otherwise equals 0

$\exp(x)$ Exponential function e^x

$\Gamma(x)$ Gamma function, $\Gamma(x) = \int_0^\infty u^{x-1}e^{-u}du$

$\Psi(x)$ Digamma function, $\Psi(x) = \frac{d}{dx}\log\Gamma(x)$

$\mathcal{X}$ A set from which values are drawn (e.g., $\mathcal{X} = \mathbb{R}^D$)

Linear algebra notation

We use boldface lowercase to denote vectors, such as $\mathbf{a}$, and boldface uppercase to denote matrices, such as $\mathbf{A}$. Vectors are assumed to be column vectors, unless noted otherwise.

Symbol	Meaning
$\mathbf{A} \succ 0$	$\mathbf{A}$ is a positive definite matrix
$\text{tr}(\mathbf{A})$	Trace of a matrix
$\det(\mathbf{A})$	Determinant of matrix $\mathbf{A}$
$|\mathbf{A}|$	Determinant of matrix $\mathbf{A}$
$\mathbf{A}^{-1}$	Inverse of a matrix
$\mathbf{A}^\dagger$	Pseudo-inverse of a matrix
$\mathbf{A}^T$	Transpose of a matrix
$\mathbf{a}^T$	Transpose of a vector
$\text{diag}(\mathbf{a})$	Diagonal matrix made from vector $\mathbf{a}$
$\text{diag}(\mathbf{A})$	Diagonal vector extracted from matrix $\mathbf{A}$
$\mathbf{I}$ or $\mathbf{I}_d$	Identity matrix of size $d \times d$ (ones on diagonal, zeros off)
$\mathbf{1}$ or $\mathbf{1}_d$	Vector of ones (of length d)
$\mathbf{0}$ or $\mathbf{0}_d$	Vector of zeros (of length d)
$||\mathbf{x}|| = ||\mathbf{x}||_2$	Euclidean or ℓ_2 norm $\sqrt{\sum_{j=1}^d x_j^2}$
$||\mathbf{x}||_1$	ℓ_1 norm $\sum_{j=1}^d |x_j|$
$\mathbf{A}_{:,j}$	j'th column of matrix
$\mathbf{A}_{i,:}$	transpose of i'th row of matrix (a column vector)
A_{ij}	Element (i,j) of matrix $\mathbf{A}$
$\mathbf{x} \otimes \mathbf{y}$	Tensor product of $\mathbf{x}$ and $\mathbf{y}$

Probability notation

We denote random and fixed scalars by lower case, random and fixed vectors by bold lower case, and random and fixed matrices by bold upper case. Occasionally we use non-bold upper case to denote scalar random variables. Also, we use $p()$ for both discrete and continuous random variables.

Symbol	Meaning
$X \perp Y$	X is independent of Y
$X \not\perp Y$	X is not independent of Y
$X \perp Y \vert Z$	X is conditionally independent of Y given Z
$X \not\perp Y \vert Z$	X is not conditionally independent of Y given Z
$X \sim p$	X is distributed according to distribution p
$\boldsymbol{\alpha}$	Parameters of a Beta or Dirichlet distribution
$\mathrm{cov}\,[\mathbf{x}]$	Covariance of $\mathbf{x}$
$\mathbb{E}\,[X]$	Expected value of X
$\mathbb{E}_q\,[X]$	Expected value of X wrt distribution q
$\mathbb{H}\,(X)$ or $\mathbb{H}\,(p)$	Entropy of distribution $p(X)$
$\mathbb{I}\,(X;Y)$	Mutual information between X and Y
$\mathbb{KL}\,(p\vert\vert q)$	KL divergence from distribution p to q
$\ell(\boldsymbol{\theta})$	Log-likelihood function
$L(\theta, a)$	Loss function for taking action a when true state of nature is θ
λ	Precision (inverse variance) $\lambda = 1/\sigma^2$
$\boldsymbol{\Lambda}$	Precision matrix $\boldsymbol{\Lambda} = \boldsymbol{\Sigma}^{-1}$
$\mathrm{mode}\,[X]$	Most probable value of X
μ	Mean of a scalar distribution
$\boldsymbol{\mu}$	Mean of a multivariate distribution
$p(x)$	Probability density or mass function
$p(x\vert y)$	Conditional probability density of x given y
Φ	cdf of standard normal
ϕ	pdf of standard normal
$\boldsymbol{\pi}$	multinomial parameter vector, Stationary distribution of Markov chain
ρ	Correlation coefficient
$\mathrm{sigm}(x)$	Sigmoid (logistic) function, $\frac{1}{1+e^{-x}}$
σ^2	Variance
$\boldsymbol{\Sigma}$	Covariance matrix
$\mathrm{var}\,[x]$	Variance of x
ν	Degrees of freedom parameter
Z	Normalization constant of a probability distribution

Machine learning/statistics notation

In general, we use upper case letters to denote constants, such as C, D, K, N, S, T, etc. We use lower case letters as dummy indexes of the appropriate range, such as $c = 1 : C$ to index classes, $j = 1 : D$ to index input features, $k = 1 : K$ to index states or clusters, $s = 1 : S$ to index samples, $t = 1 : T$ to index time, etc. To index data cases, we use the notation $i = 1 : N$, although the notation $n = 1 : N$ is also widely used.

We use $\mathbf{x}$ to represent an observed data vector. In a supervised problem, we use y or $\mathbf{y}$ to represent the desired output label. We use $\mathbf{z}$ to represent a hidden variable. Sometimes we also use q to represent a hidden discrete variable.

Symbol	Meaning
C	Number of classes
D	Dimensionality of data vector (number of features)
R	Number of outputs (response variables)
$\mathcal{D}$	Training data $\mathcal{D} = \{\mathbf{x}_i | i = 1 : N\}$ or $\mathcal{D} = \{(\mathbf{x}_i, y_i) | i = 1 : N\}$
$\mathcal{D}_{\text{test}}$	Test data
$J(\boldsymbol{\theta})$	Cost function
K	Number of states or dimensions of a variable (often latent)
$\kappa(\mathbf{x}, \mathbf{y})$	Kernel function
$\mathbf{K}$	Kernel matrix
λ	Strength of ℓ_2 or ℓ_1 regularizer
N	Number of data cases
N_c	Number of examples of class c, $N_c = \sum_{n=1}^{N} \mathbb{I}(y_n = c)$
$\phi(\mathbf{x})$	Basis function expansion of feature vector $\mathbf{x}$
$\boldsymbol{\Phi}$	Basis function expansion of design matrix X
$q()$	Approximate or proposal distribution
$Q(\boldsymbol{\theta}, \boldsymbol{\theta}_{old})$	Auxiliary function in EM
S	Number of samples
T	Length of a sequence
$T(\mathcal{D})$	Test statistic for data
$\mathbf{T}$	Transition matrix of Markov chain
$\boldsymbol{\theta}$	Parameter vector
$\boldsymbol{\theta}^{(s)}$	s'th sample of parameter vector
$\hat{\boldsymbol{\theta}}$	Estimate (usually MLE or MAP) of $\boldsymbol{\theta}$
$\hat{\boldsymbol{\theta}}_{ML}$	Maximum likelihood estimate of $\boldsymbol{\theta}$
$\hat{\boldsymbol{\theta}}_{MAP}$	MAP estimate of $\boldsymbol{\theta}$
$\overline{\boldsymbol{\theta}}$	Estimate (usually posterior mean) of $\boldsymbol{\theta}$
$\mathbf{w}$	Vector of regression weights (called $\boldsymbol{\beta}$ in statistics)
$\mathbf{W}$	Matrix of regression weights
x_{ij}	Component (i.e., feature) j of data case i, for $i = 1 : N, j = 1 : D$
$\mathbf{x}_i$	Training case, $i = 1 : N$
$\mathbf{X}$	Design matrix of size $N \times D$
$\overline{\mathbf{x}}$	Empirical mean $\overline{\mathbf{x}} = \frac{1}{N} \sum_{i=1}^{N} \mathbf{x}_i$
$\tilde{\mathbf{x}}$	Future test case
$\mathbf{x}_*$	Future test case
$\mathbf{y}$	Vector of all training labels $\mathbf{y} = (y_1, \ldots, y_N)$
z_{ij}	Latent component j for case i

Graphical model notation

In graphical models, we index nodes by $s, t, u \in \mathcal{V}$, and states by $i, j, k \in \mathcal{X}$.

Symbol	Meaning
$s \sim t$	Node s is connected to node t
bel	Belief function
$\mathcal{C}$	Cliques of a graph
ch_j	Child of node j in a DAG
desc_j	Descendants of node j in a DAG
G	A graph
$\mathcal{E}$	Edges of a graph
mb_t	Markov blanket of node t
nbd_t	Neighborhood of node t
pa_t	Parents of node t in a DAG
pred_t	Predecessors of node t in a DAG wrt some ordering
$\psi_c(\mathbf{x}_c)$	Potential function for clique c
$\mathcal{S}$	Separators of a graph
θ_{sjk}	prob. node s is in state k given its parents are in states j
$\mathcal{V}$	Nodes of a graph

List of commonly used abbreviations

Abbreviation	Meaning
cdf	Cumulative distribution function
CPD	Conditional probability distribution
CPT	Conditional probability table
CRF	Conditional random field
DAG	Directed acyclic graphic
DGM	Directed graphical model
EB	Empirical Bayes
EM	Expectation maximization algorithm
EP	Expectation propagation
GLM	Generalized linear model
GMM	Gaussian mixture model
HMM	Hidden Markov model
iid	Independent and identically distributed
iff	If and only if
KL	Kullback Leibler divergence
LDS	Linear dynamical system
LHS	Left hand side (of an equation)
MAP	Maximum A Posterior estimate
MCMC	Markov chain Monte Carlo
MH	Metropolis Hastings
MLE	Maximum likelihood estimate
MPM	Maximum of Posterior Marginals
MRF	Markov random field
MSE	Mean squared error
NLL	Negative log likelihood
OLS	Ordinary least squares
pd	Positive definite (matrix)
pdf	Probability density function
pmf	Probability mass function
RBPF	Rao-Blackwellised particle filter
RHS	Right hand side (of an equation)
RJMCMC	Reversible jump MCMC
RSS	Residual sum of squares
SLDS	Switching linear dynamical system
SSE	Sum of squared errors
UGM	Undirected graphical model
VB	Variational Bayes
wrt	With respect to

Bibliography

Abend, K., T. J. Harley, and L. N. Kanal (1965). Classification of Binary Random Patterns. *IEEE Transactions on Information Theory 11(4)*, 538–544.

Ackley, D., G. Hinton, and T. Sejnowski (1985). A learning algorithm for boltzmann machines. *Cognitive Science 9*, 147–169.

Adams, R. P., H. Wallach, and Z. Ghahramani (2010). Learning the structure of deep sparse graphical models. In *AI/Statistics*.

Agarwal, A., O. Chapelle, M. Dudik, and J. Langford (2011). A reliable effective terrascale linear learning system. Technical report.

Aggarwal, D. and S. Merugu (2007). Predictive discrete latent factor models for large scale dyadic data. In *Proc. of the Int'l Conf. on Knowledge Discovery and Data Mining*.

Ahmed, A. and E. Xing (2007). On tight approximate inference of the logistic-normal topic admixture model. In *AI/Statistics*.

Ahn, J.-H. and J.-H. Oh (2003). A Constrained EM Algorithm for Principal Component Analysis. *Neural Computation 15*, 57–65.

Ahn, S., A. Korattikara, and M. Welling (2012). Bayesian Posterior Sampling via Stochastic Gradient Fisher Scoring. In *Intl. Conf. on Machine Learning*.

Airoldi, E., D. Blei, S. Fienberg, and E. Xing (2008). Mixed-membership stochastic blockmodels. *J. of Machine Learning Research 9*, 1981–2014.

Aitchison, J. (1982). The statistical analysis of compositional data. *J. of Royal Stat. Soc. Series B 44*(2), 139–177.

Aji, S. M. and R. J. McEliece (2000, March). The generalized distributive law. *IEEE Trans. Info. Theory 46*(2), 325–343.

Alag, S. and A. Agogino (1996). Inference using message propagation and topology transformation in vector Gaussian continuous networks. In *UAI*.

Albers, C., M. Leisink, and H. Kappen (2006). The Cluster Variation Method for Efficient Linkage Analysis on Extended Pedigrees. *BMC Bioinformatics 7*.

Albert, J. and S. Chib (1993). Bayesian analysis of binary and polychotomous response data. *J. of the Am. Stat. Assoc. 88*(422), 669–679.

Allwein, E., R. Schapire, and Y. Singer (2000). Reducing multiclass to binary: A unifying approach for margin classifiers. *J. of Machine Learning Research*, 113–141.

Aloise, D., A. Deshpande, P. Hansen, and P. Popat (2009). NP-hardness of Euclidean sum-of-squares clustering. *Machine Learning 75*, 245–249.

Alpaydin, E. (2004). *Introduction to machine learning*. MIT Press.

Altun, Y., T. Hofmann, and I. Tsochantaridis (2007). Support Vector Machine Learning for Interdependent and Structured Output Spaces. In G. Bakir, T. Hofmann, B. Scholkopf, A. Smola, B. Taskar, and S. Vishwanathan (Eds.), *Predicting Structured Data*. MIT Press.

Amir, E. (2010). Approximation Algorithms for Treewidth. *Algorithmica 56*(4), 448.

Amir, E. and S. McIlraith (2005). Partition-based logical reasoning for first-order and propositional theories. *Artificial Intelligence 162*(1), 49–88.

Ando, R. and T. Zhang (2005). A framework for learning predictive structures from multiple tasks and unlabeled data. *J. of Machine Learning Research 6*, 1817–1853.

Andrews, D. and C. Mallows (1974). Scale mixtures of Normal distributions. *J. of Royal Stat. Soc. Series B 36*, 99–102.

Andrieu, C., N. de Freitas, and A. Doucet (2000). Sequential Bayesian estimation and model selection for dynamic kernel machines. Technical report, Cambridge Univ.

Andrieu, C., N. de Freitas, and A. Doucet (2001). Robust Full Bayesian Learning for Radial Basis Networks. *Neural Computation 13*(10), 2359–2407.

Andrieu, C., N. de Freitas, A. Doucet, and M. Jordan (2003). An introduction to MCMC for machine learning. *Machine Learning 50*, 5–43.

Andrieu, C., A. Doucet, and V. Tadic (2005). Online EM for parameter estimation in nonlinear-non Gaussian state-space models. In *Proc. IEEE CDC*.

Andrieu, C. and J. Thoms (2008). A tutorial on adaptive MCMC. *Statistical Computing 18*, 343–373.

Aoki, M. (1971). *Introduction to optimization techniques*. Macmillan.

Aoki, M. (1987). *State space modeling of time series*. Springer.

Archambeau, C. and F. Bach (2008). Sparse probabilistic projections. In *NIPS*.

Argyriou, A., T. Evgeniou, and M. Pontil (2008). Convex multi-task feature learning. *Machine Learning 73*(3), 243–272.

Armagan, A., D. Dunson, and J. Lee (2011). Generalized double pareto shrinkage. Technical report, Duke.

Armstrong, H. (2005). *Bayesian estimation of decomposable Gaussian graphical models*. Ph.D. thesis, UNSW.

Armstrong, H., C. Carter, K. Wong, and R. Kohn (2008). Bayesian Covariance Matrix Estimation using a Mixture of Decomposable Graphical Models. *Statistics and Computing*, 1573–1375.

Arnborg, S., D. G. Corneil, and A. Proskurowski (1987). Complexity of finding embeddings in a k-tree. *SIAM J. on Algebraic and Discrete Methods 8*, 277–284.

Arora, S. and B. Barak (2009). *Complexity Theory: A Modern Approach.* Cambridge.

Arthur, D. and S. Vassilvitskii (2007). k-means++: the advantages of careful seeding. In *Proc. 18th ACM-SIAM symp. on Discrete algorithms*, pp. 1027âĂŞ1035.

Arulampalam, M., S. Maskell, N. Gordon, and T. Clapp (2002, February). A Tutorial on Particle Filters for Online Nonlinear/Non-Gaussian Bayesian Tracking. *IEEE Trans. on Signal Processing 50*(2), 174–189.

Asavathiratham, C. (2000). *The Influence Model: A Tractable Representation for the Dynamics of Networked Markov Chains.* Ph.D. thesis, MIT, Dept. EECS.

Atay-Kayis, A. and H. Massam (2005). A Monte Carlo method for computing the marginal likelihood in non-decomposable Gaussian graphical models. *Biometrika 92*, 317–335.

Attenberg, J., K. Weinberger, A. Smola, A. Dasgupta, and M. Zinkevich (2009). Collaborative spam filtering with the hashing trick. In *Virus Bulletin.*

Attias, H. (1999). Independent factor analysis. *Neural Computation 11*, 803–851.

Attias, H. (2000). A variational Bayesian framework for graphical models. In *NIPS-12.*

Bach, F. (2008). Bolasso: Model Consistent Lasso Estimation through the Bootstrap. In *Intl. Conf. on Machine Learning.*

Bach, F. and M. Jordan (2001). Thin junction trees. In *NIPS.*

Bach, F. and M. Jordan (2005). A probabilistic interpretation of canonical correlation analysis. Technical Report 688, U. C. Berkeley.

Bach, F. and E. Moulines (2011). Non-asymptotic analysis of stochastic approximation algorithms for machine learning. In *NIPS.*

Bahmani, B., B. Moseley, A. Vattani, R. Kumar, and S. Vassilvitskii (2012). Scalable k-Means++. In *VLDB.*

Bakker, B. and T. Heskes (2003). Task Clustering and Gating for Bayesian Multitask Learning. *J. of Machine Learning Research 4*, 83–99.

Baldi, P. and Y. Chauvin (1994). Smooth online learning algorithms for hidden Markov models. *Neural Computation 6*, 305–316.

Balding, D. (2006). A tutorial on statistical methods for population association studies. *Nature Reviews Genetics 7*, 81–91.

Banerjee, A., S. Basu, and S. Merugu (2007). Multi-way clustering on relation graphs. In *Proc. SIAM Intl. Conf. on Data Mining (SDM).*

Banerjee, A., H. Wang, Q. Fu, S. Liess, and P. Snyder (2012). MAP Inference on Million Node Graphical Models: KL-divergence based Alternating Directions Method. Technical report, U. Minnesota.

Banerjee, O., L. E. Ghaoui, and A. d'Aspremont (2008). Model selection through sparse maximum likelihood estimation for multivariate gaussian or binary data. *J. of Machine Learning Research 9*, 485–516.

Bar-Shalom, Y. and T. Fortmann (1988). *Tracking and data association.* Academic Press.

Bar-Shalom, Y. and X. Li (1993). *Estimation and Tracking: Principles, Techniques and Software.* Artech House.

Barash, Y. and N. Friedman (2002). Context-specific Bayesian clustering for gene expression data. *J. Comp. Bio. 9*, 169–191.

Barber, D. (2006). Expectation Correction for Smoothed Inference in Switching Linear Dynamical Systems. *J. of Machine Learning Research 7*, 2515–2540.

Barber, D. and C. Bishop (1998). Ensemble Learning in Bayesian Neural Networks. In C. Bishop (Ed.), *Neural Networks and Machine Learning*, pp. 215–237. Springer.

Barber, D. and S. Chiappa (2007). Unified inference for variational bayesian linear gaussian state space models. In *NIPS.*

Barbieri, M. and J. Berger (2004). Optimal predictive model selection. *Annals of Statistics 32*, 870–897.

Bartlett, P., M. Jordan, and J. McAuliffe (2006). Convexity, Classification, and Risk Bounds. *J. of the Am. Stat. Assoc. 101*(473), 138–156.

Baruniak, R. (2007). Compressive sensing. *IEEE Signal Processing Magazine.*

Barzilai, J. and J. Borwein (1988). Two point step size gradient methods. *IMA J. of Numerical Analysis 8*, 141–148.

Basu, S., T. Choudhury, B. Clarkson, and A. Pentland (2001). Learning human interactions with the influence model. Technical Report 539, MIT Media Lab. ftp://whitechapel.media.mit.edu/pub/tech-reports/TR-539-ABSTRACT.html.

Baum, L. E., T. Petrie, G. Soules, and N. Weiss (1970). A maximization technique occuring in the statistical analysis of probabalistic functions in markov chains. *The Annals of Mathematical Statistics 41*, 164–171.

Beal, M. (2003). *Variational Algorithms for Approximate Bayesian Inference.* Ph.D. thesis, Gatsby Unit.

Beal, M. and Z. Ghahramani (2006). Variational Bayesian Learning of Directed Graphical Models with Hidden Variables. *Bayesian Analysis 1*(4).

Beal, M. J., Z. Ghahramani, and C. E. Rasmussen (2002). The infinite hidden Markov model. In *NIPS-14.*

Beck, A. and M. Teboulle (2009). A fast iterative shrinkage-thresholding algorithm for linear inverse problems. *SIAM J. on Imaging Sciences 2*(1), 183–202.

Beck, A. and M. Teoulle (2003). Mirror descent and nonlinear projected subgradient methods for convex optimization. *Operations Research Letters 31*(3), 167–175.

Beinlich, I., H. Suermondt, R. Chavez, and G. Cooper (1989). The ALARM monitoring system: A case study with two probabilistic inference techniques for belief networks. In *Proc. of the Second European Conf. on AI in Medicine*, pp. 247–256.

Bekkerman, R., M. Bilenko, and J. Langford (Eds.) (2011). *Scaling Up Machine Learning.* Cambridge.

Bell, A. J. and T. J. Sejnowski (1995). An information maximisation approach to blind separation and blind deconvolution. *Neural Computation 7*(6), 1129–1159.

Bengio, Y. (2009). Learning deep architectures for AI. *Foundations and Trends in Machine Learning 2*(1), 1–127.

Bengio, Y. (2012). Practical recommendations for gradient-based training of deep architectures. In G. Montavon, G. Orr, and K.-R. Mueller (Eds.), *Neural Networks: Tricks of the Trade Second Edition.*

Bengio, Y. and S. Bengio (2000). Modeling high-dimensional discrete data with multi-layer neural networks. In *NIPS.*

Bengio, Y., O. Delalleau, N. Roux, J. Paiement, P. Vincent, and M. Ouimet (2004). Learning eigenfunctions links spectral embedding and kernel PCA. *Neural Computation 16*, 2197–2219.

Bengio, Y. and P. Frasconi (1995). Diffusion of context and credit information in markovian models. *J. of AI Research 3*, 249–270.

Bengio, Y. and P. Frasconi (1996). Input/output HMMs for sequence processing. *IEEE Trans. on Neural Networks 7*(5), 1231–1249.

Bengio, Y., P. Lamblin, D. Popovici, and H. Larochelle (2007). Greedy layer-wise training of deep networks. In *NIPS.*

Berchtold, A. (1999). The double chain markov model. *Comm. Stat. Theor. Methods 28*, 2569–2589.

Berger, J. (1985). Bayesian salesmanship. In P. K. Goel and A. Zellner (Eds.), *Bayesian Inference and Decision Techniques with Applications: Essays in Honor of Bruno deFinetti.* North-Holland.

Berger, J. and R. Wolpert (1988). *The Likelihood Principle.* The Institute of Mathematical Statistics. 2nd edition.

Berkhin, P. (2006). A survey of clustering datamining techniques. In J. Kogan, C. Nicholas, and M. Teboulle (Eds.), *Grouping Multidimensional Data: Recent Advances in Clustering*, pp. 25–71. Springer.

Bernardo, J. and A. Smith (1994). *Bayesian Theory.* John Wiley.

Berrou, C., A. Glavieux, and P. Thitimajashima (1993). Near Shannon limit error-correcting coding and decoding: Turbo codes. *Proc. IEEE Intl. Comm. Conf..*

Berry, D. and Y. Hochberg (1999). Bayesian perspectives on multiple comparisons. *J. Statist. Planning and Inference 82*, 215–227.

Bertele, U. and F. Brioschi (1972). *Nonserial Dynamic Programming.* Academic Press.

Bertsekas, D. (1997). *Parallel and Distribution Computation: Numerical Methods.* Athena Scientific.

Bertsekas, D. (1999). *Nonlinear Programming* (Second ed.). Athena Scientific.

Bertsekas, D. and J. Tsitsiklis (2008). *Introduction to Probability.* Athena Scientific. 2nd Edition.

Besag, J. (1975). Statistical analysis of non-lattice data. *The Statistician 24*, 179–196.

Bhatnagar, N., C. Bogdanov, and E. Mossel (2010). The computational complexity of estimating convergence time. Technical report, .

Bhattacharya, A. and D. B. Dunson (2011). Simplex factor models for multivariate unordered categorical data. *J. of the Am. Stat. Assoc..* To appear.

Bickel, P. and E. Levina (2004). Some theory for Fisher's linear discriminant function, "Naive Bayes", and some alternatives when there are many more variables than observations. *Bernoulli 10*, 989–1010.

Bickson, D. (2009). *Gaussian Belief Propagation: Theory and Application.* Ph.D. thesis, Hebrew University of Jerusalem.

Bilmes, J. (2000). Dynamic Bayesian multinets. In *UAI.*

Bilmes, J. A. (2001). Graphical models and automatic speech recognition. Technical Report UWEETR-2001-0005, Univ. Washington, Dept. of Elec. Eng.

Binder, J., D. Koller, S. J. Russell, and K. Kanazawa (1997). Adaptive probabilistic networks with hidden variables. *Machine Learning 29*, 213–244.

Binder, J., K. Murphy, and S. Russell (1997). Space-efficient inference in dynamic probabilistic networks. In *Intl. Joint Conf. on AI.*

Birnbaum, A. (1962). On the foundations of statistical infernece. *J. of the Am. Stat. Assoc. 57*, 269–326.

Bishop, C. (1999). Bayesian PCA. In *NIPS.*

Bishop, C. (2006). *Pattern recognition and machine learning.* Springer.

Bishop, C. and G. James (1993). Analysis of multiphase flows using dual-energy densitometry and neural networks. *Nuclear Instruments and Methods in Physics Research A327*, 580–593.

Bishop, C. and M. Svensén (2003). Bayesian hierarchical mixtures of experts. In *UAI.*

Bishop, C. and M. Tipping (2000). Variational relevance vector machines. In *UAI.*

Bishop, C. M. (1994). Mixture density networks. Technical Report NCRG 4288, Neural Computing Research Group, Department of Computer Science, Aston University.

Bishop, C. M. (1995). *Neural Networks for Pattern Recognition.* Clarendon Press.

Bishop, Y., S. Fienberg, and P. Holland (1975). *Discrete Multivariate Analysis: Theory and Practice.* MIT Press.

Bistarelli, S., U. Montanari, and F. Rossi (1997). Semiring-based constraint satisfaction and optimization. *J. of the ACM 44*(2), 201–236.

Blake, A., P. Kohli, and C. Rother (Eds.) (2011). *Advances in Markov Random Fields for Vision and Image Processing.* MIT Press.

Blei, D. and J. Lafferty (2006a). Correlated topic models. In *NIPS.*

Blei, D. and J. Lafferty (2006b). Dynamic topic models. In *Intl. Conf. on Machine Learning*, pp. 113–120.

Blei, D. and J. Lafferty (2007). A Correlated Topic Model of "Science". *Annals of Applied Stat.* 1(1), 17–35.

Blei, D. and J. McAuliffe (2010, March). Supervised topic models. Technical report, Princeton.

Blei, D., A. Ng, and M. Jordan (2003). Latent dirichlet allocation. *J. of Machine Learning Research 3*, 993–1022.

Blumensath, T. and M. Davies (2007). On the difference between Orthogonal Matching Pursuit and Orthogonal Least Squares. Technical report, U. Edinburgh.

Bo, L., C. Sminchisescu, A. Kanaujia, and D. Metaxas (2008). Fast Algorithms for Large Scale Conditional 3D Prediction. In *CVPR*.

Bohning, D. (1992). Multinomial logistic regression algorithm. *Annals of the Inst. of Statistical Math. 44*, 197–200.

Bollen, K. (1989). *Structural Equation Models with Latent Variables*. John Wiley & Sons.

Bordes, A., L. Bottou, and P. Gallinari (2009, July). Sgd-qn: Careful quasi-newton stochastic gradient descent. *J. of Machine Learning Research 10*, 1737–1754.

Bordes, A., L. Bottou, P. Gallinari, J. Chang, and S. A. Smith (2010). Erratum: SGDQN is Less Careful than Expected. *J. of Machine Learning Research 11*, 2229–2240.

Boser, B. E., I. M. Guyon, and V. N. Vapnik (1992). A training algorithm for optimal margin classifiers. In *Proc. of the Workshop on Computational Learning Theory*.

Bottcher, S. G. and C. Dethlefsen (2003). deal: A package for learning bayesian networks. *J. of Statistical Software 8*(20).

Bottolo, L. and S. Richardson (2010). Evolutionary stochastic search. *Bayesian Analysis 5*(3), 583–618.

Bottou, L. (1998). Online algorithms and stochastic approximations. In D. Saad (Ed.), *Online Learning and Neural Networks*. Cambridge.

Bottou, L. (2007). Learning with large datasets (nips tutorial).

Bottou, L., O. Chapelle, D. DeCoste, and J. Weston (Eds.) (2007). *Large Scale Kernel Machines*. MIT Press.

Bouchard, G. (2007). Efficient bounds for the softmax and applications to approximate inference in hybrid models. In *NIPS 2007 Workshop on Approximate Inference in Hybrid Models*.

Bouchard-Cote, A. and M. Jordan (2009). Optimization of structured mean field objectives. In *UAI*.

Bowman, A. and A. Azzalini (1997). *Applied Smoothing Techniques for Data Analysis*. Oxford.

Box, G. and N. Draper (1987). *Empirical Model-Building and Response Surfaces*. Wiley.

Box, G. and G. Tiao (1973). *Bayesian inference in statistical analysis*. Addison-Wesley.

Boyd, S. and L. Vandenberghe (2004). *Convex optimization*. Cambridge.

Boyen, X. and D. Koller (1998). Tractable inference for complex stochastic processes. In *UAI*.

Boykov, Y., O. Veksler, and R. Zabih (2001). Fast approximate energy minimization via graph cuts. *IEEE Trans. on Pattern Analysis and Machine Intelligence 23*(11).

Brand, M. (1996). Coupled hidden Markov models for modeling interacting processes. Technical Report 405, MIT Lab for Perceptual Computing.

Brand, M. (1999). Structure learning in conditional probability models via an entropic prior and parameter extinction. *Neural Computation 11*, 1155–1182.

Braun, M. and J. McAuliffe (2010). Variational Inference for Large-Scale Models of Discrete Choice. *J. of the Am. Stat. Assoc. 105*(489), 324–335.

Breiman, L. (1996). Bagging predictors. *Machine Learning 24*, 123–140.

Breiman, L. (1998). Arcing classifiers. *Annals of Statistics 26*, 801–849.

Breiman, L. (2001a). Random forests. *Machine Learning 45*(1), 5–32.

Breiman, L. (2001b). Statistical modeling: the two cultures. *Statistical Science 16*(3), 199–231.

Breiman, L., J. Friedman, and R. Olshen (1984). *Classification and regression trees*. Wadsworth.

Breslow, N. E. and D. G. Clayton (1993). Approximate inference in generalized linear mixed models. *J. of the Am. Stat. Assoc. 88*(421), 9–25.

Briers, M., A. Doucet, and S. Maskel (2010). Smoothing algorithms for state-space models. *Annals of the Institute of Statistical Mathematics 62*(1), 61–89.

Brochu, E., M. Cora, and N. de Freitas (2009, November). A tutorial on Bayesian optimization of expensive cost functions, with application to active user modeling and hierarchical reinforcement learning. Technical Report TR-2009-23, Department of Computer Science, University of British Columbia.

Brooks, S. and G. Roberts (1998). Assessing convergence of Markov Chain Monte Carlo algorithms. *Statistics and Computing 8*, 319–335.

Brown, L., T. Cai, and A. DasGupta (2001). Interval estimation for a binomial proportion. *Statistical Science 16*(2), 101–133.

Brown, M. P., R. Hughey, A. Krogh, I. S. Mian, K. Sjölander, and D. Haussler (1993). Using dirichlet mixtures priors to derive hidden Markov models for protein families. In *Intl. Conf. on Intelligent Systems for Molecular Biology*, pp. 47–55.

Brown, P., M. Vannucci, and T. Fearn (1998). Multivariate Bayesian variable selection and prediction. *J. of the Royal Statistical Society B 60*(3), 627–641.

Bruckstein, A., D. Donoho, and M. Elad (2009). From sparse solutions of systems of equations to sparse modeling of signals and images. *SIAM Review 51*(1), 34–81.

Bryan, K. and T. Leise (2006). The $25,000,000,000 Eigenvector: The Linear Algebra behind Google. *SIAM Review 48*(3).

Bryson, A. and Y.-C. Ho (1969). *Applied optimal control: optimization, estimation, and control*. Blaisdell Publishing Company.

Buhlmann, P. and T. Hothorn (2007). Boosting Algorithms: Regularization, Prediction and Model Fitting. *Statistical Science 22*(4), 477–505.

Buhlmann, P. and S. van de Geer (2011). *Statistics for High-Dimensional Data: Methodology, Theory and Applications*. Springer.

Buhlmann, P. and B. Yu (2003). Boosting with the L2 loss: Regression and classification. *J. of the Am. Stat. Assoc. 98*(462), 324–339.

Buhlmann, P. and B. Yu (2006). Sparse boosting. *J. of Machine Learning Research 7*, 1001–1024.

Bui, H., S. Venkatesh, and G. West (2002). Policy Recognition in the Abstract Hidden Markov Model. *J. of AI Research 17*, 451–499.

Buntine, W. (2002). Variational Extensions to EM and Multinomial PCA. In *Intl. Conf. on Machine Learning*.

Buntine, W. and A. Jakulin (2004). Applying Discrete PCA in Data Analysis. In *UAI*.

Buntine, W. and A. Jakulin (2006). Discrete Component Analysis. In *Subspace, Latent Structure and Feature Selection: Statistical and Optimization Perspectives Workshop*.

Buntine, W. and A. Weigend (1991). Bayesian backpropagation. *Complex Systems 5*, 603–643.

Burges, C. J., T. Shaked, E. Renshaw, A. Lazier, M. Deeds, N. Hamilton, and G. Hullender (2005). Learning to rank using gradient descent. In *Intl. Conf. on Machine Learning*, pp. 89–96.

Burkard, R., M. Dell'Amico, and S. Martello (2009). *Assignment Problems*. SIAM.

Calvetti, D. and E. Somersalo (2007). *Introduction to Bayesian Scientific Computing*. Springer.

Candes, E., J. Romberg, and T. Tao (2006). Robust uncertainty principles: Exact signal reconstruction from highly incomplete frequency information. *IEEE. Trans. Inform. Theory 52*(2), 489–509.

Candes, E. and M. Wakin (2008, March). An introduction to compressive sampling. *IEEE Signal Processing Magazine 21*.

Candes, E., M. Wakin, and S. Boyd (2008). Enhancing sparsity by reweighted l1 minimization. *J. of Fourier Analysis and Applications 1*, 877–905.

Cannings, C., E. A. Thompson, and M. H. Skolnick (1978). Probability functions in complex pedigrees. *Advances in Applied Probability 10*, 26–61.

Canny, J. (2004). Gap: a factor model for discrete data. In *Proc. Annual Intl. ACM SIGIR Conference*, pp. 122–129.

Cao, Z., T. Qin, T.-Y. Liu, M.-F. Tsai, and H. Li (2007). Learning to rank: From pairwise approach to listwise approach. In *Intl. Conf. on Machine Learning*, pp. 129–136.

Cappe, O. (2010). Online Expectation Maximisation. In K. Mengersen, M. Titterington, and C. Robert (Eds.), *Mixtures*. Wiley.

Cappe, O. and E. Mouline (2009, June). Online EM Algorithm for Latent Data Models. *J. of Royal Stat. Soc. Series B 71*(3), 593–613.

Cappe, O., E. Moulines, and T. Ryden (2005). *Inference in Hidden Markov Models*. Springer.

Carbonetto, P. (2003). Unsupervised statistical models for general object recognition. Master's thesis, University of British Columbia.

Carbonetto, P. and M. Stephens (2012). Scalable variational inference for bayesian variable selection in regression, and its accuracy in genetic association studies. *Bayesian Analysis 7*, 73–108.

Carlin, B. P. and T. A. Louis (1996). *Bayes and Empirical Bayes Methods for Data Analysis*. Chapman and Hall.

Caron, F. and A. Doucet (2008). Sparse Bayesian nonparametric regression. In *Intl. Conf. on Machine Learning*.

Carreira-Perpinan, M. and C. Williams (2003). An isotropic gaussian mixture can have more modes than components. Technical Report EDI-INF-RR-0185, School of Informatics, U. Edinburgh.

Carter, C. and R. Kohn (1994). On Gibbs sampling for state space models. *Biometrika 81*(3), 541–553.

Carterette, B., P. Bennett, D. Chickering, and S. Dumais (2008). Here or There: Preference Judgments for Relevance. In *Proc. ECIR*.

Caruana, R. (1998). A dozen tricks with multitask learning. In G. Orr and K.-R. Mueller (Eds.), *Neural Networks: Tricks of the Trade*. Springer-Verlag.

Caruana, R. and A. Niculescu-Mizil (2006). An empirical comparison of supervised learning algorithms. In *Intl. Conf. on Machine Learning*.

Carvahlo, C., N. Polson, and J. Scott (2010). The horseshoe estimator for sparse signals. *Biometrika 97*(2), 465.

Carvahlo, C. M. and M. West (2007). Dynamic matrix-variate graphical models. *Bayesian Analysis 2*(1), 69–98.

Carvahlo, L. and C. Lawrence (2007). Centroid estimation in discrete high-dimensional spaces with applications in biology. *Proc. of the National Academy of Science, USA 105*(4).

Casella, G. and R. Berger (2002). *Statistical inference*. Duxbury. 2nd edition.

Castro, M., M. Coates, and R. D. Nowak (2004). Likelihood based hierarchical clustering. *IEEE Trans. in Signal Processing 52*(8), 230.

Celeux, G. and J. Diebolt (1985). The SEM algorithm: A probabilistic teacher derive from the EM algorithm for the mixture problem. *Computational Statistics Quarterly 2*, 73–82.

Cemgil, A. T. (2001). A technique for painless derivation of kalman filtering recursions. Technical report, U. Nijmegen.

Cesa-Bianchi, N. and G. Lugosi (2006). *Prediction, learning, and games*. Cambridge University Press.

Cevher, V. (2009). Learning with compressible priors. In *NIPS*.

Chai, K. M. A. (2010). *Multi-task learning with Gaussian processes*. Ph.D. thesis, U. Edinburgh.

Chan, A. and D. Dong (2011). Generalized gaussian process models. In *CVPR*.

Chang, H., Y. Weiss, and W. Freeman (2009). Informative Sensing. Technical report, Hebrew U. Submitted to IEEE Transactions on Info. Theory.

Chang, J. and D. Blei (2010). Hierarchical relational models for document networks. *The Annals of Applied Statistics 4*(1), 124–150.

Chang, J., J. Boyd-Graber, S. Gerrish, C. Wang, and D. Blei (2009). Reading tea leaves: How humans interpret topic models. In *NIPS*.

Chapelle, O. and L. Li (2011). An empirical evaluation of Thompson sampling. In *NIPS*.

Chartrand, R. and W. Yin (2008). Iteratively reweighted algorithms for compressive sensing. In *Intl. Conf. on Acoustics, Speech and Signal Proc.*

Chechik, G., A. G. N. Tishby, and Y. Weiss (2005). Information bottleneck for gaussian variables. *J. of Machine Learning Research 6*, 165âĂŞ188.

Cheeseman, P., J. Kelly, M. Self, J. Stutz, W. Taylor, and D. Freeman (1988). Autoclass: A Bayesian classification system. In *Proc. of the Fifth Intl. Workshop on Machine Learning*.

Cheeseman, P. and J. Stutz (1996). Bayesian classification (autoclass): Theory and results. In Fayyad, Pratetsky-Shapiro, Smyth, and Uthurasamy (Eds.), *Advances in Knowledge Discovery and Data Mining*. MIT Press.

Chen, B., K. Swersky, B. Marlin, and N. de Freitas (2010). Sparsity priors and boosting for learning localized distributed feature representations. Technical report, UBC.

Chen, B., J.-A. Ting, B. Marlin, and N. de Freitas (2010). Deep learning of invariant spatio-temporal features from video. In *NIPS Workshop on Deep Learning*.

Chen, M., D. Carlson, A. Zaas, C. Woods, G. Ginsburg, A. Hero, J. Lucas, and L. Carin (2011, March). The Bayesian Elastic Net: Classifying Multi-Task Gene-Expression Data. *IEEE Trans. Biomed. Eng. 58*(3), 468–79.

Chen, R. and S. Liu (2000). Mixture Kalman filters. *J. Royal Stat. Soc. B*.

Chen, S. and J. Goodman (1996). An empirical study of smoothing techniques for language modeling. In *Proc. 34th ACL*, pp. 310–318.

Chen, S. and J. Goodman (1998). An empirical study of smoothing techniques for language modeling. Technical Report TR-10-98, Dept. Comp. Sci., Harvard.

Chen, S. and J. Wigger (1995, July). Fast orthogonal least squares algorithm for efficient subset model selection. *IEEE Trans. Signal Processing 3*(7), 1713–1715.

Chen, S. S., D. L. Donoho, and M. A. Saunders (1998). Atomic decomposition by basis pursuit. *SIAM Journal on Scientific Computing 20*(1), 33–61.

Chen, X., S. Kim, Q. Lin, J. G. Carbonell, and E. P. Xing (2010). Graph-Structured Multi-task Regression and an Efficient Optimization Method for General Fused Lasso. Technical report, CMU.

Chib, S. (1995). Marginal likelihood from the Gibbs output. *J. of the Am. Stat. Assoc. 90*, 1313–1321.

Chickering, D. (1996). Learning Bayesian networks is NP-Complete. In *AI/Stats V*.

Chickering, D. and D. Heckerman (1997). Efficient approximations for the marginal likelihood of incomplete data given a Bayesian network. *Machine Learning 29*, 181–212.

Chickering, D. M. (2002). Optimal structure identification with greedy search. *Journal of Machine Learning Research 3*, 507–554.

Chipman, H., E. George, and R. McCulloch (1998). Bayesian CART model search. *J. of the Am. Stat. Assoc. 93*, 935–960.

Chipman, H., E. George, and R. McCulloch (2001). The practical implementation of Bayesian Model Selection. *Model Selection*. IMS Lecture Notes.

Chipman, H., E. George, and R. McCulloch (2006). Bayesian Ensemble Learning. In *NIPS*.

Chipman, H., E. George, and R. McCulloch (2010). BART: Bayesian additive regression trees. *Ann. Appl. Stat. 4*(1), 266–298.

Choi, M., V. Tan, A. Anandkumar, and A. Willsky (2011). Learning latent tree graphical models. *J. of Machine Learning Research*.

Choi, M. J. (2011). *Trees and Beyond: Exploiting and Improving Tree-Structured Graphical Models*. Ph.D. thesis, MIT.

Choset, H. and K. Nagatani (2001). Topological simultaneous localization and mapping (SLAM): toward exact localization without explicit localization. *IEEE Trans. Robotics and Automation 17*(2).

Chow, C. K. and C. N. Liu (1968). Approximating discrete probability distributions with dependence trees. *IEEE Trans. on Info. Theory 14*, 462–67.

Christensen, O., G. Roberts, and M. Skőld (2006). Robust Markov chain Monte Carlo methods for spatial generalized linear mixed models. *J. of Computational and Graphical Statistics 15*, 1–17.

Chung, F. (1997). *Spectral Graph Theory*. AMS.

Cimiano, P., A. Schultz, S. Sizov, P. Sorg, and S. Staab (2009). Explicit versus latent concept models for cross-language information retrieval. In *Intl. Joint Conf. on AI*.

Cipra, B. (2000). The Ising Model Is NP-Complete. *SIAM News 33*(6).

Ciresan, D. C., U. Meier, L. M. Gambardella, and J. Schmidhuber (2010). Deep big simple neural nets for handwritten digit recognition. *Neural Computation 22*(12), 3207–3220.

Clarke, B. (2003). Bayes model averaging and stacking when model approximation error cannot be ignored. *J. of Machine Learning Research*, 683–712.

Clarke, B., E. Fokoue, and H. H. Zhang (2009). *Principles and Theory for Data Mining and Machine Learning*. Springer.

Cleveland, W. and S. Devlin (1988). Locally-weighted regression: An approach to regression analysis by local fitting. *J. of the Am. Stat. Assoc. 83*(403), 596–610.

Collins, M. (2002). Discriminative Training Methods for Hidden Markov Models: Theory and Experiments with Perceptron Algorithms. In *EMNLP*.

Collins, M., S. Dasgupta, and R. E. Schapire (2002). A generalization of principal components analysis to the exponential family. In *NIPS-14*.

Collins, M. and N. Duffy (2002). Convolution kernels for natural language. In *NIPS*.

Collobert, R. and J. Weston (2008). A Unified Architecture for Natural Language Processing: Deep Neural Networks with Multitask Learning. In *Intl. Conf. on Machine Learning*.

Combettes, P. and V. Wajs (2005). Signal recovery by proximal forward-backward splitting. *SIAM J. Multiscale Model. Simul. 4*(4), 1168–1200.

Cook, J. (2005). Exact Calculation of Beta Inequalities. Technical report, M. D. Anderson Cancer Center, Dept. Biostatistics.

Cooper, G. and E. Herskovits (1992). A Bayesian method for the induction of probabilistic networks from data. *Machine Learning 9*, 309–347.

Cooper, G. and C. Yoo (1999). Causal discovery from a mixture of experimental and observational data. In *UAI*.

Cortez, P. and M. J. Embrechts (2012). Using sensitivity analysis and visualization techniques to open black box data mining models. *Information Sciences*.

Cover, T. and P. Hart (1967). Nearest neighbor pattern classification. *IEEE Trans. Inform. Theory 13*(1), 21–27.

Cover, T. M. and J. A. Thomas (1991). *Elements of Information Theory*. John Wiley.

Cover, T. M. and J. A. Thomas (2006). *Elements of Information Theory*. John Wiley. 2nd edition.

Cowell, R. G., A. P. Dawid, S. L. Lauritzen, and D. J. Spiegelhalter (1999). *Probabilistic Networks and Expert Systems*. Springer.

Cowles, M. and B. Carlin (1996). Markov chain monte carlo convergence diagnostics: A comparative review. *J. of the Am. Stat. Assoc. 91*, 883–904.

Crisan, D., P. D. Moral, and T. Lyons (1999). Discrete filtering using branching and interacting particle systems. *Markov Processes and Related Fields 5*(3), 293–318.

Cui, Y., X. Z. Fern, and J. G. Dy (2010). Learning multiple nonredundant clusterings. *ACM Transactions on Knowledge Discovery from Data 4*(3).

Cukier, K. (2010, February). Data, data everywhere.

Dagum, P. and M. Luby (1993). Approximating probabilistic inference in Bayesian belief networks is NP-hard. *Artificial Intelligence 60*, 141–153.

Dahl, J., L. Vandenberghe, and V. Roychowdhury (2008, August). Covariance selection for non-chordal graphs via chordal embedding. *Optimization Methods and Software 23*(4), 501–502.

Dahlhaus, R. and M. Eichler (2000). Causality and graphical models for time series. In P. Green, N. Hjort, and S. Richardson (Eds.), *Highly structured stochastic systems*. Oxford University Press.

Dallal, S. and W. Hall (1983). Approximating priors by mixtures of natural conjugate priors. *J. of Royal Stat. Soc. Series B 45*, 278–286.

Darwiche, A. (2009). *Modeling and Reasoning with Bayesian Networks*. Cambridge.

Daume, H. (2007a). Fast search for Dirichlet process mixture models. In *AI/Statistics*.

Daume, H. (2007b). Frustratingly easy domain adaptation. In *Proc. the Assoc. for Comp. Ling.*

Dawid, A. P. (1992). Applications of a general propagation algorithm for probabilistic expert systems. *Statistics and Computing 2*, 25–36.

Dawid, A. P. (2002). Influence diagrams for causal modelling and inference. *Intl. Stat. Review 70*, 161–189. Corrections p437.

Dawid, A. P. (2010). Beware of the DAG! *J. of Machine Learning Research 6*, 59–86.

Dawid, A. P. and S. L. Lauritzen (1993). Hyper-markov laws in the statistical analysis of decomposable graphical models. *The Annals of Statistics 3*, 1272–1317.

de Freitas, N., R. Dearden, F. Hutter, R. Morales-Menendez, J. Mutch, and D. Poole (2004). Diagnosis by a waiter and a mars explorer. *Proc. IEEE 92*(3).

de Freitas, N., M. Niranjan, and A. Gee (2000). Hierarchical Bayesian models for regularisation in sequential learning. *Neural Computation 12*(4), 955–993.

Dean, J., G. S. Corrado, R. Monga, K. Chen, M. Devin, Q. V. Le, M. Z. Mao, M. A. Ranzato, A. Senior, P. Tucker, K. Yang, and A. Y. Ng (2012). Large scale distributed deep networks. In *NIPS*.

Dechter, R. (1996). Bucket elimination: a unifying framework for probabilistic inference. In *UAI*.

Dechter, R. (2003). *Constraint Processing*. Morgan Kaufmann.

Decoste, D. and B. Schoelkopf (2002). Training invariant support vector machines. *Machine learnng 41*, 161–190.

Deerwester, S., S. Dumais, G. Furnas, T. Landauer, and R. Harshman (1990). Indexing by latent semantic analysis. *J. of the American Society for Information Science 41*(6), 391–407.

DeGroot, M. (1970). *Optimal Statistical Decisions*. McGraw-Hill.

Deisenroth, M., C. Rasmussen, and J. Peters (2009). Gaussian Process Dynamic Programming. *Neurocomputing 72*(7), 1508–1524.

Dellaportas, P., P. Giudici, and G. Roberts (2003). Bayesian inference for nondecomposable graphical gaussian models. *Sankhya, Ser. A 65*, 43–55.

Dellaportas, P. and A. F. M. Smith (1993). Bayesian Inference for Generalized Linear and Proportional Hazards Models via Gibbs Sampling. *J. of the Royal Statistical Society. Series C (Applied Statistics) 42*(3), 443–459.

Delyon, B., M. Lavielle, and E. Moulines (1999). Convergence of a stochastic approximation version of the EM algorithm. *Annals of Statistics 27*(1), 94–128.

Dempster, A. (1972). Covariance selection. *Biometrics 28*(1).

Dempster, A. P., N. M. Laird, and D. B. Rubin (1977). Maximum likelihood from incomplete data via the EM algorithm. *J. of the Royal Statistical Society, Series B 34*, 1–38.

Denison, D., C. Holmes, B. Mallick, and A. Smith (2002). *Bayesian methods for nonlinear classification and regression.* Wiley.

Denison, D., B. Mallick, and A. Smith (1998). A Bayesian CART algorithm. *Biometrika 85*, 363–377.

Desjardins, G. and Y. Bengio (2008). Empirical evaluation of convolutional RBMs for vision. Technical Report 1327, U. Montreal.

Dey, D., S. Ghosh, and B. Mallick (Eds.) (2000). *Generalized Linear Models: A Bayesian Perspective.* Chapman & Hall/CRC Biostatistics Series.

Diaconis, P., S. Holmes, and R. Montgomery (2007). Dynamical Bias in the Coin Toss. *SIAM Review 49*(2), 211–235.

Diaconis, P. and D. Ylvisaker (1985). Quantifying prior opinion. In *Bayesian Statistics 2.*

Dietterich, T. G. and G. Bakiri (1995). Solving multiclass learning problems via ECOCs. *J. of AI Research 2*, 263–286.

Diggle, P. and P. Ribeiro (2007). *Model-based Geostatistics.* Springer.

Ding, Y. and R. Harrison (2010). A sparse multinomial probit model for classification. *Pattern Analysis and Applications*, 1–9.

Dobra, A. (2009). Dependency networks for genome-wide data. Technical report, U. Washington.

Dobra, A. and H. Massam (2010). The mode oriented stochastic search (MOSS) algorithm for log-linear models with conjugate priors. *Statistical Methodology 7*, 240–253.

Domingos, P., S. Kok, H. Poon, M. Richardson, and P. Singla (2006). Unifying Logical and Statistical AI. In *Intl. Joint Conf. on AI.*

Domingos, P. and D. Lowd (2009). *Markov Logic: An Interface Layer for AI.* Morgan & Claypool.

Domingos, P. and M. Pazzani (1997). On the optimality of the simple bayesian classifier under zero-one loss. *Machine Learning 29*, 103–130.

Domke, J., A. Karapurkar, and Y. Aloimonos (2008). Who killed the directed model? In *CVPR.*

Doucet, A., N. de Freitas, and N. J. Gordon (2001). *Sequential Monte Carlo Methods in Practice.* Springer Verlag.

Doucet, A., N. Gordon, and V. Krishnamurthy (2001). Particle Filters for State Estimation of Jump Markov Linear Systems. *IEEE Trans. on Signal Processing 49*(3), 613–624.

Dow, J. and J. Endersby (2004). Multinomial probit and multinomial logit: a comparison of choice models for voting research. *Electoral Studies 23*(1), 107–122.

Drineas, P., A. Frieze, R. Kannan, S. Vempala, and V. Vinay (2004). Clustering large graphs via the singular value decomposition. *Machine Learning 56*, 9–33.

Drugowitsch, J. (2008). Bayesian linear regression. Technical report, U. Rochester.

Druilhet, P. and J.-M. Marin (2007). Invariant HPD credible sets and MAP estimators. *Bayesian Analysis 2*(4), 681–692.

Duane, S., A. Kennedy, B. Pendleton, and D. Roweth (1987). Hybrid Monte Carlo. *Physics Letters B 195*(2), 216–222.

Duchi, J., S. Gould, and D. Koller (2008). Projected subgradient methods for learning sparse gaussians. In *UAI.*

Duchi, J., E. Hazan, and Y. Singer (2010). Adaptive Subgradient Methods for Online Learning and Stochastic Optimization. In *Proc. of the Workshop on Computational Learning Theory.*

Duchi, J., S. Shalev-Shwartz, Y. Singer, and T. Chandra (2008). Efficient projections onto the L1-ball for learning in high dimensions. In *Intl. Conf. on Machine Learning.*

Duchi, J. and Y. Singer (2009). Boosting with structural sparsity. In *Intl. Conf. on Machine Learning.*

Duchi, J., D. Tarlow, G. Elidan, and D. Koller (2007). Using combinatorial optimization within max-product belief propagation. In *NIPS.*

Duda, R. O., P. E. Hart, and D. G. Stork (2001). *Pattern Classification.* Wiley Interscience. 2nd edition.

Dumais, S. and T. Landauer (1997). A solution to Plato's problem: The latent semantic analysis theory of acquisition, induction and representation of knowledge. *Psychological Review 104*, 211–240.

Dunson, D., J. Palomo, and K. Bollen (2005). Bayesian Structural Equation Modeling. Technical Report 2005-5, SAMSI.

Durbin, J. and S. J. Koopman (2001). *Time Series Analysis by State Space Methods.* Oxford University Press.

Durbin, R., S. Eddy, A. Krogh, and G. Mitchison (1998). *Biological Sequence Analysis: Probabilistic Models of Proteins and Nucleic Acids.* Cambridge: Cambridge University Press.

Earl, D. and M. Deem (2005). Parallel tempering: Theory, applications, and new perspectives. *Phys. Chem. Chem. Phys. 7*, 3910.

Eaton, D. and K. Murphy (2007). Exact Bayesian structure learning from uncertain interventions. In *AI/Statistics.*

Edakunni, N., S. Schaal, and S. Vijayakumar (2010). Probabilistic incremental locally weighted learning using randomly varying coefficient model. Technical report, USC.

Edwards, D., G. de Abreu, and R. Labouriau (2010). Selecting high-dimensional mixed graphical models using minimal AIC or BIC forests. *BMC Bioinformatics 11*(18).

Efron, B. (1986). Why Isn't Everyone a Bayesian? *The American Statistician 40*(1).

Efron, B. (2010). *Large-Scale Inference: Empirical Bayes Methods for Estimation, Testing, and Prediction.* Cambridge.

Efron, B., I. Johnstone, T. Hastie, and R. Tibshirani (2004). Least angle regression. *Annals of Statistics 32*(2), 407–499.

Efron, B. and C. Morris (1975). Data analysis using stein's estimator and its generalizations. *J. of the Am. Stat. Assoc. 70*(350), 311–319.

Elad, M. and I. Yavnch (2009). A plurality of sparse representations is better than the sparsest one alone. *IEEE Trans. on Info. Theory 55*(10), 4701–4714.

Elidan, G. and S. Gould (2008). Learning Bounded Treewidth Bayesian Networks. *J. of Machine Learning Research*, 2699–2731.

Elidan, G., N. Lotner, N. Friedman, and D. Koller (2000). Discovering hidden variables: A structure-based approach. In *NIPS*.

Elidan, G., I. McGraw, and D. Koller (2006). Residual belief propagation: Informed scheduling for asynchronous message passing. In *UAI*.

Elkan, C. (2003). Using the triangle inequality to accelerate k-means. In *Intl. Conf. on Machine Learning*.

Elkan, C. (2005). Deriving TF-IDF as a Fisher kernel. In *Proc. Intl. Symp. on String Processing and Information Retrieval (SPIRE)*, pp. 296–301.

Elkan, C. (2006). Clustering documents with an exponential family approximation of the Dirichlet compound multinomial model. In *Intl. Conf. on Machine Learning*.

Ellis, B. and W. H. Wong (2008). Learning causal bayesian network structures from experimental data. *J. of the Am. Stat. Assoc. 103*(482), 778–789.

Engel, Y., S. Mannor, and R. Meir (2005). Reinforcement Learning with Gaussian Processes. In *Intl. Conf. on Machine Learning*.

Erhan, D., Y. Bengio, A. Courville, P.-A. Manzagol, P. Vincent, and S. Bengio (2010). Why Does Unsupervised Pre-training Help Deep Learning? *J. of Machine Learning Research 11*, 625–660.

Erosheva, E. and S. M. Curtis (2011). Dealing with rotational invariance in Bayesian confirmatory factor analysis. Technical Report 589, U. Washington Dept. Statistics.

Erosheva, E., S. Fienberg, and C. Joutard (2007). Describing disability through individual-level mixture models for multivariate binary data. *Annals of Applied Statistics*.

Erosheva, E., S. Fienberg, and J. Lafferty (2004). Mixed-membership models of scientific publications. *Proc. of the National Academy of Science, USA 101*, 5220–2227.

Escobar, M. D. and M. West (1995). Bayesian density estimation and inference using mixtures. *J. of the Am. Stat. Assoc. 90*(430), 577–588.

Ewens, W. (1990). Population genetics theory - the past and the future. In S.Lessard (Ed.), *Mathemetical and Statistica Developments of Evolutionary Theory*, pp. 177–227. Reidel.

Fan, J. and R. Z. Li (2001). Variable selection via non-concave penalized likelihood and its oracle properties. *J. of the Am. Stat. Assoc. 96*(456), 1348–1360.

Fawcett, T. (2003). ROC Graphs: Notes and Practical Considerations for Data Mining Researchers. Technical report, HP labs.

Fearnhead, P. (2004). Exact bayesian curve fitting and signal segmentation. *IEEE Trans. Signal Processing 53*, 2160–2166.

Felzenszwalb, P., R. Girshick, D. McAllester, and D. Ramanan (2010, September). Object Detection with Discriminatively Trained Part Based Models. *IEEE Trans. on Pattern Analysis and Machine Intelligence 32*(9).

Felzenszwalb, P. and D. Huttenlocher (2006). Efficient belief propagation for early vision. *Intl. J. Computer Vision 70*(1), 41–54.

Ferrucci, D., E. Brown, J. Chu-Carroll, J. Fan, D. Gondek, A. Kalyanpur, A. Lally, J. W. Murdock, E. N. amd J. Prager, N. Schlaefter, and C. Welty (2010). Building Watson: An Overview of the DeepQA Project. *AI Magazine*, 59–79.

Fienberg, S. (1970). An iterative procedure for estimation in contingency tables. *Annals of Mathematical Statistics 41*(3), 907–917.

Figueiredo, M. (2003). Adaptive sparseness for supervised learning. *IEEE Trans. on Pattern Analysis and Machine Intelligence 25*(9), 1150–1159.

Figueiredo, M., R. Nowak, and S. Wright (2007). Gradient projection for sparse reconstruction: application to compressed sensing and other inverse problems. *IEEE. J. on Selected Topics in Signal Processing*.

Figueiredo, M. A. T. and A. K. Jain (2002). Unsupervised learning of finite mixture models. *IEEE Trans. on Pattern Analysis and Machine Intelligence 24*(3), 381–396. Matlab code at http://www.lx.it.pt/ mtf/mixture-code.zip.

Fine, S., Y. Singer, and N. Tishby (1998). The hierarchical Hidden Markov Model: Analysis and applications. *Machine Learning 32*, 41.

Finkel, J. and C. Manning (2009). Hierarchical bayesian domain adaptation. In *NAACL*, pp. 602–610.

Fischer, B. and J. Schumann (2003). Autobayes: A system for generating data analysis programs from statistical models. *J. Functional Programming 13*(3), 483–508.

Fishelson, M. and D. Geiger (2002). Exact genetic linkage computations for general pedigrees. *BMC Bioinformatics 18*.

Fletcher, R. (2005). On the Barzilai-Borwein Method. *Applied Optimization 96*, 235–256.

Fokoue, E. (2005). Mixtures of factor analyzers: an extension with covariates. *J. Multivariate Analysis 95*, 370–384.

Forbes, J., T. Huang, K. Kanazawa, and S. Russell (1995). The BATmobile: Towards a Bayesian automated taxi. In *Intl. Joint Conf. on AI*.

Forsyth, D. and J. Ponce (2002). *Computer vision: a modern approach*. Prentice Hall.

Fraley, C. and A. Raftery (2002). Model-based clustering, discriminant analysis, and density estimation. *J. of the Am. Stat. Assoc. (97)*, 611–631.

Fraley, C. and A. Raftery (2007). Bayesian Regularization for Normal Mixture Estimation and Model-Based Clustering. *J. of Classification 24*, 155–181.

Franc, V., A. Zien, and B. Schoelkopf (2011). Support vector machines as probabilistic models. In *Intl. Conf. on Machine Learning.*

Frank, I. and J. Friedman (1993). A statistical view of some chemometrics regression tools. *Technometrics 35*(2), 109–135.

Fraser, A. (2008). *Hidden Markov Models and Dynamical Systems.* SIAM Press.

Freund, Y. and R. R. Schapire (1996). Experiments with a new boosting algorithm. In *Intl. Conf. on Machine Learning.*

Frey, B. (1998). *Graphical Models for Machine Learning and Digital Communication.* MIT Press.

Frey, B. (2003). Extending factor graphs so as to unify directed and undirected graphical models. In *UAI.*

Frey, B. and D. Dueck (2007, February). Clustering by Passing Messages Between Data Points. *Science 315*, 972âĂŞ976.

Friedman, J. (1991). Multivariate adaptive regression splines. *Ann. Statist. 19*, 1–67.

Friedman, J. (1997a). On bias, variance, 0-1 loss and the curse of dimensionality. *J. Data Mining and Knowledge Discovery 1*, 55–77.

Friedman, J. (2001). Greedy function approximation: a gradient boosting machine. *Annals of Statistics 29*, 1189–1232.

Friedman, J., T. Hastie, and R. Tibshirani (2000). Additive logistic regression: a statistical view of boosting. *Annals of statistics 28*(2), 337–374.

Friedman, J., T. Hastie, and R. Tibshirani (2008). Sparse inverse covariance estimation the graphical lasso. *Biostatistics 9*(3), 432–441.

Friedman, J., T. Hastie, and R. Tibshirani (2010, Februrary). Regularization Paths for Generalized Linear Models via Coordinate Descent. *J. of Statistical Software 33*(1).

Friedman, N. (1997b). Learning Bayesian networks in the presence of missing values and hidden variables. In *UAI.*

Friedman, N., D. Geiger, and M. Goldszmidt (1997). Bayesian network classifiers. *Machine Learning J. 29*, 131–163.

Friedman, N., D. Geiger, and N. Lotner (2000). Likelihood computation with value abstraction. In *UAI.*

Friedman, N. and D. Koller (2003). Being Bayesian about Network Structure: A Bayesian Approach to Structure Discovery in Bayesian Networks. *Machine Learning 50*, 95–126.

Friedman, N., M. Ninion, I. Pe'er, and T. Pupko (2002). A Structural EM Algorithm for Phylogenetic Inference. *J. Comp. Bio. 9*, 331–353.

Friedman, N. and Y. Singer (1999). Efficient Bayesian parameter estimation in large discrete domains. In *NIPS-11.*

Fruhwirth-Schnatter, S. (2007). *Finite Mixture and Markov Switching Models.* Springer.

Fruhwirth-Schnatter, S. and R. Fruhwirth (2010). Data Augmentation and MCMC for Binary and Multinomial Logit Models. In T. Kneib and G. Tutz (Eds.), *Statistical Modelling and Regression Structures*, pp. 111–132. Springer.

Fu, W. (1998). Penalized regressions: the bridge versus the lasso. *J. Computational and graphical statistics 7*, 397âĂŞ 416.

Fukunaga, K. (1990). *Introduction to Statistical Pattern Recognition.* Academic Press. 2nd edition.

Fukushima, K. (1975). Cognitron: a self-organizing multilayered neural network. *Biological Cybernetics 20*(6), 121–136.

Fung, R. and K. Chang (1989). Weighting and integrating evidence for stochastic simulation in Bayesian networks. In *UAI.*

Gabow, H., Z. Galil, and T. Spencer (1984). Efficient implementation of graph algorithms using contraction. In *IEEE Symposium on the Foundations of Computer Science.*

Gales, M. (2002). Maximum likelihood multiple subspace projections for hidden Markov models. *IEEE. Trans. on Speech and Audio Processing 10*(2), 37–47.

Gales, M. J. F. (1999). Semi-tied covariance matrices for hidden Markov models. *IEEE Trans. on Speech and Audio Processing 7*(3), 272–281.

Gamerman, D. (1997). Efficient sampling from the posterior distribution in generalized linear mixed models. *Statistics and Computing 7*, 57–68.

Geiger, D. and D. Heckerman (1994). Learning Gaussian networks. In *UAI*, Volume 10, pp. 235–243.

Geiger, D. and D. Heckerman (1997). A characterization of Dirchlet distributions through local and global independence. *Annals of Statistics 25*, 1344–1368.

Gelfand, A. (1996). Model determination using sampling-based methods. In Gilks, Richardson, and Spiegelhalter (Eds.), *Markov Chain Monte Carlo in Practice.* Chapman & Hall.

Gelfand, A. and A. Smith (1990). Sampling-based approaches to calculating marginal densities. *J. of the Am. Stat. Assoc. 85*, 385–409.

Gelman, A., J. Carlin, H. Stern, and D. Rubin (2004). *Bayesian data analysis.* Chapman and Hall. 2nd edition.

Gelman, A. and J. Hill (2007). *Data analysis using regression and multilevel/ hierarchical models.* Cambridge.

Gelman, A. and X.-L. Meng (1998). Simulating normalizing constants: from importance sampling to bridge sampling to path sampling. *Statisical Science 13*, 163–185.

Gelman, A. and T. Raghunathan (2001). Using conditional distributions for missing-data imputation. *Statistical Science.*

Gelman, A. and D. Rubin (1992). Inference from iterative simulation using multiple sequences. *Statistical Science 7*, 457–511.

Geman, S., E. Bienenstock, and R. Doursat (1992). Neural networks and the bias-variance dilemma. *Neural Computing 4*, 1–58.

Geman, S. and D. Geman (1984). Stochastic relaxation, Gibbs distributions, and the Bayesian restoration of images. *IEEE Trans. on Pattern Analysis and Machine Intelligence 6*(6).

Geoffrion, A. (1974). Lagrangian relaxation for integer programming. *Mathematical Programming Study 2*, 82–114.

George, E. and D. Foster (2000). Calibration and empirical bayes variable selection. *Biometrika 87*(4), 731–747.

Getoor, L. and B. Taskar (Eds.) (2007). *Introduction to Relational Statistical Learning.* MIT Press.

Geyer, C. (1992). Practical markov chain monte carlo. *Statistical Science 7*, 473–483.

Ghahramani, Z. and M. Beal (1999). Variational inference for bayesian mixtures of factor analysers. In *NIPS*.

Ghahramani, Z. and M. Beal (2000). Variational inference for Bayesian mixtures of factor analysers. In *NIPS-12*.

Ghahramani, Z. and M. Beal (2001). Propagation algorithms for variational Bayesian learning. In *NIPS-13*.

Ghahramani, Z. and G. Hinton (1996a). The EM algorithm for mixtures of factor analyzers. Technical report, Dept. of Comp. Sci., Uni. Toronto.

Ghahramani, Z. and G. Hinton (1996b). Parameter estimation for linear dynamical systems. Technical Report CRG-TR-96-2, Dept. Comp. Sci., Univ. Toronto.

Ghahramani, Z. and M. Jordan (1997). Factorial hidden Markov models. *Machine Learning 29*, 245–273.

Gilks, W. and C. Berzuini (2001). Following a moving target – Monte Carlo infernece for dynamic Bayesian models. *J. of Royal Stat. Soc. Series B 63*, 127–146.

Gilks, W., N. Best, and K. Tan (1995). Adaptive rejection Metropolis sampling. *Applied Statistics 44*, 455–472.

Gilks, W. and P. Wild (1992). Adaptive rejection sampling for Gibbs sampling. *Applied Statistics 41*, 337–348.

Girolami, M., B. Calderhead, and S. Chin (2010). Riemannian Manifold Hamiltonian Monte Carlo. *J. of Royal Stat. Soc. Series B*. To appear.

Girolami, M. and S. Rogers (2005). Hierarchic bayesian models for kernel learning. In *Intl. Conf. on Machine Learning*, pp. 241–248.

Girolami, M. and S. Rogers (2006). Variational Bayesian multinomial probit regression with Gaussian process priors. *Neural Comptuation 18*(8), 1790 – 1817.

Girshick, R., P. Felzenszwalb, and D. McAllester (2011). Object detection with grammar models. In *NIPS*.

Gittins, J. (1989). *Multi-armed Bandit Allocation Indices.* Wiley.

Giudici, P. and P. Green (1999). Decomposable graphical gaussian model determination. *Biometrika 86*(4), 785–801.

Givoni, I. E. and B. J. Frey (2009, June). A binary variable model for affinity propagation. *Neural Computation 21*(6), 1589–1600.

Globerson, A. and T. Jaakkola (2008). Fixing max-product: Convergent message passing algorithms for MAP LP-relaxations. In *NIPS*.

Glorot, X. and Y. Bengio (2010, May). Understanding the difficulty of training deep feedforward neural networks. In *AI/Statistics*, Volume 9, pp. 249–256.

Glorot, X., A. Bordes, and Y. Bengio (2011). Deep sparse rectifier neural networks. In *AI/Statistics*.

Gogate, V., W. A. Webb, and P. Domingos (2010). Learning efficient Markov networks. In *NIPS*.

Goldenberg, A., A. X. Zheng, S. E. Fienberg, and E. M. Airoldi (2009). A Survey of Statistical Network Models. *Foundations and Trends in Machine Learning*, 129–233.

Golub, G. and C. F. van Loan (1996). *Matrix computations.* Johns Hopkins University Press.

Gonen, M., W. Johnson, Y. Lu, and P. Westfall (2005, August). The Bayesian Two-Sample t Test. *The American Statistician 59*(3), 252–257.

Gonzales, T. (1985). Clustering to minimize the maximum intercluster distance. *Theor. Comp. Sci. 38*, 293–306.

Gonzalez, J., Y. Low, C. Guestrin, and D. O'Hallaron (2009). Distributed parallel inference on large factor graphs. In *UAI*.

Gorder, P. F. (2006, Nov/Dec). Neural networks show new promise for machine vision. *Computing in science & engineering 8*(6), 4–8.

Gordon, N. (1993). Novel approach to nonlinear/non-Gaussian Bayesian state estimation. *IEE Proceedings (F) 140*(2), 107–113.

Graepel, T., J. Quinonero-Candela, T. Borchert, and R. Herbrich (2010). Web-Scale Bayesian Click-Through Rate Prediction for Sponsored Search Advertising in Microsoft's Bing Search Engine. In *Intl. Conf. on Machine Learning*.

Grauman, K. and T. Darrell (2007, April). The Pyramid Match Kernel: Efficient Learning with Sets of Features. *J. of Machine Learning Research 8*, 725–760.

Green, P. (1998). Reversible jump Markov chain Monte Carlo computation and Bayesian model determination. *Biometrika 82*, 711–732.

Green, P. (2003). Tutorial on transdimensional MCMC. In P. Green, N. Hjort, and S. Richardson (Eds.), *Highly Structured Stochastic Systems.* OUP.

Green, P. and B. Silverman (1994). *Nonparametric regression and generalized linear models.* Chapman and Hall.

Greenshtein, E. and J. Park (2009). Application of Non Parametric Empirical Bayes Estimation to High Dimensional Classification. *J. of Machine Learning Research 10*, 1687–1704.

Greig, D., B. Porteous, and A. Seheult (1989). Exact maximum a posteriori estimation for binary images. *J. of Royal Stat. Soc. Series B 51*(2), 271–279.

Griffin, J. and P. Brown (2007). Bayesian adaptive lassos with non-convex penalization. Technical report, U. Kent.

Griffin, J. and P. Brown (2010). Inference with normal-gamma prior distributions in regression problems. *Bayesian Analysis 5*(1), 171–188.

Griffiths, T. and M. Steyvers (2004). Finding scientific topics. *Proc. of the National Academy of Science, USA 101*, 5228–5235.

Griffiths, T., M. Steyvers, D. Blei, and J. Tenenbaum (2004). Integrating topics and syntax. In *NIPS*.

Griffiths, T. and J. Tenenbaum (2001). Using vocabulary knowledge in bayesian multinomial estimation. In *NIPS*, pp. 1385–1392.

Griffiths, T. and J. Tenenbaum (2005). Structure and strength in causal induction. *Cognitive Psychology 51*, 334–384.

Griffiths, T. and J. Tenenbaum (2009). Theory-Based Causal Induction. *Psychological Review 116*(4), 661–716.

Grimmett, G. and D. Stirzaker (1992). *Probability and Random Processes*. Oxford.

Gu, Q., Z. Li, and J. Han (2011). Generalized Fisher Score for Feature Selection. In *UAI*.

Guan, Y., J. Dy, D. Niu, and Z. Ghahramani (2010). Variational Inference for Nonparametric Multiple Clustering. In *1st Intl. Workshop on Discovering, Summarizing and Using Multiple Clustering (MultiClust)*.

Guedon, Y. (2003). Estimating hidden semi-markov chains from discrete sequences. *J. of Computational and Graphical Statistics 12*, 604–639.

Guo, Y. (2009). Supervised exponential family principal component analysis via convex optimization. In *NIPS*.

Gustafsson, M. (2001). A probabilistic derivation of the partial least-squares algorithm. *Journal of Chemical Information and Modeling 41*, 288–294.

Guyon, I., S. Gunn, M. Nikravesh, and L. Zadeh (Eds.) (2006). *Feature Extraction: Foundations and Applications*. Springer.

Hacker, J. and P. Pierson (2010). *Winner-Take-All Politics: How Washington Made the Rich Richer-and Turned Its Back on the Middle Class*. Simon & Schuster.

Halevy, A., P. Norvig, and F. Pereira (2009). The unreasonable effectiveness of data. *IEEE Intelligent Systems 24*(2), 8–12.

Hall, P., J. T. Ormerod, and M. P. Wand (2011). Theory of Gaussian Variational Approximation for a Generalised Linear Mixed Model. *Statistica Sinica 21*, 269–389.

Hamilton, J. (1990). Analysis of time series subject to changes in regime. *J. Econometrics 45*, 39–70.

Hans, C. (2009). Bayesian Lasso regression. *Biometrika 96*(4), 835–845.

Hansen, M. and B. Yu (2001). Model selection and the principle of minimum description length. *J. of the Am. Stat. Assoc.*.

Hara, H. and A. Takimura (2008). A Localization Approach to Improve Iterative Proportional Scaling in Gaussian Graphical Models. *Communications in Statistics - Theory and Method*. to appear.

Hardin, J. and J. Hilbe (2003). *Generalized Estimating Equations*. Chapman and Hall/CRC.

Harmeling, S. and C. K. I. Williams (2011). Greedy learning of binary latent trees. *IEEE Trans. on Pattern Analysis and Machine Intelligence 33*(6), 1087–1097.

Harnard, S. (1990). The symbol grounding problem. *Physica D 42*, 335–346.

Harvey, A. C. (1990). *Forecasting, Structural Time Series Models, and the Kalman Filter*. Cambridge University Press.

Hastie, T., S. Rosset, R. Tibshirani, and J. Zhu (2004). The entire regularization path for the support vector machine. *J. of Machine Learning Research 5*, 1391–1415.

Hastie, T. and R. Tibshirani (1990). *Generalized additive models*. Chapman and Hall.

Hastie, T., R. Tibshirani, and J. Friedman (2001). *The Elements of Statistical Learning*. Springer.

Hastie, T., R. Tibshirani, and J. Friedman (2009). *The Elements of Statistical Learning*. Springer. 2nd edition.

Hastings, W. (1970). Monte carlo sampling methods using markov chains and their applications. *Biometrika 57*(1), 97–109.

Haykin, S. (1998). *Neural Networks: A Comprehensive Foundation*. Prentice Hall. 2nd Edition.

Haykin, S. (Ed.) (2001). *Kalman Filtering and Neural Networks*. Wiley.

Hazan, T. and A. Shashua (2008). Convergent message-passing algorithms for inference over general graphs with convex free energy. In *UAI*.

Hazan, T. and A. Shashua (2010). Norm-product belief propagation: primal-dual message passing for approximate inference. *IEEE Trans. on Info. Theory 56*(12), 6294–6316.

He, Y.-B. and Z. Geng (2009). Active learning of causal networks with intervention experiments and optimal designs. *J. of Machine Learning Research 10*, 2523–2547.

Heaton, M. and J. Scott (2009). Bayesian computation and the linear model. Technical report, Duke.

Heckerman, D., D. Chickering, C. Meek, R. Rounthwaite, and C. Kadie (2000). Dependency networks for density estimation, collaborative filtering, and data visualization. *J. of Machine Learning Research 1*, 49–75.

Heckerman, D., D. Geiger, and M. Chickering (1995). Learning Bayesian networks: the combination of knowledge and statistical data. *Machine Learning 20*(3), 197–243.

Heckerman, D., C. Meek, and G. Cooper (1997, February). A Bayesian approach to causal discovery. Technical Report MSR-TR-97-05, Microsoft Research.

Heckerman, D., C. Meek, and D. Koller (2004). Probabilistic models for relational data. Technical Report MSR-TR-2004-30, Microsoft Research.

Heller, K. and Z. Ghahramani (2005). Bayesian Hierarchical Clustering. In *Intl. Conf. on Machine Learning*.

Henrion, M. (1988). Propagation of uncertainty by logic sampling in Bayes' networks. In *UAI*, pp. 149–164.

Herbrich, R. (2005). On Gaussian Expectation Propagation. Technical report, MSR.

Herbrich, R., T. Minka, and T. Graepel (2007). TrueSkill: A Bayesian skill rating system. In *NIPS*.

Hertz, J., A. Krogh, and R. G. Palmer (1991). *An Introduction to the Theory of Neural Comptuation*. Addison-Wesley.

Hillar, C., J. Sohl-Dickstein, and K. Koepsell (2012, April). Efficient and optimal binary hopfield associative memory storage using minimum probability flow. Technical report.

Hinton, G. (1999). Products of experts. In *Proc. 9th Intl. Conf. on Artif. Neural Networks (ICANN)*, Volume 1, pp. 1–6.

Hinton, G. (2002). Training products of experts by minimizing contrastive divergence. *Neural Computation 14*, 1771–1800.

Hinton, G. (2010). A Practical Guide to Training Restricted Boltzmann Machines. Technical report, U. Toronto.

Hinton, G. and D. V. Camp (1993). Keeping neural networks simple by minimizing the description length of the weights. In *in Proc. of the 6th Ann. ACM Conf. on Computational Learning Theory*, pp. 5–13. ACM Press.

Hinton, G., L. Deng, D. Yu, G. Dahl, A.Mohamed, N. Jaitly, A. Senior, V. Vanhoucke, P. Nguyen, T. Sainath, and B. Kingsbury (2012, November). Deep neural networks for acoustic modeling in speech recognition. *IEEE Signal Proc. Magazine 29*(6).

Hinton, G., S. Osindero, and Y. Teh (2006). A fast learning algorithm for deep belief nets. *Neural Computation 18*, 1527–1554.

Hinton, G. and R. Salakhutdinov (2006, July). Reducing the dimensionality of data with neural networks. *Science 313*(5786), 504–507.

Hinton, G. E. and Y. Teh (2001). Discovering multiple constraints that are frequently approximately satisïñĄed. In *UAI*.

Hjort, N., C. Holmes, P. Muller, and S. Walker (Eds.) (2010). *Bayesian Nonparametrics*. Cambridge.

Hochreiter, S. and J. Schmidhuber (1997). Long short-term memory. *Neural Computation 9*(8), 1735âĂŞ1780.

Hoefling, H. (2010). A Path Algorithm for the Fused Lasso Signal Approximator. Technical report, Stanford.

Hoefling, H. and R. Tibshirani (2009). Estimation of Sparse Binary Pairwise Markov Networks using Pseudo-likelihoods. *J. of Machine Learning Research 10*.

Hoeting, J., D. Madigan, A. Raftery, and C. Volinsky (1999). Bayesian model averaging: A tutorial. *Statistical Science 4*(4).

Hoff, P. D. (2009, July). *A First Course in Bayesian Statistical Methods*. Springer.

Hoffman, M., D. Blei, and F. Bach (2010). Online learning for latent dirichlet allocation. In *NIPS*.

Hoffman, M. and A. Gelman (2011). The no-U-turn sampler: Adaptively setting path lengths in Hamiltonian Monte Carlo. Technical report, Columbia U.

Hofmann, T. (1999). Probabilistic latent semantic indexing. *Research and Development in Information Retrieval*, 50–57.

Holmes, C. and L. Held (2006). Bayesian auxiliary variable models for binary and multinomial regression. *Bayesian Analysis 1*(1), 145–168.

Honkela, A. and H. Valpola (2004). Variational Learning and Bits-Back Coding: An Information-Theoretic View to Bayesian Learning. *IEEE. Trans. on Neural Networks 15*(4).

Honkela, A., H. Valpola, and J. Karhunen (2003). Accelerating Cyclic Update Algorithms for Parameter Estimation by Pattern Searches. *Neural Processing Letters 17*, 191–203.

Hopfield, J. J. (1982, April). Neural networks and physical systems with emergent collective computational abilities. *Proc. of the National Academy of Science, USA 79*(8), 2554âĂŞ2558.

Hornik, K. (1991). Approximation capabilities of multilayer feedforward networks. *Neural Networks 4*(2), 251–257.

Horvitz, E., J. Apacible, R. Sarin, and L. Liao (2005). Prediction, Expectation, and Surprise: Methods, Designs, and Study of a Deployed Traffic Forecasting Service. In *UAI*.

Howard, R. and J. Matheson (1981). Influence diagrams. In R. Howard and J. Matheson (Eds.), *Readings on the Principles and Applications of Decision Analysis, volume II*. Strategic Decisions Group.

Hoyer, P. (2004). Non-negative matrix factorizaton with sparseness constraints. *J. of Machine Learning Research 5*, 1457–1469.

Hsu, C.-W., C.-C. Chang, and C.-J. Lin (2003). A practical guide to support vector classification.

Hu, D., L. van der Maaten, Y. Cho, L. Saul, and S. Lerner (2010). Latent Variable Models for Predicting File Dependencies in Large-Scale Software Development. In *NIPS*.

Hu, M., C. Ingram, M.Sirski, C. Pal, S. Swamy, and C. Patten (2000). A Hierarchical HMM Implementation for Vertebrate Gene Splice Site Prediction. Technical report, Dept. Computer Science, Univ. Waterloo.

Huang, J., Q. Morris, and B. Frey (2007). Bayesian inference of MicroRNA targets from sequence and expression data. *J. Comp. Bio.*.

Hubel, D. and T. Wiesel (1962). Receptive fields, binocular itneraction, and functional architecture in the cat's visual cortex. *J. Physiology 160*, 106–154.

Huber, P. (1964). Robust estimation of a location parameter. *Annals of Statistics 53*, 73âĂŞ101.

Hubert, L. and P. Arabie (1985). Comparing partitions. *J. of Classification 2*, 193–218.

Hunter, D. and R. Li (2005). Variable selection using MM algorithms. *Annals of Statistics 33*, 1617–1642.

Hunter, D. R. and K. Lange (2004). A Tutorial on MM Algorithms. *The American Statistician 58*, 30–37.

Hyafil, L. and R. Rivest (1976). Constructing Optimal Binary Decision Trees is NP-complete. *Information Processing Letters 5*(1), 15–17.

Hyvarinen, A., J. Hurri, and P. Hoyer (2009). *Natural Image Statistics: a probabilistic approach to early computational vision*. Springer.

Hyvarinen, A. and E. Oja (2000). Independent component analysis: algorithms and applications. *Neural Networks 13*, 411–430.

Ilin, A. and T. Raiko (2010). Practical Approaches to Principal Component Analysis in the Presence of Missing Values. *J. of Machine Learning Research 11*, 1957–2000.

Insua, D. R. and F. Ruggeri (Eds.) (2000). *Robust Bayesian Analysis*. Springer.

Isard, M. (2003). PAMPAS: Real-Valued Graphical Models for Computer Vision. In *CVPR*, Volume 1, pp. 613.

Isard, M. and A. Blake (1998). CONDENSATION - conditional density propagation for visual tracking. *Intl. J. of Computer Vision 29*(1), 5–18.

Jaakkola, T. (2001). Tutorial on variational approximation methods. In M. Opper and D. Saad (Eds.), *Advanced mean field methods*. MIT Press.

Jaakkola, T. and D. Haussler (1998). Exploiting generative models in discriminative classifiers. In *NIPS*, pp. 487–493.

Jaakkola, T. and M. Jordan (1996a). Computing upper and lower bounds on likelihoods in intractable networks. In *UAI*.

Jaakkola, T. and M. Jordan (1996b). A variational approach to Bayesian logistic regression problems and their extensions. In *AI + Statistics*.

Jaakkola, T. S. and M. I. Jordan (2000). Bayesian parameter estimation via variational methods. *Statistics and Computing 10*, 25–37.

Jacob, L., F. Bach, and J.-P. Vert (2008). Clustered Multi-Task Learning: a Convex Formulation. In *NIPS*.

Jain, A. and R. Dubes (1988). *Algorithms for Clustering Data*. Prentice Hall.

James, G. and T. Hastie (1998). The error coding method and PICTS. *J. of Computational and Graphical Statistics 7*(3), 377–387.

Japkowicz, N., S. Hanson, and M. Gluck (2000). Nonlinear autoassociation is not equivalent to PCA. *Neural Computation 12*, 531–545.

Jaynes, E. T. (2003). *Probability theory: the logic of science*. Cambridge university press.

Jebara, T., R. Kondor, and A. Howard (2004). Probability product kernels. *J. of Machine Learning Research 5*, 819–844.

Jeffreys, H. (1961). *Theory of Probability*. Oxford.

Jeffreys, H. (1973). *Scientific Inference*. Cambridge. Third edition.

Jelinek, F. (1997). *Statistical methods for speech recognition*. MIT Press.

Jensen, C. S., A. Kong, and U. Kjaerulff (1995). Blocking-gibbs sampling in very large probabilistic expert systems. *Intl. J. Human-Computer Studies*, 647–666.

Jensen, D., J. Neville, and B. Gallagher (2004). Why collective inference improves relational classification. In *Proc. of the Int'l Conf. on Knowledge Discovery and Data Mining*.

Jermyn, I. (2005). Invariant bayesian estimation on manifolds. *Annals of Statistics 33*(2), 583–605.

Jerrum, M. and A. Sinclair (1993). Polynomial-time approximation algorithms for the Ising model. *SIAM J. on Computing 22*, 1087–1116.

Jerrum, M. and A. Sinclair (1996). The markov chain monte carlo method: an approach to approximate counting and integration. In D. S. Hochbaum (Ed.), *Approximation Algorithms for NP-hard problems*. PWS Publishing.

Jerrum, M., A. Sinclair, and E. Vigoda (2004). A polynomial-time approximation algorithm for the permanent of a matrix with non-negative entries. *Journal of the ACM*, 671–697.

Ji, S., D. Dunson, and L. Carin (2009). Multi-task compressive sensing. *IEEE Trans. Signal Processing 57*(1).

Ji, S., L. Tang, S. Yu, and J. Ye (2010). A shared-subspace learning framework for multi-label classification. *ACM Trans. on Knowledge Discovery from Data 4*(2).

Jirousek, R. and S. Preucil (1995). On the effective implementation of the iterative proportional fitting procedure. *Computational Statistics & Data Analysis 19*, 177–189.

Joachims, T. (2006). Training Linear SVMs in Linear Time. In *Proc. of the Int'l Conf. on Knowledge Discovery and Data Mining*.

Joachims, T., T. Finley, and C.-N. Yu (2009). Cutting-Plane Training of Structural SVMs. *Machine Learning 77*(1), 27–59.

Johnson, J. K., D. M. Malioutov, and A. S. Willsky (2006). Walk-sum interpretation and analysis of gaussian belief propagation. In *NIPS*, pp. 579–586.

Johnson, M. (2005). Capacity and complexity of HMM duration modeling techniques. *Signal Processing Letters 12*(5), 407–410.

Johnson, N. (2009). A study of the NIPS feature selection challenge. Technical report, Stanford.

Johnson, V. and J. Albert (1999). *Ordinal data modeling*. Springer.

Jones, B., A. Dobra, C. Carvalho, C. Hans, C. Carter, and M. West (2005). Experiments in stochastic computation for high-dimensional graphical models. *Statistical Science 20*, 388–400.

Jordan, M. I. (2007). An introduction to probabilistic graphical models. In preparation.

Jordan, M. I. (2011). The era of big data. In *ISBA Bulletin*, Volume 18, pp. 1–3.

Jordan, M. I., Z. Ghahramani, T. S. Jaakkola, and L. K. Saul (1998). An introduction to variational methods for graphical models. In M. Jordan (Ed.), *Learning in Graphical Models*. MIT Press.

Jordan, M. I. and R. A. Jacobs (1994). Hierarchical mixtures of experts and the EM algorithm. *Neural Computation 6*, 181–214.

Journee, M., Y. Nesterov, P. Richtarik, and R. Sepulchre (2010). Generalized power method for sparse principal components analysis. *J. of Machine Learning Research 11*, 517–553.

Julier, S. and J. Uhlmann (1997). A new extension of the Kalman filter to nonlinear systems. In *Proc. of AeroSense: The 11th Intl. Symp. on Aerospace/Defence Sensing, Simulation and Controls*.

Jurafsky, D. and J. H. Martin (2000). *Speech and language processing: An Introduction to Natural Language Processing, Computational Linguistics, and Speech Recognition*. Prentice-Hall.

Jurafsky, D. and J. H. Martin (2008). *Speech and language processing: An Introduction to Natural Language Processing, Computational Linguistics, and Speech Recognition*. Prentice-Hall. 2nd edition.

Kaariainen, M. and J. Langford (2005). A Comparison of Tight Generalization Bounds. In *Intl. Conf. on Machine Learning*.

Kaelbling, L., M. Littman, and A. Moore (1996). Reinforcement learning: A survey. *J. of AI Research 4*, 237–285.

Kaelbling, L. P., M. Littman, and A. Cassandra (1998). Planning and acting in partially observable stochastic domains. *Artificial Intelligence 101*.

Kaiser, H. (1958). The varimax criterion for analytic rotation in factor analysis. *Psychometrika 23*(3).

Kakade, S., Y. W. Teh, and S. Roweis (2002). An alternate objective function for markovian fields. In *Intl. Conf. on Machine Learning*.

Kanazawa, K., D. Koller, and S. Russell (1995). Stochastic simulation algorithms for dynamic probabilistic networks. In *UAI*.

Kandel, E., J. Schwarts, and T. Jessell (2000). *Principles of Neural Science*. McGraw-Hill.

Kappen, H. and F. Rodriguez (1998). Boltzmann machine learning using mean field theory and linear response correction. In *NIPS*.

Kappes, J., B. Andres, F. Hamprecht, C. Schnoerr, S. Nowozin, D. Batra, S. Kim, B. Kausler, J. Lellmann, N. Komodakis, , and C. Rother (2013). A comparative study of modern inference techniques for discrete energy minimization problems. In *CVPR*.

Karhunen, J. and J. Joutsensalo (1995). Generalizations of principal component analysis, optimization problems, and neural networks. *Neural Networks 8*(4), 549–562.

Kass, R. and L. Wasserman (1995). A reference bayesian test for nested hypotheses and its relationship to the schwarz criterio. *J. of the Am. Stat. Assoc. 90*(431), 928–934.

Katayama, T. (2005). *Subspace Methods for Systems Identification*. Springer Verlag.

Kaufman, L. and P. Rousseeuw (1990). *Finding Groups in Data: An Introduction to Cluster Analysis*. Wiley.

Kawakatsu, H. and A. Largey (2009). EM algorithms for ordered probit models with endogenous regressors. *The Econometrics Journal 12*(1), 164–186.

Kearns, M. J. and U. V. Vazirani (1994). *An Introduction to Computational Learning Theory*. MIT Press.

Keerthi, S. and S. Sundararajan (2007). CRF versus SVM-Struct for Sequence Labeling. Technical report, Yahoo research.

Kelley, J. E. (1960). The cutting-plane method for solving convex programs. *J. of the Soc. for Industrial and Applied Math. 8*, 703–712.

Kemp, C., J. Tenenbaum, S. Niyogi, and T. Griffiths (2010). A probabilistic model of theory formation. *Cognition 114*, 165–196.

Kemp, C., J. Tenenbaum, T. Y. T. Griffiths and, and N. Ueda (2006). Learning systems of concepts with an infinite relational model. In *AAAI*.

Kersting, K., S. Natarajan, and D. Poole (2011). Statistical Relational AI: Logic, Probability and Computation. Technical report, UBC.

Khan, M. E., B. Marlin, G. Bouchard, and K. P. Murphy (2010). Variational bounds for mixed-data factor analysis. In *NIPS*.

Khan, Z., T. Balch, and F. Dellaert (2006). MCMC Data Association and Sparse Factorization Updating for Real Time Multitarget Tracking with Merged and Multiple Measurements. *IEEE Trans. on Pattern Analysis and Machine Intelligence 28*(12).

Kirkpatrick, S., C. G. Jr., and M. Vecchi (1983). Optimization by simulated annealing. *Science 220*, 671–680.

Kitagawa, G. (2004). The two-filter formula for smoothing and an implementation of the Gaussian-sum smoother. *Annals of the Institute of Statistical Mathematics 46*(4), 605–623.

Kivinen, J. and M. Warmuth (1997). Additive versus exponentiated gradient updates for linear prediction. *Info. and Computation 132*(1), 1–64.

Kjaerulff, U. (1990). Triangulation of graphs – algorithms giving small total state space. Technical Report R-90-09, Dept. of Math. and Comp. Sci., Aalborg Univ., Denmark.

Kjaerulff, U. and A. Madsen (2008). *Bayesian Networks and Influence Diagrams: A Guide to Construction and Analysis*. Springer.

Klaassen, C. and J. A. Wellner (1997). Efficient estimation in the bivariate noramal copula model: Normal margins are least favorable. *Bernoulli 3*(1), 55–77.

Klami, A. and S. Kaski (2008). Probabilistic approach to detecting dependencies between data sets. *Neurocomputing 72*, 39–46.

Klami, A., S. Virtanen, and S. Kaski (2010). Bayesian exponential family projections for coupled data sources. In *UAI*.

Kleiner, A., A. Talwalkar, P. Sarkar, and M. I. Jordan (2011). A scalable bootstrap for massive data. Technical report, UC Berkeley.

Kneser, R. and H. Ney (1995). Improved backing-off for n-gram language modeling. In *Intl. Conf. on Acoustics, Speech and Signal Proc.*, Volume 1, pp. 181–184.

Knowles, D. and T. Minka (2011). Non-conjugate message passing for multinomial and binary regression. In *NIPS*.

Ko, J. and D. Fox (2009). GP-BayesFilters: Bayesian Filtering Using Gaussian Process Prediction and Observation Models. *Autonomous Robots Journal*.

Kohn, R., M. Smith, and D. Chan (2001). Nonparametric regression using linear combinations of basis functions. *Statistical Computing 11*, 313–322.

Koivisto, M. (2006). Advances in exact Bayesian structure discovery in Bayesian networks. In *UAI*.

Koivisto, M. and K. Sood (2004). Exact Bayesian structure discovery in Bayesian networks. *J. of Machine Learning Research 5*, 549–573.

Koller, D. and N. Friedman (2009). *Probabilistic Graphical Models: Principles and Techniques.* MIT Press.

Koller, D. and U. Lerner (2001). Sampling in Factored Dynamic Systems. In A. Doucet, N. de Freitas, and N. Gordon (Eds.), *Sequential Monte Carlo Methods in Practice.* Springer.

Kolmogorov, V. (2006, October). Convergent Tree-reweighted Message Passing for Energy Minimization. *IEEE Trans. on Pattern Analysis and Machine Intelligence 28*(10), 1568–1583.

Kolmogorov, V. and M. Wainwright (2005). On optimality properties of tree-reweighted message passing. In *UAI*, pp. 316–322.

Kolmogorov, V. and R. Zabin (2004). What energy functions can be minimized via graph cuts? *IEEE Trans. on Pattern Analysis and Machine Intelligence 26*(2), 147–159.

Komodakis, N., N. Paragios, and G. Tziritas (2011). MRF Energy Minimization and Beyond via Dual Decomposition. *IEEE Trans. on Pattern Analysis and Machine Intelligence 33*(3), 531–552.

Koo, T., A. M. Rush, M. Collins, T. Jaakkola, and D. Sontag (2010). Dual Decomposition for Parsing with Non-Projective Head Automata. In *Proc. EMNLP*, pp. 1288–1298.

Koren, Y. (2009a). The bellkor solution to the netflix grand prize. Technical report, Yahoo! Research.

Koren, Y. (2009b). Collaborative filtering with temporal dynamics. In *Proc. of the Int'l Conf. on Knowledge Discovery and Data Mining*.

Koren, Y., R. Bell, and C. Volinsky (2009). Matrix factorization techniques for recommender systems. *IEEE Computer 42*(8), 30–37.

Krishnapuram, B., L. Carin, M. Figueiredo, and A. Hartemink (2005). Learning sparse bayesian classifiers: multi-class formulation, fast algorithms, and generalization bounds. *IEEE Transaction on Pattern Analysis and Machine Intelligence*.

Krizhevsky, A. and G. Hinton (2010). Using Very Deep Autoencoders for Content-Based Image Retrieval. Submitted.

Kschischang, F., B. Frey, and H.-A. Loeliger (2001, February). Factor graphs and the sum-product algorithm. *IEEE Trans Info. Theory*.

Kuan, P., G. Pan, J. A. Thomson, R. Stewart, and S. Keles (2009). A hierarchical semi-Markov model for detecting enrichment with application to ChIP-Seq experiments. Technical report, U. Wisconsin.

Kulesza, A. and F. Pereira (2007). Structured learning with approximate inference. In *NIPS*.

Kulesza, A. and B. Taskar (2011). Learning Determinantal Point Processes. In *UAI*.

Kumar, N. and A. Andreo (1998). Heteroscedastic discriminant analysis and reduced rank HMMs for improved speech recognition. *Speech Communication 26*, 283–297.

Kumar, S. and M. Hebert (2003). Discriminative random fields: A discriminative framework for contextual interaction in classification. In *Intl. Conf. on Computer Vision*.

Kuo, L. and B. Mallick (1998). Variable selection for regression models. *Sankhya Series B 60*, 65–81.

Kurihara, K., M. Welling, and N. Vlassis (2006). Accelerated variational DP mixture models. In *NIPS*.

Kushner, H. and G. Yin (2003). *Stochastic approximation and recursive algorithms and applications.* Springer.

Kuss, M. and C. Rasmussen (2005). Assessing approximate inference for binary gaussian process classification. *J. of Machine Learning Research 6*, 1679–1704.

Kwon, J. and K. Murphy (2000). Modeling freeway traffic with coupled HMMs. Technical report, Univ. California, Berkeley.

Kyung, M., J. Gill, M. Ghosh, and G. Casella (2010). Penalized Regression, Standard Errors and Bayesian Lassos. *Bayesian Analysis 5*(2), 369–412.

Lacoste-Julien, S., F. Huszar, and Z. Ghahramani (2011). Approximate inference for the loss-calibrated Bayesian. In *AI/Statistics*.

Lacoste-Julien, S., F. Sha, and M. I. Jordan (2009). DiscLDA: Discriminative learning for dimensionality reduction and classification. In *NIPS*.

Lafferty, J., A. McCallum, and F. Pereira (2001). Conditional random fields: Probabilistic models for segmenting and labeling sequence data. In *Intl. Conf. on Machine Learning*.

Lange, K., R. Little, and J. Taylor (1989). Robust statistical modeling using the t disribution. *J. of the Am. Stat. Assoc. 84*(408), 881–896.

Langville, A. and C. Meyer (2006). Updating Markov chains with an eye on Google's PageRank. *SIAM J. on Matrix Analysis and Applications 27*(4), 968–987.

Larranaga, P., C. M. H. Kuijpers, M. Poza, and R. H. Murga (1997). Decomposing bayesian networks: triangulation of the moral graph with genetic algorithms. *Statistics and Computing (UK) 7*(1), 19–34.

Lashkari, D. and P. Golland (2007). Convex clustering with examplar-based models. In *NIPS*.

Lasserre, J., C. Bishop, and T. Minka (2006). Principled hybrids of generative and discriminative models. In *CVPR*.

Lau, J. and P. Green (2006). Bayesian model-based clustering procedures. *Journal of Computational and Graphical Statistics 12*, 351–357.

Lauritzen, S. (1996). *Graphical Models*. OUP.

Lauritzen, S. (2000). Causal inference from graphical models. In D. R. C. O. E. Barndoff-Nielsen and C. Klueppelberg (Eds.), *Complex stochastic systems*. Chapman and Hall.

Lauritzen, S. and D. Nilsson (2001). Representing and solving decision problems with limited information. *Management Science 47*, 1238–1251.

Lauritzen, S. L. (1992, December). Propagation of probabilities, means and variances in mixed graphical association models. *J. of the Am. Stat. Assoc. 87*(420), 1098–1108.

Lauritzen, S. L. (1995). The EM algorithm for graphical association models with missing data. *Computational Statistics and Data Analysis 19*, 191–201.

Lauritzen, S. L. and D. J. Spiegelhalter (1988). Local computations with probabilities on graphical structures and their applications to expert systems. *J. R. Stat. Soc. B B*(50), 127–224.

Law, E., B. Settles, and T. Mitchell (2010). Learning to tag from open vocabulary labels. In *Proc. European Conf. on Machine Learning*.

Law, M., M. Figueiredo, and A. Jain (2004). Simultaneous Feature Selection and Clustering Using Mixture Models. *IEEE Trans. on Pattern Analysis and Machine Intelligence 26*(4).

Lawrence, N. D. (2005). Probabilistic non-linear principal component analysis with gaussian process latent variable models. *J. of Machine Learning Research 6*, 1783–1816.

Lawrence, N. D. (2012). A unifying probabilistic perspective for spectral dimensionality reduction: insights and new models. *J. of Machine Learning Research 13*, 1609–1638.

Learned-Miller, E. (2004). Hyperspacings and the estimation of information theoretic quantities. Technical Report 04-104, U. Mass. Amherst Comp. Sci. Dept.

LeCun, Y., B. Boser, J. S. Denker, D. Henderson, R. E. Howard, W. Hubbard, and L. D. Jackel (1989, Winter). Backpropagation applied to handwritten zip code recognition. *Neural Computation 1*(4), 541–551.

LeCun, Y., L. Bottou, Y. Bengio, and P. Haffner (1998, November). Gradient-based learning applied to document recognition. *Proceedings of the IEEE 86*(11), 2278–2324.

LeCun, Y., S. Chopra, R. Hadsell, F.-J. Huang, and M.-A. Ranzato (2006). A tutorial on energy-based learning. In B. et al. (Ed.), *Predicting Structured Outputs*. MIT press.

Ledoit, O. and M. Wolf (2004a). Honey, I Shrunk the Sample Covariance Matrix. *J. of Portfolio Management 31*(1).

Ledoit, O. and M. Wolf (2004b). A well-conditioned estimator for large-dimensional covariance matrices. *J. of Multivariate Analysis 88*(2), 365–411.

Lee, A., F. Caron, A. Doucet, and C. Holmes (2010). A hierarchical bayesian framework for constructing sparsity-inducing priors. Technical report, U. Oxford.

Lee, A., F. Caron, A. Doucet, and C. Holmes (2011). Bayesian Sparsity-Path-Analysis of Genetic Association Signal using Generalized t Prior. Technical report, U. Oxford.

Lee, D. and S. Seung (2001). Algorithms for non-negative matrix factorization. In *NIPS*.

Lee, H., R. Grosse, R. Ranganath, and A. Ng (2009). Convolutional deep belief networks for scalable unsupervised learning of hierarchical representations. In *Intl. Conf. on Machine Learning*.

Lee, H., Y. Largman, P. Pham, and A. Ng (2009). Unsupervised feature learning for audio classification using convolutional deep belief networks. In *NIPS*.

Lee, S.-I., V. Ganapathi, and D. Koller (2006). Efficient structure learning of Markov networks using L1-regularization. In *NIPS*.

Lee, T. S. and D. Mumford (2003). Hierarchical Bayesian inference in the visual cortex. *J. of Optical Society of America A 20*(7), 1434–1448.

Lenk, P., W. S. DeSarbo, P. Green, and M. Young (1996). Hierarchical Bayes Conjoint Analysis: Recovery of Partworth Heterogeneity from Reduced Experimental Designs. *Marketing Science 15*(2), 173–191.

Lenkoski, A. and A. Dobra (2008). Bayesian structural learning and estimation in Gaussian graphical models. Technical Report 545, Department of Statistics, University of Washington.

Lepar, V. and P. P. Shenoy (1998). A Comparison of Lauritzen-Spiegelhalter, Hugin and Shenoy-Shafer Architectures for Computing Marginals of Probability Distributions. In G. Cooper and S. Moral (Eds.), *UAI*, pp. 328–337. Morgan Kaufmann.

Lerner, U. and R. Parr (2001). Inference in hybrid networks: Theoretical limits and practical algorithms. In *UAI*.

Leslie, C., E. Eskin, A. Cohen, J. Weston, and W. Noble (2003). Mismatch string kernels for discriminative protein classification. *Bioinformatics 1*, 1–10.

Levy, S. (2011). *In The Plex: How Google Thinks, Works, and Shapes Our Lives*. Simon & Schuster.

Li, L., W. Chu, J. Langford, and X. Wang (2011). Unbiased offline evaluation of contextual-bandit-based news article recommendation algorithms. In *WSDM*.

Liang, F., S. Mukherjee, and M. West (2007). Understanding the use of unlabelled data in predictive modelling. *Statistical Science 22*, 189–205.

Liang, F., R. Paulo, G. Molina, M. Clyde, and J. Berger (2008). Mixtures of g-priors for Bayesian Variable Selection. *J. of the Am. Stat. Assoc. 103*(481), 410–423.

Liang, P. and M. I. Jordan (2008). An asymptotic analysis of generative, discriminative, and pseudolikelihood estimators. In *International Conference on Machine Learning (ICML)*.

Liang, P. and D. Klein (2009). Online EM for Unsupervised Models. In *Proc. NAACL Conference*.

Liao, L., D. J. Patterson, D. Fox, and H. Kautz (2007). Learning and Inferring Transportation Routines. *Artificial Intelligence 171*(5), 311–331.

Lindley, D. (1982). Scoring rules and the inevetability of probability. *ISI Review 50*, 1–26.

Lindley, D. (2006). *Understanding Uncertainty*. Wiley.

Lindley, D. V. (1972). *Bayesian Statistics: A Review*. SIAM.

Lindley, D. V. and L. D. Phillips (1976). Inference for a Bernoulli Process (A Bayesian View). *The American Statistician 30*(3), 112–119.

Lindsay, B. (1988). Composite likelihood methods. *Contemporary Mathematics 80*(1), 221–239.

Lipton, R. J. and R. E. Tarjan (1979). A separator theorem for planar graphs. *SIAM Journal of Applied Math 36*, 177–189.

Little, R. J. and D. B. Rubin (1987). *Statistical Analysis with Missing Data*. New York: Wiley and Son.

Liu, C. and D. Rubin (1995). ML Estimation of the T distribution using EM and its extensions, ECM and ECME. *Statistica Sinica 5*, 19–39.

Liu, H., J. Lafferty, and L. Wasserman (2009). The nonparanormal: Semiparametric estimation of high dimensional undirected graphs. *J. of Machine Learning Research 10*, 2295–2328.

Liu, J. (2001). *Monte Carlo Strategies in Scientific Computation*. Springer.

Liu, J. S., W. H. Wong, and A. Kong (1994). Covariance structure of the gibbs sampler with applications to the comparisons of estimators and augmentation schemes. *Biometrika 81*(1), 27–40.

Liu, T.-Y. (2009). Learning to rank for information retrieval. *Foundations and Trends in Information Retrieval 3*(3), 225–331.

Lizotte, D. (2008). *Practical Bayesian optimization*. Ph.D. thesis, U. Alberta.

Ljung, L. (1987). *System Identification: Theory for the User*. Prentice Hall.

Lo, C. H. (2009). *Statistical methods for high throughput genomics*. Ph.D. thesis, UBC.

Lo, K., F. Hahne, R. Brinkman, R. Ryan, and R. Gottardo (2009, May). flowclust: a bioconductor package for automated gating of flow cytometry data. *BMC Bioinformatics 10*, 145+.

Lopes, H. and M. West (2004). Bayesian model assessment in factor analysis. *Statisica Sinica 14*, 41–67.

Lowe, D. G. (1999). Object recognition from local scale-invariant features. In *Proc. of the International Conference on Computer Vision ICCV, Corfu*, pp. 1150–1157.

Luce, R. (1959). *Individual choice behavior: A theoretical analysis*. Wiley.

Lunn, D., N. Best, and J. Whittaker (2009). Generic reversible jump MCMC using graphical models. *Statistics and Computing 19*(4), 395–408.

Lunn, D., C. Jackson, N. Best, A. Thomas, and D. Spiegelhalter (2012). *The BUGS Book: A practical Introduction to Bayesian Analysis*. CRC Press.

Lunn, D., A. Thomas, N. Best, and D. Spiegelhalter (2000). WinBUGS – a Bayesian modelling framework: concepts, structure, and extensibility. *Statistics and Computing 10*, 325–337.

Ma, H., H. Yang, M. Lyu, and I. King (2008). SoRec: Social recommendation using probabilistic matrix factorization. In *Proc. of 17th Conf. on Information and Knowledge Management*.

Ma, S., C. Ji, and J. Farmer (1997). An efficient EM-based training algorithm for feedforward neural networks. *Neural Networks 10*(2), 243–256.

Maathuis, M., D. Colombo, M. Kalisch, and P. BÃijhlmann (2010). Predicting causal effects in large-scale systems from observational data. *Nature Methods 7*, 247–248.

Maathuis, M., M. Kalisch, and P. BÃijhlmann (2009). Estimating high-dimensional intervention effects from observational data. *Annals of Statistics 37*, 3133–3164.

MacKay, D. (1992). Bayesian interpolation. *Neural Computation 4*, 415–447.

MacKay, D. (1995a). Developments in probabilistic modeling with neural networks — ensemble learning. In *Proc. 3rd Ann. Symp. Neural Networks*.

MacKay, D. (1995b). Probable networks and plausible predictions — a review of practical Bayesian methods for supervised neural networks. *Network*.

MacKay, D. (1997). Ensemble learning for Hidden Markov Models. Technical report, U. Cambridge.

MacKay, D. (1999). Comparison of approximate methods for handling hyperparameters. *Neural Computation 11*(5), 1035–1068.

MacKay, D. (2003). *Information Theory, Inference, and Learning Algorithms*. Cambridge University Press.

MacKay, D. and L. Peto (1995). A hierarchical dirichlet language model. *Natural Language Engineering 1*(3), 289–307.

Macnaughton-Smith, P., W. T. Williams, M. B. Dale, and G. Mockett (1964). Dissimilarity analysis: a new technique of hierarchical sub-division. *Nature 202*, 1034 – 1035.

Madeira, S. C. and A. L. Oliveira (2004). Biclustering algorithms for biological data analysis: A survey. *IEEE/ACM Transactions on Computational Biology and Bioinformatics 1*(1), 24–45.

Madigan, D. and A. Raftery (1994). Model selection and accounting for model uncertainty in graphical models using Occam's window. *J. of the Am. Stat. Assoc. 89*, 1535–1546.

Madsen, R., D. Kauchak, and C. Elkan (2005). Modeling word burstiness using the Dirichlet distribution. In *Intl. Conf. on Machine Learning*.

Mairal, J., F. Bach, J. Ponce, and G. Sapiro (2010). Online learning for matrix factorization and sparse coding. *J. of Machine Learning Research 11*, 19–60.

Mairal, J., M. Elad, and G. Sapiro (2008). Sparse representation for color image restoration. *IEEE Trans. on Image Processing 17*(1), 53–69.

Malioutov, D., J. Johnson, and A. Willsky (2006). Walk-sums and belief propagation in gaussian graphical models. *J. of Machine Learning Research 7*, 2003–2030.

Mallat, S., G. Davis, and Z. Zhang (1994, July). Adaptive time-frequency decompositions. *SPIE Journal of Optical Engineering 33*, 2183–2919.

Mallat, S. and Z. Zhang (1993). Matching pursuits with time-frequency dictionaries. *IEEE Transactions on Signal Processing 41*(12), 3397–3415.

Malouf, R. (2002). A comparison of algorithms for maximum entropy parameter estimation. In *Proc. Sixth Conference on Natural Language Learning (CoNLL-2002)*, pp. 49–55.

Manning, C., P. Raghavan, and H. Schuetze (2008). *Introduction to Information Retrieval*. Cambridge University Press.

Manning, C. and H. Schuetze (1999). *Foundations of statistical natural language processing*. MIT Press.

Mansinghka, V., D. Roy, R. Rifkin, and J. Tenenbaum (2007). AClass: An online algorithm for generative classification. In *AI/Statistics*.

Mansinghka, V., P. Shafto, E. Jonas, C. Petschulat, and J. Tenenbaum (2011). Cross-Categorization: A Nonparametric Bayesian Method for Modeling Heterogeneous, High Dimensional Data. Technical report, MIT.

Margolin, A., I. Nemenman, K. Basso, C. Wiggins, G. Stolovitzky, and R. F. abd A. Califano (2006). ARACNE: An Algorithm for the Reconstruction of Gene Regulatory Networks in a Mammalian Cellular Context. *BMC Bioinformatics 7*.

Marin, J.-M. and C. Robert (2007). *Bayesian Core: a practical approach to computational Bayesian statistics*. Springer.

Marks, T. K. and J. R. Movellan (2001). Diffusion networks, products of experts, and factor analysis. Technical report, University of California San Diego.

Marlin, B. (2003). Modeling user rating profiles for collaborative filtering. In *NIPS*.

Marlin, B. (2008). *Missing Data Problems in Machine Learning*. Ph.D. thesis, U. Toronto.

Marlin, B., E. Khan, and K. Murphy (2011). Piecewise Bounds for Estimating Bernoulli-Logistic Latent Gaussian Models. In *Intl. Conf. on Machine Learning*.

Marlin, B. and R. Zemel (2009). Collaborative prediction and ranking with non-random missing data. In *Proc. of the 3rd ACM Conference on Recommender Systems*.

Marlin, B. M., K. Swersky, B. Chen, and N. de Freitas (2010). Inductive principles for restricted boltzmann machine learning. In *AI/Statistics*.

Marroquin, J., S. Mitter, and T. Poggio (1987). Probabilistic solution of ill-posed problems in computational vision. *J. of the Am. Stat. Assoc. 82*(297), 76–89.

Martens, J. (2010). Deep learning via hessian-free optimization. In *Intl. Conf. on Machine Learning*.

Maruyama, Y. and E. George (2008). A g-prior extension for $p > n$. Technical report, U. Tokyo.

Mason, L., J. Baxter, P. Bartlett, and M. Frean (2000). Boosting algorithms as gradient descent. In *NIPS*, Volume 12, pp. 512–518.

Matthews, R. (1998). Bayesian Critique of Statistics in Health: The Great Health Hoax.

Maybeck, P. (1979). *Stochastic models, estimation, and control*. Academic Press.

Mazumder, R. and T. Hastie (2012). The Graphical Lasso: New Insights and Alternatives. Technical report, Stanford Dept. Statistics.

McAuliffe, J., D. Blei, and M. Jordan (2006). Nonparametric empirical bayes for the dirichlet process mixture model. *Statistics and Computing 16*(1), 5–14.

McCallum, A. (2003). Efficiently inducing features of conditional random fields. In *UAI*.

McCallum, A., D. Freitag, and F. Pereira (2000). Maximum Entropy Markov Models for Information Extraction and Segmentation. In *Intl. Conf. on Machine Learning*.

McCallum, A. and K. Nigam (1998). A comparison of event models for naive Bayes text classification. In *AAAI/ICML workshop on Learning for Text Categorization*.

McCray, A. (2003). An upper level ontology for the biomedical domain. *Comparative and Functional Genomics 4*, 80–84.

McCullagh, P. and J. Nelder (1989). *Generalized linear models*. Chapman and Hall. 2nd edition.

McCulloch, W. and W. Pitts (1943). A logical calculus of the ideas immanent in nervous activity. *Bulletin of Mathematical Biophysics 5*, 115–137.

McDonald, J. and W. Newey (1988). Partially Adaptive Estimation of Regression Models via the Generalized t Distribution. *Econometric Theory 4*(3), 428–445.

McEliece, R. J., D. J. C. MacKay, and J. F. Cheng (1998). Turbo decoding as an instance of Pearl's 'belief propagation' algorithm. *IEEE J. on Selected Areas in Comm. 16*(2), 140–152.

McFadden, D. (1974). Conditional logit analysis of qualitative choice behavior. In P. Zarembka (Ed.), *Frontiers in econometrics*, pp. 105–142. Academic Press.

McGrayne, S. B. (2011). *The theory that would not die: how Bayes' rule cracked the enigma code, hunted down Russian submarines, and emerged triumphant from two centuries of controversy*. Yale University Press.

McKay, B. D., F. E. Oggier, G. F. Royle, N. J. A. Sloane, I. M. Wanless, and H. S. Wilf (2004). Acyclic digraphs and eigenvalues of (0,1)-matrices. *J. Integer Sequences 7*(04.3.3).

McLachlan, G. J. and T. Krishnan (1997). *The EM Algorithm and Extensions*. Wiley.

Meek, C. and D. Heckerman (1997). Structure and parameter learning for causal independence and causal interaction models. In *UAI*, pp. 366–375.

Meek, C., B. Thiesson, and D. Heckerman (2002). Staged mixture modelling and boosting. In *UAI*, San Francisco, CA, pp. 335–343. Morgan Kaufmann.

Meila, M. (2001). A random walks view of spectral segmentation. In *AI/Statistics*.

Meila, M. (2005). Comparing clusterings: an axiomatic view. In *Intl. Conf. on Machine Learning*.

Meila, M. and T. Jaakkola (2006). Tractable Bayesian learning of tree belief networks. *Statistics and Computing 16*, 77–92.

Meila, M. and M. I. Jordan (2000). Learning with mixtures of trees. *J. of Machine Learning Research 1*, 1–48.

Meinshausen, N. (2005). A note on the lasso for gaussian graphical model selection. Technical report, ETH Seminar fur Statistik.

Meinshausen, N. and P. Buehlmann (2010). Stability selection. *J. of Royal Stat. Soc. Series B 72*, 417–473.

Meinshausen, N. and P. Buhlmann (2006). High dimensional graphs and variable selection with the lasso. *The Annals of Statistics 34*, 1436–1462.

Meltzer, T., C. Yanover, and Y. Weiss (2005). Globally optimal solutions for energy minimization in stereo vision using reweighted belief propagation. In *ICCV*, pp. 428–435.

Meng, X. L. and D. van Dyk (1997). The EM algorithm — an old folk song sung to a fast new tune (with Discussion). *J. Royal Stat. Soc. B 59*, 511–567.

Mesot, B. and D. Barber (2009). A Simple Alternative Derivation of the Expectation Correction Algorithm. *IEEE Signal Processing Letters 16*(1), 121–124.

Metropolis, N., A. Rosenbluth, M. Rosenbluth, A. Teller, and E. Teller (1953). Equation of state calculations by fast computing machines. *J. of Chemical Physics 21*, 1087–1092.

Metz, C. (2010). Google behavioral ad targeter is a Smart Ass. *The Register*.

Mikolov, T., D. Anoop, K. Stefan, B. Lukas, and C. Jan (2011). Empirical evaluation and combination of advanced language modeling techniques. In *Proc. 12th Annual Conf. of the Intl. Speech Communication Association (INTERSPEECH)*.

Miller, A. (2002). *Subset selection in regression*. Chapman and Hall. 2nd edition.

Mimno, D. and A. McCallum (2008). Topic models conditioned on arbitrary features with dirichlet-multinomial regression. In *UAI*.

Minka, T. (1999). Pathologies of orthodox statisics. Technical report, MIT Media Lab.

Minka, T. (2000a). Automatical choice of dimensionality for PCA. Technical report, MIT.

Minka, T. (2000b). Bayesian linear regression. Technical report, MIT.

Minka, T. (2000c). Bayesian model averaging is not model combination. Technical report, MIT Media Lab.

Minka, T. (2000d). Empirical risk minimization is an incomplete inductive principle. Technical report, MIT.

Minka, T. (2000e). Estimating a Dirichlet distribution. Technical report, MIT.

Minka, T. (2000f). Inferring a Gaussian distribution. Technical report, MIT.

Minka, T. (2001a). Bayesian inference of a uniform distribution. Technical report, MIT.

Minka, T. (2001b). Empirical Risk Minimization is an incomplete inductive principle. Technical report, MIT.

Minka, T. (2001c). Expectation propagation for approximate Bayesian inference. In *UAI*.

Minka, T. (2001d). *A family of algorithms for approximate Bayesian inference*. Ph.D. thesis, MIT.

Minka, T. (2001e). Statistical approaches to learning and discovery 10-602: Homework assignment 2, question 5. Technical report, CMU.

Minka, T. (2003). A comparison of numerical optimizers for logistic regression. Technical report, MSR.

Minka, T. (2005). Divergence measures and message passing. Technical report, MSR Cambridge.

Minka, T. and Y. Qi (2003). Tree-structured approximations by expectation propagation. In *NIPS*.

Minka, T., J. Winn, J. Guiver, and D. Knowles (2010). Infer.NET 2.4. Microsoft Research Cambridge. http://research.microsoft.com/infernet.

Minsky, M. and S. Papert (1969). *Perceptrons*. MIT Press.

Mitchell, T. (1997). *Machine Learning*. McGraw Hill.

Mitchell, T. and J. Beauchamp (1988). Bayesian Variable Selection in Linear Regression. *J. of the Am. Stat. Assoc. 83*, 1023–1036.

Mobahi, H., R. Collobert, and J. Weston (2009). Deep learning from temporal coherence in video. In *Intl. Conf. on Machine Learning*.

Mockus, J., W. Eddy, A. Mockus, L. Mockus, and G. Reklaitis (1996). *Bayesian Heuristic Approach to Discrete and Global Optimization: Algorithms, Visualization, Software, and Applications*. Kluwer.

Moghaddam, B., A. Gruber, Y. Weiss, and S. Avidan (2008). Sparse regression as a sparse eigenvalue problem. In *Information Theory & Applications Workshop (ITA'08)*.

Moghaddam, B., B. Marlin, E. Khan, and K. Murphy (2009). Accelerating Bayesian Structural Inference for Non-Decomposable Gaussian Graphical Models. In *NIPS*.

Moghaddam, B. and A. Pentland (1995). Probabilistic visual learning for object detection. In *Intl. Conf. on Computer Vision*.

Mohamed, S., K. Heller, and Z. Ghahramani (2008). Bayesian Exponential Family PCA. In *NIPS*.

Moler, C. (2004). *Numerical Computing with MATLAB*. SIAM.

Mooij, J. and H. Kappen (2005). Sufficient conditions for convergence of loopy belief propagation. In *UAI*.

Morris, R. D., X. Descombes, and J. Zerubia (1996). The Ising/Potts model is not well suited to segmentation tasks. In *IEEE DSP Workshop*.

Mosterman, P. J. and G. Biswas (1999). Diagnosis of continuous valued systems in transient operating regions. *IEEE Trans. on Systems, Man, and Cybernetics, Part A 29*(6), 554–565.

Moulines, E., J.-F. Cardoso, and E. Gassiat (1997). Maximum likelihood for blind separation and deconvolution of noisy signals using mixture models. In *Proc. IEEE Int. Conf. on Acoustics, Speech and Signal Processing (ICASSP'97)*, Munich, Germany, pp. 3617–3620.

Muller, P., G. Parmigiani, C. Robert, and J. Rousseau (2004). Optimal sample size for multiple testing: the case of gene expression microarrays. *J. of the Am. Stat. Assoc. 99*, 990–1001.

Mumford, D. (1994). Neuronal architectures for pattern-theoretic problems. In C. Koch and J. Davis (Eds.), *Large Scale Neuronal Theories of the Brain*. MIT Press.

Murphy, K. (2000). Bayesian map learning in dynamic environments. In *NIPS*, Volume 12.

Murphy, K. and M. Paskin (2001). Linear time inference in hierarchical HMMs. In *NIPS*.

Murphy, K., Y. Weiss, and M. Jordan (1999). Loopy belief propagation for approximate inference: an empirical study. In *UAI*.

Murphy, K. P. (1998). Filtering and smoothing in linear dynamical systems using the junction tree algorithm. Technical report, U.C. Berkeley, Dept. Comp. Sci.

Murray, I. and Z. Ghahramani (2005). A note on the evidence and bayesian occam's razor. Technical report, Gatsby.

Musso, C., N. Oudjane, and F. LeGland (2001). Improving regularized particle filters. In A. Doucet, J. F. G. de Freitas, and N. Gordon (Eds.), *Sequential Monte Carlo Methods in Practice*. Springer.

Nabney, I. (2001). *NETLAB: algorithms for pattern recognition*. Springer.

Neal, R. (1992). Connectionist learning of belief networks. *Artificial Intelligence 56*, 71–113.

Neal, R. (1993). Probabilistic Inference Using Markov Chain Monte Carlo Methods. Technical report, Univ. Toronto.

Neal, R. (1996). *Bayesian learning for neural networks*. Springer.

Neal, R. (1997). Monte Carlo Implementation of Gaussian Process Models for Bayesian Regression and Classification. Technical Report 9702, U. Toronto.

Neal, R. (1998). Erroneous Results in 'Marginal Likelihood from the Gibbs Output'. Technical report, U. Toronto.

Neal, R. (2000). Markov Chain Sampling Methods for Dirichlet Process Mixture Models. *J. of Computational and Graphical Statistics 9*(2), 249–265.

Neal, R. (2003a). Slice sampling. *Annals of Statistics 31*(3), 7–5–767.

Neal, R. (2010). MCMC using Hamiltonian Dynamics. In S. Brooks, A. Gelman, G. Jones, and X.-L. Meng (Eds.), *Handbook of Markov Chain Monte Carlo*. Chapman & Hall.

Neal, R. and D. MacKay (1998). Likelihood-based boosting. Technical report, U. Toronto.

Neal, R. and J. Zhang (2006). High dimensional classification Bayesian neural networks and Dirichlet diffusion trees. In I. Guyon, S. Gunn, M. Nikravesh, and L. Zadeh (Eds.), *Feature Extraction*. Springer.

Neal, R. M. (2001). Annealed importance sampling. *Statistics and Computing 11*, 125–139.

Neal, R. M. (2003b). Density Modeling and Clustering using Dirichlet Diffusion Trees. In J. M. Bernardo et al. (Eds.), *Bayesian Statistics 7*, pp. 619–629. Oxford University Press.

Neal, R. M. and G. E. Hinton (1998). A new view of the EM algorithm that justifies incremental and other variants. In M. Jordan (Ed.), *Learning in Graphical Models*. MIT Press.

Neapolitan, R. (2003). *Learning Bayesian Networks*. Prentice Hall.

Nefian, A., L. Liang, X. Pi, X. Liu, and K. Murphy (2002). Dynamic Bayesian Networks for Audio-Visual Speech Recognition. *J. Applied Signal Processing*.

Nemirovski, A. and D. Yudin (1978). On Cezari's convergence of the steepest descent method for approximating saddle points of convex-concave functions. *Soviet Math. Dokl. 19*.

Nesterov, Y. (2004). *Introductory Lectures on Convex Optimization. A basic course*. Kluwer.

Newton, M., D. Noueiry, D. Sarkar, and P. Ahlquist (2004). Detecting differential gene expression with a semiparametric hierarchical mixture method. *Biostatistics 5*, 155–176.

Newton, M. and A. Raftery (1994). Approximate Bayesian Inference with the Weighted Likelihood Bootstrap. *J. of Royal Stat. Soc. Series B 56*(1), 3–48.

Ng, A., M. Jordan, and Y. Weiss (2001). On Spectral Clustering: Analysis and an algorithm. In *NIPS*.

Ng, A. Y. and M. I. Jordan (2002). On discriminative vs. generative classifiers: A comparison of logistic regression and naive bayes. In *NIPS-14*.

Nickisch, H. and C. Rasmussen (2008). Approximations for binary gaussian process classification. *J. of Machine Learning Research 9*, 2035–2078.

Nilsson, D. (1998). An efficient algorithm for finding the M most probable configurations in a probabilistic expert system. *Statistics and Computing 8*, 159–173.

Nilsson, D. and J. Goldberger (2001). Sequentially finding the N-Best List in Hidden Markov Models. In *Intl. Joint Conf. on AI*, pp. 1280–1285.

Nocedal, J. and S. Wright (2006). *Numerical Optimization*. Springer.

Nowicki, K. and T. A. B. Snijders (2001). Estimation and prediction for stochastic blockstructures. *Journal of the American Statistical Association 96*(455), 1077–1087.

Nowlan, S. and G. Hinton (1992). Simplifying neural networks by soft weight sharing. *Neural Computation 4*(4), 473–493.

Nowozin, S. and C. Lampert (2011). Structured learning and prediction in computer vision. *Foundations and Trends in Computer Graphics and Vision 6*(3-4).

Nummiaro, K., E. Koller-Meier, and L. V. Gool (2003). An adaptive color-based particle filter. *Image and Vision Computing 21*(1), 99–110.

Obozinski, G., B. Taskar, and M. I. Jordan (2007). Joint covariate selection for grouped classification. Technical report, UC Berkeley.

Oh, M.-S. and J. Berger (1992). Adaptive importance sampling in Monte Carlo integration. *J. of Statistical Computation and Simulation 41*(3), 143 – 168.

Oh, S., S. Russell, and S. Sastry (2009). Markov Chain Monte Carlo Data Association for Multi-Target Tracking. *IEEE Trans. on Automatic Control 54*(3), 481–497.

O'Hagan, A. (1978). Curve fitting and optimal design for prediction. *J. of Royal Stat. Soc. Series B 40*, 1–42.

O'Hara, R. and M. Sillanpaa (2009). A Review of Bayesian Variable Selection Methods: What, How and Which. *Bayesian Analysis 4*(1), 85–118.

Olshausen, B. A. and D. J. Field (1996). Emergence of simple cell receptive field properties by learning a sparse code for natural images. *Nature 381*, 607–609.

Opper, M. (1998). A Bayesian approach to online learning. In D. Saad (Ed.), *On-line learning in neural networks*. Cambridge.

Opper, M. and C. Archambeau (2009). The variational Gaussian approximation revisited. *Neural Computation 21*(3), 786–792.

Opper, M. and D. Saad (Eds.) (2001). *Advanced mean field methods: theory and practice*. MIT Press.

Osborne, M. R., B. Presnell, and B. A. Turlach (2000a). A new approach to variable selection in least squares problems. *IMA Journal of Numerical Analysis 20*(3), 389–403.

Osborne, M. R., B. Presnell, and B. A. Turlach (2000b). On the lasso and its dual. *J. Computational and graphical statistics 9*, 319–337.

Ostendorf, M., V. Digalakis, and O. Kimball (1996). From HMMs to segment models: a unified view of stochastic modeling for speech recognition. *IEEE Trans. on Speech and Audio Processing 4*(5), 360–378.

Overschee, P. V. and B. D. Moor (1996). *Subspace Identification for Linear Systems: Theory, Implementation, Applications*. Kluwer Academic Publishers.

Paatero, P. and U. Tapper (1994). Positive matrix factorization: A nonnegative factor model with optimal utilization of error estimates of data values. *Environmetrics 5*, 111–126.

Padadimitriou, C. and K. Steiglitz (1982). *Combinatorial optimization: Algorithms and Complexity*. Prentice Hall.

Paisley, J. and L. Carin (2009). Nonparametric factor analysis with beta process priors. In *Intl. Conf. on Machine Learning*.

Palmer, S. (1999). *Vision Science: Photons to Phenomenology*. MIT Press.

Parise, S. and M. Welling (2005). Learning in Markov Random Fields: An Empirical Study. In *Joint Statistical Meeting*.

Park, T. and G. Casella (2008). The Bayesian Lasso. *J. of the Am. Stat. Assoc. 103*(482), 681–686.

Parviainen, P. and M. Koivisto (2011). Ancestor relations in the presence of unobserved variables. In *Proc. European Conf. on Machine Learning*.

Paskin, M. (2003). Thin junction tree filters for simultaneous localization and mapping. In *Intl. Joint Conf. on AI*.

Pearl, J. (1988). *Probabilistic Reasoning in Intelligent Systems: Networks of Plausible Inference*. Morgan Kaufmann.

Pearl, J. (2000). *Causality: Models, Reasoning and Inference*. Cambridge Univ. Press.

Pearl, J. and T. Verma (1991). A theory of inferred causation. In *Knowledge Representation*, pp. 441–452.

Pe'er, D. (2005, April). Bayesian network analysis of signaling networks: a primer. *Science STKE 281*, 14.

Pena, J. (2013). Reading dependencies from covariance graphs. *Intl. J. of Approximate Reasoning 54*(1).

Peng, F., R. Jacobs, and M. Tanner (1996). Bayesian Inference in Mixtures-of-Experts and Hierarchical Mixtures-of-Experts Models With an Application to Speech Recognition. *J. of the Am. Stat. Assoc. 91*(435), 953–960.

Petris, G., S. Petrone, and P. Campagnoli (2009). *Dynamic linear models with R*. Springer.

Pham, D.-T. and P. Garrat (1997). Blind separation of mixture of independent sources through a quasi-maximum likelihood approach. *IEEE Trans. on Signal Processing 45*(7), 1712–1725.

Pietra, S. D., V. D. Pietra, and J. Lafferty (1997). Inducing features of random fields. *IEEE Trans. on Pattern Analysis and Machine Intelligence 19*(4).

Plackett, R. (1975). The analysis of permutations. *Applied Stat. 24*, 193–202.

Platt, J. (1998). Using analytic QP and sparseness to speed training of support vector machines. In *NIPS*.

Platt, J. (2000). Probabilities for sv machines. In A. Smola, P. Bartlett, B. Schoelkopf, and D. Schuurmans (Eds.), *Advances in Large Margin Classifiers*. MIT Press.

Platt, J., N. Cristianini, and J. Shawe-Taylor (2000). Large margin DAGs for multiclass classification. In *NIPS*, Volume 12, pp. 547–553.

Plummer, M. (2003). JAGS: A Program for Analysis of Bayesian Graphical Models Using Gibbs Sampling. In *Proc. 3rd Intl. Workshop on Distributed Statistical Computing*.

Polson, N. and S. Scott (2011). Data augmentation for support vector machines. *Bayesian Analysis 6*(1), 1–124.

Pontil, M., S. Mukherjee, and F. Girosi (1998). On the Noise Model of Support Vector Machine Regression. Technical report, MIT AI Lab.

Poon, H. and P. Domingos (2011). Sum-product networks: A new deep architecture. In *UAI*. Java code at http://alchemy.cs.washington.edu/spn/. Short intro at http://lessoned.blogspot.com/2011/10/intro-to-sum-product-networks.html.

Pourahmadi, M. (2004). Simultaneous Modelling of Covariance Matrices: GLM, Bayesian and Nonparametric Perspectives. Technical report, Northern Illinois University.

Prado, R. and M. West (2010). *Time Series: Modelling, Computation and Inference.* CRC Press.

Press, S. J. (2005). *Applied multivariate analysis, using Bayesian and frequentist methods of inference.* Dover. Second edition.

Press, W., W. Vetterling, S. Teukolosky, and B. Flannery (1988). *Numerical Recipes in C: The Art of Scientific Computing* (Second ed.). Cambridge University Press.

Prince, S. (2012). *Computer Vision: Models, Learning and Inference.* Cambridge.

Pritchard, J., M. M. Stephens, and P. Donnelly (2000). Inference of population structure using multilocus genotype data. *Genetics 155,* 945–959.

Qi, Y. and T. Jaakkola (2008). Parameter Expanded Variational Bayesian Methods. In *NIPS.*

Qi, Y., M. Szummer, and T. Minka (2005). Bayesian Conditional Random Fields. In *10th Intl. Workshop on AI/Statistics.*

Quinlan, J. (1990). Learning logical definitions from relations. *Machine Learning 5,* 239–266.

Quinlan, J. R. (1986). Induction of decision trees. *Machine Learning 1,* 81–106.

Quinlan, J. R. (1993). *C4.5 Programs for Machine Learning.* Morgan Kauffman.

Quinonero-Candela, J., C. Rasmussen, and C. Williams (2007). Approximation methods for gaussian process regression. In L. Bottou, O. Chapelle, D. DeCoste, and J. Weston (Eds.), *Large Scale Kernel Machines,* pp. 203–223. MIT Press.

Rabiner, L. R. (1989). A tutorial on Hidden Markov Models and selected applications in speech recognition. *Proc. of the IEEE 77*(2), 257–286.

Rai, P. and H. Daume (2009). Multilabel prediction via sparse infinite CCA. In *NIPS.*

Raiffa, H. (1968). *Decision Analysis.* Addison Wesley.

Raina, R., A. Madhavan, and A. Ng (2009). Large-scale deep unsupervised learning using graphics processors. In *Intl. Conf. on Machine Learning.*

Raina, R., A. Ng, and D. Koller (2005). Transfer learning by constructing informative priors. In *NIPS.*

Rajaraman, A. and J. Ullman (2010). *Mining of massive datasets.* Self-published.

Rajaraman, A. and J. Ullman (2011). *Mining of massive datasets.* Cambridge.

Rakotomamonjy, A., F. Bach, S. Canu, and Y. Grandvalet (2008). SimpleMKL. *J. of Machine Learning Research 9,* 2491–2521.

Ramage, D., D. Hall, R. Nallapati, and C. Manning (2009). Labeled LDA: A supervised topic model for credit attribution in multi-labeled corpora. In *EMNLP.*

Ramage, D., C. Manning, and S. Dumais (2011). Partially Labeled Topic Models for Interpretable Text Mining. In *Proc. of the Int'l Conf. on Knowledge Discovery and Data Mining.*

Ramaswamy, S., P. Tamayo, R. Rifkin, S. Mukherjee, C. Yeang, M. Angelo, C. Ladd, M. Reich, E. Latulippe, J. Mesirov, T. Poggio, W. Gerald, M. Loda, E. Lander, and T. Golub (2001). Multiclass cancer diagnosis using tumor gene expression signature. *Proc. of the National Academy of Science, USA 98,* 15149–15154.

Ranzato, M. and G. Hinton (2010). Modeling pixel means and covariances using factored third-order Boltzmann machines. In *CVPR.*

Ranzato, M., F.-J. Huang, Y.-L. Boureau, and Y. LeCun (2007). Unsupervised Learning of Invariant Feature Hierarchies with Applications to Object Recognition. In *CVPR.*

Ranzato, M., C. Poultney, S. Chopra, and Y. LeCun (2006). Efficient learning of sparse representations with an energy-based model. In *NIPS.*

Rao, A. and K. Rose (2001, February). Deterministically Annealed Design of Hidden Markov Model Speech Recognizers. *IEEE Trans. on Speech and Audio Proc. 9*(2), 111–126.

Rasmussen, C. (2000). The infinite gaussian mixture model. In *NIPS.*

Rasmussen, C. E. and J. Quiñonero-Candela (2005). Healing the relevance vector machine by augmentation. In *Intl. Conf. on Machine Learning,* pp. 689–696.

Rasmussen, C. E. and C. K. I. Williams (2006). *Gaussian Processes for Machine Learning.* MIT Press.

Ratsch, G., T. Onoda, and K. Muller (2001). Soft margins for adaboost. *Machine Learning 42,* 287–320.

Rattray, M., O. Stegle, K. Sharp, and J. Winn (2009). Inference algorithms and learning theory for Bayesian sparse factor analysis. In *Proc. Intl. Workshop on Statistical-Mechanical Informatics.*

Rauch, H. E., F. Tung, and C. T. Striebel (1965). Maximum likelihood estimates of linear dynamic systems. *AIAA Journal 3*(8), 1445–1450.

Ravikumar, P., J. Lafferty, H. Liu, and L. Wasserman (2009). Sparse Additive Models. *J. of Royal Stat. Soc. Series B 71*(5), 1009–1030.

Raydan, M. (1997). The barzilai and borwein gradient method for the large scale unconstrained minimization problem. *SIAM J. on Optimization 7*(1), 26–33.

Rendle, S., C. Freudenthaler, Z. Gantner, and L. Schmidt-Thieme (2009). BPR: Bayesian Personalized Ranking from Implicit Feedback. In *UAI.*

Rennie, J. (2004). Why sums are bad. Technical report, MIT.

Rennie, J., L. Shih, J. Teevan, and D. Karger (2003). Tackling the poor assumptions of naive Bayes text classifiers. In *Intl. Conf. on Machine Learning.*

Reshef, D., Y. Reshef, H. Finucane, S. Grossman, G. McVean, P. Turnbaugh, E. Lander, M. Mitzenmacher, and P. Sabeti (2011, December). Detecting novel associations in large data sets. *Science 334,* 1518–1524.

Resnick, S. I. (1992). *Adventures in Stochastic Processes.* Birkhauser.

Rice, J. (1995). *Mathematical statistics and data analysis*. Duxbury. 2nd edition.

Richardson, M. and P. Domingos (2006). Markov logic networks. *Machine Learning 62*, 107–136.

Richardson, S. and P. Green (1997). On Bayesian Analysis of Mixtures With an Unknown Number of Components. *J. of Royal Stat. Soc. Series B 59*, 731–758.

Riesenhuber, M. and T. Poggio (1999). Hierarchical models of object recognition in cortex. *Nature Neuroscience 2*, 1019–1025.

Rish, I., G. Grabarnik, G. Cecchi, F. Pereira, and G. Gordon (2008). Closed-form supervised dimensionality reduction with generalized linear models. In *Intl. Conf. on Machine Learning*.

Ristic, B., S. Arulampalam, and N. Gordon (2004). *Beyond the Kalman Filter: Particle Filters for Tracking Applications*. Artech House Radar Library.

Robert, C. (1995). Simulation of truncated normal distributions. *Statistics and computing 5*, 121–125.

Robert, C. and G. Casella (2004). *Monte Carlo Statisical Methods*. Springer. 2nd edition.

Roberts, G. and J. Rosenthal (2001). Optimal scaling for various Metropolis-Hastings algorithms. *Statistical Science 16*, 351–367.

Roberts, G. O. and S. K. Sahu (1997). Updating schemes, correlation structure, blocking and parameterization for the gibbs sampler. *J. of Royal Stat. Soc. Series B 59*(2), 291–317.

Robinson, R. W. (1973). Counting labeled acyclic digraphs. In F. Harary (Ed.), *New Directions in the Theory of Graphs*, pp. 239–273. Academic Press.

Roch, S. (2006). A short proof that phylogenetic tree reconstruction by maximum likelihood is hard. *IEEE/ACM Trans. Comp. Bio. Bioinformatics 31*(1).

Rodriguez, A. and K. Ghosh (2011). Modeling relational data through nested partition models. *Biometrika*. To appear.

Rose, K. (1998, November). Deterministic annealing for clustering, compression, classification, regression, and related optimization problems. *Proc. IEEE 80*, 2210–2239.

Rosenblatt, F. (1958). The perceptron: A probabilistic model for information storage and organization in the brain. *Psychological Review 65*(6), 386–408.

Ross, S. (1989). *Introduction to Probability Models*. Academic Press.

Rosset, S., J. Zhu, and T. Hastie (2004). Boosting as a regularized path to a maximum margin classifier. *J. of Machine Learning Research 5*, 941–973.

Rossi, P., G. Allenby, and R. McCulloch (2006). *Bayesian Statistics and Marketing*. Wiley.

Roth, D. (1996, Apr). On the hardness of approximate reasoning. *Artificial Intelligence 82*(1-2), 273–302.

Roth, S. and M. Black (2009, April). Fields of experts. *Intl. J. Computer Vision 82*(2), 205–229.

Rother, C., P. Kohli, W. Feng, and J. Jia (2009). Minimizing sparse higher order energy functions of discrete variables. In *CVPR*, pp. 1382–1389.

Rouder, J., P. Speckman, D. Sun, and R. Morey (2009). Bayesian t tests for accepting and rejecting the null hypothesis. *Psychonomic Bulletin & Review 16*(2), 225–237.

Roverato, A. (2002). Hyper inverse Wishart distribution for non-decomposable graphs and its application to Bayesian inference for Gaussian graphical models. *Scand. J. Statistics 29*, 391–411.

Roweis, S. (1997). EM algorithms for PCA and SPCA. In *NIPS*.

Rubin, D. (1998). Using the SIR algorithm to simulate posterior distributions. In *Bayesian Statistics 3*.

Rue, H. and L. Held (2005). *Gaussian Markov Random Fields: Theory and Applications*, Volume 104 of *Monographs on Statistics and Applied Probability*. London: Chapman & Hall.

Rue, H., S. Martino, and N. Chopin (2009). Approximate Bayesian Inference for Latent Gaussian Models Using Integrated Nested Laplace Approximations. *J. of Royal Stat. Soc. Series B 71*, 319–392.

Rumelhart, D., G. Hinton, and R. Williams (1986). Learning internal representations by error propagation. In D. Rumelhart, J. McClelland, and the PDD Research Group (Eds.), *Parallel Distributed Processing: Explorations in the Microstructure of Cognition*. MIT Press.

Ruppert, D., M. Wand, and R. Carroll (2003). *Semiparametric Regression*. Cambridge University Press.

Rush, A. M. and M. Collins (2012). A tutorial on Lagrangian relaxation and dual decomposition for NLP. Technical report, Columbia U.

Russell, S., J. Binder, D. Koller, and K. Kanazawa (1995). Local learning in probabilistic networks with hidden variables. In *Intl. Joint Conf. on AI*.

Russell, S. and P. Norvig (1995). *Artificial Intelligence: A Modern Approach*. Englewood Cliffs, NJ: Prentice Hall.

Russell, S. and P. Norvig (2002). *Artificial Intelligence: A Modern Approach*. Prentice Hall. 2nd edition.

Russell, S. and P. Norvig (2010). *Artificial Intelligence: A Modern Approach*. Prentice Hall. 3rd edition.

Sachs, K., O. Perez, D. Pe'er, D. Lauffenburger, and G. Nolan (2005). Causal protein-signaling networks derived from multiparameter single-cell data. *Science 308*.

Sahami, M. and T. Heilman (2006). A Web-based Kernel Function for Measuring the Similarity of Short Text Snippets. In *WWW conference*.

Salakhutdinov, R. (2009). *Deep Generative Models*. Ph.D. thesis, U. Toronto.

Salakhutdinov, R. and G. Hinton (2009). Deep Boltzmann machines. In *AI/Statistics*, Volume 5, pp. 448–455.

Salakhutdinov, R. and G. Hinton (2010). Replicated Softmax: an Undirected Topic Model. In *NIPS*.

Salakhutdinov, R. and H. Larochelle (2010). Efficient Learning of Deep Boltzmann Machines. In *AI/Statistics*.

Salakhutdinov, R. and A. Mnih (2008). Probabilistic matrix factorization. In *NIPS*, Volume 20.

Salakhutdinov, R. and S. Roweis (2003). Adaptive overrelaxed bound optimization methods. In *Proceedings of the International Conference on Machine Learning*, Volume 20, pp. 664–671.

Salakhutdinov, R., J. Tenenbaum, and A. Torralba (2011). Learning To Learn with Compound HD Models. In *NIPS*.

Salakhutdinov, R. R., A. Mnih, and G. E. Hinton (2007). Restricted boltzmann machines for collaborative filtering. In *Intl. Conf. on Machine Learning*, Volume 24, pp. 791–798.

Salojarvi, J., K. Puolamaki, and S. Klaski (2005). On discriminative joint density modeling. In *Proc. European Conf. on Machine Learning*.

Sampson, F. (1968). *A Novitiate in a Period of Change: An Experimental and Case Study of Social Relationships*. Ph.D. thesis, Cornell.

Santner, T., B. Williams, and W. Notz (2003). *The Design and Analysis of Computer Experiments*. Springer.

Sarkar, J. (1991). One-armed bandit problems with covariates. *The Annals of Statistics 19*(4), 1978–2002.

Sato, M. and S. Ishii (2000). On-line EM algorithm for the normalized Gaussian network. *Neural Computation 12*, 407–432.

Saul, L., T. Jaakkola, and M. Jordan (1996). Mean Field Theory for Sigmoid Belief Networks. *J. of AI Research 4*, 61–76.

Saul, L. and M. Jordan (1995). Exploiting tractable substructures in intractable networks. In *NIPS*, Volume 8.

Saul, L. and M. Jordan (2000). Attractor dynamics in feedforward neural networks. *Neural Computation 12*, 1313–1335.

Saunders, C., J. Shawe-Taylor, and A. Vinokourov (2003). String Kernels, Fisher Kernels and Finite State Automata. In *NIPS*.

Savage, R., K. Heller, Y. Xi, Z. Ghahramani, W. Truman, M. Grant, K. Denby, and D. Wild (2009). R/BHC: fast Bayesian hierarchical clustering for microarray data. *BMC Bioinformatics 10*(242).

Schaefer, J. and K. Strimmer (2005). A shrinkage approach to large-scale covariance matrix estimation and implications for functional genomics. *Statist. Appl. Genet. Mol. Biol 4*(32).

Schapire, R. (1990). The strength of weak learnability. *Machine Learning 5*, 197–227.

Schapire, R. and Y. Freund (2012). *Boosting: Foundations and Algorithms*. MIT Press.

Schapire, R., Y. Freund, P. Bartlett, and W. Lee (1998). Boosting the margin: a new explanation for the effectiveness of voting methods. *Annals of Statistics 5*, 1651–1686.

Scharstein, D. and R. Szeliski (2002). A taxonomy and evaluation of dense two-frame stereo correspondence algorithms. *Intl. J. Computer Vision 47*(1), 7–42.

Schaul, T., S. Zhang, and Y. LeCun (2012). No more pesky learning rates. Technical report, Courant Instite of Mathematical Sciences.

Schmee, J. and G. Hahn (1979). A simple method for regresssion analysis with censored data. *Technometrics 21*, 417–432.

Schmidt, M. (2010). *Graphical model structure learning with L1 regularization*. Ph.D. thesis, UBC.

Schmidt, M., G. Fung, and R. Rosales (2009). Optimization methods for ℓ − 1 regularization. Technical report, U. British Columbia.

Schmidt, M. and K. Murphy (2009). Modeling Discrete Interventional Data using Directed Cyclic Graphical Models. In *UAI*.

Schmidt, M., K. Murphy, G. Fung, and R. Rosales (2008). Structure Learning in Random Fields for Heart Motion Abnormality Detection. In *CVPR*.

Schmidt, M., A. Niculescu-Mizil, and K. Murphy (2007). Learning Graphical Model Structure using L1-Regularization Paths. In *AAAI*.

Schmidt, M., E. van den Berg, M. Friedlander, and K. Murphy (2009). Optimizing Costly Functions with Simple Constraints: A Limited-Memory Projected Quasi-Newton Algorithm. In *AI & Statistics*.

Schniter, P., L. C. Potter, and J. Ziniel (2008). Fast Bayesian Matching Pursuit: Model Uncertainty and Parameter Estimation for Sparse Linear Models. Technical report, U. Ohio. Submitted to IEEE Trans. on Signal Processing.

Schnitzspan, P., S. Roth, and B. Schiele (2010). Automatic discovery of meaningful object parts with latent CRFs. In *CVPR*.

Schoelkopf, B. and A. Smola (2002). *Learning with Kernels: Support Vector Machines, Regularization, Optimization, and Beyond*. MIT Press.

Schoelkopf, B., A. Smola, and K.-R. Mueller (1998). Nonlinear component analysis as a kernel eigenvalue problem. *Neural Computation 10*, 1299 - 1319.

Schraudolph, N. N., J. Yu, and S. Günter (2007). A Stochastic Quasi-Newton Method for Online Convex Optimization. In *AI/Statistics*, pp. 436–443.

Schwarz, G. (1978). Estimating the dimension of a model. *Annals of Statistics 6*(2), 461–464.

Schwarz, R. and Y. Chow (1990). The n-best algorithm: an efficient and exact procedure for finding the n most likely hypotheses. In *Intl. Conf. on Acoustics, Speech and Signal Proc.*

Schweikerta, G., A. Zien, G. Zeller, J. Behr, C. Dieterich, C. Ong, P. Philips, F. D. Bona, L. Hartmann, A. Bohlen, N. KrÄijger, S. Sonnenburg, and G. RÄd'tsch (2009). mGene: Accurate SVM-based Gene Finding with an Application to Nematode Genomes. *Genome Research, 19*, 2133–2143.

Schwing, A., T. Hazan, M. Pollefeys, and R. Urtasun (2011). Distributed message passing for large scale graphical models. In *CVPR*. C++ MPI/EC2 implementation is available here: http://www.alexander-schwing.de/projects.php.

Scott, D. (1979). On optimal and data-based histograms. *Biometrika 66*(3), 605–610.

Scott, J. G. and C. M. Carvalho (2008). Feature-inclusion stochastic search for gaussian graphical models. *J. of Computational and Graphical Statistics 17*(4), 790–808.

Scott, S. (2009). Data augmentation, frequentist estimation, and the bayesian analysis of multinomial logit models. *Statistical Papers.*

Scott, S. (2010). A modern Bayesian look at the multi-armed bandit. *Applied Stochastic Models in Business and Industry 26*, 639–658.

Sedgewick, R. and K. Wayne (2011). *Algorithms.* Addison Wesley.

Seeger, M. (2008). Bayesian Inference and Optimal Design in the Sparse Linear Model. *J. of Machine Learning Research 9*, 759–813.

Seeger, M. and H. Nickish (2008). Compressed sensing and bayesian experimental design. In *Intl. Conf. on Machine Learning.*

Segal, D. (2011, 12 February). The dirty little secrets of search. *New York Times.*

Seide, F., G. Li, and D. Yu (2011). Conversational Speech Transcription Using Context-Dependent Deep Neural Networks. In *Interspeech.*

Sejnowski, T. and C. Rosenberg (1987). Parallel networks that learn to pronounce english text. *Complex Systems 1*, 145–168.

Sellke, T., M. J. Bayarri, and J. Berger (2001). Calibration of p Values for Testing Precise Null Hypotheses. *The American Statistician 55*(1), 62–71.

Serre, T., L. Wolf, and T. Poggio (2005). Object recognition with features inspired by visual cortex. In *CVPR,* pp. 994–1000.

Shachter, R. (1998). Bayes-ball: The rational pastime (for determining irrelevance and requisite information in belief networks and influence diagrams). In *UAI.*

Shachter, R. and C. R. Kenley (1989). Gaussian influence diagrams. *Managment Science 35*(5), 527–550.

Shachter, R. D. and M. A. Peot (1989). Simulation approaches to general probabilistic inference on belief networks. In *UAI,* Volume 5.

Shafer, G. R. and P. P. Shenoy (1990). Probability propagation. *Annals of Mathematics and AI 2*, 327–352.

Shafto, P., C. Kemp, V. Mansinghka, M. Gordon, and J. B. Tenenbaum (2006). Learning cross-cutting systems of categories. In *Cognitive Science Conference.*

Shahaf, D., A. Chechetka, and C. Guestrin (2009). Learning Thin Junction Trees via Graph Cuts. In *AISTATS.*

Shalev-Shwartz, S., Y. Singer, and N. Srebro (2007). Pegasos: primal estimated sub-gradient solver for svm. In *Intl. Conf. on Machine Learning.*

Shalizi, C. (2009). Cs 36-350 lecture 10: Principal components: mathematics, example, interpretation.

Shan, H. and A. Banerjee (2010). Residual Bayesian co-clustering for matrix approximation. In *SIAM Intl. Conf. on Data Mining.*

Shawe-Taylor, J. and N. Cristianini (2004). *Kernel Methods for Pattern Analysis.* Cambridge.

Sheng, Q., Y. Moreau, and B. D. Moor (2003). Biclustering Microarray data by Gibbs sampling. *Bioinformatics 19*, ii196–ii205.

Shi, J. and J. Malik (2000). Normalized cuts and image segmentation. *IEEE Trans. on Pattern Analysis and Machine Intelligence.*

Shoham, Y. and K. Leyton-Brown (2009). *Multiagent Systems: Algorithmic, Game- Theoretic, and Logical Foundations.* Cambridge University Press.

Shotton, J., A. Fitzgibbon, M. Cook, T. Sharp, M. Finocchio, R. Moore, A. Kipman, and A. Blake (2011). Real-time human pose recognition in parts from a single depth image. In *CVPR.*

Shwe, M., B. Middleton, D. Heckerman, M. Henrion, E. Horvitz, H. Lehmann, and G. Cooper (1991). Probabilistic diagnosis using a reformulation of the internist-1/qmr knowledge base. *Methods. Inf. Med 30*(4), 241–255.

Siddiqi, S., B. Boots, and G. Gordon (2007). A constraint generation approach to learning stable linear dynamical systems. In *NIPS.*

Siepel, A. and D. Haussler (2003). Combining phylogenetic and hidden markov models in biosequence analysis. In *Proc. 7th Intl. Conf. on Computational Molecular Biology (RECOMB).*

Silander, T., P. Kontkanen, and P. Myllymaki (2007). On Sensitivity of the MAP Bayesian Network Structure to the Equivalent Sample Size Parameter. In *UAI,* pp. 360–367.

Silander, T. and P. Myllymaki (2006). A simple approach for finding the globally optimal Bayesian network structure. In *UAI.*

Sill, J., G. Takacs, L. Mackey, and D. Lin (2009). Feature-weighted linear stacking. Technical report, .

Silver, N. (2012). *The Signal and the Noise.* Penguin Press.

Silverman, B. W. (1984). Spline smoothing: the equivalent variable kernel method. *Annals of Statistics 12*(3), 898–916.

Simard, P., D. Steinkraus, and J. Platt (2003). Best practices for convolutional neural networks applied to visual document analysis. In *Intl. Conf. on Document Analysis and Recognition (ICDAR).*

Simon, D. (2006). *Optimal State Estimation: Kalman, H Infinity, and Nonlinear Approaches.* Wiley.

Singliar, T. and M. Hauskrecht (2006). Noisy-OR Component Analysis and its Application to Link Analysis. *J. of Machine Learning Research 7.*

Smidl, V. and A. Quinn (2005). *The Variational Bayes Method in Signal Processing.* Springer.

Smith, A. F. M. and A. E. Gelfand (1992). Bayesian statistics without tears: A sampling-resampling perspective. *The American Statistician 46*(2), 84–88.

Smith, R. and P. Cheeseman (1986). On the representation and estimation of spatial uncertainty. *Intl. J. Robotics Research 5*(4), 56–68.

Smith, V., J. Yu, T. Smulders, A. Hartemink, and E. Jarvis (2006). Computational Inference of Neural Information Flow Networks. *PLOS Computational Biology 2*, 1436–1439.

Smolensky, P. (1986). Information processing in dynamical systems: foundations of harmony theory. In D. Rumehart and J. McClelland (Eds.), *Parallel Distributed Processing: Explorations in the Microstructure of Cognition. Volume 1.* McGraw-Hill.

Smyth, P., D. Heckerman, and M. I. Jordan (1997). Probabilistic independence networks for hidden Markov probability models. *Neural Computation 9*(2), 227–269.

Sohl-Dickstein, J., P. Battaglino, and M. DeWeese (2011). Minimum probability flow learning. In *Intl. Conf. on Machine Learning.*

Sollich, P. (2002). Bayesian methods for support vector machines: evidence and predictive class probabilities. *Machine Learning 46*, 21–52.

Sontag, D., A. Globerson, and T. Jaakkola (2011). Introduction to dual decomposition for inference. In S. Sra, S. Nowozin, and S. J. Wright (Eds.), *Optimization for Machine Learning.* MIT Press.

Sorenson, H. and D. Alspach (1971). Recursive Bayesian estimation using Gaussian sums. *Automatica 7*, 465âĂŞ 479.

Soussen, C., J. Iier, D. Brie, and J. Duan (2010). From Bernoulli-Gaussian deconvolution to sparse signal restoration. Technical report, Centre de Recherche en Automatique de Nancy.

Spaan, M. and N. Vlassis (2005). Perseus: Randomized Point-based Value Iteration for POMDPs. *J. of AI Research 24*, 195–220.

Spall, J. (2003). *Introduction to Stochastic Search and Optimization: Estimation, Simulation, and Control.* Wiley.

Speed, T. (2011, December). A correlation for the 21st century. *Science 334*, 152–1503.

Speed, T. and H. Kiiveri (1986). Gaussian Markov distributions over finite graphs. *Annals of Statistics 14*(1), 138–150.

Spiegelhalter, D. J. and S. L. Lauritzen (1990). Sequential updating of conditional probabilities on directed graphical structures. *Networks 20*.

Spirtes, P., C. Glymour, and R. Scheines (2000). *Causation, Prediction, and Search.* MIT Press. 2nd edition.

Srebro, N. (2001). Maximum Likelihood Bounded Tree-Width Markov Networks. In *UAI.*

Srebro, N. and T. Jaakkola (2003). Weighted low-rank approximations. In *Intl. Conf. on Machine Learning.*

Steinbach, M., G. Karypis, and V. Kumar (2000). A comparison of document clustering techniques. In *KDD Workshop on Text Mining.*

Stephens, M. (2000). Dealing with label-switching in mixture models. *J. Royal Statistical Society, Series B 62*, 795–809.

Stern, D., R. Herbrich, and T. Graepel (2009). Matchbox: Large Scale Bayesian Recommendations. In *Proc. 18th. Intl. World Wide Web Conference.*

Steyvers, M. and T. Griffiths (2007). Probabilistic topic models. In T. Landauer, D. McNamara, S. Dennis, and W. Kintsch (Eds.), *Latent Semantic Analysis: A Road to Meaning.* Laurence Erlbaum.

Stigler, S. (1986). *The history of statistics.* Harvard University press.

Stolcke, A. and S. M. Omohundro (1992). Hidden Markov Model Induction by Bayesian Model Merging. In *NIPS-5.*

Stoyanov, V., A. Ropson, and J. Eisner (2011). Empirical risk minimization of graphical model parameters given approximate inference, decoding, and model structure. In *AI/Statistics.*

Sudderth, E. (2006). *Graphical Models for Visual Object Recognition and Tracking.* Ph.D. thesis, MIT.

Sudderth, E. and W. Freeman (2008, March). Signal and Image Processing with Belief Propagation. *IEEE Signal Processing Magazine.*

Sudderth, E., A. Ihler, W. Freeman, and A. Willsky (2003). Nonparametric Belief Propagation. In *CVPR.*

Sudderth, E., A. Ihler, M. Isard, W. Freeman, and A. Willsky (2010). Nonparametric Belief Propagation. *Comm. of the ACM 53*(10).

Sudderth, E. and M. Jordan (2008). Shared Segmentation of Natural Scenes Using Dependent Pitman-Yor Processes. In *NIPS.*

Sudderth, E., M. Wainwright, and A. Willsky (2008). Loop series and bethe variational bounds for attractive graphical models. In *NIPS.*

Sun, J., N. Zheng, and H. Shum (2003). Stereo matching using belief propagation. *IEEE Trans. on Pattern Analysis and Machine Intelligence 25*(7), 787–800.

Sun, L., S. Ji, S. Yu, and J. Ye (2009). On the equivalence between canonical correlation analysis and orthonormalized partial least squares. In *Intl. Joint Conf. on AI.*

Sunehag, P., J. Trumpf, S. V. N. Vishwanathan, and N. N. Schraudolph (2009). Variable Metric Stochastic Approximation Theory. In *AI/Statistics*, pp. 560–566.

Sutton, C. and A. McCallum (2007). Improved Dynamic Schedules for Belief Propagation. In *UAI.*

Sutton, R. and A. Barto (1998). *Reinforcment Learning: An Introduction.* MIT Press.

Swendsen, R. and J.-S. Wang (1987). Nonuniversal critical dynamics in Monte Carlo simulations. *Physical Review Letters 58*, 86–88.

Swersky, K., B. Chen, B. Marlin, and N. de Freitas (2010). A Tutorial on Stochastic Approximation Algorithms for Training Restricted Boltzmann Machines and Deep Belief Nets. In *Information Theory and Applications (ITA) Workshop.*

Szeliski, R. (2010). *Computer Vision: Algorithms and Applications.* Springer.

Szeliski, R., R. Zabih, D. Scharstein, O. Veksler, V. Kolmogorov, A. Agarwala, M. Tappen, and C. Rother (2008). A Comparative Study of Energy Minimization Methods for Markov Random Fields with Smoothness-Based Priors. *IEEE Trans. on Pattern Analysis and Machine Intelligence 30*(6), 1068–1080.

Szepesvari, C. (2010). *Algorithms for Reinforcement Learning.* Morgan Claypool.

Taleb, N. (2007). *The Black Swan: The Impact of the Highly Improbable.* Random House.

Talhouk, A., K. Murphy, and A. Doucet (2012). Efficient Bayesian Inference for Multivariate Probit Models with Sparse Inverse Correlation Matrices. *J. Comp. Graph. Statist. 21*(3), 739–757.

Tanner, M. (1996). *Tools for statistical inference*. Springer.

Tanner, M. and W. Wong (1987). The calculation of posterior distributions by data augmentation. *J. of the Am. Stat. Assoc. 82*(398), 528–540.

Tarlow, D., I. Givoni, and R. Zemel (2010). Hop-map: efficient message passing with high order potentials. In *AI/Statistics*.

Taskar, B., C. Guestrin, and D. Koller (2003). Max-margin markov networks. In *NIPS*.

Taskar, B., D. Klein, M. Collins, D. Koller, and C. Manning (2004). Max-margin parsing. In *Proc. Empirical Methods in Natural Language Processing*.

Teh, Y. W. (2006). A hierarchical Bayesian language model based on Pitman-Yor processes. In *Proc. of the Assoc. for Computational Linguistics*, pp. 985=992.

Teh, Y.-W., M. Jordan, M. Beal, and D. Blei (2006). Hierarchical Dirichlet processes. *J. of the Am. Stat. Assoc. 101*(476), 1566–1581.

Tenenbaum, J. (1999). *A Bayesian framework for concept learning*. Ph.D. thesis, MIT.

Tenenbaum, J., C. Kemp, T. Griffiths, and N. Goodman (2011). How to grow a mind: Statistics, structure, and abstraction. *Science 6022*, 1279–1285.

Tenenbaum, J. B. and F. Xu (2000). Word learning as bayesian inference. In *Proc. 22nd Annual Conf.of the Cognitive Science Society*.

Theocharous, G., K. Murphy, and L. Kaelbling (2004). Representing hierarchical POMDPs as DBNs for multi-scale robot localization. In *IEEE Intl. Conf. on Robotics and Automation*.

Thiesson, B., C. Meek, D. Chickering, and D. Heckerman (1998). Learning mixtures of DAG models. In *UAI*.

Thomas, A. and P. Green (2009). Enumerating the decomposable neighbours of a decomposable graph under a simple perturbation scheme. *Comp. Statistics and Data Analysis 53*, 1232–1238.

Thrun, S., W. Burgard, and D. Fox (2006). *Probabilistic Robotics*. MIT Press.

Thrun, S., M. Montemerlo, D. Koller, B. Wegbreit, J. Nieto, and E. Nebot (2004). Fastslam: An efficient solution to the simultaneous localization and mapping problem with unknown data association. *J. of Machine Learning Research 2004*.

Thrun, S. and L. Pratt (Eds.) (1997). *Learning to learn*. Kluwer.

Tibshirani, R. (1996). Regression shrinkage and selection via the lasso. *J. Royal. Statist. Soc B 58*(1), 267–288.

Tibshirani, R., G. Walther, and T. Hastie (2001). Estimating the number of clusters in a dataset via the gap statistic. *J. of Royal Stat. Soc. Series B 32*(2), 411–423.

Tieleman, T. (2008). Training restricted Boltzmann machines using approximations to the likelihood gradient. In *Proceedings of the 25th international conference on Machine learning*, pp. 1064–1071. ACM New York, NY, USA.

Ting, J., A. D'Souza, S. Vijayakumar, and S. Schaal (2010). Efficient learning and feature selection in high-dimensional regression. *Neural Computation 22*(4), 831–886.

Tipping, M. (1998). Probabilistic visualization of high-dimensional binary data. In *NIPS*.

Tipping, M. (2001). Sparse bayesian learning and the relevance vector machine. *J. of Machine Learning Research 1*, 211–244.

Tipping, M. and C. Bishop (1999). Probabilistic principal component analysis. *J. of Royal Stat. Soc. Series B 21*(3), 611–622.

Tipping, M. and A. Faul (2003). Fast marginal likelihood maximisation for sparse bayesian models. In *AI/Stats*.

Tishby, N., F. Pereira, and W. Biale (1999). The information bottleneck method. In *The 37th annual Allerton Conf. on Communication, Control, and Computing*, pp. 368–377.

Torralba, A., R. Fergus, and Y. Weiss (2008). Small codes and large image databases for recognition. In *CVPR*.

Train, K. (2009). *Discrete choice methods with simulation*. Cambridge University Press. Second edition.

Tseng, P. (2008). On Accelerated Proximal Gradient Methods for Convex-Concave Optimization. Technical report, U. Washington.

Tsochantaridis, I., T. Joachims, T. Hofmann, and Y. Altun (2005, September). Large margin methods for structured and interdependent output variables. *J. of Machine Learning Research 6*, 1453–1484.

Tu, Z. and S. Zhu (2002). Image Segmentation by Data-Driven Markov Chain Monte Carlo. *IEEE Trans. on Pattern Analysis and Machine Intelligence 24*(5), 657–673.

Turian, J., L. Ratinov, and Y. Bengio (2010). Word representations: a simple and general method for semi-supervised learning. In *Proc. ACL*.

Turlach, B., W. Venables, and S. Wright (2005). Simultaneous variable selection. *Technometrics 47*(3), 349–363.

Turner, R., P. Berkes, M. Sahani, and D. Mackay (2008). Counterexamples to variational free energy compactness folk theorems. Technical report, U. Cambridge.

Ueda, N. and R. Nakano (1998). Deterministic annealing EM algorithm. *Neural Networks 11*, 271–282.

Usunier, N., D. Buffoni, and P. Gallinari (2009). Ranking with ordered weighted pairwise classification. In *Intl. Conf. on Machine Learning*.

Vaithyanathan, S. and B. Dom (1999). Model selection in unsupervised learning with applications to document clustering. In *Intl. Conf. on Machine Learning*.

van der Merwe, R., A. Doucet, N. de Freitas, and E. Wan (2000). The unscented particle filter. In *NIPS-13*.

van Dyk, D. and X.-L. Meng (2001). The Art of Data Augmentation. *J. Computational and Graphical Statistics 10*(1), 1–50.

van Iterson, M., H. van Haagen, and J. Goeman (2012). Resolving confusion of tongues in statistics and machine learning: A primer for biologists and bioinformaticians. *Proteomics 12*, 543–549.

Vandenberghe, L. (2006). Applied numerical computing: Lecture notes.

Vandenberghe, L. (2011). Optimization Methods for Large-Scale Systems (UCLA EE236C Lecture Notes).

Vanhatalo, J. (2010). *Speeding up the inference in Gaussian process models*. Ph.D. thesis, Helsinki Univ. Technology.

Vanhatalo, J., V. Pietilainen, and A. Vehtari (2010). Approximate inference for disease mapping with sparse gaussian processes. *Statistics in Medicine 29*(15), 1580–1607.

Vapnik, V. (1998). *Statistical Learning Theory*. Wiley.

Vapnik, V., S. Golowich, and A. Smola (1997). Support vector method for function approximation, regression estimation, and signal processing. In *NIPS*.

Varian, H. (2011). Structural time series in R: a Tutorial. Technical report, Google.

Verma, T. and J. Pearl (1990). Equivalence and synthesis of causal models. In *UAI*.

Viinikanoja, J., A. Klami, and S. Kaski (2010). Variational Bayesian Mixture of Robust CCA Models. In *Proc. European Conf. on Machine Learning*.

Vincent, P. (2011). A Connection between Score Matching and Denoising Autoencoders. *Neural Computation 23*(7), 1661–1674.

Vincent, P., H. Larochelle, I. Lajoie, Y. Bengio, and P.-A. Manzagol (2010). Stacked Denoising Autoencoders: Learning Useful Representations in a Deep Network with a Local Denoising Criterion. *J. of Machine Learning Research 11*, 3371–3408.

Vinh, N., J. Epps, and J. Bailey (2009). Information theoretic measures for clusterings comparison: Is a correction for chance necessary? In *Intl. Conf. on Machine Learning*.

Vinyals, M., J. Cerquides, J. Rodriguez-Aguilar, and A. Farinelli (2010). Worst-case bounds on the quality of max-product fixed-points. In *NIPS*.

Viola, P. and M. Jones (2001). Rapid object detection using a boosted cascade of simple classifiers. In *CVPR*.

Virtanen, S. (2010). Bayesian exponential family projections. Master's thesis, Aalto University.

Vishwanathan, S. V. N. and A. Smola (2003). Fast kernels for string and tree matching. In *NIPS*.

Viterbi, A. (1967). Error bounds for convolutional codes and an asymptotically optimal decoding algorithm. *IEEE Trans. on Information Theory 13*(2), 260–269.

von Luxburg, U. (2007). A tutorial on spectral clustering. *Statistics and Computing 17*(4), 395–416.

Wagenmakers, E.-J., R. Wetzels, D. Borsboom, and H. van der Maas (2011). Why Psychologists Must Change the Way They Analyze Their Data: The Case of Psi. *Journal of Personality and Social Psychology*.

Wagner, D. and F. Wagner (1993). Between min cut and graph bisection. In *Proc. 18th Intl. Symp. on Math. Found. of Comp. Sci.*, pp. 744–750.

Wainwright, M., T. Jaakkola, and A. Willsky (2001). Tree-based reparameterization for approximate estimation on loopy graphs. In *NIPS-14*.

Wainwright, M., T. Jaakkola, and A. Willsky (2005). A new class of upper bounds on the log partition function. *IEEE Trans. Info. Theory 51*(7), 2313–2335.

Wainwright, M., P. Ravikumar, and J. Lafferty (2006). Inferring graphical model structure using $\ell - 1$-regularized pseudo-likelihood. In *NIPS*.

Wainwright, M. J., T. S. Jaakkola, and A. S. Willsky (2003). Tree-based reparameterization framework for analysis of sum-product and related algorithms. *IEEE Trans. on Information Theory 49*(5), 1120–1146.

Wainwright, M. J. and M. I. Jordan (2008). Graphical models, exponential families, and variational inference. *Foundations and Trends in Machine Learning 1–2*, 1–305.

Wallach, H., I. Murray, R. Salakhutdinov, and D. Mimno (2009). Evaluation methods for topic models. In *Intl. Conf. on Machine Learning*.

Wan, E. A. and R. V. der Merwe (2001). The Unscented Kalman Filter. In S. Haykin (Ed.), *Kalman Filtering and Neural Networks*. Wiley.

Wand, M. (2009). Semiparametric regression and graphical models. *Aust. N. Z. J. Stat. 51*(1), 9–41.

Wand, M. P., J. T. Ormerod, S. A. Padoan, and R. Fruhrwirth (2011). Mean Field Variational Bayes for Elaborate Distributions. *Bayesian Analysis 6*(4), 847 – 900.

Wang, C. (2007). Variational Bayesian Approach to Canonical Correlation Analysis. *IEEE Trans. on Neural Networks 18*(3), 905–910.

Wasserman, L. (2004). *All of statistics. A concise course in statistical inference*. Springer.

Wei, G. and M. Tanner (1990). A Monte Carlo implementation of the EM algorithm and the poor man's data augmentation algorithms. *J. of the Am. Stat. Assoc. 85*(411), 699–704.

Weinberger, K., A. Dasgupta, J. Attenberg, J. Langford, and A. Smola (2009). Feature hashing for large scale multitask learning. In *Intl. Conf. on Machine Learning*.

Weiss, D., B. Sapp, and B. Taskar (2010). Sidestepping intractable inference with structured ensemble cascades. In *NIPS*.

Weiss, Y. (2000). Correctness of local probability propagation in graphical models with loops. *Neural Computation 12*, 1–41.

Weiss, Y. (2001). Comparing the mean field method and belief propagation for approximate inference in MRFs. In Saad and Opper (Eds.), *Advanced Mean Field Methods*. MIT Press.

Weiss, Y. and W. T. Freeman (1999). Correctness of belief propagation in Gaussian graphical models of arbitrary topology. In *NIPS-12*.

Weiss, Y. and W. T. Freeman (2001a). Correctness of belief propagation in Gaussian graphical models of arbitrary topology. *Neural Computation 13*(10), 2173–2200.

Weiss, Y. and W. T. Freeman (2001b). On the optimality of solutions of the max-product belief propagation algorithm in arbitrary graphs. *IEEE Trans. Information Theory, Special Issue on Codes on Graphs and Iterative Algorithms 47*(2), 723–735.

Weiss, Y., A. Torralba, and R. Fergus (2008). Spectral hashing. In *NIPS*.

Welling, M., C. Chemudugunta, and N. Sutter (2008). Deterministic latent variable models and their pitfalls. In *Intl. Conf. on Data Mining*.

Welling, M., T. Minka, and Y. W. Teh (2005). Structured region graphs: Morphing EP into GBP. In *UAI*.

Welling, M., M. Rosen-Zvi, and G. Hinton (2004). Exponential family harmoniums with an application to information retrieval. In *NIPS-14*.

Welling, M. and C. Sutton (2005). Learning in Markov random fields with contrastive free energies. In *Tenth International Workshop on Artificial Intelligence and Statistics (AISTATS)*.

Welling, M. and Y.-W. Teh (2001). Belief optimization for binary networks: a stable alternative to loopy belief propagation. In *UAI*.

Weng, R. and C.-J. Lin (2011). A Bayesian Approximation Method for Online Ranking. *J. of Machine Learning Research 12*, 267–300.

Werbos, P. (1974). *Beyond regression: New Tools for Prediction and Analysis in the Behavioral Sciences*. Ph.D. thesis, Harvard.

West, M. (1987). On scale mixtures of normal distributions. *Biometrika 74*, 646–648.

West, M. (2003). Bayesian Factor Regression Models in the "Large p, Small n" Paradigm. *Bayesian Statistics 7*.

West, M. and J. Harrison (1997). *Bayesian forecasting and dynamic models*. Springer.

Weston, J., S. Bengio, and N. Usunier (2010). Large Scale Image Annotation: Learning to Rank with Joint Word-Image Embeddings. In *Proc. European Conf. on Machine Learning*.

Weston, J., F. Ratle, and R. Collobert (2008). Deep Learning via Semi-Supervised Embedding. In *Intl. Conf. on Machine Learning*.

Weston, J. and C. Watkins (1999). Multi-class support vector machines. In *ESANN*.

Wick, M., K. Rohanimanesh, K. Bellare, A. Culotta, and A. McCallum (2011). SampleRank: Training Factor Graphs with Atomic Gradients. In *UAI*.

Wiering, M. and M. van Otterlo (Eds.) (2012). *Reinforcement learning: State-of-the-art*. Springer.

Wilkinson, D. and S. Yeung (2002). Conditional simulation from highly structured gaussian systems with application to blocking-mcmc for the bayesian analysis of very large linear models. *Statistics and Computing 12*, 287–300.

Williams, C. (1998). Computation with infinite networks. *Neural Computation 10*(5), 1203–1216.

Williams, C. (2000). A MCMC approach to Hierarchical Mixture Modelling . In S. A. Solla, T. K. Leen, and K.-R. Müller (Eds.), *NIPS*. MIT Press.

Williams, C. (2002). On a Connection between Kernel PCA and Metric Multidimensional Scaling. *Machine Learning J. 46*(1).

Williams, O. and A. Fitzgibbon (2006). Gaussian process implicit surfaces. In *Gaussian processes in practice*.

Williamson, S. and Z. Ghahramani (2008). Probabilistic models for data combination in recommender systems. In *NIPS Workshop on Learning from Multiple Sources*.

Winn, J. and C. Bishop (2005). Variational message passing. *J. of Machine Learning Research 6*, 661–694.

Wipf, D. and S. Nagarajan (2007). A new view of automatic relevancy determination. In *NIPS*.

Wipf, D. and S. Nagarajan (2010, April). Iterative Reweighted $\ell-1$ and $\ell-2$ Methods for Finding Sparse Solutions. *J. of Selected Topics in Signal Processing (Special Issue on Compressive Sensing) 4*(2).

Wipf, D., B. Rao, and S. Nagarajan (2010). Latent variable bayesian models for promoting sparsity. *IEEE Transactions on Information Theory*.

Witten, D., R. Tibshirani, and T. Hastie (2009). A penalized matrix decomposition, with applications to sparse principal components and canonical correlation analysis. *Biostatistics 10*(3), 515–534.

Witten, I. and E. Frank (2005). *Data Mining: Practical Machine Learning Tools and Techniques*. Morgan-Kaufman. Second Edition.

Wolpert, D. (1992). Stacked generalization. *Neural Networks 5*(2), 241–259.

Wolpert, D. (1996). The lack of a priori distinctions between learning algorithms. *Neural Computation 8*(7), 1341–1390.

Wong, F., C. Carter, and R. Kohn (2003). Efficient estimation of covariance selection models. *Biometrika 90*(4), 809–830.

Wood, F., C. Archambeau, J. Gasthaus, L. James, and Y. W. Teh (2009). A stochastic memoizer for sequence data. In *Intl. Conf. on Machine Learning*.

Wright, S., R. Nowak, and M. Figueiredo (2009). Sparse reconstruction by separable approximation. *IEEE Trans. on Signal Processing 57*(7), 2479–2493.

Wu, T. T. and K. Lange (2008). Coordinate descent algorithms for lasso penalized regression. *Ann. Appl. Stat 2*(1), 224–244.

Wu, Y., H. Tjelmeland, and M. West (2007). Bayesian CART: Prior structure and MCMC computations. *J. of Computational and Graphical Statistics 16*(1), 44–66.

Xu, F. and J. Tenenbaum (2007). Word learning as Bayesian inference. *Psychological Review 114*(2).

Xu, Z., V. Tresp, A. Rettinger, and K. Kersting (2008). Social network mining with nonparametric relational models. In *ACM Workshop on Social Network Mining and Analysis (SNA-KDD 2008)*.

Xu, Z., V. Tresp, K. Yu, and H.-P. Kriegel (2006). Infinite hidden relational models. In *UAI*.

Xu, Z., V. Tresp, S. Yu, K. Yu, and H.-P. Kriegel (2007). Fast inference in infinite hidden relational models. In *Workshop on Mining and Learning with Graphs*.

Xue, Y., X. Liao, L. Carin, and B. Krishnapuram (2007). Multi-task learning for classification with dirichlet process priors. *J. of Machine Learning Research 8*, 2007.

Yadollahpour, P., D. Batra, and G. Shakhnarovich (2011). Diverse M-best Solutions in MRFs. In *NIPS workshop on Disrete Optimization in Machine Learning*.

Yan, D., L. Huang, and M. I. Jordan (2009). Fast approximate spectral clustering. In *15th ACM Conf. on Knowledge Discovery and Data Mining*.

Yang, A., A. Ganesh, S. Sastry, and Y. Ma (2010, Feb). Fast l1-minimization algorithms and an application in robust face recognition: A review. Technical Report UCB/EECS-2010-13, EECS Department, University of California, Berkeley.

Yang, C., R. Duraiswami, and L. David (2005). Efficient kernel machines using the improved fast Gauss transform. In *NIPS*.

Yang, S., B. Long, A. Smola, H. Zha, and Z. Zheng (2011). Collaborative competitive filtering: learning recommender using context of user choice. In *Proc. Annual Intl. ACM SIGIR Conference*.

Yanover, C., O. Schueler-Furman, and Y. Weiss (2007). Minimizing and Learning Energy Functions for Side-Chain Prediction. In *Recomb*.

Yaun, G.-X., K.-W. Chang, C.-J. Hsieh, and C.-J. Lin (2010). A Comparison of Optimization Methods and Software for Large-scale L1-regularized Linear Classification. *J. of Machine Learning Research 11*, 3183–3234.

Yedidia, J., W. T. Freeman, and Y. Weiss (2001). Understanding belief propagation and its generalizations. In *Intl. Joint Conf. on AI*.

Yoshida, R. and M. West (2010). Bayesian learning in sparse graphical factor models via annealed entropy. *J. of Machine Learning Research 11*, 1771–1798.

Younes, L. (1989). Parameter estimation for imperfectly observed Gibbsian fields. *Probab. Theory and Related Fields 82*, 625–645.

Yu, C. and T. Joachims (2009). Learning structural SVMs with latent variables. In *Intl. Conf. on Machine Learning*.

Yu, S., K. Yu, V. Tresp, K. H-P., and M. Wu (2006). Supervised probabilistic principal component analysis. In *Proc. of the Int'l Conf. on Knowledge Discovery and Data Mining*.

Yu, S.-Z. and H. Kobayashi (2006). Practical implementation of an efficient forward-backward algorithm for an explicit-duration hidden Markov model. *IEEE Trans. on Signal Processing 54*(5), 1947–1951.

Yuan, M. and Y. Lin (2006). Model selection and estimation in regression with grouped variables. *J. Royal Statistical Society, Series B 68*(1), 49–67.

Yuan, M. and Y. Lin (2007). Model selection and estimation in the gaussian graphical model. *Biometrika 94*(1), 19–35.

Yuille, A. (2001). CCCP algorithms to minimze the Bethe and Kikuchi free energies: convergent alternatives to belief propagation. *Neural Computation 14*, 1691–1722.

Yuille, A. and A. Rangarajan (2003). The concave-convex procedure. *Neural Computation 15*, 915.

Yuille, A. and S. Zheng (2009). Compositional noisy-logical learning. In *Intl. Conf. on Machine Learning*.

Yuille, A. L. and X. He (2012). Probabilistic models of vision and max-margin methods. *Frontiers of Electrical and Electronic Engineering 7*(1).

Zellner, A. (1986). On assessing prior distributions and bayesian regression analysis with g-prior distributions. In *Bayesian inference and decision techniques, Studies of Bayesian and Econometrics and Statistics volume 6*. North Holland.

Zhai, C. and J. Lafferty (2004). A study of smoothing methods for language models applied to information retrieval. *ACM Trans. on Information Systems 22*(2), 179–214.

Zhang, N. (2004). Hierarchical latent class models for cluster analysis. *J. of Machine Learning Research*, 301–308.

Zhang, N. and D. Poole (1996). Exploiting causal independence in Bayesian network inference. *J. of AI Research*, 301–328.

Zhang, T. (2008). Adaptive Forward-Backward Greedy Algorithm for Sparse Learning with Linear Models. In *NIPS*.

Zhang, X., T. Graepel, and R. Herbrich (2010). Bayesian Online Learning for Multi-label and Multi-variate Performance Measures. In *AI/Statistics*.

Zhao, J.-H. and P. L. H. Yu (2008, November). Fast ML Estimation for the Mixture of Factor Analyzers via an ECM Algorithm. *IEEE. Trans. on Neural Networks 19*(11).

Zhao, P., G. Rocha, and B. Yu (2005). Grouped and Hierarchical Model Selection through Composite Absolute Penalties. Technical report, UC Berkeley.

Zhao, P. and B. Yu (2007). Stagewise Lasso. *J. of Machine Learning Research 8*, 2701–2726.

Zhou, H., D. Karakos, S. Khudanpur, A. Andreou, and C. Priebe (2009). On Projections of Gaussian Distributions using Maximum Likelihood Criteria. In *Proc. of the Workshop on Information Theory and its Applications*.

Zhou, M., H. Chen, J. Paisley, L. Ren, G. Sapiro, and L. Carin (2009). Non-parametric Bayesian Dictionary Learning for Sparse Image Representations. In *NIPS*.

Zhou, X. and X. Liu (2008). The EM algorithm for the extended finite mixture of the factor analyzers model. *Computational Statistics and Data Analysis 52*, 3939–3953.

Zhu, C. S., N. Y. Wu, and D. Mumford (1997, November). Minimax entropy principle and its application to texture modeling. *Neural Computation 9*(8).

Zhu, J. and E. Xing (2010). Conditional topic random fields. In *Intl. Conf. on Machine Learning*.

Zhu, L., Y. Chen, A.Yuille, and W. Freeman (2010). Latent hierarchical structure learning for object detection. In *CVPR*.

Zhu, M. and A. Ghodsi (2006). Automatic dimensionality selection from the scree plot via the use of profile likelihood. *Computational Statistics & Data Analysis 51*, 918–930.

Zhu, M. and A. Lu (2004). The counterintuitive non-informative prior for the bernoulli family. *J. Statistics Education*.

Zinkevich, M. (2003). Online convex programming and generalized infinitesimal gradient ascent. In *Intl. Conf. on Machine Learning*, pp. 928–936.

Zobay, O. (2009). Mean field inference for the Dirichlet process mixture model. *Electronic J. of Statistics 3*, 507–545.

Zoeter, O. (2007). Bayesian generalized linear models in a terabyte world. In *Proc. 5th International Symposium on image and Signal Processing and Analysis*.

Zou, H. (2006). The adaptive Lasso and its oracle properties. *J. of the Am. Stat. Assoc.*, 1418–1429.

Zou, H. and T. Hastie (2005). Regularization and variable selection via the elastic net. *J. of Royal Stat. Soc. Series B 67*(2), 301–320.

Zou, H., T. Hastie, and R. Tibshirani (2006). Sparse principal component analysis. *J. of Computational and Graphical Statistics 15*(2), 262–286.

Zou, H., T. Hastie, and R. Tibshirani (2007). On the "Degrees of Freedom" of the Lasso. *Annals of Statistics 35*(5), 2173–2192.

Zou, H. and R. Li (2008). One-step sparse estimates in nonconcave penalized likelihood models. *Annals of Statistics 36*(4), 1509–1533.

Zou, J. and R. Adams (2012). Priors for diversity in generative latent variable models. In *NIPS*.

Zweig, G. and M. Padmanabhan (2000). Exact alpha-beta computation in logarithmic space with application to map word graph construction. In *Proc. Intl. Conf. Spoken Lang.*

Index to code

Index to keywords